A Brief Table of Integrals

(An arbitrary constant may be added to each integral.)
(The notation arc sin means $\sin^{-1}$*, etc.)*

1. $\displaystyle \int x^n \, dx = \frac{1}{n+1} x^{n+1} \quad (n \neq -1)$

2. $\displaystyle \int \frac{1}{x} \, dx = \ln|x|$

3. $\displaystyle \int e^x \, dx = e^x$

4. $\displaystyle \int a^x \, dx = \frac{a^x}{\ln a}$

5. $\displaystyle \int \sin x \, dx = -\cos x$

6. $\displaystyle \int \cos x \, dx = \sin x$

7. $\displaystyle \int \tan x \, dx = -\ln|\cos x|$

8. $\displaystyle \int \cot x \, dx = \ln|\sin x|$

9. $\displaystyle \int \sec x \, dx = \ln|\sec x + \tan x|$
$\qquad\qquad = \ln|\tan(\tfrac{1}{2}x + \tfrac{1}{4}\pi)|$

10. $\displaystyle \int \csc x \, dx = \ln|\csc x - \cot x|$
$\qquad\qquad = \ln|\tan \tfrac{1}{2}x|$

11. $\displaystyle \int \arcsin \frac{x}{a} \, dx = x \arcsin \frac{x}{a} + \sqrt{a^2 - x^2} \quad (a > 0)$

12. $\displaystyle \int \arccos \frac{x}{a} \, dx = x \arccos \frac{x}{a} - \sqrt{a^2 - x^2} \quad (a > 0)$

13. $\displaystyle \int \arctan \frac{x}{a} \, dx = x \arctan \frac{x}{a} - \frac{a}{2} \ln(a^2 + x^2) \quad (a > 0)$

14. $\displaystyle \int \sin^2 mx \, dx = \frac{1}{2m} (mx - \sin mx \cos mx)$

15. $\displaystyle \int \cos^2 mx \, dx = \frac{1}{2m} (mx + \sin mx \cos mx)$

16. $\displaystyle \int \sec^2 x \, dx = \tan x$

17. $\displaystyle \int \csc^2 x \, dx = -\cot x$

18. $\displaystyle \int \sin^n x \, dx = -\frac{\sin^{n-1} x \cos x}{n} + \frac{n-1}{n} \int \sin^{n-2} x \, dx$

19. $\displaystyle \int \cos^n x \, dx = \frac{\cos^{n-1} x \sin x}{n} + \frac{n-1}{n} \int \cos^{n-2} x \, dx$

20. $\displaystyle \int \tan^n x \, dx = \frac{\tan^{n-1} x}{n-1} - \int \tan^{n-2} x \, dx \quad (n \neq 1)$

21. $\displaystyle \int \cot^n x \, dx = -\frac{\cot^{n-1} x}{n-1} - \int \cot^{n-2} x \, dx \quad (n \neq 1)$

22. $\displaystyle \int \sec^n x \, dx = \frac{\tan x \sec^{n-2} x}{n-1} + \frac{n-2}{n-1} \int \sec^{n-2} x \, dx \quad (n \neq 1)$

23. $\displaystyle \int \csc^n x \, dx = -\frac{\cot x \csc^{n-2} x}{n-1} + \frac{n-2}{n-1} \int \csc^{n-2} x \, dx \quad (n \neq 1)$

Continued on overleaf.

24. $\displaystyle\int \sinh x\, dx = \cosh x$

25. $\displaystyle\int \cosh x\, dx = \sinh x$

26. $\displaystyle\int \tanh x\, dx = \ln|\cosh x|$

27. $\displaystyle\int \coth x\, dx = \ln|\sinh x|$

28. $\displaystyle\int \operatorname{sech} x\, dx = \arctan(\sinh x)$

29. $\displaystyle\int \operatorname{csch} x\, dx = \ln\left|\tanh\frac{x}{2}\right| = -\frac{1}{2}\ln\frac{\cosh x + 1}{\cosh x - 1}$

30. $\displaystyle\int \sinh^2 x\, dx = \tfrac{1}{4}\sinh 2x - \tfrac{1}{2}x$

31. $\displaystyle\int \cosh^2 x\, dx = \tfrac{1}{4}\sinh 2x + \tfrac{1}{2}x$

32. $\displaystyle\int \operatorname{sech}^2 x\, dx = \tanh x$

33. $\displaystyle\int \sinh^{-1}\frac{x}{a}\, dx = x\sinh^{-1}\frac{x}{a} - \sqrt{x^2 + a^2} \quad (a > 0)$

34. $\displaystyle\int \cosh^{-1}\frac{x}{a}\, dx = \begin{cases} x\cosh^{-1}\dfrac{x}{a} - \sqrt{x^2 - a^2} & \left[\cosh^{-1}\left(\dfrac{x}{a}\right) > 0, a > 0\right] \\[2ex] x\cosh^{-1}\dfrac{x}{a} + \sqrt{x^2 - a^2} & \left[\cosh^{-1}\left(\dfrac{x}{a}\right) < 0, a > 0\right] \end{cases}$

35. $\displaystyle\int \tanh^{-1}\frac{x}{a}\, dx = x\tanh^{-1}\frac{x}{a} + \frac{a}{2}\ln|a^2 - x^2|$

36. $\displaystyle\int \frac{1}{\sqrt{a^2 + x^2}}\, dx = \ln(x + \sqrt{a^2 + x^2}) = \sinh^{-1}\frac{x}{a} \quad (a > 0)$

37. $\displaystyle\int \frac{1}{a^2 + x^2}\, dx = \frac{1}{a}\arctan\frac{x}{a} \quad (a > 0)$

38. $\displaystyle\int \sqrt{a^2 - x^2}\, dx = \frac{x}{2}\sqrt{a^2 - x^2} + \frac{a^2}{2}\arcsin\frac{x}{a} \quad (a > 0)$

39. $\displaystyle\int (a^2 - x^2)^{3/2}\, dx = \frac{x}{8}(5a^2 - 2x^2)\sqrt{a^2 - x^2} + \frac{3a^4}{8}\arcsin\frac{x}{a} \quad (a > 0)$

40. $\displaystyle\int \frac{1}{\sqrt{a^2 - x^2}}\, dx = \arcsin\frac{x}{a} \quad (a > 0)$

41. $\displaystyle\int \frac{1}{a^2 - x^2}\, dx = \frac{1}{2a}\ln\left|\frac{a + x}{a - x}\right|$

42. $\displaystyle\int \frac{1}{(a^2 - x^2)^{3/2}}\, dx = \frac{x}{a^2\sqrt{a^2 - x^2}}$

43. $\displaystyle\int \sqrt{x^2 \pm a^2}\, dx = \frac{x}{2}\sqrt{x^2 \pm a^2} \pm \frac{a^2}{2}\ln|x + \sqrt{x^2 \pm a^2}|$

44. $\displaystyle\int \frac{1}{\sqrt{x^2 - a^2}}\, dx = \ln|x + \sqrt{x^2 - a^2}| = \cosh^{-1}\frac{x}{a} \quad (a > 0)$

45. $\displaystyle\int \frac{1}{x(a + bx)}\, dx = \frac{1}{a}\ln\left|\frac{x}{a + bx}\right|$

46. $\displaystyle\int x\sqrt{a + bx}\, dx = \frac{2(3bx - 2a)(a + bx)^{3/2}}{15b^2}$

47. $\displaystyle\int \frac{\sqrt{a + bx}}{x}\, dx = 2\sqrt{a + bx} + a\int \frac{1}{x\sqrt{a + bx}}\, dx$

This table is continued on the endpapers at the back.

CALCULUS

Jerrold Marsden
University of California, Berkeley

Alan Weinstein
University of California, Berkeley
California Institute of Technology

The Benjamin/Cummings Publishing Company, Inc.
Menlo Park, California • Reading, Massachusetts
London • Amsterdam • Don Mills, Ontario • Sydney

TO NANCY AND MARGO

Sponsoring editors: James W. Behnke, Susan A. Newman

Production editor: Margaret Moore

Book and cover designer: James Stockton

Artists: Georg Klatt, Michael Fornalski

Cover photo: Tom Tracy

Computer graphics: Don Long and Jim Consul of Lockheed Missiles and Space Co., Inc., Sunnyvale, California; J. Kazdan, H. Ferguson, A. Cook. See the preface for further details.

Chapter opening design: Chris Walker

Manuscript acquisition: John Staples

About the cover: Solar energy fuels the air currents that keep the Rogallo wing hang-glider aloft. Energy and motion are recurrent themes in calculus.

Library of Congress Cataloging in Publication Data
Marsden, Jerrold E
    Calculus.

    Includes index.
    1. Calculus.   I.   Weinstein, Alan, 1943–   joint
author. II. Title.
QA303.M337
ISBN 0-8053-6930-9

abcdefghij–MU–8210

The Benjamin/Cummings Publishing Company, Inc.
2727 Sand Hill Road
Menlo Park, California 94025

Contents

Acknowledgments

Special Reviewer, Consultant, and Contributor
   Grant B. Gustafson, University of Utah

Reviewers
   Thomas Banchoff, Brown University
   Fred Borges, San Diego Mesa College
   Stephen H. Brown, Auburn University
   Donald Chakerian, University of California, Davis
   Bruce Edwards, University of Florida
   Ray Edwards, Chabot College
   Marshall Fraser
   W. R. Fuller, Purdue University
   L. J. Gerstein, University of California, Santa Barbara
   Kent Goodrich, University of Colorado
   Mark P. Hale, University of Florida
   D. W. Hall, Michigan State University
   Alan Hatcher, University of California, Los Angeles
   Baxter Johns, Baylor University
   Kenneth Millett, University of California, Santa Barbara
   Edward R. Millman
   Jack Milton, University of California, Davis
   Peter Renz
   Donald Sherbert, University of Illinois
   Anthony Tromba, University of California, Santa Cruz

Assistants and Proofreaders
   Yilmaz Akyildiz, Technical University of Ankara
   Roger Apodaca, University of California, Berkeley
   John Jacob, Pepperdine University
   Dana Kwong, University of Maryland
   Teresa Ling, University of California, Berkeley
   Michael Hoffman, California Institute of Technology
   Richard McIntosh, University of Calgary
   Katie Perkins, Princeton University
   Frederick Soon, University of California, Berkeley
   Tudor Ratiu, University of California, Berkeley
   Allan E. Crawford, University of Utah
   Adeline M. Gustafson, University of Utah

Typists
   Ikuko Workman
   Connie Calica
   Marnie McElhiny

We wish to thank the following institutions
   University of California, Berkeley
   University of Toronto
   I.H.E.S., Bures-Sur-Yvette
   Rice University
   University of Paris VI
   California Institute of Technology
   University of Calgary

Preface and Notes to the Teacher

The goal of this text is to help students learn to use calculus intelligently for solving a wide variety of mathematical and physical problems. The major features that contribute toward this goal may be grouped under three headings: *applications*, *organization* and *format*.

APPLICATIONS

Motivating Examples

To enhance the student's understanding of calculus, we emphasize the geometric and physical side of the subject and include many applications to show how calculus can be used. A special feature is that the applications are used not only to illustrate the mathematics, but as motivation for the introduction of each new topic and to make these topics natural and plausible. For instance, the discussion of inverse functions is introduced by yogurt making (see p. 218) and partial differentiation is motivated by analysis of water waves (see p. 686).

Down-to-Earth Applications

Calculus students should not be treated as if they are already the engineers, physicists, biologists, mathematicians, physicians or business executives they are preparing to become. The applications used in this text are balanced and include examples that are directed toward students as human beings who eat, run, bathe and watch the sun rise and set.

In view of the increasing attention we all pay to problems of energy supply and use, we have included many examples in this area. In particular, the question "how much solar energy is received in a given time and area at a given latitude?" is dealt with at pertinent points in the text using calculus techniques of increasing sophistication to piece together an answer to the question.

Specialized applications to a variety of disciplines occur in virtually every section; they may be selected according to the tastes of student and instructor. In each case, no extensive background in the specialized area is needed.

Sample Applications

	Motivation	*Worked Example*	*Problem*
Engineering	estimating accumulated energy, p. 178	circuits, p. 396	membrane deflection, p. 551
Physics/Chemistry	gyroscope, p. 620	pressure/temperature, p. 58	reaction rates, p. 405
Biology/Medicine	ear popping, p. 231	predator-prey, p. 397	blood pressure, p. 336
Business/Economics	cost of gasoline, p. 84	maximizing output, p. 731	marginal costs and profit, p. 96
General interest	sections of a sailing ship, p. 648	solar energy, p. 438	topography, p. 681

Calculators

Calculator applications are used for motivation (such as for composition on p. 233) and to illustrate the numerical content of calculus (see, for instance, p. 530). Special calculator discussions tell how to use a calculator and recognize its advantages and shortcomings.

Computer-Generated Graphics

Computer-generated graphics are becoming increasingly important as a tool for the study of calculus. High-resolution plotters were used to plot the graphs of curves and surfaces which arose in the study of Taylor polynomial approximation, maxima and minima for several variables, and three-dimensional surface geometry. These accurate perspective reproductions are possible because of recent advances in computer technology.

Lockheed, Inc., used CADAM (Computer Graphics Augmented Design and Manufacturing) computer graphic techniques to produce a sequence of 19 chapter opening illustrations expressly for this book. The sequence shows a hangglider moving along its descending path while the setting sun moves toward the horizon. Calculus makes it possible to describe accurately such continuous phenomena.

ORGANIZATION

The book is naturally divided into segments on one-variable calculus (Chapters 1–13) and several-variable calculus (Chapters 14–18).

Flexibility

Maximum flexibility and minimum interdependence of chapters allow the book to be used for a wide variety of courses. For example, differentiation and integration can be interwoven or can be studied separately, and the transcendental functions can be treated before or after integration. Many topics can be omitted without affecting other sections.

Sample courses for three-semester or three-quarter programs

Suggested chapters	Features	Notes
I. 1, 2, 3, 4, 5/5S, 6 II. 7, 8, 9–1, 10, 11, 12–3, 12–4 III. 13, 14, 15, 16–1, 16–2, 17–1	Basic calculus to be followed by a second year of linear algebra, differential equations and vector analysis. Accelerated course. Science-engineering emphasis.	(a) More on limits can be done as time allows by moving 12–1 and 12–2 into term I. (b) Portions of Chapter 11 can be omitted and replaced by Chapter 9 if less integration is desired.
I. 1, 2, 3, 4, 5, 7, 5S II. 6, 8, 10, 11, 12–3, 12–4 III. 13, 14, 15, 16–1, 16–2, 17–1	First term free of trigonometric functions for business, architecture audiences. Accelerated course to be followed by one year of linear algebra, differential equations and vector analysis.	(a) Sections 12–1 and 12–2 can be added as time allows in any term. (b) Chapter 9 can be added as soon as Chapter 8 is completed.

Suggested chapters	Features	Notes
I. 1, 2, 3, 4, 5, 5S II. 6, 7, 8, 9, 10 III. 11, 12, 13, 14,	Completes Calculus AB in terms I, II and Calculus BC in terms I, II, III for easy placement of transfer students. One-variable calculus only, plus vectors.	Chapters 9 and 12 may be reordered in terms II and III to place more theory in term II.
I. 1, 2, 3, 4, 5, 5S II. 6, 7, 8, 10, 11 III. 9, 12, 13, 14, 15	Accelerated one-variable calculus with three-dimensional geometry and an introduction to several-variables and vectors.	(a) Chapter 9 and Sections 12–1, 12–2 and 14–3 can be omitted from term III without loss of continuity. (b) Enrichment from appendices may be desirable in terms I and II.

Sample courses for four-semester or four-quarter programs

Suggested chapters	Features	Notes
I. 1, 2, 3, 4, 5/5S II. 6, 7, 8, 9, 10 III. 11, 12, 13, 14 IV. 15, 16, 17, 18	Integration in term I, differential equations in term II. Transcendental function exposure in term I. Science-engineering emphasis.	(a) A preview of term II is given in 5S. (b) Chapters 6, 7 can be done in any order. (c) Sections 12–1, 12–2 can be omitted or covered quickly to allow parts of term II to be moved to term III.
I. 1, 2, 3, 5, 6, II. 7, 4, 8, 9, 10 III. 11, 12, 13, 14 IV. 15, 16, 17, 18	Integration done as a separate topic in terms II and III. Trigonometric functions in term I. Early differential equations.	(a) Chapters 6, 7 can be interchanged in terms I and II. (b) Chapter 5S can be added to term I to preview Chapter 7 and to review Chapter 6.
I. 1, 2, 3, 5, 6, 5S II. 7, 4, 8, 10, 11 III. 9, 12, 13, 14 IV. 15, 16, 17, 18	Differentiation and integration done separately in terms I, II, which parallels high-school Calculus AB. Terms I, II, III cover Calculus BC.	(a) Chapter 12 can be split between two or more terms. (b) Placement of transfer students is easiest with this program.

Sample course for intensive calculus of one variable (summer school)

First session   Chapters 1, 2, 3, 5, 5S, 4, 6, 7
Second session   Chapters 8, 9, 10, 11, 12, 13

Summary of chapter interdependence*

Chapter	Prerequisite chapter(s)	Chapter	Prerequisite chapter(s)	Chapter	Prerequisite chapter(s)
1	R–1, R–2, R–3	7	5	14	R
2	1	8	4, 6–1, 6–2, 7	15	11, 14–1, 14–2
3	2	9	8	16–1, 16–2	15
4	2	10	8	16–3	3, 6–3, 16–1, 16–2
5	2	11	8	16–4	14, 16–1, 16–2
5S	5	12–1, 12–2	2	17	10, 16–1, 16–2
6–1, 6–2	5	12–3 and 12–4	8	18–1	15
6–3	3, 6–1, 6–2	13	8, 12–1	18–2	17, 18–1

*See the instructor's guide for more information.

Prerequisites and Preliminaries

An historical introduction to calculus is designed to orient students before technical material begins.

Prerequisite material from algebra, trigonometry and analytic geometry appears in Chapters R, 6, 14 and 15. These topics are treated completely; however, high school algebra is only lightly reviewed and knowledge of some plane geometry, such as the study of similar triangles, is assumed.

Analytic geometry and trigonometry are treated in enough detail to serve as a first introduction to the subjects.

Placement of Differentiation and Integration

Differentiation is introduced in Chapter 1 and is developed in Chapters 2, 3, 5, 5S, 6 and 7. Integration is introduced in Chapter 4 and is developed in Chapters 8, 10 and 11. Antidifferentiation is studied in Chapter 1. Chapter 4 can be placed anywhere between Chapters 2 and 8; the other chapters are written so this can be done with no modifications necessary. Continuity is maintained by the presence of antiderivatives in Chapters 5, 6 and 7.

Placement of the Transcendental Functions

If you treat integration late, there is normally no difficulty treating the trigonometric functions in the first semester. If you treat integration early, a supplement to Chapter 5 enables you to give the students a working knowledge of the transcendental functions in the first semester. This can also serve as a bridge to the second semester, and can give students needed practice with the differentiation rules, especially the chain rule.

Early Differential Equations

The occurrence of differential equations in Chapter 9 enables one to treat many interesting applications in the second semester. The topics in Section 9–1 are simple harmonic motion and growth and decay, topics that most people will wish to treat in detail. Section 9–2 treats separable and linear equations and applications. This chapter can be entirely or partially postponed without affecting the subsequent chapters. Exact equations appear in Chapter 18.

Limits

Limits first appear in Section 1–2 to introduce the derivative. This first encounter with limits is presented at an intuitive level. Notational precision is maintained, but we have placed the definition and further theory in Chapter 12. A deep study of limits can be undertaken any time after Chapter 2 by covering Sections 12–1, 12–2 and a portion of 12–3.

Matrix Algebra

Linear algebra does not appear as a separate topic. Instead, we include facts about 2×2 and 3×3 determinants (Chapter 14) and develop matrix multiplication separately (Chapter 16). This approach entirely avoids abstract linear algebra.

FORMAT AND SPECIAL FEATURES

Chapter Structure

The structure of the chapters is designed to guide the student through the material and to encourage "learning by doing."

The chapters are divided into numbered sections. The opening page of each chapter is intended to orient the student. After the chapter title, there is a one-sentence "headline" which presents some essential point in the chapter. This is followed by a paragraph describing the chapter and its relationship to other chapters. Following a detailed table of contents there is a paragraph describing the sections comprising the chapter. Each chapter ends with a set of *Review Problems.*

Section Structure

Each section begins with another "headline" followed by a summary of the contents of the section and its relation to other sections. Then there is a list of goals, containing the most important skills to be mastered in the section. Many instructors will wish to demand more than the contents of our goals.

Each section is divided into subsections, each of which is short enough so that it should never be necessary to split a subsection for a given day's assignment. Each section ends with a set of problems suitable for homework assignment.

Subsection Structure

A subsection normally begins with the statement of a problem to be solved or a "real-world" example. Then comes the mathematical development, including *Worked Examples.* Summary displays occur throughout the subsection, highlighting the main points. These are useful for review and help students find relevant material for problem solving. We give plenty of *Exercises* and *Problems* that can be solved by the same methods used in the examples, but we have avoided a "cook-book" attitude.

At the end of each subsection are *Solved Exercises* and *Exercises.* These are intended to help the student master the material in small pieces and to "warm-up" for doing the problems. These exercises can be used for individual study, daily homework, or quizzes.

Answers and Solutions

Complete solutions to the Solved Exercises are given in Appendix A. They are physically separated from the statements of the Exercises to encourage the student to try to solve the Exercises before checking the solutions.

Answers and hints to all the exercises are found in the Student Guide.

Answers to all the odd-numbered problems are found in Appendix C.

Special Symbols

A star (★) indicates the most demanding homework problems. *Calculator problems* and general calculator discussions are marked with the special calculator symbol ▦. Theorem statements and the ends of proofs are signaled by horizontal lines. Small type and vertical lines in the margin indicate optional material.

Precalculus Quizzes

Several *orientation quizzes* with answers and a *review section* (Chapter R) contribute to bridging the gap between previous training and this book. Students are advised to assess themselves and to take a precalculus course if they lack the background.

SUPPLEMENTS

Student Guide

Contains
  • Answers and hints to the Exercises
  • Sample exams
  • Proofs and technical appendices

Instructor's Guide

Contains
  • Suggestions for the instructor, section by section
  • Sample exams
  • Supplementary answers

Calculus Unlimited

Companion paperback primarily for honors students and instructors who wish to see calculus developed from a non-traditional point of view without using limits.

Misprints are a plague to authors (and readers) of mathematical textbooks. We have made a special effort to weed them out and we will be grateful to the readers who help us eliminate any that remain.

Jerry Marsden
Berkeley, California

Alan Weinstein
Pasadena, California

How to Use This Book: A Note to the Student

Begin by orienting yourself. Try to get a rough feel for what we are trying to accomplish in calculus by rapidly reading the introduction and by looking at some of the chapter headings.

Next, make a preliminary assessment of your own preparation for calculus by taking the self-assessment tests at the beginning of Chapter R. If you need to, study the first three sections of Chapter R in detail and begin reviewing trigonometry (Section 6–1) as soon as possible.

You can learn a little bit about calculus by reading this book, but you can learn to use calculus only by practicing it yourself. The book is organized with the intention of encouraging and facilitating practice. At several points in each section of each chapter, the text is interrupted by *Worked Examples, Solved Exercises* and *Exercises.*

The Worked Examples and Solved Exercises, together with their solutions, are an essential part of the book—you should read them all. Each Worked Example is followed immediately by a solution and is intended to reinforce an important textual point.

The solutions to all the Solved Exercises are collected at the end of the book in Appendix A. We suggest that you first try to work each Solved Exercise without consulting the solution, but that you then read the solution provided, as it may contain important comments on the material.

The Solved Exercises are followed by Exercises, which are similar in nature to the Solved Exercises, but for which solutions are not given in the text. You should work as many Exercises as you need in order to gain confidence with the subject matter. Answers to all the exercises are contained in the Student Guide.

At the end of each section is a list of *Problems.* These cover material in the entire section and range in difficulty from easy to challenging. Many teachers will make up homework assignments from the Problems. Finally, there are Review Problems at the end of each chapter, which cover material in the entire chapter. Answers to all the odd-numbered problems are found at the back of the book.

In order to prepare for examinations, try reworking the Worked Examples and Solved Exercises in the text and the Sample Examinations in the Student Guide without looking at the solutions. Be sure that you can do all of the assigned homework problems.

When writing solutions to homework or exam problems, you should use the English language liberally and correctly. A page of disconnected formulas with no explanatory words is impossible to comprehend.

We have written the book with your needs in mind. Please inform us of shortcomings you have found so we can correct them for future students. We wish you luck in the course and hope you find the study of calculus stimulating, enjoyable, and useful.

Jerry Marsden
Alan Weinstein

Introduction

Calculus has earned a reputation for being an essential tool in the sciences. Our aim in this introduction is to give the reader an idea of what calculus is all about and why it is useful.

Calculus has two main divisions, called differential calculus and integral calculus. We shall give a sample application of each of these divisions, followed by a discussion of the history and theory of calculus.

DIFFERENTIAL CALCULUS

The graph in Fig. I-1 portrays the variation of the temperature y (in degrees Celsius) with the time x (in hours from midnight) on an October day in New Orleans.

Fig. I-1 Temperature in °C as a function of time.

Each point on the graph indicates the temperature at a particular time. For example, at $x = 12$ (noon), the temperature was 15°C. The fact that for each x there is an associated y means that y is a *function of x.*

The graph as a whole can reveal information more readily than a table. For example, we can see at a glance that, from about 5 A.M. to 2 P.M., the temperature was rising, and that at the end of this period the maximum temperature for the day was reached. At 2 P.M. the air cooled (perhaps due to a brief shower), although the temperature rose again later in the afternoon. We also see that the lowest temperature occurred at about 5 A.M.

We know that the sun is highest at noon, but the highest temperature did not occur until 2 hours later. How, then, is the high position of the sun at noon reflected in the shape of the graph? The answer lies in the concept of *rate of change*, which is the central idea of differential calculus.

At any given moment of time, we can consider the rate at which temperature is changing with respect to time. What is this rate? If the graph of temperature against time were a segment of straight line, as it is in Fig. I-2, the answer would be easy. If we compare the temperature measurement at times x_1 and y_2, the ratio $(y_2 - y_1)/(x_2 - x_1)$ of change in temperature to change in time, measured in degrees per hour, is the rate of change. It is a basic property of straight lines that this ratio, called the *slope* of the line, does not depend upon which two points are used to form the ratio.

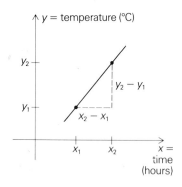

Fig. I-2 The ratio $(y_2 - y_1)/(x_2 - x_1)$ is the ratio of change of temperature with respect to time.

Returning to Fig. I-1, we may ask what the rate of change of temperature was with respect to time at noon. We cannot simply take a ratio $(y_2 - y_1)/(x_2 - x_1)$, since the graph is no longer a straight line, and the answer would depend on which points on the graph we chose. One solution to our problem is to draw the line l which best fits the graph at the point $(x, y) = (12, 15)$, and to take the slope of this line (see Fig. I-3). The line l is called the *tangent line* to the temperature curve at $(12, 15)$; its slope can be measured with a ruler to be about 2°C per hour. By drawing tangent lines to the curve at other points, the reader will find that for no other point is the slope of the tangent line as great as 2°C per hour. Thus, the high position of the sun at noon is reflected by the fact that the rate of change of temperature with respect to time was greatest then.

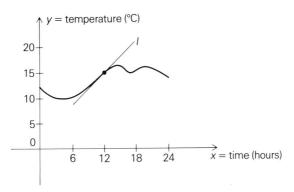

Fig. I-3 The rate of change of temperature with respect to time when $x = 12$ is the slope of the line l.

The example just given shows the importance of rates of change and tangent lines, but it leaves open the question of just what the tangent line *is*. Our definition of the tangent line as the one which "best fits" the curve leaves much to be desired, since it appears to depend on the judgment of the person drawing the line. Giving a mathematically precise definition of the tangent line to the graph of a function in the xy plane is the first step in the development of *differential calculus*. The slope of the tangent line, which represents the rate of change of y with respect to x, is called the *derivative* of the function. The process of determining the derivative is called *differentiation*.

The principal tool of differential calculus consists of a series of rules which enable us to compute a formula for the rate of change of y with respect to x, given a formula for y in terms of x. For instance, if $y = x^2 + 3x$, the derivative at x turns out to be $2x + 3$. These rules were discovered by Isaac Newton (1642–1727) in England and, independently, by Gottfried Leibniz (1642–1716), a German working in France. Newton and Leibniz had many precursors. The ancient Greeks, notably Archimedes of Syracuse (287–212 B.C.), knew how to construct the tangent lines to parabolas, hyperbolas, and certain spirals. They were, in effect, computing derivatives. After a long period with little progress, development of Archimedes' ideas were revived around 1600. By the middle of the 17th century, mathematicians could differentiate powers (i.e., the functions $y = x, x^2, x^3$, and so on) and some other functions, but a *general* method, which could be used by anyone with a little training, was first developed by Newton and Leibniz in the 1670s. Thanks to their work, it is no longer necessary to be a gifted mathematician, with a lot of time on your hands, to differentiate functions.

INTEGRAL CALCULUS AND THE FUNDAMENTAL THEOREM

The second fundamental operation of calculus is called *integration.* To illustrate this operation, we consider another question about Fig. I-1: what was the *average* temperature on this day?

We know that the average of a list of numbers is found by adding the entries in the list and then dividing by the number of entries. In the problem at hand, though, we do not have a finite list of numbers, but rather a continuous graph.

As we did with rates of change, let us look at a simpler example. Suppose that the temperature changed by jumps every two hours, as in Fig. I-4. Then we could simply add the 12 temperature readings and divide by 12 to get the average.

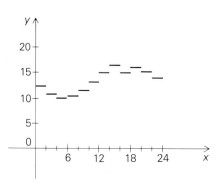

Fig. I-4 If the temperature jumps, the average is easy to find.

We can interpret this averaging process graphically in the following way. Let $y_1, \ldots, y_{12}$ be the 12 temperature readings, so that their average is $y_{\text{ave}} = (1/12)(y_1 + \cdots + y_{12})$. The region under the graph, shaded in Fig. I-5, is composed of 12 rectangles. The area of the ith rectangle is (base) $\times$ (height) $= 2y_i$), so the total area is $A = 2y_1 + 2y_2 + \cdots + 2y_{12} = 2(y_1 + \cdots + 2y_{12})$. Comparing this with the formula for the average, we find that $y_{\text{ave}} = A/24$. In other words, the average temperature is equal to the area under the graph, divided by the length of the time interval. Now we can guess how to define the average temperature for Fig. I-1. It is simply the area of the region under the graph (shaded in Fig. I-6) divided by 24.

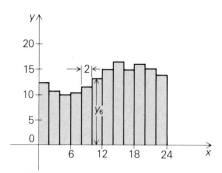

Fig. I-5 The area of the ith rectangle is $2y_i$.

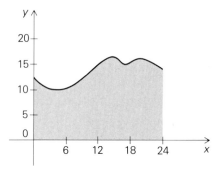

Fig. I-6 The average temperature is $1/24$ times the shaded area.

The area under the graph of a function on an interval is called the *integral* of the function over the interval. Finding integrals, or *integrating,* is the subject of the *integral calculus.*

Progress in integration was parallel to that in differentiation, and eventually the two problems became linked. The ancient Greeks knew the area of simple geometric figures, like circles and segments of parabolas. By the middle of the seventeenth century, areas under the graph of x, x^2, x^3, and so on, and other functions could be calculated. Mathematicians at that time realized that the slope and area problems were related. Newton and Leibniz formulated this relation precisely in the form of the *fundamental theorem of calculus,* which states that integration and differentiation are inverse operations. To give the reader the idea behind this theorem, we simply observe that if a list of numbers $b_1, b_2, \ldots, b_n$ is given, and the differences $d_1 = b_2 - b_1, d_2 = b_3 - b_2, \ldots,$ $d_{n-1} = b_n - b_{n-1}$ are taken (this corresponds to differentiation), then we can recover the original list from the d_i's and the initial entry b_1 by adding (this corresponds to integration): $b_2 = b_1 + d_1, b_3 = b_1 + d_1 + d_2, \ldots, b_n = b_1 +$ $d_1 + d_2 + \cdots + d_{n-1}$.

The fundamental theorem of calculus, together with the rules of differentiation, bring the solution of many integration problems within reach of anyone who has learned the differential calculus.

The importance and applicability of calculus lies in the fact that a wide variety of quantities are related by the operations of differentiation and integration. Some examples are listed in Fig. I-7.

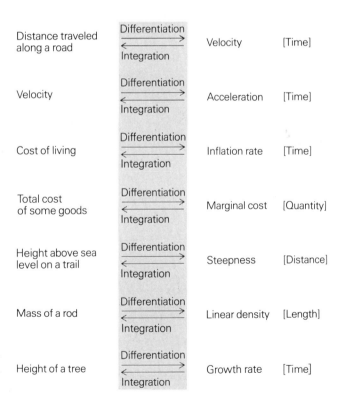

Fig. I-7 Quantities related by the operations of calculus. (The independent variable is in brackets.)

The primary aim of this book is to help you learn *how* to carry out the operations of differentiation and integration and *when* to use them in the solution of many types of problems.

THE THEORY OF CALCULUS

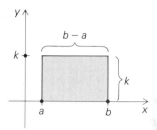

Fig. I-8 The integral of the constant function, $y = k$ over the interval $[a, b]$, is just the area $k(b - a)$ of this rectangle.

We shall describe three approaches to the theory of calculus. It will be simpler, as well as more faithful to history, if we begin with integration.

The simplest function to integrate is a constant $y = k$. Its integral over the interval $[a, b]$ is simply the area $k(b - a)$ of the rectangle under its graph (see Fig. I-8). Next in simplicity are the functions whose graphs are composed of several horizontal straight lines, as in Fig. I-9. Such functions are called *piecewise constant*. The integral of such a function is the sum of the areas of the constituent rectangles, and so it can be computed easily.

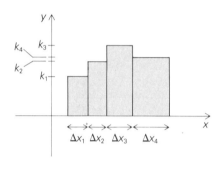

Fig. I-9 The integral over $[a, b]$ of this piecewise constant function is

$$k_1\Delta x_1 + k_2\Delta x_2 + k_3\Delta x_3 + k_4\Delta x_4,$$

where k_i is the value of y on the ith interval, and Δx_i is the length of that interval.

There are three ways to pass from the simple problem of integrating piecewise constant functions to the interesting problem of integrating more general functions, like $y = x^2$ or the function in Fig. I-1. These three ways are as follows:

1. **The method of exhaustion** This method was invented by Eudoxus of Cnidus (408–355 B.C.) and was exploited by Archimedes of Syracuse (287–212 B.C.) to calculate the areas of circles, parabolic segments, and other figures. In terms of functions, the basic idea is to *compare* the function to be integrated with piecewise constant functions. In Fig. I-10, we show the graph of $y = x^2$ on $[0, 1]$, and piecewise constant functions whose graphs lie below and above it. Since a figure inside another figure has a smaller area, we may conclude that the integral of $y = x^2$ on $[0, 1]$ lies between the integrals of these two piecewise constant functions. In this way, we can get lower and upper estimates for the integral. By choosing piecewise constant functions with shorter and shorter "steps," it is reasonable to expect that we can *exhaust* the area between the rectangles and the curve and, thereby, calculate the area to any accuracy desired. By reasoning with arbitrarily small steps, we can in some cases determine the exact area—that is just what Archimedes did.

Fig. I-10 The integral of $y = x^2$ on the interval $[0, 1]$ lies between the integrals of the two piecewise constant functions.

 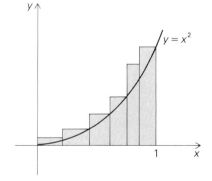

2. **The method of limits**   This method was fundamental in the seventeenth-century development of calculus and is the one which is most important today. Instead of comparing the function to be integrated with piecewise constant functions, we *approximate* it by piecewise constant functions, as in Fig. I-11. If, as we allow the steps to get shorter and shorter, the approximation gets better and better, we say that the integral of the given function is the *limit* of these approximations.

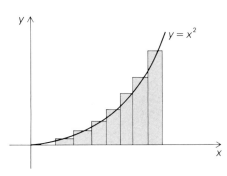

Fig. I-11 The integral of the piecewise constant function is an approximation for the integral of x^2.

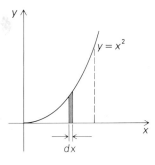

Fig. I-12 The integral of x^2 on [0, 1] may be thought of as the sum of the areas of infinitely many rectangles of infinitesimal width dx.

3. **The method of infinitesimals**   This method, too, was invented by Archimedes, but he kept it for his personal use since it did not meet the standards of rigor demanded at that time. (Archimedes' use of infinitesimals was not discovered until 1906. It was found as a *palimpsest*, a parchment which had been washed and reused for some religious writing.)* The infinitesimal method was also used in the seventeenth century, especially by Leibniz. The idea behind the infinitesimal method is to consider any function as being a piecewise constant function whose graph has infinitely many steps, each of them infinitely small, or *infinitesimal*, in length. It is impossible to represent this idea faithfully by a drawing, but Fig. I-12 suggests what is going on.

Each of these three methods—exhaustion, limits, and infinitesimals—has its advantages and disadvantages. The method of exhaustion is the easiest to comprehend and to make rigorous, but it is usually cumbersome in applications. Limits are much more efficient for calculation, but their theory is considerably harder to understand; indeed, it was not until the middle of the nineteenth century with the work of Augustin-Louis Cauchy (1789–1857) and Karl Weierstrass (1815–1897), among others, that limits were given a firm mathematical foundation. Infinitesimals lead most quickly to answers to many problems, but the idea of an "infinitely small" quantity is hard to comprehend fully,[†] and the method can lead to wrong answers as well as correct ones if it is not used carefully. The mathematical foundations of the method of infinitesimals were not established until the twentieth century, with the work of the logician, Abraham Robinson (1918–1974).[‡]

---

*See S. H. Gould, The Method of Archimedes, *American Mathematical Monthly* 62 (1955), 973–976.

[†]*An early critic of infinitesimals was Bishop George Berkeley, who referred to them as "ghosts of departed quantities" in his anticalculus book,* The Analyst *(1734). The city in which this calculus book has been written is named after him.*

[‡]*A calculus textbook based upon this work is H. J. Keisler,* Elementary Calculus, *Prindle, Weber, and Schmidt, Boston (1976).*

The three methods used to define the integral can be applied to differentiation as well. In this case, we replace the piecewise constant functions by the linear functions $y = mx + b$. For a function of this form, a change of Δx in x produces a change $\Delta y = m\Delta x$ in y, so the rate of change, given by the ratio $\Delta y/\Delta x$, is equal to m, independent of x and of Δx (see Fig. I-13).

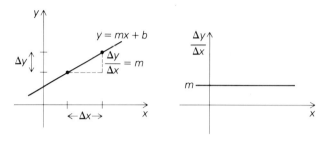

Fig. I-13 If $y = mx + b$, the rate of change of y with respect to x is constant and equal to m.

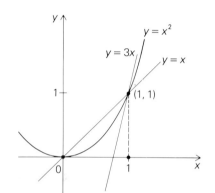

Fig. I-14 The rate of change of $y = x^2$ at $x = 1$ lies between 1 and 3.

1. **The method of exhaustion** To find the rate of change of a general function, we may compare the function with linear functions by seeing how straight lines with various slopes cross the graph at a given point. In Fig. I-14, we show the graph of $y = x^2$, together with lines which are more and less steep at the point $x = 1$, $y = 1$. By bringing our comparison lines closer and closer together, we can calculate the rate of change to any accuracy desired; if the algebra is simple enough, we can even calculate the rate of change exactly.

The historical origin of this method can be found in the following definition of tangency used by the ancient Greeks: "the tangent line touches the curve, and in the space between the line and curve, no other straight line can be interposed."*

2. **The method of limits** To approximate the tangent line to a curve we draw the *secant* line through two nearby points. As the two points become closer and closer, the slope of the secant approaches a limiting value which is the rate of change of the function (see Fig. I-15).

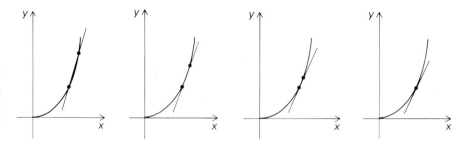

Fig. I-15 The rate of change of a function is the limit of the slopes of secant lines drawn through two points on the graph.

---

*See C. Boyer, The History of the Calculus and Its Conceptual Development, *Dover, N.Y. p. 57.*
The method of exhaustion is not normally used in calculus courses for differentiation and this book is no exception. However, it could be used and it is intellectually satisfying to do so; see Calculus Unlimited, *Benjamin/Cummings (1980) written by the present authors.*

This approach to rates of change derives from the work of Pierre de Fermat* (1601–1665), whose interest in tangents arose from the idea, due originally to Kepler, that the slope of the tangent line should be zero at a maximum or minimum point (Fig. I-16).

Fig. I-16 The slope of the tangent line is zero at a maximum or minimum point.

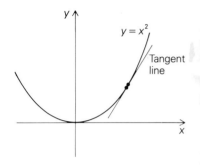

Fig. I-17 The tangent line may be thought of as the secant line through a pair of infinitesimally near points.

3. **The method of infinitesimals** In this method, we simply think of the tangent line to a curve as a secant line drawn through two infinitesimally close points on the curve, as suggested by Fig. I-17. This idea seems to go back to Galileo† (1564–1642) and his student Cavalieri (1598–1647), who defined instantaneous velocity as the ratio of an infinitely small distance to an infinitely short time.

As with integration, infinitesimals lead most quickly to answers (but not always the right ones), and the method of exhaustion is conceptually simplest. Because of its computational power, the method of limits has become the most widely used of the approaches to differential calculus. It is this method which we shall use in this book.

THE POWER OF CALCULUS (THE CALCULUS OF POWER)

To end this introduction, we shall give an example of a practical problem which calculus can help us to solve.

The sun, which is the ultimate source of nearly all of the earth's energy, has always been an object of fascination to mankind. The relation between the sun's position and the seasons was predicted by early agricultural societies, some of which developed quite sophisticated astronomical techniques. Today, as the earth's resources of fossil fuels dwindle, the sun has new importance as a *direct* source of energy. To use this energy efficiently, it is useful to know just how much solar radiation is available at various locations at different times of year.

From basic astronomy we know that the earth revolves about the sun while rotating about an axis inclined at 23.5° to the plane of its orbit (see

---

*Fermat is also famous for his work in number theory. Fermat's last theorem: "If n is an integer greater than 2, there are no positive integers x, y, and z such that $x^n + y^n = z^n$," remains unproven today. Fermat claimed to have proved it, but his proof has not been found, and most mathematicians now doubt that it could have been correct.

†Newton's acknowledgment, "If I have seen further than others, it is because I have stood on the shoulders of giants," probably refers chiefly to Galileo, who died the year Newton was born. (A similar quotation from Lucan (39–65 A.D.) was cited by Robert Burton in the early 1600s—"Pygmies see further than the giants on whose shoulders they stand.")

Fig. I-18). Even assuming idealized conditions, such as a perfectly spherical earth revolving in a circle about the sun, it is not a simple matter to predict the length of the day or the exact time of sunset at a given latitude on the earth on a given day of the year.

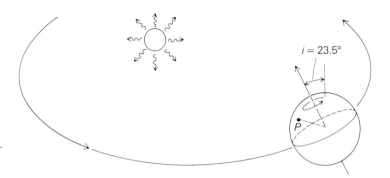

Fig. I-18 The earth revolving about the sun.

In 1857, an American scientist named L. W. Meech published in the *Smithsonian Contributions to Knowledge* (Volume 9, Article II) a paper entitled "On the relative intensity of the heat and light of the sun upon different latitudes of the earth." Meech was interested in determining the extent to which the variation of temperature on the surface of the earth could be correlated with the variations of the amount of sunlight impinging on different latitudes at different times. One of Meech's ultimate goals was to predict whether or not there was an open sea near the north pole—a region then unexplored. He used the integral calculus to sum the total amount of sunlight arriving at a given latitude on a given day of the year, and then he summed this quantity over the entire year. Meech found that the amount of sunlight reaching the atmosphere above polar regions was surprisingly large during the summer due to the long days (see Fig. I-19). The differential calculus is used to predict the shape of graphs like those in Fig. I-19 by calculating the slopes of their tangent lines.

Meech realized that, since the sunlight reaching the polar regions arrives at such a low angle, much of it is absorbed by the atmosphere, so one cannot conclude the existence of "a brief tropical summer with teeming forms of vegetable and animal life in the centre of the frozen zone." Thus, Meech's calculations fell short of permitting a firm conclusion as to the existence or not of an open sea at the North Pole, but his work has recently taken on new importance. Graphs like Fig. I-19 have appeared in books devoted to meteorology, geology, ecology (with regard to the biological energy balance), and solar energy engineering.

Even if one takes into account the absorption of energy by the atmosphere, on a summer's day the middle latitudes still receive more energy at the earth's surface than does the equator. In fact, the hottest places on earth are not at the equator but in a band north and south of the equator. (This is enhanced by climate: the low-middle latitudes are much freer of clouds than the equatorial zone.)*

As we carry out our study of calculus in this book, we will from time to time reproduce parts of Meech's calculations (slightly simplified) to show how the material being learned may be applied to a substantial problem. By the time you have finished this book, you should be able to read Meech's article by yourself.

---

*According to the "Guinness Book of World Records," the world's highest temperatures have occurred at Ouargla, Algeria (latitude 32°N), Death Valley, California (latitude 36°N), and Al'Aziziyah, Libya (latitude 32°N). Locations in Chile, Southern Africa, and Australia approach these records.

Fig. I-19 The sun's diurnal intensity along the meridian, at intervals of thirty days.

R

Review of Fundamentals

Functions are to calculus what numbers are to algebra.

Success in the study of calculus depends upon a solid understanding of algebra and analytic geometry. In this chapter, we review those topics from these preparatory subjects which are particularly important for calculus.

CONTENTS

This chapter begins with a discussion of the algebra of real numbers. We emphasize fundamental manipulative skills important in calculus and in analytic geometry, such as completing the square and transforming inequalities. We go on in Sections R-2 and R-3 to review the basic analytic geometry of lines, circles, parabolas, and graphs of functions. The conic sections are discussed in Section R-4—this section is not immediately essential for calculus and may be postponed. Students with good preparation may skip directly to Chapter 1, referring back to this chapter when necessary.

Orientation Quizzes

To test your preparation, take Quiz A below. If you score at least 8/10, you are probably ready for Chapter 1. If you score less than 8/10, take Quizzes B and C. If you pass them both, a thorough review of the first three sections of this chapter should suffice. If you fail Quiz B, you probably need additional work in algebra before taking calculus. If you fail Quiz C, you may need additional work in geometry and trigonometry.

Quiz A   (Passing score is 8/10. Answers are on p. C-1.)

1. What is the slope of the line $3y + 4x = 2$?

2. For which values of x is $3x + 2 > 0$?

3. For which values of x is $3x^2 - 2x - 1 > 0$?

4. Solve for x: $2x^2 + 8x - 11 = 0$.

5. Sketch the graph of $f(x) = x^2 - x - 2$.

6. Let $g(x) = (3x^2 + 2x - 8)/(2x^2 - x)$. Compute $g(2)$ and state the domain of g.

7. For what values of x is $3x + 2 > 2x - 8$?

8. Where does the graph of $f(x) = 3x - 2$ intersect the graph of $g(x) = x^2$?

9. Sketch the curve $x^2 + y^2 - 2x - 4y + 1 = 0$.

10. Find the distance between the point $(1, 1)$ and the intersection point of the lines $y = -2x + 1$ and $y = 4x - 5$.

Quiz B   (Passing score is 8/10. Answers are on p. C-1.)

1. $\frac{3}{2} + \frac{5}{4} =$ _____.

2. Factor: $x^2 + 3x =$ _____.

3. $(-6)(-3) + 8(-1) =$ _____.

4. If a bag of sand weighs 80 kilograms, 5% of the bag weighs _____.

5. $[(3x - 1)/2x] - \frac{1}{2} =$ _____. (Bring to a common denominator.)

6. $x^3 \cdot x^4 =$ _____.

7. Arrange these numbers in ascending order (smallest to largest): 8, −6, $\frac{1}{2}$, 0, −4.

8. Solve for x: $6x + 2 = -x - 1$.

9. If $x = 3$ and $y = 9$, then $\sqrt{y}/3x =$ _____.

10. $(x^2 - 16)/(x - 4) =$ _____. (Simplify.)

Quiz C   (Passing score is 7/10. Answers are on p. C-1.)

1. Find the coordinates of the point P:

2. Find x:

3. Find θ:

4. Find $\sin \theta$:

5. Find y:

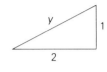

6. What is the circumference of a circle whose radius is 4 centimeters?

7. A rectangle has area 10 square meters and one side of length 5 meters. What is its perimeter?

8. The volume of a cylindrical can with radius 3 centimeters and height 2 centimeters is _____.

9. $\cos 60° = $ _____.

10. $2 \sin^2 x + 2 \cos^2 x = $ _____.

R-1 Real Numbers and Inequalities

The real numbers are ordered like the points on a line.

The most important facts about the real numbers concern algebraic operations (addition, multiplication, subtraction, and division) and order (greater and less). In this section, we review some of these facts.

Goals

After studying this section, you should be able to:

Manipulate algebraic equations and inequalities.

Identify intervals of real numbers.

Solve quadratic equations.

Use the absolute value notation.

REAL NUMBERS

The positive whole numbers 1, 2, 3, 4, ... (... means "and so on") that arise from the counting process are called the *natural numbers* by mathematicians. The arithmetic operations of addition and multiplication can be performed within the natural numbers, but the "inverse" operations of subtraction and division lead to the introduction of zero $(3 - 3 = 0)$, negative numbers $(2 - 6 = -4)$, and fractions $(3 \div 5 = \frac{3}{5})$. The whole numbers, positive, zero, and negative, are called *integers*. All numbers which can be put in the form m/n, where m and n are integers, are called *rational numbers*.

Worked Example 1   Fill in each box in the following table with T (true) or F (false):

	Is a natural number	Is an integer	Is a rational number
0			
$3 - [(4 + 5)/2]$			
$7 - 6$			
$(4 + 5) / (-3)$			

Solution   The correct entries are:

F	T	T
F	F	T
T	T	T
F	T	T

For example, $(4 + 5) \div (-3)$ is equal to -3, which is an integer but not a natural number. Every integer m is a rational number as well, since it can be written as $m/1$.

The ancient Greeks already knew that lines in simple geometric figures could have lengths which did not correspond to ratios of whole numbers. For instance, neither the length $\sqrt{2}$ of the diagonal of a square with sides of unit length, nor the area π of a circle with unit radius, can be expressed in the form m/n, with m and n integers.* Numbers which are not ratios of integers are called *irrational numbers*. These, together with the rational numbers, comprise the *real numbers*.

The usual arithmetic operations of addition, multiplication, subtraction, and division (except by zero) may be performed on real numbers, and these operations satisfy the usual algebraic rules. You are familiar with these rules; some examples are "if equals are added to equals, the results are equal," "$a + b = b + a$," and "if $ab = ac$ and $a \neq 0$, we can divide both sides by a to conclude that $b = c$."

The worked examples and exercises below review some of the algebra skills which we will use.

Worked Example 2   Solve the equation $3x + 2 = 8$ for x.

Solution   Subtracting 2 from both sides of the equation gives

$$3x + 2 - 2 = 8 - 2$$
$$3x = 6$$

Dividing both sides by 3 gives $x = 2$.

The irrationality of $\sqrt{2}$ was proved by Euclid around 300 B.C.. The irrationality of π was not proved until around 1880.

Worked Example 3 Solve for x: $(3x + 1) + 8(x + 9) = 4$.

Solution We write down a string of equations, each equivalent to the next:

$$(3x + 1) + 8(x + 9) = 4$$
$$3x + 1 + 8x + 72 = 4$$
$$11x + 73 = 4$$
$$11x = -69$$
$$x = -\frac{69}{11}$$

Thus $x = -69/11$ is the solution.

Worked Example 4 Simplify: $(a + b)(a - b) + b^2$.

Solution Since $(a + b)(a - b) = a^2 - b^2$ (this fact will be used often), we have $(a + b)(a - b) + b^2 = a^2 - b^2 + b^2 = a^2$.

Worked Example 5 Expand: $(a + b)^3$.

Solution By the rule for the square of a binomial, we have $(a + b)^2 = a^2 + 2ab + b^2$. Therefore

$$(a + b)^3 = (a + b)^2(a + b)$$
$$= (a^2 + 2ab + b^2)(a + b)$$
$$= (a^2 + 2ab + b^2)a + (a^2 + 2ab + b^2)b$$
$$= a^3 + 2a^2b + ab^2 + a^2b + 2ab^2 + b^3$$
$$= a^3 + 3a^2b + 3ab^2 + b^3$$

Worked Example 6 Factor: $2x^2 + 4x - 6$.

Solution We notice first that $2x^2 + 4x - 6 = 2(x^2 + 2x - 3)$. Using the fact that the only integer factors of -3 are ± 1 and ± 3, we find by trial and error that $x^2 + 2x - 3 = (x + 3)(x - 1)$, so we have $2x^2 + 4x - 6 = 2(x + 3)(x - 1)$.

Exercises*
1. State whether or not each of the following is a natural number, an integer, or a rational number:

   (a) $\dfrac{8}{6} - \dfrac{9}{4}$ (b) $(-1) \div (-1)$

   (c) $\left(\dfrac{1}{\sqrt{2}} + \dfrac{1}{\sqrt{3}}\right)\left(\dfrac{1}{\sqrt{2}} - \dfrac{1}{\sqrt{3}}\right)$ (d) $\pi - \dfrac{1}{2}$

2. Simplify:
   (a) $(a - 3)(b + c) - (ac + 2b)$ (b) $(b^2 - a)^2 - a(2b^2 - a)$
   (c) $a^2c + (a - b)(b - c)(a + b)$ (d) $(3a + 2)^2 - (4a + b)(2a - 1)$

3. Expand:
   (a) $(a - b)^3$ (b) $(3a + b^2 + c)^2$ (c) $(b + c)^4$

Answers to the exercises are found in the Student Guide.

4. Factor:

(a) $x^2 + 5x + 6$ (b) $x^2 - 5x + 6$

(c) $x^2 - 5x - 6$ (d) $x^2 + 5x - 6$

(e) $3x^2 - 6x - 24$ (f) $-5x^2 + 15x - 10$

(g) $x^2 - 1$ (h) $4x^2 - 9$

5. Solve for x:

(a) $2(3x - 7) - (4x - 10) = 0$ (b) $3(3 + 2x) + (2x - 1) = 8$

(c) $(2x + 1)^2 + (9 - 4x^2) + (x - 5) = 10$

6. (a) Verify that $x^3 - 1 = (x - 1)(x^2 + x + 1)$.

(b) Factor $x^3 + 1$ into linear and quadratic factors.

(c) Factor $x^3 + x^2 - 2x$ into linear factors.

(d) Factor $x^4 - 2x^2 + 1$ into linear factors. [*Hint:* First consider x^2 as the variable.]

INEQUALITIES

The real numbers have a relation of order: if two real numbers are unequal, one of them is less than the other. We may represent the real numbers as points on a line, with larger numbers to the right, as shown in Fig. R-1-1. If the number a is less than b, we write $a < b$. In this case, we also say that b is greater than a and write $b > a$.

Fig. R-1-1   The real-number line.

Given any two numbers, a and b, exactly one of the following three possibilities holds:

(1) $a < b$

(2) $a = b$

(3) $a > b$

Combinations of these possibilities have special names and notations.

If (1) or (2) holds, we write $a \leq b$ and say that "a is less than or equal to b."

If (2) or (3) holds, we write $a \geq b$ and say that "a is greater than or equal to b."

If (1) or (3) holds, we write $a \neq b$ and say that "a is unequal to b."

For instance, if x is any real number, we know that $x^2 \geq 0$. If $x \neq 0$, we can make the stronger statement that $x^2 > 0$.

Worked Example 7   Tell whether each of the following statements is true or false:

(a) $3 \leq 3$ (b) $(-2)^2 \leq 0$ (c) $-\pi < -\frac{1}{2}\pi$

Solution   (a) true because $3 = 3$; that is, (2) holds;

(b) false because $(-2)^2 = 4$, which is greater than zero;

(c) true because $-\pi$ and $-\frac{1}{2}\pi$ both lie to the left of zero on the number line and since $-\frac{1}{2}\pi$ is only half as far from zero as $-\pi$, it lies to the right of $-\pi$.

Worked Example 8 Write the proper inequality sign between each of the following pairs of numbers:

(a) 0.0000025 and $-100,000$ (b) $\frac{3}{4}$ and $\frac{6}{7}$ (c) $\sqrt{12}$ and 4

Solution (a) $0.0000025 > -100,000$ since a positive number is always to the right of a negative number.

(b) $\frac{3}{4} < \frac{6}{7}$ since $\frac{3}{4} = \frac{21}{28}$ and $\frac{6}{7} = \frac{24}{28}$.

(c) $\sqrt{12} < 4$ since $12 < 4^2$.

We can summarize the most important properties of inequalities as follows:

1. If $a < b$ and $b < c$, then $a < c$.

2. If $a < b$, then $a + c < b + c$ for any c, and $ac < bc$ if $c > 0$, while $ac > bc$ if $c < 0$. (Multiplication by a negative number reverses the sign of inequality. For instance, $3 < 4$, and multiplication by -2 gives $-6 > -8$.)

3. $ab > 0$ when a and b have the same sign; $ab < 0$ when a and b have opposite signs. (See Fig. R-1-2.)

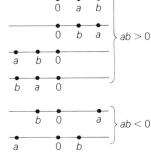

Fig. R-1-2 Possible positions of a and b when they have the same sign ($ab > 0$) or the opposite sign ($ab < 0$).

4. If a and b are any two numbers, then $a < b$ when $a - b < 0$ and $a > b$ when $a - b > 0$.

Worked Example 9 Find all numbers x for which $x^2 < 9$.

Solution We transform the inequality as follows (all steps are reversible):

$$x^2 < 9$$
$$x^2 - 9 < 0 \quad \text{(add } -9 \text{ to both sides)}$$
$$(x + 3)(x - 3) < 0 \quad \text{(factor)}$$

Since the product $(x + 3)(x - 3)$ is negative, the factors $x + 3$ and $x - 3$ must have opposite signs. Thus, either $x + 3 > 0$ and $x - 3 < 0$, in which case $x > -3$ and $x < 3$ (that is, $-3 < x < 3$); or $x + 3 < 0$ and $x - 3 > 0$, in which case $x < -3$ and $x > 3$, which is impossible. We conclude that $x^2 < 9$ if and only if $-3 < x < 3$.

Worked Example 10   Transform $a + (b - c) > b - a$ to an inequality with a alone on one side.

Solution   We transform by reversible steps:

$$a + b - c > b - a$$
$$2a + b - c > b \qquad \text{(add } a \text{ to both sides)}$$
$$2a - c > 0 \qquad \text{(add } -b \text{ to both sides)}$$
$$2a > c \qquad \text{(add } c \text{ to both sides)}$$
$$a > \tfrac{1}{2}c \qquad \text{(multiply both sides by } \tfrac{1}{2}\text{)}$$

Worked Example 11   Find all numbers x such that $x^2 - 2x - 3 > 0$.

Solution   The inequality $x^2 - 2x - 3 > 0$ is the same as $(x - 3)(x + 1) > 0$. That is, $x - 3$ and $x + 1$ have the same sign. There are two cases to consider:

Case 1: $x - 3$ and $x + 1$ are both positive; that is, $x - 3 > 0$ and $x + 1 > 0$; that is, $x > 3$ and $x > -1$, which is the same as $x > 3$ (since any number greater than 3 is certainly greater than -1).

Case 2: $x - 3 < 0$ and $x + 1 < 0$; that is, $x < 3$ and $x < -1$, which is the same as $x < -1$.

Thus $x^2 - 2x - 3 > 0$ whenever $x > 3$ or $x < -1$. These numbers x are illustrated in Fig. R-1-3.

Fig. R-1-3   Solution of an inequality.

Exercises   7. Simplify:
   (a) $(a - b) + c > 2c - b$ (b) $(a + c^2) + c(a - c) \geq ac + 1$
   (c) $ab - (a - 2b)b < b^2 + c$ (d) $2(a + ac) - 4ac > 2a - c$
   (e) $b(b + 2) > (b + 1)(b + 2)$

8. Put the following list of numbers in ascending order. (Try to do it without finding decimal equivalents for the numbers.)
   $-\frac{5}{3}, \quad -\sqrt{2}, \quad -\frac{7}{5}, \quad \frac{22}{7}, \quad 3, \quad \frac{23}{8}, \quad 0, \quad \frac{9}{5}, \quad -\sqrt{3}$

9. Find all numbers x such that: (a) $4x - 13 < 3$ (b) $2(7 - x) \geq x + 1$ (c) $5(x - 3) - 2x + 6 > 0$. Sketch your solutions on a number line.

10. Find all numbers x such that: (a) $2(x^2 - x) > 0$ (b) $3x^2 + 2x - 1 \geq 0$ (c) $x^2 - 5x + 6 < 0$. Sketch your solutions on a number line.

INTERVALS AND OTHER SETS OF REAL NUMBERS

Special notations for certain parts of the real numbers called *intervals* will be important throughout this book:

Fig. R-1-4 Various types of intervals.

(a, b)	means all x such that	$a < x < b$	open interval
$[a, b]$	means all x such that	$a \le x \le b$	closed interval
$(a, b]$	means all x such that	$a < x \le b$	half-open interval
$[a, b)$	means all x such that	$a \le x < b$	half-open interval
$[a, \infty)$	means all x such that	$a \le x$	half-open interval
(a, ∞)	means all x such that	$a < x$	open interval
$(-\infty, b]$	means all x such that	$x \le b$	half-open interval
$(-\infty, b)$	means all x such that	$x < b$	open interval
$(-\infty, \infty)$	means all real numbers		open interval

These collections of real numbers are illustrated in Fig. R-1-4. A black dot indicates that the corresponding endpoint is included in the interval; a white circle indicates that the endpoint is not included in the interval. Notice that a closed interval contains both its endpoints, a half-open interval contains one endpoint, and an open interval contains none.

Warning The symbol ∞ ("infinity") does not denote a real number. It is merely a placeholder to indicate that an interval extends without limit.

In the formation of intervals, we can allow $a = b$. Thus the interval $[a, a]$ consists of the number a alone (if $a \le x \le a$, then $x = a$), while (a, a), $(a, a]$, and $[a, a)$ contain no numbers at all.

Many collections of real numbers are not intervals. For example, the integers, ... $-3, -2, -1, 0, 1, 2, 3, ...$, form a collection of real numbers which cannot be designated as a single interval. The same goes for the rational numbers, as well as the collection of all x for which $x^2 - 2x - 3 > 0$. (See Fig. R-1-3.)

A collection of real numbers is also called a *set* of real numbers. Intervals are examples of sets of real numbers, but, as the preceding examples show, not every set is an interval. We will often use capital letters to denote sets of numbers. If A is a set and x is a number, we write $x \in A$ and say that "x is an element of A" if x belongs to the collection A. For example, if we write $x \in [a, b]$ (read "x is an element of $[a, b]$"), we mean that x is a member of the collection $[a, b]$; that is, $a \le x \le b$. Similar notation is used for the other types of intervals.

Worked Example 12   True or false: (a) $3 \in [1, 8]$ (b) $-1 \in (-\infty, 2)$ (c) $1 \in [0, 1)$
(d) $8 \in (-8, \infty)$?

Solution   (a) True, because $1 \le 3 \le 8$ is true;
(b) true, because $-1 < 2$ is true;
(c) false, because $0 \le 1 < 1$ is false ($1 < 1$ is false);
(d) true, because $-8 < 8$ is true.

Worked Example 13   Prove: if $a < b$, then $(a + b)/2 \in (a, b)$.

Solution   We must show that $a < (a + b)/2 < b$. For the first inequality:

$$a < b$$
$$2a < a + b \qquad \text{(adding } a\text{)}$$
$$a < \frac{a + b}{2} \qquad \text{(multiplying by } \tfrac{1}{2}\text{)}$$

The proof that $(a + b)/2 < b$ is done similarly; add b to $(a < b)$ and divide by 2. Thus $a < (a + b)/2 < b$; i.e., the average of two numbers lies between them.

Worked Example 14   Let A be the set consisting of those x for which $x^2 - 2x - 3 > 0$. Describe A in terms of intervals.

Solution   See Worked Example 11. A consists of the intervals $(-\infty, -1)$ and $(3, \infty)$.

Exercises   11. True or false:

(a) $-7 \in [-8, 1]$ (b) $5 \in (\frac{11}{2}, 6]$ (c) $4 \in (-4, 6]$ (d) $4 \in (4, 6)$?

12. Which numbers in the list -54, -9, $-\frac{2}{3}$, 0, $\frac{5}{16}$, $\frac{1}{2}$, 8, 32, 100 belong to which of the following intervals?

(a) $[-10, 1)$ (b) $(-\infty, 44)$ (c) $(75, 500)$ (d) $(-20, -\frac{2}{3})$ (e) $(-9, \frac{5}{16}]$

13. Describe the solutions of each of the following inequalities in terms of intervals:

(a) $x + 4 \geq 7$ (b) $2x + 5 \leq -x + 1$

(c) $x > 4x - 6$ (d) $5 - x > 4 - 2x$

(e) $x^2 + 2x - 3 > 0$ (f) $2x^2 - 6 \leq 0$

(g) $x^2 - x \geq 0$ (h) $(2x + 1)(x - 5) \leq 0$

THE QUADRATIC FORMULA

The quadratic formula is used to solve for x in equations of the form $ax^2 + bx + c = 0$ when the left-hand side cannot be readily factored. The method of *completing the square*, by which the quadratic formula may be derived, is often more important than the formula itself.

COMPLETING THE SQUARE

To complete the square in the expression $ax^2 + bx + c$, factor out a and then add and subtract $(b/2a)^2$. The result is

$$ax^2 + bx + c = a\left[\left(x + \frac{b}{2a}\right)^2 + \left(\frac{c}{a} - \frac{b^2}{4a^2}\right)\right]$$

Worked Example 15   Solve the equation $x^2 - 5x + 3 = 0$ by completing the square.

Solution   We transform the equation by adding and subtracting $(\frac{5}{2})^2$ on the left-hand side:

$$x^2 - 5x + \left(\frac{5}{2}\right)^2 - \left(\frac{5}{2}\right)^2 + 3 = 0$$

$$\left(x - \frac{5}{2}\right)^2 - \frac{13}{4} = 0$$

$$\left(x - \frac{5}{2}\right)^2 = \frac{13}{4}$$

$$x - \frac{5}{2} = \pm \frac{\sqrt{13}}{2}$$

$$x = \frac{5}{2} \pm \frac{\sqrt{13}}{2}$$

When the method of completing the square is applied to the *general* quadratic equation $ax^2 + bx + c = 0$, one obtains the following general formula for the solution of the equation.

QUADRATIC FORMULA

To solve $ax^2 + bx + c = 0$, where $a \neq 0$,

$$\text{put } x = \frac{-b \pm \sqrt{b^2 - 4ac}}{2a}.$$

If $b^2 - 4ac > 0$, there are two solutions.

If $b^2 - 4ac = 0$, there is one solution.

If $b^2 - 4ac < 0$, there are no solutions.

The expression $b^2 - 4ac$ is called the *discriminant*.

In case $b^2 - 4ac < 0$, there is no real number $\sqrt{b^2 - 4ac}$, because the square of every real number is greater than or equal to zero. (Square roots of negative numbers can be found if we extend the real-number system to encompass the so-called *imaginary numbers*.)* Thus the symbol $\sqrt{r}$ represents a real number only when $r \geq 0$, in which case we always take $\sqrt{r}$ to mean the *nonnegative* number whose square is r.

Worked Example 16 Solve for x: $4x^2 = 2x + 5$.

Solution Subtracting $2x + 5$ from both sides of the equation, we have $4x^2 - 2x - 5 = 0$, which is in the form $ax^2 + bx + c = 0$ with $a = 4$, $b = -2$, and $c = -5$. The quadratic formula gives

$$x = \frac{-(-2) \pm \sqrt{(-2)^2 - 4(4)(-5)}}{2(4)}$$

$$= \frac{2 \pm \sqrt{4 + 80}}{8} = \frac{2 \pm \sqrt{84}}{8} = \frac{1}{4} \pm \frac{\sqrt{21 \cdot 4}}{8} = \frac{1}{4} \pm \frac{\sqrt{21}}{4}$$

Worked Example 17 Solve for x: $x^2 - 5x + 20 = 0$.

*Imaginary numbers are discussed in the appendix to Chapter 13.

Solution   We use the quadratic formula:

$$x = \frac{5 \pm \sqrt{25 - 4 \cdot 20}}{2}$$

The discriminant is negative, so there are no real solutions.

Exercises   14. Solve the equation $x^2 + 5x + 4 = 0$ in three ways: (a) by factoring;   (b) by completing the square;   (c) by using the quadratic formula.

15. Solve for x:

(a) $x^2 + \frac{1}{2}x - \frac{1}{2} = 0$ (b) $4x^2 - 18x + 20 = 0$

(c) $-x^2 + 5x + 0.3 = 0$ (d) $5x^2 + 2x - 1 = 0$

(e) $x^2 - 5x + 7 = 0$ (f) $0.1x^2 - 1.3x + 0.7 = 0$

16. Solve for x:

(a) $x^2 + 4 = 3x^2 - x$ (b) $4x = 3x^2 + 7$

(c) $2x + x^2 = 9 + x^2$ (d) $(5 - x)(2 - x) = 1$

(e) $2x^2 - 2\sqrt{7}x + 7/2 = 0$ (f) $x^2 + 9x = 0$

ABSOLUTE VALUE

If a real number x is considered as a point on the number line, the distance between this point and zero is called the *absolute value* of x. If x is positive or zero, the absolute value of x is equal to x itself. If x is negative, however, the absolute value of x is equal to the positive number $-x$ (see Fig. R-1-5). The absolute value of x is denoted by $|x|$. For instance, $|8| = 8$, $|-7| = 7$, $|-10^8| = 10^8$.

Fig. R-1-5 The absolute value measures the distance to the origin.

ABSOLUTE VALUE

The absolute value $|x|$ of a real number x is equal to

$$\begin{cases} x & \text{if } x \geq 0 \\ -x & \text{if } x < 0 \end{cases}$$

That is, change the sign of x, if necessary, to make a positive number (or zero).

Worked Example 18   Find all x such that $|x| = 2$.

Solution   If $|x| = 2$ and $x \geq 0$, we must have $x = 2$. If $|x| = 2$ and $x < 0$, we must have $-x = 2$; that is, $x = -2$. Thus $|x| = 2$ if and only if $x = \pm 2$.

For any real number x, $|x| \geq 0$, and $|x| = 0$ exactly when $x = 0$. If b is a positive number, there are two numbers having b as their absolute value: b and $-b$. Geometrically, if $x < 0$, $|x|$ is the "mirror image" point which is obtained from x by flipping the line over, keeping zero fixed.

If x_1 and x_2 are any two real numbers, the distance between x_1 and x_2 is $x_1 - x_2$ if $x_1 > x_2$ and $x_2 - x_1$ if $x_1 < x_2$. (See Fig. R-1-6 and note that the position of zero in this figure is unimportant.) Since $x_1 \geq x_2$ if and only if $x_1 - x_2 \geq 0$, and $x_2 - x_1 = -(x_1 - x_2)$, we have the result shown in the next display.

Fig. R-1-6 The distance between x_1 and x_2 is $|x_1 - x_2|$.

DISTANCE FORMULA ON THE LINE

If x_1 and x_2 are points on the number line, the distance between x_1 and x_2 is equal to $|x_1 - x_2|$.

Worked Example 19   Describe as an interval the set of real numbers x for which $|x - 8| \leq 3$.

Solution $|x - 8| \leq 3$ means that either $x - 8 \geq 0$ and $x - 8 \leq 3$, or $x - 8 < 0$ and $-(x - 8) \leq 3$. In the first case, we have $x \geq 8$ and $x \leq 11$. In the second case, we have $x < 8$ and $x \geq 5$. Thus $|x - 8| \leq 3$ if and only if $x \in [5, 11]$.

Worked Example 20   Describe the interval $(4, 9)$ by a single inequality involving absolute values.

Solution   Let m be the midpoint of the interval $(4, 9)$; that is, $m = \frac{1}{2}(4 + 9) = \frac{13}{2}$. A number x belongs to $(4, 9)$ if and only if its distance from m is less than the distance from 9 to m. This distance is equal to $|9 - \frac{13}{2}| = \frac{5}{2}$. (Note that the distance from 4 to m is $|4 - \frac{13}{2}| = |-\frac{5}{2}| = \frac{5}{2}$ as well.) So we have $x \in (4, 9)$ if and only if $|x - \frac{13}{2}| < \frac{5}{2}$. (See Fig. R-1-7.)

Fig. R-1-7 The interval $(4, 9)$ may be described by the inequality $|x - \frac{13}{2}| < \frac{5}{2}$.

The most important algebraic properties of absolute values are listed below.

PROPERTIES OF ABSOLUTE VALUES

If x and y are any real numbers:

1. $|x + y| \leq |x| + |y|$
2. $|xy| = |x||y|$
3. $|x| = \sqrt{x^2}$

Worked Example 21   Show by example that $|x + y|$ is not always equal to $|x| + |y|$.

Solution   Let $x = 3$ and $y = -5$. Then $|x + y| = |3 - 5| = 2$, while $|x| + |y| = 3 + 5 = 8$. (Many other numbers will work as well. In fact, if $x > 0$ and $y < 0$, then $|x + y|$ will be less than $|x| + |y|$.)

Worked Example 22   Prove that $|x| = \sqrt{x^2}$.

Solution   For any number x, we have $(-x)^2 = x^2$, so $|x|^2 = x^2$ whatever the sign of x. Thus $|x|$ is a number such that $|x| \geq 0$ and $|x|^2 = x^2$, so it is the square root of x^2.

Exercises

17. Find the following absolute values:
    (a) $|3 - 5|$ (b) $|3 + 5|$ (c) $|-3 - 5|$ (d) $|-3 + 5|$
    (e) $|3 \cdot 5|$ (f) $|(-3)(-5)|$ (g) $|(-3) \cdot 5|$ (h) $|3 \cdot (-5)|$

18. Describe in terms of absolute values the set of x such that $x^2 + 5x > 0$.

19. Is the formula $|x - y| \leq |x| - |y|$ always true?

20. Express each of the following inequalities in the form "x belongs to the interval . . .":
    (a) $3 < x \leq 4$ (b) $x > 5$
    (c) $|x| < 5$ (d) $|x - 3| \leq 6$
    (e) $|3x + 1| < 2$

21. Express each of the following statements in terms of an inequality involving absolute values:
    (a) $x \in (-3, 3)$ (b) $-x \in (-4, 4)$
    (c) $x \in (-6, 6)$ (d) $x \in (2, 6)$
    (e) $x \in [-8, 12]$

22. Using the formula $|xy| = |x||y|$, find a formula for $|a/b|$. [*Hint:* Let $x = b$ and $y = a/b$.]

R-2 Lines, Circles, and Parabolas

The simplest plane figures are described by linear and quadratic equations in two variables.

In this section, we review some basic analytic geometry. The point-slope form of the equation of a straight line will be essential for the discussion of tangent lines given in Chapter 1. Circles and parabolas will provide useful examples throughout our study of calculus.

Goals

After studying this section, you should be able to:

Find the distance between points in the plane.
Write the equation of a line in point-slope form.
Recognize the equations for circles and parabolas.

COORDINATES AND DISTANCE

Analytic geometry combines the disciplines of algebra and geometry, enabling one to describe many geometric figures algebraically and bring the forces of algebra to bear on geometric problems.

One begins the algebraic representation of the plane by drawing two perpendicular lines, called the x and y axes, and then placing the real numbers on each of these lines, as shown in Fig. R-2-1.

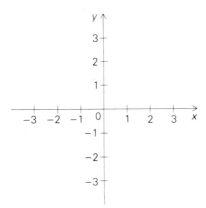

Fig. R-2-1   The x and y axes in the plane.

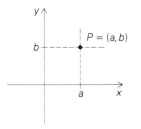

Fig. R-2-2   The point P has coordinates (a, b).

Any point P in the plane can now be described by the pair (a, b) of real numbers obtained by dropping perpendiculars to the x and y axes, as shown in Fig. R-2-2.

The numbers which describe the point P are called the *coordinates* of P: the first coordinate listed is called the x coordinate; the second is the y coordinate. We can use any letters we wish for the coordinates, including x and y themselves.

By reversing the process which produced the coordinates of a given point, we can always find a point P having a given pair as its coordinates; P is the intersection of the perpendicular lines drawn in Fig. R-2-2. Often the point with coordinates (a, b) is simply called "the point (a, b)." Drawing a point (a, b) on a graph is called *plotting* the point. Some points are plotted in Fig. R-2-3. Note that the point $(0, 0)$ is located at the intersection of the coordinate axes; it is called the *origin* of the coordinate system.

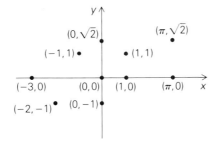

Fig. R-2-3   Examples of plotted points.

Worked Example 1   Let $a = 3$ and $b = 2$. Plot the points (a, b), (b, a), $(-a, b)$, $(a, -b)$, and $(-a, -b)$.

Solution   The points to be plotted are $(3, 2)$, $(2, 3)$, $(-3, 2)$, $(3, -2)$, and $(-3, -2)$; they are plotted in Fig. R-2-4.

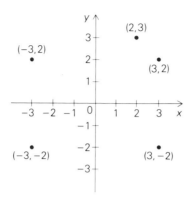

Fig. R-2-4   More plotted points.

The theorem of Pythagoras leads to a simple formula for the distance between two points (see Fig. R-2-5):

DISTANCE FORMULA

If P_1 has coordinates (x_1, y_1) and P_2 has coordinates (x_2, y_2),   the distance from P_1 to P_2 is

$$\sqrt{(x_1 - x_2)^2 + (y_1 - y_2)^2}$$

The distance between P_1 and P_2 is denoted $|P_1 P_2|$.

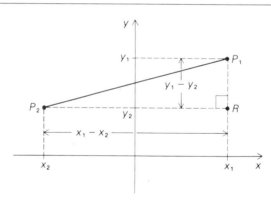

Fig. R-2-5   By the Pythagorean theorem,

$$|P_1 P_2| = \sqrt{|P_2 R|^2 + |P_1 R|^2}$$
$$= \sqrt{(x_1 - x_2)^2 + (y_1 - y_2)^2}$$

Worked Example 2   Find the distance from $(6, -10)$ to $(2, -1)$.

Solution   The distance is

$$\sqrt{(6 - 2)^2 + [-10 - (-1)]^2} = \sqrt{4^2 + (-9)^2}$$
$$= \sqrt{16 + 81} = \sqrt{97} \approx 9.85$$

If we have two points on the x axis, $(x_1, 0)$ and $(x_2, 0)$, the distance between them is $\sqrt{(x_1 - x_2)^2 + (0 - 0)^2} = \sqrt{(x_1 - x_2)^2}$. By property 3 of absolute values (see the second display on p. 15), this distance can be written as $|x_1 - x_2|$. Comparing this with the first display on p. 15, we see that *the distance formula in the plane includes the distance formula on the line* as a special case.

Exercises
1. Plot: $(0, 0)$, $(1, -1)$, $(-1, 1)$, $(2, -8)$, $(-2, 8)$, $(3, -27)$, $(-3, 27)$. Try to draw a smooth curve passing through all these points.

2. Plot the points $(-1, 2)$, $(-1, -2)$, $(1, -2)$, and $(1, 2)$.

3. Plot the points $(x, x^4 - x^2)$ for $x = -2, -\frac{3}{2}, -1, -\frac{1}{2}, 0, \frac{1}{2}, 1, \frac{3}{2}, 2$. Draw a smooth curve through the points.

4. Find the distance between each of the following pairs of points:

   (a) $(1, 1)$, $(1, -1)$ (b) $(-1, 1)$, $(-1, -1)$

   (c) $(-3, 9)$, $(2, -8)$ (d) $(0, 0)$, $(-3, 27)$

   (e) $(43721, 56841)$, $(3, 56841)$

5. Find the distance or a formula for the distance between each pair of points:

   (a) $(2, 1)$; $(3, 2)$ (b) $(a, 2)$; $(3 + a, 6)$

   (c) (x, y); $(3x, y + 10)$ (d) $(a, 0)$; $(a + b, b)$

   (e) (a, a); $(-a, -a)$

6. Find the coordinates of a point whose distance from $(0, 0)$ is $2\sqrt{2}$ and whose distance from $(4, 4)$ is $2\sqrt{2}$.

STRAIGHT LINES

In this book we will always use the term *line* in the sense of *straight line*. Rather than referring to "curved lines," we will use the term *curve*.

Draw a line l in the plane and pick two distinct points P_1 and P_2 on l. Let P_1 have coordinates (x_1, y_1) and P_2 have coordinates (x_2, y_2). The ratio $(y_2 - y_1)/(x_2 - x_1)$ (assuming that $x_2 \neq x_1$) is called the *slope* of the line l and is often denoted by the letter m. See Fig. R-2-6.

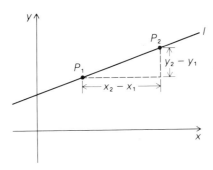

Fig. R-2-6 The slope of this line is $(y_2 - y_1)/(x_2 - x_1)$.

SLOPE FORMULA

If (x_1, y_1) and (x_2, y_2) lie on the line l,   the slope of l is

$$m = \frac{y_2 - y_1}{x_2 - x_1}.$$

An important feature of the slope m is that it does not depend upon which two points we pick, so long as they lie on the line l. To see that this is true, we observe (see Fig. R-2-7) that the right triangles $P_1 P_2 R$ and $P_1' P_2' R'$ are similar, since corresponding angles are equal, so $P_2 R / P_1 R = P_2' R' / P_1' R'$. In other words, the slope calculated using P_1 and P_2 is the same as the slope calculated using P_1' and P_2'. The slopes of some lines through the origin are shown in Fig. R-2-8.

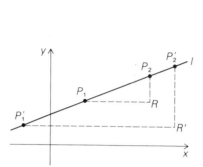

Fig. R-2-7 The slope does not depend on which two points on l are used.

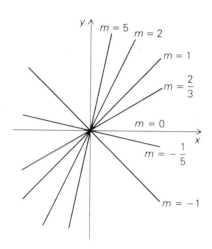

Fig. R-2-8 Slopes of some lines through the origin.

Worked Example 3   What is the slope of the line which passes through the points $(0, 1)$ and $(1, 0)$?

Solution   By the slope formula, with $x_1 = 0$, $y_1 = 1$, $x_2 = 1$, and $y_2 = 0$, the slope is $(0 - 1)/(1 - 0) = -1$.

> **Warning**   A line which is parallel to the y axis does not have a slope. In fact, any two points on such a line have the same x coordinates, so when we form the ratio $(y_1 - y_2)/(x_1 - x_2)$, the denominator becomes zero, which makes the expression meaningless.

To find the equation satisfied by the coordinates of the points on a line, we consider a line l with slope m and which passes through the point (x_1, y_1). If (x, y) is any *other* point on l, the slope formula gives

$$\frac{y - y_1}{x - x_1} = m$$

That is,

$$y = y_1 + m(x - x_1)$$

This is called the *point-slope form* of the equation of l; a general point (x, y) lies on l exactly when the equation holds.

If, for the point in the point-slope form of the equation, we take the point $(0, b)$ where l intersects the y axis (the number b is called the y intercept of l), we have $x_1 = 0$ and $y_1 = b$ and obtain the *slope-intercept form* $y = mx + b$. A vertical line has the equation $x = x_1$; y can take any value.

If we are given two points (x_1, y_1) and (x_2, y_2) on a line, we know that the slope is $(y_2 - y_1)/(x_2 - x_1)$. Substituting this term for m in the point-slope form of the equation gives the *point-point form:*

$$y = y_1 + \left(\frac{y_2 - y_1}{x_2 - x_1}\right)(x - x_1)$$

STRAIGHT LINES

Name	Data needed	Formula
point-slope	one point (x_1, y_1) on the line and the slope m	$y = y_1 + m(x - x_1)$
slope-intercept	the slope m of the line and the y-intercept b	$y = mx + b$
point-point	two points (x_1, y_1) and (x_2, y_2) on the line	$y = y_1 + \left(\dfrac{y_2 - y_1}{x_2 - x_1}\right)(x - x_1)$

For calculus, the point-slope form will turn out to be the most important of the three forms of the equation of a line, illustrated in Fig. R-2-9(a).

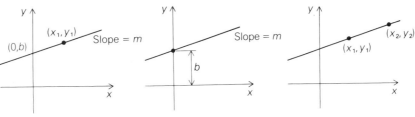

Fig. R-2-9 Three forms for the equation of a line.

(a) Point-Slope: $y = y_1 + m(x - x_1)$

(b) Slope-Intercept: $y = mx + b$

(c) Point-Point: $y = y_1 + \left(\dfrac{y_2 - y_1}{x_2 - x_1}\right)(x - x_1)$

Worked Example 4 Find the equation of the line through $(1, 1)$ with slope 5. Put the equation into slope-intercept form.

Solution Using the point-slope form, with $x_1 = 1$, $y_1 = 1$, and $m = 5$, we get $y = 1 + 5(x - 1)$. This simplifies to $y = 5x - 4$, which is the slope-intercept form.

Worked Example 5 Let l be the line through the points $(3, 2)$ and $(4, -1)$. Find the point where this line intersects the x axis.

Solution The equation of the line, in point-point form, with $x_1 = 3$, $y_1 = 2$, $x_2 = 4$, and $y_2 = -1$, is

$$y = 2 + \frac{-1 - 2}{4 - 3}(x - 3)$$

$$= 2 - 3(x - 3)$$

The line intersects the x axis at the point where $y = 0$, that is, where

$$0 = 2 - 3(x - 3)$$

Solving this equation for x, we get $x = \frac{11}{3}$, so the point of intersection is $(\frac{11}{3}, 0)$. (See Fig. R-2-10.)

Fig. R-2-10 Finding where the line through $(3, 2)$ and $(4, -1)$ meets the x axis.

Worked Example 6   Find the slope and y intercept of the line $3y + 8x + 5 = 0$.

Solution   The following equations are equivalent:

$$3y + 8x + 5 = 0$$
$$3y = -8x - 5$$
$$y = -\tfrac{8}{3}x - \tfrac{5}{3}$$

The last equation is in slope-intercept form, with slope $-\frac{8}{3}$ and y intercept $-\frac{5}{3}$.

Using the method of Worked Example 6, one can show that any equation of the form $Ax + By + C = 0$ describes a straight line, as long as A and B are not both zero. If $B \neq 0$, the slope of the line is $-A/B$; if $B = 0$, the line is vertical; and if $A = 0$, it is horizontal.

Finally, we recall without proof the fact that lines with slopes m_1 and m_2 are perpendicular if and only if $m_1 m_2 = -1$. In other words, the slopes of perpendicular lines are negative reciprocals of each other.

Worked Example 7   Find the equation of the line through $(0, 0)$ which is perpendicular to the line $3y - 2x + 8 = 0$.

Solution   The given equation has the form $Ax + By + C = 0$, with $A = -2$, $B = 3$, and $C = 8$; the slope of the line it describes is $-A/B = 2/3 = m_1$. The slope m_2 of the perpendicular line must satisfy $m_1 m_2 = -1$, so $m_2 = -3/2$. The line through the origin with this slope has the equation $y = -\frac{3}{2}x$.

Exercises

7. In each case, find the equation of the line through the point P with slope m, and sketch a graph of the line. Which two lines are perpendicular?

   (a) $P = (2, 3)$; $m = 2$ (b) $P = (-2, 6)$; $m = -\frac{1}{2}$

   (c) $P = (-1, 7)$; $m = 0$

8. Find the equation of the line through each of the following pairs of points:

   (a) $(5, 7)$; $(-1, 4)$ (b) $(1, 1)$; $(3, 2)$ (c) $(1, 4)$; $(3, 4)$ (d) $(1, 4)$; $(1, 6)$

9. Find the slope and y intercept of each of the following lines:

   (a) $x + 2y + 4 = 0$ (b) $\frac{1}{2}x - 3y + \frac{1}{3} = 0$ (c) $4y = 17$

10. Find the equation of the line with the given data:

    (a) slope $= 5$; y intercept $= 14$
    (b) y intercept $= 6$; passes through $(7, 8)$
    (c) passes through $(4, 2)$ and $(2, 4)$
    (d) passes through $(-1, -1)$; slope $= -10$

11. Find the slope and y intercept of the following lines:

    (a) $13 - 4x = 7(x + y)$ (b) $x - y = 14(x + 2y)$
    (c) $y = 17$ (d) $x = 60$

12. (a) Find the slope of the line $4x + 5y - 9 = 0$.

    (b) Find the equation of the line through $(1, 1)$ which is perpendicular to the line in part (a).

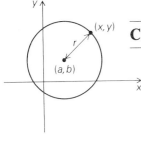

Fig. R-2-11 The point (x, y) is a typical point on the circle with radius r and center (a, b).

CIRCLES AND PARABOLAS

We now consider two more geometric figures which can be described by simple algebraic formulas: the circle and the parabola.

The circle C with radius $r > 0$ and center at (a, b) consists of those points (x, y) for which the distance from (x, y) to (a, b) is equal to r. (See Fig. R-2-11.)

The distance formula yields $\sqrt{(x - a)^2 + (y - b)^2} = r$ or, equivalently, $(x - a)^2 + (y - b)^2 = r^2$. If the center of the circle is at the origin, this equation takes the simpler form $x^2 + y^2 = r^2$.

Worked Example 8   Find the equation of the circle with center $(1, 0)$ and radius 5.

Solution   Here $a = 1$, $b = 0$, and $r = 5$, so $(x - a)^2 + (y - b)^2 = r^2$ becomes $(x - 1)^2 + y^2 = 25$ or $x^2 - 2x + y^2 = 24$.

Worked Example 9   Find the equation of the circle whose center is $(2, 1)$ and which passes through the point $(5, 6)$.

Solution   The equation must be of the form $(x - 2)^2 + (y - 1)^2 = r^2$; the problem is to determine r^2. Since the point $(5, 6)$ lies on the circle, it must satisfy the equation. That is,

$$r^2 = (5 - 2)^2 + (6 - 1)^2 = 3^2 + 5^2 = 34$$

so the correct equation is $(x - 2)^2 + (y - 1)^2 = 34$.

Worked Example 10   Prove that the graph of $x^2 + y^2 - 6x - 16y + 8 = 0$ is a circle. Find its center.

Solution   Complete the squares:

$$\begin{aligned}
0 = x^2 + y^2 - 6x - 16y + 8 &= x^2 - 6x \qquad + y^2 - 16y \qquad\qquad + 8 \\
&= x^2 - 6x + 9 + y^2 - 16y + 64 - 9 - 64 + 8 \\
&= (x - 3)^2 + (y - 8)^2 - 65
\end{aligned}$$

Thus the equation becomes $(x - 3)^2 + (y - 8)^2 = 65$, whose graph is a circle with center $(3, 8)$ and radius $\sqrt{65} \approx 8.06$.

x	2	1.5	1	0.5	0
y	4	2.25	1	0.25	0

Axis of symmetry

Fig. R-2-12 The graph of the parabola $y = x^2$.

Consider next the equation $y = x^2$. If we plot a number of points whose coordinates satisfy this equation, by choosing values for x and computing y, we find that these points may be joined by a smooth curve as in Fig. R-2-12. This curve is called a *parabola*. It is also possible to give a purely geometric definition of a parabola and derive the equation from geometry as was done for the line and circle. In fact, we will do so in Section R-4.

There is no need to compute y for negative values of x. In fact, if x is replaced by $-x$, the value of y is unchanged, so the graph is symmetric about the y axis. Similarly, we can plot $y = 3x^2$, $y = 10x^2$, $y = -\frac{1}{2}x^2$, $y = -8x^2$, and so on. (See Fig. R-2-13.) These graphs are also parabolas. The general parabola of this type has the equation $y = ax^2$, where a is a nonzero constant; these parabolas all have their *vertex* at the origin. If $a > 0$ the parabola opens upwards and if $a < 0$ it opens downwards.

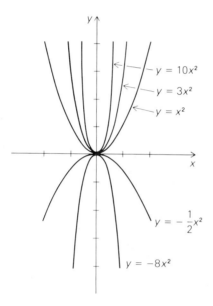

Fig. R-2-13 Parabolas $y = ax^2$ for various values of a.

Worked Example 11 Let C be the parabola with vertex at the origin and passing through the point $(2, 8)$. Find the point on C whose x coordinate is 10.

Solution The equation of C is of the form $y = ax^2$. To find a, we use the fact that $(2, 8)$ lies on C. Thus $8 = a \cdot 2^2 = 4a$, so $a = 2$ and the equation is $y = 2x^2$. If the x coordinate of a point on C is 10, the y coordinate is $2 \cdot 10^2 = 200$, so the point is $(10, 200)$.

A special focusing property of parabolas is of practical interest: a parallel beam of light rays (as from a star) impinging upon a parabola in the direction of its axis of symmetry will focus at a single point as shown in Fig. R-2-14. This property follows from the law that the angle of incidence equals the angle of reflection, together with some geometry or calculus.* (See Review Problems 30 and 31 at the end of Chapter 1.)

Just as we considered circles with center at an arbitrary point (a, b), we can consider parabolas with vertex at any point (p, q). The equation

---

This problem was historically important in the development of calculus, for it required knowing the tangent to the mirror, as in Fig. R-2-14.

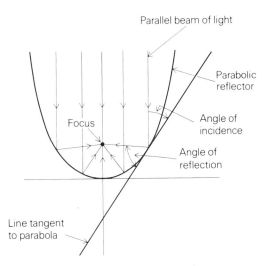

Fig. R-2-14 The focusing property of a parabolic reflector.

of such a *shifted parabola* is $y = a(x - p)^2 + q$. We have started with $y = ax^2$, then replaced x by $x - p$ and y by $y - q$ to get

$$y - q = a(x - p)^2$$

or $y = a(x - p)^2 + q$. This process is illustrated in Fig. R-2-15. Notice that if (x, y) lies on the shifted parabola, then the corresponding point on the original parabola is $(x - p, y - q)$, which must therefore satisfy the equation of the original parabola; i.e., $y - q$ must equal $a(x - p)^2$.

Given an equation of the form $y = ax^2 + bx + c$, we can complete the square on the right-hand side to put it in the form $y = a(x - p)^2 + q$. Thus the graph of any equation $y = ax^2 + bx + c$ is a shifted parabola. From now on, we will refer to a shifted parabola simply as a parabola.

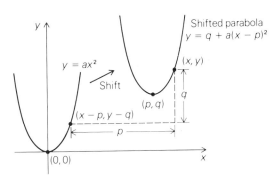

Fig. R-2-15 $y = q + a(x - p)^2$ is the parabola $y = ax^2$ shifted from $(0, 0)$ to (p, q).

Worked Example 12   Graph $y = -2x^2 + 4x + 1$.

Solution   Completing the square gives

$$\begin{aligned}
y &= -2x^2 + 4x + 1 \\
&= -2(x^2 - 2x - \tfrac{1}{2}) \\
&= -2(x^2 - 2x + 1 - 1 - \tfrac{1}{2}) \\
&= -2(x^2 - 2x + 1 - \tfrac{3}{2}) \\
&= -2(x - 1)^2 + 3
\end{aligned}$$

Fig. R-2-16 A parabola with vertex at $(1, 3)$.

The vertex is thus at $(1, 3)$ and the parabola opens downward like $y = -2x^2$ (see Fig. R-2-16).

EQUATIONS OF CIRCLES AND PARABOLAS

The equation of the circle with radius r and center at (a, b) is

$$(x - a)^2 + (y - b)^2 = r^2$$

The equation of a parabola with vertex at (p, q) is

$$y = a(x - p)^2 + q$$

Exercises

13. Find the equation of the circle with center at P and radius r. Sketch.
    (a) $P = (1, 1)$; $r = 3$ (b) $P = (-1, 7)$; $r = 5$ (c) $P = (0, 5)$; $r = 5$

14. Find the equation of the circle whose center is at $(0, 1)$ and which passes through the point $(-1, 4)$. Sketch.

15. Find the center and radius of each of the following circles. Sketch.
    (a) $x^2 + y^2 - 2x + y - \frac{3}{4} = 0$
    (b) $2x^2 + 2y^2 + 8x + 4y + 3 = 0$
    (c) $-x^2 - y^2 + 8x - 4y - 11 = 0$

16. Find the equation of the parabola whose vertex is at V and which passes through the point P.
    (a) $V = (1, 2)$; $P = (0, 1)$ (b) $V = (0, 1)$; $P = (1, 2)$
    (c) $V = (5, 5)$; $P = (0, 0)$

17. Sketch the graph of each of the following, marking the vertex in each case.
    (a) $y = x^2 - 4x + 7$ (b) $y = -x^2 + 4x - 1$
    (c) $y = -2x^2 + 8x - 5$ (d) $y = 3x^2 + 6x + 2$

18. Graph each of the following equations:
    (a) $y = -3x^2$ (b) $y = -3x^2 + 4$
    (c) $y = -6x^2 + 8$ (d) $y = -3(x + 4)^2 + 4$
    (e) $y = 4x^2 + 4x + 1$

INTERSECTIONS OF PLANE FIGURES

Analytic geometry provides an algebraic technique for finding the points where two geometric figures intersect. If each figure is given by an equation in x and y, we solve for those pairs (x, y) which satisfy both equations.

For two lines, we know that they will have either zero, one, or infinitely many intersection points. (There are none if the two lines are parallel and different, infinitely many if the two lines are the same.) For a line and a circle or parabola, there may be zero, one, or two intersection points. (See Fig. R-2-17.)

Fig. R-2-17 Intersections of some geometric figures.

Worked Example 13   Where do the lines $x + 3y + 8 = 0$ and $y = 3x + 4$ intersect?

Solution   To find the intersection point, we solve the simultaneous equations

$$x + 3y + 8 = 0$$
$$-3x + y - 4 = 0$$

Multiply the first equation by 3 and add to the second to get $0 + 10y + 20 = 0$, or $y = -2$. Substituting $y = -2$ into the first equation gives $x - 6 + 8 = 0$, or $x = -2$. The intersection point is $(-2, -2)$.

Worked Example 14   Where does the line $x + y = 1$ meet the parabola $y = 2x^2 + 4x + 1$?

Solution   We look for pairs (x, y) which satisfy both equations. We may substitute $2x^2 + 4x + 1$ for y in the equation of the line to obtain

$$x + 2x^2 + 4x + 1 = 1$$
$$2x^2 + 5x = 0$$
$$x(2x + 5) = 0$$

So $x = 0$ or $-\frac{5}{2}$. We may use either equation to find the corresponding values of y. The linear equation $x + y = 1$ is simpler; it gives $y = 1 - x$, so $y = 1$ when $x = 0$ and $y = \frac{7}{2}$ when $x = -\frac{5}{2}$. Thus the points of intersection are $(0, 1)$ and $(-\frac{5}{2}, \frac{7}{2})$. (See Fig. R-2-18.)

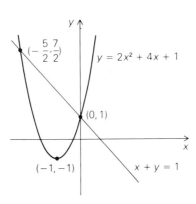

Fig. R-2-18   Intersections of the line $x + y = 1$ and the parabola $y = 2x^2 + 4x + 1$ occur at $(0, 1)$ and $(-\frac{5}{2}, \frac{7}{2})$.

Worked Example 15   (a) Where does the line $y = 3x + 4$ intersect the parabola $y = 8x^2$?
(b) For which values of x is $8x^2 < 3x + 4$? Explain your answer geometrically.

Solution   (a) We solve these equations simultaneously:

$$-3x + y - 4 = 0$$
$$y = 8x^2$$

Substituting the second into the first gives $-3x + 8x^2 - 4 = 0$, so $8x^2 - 3x - 4 = 0$.
   By the quadratic formula,

$$x = \frac{3 \pm \sqrt{9 + 4 \cdot 8 \cdot 4}}{16} = \frac{3 \pm \sqrt{137}}{16} \approx \frac{3 \pm 11.705}{16}$$
$$\approx 0.919 \text{ and } -0.544$$

Fig. R-2-19 The line $y = 3x + 4$ intersects the parabola $y = 8x^2$ at two points.

When $x \approx 0.919$, $y \approx 8(0.919)^2 \approx 6.76$. Similarly, when $x \approx -0.544$, $y \approx 2.37$, so the two points of intersection are approximately $(0.919, 6.76)$ and $(-0.544, 2.37)$. (See Fig. R-2-19.) As a check, you may substitute these pairs into the equation $-3x + y - 4 = 0$.

If the final quadratic equation had just one root—a double root—there would have been just one point of intersection; if no (real) roots, then no points of intersection.

(b) The inequality is satisfied where the parabola lies below the line, that is, for x in the interval between the x-values of the intersection points. Thus we have $8x^2 < 3x + 4$ whenever

$$x \in \left(\frac{3 - \sqrt{137}}{16}, \frac{3 + \sqrt{137}}{16} \right) \approx (-0.544, 0.919)$$

INTERSECTION POINTS

To find the intersection points of two figures,   find pairs (x, y) which simultaneously satisfy the equations describing the two figures.

Exercises

19. Find the points where the following pairs of figures intersect. Sketch a graph.
   (a) $y = -2x + 7$ and $y = 5x + 1$
   (b) $y = \frac{1}{3}x - 4$ and $y = 2x^2$
   (c) $y = 5x^2$ and $y = -6x + 7$
   (d) $x^2 + y^2 - 2y - 3 = 0$ and $y = 3x + 1$
   (e) $y = 3x^2$ and $y - x + 1 = 0$

20. What are the possible numbers of points of intersection between two circles? Make a drawing similar to Fig. R-2-17.

21. What are the possible numbers of points of intersection between a circle and a parabola? Make a drawing similar to Fig. R-2-17.

22. Find the points of intersection between the graphs of each of the following pairs of equations. Sketch your answers.
   (a) $y = 4x^2$ and $x^2 + 2y + y^2 - 3 = 0$
   (b) $x^2 + 2x + y^2 = 0$ and $x^2 - 2x + y^2 = 0$
   (c) $y = x^2 + 4x + 5$ and $y = x^2 - 1$

R-3 Functions and Graphs

A curve which intersects each vertical line at most once is the graph of a function.

In arithmetic and algebra, we operate with *numbers* (and letters which represent them). The mathematical objects of central interest in calculus are *functions*. In this section, we review some basic material concerning functions in preparation for their appearance in calculus.

Goals

After studying this section, you should be able to:

Evaluate a function at a given value of its argument.

Sketch the graph of a function by plotting points.

Tell whether a curve is the graph of a function.

EVALUATING FUNCTIONS

> ***Definition*** A *function* f on the real-number line is a rule which associates to each real number x a uniquely specified real number called $f(x)$ and pronounced "f of x."

Very often, $f(x)$ is given by a formula (such as $f(x) = x^3 + 3x + 2$) which tells us how to compute $f(x)$ when x is given. The process of calculating $f(x)$ is called *evaluating f* at x. We speak of x as the *independent variable* or the *argument* of f.

Worked Example 1   If $f(x) = x^3 + 3x + 2$, what is $f(-2), f(2.9), f(q)$?

Solution   Substituting -2 for x in the formula defining f, we have $f(-2) = (-2)^3 + 3(-2) + 2 = -8 - 6 + 2 = -12$. Similarly, $f(2.9) = (2.9)^3 + 3(2.9) + 2 = 24.389 + 8.7 + 2 = 35.089$. Finally, substituting q for x in the formula for f gives $f(q) = q^3 + 3q + 2$.

A function need not be given by a single formula, nor does it have to be denoted by the letter f. For instance, we can define a function H as follows:

$$H(x) = 0 \quad \text{if } x < 0 \quad \text{and} \quad H(x) = 1 \quad \text{if } x \geq 0$$

We have a uniquely specified value $H(x)$ for any given x, so H is a function.

Worked Example 2   Find $H(3), H(-5), H(0)$, and $H(|x|)$ for the function defined in the preceding paragraph.

Solution $H(3) = 1$ since $3 \geq 0$; $H(-5) = 0$ since $-5 < 0$; $H(0) = 1$ since $0 \geq 0$. Finally, since $|x| \geq 0$ no matter what the value of x, we may write $H(|x|) = 1$.

Calculator Discussion

We can think of a function as a *machine* or as a *program* in a calculator or computer which feeds us the output $f(x)$ when we feed in the number x. (See Fig. R-3-1.)

$x \xrightarrow{\text{Input}}$ | Programmed Calculator | $\xrightarrow{\text{Output}} f(x)$

Fig. R-3-1   A function on a calculator.

Many pocket calculators have functions built into them. Take, for example, the key labeled x^2. Enter a number x, say, 3.814. Now press the x^2 key and read: 14.546596. The x^2 key thus represents a function, the "squaring function." Whatever number x is fed in, pressing this key causes the calculator to give x^2 as an output.

We remark that the functions computed by calculators are often only *approximately* equal to the idealized mathematical functions indicated on the keys. For instance, entering 2.000003 and pressing the x^2 key gives the result 4.000012, while squaring 2.000003 by hand gives 4.000012000009.

Worked Example 3   Let $f(x) = [x^2 - 4.000012] \cdot 10^{13} + 2$.
(a) What is $f(2.000003)$?

(b) What result would you obtain in part (*a*) if you calculated $f(2.000003)$ by using the calculator described above to square 2.000003?

Solution   (a) $f(2.000003) = [(2.000003)^2 - 4.000012] \cdot 10^{13} + 2$
$= [4.000012000009 - 4.000012] \cdot 10^{13} + 2$
$= 0.000000000009 \cdot 10^{13} + 2$
$= 90 + 2 = 92$

(b) If we used the calculator to square 2.000003, we would obtain $f(2.000003) = 0 \cdot 10^{13} + 2 = 2$, which is nowhere near the correct answer.

Some very simple functions turn out to be quite useful:

$$f(x) = x$$

defines a perfectly respectable function called the *identity function* ("identity" because if we feed in x we get back the identical number x). Similarly,

$$f(x) = 0$$

is the *zero function* and

$$f(x) = 1$$

is the *unit function* whose value is always 1, no matter what x is fed in.

Some formulas are not defined for all x. For instance, $1/x$ is defined only if $x \neq 0$. With a slightly more general definition, we can still consider $f(x) = 1/x$ as a function.

Definition   Let D be a set of real numbers. A *function f* with *domain D* is a rule which assigns a unique real number $f(x)$ to each number x in D.

This symbol denotes problems or discussions that may require use of a hand-held calculator.

If we specify a function by a formula like $f(x) = (x - 2)/(x - 3)$, its domain may be assumed to consist of all x for which the formula is defined (in this case all $x \neq 3$), unless another domain is explicitly mentioned. If we wish, for example, to consider the squaring function applied only to positive numbers, we would write: "Let f be defined by $f(x) = x^2$ for $x > 0$."

Worked Example 4 (a) What is the domain of $f(x) = 3x/(x^2 - 2x - 3)$? (b) Evaluate $f(1.6)$.

Solution (a) The domain of f consists of all x for which the denominator is not zero. But $x^2 - 2x - 3 = (x - 3)(x + 1)$ is zero just at $x = 3$ and $x = -1$. Thus the domain consists of all real numbers except 3 and -1.

(b) $f(1.6) = 3(1.6)/[(1.6)^2 - 2(1.6) - 3]$
$= 4.8/[2.56 - 3.2 - 3]$
$= 4.8/(-3.64) \approx -1.32$

To visualize a function, we can draw its *graph*.

Definition Let f be a function with domain D. The set of all points (x, y) in the plane with x in D and $y = f(x)$ is called the *graph* of f.

Worked Example 5 (a) Let $f(x) = 3x + 2$. Evaluate $f(-1), f(0), f(1)$, and $f(2.3)$.
(b) Draw the graph of f.

Solution (a) $f(-1) = 3(-1) + 2 = -1$; $f(0) = 3 \cdot 0 + 2 = 2$; $f(1) = 3 \cdot 1 + 2 = 5$; $f(2.3) = 3(2.3) + 2 = 8.9$.
(b) The graph is the set of all (x, y) such that $y = 3x + 2$. This is then just the straight line $y = 3x + 2$. It has y intercept 2 and slope 3, so we can plot it directly (Fig. R-3-2).

Fig. R-3-2 The graph of $f(x) = 3x + 2$ is a line.

Worked Example 6 Draw the graph of $f(x) = 3x^2$.

Solution The graph of $f(x) = 3x^2$ is just the parabola $y = 3x^2$, drawn in Fig. R-3-3 (see Section R-2).

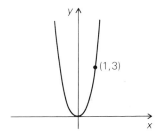

Fig. R-3-3 The graph of $f(x) = 3x^2$ is a parabola.

Worked Example 7   Let g be the *absolute value function* defined by $g(x) = |x|$. (The domain consists of all real numbers.) Draw the graph of g.

Solution   We begin by choosing various values of x in the domain, computing $g(x)$, and plotting the points $(x, g(x))$. Connecting these points results in the graph shown in Fig. R-3-4. Another approach is to use the definition

$$g(x) = |x| = \begin{cases} x & \text{if } x \geq 0 \\ -x & \text{if } x < 0 \end{cases}$$

We observe that the part of the graph of g for $x \geq 0$ is a line through $(0,0)$ with slope 1, while the part for $x < 0$ is a line through $(0,0)$ with slope -1. It follows that the graph of g is as drawn in Fig. R-3-4.

x	-3	-2	-1	0	$\frac{1}{2}$	1	4
$g(x)$	3	2	1	0	$\frac{1}{2}$	1	4

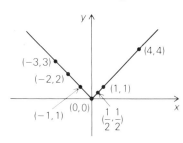

Fig. R-3-4   Some points on the graph of $y = |x|$.

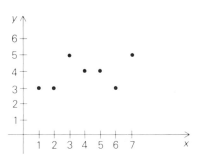

Fig. R-3-5   The graph of "one, two, three, four, five, six, seven."

If the domain of a function consists of finitely many points, then the graph consists of isolated dots; there is no line to be filled in. For instance, Fig. R-3-5 shows the graph of a function l whose domain is $\{1, 2, 3, 4, 5, 6, 7\}$ and for which $l(x)$ is the number of letters in the English name for x.

Exercises   1. Evaluate each of the following functions at $x = -1$ and $x = 1$:
   (a) $f(x) = 5x^2 - 2x$ (b) $f(x) = -x^2 + 3x - 5$
   (c) $f(x) = x^3 - 2x^2 + 1$ (d) $f(x) = 4x^2 + x - 2$
   (e) $f(x) = -x^3 + x^2 - x + 1$

2. Draw graphs of the functions of parts (a), (b), and (d) in Exercise 1.

3. Find the domain of each of the following functions, and evaluate each function at $x = 10$.

   (a) $f(x) = \dfrac{x^2}{x - 1}$ (b) $f(x) = \dfrac{x^2}{-x^2 + 2x - 1}$

   (c) $f(x) = 5x\sqrt{1 - x^2}$ (d) $f(x) = \dfrac{x^2 - 1}{\sqrt{x - 4}}$

   (e) $f(x) = \dfrac{5x + 2}{x^2 - x - 6}$

4. Plot 10 points on the graph of each function of parts (a) and (d) in Exercise 3 and connect with a smooth curve.

5. Sketch the graph of each function:

(a) $f(x) = (x - 1)^2 + 3$ (b) $f(x) = x^2 - 9$

(c) $f(x) = 3x + 10$ (d) $f(x) = x^2 + 4x + 2$

THE SHAPE OF A GRAPH

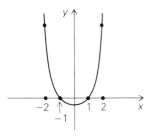

Fig. R-3-6 Correct appearance of graph?

In plotting a complicated function such as $f(x) = 0.3x^4 - 0.2x^2 - 0.1$, we must be sure to take enough values of x, for we might otherwise miss some important details. Choosing $x = -2, -1, 0, 1, 2$ gives the points $(-2, 3.9)$, $(-1, 0)$, $(0, -0.1)$, $(1, 0)$, $(2, 3.9)$ on the graph. See Fig. R-3-6. Should we draw a smooth curve through these points? How can we be sure there are no other little bumps in the graph?

To answer this question, we can do some serious calculating: let us plot points on the graph of $f(x) = 0.3x^4 - 0.2x^2 - 0.1$ for values of x at intervals of 0.1 between -2 and 2. If we notice that $f(x)$ is unchanged if x is replaced by $-x$, we can cut the work in half. It is only necessary to calculate $f(x)$ for $x \geq 0$, since the values for negative x are the same. The graph of f is therefore symmetric about the y axis. The results of this calculation are given here and plotted in Fig. R-3-7:

x	$f(x)$	x	$f(x)$	x	$f(x)$	x	$f(x)$
0	−0.10000	0.5	−0.13125	1.0	0.00000	1.5	0.96875
0.1	−0.10197	0.6	−0.13312	1.1	0.09723	1.6	1.35408
0.2	−0.10752	0.7	−0.12597	1.2	0.23408	1.7	1.82763
0.3	−0.11557	0.8	−0.10512	1.3	0.41883	1.8	2.40128
0.4	−0.12432	0.9	−0.06517	1.4	0.66048	1.9	3.08763
						2.0	3.90000

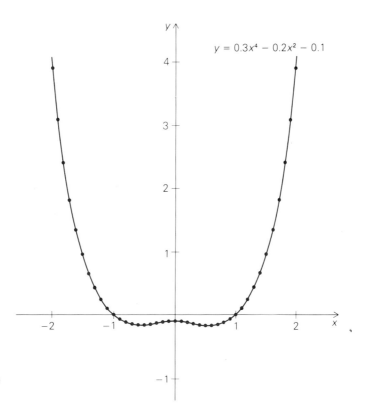

Fig. R-3-7 The graph more carefully plotted.

Thus we see that indeed our original guess (Fig. R-3-6) was wrong and that a more refined calculation gives Fig. R-3-7. How can we be sure not to have missed still more bumps and wiggles? By plotting many points we can make good guesses but can never know for sure. The calculus we will develop in the first two chapters of this book can tell us exactly how many wiggles the graph of a function can have and so will greatly facilitate plotting.

(optional)

Calculator Discussion

There is a trick called Horner's recurrence (it may be due to Newton) which simplifies the numerical evaluation of polynomial functions. Given the polynomial $0.3x^4 - 0.2x^2 - 0.1$, you may rewrite it as

$$(0.3x^2 - 0.2)x^2 - 0.1 \qquad \text{or} \qquad (0.3 \cdot x \cdot x - 0.2) \cdot x \cdot x - 0.1.$$

Given any value of x, you can evaluate this expression from left to right without having to store any intermediate numbers for reentry into the calculator.

This procedure works for any polynomial. The general fourth-order polynomial, for instance, usually written as $ax^4 + bx^3 + cx^2 + dx + e$, can be rewritten as $\{[(ax + b)x + c]x + d\}x + e$. Although the parentheses, brackets, and braces make this expression look complicated, in fact it can be read as "a times x plus b times x plus c times x plus d times x plus e," and the calculations can be done in just that order. (If a coefficient, say d, is negative, replace "plus d" by "minus $-d$.")

If your calculator has a memory, it is useful to enter the value of x in the memory at the beginning of the computation so that you can recall it at the press of a single key. (Notice that in the case of the polynomial $0.3x^4 - 0.2x^2 - 0.1 = (0.3x^2 - 0.2)x^2 - 0.1$, you can do even better by computing x^2 first and entering that in the memory.) Incidentally, the procedure described here is also a good one to use in programming a computer to evaluate polynomials. (It is not necessarily the most efficient procedure, depending on the cost of multiplications, additions, and accuracy desired. For example, $0.3x^4 - 0.2x^2 - 0.1$ is evaluated more accurately on a calculator for x near 1 if one uses $0.3x^4 - 0.2x^2 - 0.1 = (0.3x^2 + 0.1)(x - 1)(x + 1)$.)

Graphs of functions can have various shapes, but not every set of points in the plane is the graph of a function. Consider, for example, the circle $x^2 + y^2 = 5$ (Fig. R-3-8). If this circle were the graph of a function f, what would $f(2)$ be? Since $(2, 1)$ lies on the circle, we must have $f(2) = 1$. Since $(2, -1)$ also lies on the circle, $f(2)$ should also be equal to -1. But our definition of a function requires that $f(2)$ should have a definite value. Our only escape from this apparent contradiction is the conclusion that *the circle is not the graph of any function*. However, the upper semicircle alone is the graph of $y = \sqrt{5 - x^2}$ and the lower semicircle is the graph of $y = -\sqrt{5 - x^2}$, each with domain $[-\sqrt{5}, \sqrt{5}]$. Thus, while the circle is not a graph, it can be broken into two graphs.

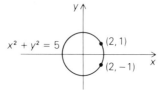

Fig. R-3-8 The circle is not the graph of a function.

Worked Example 8   Which straight lines in the plane are graphs of functions?

Solution   If a line is not vertical, it has the form $y = mx + b$, so it is the graph of the function $f(x) = mx + b$. (If $m = 0$, the function is a constant function.) A vertical line is not the graph of a function—if the line is $x = a$, then $f(a)$ is not determined since y can take on any value.

There is a simple test for determining whether a set of points in the plane is the graph of a function. If the number x_0 belongs to the domain of a function f, the vertical line $x = x_0$ intersects the graph of f at the point $(x_0, f(x_0))$ and at no other point. If x_0 does not belong to the domain, (x_0, y) is not on the graph for any value of y, so the vertical line does not intersect the graph at all. Thus we have the following criterion:

RECOGNIZING GRAPHS OF FUNCTIONS

A set of points in the plane is the graph of a function	if and only if	every vertical line intersects the set in *at most* one point.
The domain of the function	is	the set of x_0 such that the vertical line $x = x_0$ meets the graph.

If C is a set of points satisfying this criterion, we can reconstruct the function f of which C is the graph. For each value x_0 of x, look for a point where the line $x = x_0$ meets C. The y coordinate of this point is $f(x_0)$. If there is no such point, x_0 is not in the domain of f.

Worked Example 9   For each of the sets in Fig. R-3-9:
  (*i*) Tell whether it is the graph of a function.
  (*ii*) If the answer to part (*i*) is yes, tell whether $x = 3$ is in the domain of the function.
  (*iii*) If the answer to part (*ii*) is yes, evaluate the function at $x = 3$.

(a)   (b)   (c)

Fig. R-3-9 Which curves are graphs of functions?

Solution   (a) (*i*) yes; (*ii*) no; this is indicated by the white dot.
  (b) (*i*) no; for example, the line $x = -3$ cuts the curve in two points.
  (c) (*i*) yes;   (*ii*) yes;   (*iii*) 1.

Worked Example 10   Which of the curves $y^2 = x$ and $y^3 = x$ is the graph of a function?

Solution   We begin with $y^2 = x$. Note that each value of y determines a unique value of x; we plot a few points according to the following table (see Fig. R-3-10):

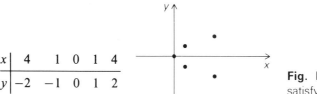

x	4	1	0	1	4
y	−2	−1	0	1	2

Fig. R-3-10 Five points satisfying $y^2 = x$.

We see immediately that the vertical line $x = 4$ meets $y^2 = x$ in two points, so $y^2 = x$ cannot be the graph of a function. (You may convince youself that the curve $y^2 = x$ is a parabola whose axis of symmetry is horizontal.) Now look at $y^3 = x$. We begin by plotting a few points (Fig. R-3-11):

x	−8	−1	0	1	8
y	−2	−1	0	1	2

Fig. R-3-11 Five points on $y^3 = x$.

Fig. R-3-12 The curve $y^3 = x$.

These points could all lie on the graph of a function. In fact, the full curve $y^3 = x$ looks as in Fig. R-3-12. We see by inspection that the curve intersects each vertical line exactly once, so there is a function f whose graph is the given curve. Since $f(x)$ is a number whose cube is x, f is called the *cube root* function.

Exercises   Plot the graphs of the functions in Exercises 6 and 7 using the two given sets of values for x. (See the discussion on p. 33.)

   6. $f(x) = 3x^3 - x^2 + 1$:
     (a) at $x = -2, -1, 0, 1, 2$ (b) at 0.2 intervals in $[-2, 2]$

   7. $f(x) = 2x^4 - x^3$:
     (a) at $x = -2, -1, 0, 1, 2$ (b) at 0.1 intervals in $[-2, 2]$

   8. Which of the curves in Fig. R-3-13 are graphs of functions?

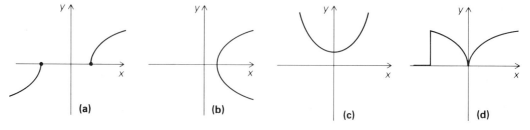

(a)   (b)   (c)   (d)

Fig. R-3-13 Which curves are graphs?

9. Match the following formulas with the curves in Fig. R-3-14:

   (a) $x + y \geq 1$ (b) $x - y \leq 1$

   (c) $y = (x - 1)^2$ (d) $y = \left(\dfrac{x + |x|}{2} \right)^2$

   (e) Not the graph of a function   (f) None of the above

    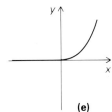

Fig. R-3-14   Match the curve with the formula.

10. Tell whether the curve defined by each of the following equations is the graph of a function. If the answer is yes, find the domain of the function.

   (a) $xy = 1$ (b) $x^2 - y^2 = 1$

   (c) $y = \sqrt{x^2 - 1}$ (d) $x + y^2 = 3$

   (e) $y + x^2 = 3$

(optional) # R-4 The Conic Sections*

All the curves described by quadratic equations in two variables can be obtained by cutting a cone with planes.

In this section, we discuss the analytic geometry of ellipses and hyperbolas, together with the circles and parabolas already studied. These curves, called the conic sections, will be used occasionally as examples throughout this book, but an understanding of the material in this section is *not* essential for the further study of calculus.

Goals

After studying this section, you should be able to:

   Find the equation of a conic, given its geometric description.

   Identify and plot a conic, given its equation.

THE ELLIPSE

The ellipse, hyperbola, parabola, and circle are called *conic sections* because they can all be obtained by slicing a cone with a plane (see Fig. R-4-1). The theory of these curves, developed by Apollonius of Perga (262–200 BC), is a masterwork of Greek geometry. We will return to the three-dimensional origin of the conics in Chapter 15, when we have studied some analytic geometry in space. For now,

---

This section is not urgently needed for calculus, and its study may be deferred. It is needed later for occasional problems and is a prerequisite for Chapter 15.

(optional)

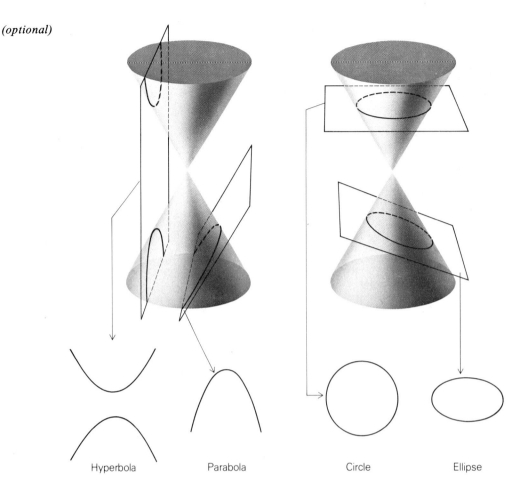

Fig. R-4-1   Conic sections are obtained by slicing a cone with a plane; which conic section is obtained depends on the direction of the slicing plane.

we will treat these curves, beginning with the ellipse, purely as objects in the plane.

Definition   An *ellipse* is the set of points in the plane for which the sum of the distances from two fixed points is constant. These two points are called the *foci* (plural of *focus*).

An ellipse can be drawn with the aid of a string tacked at the foci, as shown in Fig. R-4-2.

Fig. R-4-2   Mechanical construction of an ellipse.

To find an equation for the ellipse, we locate the foci on the x axis at the points $F' = (-c, 0)$ and $F = (c, 0)$. Let $2a > 0$ be the sum of the distances from a point on the ellipse to the foci. Since the distance between the foci is $2c$ and the length of a side of a triangle is less than the sum of the lengths of *(optional)* | the other sides, we must have $2c < 2a$; i.e., $c < a$. Referring to Fig. R-4-3, we see that a point $P = (x, y)$ is on the ellipse precisely when

$$|FP| + |F'P| = 2a$$

That is,

$$\sqrt{(x + c)^2 + y^2} + \sqrt{(x - c)^2 + y^2} = 2a.$$

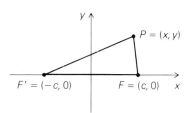

Fig. R-4-3 P is on the ellipse when $|FP| + |F'P| = 2a$.

Transposing $\sqrt{(x - c^2 + y^2}$, squaring, simplifying and squaring again yields

$$(a^2 - c^2)x^2 + a^2y^2 = a^2(a^2 - c^2)$$

Let $a^2 - c^2 = b^2$ (remember that $a > c > 0$ and so $a^2 - c^2 > 0$). Then, after division by a^2b^2, the equation becomes

$$\frac{x^2}{a^2} + \frac{y^2}{b^2} = 1$$

This is the *equation of an ellipse in standard form.*

Since $b^2 = a^2 - c^2 < a^2$, we have $b < a$. If we had put the foci on the y axis, we would have obtained an equation of the same form with $b > a$; the length of the "string" would now be $2b$ rather than $2a$. (See Fig. R-4-4.) The length of the long axis of the ellipse [$2a$ in Fig. R-4-4(a) and $2b$ in Fig. R-4-4(b)] is called the *major axis* and the length of the short axis [$2b$ in Fig. R-4-4(a) and $2a$ in Fig. R-4-4(b)] is the *minor axis.*

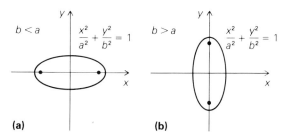

Fig. R-4-4   The appearance of an ellipse in the two cases $b < a$ and $b > a$.

(a)   **(b)**

Worked Example 1   Sketch the graph of $4x^2 + 9y^2 = 36$. Where are the foci?

Solution   Dividing both members of the equation by 36, we obtain the standard form

$$\frac{x^2}{9} + \frac{y^2}{4} = 1$$

(optional) | Hence $a = 3$, $b = 2$, and $c = \sqrt{a^2 - b^2} = \sqrt{5}$. The foci are $(\pm\sqrt{5}, 0)$ and the x intercepts are $(\pm 3, 0)$. The graph is shown in Fig. R-4-5.

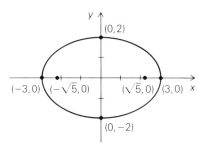

Fig. R-4-5   The graph of $4x^2 + 9y^2 = 36$.

Worked Example 2   Sketch the graph of $9x^2 + y^2 = 81$. Where are the foci?

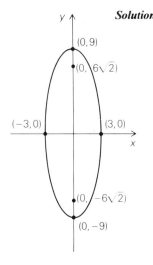

Fig. R-4-6   The ellipse $9x^2 + y^2 = 81$.

Solution   Dividing by 81, we obtain the standard form $x^2/3^2 + y^2/9^2 = 1$. The graph is sketched in Fig. R-4-6. The foci are at $(0, \pm 6\sqrt{2})$.

ELLIPSE

Equation: $\dfrac{x^2}{a^2} + \dfrac{y^2}{b^2} = 1$ (standard form)

Foci: $(\pm c, 0)$ where $\sqrt{a^2 - b^2} = c$ if $b < a$ or $(0, \pm c)$ where $\sqrt{b^2 - a^2} = c$ if $b > a$. If $b = a$, the ellipse is a circle.

x intercepts: $(a, 0)$ and $(-a, 0)$

y intercepts: $(0, b)$ and $(0, -b)$

If P is any point on the ellipse, the sum of its distances from the foci is $2a$ if $b < a$ or $2b$ if $b > a$.

Exercises

1. Sketch the graph of $x^2 + 9y^2 = 36$. Where are the foci?

2. Sketch the graph of $x^2 + \frac{1}{9}y^2 = 1$. Where are the foci?

3. Sketch the graphs of $x^2 + 4y^2 = 4$, $x^2 + y^2 = 4$, and $4x^2 + 4y^2 = 4$ on the same set of axes.

THE HYPERBOLA

Definition   A hyperbola is the set of points in the plane for which the *difference* of the distances from two fixed points is constant. These two points are called the *foci*.

To draw a hyperbola requires a mechanical device more elaborate than the one for the ellipse (see Fig. R-4-7); however, we can obtain the equation in the same way as we did for the ellipse. Again let the foci be placed at $F' = (-c, 0)$ and $F = (c, 0)$, and let the difference in question be $2a$, $a > 0$. Since the difference of the distances from the two foci is $2a$ and we must have $|F'P| < |FP| +$

(optional)

Fig. **R-4-7** Mechanical construction of a hyperbola.

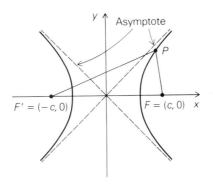

Fig. R-4-8 P is on the hyperbola when
$|PF'| - |PF| = \pm 2a$

$|F'F|$, it follows that $|F'P| - |FP| < |F'F|$, and so $2a < 2c$. Thus we must have $a < c$ (see Fig. R-4-8).

The point $P = (x, y)$ lies on the hyperbola exactly when

$$\sqrt{(x + c)^2 + y^2} - \sqrt{(x - c)^2 + y^2} = \pm 2a$$

After some calculation, we get

$$(a^2 - c^2)x^2 + a^2 y^2 = a^2(a^2 - c^2)$$

If we let $c^2 - a^2 = b^2$ (since $a < c$), we get

$$\frac{x^2}{a^2} - \frac{y^2}{b^2} = 1$$

which is the equation of a hyperbola in standard form.

For x large in magnitude, the hyperbola approaches the lines $y = \pm(b/a)x$, which are called the *asymptotes* of the hyperbola. To see this, for x and y positive we first solve for y in the equation of the hyperbola, obtaining $y = (b/a)\sqrt{x^2 - a^2}$. Subtracting this from the linear function $(b/a)x$, we find that the vertical distance from the hyperbola to the line $y = (b/a)x$ is given by

$$d = \frac{b}{a}(x - \sqrt{x^2 - a^2})$$

To study the behavior of this expression as x becomes large, we multiply by $(x + \sqrt{x^2 - a^2})/(x + \sqrt{x^2 - a^2})$ and simplify to obtain $ab/(x + \sqrt{x^2 - a^2})$. As x becomes larger and larger, the denominator gets larger as well, so the quantity *(optional)* | d comes closer and closer to zero. Thus the hyperbola comes closer and closer to the line. The other quadrants are treated similarly. See Fig. R-4-9.

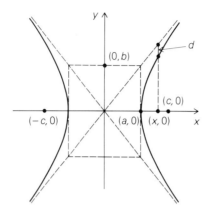

Fig. R-4-9 The vertical distance d from the hyperbola

to its asymptote $y = \dfrac{b}{a}x$ is:

$\dfrac{b}{a}(x - \sqrt{x^2 - a^2}) = \dfrac{ab}{x + \sqrt{x^2 - a^2}}$

Worked Example 3   Sketch the curve $25x^2 - 16y^2 = 400$.

Solution   Dividing by 400, we get the standard form $x^2/16 - y^2/25 = 1$, so $a = 4$ and $b = 5$. The asymptotes are $y = \pm\frac{5}{4}x$, and the curve intersects the x axis at $(\pm 4, 0)$ (see Fig. R-4-10).

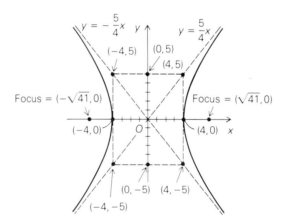

Fig. R-4-10 The hyperbola $25x^2 - 16y^2 = 400$.

If the foci are located on the y axis, the equation of the hyperbola becomes $y^2/b^2 - x^2/a^2 = 1$ (see Fig. R-4-11).

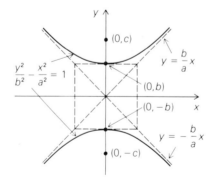

Fig. R-4-11 A hyperbola with foci on the y axis.

Notice that if we draw the rectangle with $(\pm a, 0)$ and $(0, \pm b)$ at the midpoints of its sides, then the asymptotes are the lines through opposite corners, as shown in Figs. R-4-10 and R-4-11.

Worked Example 4   Sketch the graph of $4y^2 - x^2 = 4$.

Solution Dividing by 4, we get $y^2 - x^2/2^2 = 1$, which is in the second standard form
(optional) with $a = 2$ and $b = 1$. The hyperbola and its asymptotes are sketched in Fig.
R-4-12.

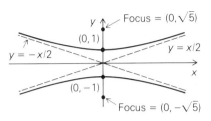

Fig. R-4-12   The hyperbola
$4y^2 - x^2 = 4$.

HYPERBOLA

Case 1:   Foci on x axis

Case 2:   Foci on y axis

Equation: $\dfrac{x^2}{a^2} - \dfrac{y^2}{b^2} = 1$ $\dfrac{y^2}{b^2} - \dfrac{x^2}{a^2} = 1$

Foci: $(\pm c, 0), c = \sqrt{a^2 + b^2}$ $(0, \pm c), c = \sqrt{a^2 + b^2}$

x intercepts: $(\pm a, 0)$ none

y intercepts: none $(0, \pm b)$

Asymptotes: $y = \pm \dfrac{b}{a} x$ $y = \pm \dfrac{b}{a} x$

If P is any point on the hyperbola, the difference between its distances from
the two foci is $2a$ in case 1 and $2b$ in case 2.

Exercises

4. Sketch the graph of $y^2 - x^2 = 2$, showing asymptotes and foci.

5. Sketch the graph of $3x^2 = 2 + y^2$, showing asymptotes and foci.

6. Sketch the graphs of $x^2 + 4y^2 = 4$ and $x^2 - 4y^2 = 4$ on the same set of
   axes.

THE PARABOLA AND THE SHIFTED CONIC SECTIONS

We are already familiar with the circle and parabola from Section R-2. The circle
is a special case of an ellipse in which $a = b$; that is, the foci coincide.

The parabola can be thought of as a special case of the ellipse or hyperbola,
in which one of the foci has moved to infinity. It can also be described as follows:

Definition   A parabola is the set of points in the plane for which the distances
from a fixed point, the *focus*, and a fixed line, the *directrix*, are equal.

Placing the focus at $(0, c)$ and the directrix at the line $y = -c$, we are
led, as above, to an equation relating x and y. Here we have (see Fig. R-4-13)
$|PF| = |PG|$. That is, $\sqrt{x^2 + (y - c)^2} = |y + c|$ and so

$$x^2 + (y - c)^2 = (y + c)^2$$
$$x^2 - 4cy = 0$$
$$y = \frac{x^2}{4c}$$

which is the form of a parabola as given in Section R-2.

(optional) | If we place the focus on the x axis, and use $x = -c$ as the directrix, we get the horizontal parabola $x = y^2/4c$.

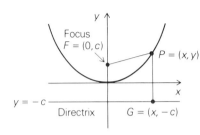

Fig. R-4-13 P is on the parabola when $|PF| = |PG|$.

PARABOLA

Case 1: Focus on y axis   **Case 2:** Focus on x axis

Equation: $y = ax^2$ $\left(a = \dfrac{1}{4c}\right)$ $x = by^2$ $\left(b = \dfrac{1}{4c}\right)$

Focus: $(0, c)$ $(c, 0)$

Directrix: $y = -c$ $x = -c$

If P is any point on the parabola, its distances from the focus and directrix are equal.

In Section R-2 we studied the shifted parabola: $y = ax^2$ became $(y - q) = a(x - p)^2$ if we moved the origin to (p, q). We can do the same for the other conic sections:

SHIFTED CONIC SECTIONS

Shifted ellipse: $\dfrac{(x - p)^2}{a^2} + \dfrac{(y - q)^2}{b^2} = 1$ (shifted circle if $a = b$)

Shifted hyperbola: $\dfrac{(x - p)^2}{a^2} - \dfrac{(y - q)^2}{b^2} = a^2$ (horizontal hyperbola)

                 $\dfrac{(y - q)^2}{b^2} - \dfrac{(x - p)^2}{a^2} = 1$ (vertical hyperbola)

Shifted parabola: $y - q = a(x - p)^2$ (vertical)

                 $x - p = b(y - q)^2$ (horizontal)

We can recognize shifted conic sections by completing the square, as we did with parabolas and circles in Section R-2.

Worked Example 5   Sketch the graph of $x^2 - 4y^2 - 2x + 16y = 19$.

Solution   We complete the square:

$$x^2 - 2x = (x - 1)^2 - 1$$
$$-4y^2 + 16y = -4(y^2 - 4y)$$
$$= -4[(y - 2)^2 - 4]$$

(optional)   Thus

$$0 = x^2 - 4y^2 - 2x + 16y - 19 = (x - 1)^2 - 1 - 4[(y - 2)^2 - 4] - 19$$
$$= (x - 1)^2 - 4(y - 2)^2 - 4$$

Hence our equation is

$$\frac{(x - 1)^2}{4} - (y - 2)^2 = 1$$

which is the hyperbola $x^2/4 - y^2 = 1$ shifted over to $(1, 2)$. See Fig. R-4-14.

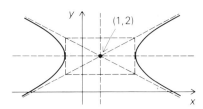

Fig. R-4-14   The hyperbola
$x^2 - 4y^2 - 2x + 16y = 19$.

Similarly, any equation of the form

$$Ax^2 + Cy^2 + Dx + Ey + F = 0$$

describes a conic section (or one or two straight lines, a point, or the empty set). To bring it into standard form, one must first complete the square to find the center of the conic.

Worked Example 6   Sketch the curve $y^2 + x + 3y - 8 = 0$.

Solution   Completing the square, we get $y^2 + 3y = (y + \frac{3}{2})^2 - \frac{9}{4}$, so that $y^2 + x + 3y - 8 = 0$ becomes $(y + \frac{3}{2})^2 + x - \frac{41}{4} = 0$; that is, $x - \frac{41}{4} = -(y + \frac{3}{2})^2$. This is a shifted parabola opening to the left, as in Fig. R-4-15.

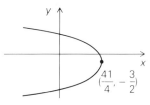

Fig. R-4-15   The parabola
$y^2 + x + 3y - 8 = 0$.

Exercises   7. Put each of the following quadratic expressions into one of the standard forms for a conic section. Identify the resulting curve and sketch it. Indicate the location of important features such as foci, center, directrix, vertex, and asymptotes.

(a) $x^2 + y^2 = 4$ (b) $x^2 + 2x + y^2 - 2y = 2$

(c) $x^2 - y^2 = 4$ (d) $x^2 + 2x - y^2 - 2y = 1$

(e) $2x^2 + 4y^2 - 6y = 8$ (f) $3x^2 - 6x + y - 7 = 0$

8. Find the equations of the curves described as follows:

(a) The circle with center $(2, 3)$ and radius 5

(b) The ellipse consisting of those points whose distances from $(0, 0)$ and $(2, 0)$ sum to 8

(c) The parabola with vertex at $(1, 0)$ and passing through $(0, 1)$ and $(2, 1)$

(d) The circle passing through $(0, 0)$, $(1, \frac{1}{2})$, and $(2, 0)$

(e) The hyperbola with foci at $(0, 2)$ and $(0, -2)$ and passing through $(2, 6)$

1

The Derivative

Differentiation is one of the two fundamental operations of calculus.

Differential calculus is a technique for describing and analyzing *change* in very general situations. The position of a moving object, the population of a city or a bacterial colony, the height of the sun in the sky, the rate of traffic flow on a road, and the price of cheese all change with time. The altitude of a road changes with position along the road; the pressure inside a balloon changes with temperature; the temperature in the atmosphere changes with altitude. The *derivative,* which we will define in this chapter, is a measure of the rate of change of one quantity with respect to another.

CONTENTS

Our calculus course begins in earnest with this, the first of six chapters on differential calculus. In Section 1-1, the derivative is motivated by the problem of finding the velocity of a moving object and the tangent line to a graph. Section 1-2 begins with the definition and basic properties of limits. The derivative is expressed as a limit, and limits are used in the calculation of derivatives. The basic algebraic rules for differentiation are presented in Section 1-3; these rules are essential for everything which follows. Section 1-4 is devoted to inverting the operation of differentiation; this new operation is closely related to integration and is important in many applications of calculus.

1-1 Introduction to the Derivative

Velocities and slopes are both derivatives.

This section introduces the basic idea of the derivative by studying two problems. The first is the problem of finding the velocity of a moving object, and the second is the problem of finding the equation of the line tangent to a graph. The next section will take up the definition of the derivative in a more formal manner.

Goals

After studying this section you should be able to:

Calculate velocities of objects moving according to given formulas in simple cases.

Calculate the equation of the line tangent to the graph of a simple function.

Differentiate quadratic functions.

VELOCITY

To illustrate the notion of the derivative, which will occupy us throughout the first portion of this book, we begin with an example.

Consider a bus whose motion along a straight highway is described by a function $y = f(x)$ where

x = time, measured in seconds from some designated starting position

y = the bus's position, measured in meters from the designated starting position (to the front of the bus)

See Fig. 1-1-1. The velocity of the bus at any given moment, measured in meters per second, is a definite physical quantity; it can be measured by a speedometer on the bus or by a stationary radar device.

Fig. 1-1-1 What is the velocity of the bus in terms of its position?

We wish to study this problem: given a formula for the position of the bus, such as $f(x) = x + 2x^2$, how can we calculate the velocity of the bus at any specific time, say $x_0 = 3$? We may call this the *velocity problem*.

Before the calculus was invented, velocity problems were solved by various geometric techniques. What Newton and Leibniz did was to discover a general method which led to efficient rules for computing velocities. We shall develop most of these rules in Section 1-3.

There is one kind of motion, called *uniform motion,* for which the velocity problem has an easy solution. We shall describe it in some detail, since the solution for general functions is based on it.

Suppose that the position of the bus is measured at time x_0 and again at a later time x. These positions are given by $y_0 = f(x_0)$ and $y = f(x)$. Let $\Delta x = x - x_0$ denote the elapsed time between x_0 and x. Here Δ is the capital Greek letter "delta," which corresponds to the Roman D and stands for *d*ifference. The combination "Δx" read "delta-x" is not the product of Δ and x but rather a single entity denoting the difference between two values of x. Similarly we write $\Delta y = y - y_0$ for the difference in the positions. Notice that $x = x_0 + \Delta x$ so that

$$\Delta y = y - y_0$$
$$= f(x) - f(x_0)$$

i.e.,

$$\Delta y = f(x_0 + \Delta x) - f(x_0)$$

The motion is called *uniform* if the ratio

$$\frac{\text{distance travelled}}{\text{elapsed time}} = \frac{\Delta y}{\Delta x} = \frac{f(x_0 + \Delta x) - f(x_0)}{\Delta x}$$

is the same for all choices of x_0 and Δx. This ratio is the *velocity* v of the bus. We thus have the equation: $\Delta y / \Delta x = v$, i.e., $(y = y_0)/(x - x_0) = v$, which we can rewrite as $y = y_0 + v(x - x_0)$. We recognize this as the equation of a straight line with slope v. Letting $m = v$ and $b = y_0 - vx_0$, we can rewrite it in slope-intercept form: $y = mx + b$. (See Fig. 1-1-2.)

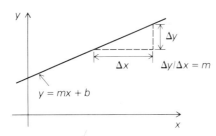

Fig. 1-1-2   If $\Delta y / \Delta x$ is constant, then the graph of y against x is a straight line.

Conversely, if y is a linear function of x, i.e., if $y = mx + b$, then for any x_0 and Δx we have

$$\frac{\Delta y}{\Delta x} = \frac{f(x_0 + \Delta x) - f(x_0)}{\Delta x}$$
$$= \frac{m(x_0 + \Delta x) - mx_0}{\Delta x}$$
$$= \frac{m \, \Delta x}{\Delta x} = m$$

Thus the motion is uniform with velocity $v = m$, the slope of the line.

UNIFORM MOTION

If the position of an object is given by	then its velocity is
$$y = mx + b$$	$$v = m = \frac{\Delta y}{\Delta x} \quad \text{for any } \Delta x$$

Worked Example 1   A particle moving on a line has position $y = 3x - 5$ at time x. What is its velocity at time $x = 8$?

Solution   Comparison of $y = 3x - 5$ with $y = mx + b$ gives $m = 3$ and $b = -5$. Thus the velocity is $v = 3$ at all times, and in particular, at $x = 8$. (The graph of position versus time is shown in Fig. 1-1-3.)

Fig. 1-1-3   The graph of $y = 3x - 5$.

Now suppose that the motion of the bus is not uniform. Then the ratio of distance travelled to elapsed time

$$\frac{\Delta y}{\Delta x} = \frac{f(x) - f(x_0)}{x - x_0}$$

depends on which x_0 and x are chosen. We call $\Delta y / \Delta x$ the *average velocity* during the time interval Δx. On the graph, the average velocity is the slope of the line through the points $(x_0, f(x_0))$ and $(x, f(x))$ as in Fig. 1-1-4. But the speedometer does not measure the average velocity; it measures the instantaneous velocity, that is, the velocity at a certain moment. Nevertheless, if Δx is very small, we expect the average velocity $\Delta y / \Delta x$ to be close to the instantaneous velocity v; the error $(\Delta y / \Delta x) - v$ should become arbitrarily small as Δx approaches zero. We do *not* make $\Delta x = 0$ for then $\Delta y = 0$ as well, and $\Delta y / \Delta x = 0/0$ is meaningless. Instead, we calculate $\Delta y / \Delta x$ for small Δx and see if $\Delta y / \Delta x$ becomes close to some number if Δx is made arbitrarily small; that number is the instantaneous velocity.* The procedure is illustrated in the next Worked Example.

Fig. 1-1-4   The average velocity over the interval $[x_0, x]$ equals the slope of the straight line through $(x_0, f(x_0))$ and $(x, f(x))$.

Worked Example 2   The position of a bus at time x (in seconds) is $f(x) = 1 + 2x^2$ meters from a reference point, measured along a straight road. Calculate its (instantaneous) velocity at $x_0 = 3$.

---

*One can legitimately debate the question: "*just what* is *velocity, a pre-existing entity we are measuring or a quantity we are* defining *to be the value $\Delta y / \Delta x$ approaches as Δx approaches zero?" Our view is that velocity is a physical quantity that can be measured; mathematical definitions are arranged to* correspond *to the physical situation so that the consequent mathematical deductions can yield physical conclusions.*

Solution   We choose Δx arbitrarily and calculate the average velocity for a time interval Δx starting at time $x_0 = 3$:

$$\frac{\Delta y}{\Delta x} = \frac{f(3 + \Delta x) - f(3)}{\Delta x}$$

$$= \frac{[1 + 2(3 + \Delta x)^2] + [1 + 2 \cdot 3^2]}{\Delta x}$$

$$= \frac{[1 + 2(9 + 6\Delta x + (\Delta x)^2)] - [1 + 2 \cdot 9]}{\Delta x}$$

$$= \frac{12\Delta x + 2(\Delta x)^2}{\Delta x} = 12 + 2\Delta x$$

If we let Δx get arbitrarily small in this last expression, $2\Delta x$ gets arbitrarily small as well, and so $\Delta y / \Delta x = 12 + 2\Delta x$ gets close to 12. Thus the required (instantaneous) velocity at $x_0 = 3$ is 12 meters per second.

In this book, we shall use "velocity" to mean "instantaneous velocity."

NONUNIFORM MOTION

To calculate the velocity v at time x_0 of an object whose position at time x is $y = f(x)$:

1. Form the average velocity

$$\frac{\Delta y}{\Delta x} = \frac{f(x_0 + \Delta x) - f(x_0)}{\Delta x}$$

2. Simplify your expression for $\Delta y / \Delta x$ as much as possible, cancelling Δx from numerator and denominator, if you can.

3. Find the number v that is approached by $\Delta y / \Delta x$ as Δx is taken closer and closer to zero.

This method works well for simple functions but becomes unwieldy even for moderately complicated functions like $f(x) = (1 + 2x^2)/(1 + x^4)$. The rules of calculus developed in Section 1-3 will enable us to solve such problems easily, completely bypassing the method above.

Solved Exercises*   1. The position y in meters (from a fixed reference point on a straight road) at time x in seconds (from a fixed starting time) is given by $y = 8x + 8$. Find the velocity at time $x_0 = 2$.

2. Repeat Solved Exercise 1 for $y = 3x^2 + 8x$ and $x_0 = 1$.

Exercises†   In the following exercises, y represents the position in meters (from a fixed reference point on a straight road) and x represents the time in

---

*Solutions are found in Appendix A.
†Answers and partial solutions to exercises are found in the Student Guide.

seconds (from a fixed starting time). Find the velocity at the specified time x_0:

1. $y = 2x - 3$; $x_0 = 3$ 2. $y = 10x + 18$; $x_0 = 5$

3. $y = x^2 + 1$; $x_0 = 2$ 4. $y = 3x^2 + 18x - 8$; $x_0 = 2$

TANGENT LINES AND THE DERIVATIVE

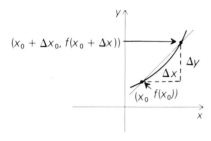

Fig. 1-1-5 $\Delta y / \Delta x$ is the slope of the secant line.

Let us now consider the geometric problem of finding the equation of the line tangent to the graph of a function $y = f(x)$ at a given point $(x_0, f(x_0))$ on the graph. We may call this the *slope problem*.

To solve this problem, let $\Delta x \neq 0$ be a number, and draw the line through the points $(x_0, f(x_0))$ and $(x_0 + \Delta x, f(x_0 + \Delta x))$; see Fig. 1-1-5. This line is called a secant line and $\Delta y / \Delta x = (f(x_0 + \Delta x) - f(x_0))/\Delta x$ is its slope.

As Δx becomes small, x_0 being fixed, it appears that the secant line comes close to the tangent line, so that the slope $\Delta y / \Delta x$ of the secant line comes close to the slope of the tangent line. (See Fig. 1-1-6.)

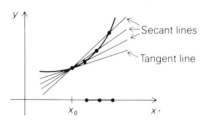

Fig. 1-1-6 The secant line comes close to the tangent line as the second point moves close to x_0.

Worked Example 3   Let $f(x) = x^2$ and $x_0 = 3$. What is the value of $\Delta y / \Delta x$ for $\Delta x = 1$, 0.1, 0.01, 0.0001? What number does $\Delta y / \Delta x$ come close to as Δx becomes small?

Solution   By definition, $\Delta y = f(x_0 + \Delta x) - f(x_0)$. When $f(x) = x^2$ and $x_0 = 3$, we find that $\Delta y = (3 + \Delta x)^2 - 3^2 = 9 + 6\Delta x + (\Delta x)^2 - 9 = 6\Delta x + (\Delta x)^2$, and so

$$\frac{\Delta y}{\Delta x} = 6 + \Delta x \qquad \text{for } \Delta x \neq 0$$

Substituting the given values of Δx, we obtain:

Δx	1	0.1	0.01	0.0001
$\Delta y / \Delta x$	7	6.1	6.01	6.0001

As Δx comes close to zero, $\Delta y / \Delta x = 6 + \Delta x$ comes close to 6.

The number m that $\Delta y / \Delta x$ gets close to as Δx gets near zero will be taken to be the slope of the line tangent to the graph of f at $(x_0, f(x_0))$. Thus, from Worked Example 3, the slope of the line tangent to the graph of $y = x^2$ at $x_0 = 3$ is $m = 6$.

Notice that the procedure for calculating the slope of the tangent line is identical to that for calculating velocity. The slope of the line tangent to the graph of f at $(x_0, f(x_0))$ thus equals the velocity of an object whose position at time x is $y = f(x)$. This common value is denoted $f'(x_0)$ and is called *the derivative* of f at x_0.

THE DERIVATIVE

The derivative of $y = f(x)$ at x_0 is denoted $f'(x_0)$ and is

the value, if there is one, that

$$\frac{\Delta y}{\Delta x} = \frac{f(x_0 + \Delta x) - f(x_0)}{\Delta x}$$

gets close to as Δx gets near zero.

Since the tangent line to the graph of $y = f(x)$ at $(x_0, f(x_0))$ has slope $f'(x_0)$ and passes through the point $(x_0, f(x_0))$, the point-slope form of the equation of a straight line (see Section R-2) yields the following:

EQUATION OF THE TANGENT LINE

The equation of the line tangent to the graph of $y = f(x)$ at $(x_0, f(x_0))$ is

$y = y_0 + f'(x_0)(x - x_0)$,
where $y_0 = f(x_0)$.

Worked Example 4 Find the equation of the line tangent to the graph of $y = 3x^2 + 2$ at $x_0 = 2$. Sketch.

Solution First we compute the derivative $f'(2)$ at $x_0 = 2$, where $f(x) = 3x^2 + 2$. To do so, we form

$$\frac{\Delta y}{\Delta x} = \frac{f(2 + \Delta x) - f(2)}{\Delta x}$$

$$= \frac{[3(2 + \Delta x)^2 + 2] - [3 \cdot 2^2 + 2]}{\Delta x}$$

$$= \frac{[3(4 + 4\Delta x + (\Delta x)^2) + 2] - [3 \cdot 4 + 2]}{\Delta x}$$

$$= 12 + 3\Delta x$$

which approaches 12 as Δx approaches zero. Thus $f'(2) = 12$. Also, $y_0 = f(2) = 14$ so substitution into the equation of the tangent line in the preceding display gives

$$y = 14 + 12(x - 2)$$

or

$$y = 12x - 10$$

The parabola and its tangent line are sketched in Fig. 1-1-7.

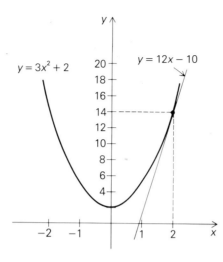

$y = 3x^2 + 2$

$y = 12x - 10$

Fig. 1-1-7 The parabola $y = 3x^2 + 2$ and its tangent line at $(2, 14)$.

Solved Exercises

3. Find the derivative of $f(x) = 12x + 2$ at $x_0 = 10$.

4. Find the equation of the line tangent to the graph of $y = 10x^2 + 2x$ at $x_0 = 1$.

Exercises

5. Find the derivative of $f(x) = x + 2$ at $x_0 = 0$.

6. If $f(x) = 2x + 8$, find $f'(0)$.

7. Find the equation of the line tangent to the graph of $y = 8x^2 - x + 1$ at $x_0 = 2$.

8. Where does the line tangent to the graph of $y = x^2$ at $x_0 = 2$ intersect the x-axis?

THE DERIVATIVE AS A FUNCTION

The preceding examples show how derivatives may be calculated directly from the definition. Usually, we will not use this cumbersome method; instead, we will use *differentiation rules*. These rules, once derived, enable us to differentiate many functions quite simply. In this section, we will content ourselves with deriving the rules for differentiating linear and quadratic functions. General rules will be introduced in Section 1-3 after we have studied the method of limits more thoroughly.

The following rule will enable us to find the tangent line to any parabola at any point.

QUADRATIC FUNCTION RULE

Let $f(x) = ax^2 + bx + c$,
where a, b, and c are constants
and let x_0 be any real number.

Then
$f'(x_0) = 2ax_0 + b$.

To justify the quadratic function rule, we form the quotient

$$\frac{\Delta y}{\Delta x} = \frac{f(x_0 + \Delta x) - f(x_0)}{\Delta x}$$

$$= \frac{a(x_0 + \Delta x)^2 + b(x_0 + \Delta x) + c - ax_0^2 - bx_0 - c}{\Delta x}$$

$$= \frac{ax_0^2 + 2ax_0\Delta x + a(\Delta x)^2 + bx_0 + b\Delta x + c - ax_0^2 - bx_0 - c}{\Delta x}$$

$$= \frac{2ax_0\Delta x + a(\Delta x)^2 + b\Delta x}{\Delta x}$$

$$= 2ax_0 + b + a(\Delta x)$$

As Δx gets near zero, $a(\Delta x)$ does as well, and so $\Delta y/\Delta x$ gets close to $2ax_0 + b$. Therefore $2ax_0 + b$ is the derivative of $ax^2 + bx + c$ at $x = x_0$.

Worked Example 5   Find the derivative at -2 of $f(x) = 3x^2 + 2x - 1$.

Solution   Applying the quadratic function rule with $a = 3$, $b = 2$, $c = -1$, and $x_0 = -2$, we find $f'(-2) = 2(3)(-2) + 2 = -10$.

We can use the quadratic function rule to obtain quickly a fact which may be known to you from analytic geometry.

Worked Example 6   Suppose that $a \neq 0$. At which point does the parabola $y = ax^2 + bx + c$ have a horizontal tangent line?

Solution   The slope of the tangent line through the point $(x_0, ax_0^2 + bx_0 + c)$ is $2ax_0 + b$. This line is horizontal when its slope is zero—that is, when $2ax_0 + b = 0$, so $x_0 = -b/2a$. The y value here is $a(-b/2a)^2 + b(-b/2a) + c = b^2/4a - b^2/2a + c = -(b^2/4a) + c$. The point $(-b/2a, -b^2/4a + c)$ is called the *vertex* of the parabola $y = ax^2 + bx + c$. (See Fig. 1-1-8.)

In the quadratic function rule we did not require that $a \neq 0$. When $a = 0$, the function $f(x) = ax^2 + bx + c$ is linear, and when $a = 0$ and $b = 0$, the function $f(x) = ax^2 + bx + c$ is constant. The following display summarizes these two special cases.

Fig. 1-1-8 The vertex of the parabola $y = ax^2 + bx + c$ is at $(-b/2a, (-b^2/4a) + c)$. In this graph, $a > 0$.

LINEAR FUNCTION RULE	
If $f(x) = bx + c$ where b and c are constants and if x_0 is any real number,	then $f'(x_0) = b$.
If $f(x) = c$, where c is a constant and x_0 is any real number,	then $f'(x_0) = 0$.

The linear function rule tells us that the derivative of a linear function is the slope of its graph. Note that it does not depend on x_0: the rate of change of a linear function is *constant*. For instance, if $f(x) = 3x + 4$ then $f'(x_0) = 3$ for any x_0, and if $g(x) = 4$ then $g'(x_0) = 0$ for any x_0. For a general quadratic function, though, the derivative $f'(x_0)$ does depend upon the point x_0 at which the derivative is taken. In fact, we can consider f' as a *new function;* writing the letter x instead of x_0, we have $f'(x) = 2ax + b$.

Definition Let f be a given function. The *derivative* of f is the new function f' whose value $f'(x)$ at x (if such a value exists) is the derivative of f at x, as defined by the display on p. 53 with x substituted for x_0.

Worked Example 7 What is the derivative of $f(x) = 3x^2 - 2x + 1$?

Solution By the quadratic function rule, $f'(x_0) = 2 \cdot 3x_0 - 2 = 6x_0 - 2$. Writing x instead of x_0, we find that the derivative of $f(x) = 3x^2 - 2x + 1$ is $f'(x) = 6x - 2$.

When we are dealing with functions given by specific formulas, we often omit the function names. For example, we could state the result of Worked Example 7 as "the derivative of $3x^2 - 2x + 1$ is $6x - 2$."

Since the derivative of a function f is another function f', we can go on to differentiate f' again. The result is yet another function, called the *second derivative* of f and denoted by f''.

Worked Example 8 Find the second derivatives of (a) $f(x) = 3x^2 - 2x + 1$ and (b) $f(x) = 10x + 2$.

Solution (a) We must differentiate $f'(x) = 6x - 2$. This is a linear function; applying the formula for the derivative of a linear function, we find $f''(x) = 6$. The second derivative of $3x^2 - 2x + 1$ is thus the constant function whose value for every x is equal to 6.
(b) The first derivative is $f'(x) = 10$. Differentiating again gives zero. Thus, $f''(x) = 0$.

SECOND DERIVATIVES

To compute the second derivative $f''(x)$:

1. Compute the first derivative $f'(x)$,
2. Calculate the derivative of $f'(x)$.

We end with a remark on notation. It is not necessary to represent functions by f and independent and dependent variables by x and y; as long as we say what we are doing, we can use any letters we wish.

Worked Example 9 (a) Let $g(a) = 4a^2 + 3a - 2$. What is $g'(a)$? What is $g'(2)$?

(b) A ball thrown downwards from a balloon has height $100 - 10t - 4.9t^2$ meters after t seconds. Finds its velocity at $t = 2$.

Solution (a) If $f(x) = 4x^2 + 3x - 2$, we know $f'(x) = 8x + 3$. Using g instead of f and a instead of x, we have $g'(a) = 8a + 3$. Finally, $g'(2) = 8 \cdot 2 + 3 = 19$.

(b) The height of time t is $h(t) = 100 - 10t - 4.9t^2$. By the quadratic function rule, $h'(t) = -10 - 9.8t$. Thus, $h'(2) = -10 - (9.8)(2) = -29.6$ and so the velocity is 29.6 meters per second *downward*.

DIFFERENTIATING THE SIMPLEST FUNCTIONS

The derivative of the quadratic function $f(x) = ax^2 + bx + c$ is the linear function $f'(x) = 2ax + b$.

The derivative of the linear function $f(x) = bx + c$ is the constant function $f'(x) = b$.

The derivative of the constant function $f(x) = c$ is the zero function $f'(x) = 0$.

Solved Exercises

5. Let $f(x) = 3x + 1$. What is $f'(8)$?

6. An apple falls from a tall tree toward the earth. After t seconds, it has fallen $4.9t^2$ meters. What is the velocity of the apple when $t = 3$?

7. Find the equation of the line tangent to the graph of $f(x) = 3x^2 + 4x + 2$ at the point where $x_0 = 1$.

8. For which functions $f(x) = ax^2 + bx + c$ is the second derivative equal to the zero function?

Exercises

9. Differentiate the following functions:

   (a) $f(x) = x^2 + 3x - 1$ (b) $f(x) = (x - 1)(x + 1)$

   (c) $f(x) = -3x + 4$ (d) $g(t) = -4t^2 + 3t + 6$

10. A ball is thrown upward at $t = 0$; its height in meters until it strikes the ground is $24.5t - 4.9t^2$ where the time is t seconds. Find:

    (a) The velocity at $t = 0, 1, 2, 3, 4, 5$.

    (b) The time when the ball is at its highest point.

    (c) The time when the velocity is zero.

    (d) The time when the ball strikes the ground.

11. Find the tangent line to the parabola $y = x^2 - 3x + 1$ when $x_0 = 2$. Sketch.

12. Find the second derivative of each of the following:

    (a) $f(x) = x^2 - 5$ (b) $f(x) = x - 2$

    (c) A function whose derivative is $3x^2 - 7$

RATES OF CHANGE

The derivative concept applies to more situations than just velocities and slopes. It leads to the idea of rates of change in general. Let us begin with a study of the general situation when two quantities are related linearly, which is the analog of uniform motion.

Suppose that two quantities x and y are related in such a way that a change Δx in x always produces a change Δy in y which is proportional to Δx; that is, the ratio $\Delta y / \Delta x$ equals a constant, m. We say that y changes *proportionally* or *linearly* with x.

For instance, consider a hanging spring to which objects may be attached. Let x be the weight in grams of the object, and let y be the resulting length in centimeters of the spring. It is an experimental fact, called Hooke's law, that (for values of Δx which are not too large) a change Δx in the weight of the object produces a proportional change Δy in the length of the spring. (See Fig. 1-1-9.)

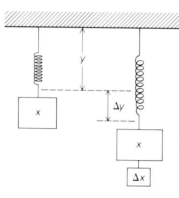

Fig. 1-1-9 Hooke's law states that the change in length Δy is proportional to the change in weight Δx.

If we graph y against x, we get a segment of a straight line with slope

$$m = \frac{\Delta y}{\Delta x}$$

as shown in Fig. 1-1-2. The equation of the line is $y = mx + b$, and the function $f(x) = mx + b$ is a *linear function*. The slope m of a straight line thus represents *the rate of change of y with respect to x*. (The quantity b is the length of the spring with no weight.)

LINEAR OR PROPORTIONAL CHANGE

y is related to x by a linear function: $y = mx + b$ $\dfrac{\Delta y}{\Delta x} = m$

Worked Example 10   For temperatures in the range $[-50, 150]$ (degrees Celsius), the pressure in a certain closed container of gas changes linearly with the temperature. Suppose that a $40°$ increase in temperature causes the pressure to increase by 30 millibars (a millibar is one-thousandth of the average atmospheric pressure at sea level). (a) What is the rate of change of pressure with respect to temperature? (b) What change of temperature would cause the pressure to drop by 9 millibars?

Solution   (a) Let T be the temperature and P be the pressure. When $\Delta T = 40$, then $\Delta P = 30$, so the rate of change of pressure with respect to temperature is $\Delta P / \Delta T = 30/40 = \frac{3}{4}$ (millibars per degree). Note that, in this example, the role of x in the general discussion is played by T; P plays the role of y, and $m = \frac{3}{4}$. (b) If the pressure is to drop 9 millibars, we must have $\Delta P = -9$. From $\Delta P / \Delta T = \frac{3}{4}$, we find $\Delta T = \frac{4}{3}\Delta P = -\frac{4}{3}(9) = -12$. Hence a temperature drop of 12°C causes a pressure drop of 9 millibars.

If two quantities x and y are related by a function $y = f(x)$ then the value that $\Delta y / \Delta x = (f(x_0 + \Delta x) - f(x_0))/\Delta x$ approaches as Δx gets small is the *rate of change of y with respect to x at x_0*; it equals the derivative $f'(x_0)$.

RATE OF CHANGE	
If two quantities x and y are related by $y = f(x)$,	$f'(x_0)$ measures the number of units of increase of y per unit increase of x at the value x_0.

Worked Example 11   The area of a circle of radius r is $A = \pi r^2$. Find the rate of change of A with respect to r at $r_0 = 5$.

Solution   Here $y = A$ and $x = r$ and $f(x) = \pi x^2$. Since π is a constant, the quadratic function rule gives $f'(x_0) = 2\pi x_0$. At $x_0 = 5$ we get $f'(x_0) = 10\pi$; thus the area is increasing at a rate of 10π square units per unit increase in radius when the radius is 5.

If $f(x)$ is the position (in meters on a straight road from a designated starting position) of a bus at time x (in seconds), we know that $f'(x)$, the rate of change of position with respect to time, is the velocity of the bus. The rate of change of velocity with respect to time is called the *acceleration*. Thus the acceleration at time x equals $f''(x)$. It is measured in meters per second per second, i.e., meters per second^2. If distance is measured in feet, the acceleration is measured in feet per second^2.

ACCELERATION	
If the position of an object is $f(x)$ at time x,	
the velocity at time x is	$v = f'(x)$
and the acceleration at time x is	$a = f''(x)$

Worked Example 12   A dragster travels the $\frac{1}{4}$ mile strip in 6.00 seconds, its distance from the start in feet after x seconds being $f(x) = 44x^2/3 + 132x$
(a) Find its velocity and acceleration as it crosses the finish line.
(b) How fast was it going halfway down the strip?

Solution   (a) The velocity at time x is $v = f'(x) = 88x/3 + 132$, and the acceleration is $a = f''(x) = 88/3$. Substituting $x = 6.00$, we get $v = 308$ feet per second ($= 210$ miles per hour) and $a = 29.3$ feet per second^2.

(b) To find the velocity halfway down, we do *not* substitute $x = 3.00$ in $v = f'(x)$ . . . that would be its velocity at half time. The total distance covered is $f(6) = (44)(36)/3 + (132)(6) = 1320$ feet ($= \frac{1}{4}$ mile). Thus, half the distance is 660 feet. To find the time x corresponding to the distance $y = 660$, we write $f(x) = 660$ and solve for x using the quadratic formula:

$$\frac{44x^2}{3} + 132x = 660$$

$$x^2 + 9x - 45 = 0 \qquad \text{(multiply by } \tfrac{3}{44})$$

$$x = \frac{-9 \pm \sqrt{81 + 180}}{2} = -12.58, \; 3.58$$

Since the time during the race is positive, we discard the negative root and retain $x = 3.58$. Substituting into $v = f'(x) = 88x/3 + 132$ gives $v \approx 237$ feet per second (≈ 162 miles per hour).

Rates of change also arise in many situations in economics. The following is an illustration.

Worked Example 13 Consider a factory in which, if x man-hours of labor are employed, then $y = f(x)$ dollars worth of output are produced.
(optional)
(a) If y changes proportionally with x, express $\Delta y / \Delta x$ in words.
(b) Express $f'(x_0)$ in words if f is not necessarily linear.

Solution (a) $\Delta y = f(x_0 + \Delta x) - f(x_0)$ represents the amount of extra output produced if Δx extra man-hours of labor are employed. Thus, $\Delta y / \Delta x$ is the output per man-hour. This number is called the *productivity* of labor.
(b) $\Delta y / \Delta x$ is the extra output per extra man-hour of extra labor when Δx extra man-hours are employed. The limiting value as Δx gets very small is $f'(x_0)$. Thus, if x_0 man-hours are presently employed, the value of $f'(x_0)$ represents the additional output per additional man-hour of extra labor which is employed. One calls $f'(x_0)$ the *marginal productivity* at level x_0.

In Fig. 1-1-10 we sketch a possible productivity curve $y = f(x)$. Notice that as x_0 becomes larger and larger, the marginal productivity $f'(x_0) = $ (dollars of output per man-hour at level x_0) gets smaller. One says that *the law of diminishing returns* applies.

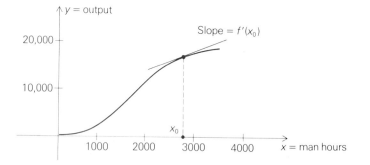

Fig. 1-1-10 A possible productivity curve; the slope of the tangent line is the marginal productivity.

Solved Exercises

9. Suppose that y changes proportionally with x and the rate of change is 3. If $y = 2$ when $x = 0$, find the equation relating y to x.

10. Let S denote the supply of hogs in Chicago, measured in thousands, and let P denote the price of pork in cents per pound. Suppose that, for S between 0 and 100, P changes linearly with S. On April 1, $S = 50$ and $P = 163$; on April 3, a rise in S of 10 leads to a decline in P to 161. What happens if S falls to 30?

11. Suppose that the price of pork P depends on the supply S by the formula $P = 160 - (.03)S + (.01)S^2$. Find the rate of change of P with respect to S when $S = 50$.

12. Find the acceleration of the apple in Solved Exercise 6 when $t = 3$.

Exercises

13. It will take a certain woman seven bags of cement to build a 6-meter-long sidewalk of uniform width and thickness. Her husband offers to contribute enough of his own labor to extend the sidewalk to 7 meters. How much more cement do they need?

14. A rock is thrown straight down the face of a vertical cliff with an initial velocity of 3 meters per second. Two seconds later, the rock is falling at a velocity of 22.6 meters per second. Assuming that the velocity v changes proportionally with time t, find the equation relating v to t. How fast is the rock falling after 15 seconds?

15. In November 1974, Mr. A used 302 kilowatt-hours of electricity and paid \$8.95 to do so. In December 1974, he paid \$10.23 for 366 kilowatt-hours. Assuming that the cost of electricity changes linearly with the amount used, how much would Mr. A have to pay if he used no electricity at all? Assuming that he could reduce his bill to zero by giving some electricity back to the power company, how much would he have to give them?

16. If the height H in feet of a certain species of tree depends on its base diameter d in feet through the formula $H = 56d - 3d^2$, find the rate of change of H with respect to d at $d = 0.5$.

17. The height of a pebble dropped off a building at time $t = 0$ is $h(t) = 44.1 - 4.9t^2$ meters at time t. The pebble strikes the ground at $t = 3.00$ seconds.

   (a) What is its velocity and acceleration when it strikes the ground?

   (b) What is its velocity when it is half-way down the building?

Problems for Section 1-1

1. Find the velocities and accelerations at the indicated times of the objects whose positions y (in meters) as a function of time x (in seconds) are given by the following formulas:

   (a) $y = 2x - 1$; $x_0 = 1$

   (b) $y = 8x + 2$; $x_0 = 2$

   (c) $y = -5x^2$; $x_0 = 1$ (d) $y = 5x^2$; $x_0 = 0$

   (e) $y = 8x^2 - 2x + 1$; $x_0 = 2$

   (f) $y = 3x^2 - x$; $x_0 = 1$

2. Find the velocities and accelerations at the indicated times of the particles whose positions y (in meters) on a line are given by the following functions of time t (in seconds):

   (a) $y = 3t + 2$; $t_0 = 1$

   (b) $y = 5t - 1$; $t_0 = 0$

   (c) $y = 8t^2 + 1$; $t_0 = 0$

   (d) $y = 18t^2 - 2t + 5$; $t_0 = 2$

   (e) $y = 5t^2 - t$; $t_0 = 0$

   (f) $y = t - 5t^2$; $t_0 = 1$

3. Find the equation of the line tangent to the graphs of the following functions at the indicated points:

   (a) $y = x^2$; $x_0 = 1$ (b) $y = -x^2$; $x_0 = 2$

   (c) $y = 5x^2 - 3x + 1$; $x_0 = 0$

   (d) $y = x + 1 - x^2$; $x_0 = 2$

4. Find the equation of the line tangent to the graphs of the following functions at the indicated points and sketch:

   (a) $y = 1 - x^2$; $x_0 = 1$

   (b) $y = 1 - x$; $x_0 = 0$

   (c) $y = x^2 - 2x + 1$; $x_0 = 2$

   (d) $y = 3x^2 + 1 - x$; $x_0 = 5$

5. Find the derivative of each of the following functions by finding the value $\Delta y / \Delta x$ approaches for Δx small:

   (a) $4x^2 + 3x + 2$ (b) $(x - 3)(x + 1)$

   (c) $1 - x^2$ (d) $-x^2$

   (e) $-2x^2 + 5x$ (f) $1 - x$

6. Find the derivatives in Problem 5 using the quadratic function rule.

7. Find the following derivatives:

   (a) $f(x) = x^2 - 2$; find $f'(3)$.

   (b) $f(x) = 1$; find $f'(7)$.

   (c) $f(x) = -13x^2 - 9x + 5$; find $f'(1)$.

   (d) $g(s) = 0$; find $g'(3)$.

   (e) $k(y) = (y + 4)(y - 7)$; find $k'(-1)$.

   (f) $x(f) = 1 - f^2$; find $x'(0)$.

   (g) $f(x) = -x + 2$; find $f'(3.752764)$.

8. Assume that y changes proportionally with x and the rate of change is r. In each case, find y as a function of x.

   (a) $r = 5$
       $y = 1$ when $x = 4$

   (b) $r = -2$
       $y = 10$ when $x = 15$

   (c) $r = \frac{1}{2}$
       $y = 1$ when $x = 3$

9. In each of the following situations, assume that y is a linear function of x. What name would you give to the rate of change of y with respect to x? In what units could this rate be expressed?

   (a) $y =$ distance driven in an automobile; $x =$ amount of fuel used.

   (b) $x =$ distance driven in an automobile; $y =$ amount of fuel used.

   (c) $x =$ amount of fuel purchased; $y =$ amount of money paid for fuel.

   (d) $x =$ distance driven in an automobile; $y =$ amount of money paid for fuel.

10. If the price of electricity changes proportionally with time, and if the price goes from 2 cents per kilowatt-hour in 1978 to 3.2 cents per kilowatt-hour in 1980, what is the rate of change of price with respect to time? When will the price be 5 cents per kilowatt-hour? What will the price be in 1987?

11. Let $f(x) = 2x^2 - 5x + 2$, $g(x) = \frac{3}{4}x^2 + 2x$, and $h(x) = -3x^2 + x + 3$.

    (a) Find the derivative of $f(x) + g(x)$ at $x = 1$.

    (b) Find the derivative of $3f(x) - 2h(x)$ at $x = 0$.

    (c) Find the equation of the tangent line to the graph of $f(x)$ at $x = 1$.

    (d) Find the second derivative of $f(x)$ at $x = 2$.

    (e) Find the second derivative of $3f(x) + g(x)$ at $x = 1$.

12. Two trains, A and B, are moving on adjacent tracks with positions given by the functions $A(t) = t^2 + t + 5$ and $B(t) = 3t + 4$. What is the best time for a hobo on train B to make a moving transfer to train A? (You may assume that the trains are long enough so that a part of train A will be alongside the hobo at the time you select.)

13. Suppose that tension T of a muscle is related to the time t of exertion by $T = 5 + 3t - t^2$, $0 \le t \le 3/2$. Find the rate of change of T with respect to t at $t = 1$.

14. Find the lines through the point $(4, 7)$ which are tangent to the graph of $y = x^2$. Sketch.

★15. Given a point $(\bar{x}, \bar{y})$, find a general rule for determining how many lines through the point are tangent to the parabola $y = x^2$.

16. Let $f(t) = 2t^2 - 5t + 2$ be the position of object A and let $h(t) = -3t^2 + t + 3$ be the position of object B.

    (a) When is A moving faster than B?

    (b) How fast is B going when A stops?

    (c) When does B change direction?

17. How do the graphs of functions $ax^2 + bx + c$ whose second derivative is positive compare with those for which the second derivative is negative and those for which the second derivative is zero?

★ *A star denotes a somewhat more challenging problem.*

18. In Fig. 1-1-11, show that P is the midpoint of OQ.

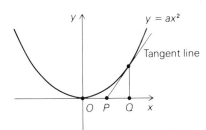

Fig. 1-1-11 P lies halfway between O and Q.

★ 19. Let R be any point on the parabola $y = x^2$. (a) Draw the horizontal line through R. (b) Draw the perpendicular to the tangent line at R. Show that the distance between the points where these lines cross the y axis is equal to $\frac{1}{2}$, regardless of the value of x. (Assume, however, that $x \neq 0$.)

★ 20. If $f(x) = ax^2 + bx + c = a(x - r)(x - s)$ (r and s are the roots of f), show that the values of $f'(x)$ at r and s are negatives of one another. Explain this by appeal to the symmetry of the graph.

21. Let $f(x) = 2x^2 + 3x + 1$.
    (a) For which values of x is $f'(x)$ negative, positive, and zero?
    (b) Identify these points on a graph of f.

22. Differentiate:
    (a) $f(x) = 3x + 4$
    (b) $g(s) = 1 - s^2$
    (c) $h(t) = 3t^2 - 5t + 9$
    (d) $f(x) = (9 - x)(1 - x)$

23. A manufacturer of hand calculators can produce up to 50,000 units with a wholesale price of $16 for a fixed cost of $9000 plus $11.50 for each unit produced. Let x stand for the number of units produced.
    (a) Explain why the *total revenue R, total cost C,* and *profit P* must satisfy the equations
    $$R = 16x, \quad C = 9000 + (11.5)x,$$
    $$P = R - C = (4.5)x - 9000$$
    (b) The *break-even point* is the production level x for which the profit is zero. Find it.
    (c) Determine the production level x which corresponds to a $4500 profit.
    (d) Express $\Delta P/\Delta x$ in words.

24. Let $P be invested at $r\%$ for t years. Then the amount A due is given by $A = Prt/100 + P$ (simple interest). If $100 is invested at 7%, then $A = 7t + 100$.
    (a) What will $100 amount to after six years?
    (b) The slope of the graph of A versus t indicates the increase in the amount A for each additional year of investment. Find it.

25. The stopping distance y, in feet, for an automobile traveling on a level road at x mph is given by $y = (0.044)x^2 + (1.1)x$.
    (a) What speed corresponds to stopping in 600 feet? (A hand calculator may be useful.)
    (b) Find the derivative of the quadratic function given above, and interpret it as the rate of change of one quantity with respect to another.

26. A toolbox falls from a building, its height y in feet from the ground after t seconds being given by $y = 100 - 16t^2$. This formula does not depend on the weight W of the toolbox.
    (a) Find the *impact time* t^*, i.e., the positive time for which $y = 0$.
    (b) Find the *impact velocity*, i.e., the velocity at t^*.
    (c) The momentum p is defined by $p = Wv/32$, where W is the weight in pounds, and v is the velocity in feet per second. Find the impact momentum for a 20-lb toolbox.

27. A biologist prepares a special diet for white rats from Purina Formulab and red wheat. Assume nominal protein levels of 16% and 10%, respectively. Find a linear equation which relates the amount y of Formulab and the amount x of red wheat producing a diet with 12% protein.

28. Straight-line depreciation assumes that the difference between current value and original value is directly proportional to the time t. Suppose a home office is presently furnished for $4000 and salvaged for $500 after ten years. Assume straight line depreciation.
    (a) Find a linear equation for the value V of the office furniture after t years, for tax purposes.
    (b) The slope of the line indicates the decrease in value each year of the office furniture, to be used in preparing a tax return. Find it.

29. One summer day in Los Angeles, the pollution index at 7:00 AM was 20 parts per million, increasing linearly 15 parts per million each hour until 5:00 PM. Let y be the amount of pollutants in the air x hours after 7:00 AM.

    (a) Find a linear equation relating y and x.

    (b) The slope is the increase in pollution for each hour increase in time. Find it.

    (c) Find the pollution level at 5:00 PM.

30. The amount of rain y in inches at time x in hours from the start of the September 3, 1975 Owens Valley thunderstorm was given by $y = 2x - x^2$, $0 \leq x \leq 1$.

    (a) Find how many inches of rain per hour are falling half-way through the storm.

    (b) Find how many inches of rain per hour are falling after half an inch of rain has fallen.

1-2 The Limit Method

The derivative of a function can be expressed as a limit of difference quotients.

We begin this section with a general discussion of limits. Limits are needed to formulate and compute with the idea of "gets close to" that was used in Section 1-1. The method of limits is then applied to differential calculus, where it enables us to calculate derivatives. The idea of the derivative as a limit also leads to the useful Leibniz notation.

Goals

After studying this section, you should be able to:

Guess limits by numerical calculation.

Use the limit rules to evaluate limits.

Find derivatives by the method of limits.

Express derivatives in the Leibniz notation.

INTRODUCTION TO LIMITS

We illustrate the idea of a limit by looking at the function

$$g(x) = \frac{2x^2 - 7x + 3}{x - 3}$$

which is defined for all real numbers except 3. Computing values of $g(x)$ for some values of x near 3, we obtain the following tables:

x	3.5	3.1	3.01	3.0001	3.000001
$g(x)$	6	5.2	5.02	5.0002	5.000002

x	2.5	2.9	2.99	2.9999	2.999999
$g(x)$	4	4.8	4.98	4.9998	4.999998

It appears that, as x gets closer and closer to 3, $g(x)$ gets closer and closer to 5.

In general, if the value $f(x)$ of a function f gets close to a certain number l as x gets close to a number a (which may or may not be in the domain of f), we say that "l is the limit of $f(x)$ as x approaches a," or "$f(x)$ approaches l as x approaches a." See Fig. 1-2-1. The usual notations for this are

$$f(x) \to l \quad \text{as} \quad x \to a$$

or

$$\lim_{x \to a} f(x) = l$$

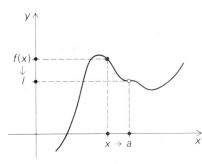

Fig. 1-2-1 The notion of limit: as x gets near to a, $f(x)$ gets near to l.

For example, the preceding tables suggest that

$$\frac{2x^2 - 7x + 3}{x - 3} \to 5 \quad \text{as} \quad x \to 3$$

or

$$\lim_{x \to 3} \frac{2x^2 - 7x + 3}{x - 3} = 5$$

Worked Example 1 Using numerical computations, guess the value of $\lim_{x \to 4} [1/(4x - 2)]$.

Solution We make a table using a calculator and round off to three significant figures:

x	4.1	4.01	4.001	3.9	3.99	3.999
$1/(4x - 2)$	0.0694	0.0712	0.0714	0.0735	0.0716	0.0714

It appears that the limit is a number which, when rounded to three decimal places, is 0.071. Repeating our calculations, we may notice that as $x \to 4$, the expression $4x - 2$ in the denominator of our fraction is approaching 14. The decimal expansion of $\frac{1}{14}$ is 0.071428..., so we may guess that

$$\lim_{x \to 4} \frac{1}{4x - 2} = \frac{1}{14}$$

We summarize in the following display an intuitive formulation of the limit idea, to be followed by some remarks regarding its application.

THE NOTION OF LIMIT

If, as the value of the variable x comes close to a, the value of the function $f(x)$ comes close to l,

then we say that f approaches the limit l as x approaches a and we write

$$f(x) \to l \quad \text{as} \quad x \to a \qquad \text{or} \qquad \lim_{x \to a} f(x) = l$$

Remarks

1. The limit $\lim_{x \to a} f(x)$ depends upon the values of $f(x)$ for x *near* a, but not for x equal to a. Indeed, $f(a)$ might not be defined, but even if it is, it may not be equal to the limit.

2. In determining $\lim_{x \to a} f(x)$, we must consider values of x on both sides of a.

3. As x gets nearer and nearer to a, the values of $f(x)$ might not approach any fixed number. In this case, we say that $f(x)$ *has no limit as $x \to a$* or that $\lim_{x \to a} f(x)$ *does not exist.*

Worked Example 2   Reading the graph in Fig. 1-2-2, find $\lim_{t \to b} g(t)$ if it exists, for $b = 1, 2, 3, 4,$ and 5.

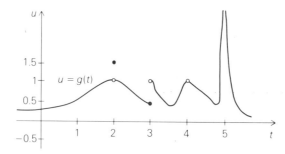

Fig. 1-2-2   Find the limits of g at the indicated points. A small circle means that the indicated point does not belong to the graph.

Solution   Notice first of all that we have introduced new letters; $\lim_{t \to b} g(t)$ means the value approached by $g(t)$ as t approaches b.

$b = 1$: $\lim_{t \to 1} g(t) = 0.5$. In this case, $g(b)$ is defined and happens to be equal to the limit.

$b = 2$: $\lim_{t \to 2} g(t) = 1$. In this case, $g(b)$ is defined and equals 1.5, which is not the same as the limit.

$b = 3$: $\lim_{t \to 3} g(t)$ does not exist. For t near 3, $g(t)$ has values near 0.5 (for $t < 3$) and near 1 (for $t > 3$). There is no *single* number approached by $g(t)$ as t approaches 3.

$b = 4$: $\lim_{t \to 4} g(t) = 1$. In this case, $g(b)$ is not defined.

$b = 5$: $\lim_{t \to 5} g(t)$ does not exist. As t approaches 5, $g(t)$ grows larger and larger and does not approach any limit.

(optional)   We give next a formal definition of limits to show that the concept has no ambiguities. *It is not necessary to master the definition at this stage;* the detailed theory of limits will be given in Chapter 12.

(optional)

Definition Let f be a function and a a real number such that $f(x)$ is defined for all x near a (but possibly not at a itself). A number l is called the *limit* of f at a provided that:

1. For every $y_1 < l$, there is an open interval (c, d) about a such that $y_1 < f(x)$ if $c < x < d$ and $x \neq a$.

2. For every $y_2 > l$, there is an open interval (c, d) about a such that $y_2 > f(x)$ if $c < x < d$ and $x \neq a$.

We write $\lim_{x \to a} f(x) = l$. See Fig. 1-2-3.

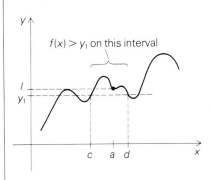

Fig. 1-2-3 The definition of limit; condition 1 for $y_1 < l$.

This definition makes precise our intuitive idea that to bring $f(x)$ close to its limit l it suffices to make x close to a.

The computation of limits is facilitated by certain rules, which we list in the following display. We will make no attempt to prove the rules at this point. (We give proofs in the appendix to Chapter 12.)* Instead, we will simply make some remarks and give examples which suggest that the rules are reasonable.

BASIC PROPERTIES OF LIMITS

Assume that $\lim_{x \to a} f(x)$ and $\lim_{x \to a} g(x)$ exist:

Sum rule: $\quad \lim_{x \to a} [f(x) + g(x)] = \lim_{x \to a} f(x) + \lim_{x \to a} g(x)$

Product rule: $\quad \lim_{x \to a} [f(x)g(x)] = \lim_{x \to a} f(x) \lim_{x \to a} g(x)$

Reciprocal rule: $\quad \lim_{x \to a} \left[\dfrac{1}{f(x)} \right] = \dfrac{1}{\lim_{x \to a} f(x)}$ if $\lim_{x \to a} f(x) \neq 0$

Constant function rule: $\quad \lim_{x \to a} c = c$

Identity function rule: $\quad \lim_{x \to a} x = a$

Replacement rule: If the functions f and g agree for all x near a (not necessarily including $x = a$), then $\lim_{x \to a} f(x) = \lim_{x \to a} g(x)$

This appendix is found in the Student Guide.

The sum, product, and reciprocal rules are based on the following observation: If we replace the numbers y_1 and y_2 by numbers z_1 and z_2 which are close to y_1 and y_2, then $z_1 + z_2$, $z_1 z_2$, and $1/z_1$ will be close to $y_1 + y_2$, $y_1 y_2$, and $1/y_1$, respectively.

The constant function rule says that if $f(x)$ is identically equal to c, then $f(x)$ is near c for all x near a. This is true because c is near c. (Notice a certain asymmetry in the definition of the limit $\lim_{x \to a} f(x) = l$: x is to be near a but unequal to a, while $f(x)$ is to be near l and can be equal to l.) The identity function rule is true since it merely says that x is near a if x is near a. Finally, the replacement rule follows from the fact that $\lim_{x \to a} f(x)$ depends only on the values of $f(x)$ for x *near a*, and not *at a*, nor on values of x far away from a.

Worked Example 3   Use the limit rules to find $\lim_{x \to 3}(x^2 + 2x + 5)$.

Solution   By the product and identity function rules,

$$\lim_{x \to 3} x^2 = \lim_{x \to 3}(x \cdot x) = (\lim_{x \to 3} x)(\lim_{x \to 3} x) = 3 \cdot 3 = 9$$

By the product, constant function, and identity function rules,

$$\lim_{x \to 3} 2x = (\lim_{x \to 3} 2)(\lim_{x \to 3} x) = 2 \cdot 3 = 6$$

By the sum rule,

$$\lim_{x \to 3}(x^2 + 2x) = \lim_{x \to 3} x^2 + \lim_{x \to 3} 2x = 9 + 6 = 15$$

Finally, by the sum and constant function rules,

$$\lim_{x \to 3}(x^2 + 2x + 5) = \lim_{x \to 3}(x^2 + 2x) + \lim_{x \to 3} 5 = 15 + 5 = 20$$

As you gain experience with limits, you can eliminate some of the steps we used to solve Worked Example 3. Moreover, you can use some further rules which can be derived from the basic properties.

DERIVED PROPERTIES OF LIMITS

Extended sum rule: $\lim_{x \to a}[f_1(x) + \cdots + f_n(x)] = \lim_{x \to a} f_1(x) + \cdots + \lim_{x \to a} f_n(x)$

Extended product rule: $\lim_{x \to a}[f_1(x) \cdots f_n(x)] = \lim_{x \to a} f_1(x) \cdots \lim_{x \to a} f_n(x)$

Constant multiple rule: $\lim_{x \to a} cf(x) = c \lim_{x \to a} f(x)$

Quotient rule: $\lim_{x \to a} \dfrac{f(x)}{g(x)} = \dfrac{\lim_{x \to a} f(x)}{\lim_{x \to a} g(x)}$ if $\lim_{x \to a} g(x) \neq 0$

Power rule: $\lim_{x \to a} x^n = a^n$ ($n = 0, \pm 1, \pm 2, \pm 3, \ldots$ and $a \neq 0$ if n is negative)

The extended sum and product rules are obtained by applying the basic sum and product rules repeatedly. The constant multiple rule comes from applying the product rule with one function equal to a constant. The power rule for $n > 0$ comes from the identity function rule and the extended product rule; for $n < 0$, the reciprocal rule is used. The quotient rule is obtained by combining the product and reciprocal rules. Details of the derivations are presented in the solved exercises and exercises.

The next example illustrates the use of the replacement rule.

Worked Example 4 Use the properties of limits to show that

$$\lim_{x \to 3} \frac{2x^2 - 7x + 3}{x - 3} = 5$$

as was suggested by our earlier numerical calculations.

Solution We cannot use the quotient rule, since $\lim_{x \to 3}(x - 3) = \lim_{x \to 3} x - \lim_{x \to 3} 3 = 3 - 3 = 0$. Since substituting $x = 3$ into the numerator yields zero, $x - 3$ must be a factor; in fact, $2x^2 - 7x + 3 = (2x - 1)(x - 3)$, and we have

$$\frac{2x^2 - 7x + 3}{x - 3} = \frac{(2x - 1)(x - 3)}{x - 3}$$

For $x \neq 3$, we can divide numerator and denominator by $x - 3$ to obtain $2x - 1$. Now we apply the replacement rule, with

$$f(x) = \frac{2x^2 - 7x + 3}{x - 3} \quad \text{and} \quad g(x) = 2x - 1$$

since these two functions agree for $x \neq 3$. Therefore

$$\lim_{x \to 3} \frac{2x^2 - 7x + 3}{x - 3} = \lim_{x \to 3}(2x - 1) = 2(\lim_{x \to 3} x) - 1 = 2 \cdot 3 - 1 = 5$$

You may have noticed that certain limits of the form $\lim_{x \to a} f(x)$ can be evaluated by setting x equal to a. In fact, combining the power, constant multiple, extended sum, and quotient rules yields the result that, if

$$f(x) = \frac{c_m x^m + c_{m-1} x^{m-1} + \cdots + c_0}{d_n x^n + d_{n-1} x^{n-1} + \cdots + d_0}$$

is a quotient of polynomials, then

$$\lim_{x \to a} f(x) = \frac{c_m a^m + c_{m-1} a^{m-1} + \cdots + c_0}{d_n a^n + d_{n-1} a^{n-1} + \cdots + d_0}$$

provided the denominator is unequal to zero. In other words, for these functions $\lim_{x \to a} f(x) = f(a)$. Just as a quotient of integers is called a rational number, a quotient of polynomials is called a *rational function*.

RATIONAL FUNCTION RULE

If $f(x)$ is a polynomial or a quotient of polynomials and $f(a)$ is defined, then

$$\lim_{x \to a} f(x) = f(a)$$

Warning Students seduced by the simplicity of the rational function rule often believe that a *limit* is nothing more than a *value*. Reread Worked Examples 2 and 4 whenever you feel yourself falling into this trap.

It is often useful to consider limits of the form $\lim_{x \to \infty} f(x)$. This symbol refers to the value approached by $f(x)$ as x becomes arbitrarily large. For instance, $\lim_{x \to \infty} 1/x = 0$, and

$$\lim_{x \to \infty} \frac{2x+1}{3x+1} = \lim_{x \to \infty} \frac{2 + \dfrac{1}{x}}{3 + \dfrac{1}{x}} = \frac{2+0}{3+0} = \frac{2}{3}$$

Limits as $x \to \infty$ obey the same rules as those with $x \to a$.

(optional) | The following is a "word problem" involving limits. Problems of this type require some interpretation and translation into mathematics together with the mathematical solution. (Examples and problems of this type are scattered throughout the book to help you apply the mathematics you are learning.)

Worked Example 5 | Population equilibrium can be modeled via the logistic equation, giving the formula $kN^{-1} - \beta = (kN_0^{-1} - \beta)x$. Here, $N = F(t)$ is the number of individuals at a time t, $N_0 = F(0)$, $k > 0$, $\beta > 0$, and $x = f(t)$ is a positive time-dependent factor satisfying $0 < x \le 1$ and $\lim_{t \to \infty} f(t) = 0$.

(a) Suppose the *vital factors* k and β satisfy $kN_0^{-1} - \beta > 0$. Show that $\lim_{t \to \infty} F(t) = k/\beta$.

(b) Examine the case $kN_0^{-1} - \beta < 0$.

(c) In which of the above cases does the population decay to the equilibrium level k/β?

(d) Sketch a possible graph of N versus t for the two cases cited above.

Solution | (a) The number of individuals is given to satisfy $kN^{-1} - \beta = (kN_0^{-1} - \beta)x$. Therefore, solving for N,

$$N = F(t) = \frac{k}{(kN_0^{-1} - \beta)x + \beta}$$

By the rules for limits,

$$\lim_{t \to \infty} F(t) = \frac{k}{[(kN_0^{-1} - \beta) \lim_{t \to \infty} f(t)] + \beta} = \frac{k}{\beta}$$

Thus the population level approaches k/β for large t. The condition $kN_0^{-1} - \beta > 0$, i.e., $k/\beta > N_0$, means that the limiting population is greater than the initial population.

(optional)

(b) In this case, reasoning as in (a) also shows that $\lim_{t \to \infty} F(t) = k/\beta$. The difference with (a) is that in (a), $(kN_0^{-1} - \beta)x > 0$ implies $(kN_0^{-1} - \beta)x + \beta > \beta$ and so $F(t) < k/\beta$ and thus $F(t)$ approaches k/β from below. Here $F(t)$ approaches k/β from above.

(c) Our analysis in (b) suggests that *decay* occurs in case $kN_0^{-1} - \beta < 0$.

(d) See Figure 1-2-4.

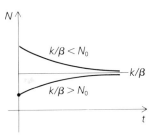

Fig. 1-2-4 Possible graphs for $N = F(t)$ in Worked Example 5.

Solved Exercises

1. Using only the basic properties of limits, find $\lim_{x \to 4} 1/(4x - 2)$. Check your answer by using the rational function rule. (See Worked Example 1.)

2. Guess $\lim_{x \to 1} [(x^3 - 3x^2 + 5x - 3)/(x - 1)]$ by doing numerical calculations. Verify your guess by using the properties of limits.

3. (a) Prove the constant multiple rule from the basic properties of limits.

   (b) Prove the quotient rule using the product and reciprocal rules.

Exercises

1. Find $\lim_{x \to -1} 2x/(4x^2 + 5)$, first by numerical calculation and guesswork, then by the basic properties of limits, and finally by the rational function rule.

2. For the function in Fig. 1-2-5, find $\lim_{x \to a} f(x)$ for $a = 0, 1, 2, 3, 4$, if it exists. In each case, tell whether $\lim_{x \to a} f(x) = f(a)$.

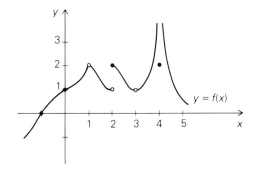

Fig. 1-2-5 Find the limits at 0, 1, 2, 3, 4.

3. Prove the extended sum rule for $n = 3$.

4. Find $\lim_{x \to 2} [(x^2 + 3x - 10)/(x - 2)]$.

THE DERIVATIVE AS A LIMIT

When $y = f(x)$ varies linearly with x—that is, when $y = mx + b$—the rate of change $f'(x) = m$ of y with respect to x is just the ratio $\Delta y / \Delta x$.

(See Fig. 1-1-2.) Since the graph is a straight line, the geometry of similar triangles tells us that the ratio $\Delta y / \Delta x$ does not depend upon which pairs of points we choose to compute it. Conversely, if we know the rate of change m in advance, we can compute Δy given Δx, since $\Delta y = m \Delta x$. This formula is useful in many situations: we can find distance traveled in a given time if we know the velocity; we can find the cost of an extra Δp pounds of sugar if we know the price per pound.

If $y = f(x)$ is not necessarily a linear function, we introduced the following notation in Section 1-1: let $(x_0, f(x_0))$ and $(x_0 + \Delta x, f(x_0 + \Delta x))$ be two points on the graph. (See Fig. 1-1-5.) Let $\Delta y = f(x_0 + \Delta x) - f(x_0)$. The difference quotient $\Delta y / \Delta x$ is the *average rate of change of y with respect to x on the interval from x_0 to $x_0 + \Delta x$*. Recall that if y is the position of a moving object at time x, the ratio $\Delta y / \Delta x$ of distance traveled to time elapsed is called the *average velocity* over the period from x_0 to $x_0 + \Delta x$. As Δx gets small, with x_0 being fixed, we expect the average velocity over the period from x_0 to $x_0 + \Delta x$ to come close to the instantaneous velocity $f'(x_0)$.

Geometrically, the ratio $\Delta y / \Delta x$ is the slope of the secant line drawn through the points $(x_0, f(x_0))$ and $(x_0 + \Delta x, f(x_0 + \Delta x))$. As Δx becomes smaller and smaller, x_0 being fixed, the secant line comes close to the tangent line, so that the slope $\Delta y / \Delta x$ of the secant line comes close to the slope $f'(x_0)$ of the tangent line. (See Fig. 1-1-6.)

Our work in Section 1-1 illustrates the general principle that *the derivative is a limit of difference quotients*. We may formulate this principle as follows.

Let $f(x)$ be a function whose domain contains an open interval about x_0. Considering Δx as a new variable, we define the difference quotient function $q(\Delta x)$ by the formula

$$q(\Delta x) = \frac{f(x_0 + \Delta x) - f(x_0)}{\Delta x}$$

That is, $q(\Delta x) = \Delta y / \Delta x$. The domain of q consists of those Δx which are near enough to zero so that $x_0 + \Delta x$ is in the domain of f; *zero itself is not in the domain of q*. Still, we can look at the limit $\lim_{\Delta x \to 0} q(\Delta x)$, which we may also write as

$$\lim_{\Delta x \to 0} \frac{f(x_0 + \Delta x) - f(x_0)}{\Delta x}$$

The concept of limit may now be used to formulate the definition of the derivative.

Definition   Let $f(x)$ be a function whose domain contains an open interval about x_0. We say that f is *differentiable at x_0* when the following limit exists:

$$f'(x_0) = \lim_{\Delta x \to 0} \frac{f(x_0 + \Delta x) - f(x_0)}{\Delta x}$$

Let us now re-examine the justification of the quadratic function rule given in Section 1-1 by verifying the steps in the language of limits.

Worked Example 6 Using the formula for the derivative as a limit, find $f'(x_0)$ when $f(x) = ax^2 + bx + c$.

Solution We write down the difference quotient and simplify:

$$q(\Delta x) = \frac{f(x_0 + \Delta x) - f(x_0)}{\Delta x}$$

$$= \frac{a(x_0 + \Delta x)^2 + b(x_0 + \Delta x) + c - ax_0^2 - bx_0 - c}{\Delta x}$$

$$= \frac{ax_0^2 + 2ax_0\Delta x + a(\Delta x)^2 + bx_0 + b\Delta x + c - ax_0^2 - bx_0 - c}{\Delta x}$$

$$= \frac{2ax_0\Delta x + a(\Delta x)^2 + b\Delta x}{\Delta x}$$

Now we observe that, for $\Delta x \neq 0$, we have $q(\Delta x) = 2ax_0 + a\Delta x + b$. The replacement rule gives

$$\lim_{\Delta x \to 0} q(\Delta x) = \lim_{\Delta x \to 0} (2ax_0 + a\Delta x + b)$$

Since $2ax_0 + a\Delta x + b$ is a polynomial (in fact, a linear function) in the variable Δx, we can evaluate the last limit by setting Δx equal to zero, obtaining $2ax_0 + b$. (Note that we could not have done this evaluation before we used the replacement rule.) We now conclude that $f'(x_0) = 2ax_0 + b$, as in the quadratic function rule.

DIFFERENTIATION BY THE METHOD OF LIMITS

To find the derivative of f at x_0:

1. Write down the difference quotient

$$q(\Delta x) = \frac{f(x_0 + \Delta x) - f(x_0)}{\Delta x}$$

2. Find $\lim_{\Delta x \to 0} q(\Delta x)$. (Try to factor Δx out of the numerator, and use the replacement rule.)

The average rate of change of $f(x)$ with respect to x for x between x_0 and $x_0 + \Delta x$ approaches the instantaneous rate of change at x_0, as Δx approaches zero.

$$f'(x_0) = \lim_{\Delta x \to 0} \frac{f(x_0 + \Delta x) - f(x_0)}{\Delta x}$$

Solved Exercises

4. Suppose that a particle moves according to the rule $f(t) = 3t^2 + t$. Find the average velocity on the interval between $t = 0$ and $t = 1$. At what time is the instantaneous velocity equal to this average?

5. Find the average velocity in Solved Exercise 4 between
   (a) $t = 0$ and $t = 0.5$
   (b) $t = 0$ and $t = 0.1$
   (c) $t = 0$ and $t = 0.01$
   (d) $t = 0$ and $t = 0.001$
   Compare with the instantaneous velocity at $t = 0$.

6. Use the method of limits to differentiate x^3.

5. Find the average rate of change of each of the following functions on the specified interval. Also find the derivative at the endpoints.
   (a) $f(t) = 400 - 20t - 16t^2$; between $t_0 = 1$ and $t_1 = \frac{3}{2}$
   (b) $f(x) = (x + 1)(x + 2)$; $x_0 = 2$, $\Delta x = 0.5$
   (c) $g(s) = (3s + 2)(s - 1) - 3s^2$; $s_0 = 0$, $s_1 = 6$

6. Let $y = 4x^2 - 2x + 7$. Compute the average rate of change of y with respect to x over the interval from $x_0 = 0$ to $x_1 = \Delta x$ for the following values of Δx: 0.1, 0.001, 0.000001. Compare with the derivative at $x_0 = 0$.

7. Using the limit method, find the derivative of $2x^3 + x^2 - 3$ at $x_0 = 1$.

8. (a) Expand $(a + b)^4$. (b) Use the limit method to differentiate x^4.

THE LINEAR APPROXIMATION

We have seen that average rates of change and derivatives are close to one another, and we have used the limit idea to pass from average rates to derivatives. We can also go in the other direction: given $f(x_0)$ and $f'(x_0)$, we can use the derivative to get an approximate value for $f(x)$ when x is near x_0.

According to the definition of the derivative, the difference quotient $\Delta y / \Delta x = [f(x_0 + \Delta x) - f(x_0)] / \Delta x$ is close to $f'(x_0)$ when Δx is small. That is, the difference

$$\frac{\Delta y}{\Delta x} - f'(x_0) = \frac{f(x_0 + \Delta x) - f(x_0)}{\Delta x} - f'(x_0) = e \qquad (1)$$

is small when Δx is small.

If we rewrite equation (1) in the form

$$f(x_0 + \Delta x) = f(x_0) + f'(x_0)\Delta x + e\Delta x \qquad (2)$$

we find that $f(x_0) + f'(x_0)\Delta x$ is a good approximation to $f(x_0 + \Delta x)$ and that the error $e\Delta x$ in this approximation becomes arbitrarily small—even compared to Δx itself—as Δx approaches zero. Writing x for $x_0 + \Delta x$, we see that a good approximation for $f(x)$ is $f(x_0) + f'(x_0)(x - x_0)$, when x is near x_0. The function $g(x) = f(x_0) + f'(x_0)(x - x_0)$ is linear in x; it is called the *best linear approximation to f at x_0*. Notice that its graph is just the tangent line to the graph of f at $(x_0, f(x_0))$. (See Fig. 1-2-6.)

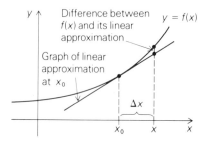

Fig. 1-2-6 As Δx approaches zero, the difference between $f(x)$ and the approximation $f(x_0) + f'(x_0) (x - x_0)$ becomes arbitrarily small compared to $\Delta x = x - x_0$.

Worked Example 7 It will be shown in Chapter 5 that the derivative of $\sqrt{x}$ is $1/(2\sqrt{x})$. Using this fact, write down the linear approximation to $\sqrt{x}$ for x near 9, and use it to find an approximate value for $\sqrt{8.92}$.

Solution Let $f(x) = \sqrt{x}$. Then $f(9) = \sqrt{9} = 3$ and $f'(9) = 1/(2\sqrt{9}) = \frac{1}{6}$, so the linear approximation to $f(x)$ at x_0 is $3 + \frac{1}{6}(x - 9)$. When $x = 8.92$, the value of this approximation is $3 + \frac{1}{6}(8.92 - 9) = 3 - \frac{1}{6}\frac{8}{100} = 2\frac{74}{75}$.

The decimal equivalent of $2\frac{74}{75}$ is 2.986666...; that of $\sqrt{8.92}$ is 2.986637.... The error in the approximation is about 0.000029, which is small compared to $\Delta x = 0.08$.

The linear approximation $f(x) \approx f(x_0) + f'(x_0)(x - x_0)$ (x near x_0) is sometimes called the *first-order approximation*. The naive approximation $f(x) \approx f(x_0)$ (x near x_0) is called the *zeroth-order approximation*. Approximations of second and higher orders also exist; we will study them in Section 13-3.

THE LINEAR (FIRST-ORDER) APPROXIMATION

For x near x_0, $f(x_0) + f'(x_0)(x - x_0)$ is a good approximation for $f(x)$.

$$f(x_0 + \Delta x) \approx f(x_0) + f'(x_0)\Delta x \quad \text{or} \quad \Delta y \approx f'(x_0)\Delta x$$

The error becomes arbitrarily small, compared with Δx, as $\Delta x \to 0$.

Solved Exercises

7. Show that the first-order approximation to $(x_0 + \Delta x)^2$ is $x_0^2 + 2x_0\Delta x$.

8. Calculate an approximate value for $(1.03)^2$. Compare with the actual value. Do the same for $(1.0003)^2$ and $(1.0000003)^2$.

9. Calculate the first-order approximation to the area of a square whose side is 2.01. Draw a geometric figure, obtained from a square of side 2, whose area is exactly that given by the first-order approximation.

Exercises

9. Find an approximate value for $(2.98)^2$. (Note that Δx is negative if you choose $x_0 = 3$.)

10. The radius of a circle is increased from 3 to 3.04. Using the linear approximation, what do you find to be the increase in the area of the circle? Compare with the actual increase.

11. Let $f(x) = 3x^2 - 4x + 7$. Using the first-order approximation, find approximate values for $f(2.02)$, $f(1.98)$, and $f(2.004)$. Compute the actual values and compare with the approximations. Compare the amount of time you spend in computing the approximations as opposed to the actual values.

12. Let $f(x) = 3x^2 - 4x + 7$. Show that the first-order approximation to $f(2 + \Delta x)$ always gives an answer which is too small, regardless of whether Δx is positive or negative. Interpret your answer geometrically by drawing a graph of f and its tangent line at $x_0 = 2$.

THE LEIBNIZ NOTATION: dy/dx

We have seen that the rate of change $f'(x_0)$ of $y = f(x)$ with respect to x at x_0 is closely approximated by the average rate of change $\Delta y / \Delta x$ over the interval from x_0 to $x_0 + \Delta x$. In the view of Gottfried Wilhelm von Leibniz (1646–1716), one of the founders of calculus, one could think of Δx as becoming "infinitely small" or "infinitesimal." The resulting quantity he denoted as dx, the letters d and Δ being the Roman and Greek equivalents of one another. When Δx became the infinitesimal dx, Δy simultaneously became the infinitesimal dy and the ratio $\Delta y / \Delta x$ became dy/dx, which was no longer an approximation to the derivative but exactly equal to it.

The infinitesimal quantities, being smaller than any real number but still not zero, were the cause of much controversy, leading to much criticism of the developing subject of calculus. The problems were not cleared up until the development of the theory of limits in the nineteenth century. Recent advances in logic by Abraham Robinson (1918–1974) have also made possible a mathematically respectable treatment of infinitesimals.

Despite its troubled history, the notation dy/dx has proved to be extremely convenient—not as a ratio of infinitesimal quantities but as a *synonym* for $f'(x)$. In other words, dy/dx simply denotes the derivative of y with respect to x (or the limit of $\Delta y / \Delta x$), while the numerator dy and the denominator dx are not taken to have any independent meaning.

LEIBNIZ NOTATION

If $y = f(x)$, the derivative of $f'(x)$ may be written $\dfrac{dy}{dx}$, dy/dx, $\dfrac{df(x)}{dx}$, $(d/dx)f(x)$ or $d(f(x))/dx$. This is just a notation and does not represent division. If we wish to denote the value $f'(x_0)$ of f' at a specific point x_0, we may write

$$\left. \frac{dy}{dx} \right|_{x_0} \quad \text{or} \quad \left. \frac{df(x)}{dx} \right|_{x_0}$$

$\dfrac{dy}{dx}$ is read "the derivative of y with respect to x" or "dy by dx."

Of course, we can use this notation if the variables are named other than x and y. For instance, the area of a circle of radius r is $A = \pi r^2$, and we can write the rate of change of A with respect to r as $dA/dr = 2\pi r$.

This notation is particularly useful when functions are given by explicit formulas and the function names would be superfluous. For instance, instead of writing

> If $y = f(x)$, where $f(x) = 3x^2 + 2x$, then $f'(x) = 6x + 2$

we may simply write

> If $y = 3x^2 + 2x$, then $\dfrac{dy}{dx} = 6x + 2$

In fact, the name of the dependent variable itself is often superfluous. Since y is equal to $3x^2 + 2x$, we may substitute $3x^2 + 2x$ for y and write

$$\frac{d(3x^2 + 2x)}{dx} = 6x + 2$$

This is the shortest way of writing the result of a differentiation. To avoid putting a complicated expression in the numerator of a fraction, we may also write the preceding formula as

$$\frac{d}{dx}(3x^2 + 2x) = 6x + 2$$

Here the d/dx may be thought of as a symbol for the *operation* of differentiation. It takes the place of the prime $(')$ in the functional notation.

Worked Example 8   Write the differentiation rule for quadratic functions in Leibniz notation.

Solution $(d/dx)(ax^2 + bx + c) = 2ax + b$.

Since the second derivative $f''(x)$ is obtained by applying the operation d/dx twice—that is, $(d/dx)(d/dx)f(x)$—we denote it by $(d^2/dx^2)(f(x))$, or d^2y/dx^2.

Worked Example 9   Find $(d^2/dr^2)(8r^2 + 2r + 10)$.

Solution $(d^2/dr^2)(8r^2 + 2r + 10) = (d/dr)[(d/dr)(8r^2 + 2r + 10)] = (d/dr)[16r + 2] = 16$.

To summarize this discussion, we may say that the dy/dx (or Leibniz) notation is most useful when we are dealing with variables, especially when they are related by formulas. The quantity dy/dx represents the *rate of change* of the variable y with respect to the variable x. The f' (or functional) notation is still valuable if we wish to emphasize the role of the functional relationship or if we must deal with functions not defined by simple formulas. It is also the most convenient notation when we want to evaluate derivatives at particular values of the independent variable.

Solved Exercises   10. Find dy/dx if $y = 8x^2 + 2x + 7$.

11. Let V represent the volume of a box with fixed height h and variable square base of side a. Find the rate of change of V with respect to a.

12. Find $(d/dm)(4m^2 + 7m + 6)$.

13. Calculate d^2y/dx^2 if $y = 3x^2 + 5x + 2$.

14. Find $(d^2/dx^2)(x^3)$.

Exercises   13. Compute dy/dx in the following cases:
   (a) $y = 3x^2 + 5x$
   (b) $y = 5 + \frac{3}{2}x - 4x^2$
   (c) $y = \frac{1}{4}(x - 2)^2$
   (d) $y = 11x - 8$

14. Find:

    (a) $(d/du)(8u^2 + 10u)$ (b) $(d/d\alpha)(\alpha^2)$

    (c) $d(x)/dx$ (d) $d(2)/dx$

15. Let V be the volume of a cylindrical can with height h and radius r.

    (a) Compute dV/dr, assuming that h is fixed.

    (b) Compute dV/dh, assuming that r is fixed.

16. Suppose that $z = 2y^2 + 3y$ and $y = 5x + 1$.

    (a) Find dz/dy and dy/dx.

    (b) Express z in terms of x and find dz/dx.

    (c) Compare dz/dx with $(dz/dy) \cdot (dy/dx)$. (Write everything in terms of x.)

    (d) Solve for x in terms of y and find dx/dy.

    (e) Compare dx/dy with dy/dx.

17. Find:

    (a) $(d^2/du^2)(-u^2 + 2u + 6)$ (b) $(d^2/du^2)(u)$

Problems for Section 1-2

1. Find the following limits, if they exist.

    (a) $\displaystyle\lim_{x\to3} \frac{x^2 - 9}{x - 3}$ (b) $\displaystyle\lim_{x\to\sqrt{5}} \frac{2}{x^2 + 5}$

    (c) $\displaystyle\lim_{x\to\sqrt{5}} \frac{2}{x^2 - 5}$ (d) $\displaystyle\lim_{x\to-1} \frac{x + 1}{x - 1}$

    (e) $\displaystyle\lim_{x\to0} \frac{x^2 + 5x}{x^2}$

    (f) $\displaystyle\lim_{x\to1} \frac{x^{10} + 8x^3 - 7x^2 - 2}{x + 1}$

2. Find $\lim_{x\to a} f(x)$, where $a = -2, 0,$ and 1, for f sketched in Fig. 1-2-7.

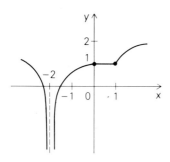

Fig. 1-2-7 Find $\lim_{x\to a} f(x)$ at the indicated points.

3. Use limits to find the derivatives of

    (a) $f(x) = x^2 - x$ (b) $f(x) = 3x^3 + x$

    (c) $f(x) = (x^2 + x)/2x$ (d) $f(x) = 1/x$

    (e) $f(x) = x/(1 + x^2)$

4. How should $f(x) = (x^5 - 1)/(x - 1)$ be defined at $x = 1$ in order that $\lim_{x\to1} f(x) = f(1)$?

5. Write the following limits in the form

    $$\lim_{\Delta x \to 0} \frac{f(x_0 + \Delta x) - f(x_0)}{\Delta x}$$

    and use your knowledge of derivatives to compute them.

    (a) $\displaystyle\lim_{s\to0} \frac{(s + 1)^2 - 1}{s}$ (b) $\displaystyle\lim_{s\to0} \frac{(s + 1)^4 - 1}{s}$

    (c) $\displaystyle\lim_{r\to0} \frac{(2 + r)^2 + 3(2 + r) - 10}{r}$

    (d) $\displaystyle\lim_{h\to0} \frac{(3 + h)[(3 + h)^2 - (3 + h) + 1] - 21}{h}$

6. During a beet strike, the price of beets was $p(t) = 26(1 + t^3/300)$ cents per pound, where t is the time in months from the beginning of the strike. What is the average rate of increase of price (cents per pound per month) during the first 6 months of the strike? What is the instantaneous rate at the beginning of the third month?

7. Let A represent the area of the shapes in Fig. 1-2-8.

(a) Find dA/dx and d^2A/dx^2. **(b)** Find dA/dr and d^2A/dr^2

(Rectangle with $\frac{1}{4}$ of disk removed)

(c) Find dA/dy and d^2A/dy^2. **(d)** Find dA/dx and d^2A/dx^2

Fig. 1-2-8 Find the indicated rates of change of the areas.

8. Let P represent the perimeter of the relevant figure in Problem 7.

   (a) For Problem 7(a), find dA/dP and dP/dx.

   (b) For Problem 7(b), find dA/dP and dP/dr.

   (c) For Problem 7(c), find dA/dP and dP/dy.

   (d) For Problem 7(d), find dA/dP and dP/dx.

9. Suppose that the population of a city at time t (in years) is given by a function $p(t)$. Write the rate of increase of population with respect to time as a limit.

10. Find the rate of change of the surface area of a cube with respect to the length of an edge.

11. In each case, express z in terms of x and find the derivatives:

$$\frac{dz}{dx}, \quad \frac{dy}{dx}, \quad \frac{dz}{dy}, \quad \frac{d^2z}{dx^2}, \quad \frac{d^2y}{dx^2}$$

   Check that $dz/dx = (dz/dy) \cdot (dy/dx)$.

   (a) $z = 4y^2 - 2y + 1$; $y = 2x - 7$.

   (b) $z = -3y^2 + y - 7$; $y = x + 2$.

   (c) $z = 5y + 6$; $y = 2x^2 + 8x - 3$.

12. In Problem 11(a), how fast is z changing with respect to y when $y = 4$; $y = \frac{1}{2}$; $y = -8$?

13. In Problem 11(b), how fast is z changing with respect to y when $y = -2$; $x = 3$ (careful!); $y = 0.2$?

★14. (a) Let $y = ax^2 + bx + c$, where a, b, and c are constant. Show that the average rate of change of y with respect to x on any interval $[x_1, x_2]$ equals the instantaneous rate of change at the midpoint—that is, at $(x_1 + x_2)/2$.

   (b) Let $f(x) = ax^2 + bx + c$, where a, b, and c are constant. Prove that, for any x_0,

   $$f(x) = f(x_0) + f'(m)(x - x_0)$$

   where $m = \dfrac{x + x_0}{2}$

15. Find the average rate of change of each of the following functions on the given interval. Compare with the derivative at the midpoint.

   (a) $f(x) = (x - \frac{1}{2})(x + 1)$ between $x = -\frac{1}{2}$ and $x = 0$.

   (b) $g(t) = 3(t + 5)(t - 3)$ on $[2, 6]$.

   (c) $h(r) = 10r^2 - 3r + 6$ on $[-0.1, 0.4]$.

   (d) $r(t) = (2 - t)(t + 4)$; t in $[3, 7]$.

16. Let $f(x) = x^3 + 2x$. Use limits to compute $df(x)/dx$ and $(d/dx)[df(x)/dx]$.

17. Calculate an approximate value for each of the following squares and compare with the exact values:

   (a) $(2.02)^2$ (b) $(199)^2$

   (c) $(4.999)^2$ (d) $(-1.002)^2$

18. Let $h(t) = -4x^2 + 7x + \frac{3}{4}$. Using the first-order approximation, find approximate values for $h(3.001)$, $h(1.97)$, and $h(4.03)$.

19. Find the first-order approximations to the following functions at the specified points:

   (a) $h(x) = x^2 - 8x + 17$; $x_0 = 2.002$, $x_0 = 1.98$.

   (b) $f(r) = -2r^2 + 9r$; $r_0 = 5.005$, $r_0 = 5.01$.

   (c) $g(t) = 25 - t - 7t^2$; $t_0 = 3.02$, $t_0 = 3.0002$.

   (d) $f(x) = 4x^2 + 4x + 4$; $x_0 = 9.989$, $x_0 = 10.021$.

20. The radius r of the base of a right circular cylinder of fixed height h is changed from 4 to 3.96. Using the linear approximation, approximate the change in volume V.

★21. Let $g(x) = -4x^2 + 8x + 13$. Prove that the first-order approximation to $g(3 + \Delta x)$ always gives an answer which is too large, regardless of whether Δx is positive or negative. Interpret your answer geometrically by drawing a graph of g and its tangent line when $x_0 = 3$. Find $g'(3)$ and $g''(3)$.

22. Let $h(t) = 2t^3$ be the position of an object moving along a straight line at time t. What are the velocity and acceleration at $t = 3$?

23. A block of ice melts in a room held at 75°F. Let $f(t)$ be the base area of the block and $g(t)$ the height of the block, measured with a ruler at time t.

    (a) Assume the block of ice melts completely at time T. What values would you assign to $f(T)$ and $g(T)$?

    (b) Give physical reasons why $\lim_{t \to T} f(t) = f(T)$ and $\lim_{t \to T} g(t) = g(T)$ need not both hold; what are the limits?

    (c) The limiting volume of the ice block at time T is zero. Write this statement as a limit formula.

    (d) Using (c), illustrate the product rule for limits.

24. The monitor switch on a certain stereo receiver leaks, so that when the volume is turned up full, the FM signal can be heard faintly through the headphones. Let $f(x)$ be the dB output at the stereo headphones corresponding to the volume dial rotation of x degrees, measured from the zero volume level.

    (a) Explain in layman terms why $f(0) = 0$.

    (b) The volume dial has a range of 270°. Explain in layman terms why $\lim_{x \to 270-} f(x) = f(270)$ (the expression $\lim_{x \to 270-} f(x)$ means the limit of $f(x)$ as x approaches 270 from the left).

    (c) The symbol $\lim_{x \to 270-} f(x)$ implicitly involves the examination of the dB output levels at the headphones at all sufficiently high volume levels, whereas $f(270)$ is obtained by a single meter reading. Explain.

25. A thermometer is stationed at x centimeters from a candle flame. Let $f(x)$ be the Celsius scale reading on the thermometer. Assume the glass in the thermometer will crack upon contact with the flame.

    (a) Explain physically why $f(0)$ doesn't make any sense.

    (b) Describe in terms of the thermometer scale the meaning of $\lim_{x \to 0+} f(x)$ (i.e., the limit of $f(x)$ as x approaches zero through positive values).

    (c) Draw a realistic graph of $f(x)$ for a scale with maximum value 200°C. (Assume that the flame temperature is 400°C).

    (d) Repeat (c) for a maximum scale value of 500°C.

★26. Suppose that a function f is defined on an open interval I containing x_0 and that there are numbers m and K such that

$$|f(x) - f(x_0) - m(x - x_0)| \le K|x - x_0|^2$$

for all x in I. Prove that f is differentiable at x_0 with derivative $f'(x_0) = m$.

★27. Prove that

$$\lim_{x \to 0} \frac{(a + x)^n - a^n}{x} = na^{n-1}$$

if n is any positive integer.

1-3 Algebraic Rules for Differentiation

The Newton-Leibniz "machine" makes it easy to differentiate sums, products, and quotients.

Suppose that you want to differentiate

$$f(x) = \frac{(x^2 + 5)^8}{(2x + 1)(x^3 + 7)}$$

Using the method of limits, you might eventually obtain an answer, but the prospect of simplifying $[f(x_0 + \Delta x) - f(x_0)]/\Delta x$ is frightful to contemplate. The power of calculus is provided by systematic rules for differentiation that enable us to bypass the complexities of limit computations.

Goals

After studying this section, you should be able to:

Reproduce from memory the sum, product, quotient, and power of a function rule.

Differentiate any polynomial or rational function.

THE SUM RULE

If f and g are two functions, we can define a new function, $f + g$, by the formula $(f + g)(x) = f(x) + g(x)$.

Worked Example 1 If $f(x) = x^2$ and $g(x) = 3x^2 + 2x + 1$, find $(f + g)(2)$. Find a formula for $(f + g)(x)$.

Solution By definition, $(f + g)(2) = f(2) + g(2)$. Now $f(2) = 2^2 = 4$ and $g(2) = 3 \cdot 2^2 + 2 \cdot 2 + 1 = 17$, so $(f + g)(2) = 4 + 17 = 21$. For general x, $(f + g)(x) = f(x) + g(x) = x^2 + 3x^2 + 2x + 1 = 4x^2 + 2x + 1$. Substituting 2 for x in the last expression gives 21, as it should.

If we differentiate the functions f, g, and $f + g$ in Worked Example 1, we find that

$$f'(x) = 2x$$
$$g'(x) = 6x + 2$$

and

$$(f + g)'(x) = 8x + 2$$

Note that differentiating the sum $f + g$ gives the same result as differentiating f and g separately and then adding. This is a special case of a general fact called the *sum rule*.

SUM RULE

To differentiate a sum $f(x) + g(x)$, take the sum of the derivatives $f'(x) + g'(x)$:

$$(f + g)'(x) = f'(x) + g'(x)$$

or

$$\frac{d}{dx}(u + v) = \frac{du}{dx} + \frac{dv}{dx}$$

Worked Example 2 Use the sum rule to find the derivative of $x^3 + 5x^2 + 9x + 2$.

Solution By Solved Exercise 6 of Section 1-2, we have $(d/dx)(x^3) = 3x^2$. By the quadratic function rule, $(d/dx)(5x^2 + 9x + 2) = 10x + 9$. By the sum rule, the derivative of $x^3 + 5x^2 + 9x + 2$ is the sum of these two derivatives, $3x^2 + 10x + 9$.

We can motivate the sum rule by example. Imagine a train, on a straight track, whose distance at time x from a fixed reference point on the ground is $f(x)$. There is a runner on the train whose distance from a reference point on the train is $g(x)$. Then the distance of the runner from the fixed reference point on the ground is $f(x) + g(x)$. (See Fig. 1-3-1.) Suppose that, at a certain time x_0, the runner is going at 20 kilometers per hour with respect to the train while the train is going at 140 kilometers per hour—that is, $f'(x_0) = 140$ and $g'(x_0) = 20$. What is the velocity of the runner as seen from an observer on the ground? It is the sum of 140 and 20—that is, 160 kilometers per hour.* Considered as the sum of two velocities, the number 160 is $f'(x_0) + g'(x_0)$; considered as the velocity of the runner with respect to the ground, the number 160 is $(f + g)'(x_0)$. Thus we have $f'(x_0) + g'(x_0) = (f + g)'(x_0)$.

$f(x)$ $g(x)$

$f(x) + g(x)$

Fig. 1-3-1   The sum rule illustrated in terms of velocities.

We can *justify* the sum rule by the method of limits. By the formula for the derivative as a limit, $(f + g)'(x_0)$ is equal to

$$\lim_{\Delta x \to 0} \frac{(f + g)(x_0 + \Delta x) - (f + g)(x_0)}{\Delta x}$$

(if this limit exists). We can rewrite the limit as

$$\lim_{\Delta x \to 0} \frac{f(x_0 + \Delta x) + g(x_0 + \Delta x) - f(x_0) - g(x_0)}{\Delta x}$$

$$= \lim_{\Delta x \to 0} \left[\frac{f(x_0 + \Delta x) - f(x_0)}{\Delta x} + \frac{g(x_0 + \Delta x) - g(x_0)}{\Delta x} \right]$$

By the sum rule for *limits*, this is

$$\lim_{\Delta x \to 0} \frac{f(x_0 + \Delta x) - f(x_0)}{\Delta x} + \lim_{\Delta x \to 0} \frac{g(x_0 + \Delta x) - g(x_0)}{\Delta x}$$

If f and g are differentiable at x_0, these two limits are just $f'(x_0)$ and $g'(x_0)$. We conclude that $f + g$ is differentiable at x_0 and $(f + g)'(x_0) = f'(x_0) + g'(x_0)$.

Solved Exercises

1. Verify the sum rule for $u = 3x^2 + 5x + 9$ and $v = 2x^2 + 5x$.

2. Find a formula for the derivative of $f(x) + g(x) + h(x)$.

The fact that one does not add velocities this way in the theory of special relativity does not violate the sum rule. In classical mechanics, velocities are derivatives; but in relativity, velocities are not simply derivatives, so the formula for their combination is more complicated.

Exercises   1. Verify the sum rule for $f(x) = 3x^2 + 6$ and $g(x) = x + 7$.

2. Let $g(x) = 2f(x)$. Express $g'(x)$ in terms of $f'(x)$.

3. If a train is moving at 20 kilometers per hour, what velocity must a runner on the train have in order to remain fixed with respect to the ground?

THE PRODUCT RULE

The product fg of two functions is defined by $(fg)(x) = f(x)g(x)$.

Worked Example 3   Let $f(x) = 3x + 3$ and $g(x) = 2x + 7$. Is $(fg)'$ equal to $f'g'$?

Solution   We find that $(fg)(x) = (3x + 3)(2x + 7) = 6x^2 + 27x + 21$, so $(fg)'(x) = 12x + 27$. On the other hand, $f'(x) = 3$ and $g'(x) = 2$, so $f'(x)g'(x) = 6$. Since $12x + 27$ and 6 are not the same function, $(fg)'$ is *not* equal to $f'g'$.

Worked Example 3 shows that the derivative of the product of two functions is not the product of their derivatives. The correct rule for products is a bit more complicated.

PRODUCT RULE

To differentiate a product $f(x)g(x)$,   differentiate each factor and multiply it by the other one, then add the two products:

$$(fg)'(x) = f(x)g'(x) + f'(x)g(x)$$

or

$$\frac{d}{dx}(uv) = \frac{du}{dx}v + u\frac{dv}{dx}$$

Worked Example 4   Verify the product rule for f and g in Worked Example 3.

Solution   We know that $(fg)'(x) = 12x + 27$. On the other hand, $f(x)g'(x) + f'(x)g(x) = (3x + 3)2 + 3(2x + 7) = 6x + 6 + 6x + 21 = 12x + 27$, so the product rule gives the right answer.

Of course, the best use of the product rule is to find derivatives which we don't already know.

Worked Example 5   Find $(d/dx)(x^4)$.

Solution   We write $x^4 = x^2 \cdot x^2$. Then

$$\frac{d}{dx}(x^4) = \frac{d}{dx}(x^2 \cdot x^2) = 2x \cdot x^2 + x^2 \cdot 2x = 2x^3 + 2x^3 = 4x^3$$

The form of the product rule may be a surprise to you. Why should that strange combination of f, g, and their derivatives be the derivative of fg? The following motivation should help you to understand.

The cost of fuel for driving, measured in cents per kilometer, can increase with time for two reasons. Inflation causes the price of gasoline to rise, while the deterioration of your car causes its fuel consumption rate to increase. To represent the situation mathematically, we write $c(t)$ for the fuel cost at time t in cents per kilometer. If $g(t)$ is the price of gasoline at time t, in cents per liter, and $m(t)$ is the fuel consumption rate at time t in liters per kilometer, then $c(t) = g(t)m(t)$.

We wish to find the rate of change, $c'(t_0)$, of cost with respect to time. It turns out that this cannot be expressed in terms of $g'(t_0)$ and $m'(t_0)$ alone. In fact, the effect of the increase in gasoline price is proportional to how much gasoline you use: the part of $c'(t_0)$ due to inflation is $g'(t_0)m(t_0)$. Similarly, the effect of the increase in fuel consumption is proportional to the price of gasoline: the part of $c'(t_0)$ due to deterioration is $g(t_0)m'(t_0)$. Adding these two contributions together, we come up with $c'(t_0) = g'(t_0)m(t_0) + g(t_0)m'(t_0)$, which is exactly the product rule.

(optional)

The method of limits may be used to prove the formula for the derivative of a product. To find $(fg)'(x_0)$, we take the limit

$$\lim_{\Delta x \to 0} \frac{(fg)(x_0 + \Delta x) - (fg)(x_0)}{\Delta x}$$
$$= \lim_{\Delta x \to 0} \frac{f(x_0 + \Delta x)g(x_0 + \Delta x) - f(x_0)g(x_0)}{\Delta x}$$

Simplifying this expression is not as straightforward as for the sum rule. We may make use of a geometric device: think of $f(x)$ and $g(x)$ as the lengths of the sides of a rectangle; then $f(x)g(x)$ is its area. The rectangles for $x = x_0$ and $x = x_0 + \Delta x$ are shown in Fig. 1-3-2. The area of the large rectangle is $f(x_0 + \Delta x)g(x_0 + \Delta x)$; that of the darker rectangle is $f(x_0)g(x_0)$. The difference $f(x_0 + \Delta x)g(x_0 + \Delta x) - f(x_0)g(x_0)$ is the area of the lighter region, which can be decomposed into three rectangles having areas $[f(x_0 + \Delta x) - f(x_0)]g(x_0)$, $[f(x_0 + \Delta x) - f(x_0)][g(x_0 + \Delta x) - g(x)]$, and $f(x_0)[g(x_0 + \Delta x) - g(x)]$. Thus we have the identity:

$$f(x_0 + \Delta x)g(x_0 + \Delta x) - f(x_0)g(x_0) = [f(x_0 + \Delta x) - f(x_0)]g(x_0) +$$
$$f(x_0)[g(x_0 + \Delta x) - g(x_0)] + [f(x_0 + \Delta x) - f(x_0)][g(x_0 + \Delta x) - g(x_0)]$$

(The reader who is suspicious of geometric arguments can verify this identity algebraically.)

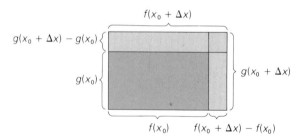

Fig. 1-3-2 The geometry behind the proof of the product rule.

(optional)

Applying this identity to our limit expression, we obtain

$$\lim_{\Delta x \to 0} \left\{ \frac{f(x_0 + \Delta x) - f(x_0)}{\Delta x} g(x_0) + f(x_0) \frac{g(x_0 + \Delta x) - g(x_0)}{\Delta x} + \frac{[f(x_0 + \Delta x) - f(x_0)][g(x_0 + \Delta x) - g(x_0)]}{\Delta x} \right\}$$

By the sum and constant multiple rules for limits, this equals

$$\left[\lim_{\Delta x \to 0} \frac{f(x_0 + \Delta x) - f(x_0)}{\Delta x} \right] g(x_0) + f(x_0) \left[\lim_{\Delta x \to 0} \frac{g(x_0 + \Delta x) - g(x_0)}{\Delta x} \right] + \lim_{\Delta x \to 0} \frac{[f(x_0 + \Delta x) - f(x_0)][g(x_0 + \Delta x) - g(x_0)]}{\Delta x}$$

We recognize the first two limits as $f'(x_0)$ and $g'(x_0)$, so the first two terms give $f'(x_0)g(x_0) + f(x_0)g'(x_0)$—precisely the product rule. To show that the third limit is zero, we can multiply and divide by Δx. Since $\Delta x \neq 0$, the product rule for *limits* yields

$$\lim_{\Delta x \to 0} \frac{f(x_0 + \Delta x) - f(x_0)}{\Delta x} \cdot \lim_{\Delta x \to 0} \frac{g(x_0 + \Delta x) - g(x_0)}{\Delta x} \cdot \lim_{\Delta x \to 0} \Delta x = f'(x_0) \cdot g'(x_0) \cdot 0 = 0$$

(This last term is represented geometrically by the small rectangle in the upper right-hand corner of Fig. 1-3-2.)

Using the product rule, we can go on to differentiate all the power functions: x^5, x^6, x^7, x^8, and so forth. The general formula for the derivative of x^n is nx^{n-1}. We have checked it for $n = 1, 2, 3$, and 4. Check it yourself for $n = 5$. The proof for arbitrary n uses mathematical induction (see Problem 18).

POWER RULE

To differentiate a power x^n ($n = 1, 2, 3, \ldots$), take out the exponent as a factor and then reduce the exponent by 1.

If $f(x) = x^n$, then $f'(x) = nx^{n-1}$, that is, $\dfrac{d}{dx}(x^n) = nx^{n-1}$.

Worked Example 6 Find $(d/dx)(x^{95} + x^{23} + 2x^2 + 4x + 1)$.

Solution
$$\frac{d}{dx}(x^{95} + x^{23} + 2x^2 + 4x + 1)$$

$$= \frac{d}{dx}(x^{95}) + \frac{d}{dx}(x^{23}) + \frac{d}{dx}(2x^2 + 4x + 1) \quad \text{(sum rule)}$$

$$= 95x^{94} + 23x^{22} + 4x + 4 \quad \text{(power rule)}$$

We know how to differentiate ax^2, where a is a constant. What about ax^{23}? The following rule will enable us to do this and much more.

CONSTANT MULTIPLE RULE

To differentiate the product of a number a with $f(x)$, multiply the number a by the derivative $f'(x)$:

$$(af)'(x) = af'(x)$$

$$\frac{d}{dx}(au) = a\frac{du}{dx}$$

The constant multiple rule follows from the product rule. We consider a as a constant *function;* then $da/dx = 0$ and

$$\frac{d}{dx}[af(x)] = \frac{da}{dx}\cdot f(x) + a\frac{d}{dx}f(x)$$

$$= 0\cdot f(x) + a\frac{d}{dx}f(x)$$

$$= a\frac{d}{dx}f(x)$$

Taken together, the sum rule, power rule, and constant multiple rule enable us to differentiate any polynomial.

Worked Example 7 Differentiate $4x^9 + 6x^5 + 3x$.

Solution

$$\frac{d}{dx}(4x^9 + 6x^5 + 3x) = 4\frac{d}{dx}(x^9) + 6\frac{d}{dx}(x^5) + 3\frac{d}{dx}(x)$$

$$= 4\cdot 9x^8 + 6\cdot 5x^4 + 3\cdot 1$$

$$= 36x^8 + 30x^4 + 3$$

Here, for reference, is a general rule, but you need not memorize it, since you can do any example by using the sum, power, and constant multiple rules.

DERIVATIVE OF A POLYNOMIAL

$$\frac{d}{dx}(c_n x^n + \cdots + c_2 x^2 + c_1 x + c_0) = nc_n x^{n-1} + (n-1)c_{n-1}x^{n-2} + \cdots + 2c_2 x + c_1$$

Worked Example 8 *(Marginal Cost and Marginal Revenue)*

(optional) *Preamble.* Consider a company that makes calculators. Suppose that x calculators are produced per week and that the management is free to adjust x. Define the following quantities:

$C(x)$ = the *cost* of making x calculators (labor, supplies, etc).

$R(x)$ = the *revenue* obtained by producing x calculators (sales).

$P(x) = R(x) - C(x) =$ the *profit.*

(optional) Even though $C(x)$, $R(x)$, and $P(x)$ are defined only for integers x, economists find it useful to imagine them defined for all real x. The cost and revenue per calculator can vary with x for many reasons; for example, mass production may lower the per unit cost but may also lower the price.

The derivative $C'(x)$ is called the *marginal cost* and $R'(x)$ is the *marginal revenue*:

$$C'(x) = \text{marginal cost} = \begin{cases} \text{the cost per calculator for producing addi-} \\ \text{tional calculators at production level } x. \end{cases}$$

$$R'(x) = \text{marginal revenue} = \begin{cases} \text{the revenue per calculator obtained by} \\ \text{producing additional calculators at produc-} \\ \text{tion level } x. \end{cases}$$

Problem. (a) What is $P'(x)$?

(b) If calculators are priced at $f(x)$ dollars per calculator and all x calculators are sold, what is the marginal revenue?

Solution (a) Since $P(x) = R(x) - C(x)$ we get $P'(x) = R'(x) - C'(x)$, the profit per additional calculator at production level x.

(b) If the price per unit is $f(x)$ and x calculators are sold, then $R(x) = xf(x)$. By the product rule, the marginal revenue is $R'(x) = xf'(x) + f(x)$.

Solved Exercises

3. Differentiate the following functions:

(a) $x^5 + 8x$ (b) $5x^3$

(c) $x^5 + 6x^2 + 8x + 2$ (d) $s^{10} + 8s^9 + 5s^8 + 2$

4. Use the product rule to differentiate $(x^2 + 2)(x + 8)$, and check the answer by multiplying out before differentiating.

5. A particle moves on a line with position $f(t) = 16t^2 + (0.03)t^4$ at time t. Find the velocity at $t = 8$.

6. Find the equation of the line tangent to the graph of $f(x) = x^8 + 2x^2 + 1$ at $(1,4)$.

7. Using the linear approximation, find an approximate value for $(2.94)^4$.

Exercises

4. Differentiate the following functions:

(a) $f(x) = x^4 - 7x^2 - 3x + 1$ (b) $h(x) = 3x^{11} + 8x^5 - 9x^3 - x$

(c) $f(t) = (t^3 - 17t + 9)(3t^5 - t^2 - 1)$

(d) $g(s) = s^{13} + 12s^8 - \frac{3}{8}s^7 + s^4 + \frac{4}{3}s^3$

(e) $f(y) = -y^3 - 8y^2 - 14y - \frac{1}{3}$ (f) $p(x) = (x^2 + 1)^3$

(g) $r(t) = (t^4 + 2t^2)^2$

5. A sphere of radius r has volume $V = \frac{4}{3}\pi r^3$. What is the rate of change of the volume of the sphere with respect to its radius? Give a geometric interpretation of the answer.

6. Differentiate:

(a) $f(x) = x^4 - 3x^3 + 2x^2$ (b) $f(t) = t^4 + 4t^3$

(c) $g(h) = 8h^{10} + h^9 - 56.5h^2$ (d) $h(\gamma) = \pi\gamma^{10} + \frac{21}{7}\gamma^9 - \sqrt{2}\,\gamma^2$

7. Find first-order approximations for: (a) $(1.03)^4$ (b) $(3.99)^3$ (c) $(101)^8$.

8. Find the equation of the tangent line to the graph of $x^4 - x^2 + 3x$ at $x = 1$.

THE POWER OF A FUNCTION; A SPECIAL CASE OF THE CHAIN RULE

If $f(x)$ is any function, we can use the product rule to differentiate $[f(x)]^2$:

$$\frac{d}{dx}[f(x)]^2 = \frac{d}{dx}[f(x)f(x)] = f'(x)f(x) + f(x)f'(x) = 2f(x)f'(x)$$

If we write $u = f(x)$, this can be expressed in Leibniz notation as

$$\frac{d}{dx}(u^2) = 2u\frac{du}{dx}$$

In the same way, we may differentiate u^3:

$$\frac{d}{dx}(u^3) = \frac{d}{dx}(u^2 \cdot u) = \frac{d}{dx}(u^2) \cdot u + u^2\frac{du}{dx}$$

$$= 2u\frac{du}{dx} \cdot u + u^2\frac{du}{dx} = 3u^2\frac{du}{dx}$$

Similarly, $(d/dx)(u^4) = 4u^3(du/dx)$ (check it yourself). And, for a general positive integer n, we have $(d/dx)u^n = nu^{n-1}(du/dx)$. (See Problem 18(b) at the end of this section.)

POWER OF A FUNCTION RULE

To differentiate the nth power $[f(x)]^n$ of a function $f(x)$ ($n = 1, 2, 3, \ldots$), take out the exponent as a factor, reduce the exponent by 1, and multiply by the derivative of $f(x)$:

$$(f^n)'(x) = n[f(x)]^{n-1}f'(x)$$

$$\frac{d}{dx}(u^n) = nu^{n-1}\frac{du}{dx}$$

If $u = x$, then $du/dx = 1$ and the power of a function rule reduces to the ordinary power rule.

Worked Example 9 Find $(d/ds)(s^4 + 2s^3 + 3)^8$.

Solution We apply the power of a function rule, with $u = s^4 + 2s^3 + 3$ (and the variable x replaced by s):

$$\frac{d}{ds}(s^4 + 2s^3 + 3)^8 = 8(s^4 + 2s^3 + 3)^7\frac{d}{ds}(s^4 + 2s^3 + 3)$$

$$= 8(s^4 + 2s^3 + 3)^7(4s^3 + 6s^2)$$

(You could also do this problem by multiplying out the eighth power and then differentiating; obviously, this practice is not recommended.)

Warning The most common mistake made by students in applying the power of a function rule is to forget the extra factor of $f'(x)$—that is, du/dx—at the end.

Solved Exercises

8. Find the derivative of $[f(x)]^3$, where $f(x) = x^4 + 2x^2$, first by using the power of a function rule and then by expanding out the cube and differentiating directly. Compare the answers. Finally, evaluate $3[f(x)]^2$ and compare this result with your previous answers.

9. If $f(x) = (x^2 + 1)^{27}(x^4 + 3x + 1)^8$, find $f'(x)$.

10. Find a general formula for $(d^2/dx^2)(u^n)$, where $u = f(x)$ is any function of x.

Exercises

9. Find the derivatives of:

    (a) $(x + 3)^4$
    (b) $(x^2 + 3x + 1)^5$
    (c) $(x^3 + 10x)^{100}$
    (d) $(x^2 + 8x)^3 \cdot x$
    (e) $(x^2 + 2)^3(x^9 + 8)$
    (f) $(x^2 + 2)(x^9 + 8)^3$
    (g) $(y + 1)^3(y + 2)^2(y + 3)$
    (h) $(s^4 + 4s^3 + 3s^2 + 2s + 1)^8$

10. If $f(x) = (x^3 + 2)^9(x^8 + 2x + 1)^{16}$, find $f'(x)$.

11. Find the second derivatives of:

    (a) $(x^4 + 10x^2 + 1)^{98}$ (b) $(x + 1)^{13}$ (c) $(x^2 + 1)^3(x^3 + 1)^2$

THE QUOTIENT RULE

We come now to quotients of functions. Let $h(x) = f(x)/g(x)$, where f and g are differentiable at x_0, and suppose that $g(x_0) \neq 0$ so that the quotient is defined at x_0. We want a formula for $h'(x_0)$. If we assume the existence of $h'(x_0)$, it is easy to compute its value from the product rule, as we now show.

Since $h(x) = f(x)/g(x)$, we have $f(x) = g(x)h(x)$. Apply the product rule to obtain

$$f'(x_0) = g'(x_0)h(x_0) + g(x_0)h'(x_0)$$

Solving for $h'(x_0)$, we get

$$h'(x_0) = \frac{f'(x_0) - g'(x_0)h(x_0)}{g(x_0)}$$

$$= \frac{f'(x_0) - g'(x_0)[f(x_0)/g(x_0)]}{g(x_0)}$$

$$= \frac{f'(x_0)g(x_0) - f(x_0)g'(x_0)}{[g(x_0)]^2}$$

This is the quotient rule.

QUOTIENT RULE

To differentiate a quotient $f(x)/g(x)$ (where $g(x) \neq 0$),

take the derivative of the numerator times the denominator, subtract the numerator times the derivative of the denominator, and divide the result by the square of the denominator:

$$\left(\frac{f}{g}\right)'(x) = \frac{f'(x)g(x) - f(x)g'(x)}{[g(x)]^2} \quad \text{or} \quad \frac{d}{dx}\left(\frac{u}{v}\right) = \frac{\dfrac{du}{dx}v - u\dfrac{dv}{dx}}{v^2}$$

When you use the quotient rule, it is important to remember which term in the numerator comes first. (In the product rule, both terms occur with a plus sign, so the order doesn't matter.) One memory aid goes as follows: Write down your guess for the right formula and set $g = 1$ and $g' = 0$. Your formula should reduce to f'. If it comes out as $-f'$ instead, you have the terms in the wrong order.

Worked Example 9   Find the derivative of $h(x) = (2x + 1)/(x^2 - 2)$.

Solution   We apply the quotient rule with $f(x) = 2x + 1$, $g(x) = x^2 - 2$, $f'(x) = 2$, and $g'(x) = 2x$:

$$h'(x) = \frac{2(x^2 - 2) - (2x + 1)2x}{(x^2 - 2)^2}$$
$$= \frac{2x^2 - 4 - 4x^2 - 2x}{(x^2 - 2)^2}$$
$$= -\frac{2x^2 + 2x + 4}{(x^2 - 2)^2}$$

(optional)   We can prove the quotient rule by the method of limits. If $h(x) = f(x)/g(x)$,

$$h'(x_0) = \lim_{\Delta x \to 0} \frac{h(x_0 + \Delta x) - h(x_0)}{\Delta x}$$
$$= \lim_{\Delta x \to 0} \frac{\dfrac{f(x_0 + \Delta x)}{g(x_0 + \Delta x)} - \dfrac{f(x_0)}{g(x_0)}}{\Delta x}$$
$$= \lim_{\Delta x \to 0} \frac{f(x_0 + \Delta x)g(x_0) - f(x_0)g(x_0 + \Delta x)}{g(x_0)g(x_0 + \Delta x)\Delta x}$$

A look at the calculations in the limit derivation of the product rule suggests that we add $-f(x_0)g(x_0) + f(x_0)g(x_0) = 0$ to the numerator. We get

(optional)

$$h'(x_0) = \lim_{\Delta x \to 0} \frac{f(x_0 + \Delta x)g(x_0) - f(x_0)g(x_0) + f(x_0)g(x_0) - f(x_0)g(x_0 + \Delta x)}{g(x_0)g(x_0 + \Delta x)\Delta x}$$

$$= \lim_{\Delta x \to 0} \frac{1}{g(x_0)g(x_0 + \Delta x)} \left[\frac{f(x_0 + \Delta x) - f(x_0)}{\Delta x} g(x_0) - f(x_0) \frac{g(x_0 + \Delta x) - g(x_0)}{\Delta x} \right]$$

$$= \frac{1}{\lim_{\Delta x \to 0} g(x_0 + \Delta x)} \frac{1}{g(x_0)} [f'(x_0)g(x_0) - f(x_0)g'(x_0)] \qquad (1)$$

What is $\lim_{\Delta x \to 0} g(x_0 + \Delta x)$? We have $g'(x_0) = \lim_{\Delta x \to 0} [g(x_0 + \Delta x) - g(x_0)]/\Delta x$. Thus $\lim_{\Delta x \to 0} [g(x_0 + \Delta x) - g(x_0)] = \lim_{\Delta x \to 0} [\Delta x \cdot (g(x_0 + \Delta x) - g(x_0))/\Delta x] = [\lim_{\Delta x \to 0} \Delta x][\lim_{\Delta x \to 0} (g(x_0 + \Delta x) - g(x_0))/\Delta x] = 0 \cdot g'(x_0) = 0$ and so

$$\lim_{\Delta x \to 0} g(x_0 + \Delta x) = g(x_0) \qquad (2)$$

Substituting (2) into (1) gives the quotient rule.

Solved Exercises

11. Differentiate $x^2/(x^3 + 5)$.

12. Find the equation of the line tangent to the graph of $f(x) = (2x + 1)/(3x + 1)$ at $x = 1$.

13. Find a first-order approximation for $1/0.98$.

Exercises

12. Differentiate the following functions:
   (a) $(x - 2)/(x^2 + 3)$
   (b) $(x^3 - 3x + 5)/(x^4 - 1)$
   (c) $[1/x^2] + [x/(x^2 + 1)]$
   (d) $[s/(1 - s)]^5$
   (e) $(r^2 + 2)/r^8$
   (f) $1/t^9$

13. Find $\dfrac{d^2}{dt^2}\left(\dfrac{t^2 + 1}{t^2 - 1}\right)$.

14. Find the equation of the line tangent to the graph of $f(x) = 1/x$ at $x = 2$.

15. Find a first-order approximation for $1/1.98$.

RECIPROCALS AND NEGATIVE POWERS

Certain special cases of the quotient rule are particularly useful. If $f(x) = 1$, then $h(x) = 1/g(x)$ and we get the reciprocal rule:

RECIPROCAL RULE

To differentiate the reciprocal $1/g(x)$ of a function (where $g(x) \neq 0$), take the negative of the derivative of the function and divide by the square of the function:

$$\left(\frac{1}{g}\right)'(x) = \frac{-g'(x)}{[g(x)]^2} \qquad \text{or} \qquad \frac{d}{dx}\left(\frac{1}{u}\right) = -\frac{1}{u^2}\frac{du}{dx}$$

Worked Example 11   Differentiate $1/(x^3 + 3x^2)$.

Solution

$$\frac{d}{dx}\left(\frac{1}{x^3 + 3x^2}\right) = -\frac{1}{(x^3 + 3x^2)^2}\frac{d}{dx}(x^3 + 3x^2)$$

$$= -\frac{3x^2 + 6x}{(x^3 + 3x^2)^2}$$

Using the reciprocal rule and the power of a function rule, we can differentiate negative powers of a function.* If n is a positive integer,

$$\frac{d}{dx}[f(x)]^{-n} = \frac{d}{dx}\left(\frac{1}{[f(x)]^n}\right)$$

$$= -\frac{d[f(x)]^n}{dx} \cdot \frac{1}{[f(x)]^{2n}}$$

$$= -n[f(x)]^{n-1} \cdot \frac{df(x)}{dx} \cdot [f(x)]^{-2n}$$

$$= -n[f(x)]^{-n-1}\frac{df(x)}{dx}$$

If we write m for $-n$, this is exactly

$$\frac{d}{dx}[f(x)]^m = m[f(x)]^{m-1}\frac{df(x)}{dx}$$

In other words, the power of a function rule works just as well for negative integers as for positive ones. In particular, so does the power rule:

$$\frac{d}{dx}(x^m) = mx^{m-1}$$

for m positive or negative.

If $m = 0$, then $[f(x)]^m = 1$ and the derivative is zero, so the power rule works for zero as well and, thus, for all integer powers.

Worked Example 12   Let $f(x) = 1/[(3x^2 - 2x + 1)^{100}]$. Find $f'(x)$.

Solution   We write $f(x)$ as $(3x^2 - 2x + 1)^{-100}$. Thus

$$f'(x) = -100(3x^2 - 2x + 1)^{-101}(6x - 2)$$

(Don't forget the last factor.)

Solved Exercises   14. Differentiate $1/[(x^3 + 5x)^4]$.

15. In the text, we derived the reciprocal rule from the quotient rule. By writing $f(x)/g(x) = f(x) \cdot [1/g(x)]$, show that the quotient rule also follows from the product rule and the reciprocal rule.

16. Differentiate $1/[(x^2 + 3)(x^2 + 4)]$.

---

*Students requiring review of negative exponents should read the subsection "The Laws of Exponents" found at the beginning of Section 7-1.

Exercises 16. Find the following derivatives:

(a) $\dfrac{d}{dx}\left(\dfrac{1}{x^4}\right)$ (b) $\dfrac{d}{dx}\left(\dfrac{1}{(x+1)^2}\right)$ (c) $\dfrac{d}{dx}\left(\dfrac{1}{x^5+5x^2}\right)$

(d) $\dfrac{d}{dx}\left(\dfrac{1}{(x^2+9)^{100}}\right)$ (e) $\dfrac{d^2}{ds^2}\left(\dfrac{1}{s^2+2}\right)$

17. Use the reciprocal rule twice to differentiate $1/[1/g(x)]$ and show that the answer comes out to $g'(x)$.

SUMMARY OF DIFFERENTIATION RULES

We conclude this section with a summary of the differentiation rules obtained so far. (More rules will be given in Chapters 5, 6 and 7.) Some of these rules are special cases of the others. For instance, the linear and quadratic function rules are special cases of the polynomial rule, and the reciprocal rule is the quotient rule for $f(x) = 1$.

Remember that the basic idea for differentiating a complicated function is to break it into its component parts and combine the derivatives of the parts according to the rules. The problems which follow (together with the ones at the end of the chapter) will provide you with material for the practice which is necessary for mastering the use of the rules. Your efforts will be rewarded in Chapters 2 and 3, where the operation of differentiation is used to solve some problems of mathematical and practical interest.

DIFFERENTIATION RULES

	The derivative of	is	In Leibniz notation	See page
Linear function	$mx + b$	m	$\dfrac{d}{dx}(mx + b) = m$	57
Quadratic function	$ax^2 + bx + c$	$2ax + b$	$\dfrac{d}{dx}(ax^2 + bx + c) = 2ax + b$	57
Sum	$f(x) + g(x)$	$f'(x) + g'(x)$	$\dfrac{d}{dx}(u + v) = \dfrac{du}{dx} + \dfrac{dv}{dx}$	81
Constant multiple	$af(x)$	$af'(x)$	$\dfrac{d}{dx}(au) = a\dfrac{du}{dx}$	86
Product	$f(x)g(x)$	$f'(x)g(x) + f(x)g'(x)$	$\dfrac{d}{dx}(uv) = \dfrac{du}{dx}v + u\dfrac{dv}{dx}$	83
Power	x^n {n any integer}	nx^{n-1}	$\dfrac{d}{dx}(x^n) = nx^{n-1}$	85
Power of a function	$[f(x)]^n$ {n any integer}	$n[f(x)]^{n-1}f'(x)$	$\dfrac{d}{dx}(u^n) = nu^{n-1}\dfrac{du}{dx}$	88
Polynomial	$c_n x^n + \cdots + c_2 x^2 + c_1 x + c_0$	$nc_n x^{n-1} + \cdots + 2c_2 x + c_1$	$\dfrac{d}{dx}(c_n x^n + \cdots + c_2 x^2 + c_1 x + c_0)$ $= nc_n x^{n-1} + \cdots + 2c_2 x + c_1$	86
Quotient	$f(x)/g(x)$ {$g(x) \neq 0$}	$\dfrac{f'(x)g(x) - f(x)g'(x)}{[g(x)]^2}$	$\dfrac{d}{dx}\left(\dfrac{u}{v}\right) = \dfrac{\dfrac{du}{dx}v - u\dfrac{dv}{dx}}{v^2}$	90
Reciprocal	$1/g(x)$ {$g(x) \neq 0$}	$-g'(x)/[g(x)]^2$	$\dfrac{d}{dx}\left(\dfrac{1}{v}\right) = -\dfrac{1}{v^2}\dfrac{dv}{dx}$	91

Problems for Section 1-3

1. (a) Find $f'(r)$ if $f(r) = -5r^6 + 5r^4 - 13r^2 + 15$.
   (b) Find $g'(s)$ if $g(s) = s^7 + 13s^6 - 18s^3 + \frac{3}{2}s^2$.
   (c) Find $h'(t)$ if $h(t) = (t^4 + 9)(t^3 - t)$.

2. (a) Find $(d/dx)(x^5 + 2x^4 + 7)$.
   (b) Find $(d/du)[(u^4 + 5)(u^3 + 7u^2 + 19)]$.
   (c) Find $(d/dt)(3t^5 + 9t^3 + 5t)^8$.

3. Differentiate the following functions:
   (a) $(x^7 - x^2)/(x^3 + 1)$
   (b) $4/[(x^2 - 1)^3(x + 7)^2]$
   (c) $(5x^3 + x - 10)/(3x^4 + 2)$
   (d) $[(x - 1)/(x + 1)]^{300}$
   (e) $(x^2 + 2)/(x^2 - 2)$

4. Let $f(x) = 4x^5 - 13x$ and $g(x) = x^3 + 2x - 1$. Find the derivatives of the following functions:
   (a) $f(x)g(x)$ (b) $f(x) + x^3 - 7x$
   (c) $xf(x) + g(x)$
   (d) $[f(x)/g(x)] + (x^3 - 3x) - 7$
   (e) $g(x)/f(x)$ (f) $[f(x)]^8[g(x)]^{-4}$

5. Find the equation of the tangent line to the graph of $f(x)$ at $(x_0, f(x_0))$ in each of the following cases:
   (a) $f(x) = (x^2 - 7)[3x/(x + 2)]$; $x_0 = 0$.
   (b) $f(x) = [(1/x) - 2x](x^2 + 2)$; $x_0 = \frac{1}{2}$.
   (c) $f(x) = 1/(x^2 + 4)^8$; $x_0 = 2$.

6. Find d^2y/dx^2 if:
   (a) $y = x^5 + 7x^4 - 2x + 3$
   (b) $y = (x - 1)^3 + x^2$
   (c) $y = x^2/(x - 1)$
   (d) $y = x^5 + 1/x + 2/x^3$
   (e) $y = (x - a)^n$ (a constant; n positive integer)
   (f) $y = 1/(x - a)^n$ (a constant; n positive integer)

★7. Let $P(x)$ be a quadratic polynomial. Show that $(d/dx)(1/P(x))$ is zero for at most one value of x in its domain. Find an example of a $P(x)$ for which $(d/dx)(1/P(x))$ is never zero on its domain.

8. What is the rate of change of the volume of a right circular cylinder with respect to its radius if the height is equal to the radius?

9. If an object has position $(t^2 + 4)^5$ at time t, what is its velocity when $t = -1$?

10. Calculate approximate values for each of the following:
    (a) $s^4 - 5s^3 + 3s - 4$; if $s = 0.9997$
    (b) $x^4/(x^5 - 2x^2 - 1)$; if $x = 2.0041$
    (c) $(2.01)^{20}$ (d) $1/(1.99)^2$

11. Show that a good approximation to $1/(1 + x)$ when x is small is $1 - x$.

12. If you travel 1 mile in $60 + x$ seconds, show that a good approximation to your average speed, for x small, is $60 - x$ miles per hour. (This works quite well on roads which have mileposts.) Find the error in this approximation if $x = 1, -1, 5, -5, 10, -10$.

13. Return to Problem 11. Show experimentally that a better approximation to $1/(1 + x)$ is $1 - x + x^2$. Use this result to refine the speedometer checking rule in Problem 12.

★14. Devise a speedometer checking rule for metric units which works for speeds in the vicinity of about 90 or 100 kilometers per hour.

15. For each of the following functions, find those values of x for which the derivative is zero:
    (a) $f(x) = x^2 - x + 5$
    (b) $f(x) = (3x - 1)/(x^2 + 5)$
    (c) $f(x) = 4x^2/(x - 1)$

16. Suppose that the position x of a car at time t is $(t - 2)^3$.
    (a) What is the velocity at $t = -1, 0, 1$?
    (b) Show that the average velocity over every interval of time is strictly positive.
    (c) There is a stop sign at $x = 0$. A police officer gives the driver a ticket because there was no period of time during which the car was stopped. The driver argues that, since his velocity was zero at $t = 2$, he obeyed the stop sign. Who is right?

★17. Show that $f(x) = x$ and $g(x) = 1/(1 - x)$ obey the "false product rule" $(fg)'(x) = f'(x)g'(x)$.

★18. This problem requires a knowledge of mathematical induction.
    (a) Prove that $(d/dx)x^n = nx^{n-1}$ for all $n = 1, 2, 3, \ldots$ by
        1. Verifying that it is true for $n = 1$ and $n = 2$.
        2. Assuming that it holds for n and applying the product rule to
        $$\frac{d}{dx}x^{n+1} = \frac{d}{dx}x \cdot x^n$$
        to deduce it for $n + 1$.
    (b) By a similar procedure prove that
        $$\frac{d}{dx}[f(x)]^n = n[f(x)]^{n-1}f'(x)$$
        for all $n = 1, 2, 3, \ldots$.

★19. (a) Find a formula for the second and third derivatives of x^n.
    (b) Find a formula for the rth derivative of x^n if $n > r$.
    (c) Find a formula for the derivative of the product $f(x)g(x)h(x)$ of three functions.

This symbol denotes problems or discussions that may require use of a hand-held calculator.

★20. For reasons which will become clear in Chapter 7, the quotient $f'(x)/f(x)$ is called the *logarithmic derivative* of $f(x)$.

(a) Show that the logarithmic derivative of the product of two functions is the sum of the logarithmic derivatives of the functions.

(b) Show that the logarithmic derivative of the quotient of two functions is the difference of their logarithmic derivatives.

(c) Show that the logarithmic derivative of the nth power of a function is n times the logarithmic derivative of the function.

(d) Develop a formula for the logarithmic derivative of

$$[f_1(x)]^{n_1} [f_2(x)]^{n_2} \cdots [f_k(x)]^{n_k}$$

in terms of the logarithmic derivatives of f_1 through f_k.

(e) Using your formula in part (d), find the ordinary (*not* logarithmic) derivative of

$$f(x) = \frac{(x^2 + 3)(x + 4)^8 (x + 7)^9}{(x^4 + 3)^{17} (x^4 + 2x + 1)^5}$$

If you have enough courage, compute $f'(x)$ without using the formula in part (d).

★21. Differentiate $(1 + (1 + (1 + x^2)^8)^8)^8$.

22. For each of the following functions, find a function whose derivative is $f(x)$. (Do not find $f'(x)$.)

(a) $f(x) = x^2$ (b) $f(x) = x^2 + 2x + 3$

(c) $f(x) = x^n$ (n any integer except -1)

(d) $f(x) = (x + 3)(x^2 + 1)$

★23. Calculate the following limits by expressing each one as the derivative of some function:

(a) $\displaystyle\lim_{x \to 1} \frac{x^8 - x^7 + 3x^2 - 3}{x - 1}$

(b) $\displaystyle\lim_{x \to 2} \frac{1/x^3 - 1/2^3}{x - 2}$

(c) $\displaystyle\lim_{x \to -1} \frac{x^2 + x}{(x + 2)(x + 1)}$

24. The total cost C in dollars for producing x cases of solvent is given by $C(x) = 20 + 5x - (0.01)x^2$. The number 20 in the formula represents the fixed cost for placing the order, regardless of size. The other terms represent the variable costs.

(a) Find the marginal cost.

(b) Find the cost for the 85th case of solvent, i.e., the marginal cost for a purchase of 84 cases.

(c) Explain in the language of marginal cost the statement "the more you buy, the cheaper it gets."

(d) Find a large value of x, beyond which it is unreasonable for the given formula for $C(x)$ to be applicable.

25. In problem 24, suppose that the solvent is priced at $8 - [(8x + 100)/(x + 300)]$ dollars per case at production level x. Calculate the marginal revenue and the marginal profit.

1-4 Antiderivatives

An antiderivative of f is a function whose derivative is f.

Many applications of calculus require one to find a function whose derivative is given. In this section, we show how to solve simple problems of this type. Our study of antidifferentiation will continue in Sections 3-3 and 4-2.

Goals

After studying this section, you should be able to:

Find the antiderivative of any polynomial.

Find the position function for a moving object, given its velocity function and its position at one time.

Use the indefinite integral notation.

ANTIDIFFERENTIATION AS AN INVERSE PROBLEM

A child learning arithmetic is first presented with problems like "how much is two plus five?" After some experience with this kind of problem, the child can be presented with "what number must you add to three to get eight?" In this second type of problem, what was formerly the answer is presented as part of the data; it is now the "question" which is to be determined. Such a problem is called an *inverse* problem. Calculus has its own inverse problems; here is an example.

Worked Example 1   Find a function whose derivative is $2x + 3$.

Solution   We recall that the derivative of x^2 is $2x$ and that the derivative of $3x$ is 3, so the unknown function could be $x^2 + 3x$. We may check our answer by differentiating: $(d/dx)(x^2 + 3x) = 2x + 3$.

The function $x^2 + 3x$ is not the only possible solution to Worked Example 1; so are $x^2 + 3x + 1$, $x^2 + 3x + 2$, etc. In fact, since the derivative of a constant function is zero, $x^2 + 3x + C$ solves the problem for any number C. Just like the arithmetic inverse problem "find all pairs of numbers which add up to eight," our calculus inverse problem has more than one solution.

Definition   If $F' = f$, then F is called an *antiderivative* of f.

The antiderivative of a function, unlike the derivative, is never unique. Indeed, if F is an antiderivative of f, so is $F + C$ for an arbitrary constant C. We will show in Section 3-3 that *all* the antiderivatives are of this form. For now, we take this fact for granted.

We can make the solution of an antidifferentiation problem unique by imposing an extra condition on the unknown function. The following example is a typical application of antidifferentiation to physics.

Worked Example 2   The velocity of a particle moving along a line is $3t + 5$ at time t. At time 1, the particle is at position 4. Where is it at time 10?

Solution   Let $F(t)$ denote the position of the particle at time t. We will determine the function F. Since velocity is the rate of change of position with respect to time, we must have $F'(t) = 3t + 5$; that is, F is an antiderivative of $f(t) = 3t + 5$. A function whose derivative is $3t$ is $\frac{3}{2}t^2$, since $(d/dt)\frac{3}{2}t^2 = \frac{3}{2}2t = 3t$. Similarly, a function whose derivative is 5 is $5t$. Therefore, we take

$$F(t) = \tfrac{3}{2}t^2 + 5t + C \tag{1}$$

where C is a constant to be determined. To find the value of C, we use the information that the particle is at position 4 at time 1; that is, $F(1) = 4$. Substituting 1 for t and 4 for $F(t)$ in equation (1) gives

$$4 = \tfrac{3}{2} + 5 + C = \tfrac{13}{2} + C$$

or $C = -\frac{5}{2}$, and so $F(t) = \frac{3}{2}t^2 + 5t - \frac{5}{2}$. Finally, we substitute 10 for t, obtaining the position at time 10: $F(10) = \frac{3}{2} \cdot 100 + 5 \cdot 10 - \frac{5}{2} = 197\frac{1}{2}$. See Fig. 1-4-1.

Fig. 1-4-1 The graph of the solution to Worked Example 2.

In figure: $y = \frac{3}{2}t^2 + 5t - \frac{5}{2}$

At this point in our study of calculus, we must solve antidifferentiation problems by guessing the answer and then checking and refining our guesses if necessary. More systematic methods will be given shortly.

Solved Exercises

1. Find all the antiderivatives you can for the function $f(x) = x^4 + 5$.

2. The acceleration of a falling body near the earth's surface is 9.8 meters per second per second. If the body has downward velocity v_0 at $t = 0$, what is its velocity at time t? If the position is x_0 at time 0, what is the position at time t? See Fig. 1-4-2.

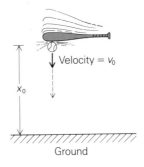

Fig. 1-4-2   The body is projected downward at $t = 0$ with velocity v_0.

Exercises

1. Find antiderivatives of each of the following functions:
   (a) $x + 2$ (b) $x^5 + [2/x^4]$ (c) $x^6 + 9$
   (d) $4x^8 + 3x^2$ (e) $s(s + 1)(s + 2)$ (f) $1/t^3$

2. A ball is thrown downward with a velocity of 10 meters per second. How long does it take the ball to fall 150 meters?

3. A particle moves along a line with velocity $v(t) = \frac{1}{2}t^2 + t$. If it is at $x = 0$ when $t = 0$, find its position as a function of t.

LEIBNIZ NOTATION AND THE INDEFINITE INTEGRAL

The most widely used notation for the antiderivative is due to Leibniz. If F is an antiderivative of f, we write

$$F(x) = \int f(x)\,dx \qquad \qquad (2)$$

The elongated S is called an *integral sign,* and the expression on the right side of equation (2) is read "the antiderivative of f of x dx." The antiderivative is also called an *indefinite integral.* Leibniz used an S-like symbol because the operation of antidifferentiation (also called *integration*) turns out to be a form of $\int$ummation.

We will study integration as a summation process in Chapter 4. For the moment, we will simply consider the symbol $\int$ as meaning "antiderivative." The dx has no meaning by itself except to tell us that the independent variable is x.

Worked Example 3   Express the result of Solved Exercise 1 in Leibniz notation.

Solution $\int (x^4 + 5)\,dx = \frac{1}{5}x^5 + 5x + C$

ANTIDIFFERENTIATION AND INDEFINITE INTEGRALS

An antiderivative for f is a function F such that $F'(x) = f(x)$. We write $F(x) = \int f(x)\,dx$. The function $\int f(x)\,dx$ is also called the indefinite integral of f and f is called the integrand.

If $F(x)$ is an antiderivative of $f(x)$, the general antiderivative has the form $F(x) + C$ for an arbitrary constant C.

Some of the differentiation rules lead directly to systematic rules for antidifferentiation. The rules in the following display can be proved by differentiation of their right-hand side (see Worked Example 5 below).

ANTIDIFFERENTIATION RULES

Sum:
$$\int [f(x) + g(x)]\,dx = \int f(x)\,dx + \int g(x)\,dx$$

Constant multiple:
$$\int af(x)\,dx = a\int f(x)\,dx$$

Power:
$$\int x^n\,dx = \frac{x^{n+1}}{n+1} + C \quad (n \text{ an integer}; n \neq -1)$$

Polynomial:
$$\int (c_n x^n + \cdots + c_1 x + c_0)\,dx = \frac{c_n}{n+1}x^{n+1} + \cdots + \frac{c_1}{2}x^2 + c_0 x + C$$

Worked Example 4 Find $\int [(1/x^2) + 3x + 2]\,dx$.

Solution
$$\int \left(\frac{1}{x^2} + 3x + 2\right)dx = \int \frac{1}{x^2}\,dx + 3\int x\,dx + 2\int 1\,dx$$
$$= \int x^{-2}\,dx + 3\int x^1\,dx + 2\int x^0\,dx$$
$$= -1x^{-1} + \frac{3}{2}x^2 + 2x^1 + C$$
$$= -\frac{1}{x} + \frac{3}{2}x^2 + 2x + C$$

We write C only once because the sum of three constants is a constant.

Worked Example 5 Prove the power rule
$$\int x^n\,dx = \frac{x^{n+1}}{n+1} + C \quad (n \neq -1)$$

Solution By definition, $F(x) = \int x^n\,dx$ is a function such that $F'(x) = f(x) = x^n$. However, $F(x) = [x^{n+1}/(n+1)] + C$ is such a function since $F'(x) = (n+1)\cdot x^{n+1-1}/(n+1) = x^n$, by the power rule for derivatives.

Solved Exercises

3. Explain the exclusion $n \neq -1$ in the power rule.

4. Find $\int dx/(3x + 1)^5$.

5. Let x = position, v = velocity, a = acceleration, t = time. Express the relations between these variables by using the indefinite integral notation.

Exercises

4. Find the following indefinite integrals:
   (a) $\int (x^2 + 3x + 2)\,dx$ (b) $\int 4\pi r^2\,dr$
   (c) $\int 4db/(3b + 2)^9$ (d) $\int [(1/x^4) + x^4]\,dx$
   (e) $\int 3t(t + 1)(t + 2)\,dt$

5. Is it true that $\int f(x)g(x)\,dx = [\int f(x)\,dx]g(x) + f(x)[\int g(x)\,dx]$?

6. Prove the constant multiple rule for antidifferentiation.

Problems for Section 1-4

1. Find functions which have the following functions as their derivatives:
   (a) $f(x) = 3x$ (b) $f(t) = (t + 1)^2$
   (c) $f(x) = (x + 1)/x^3$ (d) $f(x) = 3x^4 + 4x^3$

2. Find the indicated antiderivatives:
   (a) $\int (x^3 + 3x)\,dx$ (b) $\int (t^3 + t^{-2})\,dt$
   (c) $\int [1/(t + 1)^2]\,dt$
   (d) $\int [(w^2 + 2)/w^5]\,dw$

3. (a) Find $(d/dx)[(x^3 + 1)/(x^3 - 1)]$.
   (b) Find $\int [-6x^2/(x^3 - 1)^2]\,dx$.

4. The population of Booneville increases at a rate of $r(t) = (3.62)(1 + 0.8t^2)$ people per year, where t is the time in years from 1970. The population in 1976 was 726. What will it be in 1984?

5. Water is being poured into a tub at $g(t)$ gallons per minute at time t. If $G(t)$ is the number of gallons in the tub at time t, write G in terms of g using integral notation. How can you determine the arbitrary constant?

6. Prove the sum rule for antiderivatives.

7. Find:
   (a) $\int (3t^2 + 2t + 1)\,dt$
   (b) $\int [(t^3 + t + 1)/t^5]\,dt$
   (c) $\int (8t + 1)^{-2}\,dt$ (d) $\int (u^4 - 6u)\,du$

8. Find a function $F(x)$ such that
   $x^5 F'(x) + x^3 + 2x = 3$

9. Find a function $f(x)$ whose graph passes through $(1, 1)$ and such that the slope of its tangent line at $(x, f(x))$ is $3x + 1$.

10. Find the antiderivative $F(x)$ of
    $f(x) = x^3 + 3x^2 + 2$
    which satisfies $F(0) = 1$.

11. A rock is thrown vertically upward with velocity 19.6 meters per second. After how long does it return to the thrower? (The acceleration due to gravity is 9.8 meters per second per second; see Solved Exercise 2.) See Fig. 1-4-3.

Fig. 1-4-3 The path of a rock thrown upwards from the earth.

Ground

Review Problems for Chapter 1

1. Differentiate each of the following functions:
   (a) $6x$
   (b) $x^2 + 9x + 10$
   (c) $x^3 + x^2 - 1$
   (d) $6/x$
   (e) $(x^2 + 1)^{13}$

2. Differentiate the following functions:
   (a) $(x + 1)/(x - 1)$
   (b) $(x^2 - 1)/(x^2 - 2)$
   (c) $(x^2 + 1)^{13}/(x^2 - 1)^{14}$
   (d) $[(x^3 + 6)^2 - (2x^4 + 1)^3]/(x^5 + 8)$
   (e) $(9x^9 - x^8 + 14x^7 + x^6 + 5x^4 + x^2 + 2)/x^2$

3. Let $A(x) = x^3 - x^2 - 2x$
   $B(x) = x^2 + \frac{1}{4}x - \frac{3}{8}$
   $C(x) = 2x^3 - 5x^2 + x + 2$
   $D(x) = x^2 + 8x + 16$

   Differentiate:
   (a) $3A(x) + 2B(x)$
   (b) $[A(x)]^2$
   (c) $A(x)/D(x)$
   (d) $C(x) - 3[D(x)/A(x)]$
   (e) $A(2x)$

4. Find the equation of the tangent line to the graph of:
   (a) $A(x)$ at $x = 1$
   (b) $B(x)$ at $x = 0$
   (c) $C(x)$ at $x = -2$
   (d) $D(x)$ at $x = -1$
   where A, B, C, D are defined in problem 3.

5. Let $x(t) = t^3 + ct$ be the position of a particle at time t. For which values of c does the particle reverse direction, and at what times does the reversal take place for each such value of c? Does the value of c affect the particle's acceleration?

6. An evasive moth has position $t^3 - t + 2$ at time t. A hungry bat has position $y(t) = -\frac{2}{3}t^2 + t + 2$ at time t. How many chances does the bat have to catch the moth? How fast are they going and what are their accelerations at these times?

7. Find the equation of the line tangent to the graph of the function at the indicated point:
   (a) $f(x) = x^3 - 6x;\ (0,0)$
   (b) $f(x) = (x^4 - 1)/(6x^2 + 1);\ (1,0)$
   (c) $f(x) = (x^5 - 6x^4 + 2x^3 - x)/(x^2 + 1);$ $(1,-2)$
   (d) $f(x) = (x^3 - 7)/(x^3 + 11);\ (2,\frac{1}{19})$
   (e) $f(x) = (3x^2 + 2x + 1)/(9x^2 - x - 2);\ (-1,\frac{1}{4})$

8. Use the definition of the derivative directly to find $f'(1)$, where $f(x) = 3x^3 + 8x$.

9. Find:
   (a) $\lim_{x \to 1} x^2 + 1$
   (b) $\lim_{x \to 1} \dfrac{x^3 - 1}{x - 1}$
   (c) $\lim_{x \to 1} \dfrac{x^3 + 1}{x + 1}$
   (d) $\lim_{h \to 0} [(h - 2)^6 - 64]/h$
   (e) $\lim_{h \to 0} [(h - 2)^6 + 64]/h$

10. Find:
    (a) $\lim_{x \to 3} (3x^2 + 2x)/x$
    (b) $\lim_{x \to 0} (3x^2 + 2x)/x$
    (c) $\lim_{x \to 3} (x^2 - 9)/(x - 3)$
    (d) $\lim_{x \to 1} (x^5 - 1)/(x - 1)$
    (e) $\lim_{s \to 0} [(s + 3)^9 - 3^9]/s$

11. For the function sketched in Fig. 1-R-1, find $\lim_{x \to x_0} f(x)$ for $x_0 = -3, -2, -1, 0, 1, 2, 3$. If the limit is not defined or does not exist, say so.

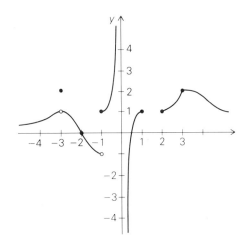

Fig. 1-R-1 Find $\lim_{x \to x_0} f(x)$ at the indicated points.

12. (a) Find the linear approximation to $(x^{40} - 1)/(x^{29} + 1)$ at $x_0 = 1$.
    (b) Calculate $[(1.021)^{40} - 1]/[(1.021)^{29} + 1]$ approximately.

13. Find:
    (a) $(d/ds)[(s^4 - 1)/(\pi s^3 - 1)]$
    (b) $(d^2/du^2)(u^5 + 8)$
    (c) The rate of change of the surface area of a sphere with respect to its radius. (The surface area of a sphere is $A = 4\pi r^2$.)
    (d) The rate of change of the volume of a cube with respect to the length of one of its sides.

14. A particle is said to be *accelerating* (decelerating) if the sign of its acceleration is equal (opposite) to the sign of its velocity. (a) Let $f(t) = -t^3$ be the position of a particle on a straight line at time t. When is the particle accelerating and when is it decelerating? (b) If the position of a particle on a line is given as a quadratic function of time and the particle is accelerating at time t_0, does the particle ever decelerate?

15. Find the first and second derivatives of the following functions:

   (a) $f(x) = (x - a)/(x^2 + 2bx + c)$ (a, b, c constants)

   (b) $g(t) = (t^5 - 3t^4)/(t^3 - 1)$

   (c) $h(r) = r^{13} - \sqrt{2}\,r^4 - [r/(r^2 + 3)]$

   (d) $h(x) = (x - 2)^4(x^2 + 2)$

   (e) $q(s) = s^5 [(1 - 2s)/(1 - s)]$

   (f) $f(z) = (az + b)/(cz + d)$ (a, b, c, d constants)

16. Find the first and second derivatives of the following functions.

   (a) $x(t) = [A/(1 - t)] + [B/(1 - t^2)] + [C/(1 - t^3)]$ (A, B, C constants)

   (b) $f(x) = (x - 1)^3 g(x)$ ($g(x)$ is some differentiable function)

   (c) $s(t) = (t - a)^n/(t - b)^m$ (a, b constants; n, m positive integers)

   (d) $k(s) = s^{15} + [(s^{12} - 2)/(s - 1)] + s^2 - 1$

   (e) $V(r) = \frac{4}{3}\pi r^2 + 2\pi rh(r)$, where $h(r) = 2r - 1$

   (f) $f(x) = 1 - 1/x - 1/x^2 + 1/x^3 + 1/x^{10}$

17. Find the following antiderivatives:

   (a) $\int (4.9t + 15)\,dt$

   (b) $\int (s^5 + 4s^4 + 9)\,ds$

   (c) $\int (1/z^2 + 4/z^3)\,dz$

   (d) $\int [2x(x^2 + 1) + x^2(2x)]\,dx$ (*Hint:* Recognize the integrand as the derivative of a product.)

   (e) $\int 1/(x - 1)^2\,dx$

   (f) $\int (ax^2 + bx + c)\,dx$ (a, b, c constants)

   (g) $\int [(t^2 + t)5t^4 - (2t + 1)t^5]/(t^2 + t)^2\,dt$ (*Hint:* Recognize the integrand as the derivative of a quotient.)

18. Find antiderivatives for each of the following functions:

   (a) $x^2 + 1$ (b) $x^3 + 3x^2 + 9$

   (c) $(x^4 + 3x^2 + 2)/x^2$ (d) $(x^2 + 1)^3$

   (e) $(x^3 - 1)^2 - (x^3 + 1)^2$

★19. Find the equation of a line through the origin, with positive slope, which is tangent to the parabola $y = x^2 - 2x + 2$.

20. Differentiate both sides of the equation

$$\frac{f(x)}{g(x)} = \frac{1}{g(x)/f(x)}$$

and show that you get the same result on each side.

★21. The polynomial $a_n x^n + a_{n-1}x^{n-1} + \cdots + a_0$ is said to have *degree n* if $a_n \neq 0$. For example: $\deg(x^3 - 2x + 3) = 3$, $\deg(x^4 + 5) = 4$, $\deg(0x^2 + 3x + 1) = 1$. The degree of the rational function $f(x)/g(x)$, where $f(x)$ and $g(x)$ are polynomials, is defined to be the degree of f minus the degree of g.

   (a) Prove that, if $f(x)$ and $g(x)$ are polynomials, then $\deg f(x)g(x) = \deg f(x) + \deg g(x)$.

   (b) Prove the result in part (a) when $f(x)$ and $g(x)$ are rational functions.

   (c) Prove that, if $f(x)$ is a rational function with nonzero degree, then $\deg f'(x) = \deg f(x) - 1$. What if $\deg f(x) = 0$?

22. Using the formula $\sqrt{x} \cdot \sqrt{x} = x$ and the product rule, find what $(d/dx)\sqrt{x}$ should be.

23. Find the rate of change of the area of an equilateral triangle with respect to the length of one of its sides.

24. Find d^2u/dx^2, where $u = (x^4 + 3)(2x^2 + 7x)$.

25. Find a formula for $(d^2/dx^2)[f(x)g(x)]$.

26. If the position of an object at time t is $(t^5 + 1)(t + 2)$, find its velocity and acceleration when $t = 0.1$.

27. Let $y = x^2 + 3x + 1$ and $z = y^2 + 4y$.

   (a) Find dy/dx and dz/dy.

   (b) Express z in terms of x and find dz/dx.

   (c) Express $(dz/dy) \cdot (dy/dx)$ in terms of x and compare with the answer to part (b).

28. Find a formula for $(d/dx)[f(x)^m g(x)^n]$, where f and g are differentiable functions and m and n are positive integers.

29. If $y = f(x)$ and $z = y^n$, find dz/dx. Compare the result with $(dz/dy) \cdot (dy/dx)$.

30. Prove that the parabola $y = x^2$ has the optical focusing property mentioned on p. 24. (This problem requires trigonometry; consult Section 6-1 for a review.) *Hint:* Refer to Fig. 1-R-2 and carry out the following program:

    (a) Express $\tan \phi$ and $\tan \theta$ in terms of x.

    (b) Prove that $90° - \theta = \theta - \phi$ by using the trigonometric identities:

    $$\tan 2\theta = \frac{2 \tan \theta}{1 - \tan^2 \theta}$$

    and

    $$\tan(\phi + 90°) = -\frac{1}{\tan \phi}$$

★ 31. Prove that the parabola $y = ax^2$ has the optical focusing property. (You should start by figuring out where the focal point will be.)

32. The Stefan-Boltzman constant σ appears in *Stefan's law* $R = \sigma T^4$. The symbol R is the rate of change of radiant energy emitted per unit area from a body's surface held at temperature T (in degrees Kelvin). Find dR/dT for $500°$ Kelvin, assuming $\sigma = 5.67 \times 10^{-8}$ watts/(meter)^2 (degrees Kelvin)^4. Interpret your answer as a rate of change.

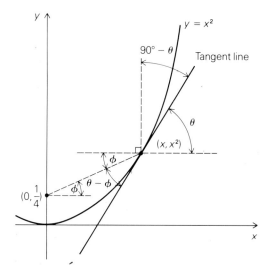

Fig. 1-R-2 The geometry needed to prove that the parabola has the optical focusing property.

2

The Shape of Graphs

Differential calculus provides tests for locating the key features of graphs.

Now that we know how to differentiate rational functions, we can use this information to assist us in plotting their graphs. The signs of the derivative and the second derivative of a function will tell us which way the graph of that function is "leaning" and "bending"; changes of sign of the derivatives are thus of particular importance.

In Chapters 5, 6, and 7 we will learn how to differentiate new functions. The methods learned in this chapter will help us study them.

CONTENTS

Section 2-1 covers the concept of continuity, a property more general than differentiability. The main result in Section 2-2 states that a function f is increasing (or decreasing) at points where f' is positive (or negative). Similarly, the sign of the second derivative gives information about the concavity of a function, the subject of Section 2-3. In Section 2-4 we lay out a systematic procedure for drawing the graph of a given function.

This apppendix is found in the Student Guide.

2-1 Continuity and the Intermediate Value Theorem

If a continuous function f is negative at a and is positive at b, then f must be zero at some point between a and b.

Before embarking on our study of graphing, we have to develop some tools. The first of these is the property of continuity, which has many important consequences, including the intermediate value theorem, which is useful for locating the zeros of functions. In the following sections, we develop tests directly useful in graphing; these rely on the concept of continuity as well.

Goals

After studying this section, you should be able to:

Determine where a given function is continuous.

Use the intermediate value theorem to estimate the location of roots of continuous functions.

THE DEFINITION OF CONTINUITY

A film editor splices footage of two scenes to produce a film in which a magician disappears from one side of a screen and instantly reappears on the other side. To the viewer of the film, it seems miraculous that the magician was able to do this without passing through the screen. The magician's instantaneous relocation is interpreted by the viewer as *discontinuous motion*.

Let us formulate this event mathematically. Let $x = f(t)$ denote the position of the magician at time t; see Fig. 2-1-1. Let t_0 be the instant of relocation. If the screen is at $x = c$ then $f(t_0) < c$ (the magician is to the left of the screen), yet $f(t) > c$ if $t > t_0$ (the magician is suddenly to the right of the screen). When this happens, we call f *discontinuous* at t_0. (We would use the same terminology if the magician suddenly relocated going right to left.) If a function f is not discontinuous, we call it *continuous*.

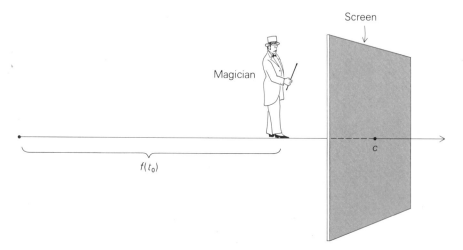

Fig. 2-1-1 The position $f(t)$ of a magician is discontinuous since he can suddenly disappear and reappear on the other side of a screen.

The notion of a continuous curve can be illustrated geometrically. Naively, we think of a curve as being continuous if we can draw it "without removing the pencil from the paper." Let (x_0, y_0) be a point on the curve, and draw the lines $y = c_1$ and $y = c_2$ with $c_1 < y_0 < c_2$. If the curve is continuous, at least a "piece" of the curve on each side of (x_0, y_0) should be between these lines, as in Fig. 2-1-2 (left). Compare this with the behavior of the discontinuous curve in Fig. 2-1-2 (right).

Fig. 2-1-2 Illustrating a continuous curve (left) and a discontinuous curve (right).

(optional)

The following definition is a precise formulation, for functions, of the ideas we have just described in terms of motions and curves. It is not necessary to master this theory in order to work with the idea of continuity.

Definition Let f be a function whose domain contains an open interval about x_0. We say that f is *continuous* at x_0 if the following two conditions are met:

1. For every number c_1 such that $c_1 < f(x_0)$, there is an open interval I about x_0 such that $c_1 < f(x)$ for all $x \in I$.

2. For every number c_2 such that $f(x_0) < c_2$, there is an open interval J about x_0 such that $f(x) < c_2$ for all $x \in J$.

If f is continuous at each point of its domain, we just say that f is *continuous*. (Notice that f can be continuous at x_0 only if $f(x_0)$ is defined.)

Worked Example 1 Suppose that f is a continuous function and that $f(1) = 10$. Show that if x is close enough to 1, $f(x)$ is less than 10.5.

Solution Since f is continuous, it has both properties 1 and 2 of the definition. In property 2, choose $x_0 = 1$ and $c_2 = 10.5$. Thus, $f(x_0) < c_2$, since $10 < 10.5$. Thus there is an open interval J, such that when x is in J, $f(x) < 10.5$. We take $x \in J$ as a measure of how close x has to be to x_0. (We cannot compute J explicitly without more information on f.)

Worked Example 2 Let $g(x)$ be the *step function* defined by

$$g(x) = \begin{cases} 0 & \text{if } x \le 0 \\ 1 & \text{if } x > 0 \end{cases}$$

Show that g is not continuous at $x_0 = 0$.

Solution

Fig. 2-1-3 This step function is discontinuous at $x_0 = 0$.

First we sketch the graph of g (Fig. 2-1-3). Take $\frac{1}{2}$ for c_2. The inequality $g(0) < \frac{1}{2}$ is satisfied by g at $x_0 = 0$, since $g(0) = 0$, but no matter what open interval I we take about zero, there are positive numbers x in I for which $g(x) = 1$, which is greater than $\frac{1}{2}$. Since it is not possible to choose I such that condition 1 in the definition of continuity is satisfied, with $x_0 = 0$, $c_2 = \frac{1}{2}$, it follows that g is not continuous at zero.

If f is continuous at x_0, then $f(x)$ must become closer and closer to $f(x_0)$ as x approaches x_0; and this condition is also sufficient for continuity: i.e., f is continuous at x_0 if and only if $\lim_{x \to x_0} f(x) = f(x_0)$. Equivalently, we can say $\lim_{x \to x_0} [f(x) - f(x_0)] = 0$ or we can say $\lim_{x \to 0} \Delta y = 0$ where $\Delta x = x - x_0$ and $\Delta y = f(x) - f(x_0) = f(x_0 + \Delta x) - f(x_0)$.

CONTINUITY AND LIMITS

If f is defined in an open interval containing x_0, then f is continuous at x_0 if and only if

$$\lim_{x \to x_0} f(x) = f(x_0).$$

This condition is equivalent to $\lim_{\Delta x \to 0} \Delta y = 0$ where $\Delta y = f(x_0 + \Delta x) - f(x_0)$.

In other words, continuity at x_0 means tha the *limit* of f at x_0 exists *and* is equal to the *value* of f at x_0. For example, the rational function rule for limits (p. 70) says precisely that *all rational functions are continuous.*

Worked Example 3   Using the laws of limits, show that if f and g are continuous at x_0, so is fg.

Solution   We must show that $\lim_{x \to x_0} (fg)(x) = (fg)(x_0)$. By the product rule for limits, $\lim_{x \to x_0} f(x)g(x) = \lim_{x \to x_0} f(x) \lim_{x \to x_0} g(x) = f(x_0)g(x_0)$, since f and g are continuous at x_0. But $f(x_0)g(x_0) = (fg)(x_0)$, and so $\lim_{x \to x_0} (fg)(x) = (fg)(x_0)$, as required.

(optional)   The statement in the display above can be proven rigorously by a comparison of the definitions on p. 67 and p. 107. Alternatively, it can be taken as a *definition* of continuity in terms of limits.

It is sometimes necessary to speak of the continuity of a function defined on a *closed* interval $[a, b]$. For x_0 in (a, b), the preceding definition applies, but if x_0 is a or b, we must use a slightly modified version as follows.

(optional)

Definition If x_0 is an element of the domain D of a function f, we say that f is continuous at x_0 if:

1. For each $c_1 < f(x_0)$ there is an open interval I about x_0 such that, for those x in I *which also lie in D*, $c_1 < f(x)$.

2. For each $c_2 > f(x_0)$ there is an open interval J about x_0 such that, for those x in J *which also lie in D*, $f(x) < c_2$.

If f is continuous at every point of its domain, we simply say that f is continuous or f is continuous on D.

In practice, when we discuss a function on an interval $[a, b]$ we are often dealing with a function which is really definable on some larger open interval; we are simply restricting attention to the behavior of the function on $[a, b]$. If the function is continuous on the larger open interval, it is certainly continuous on a smaller closed interval.

Solved Exercises* ★1. Let f be continuous at x_0 and defined in an open interval containing x_0. Assume that $f(x_0) > 0$. Prove that there is an open interval I containing x_0 on which f is positive.

★ 2. Let $f(x)$ be the *absolute value function:* $f(x) = |x|$; that is,

$$f(x) = \begin{cases} x & \text{if } x \geq 0 \\ -x & \text{if } x < 0 \end{cases}$$

Show that f is continuous at $x_0 = 0$.

3. Decide whether each of the functions whose graphs appear in Fig. 2-1-4 is continuous. Explain your answers.

(a)   **(b)**   **(c)**   **(d)**   **Fig. 2-1-4** Which functions are continuous?

Exercises ★1. Let $f(x)$ be the step function defined by

$$f(x) = \begin{cases} -1 & \text{if } x < 0 \\ -2 & \text{if } x \geq 0 \end{cases}$$

Show that f is discontinuous at zero.

★ 2. Show directly from the definition of continuity that the linear function $f(x) = 5x - 10$ is continuous at $x_0 = 2$ (that is, do not use limits).

Solutions are found in Appendix A. The Exercises and Solved Exercises marked with a ★ refer to (optional) text material in small print.

3. Let $f(x)$ be defined by

$$f(x) = \begin{cases} x^2 + 1 & \text{if } x < 1 \\ ? & \text{if } 1 \leq x \leq 3 \\ x - 6 & \text{if } 3 < x \end{cases}$$

How can you define $f(x)$ on the interval $[1, 3]$ in order to make f continuous on $(-\infty, \infty)$? (A geometric argument will suffice.)

4. Let $f(x)$ be defined by $f(x) = (x^2 - 1)/(x - 1)$ for $x \neq 1$. How should you define $f(1)$ to make the resulting function continuous? [*Hint:* Plot a graph of $f(x)$ for x near 1 by factoring the numerator.]

5. Let $f(x)$ be defined by $f(x) = 1/x$ for $x \neq 0$. Is there any way to define $f(0)$ so that the resulting function will be continuous?

DIFFERENTIABILITY AND CONTINUITY

If a function $f(x)$ is differentiable at $x = x_0$, then the graph of f has a tangent line at $(x_0, f(x_0))$. Our intuition suggests that if a curve is smooth enough to have a tangent line then the curve should have no breaks—that is, a differentiable function should be continuous. The following theorem says just that.

Theorem 1   *If the function f is differentiable at x_0, then f is continuous at x_0.*

Proof   If f is differentiable at x_0, then

$$\lim_{\Delta x \to 0} \frac{f(x_0 + \Delta x) - f(x_0)}{\Delta x} = f'(x_0)$$

Multiplying by $\lim_{\Delta x \to 0} \Delta x = 0$ gives

$$\left[\lim_{\Delta x \to 0} \Delta x \right] \left[\lim_{\Delta x \to 0} \frac{f(x_0 + \Delta x) - f(x_0)}{\Delta x} \right] = 0 \cdot f'(x_0) = 0$$

By the product rule for limits, this becomes

$$\lim_{\Delta x \to 0} [f(x_0 + \Delta x) - f(x_0)] = 0$$

or

$$\lim_{\Delta x \to 0} \Delta y = 0$$

Thus, by the display on p. 108, f is continuous at x_0.

Worked Example 4   Show that the function $f(x) = (x - 1)/3x^2$ is continuous at $x_0 = 4$.

Solution   *Method 1:*   We know from Section 1-3 that x, $x - 1$, x^2, $3x^2$, and hence $(x - 1)/3x^2$ are differentiable (when $x \neq 0$). Since $4 \neq 0$, Theorem 1 implies that f is continuous at $x = 4$.

Method 2:   Since $3x^2$ vanishes only at $x = 0$, the quotient rule (or rational function rule) for limits gives

$$\lim_{x \to 4} \frac{x-1}{3x^2} = \frac{\lim_{x \to 4} x - 1}{\lim_{x \to 4} 3x^2} = \frac{3}{3 \cdot 16} = \frac{1}{16} = f(4)$$

where $f(x) = (x - 1)/3x^2$. Thus f is continuous at $x_0 = 4$.

(These methods are certainly much easier than attempting to verify directly the conditions in the definition of continuity.)

The argument used in Worked Example 4 leads to the following general result.

Corollary

1. *Any polynomial $P(x)$ is continuous.*

2. *Let $P(x)$ and $Q(x)$ be polynomials, with $Q(x)$ not identically zero. Then the rational function $R(x) = P(x)/Q(x)$ is continuous at all points of its domain; that is, R is continuous at all x_0 such that $Q(x_0) \neq 0$.*

The corollary follows from Theorem 1 because we know from Section 1-3 that polynomials and rational functions are differentiable; or, alternatively, the corollary follows from the rational function rule for limits.

Solved Exercises

4. Prove that $(x^2 - 1)/(x^3 + 3x)$ is continuous at $x = 1$.

5. Is the converse of Theorem 1 true? That is, is a function which is continuous at x_0 necessarily differentiable there? Prove or give a counterexample.

6. Where is $(x^4 + 1)/(x^3 - 8)$ continuous?

7. Let $f(x) = (x^3 + 2)/(x^2 - 1)$. Show that f is continuous on $[-\frac{1}{2}, \frac{1}{2}]$.

Exercises

6. Is the function $(x^3 - 1)/(x^2 - 1)$ continuous at 1?

7. Let $f(x) = [1/x] + [(x^2 - 1)/x]$. Can you define $f(0)$ so that the resulting function is continuous at all x?

8. Find a function which is continuous on the whole real line and differentiable for all x except 1, 2, and 3. (A sketch will do.)

9. Where is $(x^2 - 1)/(x^4 + x^2 + 1)$ continuous?

THE INTERMEDIATE VALUE THEOREM

According to our previous discussion, a continuous function is one whose graph never "jumps" across horizontal lines. This is a *local* property of the function, since the definition involves an indefinitely small interval about the point of continuity. There is a corresponding *global* statement, called the intermediate value theorem, which applies to the function on an entire interval $[a, b]$.

> **Theorem 2** **Intermediate Value Theorem** *(first version). Let f be continuous on $[a, b]$ and suppose that, for some number c, $f(a) < c < f(b)$ or $f(a) > c > f(b)$. Then there is some point x_0 in (a, b) such that $f(x_0) = c$.*

Geometrically interpreted, this theorem says that for the graph of a continuous function to get from one side of a horizontal line to the other, the graph must meet the line somewhere (see Fig. 2-1-5). The proof of the theorem depends on a careful study of properties of the real numbers and is given in the appendix to this chapter (see the Student Guide).

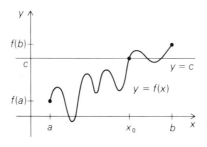

Fig. 2-1-5 The graph of f must pierce the horizontal line $y = c$ if it is to get across.

Worked Example 5 Show that there is a number x_0 such that $x_0^5 - x_0 = 3$.

Solution Let $f(x) = x^5 - x$. Then $f(0) = 0$ and $f(2) = 30$. Since $0 < 3 < 30$, the intermediate value theorem ensures the existence of a number x_0 in $(0, 2)$ such that $f(x_0) = 3$. (The function f is continuous on $[0, 2]$ because it is a polynomial.)

Notice that the intermediate value theorem does not tell us how to find the number x_0 but merely that it exists. (A look at Fig. 2-1-5 should convince you that there may be more than one possible choice for x_0.) Nevertheless, by repeatedly dividing an interval into two or more parts and evaluating $f(x)$ at the dividing points, we can solve the equation $f(x_0) = c$ as accurately as we wish. This *method of bisection* is illustrated in Solved Exercise 8 below.

There is a second way of stating the intermediate value theorem (the *contrapositive* statement) which is useful in its own right.

> **Theorem 2′** **Intermediate Value Theorem** *(second version). Let f be a function which is continuous on $[a, b]$ and suppose that $f(x) \neq c$ for all x in $[a, b]$. If $f(a) < c$, then $f(b) < c$ as well. (See Fig. 2-1-6.) Similarly, if $f(a) > c$, then $f(b) > c$ as well.*

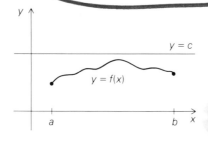

Fig. 2-1-6 The graph starts below the $y = c$ line and never pierces the line, so it stays below the line.

In geometric language, this second version of the theorem says: "The graph of a continuous function which never meets a horizontal line must remain on one side of it." The first version says: "If the graph of a continuous function gets from one side of a horizontal line to the other, the graph must meet the line somewhere." You should convince yourself that these two statements are really saying the same thing.

In practice, the second version of the intermediate value theorem is useful for determining the sign of a function on intervals where it has no roots.

Worked Example 6 Suppose that f is continuous on $[0,3]$, that f has no roots on the interval, and that $f(0) = 1$. Prove that $f(x) > 0$ for *all* x in $[0,3]$.

Solution Apply the intermediate value theorem (version 2), with $c = 0$, to f on $[0, \bar{x}]$ for each $\bar{x}$ in $(0,3]$. Since f is continuous on $[0,3]$, it is continuous on $[0, \bar{x}]$; since $f(0) = 1 > 0$, we have $f(\bar{x}) > 0$. But $\bar{x}$ was anything in $(0,3]$, so we have proved what we want.

Solved Exercises

8. (The method of bisection) Find a solution of the equation $x^5 - x = 3$ in $[0, 2]$ to within an accuracy of 0.1 by repeatedly chopping intervals in half and testing each half for a root.

9. The function $f(x) = 1/(x - 1)$ never takes the value zero, yet $f(0) = -1$ is negative and $f(2) = 1$ is positive. Why isn't this a counterexample to the intermediate value theorem?

10. "Prove" that you were once exactly one meter tall.

Exercises

10. Prove that the equation $x^3 + 2x - 1 = 7$ has a real solution.

11. The roots of $f(x) = x^3 - 2x - x^2 + 2$ are $\sqrt{2}$, $-\sqrt{2}$, and 1. By evaluating $f(-3)$, $f(0)$, $f(1.3)$, and $f(2)$, determine the sign of $f(x)$ on each of the intervals between its roots.

12. Find a solution of the equation $x^5 - x = 5$ to an accuracy of 0.1.

Problems for Section 2-1

1. Let $f(x) = x + (4/x)$ for $x \le -\frac{1}{2}$ and $x \ge 2$. Define $f(x)$ for x in $(-\frac{1}{2}, 2)$ in such a way that the resulting function is continuous on the whole real line.

2. Sketch the graph of a function which is continuous on the whole real line and differentiable everywhere except at $x = 0, 1, 2, 3, 4, 5, 6, 7, 8, 9, 10$.

3. Explain why the function h given by
$$h(x) = \begin{cases} -x^3 + 5 & \text{for } x \le 2 \\ x^2 - 7 & \text{for } x > 2 \end{cases}$$
is continuous at 2. Is h continuous on the whole real line?

4. An empty bucket with a capacity of 10 liters is placed beneath a faucet. At time $t = 0$, the faucet is turned on, and water flows from the faucet at the rate of 5 liters per minute. Let $V(t)$ be the volume of the water in the bucket at time t. Present a plausible argument showing that V is a continuous function on $(-\infty, \infty)$. Sketch a graph of V. Is V differentiable on $(-\infty, \infty)$?

5. Let $f(x) = (x^2 - 4)/(x - 2)$, $x \ne 2$. Define $f(2)$ so that the resulting function is continuous at $x = 2$.

6. Let $f(x) = (x^3 - 1)/(x - 1)$, $x \ne 1$. How should $f(1)$ be defined in order that f be continuous at each point?

★7. The function $f(x) = x^2 + 4x + 1$ is continuous at 1, and $f(1) = 6$. Therefore, for every $c > 6$, there is an open interval I containing 1 such that $f(x) < c$ for all $x \in I$.

(a) Find such an interval for $c = 6.1$.

(b) Find such an interval for $c = 6.0001$.

(c) Find such an interval for a general $c > 6$. What happens to the interval as c gets nearer to 6?

★8. Prove that if f is continuous on an interval I (not necessarily closed) and $f(x) \neq 0$ for all x in I, then the sign of $f(x)$ is the same for all x in I.

★9. Suppose that f is continuous at x_0 and that in any open interval I containing x_0 there are points x_1 and x_2 such that $f(x_1) < 0$ and $f(x_2) > 0$. Prove that $f(x_0) = 0$.

10. Show that the equation $-s^5 + s^2 = 2s - 6$ has a solution.

11. Prove that $f(x) = x^8 + 3x^{4'} - 1$ has at least two distinct real roots.

12. Find a solution of the equation $x^5 - x = 3$ to an accuracy of 0.01.

★13. (a) Suppose that f and g are continuous on the real line. Show that $f - g$ is continuous.

(b) Suppose that f and g are continuous functions on the whole real line. Prove that if $f(x) \neq g(x)$ for all x, and $f(0) > g(0)$, then $f(x) > g(x)$ for all x. Interpret your answer geometrically.

★14. In the method of bisection (see Solved Exercise 8), each estimate of the solution of $f(x) = c$ is approximately twice as accurate as the previous one. Examining the list of powers of 2— 2, 4, 8, *16*, 32, 64, *128*, 256, 512, *1024*, 2048, 4096, 8192, *16384*, 32768, 65536, *131072*, ...—suggests that we get one more decimal place of accuracy for every three or four repetitions of the procedure. Explain.

★15. How many times must we apply the bisection procedure to guarantee the accuracy A for the interval $[a, b]$ if:

(a) $A = \frac{1}{100}$; $[a,b] = [3, 4]$

(b) $A = \frac{1}{1000}$; $[a, b] = [-1, 3]$

(c) $A = \frac{1}{700}$; $[a, b] = [11, 23]$

(d) $A = \frac{1}{15}$; $[a, b] = [0.1, 0.2]$

16. Use the method of bisection to find a root of the given function, on the given interval, to the given accuracy.

(a) $x^3 + 7$ on $[-3, -1]$; accuracy $\frac{1}{100}$

(b) $x^5 + x^2 + 1$ on $[-2, -1]$; accuracy $\frac{1}{1000}$

(c) $x^4 - 3x - 2$ on $[1, 2]$; accuracy $\frac{1}{500}$

17. Use the method of bisection to approximate $\sqrt{7}$ to within two decimal places. [*Hint:* Let $f(x) = x^2 - 7$. What should you use for a and b?]

★18. Can you improve the method of bisection by choosing a better point than the one halfway between the two previous points? [*Hint:* If $f(a)$ and $f(b)$ have opposite signs, choose the point where the line through $(a, f(a))$ and $(b, f(b))$ crosses the x axis. Is there a method of division more appropriate to the decimal system?]

Do some experiments to see whether your method gives more accurate answers than the bisection method in the same number of steps. If so, does the extra accuracy justify the extra time involved in carrying out each step? You might wish to compare various methods on a competitive basis, either with friends or with yourself by timing the calculations.

19. Let $f(x)$ be 1 if a certain sample of lead is in the solid state at temperature x, let $f(x)$ be 0 if it is in the liquid state. Define x_0 to be the melting point of this lead sample. Is there any way to define $f(x_0)$ so as to make f continuous? Give reasons for your answer, and supply a graph for f.

20. The performance of stereo cassette tape is often judged by a graph which is the plot of relative output y (in dB) versus signal frequency x (in Hz). The range of interest is 20 Hz to 10,000 Hz, in the case of 0 dB settings on the VU meters.

The desired graph is the x-axis: $y = 0$. In practice, this does not happen at all, and the graph resembles a piece of crinkled-up wire, which is close to the x-axis from 100 Hz to 6000 Hz.

(a) Using the intermediate value theorem and other properties of continuous functions, supply a continuous graph of $y = f(x)$ for the following data (Maxell UD-XLII): $f(100) = 3.6$, $f(400) = 3.3$, $f(8400) = -3$, $f(3000) = -0.2$, $f(10,000) = -6.5$, $f(50) = 0$, $f(25) = -5$, $f(x) > 0$ for $50 < x < 1000$, $f(x) < 0$ for $20 < x < 50$, $f(x) < 0$ for $1000 < x < 10,000$.

(b) Recording engineers demand $|f(x)| < 3$. For which values of x does your graph satisfy this condition?

21. Let x and y be the true theoretical values of two chemical concentrations and let a and b be decimal approximations to x and y, respectively, obtained by experimental methods. Then there are remainder decimals h, k, l satisfying $x = a + h$, $y = b + k$, $xy = ab + l$. Suppose that it is known by experimental means that h and k are zero through the 20th decimal place (the other digits are undetermined).

    (a) Does l have to be zero through the 20th decimal place? If yes, prove it. If no, give examples to support your claim.

    (b) Suppose $|a| \leq 1$, $|b| \leq 1$. Through what decimal place must the remainder l be zero (in all cases)? Justify your answer with algebra.

★ 22. Prove that any odd-degree polynomial has a root by following these steps.

   1. Reduce the problem to showing that $f(x) = x^n + a_{n-1}x^{n-1} + \cdots + a_1 x + a_0$ has a root when a_i are constants and n is odd.
   2. Show that if $|x| > 1$ and $|x| > 2\{|a_0| + \cdots + |a_{n-1}|\}$, then $f(x)/x^n$ is positive.
   3. Conclude that $f(x) < 0$ if x is large negative and $f(x) > 0$ if x is large positive.
   4. Apply the intermediate value theorem.

2-2 Increasing and Decreasing Functions

The sign of the derivative indicates whether a function is increasing or decreasing.

We begin by defining what it means for a function to be increasing or decreasing at a point. Change of sign is then defined. Points where the derivative of a function changes sign are particularly important; called turning points, they separate the intervals on which the function increases and decreases.

Goals

After studying this section, you should be able to:

Find the intervals on which a function is increasing or decreasing.

Find the turning points of a function and identify them as local maxima or minima.

INCREASING OR DECREASING AT A POINT

We make a precise definition out of the vague notion that an increasing function is one which is "getting bigger."

Definition Let f be a function whose domain contains an open interval about x_0. We say that f is *increasing* at x_0 if there is an interval (a, b) containing x_0 such that:

   1. If $a < x < x_0$, then $f(x) < f(x_0)$.

   2. If $x_0 < x < b$, then $f(x) > f(x_0)$.

Similarly, f is *decreasing* at x_0 if there is an interval (a, b) containing x_0 such that:

   1. If $a < x < x_0$, then $f(x) > f(x_0)$.

   2. If $x_0 < x < b$, then $f(x) < f(x_0)$.

We can tell whether a function is increasing or decreasing at x_0 by seeing how its graph crosses the horizontal line $y = f(x_0)$ at x_0 (see Fig. 2-2-1).

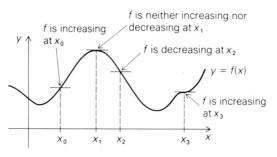

Fig. 2-2-1   The graphic significance of increasing and decreasing.

Worked Example 1   Show that $f(x) = x^2$ is increasing at $x_0 = 2$.

Solution   Choose (a, b) to be $(1, 3)$. If $1 < x < 2$, we have $f(x) = x^2 < 4 = x_0^2$. If $2 < x < 3$, then $f(x) = x^2 > 4 = x_0^2$. We have verified conditions 1 and 2 of the definition, so f is increasing at 2.

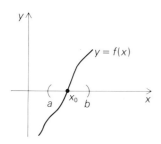

Fig. 2-2-2   This function changes sign from negative to positive at x_0.

Of special interest is the case $f(x_0) = 0$. If f is increasing at such an x_0 we say that *f changes sign from negative to positive at x_0*. By definition, this occurs when $f(x_0) = 0$ and there is an open interval (a, b) containing x_0 such that $f(x) < 0$ when $a < x < x_0$ and $f(x) > 0$ when $x_0 < x < b$. (See Figure 2-2-2.)

Similarly if $f(x_0) = 0$ and f is decreasing at x_0 we say that *f changes sign from positive to negative at x_0*. This occurs when $f(x_0) = 0$ and there is an open interval (a, b) containing x_0 such that $f(x) > 0$ when $a < x < x_0$ and $f(x) < 0$ when $x_0 < x < b$. (See Figure 2-2-3.)

Notice that the interval (a, b) may have to be chosen small, since a function which changes sign from negative to positive may later change back from positive to negative (see Fig. 2-2-4).

Fig. 2-2-3   This function changes sign from positive to negative at x_0.

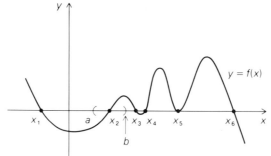

Fig. 2-2-4   This function changes sign from negative to positive at x_2 and x_4 and from positive to negative at x_1, x_3 and x_6; it does neither at x_5.

Changes of sign can be significant in everyday life as, for instance, when the function $b = f(t)$ representing your bank balance changes from positive to negative. Changes of sign will be important for mathematical reasons in the next few sections of this chapter; it is useful to be able to determine when a function changes sign by looking at its formula.

Worked Example 2   Where does $f(x) = 3x - 5$ change sign?

Solution   We begin by finding those x for which $f(x)$ is negative and those for which it is positive. First, f is negative when

$$3x - 5 < 0$$
$$3x < 5$$
$$x < \tfrac{5}{3}$$

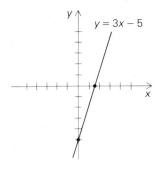

Fig. 2-2-5 This function changes sign from negative to positive at $x = 5/3$.

(If you had difficulty following this argument, you may wish to review the material on inequalities in Section R-1.) Similarly, f is positive when $x > \tfrac{5}{3}$. So f changes sign from negative to positive at $x = \tfrac{5}{3}$. (See Fig. 2-2-5.) Here the interval (a, b) can be chosen arbitrarily large.

If a function is given by a formula which factors, this often helps us to find the changes of sign.

Worked Example 3   Where does $f(x) = x^2 - 5x + 6$ change sign?

Solution   We write $f(x) = (x - 3)(x - 2)$. The function f changes sign whenever one of its factors does. This occurs at $x = 2$ and $x = 3$. The factors have opposite signs for x between 2 and 3, and the same sign otherwise, so f changes from positive to negative at $x = 2$ and from negative to positive at $x = 3$. (See Fig. 2-2-6.)

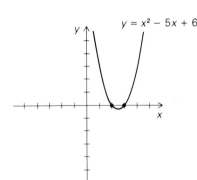

Fig. 2-2-6 This function changes sign from positive to negative at $x = 2$ and from negative to positive at $x = 3$.

Given a more complicated function, such as $x^5 - x^3 - 2x^2$, it may be difficult to tell directly whether it is increasing or decreasing at a given point. The derivative is a very effective tool for helping us answer such questions—if we use the following theorem.*

Theorem 3   *Let f be differentiable at x_0.*

1.   *If $f'(x_0) > 0$, then f is increasing at x_0.*

2.   *If $f'(x_0) < 0$, then f is decreasing at x_0.*

3.   *If $f'(x_0) = 0$, then f may be increasing at x_0, decreasing at x_0, or neither.*

It is possible to turn the tables and base the definition of the derivative on the concept of change of sign. See Calculus Unlimited *by the authors (Benjamin/Cummings, 1980).*

Proof   1. If $f'(x_0) > 0$ then

$$\frac{\Delta y}{\Delta x} = \frac{f(x_0 + \Delta x) - f(x_0)}{\Delta x} > 0$$

for Δx sufficiently small, since $\Delta y / \Delta x$ approaches $f'(x_0)$ as Δx approaches zero. Thus, there is an interval (a,b) about x_0 in which $\Delta y / \Delta x > 0$. Therefore if $a < x < x_0$, $\Delta x = x - x_0 < 0$ and so $\Delta y = f(x) - f(x_0) < 0$ as well; i.e., $f(x) < f(x_0)$. Similarly, if $x_0 < x < b$ then $\Delta x > 0$ and so $\Delta y > 0$ as well; i.e., $f(x) > f(x_0)$. Thus, by definition f is increasing at x_0.

2. This proof is similar to 1.

3. This is established in Solved Exercise 5 below using the examples $y = x^3$, $y = -x^3$, and $y = x^2$.

Worked Example 4   Is $x^5 - x^3 - 2x^2$ increasing or decreasing at -2?

Solution   Letting $f(x) = x^5 - x^3 - 2x^2$, we have $f'(x) = 5x^4 - 3x^2 - 4x$ and $f'(-2) = 5(-2)^4 - 3(-2)^2 - 4(-2) = 80 - 12 + 8 = 76$, which is positive. By Theorem 3 (part 1), $x^5 - x^3 - 2x^2$ is increasing at -2.

Theorem 3 can be interpreted geometrically: if the linear approximation to f at x_0 (that is, the tangent line) is an increasing or decreasing function, then f itself is increasing or decreasing at x_0. If the tangent line is horizontal, the behavior of f at x_0 is not determined by the tangent line. (See Fig. 2-2-7.)

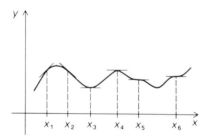

Fig. 2-2-7 $f'(x_1) > 0$; f is increasing at x_1
$f'(x_2) < 0$; f is decreasing at x_2
$f'(x_3) = f'(x_4) = 0$; f is neither increasing nor decreasing at x_3 and x_4
$f'(x_5) = 0$; f is decreasing at x_5
$f'(x_6) = 0$; f is increasing at x_6

We can also interpret Theorem 3 in terms of velocities. If $f(t)$ is the position of a particle on the real-number line at time t, and $f'(t_0) > 0$, then the particle is moving to the right at time t_0; if $f'(t_0) < 0$, the particle is moving to the left (see Fig. 2-2-8.)

Fig. 2-2-8 Positive velocity means that the motion is to the right, and negative velocity means that the motion is to the left.

Let us summarize the results we have obtained.

INCREASING-DECREASING TEST

1. If $f'(x_0) > 0$, f is increasing at x_0.
2. If $f'(x_0) < 0$, f is decreasing at x_0.
3. If $f'(x_0) = 0$, the test is inconclusive.

Combined with the techniques for differentiation in Chapter 1, this test provides an effective means for deciding where a function is increasing or decreasing.

Solved Exercises

1. Using only algebra, determine the sign change of $f(x) = (x - r_1) \times (x - r_2)$ at $x = r_1$, where $r_1 < r_2$.

2. The temperature at time x is given by $f(x) = (x + 1)/(x - 1)$ for $x \neq 1$. Is it getting warmer or colder at $x = 0$?

3. Using Theorem 3, find the points at which $f(x) = 2x^3 - 9x^2 + 12x + 5$ is increasing or decreasing.

4. (a) For which positive integers n does $y = x^n$ change sign at $x = 0$?

   (b) Decide whether each of the functions x^3, $-x^3$, and x^2 is increasing, decreasing, or neither at $x = 0$.

Exercises

1. Using only algebra, find the sign changes of each of the following functions at the indicated point.

   (a) $f(x) = 2x - 1$; $x = 1/2$ (b) $f(x) = x^2 - 1$; $x = -1$

   (c) $f(x) = x^5$; $x = 0$ (d) $h(z) = z(z - 1)(z - 2)$; $z = 1$

   (e) $f(x) = (x - r_1)(x - r_2)$; $x = r_1$, $r_1 > r_2$

2. If $f(t) = t^5 - t^4 + 2t^2$ is the position of a particle on the real-number line at time t, is it moving to the left or right at $t = 1$?

3. (a) Find the points at which $f(x) = x^2 - 1$ is increasing or decreasing.

   (b) Find the points at which $x^3 - 3x^2 + 2x$ is increasing or decreasing.

4. Is $f(x) = 1/(x^2 + 1)$ increasing or decreasing at $x = 1$, -3, $\frac{3}{4}$, 25, -36?

5. A ball is thrown upward with an initial velocity of 30 meters per second. The ball's height above the ground at time t is $h(t) = 30t - 4.9t^2$. When is the ball rising? When is it falling?

INCREASING OR DECREASING ON AN INTERVAL

If f is increasing at every point of an interval $[a, b]$, one would expect $f(b)$ to be larger than $f(a)$. In fact, we have the following useful result.

Theorem 4 *Let f be continuous on $[a, b]$, where $a < b$, and suppose that f is increasing [decreasing] at all points of (a, b). Then*

$$f(b) > f(a) \quad [f(b) < f(a)]$$

The statement of Theorem 4 may appear to be tautological—that is, "trivially true"—but in fact it requires a proof, which is given in the appendix to this chapter (see the Student Guide). Like the intermediate value theorem, Theorem 4 connects a *local* property of functions (increasing at each point of an interval) with a *global* property (relation between values of the function at endpoints). We do not insist that f be increasing or decreasing at a or b because we wish the theorem to apply in cases of the type illustrated in Fig. 2-2-9.

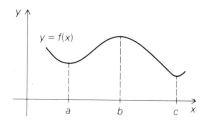

Fig. 2-2-9 f is increasing at each point of (a, b): $f(b) > f(a)$; f is decreasing at each point of (b, c): $f(c) < f(b)$; f is neither increasing nor decreasing at a, b, c.

Worked Example 5   Show that if $a > 1$, then $a^4/(1 + a^2) > \frac{1}{2}$.

Solution   We consider the function $f(x) = x^4/(1 + x^2)$ on the interval $[1, a]$. By the quotient rule,

$$f'(x) = \frac{4x^3(1 + x^2) - x^4(2x)}{(1 + x^2)^2} = \frac{4x^3 + 2x^5}{(1 + x^2)^2} = \frac{2x^3(2 + x^2)}{(1 + x^2)^2}$$

which is positive for all $x > 0$ and, hence, for all x in $(1, a)$. By Theorem 3 (part 1), f is increasing at all points of $(1, a)$; since f is continuous on $[1, a]$, we can apply Theorem 4 to conclude that $f(a) > f(1)$. Since $f(a) = a^4/(1 + a^2)$ and $f(1) = \frac{1}{2}$, we have proved the required inequality.

Note that if the hypotheses of the "increasing" version of Theorem 4 hold for f on $[a, b]$, they also hold for f on any interval $[x_1, x_2]$ where $a \leq x_1 < x_2 \leq b$. We conclude that $f(x_1) < f(x_2)$ for any x_1 and x_2 such that $a < x_1 \leq x_2 < b$. This suggests that we should make the following definition.

Definition   Let f be a function defined on an interval I. If $f(x_1) < f(x_2)$ for all $x_1 < x_2$ in I, we say that f is *increasing* on I. If $f(x_1) > f(x_2)$ for all $x_1 < x_2$ in I, we say that f is *decreasing* on I. If f is either increasing on I or decreasing on I, we say that f is *monotonic* on I.

We can summarize this discussion as follows.

Theorem 4′   *Let f be continuous on $[a, b]$ and increasing [decreasing] at all points of (a, b). Then f is increasing [decreasing] on $[a, b]$.*

For example, the function in Fig. 2-2-9 is monotonic on $[a, b]$ and monotonic on $[b, c]$, but it is not monotonic on $[a, c]$. The function $f(x) = x^2$ is monotonic on $(-\infty, 0)$ and $(0, \infty)$, but not on $(-\infty, \infty)$. (Draw a sketch to convince yourself.)

Combining Theorems 3 and 4′ with the intermediate value theorem gives a result which is useful for graphing.

Theorem 5 *Suppose that:*

1. *f is continuous on* $[a, b]$.

2. *f is differentiable on* (a, b) *and* f' *is continuous on* (a, b).

3. f' *is never zero on* (a, b).

Then f is monotonic on $[a, b]$. *To check whether f is increasing or decreasing on* $[a, b]$, *it suffices to compute the value of* f' *at any one point of* (a, b) *and see whether it is positive or negative.*

Proof* By the intermediate value theorem, f' must be either positive everywhere on (a, b) or negative everywhere on (a, b). (If f' took values with both signs, it would have to be zero somewhere in between.) If f' is positive everywhere, f is increasing at each point of (a, b) by Theorem 3. By Theorem 4′, f is increasing on $[a, b]$. If f' is negative everywhere, f is decreasing on $[a, b]$.

Worked Example 6 Show that the function $f(x) = x^3 + 3x^2 - 6x$ is monotonic on the interval $[-2, \frac{1}{2}]$. Is it increasing or decreasing there?

Solution The derivative is $3x^2 + 6x - 6$, whose roots are

$$\frac{-6 \pm \sqrt{36 + 72}}{6} = -1 \pm \sqrt{3} \approx -2.73, 0.73$$

Since the derivative is continuous and has no roots on $(-2, \frac{1}{2})$, and f is continuous on $[-2, \frac{1}{2}]$, we conclude from Theorem 5 that f is monotonic on $[-2, \frac{1}{2}]$. To check whether f is increasing or decreasing, we compute $f'(0)$. It equals -6, so f is decreasing on the interval.

Corollary *Suppose that:*

1. *f is continuous on an open interval* (a, b).

2. *f is differentiable and* f' *is continuous at each point of* (a, b).

3. f' *is not zero at any point of* (a, b).

Then f is monotonic on (a, b) *(increasing if* $f' > 0$, *decreasing if* $f' < 0$).

Proof By the intermediate value theorem, f' is everywhere positive or everywhere negative on (a, b). Suppose it to be everywhere positive. Let $x_1 < x_2$ be in (a, b). We may apply Theorems 3 and 4 to f on $[x_1, x_2]$ and conclude that $f(x_1) < f(x_2)$. Thus f is increasing on (a, b). Similarly, if f' is everywhere negative on (a, b), f is decreasing on (a, b).

We can also apply Theorem 5 to intervals which are not closed.

---

* *Theorem 5 is still true if* f' *exists and is nonzero on* $[a, b]$, *even if it is not continuous. This technical point may be dealt with by use of the mean value theorem; see Section 3-3, problem 12.*

Similar statements hold for half-open intervals $[a, b)$ or $(a, b]$.

INCREASING-DECREASING TEST

If f has a continuous derivative,	then f is monotonic on each of the intervals between successive points where $f'(x) = 0$.
To test whether f is increasing or decreasing on each such interval,	it suffices to determine the sign of $f'(x)$ at any point in the interval.

Solved Exercises

5. On what intervals is $f(x) = x^3 - 2x + 6$ monotonic? Is f increasing or decreasing on each of these intervals?

6. Let

$$f(x) = \begin{cases} x & \text{for } x \leq 0 \\ 2x & \text{for } x \geq 0 \end{cases}$$

On what intervals is f monotonic?

7. Match each of the functions in the left-hand column of Fig. 2-2-10 with its derivative in the right-hand column.

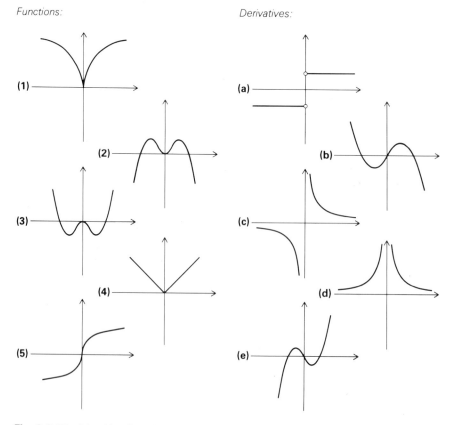

Fig. 2-2-10   Matching functions and their derivatives.

Exercises 6. On what intervals is each of the following functions increasing or decreasing?

  (a) $x^3 + x - 6$ (b) $\frac{1}{3}x^3 - 2x + 1$ (c) $-(x + 1)/(x - 1)$

7. Sketch the derivative of each of the functions in Fig. 2-2-11.

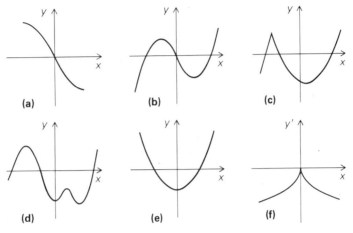

Fig. 2-2-11 Sketch the derivatives of these functions.

8. Show that if $b > a > 0$, then $b/(1 + b) > a/(1 + a)$. [*Hint:* Consider the function $f(x) = x/(1 + x)$.]

TURNING POINTS

Points where the derivative f' changes sign are called *turning points* of f; they separate the intervals on which f is increasing and decreasing. A simple example will illustrate the importance of turning points. Let $g(t)$ be the total weight of the fish in Lake Erie at time t. When $g'(t) < 0$, the total weight of the fish is decreasing and so the lake is "dying"; during part of the 1960s and 1970s, this was the case. At a certain time t_0 in the early 1970s, the effect of strict pollution controls was to cause g' to change sign from negative to positive—at this "turning point" in time, the lake became alive again; that is, the fish population began growing instead of shrinking. Lake Erie's fish population was not always shrinking before t_0, nor will it continue growing forever—thus it is only the behavior of g *near* t_0 which is relevant for t_0 to be a turning point.

> **Definition** If f is differentiable at x_0 and $f'(x_0) = 0$, we call x_0 a *critical point* of f.
>
> If f is defined and differentiable throughout an open interval containing x_0, we call x_0 a *turning point* of f if f' changes sign at x_0.

Since $f'(x_0) = 0$ if f' changes sign at x_0, a turning point is also a critical point. A critical point *need not* be a turning point, however, as part (*b*) of the following example shows.

Worked Example 7 Find the critical points and turning points of (a) $f(x) = x^2 - 2x + 5$; (b) $f(x) = x^3$; and (c) $f(x) = x^3 - 6x^2 + 10$.

Solution (a) To find the critical points we set $f'(x) = 0$. Here $f'(x) = 2x - 2 = 2(x - 1)$, which is zero at $x_0 = 1$. Thus $x_0 = 1$ is the only critical point. Since $f'(x)$ changes sign from negative to positive at $x_0 = 1$, 1 is also a turning point of f.

(b) For the critical points we set $f'(x) = 0$; that is, $3x^2 = 0$, so $x_0 = 0$ is the only critical point. It is *not* a turning point, since $f'(x) = 3x^2$ does not change sign at $x_0 = 0$.

(c) Again for the critical points we set $f'(x) = 0$; that is, $3x^2 - 12x = 0$; that is, $x_0 = 0$ or $x_0 = 4$. Thus the critical points are $x_0 = 0$ and $x_0 = 4$. Since $f'(x) = 3x(x - 4)$, f' changes sign from positive to negative at $x_0 = 0$ and from negative to positive at $x_0 = 4$; both are turning points.

To investigate the behavior of the graph of f at a turning point, it is useful to treat separately the two possible kinds of sign change. Suppose first that f' changes sign from negative to positive at x_0. Then there is an open interval (a, b) about x_0 such that $f'(x) < 0$ on (a, x_0) and $f'(x) > 0$ on (x_0, b). According to the display on p. 122, we find that f is decreasing on $(a, x_0]$ and increasing on $[x_0, b)$. It follows that $f(x) > f(x_0)$ for x in (a, x_0) and $f(x) > f(x_0)$ for x in (x_0, b). (See Fig. 2-2-12.)

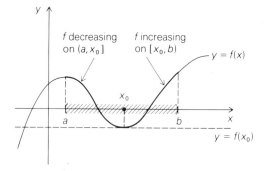

Fig. 2-2-12 At the local minimum point x_0, f' changes sign from negative to positive.

Notice that for all $x \neq x_0$ in (a, b), $f(x_0) < f(x)$; that is, the smallest value taken on by f in (a, b) is achieved at x_0. For this reason, we call x_0 a *local minimum point** for f.

In case f' changes from positive to negative at x_0, a similar argument shows that f behaves as shown in Fig. 2-2-13 with $f(x_0) > f(x)$ for all $x \neq x_0$ in an open interval (a, b) around x_0. In this case, x_0 is called a *local maximum point* for f.

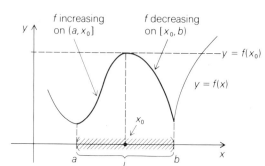

Fig. 2-2-13 At the local maximum point x_0, f' changes sign from positive to negative.

Sometimes the phrase ''strict local minimum'' is used for this.

Worked Example 8.   Find and classify the turning points of the function $f(x) = x^3 + 3x^2 - 6x$ in Worked Example 6.

Solution   The derivative $f'(x) = 3x^2 + 6x - 6$ has roots at $-1 \pm \sqrt{3}$; it is positive on $(-\infty, -1 - \sqrt{3})$ and $(-1 + \sqrt{3}, \infty)$ and is negative on $(-1 - \sqrt{3}, -1 + \sqrt{3})$. Changes of sign occur at $-1 - \sqrt{3}$ (positive to negative) and $-1 + \sqrt{3}$ (negative to positive), so $-1 - \sqrt{3}$ is a local maximum point and $-1 + \sqrt{3}$ is a local minimum point.

CRITICAL POINTS AND TURNING POINTS

If $f'(x_0) = 0$,	x_0 is called a *critical point* for f.
If f' changes sign at x_0,	x_0 is called a *turning point* for f.
If the sign change of f' is from negative to positive,	x_0 is a *local minimum point;* i.e., for x in some open interval about x_0 and $x \neq x_0$, we have $f(x) > f(x_0)$.
If the sign change of f' is from positive to negative,	x_0 is a *local maximum point;* i.e., for x in some open interval about x_0 and $x \neq x_0$, we have $f(x) < f(x_0)$.

Every turning point is a critical point, but a critical point is not necessarily a turning point.

Solved Exercises

8. Find the turning points of the function $f(x) = 3x^4 - 8x^3 + 6x^2 - 1$. Are they local maximum or minimum points?

9. Find the turning points of x^n for $n = 0, 1, 2, \ldots$.

10. Does $g(x) = 1/f(x)$ have the same turning points as $f(x)$?

Exercises

9. For each of the functions in Fig. 2-2-14, tell whether x_0 is a turning point, a local minimum point, or a local maximum point.

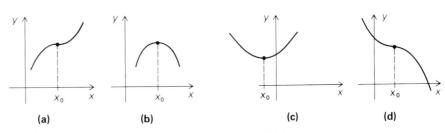

Fig. 2-2-14   Is x_0 a turning point? A local maximum? A local minimum?

10. Find the turning points of each of the following functions and tell whether they are local maximum or minimum points:
    (a) $f(x) = \frac{1}{3}x^3 - \frac{5}{2}x^2 + 4x + 1$ (b) $f(x) = 1/x^2$
    (c) $f(x) = ax^2 + bx + c$ $(a \neq 0)$ (d) $f(x) = 1/(x^4 - 2x^2 - 5)$

11. Find a function f with turning points at $x = -1$ and $x = \frac{1}{2}$.

12. Can a function be increasing at a turning point? Explain.

Problems for Section 2-2

1. Determine whether each of the functions below is increasing, decreasing, or neither at the indicated point:

   (a) $x^3 + x + 1$; $x_0 = 0$

   (b) $(x^2 - 1)/(x^2 + 1)$; $x_0 = 0$

   (c) $x^4 + 5$; $x_0 = 0$

   (d) $x^4 + x + 5$; $x_0 = 0$

   (e) $1/(x^4 - x^3 + 1)$; $x_0 = 1$

2. Find the turning points of the following functions and decide whether they are maxima or minima:

   (a) $f(x) = x^2 - 2$

   (b) $f(x) = x^3 + x^2 - 2$

   (c) $f(x) = 3x^4$

   (d) $f(x) = (x^2 + 1)/(x^2 - 1)$

3. At what points do the functions in Problem 2 change sign?

4. (a) Describe the change of sign at $x = 0$ of the function $f(x) = mx$ for $m = -2, 0, 2$.

   (b) Describe the change of sign at $x = 0$ of the function $f(x) = mx - x^2$ for $m = -1, -\frac{1}{2}, 0, \frac{1}{2}, 1$.

5. Let $f(t)$ denote the angle of the sun above the horizon at time t. When does $f(t)$ change sign?

6. Find the sign changes (if any) of each of the following using Theorem 3:

   (a) $x^2 + 3x + 2$ (b) $x^2 + x - 1$

   (c) $x^2 - 4x + 4$ (d) $(x^3 - 1)/(x^2 + 1)$

7. Let $f(x) = 4x^2 + (1/x)$. Determine whether f is increasing or decreasing at each of the following points: (a) 1; (b) $-\frac{1}{2}$; (c) -5; (d) $2\frac{1}{3}$.

8. Find all points at which $f(x) = x^2 - 3x + 2$ is increasing, and at which it changes sign.

9. The annual inflation rate in Uland during 1968 was approximately $r(t) = 20[1 + (t^2 - 6t)/500]$ percent *per year*, where t is the time *in months* from the beginning of the year. During what months was the inflation rate decreasing? What are the turning points of $r(t)$? Explain their (political) significance.

10. Prove the following assertions concerning the function $f(x) = (x^3 - 1)/(x^2 - 1)$:

    (a) f can be defined at $x = 1$ so that f becomes continuous and differentiable there, but cannot be so defined at $x = -1$.

    (b) f is increasing on $(-\infty, -2]$ and decreasing on $[-2, -1)$.

    (c) If $a < -2$, then $(a^3 - 1)/(a^2 - 1) < -3$.

    (d) f is increasing on $[0, \infty)$ and decreasing on $(-1, 0]$. Make up an inequality based on the result for part (d).

11. Find a quadratic polynomial which is zero at $x = 1$, is decreasing if $x < 2$, and is increasing if $x > 2$.

12. Match each derivative on the left in Fig. 2-2-15 with the function on the right.

Derivatives: *Functions:*

(a)   (1)

(b)   (2)

(c)   (3)

(d)   (4)

(e)   (5)

Fig. 2-2-15 Matching derivatives and functions.

13. Sketch functions whose derivatives are shown in Fig. 2-2-16.

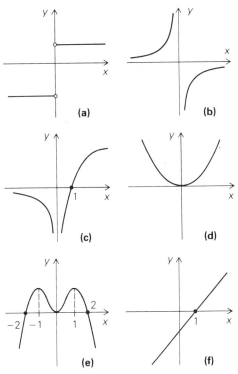

(a)   **(b)**

(c)   **(d)**

(e)   **(f)**

Fig. 2-2-16 Sketch functions that have these derivatives.

14. For each of the functions shown in Fig. 2-2-16, state: (a) where it is increasing; (b) where it is decreasing; (c) what its turning points are; (d) where it changes sign.

15. For each of the functions in Fig. 2-2-17, tell whether the function is increasing or decreasing and whether the derivative of the function is increasing or decreasing.

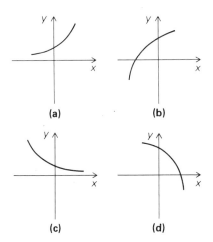

(a)   **(b)**

(c)   **(d)**

Fig. 2-2-17 Are f and f' increasing or decreasing?

★16. Using the definition of an increasing function, prove that if f and g are increasing at x_0, then so is $f + g$.

★17. Prove that if f and g are increasing and positive on an interval I, then fg is increasing on I.

★18. Let $f(x) = a_0 + a_1 x + a_2 x^2 + \cdots + a_n x^n$. Under what conditions on the a_i's is f increasing at $x_0 = 0$?

19. Show that the parabola $f(x) = ax^2 + bx + c$ ($a \neq 0$) has a turning point at its vertex.

20. Find the intervals on which each of the functions below is increasing or decreasing:

(a) $2x^3 - 5x + 7$ (b) $x^5 - x^3$

(c) $(x^2 + 1)/(x - 3)$ (d) $x^4 - 2x^2 + 1$

(e) $f(x) = \begin{cases} \dfrac{x^3}{x^2 + 5} & \text{for } x \leq 0 \\ -x^2 & \text{for } x \geq 0 \end{cases}$

(f is continuous on $(-\infty, \infty)$)

★21. Under what conditions on a, b, c, and d is the cubic polynomial $ax^3 + bx^2 + cx + d$ monotonic on $(-\infty, \infty)$? (Assume $a \neq 0$.)

★22. If g and h are positive functions, find criteria involving $g'(x)/g(x)$ and $h'(x)/h(x)$ to tell when (a) the product $g(x)h(x)$ and (b) the quotient $g(x)/h(x)$ are increasing or decreasing.

23. Herring production T (in grams) is related to the number N of fish stocked in a storage tank by the equation $T = 500N - 50N^2$.

(a) Find dT/dN.

(b) Unless too many fish are stocked, an increase in the number of fish stocked will cause an increase in production at the expense of a reduction in the growth of each fish. (The weight of each fish is T/N.) Explain this statement mathematically in terms of derivatives and the level N^* of stocking which corresponds to maximum production.

24. The cost, in dollars, for manufacturing x electronic parts is given by $C(x) = 3 + 10x - x^2$.

(a) The marginal cost is the cost of producing one additional part, or, alternatively, marginal cost $= C'(x)$. Find it.

(b) How many parts must be produced to get the cost for one additional part down to at least $1.00?

(c) At what production level does the total cost begin to decrease? Is this a turning point? Classify further as a maximum or minimum.

★25. Find a cubic polynomial with a graph like the one shown in Fig. 2-2-18.

Fig. 2-2-18 This is the graph of what cubic polynomial?

★26. (a) Show that there is no quartic polynomial whose graph is consistent with the information shown in Fig. 2-2-16(e).

(b) Show that if -2 is replaced by $-\sqrt{2}$, then there is a quartic polynomial consistent with the information in Fig. 2-2-16(e).

2-3 The Second Derivative and Concavity

The sign of the second derivative indicates which way the graph of a function is bending.

The second derivative $f''(x)$ is the rate of change of the first derivative $f'(x)$. If $f''(x)$ is positive, $f'(x)$ is increasing; if it is negative, $f'(x)$ is decreasing. Thus the sign of $f''(x)$ indicates whether the tangent line to the graph of f rotates counterclockwise or clockwise as x increases, and so it tells us whether the graph curves up or down. In particular, if $f'(x_0) = 0$ and $f''(x_0) > 0$, the graph curves up at x_0, so we have a local minimum; if $f'(x_0) = 0$ and $f''(x_0) < 0$, the graph curves down, so we have a local maximum.

Goals

After reading this section, you should be able to:

Use the second derivative to test whether a critical point is a local maximum or minimum.

Find the intervals on which the graph of a function is concave upward or downward.

Find the inflection points of a function and determine how the tangent line crosses the graph at these points.

A TEST FOR TURNING POINTS

If g is a function such that $g(x_0) = 0$ and $g'(x_0) > 0$, then g is increasing at x_0, so g changes sign from negative to positive at x_0. Similarly, if $g(x_0) = 0$ and $g'(x_0) < 0$, then g changes sign from positive to negative at x_0. Applying these assertions to $g = f'$, we obtain the following useful result.

Theorem 6   Second derivative test.   *Let f be a differentiable function, let x_0 be a critical point of f (that is, $f'(x_0) = 0$), and assume that $f''(x_0)$ exists.*

1. *If $f''(x_0) > 0$, then x_0 is a local minimum point for f.*

2. *If $f''(x_0) < 0$, then x_0 is a local maximum point for f.*

3. If $f''(x_0) = 0$, then x_0 may be a local maximum or minimum point, or it may not be a turning point at all.

Proof For parts 1 and 2, we use the definition of turning point as a point where the derivative changes sign (see the display on p. 125). For part 3, we observe that the functions x^4, $-x^4$, and x^3 all have $f'(0) = f''(0) = 0$, but zero is a local minimum point for x^4, a local maximum point for $-x^4$, and not a turning point for x^3.

Worked Example 1 Use the second derivative test to analyze the critical points of $f(x) = x^3 - 6x^2 + 10$.

Solution Since $f'(x) = 3x^2 - 12x = 3x(x - 4)$, the critical points are at 0 and 4. Since $f''(x) = 6x - 12$, we find that $f''(0) = -12 < 0$ and $f''(4) = 12 > 0$. By Theorem 6, zero is a local maximum point and 4 is a local minimum point.

When $f''(x_0) = 0$, the second derivative test is inconclusive. We may still use the first derivative test to analyze the critical points, however.

Worked Example 2 Analyze the critical point $x_0 = -1$ of $f(x) = 2x^4 + 8x^3 + 12x^2 + 8x + 7$.

Solution The derivative is $f'(x) = 8x^3 + 24x^2 + 24x + 8$, and $f'(-1) = -8 + 24 - 24 + 8 = 0$, so -1 is a critical point. Now $f''(x) = 24x^2 + 48x + 24$, so $f''(-1) = 24 - 48 + 24 = 0$, and the second derivative test is inconclusive. If we factor f', we find $f'(x) = 8(x^3 + 3x^2 + 3x + 1) = 8(x + 1)^3$. Thus -1 is the only root of f', $f'(-2) = -8$, and $f'(0) = 8$, so f' changes sign from negative to positive at -1; hence -1 is a local minimum point for f.

SECOND DERIVATIVE TEST

Suppose that $f'(x_0) = 0$.
1. If $f''(x_0) > 0$, then x_0 is a local minimum point.
2. If $f''(x_0) < 0$, then x_0 is a local maximum point.
3. If $f''(x_0) = 0$, the test is inconclusive.

Solved Exercises 1. Analyze the critical points of $f(x) = x^3 - x$.

2. Show that if $f'(x_0) = f''(x_0) = 0$ and $f'''(x_0) > 0$, then x_0 is *not* a turning point of f.

Exercises 1. Analyze the critical points of the following functions:
   (a) $f(x) = x^4 - x^2$
   (b) $g(s) = s/(1 + s^2)$
   (c) $h(p) = p + (1/p)$

2. Show that if $f'(x_0) = f''(x_0) = 0$ and $f'''(x_0) < 0$, then x_0 is *not* a turning point for f.

CONCAVITY

Whether or not $f'(x_0)$ is zero, the sign of $f''(x_0)$ has an important geometric interpretation: it tells us which way the tangent line to the graph of f turns as the point of tangency moves to the right along the graph (see Fig. 2-3-1). The two graphs in Fig. 2-3-1 are bent in opposite ways. The graph in part (a) is said to be *concave upward;* the graph in part (b) is said to be *concave downward.*

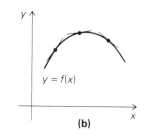

Fig. 2-3-1 (a) The slope of the tangent line is increasing; $f''(x) > 0$. (b) The slope of the tangent line is decreasing; $f''(x) < 0$.

We can give precise definitions of upward and downward concavity by considering how the graph of f lies in relation to one of its tangent lines. To do this, we compare $f(x)$ with its linear approximation (see p. 75).

Definition Let f be differentiable at x_0 and let $l(x) = f(x_0) + f'(x_0)(x - x_0)$.

1. If there is an open interval (a, b) about x_0 such that $f(x) > l(x)$ whenever $a < x < x_0$ or $x_0 < x < b$, then f is called *concave upward* at x_0.

2. If there is an open interval (a, b) about x_0 such that $f(x) < l(x)$ whenever $a < x < x_0$ or $x_0 < x < b$, then f is called *concave downward* at x_0.

Geometrically, f is concave upward (downward) at x_0 if the graph of f lies locally above (below) its tangent line at x_0, as in Fig. 2-3-2.

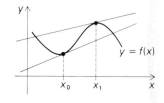

Fig. 2-3-2 The function f is concave upward at x_0 and concave downward at x_1.

Theorem 7 *Let f be a differentiable function and suppose that $f''(x_0)$ exists.*

1. *If $f''(x_0) > 0$, then f is concave upward at x_0.*

2. *If $f''(x_0) < 0$, then f is concave downward at x_0.*

3. *If $f''(x_0) = 0$, then f at x_0 may be concave upward, concave downward, or neither.*

Proof Let $h(x) = f(x) - [f(x_0) + f'(x_0)(x - x_0)]$ be the difference between $f(x)$ and its linear approximation at x_0 (see Fig. 2-3-3). Then $h'(x) = f'(x) - f'(x_0)$ and $h''(x) = f''(x)$. We have $h(x_0) = 0$, $h'(x_0) = 0$, and $h''(x_0) = f''(x_0)$. In particular, x_0 is a critical point of h. Now suppose that $f''(x_0) > 0$. Then $h''(x_0) > 0$ and (by Theorem 6) x_0 is a local minimum point for h. In other words

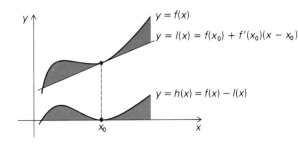

Fig. 2-3-3 $h(x)$ is the difference between $f(x)$ and its linear approximation at x_0.

(see p. 125), there is an interval (a, b) about x_0 such that $h(x) > h(x_0)$ whenever $a < x < x_0$ or $x_0 < x < b$. Substituting $f(x) - [f(x_0) + f'(x_0)(x - x_0)]$ for $h(x)$, we have $f(x) - [f(x_0) + f'(x_0)(x - x_0)] > 0$ or $f(x) > f(x_0) + f'(x_0)(x - x_0)$ whenever $a < x < x_0$ or $x_0 < x < b$; that is, f is concave upward at x_0. This proves part 1 of the theorem; the proof of part 2 is similar. The functions x^4, $-x^4$, and x^3 illustrate all the possibilities in part 3 (see Solved Exercise 3).

Worked Example 3   Discuss the concavity of $f(x) = 4x^3$ at the points $x = -1$, $x = 0$, and $x = 1$.

Solution   We have $f'(x) = 12x^2$ and $f''(x) = 24x$, so $f''(-1) = -24$, $f''(0) = 0$, and $f''(1) = 24$. Therefore f is concave downward at -1 and concave upward at 1. At zero, Theorem 7 is inconclusive; we can see, however, that f is neither concave upward nor downward by noticing that f is increasing at zero, so that it crosses its tangent line at zero (the x axis). That is, f is neither above nor below its tangent line near zero, so f is neither concave up nor concave down at zero.

THE SECOND DERIVATIVE AND CONCAVITY

1. If $f''(x_0) > 0$, then f is concave upward at x_0.

2. If $f''(x_0) < 0$, then f is concave downward at x_0.

3. If $f''(x_0) = 0$, then f may be either concave upward at x_0, concave downward at x_0, or neither.

Solved Exercises   3. Show: (a) x^4 is concave upward at zero; (b) $-x^4$ is concave downward at zero; (c) x^3 is neither concave upward nor concave downward at zero.

4. Find the points at which $f(x) = 3x^3 - 8x + 12$ is concave upward and those at which it is concave downward. Make a rough sketch of the graph.

5. (a) Relate the sign of the error made in the linear approximation to f with the second derivative of f. (b) Apply your conclusion to the linear approximation of $1/x$ at $x_0 = 1$.

Exercises   3. Find the intervals on which each of the following functions is concave upward or downward. Sketch their graphs.

(a) $1/x$ (b) $1/(1 + x^2)$ (c) $x^3 + 4x^2 - 8x + 1$

4. (a) Use the second derivative to compare x^2 with $9 + 6(x - 3)$ for x near 3. (b) Show by algebra that $x^2 > 9 + 6(x - 3)$ for all $x \neq$ to 3.

INFLECTION POINTS

We just saw that a function f is concave upward where $f''(x) > 0$ and concave downward where $f''(x) < 0$. Points which separate intervals where f has the two types of concavity are of special interest and are called *inflection points*. An example is shown in Fig. 2-3-4. The following is the most useful technical definition.

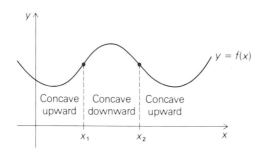

Fig. 2-3-4   The inflection points of f are x_1 and x_2.

Definition   The point x_0 is called an *inflection point* for the function f if $f''(x)$ is defined for x in an open interval containing x_0 and f'' changes sign at x_0.

Comparing the preceding definition with the one on p. 123, we find that *inflection points for f* are just *turning points for f'*; therefore we can use all our techniques for finding turning points in order to locate inflection points. We summarize these tests in the following display.

> ### INFLECTION POINTS
>
> An inflection point for f is a point where f'' changes sign.
>
> If x_0 is an inflection point for f, $f''(x_0) = 0$.
>
> If $f''(x_0) = 0$ and $f'''(x_0) \neq 0$, then x_0 is an inflection point for f.

Worked Example 4   Find the inflection points of $f(x) = x^2 + (1/x)$.

Solution   The first derivative is $f'(x) = 2x - (1/x^2)$, and so the second derivative is $f''(x) = 2 + (2/x^3)$. The only possible inflection points occur where

$$0 = f''(x) = 2 + \frac{2}{x^3}$$

That is, $x^3 = -1$; hence $x = -1$. To test whether this is an inflection point, we calculate the third derivative: $f'''(x) = 6/x^4$, so $f'''(-1) = -6 \neq 0$; hence -1 is an inflection point.

Some additional insight into the meaning of inflection points can be obtained by considering the motion of a moving object. If $x = f(t)$ is its position at time t, then the second derivative $d^2x/dt^2 = f''(t)$ is the rate of change of the velocity $dx/dt = f'(t)$ with respect to time—the *acceleration*. Assume that $dx/dt > 0$, so that the object is moving to the right on the number line. If $d^2x/dt^2 > 0$, the velocity is increasing; that is, the object is *accelerating*. If $d^2x/dt^2 < 0$, the velocity is decreasing; the object is *decelerating*. A point of inflection therefore occurs when the object switches from accelerating to decelerating or vice versa.

If you draw the tangent line at one of the points of inflection in Fig. 2-3-4, you will find that it crosses the graph at the point of tangency. The following theorem shows that this occurs in general.

Theorem 8   Let x_0 be an inflection point for f and let $h(x) = f(x) - [f(x_0) + f'(x_0)(x - x_0)]$ be the difference between f and its linear approximation at x_0.
 If f'' changes sign from negative to positive [positive to negative] at x_0 (for example, if $f'''(x_0) > 0$ [<0]), then h changes sign from negative to positive [positive to negative] at x_0.

We invite you to try writing out the proof of Theorem 8 (see Solved Exercise 7). The two cases are illustrated in Fig. 2-3-5.

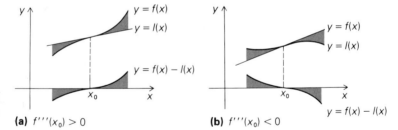

Fig. 2-3-5 The graph of f crosses its tangent line at a point of inflection.

(a) $f'''(x_0) > 0$ **(b)** $f'''(x_0) < 0$

Solved Exercises

6. Discuss the behavior of $f(x)$ at $x = 0$ for $f(x) = x^4, -x^4, x^5, -x^5$.

7. Prove Theorem 8. [*Hint:* See Solved Exercise 2.]

8. Find the inflection points of the function $f(x) = 24x^4 - 32x^3 + 9x^2 + 1$.

Exercises

5. Find the inflection points for x^n, n a positive integer. How does the answer depend upon n?

6. Find the inflection points for the following functions. In each case, tell how the tangent line behaves with respect to the graph.
    (a) $x^3 - x$ (b) $x^4 - x^2 + 1$
    (c) $(x - 1)^4$ (d) $1/(1 + x^2)$

Problems for Section 2-3

1. Find the local maximum and minimum points of the following functions:
   (a) $f(x) = 3x^2 + 2$
   (b) $f(x) = (x^2 - 1)/(x^2 + 1)$
   (c) $f(x) = (x^4 - 1)/(x^2 + 1)$
   (d) $f(x) = x^3 + 6x - 3$
   (e) $f(x) = 6x^5 + x + 20$

2. Find the intervals on which the following functions are concave upward and those on which they are concave downward:
   (a) $f(x) = 3x^2 + 8x + 10$
   (b) $f(x) = x^3 + 3x + 8$
   (c) $f(x) = x/(x - 1)$
   (d) $f(x) = x^4$

3. Find the inflection points of each of the following:
   (a) $f(x) = 2x^3 + 3x$
   (b) $f(x) = x^7$
   (c) $f(x) = (x^2 - 1)/(x^2 + 1)$

4. In each of the graphs of Fig. 2-3-6, tell whether x_0 is a turning point, a local maximum point, a local minimum point, or an inflection point of f.

5. Find the local maxima, local minima, and inflection points of each of the following functions. Also find the intervals on which each function is increasing, decreasing, and concave upward and downward.
   (a) $\frac{1}{4}x^2 - 1$ (b) $1/[x(x-1)]$
   (c) $x^3 + 2x^2 - 4x + \frac{3}{2}$ (d) $x^2 - x^4$
   (e) $f(x) = \begin{cases} 3 - x^2 & \text{for } x \le 2 \\ x^2 - 5 & \text{for } x \ge 2 \end{cases}$

★6. If $f(x)$ is positive for all x, do $f(x)$ and $1/f(x)$ have the same inflection points?

7. Find a function with inflection points at 1 and 2. [*Hint:* Start by writing down $f''(x)$. Then figure out what $f'(x)$ and $f(x)$ should be.]

★8. Find a function with turning points at 1 and 3 and an inflection point at 2.

★9. Sketch the graphs of continuous functions defined on $(-\infty, \infty)$ with the following properties. (If you think no such function can exist, state so as your answer.)
   (a) Three turning points and two points of inflection.
   (b) Two turning points and three points of inflection.
   (c) Four turning points and no points of inflection.
   (d) Two local maxima and no local minima.

★10. Suppose that $f'(x_0) = f''(x_0) = f'''(x_0) = 0$, but $f''''(x_0) \ne 0$. Is x_0 a turning point or an inflection point? Give examples to show that anything can happen if $f''''(x_0) = 0$.

★11. If $f''(x_0) > 0$ and $g''(x_0) > 0$, is $(fg)''(x_0)$ necessarily positive? Prove or give a counterexample.

12. Let $f(x) = x^3 - x$.
   (a) Find the linear approximations to f at $x_0 = -1, 0,$ and 1.
   (b) For each such x_0, compare the value of $f(x_0 + \Delta x)$ with the linear approximation for $\Delta x = \pm 1, \pm 0.1, \pm 0.01$. How does the error depend upon $f''(x_0)$?

(a)

(b)

(c)

(d)

(e)

(f)

(g)

(h)

Fig. 2-3-6 Is x_0 a turning point? A local maximum? A local minimum? An inflection point?

★13. Show that if a function f is concave upward at x_0, it cannot be concave downward at x_0 as well.

14. The power output of a battery is given by $P = EI - RI^2$ where E and R are positive constants.

   (a) For which current I is the power P a maximum? Justify using the second derivative test.

   (b) What is the maximum power?

15. A rock thrown upward attains a height $s = 3 + 40t - 16t^2$ feet in t seconds. Using the second derivative test, find the maximum height of the rock.

16. A generator of E volts is connected to an inductor of L henrys, a resistor of R ohms, and a second resistor of x ohms. Heat is dissipated from the second resistor, the power P being given by

$$P = \frac{E^2 x}{(2\pi L)^2 + (x + R)^2}$$

   (a) Find the resistance value x^* which makes the power as large as possible. Justify with the second derivative test.

   (b) Find the maximum power which can be achieved by adjustment of the resistance x.

17. An Idaho cattle rancher owns 1600 acres adjacent to the Snake River. He wishes to make a three-sided fence from 2 miles of surplus fencing, the enclosure being set against the river to make a rectangular corral. If x is the length of the short side of the fence, then $A = x(2 - 2x)$ is the area enclosed by the fence (assuming the river is straight).

   (a) Show that the maximum area occurs when $x = \dfrac{1}{2}$, using the second derivative test.

   (b) Verify that the maximum area enclosed is 0.5 square mile.

   (c) Verify that the fence dimensions are $\frac{1}{2}$, 1, $\frac{1}{2}$ miles, when the area enclosed is a maximum.

★18. Prove that no odd-degree polynomial can be everywhere concave upward. (As part of your solution, give a few simple examples and include a brief discussion of the possibilities for even-degree polynomials.)

2-4 Drawing Graphs

Using calculus to determine the principal features of a graph often produces better results than simple plotting.

One of the best ways to understand the behavior of a function is to draw its graph. The simplest way to draw a graph is by plotting some points and connecting them with a smooth curve, but this method can lead to serious errors unless we are sure that we have plotted enough points. The methods described in the first three sections of this chapter, combined with the techniques of differentiation in Chapter 1, enable us to make a good choice of which points to plot and permit us to connect the points by a curve of the proper shape.

Goals

After studying this section, you should be able to:

Tell whether a function is even or odd.

Describe the asymptotic behavior of a function.

Plot the graph of a polynomial or rational function.

A SYSTEMATIC PROCEDURE

We begin by outlining a systematic procedure to follow in graphing any function. The details of the procedure are explained below.

GRAPHING PROCEDURE

To sketch the graph of a function f:

1. Note any *symmetries* of f. Is $f(x) = f(-x)$, or $f(x) = -f(-x)$, or neither? In the first case, f is called *even*; in the second case, f is called *odd*. (See Fig. 2-4-1 and the remarks below.)

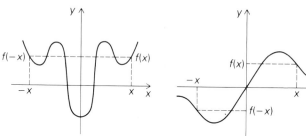

Fig. 2-4-1 f is even when $f(-x) = f(x)$ and odd when $f(-x) = -f(x)$.

(a) Even function   **(b)** Odd function

2. Locate any points where f is not defined and determine the behavior of f near these points. Also determine, if you can, the behavior of $f(x)$ for x very large positive and negative.

3. Locate the local maxima and minima of f, and determine the intervals on which f is increasing and decreasing.

4. Locate the inflection points of f, and determine the intervals on which f is concave upward and downward.

5. Plot a few other key points, such as x and y intercepts, and draw a small piece of the tangent line to the graph at each of the points you have plotted. (To do this, you must evaluate f' at each point.)

6. Fill in the graph consistent with the information gathered in steps 1 through 5.

Let us examine the foregoing instructions, beginning with step 1. If f is even—that is, $f(x) = f(-x)$—we may plot the graph for $x \geq 0$ and then reflect the result across the y-axis to obtain the graph for $x \leq 0$. (See Fig. 2-4-1 (left).) If f is odd, that is, $f(x) = -f(-x)$ then, having plotted f for $x \geq 0$, we may reflect the graph in the y-axis and then in the x-axis to obtain the graph for $x \leq 0$. (See Fig. 2-4-1 (right).)

To decide whether a function is even or odd, substitute $-x$ for x in the expression for $f(x)$ and see if the resulting expression is the same as $f(x)$, the negative of $f(x)$, or neither. For example, $x^4 + 3x^2 + 12$ is even, $x/(1 + x^2)$ is odd, and $x/(1 + x)$ is neither even nor odd.

Step 2 is concerned with what is known as the *asymptotic* behavior of the function f and is best explained through an example.

Worked Example 1   Find the asymptotic behavior of $f(x) = x/(1 - x)$ (f is not defined for $x = 1$).

Solution For x near 1 and $x > 1$, $1 - x$ is a small negative number, so $f(x) = x/(1 - x)$ is large and negative; for x near 1 and $x < 1$, $1 - x$ is small and positive, so $x/(1 - x)$ is large and positive. Thus we could sketch the part of the graph of f near $x = 1$ as in Fig. 2-4-2. The line $x = 1$ is called a *vertical asymptote* for $x/(1 - x)$.

Fig. 2-4-2 Pieces of the graph of $y = x/(1 - x)$ are plotted near the vertical asymptote $x = 1$.

Next we examine the behavior of $x/(1 - x)$ when x is large positive and negative. Since both the numerator and the denominator also get large, it is not clear what the ratio does. We may note, however, that

$$\frac{x}{1 - x} = \frac{1}{(1/x) - 1}$$

for $x \neq 0$. As x gets large (positive or negative), $1/x$ gets small, and $(1/x) - 1$ gets close to -1, so $1/[(1/x) - 1]$ gets close to $1/-1 = -1$. Furthermore, when x is large positive, $1/x > 0$, $(1/x) - 1 > -1$, and $1/[(1/x) - 1] < 1/-1 = -1$, so the graph lies near to, but below, the line $y = -1$. Similarly, for x large negative, $1/x < 0$, $(1/x) - 1 < -1$, and $1/[(1/x) - 1] > -1$, so the graph lies near to, but above, the line $y = -1$. Thus we could sketch the part of the graph for x large as in Fig. 2-4-3. The line $y = -1$ is called a *horizontal asymptote* for f. The location of horizontal and vertical asymptotes will be dealt with in more detail in Chapter 12, but this procedure suffices for now.

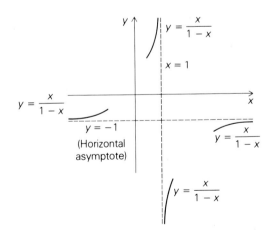

Fig. 2-4-3 Pieces of the graph of $y = x/(1 - x)$ are plotted near the horizontal asymptote $y = -1$.

Steps 3 and 4 were described in detail in Sections 2-2 and 2-3; step 5 increases the accuracy of plotting, and step 6 completes the job.

Some words of advice: It is important to be systematic; follow the procedure step by step, and introduce the information on the graph as you proceed. A haphazard attack on a graph often leads to confusion and sometimes to desperation. Just knowing steps 1 through 6 is not enough—you must be able to employ them effectively. The only way to develop this ability is through practice.

Worked Example 2   Sketch the graph of $f(x) = x - (1/x)$.

Solution   We carry out the six-step procedure:

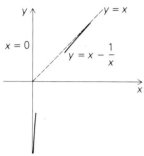

Fig. 2-4-4   The lines $x = 0$ and $y = x$ are asymptotes.

1. $f(-x) = -x + (1/x) = -f(x)$; f is odd, so we need only study $f(x)$ for $x \geq 0$.

2. f is not defined for $x = 0$. For x small and positive, $-(1/x)$ is large in magnitude and negative in sign, so $x - (1/x)$ is large and negative as well; $x = 0$ is a vertical asymptote. For x large and positive, $-(1/x)$ is small and negative; thus the graph of $f(x) = x - (1/x)$ lies below the line $y = x$, coming closer to the line as x becomes larger. The line $y = x$ is again called an *asymptote* (see Fig. 2-4-4).

3. $f'(x) = 1 + (1/x^2)$, which is positive for all $x \neq 0$. Thus f is always increasing and there are no maxima or minima.

4. $f''(x) = -(2/x^3)$, which is negative for all $x > 0$; f is concave downward on $(0, \infty)$.

5. The x intercept occurs where $x - (1/x) = 0$; that is, $x = 1$. We have $f'(1) = 2$, $f(2) = \frac{3}{2}$, $f'(2) = \frac{5}{4}$.

The information obtained in steps 1 through 5 is placed on the graph in Fig. 2-4-5.

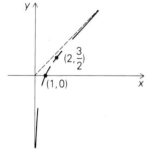

Fig. 2-4-5   The information obtained from steps 1 to 5.

6. We fill in the graph for $x > 0$ (Fig. 2-4-6). Finally, we use the fact that f is odd to obtain the other half of the graph by reflecting through the x and y axes. See Fig. 2-4-7.

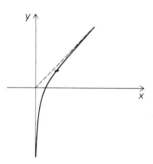

Fig. 2-4-6   The graph is filled in (step 6).

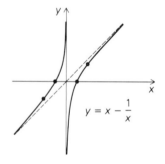

Fig. 2-4-7   The complete graph is obtained by using the fact that f is odd.

Calculator Remarks

While calculators enable one to plot points relatively quickly, and computers will plot graphs from formulas, the use of calculus is still essential. A calculator can be deceptive if used alone, as we saw on pg. 33. In Chapter 15 we will see how the computer can help us graph complicated surfaces in space, but it is inefficient and unwise to begin expensive computation before a thorough analysis using calculus.

Solved Exercises

1. Sketch the graph of $f(x) = x/(1 - x)$.

2. Sketch the graph of $f(x) = x/(1 + x^2)$.

3. Sketch the graph of $f(x) = 2x^3 + 8x + 1$.

4. Sketch the graph of $f(x) = 2x^3 - 8x + 1$.

Exercises

1. Sketch the graph of $f(x) = x^2/(1 - x^2)$.

2. Sketch the graph of $f(x) = x^3 + \frac{1}{4}x^2 - \frac{7}{8}x - \frac{3}{8}$.

3. Sketch the graph of $f(x) = -x^3 + x + 1$.

4. Sketch the graph of $f(x) = x^2/(2 + x)$.

5. Let $f(x)$ be a polynomial. Show that $f(x)$ is an even (odd) function if only even (odd) powers of x occur with nonzero coefficients in $f(x)$.

THE GENERAL CUBIC

(optional)

In Section R-2, we saw how to plot the general linear function, $f(x) = ax + b$, and the general quadratic function $f(x) = ax^2 + bx + c$ ($a \neq 0$). The methods of this section yield the same results: the graph of $ax + b$ is a straight line; while the graph of $f(x) = ax^2 + bx + c$ is a parabola, concave upward if $a > 0$ and concave downward if $a < 0$. Moreover, the turning point of the parabola occurs when $f'(x) = 2ax + b = 0$, that is, $x = -(b/2a)$, which is the same result as was obtained by completing the square.

A more ambitious task is to determine the shape of the graph of the general cubic $f(x) = ax^3 + bx^2 + cx + d$. (We assume that $a \neq 0$; otherwise, we are dealing with a quadratic or linear function.) Of course, any specific cubic can be plotted by techniques already developed, but we wish to get an idea of what *all possible* cubics look like and how their shapes depend on a, b, c, and d.

We begin our analysis with some simplifying transformations. First of all, we can factor out a and obtain a new polynomial $f_1(x)$ as follows:

$$f(x) = a\left(x^3 + \frac{b}{a}x^2 + \frac{c}{a}x + \frac{d}{a}\right) = af_1(x)$$

The graphs of f and f_1 have the same basic shape; if $a > 0$, the y axis is just rescaled by multiplying all y values by a; if $a < 0$, the y axis is rescaled and the graph is flipped about the x axis. It follows that we do not lose any generality by assuming that the coefficient of x^3 is 1.

Consider therefore the simpler form

$$f_1(x) = x^3 + b_1x^2 + c_1x + d_1$$

where $b_1 = b/a$, $c_1 = c/a$ and $d_1 = d/a$. In trying to solve cubic equations, mathematicians of the early Renaissance noticed a useful trick: if we replace x by $x - (b_1/3)$, then the quadratic term drops out; that is,

$$f_1\left(x - \frac{b_1}{3}\right) = x^3 + c_2x + d_2$$

where c_2 and d_2 are new constants, depending on b_1, c_1, and d_1. (We leave to the reader the task of verifying this last statement and expressing c_2 and d_2 in terms of b_1, c_1, and d_1.)

The graph of $f_1(x - b_1/3) = f_2(x)$ is the same as that of $f_1(x)$ except that it is shifted by $b_1/3$ units along the x axis. This means that we lose no generality by assuming that the coefficient of x^2 is zero—that is, we only need to graph *(optional)* $f_2(x) = x^3 + c_2 x + d_2$. Finally, replacing $f_2(x)$ by $f_3(x) = f_2(x) - d_2$ just corresponds to shifting the graph d_2 units parallel to the y axis.

We have now reduced the graphing of the general cubic to the case of graphing $f_3(x) = x^3 + c_2 x$.* For simplicity let us write $f(x)$ for $f_3(x)$ and c for c_2. To plot $f(x) = x^3 + cx$, we go through steps 1 to 6:

1. f is odd.

2. f is defined everywhere. Since $f(x) = x^3 (1 + c/x^2)$, $f(x)$ is large positive (negative) when x is large positive (negative); there are no horizontal or vertical asymptotes.

3. $f'(x) = 3x^2 + c$. If $c > 0$, $f'(x) > 0$ for all x, and f is increasing everywhere. If $c = 0$, $f'(x) > 0$ except at $x = 0$, so f is increasing everywhere even though the graph has a horizontal tangent at $x = 0$. If $c < 0$, $f'(x)$ has roots at $\pm\sqrt{-c/3}$; f is increasing on $(-\infty, -\sqrt{-c/3}\,]$ and $[\sqrt{-c/3}, \infty)$ and decreasing on $[-\sqrt{-c/3}, \sqrt{-c/3}\,]$. Thus $-\sqrt{-c/3}$ is a local maximum point and $\sqrt{-c/3}$ is a local minimum point.

4. $f''(x) = 6x$, so f is concave downward for $x < 0$ and concave upward for $x > 0$. Zero is an inflection point.

5. $f(0) = 0, f'(0) = c$

$$f(\pm\sqrt{-c}\,) = 0, f'(\pm\sqrt{-c}\,) = -2c \qquad \text{(if } c < 0)$$

$$f\left(\pm\sqrt{\frac{-c}{3}}\right) = \pm\frac{2}{3}c\sqrt{\frac{-c}{3}}, f'\left(\pm\sqrt{\frac{-c}{3}}\right) = 0 \qquad \text{(if } c < 0)$$

6. We skip the preliminary sketch and draw the final graphs (Fig. 2-4-8).

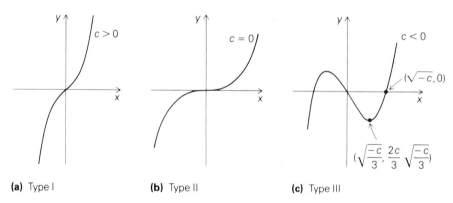

(a) Type I   **(b)** Type II   **(c)** Type III

Fig. 2-4-8   The graph of $y = x^3 + cx$ for $c > 0$ (I), $c = 0$ (II), and $c < 0$ (III).

---

* These algebraic ideas are related to a formula for the roots of a general cubic, discovered by Niccolo Tartaglia (1506–1559) but published (without Tartaglia's permission) by and often credited to Girolamo Cardano (1501–1576). Namely, the solutions of the equation $x^3 + bx^2 + cx + d = 0$ are

$$x_1 = S + T - \frac{b}{3} \qquad \text{where} \qquad S = \sqrt[3]{R + \sqrt{Q^3 + R^2}},$$

$$x_2 = -\tfrac{1}{2}(S + T) - \frac{b}{3} + \tfrac{1}{2}\sqrt{-3}\,(S - T) \qquad T = \sqrt[3]{R - \sqrt{Q^3 + R^2}}$$

$$x_3 = -\tfrac{1}{2}(S + T) - \frac{b}{3} - \tfrac{1}{2}\sqrt{-3}\,(S - T) \qquad Q = \frac{3c - b^2}{9}$$

$$R = \frac{9bc - 27d - 2b^3}{54}$$

There is also a formula for the roots of a quartic equation, but a famous theorem (due to Abel and Ruffini in the nineteenth century) states that there can be no such algebraic formula for the general equation of degree ≥ 5. Modern proofs of this theorem can be found in advanced textbooks on algebra (such as L. Goldstein, Abstract Algebra, Prentice-Hall, (1973). These proofs are closely related to the proof of the impossibility of trisecting angles with ruler and compass.

(optional)

Thus there are three types of cubics (with $a > 0$):

Type I: f is increasing at all points; $f' > 0$ everywhere.

Type II: f is increasing at all points; $f' = 0$ at one point.

Type III: f has two turning points.

Type II is the transition type between types I and III. You may imagine the graph changing as c begins with a negative value and then moves toward zero. As c gets smaller and smaller, the turning points move in toward the origin and the bumps in the graph become smaller and smaller until, when $c = 0$, the bumps merge at the point where $f'(x) = 0$. As c passes zero to become positive, the bumps disappear completely.

Worked Example 3   A drug is injected into a person's bloodstream and the temperature increase T recorded one hour later. If x milligrams are injected, then

$$T(x) = \frac{x^2}{8}\left(1 - \frac{x}{16}\right), \quad 0 \leq x \leq 16$$

(a) The rate of change of T with respect to dosage x is called the *sensitivity* of the body to the dosage. The sensitivity is therefore $T'(x)$. Find it.

(b) Use the techniques for graphing cubics to graph T versus x.

(c) Find the dosage producing maximum sensitivity.

Solution   (a) $T(x) = (x^2/8) - (x^3/128)$ and so $T'(x) = (x/4) - 3x^2/128 = x[(1/4) - (3x/128)]$.

(b) T is neither even nor odd and is everywhere defined on its domain $[0, 16]$. The graph intersects the x-axis when $x = 0$ and $x = 16$. The critical points, where T' vanishes, are $x = 0$ and $x = 32/3$. We have $T'' = (1/4) - (3x/64)$ which is positive at $x = 0$ and negative at $x = 32/3$ so $x = 0$ is a local minimum and $x = 32/3$ is a local maximum. We plot a few more points to obtain the graph in Fig. 2-4-9.

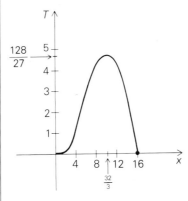

Fig. 2-4-9 A graph of temperature increase T versus dosage x.

(c) We are to maximize $f(x) = T'(x) = (x/4) - 3x^2/128$. Now $f'(x) = (1/4) - 3x/64$ which vanishes at $x = 16/3$. Since $f''(x) = -3/64 < 0$, this is a maximum. Thus $x = 16/3$ is the required dosage.

Solved Exercises ★5. Convert $2x^3 + 3x^2 + x + 1$ to the form $x^3 + cx$ and determine whether the cubic is of type I, II, or III.

Exercises ★6. Convert $x^3 + x^2 + 3x + 1$ to the form $x^3 + cx$ and determine whether the cubic is of type I, II, or III.

★7. Convert $x^3 - 3x^2 + 3x + 1$ to the form $x^3 + cx$ and determine whether the cubic is of type I, II, or III.

★8. A simple model for the voting population in a district is given by $N(t) = 30 + 12t^2 - t^3$, $0 \le t \le 8$, where t is the time in years, N the population in thousands.

(a) Graph N versus t on $0 \le t \le 8$, using the techniques for graphing cubics.

(b) At what time t will the population of voters increase most rapidly?

(c) Explain the significance of the points $t = 0$ and $t = 8$.

Problems for Section 2-4

1. Which of the following functions are even and which are odd?

    (a) $f(x) = (x^3 + 6x)/(x^2 + 1)$

    (b) $f(x) = x$

    (c) $f(x) = x/(x^3 + 1)$

    (d) $f(x) = x^6 + 8x^2 + 3$

    (e) $f(x) = xg(x)$, where g is even

2. Describe the behavior of the following functions near their vertical asymptotes:

    (a) $f(x) = x/(3x + 1)$

    (b) $f(x) = (x^3 + 1)/(x^2 - 1)$

    (c) $f(x) = x/(x^3 - 1)$

    (d) $f(x) = (x^3 - 1)/(x^2 - 1)$

3. Match the following functions with the graphs in Fig. 2-4-10:

    (a) $(x^2 + 1)/(x^2 - 1)$

    (b) $x + (1/x)$

    (c) $x^3 - 3x^2 - 9x + 1$

    (d) $(x^2 - 1)/(x^2 + 1)$

4. Sketch and discuss the graphs of the following functions:

    (a) $[x^3/(2 - x^2)] + x$

    (b) $x^3 + 7x^2 - 2x + 10$

    (c) $8x^3 - 3x^2 + 2x$ (d) $x^2 + [1/(1 - x)]$

5. Sketch and discuss the graphs of the following functions:

    (a) $x(x - 2)/(x - 1)$ *Hint:* Let $u = x - 1$.

    (b) $-x^3 + \frac{3}{2}x^2 + x + 1$

    (c) $(1 - x^2)/(1 + x^2)$

    (d) $x^4 - x^2$

★6. What does the graph of $[ax/(bx + c)] + d$ look like if a, b, c, and d are positive constants?

★7. Prove that the graph of any cubic $f(x) = ax^3 + bx^2 + cx + d$ ($a \ne 0$) is symmetric about its inflection point in the sense that the function

$$g(x) = f\left(x - \frac{b}{3a}\right) - f\left(-\frac{b}{3a}\right)$$

is odd. [*Hint:* $g(x)$ is again a cubic; where is its inflection point?]

8. Find a criterion for telling when the quotient $f(x)/g(x)$ of two polynomials is even, odd, or neither.

★9. Suppose that $f(x)$ is defined on all of $(-\infty, \infty)$. Show that

$$f(x) = e(x) + o(x)$$

where e is an even function and o is an odd function. [*Hint:* Substitute $-x$ for x, use the fact that e is even and o is odd, and solve for $e(x)$ and $o(x)$.]

10. Sketch the graphs of the following functions:

    (a) $x^3 + 3x^2 + 2x$ (b) $x^4 - x^3$

    (c) $x^4 - 6x^2 + 5x + 2$ [*Hint:* Sketch the derivative first.]

    (d) $(3x^2 + 4)/(x^2 - 9)$

(A)   (B)

(C) 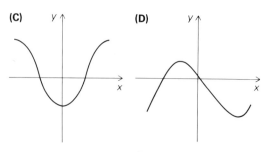 (D)

Fig. 2-4-10   Matching.

11. Sketch the graphs of the following functions:

    (a) $3x^3 + x^2 + 1$ (b) $x^4 + x^3$

    (c) $x^4 - 3x^2 + 2\sqrt{2}$ (d) $(x^3 + 1)/(x^2 - 1)$

    (e) $(x^2 + 1)/[(x - 1)(x + 1)^2]$

★12. If f is twice differentiable and x_0 is a critical point of f, must x_0 be either a local maximum, local minimum, or inflection point?

13. The population P of mice in a wood varies with the number x of owls in the wood according to the formula $P = 30 + 10x^2 - x^3$, $0 \le x \le 10$. Graph P versus x.

★14. This problem concerns the graph of the general cubic:

    (a) Find an explicit formula for the coefficient c_2 in $f(x_1 - b_1/3)$ (p. 139) in terms of b_1, c_1, and d_1 and thereby give a simple rule for determining whether the cubic

    $$x^3 + b_1 x^2 + c_1 x + d_1$$

    is of type I, II, or III.

    (b) Give a rule, in terms of a, b, c, d, for determining the type of the general cubic $ax^3 + bx^2 + cx + d$.

    (c) Use the quadratic formula on the derivative of $ax^3 + bx^2 + cx + d$ to determine, in terms of a, b, c, and d, how many turning points there are. Compare with the result in part (b).

Problems 15 to 20 concern the graph of the general quartic:

$$f(x) = ax^4 + bx^3 + cx^2 + dx + e \quad a \ne 0$$

★15. Using the substitution $x - b/4a$ for x, show that one can reduce to the case

    $$f(x) = x^4 + cx^2 + dx + e$$

    (with a new c, d and e!).

★16. According to the classification of cubics, $f'(x) = 4x^3 + 2cx + d$ can be classified into three types: I ($c > 0$), II ($c = 0$), and III ($c < 0$), so we may name each quartic by the type of its derivative. Sketch the graph of a typical quartic of type I.

★17. Divide type II quartics into three cases: II_1 ($d > 0$), II_2 ($d = 0$), II_3 ($d < 0$). Sketch their graphs with $e = 0$.

★18. In case III ($c < 0$), $f'(x)$ has two turning points and can have one, two, or three roots. By considering Fig. 2-4-11, show that the sign of $f'(x)$ at its turning points determines the number of roots. Obtain thereby a classification of type III quartics into five subtypes: III_1, III_2, III_3, III_4, and III_5. Sketch graphs for each case and determine the conditions on c and d which govern the cases.

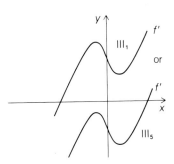

(a) One root: (f' has the same sign at its two turning points)

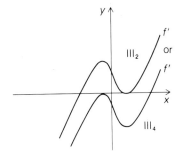

(b) Two roots: (f' is zero at one of its turning points)

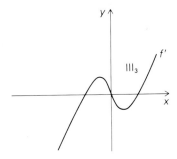

(c) Three roots: (f' has opposite signs at its turning points)

Fig. 2-4-11 The five possible positions of the graph of f' in case III.

★19. Using your results from Problems 17 and 18, show that the the (c, d) plane may be divided into regions, as shown in Fig. 2-4-12, which determine the type of quartic.*

★20. Classify and sketch the graphs of the following quartics: (a) $x^4 - 3x^2 - 4x$; (b) $x^4 + 4x^2 + 6x$; (c) $x^4 + 7x$.

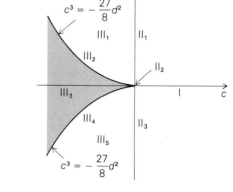

Fig. 2-4-12 Locating the value of (c, d) in this graph tells the type of quartic.

---

*For advanced (and sometimes controversial) applications of this figure, see T. Poston and I. Stuart, Catastrophe Theory and its Applications, *Pitman* (1978).

Review Problems for Chapter 2

1. In what intervals are each of the following functions increasing? Decreasing?
   (a) $x^3 + 3x + 2$ (b) $x^5 - 3x^4$
   (c) $1/(x + 3)$ (d) $1/(1 + x^4)$
   (e) $(1 - x^3)/(1 + 2x^3)$

2. For each of the functions in Problem 1, find the points of inflection and the points where the second derivative vanishes.

3. Sketch the graph of each of the functions in Problem 1.

4. Suppose that the British inflation rate (in percent per year) from 1980 to 1990 is given by the function $B(t) = 10(t^3/100 - t^2/20 - t/5 + 1)$, where t is the time in years from January 1, 1983. When is the inflation rate increasing? How might a politician react to turning points? Inflection points?

5. Tell whether each statement below is true or false:
   (a) Every continuous function is differentiable.
   (b) If f is a continuous function on $[0, 2]$, and $f(1) = -1$ and $f(2) = 1$, then there must be a point x in $[0, 1]$ where $f(x) = 0$.
   (c) If a function is increasing at $x = 1$ and at $x = 2$, it must be increasing at every point between 1 and 2.
   (d) If a differentiable function f on $(-\infty, \infty)$ has a local maximum point at $x = 0$, then $f'(0) = 0$.
   (e) $f(x) = x^4 - x^3$ is increasing at zero.
   (f) $f(x) = x^3 + 3x$ takes its largest value on $[-1, 1]$ at an endpoint.
   (g) Parabolas never have inflection points.

   (h) All cubic functions $y = ax^3 + bx^2 + cx + d$, $a \neq 0$ have exactly one inflection point.
   (i) $f(x) = 3/(5x^2 + 1)$ on $(-\infty, \infty)$ has a local maximum at $x = 0$.

★6. A function defined on (a, b) such that

   $$f(tx + (1 - t)y) \leq tf(x) + (1 - t)f(y)$$

   for all x, y in (a, b) and all t in $[0, 1]$ is called *convex*. Prove that if f'' is continuous and positive on (a, b), then f is convex. [*Hint:* Show first of all that f' is increasing so that, for $0 \leq t \leq 1$,

   $$f'(x) \leq f'(tx + (1 - t)y) \leq f'(y)$$

   Second, let

   $$g(t) = tf(x) + (1 - t)f(y) - f(tx + (1 - t)y)$$

   on $[0, 1]$ and show that the minima of g are endpoints.]

7. For each of the functions (a) through (d), answer the questions below and draw a graph:
   (i) Where is f continuous?
   (ii) Where is f differentiable?
   (iii) On which intervals is f monotonic?
   (iv) Where is f concave upward or downward?
   (v) What are the critical points, endpoints, turning points, local maximum and minimum points, and inflection points for f?
   (a) $f(x) = -7x^3 + 2x^2 + 15$ on $(-\infty, \infty)$
   (b) $f(x) = 6x^2 + 3x + 4$ on $(-\infty, \infty)$
   (c) $f(x) = 3x/(2 - 5x)$ on its domain
   (d) $f(x) = x^3 - 2x + 1$ on $[-1, 2]$

8. For each of the functions (a) through (d), answer the questions below and draw a graph:

   (i) Where is f continuous?

   (ii) Where is f differentiable?

   (iii) On which intervals is f monotonic?

   (iv) Where is f concave upward or downward?

   (v) What are the critical points, endpoints, singular points, turning points, local maximum and minimum points, and inflection points for f?

   (a) $f(x) = x^4 - 3x^3 + x^2$ on $[-6, 6]$

   (b) $f(x) = \begin{cases} x^2 & x \ge 0 \\ -x^2 & x \le 0 \end{cases}$ on $(-1, 1)$

   (c) $f(x) = x/(1 + x^{2n})$ on $(-\infty, \infty)$

   (d) $f(x) = \begin{cases} x^2 & x \ge 0 \\ x^2 + 1 & x < 0 \end{cases}$ on $(-2, 2]$

9. A bicycle is moving at $\frac{1}{16} t^3 + 4t^2 - 3t + 2$ miles per hour, where t is the time in hours, $0 \le t \le 2$. Does the bicycle turn around during this period?

10. Find the local maxima and minima of the function $(x^3 - 12x + 1)^6$.

★11. Let $f(x) = 1/(1 + x^2)$.

   (a) For which values of c is the function $f(x) + cx$ increasing on the whole real line? Sketch the graph of $f(x) + cx$ for one such c.

   (b) For which values of c is the function $f(x) - cx$ decreasing on the whole real line? Sketch the graph of $f(x) - cx$ for one such c.

   (c) How are your answers in parts (a) and (b) related to the inflection points of f?

★12. Let f be a nonconstant polynomial such that $f(0) = f(1)$. Prove that f has a turning point somewhere in the interval $(0, 1)$.

13. A manufacturer sells x hole punchers per week. The weekly cost and revenue equations are

$$C(x) = 5000 + 2x, \quad R(x) = 10x - \frac{x^2}{1000},$$

$$0 \le x \le 8000$$

   (a) Find the minimum cost.

   (b) Find the maximum revenue.

   (c) Define the profit P by the formula $P(x) = R(x) - C(x)$. Find the maximum profit.

14. A paint can is gently kicked off the roof of a 156-foot building, its distance S from the ground being given by $S = 156 + 4t - 16t^2$ (S in feet, t in seconds).

   (a) How high up does the can go before it falls downward?

   (b) At what time does the can return to roof level?

   (c) Find the velocity of the can when it collides with the ground.

15. The bloodstream drug concentration $C(t)$ in a certain patient's bloodstream t hours after injection is given by

$$C(t) = \frac{16t}{(10t + 20)^2}$$

   Find the maximum concentration and the number of hours after injection at which it occurs.

16. The power P developed by the engine of an aircraft flying at a constant (subsonic) speed v given by

$$P = \left(cv^2 + \frac{d}{v^2} \right) v$$

   where $c > 0$, $d > 0$.

   (a) Find the speed v_0 which minimizes the power.

   (b) Let $Q(t)$ denote the amount of fuel (in gallons) the aircraft has at time t. Assume the power is proportional to the rate of fuel consumption. At what speed will the flight time from takeoff to fuel exhaustion be maximized?

17. Let x and y be the horizontal and vertical velocities of a 22-caliber-long rifle shell, and assume a and b, respectively, are decimal approximations to these velocities, correct to the 12th decimal place. Suppose $0 \le x \le 1365$, $0 \le y \le 1365$, and $x^2 + y^2 = a^2 + b^2 + l$. Does the remainder term l have zeros in its first 12 decimal places, necessarily? Explain.

18. In a drug sensitivity problem, the change T in body temperature for an x-milligram injection is given by

$$T(x) = x^2 \left(1 - \frac{x}{4} \right), \quad 0 \le x \le 4$$

   (a) Find the *sensitivity* $T'(x)$ when $x = 2$.

   (b) Graph T versus x, using cubic graphing methods.

19. A fixed-frequency generator producing 6 volts is connected to a coil of 0.05 henry, in parallel with a resistor of 0.5 ohm and a second resistor of x ohms. The power P is given by

$$P = \frac{36x}{(\pi/10)^2 + (x + 1/2)^2}$$

Find the maximum power and the value of x which produces it.

20. A function f is said to majorize a function g on $[a, b]$ if $f(x) \geq g(x)$ for all $a \leq x \leq b$.

(a) Show by means of a graph that x majorizes x^2 on $[0, 1]$.

(b) Argue that "f majorizes g on $[a, b]$" means that the curve $y = f(x)$, $a \leq x \leq b$, lies above the curve $y = g(x)$, $a \leq x \leq b$.

(c) If $m > n > 0$, then x^n majorizes x^m on $[0, 1]$. Explain fully.

(d) If $m > n > 0$, then $(x - a)^n$ majorizes $(x - a)^m$ on $[a, a + 1]$. Why?

(e) Given that $m \neq n$, $m > 0$, and $n > 0$, determine on which intervals $(x - a)^n$ majorizes $(x - a)^m$ (or conversely).

(f) Graph $y = x - 1$, $y = (x - 1)^2$, $y = (x - 1)^3$, $y = (x - 1)^4$, $y = (x - 1)^5$ on the same set of axes for $-2 \leq x \leq 2$.

21. The geometric appearance of the graph of a factored polynomial (such as $y = x(x - 1)^2 (x - 2)^3 (x - 3)^7 (x - 4)^{12}$) near a root r is the same as that of the equation $y = c(x - r)^n$, where c and n are chosen appropriately for the root r.

(a) Let $y = x(x + 1)^2 (x - 2)^4$. For values of x near 2, the factor $x(x + 1)^2$ is nearly $2(2 + 1)^2 = 18$, so y is approximately $18(x - 2)^4$. Sketch the graph near $x = 2$.

(b) Argue that near the roots 0, 1, 3 the equation $y = 10x(x - 1)^3 (x - 3)^2$ looks geometrically like $y = -90x$, $y = 40(x - 1)^3$, $y = 240(x - 3)^2$, respectively. Use this information to help sketch the graph from $x = 0$ to $x = 4$.

3

Maxima and Minima

The global maxima and minima of functions are important in both the theory and the applications of calculus.

Differential calculus has many important applications to problems of *optimization*. Many practical problems in which one wishes to choose the policy which will have the "best" outcome may be translated into mathematical problems in which one seeks the maximum or minimum value of a function. In this chapter, we will deal with both practical and theoretical applications of maxima and minima for functions of one variable. Analogous problems for functions of several variables will be treated in Chapter 16.

CONTENTS

In Section 3-1, we show how differential calculus can be used to find maxima and minima of functions on intervals. These techniques are used in Section 3-2 to solve word problems in a variety of fields. In Section 3-3, we use the idea of maxima and minima to prove a theorem important to the further development of calculus.

---

*This appendix is found in the Student Guide.

3-1 Global Maxima and Minima

Differential calculus helps us to locate the points at which a function reaches its largest or smallest value.

The maximum and minimum points discussed in Section 2-3 were *local*, since we compared $f(x_0)$ with $f(x)$ for x near x_0. For many applications of calculus, however, it is important to find the points where $f(x)$ has the largest or smallest possible value as x ranges over a fixed interval. In this section, we show how calculus helps us to locate these *global* maximum and minimum points; in the following section, we discuss how to translate word problems into calculus problems involving maxima and minima.

Goals

After studying this section, you should be able to:

Distinguish local maxima and minima from global maxima and minima for a function on an interval.

Apply the closed interval test to locate the maxima and minima for a given function.

THE CONCEPT OF GLOBAL MAXIMA AND MINIMA

Global maxima and minima should be as familiar to you as the daily weather report. The statement on the 6 P.M. news that "today's high temperature was 26°C" means that:

1. At no time today was the temperature higher than 26°C.

2. At some time today the temperature was exactly 26°C.

If we let f be the function which assigns to each t in the interval $[0, 18]$ the temperature in degrees at time t hours after midnight, then we may say that 26 is the (global) *maximum value of f on* $[0, 18]$. Here is a formal definition.

Definition. Let f be a function which is defined on a set I of real numbers.

If M is a real number such that:

1. $f(x) \leq M$ for all x in I

2. $f(x_0) = M$ for at least one x_0 in I

then we call M the *maximum value of f on I.*

If m is a real number such that:

1. $f(x) \geq m$ for all x in I

2. $f(x_1) = m$ for at least one x_1 in I

then we call m the *minimum value of f on I.*

The numbers x_0 and x_1 in the definition of maximum and minimum values are also of interest. They represent the points at which the maximum and minimum values are attained. For the temperature function discussed above, x_0 might be 15.5, indicating that the high temperature occurred at 3:30 PM. Of course, it might be possible that the temperature rose to 26° at 2 PM, dipped due to a sudden rain shower, rose again to 26° at 3:30 PM, and finally decreased toward evening. In that case, both 14 and 15.5 would be acceptable values for x_0.

In general, those points x_0 $[x_1]$ for which $f(x_0) = M$ $[f(x_1) = m]$ are called maximum [minimum] *points* for f on I.* As we just observed, there may be more than one maximum or minimum point. Sometimes, to distinguish the points considered here from the turning points discussed in Section 2-3, we call them *global* maximum and minimum points.

We will see shortly that differential calculus provides a powerful technique for locating the maximum and minimum points of functions defined by formulas. For functions defined by other means, there are sometimes direct approaches to finding maxima and minima. For example:

1. The highest ring around a bathtub indicates the maximum water level achieved since the tub was last scrubbed. (Other rings indicate local maxima.)

2. On a maximum-minimum thermometer, one can read directly the maximum and minimum temperatures achieved since the last time the thermometer was reset.

3. If the graph of a function is available, either from experimental data or by plotting from a formula as in Section R-3, the maxima and minima can be seen as the high and low points on the graph.

4. Number the students in a class from 1 to 25. Let $I = \{1, 2, 3, ..., 25\}$ be the set consisting of just the 25 integers between 1 and 25. For each x in I, let $f(x)$ be the examination grade of the student whose number is x. If we list $f(1)$ through $f(25)$, we can find the maximum and minimum values of f on I by simply picking out the highest and lowest numbers on the list. The maximum and minimum *points* are the numbers of the students with the best and worst grades. (There may be more than one of each.)

Worked Example 1 Find the maximum and minimum points and values of the function $f(x) = 1/(1 + x^2)$ on the interval $[-2, 2]$.

Solution $f(x)$ is largest where its denominator is smallest and vice versa. The maximum point occurs, therefore, at $x = 0$; the maximum value is 1. The minimum points occur at -2 and 2; the minimum value is $\frac{1}{5}$.

We may verify these statements with the aid of calculus: $f'(x) = -2x/(1 + x^2)^2$, which is positive for $x < 0$ and negative for $x > 0$, so f is increasing on $[-2, 0]$ and decreasing on $[0, 2]$. It follows that

$$\tfrac{1}{5} = f(-2) < f(x) < f(0) = 1 \qquad \text{for } -2 < x < 0$$

and

$$1 = f(0) > f(x) > f(2) = \tfrac{1}{5} \qquad \text{for } 0 < x < 2$$

and we have $\tfrac{1}{5} \le f(x) \le 1$ for $-2 \le x \le 2$. See Fig. 3-1-1.

Fig. 3-1-1 The function $f(x) = 1/(1 + x^2)$ on $[-2, 2]$ has a maximum point at $x = 0$ and minimum points at $x = \pm 2$.

---

*Here, as after the adjectives turning, critical, and inflection, the term point refers to a point on the number line rather than in the plane.

Solved Exercises*

1. Find the maximum and minimum points and values, if they exist, for the function $f(x) = x^2 + 1$ on each of the following intervals:

 (a) $(-\infty, \infty)$ (b) $(0, \infty)$

 (c) $(-1, 1)$ (d) $[-1, 1]$

 (e) $(0, 1]$ (f) $[\frac{1}{2}, 1]$

 (g) $(-2, 1]$ (h) $[-2, 1)$

2. Fig. 3-1-2 shows the amount of solar energy received at various latitudes in the northern hemisphere on June 21 on a square meter of horizontal surface located at the top of the atmosphere. Find the maximum and minimum points and values.

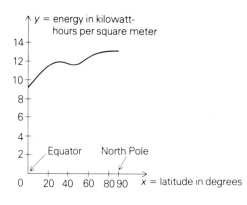

Fig. 3-1-2 The solar energy y received on June 21 at the top of the atmosphere at various latitudes x. (Reproduced from W. G. Kendrew, *Climatology*, Oxford University Press, 1949.)

3. Let the set I be contained in the set J, and suppose that f is defined on J (and so is defined on I). Show that if m_I and m_J are the minimum values of f on I and J, respectively, then $m_J \leq m_I$.

Exercises

1. Find the maximum and minimum points and values, if they exist, for the function $f(x) = x^2 - 3x + 1$ on each of the following intervals. [*Hint:* Begin by graphing f by the methods of Section 2-4.]

 (a) $(2, \infty)$ (b) $(-\infty, \frac{1}{2}]$

 (c) $(-3, 2]$ (d) $(-\frac{3}{2}, 2]$

 (e) $(-2, 2)$ (f) $(-\infty, \infty)$

 (g) $[-1, 1]$ (h) $[-8, 8]$

2. Find the maximum and minimum points and values of the function graphed in Fig. 3-1-6, p. A-12. Discuss your answer.

3. Prove that the maximum value of f on I is unique; that is, show that if M_1 and M_2 both satisfy conditions 1 and 2 of the definition, then $M_1 = M_2$. [*Hint:* Show that $M_1 \leq M_2$ and $M_2 \leq M_1$.]

CRITICAL POINTS, ENDPOINTS, AND SINGULAR POINTS

We begin the application of differential calculus to maximum-minimum problems by classifying maximum and minimum points into three types.

Solutions appear in Appendix A.

Definition Let f be a function defined on an interval I. A point x_0 in I which is not an endpoint is called:

1. A *critical point* for f if $f'(x_0) = 0$
2. A *singular point* for f if f is not differentiable at x_0

Theorem 1 *If x_0 is a maximum (or minimum) point for f on I, then x_0 is either a critical point, an endpoint of I, or a singular point.*

Proof Suppose that x_0 is neither a critical point, an endpoint, nor a singular point. Then f is defined in an open interval containing x_0, f is differentiable at x_0, and $f'(x_0) \neq 0$. By Theorem 3 (p. 117), f is either increasing or decreasing at x_0, so f takes values on I which are both larger and smaller than $f(x_0)$. Hence x_0 could not be a maximum or a minimum point.

Theorem 1 allows us to narrow down the search for maximum and minimum points to a hunt for critical points, endpoints, and singular points, which can often be found easily. (In many problems, singular points do not occur.)

Warning Theorem 1 does not say that every critical point, endpoint, or singular point is *necessarily* a maximum or minimum point; further arguments are necessary to determine whether or not they are.

Worked Example 2 Are the maximum and minimum points in Worked Example 1 critical points, endpoints, or singular points?

Solution The minimum points -2 and 2 are endpoints of $[-2, 2]$. The maximum point zero is neither an endpoint nor a singular point; by Theorem 1, it must be a critical point. Indeed, we saw that $f'(x) = -2x/(x^2 + 1)^2$, so $f'(0) = 0$.

Solved Exercises

4. Find the critical points, endpoints, singular points for $f(x) = x^2 + 1$ on each of the intervals in Solved Exercise 1 above, and compare them with the maximum and minimum points.

5. Find the critical points, endpoints, and maximum and minimum points for $f(x) = x^3$ on $[-1, \infty)$.

6. Find the critical points and maximum and minimum points for $f(x) = x/(1 + x^2)$ on $(-\infty, \infty)$. (See Solved Exercise 2, Section 2-4.)

7. Let $f(x) = |x|$ be the absolute value function. Find the critical points, endpoints, singular points, and maximum and minimum points for f on the interval $(-2, 1]$.

Exercises

4. Find the critical points and endpoints for $f(x) = x^3 - 2x + 1$ on each of the intervals in Exercise 1 above. Compare them with the maximum and minimum points.

5. Find the critical points, endpoints, singular points, and maximum and minimum points for each function on the given interval:

(a) $f(x) = x^2 - x$ on $[0, 1)$

(b) $f(x) = x/(1 - x^2)$ for $-1 < x \le 0$ and $f(x) = x^3 - x$ for $0 \le x < 2$, on $(-1, 2)$.

(c) $f(x) = \begin{cases} -x^4 - 1 & \text{for } -1 < x \le 0 \\ -1 & \text{for } 0 \le x \le 1 \\ x - 2 & \text{for } 1 \le x < \infty \end{cases}$ on $(-1, \infty)$

Is f differentiable at $x = 0$?

A TEST ON CLOSED INTERVALS

If I is a closed interval, the problem of finding maxima and minima for a function f on I is simplified by the following theoretical result.

> **Theorem 2   Extreme Value Theorem.**   Let f be continuous on the closed interval $[a, b]$. Then f has both a maximum and a minimum value on $[a, b]$.

The proof of this theorem is in the appendix to this chapter, but the statement should be understood by everyone. It says that two conditions together are *sufficient* to ensure that f has both a maximum and a minimum value on I:

1. f is continuous on I.

2. I is a closed interval.

Cases (a), (b), (c), (e), and (g) of Solved Exercise 1 in this section show that condition 1 alone is not sufficient; that is, if I is not closed, the maxima and minima may not exist. Case (h) shows that the maxima and minima might happen to exist even if I is not closed; thus these two conditions are not *necessary* for the existence of maxima and minima. (See Solved Exercise 8 for further discussion.)

Notice that, like the intermediate value theorem, the extreme value theorem is an *existence theorem* which tells you nothing about how to find the maxima and minima. Combining the extreme value theorem with Theorem 1, however, yields a practical test, described in the display which follows.

CLOSED INTERVAL TEST

To find the maxima and minima for a function f which is continuous on $[a, b]$:

1. Make a list $x_1, \ldots, x_n$ consisting of the critical points of f on (a, b), the endpoints a and b of $[a, b]$, and the singular points of f in (a, b) (points where f is not differentiable).*
2. Compute the values $f(x_1), \ldots, f(x_n)$.

Then the largest [smallest] of the $f(x_i)$ is the maximum [minimum] value of f on $[a, b]$. The maximum [minimum] points for f on $[a, b]$ are those x_i for which $f(x_i)$ equals the maximum [minimum] value.

*In bizarre cases the list may be infinite. We ignore this possibility.

To justify the closed interval test, we first note that by the extreme value theorem, the maximum and minimum points must exist; by Theorem 1, these points must appear on the list $x_1, \ldots, x_n$. It remains, therefore, to determine *which* of the points in the list are the maximum and minimum points; to do this, it suffices to evaluate f at all of them and then to compare the values.

Worked Example 3   Find the maximum and minimum points and values for the function $f(x) = (x^2 - 8x + 12)^4$ on the interval $[-10, 10]$.

Solution   There are no singular points; the list indicated by the closed interval test consists of -10, 10, and the critical points. To find the critical points we differentiate:

$$\frac{d}{dx}(x^2 - 8x + 12)^4 = 4(x^2 - 8x + 12)^3 (2x - 8)$$

$$= 8(x - 6)^3 (x - 2)^3 (x - 4)$$

which is zero when $x = 2, 4,$ or 6. We compute the value of f at each of these points and put the results in a table:

x	-10	2	4	6	10
$f(x)$	$(192)^4$	0	$(-4)^4$	0	$(32)^4$

The maximum value is $(192)^4 = 1358954496$; the maximum point is the endpoint -10. The minimum value is zero; the minimum points are the critical points 2 and 6.

Solved Exercises

8. Find a function defined on $[0, 1]$ which does not have a maximum value on $[0, 1]$.

9. Find the maximum and minimum values of $f(x) = x^4 - 4x^2 + 7$ on the interval $[-4, 2]$.

10. Let $f(x)$ be defined by

$$f(x) = \begin{cases} 5x & x \le 0 \\ -3x & x \ge 0 \end{cases}$$

Find the maximum and minimum values of f on $[-1, 1]$.

Exercises

6. Find a function defined on $[-2, 2]$ which has neither a maximum value nor a minimum value on $[-2, 2]$.

7. Find the maximum and minimum value and the maximum and minimum points for each function on the interval given:

(a) $4x^4 - 2x^2 + 1$ on $[-10, 20]$

(b) $4x^4 - 2x^2 + 1$ on $[-0.2, -0.1]$

(c) $1/(4x^4 - 2x^2 + 200,000)$ on $[-10, 20]$

(d) $(1 + x^2)/(1 - x^2)$ on $[-\frac{1}{2}, \frac{3}{4}]$

OTHER TESTS

Many maximum-minimum problems involve intervals which are not closed. A general approach to finding maxima and minima in these problems is to begin by finding the critical points, endpoints, and singular points and then test by various means to see which of these points, if any, are actually maxima and minima. Although we can give a few specific tests, it is best to use your common sense, as developed by practice, together with rough graphs. We leave it to the reader to write out formal justifications for the following tests, which are based on the ideas of Sections 2-2 and 2-3.

TURNING POINT TEST

Suppose that f is continuous on an open interval (a, b), that f' is continuous everywhere on (a, b), and that $f'(x)$ has exactly one root x_0 on (a, b). Choose points $x_1 < x_0 < x_2$ in (a, b).

1. If $f'(x_1) < 0, f'(x_2) > 0$, then x_0 is a minimum point.
2. If $f'(x_1) > 0, f'(x_2) < 0$, then x_0 is a maximum point.
3. If $f'(x_1)$ and $f'(x_2)$ have the same sign, then x_0 is neither a maximum nor a minimum point.

SECOND DERIVATIVE TEST

Suppose that f is differentiable on an open interval (a, b), that f' is continuous everywhere on (a, b), and that $f'(x)$ has exactly one root x_0 on (a, b).

1. If $f''(x_0) > 0$, then x_0 is a minimum point.
2. If $f''(x_0) < 0$, then x_0 is a maximum point.
3. If $f''(x_0) = 0$ or $f''(x_0)$ does not exist, the test is inconclusive.

There are similar statements for half-open or closed intervals; in them, we require f to be continuous at the endpoints.

Worked Example 4 Find maximum and minimum points and values for $y = x^4 - x^2$ on $[-1, \infty)$.

Solution There are no singular points, -1 is the only endpoint, and the critical points are the solutions of

$$0 = 4x^3 - 2x = 2x(2x^2 - 1)$$

The solutions of this equation are 0, $\sqrt{\frac{1}{2}}$ and $-\sqrt{\frac{1}{2}}$, all of which lie in the interval $[-1, \infty)$. We make a table of values of the function

x	-1	$-\sqrt{\frac{1}{2}}$	0	$\sqrt{\frac{1}{2}}$
$x^4 - x^2$	0	$-\frac{1}{4}$	0	$-\frac{1}{4}$

Fig. 3-1-3 The graph of $y = x^4 - x^2$ on $[-1, \infty)$.

and its derivative

x	-1	$-\frac{1}{2}$	0	$\frac{1}{2}$	1
$4x^3 - 2x$	-2	$\frac{1}{2}$	0	$-\frac{1}{2}$	2

and draw a rough graph (Fig. 3-1-3).

We see that the critical points $-\sqrt{\frac{1}{2}}$ and $\sqrt{\frac{1}{2}}$ are minimum points; the minimum value is $-\frac{1}{4}$. The endpoint -1 and the critical point zero are not maximum or minimum points since $x^4 - x^2$ takes on values both greater and less than zero on the interval $[-1, \infty)$.

Solved Exercises

11. Find maximum and minimum points and values for $-3x^2 + 2x + 1$ on $(-\infty, \infty)$.

12. Let $f(x) = px + (q/x)$, where p and q are nonzero numbers. Find the maxima and minima of f on $(0, \infty)$.

13. Find the maxima and minima of $x^3 - 3x + 5$ on $(-3, \frac{3}{2})$.

Exercises

8. Find the maximum and minimum points and values for each function on the given interval:
   (a) $2x^3 - 5x + 2$ on $[1, \infty)$ (b) $x^3 - 6x + 3$ on $(-\infty, \infty)$
   (c) $x^2 + 2/x$ on $(0, 5]$

9. Find the maximum and minimum points for each of the following:
   (a) $x^2 + 1/x$ on $(-\infty, 0)$ (b) $-x^3 + 5x + 4$ on $(-2, \frac{5}{3})$
   (c) $(x^2 + 2x + 3)/(x^2 + 5)$ on $(-2, 2]$

Problems for Section 3-1

1. Find the critical points, endpoints, singular points, global maximum and minimum points, and maximum and minimum values for each of the following functions on the designated interval:
   (a) $7x^2 + 2x + 4$ on $[-1, 1]$
   (b) $7x^2 + 2x + 4$ on $(0, \infty)$
   (c) $7x^2 + 2x + 4$ on $[-4, 2)$
   (d) $x^3 + 6x^2 - 12x + 7$ on $(-\infty, \infty)$
   (e) $x^3 + 6x^2 - 12x + 7$ on $(-2, 6]$
   (f) $x^3 + 6x^2 - 12x + 7$ on $[-2, 1)$
   (g) $x^3 + 6x^2 - 12x + 7$ on $[-4, 4]$

2. Find the critical points, endpoints, singular points, and maximum and minimum points for each function on the given interval:
   (a) $f(x) = x^3 - 3x^2 + 3x + 1$ on $(-\infty, \infty)$
   (b) $f(x) = x^3 + 3x^2 - 3x + 1$ on $[-1, 2]$
   (c) $f(x) = \begin{cases} x^2 + 1 & x \le 1 \\ -x^3 + 3 & x \ge 1 \end{cases}$ on $(-6, 1)$
   (d) f as in part (c) on $[-2, 4]$
   (e) f as in part (c) on $(-\infty, \infty)$
   (f) $x^3/(1 + x^2)$ on $(-\infty, \infty)$
   (g) $x^n/(1 + x^2)$ on $(-\infty, \infty)$ (n a positive integer)

3. For each of the functions sketched in Fig. 3-1-4 find the critical points, singular points, endpoints, and maximum and minimum points.

(a)

(b)

(c)

(d)

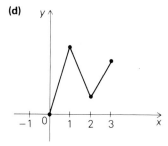

Fig. 3-1-4 Find the critical points, singular points, and maximum and minimum points.

4. Find the maximum *value* of the given function on the designated interval:
   (a) $f(x) = x^3 - x;$ $[-2, 3]$
   (b) $f(x) = x^4 + 8x^3 + 3;$ $[-1, 1]$
   (c) $f(x) = -x/(1 + x^2);$ $[-1, 6]$
   (d) $f(x) = (x^3 - 1)/(x^2 + 1);$ $[-10, 10]$
   (e) $f(x) = (x^2 + 1)/(x^4 + 1);$ $(-\infty, \infty)$

5. Find the minimum *value* of each function in Problem 4 on the designated interval.

6. Find a function defined on $[-3, -1]$ which is continuous at -3 and -1, has a maximum value on $[-3, -1]$, but has no minimum value on $[-3, -1]$.

★7. Let f and g be defined on I. Under what conditions is the maximum value of $f + g$ on I equal to the sum of the maximum values of f and g on I?

★8. Let f be a function, $a > 0$ a positive real number. Discuss the relation between the critical points of $f(x)$, $af(x)$, $a + f(x)$, $f(ax)$, and $f(a + x)$.

★9. Let I be the set consisting of the whole numbers from 1 to 1000, and let $f(x) = 45x - x^2$. Find the maximum and minimum points and values for f on I.

10. (a) Suppose that you drive from coast to coast on Interstate Route 80 and your altitude above sea level is $f(x)$ when you are x miles from San Francisco. Discuss the critical points, singular points, endpoints, global maximum and minimum points and values, and local maximum and minimum points for $f(x)$.
    (b) Do as in part (a) for a hike to the top of Mt. Whitney, where x is the distance walked from your starting point.

★11. Suppose that f is continuous on $[a, b]$ and is concave upward on the interval (a, b). Show that the maximum point of f is an endpoint.

3-2 Maximum-Minimum Word Problems

A step-by-step procedure aids in the solution of practical maximum-minimum problems.

This section is concerned with maximum-minimum problems which are presented in words rather than in formulas. Students are often overwhelmed by such "word problems," which appear to admit no systematic means of solution. Fortunately, guidelines do exist for attacking practical maximum-minimum problems; successful use of these guidelines is facilitated most of all by lots of practice.

Goals

After studying this section, you should be able to:

Convert a word problem into a maximum-minimum problem for a function.
Use calculus to solve the maximum-minimum problem.
Interpret your answer in terms of the original word problem.

THE PROBLEMS

To illustrate a general approach to maximum-minimum problems, we will go through the solution of four sample problems step by step, following the procedures described in the ensuing paragraphs:*

1. A shepherd lives on a straight coastline and has 500 meters of fencing with which to enclose his sheep. Assuming that he uses the coastline as one side of a rectangular enclosure, what dimensions should the rectangle have in order that the sheep have the largest possible area in which to graze?†

2. Illumination from a point light source is proportional to the intensity of the source and inversely proportional to the square of the distance from the source to the point of observation. Given two point sources 10 meters apart, with one source four times as intense as the other, find the darkest point on the line segment joining the sources.

3. Given four numbers, a, b, c, d, find a number x which best approximates them in the sense that the sum of the squares of the differences between x and each of the four numbers is as small as possible.

4. Suppose that it costs $(x^2/100) + 10x$ cents to run your car for x days. Once you sell your car, it will cost you 50 cents a day to take the bus. How long should you keep the car?

For a general discussion of how to attack a problem, we enthusiastically recommend How to Solve It, *by G. Polya (Princeton University Press, Second Edition, 1957).*

†*This ancient Greek problem is a variant of a famous problem ingeniously solved by Dido, the daughter of the king of Tyre and founder of Carthage (see M. Kline,* Mathematics: A Cultural Approach, *Addison-Wesley, 1962, p. 114).*

NAMING AND IDENTIFYING VARIABLES

The first thing to do is read the problem carefully. Then ask yourself: "What is given? What is required?"

Sometimes it appears that not enough data are given. In problem 2, for instance, one may think: "Illumination is proportional to the intensity, but I'm not given the constant of proportionality, so the problem isn't workable." It turns out this is a mistake; the problem *is* workable. In fact, the answer does not depend upon the proportionality constant. (If it did, you could at least express your answer in terms of this unknown constant.) On the other hand, some of the data given in the statement of a problem may be irrelevant. You should do your best at the beginning of solving a problem to decide which data are relevant and which irrelevant.

Here, in full, is the first step in attacking a max-min problem.

STEP 1: SETTING UP THE PROBLEM

Read the problem carefully, give names to any unnamed relevant variables in the problem, and note any relations among the variables.

Draw a figure, if one is appropriate.

Identify the quantity to be maximized or minimized.

Make sure that the relevant and irrelevant information is clearly distinguished.

Worked Example 1 Carry out step 1 for problem 1.

Solution We draw a picture (Fig. 3-2-1). Let l and w denote the length and width of the rectangle and let A be the area enclosed. The relations are $lw = A$ and $l + 2w = 500$ (since there are 500 meters of fencing available). We want to maximize A.

Fig. 3-2-1 For which shape is A largest?

Solved Exercise 1. Carry out step 1 for problems 2, 3, and 4.

Exercise 1. Carry out step 1 for problems (a) through (d):

(a) Of all rectangles with area 1, which has the smallest perimeter?

(b) Find the point on the arc of the parabola $y = x^2$ for $0 \le x \le 1$ which is nearest to the point $(0, 1)$. [*Hint:* Consider the *square* of the distance between points.]

(c) Two point masses which are a fixed distance apart attract one another with a force which is proportional to the product of the masses. Assuming that the sum of the two masses is M, what must the individual masses be so that the force of attraction is as large as possible?

(d) Ten miles from home you remember that you left the water running, which is costing you 10 cents an hour. Driving home at speed s miles per hour costs you $6 + (s/10)$ cents per mile. At what speed should you drive to minimize the total cost of gas and water?

TRANSLATION TO A CALCULUS PROBLEM

Having set up the problem, we are ready to apply the methods of calculus.

STEP 2: SOLVING THE PROBLEM

Write the quantity to be maximized or minimized as a function of one of the other variables in the problem. (This is usually done by expressing all other variables in terms of the one chosen.)

Note any restrictions on the chosen variable.

Find the maxima and minima by the methods of Section 3-1. Or, if the problem is not of the standard type, use any other relevant principles of calculus, such as those in Chapters 1 and 2.

State the answer in words.

The main thing to be mastered in word problems is the technique of translating words into relevant mathematical symbols to which the tools of calculus can be applied. Once the calculus work is done, the answer can then be translated back into the terms of the original word problem.

Worked Example 2   Carry out step 2 for problem 1.

Solution   We want to maximize A, so we write it as a function of l or w; we choose w. Now $l + 2w = 500$, so $l = 500 - 2w$. Thus $A = lw = (500 - 2w)w = 500w - 2w^2$.

The restriction on w is that $0 \le w \le 250$. (Clearly, only a nonnegative w can be meaningful, and w cannot be more than 250 or else l would be negative.)

To maximize $A = 500w - 2w^2$ on $[0, 250]$, we compute $dA/dw = 500 - 4w$, which is zero if $w = 125$. Since $d^2 A/dw^2 = -4$ for all w, the second derivative test (p. 156) tells us that 125 is a maximum point. Hence the maximum occurs when $w = 125$ and $l = 250$.

The rectangle should be 250 meters long in the direction parallel to the coastline and 125 meters in the direction perpendicular to the coastline in order to enclose the maximal area.

Solved Exercise   2. Carry out step 2 for problems 2, 3, and 4.

Exercise   2. Carry out step 2 for parts (a) through (d) of Exercise 1.

THE ROLE OF GUESSWORK

In the process of doing a word problem, it is useful to ask general questions like, "Can I guess any properties of the answer? Is the answer reasonable?" Sometimes a clever or educated guess can carry one surprisingly far toward the solution of a problem.

For instance, consider the problem "Find the triangle with perimeter 1 which has the greatest area." In the statement of the problem, all three sides of the triangle enter in the same way—there is nothing to single out any side as special. Therefore we guess that the answer must have the three sides equal; that is, the triangle should be equilateral. This is, in fact, correct. Such reasoning must be used with care: if we ask for the triangle with perimeter 1 and the *least* area, the answer is *not* an equilateral triangle; thus, reasoning by symmetry must often be supplemented by a more detailed analysis.

In problems 1 and 3, the answers have as much symmetry as the data, and indeed these answers might have been guessed before any calculation had been done. In problem 2, most people would have a hard time guessing the answer, but at least one can observe that the answer finally obtained has the darkest point nearer to the weaker of the two sources, which is reasonable.

Worked Example 3 Of all rectangles inscribed in a circle of radius 1, guess which has the largest area; the largest perimeter.

Solution Since the length and width enter symmetrically into the formulas for area and perimeter, we may guess that the maximum of both area and perimeter occurs when the rectangle is a square of side $\sqrt{2}$.

Solved Exercises In each example, try to guess the answer, or some part of the answer, by using some symmetry of the data.

3. Of all rectangles of area 1, which has the smallest perimeter?

4. Of all geometric figures with perimeter 1, which has the greatest area?

5. What is the answer in problem 2 if the two intensities are equal?

Exercises 3. Guess the answer in problem 2 if one of the intensities is eight times the other.

4. Of all *right* triangles of area 1, guess which one has the shortest perimeter. Which one, if any, has the longest perimeter?

5. One hundred feet of fencing is to be used to enclose two pens, one square and one triangular. What dimensions should the pens be to have the largest possible area?

6. Use calculus to show that the answer in Worked Example 3 is correct.

FURTHER MAXIMUM-MINIMUM PROBLEMS

We can now solve word problems by bringing together all of the preceding techniques.

Worked Example 4   Find the dimensions of a rectangular box of minimum cost if the manufacturing costs are 10 cents per square meter on the bottom, 5 cents per square meter on the sides, and 7 cents per square meter on the top. The volume is to be 2 cubic meters and the height is to be 1 meter.

Solution   Let the dimensions of the base be l and w and the height be 1. If the total cost is C, then

$$C = 10lw + 7lw + 2 \cdot 5 \cdot (l \cdot 1 + w \cdot 1)$$
$$= 17lw + 10l + 10w$$

Now $l \cdot w \cdot 1 = 2$ is the total volume. Eliminating w,

$$C = 34 + 10l + \frac{20}{l}$$

We are to minimize C. Let $f(l) = 34 + 10l + (20/l)$ on $(0, \infty)$. Then

$$f'(l) = 10 - \frac{20}{l^2}$$

which is zero when $l = \sqrt{2}$. (We are concerned only with $l > 0$.) Since $f''(l) = 40/l^3$ is positive at $l = \sqrt{2}$, $l = \sqrt{2}$ is a local minimum point. Since this is the only critical point, it is also the global minimum. Thus the dimensions of minimum cost are $\sqrt{2}$ by $\sqrt{2}$ by 1.

Worked Example 5   The terms marginal revenue and marginal cost were defined in Worked Example 8 of Section 1-3. Prove that at a production level x_0 that maximizes profit, the marginal revenue equals the marginal cost; see Figure 3-2-2.

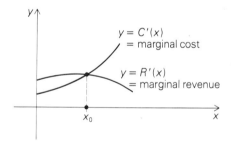

Fig. 3-2-2   The production level x_0 that maximizes profit has $C'(x_0) = R'(x_0)$.

Solution   We have $P(x) = R(x) - C(x)$ from Worked Example 8 of Section 1-3. At a maximum point x_0 for P, we must have $P'(x_0) = 0$ by the first derivative test. Therefore, $R'(x_0) - C'(x_0) = 0$ or $R'(x_0) = C'(x_0)$ as required.

Solved Exercises   6. The cost of running a boat is $10v^3$ dollars per mile where v is its speed in still water. What is the most economical speed to run the boat upstream against a current of 5 miles per hour?

7. Determine the number of units of a commodity that should be produced to maximize the profit when the cost and revenue functions are given by $C(x) = 800 + 30x + 0.02x^2$, $R(x) = 50x - 0.01x^2$.

Exercises

7. A rectangular box, open at the top, is to be constructed from a rectangular sheet of cardboard 50 centimeters by 80 centimeters by cutting out equal squares in the corners and folding up the sides. What size squares should be cut out for the container to have maximum volume?

8. The stiffness S of a wooden beam of rectangular cross-section is proportional to its breadth and the cube of its thickness. Find the stiffest rectangular beam that can be cut from a circular log of diameter d.

9. Determine the number of units of a commodity that should be produced to maximize the profit for the following cost and revenue functions: $C(x) = 360 + 80x + .002x^2 + .00001x^3$, $R(x) = 100x - .0001x^2$.

Problems for Section 3-2

1. Given n numbers, $a_1, \ldots, a_n$, find a number x which best approximates them in the sense that the sum of the squares of the differences between x and the n numbers is as small as possible.

2. (a) A can is to be made to hold 1 liter (= 1000 cubic centimeters) of oil. If the can is in the shape of a circular cylinder, what should the radius and height be in order that the surface area of the can (top and bottom and curved part) be as small as possible?

   (b) What is the answer in part (a) if the total capacity is to be V cubic centimeters?

   (c) Suppose that the surface area of a can is fixed at A square centimeters. What should the dimensions be so that the capacity is maximized?

   (d) Why aren't cans always made in the shape you find in parts (a), (b), and (c)?

3. One positive number plus the square of another equals 48. Choose the numbers so that their product is as large as possible.

4. A window in the shape of a rectangle with a semicircle on the top is to be made with a perimeter of 4 meters. What is the largest possible area for such a window?

5. A forest can support up to 10,000 rabbits. If there are x rabbits in the forest, each female can be expected to give birth to $\frac{1}{1000}(10{,}000 - x)$ bunnies in a year. What total population will give rise to the greatest number of newborn bunnies in a year? (Assume that exactly half the rabbits are female, and ignore the fact that the bunnies may themselves give birth to more young during the year, and remember that the total population including new bunnies is not to exceed 10,000).

6. (a) In Fig. 3-2-3, for which value of y does the line segment PQ have the shortest length? Express your answer in terms of a and b. [*Hint:* Minimize the square of the length.]

   (b) What is the length of the longest ladder which can be slid along the floor around the corner from a corridor of width a to a corridor of width b?

Fig. 3-2-3   The "ladder" PQ just fits into the corner of the corridor.

7. One thousand feet of fencing is to be used to surround two areas, one square and one circular. What should the size of each area be in order that the total area be (a) as large as possible and (b) as small as possible?

★8. (This minimization problem involves no calculus.) You ran out of milk the day before your weekly visit to the supermarket, and you must pick up a container at the corner grocery store. At the corner store, a quart costs Q cents and a half gallon costs G cents. At the supermarket, milk costs q cents a quart and g cents a half gallon.

   (a) If $Q = 45$, $G = 80$, $q = 38$, and $g = 65$, what size container should you buy at the corner grocery to minimize your eventual milk expense? (Assume that a quart will get you through the day.)

   (b) Under what conditions on Q, G, q, and g should you buy a half gallon today?

9. The U.S. Post Office will accept rectangular boxes only if the sum of the length and girth (twice the width plus twice the height) is less than 72 inches. What are the dimensions of the box of maximum volume the Post Office will accept? (You may assume, by symmetry, that the width and height are equal.)

10. Three equal light sources are spaced at $x = 0, 1, 2$, along a line. At what points between the sources is the total illumination least? See Problem 2, p. 159. You should get a cubic equation for the square of the position of the desired point. Solve the equation numerically by using the method of bisection; see Solved Exercise 8 and Problem 14, Section 2-1.

★11. What happens in Problem 10 if there are four light sources instead of three? Can you guess the correct answer by using a symmetry argument without doing any calculation?

12. Find the point or points on the arc of the parabola $y = x^2$ for $0 \le x \le 1$ which are nearest to the point $(0, q)$. Express your answer in terms of q. [*Hint:* Minimize the *square* of the distance between points.]

13. In Worked Example 1, suppose that we allow the fencing to assume any shape, not necessarily rectangular, but still with one side along the shore requiring no fencing. What shape gives the maximum area? [*Hint:* Read the reference in the second footnote on p. 159.] A formal proof is not required.

14. A conical dunce cap is to be made from a circular piece of paper of circumference c by cutting out a pie-shaped piece whose curved outer edge has length l. What should l be so that the resulting dunce cap has maximum volume?

15. If the cost of producing x calculators is $C(x) = 100 + 10x + 0.01x^2$ and the price per calculator at production level x is $P(x) = 26 - 0.1x$ (this is called the *demand* equation), what production level should be set in order to maximize profit?

3-3 The Mean Value Theorem

If the derivative of a function is everywhere zero, then the function is constant.

The mean value theorem is a technical result whose applications are more important than the theorem itself. We begin this section with a statement of the theorem, proceed immediately to the applications, and conclude with a proof which uses the idea of global maxima and minima.

Goals

After studying this section, you should be able to:

Estimate difference quotients of a function, given information about the derivative.

Give physical and geometric interpretations of the mean value theorem.

APPLICATIONS OF THE MEAN VALUE THEOREM

The mean value theorem is, like the intermediate value and extreme value theorems, an *existence* theorem. It asserts the existence of a point in an interval where a function has a particular behavior, but it does not tell you how to find the point.

Theorem 3 Mean Value Theorem. *Suppose that the function f is continuous on the closed interval $[a, b]$ and differentiable on the open interval (a, b). Then there is a point x_0 in the open interval (a, b) at which $f'(x_0) = [f(b) - f(a)]/(b - a)$.*

Fig. 3-3-1 The slope $[f(b) - f(a)]/(b - a)$ of the secant line l_1 is equal to the slope of the tangent line l_2.

In physical terms, the mean value theorem says that the average velocity of a moving object during an interval of time is equal to the instantaneous velocity at some moment in the interval. Geometrically, the theorem says that a secant line drawn through two points on a smooth graph is parallel to the tangent line at some intermediate point on the curve. There may be more than one such point, as in Fig. 3-3-1. Consideration of these physical and geometric interpretations should make the theorem believable.

We will prove the mean value theorem at the end of this section. For now, we will concentrate on some applications. Our first corollary tells us that if we know something about $f'(x)$ for all x in $[a, b]$, then we can conclude something about the relation between values of $f(x)$ at different points in $[a, b]$.

Corollary 1 *Let f be differentiable on (a, b) [and continuous on $[a, b]$]. Suppose that, for all x in the open interval (a, b), the derivative $f'(x)$ belongs to a certain set S of real numbers. Then, for any two distinct points x_1 and x_2 in (a, b) [in $[a, b]$], the difference quotient*

$$\frac{f(x_2) - f(x_1)}{x_2 - x_1}$$

belongs to S as well.

Proof The difference quotient stays the same if we exchange x_1 and x_2, so we may assume that $x_1 < x_2$. The interval $[x_1, x_2]$ is contained in (a, b) [in $[a, b]$]. Since f is differentiable on (a, b) [and continuous on $[a, b]$], it is continuous on $[x_1, x_2]$ and differentiable on (x_1, x_2). By the mean value theorem, applied to f on $[x_1, x_2]$, there is a number x_0 in (x_1, x_2) such that $[f(x_2) - f(x_1)]/(x_2 - x_1) = f'(x_0)$. But (x_1, x_2) is contained in (a, b), so $x_0 \in (a, b)$. By hypothesis, $f'(x_0)$ must belong to S; hence so does $[f(x_2) - f(x_1)]/(x_2 - x_1)$.

Worked Example 1 Suppose that f is differentiable on the whole real line and that $f'(x)$ is constant. Use Corollary 1 to prove that f is linear.

Solution Let m be the constant value of f' and let S be the set whose only member is m. For any x, we may apply Corollary 1 with $x_1 = 0$ and $x_2 = x$ to conclude that $[f(x) - f(0)]/(x - 0)$ belongs to S, which means that $[f(x) - f(0)]/(x - 0) = m$. But then $f(x) = mx + f(0)$ for all x, so f is linear.

Worked Example 2 Let f be continuous on $[1, 3]$ and differentiable on $(1, 3)$. Suppose that, for all x in $(1, 3)$, $1 \leq f'(x) \leq 2$. Prove that $2 \leq f(3) - f(1) \leq 4$.

Solution Apply Corollary 1, with S equal to the interval $[1, 2]$. Then we have $1 \leq [f(3) - f(1)]/(3 - 1) \leq 2$, and so $2 \leq f(3) - f(1) \leq 4$.

Corollary 2 *Suppose that $f'(x) = 0$ for all x in some open interval (a, b). Then f is constant on (a, b).*

Proof Let $x_1 < x_2$ be any two points in (a, b). Corollary 1 applies with S the set consisting only of the number zero. Thus we have $[f(x_2) - f(x_1)]/(x_2 - x_1) = 0$, or $f(x_2) = f(x_1)$, so f takes the same value at all points of (a, b).

Worked Example 3 Let $f(x) = (d/dx)|x|$. (a) Find $f'(x)$. (b) What does Corollary 2 tell you about f? What does it not tell you?

Solution (a) Since $|x|$ is linear on $(-\infty, 0)$ and $(0, \infty)$, its second derivative $d^2|x|/dx^2 = f'(x)$ is identically zero for all $x \neq 0$.

(b) By Corollary 2, f is constant *on any open interval on which it is differentiable*. It follows that f is constant on $(-\infty, 0)$ and $(0, \infty)$. The corollary does *not* say that f is constant on $(-\infty, \infty)$. In fact, $f(-2) = -1$, while $f(2) = +1$.

Finally, we can derive from Corollary 2 the fact that two antiderivatives of a function differ by a constant, a principle which was first used in Section 1-4.

Corollary 3 *Let $F(x)$ and $G(x)$ be functions such that $F'(x) = G'(x)$ for all x in an open interval (a, b). Then there is a constant C such that $F(x) = G(x) + C$ for all x in (a, b).*

Proof We apply Corollary 2 to the difference $F(x) - G(x)$. Since $(d/dx)[F(x) - G(x)] = F'(x) - G'(x) = 0$ for all x in (a, b), $F(x) - G(x)$ is equal to a constant C, so $F(x) = G(x) + C$.

Worked Example 4 Suppose that $F'(x) = x$ for all x and that $F(3) = 2$. What is $F(x)$?

Solution Let $G(x) = \frac{1}{2}x^2$. Then $G'(x) = x = F'(x)$, so $F(x) = G(x) + C = \frac{1}{2}x^2 + C$. To evaluate C, set $x = 3$: $2 = F(3) = \frac{1}{2}(3^2) + C = \frac{9}{2} + C$. Thus $C = 2 - \frac{9}{2} = -\frac{5}{2}$ and $F(x) = \frac{1}{2}x^2 - \frac{5}{2}$.

Solved Exercises

1. Let $f(x) = x^3$ on the interval $[-2, 3]$. Find explicitly the value(s) of x_0 whose existence is guaranteed by the mean value theorem.

2. If, in Corollary 1, the set S is taken to be the interval $(0, \infty)$, the result is a theorem which has already been proved. What theorem is it?

3. The velocity of a train is kept between 40 and 50 kilometers per hour during a trip of 200 kilometers. What can you say about the duration of the trip?

4. Suppose that $F'(x) = -(1/x^2)$ for all $x \neq 0$. Is $F(x) = 1/x + C$, where C is a constant?

Exercises

1. Directly verify the validity of the mean value theorem for $f(x) = x^2 - x + 1$ on $[-1, 2]$ by finding the point(s) x_0. Sketch.

2. Suppose that f is continuous on $[0, \frac{1}{2}]$ and $0.3 \leq f'(x) < 1$ for $0 < x < \frac{1}{2}$. Prove that $0.15 \leq [f(\frac{1}{2}) - f(0)] < 0.5$.

3. Suppose that $f'(x) = x^2$ and $f(1) = 0$. What is $f(x)$?

4. Suppose that an object lies at $x = 4$ when $t = 0$ and that the velocity dx/dt is 35 with a possible error of ± 1, for all t in $[0, 2]$. What can you say about the object's position when $t = 2$?

(optional) **PROOF OF THE MEAN VALUE THEOREM**

Our proof of the mean value theorem will use two results from Section 3-1 which we recall here:

1. If x_0 lies in the open interval (a, b) and is a maximum or minimum point for a function f on an interval $[a, b]$, and if f is differentiable at x_0, then $f'(x_0) = 0$ (Theorem 1, p. 153).

2. If f is continuous on a closed interval $[a, b]$, then f has a maximum and a minimum point in $[a, b]$ (Theorem 2, p. 154, extreme value theorem).

Our proof of the mean value theorem proceeds in three steps.

Fig. 3-3-2 Rolle's theorem: If f is zero at the ends of the interval, its graph must have a horizontal tangent line somewhere between.

Step 1 (Rolle's Theorem*) Let f be continuous on $[a, b]$ and differentiable on (a, b), and assume that $f(a) = f(b) = 0$. Then there is a point x_0 in (a, b) at which $f'(x_0) = 0$.

Proof If $f(x) = 0$ for all x in $[a, b]$, we can choose any x_0 in (a, b). So assume that f is not everywhere zero. By result 2 above, f has a maximum point x_1 and a minimum point x_2. Since f is zero at the ends of the interval but is not identically zero, at least one of x_1, x_2 lies in (a, b). Let x_0 be this point. By result 1, $f'(x_0) = 0$.

Rolle's theorem has a simple geometric interpretation (see Fig. 3-3-2).

Step 2 (Horserace Theorem) Suppose that f_1 and f_2 are continuous on $[a, b]$ and differentiable on (a, b), and assume that $f_1(a) = f_2(a)$ and $f_1(b) = f_2(b)$. Then there exists a point x_0 in (a, b) such that $f_1'(x_0) = f_2'(x_0)$.

Proof Let $f(x) = f_1(x) - f_2(x)$. Since f_1 and f_2 are differentiable on (a, b) and continuous on $[a, b]$, so is f. By assumption, $f(a) = f(b) = 0$, so from step 1, $f'(x_0) = 0$ for some x_0 in (a, b). Thus $f_1'(x_0) = f_2'(x_0)$ as required.

We call this the "horserace theorem" because it has the following interpretation. Suppose that two horses run a race starting together and ending in a tie. Then, at some time during the race, they must have had the same velocity.

Step 3: We apply step 2 to a given function f and the linear function l that matches f at its endpoints, namely,

$$l(x) = f(a) + (x - a) \left[\frac{f(b) - f(a)}{b - a} \right]$$

Note that $l(a) = f(a)$, $l(b) = f(b)$, and $l'(x) = [f(b) - f(a)]/(b - a)$. By step 2, $f'(x_0) = l'(x_0) = [f(b) - f(a)]/(b - a)$ for some point x_0 in (a, b).

---

*Michel Rolle (1652–1719) (pronounced "roll") was actually best known for his attacks on the calculus. He was one of the critics of the newly founded theory of Newton and Leibniz. It is an irony of history that he has become so famous for "Rolle's theorem" when he did not even prove the theorem but used it only as a remark concerning the location of roots of polynomials. (See D. E. Smith, Source Book in Mathematics, Dover, 1929, pp. 251–260, for further information.)

(optional)

Thus we have proved the mean value theorem:

Mean Value Theorem *If f is continuous on* $[a, b]$ *and is differentiable on* (a, b), *then there is a point* x_0 *in* (a, b) *at which*

$$f'(x_0) = \frac{f(b) - f(a)}{b - a}$$

Solved Exercises ★5. Let $f(x) = x^4 - 9x^3 + 26x^2 - 24x$. Note that $f(0) = 0$ and $f(2) = 0$. Show without calculating that $4x^3 - 27x^2 + 52x - 24$ has a root somewhere strictly between 0 and 2.

★6. Suppose that f is a differentiable function such that $f(0) = 0$ and $f(1) = 1$. Show that $f'(x_0) = 2x_0$ for some x_0 in $(0, 1)$.

Exercises ★5. Suppose that the horses in the race described above cross the finish line with equal velocities. Must they have had the same acceleration at some time during the race?

★6. Let $f(x) = |x| - 1$. Then $f(-1) = f(1) = 0$, but $f'(x)$ is never equal to zero on $[-1, 1]$. Does this contradict Rolle's theorem? Explain.

Problems for Section 3-3

1. Suppose that $(d/dx)[f(x) - 2g(x)] = 0$. What can you say about the relationship between f and g?

★2. Suppose that f and g are continuous on $[a, b]$ and that f' and g' are continuous on (a, b). Assume that

$$f(a) = g(a) \quad \text{and} \quad f(b) = g(b)$$

Prove that there is a number c in (a, b) such that the line tangent to the graph of f at $(c, f(c))$ is parallel to the line tangent to the graph of g at $(c, g(c))$.

3. Let $f(x) = x^7 - x^5 - x^4 + 2x + 1$. Prove that the graph of f has slope 2 somewhere between -1 and 1.

4. Find the antiderivatives of each of the following:
   (a) $f(x) = \frac{1}{2}x - 4x^2 + 21$
   (b) $f(x) = 6x^5 - 12x^3 + 15x - 11$
   (c) $f(x) = x^4 + 7x^3 + x^2 + x + 1$
   (d) $f(x) = (1/x^2) + 2x$
   (e) $f(x) = 2x(x^2 + 7)^{100}$

5. Find the antiderivative $F(x)$ for the given function $f(x)$ satisfying the given condition:
   (a) $f(x) = 2x^4$; $F(1) = 2$
   (b) $f(x) = 4 - x$; $F(2) = 1$
   (c) $f(x) = x^4 + x^3 + x^2$; $F(1) = 1$
   (d) $f(x) = 1/x^5$; $F(1) = 3$

6. If $f''(x) = 0$ on (a, b), what can you say about f?

7. (a) Let $f(x) = x^5 + 8x^4 - 5x^2 + 15$. Prove that somewhere between -1 and 0 the tangent line to the graph of f has slope -2.
   (b) Let $f(x) = 5x^4 + 9x^3 - 11x^2 + 10$. Prove that the graph of f has slope 9 somewhere between -1 and 1.

★8. Let f be a polynomial. Suppose that f has a double root at a and at b. Show that $f'(x)$ has at least three roots in $[a, b]$.

9. Let f be twice differentiable on (a, b) and suppose f vanishes at three distinct points in (a, b). Prove that there is a point x_0 in (a, b) at which $f''(x_0) = 0$.

10. The fuel consumption of an automobile varies between 17 and 23 miles per gallon, according to the conditions of driving. Let $f(x)$ be the number of gallons of fuel in the tank after x miles have been driven. If $f(100) = 15$, give upper and lower estimates for $f(200)$.

★11. The coyote population in Nevada was the same at three consecutive times t_1, t_2, t_3. Assume that the population $N(t)$ is a differentiable function of time t and is nonconstant on $[t_1, t_2]$ and on $[t_2, t_3]$. Establish by virtue of the mean value theorem the existence of two times T, T^* in (t_1, t_2) and (t_2, t_3), respectively, for which the coyote population decreased. [*Hint:* If $dN/dt > 0$ on $t_1 < t < t_2$, then dN/dt cannot equal $[N(t_2) - N(t_1)]/(t_2 - t_1)$.]

★12. Use the mean value theorem to prove theorem 5, Section 2-2, without the hypothesis that f' be continuous.

Review Problems for Chapter 3

1. The material for the top, bottom, and lateral surface of a tin can costs $\frac{1}{20}$ of a cent per square centimeter. The cost of sealing the top and bottom to the lateral surface is p cents times the total length in centimeters of the rims (see Fig. 3-R-1) which are to be sealed. Find (in terms of p) the dimensions of the cheapest can which will hold a volume of V cubic centimeters. Express your answer as the solution of a cubic equation; do not solve it. (Notice that the case $p = 0$ was Problem 2 in Section 3-2.).

Rims

Fig. 3-R-1 Minimize the cost of making the can.

2. Find the maximum and minimum *values* of each function on the designated interval:
   (a) $f(x) = 3x^3 + x^2 - x + 5$; $[-2, 2]$
   (b) $f(x) = (5x^4 + 1)/(x^2 + 1)$; $[-1, 1]$
   (c) $f(x) = x^3/(x^2 - 4)$; $[-1, 1]$
   (d) $f(x) = x^3/(x^2 - 4)$; $[-2, 2]$
   (e) $f(x) = -x^4 + 8x^2 + 2$; $(-\infty, \infty)$

3. Sketch the graph of each of the functions in Problem 2 on the interval given.

4. A box company paints its open-top square-bottom boxes white on the bottom and two sides and red on the remaining two sides. If red paint costs 50% more than white paint, what are the dimensions of the box with volume V which costs least to paint? In what sense is the "shape" of this box independent of V?

5. Find the maximum area an isosceles triangle can have if each of its equal sides has a length of 10 centimeters.

6. Find the number x which best approximates 1, 2, 3, and 5 in the sense that the sum of the *fourth* powers of the differences between x and each number is minimized. Compute x to within 0.1 by using the bisection method. (See Solved Exercise 8, Section 2-1.)

★ 7. Prove that, given any n numbers $a_1, \ldots, a_n$, there is a uniquely determined number x for which the sum of the fourth powers of the differences between x and the a_i's is minimized. [*Hint:* Use the second derivative.]

8. A wooden picture frame is to be 2 inches wide on top and bottom and 1 inch wide on the sides. Assuming the cost of a frame to be proportional to its front surface area, find the dimensions of the cheapest frame which will surround an area of 100 square inches.

9. At time t, a rectangle has sides given by $l(t) = 2 + t$ and $w(t) = 1 + t^2$, for $-1 \le t \le 1$. (a) When does this rectangle have minimum area? (b) When is the area shrinking the fastest?

10. Find the dimensions of the right circular cylinder of greatest volume that can be inscribed in a given right circular cone. Express your answer in terms of the height h of the cone and the radius r of the base of the cone.

11. We quote from V. Belevitch, *Classical Network Theory* (Holden-Day, 1968, p. 159): "When a generator of e.m.f. e and internal *positive* resistance R_i is connected to a positive load resistance R, the active power dissipated in the load

    $$w = |e|^2 R/(R + R_i)^2$$

    is maximum with respect to R for $R = R_i$, that is, when the load resistance is *matched* to the internal resistance of the generator."

    Verify this statement. (You do not need to know any electrical engineering.)

12. If $f'''(x) = 0$ on $(-\infty, \infty)$, show that there are constants A, B, and C such that $f(x) = Ax^2 + Bx + C$.

★13. Let f be differentiable function on $(0, \infty)$ such that all the tangent lines to the graph of f pass through the origin. Prove that f is linear. [*Hint:* Consider $f(x)/x$.]

★14. Let f be continuous on $[3, 5]$ and differentiable on $(3, 5)$, and suppose that $f(3) = 6$ and $f(5) = 10$. Prove that, for some x_0 in the interval $(3, 5)$, the tangent line to the graph of f at x_0 passes through the origin. [*Hint:* Consider $f(x)/x$.] Illustrate your result with a sketch.

15. A rectangular box with square bottom is to have fixed volume 648 cubic centimeters. The top and bottom are to be padded with foam and pressboard, which costs three times as much per square centimeter as the fiberboard used for the sides. Which dimensions produce the box of least cost?

16. A homeowner plans to construct a rectangular vegetable garden with a fence around it. The garden requires 800 square feet, and one edge is on the property line. Three sides of the fence will be chain-link costing $2.00 per linear foot, while the property line side will be inexpensive screening costing $.50 per linear foot. Which dimensions will cost the least?

17. A rental agency for compact cars rents 96 cars each day for $16.00 per day. Each dollar increase in the rental rate results in four fewer cars being rented.
    (a) How should the rate be adjusted to maximize the income?
    (b) What is the maximum income?

18. The Smellter steel works and the Green Copper Corporation smelter are located about 40 miles apart. Particulate matter concentrations in parts per million theoretically decrease by an inverse square law, giving, for example,

$$C(x) = \frac{k}{x^2} + \frac{3k}{(40 - x)^2} \qquad (1 \le x \le 39)$$

as the concentration x miles from Smellter. This model assumes that Green emits three times more particulate matter than Smellter.
    (a) Find $C'(x)$ and all critical points of C.
    (b) Assuming you wished to build a house between Smellter and Green at the point of least particulate concentration, how far would you be from Smellter?

19. Saveway checkers have to memorize the sale prices from the previous day's newspaper advertisement. A reasonable approximation for the percentage P of the new prices memorized after t hours of checking is $P(t) = 96t - 24t^2$, $0 \le t \le 3$, on the average.
    (a) Is a checker who memorizes the whole list in 30 minutes above or below average?
    (b) What is the maximum percentage of the list memorized after three hours, on the average?

20. Persons between 30 and 75 inches in height h have average weight $W = \frac{1}{2}(h/10)^3$ lb.
    (a) What is the average weight of a person 5 feet 2 inches in height?
    (b) A second grade child grows from 48 inches to 50 inches. Use the mean value theorem to estimate his approximate weight gain and compare with a direct calculation.

21. A storage vessel for a chemical bleach mixture is manufactured by coating the inside of a thin, hollow plastic cube with fiberglass. The cube has 12-inch sides, and the coating is $\frac{1}{5}$ inch thick. Use the volume formula $V = x^3$ and the mean value theorem to approximate the volume of the fiberglass coating. [*Hint:* The fiberglass volume is $V(12) - V(11.8)$.]

22. A restaurant refrigerator has ten semicircular shelves each of radius 21 inches, aligned vertically and spaced 8 inches apart.
    (a) Find the dimensions of the box container of largest volume that will fit on the shelf. [*Hint:* Maximize (volume)^2.]
    (b) What is the percentage of the total volume of shelf space used up by these maximal box containers?
    (c) After all ten box containers are in the refrigerator, how many upright cans of tomato juice will fit inside, each 3.85 inches high by 2.1 inches in diameter?

4

The Integral

Integration, defined as a continuous summation process, is linked to differentiation by the fundamental theorem of calculus.

In everyday language, the word *integration* refers to putting things together, while *differentiation* refers to separating, or distinguishing, things.

The simplest kind of "differentiation" in mathematics is ordinary subtraction, which tells us the difference between two given numbers. The differentiation process in calculus may be thought of as telling us the difference between the values of a function at two nearby points in its domain.

By analogy, the simplest kind of "integration" in mathematics is ordinary addition. Given two or more numbers, we can put them together to obtain the number which is their sum. The integration process in calculus operates on functions, giving us a "continuous sum" of all the values of a function on an interval. This process is applicable whenever a physical quantity is the aggregate of quantities spread out over space or time. For example, the volume of a wire of variable cross-sectional area is obtained by integrating the cross-sectional area over the length of the wire, and the total electrical energy consumed in a house during a day is obtained by integrating the time-varying power consumption over the period of the day.

CONTENTS

In the first section of this chapter, we discuss in detail the operation of ordinary summation. We present a physical example of "continuous summation" which is to be represented by the integral, and we define the integration operation.

The fundamental theorem of calculus, which provides the mathematical link between differential and integral calculus, enables us to use our knowledge of antidifferentiation to calculate integrals. We present an intuitive discussion of the fundamental theorem, followed by a proof, in Section 4-2. Section 4-3 introduces the first major application of integration, namely the problem of finding the area under and between graphs of functions.

*This appendix is found in the *Student Guide*.

4-1 Definition of the Integral

The integral of a function on an interval is defined by the method of exhaustion.

We begin with some preparatory material from algebra, namely, summation notation. A physical example is given to illustrate what it is that the integral should measure and how the integral can be estimated. Finally, the integral is defined as a dividing point between two sets of real numbers.

Goals

After studying this section, you should be able to:

Manipulate expressions involving summation notation.

Find the integral of a piecewise constant function.

Estimate the integral of a continuous function by finding upper and lower sums.

SUMMATION NOTATION

In preparation for the study of the integral, we will spend some time looking at ordinary summation and, in particular, at a useful shorthand notation for sums.

Given n numbers, a_1 through a_n, we denote the sum $a_1 + a_2 + \cdots + a_n$ by

$$\sum_{i=1}^{n} a_i$$

Here Σ is the capital Greek letter *sigma,* the equivalent of the Roman S (for *sum*). We read the expression above as "the sum of the a_i, as i runs from 1 to n."

Worked Example 1 Find $\sum_{i=1}^{4} a_i$, if $a_1 = 2$, $a_2 = 3$, $a_3 = 4$, $a_4 = 6$.

Solution $\sum_{i=1}^{4} a_i = a_1 + a_2 + a_3 + a_4 = 2 + 3 + 4 + 6 = 15.$

Worked Example 2 Find $\sum_{i=1}^{4} i^2$.

Solution Here we have $a_i = i^2$, so

$$\sum_{i=1}^{4} i^2 = 1^2 + 2^2 + 3^2 + 4^2 = 1 + 4 + 9 + 16 = 30$$

The letter i is called a *dummy index.* It is used merely as a labeling device; we can replace it everywhere by any other letter without changing the value of the expression. Thus, for instance,

$$\sum_{k=1}^{n} a_k = \sum_{i=1}^{n} a_i$$

since both are equal to $a_1 + \cdots + a_n$.

We can allow the summation to start and stop at any number; for instance,

$$\sum_{i=2}^{6} b_i = b_2 + b_3 + b_4 + b_5 + b_6$$

and

$$\sum_{j=-2}^{3} c_j = c_{-2} + c_{-1} + c_0 + c_1 + c_2 + c_3$$

Worked Example 3   Find $\sum_{k=2}^{5} (k^2 - k)$.

Solution
$$\sum_{k=2}^{5} (k^2 - k) = (2^2 - 2) + (3^2 - 3) + (4^2 - 4) + (5^2 - 5)$$
$$= 2 + 6 + 12 + 20 = 40$$

SUMMATION NOTATION

To evaluate

$$\sum_{i=m}^{n} a_i$$

where $m \le n$ are integers, and a_i are real numbers,

let i take each integer value such that $m \le i \le n$—for each such i, evaluate a_i and add the resulting numbers. (There are $n - m + 1$ of them.)

We list below some general properties of the summation operation:

PROPERTIES OF SUMMATION

1. $\sum_{i=m}^{n} (a_i + b_i) = \sum_{i=m}^{n} a_i + \sum_{i=m}^{n} b_i$.
2. $\sum_{i=m}^{n} ca_i = c \sum_{i=m}^{n} a_i$.
3. If $m \le n$ and $n + 1 \le p$, then $\sum_{i=m}^{p} a_i = \sum_{i=m}^{n} a_i + \sum_{i=n+1}^{p} a_i$.
4. If $a_i = C$ for all i with $m \le i \le n$, where C is some constant, then $\sum_{i=m}^{n} a_i = C(n - m + 1)$.
5. If $a_i \le b_i$ for all i with $m \le i \le n$, then $\sum_{i=m}^{n} a_i \le \sum_{i=m}^{n} b_i$.

These are just basic properties of addition extended to sums of many numbers at a time. For instance, property 3 says that $a_m + a_{m+1} + \cdots + a_p = (a_m + \cdots + a_n) + (a_{n+1} + \cdots + a_p)$, which is a generalization of the associative law. Property 2 is a distributive law; property 1 is a commutative law. Property 4 says that repeated addition of the same number is the same as multiplication; property 5 is a generalization of the basic law of inequalities: If $a \le b$ and $c \le d$, then $a + c \le b + d$.

Another useful formula gives the sum of the first n integers:

SUM OF FIRST n INTEGERS

$$\sum_{i=1}^{n} i = \tfrac{1}{2}n(n+1)$$

To prove this formula we let $S = \sum_{i=1}^{n} i = 1 + 2 + \cdots + n$. Writing S again with the order of the terms reversed, and then adding the two sums, we have:

$$
\begin{array}{l}
S = 1 \qquad\;\; + 2 \qquad\;\; + 3 \qquad\;\; + \cdots + (n-2) + (n-1) + n \\
S = n \qquad\;\; + (n-1) + (n-2) + \cdots + 3 \qquad\;\; + 2 \qquad\;\; + 1 \\
\hline
2S = (n+1) + (n+1) + (n+1) + \cdots + (n+1)\; + (n+1)\; + (n+1)
\end{array}
$$

Since there are n terms in the sum, the right-hand side is $n(n+1)$, so $2S = n(n+1)$, and $S = \tfrac{1}{2}n(n+1)$.

For example, the sum of the first 100 integers is $\tfrac{1}{2}100 \cdot 101 = 50 \cdot 101 = 5050$.*

Worked Example 4   Find the sum $4 + 5 + 6 + \cdots + 29$.

Solution   This sum is $\sum_{i=4}^{29} i$. By summation property 3 we have $\sum_{i=1}^{29} i = \sum_{i=1}^{3} i + \sum_{i=4}^{29} i$, or $\sum_{i=4}^{29} i = \sum_{i=1}^{29} i - \sum_{i=1}^{3} i$.

Evaluating each of the sums on the right side by the formula in the preceding display, we have

$$\sum_{i=4}^{29} i = \tfrac{1}{2} \cdot 29 \cdot 30 - \tfrac{1}{2} \cdot 3 \cdot 4 = 29 \cdot 15 - 3 \cdot 2 = 435 - 6 = 429$$

Worked Example 5   Find $\sum_{j=3}^{102} (j-2)$.

Solution   We can use summation properties 1, 3, and 4 to get

$$\sum_{j=3}^{102} (j-2) = \sum_{j=3}^{102} j - \sum_{j=3}^{102} 2 = \sum_{j=1}^{102} j - \sum_{j=1}^{2} j - 2(100)$$

$$= \tfrac{1}{2}(102)(103) - 3 - 200 = 5050$$

We can also do this problem by making the substitution $i = j - 2$. As j runs from 3 to 102, i runs from 1 to 100, and we get

$$\sum_{j=3}^{102} (j-2) = \sum_{i=1}^{100} i = \tfrac{1}{2} \cdot 100 \cdot 101 = 5050$$

---

*A famous story about the mathematical giant C. F. Gauss (1777–1855) concerns a task his class had received from a demanding teacher in elementary school. They were to add up the first 100 numbers. Gauss wrote the answer 5050 on his slate immediately; he had derived $S = \tfrac{1}{2}n(n+1)$ in his head at age 10!

The second method used in Worked Example 5 may be stated as a general rule:

SUBSTITUTION OF INDEX

With the substitution $i = j + q$,

$$\sum_{j=m}^{n} a_{j+q} = \sum_{i=m+q}^{n+q} a_i$$

Worked Example 6 illustrates another general rule.

Worked Example 6 Show that $\sum_{i=1}^{n} [(i+1)^3 - i^3] = (n+1)^3 - 1$.

Solution The easiest way to do this is by writing out the sum:

$$\sum_{i=1}^{n} [(i+1)^3 - i^3] = [2^3 - 1^3] + [3^3 - 2^3] + [4^3 - 3^3]$$
$$+ \cdots + [n^3 - (n-1)^3] + [(n+1)^3 - n^3]$$

We can cancel 2^3 with -2^3, 3^3 with -3^3, and so on up to $-n^3$ with n^3. This leaves only the outer terms

$$-1^3 + (n+1)^3 = (n+1)^3 - 1$$

This kind of sum is called a *telescoping*, or *collapsing*, sum.

TELESCOPING SUM

$$\sum_{i=m}^{n} [a_{i+1} - a_i] = a_{n+1} - a_m$$

Solved Exercises*

1. Find $\sum_{j=-2}^{2} j^3$.

2. Find a formula for $\sum_{i=m}^{n} i$, where m and n are positive integers.

3. Find $\sum_{j=1}^{102} (j + 6)$.

4. Show that $\sum_{k=1}^{1000} \sin(k^2) \leq 1000$.

Exercises

1. Find $\sum_{i=1}^{6} i(i+1)$.

2. Find $\sum_{i=1}^{12} a_i$, where a_i is the number of days in the ith month of 1983.

3. Find:

   (a) $\sum_{i=1}^{4} \sin(\pi i / 2)$ (b) $\sum_{i=1}^{10001} \sin(\pi i / 2)$

   (c) $\sum_{j=-1000}^{1000} j^5$ (d) $\sum_{k=-20}^{10} k$

   (e) $\sum_{i=1}^{100} [i^4 - (i-1)^4]$

4. Prove that if c is a constant, then $\sum_{i=m}^{n} ca_i = c \sum_{i=m}^{n} a_i$.

Solutions appear in Appendix A.

ESTIMATING ACCUMULATED ENERGY

To illustrate the sort of generalized sum which is represented by the integral, we will use a physical example. Consider a solar cell attached to an energy storage unit (such as a battery) as in Fig. 4-1-1. When light shines on the solar cell, it is converted into electrical energy which is stored in the battery (as electrical-chemical energy) for later use.

Fig. 4-1-1 The storage unit accumulates the power received by the solar cell.

We will be interested in the relation between two quantities: the amount E of energy stored and the intensity I of the sunlight. The number E can be read off a dial on the energy storage device; I can be measured with a photographer's light meter. (The units in which E and I are measured are unimportant for this discussion.)

Experiment shows that when the solar cell is exposed to a steady source of sunlight, the change ΔE in the amount of energy stored is proportional to the product of the intensity I and the length Δt of the period of exposure. Thus

$$\Delta E = \kappa I \Delta t \tag{1}$$

where κ is a constant* depending on the apparatus and on the units used to measure energy, time, and intensity. (We can imagine κ being told to us as a manufacturer's specification.)

The intensity I can change—for example, the sun can move behind a cloud. If there are two periods, Δt_1 and Δt_2, during which the intensity is, respectively, I_1 and I_2, then the total energy stored is the sum of the energies stored over each individual period. That is,

$$\Delta E = \kappa I_1 \Delta t_1 + \kappa I_2 \Delta t_2 \tag{2}$$

Likewise, if there are n periods, $\Delta t_1, \ldots, \Delta t_n$, during which the intensity is $I_1, \ldots, I_n$, the energy stored will be the sum of n terms,

$$\Delta E = \kappa I_1 \Delta t_1 + \kappa I_2 \Delta t_2 + \cdots + \kappa I_n \Delta t_n = \sum_{i=1}^{n} \kappa I_i \Delta t_i \tag{3}$$

Worked Example 7   Suppose that $\kappa = 0.026$ and that the intensity is 12.3 for the first 2 seconds, 13.5 for the next 12 seconds, and 12.9 for the last 20 seconds. What is the total change in energy stored?

Solution   By formula (3),

$$\Delta E = 0.026(12.3)2 + 0.026(13.5)12 + 0.026(12.9)20$$
$$= 0.026(24.6 + 162 + 258) = 11.5596$$

κ is the Greek letter kappa.

In practice, as the sun moves gradually behind the clouds and its elevation in the sky changes, the intensity I of sunlight does not change by jumps (Fig. 4-1-2a) but varies continuously with t (Fig. 4-1-2b). The quantity ΔE can still be measured on the energy storage meter, but it can no longer be represented as a sum in the ordinary sense. In fact, the intensity now takes on infinitely many values, but the length of time during which it stays at any value reduces to zero; thus we would have to take the sum of infinitely many zeros, which is meaningless.

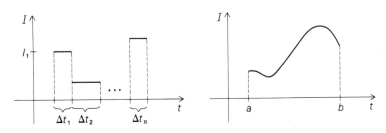

Fig. 4-1-2 The intensity of sunlight varying with time.   **(a)** Varying by jumps   **(b)** Varying continuously

Although the change in energy ΔE is not an ordinary sum of values of I, it is still some sort of "aggregate" or "generalized sum." The integral calculus will provide us with a tool for calculating such "sums" exactly; for now, we will use common-sense reasoning to *estimate* the stored energy when the intensity varies continuously.

Suppose that the intensity varies with time according to the formula $I = 9 - t^2$ during the time period $0 \le t \le 2$, and that $\kappa = 0.1$. The highest intensity during the period is 9 (at $t = 0$); the lowest intensity is $9 - 4 = 5$ (at $t = 2$). We may conclude that the actual stored energy ΔE lies between the amounts that would have been stored if the intensity were constant and equal to 5 or 9; that is,

$$\kappa \cdot 5 \cdot \Delta t \le \Delta E \le \kappa \cdot 9 \cdot \Delta t$$
$$(0.1)(5)(2) \le \Delta E \le (0.1)(9)(2)$$
$$1 \le \Delta E \le 1.8$$

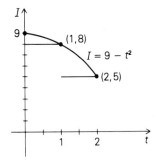

Fig. 4-1-3 The intensity is at least 8 when $0 \le t \le 1$ and it is at least 5 when $1 \le t \le 2$.

We can get a better estimate for ΔE by subdividing the interval $[0, 2]$ into two subintervals and estimating I on each subinterval. From the graph in Fig. 4-1-3, we see that the intensity is bounded below by 8 for $0 \le t \le 1$ and by 5 for $1 \le t \le 2$. Therefore the stored energy ΔE is at least as great as if we had $I = 8$ on $[0, 1]$ and $I = 5$ on $[1, 2]$; that is,

$$\kappa I_1 \Delta t_1 + \kappa I_2 \Delta t_2 \le \Delta E$$
$$(0.1)(8)(1) + (0.1)(5)(1) \le \Delta E$$
$$1.3 \le \Delta E$$

which is a somewhat better estimate than we had before.

In a similar way, we can estimate the intensity from above on each of the intervals $[0, 1]$ and $[1, 2]$ (see Fig. 4-1-4) to obtain the estimate for ΔE:

$$\Delta E \le (0.1)(9)(1) + (0.1)(8)(1)$$
$$\le 1.7$$

Thus we now have $1.3 \le E \le 1.7$ instead of $1 \le \Delta E \le 1.8$. To get more accurate estimates for ΔE, we may subdivide the interval $[0, 2]$ into still smaller pieces, as in Solved Exercise 5.

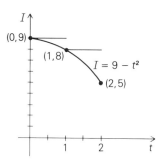

Fig. 4-1-4 The intensity is at most 9 when $0 \le t \le 1$ and it is at most 8 when $1 \le t \le 2$.

Solved Exercise

5. Obtain a better lower estimate for ΔE by dividing the interval $[0, 2]$ into subintervals of length 0.1.

Exercise

5. (a) Using the division of $[0, 2]$ into intervals of length 0.1, obtain an upper estimate for ΔE.
(b) Combining the results of part (a) and Solved Exercise 5, obtain a value of ΔE which is in error by at most 0.02.

PIECEWISE CONSTANT FUNCTIONS

You should keep the previous example in mind as we carry out the technical steps necessary for the definition of the integral. The integral of a function will be defined by comparing it with simpler functions whose integrals can be readily calculated. These comparison functions are the *piecewise constant functions*. Roughly speaking, a function f on $[a, b]$ is piecewise constant if $[a, b]$ can be broken into a finite number of subintervals such that f is constant on each subinterval.

Definition A *partition* of the interval $[a, b]$ is a sequence of numbers $(t_0, t_1, \ldots, t_n)$ such that

$$a = t_0 < t_1 < \cdots < t_{n-1} < t_n = b$$

We think of the numbers $t_0, t_1, \ldots, t_n$ as dividing $[a, b]$ into the subintervals $[t_0, t_1], [t_1, t_2], \ldots, [t_{n-1}, t_n]$. (See Fig. 4-1-5.) The number n of intervals can be as small as 1 or as large as we wish.

Fig. 4-1-5 The partition points $(t_0, t_1, \ldots, t_n)$ divide the interval $[a, b]$ into n subintervals.

Definition A function f defined on an interval $[a, b]$ is called *piecewise constant* if there is a partition $(t_0, ..., t_n)$ of $[a, b]$ and real numbers $k_1, ..., k_n$ such that

$$f(t) = k_i \quad \text{for all } t \text{ in } (t_{i-1}, t_i)$$

The partition $(t_0, ..., t_n)$ is then said to be *adapted* to the piecewise constant function f.

Notice that we put no condition on the value of f at the points $t_0, t_1, ..., t_n$, but we require that f be constant on each of the *open* intervals between successive points of the partition. As we will see in Worked Example 8, more than one partition may be adapted to a given piecewise constant function. Piecewise constant functions are sometimes called *step functions* because their graphs often resemble staircases (see Fig. 4-1-6).

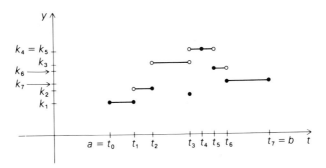

Fig. 4-1-6 The graph of a piecewise constant (step) function.

Worked Example 8 Draw a graph of the piecewise constant function f on $[0, 1]$ defined by

$$f(t) = \begin{cases} -2 & \text{if } 0 \le t < \frac{1}{3} \\ 3 & \text{if } \frac{1}{3} \le t < \frac{1}{2} \\ 3 & \text{if } \frac{1}{2} \le t \le \frac{3}{4} \\ 1 & \text{if } \frac{3}{4} < t \le 1 \end{cases}$$

Give three different partitions which are adapted to f and one partition which is not adapted to f.

Solution The graph is shown in Fig. 4-1-7 on p. 182. As usual, an open circle indicates a point which is not on the graph. (The resemblance of this graph to a staircase is rather faint.)

One partition adapted to f is $(0, \frac{1}{3}, \frac{1}{2}, \frac{3}{4}, 1)$. (That is, $t_0 = 0$, $t_1 = \frac{1}{3}$, $t_2 = \frac{1}{2}$, $t_3 = \frac{3}{4}$, $t_4 = 1$. Here $k_1 = -2$, $k_2 = k_3 = 3$, $k_4 = 1$.) If we delete $\frac{1}{2}$, the partition $(0, \frac{1}{3}, \frac{3}{4}, 1)$ is still adapted to f because f is constant on $(\frac{1}{3}, \frac{3}{4})$. Finally, we can always add extra points to an adapted partition and it will still be adapted. For example, $(0, \frac{1}{8}, \frac{1}{3}, \frac{1}{2}, \frac{3}{4}, \frac{8}{9}, 1)$ is an adapted partition.

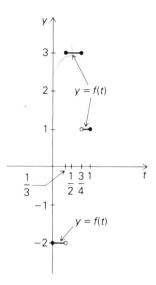

Fig. 4-1-7 The graph of the piecewise constant function in Worked Example 8.

A partition which is not adapted to f is $(0, 0.1, 0.2, 0.3, 0.4, 0.5, 0.6, 0.7, 0.8, 0.9, 1)$. Even though the intervals in this partition are very short, it is not adapted to f because f is not constant on $(0.7, 0.8)$. (Can you find another open interval in the partition on which f is not constant?)

Motivated by our physical example, we define the integral of a piecewise constant function as a sum.

Definition Let f be a piecewise constant function on $[a, b]$. Let $(t_0, \ldots, t_n)$ be a partition of $[a, b]$ which is adapted to f and let k_i be the value of f on (t_{i-1}, t_i). The sum

$$k_1(t_1 - t_0) + k_2(t_2 - t_1) + \cdots + k_n(t_n - t_{n-1})$$

is called the *integral* of f on $[a, b]$ and is denoted by $\int_a^b f(t)\, dt$; that is,

$$\int_a^b f(t)\, dt = \sum_{i=1}^n k_i \Delta t_i \qquad (4)$$

where $\Delta t_i = t_i - t_{i-1}$.

Notice that the value of the integral does not involve the values of f at the partition points $(t_0, t_1, \ldots, t_n)$. This property corresponds to the fact that in our physical example the stored energy is not affected by a blink of light with zero duration time.

It appears that the definition of the integral might depend on the choice of the partition, not just on f and $[a, b]$; however, this is not the case—the integral is *independent* of the choice of partition. We will verify this in the appendix to this chapter; see Worked Example 9 below for a special case.

If all the k_i's are nonnegative, the integral $\int_a^b f(t)\, dt$ is precisely the area "under" the graph of f—that is, the area of the set of points (x, y) such that $a \le x \le b$ and $0 \le y \le f(x)$. The region under the graph in Fig. 4-1-6 is shaded in Fig. 4-1-8.

Fig. 4-1-8 The area of the rectangle R_i is $k_i(t_i - t_{i-1})$. The area of the (shaded) region under the graph of f is $\sum_{i=1}^{7} k_i(t_i - t_{i-1})$, which equals $\int_a^b f(t)\, dt$.

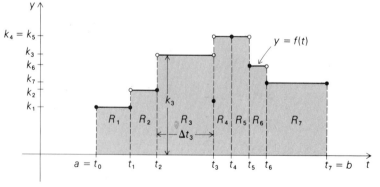

The notation $\int_a^b f(t)\, dt$ for the integral is due to Leibniz. The symbol $\int$, called an *integral sign*, is an elongated S which replaces the Greek Σ of ordinary summation. Similarly, the dt replaces the Δt_i's in the summation formula (4). The function $f(t)$ which is being integrated is called the *inte-*

grand. The endpoints a and b are also called *limits of integration* (not to be confused with the limits in Section 1-2). The relation between this notation and the one for antiderivatives will be explained in the next section.

Worked Example 9 Compute $\int_0^1 f(t)\,dt$ for the function in Worked Example 8, first using the partition $(0, \frac{1}{3}, \frac{1}{2}, \frac{3}{4}, 1)$ and then using the partition $(0, \frac{1}{3}, \frac{3}{4}, 1)$.

Solution With the first partition, we have

$$
\begin{aligned}
k_1 &= -2 & \Delta t_1 &= t_1 - t_0 = \tfrac{1}{3} - 0 = \tfrac{1}{3} \\
k_2 &= 3 & \Delta t_2 &= t_2 - t_1 = \tfrac{1}{2} - \tfrac{1}{3} = \tfrac{1}{6} \\
k_3 &= 3 & \Delta t_3 &= t_3 - t_2 = \tfrac{3}{4} - \tfrac{1}{2} = \tfrac{1}{4} \\
k_4 &= 1 & \Delta t_4 &= t_4 - t_3 = 1 - \tfrac{3}{4} = \tfrac{1}{4}
\end{aligned}
$$

Thus, by (4),

$$
\int_0^1 f(t)\,dt = \sum_{i=1}^4 k_i \Delta t_i = (-2)(\tfrac{1}{3}) + (3)(\tfrac{1}{6}) + (3)(\tfrac{1}{4}) + (1)(\tfrac{1}{4})
$$

$$
= -\tfrac{2}{3} + \tfrac{1}{2} + \tfrac{3}{4} + \tfrac{1}{4} = \tfrac{5}{6}
$$

Using the second partition, we have

$$
\begin{aligned}
k_1 &= -2 & \Delta t_1 &= t_1 - t_0 = \tfrac{1}{3} - 0 = \tfrac{1}{3} \\
k_2 &= 3 & \Delta t_2 &= t_2 - t_1 = \tfrac{3}{4} - \tfrac{1}{3} = \tfrac{5}{12} \\
k_3 &= 1 & \Delta t_3 &= t_3 - t_2 = 1 - \tfrac{3}{4} = \tfrac{1}{4}
\end{aligned}
$$

and

$$
\int_0^1 f(t)\,dt = \sum_{i=1}^3 k_i \Delta t_i = (-2)(\tfrac{1}{3}) + (3)(\tfrac{5}{12}) + (1)(\tfrac{1}{4})
$$

$$
= -\tfrac{2}{3} + \tfrac{5}{4} + \tfrac{1}{4} = \tfrac{5}{6}
$$

which is the same answer we obtained from the first partition.

Solved Exercises 6. Compute $\int_{-2}^3 f(x)\,dx$ for the function f sketched in Fig. 4-1-9. How is the integral related to the area of the shaded region in the figure?

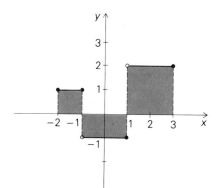

Fig. 4-1-9 What is the relation between the shaded area and the integral?

7. Let $f(t)$ be defined by

$$f(t) = \begin{cases} 2 & \text{if } 0 \le t < 1 \\ 0 & \text{if } 1 \le t < 3 \\ -1 & \text{if } 3 \le t \le 4 \end{cases}$$

For any number x in $(0, 4]$, $f(t)$ is piecewise constant on $[0, x]$.

(a) Find $\int_0^x f(t)\, dt$ as a function of x. (You will need to use different formulas on different intervals.)

(b) Let $F(x) = \int_0^x f(t)\, dt$, for $x \in (0, 4]$. Draw a graph of F.

(c) At which points is F differentiable? Find a formula for $F'(x)$.

Exercises

6. Compute the integral in Solved Exercise 6 by using the partition $(-2, -1, 0, 1, 2, 3)$.

7. Suppose that $I = f(t)$ is the intensity of light received by a solar cell, where f is the function given by

$$f(t) = \begin{cases} 1 & 0 \le t < 2 \\ 3 & 2 \le t < 3 \\ 2 & 3 \le t \le 6 \end{cases}$$

The energy E stored between time 0 and time x is

$$E = F(x) = \int_0^x \kappa f(t)\, dt$$

where κ is a constant. (We use the letter x for the time at which the stored energy is measured, so as not to confuse it with t, which varies over the interval $[0, x]$ on which the integral is taken.)

(a) Plot $F(x)$ as a function of x.

(b) What is the relation between:

    (i) The rate of change of $F(x)$ with respect to x

and   (ii) The intensity $I = f(t)$ of incoming light?

UPPER AND LOWER SUMS AND THE DEFINITION

Having defined the integral for piecewise constant functions, we will now define the integral for more general functions. We begin with a preliminary definition.

Definition Let f be a function defined on $[a, b]$. If g is any piecewise constant function on $[a, b]$ such that $g(t) \le f(t)$ for all t in the open interval (a, b), we call the number $\int_a^b g(t)\, dt$ a *lower sum* for f on $[a, b]$. If h is a piecewise constant function and $f(t) \le h(t)$ for all t in (a, b), the number $\int_a^b h(t)\, dt$ is called an *upper sum* for f on $[a, b]$.

Worked Example 10 Let $f(t) = t^2 + 1$ for $0 \le t \le 2$. Let

$$g(t) = \begin{cases} 0 & 0 \le t \le 1 \\ 2 & 1 < t \le 2 \end{cases} \quad \text{and} \quad h(t) = \begin{cases} 2 & 0 \le t \le \frac{2}{3} \\ 4 & \frac{2}{3} < t \le \frac{4}{3} \\ 5 & \frac{4}{3} < t \le 2 \end{cases}$$

Draw a graph showing $f(t)$, $g(t)$, and $h(t)$. What upper and lower sums for f can be obtained from g and h?

Solution The graph is shown in Fig. 4-1-10.

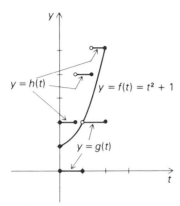

Fig. 4-1-10 The integral of h is an upper sum for f; the integral of g is a lower sum.

Since $g(t) \le f(t)$ for all t in the open interval $(0, 2)$ (the graph of g lies below that of f), we have as a lower sum

$$\int_0^2 g(t)\,dt = 0 \cdot 1 + 2 \cdot 1 = 2$$

Since the graph of h lies above that of f, $h(t) \ge f(t)$ for all t in the interval $(0, 2)$, so we have the upper sum

$$\int_0^2 h(t)\,dt = 2 \cdot \tfrac{2}{3} + 4 \cdot \tfrac{2}{3} + 5 \cdot \tfrac{2}{3} = \tfrac{22}{3} = 7\tfrac{1}{3}$$

The integral of a function should lie between the lower sums and the upper sums. For instance, the integral of the function in Worked Example 10 should lie in the interval $[2, 7\tfrac{1}{3}]$. If we could find upper and lower sums which are arbitrarily close together, then the integral would be pinned down to a single point. This idea leads to the formal definition of the integral.

Definition Let f be a function defined on $[a, b]$ and let S_0 be a real number. We say that S_0 is *the integral of f on $[a, b]$* if:

1. Every number $S < S_0$ is a lower sum for f on $[a, b]$.

2. Every number $S > S_0$ is an upper sum for f on $[a, b]$.

If such a number S_0 exists, we say that f is *integrable* on $[a, b]$, and we write $S_0 = \int_a^b f(t)\,dt$.

Remark The letter t is called the *variable of integration;* it is a dummy variable in that we can replace it by any other letter without changing the value of the integral just as we did in the summation notation. Here a and b are called the *endpoints* or *limits* for the integral.

The following two facts, which should be evident, will be proved in the appendix to this chapter:

1. Every lower sum for f is less than or equal to every upper sum.

2. Every number less than a lower sum is a lower sum, and every number greater than an upper sum is an upper sum.

In Fig. 4-1-11 we show the possible configurations for the sets of lower and upper sums. Although there appear to be many ways in which a function can be nonintegrable, in fact cases 3 through 7 occur only for unbounded functions, and case 2 is quite pathological. The appendix to this chapter contains some examples of nonintegrable functions, as well as a proof of Theorem 1 below.

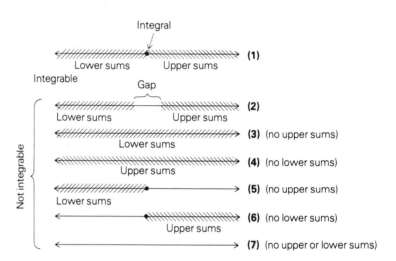

Fig. 4-1-11 Possible configurations of the sets of upper and lower sums.

Theorem 1 *If f is continuous on $[a, b]$, then f is integrable on $[a, b]$.*

Since every differentiable function is continuous, we conclude that every differentiable function is integrable; however, some noncontinuous functions can be integrable. For instance, piecewise constant functions are integrable (see Solved Exercise 8 below), but they are not continuous.

Geometrically, the integral $\int_a^b f(t)\, dt$ of an integrable function f which is nonnegative on $[a, b]$ can be interpreted as *the area under the graph of f* on $[a, b]$. See Figure 4-1-12. This relation between integrals and areas was established for piecewise constant functions on p. 182 (see Fig. 4-1-8); it extends to all integrable functions because if $0 \le g(x) \le f(x) \le h(x)$ on $[a, b]$ for piecewise constant functions g and h, then the area under the graph of f lies between the areas under the graphs of g and h. Further details will be given in Section 4-3.

The integral can also be expressed in terms of limits. If f is an integrable function on $[a, b]$ and $(t_1, t_2, ..., t_n)$ is a partition of $[a, b]$, we may choose points $s_1, ..., s_n$ such that s_i lies in the interval $[t_{i-1}, t_i]$. The sum

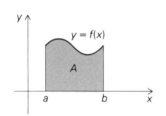

Fig. 4-1-12 The area A equals $\int_a^b f(x)\, dx$.

$$S_n = \sum_{i=1}^{n} f(s_i) \Delta t_i$$

is the integral of a piecewise constant function which approximates f (see Fig. 4-1-13); it is called a *Riemann sum* for f after the German mathematician Bernhard Riemann (1826–1866). If we choose a sequence of partitions, one for each n, such that the lengths Δt_i approach zero as n becomes larger, then the Riemann sums approach the integral $\int_a^b f(x)\, dx$ in the limit as $n \to \infty$. From this and Fig. 4-1-13, we see again the geometric interpretation of integrals in terms of areas.

Fig. 4-1-13 The area of the shaded region is a Riemann sum for f on $[a, b]$.

Just as the derivative may be defined as a limit of difference quotients, so the integral may be defined as a limit of Riemann sums; the integral is then called the *Riemann integral*. We give more details on this approach to the integral in Section 12-1.

Riemann sums provide the simplest method of *numerical integration*—that is, a technique for approximating an integral numerically. Such techniques, discussed further in Section 12-4, are particularly important since there are many integrals (such as $\int_0^1 \sqrt{1 - x^3}\, dx$) which cannot be evaluated in terms of elementary functions.

THE INTEGRAL

If f is a function on $[a, b]$, the integral $\int_a^b f(t)\, dt$ is defined by the conditions:

1. Every number $S < \int_a^b f(t)\, dt$ is a lower sum for f on $[a, b]$.

2. Every number $S > \int_a^b f(t)\, dt$ is an upper sum for f on $[a, b]$.

If f is continuous on $[a, b]$, then it has an integral.

Solved Exercises

8. Show that every piecewise constant function is integrable and that its integral as defined in the preceding box is the same as its integral as originally defined on p. 182.

9. Find $\int_1^2 (1/t)\, dt$ to within an error of no more than $\frac{1}{10}$.

10. If you used a method analogous to that in Solved Exercise 9, how many steps would it take to calculate $\int_1^2 (1/t)\, dt$ to within $\frac{1}{100}$?

11. Let f be defined on $[0, 1]$ by

$$f(t) = \begin{cases} 0 & \text{if } t = 0 \\ \dfrac{1}{t} & \text{if } 0 < t \le 1 \end{cases}$$

Is f integrable on $[0, 1]$? On $[\frac{1}{8}, 1]$?

Exercises

8. (a) Estimate $\int_1^2 (1/x^2)\, dx$ to within $\frac{1}{10}$.

   (b) Estimate $\int_1^2 (1/x^2)\, dx$ to within $\frac{1}{100}$.

9. (a) Find a function on $[0, 1]$ having upper sums but no lower sums.

   (b) Find a function on $[0, 1]$ having no upper sums and no lower sums.

10. Let

$$f(t) = \begin{cases} \ln t & t > 0 \\ 0 & t = 0 \end{cases}$$

Is f integrable on $[1, 2]$? On $[0, 1]$?

CALCULATING INTEGRALS BY HAND

(optional)

We have just given an elaborate definition of the integral and stated that continuous functions are integrable, but we have not yet computed the integral of any functions except piecewise constant functions. You may recall that in our treatment of differentiation it was also laborious to compute derivatives by using the definitions; instead, we used the algebraic rules of calculus to compute most derivatives. The situation is much the same for integration. Beginning in Section 4-2, we will develop the machinery which makes the calculation of many integrals quite simple. Before doing this, however, we will calculate one integral "by hand" in order to illustrate the definition.

After the constant functions, the simplest continuous function is $f(t) = t$. We know by Theorem 1 that the integral $\int_a^b t\, dt$ exists; to calculate this integral, we will find upper and lower sums which are closer and closer together, reducing to zero the possible error in the estimate of the integral.

Let $f(t) = t$. We divide the interval $[a, b]$ into n equal parts, using the partition

$$(t_0, \ldots, t_n) = \left(a, a + \frac{b-a}{n}, a + \frac{2(b-a)}{n}, \ldots, a + \frac{(n-1)(b-a)}{n}, b \right)$$

Note that $\Delta t_i = (b - a)/n$ for each i.

Now we define the piecewise constant function g on $[a, b]$ by setting $g(t) = t_{i-1}$ for $t_{i-1} \leq t < t_i$. (See Fig. 4-1-14.)

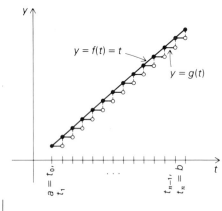

Fig. 4-1-14 The step function g gives a lower sum for f on $[a, b]$.

Note that for $t_{i-1} \leq t < t_i$, we have $g(t) = t_{i-1} \leq t = f(t)$, so $\int_a^b g(t)\, dt$ is a lower sum for f on $[a, b]$. By formula (4) on p. 182 we have

(optional)

$$\int_a^b g(t)\, dt = \sum_{i=1}^n k_i \Delta t_i = \sum_{i=1}^n t_{i-1} \Delta t_i \tag{5}$$

since t_{i-1} is the constant value of g on (t_{i-1}, t_i). We know that $\Delta t_i = (b - a)/n$ for each i. To find out what t_i is, we note that $t_0 = a$, $t_1 = a + (b - a)/n$, $t_2 = a + [2(b - a)/n]$, and so on, so the general formula is $t_i = a + [i(b - a)/n]$. Substituting for t_{i-1} and Δt_i in (5) gives

$$\int_a^b g(t)\, dt = \sum_{i=1}^n \left[a + \frac{(i - 1)(b - a)}{n} \right] \left(\frac{b - a}{n} \right)$$

$$= \sum_{i=1}^n \left[\frac{a(b - a)}{n} + \frac{i(b - a)^2}{n^2} - \frac{(b - a)^2}{n^2} \right]$$

By the properties of summation (see p. 175), we may rewrite this as:

$$\sum_{i=1}^n \frac{a(b - a)}{n} + \frac{(b - a)^2}{n^2} \left(\sum_{i=1}^n i \right) - \sum_{i=1}^n \frac{(b - a)^2}{n^2}$$

The outer terms do not involve i, so they may be summed by property 4 on p. 175; the middle term is summed by the formula on p. 176. The result is

$$a(b - a) + \left(\frac{(b - a)^2}{n^2} \right) \left(\frac{n(n + 1)}{2} \right) - \frac{(b - a)^2}{n}$$

which simplifies to

$$\frac{b^2 - a^2}{2} - \frac{1}{2n} (b - a)^2$$

(You should carry out the simplification.) We have, therefore, shown that $(b^2 - a^2)/2 - (1/2n)(b - a)^2$ is a lower sum for f.

We now move on to the upper sums. Using the same partition as before, but with the function $h(t)$ defined by $h(t) = t_i$ for $t_{i-1} \le t < t_i$, we find that $\int_a^b h(t)\, dt$ is an upper sum. Some algebra (see Solved Exercise 12 below) gives

$$\int_a^b h(t)\, dt = \frac{b^2 - a^2}{2} + \frac{(b - a)^2}{2n}$$

Our calculations of upper and lower sums therefore show that

$$\frac{b^2 - a^2}{2} - \frac{(b - a)^2}{2n} \le \int_a^b t\, dt \le \frac{b^2 - a^2}{2} + \frac{(b - a)^2}{2n}$$

These inequalities, which hold for all n, show that the integral can be neither larger nor smaller than $(b^2 - a^2)/2$, so it must be equal to $(b^2 - a^2)/2$.

INTEGRAL OF $f(t) = t$

$$\int_a^b t\, dt = \frac{b^2 - a^2}{2}$$

Worked Example 11 Find $\int_0^3 x\, dx$.

> ***Solution*** We first note that since the variable of integration is a "dummy variable,"
>
> *(optional)*
> $$\int_0^3 x\, dx = \int_0^3 t\, dt$$
>
> Using the formula just obtained, we can evaluate this integral as $\frac{1}{2}(3^2 - 0^2) = \frac{9}{2}$.
>
> In the next section you will find, to your relief, a much easier way to compute integrals. Evaluating integrals like $\int_a^b t^2\, dt$ or $\int_a^b t^3\, dt$ by hand is possible but rather tedious (see Problem 14). The methods of the next section will make these integrals simple to evaluate.

Solved Exercises ★12. Draw a graph of the function $h(t)$ used above to find an upper sum for $f(t) = t$, and show that
$$\int_a^b h(t)\, dt = \frac{b^2 - a^2}{2} + \frac{(b-a)^2}{2n}$$

★13. Using the definition of the integral, find $\int_a^b 5t\, dt$.

★14. (a) Sketch the region under the graph of $f(t) = t$ on $[a, b]$, if $0 < a < b$.
(b) Compare the area of this region with $\int_a^b t\, dt$.

Exercises ★11. Using the definition of the integral, find a formula for $\int_a^b (t + 3)\, dt$.

★12. Using the definition of the integral, find a formula for $\int_a^b (-t)\, dt$.

★13. Explain the relation between $\int_{1/2}^1 t\, dt$ and an area.

Problems for Section 4-1

1. Find:
   (a) $\sum_{k=0}^{100} (3k - 2)$ (b) $\sum_{j=2}^6 2^j$
   (c) $\sum_{k=1}^n [(k + 1)^4 - k^4]$
   (d) $\sum_{l=-75}^{75} [l^9 + 5l^7 - 13l^5 + l]$

2. Find:
   (a) $\sum_{i=0}^n (2i + 1)$
   (b) $\sum_{i=0}^n 2i$ for $n = 1, 2, 3, 4, 5$
   (c) $\sum_{i=0}^n 2^i$
   (d) $\sum_{k=1}^{300} [(k + 1)^8 - k^8]$

3. A rod 1 meter long is made of 100 segments of equal length such that the linear density of the kth segment is $30k$ grams per meter. What is the total mass of the rod?

4. The volume of a rod of uniform shape is $A\Delta x$, where A is the cross-sectional area and Δx is the length.
   (a) Suppose that the rod consists of n pieces, with the ith piece having cross-sectional area A_i and length Δx_i. Write a formula for the volume.
   (b) Suppose that the cross-sectional area is always varying, with $A = f(x)$, where f is a function on $[0, L]$, L being the total length of the rod. Write a formula for the volume of the rod, using the integral notation.
   (c) Interpret conditions 1 and 2 for the integral (p. 187) in terms of this example.

★5. (a) From Worked Example 6 we have

$$\sum_{i=1}^{n} [(i + 1)^3 - i^3] = (n + 1)^3 - 1$$

Write $(i + 1)^3 - i^3 = 3i^2 + 3i + 1$ and use properties of summation to prove that

$$\sum_{i=1}^{n} i^2 = \frac{n(n + 1)(2n + 1)}{6}$$

(b) Find a formula for

$$\sum_{i=m}^{n} i^2$$

in terms of m and n.

(c) Using the *method* and result of (a), find a formula for $\sum_{i=1}^{n} i^3$. (You may wish to try guessing an answer by experiment.)

★6. (a) Prove that $\sum_{i=1}^{n} i(i + 1) = (1/3) n(n + 1) \times (n + 2)$ by writing $i(i + 1) = (1/3) [i(i + 1) \times (i + 2) - (i - 1) i(i + 1)]$ and using a telescoping series.

(b) Find $\sum_{i=1}^{n} i(i + 1)(i + 2)$

(c) Find $\sum_{i=1}^{n} 1/i(i + 1)$

7. Let f be the function defined by

$$f(t) = \begin{cases} 2 & 1 \le t < 4 \\ 5 & 4 \le t < 7 \\ 1 & 7 \le t \le 10 \end{cases}$$

(a) Find $\int_1^{10} f(t)\, dt$.

(b) Find $\int_2^9 f(t)\, dt$.

(c) Suppose that g is a function on $[1, 10]$ such that $g(t) \le f(t)$ for all t in $[1, 10]$. What inequality can you derive for $\int_1^{10} g(t)\, dt$?

(d) With $g(t)$ as in part (c), what inequalities can you obtain for $\int_1^{10} 2g(t)\, dt$ and $\int_1^{10} -g(t)\, dt$? [*Hint:* Find functions like f with which you can compare $2g$ and $-g$.]

8. Let $f(t)$ be the "greatest integer function"; that is, $f(t)$ is the greatest integer which is less than or equal to t—for example, $f(n) = n$ for any integer, $f(5\frac{1}{2}) = 5, f(-5\frac{1}{2}) = -6$, and so on.

(a) Draw a graph of $f(t)$ on the interval $[-4, 4]$.

(b) Find $\int_0^1 f(t)\, dt$, $\int_0^6 f(t)\, dt$, $\int_{-2}^2 f(t)\, dt$, and $\int_0^{4.5} f(t)\, dt$.

(c) Find a general formula for $\int_0^n f(t)\, dt$, where n is any positive integer.

(d) Let $F(x) = \int_0^x f(t)\, dt$, where $x > 0$. Draw a graph of F for $x \in [0, 4]$, and find a formula for $F'(x)$, where it is defined.

★9. Suppose that $f(t)$ is piecewise constant on $[a, b]$. Let $g(t) = f(t) + k$, where k is a constant.

(a) Show that $g(t)$ is piecewise constant.

(b) Find $\int_a^b g(t)\, dt$ in terms of $\int_a^b f(t)\, dt$.

★10. Let $h(t) = kf(t)$, where $f(t)$ is piecewise constant on $[a, b]$.

(a) Show that $h(t)$ is piecewise constant.

(b) Find $\int_a^b h(t)\, dt$ in terms of $\int_a^b f(t)\, dt$.

★11. For $t \in [0, 1]$ let $f(t)$ be the first digit after the decimal point in the decimal expansion of t.

(a) Draw a graph of f.   (b) Find $\int_0^1 f(t)\, dt$.

12. Define the functions f and g on $[0, 3]$ as follows:

$$f(x) = \begin{cases} 4 & 0 \le x < 1 \\ -1 & 1 \le t < 2 \\ 2 & 2 \le t \le 3 \end{cases}$$

$$g(x) = \begin{cases} 2 & 0 \le x < 1\frac{1}{2} \\ 1 & 1\frac{1}{2} \le x \le 3 \end{cases}$$

(a) Draw the graph of $f(x) + g(x)$ and compute $\int_0^3 [f(t) + g(t)]\, dt$.

(b) Compute $\int_1^2 [f(t) + g(t)]\, dt$.

(c) Compare $\int_0^3 2f(t)\, dt$ with $2 \int_0^3 f(t)\, dt$.

(d) Show that

$$\int_0^3 [f(t) - g(t)]\, dt = \int_0^3 f(t)\, dt - \int_0^3 g(t)\, dt$$

(e) Is the following true?

$$\int_0^3 f(t) \cdot g(t)\, dt = \int_0^3 f(t)\, dt \cdot \int_0^3 g(t)\, dt$$

★13. Using the definition of the integral, find $\int_0^1 (1 - x)\, dx$.

★14. Using the definition of the integral, show that $\int_a^b t^2\, dt = (b^3 - a^3)/3$. (You will need the result from Problem 5(a) above.)

★15. Suppose that $f(t)$ is piecewise constant on $[a, b]$. Let $F(x) = \int_a^x f(t)\, dt$. Prove that if f is continuous at $x_0 \in [a, b]$, then F is differentiable at x_0 and $F'(x_0) = f(x_0)$. (See Solved Exercise 7.)

★16. Suppose that f is a continuous function on $[a, b]$ and that $f(x) \neq 0$ for all x in $[a, b]$. Assume that $a \neq b$ and that $f((a + b)/2) = 1$. Prove that $\int_a^b f(t)\, dt > 0$. [*Hint:* Find a lower sum.]

★17. Show that $-3 \le \int_1^2 (t^3 - 4)\, dt \le 4$.

★18. Show that $\int_0^1 t^{10}\, dt \le 1$.

★19. Compute the exact value of $\int_0^1 x^5\, dx$ by using the formula $1^5 + 2^5 + 3^5 \ldots + N^5 = N^6/6 + N^5/2 + 5N^4/12 - N^3/12$. [*Hint:* Follow the method of "calculating integrals by hand."]

4-2 The Fundamental Theorem of Calculus

The processes of integration and differentiation are inverses to one another.

The fundamental theorem of calculus states that integrating the derivative of a function gives the increment of the function which was differentiated. Knowing this, we can calculate many integrals by using techniques of antidifferentiation.

Goals

After studying this section, you should be able to:

State the fundamental theorem of calculus.

Evaluate simple integrals by using antiderivatives.

Differentiate an integral with respect to the upper limit of integration.

THE INTEGRAND AS A RATE OF CHANGE

We will illustrate the fundamental theorem of calculus by referring again to the solar cell example of the previous section. If I is the intensity of sunlight at time t, then the amount ΔE of energy stored between time a and time b can be written as an integral:

$$\Delta E = \int_a^b \kappa I \, dt \tag{1}$$

where κ is a constant. For simplicity of notation, we will combine κ and I into a single function $f(t) = \kappa I$, so that formula (1) becomes

$$\Delta E = \int_a^b f(t) \, dt \tag{2}$$

In concluding that the stored energy was an integral, we began on p. 178 with the fact that

$$\Delta E = \kappa I \, \Delta t \tag{3}$$

in case I is constant on the time interval $[a, b]$. Dividing (3) through by Δt gives the equation

$$\frac{\Delta E}{\Delta t} = \kappa I \tag{4}$$

which says that $\kappa I = f(t)$ *may be interpreted as the rate of change of E with respect to t*, at least if I is constant. In other words, the intensity of solar radiation is proportional to the rate at which the sun is delivering energy to the storage unit via the solar cell.

Our study of differential calculus has shown us that rates of change are meaningful quantities even when they are not constant; thus it is reasonable to suppose that the preceding paraphrase of equation (4) holds even when I is not constant. To express this idea mathematically, we let $E = F(t)$ denote the quantity of stored energy at time t; then the paraphrase can be written as an equation:

$$\frac{dE}{dt} = \kappa I \tag{5}$$

or

$$F'(t) = f(t) \tag{6}$$

By substituting (5) or (6) into (1) or (2), and observing that ΔE means $F(b) - F(a)$, we obtain the equation:

$$\Delta E = \int_a^b \frac{dE}{dt}\, dt \tag{7}$$

or

$$F(b) - F(a) = \int_a^b F'(t)\, dt \tag{8}$$

Equation (8) can be read, without reference to solar cells, as a simple mathematical statement about functions. We may test this statement on the quadratic function $F(t) = \frac{1}{2}t^2$.

Worked Example 1 Show that equation (8) is true when $F(t) = \frac{1}{2}t^2$.

Solution We have $F'(t) = t$, so equation (8) may be written as

$$\tfrac{1}{2}b^2 - \tfrac{1}{2}a^2 = \int_a^b t\, dt$$

But this is precisely what we found on p. 189, so (8) is verified for the case $F(t) = \frac{1}{2}t^2$.

We were led to equation (8) by physical considerations, and we verified it for a special case in Worked Example 1. The fundamental theorem of calculus states simply that equation (8) is true in general.

Theorem 2 Fundamental Theorem of Calculus. *Suppose that the function F is differentiable everywhere on $[a, b]$ and that F' is integrable on $[a, b]$. Then*

$$\int_a^b F'(t)\, dt = F(b) - F(a) \tag{9}$$

In other words, if f is integrable on $[a, b]$ and F is an antiderivative for f, then

$$\int_a^b f(t)\, dt = F(b) - F(a) \tag{10}$$

Before proving Theorem 2, we will show how easy it makes the calculation of some integrals.

Worked Example 2   Using the fundamental theorem of calculus, compute $\int_a^b t^2\,dt$.

Solution   We begin by finding an antiderivative $F(t)$ for $f(t) = t^2$; from the power rule, we may take $F(t) = \frac{1}{3}t^3$. By the fundamental theorem, we have

$$\int_a^b t^2\,dt = \int_a^b f(t)\,dt = F(b) - F(a) = \tfrac{1}{3}b^3 - \tfrac{1}{3}a^3$$

We conclude that $\int_a^b t^2\,dt = \frac{1}{3}(b^3 - a^3)$. As we already mentioned, it is possible to evaluate this integral "by hand," using partitions of $[a,b]$ and calculating upper and lower sums (see Problem 14, p. 191), but the present method is much more efficient.

We can summarize the integration method provided by the fundamental theorem as follows.

FUNDAMENTAL INTEGRATION METHOD

To integrate the function $f(t)$ over the interval $[a,b]$:   find an antiderivative $F(t)$ for $f(t)$, then evaluate F at a and b and subtract the results:

$$\int_a^b f(t)\,dt = F(b) - F(a)$$

Notice that, according to the fundamental theorem, it does not matter which antiderivative we use. In fact, we do not need the fundamental theorem to tell us that if F_1 and F_2 are both antiderivatives of f on $[a,b]$, then

$$F_1(b) - F_1(a) = F_2(b) - F_2(a) \tag{11}$$

To prove (11) we use the fact that any two antiderivatives of a function differ by a constant (see p. 167). We have, therefore, $F_1(t) = F_2(t) + C$, where C is a constant, and so

$$F_1(b) - F_1(a) = [F_2(b) + C] - [F_2(a) + C]$$

The C's cancel, and the expression on the right is just $F_2(b) - F_2(a)$.

Expressions of the form $F(b) - F(a)$ occur so often that it is useful to have a special notation for them.

Definition $F(t)\big|_a^b$ means $F(b) - F(a)$.

Worked Example 3   Find $(t^3 + 5)\big|_2^3$.

Solution   Here $F(t) = t^3 + 5$ and

$$\begin{aligned}(t^3 + 5)\big|_2^3 &= F(3) - F(2)\\ &= 3^3 + 5 - (2^3 + 5)\\ &= 32 - 13 = 19\end{aligned}$$

In terms of this new notation, we can write the formula of the fundamental theorem of calculus in the form:

$$\int_a^b f(t)\, dt = F(t) \Big|_a^b \tag{12}$$

where F is an antiderivative of f on $[a, b]$.

Worked Example 4   Find $\int_2^6 (t^2 + 1)\, dt$.

Solution   By the sum and power rules for antiderivatives, an antiderivative for $t^2 + 1$ is $\frac{1}{3}t^3 + t$. By the fundamental theorem,

$$\int_2^6 (t^2 + 1)\, dt = \left(\frac{1}{3}t^3 + t \right) \Big|_2^6$$
$$= \frac{6^3}{3} + 6 - \left(\frac{2^3}{3} + 2 \right)$$
$$= 78 - 4\tfrac{2}{3} = 73\tfrac{1}{3}$$

Solved Exercises

1. Evaluate $\int_0^1 x^4\, dx$.

2. Find $\int_0^3 (t^2 + 3t)\, dt$.

3. Light shines on a solar energy unit with intensity $I = t^2$ for the period from $t = 0$ to $t = 1$. Light shines on another, identical, unit with intensity $I = 5t^3$ over the same period. Which unit stores more energy?

4. Suppose that $v = f(t)$ is the velocity at time t of an object moving along a line. Using the fundamental theorem of calculus, interpret the integral $\int_a^b v\, dt = \int_a^b f(t)\, dt$.

5. (See Worked Example 8, Section 1-3). The marginal revenue for a company at production level x is given by $15 - 0.1x$. If $R(0) = 0$, find $R(100)$.

Exercises

1. Use the fundamental theorem to compute ΔE for the case $\kappa = 0.1$, $I = 9 - t^2$, $[a, b] = [0, 2]$ discussed on pp. 179 and 180. Verify that your answer is consistent with the estimates obtained in Solved Exercise 5 and Exercise 5 on p. 180.

2. Find $\int_a^b s^4\, ds$.

3. Find $\int_{-1}^1 (t^4 + t^{917})\, dt$.

4. Find $\int_0^{10} \left(\frac{t^4}{100} - t^2 \right) dt$

5. Suppose that the marginal revenue of a company at production level x is given by $30 - 0.02x - 0.0001x^2$. If $R(0) = 0$, find $R(300)$.

(optional)   **PROOF OF THE FUNDAMENTAL THEOREM**

We will now give a complete proof of the fundamental theorem of calculus in the form stated as Theorem 2 above. The basic idea is as follows: Letting F be an antiderivative for f on $[a, b]$, we will show that if l_f and u_f are any lower and upper sums for f on $[a, b]$, then $l_f \le F(b) - F(a) \le u_f$. Since f is assumed *(optional)* to be integrable on $[a,b]$, the only number which can separate the lower sums from the upper sums in this way is the integral $\int_a^b f(t)\, dt$. (See Fig. 4-1-11(1).) It will follow that $F(b) - F(a)$ must equal $\int_a^b f(t)\, dt$.

To show that every lower sum is less than or equal to $F(b) - F(a)$, we must take any piecewise constant g on $[a,b]$ such that $g(t) \le f(t)$ for all t in (a,b) and show that $\int_a^b g(t)\, dt \le F(b) - F(a)$. Let $(t_0, t_1, ..., t_n)$ be a partition adapted to g and let k_i be the value of g on (t_{i-1}, t_i). Since $F' = f$, we have

$$k_i = g(t) \le f(t) = F'(t)$$

Hence

$$k_i \le F'(t)$$

for all t in (t_{i-1}, t_i). (See Fig. 4-2-1.) It follows from Corollary 1 of the mean value theorem (see p. 166) that

$$k_i \le \frac{F(t_i) - F(t_{i-1})}{t_i - t_{i-1}}$$

Hence

$$k_i \Delta t_i \le F(t_i) - F(t_{i-1})$$

Summing from $i = 1$ to n, we get

$$\sum_{i=1}^{n} k_i \Delta t_i \le \sum_{i=1}^{n} [F(t_i) - F(t_{i-1})]$$

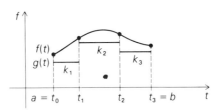

Fig. 4-2-1 The integral of g is a lower sum for f on $[a, b]$.

The left-hand side is just $\int_a^b g(t)\, dt$, by the definition of the integral of a step function. The right-hand side is a telescoping sum (see p. 177), equal to $F(t_n) - F(t_0)$. (See Fig. 4-2-2.) Thus we have

Fig. 4-2-2 The total change of F on $[a, b]$ is the sum of changes on the subintervals $[t_{i-1}, t_i]$. The change on the ith subinterval is at least $k_i \Delta t_i$.

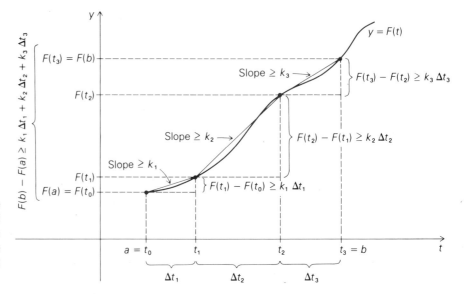

(optional)

$$\int_a^b g(t)\, dt \le F(b) - F(a)$$

which is what we wanted to prove.

In the same way (see Exercise 6), we can show that if $h(t)$ is a piecewise constant function such that $f(t) \le h(t)$ for all t in (a, b), then

$$F(b) - F(a) \le \int_a^b h(t)\, dt$$

as required. This completes the proof of the fundamental theorem.

Solved Exercise ★6. Suppose that F is continuous on $[0, 2]$, that $F'(x) < 2$ for $0 \le x \le \frac{1}{3}$, and that $F'(x) < 1$ for $\frac{1}{3} \le x \le 2$. What can you say about $F(2) - F(0)$?

Exercises ★6. Prove that if $h(t)$ is a piecewise constant function on $[a, b]$ such that $f(t) \le h(t)$ for all $t \in (a, b)$, then $F(b) - F(a) \le \int_a^b h(t)\, dt$, where F is any antiderivative for f on $[a, b]$.

★7. Let $a_0, \ldots, a_n$ be any numbers and let $\delta_i = a_i - a_{i-1}$. Let $b_k = \Sigma_{i=1}^k \delta_i$ and let $d_i = b_i - b_{i-1}$. Express the b's in terms of the a's and the d's in terms of the δ's.

DEFINITE AND INDEFINITE INTEGRALS

When we studied antiderivatives in Section 1-4, we used the notation $\int f(x)\, dx$ for an antiderivative of f, and we called it an *indefinite integral*. We can see now that this notation and the terminology come from the fundamental theorem of calculus. In fact, we can rewrite the fundamental theorem, using the indefinite integral notation, as in the following display.

DEFINITE AND INDEFINITE INTEGRALS

$$\int_a^b f(t)\, dt = \int f(t)\, dt \,\Bigg|_a^b$$

Notice that although the indefinite integral involves an arbitrary constant, the expression

$$\left(\int f(t)\, dt \right) \Bigg|_a^b$$

is unambiguously defined, since the constant cancels when we subtract the value at b from the value at a.

An expression of the form $\int_a^b f(t)\, dt$, in which the endpoints are specified, which we have been calling simply "an integral," is sometimes called a *definite* integral to distinguish it from an indefinite integral.

Note that a definite integral is a number, while an indefinite integral is a *function* (determined up to an additive constant). Given the indefinite integral of f, we can obtain any definite integral by substituting the endpoints and subtracting. You may compare this distinction with the similar situation for derivatives. We originally defined the derivative *at a point* for a function—this derivative was a number. We saw that we could consider the derivative at all possible points—this procedure yields a function, which we also called the derivative. The derivative at a point was obtained by evaluating the derivative function at the desired point.

Remember that one may check an indefinite integral formula by differentiating.

INDEFINITE INTEGRAL TEST

To check a given formula $\int f(x)\ dx = F(x) + C$, differentiate $F(x)$ and see if you get $f(x)$.

Worked Example 5 (a) Check the formula

$$\int x(1 + x)^6\ dx = \frac{1}{56}(7x - 1)(1 + x)^7 + C$$

(Do not attempt to derive the formula.)

(b) Find $\int_0^2 x(1 + x)^6\ dx$.

Solution (a) We differentiate the right-hand side using the product and power of a function rules:

$$\frac{d}{dx}\left[\frac{1}{56}(7x - 1)(1 + x)^7\right] = \frac{1}{56}[7(1 + x)^7 + (7x - 1)7(1 + x)^6]$$

$$= \frac{1}{56}(1 + x)^6[7(1 + x) + 7(7x - 1)]$$

$$= (1 + x)^6 x$$

Thus the formula checks.

(b) By the fundamental theorem and the formula we just checked, we have

$$\int_0^2 x(1 + x)^6\ dx = \frac{1}{56}(7x - 1)(1 + x)^7 \Big|_0^2$$

$$= \frac{1}{56}[13 \cdot 3^7 - (-1)]$$

$$= \frac{28,432}{56}$$

$$= \frac{3554}{7} \approx 507.7$$

We have seen that the fundamental theorem of calculus enables us to compute integrals by using antiderivatives. The inverse relationship between integration and differentiation is completed by the following alternative version of the fundamental theorem, which enables us to build up an antiderivative for a function by taking definite integrals and letting the endpoint vary.

Theorem 3 Fundamental Theorem of Calculus: Alternative Version.
Let f be continuous on the interval I and let a be a number in I. Define the function F on I by

$$F(t) = \int_a^t f(s) \, ds$$

Then $F'(t) = f(t)$; that is,

$$\frac{d}{dt}\left[\int_a^t f(s) \, ds\right] = f(t).$$

In particular, every continuous function has an antiderivative.

The full proof of Theorem 3 is given in the appendix to this chapter. However, we can prove the theorem now if we assume that f has *some* antiderivative, say F_1. For then, by Theorem 2,

$$\frac{d}{dt}\left[\int_a^t f(s) \, ds\right] = \frac{d}{dt}[F_1(t) - F_1(a)]$$

$$= \frac{d}{dt} F_1(t) \quad \text{(since } F_1(a) \text{ is constant)}$$

$$= f(t) \qquad \text{(since } F_1 \text{ is an antiderivative)}$$

as required.

Another look at our solar cell example will give further evidence. The energy stored between times a and t is

$$F(t) - F(a) = \int_a^t f(s) \, ds$$

Differentiating both sides of this equation, we get

$$F'(t) = \frac{d}{dt}\left[\int_a^t f(s) \, ds\right]$$

We saw in that example that $F'(t) = f(t)$ (see p. 193) so we arrive again at the conclusion of Theorem 3.

Worked Example 6 Verify Theorem 3 for the case $f(t) = t$.

Solution According to the formula in the display on p. 189, $\int_a^b f(t) \, dt = \int_a^b t \, dt = \frac{1}{2}(b^2 - a^2)$. Writing s for t gives

$$\int_a^b f(s) \, ds = \frac{1}{2}(b^2 - a^2)$$

and now writing t for b gives

$$\int_a^t f(s)\,ds = \tfrac{1}{2}(t^2 - a^2)$$

Differentiating both sides of the last equation with respect to t, we have

$$\frac{d}{dt}\left[\int_a^t f(s)\,ds\right] = \tfrac{1}{2}(2t) = t$$

Since $t = f(t)$, this result agrees with the conclusion of the fundamental theorem.

Worked Example 7   Let $F(t) = \int_2^t [1/(1 + s^2 + s^3)]\,ds$. Find $F'(3)$.

Solution   By Theorem 3, with $f(s) = 1/(1 + s^2 + s^3)$, we have $F'(3) = f(3) = 1/(1 + 3^2 + 3^3) = 1/37$. Notice that we did not need to differentiate or integrate $1/(1 + s^2 + s^3)$ to get the answer.

We summarize below the two forms of the fundamental theorem:

FUNDAMENTAL THEOREM OF CALCULUS

$$\int_a^b F'(t)\,dt = F(b) - F(a)$$

Integrating the derivative of F gives back F.

$$\frac{d}{dt}\int_a^t f(s)\,ds = f(t)$$

Differentiating the integral of f with respect to the upper limit gives back f.

Solved Exercises

7. Explain the meaning of the formula $\int f(t)\,dt = \int_a^t f(s)\,ds + C$.

8. Find
$$\int_{-2}^3 (x^4 + 5x^2 + 2x + 1)\,dx$$

9. If F is a differentiable function such that F' is integrable, write a formula for $F(t)$ in terms of F' and the value of F at a single number a.

10. Find
$$\int_1^2 \frac{x^2 + 2x + 2}{x^4}\,dx$$

Exercises

8. Find the following definite integrals:
   (a) $\int_{-2}^4 x^6\,dx$ (b) $\int_{-1}^1 (x^3 + 7)\,dx$ (c) $\int_1^8 [(1 + \theta^2)/\theta^4]\,d\theta$

9. Calculate:
   (a) $\int_1^2 (1 + 2t)^5\,dt$ (b) $\int_{-3}^{472} 0\,dt$ (c) $\int_2^3 dt/t^2$

10. Let $F(t) = \int_3^t \{1/[(4 - s)^2 + 8]^3\}\,ds$. Find $F'(4)$.

11. Find $\dfrac{d}{dx}\displaystyle\int_0^x \frac{t^4}{1 + t^6}\,dt$

PROPERTIES OF THE INTEGRAL

In the display on page 175 we listed five key properties of the summation process. In the following display we list the corresponding properties of the definite integral.

PROPERTIES OF INTEGRATION

1. $\int_a^b [f(t) + g(t)]\ dt = \int_a^b f(t)\ dt + \int_a^b g(t)\ dt$ (sum rule)

2. $\int_a^b c f(t)\ dt = c \int_a^b f(t)\ dt$, c a constant (constant multiple rule)

3. If $a < b < c$, then $\int_a^c f(t)\ dt = \int_a^b f(t)\ dt + \int_b^c f(t)\ dt$

4. If $f(t) = C$ is constant, then $\int_a^b f(t)\ dt = C(b - a)$

5. If $f(t) \le g(t)$ for all t satisfying $a \le t \le b$ then $\int_a^b f(t)\ dt \le \int_a^b g(t)\ dt$

These properties hold for integrable functions f and g, as will be proved in the problems for the appendix. However, it is much easier to deduce them from the sum and constant multiple rules for the indefinite integral (see p. 99) and the fundamental theorem of calculus. This approach assumes that f and g are continuous. The following example illustrates the procedure.

Worked Example 8   Prove property 1 in the display above (assuming f and g are continuous).

Solution   Let F be an antiderivative for f and G be one for g. They exist by theorem 3. Then $F + G$ is an antiderivative for $f + g$ by the sum rule for antiderivatives (Section 1–4). Thus,

$$
\begin{aligned}
\int_a^b [f(t) + g(t)]\ dt &= [F(t) + G(t)]\ \Big|_a^b \\
&= [F(b) + G(b)] - [F(a) + G(a)] \\
&= [F(b) - F(a)] + [G(b) - G(a)] \\
&= \int_a^b f(t)\ dt + \int_a^b g(t)\ dt
\end{aligned}
$$

The proofs of rules 2, 3 and 5 proceed in a similar manner (see Solved Exercise 11). Rule 4 is obvious.

We have defined the definite integral $\int_a^b f(t)\ dt$ when a is less than b; however, the right-hand side of the equation

$$
\int_a^b F'(t)\ dt = F(b) - F(a) \tag{13}
$$

makes sense even when $a \ge b$. Can we define $\int_a^b f(t)\ dt$ for the case $a \ge b$ so that equation (13) will still be true? The answer is simple.

Definition If $a > b$ and f is integrable on $[b, a]$, we define

$$\int_a^b f(t)\, dt \quad \text{to be} \quad -\int_b^a f(t)\, dt$$

If $a = b$, we define $\int_a^b f(t)\, dt$ to be zero.

Worked Example 9 Show that if F' is integrable on $[b, a]$, where $b < a$, then equation (13) holds.

Solution $\int_a^b F'(t)\, dt = -\int_b^a F'(t)\, dt = -[F(a) - F(b)] = F(b) - F(a)$.

"WRONG-WAY" INTEGRALS

$$\int_a^b f(t)\, dt = -\int_b^a f(t)\, dt \quad \text{if } b < a$$

$$\int_a^a f(t)\, dt = 0$$

Solved Exercises
11. Prove property 5 of the integral.
12. Find $\int_6^2 t^3\, dt$.
13. What property of the integral is analogous to substitution of index (see p. 177)?

Exercises
12. Prove properties 2 and 3 of the integral.
13. Find $\int_2^{-2} t^4\, dt$.
14. Verify property 3 of the integral if $b < a < c$.
15. What property of the integral is analogous to the telescoping sum (see p. 177)?

Problems for Section 4-2

1. Evaluate the following definite integrals:
   (a) $\int_1^3 t^3\, dt$
   (b) $\int_{-1}^2 (t^4 + 8t)\, dt$
   (c) $\int_0^{-4} (1 + x^2 - x^3)\, dx$
   (d) $\int_1^2 4\pi r^2\, dr$
   (e) $\int_2^{-1} (1 + t^2)^2\, dt$

2. Evaluate:
   (a) $\int_1^2 \left(s^3 + \frac{1}{s^2}\right) ds$
   (b) $\int_1^2 dt/(t + 4)^3$
   (c) $\int_{\pi/2}^{\pi} (3 + z^2)\, dz$
   (d) $\int_1^2 \frac{t^2 + 8t + 1}{t^4}\, dt$
   (e) $\int_{-1}^1 \frac{(1 + t^2)^2}{t^2}\, dt$

3. Evaluate:
   (a) $(d/dt) \int_0^t (3/(x^4 + x^3 + 1)^6)\, dx$
   (b) $(d/dt) \int_t^3 x^2 (1 + x)^5\, dx$
   (c) $\dfrac{d}{dt} \int_t^4 \dfrac{u^4}{(u^2 + 1)^3}\, du$

★4. Let f be continuous on the interval I and let a_1 and a_2 be in I. Define the functions:

$$F_1(t) = \int_{a_1}^t f(s)\, ds \quad \text{and} \quad F_2(t) = \int_{a_2}^t f(s)\, ds$$

   (a) Show that F_1 and F_2 differ by a constant.
   (b) Express the constant $F_2 - F_1$ as an integral. [*Hint:* Substitute $t = a_1$ or a_2.]

★5. Suppose that

$$f(t) = \begin{cases} t^2 & 0 \le t \le 1 \\ 1 & 1 \le t < 5 \\ (t-6)^2 & 5 \le t \le 6 \end{cases}$$

(a) Draw a graph of f on the interval $[0, 6]$.

(b) Find $\int_0^6 f(t)\, dt$.

(c) Find $\int_0^6 f(x)\, dx$.

(d) Let $F(t) = \int_0^t f(s)\, ds$. Find a formula for $F(t)$ in $[0, 6]$ and draw a graph of F.

(e) Find $F'(t)$ for t in $(0, 6)$.

6. (a) Give a formula for a function f whose graph is the broken line segment $ABCD$ in Fig. 4-2-3.

(b) Find $\int_3^{10} f(t)\, dt$.

(c) Find the area of the quadrilateral $ABCD$ by means of geometry and compare the result with the integral in part (b).

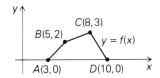

Fig. 4-2-3 Find a formula for f.

7. An object moving in a straight line has velocity $v = 6t^4 + 3t^2$ at time t. How far does the object travel between $t = 1$ and $t = 10$?

8. A tank with a faucet at the bottom has $V = F(t)$ liters of water in it at time t seconds. The faucet is manipulated so that water flows out at the rate $f(t)$ per second. Express the difference $F(b) - F(a)$ as an integral.

9. Check the following integral:

$$\int [3t^2/(1 + t^3)^2]\, dt = t^3/(1 + t^3) + C$$

10. Check the following integral formula:

$$\int \frac{x^3 + 2x + 1}{(1 - x)^5}\, dx = \frac{1}{x - 1} + \frac{3}{2}\frac{1}{(x-1)^2} + \frac{5}{3}\frac{1}{(x-1)^3} + \frac{1}{(x-1)^4} + C$$

11. Use the statement of Problems 9 and 10 to evaluate

(a) $\int_0^1 [3t^2/(1 + t^3)^2]\, dt$;

(b) $\int_2^3 [(s^3 + 2s + 1)/(1 - s)^5]\, ds$.

12. (a) Calculate the derivative of $x^3/(x^2 + 1)$.

(b) Find $\int_0^1 [(3x^2 + x^4)/(1 + x^2)^2]\, dx$.

13. Develop a formula for $\int x(1 + x)^n\, dx$ for $n \ne -1$ or -2 by studying Worked Example 5. [*Hint:* Guess the answer $(ax + b)(1 + x)^{n+1}$ and determine what a and b have to be.]

14. Prove property 3 of integration in the display on p. 201 for $c < b < a$.

15. What property of integration is analogous to the formula $\sum_{i=1}^n i = n(n + 1)/2$?

16. (a) Find the total revenue $R(x)$ from selling x units of a product, if the marginal revenue is $36 - 0.01\, x + 0.00015\, x^2$ and $R(0) = 0$.

(b) How much revenue is produced when the items from $x = 50$ to $x = 100$ are sold?

17. A small gold mine in northern Nevada was reopened in January 1979, producing 500,000 tons of ore in the first year. Let $A(t)$ be the number of tons of ore produced, in thousands, t years after 1979. Productivity $A'(t)$ is expected to decline by 20,000 tons per year until 1990.

(a) Find a formula for $A'(t)$, assuming the production decline is constant.

(b) How much ore is mined, approximately, T years after 1979 during a time period Δt?

(c) Find, by methods of definite integration, the predicted number of tons of ore to be mined from 1981 through 1986.

18. Let $W(t)$ be the number of words learned after t minutes are spent memorizing a French vocabulary list. Typically, $W(0) = 0$ and $W'(t) = 4(t/100) - 3(t/100)^2$.

(a) Apply the fundamental theorem of calculus to show that

$$W(t) = \int_0^t [4(x/100) - 3(x/100)^2]\, dx$$

(b) Evaluate the integral in part (a).

(c) How many words are learned after an hour and 40 minutes of study?

4-3 Areas

Integral calculus may be used to calculate the area of regions with curved edges.

In Section 4-1 we defined the integral with specific physical and geometric problems in mind. Now we solve a large variety of area problems.

Goals

After studying this section, you should be able to:

Compute the areas of regions under and between graphs.

Express geometric areas in terms of areas between graphs.

THE AREA UNDER A GRAPH

Fig. 4-3-1 The region under the graph of *f* on [*a*, *b*] is shaded.

In Section 4-1 (see Figures 4-1-12 and 4-1-13), we stated that the integral of a nonnegative function was the area under its graph. We now take up the subject of area in more detail.

Let *f* be a function whose domain includes [*a*, *b*] and suppose that $f(x) \geq 0$ for all *x* in [*a*, *b*]. By "the region under the graph of *f* on [*a*, *b*]," we mean the set of points (*x*, *y*) in the plane such that $a \leq x \leq b$ and $0 \leq y \leq f(x)$. (See Fig. 4-3-1.)

In other words, the region under the graph of *f* on [*a*, *b*] consists of those points which are to the right of the line *x = a*, to the left of the line *x = b*, above the line *y = 0*, and below the graph of *f*, including the bounding lines and curve.

Our problem is to determine the area of the region under the graph of *f* on [*a*, *b*]. For convenience, we will also refer to this area as "the area under the graph of *f* on [*a*, *b*]."* For the moment, we take the viewpoint that the area is a number which exists before we compute it—our task is to find the value of this number. (Later in this section, we will look at this viewpoint somewhat critically.) We will also take it for granted that areas of plane regions satisfy the following four properties:

1. Congruent figures have the same area.

2. The area of a rectangle is the product of the lengths of two adjacent sides (base times height).

3. If region *C* is made up of two regions, *A* and *B*, which do not overlap except along their common edges, then the area of *C* is the sum of the areas of *A* and *B*.

4. If region *A* is contained in region *B*, then the area of *A* is less than or equal to the area of *B*.

Using properties 1, 2, and 3 we will now show that if *f* is piecewise constant on [*a*, *b*], then the area under the graph of *f* on [*a*, *b*] is equal to the integral $\int_a^b f(x)\, dx$. In fact, if $(t_0, t_1, \ldots, t_n)$ is an adapted partition

We will consistently use the word region to denote a set of points in the plane (or in space), while the word area will denote a number which represents the "extent" of a given region. What we call a region is sometimes called an area in common usage ("this is a restricted area"), but we will never be using the word area in this way.

for f on $[a, b]$, with $f(x) = k_i$ on $[t_{i-1}, t_i]$, then the region under the graph of f on $[a, b]$ consists of n rectangles, with adjacent pairs sharing a common edge (Fig. 4-3-2). The area of the ith rectangle A_i is, by property 2, $k_i(t_i - t_{i-1}) = k_i \Delta t_i$. Applying property 3 $(n - 1)$-times, we may conclude that the area under the graph is $\Sigma_{i=1}^{n}$ area $(A_i) = \Sigma_{i=1}^{n} k_i \Delta t_i$, which is just the definition of $\int_a^b f(x)\,dx$.*

Fig. 4-3-2 The area of the shaded region is the sum of the areas of the rectangles A_i.

Now let f be *any* integrable function defined on $[a, b]$, not necessarily piecewise constant, such that $f(x) \geq 0$ for all x in $[a, b]$, and let R_f denote the region under the graph of f on $[a, b]$. If h is any piecewise constant function on $[a, b]$ such that $f(x) \leq h(x)$ for all x in $[a, b]$, the region R_h under the graph of h contains R_f (see Fig. 4-3-3). So, by property 4 of areas, we have area $(R_f) \leq$ area $(R_h) = \int_a^b h(x)\,dx$. In other words:

The area under the graph of f on $[a, b]$ is less than or equal to any upper sum for f.

Next let g be any piecewise constant function such that $0 \leq g(x) \leq f(x)$ for all x in $[a, b]$. Since R_g is contained in R_f (Fig. 4-3-3), we have $\int_a^b g(x)\,dx =$ area $(R_g) \leq$ area (R_f). It follows† that:

The area under the graph of f on $[a, b]$ is greater than or equal to any lower sum for f.

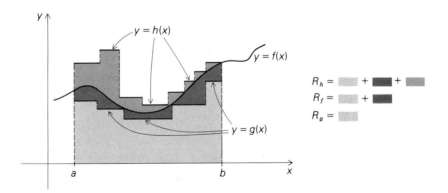

Fig. 4-3-3 The area under the graph lies between the lower and upper sums.

The reader may object that we have neglected the values of f at the partition points $t_0, ..., t_n$. Taking these into account would add some vertical line segments to our figure. We take it for granted that these line segments do not contribute to the area (or we may consider them as rectangles with base of zero length).

†*We have ignored a technical point here. Among the lower sums for x are integrals of piecewise constant functions g for which $g(x) \leq f(x)$ on $[a, b]$, but $g(x)$ may be negative for some x. In this case, we consider the intermediate function $g_+(x)$ obtained by changing $g(x)$ to zero wherever it is negative. Then we have $\int_a^b g(x)\,dx \leq \int_a^b g_+(x)\,dx$, obviously, while $\int_a^b g_+(x)\,dx \leq$ area (R_f) because we still have $g_+(x) \leq f(x)$ for all x in $[a, b]$.*

Now suppose that f is integrable on $[a, b]$. Then there is exactly one number which is less than or equal to every upper sum and greater than or equal to every lower sum—that number is the integral of f on $[a, b]$. It follows from the preceding statements that the area under the graph of f on $[a, b]$ is equal to the integral.

INTEGRALS AND AREAS

If $f(x) \geq 0$ for all x in $[a, b]$, and f is integrable on $[a, b]$,   the area under the graph of f on $[a, b]$ equals $\int_a^b f(x)\, dx$.

Since we know that any continuous function is integrable, we can express as an integral the area of any figure which is bounded below by the x axis, on the sides by two vertical lines, and above by the graph of a continuous function. Furthermore, the fundamental theorem of calculus gives us an efficient means for computing these areas.

Worked Example 1   Find the area under the graph of $f(x) = mx$ on $[0, b]$ $(m, b > 0)$. Interpret your answer geometrically.

Solution   By the fundamental principle of integrals and areas in the preceding box, the area is

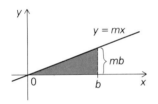

$$\int_0^b f(x)\, dx = \int_0^b mx\, dx = \tfrac{1}{2}mx^2 \Big|_0^b = \tfrac{1}{2}mb^2$$

If we draw the region under the graph (Fig. 4-3-4), we see that it is a triangle of base b and height mb. Our integral tells us that the area is $\frac{1}{2}$(base)(height), in accordance with elementary geometry.

Fig. 4-3-4   The area of the shaded triangle is equal to the integral $\int_0^b mx\, dx$.

Next we use the integral to find an area which cannot be computed by elementary geometry.

Worked Example 2   Find the area of the region bounded by the x axis, the y axis, the line $x = 2$, and the parabola $y = x^2$.

Solution   The region described is that under the graph of $f(x) = x^2$ on $[0, 2]$ (Fig. 4-3-5). The area of the region is $\int_0^2 x^2\, dx = \tfrac{1}{3}x^3 \big|_0^2 = \tfrac{8}{3}$.

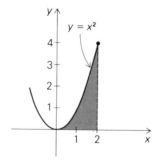

Fig. 4-3-5   The shaded area equals $\int_0^2 x^2\, dx$.

Worked Example 3 Find the area under the curve $y = x^2 + 1$ from $x = 0$ to $x = 2$.

Solution The region is the shaded portion of Fig. 4-3-6. The area is

$$\int_0^2 (x^2 + 1)\, dx = [(x^3/3) + x] \Big|_0^2 = \tfrac{8}{3} + 2 = \tfrac{14}{3}$$

From Fig. 4-3-6 we see that the area is the same as that in Fig. 4-3-5 plus the area of a rectangle with base 2 and height 1.

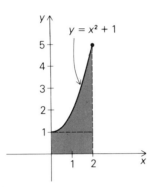

Fig. 4-3-6 The shaded area equals $\int_0^2 (x^2 + 1)\, dx$.

(optional)

Up to now, we have assumed that the area of a region is a preexisting number—all we wanted to do was to compute it. We have never given a mathematical definition of area, however, so the skeptical student may ask just what it is we are calculating. There are two approaches to this problem. The first is to construct a mathematical theory of area and show that the theory satisfies the properties 1 through 4. This can be done,* but we will not do it here. The second approach is to *define* the area of the region under the graph of *f* from *a* to *b* as the integral $\int_a^b f(x)\, dx$. Since the integral has a precise definition, we now have a precise definition of area, at least for certain regions. Although one may raise several objections to this definition, they can all be answered; furthermore, the idea of *defining* a geometric quantity as an integral has many other applications.

Possible objections:

1. How do we know that this definition of area agrees with our intuitive notion of area?

2. What about the area of regions other than those under graphs?

3. What about the area of the region under the graph of a nonintegrable function?

Responses:

1. The argument leading up to the fundamental principle on integrals and areas (p. 206) shows that if we consider area intuitively, it must be equal to the integral.

2. We will see in the next subsection how to break up many other regions into pieces, each of which is the region under a graph.

3. This is the stickiest question. Our response is this: The area of the region under the graph of a nonintegrable function is not defined. It is hard to imagine regions not having an area (as opposed to having zero area), but this is the only reasonable approach (just as we accept the existence of functions which do not have a derivative). Fortunately, the regions usually considered in geometry are bounded by continuous curves, and for those we have no problem.

The mathematical theory of area is discussed, for example, in Theory of Area, *by M. I. Knopp, Markham, 1969.*

The fundamental theorem of calculus (alternative version) states that $(d/dx) \int_a^x f(t)\,dt = f(x)$. This has a geometric interpretation: the rate of change of the area under the graph of f from a to x is $f(x)$ (assume f is positive and continuous). This idea can be used to give another justification of the connection between integrals and areas. See Problem 9.

Solved Exercises

1. (a) Find $\dfrac{d}{dx}\left[\dfrac{1}{(1 + x^2)}\right]$

   (b) Find the area under the graph of $x/(1 + x^2)^2$, from $x = 0$ to $x = 1$.

2. A parabolic doorway with base 6 feet and height 8 feet is cut out of a wall. How many square feet of wall space are removed?

Exercises

1. Find the area under the graph of $y = mx + b$ from $x = a_1$ to $x = a_2$ and verify your answer by using plane geometry. Assume that $mx + b \geq 0$ on $[a_1, a_2]$.

2. Find the area under the graph of $y = (1/x^2) + x + 1$ from $x = 1$ to $x = 2$.

3. Find the area under the graph of $y = x^2 + 1$ from -1 to 2 and sketch the region.

4. Find the area under the graph of

$$f(x) = \begin{cases} -x^3 & \text{if } x \leq 0 \\ x^3 & \text{if } x > 0 \end{cases}$$

   from $x = -1$ to $x = 1$ and sketch.

5. (a) Find $(d/dx)\,[x^3/(1 + x^3)]$.

   (b) Find the area under the graph of $x^2/(1 + x^3)^2$ from $x = 1$ to $x = 2$.

AREAS BETWEEN CURVES

By using the basic properties of area and the fundamental principle for integrals and areas, we can find the areas of many regions besides those under graphs.

If f and g are two functions defined on $[a, b]$, with $f(x) \leq g(x)$ for all x in $[a, b]$, we define *the region between the graphs* of f and g on $[a, b]$ to be the set of those points (x, y) such that $a \leq x \leq b$ and $f(x) \leq y \leq g(x)$.

Worked Example 4 Sketch and find the area of the region between the graphs of x^2 and $x + 3$ on $[-1, 1]$.

Solution The region is shaded in Fig. 4-3-7. It is not quite of the form we have been dealing with; however, we may note that if we add to it the region (dotted in Fig. 4-3-7) under the graph of x^2 on $[-1, 1]$, we obtain the region under the graph of $x + 3$ on $[-1, 1]$. Denoting the area of the shaded region by A, we have

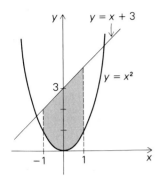

$$\int_{-1}^{1} x^2 \, dx + A = \int_{-1}^{1} (x + 3) \, dx$$

or

$$A = \int_{-1}^{1} (x + 3) \, dx - \int_{-1}^{1} x^2 \, dx = \int_{-1}^{1} (x + 3 - x^2) \, dx$$

Evaluating the integral yields $A = \frac{1}{2}x^2 + 3x - \frac{1}{3}x^3 \big|_{-1}^{1} = 5\frac{1}{3}$.

Fig. 4-3-7 The area of the shaded region is the difference between two areas under graphs.

The method of Worked Example 4 can be used in general to show that if $0 \leq f(x) \leq g(x)$ for x in $[a, b]$, then the area of the region between the graphs of f and g on $[a, b]$ is equal to $\int_a^b g(x) \, dx - \int_a^b f(x) \, dx = \int_a^b [g(x) - f(x)] \, dx$.

Worked Example 5 Find the area between the graphs of $y = x^2$ and $y = x^3$ for x between 0 and 1.

Solution Since $0 \leq x^3 \leq x^2$ on $[0, 1]$, by the principle just stated the area is $\int_0^1 (x^2 - x^3) \, dx = (x^3/3 - x^4/4) \big|_0^1 = \frac{1}{3} - \frac{1}{4} = \frac{1}{12}$.

Now let's look at a case where we do not have $f(x) \geq 0$.

Worked Example 6 Sketch and find the area of the region between the graphs of x and $x^2 + 1$ on $[-2, 2]$.

Solution The region is shaded in Fig. 4-3-8. We could find the area of this region by the method of the previous example if the region were entirely above the x axis. By adding a sufficiently large constant C to y (in this case, $C \geq 2$), we can shift the figure so that it does lie above the axis. (See Fig. 4-3-9.)

The shaded region in Fig. 4-3-9 is congruent to the one in Fig. 4-3-8 (in fact, we produced Fig. 4-3-9 by tracing Fig. 4-3-8 and moving the x axis down C units), so the two regions have the same area, by principle 1 on p. 204.

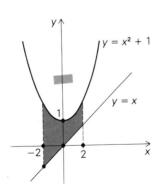

Fig. 4-3-8 What is the area of the shaded region?

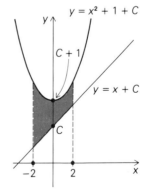

Fig. 4-3-9 The shaded region is congruent to the one in Fig. 4-3-8.

Now we can use the method of Worked Example 4 to find that the area is

$$\int_{-2}^{2} [(x^2 + 1 + C) - (x + C)]\, dx = \int_{-2}^{2} (x^2 + 1 - x)\, dx$$

Notice that the C's have canceled. We evaluate the integral as

$$\left(\frac{x^3}{3} + x - \frac{x^2}{2} \right) \Bigg|_{-2}^{2} = \left(\frac{8}{3} + 2 - \frac{4}{2} \right) - \left(-\frac{8}{3} - 2 - \frac{4}{2} \right) = \frac{28}{3}$$

The method of Worked Example 6 works whenever $f(x) \le g(x)$ on $[a, b]$ and there is some C such that $f(x) + C \ge 0$ on $[a, b]$. If f is integrable, it is bounded, so there is always such a C. We have, therefore, the following principle.

AREA BETWEEN GRAPHS

If $f(x) \le g(x)$ for all x in $[a, b]$, and f and g are integrable on $[a, b]$, then the area between the graphs of f and g on $[a, b]$ equals $\int_a^b [g(x) - f(x)]\, dx$.

As with the area under a graph, we can consider our formula for the area between two graphs as a definition of that area if we are not willing to accept area as an intuitively given quantity.

There is a heuristic argument for the formula for the area between graphs which gives a useful way of remembering the formula and deriving similar ones. We can think of the region between the graphs as being composed of infinitely many "infinitesimally wide" rectangles, of width dx, one for each x in $[a, b]$. (See Fig. 4-3-10.)

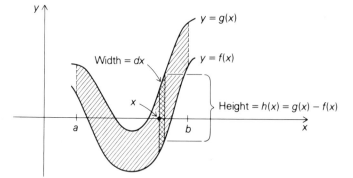

Fig. 4-3-10 We may think of the striped region as being composed of infinitely many rectangles, each of infinitesimal width.

The total area is then the "continuous sum" of the areas of these rectangles. The height of the rectangle over x is $h(x) = g(x) - f(x)$, the area of the rectangle is $[g(x) - f(x)]\, dx$, and the continuous sum of these areas is the integral $\int_a^b [g(x) - f(x)]\, dx$. This kind of infinitesimal argument was used frequently in the early days of calculus, when it was considered to be perfectly acceptable. Nowadays, we usually take the viewpoint of Archimedes, who used infinitesimals to discover results which he later proved by more rigorous, but much more tedious, arguments.

If a function $f(x)$ is *nonpositive* on the interval $[a, b]$, then we have $f(x) \leq 0$ for all x in $[a, b]$; hence $0 - f(x) \geq 0$ on $[a, b]$ and so the integral $\int_a^b [0 - f(x)] \, dx = -\int_a^b f(x) \, dx$ is the area between the graph of f and the x axis (that is, the graph of $g(x) = 0$) on $[a, b]$; thus $\int_a^b f(x) \, dx$ is the negative of this area. If the function $f(x)$ changes sign one or more times on $[a, b]$, then the integral $\int_a^b f(x) \, dx$ can be considered as a "signed" area: for every subinterval on which $f(x) \geq 0$, we take the area under the graph; for every subinterval on which $f(x) \leq 0$, we take the *negative* of the area between the graph and the x axis; adding those positive and negative areas gives $\int_a^b f(x) \, dx$ (see Fig. 4-3-11 and Solved Exercise 5).

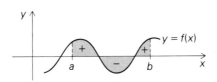

Fig. 4-3-11 The integral $\int_a^b f(x) \, dx$ is the sum of the areas marked + minus the areas marked −.

In Solved Exercise 4 below we show how integration can also be used to find the area of regions which are not "between two graphs" but can be broken into such regions.

Solved Exercises

3. Find the area between the graph of x^2 and the line $y = 4$ on $[0, 2]$.

4. The curves $x = y^2$ and $x = 1 + \frac{1}{2}y^2$ (neither of which is the graph of a function $y = f(x)$) divide the xy plane into five regions, only one of which is bounded (that is, contained in some rectangle). Sketch and find the area of this bounded region.

5. Interpret $\int_0^2 (x^2 - 1) \, dx$ in terms of areas.

Exercises

6. Find the area of the shaded region in Fig. 4-3-12.

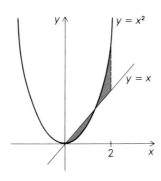

Fig. 4-3-12 Find the shaded area.

7. Find the area between the graph of $y = (2/x^2) + x^4$ and the line $y = 1$ between $x = 1$ and $x = 2$.

8. The curves $y = x^3$ and $y = x$ divide the plane into six regions, only two of which are bounded. Find their areas.

Problems for Section 4-3

1. Find the area of the region between the graph of each of the following functions and the x axis on the given interval and sketch.

   (a) $x^4 + 2$; on $[-1, 1]$ (b) x^3; on $[0, 2]$
   (c) $x^2 + 2x + 3$; on $[1, 2]$
   (d) $3x^4 - 2x + 2$; on $[-1, 1]$
   (e) $x^3 + 3x + 2$; on $[0, 2]$
   (f) $1/x^2$; on $[1, 2]$

2. Find the area under the graph of each of the following functions.

   (a) $(1/x^3) + x^2$; $1 \le x \le 3$
   (b) $(x^4 + 3x^2 + 1)$; $-2 \le x \le 1$
   (c) $(3x - 5)/x^3$; $1 \le x \le 2$
   (d) $8x^6 + 3x^4 - 2$; $1 \le x \le 2$

3. (a) Use calculus to find the area of the triangle whose vertices are $(0, 0)$, (a, h), and $(b, 0)$. (Assume $0 < a < b$ and $0 < h$.) Compare your result with a formula from geometry.

   (b) Repeat part (a) for the case $0 < b < a$.

   (c) Can you find a single calculation which handles cases (a) and (b) simultaneously?

4. Worked Example 5 in Section 4-2 showed that $\int x(1 + x)^6 \, dx = \frac{1}{56}(7x - 1)(1 + x)^7 + C$. Use this result to find the area under the graph of $1 + x(1 + x)^6$ between $x = -1$ and $x = 1$.

5. Find the area between the graphs in each of the following pairs of functions on the given interval:

   (a) x and x^4 on $[0, 1]$
   (b) $x^4 + 1$ and $1/x^2$ on $[1, 2]$
   (c) x^2 and $4x^4$ on $[2, 3]$
   (d) $3x^2$ and $x^4 + 2$ on $[-\frac{1}{2}, \frac{1}{2}]$
   (e) $4\dfrac{x^6 - 1}{x^6 + 1}$ and $\dfrac{(3x^6 - 1)(x^6 - 1)}{x^6(x^6 + 1)}$ on $[1, 2]$.

6. Find the area of each of the numbered regions in Fig. 4-3-13.

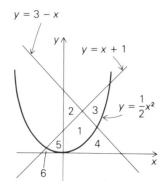

Fig. **4-3-13**   Find the area of each region.

7. Find the area of the shaded region in Fig. 4-3-14.

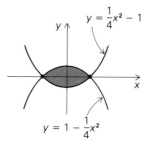

Fig. **4-3-14**   What is the area of the shaded region?

★8. Find the area in square centimeters, correct to within 1 square centimeter, of the region in Fig. 4-3-15.

Fig. **4-3-15**   Find the area of the "blob."

9. This problem gives another demonstration of the fundamental principle relating integrals to areas. Let $f(x) \ge 0$ on $[a, b]$, and define $F(x)$ to be the area under the graph of f on $[a, x]$.

   (a) Copy Fig. 4-3-16 and shade in the region whose area is $F(x + \Delta x) - F(x)$.

   (b) Compare the shaded region with a rectangle of area $f(x) \Delta x$.

   (c) Draw a figure whose area is $F(x + \Delta x) - F(x) - f(x) \Delta x$. (You may want to magnify your figure for this drawing.)

   (d) For Δx small, it appears that $f(x) \Delta x$ is a good approximation to $F(x + \Delta x) - F(x)$, so we guess that $F'(x) = f(x)$. Using this, together with the fundamental theorem of calculus, show that $F(x) = \int_a^x f(t) \, dt$.

Fig. **4-3-16**   Shade in the region whose area is $F(x + \Delta x) - F(x)$.

10. The lines $y = x$ and $y = 2x$ and the curve $y = 2/x^2$ together divide the plane into several regions, one of which is bounded. (a) How many regions are there? (b) Find the area of the bounded region.

11. The region under the graph $y = 1/x^2$ on $[1, 4]$ is to be divided into two parts of equal area by a vertical line. Where should the line be drawn?

12. Where will you draw a horizontal line to divide the region in Problem 11 into two parts of equal area?

13. Find the area of the shaded region in Fig. 4-3-17.

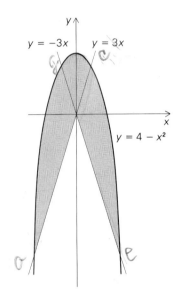

$y = -3x$ $y = 3x$

$y = 4 - x^2$

Fig. 4-3-17 What is the area of the shaded region?

14. A swimming pool has the shape of the region bounded by $y = x^2$ and $y = 2$. A swimming pool cover is estimated to cost $2.00 per square foot. If one unit along each of the x and y axes is 50 feet, then how much should the cover cost?

15. A circus tent is equipped with four exhaust fans at one end, each capable of moving 5500 cubic feet of air per minute. The rectangular base of the tent is 80 feet by 180 feet. Each corner is supported by a 20-foot-high post and there is a center beam supporting the roof, which is 32 feet off the ground and runs down the center of the tent for its 180-foot length. Canvas drapes in a parabolic shape from the center beam to the sides 20 feet off the ground. (See Fig. 4-3-18.) Determine the elapsed time for a complete change of air in the tent enclosure.

Exhaust fan
32 ft   20 ft   180 ft   80 ft

Fig. 4-3-18 The circus tent in Problem 15.

Review Problems for Chapter 4

1. Find the following sums:
   (a) $\Sigma_{i=1}^{5} 2^i/i(i+1)$ (b) $\Sigma_{j=4}^{8} (j^2 - 10)/3j$
   (c) $\Sigma_{i=0}^{100} (i^2 - 2i + 1)$
   (d) $\Sigma_{i=n}^{n+3} (i^2 - 1)/(i+1)$ (n a non-negative integer)
   (e) $\Sigma_{i=1}^{500} (3i + 7)$

★2. Find $\int_a^b (3t^2 + t)\, dt$ "by hand."

3. (a) Suppose that the velocity v of a moving object is a function $f(x)$ of its *position x*. Assume that $f(x) > 0$ for all x in $[x_1, x_2]$; express the *time* required to travel from x_1 to x_2 as an integral. [*Hint:* Consider first the case where f is piecewise constant. This would correspond to the situation where the speed limit on a highway changes occasionally, and one always drives exactly at the speed limit.]

   (b) Suppose the velocity of an object at position x is x^n, where n is some rational number $\neq 1$. Find the time required to travel from $x = \frac{1}{1000}$ to $x = 1$.

4. Let f be defined on $[0, 1]$ by

$$f(t) = \begin{cases} 1 & 0 \le t < \frac{1}{5} \\ 2 & \frac{1}{5} \le t < \frac{1}{4} \\ 3 & \frac{1}{4} \le t < \frac{1}{3} \\ 4 & \frac{1}{3} \le t < \frac{1}{2} \\ 5 & \frac{1}{2} \le t \le 1 \end{cases}$$

Find $\int_0^1 f(t)\, dt$.

5. Let

$$g(y) = \begin{cases} y & 0 \le y < 1 \\ 2 & 1 \le y < 2 \\ y & 2 \le y \le 4 \end{cases}$$

Compute $\int_0^1 g(y)\, dy + \int_3^4 g(y)\, dy$.

6. Let

$$y(t) = \begin{cases} t & 2 \le t < 3 \\ -4 & 3 \le t < 4 \\ 1 & 4 \le t \le 5 \end{cases}$$

Compute $\int_5^2 y(t)\, dt$.

7. Evaluate the definite integrals:
   (a) $\int_3^5 (-2x^3 + x^2)\, dx$
   (b) $\int_1^3 [(x^3 - 5)/x^2]\, dx$
   (c) $\int_1^2 [\frac{1}{3}s^2 - (s^4 + 1)/2s^2]\, ds$
   (d) $\int_1^2 [(x^2 + 3x + 2)/(x + 1)]\, dx$

8. Find the area under the graphs of the following functions between the indicated limits.
   (a) $y = x^3 + x^2, 0 \le x \le 1$
   (b) $y = \dfrac{x^2 + 2x + 1}{x^4}, 1 \le x \le 2$
   (c) $y = (x - 1)^3, 1 \le x \le \pi$

9. Find the area between the graphs of $y = x^3$ and $y = 5x^2 + 2x$ between $x = 0$ and $x = 2$. Sketch.

10. (a) Verify the integration formula

$$\int [x^2/(x^3 + 6)^2]\, dx = \tfrac{1}{12}(x^3 + 2)/(x^3 + 6) + C$$

   (b) Find the area under the graph of $y = x^2/(x^3 + 6)^2$ between $x = 0$ and $x = 2$.

11. (a) At time $t = 0$, a container has 1 liter of water in it. Water is poured in at the rate of $3t^2 - 2t + 3$ liters per minute (t = time in minutes). If the container has a leak which can drain 2 liters per minute, how much water is in the container at the end of 3 minutes?

   (b) What if the leak is 4 liters per minute?

   (c) What if the leak is 8 liters per minute? [*Hint:* What happens if the tank is empty for a while?]

12. An object is thrown at $t = 0$ from an airplane, and it has vertical velocity $v = -10 - 32t$ feet per second at time t. If the object is still falling after 10 seconds, what can you say about the altitude of the plane at $t = 0$?

13. (a) Find upper and lower sums for $\int_0^1 [4/(1 + x^2)]\, dx$ within 0.2 of one another.
   (b) Look at the average of these sums. Can you guess what the exact integral is?

★14. Let f be defined on $[0, 1]$ by

$$f(t) = \begin{cases} 0 & t = 0 \\ \dfrac{1}{\sqrt{t}} & 0 < t \le 1 \end{cases}$$

   (a) Show that there are no upper sums for f on $[0, 1]$, and hence that f is not integrable.

   (b) Show that every number less than 2 is a lower sum. [*Hint:* Use step functions which are zero on an interval $[0, \varepsilon)$ and approximate f very closely on $[\varepsilon, 1]$. Take ε small and use the integrability of f on $[\varepsilon, 1]$.]

   (c) Show that no number greater than or equal to 2 is a lower sum. [*Hint:* Show $\varepsilon f(\varepsilon) + \int_\varepsilon^1 f(t)\, dt < 2$ for all ε in $(0, 1)$.]

   (d) If you had to assign a value to $\int_0^1 f(t)\, dt$, what value would you assign?

★15. Modeling your discussion after Problem 14, find the upper and lower sums for each of the following functions on $[0, 1]$:

   (a) $f(t) = \begin{cases} 0 & t = 0 \\ -\dfrac{1}{\sqrt[3]{t}} & t > 0 \end{cases}$

   (b) $f(t) = \begin{cases} 0 & t = 0 \\ \dfrac{1}{t^2} & t > 0 \end{cases}$

16. Suppose that a *supply curve* $p = S(x)$ and a *demand curve* $p = D(x)$ are graphed and there is a unique point (a, b) at which supply equals demand (p = price/unit in dollars, x = number of units). The (signed) area enclosed by $x = 0$, $x = a$, $p = b$, and $p = D(x)$ is called the *consumer's surplus* or the *consumer's loss* depending on whether the sign is positive or negative, respectively. Similarly, the (signed) area enclosed by $x = 0$, $x = a$, $p = b$, and $p = S(x)$ is called the *producer's surplus* or the *producer's loss* depending on whether the sign is positive or negative, respectively.

   (a) Let $D(x) > b$. Explain why *consumer's surplus* $= \int_0^a [D(x) - b] \, dx$.

   (b) Let $S(x) < b$. Explain why *producer's surplus* $= \int_0^a [b - S(x)] \, dx$.

   (c) "If the price stabilizes at \$6/unit, then some people are still willing to pay a higher price, but benefit by paying the lower price of \$6/unit. The total of these benefits over $[0, a]$ is the consumer's surplus." Explain in the language of integration theory.

   (d) Find the consumer's and producer's surplus for the supply curve $p = x^2/8$ and the demand curve $p = -(x/4) + 1$.

17. The demand for wood products in 1975 was about 12.6 billion cubic feet. By measuring order increases, it was determined that x years after 1975, the demand increased by $9x/1000$; that is, $D'(x) = 9x/1000$, where $D(x)$ is the demand x years after 1975, in billions of cubic feet.

   (a) Use the fundamental theorem of calculus to show that

   $$D(x) = D(0) + \int_0^x (9t/1000) \, dt.$$

   (b) Find $D(x)$.

   (c) Find the demand for wood in 1982.

18. Deliveries to locations along a straight portion of highway are made by a truck from a warehouse on the highway. Let $d(t)$ be the distance from the delivery truck to its warehouse and let $v(t)$ be the velocity of the truck, after t hours of deliveries, given by $v(t) = d'(t)$.

   (a) The accumulated mileage for T hours of deliveries is generally *not* $d(T) - d(0)$. Explain.

   (b) The accumulated mileage for the first hour of deliveries is generally *not* $\int_0^1 v(t) \, dt$. Explain, using the fundamental theorem of calculus.

   (c) Suppose the velocity is positive on $(0, 1)$ and $(2, 2.5)$, negative on $(1, 2)$, and $v(0) = v(1) = v(2) = v(2.5) = 0$. Justify that the mileage after $2\frac{1}{2}$ hours is $\int_0^{2.5} |v(t)| \, dt$.

19. A rock is dropped off a bridge over a gorge. The sound of the splash is heard 5.6 seconds after the rock was dropped. (Assume the rock falls with velocity $32t$ ft/sec and sound travels at 1080 ft/sec.)

   (a) Show by integration that the rock falls $16t^2$ ft after t seconds, and that the sound of the splash travels $1080t$ feet in t seconds.

   (b) The time T required for the rock to hit the water must satisfy $16T^2 = 1080(5.6 - T)$, because the rock and the sound wave travel equal distances. Find T.

   (c) Find the height of the bridge.

   (d) Find the number of seconds required for the sound of the splash to travel from the water to the bridge.

20. The current $I(t)$ and charge $Q(t)$ at time t (in amperes and coulombs, respectively) in a circuit are related by the equation $I(t) = dQ/dt$.

   (a) Given $Q(0) = 1$, use the fundamental theorem of calculus to justify the formula $Q(t) = 1 + \int_0^t I(r) \, dr$.

   (b) The voltage drop V (in volts) across a resistor of resistance R ohms is related to the current I (in amperes) by the formula $V = RI$. Suppose in a simple circuit with a resistor made of nichrome wire, $V = 4.36$, $R = 1$, and $Q(0) = 1$. Find $Q(t)$.

   (c) Repeat (b) for a circuit with a 12-volt battery and 4-ohm resistance.

21. A ruptured sewer line caused lake contamination near a ski resort. The concentration $C(t)$ of bacteria (number per cubic centimeter) after t days is typically given by $C'(t) = 10^3(t - 7)$, $0 \le t \le 6$, after treatment of the lake at $t = 0$.

   (a) An inspector will be sent out after the bacteria concentration has dropped to half its original value $C(0)$. On which day should the inspector be sent if $C(0) = 40(10^3)$?

   (b) What is the total change in the concentration from the fourth day to the sixth day?

5

Inverse Functions and the Chain Rule

General rules for the differentiation of inverse and composite functions may be applied to fractional powers.

In this chapter we extend our techniques of differentiation to a wider class of functions, such as $\sqrt{1 + x^2}$ and $x^{5/3}$. Our expansion of techniques and applications will continue in the next two chapters, where we discuss the trigonometric, exponential, and logarithm functions. A brief introduction to these functions is given in the supplement to this chapter.

CONTENTS

Section 5-1 deals with the differentiation of inverse functions. This will be applied to the differentiation of fractional powers such as $x^{5/3}$. Section 5-2 deals with compositions; the theory there enables us to differentiate expressions such as $\sqrt{1 + x^2}$. Applications to word problems and related rate problems are also given. The theory developed in this chapter is essential in each of the succeeding two chapters.

5-1 Inverse Functions and Fractional Powers

The derivative of the inverse is the reciprocal of the derivative.

Sometimes two variable quantities are related in such a way that either one may be considered as a function of the other. The relationship between the quantities may be expressed by either of two functions, which are called *inverses* of one another. In this section, we will learn when a given function has an inverse, and we will see a useful relation between the derivative of a function and the derivative of its inverse.

Goals

After studying this section, you should be able to:

Determine whether a given function has an inverse.

Find the derivative of the inverse of f, given the derivative of f.

Differentiate fractional powers.

DEFINITION OF INVERSE FUNCTIONS

To illustrate the notion of an inverse, let us consider a certain function associated with making yogurt. A yogurt culture is added to milk which has been boiled, and the resulting mixture is set aside for 4 hours at a temperature T. At the end of the 4 hours, the "sourness" of the yogurt (specifically, the amount of lactic acid which has formed) is measured. Let S denote this sourness. The process is repeated at various temperatures between 10°C and 60°C, all other variables being kept fixed, and it is found that the sourness S is a function $f(T)$ of the fermentation temperature T. (See Fig. 5-1-1. If T is too low, the culture is dormant; if T is too high, the culture is killed.)

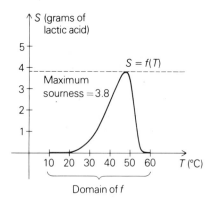

Fig. 5-1-1 The sourness of yogurt as a function of fermentation temperature.

In making yogurt to suit one's taste, one might desire a certain degree of sourness and wish to know what temperature to use. (Remember that we are fixing all other variables, including the time of fermentation.) To find the temperature which gives $S = 2$, for instance, one may draw the horizontal line $S = 2$, see if it intersects the graph of f, and read off the value of T (Fig. 5-1-2).

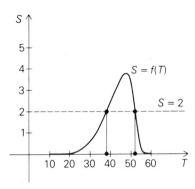

Fig. 5-1-2 Both $T = 38°C$ and $T = 52°C$ produce a sourness $S = 2$.

From the graph we see that there are two possible values of T: 38°C and 52°C. Similarly, there are two possible temperatures to achieve any value of S strictly between zero and the maximum value 3.8 of f. If, however, we restrict the allowable temperatures to the interval $[20, 47]$, then we will get a unique value of T for each S in $[0, 3.8]$. (See Fig. 5-1-3.)

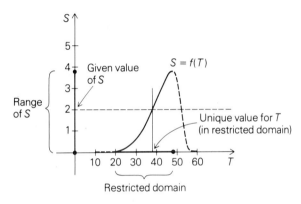

Fig. 5-1-3 If we only allow T in the range 20°C to 47°C, each value of S comes from just one value of T.

We may describe the situation by saying that there is a new function g which assigns to each S in the domain $[0, 3.8]$ a number $T = g(S)$ such that T is the unique element of $[20, 47]$ for which $S = f(T)$. While f assigns to each temperature the corresponding sourness, g assigns to each sourness the temperature which corresponds to it. We call g the *inverse function* to f, since it does just the opposite of f. The graph of g is drawn in Fig. 5-1-4; we obtain it from the graph of f by interchanging the role of the horizontal and vertical coordinates, which we can do by flipping the graph over along a line which bisects the angle between the S and T axes. (We actually drew the graph of g by tracing the graph of f from the back side of the page—try it yourself.)

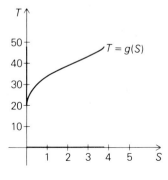

Fig. 5-1-4 The graph of T as a function of S.

Motivated by this example, we give a formal definition.

Definition Let the function f be defined on a set A. Let B be the range of values for f on A; that is, $y \in B$ means that $y = f(x)$ for some x in A. We say that f is *invertible on* A if, for every y in B, there is a *unique* x in A such that $y = f(x)$.

If f is invertible on A, then there is a function g, whose domain is B, given by this rule: $g(y)$ is that unique x in A for which $f(x) = y$. We call g the *inverse function* of f on A. The inverse function of f is denoted by f^{-1}.

Worked Example 1   Let $f(x) = x^2$. Find a suitable A such that f is invertible on A. Find the inverse function and sketch its graph.

Solution   The graph of f on $(-\infty, \infty)$ is shown in Fig. 5-1-5. For y in the range of f—that is, $y \geq 0$—there are two values of x such that $f(x) = y$: namely, $-\sqrt{y}$ and $+\sqrt{y}$. To assure only one x we may restrict f to $A = [0, \infty)$. Then $B = [0, \infty)$ and for any y in B there is exactly one x in A such that $f(x) = y$: namely, $x = \sqrt{y} = g(y)$. (Choosing $A = (-\infty, 0]$ and obtaining $x = -\sqrt{y}$ is also acceptable.) We obtain the graph of the inverse by looking at the graph of f from the back of the page.

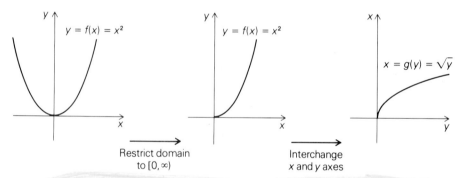

Fig. 5-1-5   The function $f(x) = x^2$ and its inverse $g(y) = \sqrt{y}$.

Warning   Notice from this example that the inverse f^{-1} is not in general the same as $1/f$.

If we are given a formula for $y = f(x)$ in terms of x, we may be able to find a formula for its inverse $x = g(y)$ by solving the equation $y = f(x)$ for x in terms of y. Sometimes, for complicated functions f, one cannot solve $y = f(x)$ to get an explicit formula for x in terms of y. In that case, one must resort to theoretical results which guarantee the existence of an inverse function. These will be discussed shortly.

INVERSE FUNCTIONS

To find the inverse function g for f,   solve the equation $y = f(x)$ for x in terms of y. If the solution x is unique, set $x = g(y)$.

It may be necessary to restrict the domain of f before there is an inverse function.

Calculator Discussion

Recall (see p. 30) that a function f may be thought of as an operational key on a calculator. The inverse of a function should be another key, which we can label f^{-1}. According to the definition, if we feed in any x, then push f to get $y = f(x)$, then push f^{-1}, we get back $x = f^{-1}(y)$, the number we started with. Likewise if we feed in a number y and push first f^{-1} and then f we get y back again. By Worked Example 1, $y = x^2$ and $x = \sqrt{y}$ (for $x \geq 0$, $y \geq 0$) are inverse functions. Try it out numerically, by pushing $x = 3.0248759$, then the x^2 key, then the $\sqrt{x}$ key. Try it also in the reverse order. (The answer may not come out exactly right because of round-off errors.)

Worked Example 2   Does $f(x) = x^3$ have an inverse on $(-\infty, \infty)$? Sketch.

Solution   From Fig. 5-1-6 we see that the range of f is $(-\infty, \infty)$ and for each $y \in (-\infty, \infty)$ there is exactly one number x such that $f(x) = y$—namely, $x = \sqrt[3]{y} = f^{-1}(y)$ (negative if $y < 0$, positive or zero if $y \geq 0$)—so the answer is yes. We can regain x as our independent variable by observing that if $f^{-1}(y) = \sqrt[3]{y}$, then $f^{-1}(x) = \sqrt[3]{x}$. Thus we can replace y by x so that the independent variable has a more familiar name. As shown in Fig. 5-1-6(b) and (c), this renaming does not affect the graph of f^{-1}, since the independent variable is always drawn on the horizontal axis.

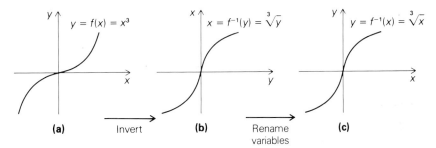

Fig. 5-1-6 The function $y = x^3$ and its inverse $x = \sqrt[3]{y}$.

Solved Exercises*

1. If $m \neq 0$, find the inverse function for $f(x) = mx + b$ on $(-\infty, \infty)$.

2. Find the inverse function for $f(x) = (ax + b)/(cx + d)$ on its domain (assume $c \neq 0$).

3. Find an inverse function g for $f(x) = x^2 + 2x + 1$ on some interval containing zero. What is $g(9)$? What is $g(x)$?

4. Sketch the graph of the inverse function for each function in Fig. 5-1-7.

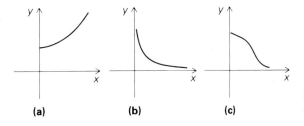

Fig. 5-1-7 Sketch the graph of the inverse.

Solutions appear in Appendix A.

5. Determine whether or not each function in Fig. 5-1-8 is invertible on its domain.

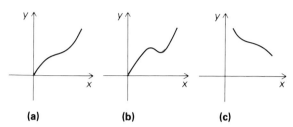

(a)   (b)   (c)

Fig. 5-1-8 Is there an inverse?

Exercises

1. Find the inverse for each of the following functions on the given interval:

(a) $f(x) = 2x + 5$ on $[-4, 4]$ (b) $f(x) = -\frac{1}{3}x + 2$ on $(-\infty, \infty)$

(c) $h(t) = t - 10$ on $[0, \pi)$

(d) $a(s) = (2s + 5)/(-s + 1)$ on $[-\frac{1}{2}, \frac{1}{2}]$

(e) $f(x) = x^5$ on $(-\infty, \infty)$ (f) $f(x) = x^8$ on $(0, 1]$

2. Determine whether each function in Fig. 5-1-9 has an inverse. Sketch the inverse if there is one.

(a)   (b)   (c)

Fig. 5-1-9 Which functions have inverses?

3. Sketch a graph of $f(x) = x/(1 + x^2)$ and find an interval on which f is invertible.

4. Enter the number 2.6 on your calculator, then push the x^2 key followed by the $\sqrt{x}$ key. Is there any round-off error? Try the $\sqrt{x}$ key, then the x^2 key. Also try a sequence such as $x^2, \sqrt{x}, x^2, \sqrt{x}, \ldots$ Do the errors build up? Try pushing the x^2 key five times, then the $\sqrt{x}$ key five times. Do you get back the original number? Try these experiments with different starting numbers.

5. If we think of a French-English dictionary as defining a function from the set of French words to the set of English words (does it really?), how is the inverse function defined? Discuss.

A TEST FOR INVERTIBILITY

A function may be invertible even though we cannot find an explicit formula for the inverse function. This fact gives us a way of obtaining "new functions." The following is a useful calculus test for finding intervals on which a function is invertible.

Theorem 1 *Suppose that f is continuous on* $[a, b]$ *and that f is increasing at each point of* (a, b). *(For instance, this holds if* $f'(x) >$ 0 *for each x in* (a, b).) *Then f is invertible on* $[a, b]$, *and the inverse* f^{-1} *is defined on* $[f(a), f(b)]$.

If f is decreasing rather than increasing at each point of (a, b), *then f is still invertible; in this case, the domain of* f^{-1} *is* $[f(b), f(a)]$.

Proof If f is continuous on $[a, b]$ and increasing at each point of (a, b), we know by the results of Section 2-2 on increasing functions that f is increasing on $[a, b]$; that is, if $a \leq x_1 < x_2 \leq b$, then $f(x_1) < f(x_2)$. In particular, $f(a) < f(b)$. If y is any number in $(f(a), f(b))$, then by the intermediate value theorem (first version, p. 112), there is an x in (a, b) such that $f(x) = y$. If $y = f(a)$ or $f(b)$, we can choose $x = a$ or $x = b$. Since f is increasing on $[a, b]$ for any y in $[f(a), f(b)]$, there can only be one x such that $y = f(x)$. In fact, if we had $f(x_1) = f(x_2) = y$, with $x_1 < x_2$, we would have $y < y$, which is impossible. Thus, by definition, f is invertible on $[a, b]$ and the domain of f^{-1} is the range $[f(a), f(b)]$ of values of f on $[a, b]$. The proof of the second assertion is similar.

Worked Example 3   Verify, using Theorem 1, that $f(x) = x^2$ has an inverse if f is defined on $[0, b]$ for a given $b > 0$.

Solution   Since f is differentiable on $(-\infty, \infty)$, it is continuous on $(-\infty, \infty)$ and hence on $[0, b]$. But $f'(x) = 2x > 0$ for $0 < x < b$. Thus f is increasing. Hence Theorem 1 guarantees that f has an inverse defined on $[0, b^2]$. Note that this example enables one to prove the existence of square roots.

In general, a function is not monotonic throughout the interval on which it is defined. Theorem 1 shows that the turning points of f divide the domain of f into subintervals on each of which f is invertible.

Solved Exercises   6. Let $f(x) = x^5 + x$.
   (a) Show that f has an inverse on $[-2, 2]$. What is the domain of this inverse?
   (b) Show that f has an inverse on $(-\infty, \infty)$.
   (c) What is $f^{-1}(2)$?
   (d) Numerically calculate $f^{-1}(3)$ to two decimal places of accuracy.

7. Find intervals on which $f(x) = x^5 - x$ is invertible.

8. Show that, if n is odd, $f(x) = x^n$ is invertible on $(-\infty, \infty)$. What is the domain of the inverse function?

9. Discuss the invertibility of $f(x) = x^n$ for n even.

Exercises   6. Show that $f(x) = -x^3 - 2x + 1$ is invertible on $[-1, 2]$. What is the domain of the inverse?

7. (a) Show that $f(x) = x^3 - 2x + 1$ is invertible on $[2, 4]$. What is the domain of the inverse? (b) Find the largest possible intervals on which f is invertible.

8. Find the largest possible intervals on which $f(x) = 1/(x^2 - 1)$ is invertible. Sketch the graphs of the inverse functions.

9. Show that $f(x) = \frac{1}{3}x^3 - x$ is not invertible on any open interval containing 1.

10. Let $f(x) = x^5 + x$. (a) Find $f^{-1}(246)$. ▦(b) Find $f^{-1}(4)$, correct to at least two decimal places.

DIFFERENTIATING INVERSE FUNCTIONS

Even though the inverse function $f^{-1}(y)$ is defined somewhat abstractly, there is a simple formula for its derivative. To motivate the formula we can proceed in several ways. First of all, note that if l is the linear function $l(x) = mx + b$, then $l^{-1}(y) = (1/m)y - (b/m)$ (see Solved Exercise 1), so $l'(x) = m$, and $(l^{-1})'(y) = 1/m$, the reciprocal. We can express this by the formula $dx/dy = 1/(dy/dx)$.

Now examine Fig. 5-1-10, where we have flipped the graph of f and its tangent line $y = mx + b$ in the drawing on the top to obtain the graph of f^{-1} together with the line $x = l^{-1}(y) = y/m - b/m$ in the drawing on the bottom. Since the line is tangent to the curve on the top, and flipping the drawing should preserve this tangency, the line $x = (1/m)y - b/m$ ought to be the tangent line to $x = f^{-1}(y)$ at (y_0, x_0), and its slope $1/m$ should be the derivative $(f^{-1})'(y_0)$. Since $m = f'(x_0)$, we conclude that

$$(f^{-1})'(y_0) = \frac{1}{f'(x_0)}$$

Since the expression $(f^{-1})'$ is awkward, we sometimes revert to the notation $g(y)$ for the inverse function and write

$$g'(y_0) = \frac{1}{f'(x_0)}$$

Fig. 5-1-10 The inverse of the tangent line is the tangent line of the inverse.

Notice that although dy/dx is not an ordinary fraction, the rule

$$\frac{dx}{dy} = \frac{1}{dy/dx}$$

is valid. (Maybe the "reciprocal" notation f^{-1} isn't so bad after all!)

Theorem 2 *Suppose that $f'(x) > 0$ or $f'(x) < 0$ for all x in an open interval I containing x_0, so that by Theorem 1 there is an inverse function g to f, defined on an open interval containing $f(x_0) = y_0$, with $g(y_0) = x_0$. Then g is differentiable at y_0 and*

$$g'(y_0) = \frac{1}{f'(x_0)} = \frac{1}{f'(g(y_0))}$$

Proof Assuming that the inverse function is continuous (see Problem 15 for a proof), then we can prove Theorem 2 by using limits. Recall that $f'(x_0) = dy/dx = \lim_{\Delta x \to 0} (\Delta y/\Delta x)$, where Δx and Δy denote changes in x and y as usual. On the other hand,

$$g'(y_0) = \frac{dx}{dy} = \lim_{\Delta y \to 0} \frac{\Delta x}{\Delta y} = \frac{1}{\lim_{\Delta y \to 0} \Delta y / \Delta x}$$

by the reciprocal rule for limits. But $\Delta x \to 0$ when $\Delta y \to 0$, since g is continuous, so

$$g'(y_0) = \frac{1}{\lim_{\Delta x \to 0} (\Delta y / \Delta x)} = \frac{1}{dy / dx}$$

Worked Example 4   Use the inverse function rule to compute the derivative of $\sqrt{x}$. Evaluate the derivative at $x = 2$.

Solution   Let us write $g(y) = \sqrt{y}$. This is the inverse function to $f(x) = x^2$. Since $f'(x) = 2x$,

$$g'(y) = \frac{1}{f'(g(y))} = \frac{1}{2g(y)} = \frac{1}{2\sqrt{y}}$$

so $(d/dy)(\sqrt{y}) = 1/(2\sqrt{y})$. We may substitute any letter for y in this result, including x, so we get the formula

$$\frac{d}{dx}\sqrt{x} = \frac{1}{2\sqrt{x}}$$

When $x = 2$, the derivative is $1/(2\sqrt{2})$.

Worked Example 5   Find the first-order approximation to $\sqrt{9.06}$.

Solution   Let $f(x) = \sqrt{x}$. Since we know that $f(9) = 3$, we use the approximation

$$f(x) \approx f(x_0) + f'(x_0)(x - x_0)$$

with $x_0 = 3$ to get

$$\sqrt{9.06} = f(9.06) \approx f(9) + f'(9)(9.06 - 9)$$

Since $f(9) = 3$ and $f'(9) = 1/(2\sqrt{9}) = \frac{1}{6}$, we have $\sqrt{9.06} \approx 3 + \frac{1}{6}(0.06) = 3.01$. (Actually, $\sqrt{9.06} = 3.00998\ldots$, so the error is less than 0.00002. Notice also that if we compute an approximation to $(3.01)^2$ by using the first-order approximation to the squaring function, we get *exactly* 9.06.)

INVERSE FUNCTION RULE

To differentiate an inverse function at y,   take the reciprocal of the derivative of the given function at $x = f^{-1}(y)$:

$$g'(y) = \frac{1}{f'(g(y))} \quad \text{if } g = f^{-1}$$

$$\frac{dx}{dy} = \frac{1}{dy / dx}$$

Solved Exercises

10. Verify the inverse function rule for $y = (ax + b)/(cx + d)$ by finding dy/dx and dx/dy directly. (See Solved Exercise 2.)

11. If $f(x) = x^3 + 2x + 1$, show that f has an inverse on $[0, 2]$. Find the derivative of the inverse function at $y = 4$.

12. Find the equation of the line tangent to the graph of $y = \sqrt{x}$ at $(2, \sqrt{2})$.

Exercises

11. Let $y = x^3 + 2$. Find dx/dy when $y = 3$.

12. If $f(x) = x^5 + x$, find the derivative of the inverse function when $y = 34$.

13. (a) Find the derivative of $\sqrt[3]{x}$ on $I = (0, \infty)$. (b) Find the first-order approximation to $\sqrt[3]{8.06}$.

14. For each function f below, find the derivative of the inverse function g at the points indicated:

   (a) $f(x) = 3x + 5$; find $g'(2)$, $g'(\frac{3}{4})$.

   (b) $f(x) = x^5 + x^3 + 2x$; find $g'(0)$, $g'(4)$.

   (c) $f(x) = \frac{1}{12}x^3 - x$ on $[-1, 1]$; find $g'(0)$, $g'(\frac{11}{12})$.

DIFFERENTIATING FRACTIONAL POWERS

We will now use the formula for the derivative of an inverse function to calculate the derivative of x^r where r is rational, not necessarily an integer. For example, what is $(d/dx)(x^{3/2})$? If you guessed $\frac{3}{2}x^{1/2}$, you were right, as we will see.

Let us begin by considering $y^{1/n}$ for n a positive integer and then differentiate with respect to y. We temporarily use the variable name "y" since we will be using the inverse function rule. As we have seen in Solved Exercises 8 and 9, $f(x) = x^n$ and $g(y) = y^{1/n}$ are inverse functions. For $x \neq 0$ (and $x > 0$ if n is even) $f'(x) = nx^{n-1} > 0$, so

$$\frac{d}{dy}y^{1/n} = \frac{1}{f'(x)} = \frac{1}{nx^{n-1}}$$

Since $x = y^{1/n}$, this equals

$$\frac{1}{ny^{(n-1)/n}} = \frac{1}{n}y^{(1-n)/n} = \frac{1}{n}y^{(1/n)-1}$$

Thus

$$\frac{d}{dy}y^{1/n} = \frac{1}{n}y^{(1/n)-1}$$

We can revert to x notation and rephrase our answer:

$$\frac{d}{dx}x^{1/n} = \frac{1}{n}x^{(1/n)-1} \quad (x \neq 0)$$

Worked Example 6 Differentiate $f(x) = 3\sqrt[5]{x}$.

Solution $\dfrac{d}{dx} 3 \sqrt[5]{x} = 3 \dfrac{d}{dx} x^{1/5} = \dfrac{3}{5} x^{1/5-1} = \dfrac{3}{5} x^{-4/5} = \dfrac{3}{5x^{4/5}}$

Next we consider the function $f(x) = x^r$, where r is a rational number. We may write r as a fraction p/q, where p and q are integers having no common factor (that is, p/q is in lowest terms) and q is positive. Thus $f(x) = x^r = (x^{1/q})^p$. The domain of this function depends on p and q as shown in the following table:

If p is	and q is	the domain of $x^{p/q}$ is
Positive	odd	$(-\infty, \infty)$
Negative	odd	all numbers except zero
Positive	even	$[0, \infty)$
Negative	even	$(0, \infty)$

We may use the power of a function rule (p. 88) to differentiate f. Let $g(x) = x^{1/q}$, so that $f(x) = [g(x)]^p$. Then

$$(d/dx)[g(x)]^p = p[g(x)]^{p-1} g'(x)$$

so by the formula just developed for $g'(x)$, with $1/n$ replaced by $1/q$, we have

$$\frac{d}{dx}(x^{p/q}) = \frac{d}{dx}(x^{1/q})^p = p(x^{1/q})^{p-1} \cdot \frac{1}{q} x^{(1/q)-1}$$

$$= \frac{p}{q} x^{(p-1)/q} x^{(1-q)/q} = \frac{p}{q} x^{(p-q)/q} = \frac{p}{q} x^{(p/q)-1}$$

The domain of this derivative is essentially the same as that of $x^{p/q}$ itself, the only difference being that zero may not belong to the domain of the derivative (for example, $x^{2/3}$ is defined at zero, but $\frac{2}{3} x^{-1/3}$ is not). We conclude that differentiation of rational powers follows the same rule as for integer powers.

RATIONAL POWER RULE

To differentiate a power x^r (r a rational number),   take out the exponent as a factor and then reduce the exponent by 1:

$$\frac{d}{dx}(x^r) = rx^{r-1}$$ (The formula is valid for all x for which the answer makes sense.)

Worked Example 7 Differentiate $f(x) = 3x^2 + (x^2 + x^{1/3})/\sqrt{x}$.

Solution $f'(x) = \dfrac{d}{dx}(3x^2 + x^{3/2} + x^{-1/6})$

$= 6x + \frac{3}{2} x^{1/2} - \frac{1}{6} x^{-7/6}$

$= 6x + \frac{3}{2}\sqrt{x} - \dfrac{1}{6x^{7/6}}$

The rational power rule can be extended to $[f(x)]^r$ as follows. Assuming differentiability of $[f(x)]^r$, we can calculate the derivative by letting $r = p/q$ and

$$g(x) = [f(x)]^{p/q}$$

so

$$g(x)^q = f(x)^p$$

By the power of a function rule for integers (p. 88),

$$qg(x)^{q-1} g'(x) = pf(x)^{p-1} f'(x)$$
$$qf(x)^{(p/q)(q-1)} g'(x) = pf(x)^{p-1} f'(x)$$

Solving for $g'(x)$:

$$g'(x) = \frac{p}{q} f(x)^{(p-1)-(p/q)(q-1)} f'(x)$$
$$= \frac{p}{q} f(x)^{(p/q)-1} f'(x)$$
$$= rf(x)^{r-1} f'(x)$$

(We could have obtained the formula for $(d/dx)\,(x^r)$ by this procedure as well as the one we used (see Exercise 18).)

Worked Example 8 Differentiate $g(x) = (9x^3 + 10)^{5/3}$.

Solution Here $f(x) = 9x^3 + 10$, $r = \frac{5}{3}$, and $f'(x) = 27x^2$. Thus

$$g'(x) = \tfrac{5}{3}(9x^3 + 10)^{2/3} \cdot 27x^2 = 45x^2(9x^3 + 10)^{2/3}$$

RATIONAL POWER OF A FUNCTION RULE

To differentiate a power $[f(x)]^r$ (r a rational number),   take out the exponent as a factor, reduce the exponent by 1, and multiply by $f'(x)$:

$$\frac{d}{dx} [f(x)]^r = r [f(x)]^{r-1} f'(x)$$

Solved Exercises 13. Differentiate:
    (a) $10x^{1/8}$ (b) $x^{3/5}$
    (c) $x^2(x^{1/3} + x^{4/3})$ (d) $1/\sqrt{x}$

14. If we had a definition of x^π (which we don't, at this point, since π is irrational), what should its derivative be?

15. Find $\int x^r\, dx$, where r is rational. Which value(s) of r must be excluded?

Exercises 15. Find the domain and the derivative of each of the following functions:
    (a) $f(x) = x^{3/11} - x^{1/5}$ (b) $g(t) = t(t^{2/3} + t^7)$
    (c) $h(y) = y^{1/8}/(y - 2)$ (d) $k(s) = 1/(s^{3/5} - s)$
    (e) $l(x) = [(x^2 + 1)/(x^2 - 1)]^{1/2}$ (f) $m(u) = (u^9 - 1)^{-6/7}$

16. Find the first-order approximations to (a) $(4.01)^{3/2}$; (b) $(16.3)^{-5/4}$.

17. Find the equation of the tangent line to the curve $y = \sqrt{x} + \sqrt[3]{x}$ at the point $(1, 2)$.

18. Assume you knew that $f(x) = x^{p/q}$ was differentiable but did not know its derivative. Differentiate both sides of the equation $[f(x)]^q = x^p$, using the power of a function rule (p. 88), and solve for $f'(x)$.

19. Find the following antiderivatives:

    (a) $\int (1/\sqrt{x}) \, dx$

    (b) $\int (2/\sqrt{x}) \, dx$

    (c) $\int [1/(s + 1)^{5/8}] \, ds$

    (d) $\int (2x + 1)^{3/2} \, dx$

Problems for Section 5-1

1. Suppose that $f(x)$ is the number of pounds of beans you can buy for x dollars. Let $g(y)$ be the inverse function. What does $g(y)$ represent?

2. Let $f(x) = x^3 - 4x^2 + 1$.

    (a) Find an interval containing 1 on which f is invertible. Denote the inverse by g.

    (b) Compute $g(-7)$ and $g'(-7)$.

    (c) What is the domain of g?

    (d) Compute $g(-5)$ and $g'(-5)$.

3. Suppose that f is concave upward and increasing on $[a, b]$.

    (a) By drawing a graph, guess whether f^{-1} is concave upward or downward on the interval $[f(a), f(b)]$.

    (b) What if f is concave upward and decreasing on $[a, b]$?

4. Show that if the inverse function to f on S is g, with domain T, then the inverse function to g on T is f, with domain S. Thus the inverse of the inverse function is the original function, that is $(f^{-1})^{-1} = f$.

5. Draw a graph of $y = f(x) = (3x + 1)/(2x - 2)$ and a graph of its inverse function.

★6. Under what conditions on a, b, c, and d is the function $f(x) = (ax + b)/(cx + d)$ equal to its own inverse function?

7. (a) What is the inverse function of $f(x) = \sqrt{x - 3}$? (b) Find $(d/dx) \sqrt{x - 3}$.

8. Find the following derivatives:

    (a) $(d/dx)(x^2 + 5)^{7/8}$

    (b) $(d^2/dx^2)(x^r)$, where r is rational

    (c) $f'(7)$, where $f(x) = 7 \sqrt[3]{x}$

    (d) $g'(0)$, where g is the inverse function to $f(x) = x^9 + x^5 + x$

    (e) $(d/dx)(x^{1/4} + 4/\sqrt{x})|_{x=81}$

9. (a) At which points is $x^{1/3}$ differentiable?

    (b) On what intervals is $x^{1/3}$ increasing or decreasing?

10. (a) Find the domain of

    $$f(x) = \sqrt{(2x + 5)/(3x + 7)}$$

    (The quantity in the square root must be ≥ 0.)

    (b) By solving for x in the equation $y = \sqrt{(2x + 5)/(3x + 7)}$, find a formula for the inverse function $g(y)$.

    (c) By using Theorem 2, find a formula for $f'(x)$.

    (d) Check your answer by using the power of a function rule.

11. Differentiate each of the following functions. [*Hint:* For part (c), differentiate $\sqrt{x^2 + 2}$ first and then use the quotient rule.]

    (a) $f(x) = 3\sqrt{x} - (1/x) + 5$

    (b) $f(x) = \sqrt{x}/(3 + x + x^3)$

    (c) $f(x) = x/\sqrt{x^2 + 2}$

    (d) $f(x) = \sqrt{x}/(1 + \sqrt{x})$

    (e) $f(x) = \sqrt{x}/\sqrt[3]{x^4 + 2}$

12. Differentiate:

    (a) $f(x) = \sqrt[3]{x}/(x^2 + 2)$

    (b) $f(x) = (3x^2 - 6x)/3\sqrt[4]{x}$

    (c) $f(x) = \sqrt{(6x^2 + 2x + 1)/(\sqrt{x} + 2x^3)}$

    (d) $f(x) = \sqrt{\sqrt[3]{x}/\sqrt{3x^2 + 1 + x}}$

13. Find the following antiderivatives:

    (a) $\int (x + \sqrt{x} + \sqrt[3]{x}) \, dx$

    (b) $\int \sqrt{x - 1} \, dx$

    (c) $\int 4x^{3/2} \, dx$

14. Find linear approximations for:

    (a) $\sqrt[4]{15.97}$ (b) $(4.02)^{3/2}$ (c) $(-26.98)^{2/3}$

★15. Suppose that, in Theorem 2, $f'(x_0) > 0$ and $f(x_0) = y_0$. If g is the inverse of f with $g(y_0) = x_0$, show that g is continuous at y_0 by filling in the details of the following argument:

(a) For Δx sufficiently small, $\frac{3}{2}f'(x_0) > \Delta y / \Delta x > \frac{1}{2}f'(x_0)$.

(b) As $\Delta y \to 0$, $\Delta x \to 0$ as well.

(c) Let $\Delta y = f(x_0 + \Delta x) - f(x_0)$. Then $\Delta x = g(y_0 + \Delta y) - g(y_0)$.

16. Determine the equation of the tangent line through $(1, 1)$ for the curve $y = x^{1/3} - x^{-2/3} + 2x/(1 + \sqrt{x})$.

17. Two positive point charges of magnitude q are fixed at $(0, a)$, $(0, -a)$. The potential V at any point on the x axis is $V = 2q/\sqrt{a^2 + x^2}$.

(a) Sketch V versus x on $-4a \le x \le 4a$.

(b) Find dV/dx.

(c) At what value of x is V equal to $\frac{1}{2}V(0)$?

18. The average pulse rate for persons 30 inches to 74 inches tall can be approximated by $y = 589/\sqrt{x}$ beats per minute for a person x inches tall.

(a) Find dy/dx, and show that it is always negative.

(b) The value of $|dy/dx|$ at $x = 65$ is the decrease in beats per minute expected for a one-inch increase in height. Explain.

(c) Do children have higher pulse rates than adults, according to this model?

19. Let $y = 24\sqrt{x}$ be the learning curve for learning y items in x hours, $0 \le x \le 5$. Apply the first-order approximation (see Worked Example 5) to approximate the increase in items learned in the 12-minute period $1 \le x \le 1.2$.

20. The daily demand function for a certain manufactured good is $p(x) = 75 + 4\sqrt{2x} - (x^2/2)$.

(a) Find dp/dx. Interpret.

(b) Find the production level x for which $dp/dx = 0$. Interpret.

21. The fundamental period of vibration P and the tension t in a certain string are related by $P = \sqrt{t}/32$ seconds. Find the rate of change of period with respect to tension when $t = 9$ lb.

22. Lions in a small district in an African game preserve defend an exclusive region of area A which depends on their body weight W by the approximate formula $A = W^{1.31}$.

(a) Find dA/dW.

(b) By what percentage should the defended area increase after a 200-lb lion undergoes a 20-lb weight gain?

23. The object distance x and image distance y satisfy the thin lens equation $1/x + 1/y = 1/f$, where f is the focal length.

(a) Solve for y as a function of x when $f = 50$ mm.

(b) Find dy/dx.

(c) Find all (x, y) such that $(d/dx)(x + y) = 0$.

5-2 The Chain Rule and Related Rates

The derivative of the composite of two functions is the product of their derivatives.

The power of a function rule stated that

$$\frac{d}{dx}[g(x)]^r = r[g(x)]^{r-1}\frac{d}{dx}[g(x)]$$

The process of forming $[g(x)]^r$ can be broken into two successive operations: first $u = g(x)$ is found, and then the power $y = u^r$ is taken. The *chain rule* tells us how to differentiate the function obtained by applying two functions in succession; it has the power of a function rule as a special case.

Goals

After studying this section, you should be able to:

Form the composite of two functions.

Differentiate composite functions.

Differentiate functions defined implicitly.

Solve word problems involving related rates.

Find the tangent line to a curve given implicitly or by parametric equations.

A PHYSICAL EXAMPLE

When the pressure of the atmosphere around you changes too rapidly, the pressure imbalance which develops between the inside and outside of your eardrums causes a popping sensation in the ears. Such rapid pressure changes usually occur, not because of weather changes, but because you are rapidly changing altitude (in an airplane or a car, for example), and pressure decreases with increasing altitude. Thus the quantity of physiological interest is the rate of change of pressure with respect to time. Our goal is to express this rate in terms of two more directly measureable quantities: the rate of change of pressure with respect to altitude and the rate of change of altitude with respect to time.

We may express the situation mathematically by saying that the pressure p is related to the altitude u by a function $p = f(u)$, while the altitude is related to the time t by some other function $u = g(t)$. By substituting the value of u from the second function into the first, we obtain

$$p = f(g(t)) \tag{1}$$

In this way, we may think of the pressure as a function of time; that is, $p = h(t)$, where the function h is defined by combining the functions f and g as in (1). The function h is called the *composition* of f and g. We sometimes say that $p = h(t)$ depends upon t through the *intermediate variable* $u = g(t)$.

Let us now consider various rates of change at time t_0. The rate of change of pressure with respect to time is $h'(t_0)$; the rate of change of altitude with respect to time is $g'(t_0)$. The rate of change of pressure with respect to altitude is $f'(u_0)$, where $u_0 = g(t_0)$ is the altitude at time t_0. (This rate will be negative, since pressure is a decreasing function of altitude.)

How is $h'(t_0)$ related to $f'(g(t_0))$ and $g'(t_0)$? To get a hint of an answer, we will look at some actual numbers, together with units of measurement. The quantity $f'(u_0)$ is meteorological data—for u_0 near sea level,

$$\frac{dp}{du} = f'(u_0) = -0.12 \, \frac{\text{gsc}}{\text{m}}$$

(gsc denotes grams per square centimeter, a unit of pressure).

If we are rolling down a steep hill at 100 kilometers per hour, we could have

$$\frac{du}{dt} = g'(t_0) = -3 \, \frac{\text{m}}{\text{sec}}$$

(Only part of our velocity is in the downward direction.)

The quantity $dp/dt = h'(t_0)$ which we are seeking should be measured in units of gsc/sec (pressure/time), and it should be positive, since the pressure increases as we descend. How can we manufacture such a quantity out of the following?

$$-0.12 \frac{\text{gsc}}{\text{m}} \quad \text{and} \quad -3 \frac{\text{m}}{\text{sec}}$$

One way is to multiply them, obtaining 0.36 gsc/sec.

We may justify this multiplication heuristically by the following argument. Over a short period of time Δt, we may assume that our velocity is approximately constant, so that we have $\Delta h \approx -3\Delta t$. Since Δt is small, so is Δh, and we have $\Delta p \approx -0.12\Delta h$. Substituting the first approximate equality into the second gives $\Delta p \approx 0.36\Delta t$, or $\Delta p/\Delta t \approx 0.36$. Applying this result to our example, we find that, in 3 seconds, a pressure increase of over 1 gsc can be built up—this small pressure can be significant when applied to a surface as sensitive as an eardrum.

In terms of derivatives, we are led to conjecture the formula

$$h'(t_0) = f'(u_0) \cdot g'(t_0) = f'(g(t_0)) \cdot g'(t_0)$$

where $h(t) = f(g(t))$. As we will learn later in this section, this formula is true in general. It is called the *chain rule*.

Solved Exercises

1. At a certain moment, an airplane is at an altitude of 1500 meters and is climbing at the rate of 5 meters per second. At this altitude, pressure decreases with altitude at the rate of 0.095 gsc per meter. What is the rate of change of pressure with respect to time?

2. Express the chain rule in Leibniz notation.

Exercises

1. Suppose that the airplane in Solved Exercise 1 is descending rather than climbing at the rate of 5 meters per second. What is the rate of change of pressure with respect to time?

2. At a certain moment, your car is consuming gasoline at the rate of 15 miles per gallon. If gasoline costs 75 cents per gallon, what is the cost per mile? Set the problem up in terms of functions and apply the chain rule.

COMPOSITION OF FUNCTIONS

The idea behind the definition of composition of functions is that one variable depends on another through an intermediate one.

Definition Let f and g be functions with domains D_f and D_g. Let D be the set consisting of those x in D_g for which $g(x)$ belongs to D_f. For x in D, we can evaluate $f(g(x))$, and we call the result $h(x)$. The resulting function $h(x) = f(g(x))$, with domain D, is called the *composition* of f and g. It is often denoted by $f \circ g$.

Worked Example 1 If $f(u) = u^3 + 2$ and $g(x) = \sqrt{x^2 + 1}$, what is $h = f \circ g$?

Solution We calculate $h(x) = f(g(x))$ by writing $u = g(x)$ and substituting in $f(u)$. We get $u = \sqrt{x^2 + 1}$ and

$$h(x) = f(u) = u^3 + 2 = (\sqrt{x^2 + 1})^3 + 2 = (x^2 + 1)^{3/2} + 2$$

Since each of f and g is defined for all numbers, the domain D of h is $(-\infty, \infty)$.

Calculator Discussion

On electronic calculators, several functions, such as $1/x$, x^2, $\sqrt{x}$, and $\sin x$, are evaluated by the push of a single key. To evaluate the composite function $f \circ g$ on x, you first enter x, then push the key for g to get $g(x)$, then push the key for f to get $f(g(x))$. For instance, let $f(x) = x^2$, $g(x) = \sin x$. To calculate $(f \circ g)(x) = f(g(x)) = (\sin x)^2$ for $x = 32$ (degrees), we enter 32, then press the sin key, then the x^2 key. The result is $32 \rightarrow 0.52991926 \rightarrow 0.28081442$. Notice that $(g \circ f)(x) = \sin(x^2)$ is quite different: entering 32 and pressing the x^2 key followed by the sin key, we get $32 \rightarrow 1024 \rightarrow -0.82903756$.

In some cases, $f \circ g$ and $g \circ f$ happen to be the same function; for example, if $f(x) = 1/x$ and $g(x) = x^2$, $(f \circ g)(x)$ and $(g \circ f)(x)$ are both equal to $1/x^2$ $(=(1/x)^2)$. Nevertheless, since a calculator computes only an approximation to the ideal mathematical function, you may get different results by doing the calculations in different orders. With $x = 9$, for example, $(1/x)^2$ and $1/x^2$ both come out to 0.01234567901, but for $x = 6$, we get (on the calculator we used) $(1/x)^2 = 0.02777777776$, while $1/x^2 = 0.02777777777$, and for $x = 397$, we get $(1/x)^2 = 6.344815331 \times 10^{-6}$, while $1/x^2 = 6.344815334 \times 10^{-6}$. (Such computations provide a useful way to check the accuracy of a calculator.)

Warning Do not confuse the composition of functions with the product. We have

$$(fg)(x) = f(x)g(x)$$

while

$$(f \circ g)(x) = f(g(x))$$

In the case of the product we evaluate $f(x)$ and $g(x)$ separately and then multiply the results; in the case of the composition, we evaluate $g(x)$ first and then apply f to the result.

Worked Example 2 Express $[(x^3 + 3)/2]^{1/2}$ as the composition of two functions from the following list: (a) x; (b) x^3; (c) $\sqrt{x}$; (d) $x^3 + 3$; (e) $(x^3 + 3)/2$.

Solution If we let $g(x) = (x^3 + 3)/2$ and $f(u) = \sqrt{u}$, then $f(g(x)) = \sqrt{g(x)} = [g(x)]^{1/2} = [(x^3 + 3)/2]^{1/2}$. Thus the given function is the composite of the functions (c) and (e).

COMPOSITION OF FUNCTIONS

The composition $f \circ g$ is obtained by writing $u = g(x)$ and evaluating $f(u)$.

To break up a given function $h(x)$ as a composition, find an intermediate expression $u = g(x)$ such that $h(x)$ can be written in terms of u.

Solved Exercises

3. Let $g(x) = x + 1$ and $f(u) = u^2$. Find $f \circ g$ and $g \circ f$.

4. Let $h(x) = x^{24} + 3x^{12} + 1$. Write $h(x)$ as a composite function $f(g(x))$.

5. Let $f(x) = x - 1$ and $g(x) = \sqrt{x}$.
   (a) What are the domains D_f and D_g?
   (b) Find $f \circ g$ and $g \circ f$. What are their domains?
   (c) Find $(f \circ g)(2)$ and $(g \circ f)(2)$.
   (d) Sketch graphs of f, g, $f \circ g$, and $g \circ f$.

6. Let i be the "identity function" $i(x) = x$. Show that $i \circ f = f$ and $f \circ i = f$ for any function f.

Exercises

3. Find $f \circ g$ and $g \circ f$ in each of the following cases:
   (a) $g(x) = x^3$; $f(x) = \sqrt{x - 2}$
   (b) $g(x) = x^r$; $f(x) = x^s$ (r, s rational)
   (c) $g(x) = 1/(1 - x)$; $f(x) = (1/2) - \sqrt{3x}$
   (d) $g(x) = (3x - 2)/(4x + 1)$; $f(x) = (2x - 7)/(9x + 3)$

4. Write the following as compositions of simpler functions:
   (a) $h(x) = 4x^2/(x^2 - 1)$
   (b) $h(r) = (r^2 + 6r + 9)^{3/2} + (1/\sqrt{r^2 + 6r + 9})$
   (c) $h(u) = \sqrt{(1 - u)/(1 + u)}$

5. Show that the inverse f^{-1} of any function f satisfies $f \circ f^{-1} = i$ and $f^{-1} \circ f = i$, where i is the identity function of Solved Exercise 6.

THE CHAIN RULE

Theorem 3 Chain Rule. *Suppose that g is differentiable at x_0 and that f is differentiable at $g(x_0)$. Then $f \circ g$ is differentiable at x_0 and its derivative there is $f'(g(x_0)) \cdot g'(x_0)$.*

Worked Example 3 Verify the chain rule for $f(u) = u^2$ and $g(x) = x^3 + 1$.

Solution Let $h(x) = f(g(x)) = [g(x)]^2 = (x^3 + 1)^2 = x^6 + 2x^3 + 1$. Thus $h'(x) = 6x^5 + 6x^2$. On the other hand, since $f'(u) = 2u$ and $g'(x) = 3x^2$,

$$f'(g(x)) \cdot g'(x) = (2 \cdot (x^3 + 1)) \, 3x^2 = 6x^5 + 6x^2$$

Hence the chain rule is verified in this case.

We have already given plausibility arguments for the chain rule at the beginning of this section. A complete proof of the chain rule can be given by using limits (See Chapter 12). We can, however, give the essence of the argument as follows.

If x is changed by a small amount Δx and the corresponding change in $u = g(x)$ is Δu, we know that

$$g'(x) = \frac{du}{dx} = \lim_{\Delta x \to 0} \frac{\Delta u}{\Delta x}$$

Corresponding to the small change Δu is a change Δy in $y = f(u)$, and

$$f'(u) = \frac{dy}{du} = \lim_{\Delta u \to 0} \frac{\Delta y}{\Delta u}$$

To calculate the rate of change dy/dx, we write

$$\frac{dy}{dx} = \lim_{\Delta x \to 0} \frac{\Delta y}{\Delta x} = \lim_{\Delta x \to 0} \frac{\Delta y}{\Delta u}\frac{\Delta u}{\Delta x} = \left(\lim_{\Delta x \to 0} \frac{\Delta y}{\Delta u}\right)\left(\lim_{\Delta x \to 0} \frac{\Delta u}{\Delta x}\right)$$

$$= \left(\lim_{\Delta u \to 0} \frac{\Delta y}{\Delta u}\right)\left(\lim_{\Delta x \to 0} \frac{\Delta u}{\Delta x}\right) = \frac{dy}{du}\frac{du}{dx} = f'(u) \cdot g'(x)$$

In going to the second line, we replaced $\Delta x \to 0$ by $\Delta u \to 0$ since the differentiable function g is continuous, and so $\Delta u \to 0$ as $\Delta x \to 0$ (see Section 2-1).

There is a flaw in this proof: the Δu determined by Δx could well be zero, and division by zero is not allowed. This difficulty is fortunately not an essential one, and the more technical proof given in Review Problem 28 in Chapter 12 avoids it.

CHAIN RULE

To differentiate a composition $f(g(x))$, differentiate g at x, differentiate f at $g(x)$, and multiply the results:

$$(f \circ g)'(x) = f'(g(x)) \cdot g'(x)$$

$$\frac{dy}{dx} = \frac{dy}{du} \cdot \frac{du}{dx}$$ if y is a function of u and u is a function of x.

Worked Example 4 If $u = x^2 + 1$ and $y = \sqrt{u}$, calculate dy/dx.

Solution By the chain rule,

$$\frac{dy}{dx} = \frac{dy}{du}\frac{du}{dx} = \left(\frac{1}{2\sqrt{u}}\right)(2x) = \frac{1}{2\sqrt{x^2+1}} \cdot 2x = \frac{x}{\sqrt{x^2+1}}$$

A special case of the chain rule may help you to understand it. Consider $h(x) = f(x + c)$, c a constant. If we let $u = g(x) = x + c$, we get $g'(x) = 1$, so

$$h'(x) = f'(g(x)) \cdot g'(x) = f'(x+c) \cdot 1 = f'(x+c)$$

Note that the graph of h is the same as that of f except that it is shifted c units to the left (see Fig. 5-2-1). It is reasonable, then, that the tangent line to the graph of h is obtained by shifting the tangent line to the graph of f. Thus, in this case, the chain rule is telling us something geometrically obvious. One might call this formula the *shifting rule*. In Leibniz notation, it reads

$$\frac{d}{dx} f(x + c) \bigg|_{x_0} = \frac{d}{dx} f(x) \bigg|_{x_0 + c}$$

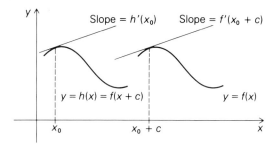

Fig. 5-2-1 The geometric interpretation of the shifting rule.

Worked Example 5 Show that the power of a function rule follows from the chain rule.

Solution If $y = [g(x)]^n$, we may write $u = g(x)$, $y = f(u) = u^n$. By the chain rule,

$$\frac{dy}{dx} = f'(g(x)) \, g'(x) = n \, [g(x)]^{n-1} g'(x)$$

This derivation explains the presence of the factor $g'(x)$, which appeared somewhat mysteriously when the power of a function rule was first derived in Section 1-3.

Solved Exercises

7. Use the chain rule to calculate dy/dx, where $y = (1 - x^3)/(1 + x^3)$, by letting $u = x^3$. Compare with a direct calculation.

8. If $h(x) = f(x^2)$, find a formula for $h'(x)$.

9. Use the shifting rule to differentiate $\sqrt{x + 3}$. Draw graphs of $y = \sqrt{x}$ and $y = \sqrt{x + 3}$.

10. Use the chain rule to differentiate $(x^4 + 2x + 1)^{5/8}$.

Exercises

6. (a) Find a "stretching rule" for the derivative of $f(cx)$, c a constant.
   (b) Draw graphs of $y = 1/(1 + x^2)$ and $y = 1/[1 + (4x)^2]$ and interpret the stretching rule geometrically.

7. Use the chain rule to differentiate the following functions:
   (a) $f(x) = (x^2 - 6x + 1)^3$ (b) $f(x) = (x - 2x^2)^{3/4}$
   (c) $y = 1/\sqrt[3]{x^2 + 2}$
   (d) $h(x) = \sqrt{f(x)}$, where $f(x)$ is an arbitrary function
   (e) $y = (9 + 2x^5)/(3 + 5x^5)$

8. Given three functions, f, g, and h:
   (a) How would you define the composition $f \circ g \circ h$?
   (b) Use the chain rule twice to obtain a formula for the derivative of $f \circ g \circ h$.

THE RATIONAL POWER OF A FUNCTION RULE

We saw in Worked Example 5 that the power of a function rule,

$$\frac{d}{dx}[f(x)]^n = n[f(x)]^{n-1} f'(x)$$

where n is an integer, can be considered as a special case of the chain rule, with $u = f(x)$ and $y = u^n$.

Now that we have the general chain rule at our disposal, we can go one step further. Let $f(x)$ be a differentiable function, r a *rational* number. Then if we let $y = u^r$, $u = f(x)$, we can apply the chain rule to obtain

$$\frac{d}{dx}[f(x)]^r = \frac{dy}{du} \cdot \frac{du}{dx} = ru^{r-1} \cdot f'(x) = r[f(x)]^{r-1} \cdot f'(x)$$

This result was obtained in the previous section under the assumption that $[f(x)]^r$ was differentiable. The method here *proves* that it is differentiable. (If the denominator of r is even, we must assume that $f(x) > 0$ so that $f(x)$ is in the domain of the rth power function.)

Solved Exercises

11. Use the power of a function rule to differentiate $(x^4 + 2x + 1)^{5/8}$.

12. Differentiate $x^2/\sqrt[3]{x^2 + 1}$.

Exercises

9. Differentiate the function $(x^3 + 2)/\sqrt{x^3 + 1}$ in at least two different ways. Be sure that the answers you get are equivalent. (Doing the same problem in several ways is a good method for checking your calculations, useful on examinations as well as in scientific work.)

10. Differentiate the following functions:
    (a) $(x^5 + 1)^{7/9}$
    (b) $1/\sqrt[4]{x^3 + 5x + 2}$
    (c) $(x^{1/3} + x^{2/3})^{1/3}$
    (d) $x^{1/7}/(3x^{5/3} + x)$
    (e) $\sqrt{x^2 + 1}/\sqrt{x^2 - 1}$
    (f) $(x + 2)^{3/2}\sqrt{x + 2}$
    (g) $\sqrt{(x + 3)^2 - 4}$

11. Find $(d^2/dx^2)(x/\sqrt{1 + x^2})$.

12. Find the slope of the tangent line to the graph of $y = \sqrt{1 - x^2}$ at the point $(\sqrt{3}/2, 1/2)$.

WORD PROBLEMS WITH THE CHAIN RULE

We now return to some examples like the one which introduced this section. As with other word problems, one must translate each problem into mathematical symbols, apply the appropriate mathematical operations, and then translate the result back into the language of the problem.

Worked Example 6   The price of eggs, in cents per dozen, is given by the formula $p = 55/(s - 1)^2$, where s is the supply of eggs, in units of 10,000 dozen, available to the wholesaler. Suppose that the supply on July 1, 1986 is $s = 2.1$ and is falling at a rate of 0.03 per month. How fast is the price rising?

Solution   By the chain rule,

$$\frac{dp}{dt} = \frac{dp}{ds} \frac{ds}{dt}$$

We calculate

$$\frac{dp}{ds} = \frac{(55)(-2)}{(s-1)^3} = \frac{-110}{(s-1)^3}$$

by the power of a function rule. When $s = 2.1$, $dp/ds = -110/(1.1)^3 = -82.64$, and we are given $ds/dt = -0.03$, so $dp/dt = (-82.64)(-0.03) = 2.48$ cents per month.

Solved Exercises

13. The population of Thin City is increasing at a rate of 10,000 people a day on March 30, 1984. The area of the city grows to keep the ratio of 1 square mile per 1000 people. How fast is the area increasing per day on this date?

14. Water is flowing into a tub at 3 gallons per minute. The level rises $\frac{1}{2}$ inch for each gallon. How fast is the level rising?

15. A dog 2 feet high trots proudly away from a 10-foot-high light post at 3 feet per second. When he is 8 feet from the light's base, how fast is the tip of his shadow moving?

Exercises

13. Suppose that Fat City occupies a circular area 10 miles in diameter and contains 500,000 inhabitants. If the population is growing now at the rate of 20,000 inhabitants per year, how fast should the diameter be increasing now in order to maintain the circular shape and the same population density (= number of people per square mile). If the population continues to grow at the rate of 20,000 per year, how fast should the diameter be increasing in 5 years? Give an intuitive explanation of the relation between the two answers.

14. The radius of a sphere S is given by $r = t^2 - 2t + 1$ at time t. How fast is the volume V of S changing at time $t = \frac{1}{2}$, 1, 2?

15. The kinetic energy K of a particle of mass m moving with speed v is $K = \frac{1}{2}mv^2$. A particle with mass 10 grams has, at a certain moment, velocity 30 centimeters per second and acceleration 5 centimeters per second per second. At what rate is the kinetic energy changing?

IMPLICIT DIFFERENTIATION

Most of the differentiation rules we have learned so far enable us to find the slope of the tangent line to a curve in the xy plane which is given by specifying y explicitly as a function of x. One exception was the inverse function rule, in which x was given as a function of y. In the rest of this section, we will consider curves given in other ways and show how to use the chain rule to find their tangent lines.

The process called *implicit differentiation* applies when a curve is specified by a relation between x and y, such as $x^2 + y^2 = 1$ or $x^4 + xy + y^5 = 2$. In general, such a curve will not be the graph of a function, but pieces of the curve will be graphs, and we can try to differentiate the functions which represent these pieces.

For example, the circle $x^2 + y^2 = 1$ is not the graph of a function, but the upper and lower semicircles are graphs (see Fig. 5-2-2).

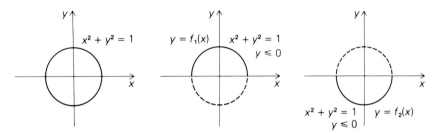

Fig. 5-2-2 Parts of the circle $x^2 + y^2 = 1$ are the graphs of functions.

Worked Example 7   If $y = f(x)$ and $x^2 + y^2 = 1$, express dy/dx in terms of x and y.

Solution   Thinking of y as a function of x, we differentiate both sides of the relation $x^2 + y^2 = 1$ with respect to x. The derivative of the left-hand side is

$$\frac{d}{dx}(x^2 + y^2) = 2x + 2y\frac{dy}{dx}$$

while the right-hand side has derivative zero. Thus

$$2x + 2y\frac{dy}{dx} = 0 \quad \text{and so} \quad \frac{dy}{dx} = -\frac{x}{y}$$

The result of Worked Example 7 can be checked, since in this case we can solve for y directly:

$$y = \pm\sqrt{1 - x^2}$$

Notice that the given relation then defines *two* functions: $f_1(x) = \sqrt{1 - x^2}$ and $f_2(x) = -\sqrt{1 - x^2}$. Taking the plus case, with $u = 1 - x^2$ and $y = \sqrt{u}$, gives

$$\frac{dy}{dx} = \frac{dy}{du}\frac{du}{dx} = \frac{1}{2\sqrt{u}}(-2x)$$

$$= \frac{1}{2\sqrt{1 - x^2}}(-2x) = \frac{-x}{\sqrt{1 - x^2}} = \frac{-x}{y}$$

so it checks. The minus case gives the same answer.

From the form of the derivative given by implicit differentiation, $dy/dx = -x/y$, we see that the tangent line to a circle at (x, y) is perpendicular to the line through (x, y) and the origin, since their slopes are negative reciprocals of one another. (See Fig. 5-2-3.) Implicit differentiation often leads directly to such striking results, and for that reason it is sometimes preferable to use this method, even when y could be expressed in terms of x.

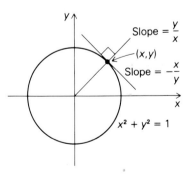

Fig. 5-2-3 If $x^2 + y^2 = 1$, the fomula $dy/dx = -x/y$ means that the tangent line to a circle at a point on the circle is perpendicular to the radius at that point.

The following is an example in which we cannot solve for y in terms of x.

Worked Example 8   Find the equation of the tangent line to the curve $2x^6 + y^4 = 9xy$ at the point $(1, 2)$.

Solution   We note first that $(1, 2)$ lies on the curve, since $2(1)^6 + 2^4 = 9(1)(2)$. Now suppose that $y = f(x)$ and differentiate both sides of the defining relation. The left-hand side gives

$$\frac{d}{dx}(2x^6 + [f(x)]^4) = 12x^5 + 4[f(x)]^3 f'(x)$$

while the right-hand side gives

$$\frac{d}{dx}(9xf(x)) = 9f(x) + 9xf'(x)$$

Equating both sides and solving for $f'(x)$, we have

$$12x^5 + 4[f(x)]^3 f'(x) = 9f(x) + 9xf'(x)$$
$$(4[f(x)]^3 - 9x)f'(x) = 9f(x) - 12x^5$$
$$f'(x) = \frac{9f(x) - 12x^5}{4[f(x)]^3 - 9x}$$

When $x = 1$ and $y = f(x) = 2$,

$$\frac{dy}{dx} = f'(x) = \frac{9(2) - 12(1)^5}{4(2)^3 - 9(1)} = \frac{18 - 12}{32 - 9} = \frac{6}{23}$$

Thus the slope of the tangent line is $\frac{6}{23}$; by the point-slope formula (p. 21), the equation of the tangent line is $y - 2 = \frac{6}{23}(x - 1)$, or $y = \frac{6}{23}x + \frac{40}{23}$.

We could have solved Worked Example 8 in the Leibniz notation without introducing the symbol $f(x)$. The functional notation has the advantage of reminding us that the factor $dy/dx = f'(x)$ must be inserted whenever we differentiate an expression involving y. If we wish to find dx/dy, the same procedure can be used with the roles of x a

---

IMPLICIT DIFFERENTIATION

To calculate dy/dx if x and y are related by an equation:

1. Differentiate both sides of the equation with respect to x, thinking of y as a function of x and using the chain rule.

2. Solve the resulting equation for dy/dx.

Solved Exercises

16. If $x^2 + y^2 = 3$, compute dy/dx when $x = 0$, $y = \sqrt{3}$.

17. If $x^3 + y^3 = xy$, compute dx/dy in terms of x and y.

18. Use implicit differentiation to find the volume of the largest right circular cylinder that can be inscribed in a sphere of radius R.

Exercises

16. Suppose that $x^4 + y^2 + y - 3 = 0$.
    (a) Compute dy/dx by implicit differentiation.
    (b) What is dy/dx when $x = 1$, $y = 1$?
    (c) Solve for y in terms of x (by the quadratic formula) and compute dy/dx directly. Compare with your answer in part (a).

17. Suppose that $xy + \sqrt{x^2 - y} = 7$.
    (a) Find dy/dx. (b) Find dx/dy.
    (c) What is the relation between dy/dx and dx/dy?

18. Suppose $x^2/(x + y^2) = y^2/2$.
    (a) Find dy/dx when $x = 2$, $y = \sqrt{2}$.
    (b) Find dy/dx when $x = 2$, $y = -\sqrt{2}$.

RELATED RATES AND PARAMETRIC CURVES

Suppose we have two quantities, x and y, each of which is a function of time t. We know that the rates of change of x and y are given by dx/dt and dy/dt. If x and y satisfy an equation, such as $x^2 + y^2 = 1$ or $x^2 + y^6 + 2y = 5$, then the *rates* dx/dt and dy/dt can be *related* by differentiating the equation with respect to t and using the chain rule.

Worked Example 9 Suppose that x and y are functions of t and $x^4 + xy + y^4 = 1$. Relate dx/dt and dy/dt.

Solution Differentiate the relation between x and y *with respect to t*, thinking of x and y as functions of t:

$$\frac{d}{dt}(x^4 + xy + y^4) = 0$$

$$4x^3 \frac{dx}{dt} + \frac{dx}{dt}y + x\frac{dy}{dt} + 4y^3 \frac{dy}{dt} = 0$$

We can simplify this to

$$\frac{dy}{dt} = -\left(\frac{4x^3 + y}{x + 4y^3}\right)\frac{dx}{dt}$$

which is the desired relation.

There is a device which may help you to remember that the chain rule must be used. First write $x = f(t)$ and $y = g(t)$, and then substitute $f(t)$ for x and $g(t)$ for y, giving

$$[f(t)]^4 + f(t)g(t) + [g(t)]^4 = 1$$

Differentiating by the power of a function rule, the product rule, and the sum rule gives

$$4[f(t)]^3 f'(t) + f'(t)g(t) + f(t)g'(t) + 4[g(t)]^3 g'(t) = 0$$

or

$$g'(t) = -\left[\frac{4[f(t)]^3 + g(t)}{f(t) + 4[g(t)]^3}\right] f'(t)$$

which is the same as the result above. Once you have done a few examples in this long-winded way, you should be ready to go back to the d/dt's.

The ear popping example at the beginning of this section may also be viewed as a problem in related rates. In that example, pressure p and altitude u are both functions of time; these functions are related since pressure is a function of altitude. When this relation is differentiated, we relate the rates dp/dt and du/dt by a direct application of the chain rule:

$$\frac{dp}{dt} = \frac{dp}{du} \frac{du}{dt}$$

RELATED RATES

To relate dx/dt and dy/dt if x and y are related by an equation:

1. Differentiate both sides of the equation with respect to t, thinking of x and y as functions of t.

2. Solve the resulting equation for dy/dt in terms of dx/dt (or vice-versa if called for).

Worked Example 10 A light L is being raised up a pole (see Fig. 5-2-4). The light shines on the object Q, casting a shadow on the ground. At a certain moment the light is 40 meters off the ground, rising at 5 meters per minute. How fast is the shadow shrinking at that instant?

Fig. 5-2-4 At what rate is the shadow shrinking?

Solution   Let the height of the light be $y = g(t)$ at time t and the length of the shadow be $x = f(t)$. By similar triangles, $x/10 = (x + 20)/y$; i.e., $xy = 10(x + 20)$. Differentiating, $x(dy/dt) + (dx/dt)y = 10(dx/dt)$. At the moment in question $y = 40$, and so $x \cdot 40 = 10(x + 20)$ or $x = 20/3$. Also, $dy/dt = 5$ and so $(20/3) \cdot 5 + 40(dx/dt) = 10(dx/dt)$. Solving for (dx/dt), we get $dx/dt = -10/9$. Thus the shadow is shrinking at $10/9$ meters per minute.

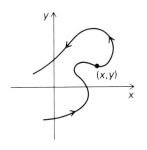

Fig. 5-2-5 If x and y are functions of t, the point (x,y) follows a curve as t varies.

There is a useful geometric interpretation of related rates. (This topic is treated in more detail in Section 11-3.) If x and y are each functions of t, say $x = f(t)$ and $y = g(t)$, we can plot the points (x, y) for various values of t. As t varies, the point (x, y) will move along a curve. When a curve is described this way, it is called a *parametric curve* (see Fig. 5-2-5). It may be possible to describe a parametric curve in other ways. Specifically, it may be described by a relation between x and y.

Worked Example 11   If $x = t^4$ and $y = t^2$, what curve does (x, y) follow for $-\infty < t < \infty$?

Solution   We notice that $y^2 = x$, so the point (x, y) lies on a parabola. As t ranges from $-\infty$ to ∞, y goes from $+\infty$ to zero and back to $+\infty$, so (x, y) stays on the half of the parabola with $y \geq 0$ and traverses it twice (see Fig. 5-2-6).

Fig. 5-2-6 As t ranges from $-\infty$ to $+\infty$, the point (t^4, t^2) traverses the parabola twice in the directions shown.

Suppose that the parametric curve $x = f(t)$, $y = g(t)$ can be described by an equation $y = h(x)$ (the case $x = k(y)$ will be similar). Then we can differentiate by the chain rule. Using Leibniz notation:

$$\frac{dy}{dt} = \frac{dy}{dx}\frac{dx}{dt} \qquad \text{so} \qquad \frac{dy}{dx} = \frac{dy/dt}{dx/dt}$$

This shows that the slope of the tangent line to a parametric curve is given by $(dy/dt)/(dx/dt)$.

Worked Example 12   Find the equation of the line tangent to the parametric curve $x = (1 + t^3)^4$, $y = t^5 + t^2 + 2$ at $t = 1$.

Solution   Here it is not clear what the relation between x and y is, but we do not need to know it. (We tacitly assume that the path followed by (x, y) can be described by a function $y = h(x)$.) We have

$$\frac{dx}{dt} = 4(1 + t^3)^3 \cdot 3t^2 = 12(1 + t^3)^3 t^2 \qquad \text{and} \qquad \frac{dy}{dt} = 5t^4 + 2t$$

so the slope of the tangent line is

$$\frac{dy}{dx} = \frac{dy/dt}{dx/dt} = \frac{5t^4 + 2t}{12t^2(1 + t^3)^3} = \frac{5t^3 + 2}{12t(1 + t^3)^3}$$

At $t = 1$, we get

$$\frac{dy}{dx} = \frac{7}{12 \cdot 2^3} = \frac{7}{96}$$

Since $x = 16$ and $y = 4$ at $t = 1$, the equation of the tangent line is given by the point-slope formula:

$$y - 4 = \tfrac{7}{96}(x - 16)$$

Solved Exercises

19. A spherical balloon is being blown up by a child. At a certain instant during inflation, air enters the balloon to make the volume increase at a rate of 50 cubic centimeters a second. At the same instant the balloon has a radius of 10 centimeters. How fast is the radius changing with time?

20. Suppose that x and y are functions of time and that (x, y) moves on the circle $x^2 + y^2 = 1$. If x is increasing at 1 centimeter per second, what is the rate of change of y when $x = 1/\sqrt{2}$ and $y = 1/\sqrt{2}$?

21. Show that the parametric equations $x = at + b$ and $y = ct + d$ describe a straight line. What is its slope?

Exercises

19. Suppose that $xy = 4$. Express dy/dt in terms of dx/dt when $x = 8$ and $y = \frac{1}{2}$.

20. If $x^2 + y^2 = x^5/y$ and $dy/dt = 3$ when $x = y = \sqrt{2}$, what is dx/dt at that point?

21. The radius and height of a circular cylinder are changing with time in such a way that the volume remains constant at 1 liter ($= 1000$ cubic centimeters). If, at a certain time, the radius is 4 centimeters and is increasing at the rate of $\frac{1}{2}$ centimeter per second, what is the rate of change of the height?

22. What curve do the parametric equations $x = t^2$ and $y = t^6$ describe?

23. Find the equation of the line tangent to the parametric curve $x = t^2 + 1$, $y = 1/(t^4 + 1)$ at $t = 2$.

Problems for Section 5-2

1. Let $h(x) = \sqrt[3]{(x^3 + 1)/(x^6 + 8)}$.
   (a) Write h as a composition of the functions $f(u) = \sqrt[3]{u}$ and $g(x) = (x^3 + 1)/(x^6 + 8)$.
   (b) Write h as a composition of $k(u) = \sqrt[3]{(u + 1)/(u^2 + 8)}$ and $l(x) = x^3$ and write k as a composition of $m(w) = \sqrt[3]{w}$ and $n(u) = (u + 1)/(u^2 + 8)$.

2. Let $f_1(x) = x$, $f_2(x) = 1/x$, $f_3(x) = 1 - x$, $f_4(x) = 1/(1 - x)$, $f_5(x) = (x - 1)/x$, and $f_6(x) = x/(x - 1)$.
   (a) Show that the composition of any two functions in this list is again in the list. Complete the "composition table" below. For example, $(f_2 \circ f_3)(x) = f_2(1 - x) = 1/(1 - x) = f_4(x)$.

$\circ$	f_1	f_2	f_3	f_4	f_5	f_6
f_1			f_3			
f_2			f_4			
f_3		f_5				
f_4			f_5			
f_5						
f_6						

   (b) Show that the inverse of any function in the preceding list is again in the list. Which of the functions equal their own inverses?

3. Differentiate the function $h(x)$ in Problem 1 according to the two compositions obtained and verify that your answers agree.

4. Verify the chain rule for the following compositions in Problem 2:
   (a) $f_4 \circ f_4 = f_5$ (b) $f_3 \circ f_2 = f_5$
   (c) $f_2 \circ f_3 = f_4$ (d) $f_1 \circ f_3 = f_3$

5. Differentiate:
   (a) $f(x) = (x^2 + 8x + 2)^{3/2}$
   (b) $f(x) = (16x^{3/2} - 4x)^{5/2}$
   (c) $f(x) = \sqrt{(\sqrt{x} + \sqrt[3]{x})/(x^2 + 1)}$
   (d) $f(x) = \sqrt{x^2 + 2}/(x^3 + 8 + \sqrt{x})$
   (e) $g(f(x))$ at $x = 1$, if $g(u) = \sqrt{u^3 + 2}$, $f(1) = 2$, and $f'(1) = -\frac{16}{23}$

6. Find the following derivatives:
   (a) $(d/dx)\sqrt{1 - x^3}$
   (b) $(d/ds)[s/(3 + s)^{5/8}]$
   (c) $f'(x)$, where $f(x) = (1 - \sqrt{x})/(1 + \sqrt{x})$
   (d) $(d^2/dx^2)(1 + x^2 + x^4)^{-1/5}$
   (e) $g'(7)$, where $g(x) = \sqrt[3]{1 + x + (x - 7)^5}$
   (f) dy/dx, where $x^3 + y^3 = (x + y)^2$
   (g) dx/dt, where x and y are functions of t, $xy + x/y = 4$, and $dy/dt = 1$

7. Find a formula for the second derivative of $f \circ g$ in terms of the first and second derivatives of f and g.

8. A hurricane is dropping 10 inches of rain per hour into a swimming pool which measures 40 feet long by 20 feet wide.

   (a) What is the rate at which the volume of water in the pool is increasing?

   (b) If the pool is 4 feet deep at the shallow end and 8 feet deep at the deep end, how fast is the water level rising after 2 hours? (Suppose that the pool was empty to begin with.) How fast after 6 hours?

9. A point in the plane moves in such a way that it is always twice as far from $(0,0)$ as it is from $(0,1)$.

   (a) Show that the point moves on a circle.

   (b) At the moment when the point crosses the segment between $(0,0)$ and $(0,1)$, what is dy/dt?

   (c) Where is the point when $dy/dt = dx/dt$? (You may assume that dx/dt and dy/dt are not simultaneously zero.)

10. Find the equation of the tangent line to the curve $x^4 + y^4 = 2$ when $x = y = 1$. Sketch the curve and the line.

11. (a) Differentiate $(1 + x^2)^r$, where r is a fixed rational number.

    (b) Find the antiderivative $\int x/\sqrt[3]{1 + x^2}\, dx$.

★12. Let $f(x) = (ax + b)/(cx + d)$, and let $g(x) = (rx + s)/(tx + u)$.

    (a) Show that $f \circ g$ and $g \circ f$ are both of the form $(kx + l)/(mx + n)$ for some $k, l, m,$ and n.

    (b) Under what conditions on a, b, c, d, r, s, t, u does $f \circ g = g \circ f$?

13. Suppose $x^2 + y^2 = t$ and that $x = 3$, $y = 4$, and $dx/dt = 7$ when $t = 25$. What is dy/dt at that moment?

14. Sketch the curve defined by the parametric equations $x = t^2$, $y = 1 - t$, $-\infty < t < \infty$.

15. (a) Find the slope of the parametric curve $y = t^4 + 2t$, $x = 8t$ at $t = 1$.

    (b) What relationship between x and y is satisfied by the points on this curve?

    (c) Verify the formula $dy/dx = (dy/dt)/(dx/dt)$ for this curve.

16. Find the equation of the tangent line to the parametric curve

$$x = \sqrt{t^4 + 6t^2 + 8t} \qquad y = \frac{t^2 + 1}{\sqrt{t - 1}}$$

    at $t = 3$.

17. Discuss and derive the formula $dx/dy = (dx/dt)/(dy/dt)$.

★18. Derive a formula for differentiating the composition $f_1 \circ f_2 \circ \ldots \circ f_n$ of n functions.

★19. Let f be differentiable with a differentiable inverse f^{-1}. Apply the chain rule to the equation $(f \circ f^{-1})(x) = x$ to deduce the inverse function rule.

20. Water is being pumped from a 20-meter square pond into a round pond with a radius 10 meters. At a certain moment, the water level in the square pond is dropping by 2 inches/minute. How fast is the water rising in the round pond?

21. A ladder 25 feet long is leaning against a vertical wall. The bottom is being shoved along the ground, towards the wall at $1\frac{1}{2}$ feet per second. How fast is the top rising when it is 15 feet off the ground?

22. A messenger service in a small resort community operates a speedboat to deliver packages and messages at various shore points around a mountain lake. Let x, y coordinates be introduced with the origin at the center of the lake. The speedboat location is then given by $x = x(t)$, $y = y(t)$ where $x(t)$ and $y(t)$ are known differentiable functions of t.

    (a) Draw a figure which details one possible path for the speedboat, in which the path crosses itself at least 12 times.

    (b) Given a point (x_0, y_0) on the speedboat path, which is non-crossing, explain the meaning of dy/dx.

    (c) Given a crossing point (x_1, y_1), explain why dy/dx is generally ill-defined.

    (d) Suppose that the boat travels the path $(x^2/a^2) + (y^2/b^2) = 1$. Identify the points on this path where dy/dx is not defined, and describe how to divide the path into two paths, each of which is representable by a function.

23. A particle travels along the path $x^4 + 4y^4 = 1$. Determine $\max \sqrt{x^2 + y^2}$ (maximum distance from $(0,0)$ to the particle).

24. Find the point (x, y) closest to $(0,0)$ on the line $y = (5 - x)/2$.

25. Let $x(t)$ be the distance traveled by a car in miles along a highway after t hours, and let $A(x)$ be the amount of gas in gallons required for the car to travel x miles.

    (a) $A(x(t))$ is the amount of gas required to drive t hours. Explain in words.

    (b) "The car's gasoline consumption in gallons per hour is the product of its gasoline consumption in gallons per mile and its velocity in miles per hour." Is this a reasonable statement? Interpret mathematically using the chain rule.

26. The mass M of a concrete beam measured x units from one end is $M = 24(2 + x^{1/3})^4$ kg. Find the density dM/dx.

Review Problems for Chapter 5 ████████████████████

1. Differentiate:
    (a) $f(x) = x^{5/3}$
    (b) $g(x) = x^{3/2}/\sqrt{1 + x^2}$
    (c) $h(x) = (1 + 2x^{1/2})^{3/2}$
    (d) $l(y) = y^2/\sqrt{1 - y^3}$

2. Differentiate:
    (a) $f(x) = \dfrac{8\sqrt{x}}{1 + \sqrt{x}} + 3x\left(\dfrac{1 + \sqrt{x}}{1 - \sqrt{x}}\right)$
    (b) $f(x) = \dfrac{1 + x^{3/2}}{1 - x^{3/2}}$
    (c) $f(y) = \dfrac{8y^4}{1 + [3/(1 + \sqrt{y})]}$
    (d) $f(y) = y^3 + \left(\dfrac{1 + y^3}{1 - y^3}\right)^{1/2}$

3. If f is a given differentiable function and $g(x) = f(\sqrt{x})$, what is $g'(x)$?

★4. If f is differentiable, has a differentiable inverse, and $g(x) = f^{-1}(\sqrt{x})$, what is $g'(x)$?

5. Find the following antiderivatives:
    (a) $\int (4x^3 + 3x^2 + 2x + 1)\,dx$
    (b) $\int 10\,dx$
    (c) $\int \left[\dfrac{-1}{x^2} + \dfrac{-2}{x^3} + \dfrac{-3}{x^4} + \dfrac{-4}{x^5}\right]dx$
    (d) $\int \frac{2}{3}x^{2/5}\,dx$

6. Find the following antiderivatives:
    (a) $\int (x^2 + \sqrt{x})\,dx$
    (b) $\int (x^{3/2} + x^{-1/2})\,dx$
    (c) $\int \dfrac{x^4 + x^6 + 1}{x^{1/2}}\,dx$
    (d) $\int \left(\sqrt{x} + \dfrac{1}{\sqrt{x}}\right)dx$

7. Let $f(x) = x^3 - 3x + 7$.
    (a) Find an interval containing zero on which f is invertible.
    (b) Denote the inverse by g. What is the domain of g?
    (c) Calculate $g'(7)$.

8. Differentiate each of the following functions, and write down the corresponding antidifferentiation formula.
    (a) $f(x) = x^{1/3}/(x^{1/2} - x^{1/3} + 1)^{1/2}$
    (b) $f(x) = \sqrt{\left(\dfrac{8x^2 + 9x + 1}{x^{1/2} + x^{-1/2}}\right)}$
    (c) $f(x) = \sqrt{x} - \sqrt{(x - 1)/(x + 1)}$
    (d) $f(x) = \sqrt{\dfrac{\sqrt{x} - 1}{\sqrt{x} + 1}}$

9. Differentiate each of the following functions, and write down the corresponding antidifferentiation formula:
    (a) $(\sqrt[4]{x} - 1)/(\sqrt[4]{x} + 1)$
    (b) $\sqrt[3]{(x - 1)/(3x + 1)}$
    (c) $\sqrt{(x^2 + 1)/(x^2 - 1)}$
    (d) $\sqrt{x^3 + 2\sqrt{x} + 1}$

10. Find the linear approximations for: (a) $\sqrt[5]{32.02}$; (b) $\sqrt[3]{27.11}$; (c) $\sqrt[3]{-63.01}$.

★11. Find a formula for the second derivative of $[g(y)]^2$ if $g(y)$ is the inverse function of $f(x)$.

12. If $x^2 + y^2 + xy^3 = 1$, find dy/dx when $x = 0$, $y = 1$.

13. If x and y are functions of t, $x^4 + xy + y^4 = 2$, and $dy/dt = 1$ at $x = 1$, $y = 1$, find dx/dt at $x = 1$, $y = 1$.

14. Let a curve be described by the parametric equations
    $$x = \sqrt{t} + t^2 + 1/t, \quad y = 1 + \sqrt[3]{t} + t, \quad 1 \le t \le 3$$
    Find the equation of the tangent line at $t = 2$.

15. Two ships, A and B, leave San Francisco together and sail due west. A sails at 20 miles per hour and B at 25 miles per hour. Ten miles out to sea, A turns due north and B continues due west. How fast are they moving away from each other 4 hours after departing San Francisco?

16. At an altitude of 2000 meters, a parachutist jumps from an airplane and falls $4.9t^2$ meters in t seconds. Suppose that the air pressure p decreases with altitude at the constant rate of 0.095 gsc per meter. The parachutist's ears pop when dp/dt reaches 2 gsc per second. At what time does this happen?

17. The *speed* of an object traveling a parametric curve is given by $v = \sqrt{(dx/dt)^2 + (dy/dt)^2}$.

    (a) Find the speed at $t = 1$ for the motion $x = t^3 - 3t^2 + 1$, $y = t^5 - t^7$.

    (b) Repeat for $x = t^2 - 3$, $y = \frac{1}{3}t^3 - t$ at $t = 1$.

18. The *lemniscate* $3(x^2 + y^2)^2 = 25(x^2 - y^2)$ is a planar curve which self-intersects at the origin.

    (a) Show by use of symmetry that the entire lemniscate can be graphed by (1) reflecting the first quadrant portion through the x-axis, and then (2) reflecting the right half-plane portion through the origin to the left half-plane.

    (b) Find by means of implicit differentiation the value of dy/dx at $(2, 1)$.

    (c) Determine the equation of the tangent line to the lemniscate through $(2, 1)$. Graph it.

    ★(d) Graph the lemniscate. Divide it into two curves, each of which is the graph of a function.

19. The *drag* on an automobile is the force opposing its motion down the highway, due largely to air resistance. The drag in pounds D can be approximated for velocities v near 50 mph by $D = kv^2$. Using $k = 0.24$, find the rate of increase of drag with respect to time at 55 mph when the automobile is undergoing acceleration of 3 mph each second.

20. The air resistance of an aircraft fuel tank is given approximately by $D = 980 + 7(v - 700)$ lbs for the velocity range of $700 \le v \le 800$ mph. Find the rate of increase in air resistance with respect to time as the aircraft accelerates past the speed of sound (740 mph) at a constant rate of 12 mph each second.

21. A physiology experiment measures the heart rate $R(x)$ in beats per minute of an athlete climbing a vertical rope of length x feet. The experiment produces two graphs: one is the heart rate R versus the length x; the other is the length x versus the time t in seconds it took to climb the rope (from a fresh start, as fast as possible).

    (a) Give a formula for the change in heart rate in going from a 12-second climb to a 13-second climb using the linear approximation.

    (b) Explain how to use the tangent lines to the two graphs and the chain rule to compute the change in part (a).

Supplement to Chapter 5: Preview of the Trigonometric and Exponential Functions

This supplement is designed for

those who wish to see a preview of some highlights in the following two chapters

those who need a working knowledge of the calculus of the trigonometric and exponential functions before the end of the first semester, but do not have time for an exhaustive study

those who need reinforcement of the chain rule (and other rules of differentiation and integration) with additional types of functions

The material in this supplement is intended to bridge the first five chapters with the following two. It is not necessary to read this section as a prerequisite for the following chapters. If it is not useful in a way such as those indicated above, we suggest proceeding directly to Chapter 6 and using this supplement for review purposes.*

THE SINE AND COSINE FUNCTIONS

The sine and cosine functions arise in the study of trigonometry, as is explained in detail in Section 6-1. For now, let us regard these functions as "God given." If you prefer, regard them as calculator buttons whose mechanism of calculation is not revealed to us. If a number x is entered and the "sin" key pushed, a new number $\sin(x)$ (or just $\sin x$) appears. Similarly, there is a key that activates the "cos" function, producing $\cos x$ for any number x which is entered.

Using a table of values of $\cos x$ and $\sin x$, or our calculator, we can plot the graphs of $y = \cos x$ and $y = \sin x$. The result is shown in Fig. 5-S-1.

We will also regard as given to us the rules for differentiation of these functions:

$$\frac{d}{dx}\sin x = \cos x \qquad \text{and} \qquad \frac{d}{dx}\cos x = -\sin x$$

The functions "sin" and "cos" (read "sine" and "cosine") are not expressible as algebraic combinations of polynomials or fractional powers in any simple, direct way. However, given x they produce definite numbers $\sin x$ and $\cos x$, so they are legitimate functions. Notice that in the general notation $f(x)$ the symbol "f" is replaced by "sin" or "cos" (and the parentheses around x are omitted).

These differentiation formulas are consistent with the graphs in Fig. 5-S-1. For example, at $x = 0$ the slope of the sine graph is $\cos 0 = 1$ (from our calculator or the graph). Likewise, the graph of $\cos x$ at $x = 0$ is horizontal, which is consistent with the value of $-\sin x$ at $x = 0$, namely 0. The reader can check consistency at other points as well.

---

*Chapters 6 and 7 do not depend on Chapter 4. If you have not yet covered Chapter 4 and wish to use this supplement, omit problems referring to definite integrals: $\int_a^b f(x)dx$.

(a) $y = \sin\theta$

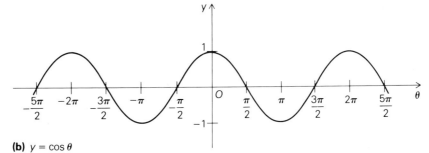

Fig. 5-S-1 The graphs of
cos *x* and sin *x*.

(b) $y = \cos\theta$

Calculator Discussion

We can check that these two differentiation formulas are consistent with the sine and cosine functions on a calculator. For instance we can check that

$$\frac{\sin(x + \Delta x) - \sin x}{\Delta x} \approx \cos x$$

for a given *x* and Δx small. The following table gives an example. (We cannot take Δx too small or else calculator errors will be significant.) Calculators usually compute $\sin x$ and $\cos x$ for *x* given either in degrees or in radian measure. Throughout the present discussion, radian measure is understood.

			$x = 1$			
Δx	1	.5	.1	.01	.001	.0001
$\dfrac{\sin(x + \Delta x) - \sin x}{\Delta x}$	.067826	.31204	.49736	.53608	.53988	.54026
$\cos x$	.54030	.54030	.54030	.54030	.54030	.54030

			$x = 2$			
Δx	1	.5	.1	.01	.001	.0001
$\dfrac{\sin(x + \Delta x) - \sin x}{\Delta x}$	−.76817	−.62165	−.46088	−.42068	−.41660	−.41619
$\cos x$	−.41614	−.41614	−.41614	−.41614	−.41614	−.41614

In this table we see that the difference quotient

$$\frac{\sin(x + \Delta x) - (\sin x)}{\Delta x}$$

is close to $\cos x$ for small Δx. This bears out our "God-given formula" $(d/dx)\sin x = \cos x$. The reader may wish to test other values of x and Δx and to verify the consistency of the formula $(d/dx)\cos x = -\sin x$ with numerical calculation.

Worked Example 1. Differentiate
(a) $(\sin x)(\cos x)$
(b) $(\sin x)^2$ (also written $\sin^2 x$)
(c) $\sin 5x$
(d) $\dfrac{\sin 3x}{(\cos x) + x^4}$

Solution (a) By the product rule,

$$\frac{d}{dx}(\sin x)(\cos x) = \left(\frac{d}{dx}\sin x\right)\cos x + \sin x\left(\frac{d}{dx}\cos x\right)$$
$$= \cos x \cos x - \sin x \sin x$$
$$= (\cos x)^2 - (\sin x)^2$$

(b) By the power of a function rule,

$$\frac{d}{dx}(\sin x)^2 = 2\sin x\frac{d}{dx}\sin x = 2\sin x \cos x$$

(c) By the chain rule,

$$\frac{d}{dx}\sin 5x = \frac{d}{du}\sin u\frac{du}{dx} \quad \text{(where } u = 5x)$$
$$= (\cos u)(5) = 5\cos 5x$$

(d) By the quotient rule and chain rule,

$$\frac{d}{dx}\frac{\sin 3x}{\cos x + x^4}$$
$$= \frac{(\cos x + x^4)\dfrac{d}{dx}\sin 3x - \sin 3x\dfrac{d}{dx}(\cos x + x^4)}{(\cos x + x^4)^2}$$
$$= \frac{(\cos x + x^4)3\cos 3x - \sin 3x(-\sin x + 4x^3)}{(\cos x + x^4)^2}$$
$$= \frac{3\cos x \cos 3x + 3x^4\cos 3x + \sin x \sin 3x - 4x^3\sin 3x}{(\cos x + x^4)^2}$$

From the equations $\cos 0 = 1$ and $\sin 0 = 0$ we can derive a further relationship. Notice that by the chain rule,

$$\frac{d}{dx}(\cos^2 x + \sin^2 x) = 2\cos x(-\sin x) + 2\sin x \cos x = 0$$

Thus $\cos^2 x + \sin^2 x$ is a constant function. But at $x = 0$ its value is 1. Thus, for all x we must have

$$\cos^2 x + \sin^2 x = 1$$

This is a useful identity that can also be derived from other facts (see Section 6-1).

From the formula $(d/dx)\sin x = \cos x$, it follows that $\sin x$ is an antiderivative of $\cos x$. Thus we have the integration formula

$$\int \cos x \, dx = \sin x + C$$

Similarly,

$$\int \sin x \, dx = -\cos x + C$$

Worked Example 2   Calculate

(a) $\int_0^1 (2 \sin x + x^3) \, dx$ (b) $\int_1^2 \cos 2x \, dx$

Solution   (a)   By the sum and power rules,

$$\int (2 \sin x + x^3) \, dx = 2 \int \sin x \, dx + \int x^3 \, dx = -2 \cos x + \frac{x^4}{4} + C$$

Thus by the Fundamental Theorem of Calculus,

$$\int_0^1 (2 \sin x + x^3) \, dx = \left(-2 \cos x + \frac{x^4}{4} \right) \Bigg|_0^1$$

$$= -2(\cos 1 - \cos 0) + \frac{1}{4}$$

$$= \frac{9}{4} - 2 \cos 1$$

(b)   An antiderivative of $\cos 2x$ may be found by first guessing $\sin 2x$; the derivative of $\sin 2x$ is $2 \cos 2x$, so we try $\frac{1}{2}\sin 2x$, which works. Thus by the Fundamental Theorem,

$$\int_1^2 \cos 2x \, dx = \frac{1}{2} \sin 2x \Bigg|_1^2 = \frac{1}{2} (\sin 4 - \sin 2)$$

Exercises   1.   Differentiate

(a) $(\cos x)^3$ (b) $\sin(20x^2)$

(c) $(\sqrt{x} + \cos x)^4$

2.   Differentiate

(a) $\dfrac{\cos \sqrt{x}}{1 + \sqrt{x}}$ (b) $\sin(x + \sqrt{x})$

(c) $\dfrac{x}{\cos x + \sin(x^2)}$

3. Find the indefinite integral:

(a) $\int (\cos x + x^4)\, dx$

(b) $\displaystyle\int \frac{x^2 \cos x + 5}{x^2}\, dx$

(c) $\int (\sin 10x + x^4)\, dx$

4. Find the definite integral:

(a) $\int_0^1 (x^3 + \sin 10x)\, dx$

(b) $\displaystyle\int_1^2 \left(x^4 + \frac{1}{x^2} + \cos 5x \right) dx$

(c) $\int_0^1 ((x+1)^2 + \cos 30x)\, dx$

5. Differentiate $\dfrac{\cos x \sin x}{1 + \cos x \sin x}$ and make up a corresponding integration formula.

6. Suppose $\varphi(x)$ is a function "appearing from the blue" with the property that

$$\frac{d\varphi}{dx} = \frac{1}{\cos x}$$

Calculate

(a) $\dfrac{d}{dx}\,(\varphi(3x)\cos x)$

(b) $\displaystyle\int_0^1 \frac{1}{\cos x}\, dx$

(c) $\dfrac{d^2}{dx^2}\,(\varphi(2x)\sin 2x)$

MORE TRIGONOMETRIC FUNCTIONS

Just as one can build up new functions from the simple building blocks x, x^2, x^3, $\sqrt{x}$, $\sqrt[3]{x}$, ... such as $(x^3 + \sqrt{x})/(x^4 + \sqrt[3]{x^2} + 2)$, one can build up new functions out of $\sin x$ and $\cos x$. For instance, $\tan x$ (read "tangent of x") is defined to be

$$\tan x = \frac{\sin x}{\cos x} \qquad \text{(tangent of } x\text{)}$$

Likewise we define

$$\cot x = \frac{1}{\tan x} = \frac{\cos x}{\sin x} \qquad \text{(cotangent of } x\text{)}$$

$$\sec x = \frac{1}{\cos x} \qquad \text{(secant of } x\text{)}$$

and

$$\csc x = \frac{1}{\sin x} \qquad \text{(cosecant of } x\text{)}$$

We can differentiate these new trigonometric functions by using the quotient rule:

$$\frac{d}{dx}\tan x = \frac{\cos x \dfrac{d}{dx}\sin x - \sin x \dfrac{d}{dx}\cos x}{\cos^2 x}$$

$$= \frac{\cos^2 x + \sin^2 x}{\cos^2 x} = \frac{1}{\cos^2 x} = \sec^2 x$$

Formulas for the derivatives of the other trigonometric functions can be similarly derived.

Worked Example 3 Differentiate
(a) $\tan(\cos\sqrt{x}\,)$ (b) $\csc\sqrt{x}$

Solution (a) By the chain rule,

$$\frac{d}{dx}\tan(\cos\sqrt{x}\,) = \frac{d}{du}\tan u \frac{du}{dx}, \quad u = \cos\sqrt{x}$$

$$= \sec^2 u \cdot \frac{d}{dx}\cos\sqrt{x}$$

$$= \sec^2(\cos\sqrt{x}\,)(-\sin\sqrt{x}\,)\frac{d}{dx}\sqrt{x}$$

$$= \frac{-\sec^2(\cos\sqrt{x}\,)\sin\sqrt{x}}{2\sqrt{x}}$$

(b) By the reciprocal rule,

$$\frac{d}{dx}\csc x = -\frac{\cos x}{\sin^2 x} = -\csc x \cot x$$

so by the chain rule,

$$\frac{d}{dx}\csc\sqrt{x} = -\frac{\csc\sqrt{x}\,\cot\sqrt{x}}{2\sqrt{x}}$$

In Section 5-1 we developed differentiation formulas for fractional powers by applying the differentiation rule for inverse functions to the power functions. We can carry out a similar program for the trigonometric functions. In Section 6-2 we shall discuss on exactly what intervals the trigonometric functions have inverses (just as the inverse of x^2, namely $\sqrt{x}$, is defined only for $x \geq 0$). For now let us ignore this issue and merely derive the formulas.

The inverse function of $\sin x$ is denoted $\sin^{-1} x$. This *does not* mean $1/\sin x$. (This is logically inconsistent with the notation $\sin^2 x$, but we are stuck with it by common usage.) By the inverse function rule (see p. 225),

$$\frac{d}{dy}\sin^{-1} y = \frac{1}{\dfrac{d}{dx}\sin x}$$

where $x = \sin^{-1}y$; that is, $\sin x = y$. Thus

$$\frac{d}{dy}\sin^{-1}y = \frac{1}{\cos x}$$

But $\cos^2 x + \sin^2 x = 1$, so $\cos x = \pm\sqrt{1 - \sin^2 x} = \pm\sqrt{1 - y^2}$. The ambiguity $\pm$ appears here because we haven't been careful in specifying $\sin^{-1}y$. The "standard" choice of $\sin^{-1}y$ works in a range of x for which $\cos x$ is positive, so $\cos x = \sqrt{1 - y^2}$. Thus

$$\frac{d}{dy}\sin^{-1}y = \frac{1}{\sqrt{1 - y^2}}$$

In x-notation,

$$\frac{d}{dx}\sin^{-1}x = \frac{1}{\sqrt{1 - x^2}}$$

One can derive formulas for the inverses of the remaining trigonometric functions in a similar way.

Worked Example 4 Differentiate $\sin^{-1}(\sqrt{x} + \cos 3x)$.

Solution By the chain rule,

$$\frac{d}{dx}\sin^{-1}(\sqrt{x} + \cos 3x)$$

$$= \frac{1}{\sqrt{1 - (\sqrt{x} + \cos 3x)^2}}\frac{d}{dx}(\sqrt{x} + \cos 3x)$$

$$= \frac{1}{\sqrt{1 - x - 2\sqrt{x}\cos 3x + \cos^2 3x}} \cdot \left(\frac{1}{2\sqrt{x}} - 3\sin 3x\right)$$

Exercises

7. Differentiate

(a) $\tan 10x$ (b) $\csc(x^2 + \sqrt{x})$ (c) $\dfrac{\cos x}{\tan x + \sqrt{x}}$

8. Differentiate
(a) $\tan(\sin\sqrt{x})$ (b) $\sin^{-1}(\sqrt{x})$ (c) $\tan(\cos x + \csc\sqrt{x})$

9. (a) Show that

$$\frac{d}{dy}\cos^{-1}y = \frac{-1}{\sqrt{1 - y^2}}$$

(on a range of x with $\sin x$ positive).
(b) Conclude that $\cos^{-1}x + \sin^{-1}x$ is a constant.

10. (a) Prove that $\dfrac{d}{dx}\tan^{-1}x = \dfrac{1}{1 + x^2}$.
(b) Differentiate $\tan^{-1}(\sqrt{1 + x^2})$.

11. Integrate

(a) $\displaystyle\int_{1}^{2} \frac{dx}{1 + x^2}$ (b) $\displaystyle\int_{0}^{1} \frac{dx}{1 + (2x)^2}$

(c) $\displaystyle\int_{1/2}^{\sqrt{2}/2} \left(\frac{x^2 + 1}{x^2} + \frac{1}{\sqrt{1 - x^2}} + \cos x \right) dx$

THE EXPONENTIAL FUNCTION

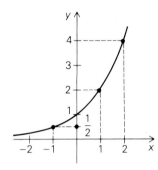

Fig. 5-S-2 The graph of $y = 2^x$.

Given a number $b > 0$, the function defined by $f(x) = b^x$ ("b to the power x") is called the *exponential function with base b*. For $b = 2$, this function is sketched in Fig. 5-S-2.

Detailed properties of exponential functions will be reviewed in Section 7-1. We accept these properties and differentiability of b^x for now. The derivative of b^x at a general point x can be obtained from the derivative at zero by the following device. For any constant x_0, the law of exponents $b^x b^y = b^{x+y}$ implies:

$$f(x + x_0) = b^{x + x_0} = b^x b^{x_0} = f(x)f(x_0)$$

Differentiating in x and using the chain rule gives

$$f'(x + x_0) = f'(x)f(x_0)$$

Setting $x = 0$ gives

$$f'(x_0) = f'(0)f(x_0)$$

Thus the derivative of f is just a constant times f itself. The constant $f'(0)$ can be adjusted if we change b. It is convenient to choose b such that $f'(0) = 1$. This can be done (as will be proved in Section 7-1) with a unique b, namely

$$b = e = 2.7182818285\ldots$$

Experiments with a calculator enable one to estimate the derivative of b^x at $x = 0$ for various b by using

$$f'(0) \approx \frac{b^{\Delta x} - 1}{\Delta x}$$

One sees that for b near 2.71, this is nearly 1 for Δx small, verifying our claim.

In summary, e is the number that has the property

$$\frac{d}{dx} e^x = e^x$$

that is, e^x reproduces itself when differentiated. (In Section 9-1 we will prove that a constant times e^x is the *only* function that reproduces itself when differentiated.) Consequently we also have

$$\int e^x \, dx = e^x + C$$

Worked Example 5 Differentiate

(a) e^{6x} (b) $e^{\cos x}$ (c) $xe^x - e^x$

Solution (a) Let $u = 6x$ and use the chain rule:

$$\frac{d}{dx}\, e^{6x} = \frac{d}{du}\, e^u \,\frac{du}{dx}$$

$$= e^u \cdot 6 = 6e^{6x}$$

(b) Again use the chain rule:

$$\frac{d}{dx}\, e^{\cos x} = e^{\cos x}\, \frac{d}{dx}\, \cos x$$

$$= -e^{\cos x}\, \sin x$$

(c) By the sum and product rules:

$$\frac{d}{dx}\, (xe^x - e^x) = e^x + xe^x - e^x = xe^x$$

Worked Example 6 Verify the formula

$$\int e^x \cos x \, dx = \frac{1}{2}\, e^x\, (\sin x + \cos x) + C$$

Solution We must check that the right-hand side is an antiderivative of the integrand. We compute, using the product rule:

$$\frac{d}{dx}\, \frac{1}{2}\, e^x\, (\sin x + \cos x)$$

$$= \frac{1}{2}\left(\frac{de^x}{dx}\right)(\sin x + \cos x) + \frac{1}{2}\, e^x \left(\frac{d}{dx}(\sin x + \cos x)\right)$$

$$= \frac{1}{2}\, e^x\, (\sin x + \cos x) + \frac{1}{2}\, e^x\, (\cos x - \sin x)$$

$$= e^x \cos x$$

Thus the formula is verified.

Exercises

12. Differentiate

   (a) $x^2 e^{10x}$ (b) $e^x \cos x$ (c) $e^{\cos 2x}$

13. Differentiate

   (a) e^{x^3} (b) $(e^x)^3$ (c) $\cos(e^x)$

14. Integrate

   (a) $\int e^{2x}\, dx$ (b) $\int (x^2 + e^x)\, dx$

   (c) $\int (\cos x + e^{4x})\, dx$

15. Integrate

   (a) $\int_0^1 (x^2 + 3e^x)\, dx$ (b) $\int_1^2 e^{-x}\, dx$

   (c) $\int_2^3 (x^3 + e^{2x})\, dx$

16. (a) Verify the integration formula

$$\int e^{ax} \sin bx \, dx = \frac{e^{ax} \, (a \sin bx - b \cos bx)}{a^2 + b^2} + C$$

(b) Find a similar formula for

$\int e^{ax} \cos bx \, dx$

(See Worked Example 6.)

THE LOGARITHM

The graph of $y = e^x$ is plotted in Fig. 5-S-3. From the graph we see that e^x has an inverse defined on $(0, \infty)$. This inverse is called the *natural logarithm function*. By definition of inverse functions, $\ln x$ is the number such that

$$e^{\ln x} = x$$

Thus $\ln 1 = 0$ and $\ln e = 1$.

Fig. 5-S-3 The graph of $y = e^x$ and $y = \ln x$.

We can differentiate $\ln x$ by using the inverse function rule from Section 5-1 with $y = e^x$,

$$\frac{d}{dy} \ln y = \frac{1}{\dfrac{d}{dx} e^x} = \frac{1}{e^x} = \frac{1}{y}$$

Thus we have the important formula

$$\frac{d}{dx} \ln x = \frac{1}{x} \quad (\text{for } x > 0)$$

and the corresponding antidifferentiation formula

$$\int \frac{1}{x} \, dx = \ln x + C$$

The latter formula now enables us to integrate all powers of x:

$$\int x^n \, dx = \begin{cases} \dfrac{x^{n+1}}{n+1} + C & \text{if } n \neq -1 \\[2ex] \ln|x| + C & \text{if } n = -1, \, x \neq 0 \end{cases}$$

(If x is negative, note that $\ln|x| = \ln(-x)$ and so by the chain rule,

$$\frac{d}{dx}\ln|x| = -\left(\frac{1}{-x}\right) = \frac{1}{x}; \quad \text{thus} \int \frac{1}{x}\,dx = \ln|x| + C$$

for $x < 0$).

Worked Example 7 Differentiate
(a) $\ln(10x^2 + 1)$ (b) $\sin(\ln x^3)\,e^{x^4}$

Solution (a) By the chain rule with $u = 10x^2 + 1$, we get

$$\frac{d}{dx}\ln(10x^2 + 1) = \frac{d}{du}\ln u \, \frac{du}{dx} = \frac{1}{u}\cdot 20x = \frac{20x}{10x^2 + 1}$$

(b) By the product rule and chain rule,

$$\frac{d}{dx}\left(\sin(\ln x^3)\,e^{x^4}\right) = \left(\frac{d}{dx}\sin(\ln x^3)\right)e^{x^4} + \sin(\ln x^3)\frac{d}{dx}e^{x^4}$$

$$= \cos(\ln x^3)\cdot \frac{3x^2}{x^3}\cdot e^{x^4} + \sin(\ln x^3)\cdot 4x^3 e^{x^4}$$

$$= e^{x^4}\left(\frac{3}{x}\cos(\ln x^3) + 4x^3\sin(\ln x^3)\right)$$

Worked Example 8 Integrate
(a) $\displaystyle\int_0^1 \frac{1}{x+1}\,dx$ (b) $\displaystyle\int_1^2 \frac{x^3 + 3x + 2}{x}\,dx$

Solution
(a) Since $\dfrac{d}{dx}\ln x = \dfrac{1}{x}$, the chain rule gives $\dfrac{d}{dx}\ln(x+1) = \dfrac{1}{x+1}$, so

$$\int_0^1 \frac{1}{x+1}\,dx = \ln(x+1)\,\bigg|_0^1 = \ln 2 - \ln 1 = \ln 2$$

(b) $\displaystyle\int_1^2 \frac{x^3 + 3x + 2}{x}\,dx = \int_1^2 \left(x^2 + 3 + \frac{2}{x}\right)dx$

$$= \left(\frac{x^3}{3} + 3x + 2\ln x\right)\bigg|_1^2$$

$$= \frac{7}{3} + 3 + 2\ln 2 = \frac{16}{3} + 2\ln 2$$

Exercises 17. Differentiate

    (a) $\ln(\cos x)$ (b) $\cos(\ln x)$

    (c) $\ln(x^4 + x^3 + 1)$

18. Differentiate

    (a) $x^3 + \ln(x^4 + 10)$ (b) $x^3 \ln(x^4)$

    (c) $\sin(e^{\cos x})$

19. Integrate

    (a) $\displaystyle\int_{-9}^{-8} \frac{dx}{x + 10}$ (b) $\displaystyle\int_{1}^{2} \frac{x^3 + 5x + 10}{x^2}\, dx$

    (c) $\displaystyle\int_{1}^{2} \left(\frac{3}{x} + \sin 5x\right) dx$

20. Verify the integration formula

$$\int \frac{1}{1 - x^2}\, dx = \frac{1}{2} \ln \left|\frac{1 + x}{1 - x}\right| + C$$

21. Verify the formula

$$\int \frac{dx}{\sqrt{x^2 - 1}} = \ln(x + \sqrt{x^2 - 1}) + C \quad (x > 1)$$

Problems for the Supplement to Chapter 5

1. Differentiate

    (a) $\cos 10x$ (b) $\sin(15(x + x^2))$

    (c) $\cos(\sqrt{1 + x^2})$ (d) $\dfrac{\sin \sqrt{x}}{1 + \sqrt{x}}$

    (e) $e^{\cos x}$ (f) $\dfrac{\sin(e^x)}{e^x + x^2}$

    (g) $\cos \sqrt{1 + e^x}$ (h) $(\tan x + x^2)^5$

    (i) $\sin^{-1} 3x$ (j) $\sqrt{1 + x^2} \sin^{-1}(x)$

2. Differentiate

    (a) $\ln(\cos x)$ (b) $\ln(\sqrt{x})$

    (c) $\dfrac{e^{\cos x}}{\cos(\sin x)}$ (d) $\tan(\sin(e^x))$

    (e) $e^{(\cos x) + x}$ (f) $e^x \cos(x^{3/2})$

    (g) $x^3 + 8x^2 + x \cos \sqrt{x}$

    (h) $\dfrac{x^2 + 2x}{1 + e^{\cos x}}$ (i) $\sin^{-1}(e^x - 1)$

    (j) $x \ln x$

3. Differentiate the following and write down corresponding integral formulas:

    (a) $\dfrac{1}{36} \sin(6x) - \dfrac{x}{6} \cos(6x)$

    (b) $x \ln x - x$ (c) $\dfrac{e^x}{x + 1}$

    (d) $\dfrac{2}{5} e^{2x} \cos x + \dfrac{1}{5} e^{2x} \sin x$

4. Integrate the following:

    (a) $\displaystyle\int \frac{x^2 + x + 2}{x}\, dx$

    (b) $\int \cos 2x\, dx$

    (c) $\displaystyle\int \frac{x + x^2 \sin 2x + 1}{x^2}\, dx$

    (d) $\int_2^3 xe^x\, dx$ (See Worked Example (5c).)

    (e) $\displaystyle\int_1^2 \frac{dx}{1 + 10x^2}$

5. Integrate

    (a) $\int_1^5 e^{3x}\, dx$

    (b) $\int_1^2 (x - \cos x - e^x)\, dx$

    (c) $\displaystyle\int_1^2 \left(\frac{1}{x} + \frac{1}{x^2} + \frac{1}{x^3}\right) dx$

    (d) $\int (\sec x \tan x + x^2)\, dx$

    (e) $\int_0^3 e^x \cos x\, dx$ (See Worked Example 6.)

6. Verify the following integration formulas:

    (a) $\displaystyle\int \frac{1}{\sqrt{1 + x^2}}\, dx = \ln(x + \sqrt{1 + x^2}) + C$

    (b) $\displaystyle\int \frac{1}{x\sqrt{1 - x^2}}\, dx = -\ln \left|\frac{1 + \sqrt{1 - x^2}}{x}\right| + C$

7. Verify the following:

    (a) $\int (x^n e^x + nx^{n-1} e^x)\, dx = x^n e^x + C$

    (b) $\int x^2 e^x\, dx = x^2 e^x - 2xe^x + 2e^x + C$

8. Use Problems 6 and 7 to evaluate

    (a) $\displaystyle\int_0^1 \frac{dx}{\sqrt{1+x^2}}$

    and

    (b) $\int_0^1 x^2 e^x \, dx$

9. Integrate

    (a) $\displaystyle\int_0^1 \frac{dx}{1+30x^2}$ (b) $\displaystyle\int_0^1 \frac{dx}{x+2}$

    (c) $\displaystyle\int_0^1 \frac{x}{x^2+2} \, dx$

    [*Hint:* differentiate $\ln(x^2+2)$.]

10. (a) By differentiating $\ln(\cos x)$, find $\int \tan x \, dx$.

    (b) Find $\int \cot x \, dx$.

11. (a) Derive a formula for the derivative of $\cot^{-1} x$.

    (b) Differentiate $\cot^{-1}(\ln(\sin 3x))$.

12. (a) Derive a formula for the derivative of $\csc^{-1} x$.

    (b) Differentiate $e^{\csc^{-1}(3x)}$.

13. Let ψ be a function such that

    $$\frac{d\psi}{dx} = \varphi$$

    where φ is described in Exercise 6. Prove that

    $$\frac{d}{dx}\left(\psi(x)\sin x + \frac{d}{dx}\left(\psi(x)\cos x - \frac{x^2}{2}\right)\right)$$
    $$= -\varphi(x)\sin x$$

14. Suppose you began by *defining* $\ln x = \displaystyle\int_1^x \frac{dt}{t}$.

    (a) Use the fundamental theorem to show that
    $$\frac{d}{dx}\ln x = \frac{1}{x}$$

    (b) Define e^x to be the inverse function of $\ln x$ and show $\dfrac{d}{dx} e^x = e^x$.

15. (a) Use the definition of $\ln x$ in Problem 14 to show that $\ln xy = \ln x + \ln y$ by showing that for a given fixed x_0,

    $$\frac{d}{dx}(\ln(x\,x_0) - \ln x - \ln x_0) = 0$$

    (b) Deduce from (a) that $e^{x+y} = e^x e^y$.

    (c) Prove $e^{x+y} = e^x e^y$ by assuming only $\dfrac{d}{dx} e^x = e^x$ and $e^0 = 1$.

Trigonometric Functions

The derivative of sin x is cos x and of cos x is −sin x; everything else follows from this.

Many problems involving angles, circles, and periodic motion lead to trigonometric functions. In this chapter, we study the differential calculus of these functions and their inverses, and we apply our knowledge to solve new problems.

CONTENTS

Section 6-1 is a review of trigonometry. Well-prepared students may skim it quickly and move on to Section 6-2, which develops the differential calculus of the trigonometric functions. In the final section, we apply the tools developed in this chapter and the previous one to solve graphing and word problems involving trigonometric functions and fractional powers.

6-1 Polar Coordinates and Trigonometry

Trigonometric functions provide the link between polar and cartesian coordinates.

This section contains a review of some trigonometry, with an emphasis on the topics which are most important for calculus. The derivatives of the trigonometric functions will be calculated in the next section.

Goals

After studying this section, you should be able to:

Plot points and sketch simple graphs in polar coordinates.

Evaluate trigonometric functions, without tables, for multiples of 30° and 45°.

Derive and use trigonometric identities.

Sketch graphs of the trigonometric functions.

DEGREES AND RADIANS

Fig. 6-1-1 The circumference and area of a circle.

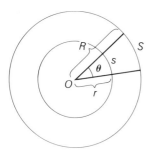

Fig. 6-1-2 $\theta = s/r = S/R$ measured in radians.

The circumference C and area A of a circle of radius r are given by

$$C = 2\pi r \qquad A = \pi r^2$$

(see Fig. 6-1-1), where π is an irrational number whose value is approximately 3.14159 ....*

The procedure for measuring an angle is given as follows:

Definition Let an angle with vertex O subtend an arc of length s on a circle with center O and radius r. (See Fig. 6-1-2.) We say that

$$\theta = \frac{s}{r}$$

is the measure of this angle in *radians*. The measure of the same angle in *degrees* is

$$\theta \cdot \frac{360°}{2\pi} = \theta \cdot \left(\frac{180°}{\pi}\right)$$

---

*For details on the fascinating history of π, see P. Beckmann, A History of π, Golem Press, 1970. To establish deeper properties of π, such as its irrationality (discovered by Hermite and Lindemann around 1880), a careful and critical examination of its definition is needed. The first explicit expression for π was given by Viète (1540–1603) as

$$\frac{2}{\pi} = \sqrt{\frac{1}{2}} \cdot \sqrt{\frac{1}{2} + \frac{1}{2}\sqrt{\frac{1}{2}}} \cdot \sqrt{\frac{1}{2} + \frac{1}{2}\sqrt{\frac{1}{2} + \frac{1}{2}\sqrt{\frac{1}{2}}}} \cdot \ldots$$

which is obtained by inscribing regular polygons in a circle. Euler's famous expression $\pi/4 = 1 - \frac{1}{3} + \frac{1}{5} - \ldots$ is discussed in Solved Exercise 9, Section 13-3. For an elementary proof of the irrationality of π, see M. Spivak, Calculus, Benjamin, 1967.*

The number θ is independent of the size of the circle used, since any two circles are similar figures; in Fig. 6-1-2 we have

$$\theta = \frac{s}{r} = \frac{S}{R}$$

Worked Example 1   An arc of length 10 meters on a circle of radius 4 meters subtends what angle at the center of the circle?

Solution   Here $s = 10$, $r = 4$, so

$$\theta = \frac{10}{4} = \frac{5}{2} = 2\frac{1}{2} \text{ radians} = \left(2\frac{1}{2}\right)\left(\frac{180°}{\pi}\right) \approx 143.24°$$

From the definition of θ and the formula $C = 2\pi r$, we see that there are 2π radians in a complete circle. In the degree system of measure (inherited from Babylonian times) a complete circle has 360°. The measures of right angles and straight angles are shown in Fig. 6-1-3.

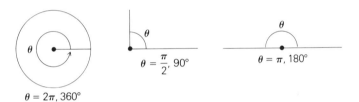

Fig. 6-1-3 A complete circle, a right angle, and a straight angle in degrees and radians.

$\theta = 2\pi, 360°$

$\theta = \frac{\pi}{2}, 90°$

$\theta = \pi, 180°$

DEGREES AND RADIANS

To convert from radians to degrees,   multiply by $180°/\pi \approx 57°18' \approx 57.296°$.

To convert from degrees to radians,   multiply by $\pi/180° \approx 0.01745$.

Degrees are used in common mensuration, but in theoretical work, including calculus, radians are much more convenient and lead to simpler differentiation formulas. Unless explicit mention is made of degrees, you should assume that radian measure is being used. The following table gives some important angles in degrees and radians:

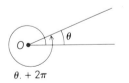

Fig. 6-1-4 θ and $\theta + 2\pi$ measure the same geometric angle.

$\theta + 2\pi$

Degrees	0°	30°	45°	60°	90°	120°	135°	150°	180°	270°	360°
Radians	0	$\frac{\pi}{6}$	$\frac{\pi}{4}$	$\frac{\pi}{3}$	$\frac{\pi}{2}$	$\frac{2\pi}{3}$	$\frac{3\pi}{4}$	$\frac{5\pi}{6}$	π	$\frac{3\pi}{2}$	2π

If θ increases beyond 2π (or 360°), we can start measuring θ over again from zero (or 0°). Thus $\theta = 2\pi$ and $\theta = 0$ measure the same angle, as do $\theta = 2\pi + \pi/4$ and $\theta = \pi/4$. In radians, θ and $\theta + 2\pi$ measure the same angle, as do $\phi° + 360°$ and $\phi°$ in degrees (see Fig. 6-1-4).

Solved Exercises*

1. Convert to radians: 36°, 160°, 280°, 300°.

2. Convert to degrees: $5\pi/18$, 2.6, 6.27, 0.2.

3. Simplify: $5\pi/2$ radians, 470°.

4. If an arc of a circle with radius 10 meters subtends an angle of 22°, how long is the arc?

Exercises

1. Convert to radians: 29°, 54°, 255°, 130°, 320°.

2. Convert to degrees: 5, $\pi/7$, 3.2, $2\pi/9$, $\frac{1}{2}$, 0.7.

3. Simplify:

   (a) Radians: $7\pi/3$, $16\pi/5$, 15π, $48\pi/11$, $13\pi + 1$.

   (b) Degrees: 520°, 1745°, 385°, 604°, 75°, 999°.

4. An arc of radius 15 feet subtends an angle of 27°. How long is the arc?

5. An arc of radius 18 meters has length 5 meters. What angle does it subtend?

POLAR COORDINATES

Fig. 6-1-5 The polar coordinates (r, θ) of a point P.

Polar coordinates were first used successfully by Isaac Newton (1671) and Jacques Bernoulli (1691). Many types of coordinate systems have been conceived, but polar coordinates have survived as one of the most useful. The definitive treatment of polar coordinates in their modern form was given by Leonhard Euler in his 1748 textbook *Introductio in analysis infinitorium*.†

Cartesian coordinates (x, y) represent points in the plane by their distances from two perpendicular lines. The polar coordinate representation labels a point in the plane by means of a pair of numbers (r, θ) which are the distance r from a point O and the angle θ measured counterclockwise in radians from a base line. When polar and cartesian systems are superimposed, the point O is taken to be the origin and the base line is the positive x axis (see Fig. 6-1-5).

POLAR COORDINATES		
If (r, θ) are the polar coordinates of a point P:	1. r is the distance of P to the origin O.	2. θ is the angle from the x axis to the line OP (measured counterclockwise in radians).
Possible values of (r, θ):	3. θ can assume any value, positive or negative (recall that θ and $\theta + 2\pi$ represent the same angle).	4. r can be negative if we agree that $(-r, \theta)$ and $(r, \theta + \pi)$ represent the same point.

*Solutions appear in Appendix A.

†See C. B. Boyer, "The Foremost Textbook of Modern Times," American Mathematical Monthly 58(1951):223–226.

With regard to points 3 and 4 in the display, we note first that positive θ means an angle measured in the counterclockwise sense, while a negative θ means an angle measured in the clockwise sense. See Fig. 6-1-6(a). Since r is the distance from P to O, it ought to be positive. It is often convenient to allow negative r, however, by measuring backwards along the ray with angle θ as shown in Fig. 6-1-6(b). Thus $(-r, \theta)$ represents the same point as $(r, \theta + \pi)$.

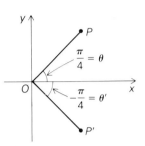

Fig. 6-1-6 Convention for plotting points with negative θ or negative r.

(a) Negative angles are measured clockwise.

(b) Both $(-r, \theta)$ and $(r, \theta + \pi)$ represent P'.

Worked Example 2 Plot the points P_1, P_2, P_3, and P_4 whose polar coordinates are $(5, \pi/6)$, $(-5, \pi/6)$, $(5, -\pi/6)$, and $(-5, -\pi/6)$, respectively.

Solution (See Fig. 6-1-7.) The point $(-5, -\pi/6)$ is obtained by rotating $\pi/6 = 30°$ clockwise to give an angle of $-\pi/6$ and then moving 5 units backwards on this line to the point P_4 shown. The other points are plotted in a similar way.

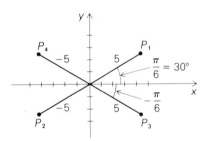

Fig. 6-1-7 Some points in polar coordinates.

Solved Exercises 5. Plot the following points in polar coordinates: $(3, \pi/2)$, $(5, -\pi/4)$, $(1, 2\pi/3)$, $(-3, \pi/2)$.

6. Describe the set of points P whose polar coordinates (r, θ) satisfy

$$0 \leq r \leq 2 \quad \text{and} \quad 0 \leq \theta < \pi$$

Exercises 6. Plot the following points in polar coordinates: $(6, 3\pi/2)$, $(-2, \pi/6)$, $(7, -2\pi/3)$, $(1, \pi/2)$, $(4, -\pi/6)$.

7. Describe the set of points whose polar coordinates (r, θ) satisfy:
   (a) $-1 \leq r \leq 2$; $\pi/3 \leq \theta < \pi/2$ (b) $0 < r < 4$; $-\pi/6 < \theta < \pi/6$
   (c) $2 \leq r < 3$; $-\pi/2 \leq \theta \leq \pi$ (d) $-2 \leq r \leq -1$; $-\pi/4 < \theta < 0$
   (e) $2 \leq r \leq 4$; $-\pi < \theta \leq \pi$

THE TRIGONOMETRIC FUNCTIONS

The following definition of the sine and cosine functions illustrates the close relation between trigonometric functions and polar coordinates.

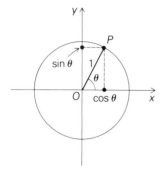

Fig. 6-1-8 The cosine and sine of θ are the x and y coordinates of the point P.

> ***Definition*** Let θ be a real number. We define $\cos \theta = x$ and $\sin \theta = y$, where (x,y) are the cartesian coordinates of the point P whose polar coordinates are $(1, \theta)$. (See Fig. 6-1-8.)

If an angle $\phi°$ is given in degrees, $\sin \phi°$ or $\cos \phi°$ means $\sin \theta$ or $\cos \theta$ where θ is the same angle measured in radians. Thus $\sin 45° = \sin \pi/4$, $\cos 60° = \cos \pi/3$, and so on. The sine and cosine functions can also be defined in terms of ratios of sides of right triangles. (See Fig. 6-1-9.) By definition, $\cos \theta = |OA'|$, and by similar triangles

$$\cos \theta = |OA'| = \left| \frac{OA'}{1} \right| = \frac{|OA'|}{|OB'|} = \frac{|OA|}{|OB|}$$

Fig. 6-1-9 The triangles OAB and $OA'B'$ are similar; $\cos \theta = |OA| / |OB|$ and $\sin \theta = |AB| / |OB|$.

In the same way, we see that

$$\sin \theta = \frac{|AB|}{|OB|}$$

Thus, in a right triangle OAB,

$$\cos \theta = \frac{\text{side adjacent}}{\text{hypotenuse}} \quad \text{that is,} \quad |OA| = |OB| \cos \theta$$

$$\sin \theta = \frac{\text{side opposite}}{\text{hypotenuse}} \quad \text{that is,} \quad |AB| = |OB| \sin \theta$$

It follows (see Fig. 6-1-10) that if the point B has cartesian coordinates (x,y) and polar coordinates (r, θ), then $\cos \theta = |OA|/|OB| = x/r$ and $\sin \theta = |AB|/|OB| = y/r$, so

Fig. 6-1-10 Converting polar to cartesian coordinates.

$$x = r \cos \theta \quad \text{and} \quad y = r \sin \theta \, ,$$

Thus the functions cos θ and sin θ provide the link between cartesian and polar coordinates. We will return to these important conversion formulas shortly.

Worked Example 3 Show that sin θ = cos $(\pi/2 - \theta)$ for $0 \leq \theta \leq \pi/2$.

Solution In Fig. 6-1-11, the angle OBA is $\pi/2 - \theta$ since the three angles must add up to 180°—that is, to π—by plane geometry. Then sin θ = (opposite/hypotenuse) = $|AB|/|OB|$ and cos $(\pi/2 - \theta)$ = (adjacent/hypotenuse) = $|AB|/|OB|$ = sin θ.

Fig. 6-1-11 If $\theta = \angle BOA$, then $\angle OBA = \pi/2 - \theta$.

The other trigonometric functions can be defined in terms of the sine and cosine:

Tangent: $\tan \theta = \dfrac{\sin \theta}{\cos \theta} = \dfrac{|AB|}{|OA|} = \dfrac{\text{side opposite}}{\text{side adjacent}}$

Cotangent: $\cot \theta = \dfrac{\cos \theta}{\sin \theta} = \dfrac{1}{\tan \theta} = \dfrac{|OA|}{|AB|}$

Secant: $\sec \theta = \dfrac{1}{\cos \theta} = \dfrac{|OB|}{|OA|}$

Cosecant: $\csc \theta = \dfrac{1}{\sin \theta} = \dfrac{|OB|}{|AB|}$

Fig. 6-1-12 Two basic examples.

Some frequently used values of the trigonometric functions can be read off the right triangles shown in Fig. 6-1-12. For example, cos $(\pi/4)$ = $1/\sqrt{2}$, sin $(\pi/4)$ = $1/\sqrt{2}$, tan $(\pi/4)$ = 1, cos $(\pi/6)$ = $\sqrt{3}/2$, tan $(\pi/3)$ = $\sqrt{3}$. (The proof that the 1, 2, $\sqrt{3}$ triangle has angles $\pi/3$, $\pi/6$, $\pi/2$ is an exercise in euclidean geometry; see Fig. 6-1-13.)

Special care should be taken with functions of angles which are not between 0 and $\pi/2$—that is, angles not in the first quadrant—to ensure that their signs are correct. For instance, we notice in Fig. 6-1-14 that sin $(2\pi/3)$ = $\sqrt{3}/2$ and cos $(2\pi/3)$ = $-\frac{1}{2}$.

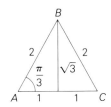

Fig. 6-1-13 The angles of an equilateral triangle are all equal to $\pi/3$.

The following table gives some commonly used values of sin, cos, and tan:

$\theta°$	0°	30°	45°	60°	90°	120°	135°	150°	180°	270°	360°
θ	0	$\dfrac{\pi}{6}$	$\dfrac{\pi}{4}$	$\dfrac{\pi}{3}$	$\dfrac{\pi}{2}$	$\dfrac{2\pi}{3}$	$\dfrac{3\pi}{4}$	$\dfrac{5\pi}{6}$	π	$\dfrac{3\pi}{2}$	2π
$\cos \theta$	1	$\dfrac{\sqrt{3}}{2}$	$\dfrac{\sqrt{2}}{2}$	$\dfrac{1}{2}$	0	$-\dfrac{1}{2}$	$-\dfrac{\sqrt{2}}{2}$	$-\dfrac{\sqrt{3}}{2}$	-1	0	1
$\sin \theta$	0	$\dfrac{1}{2}$	$\dfrac{\sqrt{2}}{2}$	$\dfrac{\sqrt{3}}{2}$	1	$\dfrac{\sqrt{3}}{2}$	$\dfrac{\sqrt{2}}{2}$	$\dfrac{1}{2}$	0	-1	0
$\tan \theta$	0	$\dfrac{\sqrt{3}}{3}$	1	$\sqrt{3}$	∞	$-\sqrt{3}$	-1	$-\dfrac{\sqrt{3}}{3}$	0	$-\infty$	0

Fig. 6-1-14 Illustrating the sine and cosine of $2\pi/3$.

To use the trigonometric functions, one must know their values for given values of θ. Over the centuries, large tables of values of the trigonometric functions have been compiled. The first such table appeared in Ptolemy's *Almagest* and was compiled by Hipparchus and Ptolemy.

Today these values are available in extensive tables as well as on many pocket calculators. Since angles as well as some lengths can be directly measured (as in surveying), the trigonometric relations can then enable us to compute lengths which may be inaccessible (see Solved Exercise 9).

Calculator Discussion

You may be curious about how pocket calculators are able to compute values of $\sin \theta$ and $\cos \theta$ with apparent ease and accuracy. Some analytic expressions are available, such as

$$\sin \theta = \theta - \frac{\theta^3}{3 \cdot 2} + \frac{\theta^5}{5 \cdot 4 \cdot 3 \cdot 2} - \frac{\theta^7}{7 \cdot 6 \cdot 5 \cdot 4 \cdot 3 \cdot 2} + \cdots$$

(as will be proved in Section 13-3), but using these is inefficient and inaccurate. Often a rational function of θ is fitted to many known values of $\sin \theta$ (or $\cos \theta$, $\tan \theta$, and so on) and this rational function is used to calculate approximate values at the remaining points. Thus when θ is entered and $\sin \theta$ pressed, a program in the calculator calculates the value of this rational function. The details depend on the calculator. If you experiment with your calculator—for example, by calculating $\tan \theta$ for θ near $\pi/2$—you might discover some inaccuracies in its method.

Worked Example 4 In Fig. 6-1-15, find x.

Fig. 6-1-15 Find x.

Solution We find that $\tan 22° = 10.3/x$, so $x = 10.3/\tan 22°$. From tables or a calculator, $\tan 22° \approx 0.404026$ so $x \approx 25.4934$.

Solved Exercises

7. Demonstrate the relationship $\tan \theta = \cot (\pi/2 - \theta)$ for $0 \le \theta \le \pi/2$.

8. Show that sin is an odd function: $\sin (-\theta) = -\sin \theta$ (assume $0 \le \theta \le \pi/2$).

9. A tree 50 meters away subtends an angle of 53° at an observer. How tall is the tree?

10. The Greek mathematician and engineer Heron discovered how to tunnel simultaneously from two sides of a mountain and have the two tunnels meet exactly. The method uses trigonometry. Try to rediscover Heron's procedure.

Exercises

8. Show that $\cos \theta = \cos (-\theta)$ for $0 \le \theta \le \pi$.

9. A mountain 3000 meters away subtends an angle of 17° at an observer. How tall is the mountain?

10. A pedestrian 100 meters from the outdoor elevator at the Fairhill Hotel at 12 noon sees the elevator at an angle of 10°. The elevator, steadily rising, makes an angle of 20° after 30 seconds has elapsed. How fast is the elevator rising?

TRIGONOMETRIC IDENTITIES

From the definition of sin and cos, the point P with cartesian coordinates $x = \cos\theta$ and $y = \sin\theta$ lies on the unit circle $x^2 + y^2 = 1$. Therefore, for any value of θ,

$$\cos^2\theta + \sin^2\theta = 1 \qquad \qquad \textbf{(1)}$$

This is an example of a trigonometric identity—a relationship among the trigonometric functions which is valid for all θ.

Relationship (1) is, in essence, a statement of Pythagoras' theorem for a right-angle triangle (triangle $OA'B'$ in Fig. 6-1-9). For a general triangle, the correct relationship is given by the *law of cosines:* with notation as in Fig. 6-1-16 we have

$$c^2 = a^2 + b^2 - 2ab\cos\theta \qquad \qquad \textbf{(2)}$$

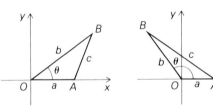

(a) θ acute   **(b)** θ obtuse

Fig. 6-1-16 Proving the law of cosines.

Proof The (x, y) coordinates of B are $x = b\cos\theta$ and $y = b\sin\theta$; those of A are $x = a$, $y = 0$. By the distance formula and equation (1),

$$c^2 = (b\cos\theta - a)^2 + (b\sin\theta)^2 = b^2\cos^2\theta - 2ab\cos\theta + a^2 + b^2\sin^2\theta$$
$$= b^2(\cos^2\theta + \sin^2\theta) + a^2 - 2ab\cos\theta = b^2 + a^2 - 2ab\cos\theta$$

Hence equation (2) is proved.

Fig. 6-1-17 Data for the law of cosines: $c^2 = a^2 + b^2 - 2ab\cos\theta$.

In Fig. 6-1-16 we situated the triangles in a particular way, but this was just a device to prove equation (2); the formula holds in the general situation of Fig. 6-1-17.

Note that $\cos\theta = 0$ when $\theta = \pi/2$; in this case, equation (2) reduces to Pythagoras' theorem: $c^2 = a^2 + b^2$.

Worked Example 5 In Fig. 6-1-18, find x.

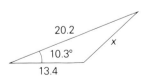

Fig. 6-1-18 Find x.

Solution By the law of cosines

$$(x)^2 = (20.2)^2 + (13.4)^2 - 2(20.2)(13.4) \cos (10.3°)$$
$$= (408.04) + (179.56) - 532.64 = 54.96$$

Taking square roots, we find $x \approx 7.41$.

Consider now the situation in Fig. 6-1-19. By the distance formula,

$$|PQ|^2 = (\cos \phi - \cos \theta)^2 + (\sin \phi - \sin \theta)^2$$
$$= \cos^2 \phi - 2 \cos \phi \cos \theta + \cos^2 \theta + \sin^2 \phi - 2 \sin \phi \sin \theta$$
$$+ \sin^2 \theta$$
$$= 2 - 2 \cos \phi \cos \theta - 2 \sin \phi \sin \theta$$

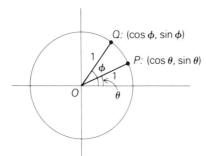

Fig. 6-1-19 Geometry for the proof of the addition formulas.

On the other hand, by the law of cosines (2) applied to ΔOPQ,

$$|PQ|^2 = 1^2 + 1^2 - 2 \cos (\phi - \theta)$$

since $\phi - \theta$ is the angle at the vertex O. Comparing these expressions gives the identity

$$\cos (\phi - \theta) = \cos \phi \cos \theta + \sin \phi \sin \theta$$

which is valid for all ϕ and θ. If we write $-\theta$ for θ, recalling that $\cos (-\theta) = \cos \theta$ and $\sin (-\theta) = -\sin \theta$, this identity yields

$$\cos (\theta + \phi) = \cos \theta \cos \phi - \sin \theta \sin \phi \tag{3}$$

Now if we write $\pi/2 - \phi$ for ϕ, recalling that $\cos(\pi/2 - \psi) = \sin \psi$, and $\sin(\pi/2 - \phi) = \cos \phi$, the same identity gives

$$\sin(\theta + \phi) = \sin \theta \cos \phi + \cos \theta \sin \phi \tag{4}$$

Identities (3) and (4) are called the *addition formulas* for sine and cosine. They are very useful for our later work. From these basic identities, we can derive many others by algebraic manipulation.

Worked Example 6 Prove that $\cos \theta \cos \phi = \frac{1}{2} \left[\cos(\theta - \phi) + \cos(\theta + \phi) \right]$.

Solution Add the identity for $\cos (\theta - \phi)$ to that for $\cos (\theta + \phi)$; you get

$$\cos (\theta - \phi) + \cos (\theta + \phi) = \cos \phi \cos \theta + \sin \phi \sin \theta + \cos \theta \cos \phi - \sin \theta \sin \phi$$
$$= 2 \cos \theta \cos \phi$$

Dividing by 2 gives the stated identity.

Worked Example 7   Prove that $1 + \tan^2 \theta = \sec^2 \theta$.

Solution   Divide both sides of $\sin^2 \theta + \cos^2 \theta = 1$ by $\cos^2 \theta$; then, using $\tan \theta = \sin \theta / \cos \theta$ and $1/\cos \theta = \sec \theta$, we get $\tan^2 \theta + 1 = \sec^2 \theta$ as required.

Some of the most important trigonometric identities are listed in Table 6-1-1. They are all useful, but you can get by quite well by memorizing only (1) through (4) above and deriving the rest when you need them.

Table 6–1–1   Trigonometric identities

Pythagorean

$$\cos^2 \theta + \sin^2 \theta = 1, \quad 1 + \tan^2 \theta = \sec^2 \theta, \quad \cot^2 \theta + 1 = \csc^2 \theta$$

Parity

$$\sin(-\theta) = -\sin \theta, \quad \cos(-\theta) = \cos \theta, \quad \tan(-\theta) = -\tan \theta$$
$$\csc(-\theta) = -\csc \theta, \quad \sec(-\theta) = \sec \theta, \quad \cot(-\theta) = -\cot \theta$$

Co-relations

$$\cos \theta = \sin(\tfrac{\pi}{2} - \theta), \quad \csc \theta = \sec(\tfrac{\pi}{2} - \theta), \quad \cot \theta = \tan(\tfrac{\pi}{2} - \theta)$$

Addition formulas

$$\sin(\theta + \phi) = \sin \theta \cos \phi + \cos \theta \sin \phi$$
$$\sin(\theta - \phi) = \sin \theta \cos \phi - \cos \theta \sin \phi$$
$$\cos(\theta + \phi) = \cos \theta \cos \phi - \sin \theta \sin \phi$$
$$\cos(\theta - \phi) = \cos \theta \cos \phi + \sin \theta \sin \phi$$

$$\tan(\theta + \phi) = \frac{(\tan \theta + \tan \phi)}{(1 - \tan \theta \tan \phi)}$$

$$\tan(\theta - \phi) = \frac{(\tan \theta - \tan \phi)}{(1 + \tan \theta \tan \phi)}$$

Double-angle formulas

$$\sin 2\theta = 2 \sin \theta \cos \theta$$
$$\cos 2\theta = \cos^2 \theta - \sin^2 \theta = 2 \cos^2 \theta - 1 = 1 - 2 \sin^2 \theta$$

$$\tan 2\theta = \frac{2 \tan \theta}{(1 - \tan^2 \theta)}$$

Half-angle formulas

$$\sin^2 \frac{\theta}{2} = \frac{1 - \cos \theta}{2} \qquad \text{or} \qquad \sin^2 \theta = \frac{1 - \cos 2\theta}{2}$$

$$\cos^2 \frac{\theta}{2} = \frac{1 + \cos \theta}{2} \qquad \text{or} \qquad \cos^2 \theta = \frac{1 + \cos 2\theta}{2}$$

$$\tan \frac{\theta}{2} = \frac{\sin \theta}{1 + \cos \theta} = \frac{1 - \cos \theta}{\sin \theta} \qquad \text{or} \qquad \tan \theta = \frac{1 - \cos 2\theta}{\sin 2\theta}$$

Product formulas

$$\sin \theta \sin \phi = \frac{1}{2}[\cos(\theta - \phi) - \cos(\theta + \phi)]$$

$$\cos \theta \cos \phi = \frac{1}{2}[\cos(\theta + \phi) + \cos(\theta - \phi)]$$

$$\sin \theta \cos \phi = \frac{1}{2}[\sin(\theta + \phi) + \sin(\theta - \phi)]$$

Solved Exercises

11. Derive the identity for $\tan(\theta + \phi)$.

12. Find $\sin 15°$ without using tables or a calculator.

13. Simplify $\sin(\theta + \pi/2)$ and $\sin(3\pi/2 + \theta)$.

14. In Fig. 6-1-20, solve for x.

15. Prove the *law of sines* using the notation in Fig. 6-1-21,

$$\frac{\sin \alpha}{a} = \frac{\sin \beta}{b} = \frac{\sin \gamma}{c}$$

Fig. 6-1-20 Find x.   **Fig. 6-1-21**   Law of sines.

Exercises

11. Derive the half-angle identity for $\cos^2(\theta/2)$.

12. Express $\sin(3\theta)$ in terms of $\sin \theta$ and $\cos \theta$.

13. Find $\cos 7\frac{1}{2}°$ and $\tan 22\frac{1}{2}°$ without using tables or a calculator.

14. Simplify:
    (a) $\cos(3\pi/2 - \theta)$ (b) $\tan(\theta + \frac{7}{2}\pi)$
    (c) $\sec(6\pi + \theta)$ (d) $\sin(\theta - \frac{9}{2}\pi)$
    (e) $\cos(\theta + \pi/2)\sin(\phi - 3\pi/2)$
    (f) $[\sin(\theta + \frac{5}{2}\pi)] / [\cos(\pi/2 - \theta)]$

15. In Fig. 6-1-22 solve for a and b.

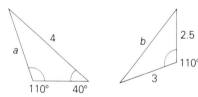

Fig. 6-1-22   Solve for a and b.

GRAPHS OF THE TRIGONOMETRIC FUNCTIONS

From the values of the trigonometric functions in tables one can draw accurate graphs of the trigonometric functions. The calculus of these functions, studied in the next section, confirms that these graphs are correct, so there are no maxima, minima, or inflection points other than those in plain view in Fig. 6-1-23. Perhaps the most important fact about all these functions is their *periodicity:* $f(\theta + 2\pi) = f(\theta)$ for any of the six functions described. Thanks to this property, the entire graph of each trigonometric function is determined by its segment on any interval of length 2π.

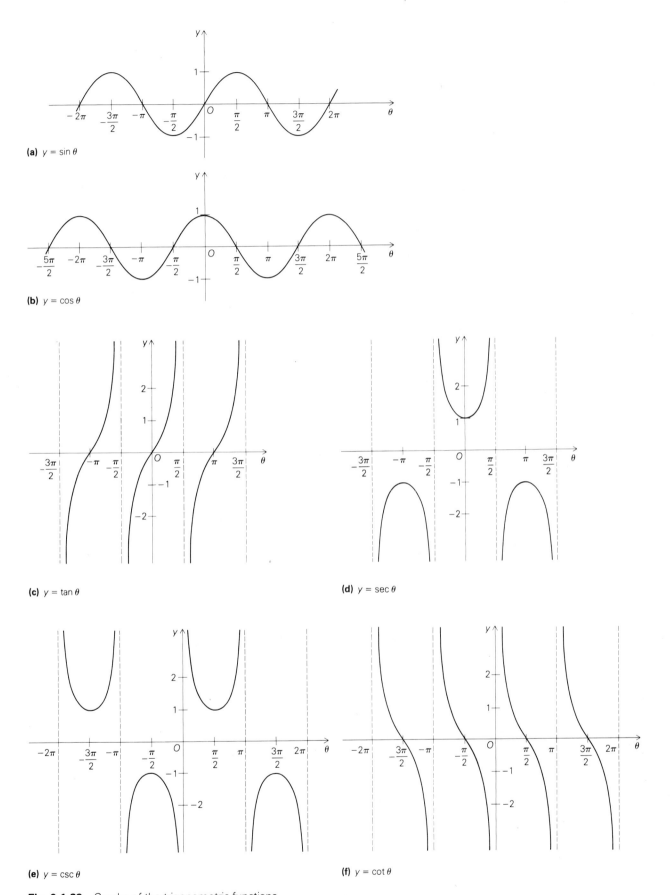

(a) $y = \sin\theta$

(b) $y = \cos\theta$

(c) $y = \tan\theta$

(d) $y = \sec\theta$

(e) $y = \csc\theta$

(f) $y = \cot\theta$

Fig. 6-1-23 Graphs of the trigonometric functions.

Worked Example 8   Sketch the graph of $y = \sin 2\theta$.

Solution   We obtain $y = \sin 2\theta$ by taking the graph of $y = \sin \theta$ and compressing the graph horizontally by a factor of 2. See Fig. 6-1-24.

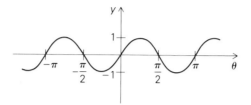

Fig. 6-1-24   The graph of $y = \sin 2\theta$.

Notice that the graph in Fig. 6-1-24 repeats itself every π units. We say that π is the *period* of the function $\sin 2\theta$. In general, if a function f satisfies $f(\theta + \tau) = f(\theta)$ for all θ and a fixed τ, we call τ the *period* of f. The reciprocal $1/\tau$ is called the *frequency*. For instance, $\sin a\theta$ and $\cos a\theta$ have period $2\pi/a$ and frequency $a/2\pi$; $\tan \theta$ has period π as well as 2π.

Solved Exercises

16. Sketch the graph of $y = 3 \cos 5\theta$.

17. Where are the inflection points of $\tan \theta$? For which values of θ do you expect $\tan \theta$ to be a differentiable function?

Exercises

16. Sketch the graph of $y = 2 \cos 3\theta$.

17. Sketch the graph of $y = \tan (3\theta/2)$.

18. Sketch the graph of $y = 4 \sin 2\theta \cos 2\theta$.

19. Locate the turning points of $\sin \theta$ by inspection of its graph.

20. For what θ do you expect $\sec \theta$ and $\cot \theta$ to be differentiable?

GRAPHS IN POLAR COORDINATES

Trigonometric functions and polar coordinates go hand in hand in view of the basic relationship between the cartesian coordinates (x, y) and polar coordinates (r, θ) of a point P (see Fig. 6-1-25).

Fig. 6-1-25
$x = r \cos \theta$ or $r = \sqrt{x^2 + y^2}$
$y = r \sin \theta$ $\cos \theta = x/r$
                            $\sin \theta = y/r$

Worked Example 9   Convert from cartesian to polar coordinates: $(2, -4)$; and from polar to cartesian coordinates: $(6, -\pi/8)$.

Solution   We plot $(2, -4)$ as in Fig. 6-1-26. Then $r = \sqrt{2^2 + (-4)^2} = \sqrt{20} = 2\sqrt{5}$ and $\cos \theta = 2/(2\sqrt{5}) = 1/\sqrt{5}$, so from tables or a calculator $\theta = 63.43°$; but we must take $\theta = -63.43°$ (or $296.57°$) since we are in the fourth quadrant. Thus the polar coordinates of $(2, -4)$ are $(2\sqrt{5}, -63.43°)$.

The cartesian coordinates of the point with polar coordinates $(6, -\pi/8)$ are

$$x = r \cos \theta = 6 \cos (-\pi/8) = (6)(0.92388) = 5.5433$$

and

$$y = r \sin \theta = 6 \sin (-\pi/8) = (6)(-0.38268) = -2.2961$$

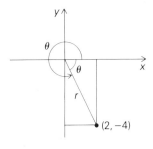

Fig. 6-1-26 Find the polar coordinates of $(2, -4)$.

That is, $(5.5433, -2.2961)$. This point is also in the fourth quadrant, as it should be.

In cartesian coordinates, we are used to graphing functions $y = f(x)$. Similarly, we can graph functions $r = f(\theta)$ in polar coordinates. Observe that if $f(\theta) = f(-\theta)$, then its graph is symmetric in the x axis; if $f(\pi - \theta) = f(\theta)$, it is symmetric in the y axis; and if $f(\theta) = f(\pi + \theta)$, it is symmetric in the origin. See Fig. 6-1-27.

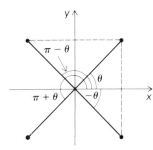

Fig. 6-1-27
Symmetry in x axis: $f(\theta) = f(-\theta)$;
Symmetry in y axis: $f(\pi - \theta) = f(\theta)$;
Symmetry in origin: $f(\theta) = f(\pi + \theta)$.

Worked Example 10   Plot the graph of $r = \cos 2\theta$ in the xy plane.

Solution   From the graph of cosine, we see that as θ increases from zero, $\cos 2\theta$ decreases to zero when $2\theta = \pi/2$ (that is, $\theta = \pi/4$) and decreases to -1 when $2\theta = \pi$ (that is, $\theta = \pi/2$). Thus (r, θ) traces out the path in Fig. 6-1-28.

We can complete the path of this point as θ goes through all values between 0 and 2π, sweeping out the four petals in Fig. 6-1-29, or else use the symmetry of $\cos 2\theta$ to reflect in the x and y axes.

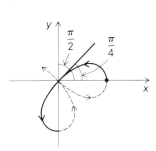

Fig. 6-1-28 Beginning of the plot of $r = \cos 2\theta$.

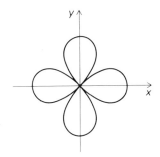

Fig. 6-1-29 $r = \cos 2\theta$, the "four-leafed rose."

In Fig. 6-1-30 the graphs of two other simple equations in polar coordinates are plotted.

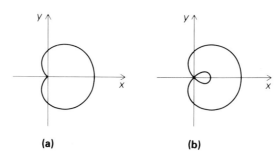

(a)   **(b)**

Fig. 6-1-30   (a) $r = 1 + \cos \theta$ ("cardioid"); (b) $r = 1 + 2 \cos \theta$ ("limaçon").

Worked Example 11   Convert the relation $r = 1 + 2 \cos \theta$ to cartesian coordinates.

Solution   We substitute $r = \sqrt{x^2 + y^2}$ and $\cos \theta = x/r = x/\sqrt{x^2 + y^2}$ to get

$$\sqrt{x^2 + y^2} = 1 + \frac{2x}{\sqrt{x^2 + y^2}}$$

That is, $x^2 + y^2 - \sqrt{x^2 + y^2} - 2x = 0$.

Solved Exercises   18. (a) What are the polar coordinates of $(x, y) = (5, -2)$?
   (b) What are the cartesian coordinates of $(r, \theta) = (2, \pi/6)$?

19. Prove that $r = f(\theta) = \cos 2\theta$ (Fig. 6-1-29) actually is symmetric in the x and y axes. What happens when θ is replaced by $\theta + \pi/2$?

20. Sketch the graph of $r = f(\theta) = \cos 3\theta$.

21. If $f(\theta + \pi/2) = f(\theta)$, what does this tell you about the graph of $r = f(\theta)$?

Exercises   21. Convert from cartesian to polar coordinates: $(1, 0)$, $(3, 4)$, $(\sqrt{3}, 1)$, $(\sqrt{3}, -1)$, $(-\sqrt{3}, 1)$.

22. Convert from polar to cartesian coordinates: $(0, \pi/8)$, $(1, 0)$, $(2, \pi/4)$, $(8, 3\pi/2)$, $(2, \pi)$.

23. Sketch the graph of $r = \cos \theta$.

24. Sketch the graph of $r = 2 \sin \theta$. Convert to cartesian coordinates.

25. If $f(\pi/2 - \theta) = f(\theta)$, what does this tell you about the graph of $r = f(\theta)$?

Problems for Section 6-1

1. Convert the following cartesian coordinates to polar coordinates:

   (a) $(1, -1)$ (b) $(0, 2)$ (c) $(\frac{1}{2}, 7)$
   (d) $(-12, -5)$ (e) $(-3, 8)$ (f) $(\frac{3}{4}, \frac{3}{4})$
   (g) $(-4, 4)$ (h) $(1, 15)$ (i) $(19, 3)$
   (j) $(-5, -6)$ (k) $(0.3, 0.9)$ (l) $(-\frac{3}{2}, \frac{1}{2})$

2. Convert the following polar coordinates to cartesian coordinates:

   (a) $(6, \pi/2)$ (b) $(12, 3\pi/4)$
   (c) $(4, -\pi)$ (d) $(2, 13\pi/2)$
   (e) $(8, -2\pi/3)$ (f) $(1, 2)$
   (g) $(-1, -1)$ (h) $(1, \pi)$
   (i) $(10, 2.7)$ (j) $(5, 7\pi/2)$
   (k) $(8, 7\pi)$ (l) $(4, -3\pi)$
   (m) $(3, \pi/4)$ (n) $(6, -\pi/4)$

3. An arc of radius 250 meters subtends an angle of 38°. How long is the arc?

4. An arc of length 110 meters subtends an angle of 24°. What is the radius of the arc?

5. An airplane flying at 5000 feet has an angle of elevation of 25° at observer A. Observer B sees the airplane directly overhead. How far apart are A and B?

6. A leaning tower in Venice tilts at 9° from the vertical directly away from an observer who is 500 meters away from its base. If the observer sees the top of the tower at an angle of elevation of 22°, how high is the tower?

7. Derive the half-angle identity for $\tan \theta/2$ in Table 6-1-1.

8. Derive the product formula identity in Table 6-1-1 for $\sin \theta \sin \phi$.

9. Find $\sec 15°$ without using tables or a calculator.

10. Sketch the graph of $y = \sin 3\theta + 1$.

11. Sketch the graph of $y = \sin \theta + \cos \theta$.

12. Sketch the graph of $y = \csc 2\theta$.

13. Sketch the graph of $y = \tan(\theta/2)$.

14. Sketch the graph of $y = \cos(3\theta + \pi/6)$.

15. Locate the inflection points of $\cot \theta$ by inspection of its graph.

16. In Fig. 6-1-31 solve for the unknown parts of the triangles.

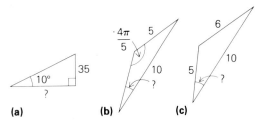

(a)   (b)   (c)

Fig. 6-1-31   Find ?

In Problems 17 through 25, sketch the graph of the given function in polar coordinates. Also convert the given equation to cartesian coordinates.

17. $r = \sin 3\theta$

18. $r = 1 - \sin \theta$

19. $r = \sin(\theta/2) + 1$

20. $r = \cos \theta - \sin \theta$

21. $r = 3$

22. $\sin \theta = 1$

23. $r^2 + 2r \cos \theta + 1 = 0$

24. $r^2 \sin 2\theta = \frac{1}{2}$

25. $\tan \theta \sec \theta = 1$

26. What is the equation of a line through the origin in polar coordinates?

Convert the relations in Problems 27 through 34 to polar coordinates.

27. $x^2 + y^2 = 1$

28. $xy = 1$

29. $x^2 + xy + y^2 = 1$

30. $y = x^5 + x^3 + 2$

31. $y = 1/(1 - x^2)$

32. $y = x/(1 + x)$

33. $y = x + 1$

34. $y = 1/x + x^2$

Demonstrate the identities in Problems 35 through 41.

35. $\sec \theta + \tan \theta = (1 + \sin \theta)/(1 - 2\sin^2 \theta/2)$

36. $8 \cos \theta + 8 \cos 2\theta = -9 + 16(\cos \theta + \frac{1}{4})^2$

37. $\sec^2(\theta/2) = (2 \sec \theta)/(\sec \theta + 1)$

38. $\csc^2(\theta/2) = (2 \sec \theta)/(\sec \theta - 1)$

39. $\csc \theta \csc \phi = [2 \sec(\theta + \phi) \sec(\theta - \phi)] / [\sec(\theta + \phi) - \sec(\theta - \phi)]$

40. $\tan \theta \tan \phi = [\sec(\theta + \phi) - \sec(\theta - \phi)] / [\sec(\theta + \phi) + \sec(\theta - \phi)]$

41. $\tan\theta = [\sin(\theta + \phi) + \sin(\theta - \phi)] / [\cos(\theta + \phi) + \cos(\theta - \phi)]$

42. Suppose that the graph of $r = f(\theta)$ is symmetric in the line $x = y$. What does this imply about f?

43. Light travels at velocity v_1 in a certain medium, enters a second medium at angle of incidence θ_1 (measured from the normal to the surface), and refracts at angle θ_2 while traveling a different velocity v_2 (in the second medium). According to Snell's law, $v_1/v_2 = \sin\theta_1 / \sin\theta_2$.

    (a) Light enters at $60°$ and refracts at $30°$. The first medium is air ($v_1 = 3 \times 10^{10}$ cm/sec). Find the velocity in the second medium.

    (b) Show that if $v_1 = v_2$, then the light travels in a straight line.

    (c) The speed halves in passing from one medium to another, the angle of incidence being $45°$. Calculate the angle of refraction.

44. A conical pendulum of cord length L and angle θ has period $T = 2\pi\sqrt{L\cos\theta/g}$ ($g = 32$ ft/sec²). Find the period for a two-foot cord length and angle $\theta = 30°$.

45. An oscillator circuit in a tape deck puts out a signal of amplitude 0.25 and frequency 400 cycles per second. Assuming that the signal is of the form $A\sin(\omega t)$, determine A and ω and then graph it.

46. The current in a wire is given by $I(t) = \sin(77t)$.

    (a) What is the *peak current*, that is, the largest value of I?

    (b) Determine the frequency; i.e., the number of cycles per second.

47. A parallelogram is formed with acute angle θ and sides ℓ, L. Find a formula for its area.

48. In Solved Exercise 15, show that the common value of $(\sin\alpha)/a$, $(\sin\beta)/b$, and $(\sin\gamma)/c$ is the reciprocal of twice the radius of the circumscribed circle.

49. Scientists and engineers often use the approximations $\sin\theta = \theta$ and $\cos\theta = 1$, valid for θ near zero. Experiment with your hand calculator to determine a region of validity for θ that guarantees eight-place accuracy for these approximations.

50. Light of wavelength λ diffracts through a single slit of width a. This light then passes through a lens and falls on a screen. A point P is on the screen, making an angle θ with the lens axis. The intensity I on the screen is

$$I = I_0\left[\frac{\sin([\pi a \sin\theta]/\lambda)}{[\pi a \sin\theta]/\lambda}\right]^2, \quad 0 < \theta < \pi$$

where I_0 is the intensity when P is on the lens axis.

    (a) Show that the intensity is zero for $\sin\theta = \lambda/a$.

    (b) Find all values of θ for which $\theta > 0$ and $I = 0$.

    (c) Verify that I is approximately I_0 when $[\pi a \sin\theta]/\lambda$ is close enough to zero. (Use $\sin\theta \approx \theta$; see Problem 49.)

    (d) In practice, $\lambda = 5 \times 10^{-5}$ cm, $a = 10^{-2}$ cm. Check by means of a calculator or table that $\sin\theta = \lambda/a$ reduces to $\theta = \lambda/a$.

51. The current I in a circuit is given by $I(t) = 20\sin(311t) + 40\cos(311t)$. Let $r = (20^2 + 40^2)^{1/2}$ and define the angle θ by $\cos\theta = 20/r$, $\sin\theta = 40/r$.

    (a) Verify by use of the sum formula for the sine function that $I(t) = r\sin(311t + \theta)$.

    (b) Show that the *peak current* is r (the maximum value of I).

    (c) Find the *period* and *frequency*. (See p. 276.)

    (d) Determine the *phase shift* (in radians), that is, the value of t which makes $311t + \theta = 0$.

52. The instantaneous power input to an AC circuit is $p = vi$ where v is the instantaneous potential difference between the circuit terminals and i is the instantaneous current. If the circuit is a pure resistor, then $v = V\sin(\omega t)$ and $i = I\sin(\omega t)$.

    (a) Verify by means of trigonometric identities that

$$p = \frac{VI}{2} - \frac{VI}{2}\cos(2\omega t)$$

    (b) Draw a graph of p as a function of t.

53. Two points are located on one bank of the Colorado River, 500 meters apart. A point on the opposite bank makes angles of $88°$ and $80°$ with the line joining the two points. Find the lengths of two cables to be stretched across the river connecting the points.

6-2 Differentiation of the Trigonometric Functions

Differentiation rules for all the trigonometric functions and their inverses follow from the differentiation formulas for sine and cosine.

In this section, we will derive differentiation formulas for the trigonometric and inverse trigonometric functions. In the course of doing so, we will use many of the techniques introduced previously, including limits (Section 1-2), algebraic rules (Section 1-3), the inverse function rule (Section 5-1), and the chain rule, related rates, and parametric curves (Section 5-2).

Goals

After studying this section, you should be able to:

Differentiate trigonometric functions and their inverses.

Differentiate combinations of trigonometric and algebraic functions.

Solve simple antidifferentiation problems involving trigonometric functions.

Evaluate inverse trigonometric functions.

Sketch the graphs of the inverse trigonometric functions.

THE DERIVATIVE OF $\sin t$ AND $\cos t$ AT $t = 0$

The unit circle $x^2 + y^2 = 1$ can be described by the parametric equations $x = \cos t$, $y = \sin t$. As t increases, the point $(x, y) = (\cos t, \sin t)$ moves along the circle in a counterclockwise direction (see Fig. 6-2-1).

Fig. 6-2-1 The point P moves at unit speed around the circle.

The length of arc on the circle between the point $(1, 0)$ (corresponding to $t = 0$) and the point $(\cos t, \sin t)$ is equal to the angle t subtended by the arc. If we think of t as time, the point $(\cos t, \sin t)$ travels a distance equal to t in time t, so it is moving with unit speed around the circle.

At $t = 0$, the tangent line to the circle is vertical, so the velocity of the point is 1 in the vertical direction; thus we expect that

$$\left.\frac{dx}{dt}\right|_{t=0} = 0 \quad \text{and} \quad \left.\frac{dy}{dt}\right|_{t=0} = 1$$

That is,

$$\cos' 0 = 0 \qquad (1)$$

and*

$$\sin' 0 = 1 \qquad (2)$$

Worked Example 1 Assuming formulas (1) and (2), calculate the derivative of $f(\theta) = \cos^2 \theta \sin \theta$ at $\theta = 0$.

Solution By the differentiation rules for products and powers,

$$f'(\theta) = 2 \cos \theta \cos' \theta \sin \theta + \cos^2 \theta \sin' \theta$$

so

$$f'(0) = 2(\cos 0)(\cos' 0)(\sin 0) + (\cos^2 0)(\sin' 0)$$
$$= 2 \cdot 1 \cdot 0 \cdot 0 + 1^2 \cdot 1 = 1$$

According to the definition of the derivative, we can rewrite formulas (1) and (2) as the following statements about limits:

$$\lim_{\theta \to 0} \frac{\cos \theta - 1}{\theta} = 0 \qquad (3)$$

and

$$\lim_{\theta \to 0} \frac{\sin \theta}{\theta} = 1 \qquad (4)$$

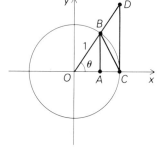

To supplement our physical argument for (1) and (2), we will prove (3) and (4) directly, using the geometry in Fig. 6-2-2. For $0 < \theta < \pi/2$,

$$\text{Area } \triangle OCB = \tfrac{1}{2} |OC| \cdot |AB| = \tfrac{1}{2} \sin \theta$$
$$\text{Area } \triangle OCB < \text{area sector } OCB = \tfrac{1}{2} \theta$$

and

$$\text{Area sector } OCB < \text{area } \triangle OCD = \tfrac{1}{2} |OC| \cdot |CD| = \tfrac{1}{2} \tan \theta = \frac{1}{2} \frac{\sin \theta}{\cos \theta}$$

Fig. 6-2-2 Geometry used to determine cos′ 0 and sin′ 0.

Thus

$$\sin \theta < \theta \qquad (5)$$

and

$$\theta < \frac{\sin \theta}{\cos \theta} \qquad (6)$$

From (5) we have

$$\sin^2 \theta < \theta^2$$
$$1 - \sin^2 \theta > 1 - \theta^2$$
$$\cos^2 \theta > 1 - \theta^2$$

---

*As usual, "′" stands for the derivative, where "cos" and "sin" play the role of the more common "f".

But $\cos \theta > \cos^2 \theta$, since $1 > \cos \theta$. Thus

$$\cos \theta > 1 - \theta^2$$
$$\cos \theta - 1 > -\theta^2$$
$$\frac{\cos \theta - 1}{\theta} > -\theta$$

From $\cos \theta < 1$, we have $0 > (\cos \theta - 1)/\theta$, and so

$$0 > \frac{\cos \theta - 1}{\theta} > -\theta \qquad\qquad (7)$$

when $0 < \theta < \pi/2$. If $-\pi/2 < \theta < 0$, then substituting $-\theta$ for θ in (7) gives

$$0 > \frac{\cos (-\theta) - 1}{-\theta} > \theta$$

That is,

$$0 < \frac{\cos \theta - 1}{\theta} < -\theta \qquad\qquad (8)$$

when $-\pi/2 < \theta < 0$. Letting θ approach zero, we find from (7) and (8) that $(\cos \theta - 1)/\theta$ is squeezed between θ and $-\theta$, so

$$(\cos \theta - 1)/\theta \to 0 \text{ as } \theta \to 0$$

and (3) is proved. As a consequence, we also have

$$\lim_{\theta \to 0} \cos \theta = 1 \qquad\qquad (9)$$

since any function is continuous wherever it is differentiable.

Now, from (5) and (6), we have

$$\cos \theta < \frac{\sin \theta}{\theta} < 1 \qquad\qquad (10)$$

when $0 < \theta < \pi/2$. Since the expressions in (10) are unchanged when θ is replaced by $-\theta$, the same inequalities hold when $-\pi/2 < \theta < 0$. Letting θ approach zero, and using equation (9), we find that $(\sin \theta)/\theta$ is squeezed between 1 and a quantity which is approaching 1, so that $(\sin \theta)/\theta \to 1$ as $\theta \to 0$, and (4) is proved.

Solved Exercises   1. Calculate

$$\frac{1 - \cos \theta}{\theta} \quad \text{and} \quad \frac{\sin \theta}{\theta}$$

for $\theta = 0.02$ and 0.001.

2. Calculate $\tan'(0)$.

3. Find $\lim\limits_{\theta \to 0} (\sin a\theta)/\theta$, where a is any constant.

Exercises   1. Prove the following inequalities by using trigonometric identities and the inequalities already established:

(a) $\theta < (\sin 2\theta)/(1 + \cos 2\theta)$ for $0 < \theta < \pi/2$

(b) $\sec \theta > \theta$ for $0 < \theta < \pi/2$

2. Find $\lim\limits_{\theta \to 0} (\tan 2\theta)/3\theta$.

3. Calculate $\sec'(0)$.

4. Calculate $(d/d\theta)[(\sin 2\theta + \theta)/(2\cos \theta)]$ at $\theta = 0$.

5. Using the half-angle formulas and the formulas

$$\sin(x + \pi/2) = \cos x, \quad \cos(x + \pi/2) = -\sin x,$$

calculate:

(a) $\dfrac{d}{d\theta} \sin\left(\theta + \dfrac{\pi}{4}\right)\Big|_{\theta = 0}$

(b) $\dfrac{d}{d\theta} \cos\left(\theta + \dfrac{\pi}{4}\right)\Big|_{\theta = 0}$

THE DERIVATIVES OF THE TRIGONOMETRIC FUNCTIONS

Formulas (1) and (2) of the preceding discussion show how to differentiate $\sin t$ and $\cos t$ at $t = 0$. To differentiate them at a general point t_0, we will use the addition formulas and the shifting rule.

We write $t = t_0 + (t - t_0)$, so

$$\sin t = \sin[t_0 + (t - t_0)]$$

That is,

$$\sin t = \sin t_0 \cos(t - t_0) + \cos t_0 \sin(t - t_0) \tag{11}$$

We can now differentiate equation (11) with respect to t and get

$$\sin' t = \sin t_0 \cos'(t - t_0) + \cos t_0 \sin'(t - t_0)$$

Setting $t = t_0$ and using (1) and (2), we get

$$\sin' t_0 = \sin t_0 \cos' 0 + \cos t_0 \sin' 0 = \cos t_0$$

In following this step, recall that, by the shifting rule (or simply by the chain rule), the derivative of $g(t) = \cos(t - t_0)$ at $t = t_0$ is $\cos'(t_0 - t_0) = \cos' 0 = 0$, and that of $h(t) = \sin(t - t_0)$ at $t = t_0$ is $\sin' 0 = 1$. (See Fig. 6-2-3 and Section 5-2.)

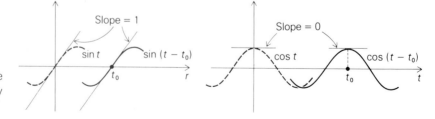

Fig. 6-2-3 Shifting the graphs of sine and cosine by t_0 units to the right.

Thus we have proved that

$$\sin' t_0 = \cos t_0$$

That is,

$$\sin' \theta = \cos \theta \qquad \qquad \textbf{(12)}$$

or

$$\frac{d}{d\theta} \sin \theta = \cos \theta$$

In the same way we can calculate the derivative of $\cos t$:

$$\cos t = \cos [t_0 + (t - t_0)] = \cos t_0 \cos (t - t_0) - \sin t_0 \sin (t - t_0) \qquad \textbf{(13)}$$

Differentiation of (13) at $t = t_0$ gives

$$\cos' t_0 = \cos t_0 \cos' 0 - \sin t_0 \sin' 0 = -\sin t_0$$

Thus

$$\cos' t_0 = -\sin t_0$$

That is,

$$\cos' \theta = -\sin \theta \qquad \qquad \textbf{(14)}$$

or

$$\frac{d}{d\theta} \cos \theta = -\sin \theta$$

In words, the derivative of the sine function is the cosine and the derivative of the cosine is *minus* the sine. Formulas (12) and (14) are worth memorizing. Study Fig. 6-2-4 to check that formulas (12) and (14) are consistent with the graphs of sine and cosine. For example, notice that on the interval $(0, \pi/2)$ $\sin \theta$ is increasing and its derivative $\cos \theta$ is positive.

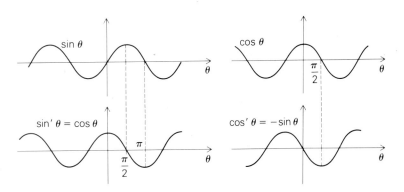

Fig. 6-2-4 Graphs of sin, cos, and their derivatives.

Worked Example 2 Differentiate $\cos \theta \sin^2 \theta$.

Solution By the product rule and the power rule,

$$\frac{d}{d\theta}(\cos \theta \sin^2 \theta) = \left(\frac{d}{d\theta} \cos \theta\right) \sin^2 \theta + \cos \theta \left(\frac{d}{d\theta} \sin^2 \theta\right)$$
$$= (-\sin \theta) \sin^2 \theta + \cos \theta \cdot 2 \sin \theta \cos \theta$$
$$= 2 \cos^2 \theta \sin \theta - \sin^3 \theta$$

Worked Example 3 Differentiate $(\sin 3x)/(1 + \cos^2 x)$.

Solution By the chain rule,

$$\frac{d}{dx} \sin 3x = 3 \cos 3x$$

so, by the quotient rule,

$$\frac{d}{dx} \frac{\sin 3x}{1 + \cos^2 x} = \frac{(1 + \cos^2 x) \cdot 3 \cos 3x - \sin 3x \cdot 2 \cos x(-\sin x)}{(1 + \cos^2 x)^2}$$
$$= \frac{3 \cos 3x(1 + \cos^2 x) + 2 \cos x \sin x \cdot \sin 3x}{(1 + \cos^2 x)^2}$$

Now that we know how to differentiate the sine and cosine functions, we can differentiate the remaining trigonometric functions by using the rules of calculus. For example, consider $\tan \theta = \sin \theta / \cos \theta$. The quotient rule gives:

$$\frac{d}{d\theta} \tan \theta = \frac{\cos \theta \, (d/d\theta) \sin \theta - \sin \theta \, (d/d\theta) \cos \theta}{\cos^2 \theta}$$
$$= \frac{\cos \theta \cdot \cos \theta + \sin \theta \cdot \sin \theta}{\cos^2 \theta}$$
$$= \frac{1}{\cos^2 \theta} = \sec^2 \theta$$

In a similar way, we see that

$$\frac{d}{d\theta} \cot \theta = -\csc^2 \theta$$

Writing $\csc \theta = 1/\sin \theta$ we get $\csc' \theta = (-\sin' \theta)/(\sin^2 \theta) = (-\cos \theta)/(\sin^2 \theta) = -\cot \theta \csc \theta$ and similarly $\sec' \theta = \tan \theta \sec \theta$.

The results we have obtained are summarized in Table 6-2-1.

Table 6–2–1   Differentiation of Trigonometric Functions

Function	Derivative	Leibniz notation
$\sin \theta$	$\cos \theta$	$d(\sin \theta)/d\theta = \cos \theta$
$\cos \theta$	$-\sin \theta$	$d(\cos \theta)/d\theta = -\sin \theta$
$\tan \theta$	$\sec^2 \theta$	$d(\tan \theta)/d\theta = \sec^2 \theta$
$\cot \theta$	$-\csc^2 \theta$	$d(\cot \theta)/d\theta = -\csc^2 \theta$
$\sec \theta$	$\tan \theta \sec \theta$	$d(\sec \theta)/d\theta = \tan \theta \sec \theta$
$\csc \theta$	$-\cot \theta \csc \theta$	$d(\csc \theta)/d\theta = -\cot \theta \csc \theta$

Worked Example 4   Differentiate $\csc x \tan 2x$.

Solution   Using the product rule and the chain rule,

$$\frac{d}{dx} \csc x \tan 2x = \left(\frac{d}{dx} \csc x\right)(\tan 2x) + \csc x \left(\frac{d}{dx} \tan 2x\right)$$
$$= -\cot x \cdot \csc x \cdot \tan 2x + \csc x \cdot 2 \cdot \sec^2 2x$$
$$= 2 \csc x \sec^2 2x - \cot x \cdot \csc x \cdot \tan 2x$$

Solved Exercises

4. Differentiate:

   (a) $\sin x \cos x$ (b) $(\tan 3x)/(1 + \sin^2 x)$ (c) $1 - \csc^2 5x$

5. Differentiate $f(\theta) = \sin (\sqrt{3\theta^2 + 1})$.

6. Find $\sin' t_0$ by using limits. [*Hint:* Use the addition formula to expand $\sin (t_0 + \Delta t)$.]

7. Discuss maxima, minima, concavity, and points of inflection for $f(x) = \sin^2 x$. Sketch its graph.

Exercises

6. Show that $(d/d\theta) \cos \theta = -\sin \theta$ can be derived from $(d/d\theta) \sin \theta = \cos \theta$ by using $\cos \theta = \sin(\pi/2 - \theta)$ and the chain rule.

7. Differentiate:

   (a) $\sin^2 x$ (b) $\tan (\theta + 1/\theta)$

   (c) $(4t^3 + 1) \sin \sqrt{t}$ (d) $\csc t \cdot \sec^2 3t$

8. Discuss maxima, minima, concavity, and points of inflection for $y = \cos 2x - 1$.

9. Use limits to find $\cos' t_0$.

ANTIDERIVATIVES OF THE TRIGONOMETRIC FUNCTIONS

By reversing the table of derivatives of trigonometric functions and multiplying through by -1 where necessary, we obtain the following indefinite integrals (antiderivatives).

| **ANTIDERIVATIVES OF SOME TRIGONOMETRIC FUNCTIONS** |

1. $\int \cos\theta\, d\theta = \sin\theta + C$ 2. $\int \sin\theta\, d\theta = -\cos\theta + C$

3. $\int \sec^2\theta\, d\theta = \tan\theta + C$ 4. $\int \csc^2\theta\, d\theta = -\cot\theta + C$

5. $\int \tan\theta \sec\theta\, d\theta = \sec\theta + C$ 6. $\int \cot\theta \csc\theta\, d\theta = -\csc\theta + C$

For instance, to check that

$$\int \sec^2\theta\, d\theta = \tan\theta + C$$

we must see if $(d/d\theta)(\tan\theta + C) = \sec^2\theta$, according to the definition of antiderivative. But this is correct since

$$\frac{d}{d\theta}(\tan\theta + C) = \frac{d}{d\theta}\tan\theta + \frac{d}{d\theta}C = \frac{d}{d\theta}\tan\theta = \sec^2\theta$$

Worked Example 5   Find $\int \sec\theta\,(\sec\theta + 3\tan\theta)\, d\theta$.

Solution   Multiplying out, we have

$$\int (\sec^2\theta + 3\sec\theta\tan\theta)\, d\theta = \int \sec^2\theta\, d\theta + 3\int \sec\theta\tan\theta\, d\theta$$
$$= \tan\theta + 3\sec\theta + C$$

Our table of antiderivatives leaves much to be desired. Where is $\int \tan\theta\, d\theta$, for instance? Solved Exercise 8 below indicates the source of the difficulty, which will be resolved in Chapter 7.

Solved Exercises

8. Suppose that $f'(x) = 1/x$. (We do not yet have such a function at our disposal, but we will see in Chapter 7 that there is one.) Show that $\int \tan\theta\, d\theta = -f(\cos\theta) + C$.

9. Find $\int \sin 4u\, du$.

Exercises

10. Find the following antiderivatives:
    (a) $\int 2\cos 4s\, ds$ (b) $\int (1 + \sec^2\theta)\, d\theta$
    (c) $\int \tan 2x \sec 2x\, dx$ (d) $\int (\sin x + \sqrt{x})\, dx$

11. Show that $\int \tan\theta\, d\theta = f(\sec\theta) + C$, where f is a function such that $f'(x) = 1/x$. (The apparent conflict with Solved Exercise 8 will be resolved in the next chapter.)

THE INVERSE SINE FUNCTION

In Section 5-1 we discussed the general concept of the inverse of a function and developed a formula for differentiating the inverse. Recall that this formula is

$$\frac{d}{dy} f^{-1}(y) = \frac{1}{(d/dx)f(x)} \tag{15}$$

where $y = f(x)$.

To apply these ideas to the sine function, we begin by using Theorem 1 (p. 223) to locate an interval on which $\sin x$ has an inverse. Since $\sin' x = \cos x > 0$ on $(-\pi/2, \pi/2)$, $\sin x$ is increasing on this interval, so $\sin x$ has an inverse on the interval $[-\pi/2, \pi/2]$. The inverse is denoted $\sin^{-1} y$.* We obtain the graph of $\sin^{-1} y$ by interchanging the x and y coordinates. (See Fig. 6-2-5.)

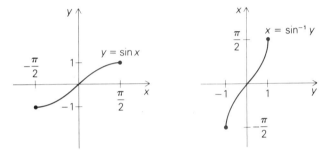

Fig. 6-2-5 The graph of $\sin x$ on $[-\pi/2, \pi/2]$ together with its inverse.

The values of $\sin^{-1} y$ may be obtained from a table for $\sin x$. (Many pocket calculators can evaluate the inverse trigonometric functions as well as the trigonometric functions.)

THE INVERSE SINE FUNCTION

The value of $\sin^{-1} y$, for $-1 \le y \le 1$, is that unique number between $[-\pi/2, \pi/2]$ whose sine is y. The number $\sin^{-1} y$ is expressed in radians unless a degree sign is explicitly shown.

Worked Example 6 Calculate $\sin^{-1} 1$, $\sin^{-1} 0$, $\sin^{-1}(-1)$, $\sin^{-1}(-\frac{1}{2})$, and $\sin^{-1}(0.342)$.

Solution Since $\sin \pi/2 = 1$, $\sin^{-1} 1 = \pi/2$. Similarly, $\sin^{-1} 0 = 0$, $\sin^{-1}(-1) = -\pi/2$. Also, $\sin(-\pi/6) = -\frac{1}{2}$, so $\sin^{-1}(-\frac{1}{2}) = -\pi/6$. Using a calculator, we find $\sin^{-1}(0.342) = 20°$.

We could have used any other interval on which $\sin x$ has an inverse, such as $[\pi/2, 3\pi/2]$, to define an inverse sine function; had we done so, the function obtained would have been different. The choice $[-\pi/2, \pi/2]$ is standard and is usually the most convenient.

Although the notation $\sin^2 y$ is commonly used to mean $(\sin y)^2$, $\sin^{-1} y$ does not mean $(\sin y)^{-1} = 1/\sin y$. Sometimes the notation $\arcsin y$ is used for the inverse sine function to avoid confusion.

Let us now calculate the derivative of $\sin^{-1}y$. By formula (15) for the derivative of an inverse,

$$\frac{d}{dy}\sin^{-1}y = \frac{1}{(d/dx)\sin x} = \frac{1}{\cos x}$$

where $y = \sin x$. However, $\cos^2 x + \sin^2 x = 1$, so $\cos x = \sqrt{1-y^2}$. (The negative root does not occur since $\cos x$ is positive on $(-\pi/2, \pi/2)$.) Thus

$$\frac{d}{dy}\sin^{-1}y = \frac{1}{\sqrt{1-y^2}} = (1-y^2)^{-1/2}, \qquad -1 < y < 1 \qquad \textbf{(16)}$$

Notice that the derivative of $\sin^{-1}y$ is not defined at $y = \pm 1$ but is "infinite" there. This is consistent with the appearance of the graph in Fig. 6-2-5.

Worked Example 7   Differentiate $h(y) = \sin^{-1}(3y^2)$.

Solution   From (16) and the chain rule, with $u = 3y^2$,

$$h'(y) = (1-u^2)^{-1/2}\frac{du}{dy} = 6y(1-9y^4)^{-1/2}$$

Worked Example 8   Differentiate $f(x) = x\sin^{-1}(2x)$.

Solution   Here we are using x for the variable name. Of course we can use any letter we please. By the product and chain rules,

$$f'(x) = \left(\frac{dx}{dx}\right)\sin^{-1}(2x) + x\frac{d}{dx}(\sin^{-1}2x)$$
$$= \sin^{-1}2x + 2x(1-4x^2)^{-1/2}$$

It is interesting to observe that, while $\sin^{-1}y$ is defined in terms of trigonometric functions, its derivative is an algebraic function, even though the derivatives of the trigonometric functions themselves are still trigonometric.

Solved Exercises

10. (a) Calculate $\sin^{-1}(\frac{1}{2})$, $\sin^{-1}(-\sqrt{3}/2)$, and $\sin^{-1}(2)$.
    (b) Simplify $\tan(\sin^{-1}x)$.

11. Calculate $(d/dx)(\sin^{-1}2x)^{3/2}$.

12. Differentiate $\sin^{-1}(\sqrt{1-x^2})$, $0 < x < 1$.

Exercises

12. What are $\sin^{-1}(0.3)$, $\sin^{-1}(2/\sqrt{3})$, $\sin^{-1}(\frac{3}{2})$, $\sin^{-1}(-\pi)$, and $\sin^{-1}(1/\sqrt{3})$?

13. Differentiate the indicated functions:
    (a) $(x^2 - 1)\sin^{-1}(x^2)$ (b) $(\sin^{-1}x)^2$
    (c) $\sin^{-1}[t/(t+1)]$ (domain $=$?)

14. What are the maxima, minima, and inflection points of $f(x) = \sin^{-1}x$?

INVERSES OF THE REMAINING TRIGONOMETRIC FUNCTIONS

The rest of the inverse trigonometric functions can be introduced in the same way as $\sin^{-1} y$.

The derivative of $\cos x$, $-\sin x$, is negative on $(0, \pi)$, so $\cos x$ on $(0, \pi)$ has an inverse $\cos^{-1} y$. Thus for $-1 \le y \le 1$, $\cos^{-1} y$ is that number (expressed in radians) in $[0, \pi]$ whose cosine is y. The graph of $\cos^{-1} y$ is shown in Fig. 6-2-6.

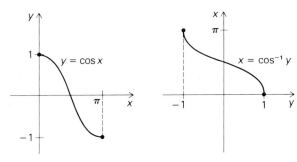

Fig. 6-2-6 The graph of cos and its inverse.

The derivative of $\cos^{-1} y$ can be calculated in the same manner as we calculated $(d/dy) \sin^{-1} y$:

$$\frac{d}{dy} \cos^{-1} y = \frac{1}{(d/dx) \cos x} = \frac{1}{-\sin x} = \frac{-1}{\sqrt{1 - y^2}} \qquad (17)$$

Worked Example 9   Differentiate $\tan(\cos^{-1} x)$.

Solution   By the chain rule and equation (17) (with x in place of y),

$$\frac{d}{dx} \tan(\cos^{-1} x) = \sec^2(\cos^{-1} x) \cdot \left(\frac{-1}{\sqrt{1 - x^2}} \right)$$

Fig. 6-2-7 $\sec(\cos^{-1} x) = 1/x$

From Fig. 6-2-7 we see that $\sec(\cos^{-1} x) = 1/x$, so

$$\frac{d}{dx} [\tan(\cos^{-1} x)] = \frac{-1}{x^2 \sqrt{1 - x^2}}$$

Another method is to use Fig. 6-2-7 directly to obtain

$$\tan(\cos^{-1} x) = \frac{\sqrt{1 - x^2}}{x}$$

Differentiating by the quotient and chain rules,

$$\frac{d}{dx} \frac{\sqrt{1 - x^2}}{x} = \frac{x \cdot (-2x) \cdot \frac{1}{2}(1 - x^2)^{-1/2} - (1 - x^2)^{1/2}}{x^2}$$

$$= -\frac{1}{x^2} \left(\frac{x^2}{\sqrt{1 - x^2}} + \sqrt{1 - x^2} \right)$$

$$= \frac{-1}{x^2 \sqrt{1 - x^2}}$$

which agrees with our previous answer.

Next let us construct the inverse tangent. Since $\tan' x = \sec^2 x$, $\tan x$ is increasing at every point of its domain. It is continuous on $(-\pi/2, \pi/2)$ and has range $(-\infty, \infty)$, so on this domain $\tan^{-1} y$ is defined as in Fig. 6-2-8. Thus for $-\infty < y < \infty$, $\tan^{-1} y$ is the unique number in $(-\pi/2, \pi/2)$ whose tangent is y.

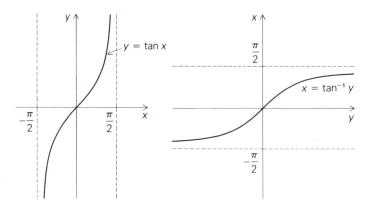

Fig. 6-2-8 tan x and its inverse.

The derivative of $\tan^{-1} y$ can be calculated as before:

$$\frac{d}{dy} \tan^{-1} y = \frac{1}{(d/dx)\tan x} = \frac{1}{\sec^2 x} = \frac{1}{1 + \tan^2 x} = \frac{1}{1 + y^2}$$

Thus

$$\frac{d}{dy} \tan^{-1} y = \frac{1}{1 + y^2} \tag{18}$$

Worked Example 10   Differentiate $f(x) = (\tan^{-1} \sqrt{x})/(\cos^{-1} x)$. Find the domain of f and f'.

Solution   By the quotient rule and the chain rule,

$$\frac{d}{dx} \frac{\tan^{-1} \sqrt{x}}{\cos^{-1} x}$$

$$= \frac{\cos^{-1} x \cdot \left[\dfrac{1}{1 + (\sqrt{x})^2} \cdot \dfrac{1}{2\sqrt{x}} \right] - \tan^{-1} \sqrt{x} \cdot \left(-\dfrac{1}{\sqrt{1 - x^2}} \right)}{(\cos^{-1} x)^2}$$

$$= \frac{\sqrt{1 - x^2} \cos^{-1} x + 2\sqrt{x}(1 + x) \tan^{-1} \sqrt{x}}{2\sqrt{x}(1 + x)\sqrt{1 - x^2}(\cos^{-1} x)^2}$$

The domain of f consists of those x for which $x \geq 0$ (so that $\sqrt{x}$ is defined) and $-1 < x < 1$ (so that $\cos^{-1} x$ is defined and not zero)—that is, the domain of f is $[0, 1)$. For f' to be defined, the denominator must be nonzero: in the derivative this requires x to belong to the interval $(0, 1)$. Thus the domain of f' is $(0, 1)$.

The remaining inverse trigonometric functions can be treated in the same way. Their graphs are shown in Fig. 6-2-9 and their properties are summarized in Table 6-2-2.

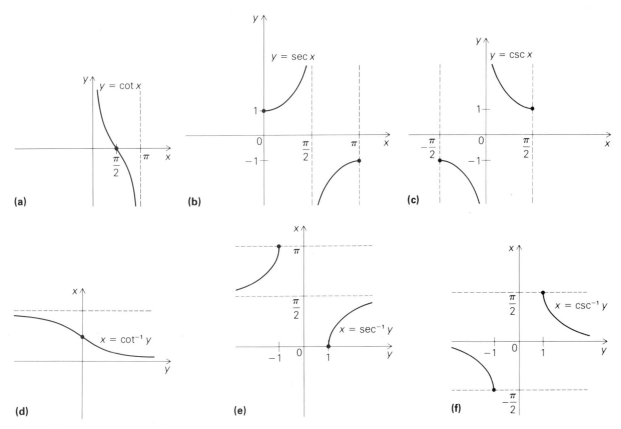

Fig. 6-2-9 cot, sec, csc, and their inverses.

Table 6–2–2   Inverse trigonometric functions

Function	Domain on which function has an inverse	Derivative of function	Inverse	Domain of inverse	Derivative of inverse	
$\sin x$	$\left[-\dfrac{\pi}{2}, \dfrac{\pi}{2}\right]$	$\cos x$	$\sin^{-1} y$	$[-1, 1]$	$\dfrac{1}{\sqrt{1 - y^2}}$	$-1 < y < 1$
$\cos x$	$[0, \pi]$	$-\sin x$	$\cos^{-1} y$	$[-1, 1]$	$-\dfrac{1}{\sqrt{1 - y^2}}$	$-1 < y < 1$
$\tan x$	$\left(-\dfrac{\pi}{2}, \dfrac{\pi}{2}\right)$	$\sec^2 x$	$\tan^{-1} y$	$(-\infty, \infty)$	$\dfrac{1}{1 + y^2}$	$-\infty < y < \infty$
$\cot x$	$(0, \pi)$	$-\csc^2 x$	$\cot^{-1} y$	$(-\infty, \infty)$	$-\dfrac{1}{1 + y^2}$	$-\infty < y < \infty$
$\sec x$	$\left[0, \dfrac{\pi}{2}\right)$ and $\left(\dfrac{\pi}{2}, \pi\right]$	$\tan x \sec x$	$\sec^{-1} y$	$(-\infty, -1]$ and $[1, \infty)$	$\dfrac{1}{\sqrt{y^2(y^2 - 1)}}$	$-\infty < y < -1$ $1 < y < \infty$
$\csc x$	$\left[-\dfrac{\pi}{2}, 0\right)$, and $\left(0, \dfrac{\pi}{2}\right]$	$-\cot x \csc x$	$\csc^{-1} y$	$(-\infty, -1]$ and $[1, \infty)$	$-\dfrac{1}{\sqrt{y^2(y^2 - 1)}}$	$-\infty < y < -1$ $1 < y < \infty$

Remembering formulas such as those in Table 6-2-2 is an unpleasant chore for most students (and professional mathematicians as well). An ideal to strive for is an understanding of the derivation of the formulas, so that any given formula can be derived efficiently. It is also useful to develop a short mental checklist: Is the sign right? Is the sign consistent with the appearance of the graph? Is the derivative undefined at the proper points?

Worked Example 11 Differentiate $f(x) = (\csc^{-1} 3x)^2$. Find the domain of f and f'.

    Solution By the chain rule,

$$\frac{d}{dx}(\csc^{-1} 3x)^2 = 2(\csc^{-1} 3x)\frac{d}{dx}\csc^{-1} 3x$$

$$= 2\csc^{-1} 3x \cdot 3 \cdot \frac{-1}{\sqrt{(3x)^2\,[(3x)^2 - 1]}}$$

$$= \frac{-2\csc^{-1}(3x)}{\sqrt{x^2(9x^2 - 1)}} = \frac{-2\csc^{-1}(3x)}{|x|\,\sqrt{9x^2 - 1}}$$

For $f(x)$ to be defined, $3x$ should lie in $[1, \infty)$ or $(-\infty, -1]$; that is, x should lie in $[\frac{1}{3}, \infty)$ or $(-\infty, -\frac{1}{3}]$. The domain of f' is $(\frac{1}{3}, \infty)$, together with $(-\infty, -\frac{1}{3})$.

Solved Exercises 13. Calculate $\cos^{-1}(\frac{1}{2})$, $\tan^{-1}(1)$, and $\csc^{-1}(2/\sqrt{3})$.

    14. Is $\cot^{-1} y = 1/(\tan^{-1} y)$?

    15. Differentiate:
        (a) $\sec^{-1}(y^2)$, $y > 0$ (b) $\cot^{-1}[(x^3 + 1)/(x^3 - 1)]$

    16. Explain why the derivative of every inverse cofunction in Table 6-2-2 is the negative of that of the inverse function.

    17. Find $\int [x^2/(1 + x^2)]\, dx$ [*Hint:* Divide first.]

Exercises 15. Calculate:
    (a) $\sec^{-1}(2/\sqrt{3})$ (b) $\sec^{-1}(2.3)$ (c) $\csc^{-1}(-5)$

    16. Calculate the derivative of each function:
        (a) $\cos^{-1}(1 - y^2)$ (b) $\sec^{-1}(y - 1/y^2)$
        (c) $\tan^{-1}[(2x^5 + x)/(1 - x^2)]$ (d) $(1/x)\csc^{-1}(1/x)$

Problems for Section 6-2

1. Differentiate each of the following functions:
    (a) $f(x) = \sqrt{x} + \cos 3x$
    (b) $f(x) = \sqrt{\cos x}$
    (c) $f(x) = (\sin^{-1} 3x)/(x^2 + 2)$
    (d) $f(x) = (x^2 \cos^{-1} x + \tan x)^{3/2}$
    (e) $f(\theta) = \cot^{-1}(\sin \theta + \sqrt{\cos^2 3\theta + \theta^2})$
    (f) $f(r) = (r^2 + \sqrt{1 - r^2})/(r \sin r)$

2. Evaluate the derivative of each of the following functions:
    (a) $(\cos^{-1} x)/(1 - \sin^{-1} x)$
    (b) $[\sin^{-1}(2t)]/(3t^2 - t + 1)$
    (c) $\sin^{-1}[(t^7 + t^4 + 1)/(t^5 + 2t)]$
    (d) $\sin(\sqrt{1 - x^3}) + \tan[x/(x^4 + 1)]$
    (e) $[\csc(\theta/\sqrt{\theta^2 + 1}) + 1]^{3/2}$
    (f) $(u^2/4) + 2\sin^{-1}[1/(u^2 + 1)^{3/2}]$

3. Calculate:
    (a) $(d/dx)[(\sin^{-1} 2x)^2 + x^2]$
    (b) $(d/dx)[\sin^{-1}(\sec 2x)/3]$
    (c) $(d^2/d\theta^2)\cos^{-1}\theta$
    (d) $(d/dv)\{(\tan\sqrt{v^2 + 1})(\sec[1/(v^2 + 1)])\}$
    (e) $(d/d\theta)\cot^{-1}(\theta^2 + 1)$
    (f) The rate of change of
        $[\sin(s\sqrt[3]{1 + s^2})]/\cos s$
        with respect to s at $s = 0$

4. Prove:
    (a) $\tan(\sin^{-1} x) = x/\sqrt{1 - x^2}$
    (b) $\csc^{-1}(1/x) = \sin^{-1} x = \pi/2 - \cos^{-1} x$

★5. (a) Evaluate $\lim_{\Delta\theta\to 0} (\tan \Delta\theta)/\Delta\theta$ using the methods of the beginning of this section.

(b) Use part (a) and the addition formula for tangent (p. 273) to prove that $\tan' \theta = \sec^2 \theta, -\pi/2 < \theta < \pi/2$, that is, prove that

$$\lim_{\Delta\theta\to 0} \frac{\tan (\theta + \Delta\theta) - \tan \theta}{\Delta\theta} = \sec^2 \theta$$

6. Differentiate $\sec [\sin^{-1}(y - 2)]$ by (a) simplifying first and (b) using the chain rule right away.

7. (a) What is the domain of $\cos^{-1}(x^2 - 3)$? Differentiate.

(b) Sketch the graph of $\cos^{-1}(x^2 - 3)$.

8. Show that $f(x) = \sec x$ satisfies the equation $f'' + f - 2f^3 = 0$.

9. Let x and y be related by the equation

$$\frac{\sin (x + y)}{xy} = 1$$

(a) Find dy/dx.

(b) If $x = t/(1 - t^2)$, find dy/dt.

(c) If $y = \sin^{-1} t$, find dx/dt.

(d) If $x = t^3 + 2t - 1$, find dy/dt.

10. Prove that $y = \tan^{-1} x$ has an inflection point at $x = 0$.

11. (a) Derive the formula for $(d/dy) \cot^{-1} y$ in Table 6-2-2.

(b) Derive the formula for $(d/dy) \csc^{-1} y$ in Table 6-2-2.

12. Find the following antiderivatives:

(a) $\int (x^3 + \sin x)\, dx$

(b) $\int (\cos 3x + 5x^{3/2})\, dx$

(c) $\int (x^4 + \sec 2x \tan 2x)\, dx$

(d) $\int [3/(1 + 4x^2)]\, dx$

13. Find the following antiderivatives:

(a) $\int [4/\sqrt{1 - x^2}]\, dx$ (b) $\int [1/\sqrt{1 - 4x^2}]\, dx$

(c) $\int [1/\sqrt{4 - x^2}]\, dx$

(d) $\int (\sin 2x + \sqrt{x})\, dx$

14. What is the equation of the line tangent to the graph of $\cos^{-1}(x^2)$ at $x = 0$?

15. A child is whirling a stone on a string 0.5 meter long in a vertical circle at 5 revolutions per second. The sun is shining directly overhead. What is the velocity of the stone's shadow when the stone is at the 10 o'clock position?

16. Is the following correct: $(d/dx) \cos^{-1} x = (-1)(\cos^{-2} x)[(d/dx) \cos x]$?

17. Find a function $f(x)$ which is differentiable and increasing for all x, yet $f(x) < \pi/2$ for all x.

18. Sketch the graph of $(\sin x)/(1 + x^2)$ for $0 \le x \le 2\pi$. (You may use a calculator to locate the critical points.)

19. Find the inflection points of $f(x) = \cos^2 3x$.

20. Where is $f(x) = x \sin x + 2 \cos x$ concave up? Concave down?

21. The displacement $x(t)$ from equilibrium of a mass m undergoing harmonic motion on a spring of Hooke's constant k is known to satisfy the equation $mx''(t) + kx(t) = 0$. Check that $x(t) = A \cos (\omega t + \theta)$ is a solution of this equation, where $\omega = \sqrt{k/m}$; A and ω are constants.

22. A slot racer travels at constant speed around a circular track, doing each lap in 3.1 seconds. The track is 3 feet in diameter.

(a) The position (x, y) of the racer can be written as $x = r \cos (\omega t)$, $y = r \sin (\omega t)$. Find the values of r and ω. Interpret these constants.

(b) The *speed* of the racer is the elapsed time divided into the distance traveled. Find its value and check that it equals $\sqrt{(dx/dt)^2 + (dy/dt)^2}$.

23. Drywall sheets weighing 6000 lb are moved across a level floor. The method is to attach a chain to the skids under the drywall stack, then pull it with a truck. The angle θ made by the chain and the floor is related to the force F along the chain by $F = 6000\, k/(k \sin \theta + \cos \theta)$, where the number k is the coefficient of friction.

(a) Compute $dF/d\theta$.

(b) Find θ for which $dF/d\theta = 0$. (This is the angle that requires the least force.)

24. Determine the equations of the tangent and normal lines to the curve $y = \cos 2x + \cos x$ at $(0, 2)$.

6-3 Old Problems with New Functions

Many interesting word problems involve fractional powers and trigonometric functions.

The graphing and word problems in Chapters 2 and 3 were limited since, at that point, we could differentiate only rational functions. Now that we have more functions at our disposal, we can solve a wider variety of problems. After giving a few illustrations of our new problem-solving capabilities, and some additional material on polar coordinates, we will study the variation in the length of the day with the latitude on the earth and the time of year.

Goals

After studying this section, you should be able to:

Solve word problems involving fractional powers and trigonometric functions.

Locate cusps.

Find tangent lines to curves described by equations in polar coordinates.

SOME PROBLEMS

In many calculus problems, square roots and trigonometric functions arise. For example, the distance $d = [(x_1 - x_2)^2 + (y_1 - y_2)^2]^{1/2}$ between two points (x_1,y_1) and (x_2,y_2) involves the square root. The trigonometric functions enter if the problem involves circular motion or the relationships between the angles and sides of a triangle.

Worked Example 1   A searchlight 10 kilometers from a straight coast makes one revolution every 30 seconds. How fast is the spot of light moving along a wall on the coast at the point P in Fig. 6-3-1?

Fig. 6-3-1 How fast is spot P moving if the beam LP revolves once every 30 seconds?

Solution   Let θ denote the angle PLQ, and let x denote the distance $|PQ|$. We know that $d\theta/dt = 2\pi/30$, since θ changes by 2π in 30 seconds. Now $x = 10 \tan\theta$, so $dx/d\theta = 10 \sec^2\theta$, and the velocity of the spot is $dx/dt = (dx/d\theta)(d\theta/dt) = 10 \sec^2\theta \ 2\pi/30 = (2\pi/3) \sec^2\theta$. When $x = 8$, $\sec\theta = |PL|/|LQ| = \sqrt{8^2 + 10^2}/10$ so the velocity is $dx/dt = (2\pi/3) \times (8^2 + 10^2)/10^2 = (82\pi/75) \approx 3.4$ kilometers per second (this is very fast!).

Worked Example 2   Given a number $a > 0$, find the minimum value of $(a + x)/\sqrt{ax}$ where $x > 0$.

Solution   We are to minimize $f(x) = (a + x)/\sqrt{ax}$ on $(0, \infty)$. This can be done straight away by the methods of Section 2.2. By the quotient rule,

$$f'(x) = \frac{\sqrt{ax} - (a + x)(a/2\sqrt{ax})}{ax}$$

$$= \frac{2ax - a(a + x)}{2(ax)^{3/2}} = \frac{a(x - a)}{2(ax)^{3/2}}$$

Thus $x = a$ is the only critical point in $(0, \infty)$. We observe that $f'(x)$ changes sign from negative to positive at $x = a$, so this is a minimum point. The minimum value is $f(a) = (a + a)/\sqrt{a^2} = 2$.

The result of this example can be rephrased by saying that

$$\frac{a + x}{\sqrt{ax}} \geq 2 \quad \text{for every } a > 0, x > 0$$

Writing b for x, this becomes

$$\frac{a + b}{2} \geq \sqrt{ab} \quad \text{for every } a > 0, b > 0$$

That is, the *arithmetic mean* is larger than the *geometric mean*. This inequality was proved by using calculus. It can also be proved from algebra alone. Indeed, since the square of any number is non-negative,

$$0 \leq (\sqrt{a} - \sqrt{b})^2$$
$$= a - 2\sqrt{ab} + b$$

so

$$2\sqrt{ab} \leq a + b$$

Hence

$$\frac{a + b}{2} \geq \sqrt{ab}$$

Sometimes this is called the *arithmetic-geometric mean inequality*.

Worked Example 3   Sketch the graph of the function $f(x) = \cos x + \cos 2x$ by using the methods of Chapter 2.

Solution   The function is defined on $(-\infty, \infty)$; there are no asymptotes. We find that $f(-x) = \cos(-x) + \cos(-2x) = \cos x + \cos 2x = f(x)$, so f is even. Furthermore, $f(x + 2\pi) = f(x)$, so the graph repeats itself every 2π units (like that of $\cos x$). It follows that we need only look for features of the graph on $[0, \pi]$, because we can obtain $[-\pi, 0]$ by reflection across the y axis and the rest of the graph by repetition of the part over $[-\pi, \pi]$.

We have

$$f'(x) = -\sin x - 2 \sin 2x$$

and

$$f''(x) = -\cos x - 4 \cos 2x$$

To find roots of the first derivative—that is, critical points—it is best to factor it, first using the formulas

$$\sin 2x = 2 \sin x \cos x \qquad \text{and} \qquad \cos 2x = 2 \cos^2 x - 1$$

We obtain

$$f'(x) = -\sin x - 4 \sin x \cos x = -\sin x (1 + 4 \cos x)$$

Also notice that

$$f''(x) = -\cos x - 4(2 \cos^2 x - 1) = -8 \cos^2 x - \cos x + 4$$

The critical points occur when $\sin x = 0$ or $1 + 4 \cos x = 0$—that is, when $x = 0$, π, or $\cos^{-1}(-\frac{1}{4}) \approx 1.82$ (radians).
We have

$$f(0) = 2 \qquad f(\pi) = 0 \qquad f[\cos^{-1}(-\tfrac{1}{4})] = -1\tfrac{1}{8}$$
$$f''(0) = -5 \qquad f''(\pi) = -3 \qquad f''[\cos^{-1}(-\tfrac{1}{4})] = 3\tfrac{3}{4}$$

Hence 0 and π are local maximum points, and $\cos^{-1}(-\frac{1}{4})$ is a local minimum point, by the second derivative test.
To find the points of inflection, we first find the roots of $f''(x) = 0$; that is,

$$-8 \cos^2 x - \cos x + 4 = 0$$

This is a quadratic equation in which $\cos x$ is the unknown, so

$$\cos x = \frac{-1 \pm \sqrt{129}}{16} \approx 0.647 \text{ and } -0.772$$

Thus our candidates for points of inflection are

$$x_1 = \cos^{-1}(0.647) \approx 0.87 \quad \text{and} \quad x_2 = \cos^{-1}(-0.772) \approx 2.45$$

We can see from the previously calculated values for $f''(x)$ that f'' change sign at these points, so they are inflection points. We calculate $f(x)$ and $f'(x)$ at the inflection points:

$$f(x_1) \approx 0.48 \qquad f'(x_1) \approx -2.74$$
$$f(x_2) \approx -0.58 \quad f'(x_2) \approx 1.33$$

Finally, the zeros of *f* may be found by writing

$$f(x) = \cos x + \cos 2x$$
$$= \cos x + 2\cos^2 x - 1$$
$$= 2(\cos x + 1)\left(\cos x - \frac{1}{2}\right)$$

Thus $f(x) = 0$ at $x = \pi$ and $x = \cos^{-1}\left(\frac{1}{2}\right) = 1.047$.

The graph on $[0, \pi]$ obtained from this information is shown in Fig. 6-3-2. Reflecting across the *y* axis and then repeating the pattern, we obtain the graph shown in Fig. 6-3-3. Such graphs, with oscillations of varying amplitudes, are typical when oscillating functions with *different* frequencies are added.

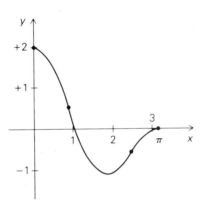

Fig. 6-3-2 The graph of $\cos x + \cos 2x$ on $[0, \pi]$.

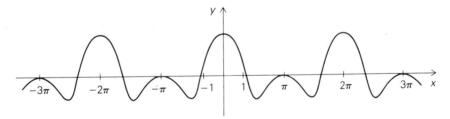

Fig. 6-3-3 The full graph of $\cos x + \cos 2x$.

Solved Exercises

Train compartment

Fig. 6-3-4 Where should you sit to get the widest view?

1. Find the point on the *x* axis for which the sum of the distances from $(0, 1)$ and (p, q) is a minimum. (Assume that *p* and *q* are positive.)

2. Two hallways, meeting at right angles, have widths *a* and *b*. Find the length of the longest pole which will go around the corner; the pole must be in a horizontal position. (In Problem 6, p. 164, you were asked to do this problem by minimizing the square of the length; redo the problem here by minimizing the length itself.)

3. (This problem was written on a train.) One normally chooses the window seat on a train to have the best view. Imagine the situation in Fig. 6-3-4 and see if this is really the best choice. (Ignore the extra advantage of the window seat which enables you to lean forward to see a special view.)

Exercises

1. The height of an object thrown straight down from an initial altitude of 1000 feet is given by $h(t) = 1000 - 40t - 16t^2$. The object is being tracked by a searchlight 200 feet from where the object will hit. How fast is the angle of elevation of the searchlight changing after 4 seconds?

2. Find the point on the curve $y = \frac{1}{2}x^2 - 4x + 5$ closest to the point $(4, 3)$.

3. Graph the following functions:

   (a) $\cos x + \sin 2x$ (b) $\cos x + 2 \cos 2x$

   (c) $\cos x + \cos 3x$ [*Hint:* Use $\sin 3x = \sin(2x + x)$ when looking for critical points.]

4. A bicycle is moving 10 feet per second. It has wheels of radius 16 inches and a reflector attached to the front spokes 12 inches from the center. If the reflector is at its lowest point at $t = 0$, how fast is the reflector accelerating vertically at $t = 5$ seconds?

(optional) **CUSPS**

Some interesting new features arise when we graph functions involving fractional powers. These first show up in the graph of $y = \sqrt{x}$ (Fig. 6-3-5). Notice that the slope $dy/dx = 1/(2\sqrt{x})$ becomes "infinite" at $x = 0$.

Next consider $y = \sqrt[3]{x}$. Recall that this is defined for every x, positive or negative, and $dy/dx = 1/(3x^{2/3})$, which is also "infinite" at $x = 0$. The graph of $y = \sqrt[3]{x}$ is obtained by flipping the graph of $y = x^3$, as we know from Section 5-1 (see Fig. 6-3-6). Sometimes one says that there is a vertical tangent line to $y = \sqrt[3]{x}$ at $x = 0$, since the inverse function has a horizontal tangent line.

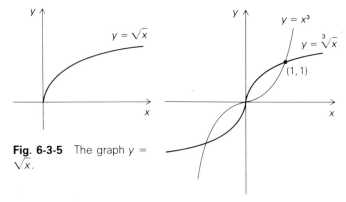

Fig. 6-3-5 The graph $y = \sqrt{x}$.

Fig. 6-3-6 The graph of $y = \sqrt[3]{x}$ has a "vertical tangent line" at the origin.

Still more interesting is the graph of $y = x^{2/3}$, which is also defined for all x. The derivative is $dy/dx = \frac{2}{3}x^{-1/3} = 2/(3\sqrt[3]{x})$. For x near zero and positive, dy/dx is large positive, whereas for x near zero and negative, dy/dx is large negative. Thus the graph has the appearance shown in Fig. 6-3-7. Again we can

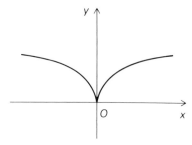

Fig. 6-3-7 The graph of $y = x^{2/3}$ has a cusp at the origin.

(optional) | say that the graph has a vertical tangent at $x = 0$. However, the shape of the graph near $x = 0$ has not been encountered before. We call $x = 0$ a *cusp*. Note that $x = 0$ is a minimum point of $f(x) = x^{2/3}$, but that $x^{2/3}$ is not differentiable there.

In general, a continuous function f is said to have a *cusp* at x_0 if $f'(x)$ has opposite signs on opposite sides of x_0 but $f'(x)$ "blows up" at x_0 in the sense that $\lim_{x \to x_0} [1/f'(x)] = 0$. In particular, f is not differentiable at x_0.

Solved Exercises ★ 4. Let $f(x) = (x^2 + 1)^{3/2}$. (a) Where is f increasing? (b) Sketch the graph of f. Are there any cusps?

★ 5. Sketch the graph of $(x + 1)^{2/3} x^2$.

Exercises ★ 5. Let $f(x) = (x^2 - 3)^{2/3}$. (a) Where is f increasing? Decreasing? (b) Sketch the graph of f, noting any cusps.

★ 6. Consider $g(x) = f(x)(x - a)^{p/q}$, where f is differentiable at $x = a$, p is even, q is odd and $p < q$. If $f(a) \neq 0$, show that g has a cusp at $x = a$. [*Hint:* Look at $g'(x)$ for x on either side of a.]

★ 7. Sketch the graph of each of the following, noting any cusps:
(a) $(x^2 + 1)^{1/3}$ (b) $(x - 1)^{4/3}(x + 1)^{2/3}$ (c) $(x - 4)^{100/99}$ (d) $x + x^{2/3}$

TANGENTS TO CURVES IN POLAR COORDINATES

In Section 6-1 we saw how to graph curves $r = f(\theta)$ in polar coordinates. To obtain further information about these graphs, we may wish to calculate the slope of the tangent line at a point (r, θ) (see Fig. 6-3-8).

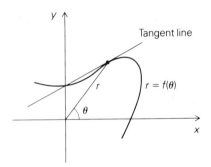

Fig. 6-3-8 What is the slope of the tangent line?

This slope is not $f'(\theta)$, since $f'(\theta)$ is the rate of change of r with respect to θ, while the slope is the rate of change of y with respect to x. Assume that y is a differentiable function of x. To calculate dy/dx, we write

$$x = r \cos \theta = f(\theta) \cos \theta \tag{1}$$

and

$$y = r \sin \theta = f(\theta) \sin \theta \tag{2}$$

This is a parametric curve with θ as the parameter. According to the formula on p. 243, with t replaced by θ,

$$\frac{dy}{dx} = \frac{dy/d\theta}{dx/d\theta} = \frac{f'(\theta)\sin\theta + f(\theta)\cos\theta}{f'(\theta)\cos\theta - f(\theta)\sin\theta}$$

Dividing numerator and denominator by $\cos\theta$ gives

$$\frac{dy}{dx} = \frac{f'(\theta)\tan\theta + f(\theta)}{f'(\theta) - f(\theta)\tan\theta} \tag{3}$$

TANGENTS TO GRAPHS IN POLAR COORDINATES

The slope of the line tangent to the graph of $r = f(\theta)$ at (r, θ) is

$$\frac{(\tan\theta)\,dr/d\theta + r}{dr/d\theta - r\tan\theta}$$

Worked Example 4 Find the slope of the line tangent to the graph of $r = 3\cos^2 2\theta$ at $\theta = \pi/6$.

Solution Here $f(\theta) = 3\cos^2 2\theta$, so $dr/d\theta = f'(\theta) = -12\cos 2\theta\sin 2\theta$ (by the chain rule). Now $f(\pi/6) = 3\cos^2(\pi/3) = \frac{3}{4}$ and $f'(\pi/6) = -12\cos(\pi/3)\sin(\pi/3) = -12\cdot\frac{1}{2}\cdot\sqrt{3}/2 = -3\sqrt{3}$. Thus formula (3) gives

$$\frac{dy}{dx} = \frac{-3\sqrt{3}\,(1/\sqrt{3}) + \frac{3}{4}}{-3\sqrt{3} - (\frac{3}{4})(1/\sqrt{3})} = \frac{3\sqrt{3}}{13}$$

Thus the slope of the tangent line is $3\sqrt{3}/13$.

Calculus can aid us when graphing in polar coordinates. As we just saw, it enables us to calculate the slope of the tangent line. As another illustration notice that a local maximum of $f(\theta)$ will be a point on the graph where the distance from the origin is a local maximum, as in Fig. 6-3-9. The methods of Chapter 2 can be used to locate these local maxima (as well as the local minima).

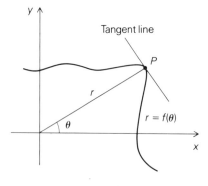

Fig. 6-3-9 The P corresponds to a local maximum point of $f(\theta)$.

Worked Example 5   Calculate the slope of the line tangent to $r = f(\theta)$ at (r, θ) if f has a local maximum there. Interpret geometrically.

Solution   At a local maximum, $f'(\theta) = 0$. Plugging this into formula (3) gives

$$\frac{dy}{dx} = \frac{f(\theta)}{-f(\theta)\tan\theta} = -\frac{1}{\tan\theta}$$

Since dy/dx is the negative reciprocal of $\tan\theta$, the tangent line is perpendicular to the line from the origin to (r, θ). (See Fig. 6-3-9.)

Solved Exercises

6. Find the slope of the line tangent to the graph of $r = \cos 3\theta$ at $(r, \theta) = (-1, \pi/3)$.

7. Find the maxima and minima of $f(\theta) = 1 + 2\cos\theta$. Sketch the graph of $r = 1 + 2\cos\theta$ in the xy plane.

Exercises

8. For each of the following, find the slope of the tangent line at the indicated point:

   (a) $r = \cos 4\theta$; $\theta = \pi/3$ (b) $r = 2\sin 5\theta$; $\theta = \pi/2$

   (c) $r = \tan\theta$; $\theta = 2\pi/3$ (d) $r = \theta^2 + 1$; $\theta = 5$

   (e) $r = 1 + 2\sin 2\theta$; $\theta = 0$ (f) $r = 2 - \sin\theta$; $\theta = \pi/6$

9. Find the maximum and minimum values of r for all the functions in Exercise 8. Sketch the graphs in the xy plane.

(optional)   **LENGTH OF DAYS**

We conclude this section with an extended application of calculus to a phenomenon which requires no specialized equipment or knowledge for its observation—the setting of the sun.

Using spherical trigonometry or vector methods, one can derive a formula relating the following variables:

A = angle of elevation of sun above horizon

l = latitude of a place on earth's surface

α = inclination of earth's axis (23.5° or 0.41 radian)

T = time of year, measured in days from first day of summer in northern hemisphere (June 21)

t = time of day, measured in hours from noon*

The formula reads:

$$\sin A = \cos l \sqrt{1 - \sin^2\alpha \cos^2\left(\frac{2\pi T}{365}\right)}\cos\left(\frac{2\pi t}{24}\right) + \sin l \sin\alpha \cos\left(\frac{2\pi T}{365}\right) \quad (4)$$

---

*By noon we mean the moment at which the sun is highest in the sky. To find out when noon occurs in your area, look in a newspaper for the times of sunrise and sunset, and take the midpoint of these times. It will probably not be 12:00, but it should change only very slowly from day to day (except when daylight savings time comes or goes).

(optional)

We will derive it in the appendix to Chapter 15.* For now, we will simply assume the formula and find some of its consequences.

At the time S of sunset, $A = 0$. That is,

$$\cos\left(\frac{2\pi S}{24}\right) = -\tan l \frac{\sin \alpha \cos (2\pi T/365)}{\sqrt{1 - \sin^2 \alpha \, \cos^2 (2\pi T/365)}} \tag{5}$$

Solving for S, and remembering that $S \geq 0$ since sunset occurs after noon, we get

$$S = \frac{12}{\pi} \cos^{-1}\left[-\tan l \frac{\sin \alpha \cos (2\pi T/365)}{\sqrt{1 - \sin^2 \alpha \, \cos^2 (2\pi T/365)}}\right] \tag{6}$$

Worked Example 6   At what time does the sun set on July 1 at 39° latitude?

Solution   We have $l = 39°$, $\alpha = 23.5°$, and $T = 11$. Substituting these values in (6), we find $\tan l = 0.8098$, $2\pi T/365 = 0.1894$ (that is, 10.85°), $\cos (2\pi T/365) = 0.9821$, and $\sin \alpha = 0.3987$. Hence

$$S = \frac{12}{\pi} \cos^{-1}(-0.3447) = \frac{12}{\pi}(1.92) = 7.344$$

Thus $S = 7.344$ (hours after noon); that is, the sun sets at 7:20:38.

For a fixed point on the earth, S may be considered a function of T. Differentiating (6) and simplifying (see Exercise 10), we find

$$\frac{dS}{dT} = -\frac{24}{365} \tan l \left\{ \frac{\sin \alpha \sin (2\pi T/365)}{[1 - \sin^2 \alpha \, \cos^2 (2\pi T/365)] \sqrt{1 - \sin^2 \alpha \, \cos^2 (2\pi T/365) \sec^2 l}} \right\} \tag{7}$$

The critical points of S occur when $2\pi T/365 = 0$, π, or 2π; that is, $T = 0$, $365/2$, or $T = 365$—the first day of summer and the first day of winter. For the northern hemisphere, $\tan l$ is positive. By the first derivative test, $T = 0$ (or 365) is a local maximum and $T = 365/2$ a minimum. Thus we get the graph shown in Fig. 6-3-10.

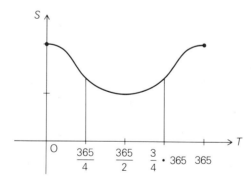

Fig. 6-3-10 Sunset time as a function of the date. The sun sets late in summer, early in winter.

Using the first-order approximation, we can also determine how much the sunset time changes from one day to the next. If we set $\Delta T = 1$, we have

$$\Delta S \approx \frac{dS}{dT} \Delta T = \frac{dS}{dT}$$

If $\pi/2 - \alpha < |l| < \pi/2$ (inside the polar circles), there will be some values of t for which the right-hand side of formula (1) does not lie in the interval $[-1, 1]$. On the days corresponding to these values of t, the sun will never set ("midnight sun").

If $l = \pm\pi/2$, $\tan l = \infty$, and the right-hand side does not make sense at all. This reflects the fact that, at the poles, it is either light all day or dark all day, depending upon the season.

(optional) and dS/dT is given by (7). Thus if the number of days after June 21 is inserted, along with the latitude l, formula (7) gives an approximation for the number of minutes later (or earlier) the sun will set the following evening. Note that the *difference* between sunset times on two days is the same whether we measure time in minutes from noon or by the clock ("standard time").

Formula (6) also tells us how long the days are as a function of latitude and day of the year. Plotting the formula on a computer (taking careful account of the polar regions) gives Fig. 6-3-11. Graphs such as the one shown in Fig. 6-3-10 result if l is fixed and only T varies.

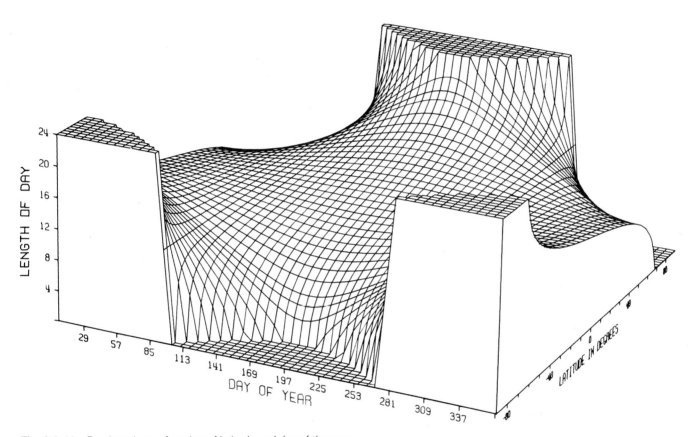

Fig. 6-3-11 Day length as a function of latitude and day of the year.

Worked Example 7
(optional)

In Worked Example 6 use formula (7) to calculate how many minutes earlier the sun will set the following evening. Compare with the values on July 1 and July 2 computed from formula (6).

Solution

Substituting $T = 11$ into (7), we obtain $\Delta S \approx -0.0055$; that is, the sun should set 0.0055 hour, or 20 seconds, earlier on July 2 than on July 1.

Computing the time of sunset on July 2 by formula (6), with $t = 12$, we obtain $S = 7.338$, or 7:20:17, which is 21 seconds earlier than the time computed in Worked Example 5 for sunset on July 2. The error in the first-order approximation to ΔS is thus about 1 part in 20, or 5%.

Differentiating formula (7), we find that the extreme values of dS/dT occur when $2\pi T/365 = \pi/2$ or $3\pi/2$. (These are the inflection points in Fig. 6-3-10; see Exercise 13.)

When $2\pi T/365 = \pi/2$ (the first day of fall), $dS/dT = -(24/365) \tan l \sin \alpha$; at this time, the days are getting shorter most rapidly.* When $2\pi T/365 = 3\pi/2$ (the first day of spring), the days are lengthening most rapidly, with $dS/dT = 24/365 \tan l \sin \alpha$.

It is interesting to note how this maximal rate, $(24/365) \tan l \sin \alpha$, depends on latitude. Near the equator, $\tan l$ is very small, so the rate is near zero, corresponding to the fact that seasons don't make much difference near the equator. Near the poles, $\tan l$ is very large, so the rate is enormous. This large rate corresponds to the sudden switch from nearly 6 months of sunlight to nearly 6 months of darkness. *At the poles, the rate is "infinite."* (Of course, in reality the change isn't quite sudden because of the sun's diameter, the fact that the earth isn't a perfect sphere, refraction by the atmosphere, and so forth.)

Solved Exercises

8. For which values of T does the sun never set if $l > \pi/2 - \alpha$ (that is, near the North Pole)? Discuss.

Exercises

10. Derive formula (7) by differentiating (6). (It may help you to use the chain rule with $u = \sin \alpha \cos(2\pi T/365)$ as the intermediate variable.)

11. According to the *Los Angeles Times* for July 12, 1975, the sun set at 8:06 PM. The latitude of Los Angeles is 33.57° North. Guess what time the July 13 paper said the sun set? What about July 14? Can you give a more accurate figure for the difference in the times of sunset?

12. Determine your latitude (approximately) by measuring the times of sunrise and sunset.

13. Calculate $d^2 S/dT^2$ to confirm that the inflection points of S occur at the first days of spring and fall.

One of the authors was stimulated to do these calculations by the observation that he was most aware of the shortening of the days at the beginning of the school year. This calculation provides one explanation for the observation; perhaps the reader can think of others.

Problems for Section 6-3

1. Two weights A and B together on the ground are joined by a 20-meter wire. The wire passes over a pulley 10 meters above the ground. Weight A is slid along the ground at 2 meters per second. How fast is the distance between the weights changing after 3 seconds? (See Fig. 6-3-12.)

Fig. 6-3-12 How fast are A and B separating?

2. Consider the two posts in Fig. 6-3-13. The light atop post A moves vertically up and down the post according to $h(t) = 55 + 5 \sin t$ (t in seconds, height in meters). How fast is the length of the shadow of the 2-meter statue changing at $t = 20$ seconds?

Fig. 6-3-13 How fast is the shadow's length changing when the light oscillates up and down?

3. Sketch the graphs of the following:
   (a) $y = 2 \cos x + \cos 2x$
   (b) $y = \cos 2x + \cos 4x$
   (c) $y = x^{2/3} \cos x$
   (d) $y = x^{2/3} + (x - 1)^{2/3}$
   (e) $y = x^{4/3}$

4. Suppose that an object has position at time t given by $x = (t - 1)^{5/3}$. Discuss its velocity and acceleration near $t = 1$.

5. Sketch the graph of (a) $y = (x^2 + 5x + 4)^{2/7} + 1$; (b) $y = (x^3 + 2x^2 + x)^{4/9}$.

6. Particle A is moving according to $x = 3 \sin 3t$ and $y = 3 \cos 3t$ in the plane; particle B is moving according to $x = 3 \cos 2t$ and $y = 3 \sin 2t$. Find the maximum distance between A and B.

7. Find the slope of the line tangent to each of the following graphs at the indicated point:
   (a) $r = \tan \theta$; $\theta = \pi/4$
   (b) $r = 3 \sin \theta + \cos(\theta^2)$; $\theta = 0$
   (c) $r = \sec \theta + 2\theta^3$; $\theta = \pi/6$
   (d) $r = \sin 3\theta \cos 2\theta$; $\theta = 0$

8. Find the maximum and minimum values of $r = \sin 3\theta \cos 2\theta$. Sketch its graph in the xy plane.

★9. Sketch the graph of $r = \sin^3 3\theta$.

10. Molasses is smeared over the upper half ($y > 0$) of the (x, y) plane. A bug crawls at 1 centimeter per minute where there is molasses and 3 centimeters per minute where there is none. Find the fastest route for the bug between each of the following pairs of points:
    (a) $(0, 1)$ and $(25, 1)$ (b) $(0, 1)$ and $(\frac{1}{8}, 1)$
    (c) $(0, 1)$ and $(1000, 1)$ (d) $(0, 1)$ and $(1, -1)$
    (e) Suppose that the bug is to travel from some point in the upper half-plane ($y > 0$) to some point in the lower half-plane ($y < 0$). The fastest route then consists of a broken line segment with a break on the x axis. Find a relation between the sines of the angles made by the two parts of the segments with the y axis.

11. Do Solved Exercise 1 without using calculus. [*Hint:* Replace (p, q) by $(p, -q)$.]

★12. Let $f(x) = \sin^{-1}[2x/(x^2 + 1)]$.
    (a) Show that $f(x)$ is defined for x in $(-\infty, \infty)$.
    (b) Compute $f'(x)$. Where is it defined?
    (c) Show that the maxima and minima occur at the singular points of f.
    (d) Sketch the graph of f. [*Hint:* What is $f'(x)$ for x near the singular points?]

13. The motion of a projectile (neglecting air friction and the curvature of the earth) is governed by the equations
$$x = v_0 t \cos \alpha \quad y = v_0 t \sin \alpha - 4.9t^2$$
where v_0 is the initial velocity and α is the initial angle of elevation. Distances are measured in meters and t is the time from launch. (See Fig. 6-3-14.)
    (a) Find the maximum height of the projectile and the distance R from launch to fall as a function of α.
    (b) Show that R is maximized when $\alpha = \pi/4$.

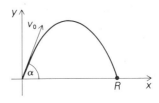

Fig. 6-3-14 Path of a projectile near the surface of the earth.

14. Consider the situation sketched in Fig. 6-3-15. At what position on the road is the angle θ maximized?

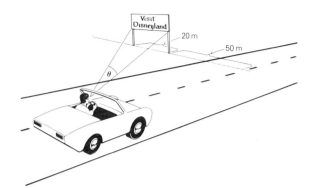

Fig. 6-3-15 Maximize θ.

★15. At latitude l on the earth, on what day of the year is the day 13 hours long? If this defines a function $T(l)$, discuss its graph.

★16. Near springtime in the temperate zone (near 45°), show that the days are getting longer at a rate of about 1.6 minutes a day (or 11 minutes a week).

★17. Planet VCH revolves about its sun once every 590 VCH "days." Each VCH "day" = 19 earth hours = 1140 earth minutes. Planet VCH's axis is inclined at 31°. What time is sunset, at a latitude of 12°, 16 VCH days after the first day of summer? (Assume that the VCH "day" is divided into 1440 "minutes.")

Review Problems for Chapter 6

1. Differentiate each of the following functions:
   (a) $f(\theta) = \theta^2 + (\theta/\sin\theta)$. What is $f'(2)$?
   (b) $g(x) = \sec x + \{1/[x\cos(x+1)]\}$
   (c) $h(y) = y^3 + 2y\tan(y^3) + 1$
   (d) $x(\theta) = [\sin(\theta/2)]^{4/7} + \theta^9 + 11\sqrt{\theta} + 4$
   (e) $y(x) = \cos(x^8 - 7x^4 - 10)$. What is $y'(0)$?

2. Differentiate:
   (a) $f(x) = \sec^{-1}[(x + \sin x)^2]$
   (b) $f(x) = \cot^{-1}(20 - \sqrt[4]{x})$
   (c) $f(y) = (2y^3 - 3\csc\sqrt{y})^{1/3}$
   (d) $r(\theta) = 6\cos^3(\theta^2 + 1) + 1$
   (e) $r(\theta) = (7\sin a\theta)/(\sin b\theta + \cos c\theta)$; a, b, c constants

3. (a) Let $f(x) = x^2 + \sin(2x + 1)$ and $x = t^3 + 1$. Find df/dx and df/dt.
   (b) Let $g(r) = 1/r^2 + (r^2 + 4)^{1/3}$ and $r = \sin 2\theta$. Find dg/dr and $dg/d\theta$.

4. (a) Let $h(x) = x\sin^{-1}(x + 1)$ and $x = y - y^3$. Find dh/dx and dh/dy.
   (b) Let $f(x) = x^{3/5} + \sqrt{2x^4 + x^2 - 6}$ and $x = y + \sin y$. Find df/dx and df/dy.
   (c) Let $f(x) = \tan^{-1}(2x^3)$ and $x = a + bt$, where a and b are constants. Find df/dx and df/dt.

5. A balloon is released from the ground 10 meters from the base of a 30-meter lamp post. The balloon rises steadily at 2 meters per second. How fast is the shadow of the balloon moving away from the base of the lamp after 4 seconds?

6. Find the following antiderivatives:
   (a) $\int \sin 3x\, dx$
   (b) $\int (x^{3/2} + \cos 2x)\, dx$
   (c) $\int -\frac{7}{5}x^{2/5}\, dx$
   (d) $\int (4\cos 4x - 4\sin 4x)\, dx$
   (e) $\int [4(x + 5)^{3/19} - 9/x^2]\, dx$
   (f) $\int (20x^{15} - 11x^7 + x^{-1/100})\, dx$

7. Evaluate:
   (a) $\int (\cos x + \sin 2x + \cos 3x)\, dx$
   (b) $\int (1.2x^{7.3} + 0.5x^{0.107} + 0.1)\, dx$
   (c) $\int nax^{n-1}\, dx$; n a rational number, a constant
   (d) $\int (14x^6 + \frac{1}{7}x^2 - 22x + 1)\, dx$
   (e) $\int (3x^2\sin x^3 + 2x)\, dx$
   (f) $\int [x^2 + (4/x^2)\sin(1 + 1/x) - 2]\, dx$

8. (a) If F and G are antiderivatives for f and g, show that $F(x)\,G(x) + C$ is an antiderivative for $f(x)\,G(x) + F(x)g(x)$.
   (b) Find the antiderivative of $x\sin x - \cos x$.
   (c) Find the antiderivative of $x\sin(x + 3) - \cos(x + 3)$.

9. Sketch the graph of $y = \theta + \sin\theta$ on $[-2\pi, 2\pi]$.

10. Sketch the graph of $r = \theta^2\cos^2\theta$ in polar coordinates. [Use your calculator to locate the zeros of $dr/d\theta$.]

11. Prove that $f(x) = x - 1 - \cos x$ is increasing on $[0, \infty)$. What inequality can you deduce?

12. Refer to Fig. 6-R-1. A girl at point G on the riverbank wishes to reach point B on the opposite side of the river as quickly as possible. She has at her disposal a rowboat which she can row 4 kilometers per hour and a bicycle which she can pedal at 16 kilometers per hour. What path should she take? (Ignore any current in the river, and assume that the bicycle can be ready where she needs it.)

Fig. 6-R-1 Find the best path from G to B.

13. Find a function $f(x)$ which satisfies $f''(x) + 4f(x) = 0$.

14. Show that $f(x) = \tan x$ satisfies $f'(x) = 1 + [f(x)]^2$ and $f(x) = \cot x$ satisfies $f'(x) = -\{1 + [f(x)]^2\}$.

15. Where is $f(x) = 1 + 2 \sin x \cos x$ concave upward? Concave downward? Find its inflection points and sketch its graph.

16. Prove that for any positive numbers a, b, c,

$$\frac{a + b + c}{3} > \sqrt[3]{abc}$$

by (a) minimizing $f(x) = (a + b + x)/\sqrt[3]{abx}$ and (b) using algebra.

17. (Refer to Review Problem 23 of Chapter 1 and Fig. 1-R-1.) Let P, with coordinates (x_1, y_1), be inside the parabola $y = x^2$; that is, $y_1 > x_1^2$. Show that the path consisting of two straight lines joining P to a point (x, x^2) on the parabola and then (x, x^2) to $(0, \frac{1}{4})$ has minimum length when the first segment is vertical.

★18. Sketch the graphs of: (a) $(x - x^2)^{3/5}$; (b) $(\sin x)^{2/3}$; (c) $x^2(\cos x)^{2/3}$.

★19. Sketch the graphs of (a) $y = x(x - 1)^{3/2}$ and (b) $x(x - 1)^{2/3}$.

20. Find the following antiderivatives:
    (a) $\int \sin(u + 1) \, du$ (b) $\int 1/(4 + y^2) \, dy$
    (c) $\int (x^3 + 3 \sec^2 x) \, dx$
    (d) $\int [1/(s^2 + 1) - 2 \cos 5s] \, ds$
    (e) $\int [2/\sqrt{r^2(r^2 - 1)}] \, dr$

21. (a) Verify that $\int \sin^{-1} x \, dx = x \sin^{-1} x + \sqrt{1 - x^2} + C$.
    (b) Differentiate $f(x) = \cos^{-1} x + \sin^{-1} x$ to conclude that f is constant. What is that constant?
    (c) Find $\int \cos^{-1} x \, dx$.
    (d) Find $\int \sin^{-1} 3x \, dx$.

22. Find a formula for the tangent line to the following graphs at the indicated point.
    (a) $r = \cos 4\theta$; $\theta = \pi/4$. (b) $r = \theta^2$; $\theta = 1$.
    (c) $r = 1/(1 + \sin^2 \theta)$; $\theta = \pi/2$.

23. A man is driving on the freeway at 50 miles per hour. He sees the sign in Fig. 6-R-2. How far from the sign is θ maximum? How fast is θ changing at this time?

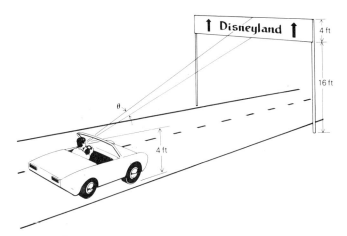

Fig. 6-R-2 At what distance is θ biggest?

★24. Consider water waves impinging on a breakwater which has two gaps as in Fig. 6-R-3. With the notation in the figure, analyze the maximum and minimum points for wave amplitude along the shore. The two wave forms emanating from P and Q can be described at any point R as $\alpha \cos (k\rho - \omega t)$, where ρ is the distance from the source P or Q; k, ω, α are constant. The net wave is described by their sum; the amplitudes do *not* add; ignore complications such as reflection of waves off the beach.*

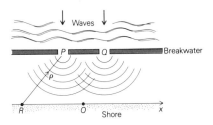

Fig. 6-R-3 Find the wave pattern on the shore.

We recommend the book Waves and Beaches by W. Bascom (Anchor Books, 1965) as a fascinating study of the mathematics, physics, engineering, and aesthetics of water waves.

★ 25. (a) When the sun rises at the equator on June 21, how fast is its angle of elevation changing? (b) Using the linear approximation, estimate how long it takes for the sun to rise 5° above the horizon.

★ 26. Let L be the latitude at which the sun has the highest noon elevation on day T. (a) Find a formula for L in terms of T. (b) Graph L as a function of T.

27. (a) Using a calculator, try to determine whether $\tan^{-1}[\tan(\pi + 10^{-20})]$ belongs to the interval $(0, 1)$. (b) Do part (a) without using a calculator. (c) Do some other calculator experiments with trigonometric functions. How else can you "fool" your calculator (or vice versa!)?

28. On a calculator, put any angle in radians on the display and successively press the buttons "sin" and "cos," alternately, until you see the numbers 0.76816 and 0.69481 appear on the display.

  (a) Try to explain this phenomenon from the graphs of $\sin x$ and $\cos x$, using composition of functions.

  (b) Can you guess the solutions x, y of the equations $\sin(\cos x) = x$, $\cos(\sin y) = y$?

  ★(c) Using the mean value and intermediate value theorems, show that the equations in (b) have exactly one solution.

29. The angle of deviation δ of a light ray entering a prism of Snell's index n and apex angle A is given by $\delta = \sin^{-1}(n \sin\rho) + \sin^{-1}(n\sin(A - \rho)) - A$ where ρ depends on the angle of incidence ϕ of the light ray. (See Fig. 6-R-4.)

  (a) Find $d\delta/d\rho$.

  (b) Show that $d\delta/d\rho = 0$ occurs for $\rho = A/2$. This is a minimum value for δ.

  (c) By Snell's law, $n = \sin\phi/\sin\rho$. Verify that $n = \sin[(A + \delta)/2]/\sin(A/2)$ when $\rho = A/2$.

Fig. 6-R-4 Light passing through a prism.

30. A pocket watch is swung counterclockwise on the end of its chain in a vertical circle. This is a circular motion, but not uniform, and the tension T in the chain is given by $T = m[(v^2/R) + g\cos\theta]$ where θ is the angle from the downward direction. Suppose the length is $R = 0.5$ meter, the watch mass is $m = 0.1$ kg, and the tangential velocity is $v = v(\theta)$.

  (a) At what points on the circular path do you expect $dv/d\theta = 0$?

  (b) Compute $dT/d\theta$ when $dv/d\theta = 0$.

  (c) If the speed v is low enough at the highest point on the path, then the chain will become slack. Find the critical speed v_c below which the chain becomes slack.

31. A wheel of unit radius rolls along the x axis, uniformly, rotating one half-turn per second. A point P on the circumference t seconds later has coordinates (x, y) given by $x = \pi t - \sin\pi t$, $y = 1 - \cos\pi t$.

  (a) Find the velocities $dx/dt, dy/dt$ and the accelerations $d^2x/dt^2, d^2y/dt^2$.

  (b) Find the *speed* $\sqrt{(dx/dt)^2 + (dy/dt)^2}$ when y is a maximum.

32. A planet travels around its sun on the polar path $r = 1/(2 + \cos\theta)$, the sun at the origin.

  (a) Verify that the path is an ellipse by changing to (x, y) coordinates.

  (b) Compute the *perihelion* distance (minimum distance from the sun to the planet).

33. Kepler's first law of planetary motion states that *the orbit of each planet is an ellipse with the sun as one focus.* The origin $(0, 0)$ is placed at the sun, and polar coordinates (r, θ) are introduced. The planet's motion is $r = r(t)$, $\theta = \theta(t)$, and these are related by $r(t) = l/[1 + e\cos\theta(t)]$, where $l = k^2/GM$ and $e^2 = 1 - (2k^2E/G^2M^2m)$; k is a constant, G is the universal gravitation constant, E is the energy of the system, M and m are the masses of the sun and planet, respectively.

  (a) Assume $e < 1$. Change to rectangular coordinates to verify that the planet's orbit is an ellipse.

  (b) Let $\mu = 1/r$. Verify the *energy equation*

  $$(d\mu/d\theta)^2 + \mu^2 = (2/k^2m)(GMm\mu - E)$$

34. The *astroid* $x^{2/3} + y^{2/3} = 1$ is a planar curve which admits no self-intersections, but it has four cusp points, and no tangent exists at these cusps.

    (a) Apply symmetry methods to graph the astroid.

    (b) Divide the astroid into two curves, each of which is a continuous function. Find equations for these functions.

    (c) Find the points where dy/dx is not defined and compare with the cusp points. Use implicit differentiation.

    (d) Explain why no tangent line exists at the cusp points.

7

Exponentials and Logarithms

The inverse of the exponential function is an antiderivative for $1/x$.

We have already defined the powers b^x of a positive number b if x is rational. We begin this chapter by defining b^x for all real x. The resulting function is written $\exp_b x = b^x$ and is called the exponential function with base b. The inverse of $\exp_b x$ is the logarithm function, denoted $\log_b x$. The calculus of these functions is simplest if we choose the special base $e = 2.7182818285 \ldots$. In fact, this choice is such that $(d/dx)e^x = e^x$, and no other b has this property. The exponential and logarithm functions arise in many important applications; we will study some of them in this chapter, and extensively in following chapters.

CONTENTS

The definitions and algebraic properties of the exponential and logarithm functions are laid out in Section 7-1. In Section 7-2, the derivatives of these functions are computed and the number e is introduced. Applications of the exponential and logarithm functions to problems such as graphing are scattered throughout the chapter. Further applications to problems involving growth and decay are given in Section 9-1.

This appendix is found in the Student Guide.

7-1 Powers and Logarithms

Any real number can be used as an exponent if the base is positive.

This section begins with a review of the laws of exponents for powers b^r, where $b > 0$ and r is rational. Since calculus deals with functions of a *real* variable, it is important to define b^r even when r is irrational. We do this by a comparison method and then go on to consider the exponential functions $\exp_b x = b^x$ and their inverses, the logarithm functions.

Goals

After studying this section, you should be able to:

Simplify exponential expressions by using the laws of exponents.

Evaluate logarithms.

Sketch graphs of exponential and logarithm functions.

THE LAWS OF EXPONENTS

The expression b^n, where b is a real number called the *base* and n is a natural number called the *exponent*, is defined as the product of b with itself n times:

$$b^n = b \cdot b \cdot \cdots \cdot b \quad (n \text{ times})$$

This operation of raising a number to a power, or *exponentiation*, has the following properties, called *laws of exponents:*

$$b^n b^m = b^{n+m} \tag{1}$$

$$(b^n)^m = b^{nm} \tag{2}$$

$$(bc)^n = b^n c^n \tag{3}$$

The first of the three laws of exponents is particularly important; it is the basis for extending the operation of exponentiation to allow rational exponents. If b^0 were defined, we ought to have $b^0 b^n = b^{0+n} = b^n$. If $b \neq 0$, then $b^n \neq 0$, and the equation $b^0 b^n = b^n$ implies that b^0 must be 1. We take this as the *definition* of b^0, noting that 0^0 is not defined (see Problem 3).

If n is a natural number, then b^{-n} is defined in order to make $b^{-n} b^n = b^{-n+n} = b^0 = 1$; that is, $b^{-n} = 1/b^n$. To define $b^{1/n}$, we observe that $b^{1/n} \cdot b^{1/n} \cdot \cdots \cdot b^{1/n} = b^{1/n + \ldots + 1/n} = b^1 = b$ ought to hold—that is, $(b^{1/n})^n$ ought to be equal to b. We saw in Section 5-1 that if $b > 0$, then there is a unique positive real number $\sqrt[n]{b}$ whose nth power is b, so we set $b^{1/n} = \sqrt[n]{b}$. (If n is odd, then $\sqrt[n]{b}$ may be defined even if b is negative, but we will reserve the notation $b^{1/n}$ for the case $b > 0$.)

Finally, if $r = m/n$ is a rational number, we define $b^r = b^{m/n} = (b^m)^{1/n}$. We leave it to you (see Exercise 4) to verify that the result is independent of the way in which r is expressed as a quotient of positive integers; for instance, $(b^4)^{1/6} = (b^6)^{1/9}$.

Having defined b^r for $b > 0$ and r rational, one can go back and prove the laws of exponents for this general case. These laws are useful for calculations with rational exponents.

Worked Example 1   Simplify $[x^{2/3}(x^{-3/2})]^{8/3}$.

Solution $(x^{2/3}x^{-3/2})^{8/3} = (x^{2/3-3/2})^{8/3} = (x^{-5/6})^{8/3} = x^{-20/9} = 1/\sqrt[9]{x^{20}}$.

To define exponential expressions with irrational powers, it will be useful for us to understand the behavior of exponentiation with respect to order. The following example contains the fact we will need.

Worked Example 2   Let $b > 1$ and p and q be rational numbers with $p < q$. Prove that $b^p < b^q$.

Solution
(optional)   By the laws of exponents, $b^q/b^p = b^{q-p}$. Let $z = q - p$; it is positive since $p < q$. We will show that $b^z > 1$, so $b^q/b^p > 1$ and thus $b^q > b^p$. Since z is rational, $z = m/n$, where m and n are positive integers. Then $b^z = (b^m)^{1/n}$. However, $b^m = b \cdot b \cdot \cdots \cdot b$ (m times) > 1 since $b > 1$, and $(b^m)^{1/n} > 1$ since $b^m > 1$. (The nth root $c^{1/n}$ of a number $c > 1$ is > 1 since, if $c^{1/n} \le 1$, $(c^{1/n})^n = c$ would also be ≤ 1.) Thus $b^z > 1$ if $z > 0$, and the solution is complete.

RATIONAL POWERS

Rational powers are defined by:

$b^n = b \cdot \cdots \cdot b$ (n times)
$b^{-n} = 1/b^n$
$b^{1/n} = \sqrt[n]{b}$ if $b > 0$ and n is a natural number
$b^{m/n} = (b^m)^{1/n}$

If $b, c > 0$ and p, q are rational, then:

$b^{p+q} = b^p b^q$

$b^{pq} = (b^p)^q$

$(bc)^p = b^p c^p$

$b^p < b^q$ if $b > 1$ and $p < q$

Solved Exercises*   1. We defined $b^{m/n}$ as $(b^m)^{1/n}$. Show that $b^{m/n} = (b^{1/n})^m$ as well.

2. Find $8^{-2/3}$ and $9^{3/2}$.

3. Simplify $(x^{2/3})^{5/2}/x^{1/4}$.

★4. How are b^p and b^q related if $0 < b < 1$ and $p < q$?

Exercises   1. Simplify by writing with rational exponents:

(a) $\left[\dfrac{\sqrt[4]{ab^3}}{\sqrt{b}}\right]^6$ (b) $\sqrt[3]{\dfrac{\sqrt{a^3 b^9}}{\sqrt[4]{a^6 b^6}}}$

2. Solve for x:
(a) $10^x = 0.001$ (b) $5^x = 1$
(c) $2^x = 0$ (d) $x - 2\sqrt{x} - 3 = 0$ (factor)

---

*Solutions appear in Appendix A.

3. Using rational exponents and the laws of exponents, verify the root formulas:

(a) $\sqrt[a]{\sqrt[b]{x}} = \sqrt[ab]{x}$ (b) $\sqrt[ac]{x^{ab}} = \sqrt[c]{x^b}$

★4. Suppose that $b > 0$ and that $p = m/n = m'/n'$. Show, using the definition of rational powers, that $b^{m/n} = b^{m'/n'}$; that is, b^p is unambiguously defined. [*Hint:* Raise both $b^{m/n}$ and $b^{m'/n'}$ to the power nn'.]

REAL EXPONENTS

If in the expression b^r we consider r as constant and b as variable, we obtain the *power function* $g(x) = x^r$ which was studied in Section 5-1. But we can also consider b as fixed and r as variable. This gives the function $f(x) = b^x$, whose domain consists of all rational numbers. The following example shows how such *exponential functions* occur naturally and suggests why we would like to have them defined for all real x.

Worked Example 3   The mass of a bacterial colony doubles after every hour. By what factor does the mass grow after: (a) 5 hours (b) 20 minutes; (c) $2\frac{1}{2}$ hours; (d) x hours, if x is rational?

Solution   (a) In 5 hours, the colony doubles five times, so it grows by a factor of $2 \cdot 2 \cdot 2 \cdot 2 \cdot 2 = 2^5 = 32$.

(b) If the colony grows by a factor of k in $\frac{1}{3}$ hour, it grows by a factor $k \cdot k \cdot k = k^3$ in 1 hour. Thus $k^3 = 2$, so $k = 2^{1/3} = \sqrt[3]{2} \approx 1.26$.

(c) In $\frac{1}{2}$ hour, the colony grows by a factor of $2^{1/2}$, so it grows by a factor of $(2^{1/2})^5 = 2^{5/2} = 5.66$ in $2\frac{1}{2}$ hours.

(d) Reasoning as in parts (a), (b), and (c) leads to the conclusion that the mass of the colony grows by a factor of 2^x in x hours.

Time is not limited to rational values; we should be able to ask how much the colony in Worked Example 3 grows after $\sqrt{3}$ hours or π hours. Let $f(x)$ be the growth factor after x hours for any real x; since the colony is increasing in size, we are led to the following mathematical problem: Find a function f defined for all real x such that f is increasing and $f(x) = 2^x$ for all rational x.

Computing some values of 2^x and plotting, we obtain the graph shown in Fig. 7-1-1. By doing more computations, we can fill in more points between those in Fig. 7-1-1, and the graph looks more and more like a smooth curve. The following theorem shows that a curve can be drawn through all these points.

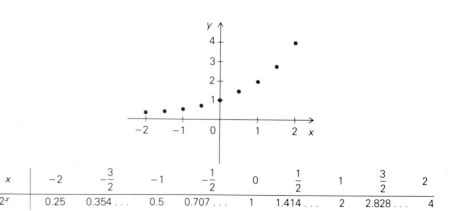

Fig. 7-1-1 Some points on the graph $y = 2^x$ for rational x.

x	-2	$-\frac{3}{2}$	-1	$-\frac{1}{2}$	0	$\frac{1}{2}$	1	$\frac{3}{2}$	2
2^x	0.25	$0.354\ldots$	0.5	$0.707\ldots$	1	$1.414\ldots$	2	$2.828\ldots$	4

Theorem 1 *Given any number $b > 0$, there is a unique function $f(x)$ defined for all real x such that:*

    1. f is increasing if $b > 1$, constant if $b = 1$, and decreasing if $0 < b < 1$.

    2. $f(x) = b^x$ for all rational x.

For $b > 1$, Worked Example 2 says b^x is increasing for rational x; Theorem 1 extends this to real x. Given Theorem 1, we can make the following definition.

Definition *If $b > 0$ and x is any real number, b^x is defined to be $f(x)$, where f is the function in Theorem 1.*

Some properties of exponentiation are summarized in the next theorem.

Theorem 2   *(a) For any $b > 0$, $f(x) = b^x$ is a continuous function.*

    (b) Let b, c, x, and y be real numbers with $b > 0$ and $c > 0$. Then:

        1. $b^{x+y} = b^x b^y$

        2. $b^{xy} = (b^x)^y$

        3. $(bc)^x = b^x c^x$

Set of numbers b^r for rational $r < x$

Fig. 7-1-2 The number b^r lies just above the set of b^r for r rational, $r < x$.

Detailed proofs of Theorems 1 and 2 are given in the appendix to this chapter; the general idea behind the definition of b^x is the following. For irrational x, the number $f(x)$ must satisfy the condition $b^r < f(x)$ whenever $r < x$ is rational, since $f(x)$ is to be an increasing function and $f(r) = b^r$. We define $f(x)$ to be the *least* number which is greater than b^r for all rational $r < x$ (see Fig. 7-1-2). Proving that this gives a function satisfying the conditions desired is a technical job that is carried out in the appendix to this chapter (see the Student Guide).

Calculator Discussion

When we compute $2^{\sqrt{3}}$ on a calculator, we are implicitly using the continuity of $f(x) = 2^x$. The calculator in fact computes a rational power of 2—namely, $2^{1.732050808}$, where 1.732050808 is a decimal approximation to $\sqrt{3}$. Continuity of $f(x)$ says precisely that if the decimal approximation to x is good, then the answer is a good approximation to $f(x)$. The fact that f is increasing gives more information. For example, since

$$\frac{1732}{1000} < \sqrt{3} < \frac{17{,}321}{10{,}000}$$

we can be sure that

$$2^{1732/1000} = 3.32188 \cdots < 2^{\sqrt{3}} < 2^{17{,}321/10{,}000} = 3.32211 \cdots$$

so $2^{\sqrt{3}} = 3.322$ is correct to three decimal places.

Worked Example 4   Simplify $(\sqrt{(3^{\pi})})(3^{-\pi/4})$.

Solution $\sqrt{3^{\pi}}\, 3^{-\pi/4} = (3^{\pi})^{1/2}\, 3^{-\pi/4} = 3^{\pi/2 - \pi/4} = 3^{\pi/4}$.

Sometimes the notation $\exp_b x$ is used for b^x, exp standing for "exponential." One reason for this is typographical: an expression like $\exp_b(x^2/2 + 3x)$ is easier on the eyes and on the printer than $b^{(x^2/2+3x)}$. Another reason is mathematical: when we write $\exp_b x$, we indicate that we are thinking of b^x as a *function of x*.

Worked Example 5   Sketch the graphs of $\exp_2$, $\exp_{3/2}$, $\exp_1$, $\exp_{2/3}$, and $\exp_{1/2}$.

Solution   From Theorem 1, $\exp_2$ and $\exp_{3/2}$ are increasing, and $\exp_{2/3}$ and $\exp_{1/2}$ are decreasing. Using these facts and a few plotted points, we sketch the graphs in Fig. 7-1-3.

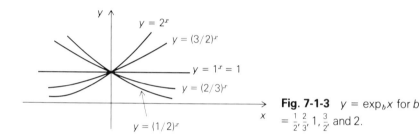

Fig. 7-1-3 $y = \exp_b x$ for $b = \frac{1}{2}, \frac{2}{3}, 1, \frac{3}{2}$, and 2.

Worked Example 6   A curve whose equation in polar coordinates has the form $r = b^{\theta}$ for some b is called an *exponential spiral*. Sketch the exponential spiral for $b = 1.1$.

Solution   We observe that $r \to \infty$ as $\theta \to \infty$ and that $r \to 0$ as $\theta \to -\infty$. To graph the spiral, we note that r increases with θ; we then plot several points (using a calculator) and connect them with a smooth curve. (See Fig. 7-1-4.) Every turn of the spiral is $(1.1)^{2\pi} \approx 1.82$ times as big as the previous one.

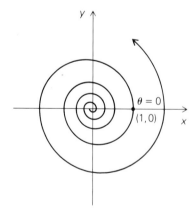

Fig. 7-1-4   The exponential spiral $r = (1.1)^{\theta}$.

Solved Exercises

5. How is the graph of $\exp_{1/b} x$ related to that of $\exp_b x$?

6. Simplify: $(2^{\sqrt{3}} + 2^{-\sqrt{3}})(2^{\sqrt{3}} - 2^{-\sqrt{3}})$.

7. Match the graphs and functions in Fig. 7-1-5.

(a)   (b)   (c)   (d)

Fig. 7-1-5 Match the graphs and functions: (A) $y = x^{\sqrt{3}}$; (B) $y = x^{1/\sqrt{3}}$; (C) $y = (\sqrt{3})^x$; (D) $y = (1/\sqrt{3})^x$.

Exercises   5. Simplify: $[(\sqrt{3})^\pi - (\sqrt{2})^{\sqrt{5}}]/[\sqrt[4]{3^\pi} + 2^{\sqrt{5}/4}]$.

6. Graph $y = 3^{x+2}$ by "shifting" the graph of $y = 3^x$ by 2 units to the left. Graph $y = 9(3^x)$ by "stretching" the graph of $y = 3^x$ by a factor of 9 in the direction of the y axis. Compare the two results. In general, how does shifting the graph of $y = 3^x$ by k units to the left compare with stretching the graph by a factor of 3^k in the direction of the y axis?

7. Carefully graph the following functions on one set of axes:

   (a) $f(x) = 2^x$ (b) $g(x) = x^2 + 1$ (c) $h(x) = x + 1$

   Can you see why $f'(x)$ should be between 1 and 2?

8. From the graph of $f(x) = 2^x$, predict the domain and range of $f'(x)$ and make a reasonable sketch of what the function $f'(x)$ might look like.

LOGARITHMS

If $b > 1$, the function $\exp_b x = b^x$ is increasing and continuous. The range of $\exp_b$ is $(0, \infty)$ since for x large positive, b^x is large positive (see Problem 9), and for x large negative, b^x is small but positive. It follows from Theorem 1 in Section 5-1 that $\exp_b$ has a unique inverse function with domain $(0, \infty)$ and range $(-\infty, \infty)$. This function is called $\log_b$. By the definition of an inverse function, $\log_b y$ is that number x such that $b^x = y$. The number b is called the *base* of the logarithm.

Worked Example 7   Find $\log_3 9$, $\log_{10} 10^a$, and $\log_9 3$.

Solution   Let $x = \log_3 9$. Then $3^x = 9$. Since $3^2 = 9$, x must be 2. Similarly, $\log_{10} 10^a$ is a and $\log_9 3 = \frac{1}{2}$ since $9^{1/2} = 3$.

The graph of $\log_b x$ for $b > 1$ is sketched in Fig. 7-1-6 and is obtained by flipping over the graph of $\exp_b x$ along the diagonal $y = x$. As usual

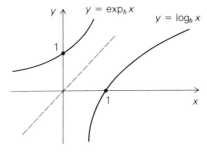

Fig. 7-1-6 The graphs of $y = \exp_b x$ and $y = \log_b x$ for $b > 1$.

with inverse functions, the label y in $\log_b y$ was only temporary to stress the fact that $\log_b y$ is the inverse of $y = \exp_b x$. From now on we will usually use the variable names x or t and write $\log_b x$ or $\log_b t$.

Worked Example 8   Compare the graphs of $\log_2 x$ and $\log_{1/2} x$.

Solution   This is done by flipping the graphs of 2^x and $(\frac{1}{2})^x$, as shown in Fig. 7-1-7. The graphs of $\log_2 x$ and $\log_{1/2} x$ are reflections of one another in the x axis.

Fig. 7-1-7 Exponential and logarithm functions with base $= 2 > 1$ and base $= \frac{1}{2} < 1$.

Notice that for $b > 1$, $\log_b x$ is increasing. If $b < 1$, $\exp_b x$ is decreasing and so is $\log_b x$. However, while $\exp_b x$ is always positive, $\log_b x$ can be either positive or negative. Since $\exp_b 0 = 1$, we can conclude that $\log_b 1 = 0$; since $\exp_b 1 = b$, $\log_b b = 1$. These properties are summarized in the following display.

PROPERTIES OF $\log_b x$

0. $\log_b x$ is that number y such that $b^y = x$

1. $\log_b x$ is defined for $x > 0$ and $b > 0$ (but $\log_b x$ can be positive or negative)

2. $\log_b 1 = 0$.

3. If $b < 1$, $\log_b x$ is decreasing; if $b > 1$, $\log_b x$ is increasing.

From the laws of exponents (p. 317) we can read off corresponding laws for $\log_b x$.

LAWS OF LOGARITHMS

1. $\log_b (xy) = \log_b x + \log_b y$, and $\log_b (x/y) = \log_b x - \log_b y$
2. $\log_b (x^y) = y \log_b x$
3. $\log_b x = (\log_b c)(\log_c x)$

To prove law 1, for instance, we remember that $\log_b x$ is the number such that $\exp_b (\log_b x) = x$. So we must check that

$$\exp_b (\log_b x + \log_b y) = \exp_b (\log_b xy)$$

But the left side is $\exp_b (\log_b x) \, \exp_b (\log_b y) = xy$, as is the right side. The other laws are proved the same way (see Exercise 13).

Worked Example 9   What is the relationship between $\log_b c$ and $\log_c b$?

Solution   Substituting b for x in law 3, we get

$$\log_b b = (\log_b c)(\log_c b)$$

But $\log_b b = 1$, so $\log_b c = 1/\log_c b$.

Worked Example 10   Evaluate $\log_{10}[(100^{3.2})\sqrt{10}]$ by hand.

Solution
$$\begin{aligned}
\log_{10}(100^{3.2}\sqrt{10}) &= \log_{10}100^{3.2} + \log_{10}\sqrt{10} && \text{(law 1)} \\
&= 3.2\log_{10}100 + \tfrac{1}{2}\log_{10}10 && \text{(law 2)} \\
&= 3.2\log_{10}10^2 + \tfrac{1}{2} \\
&= 6.4 + \tfrac{1}{2} = 6.9
\end{aligned}$$

We conclude this section with a word problem involving exponentials and logarithms.

Worked Example 11   The number N of people who contract influenza t days after a group of 1000 people are put in contact with a single person with influenza can be modeled by $N = 1000/(1 + 999 \cdot 10^{-.17t})$.

(a) How many people contract influenza after 20 days?

(b) Will everyone eventually contract the disease?

(c) In how many days will 600 people contract the disease?

Solution   (a) According to the given model, we substitute $t = 20$ into the formula to give

$$N = \frac{1000}{1 + 999 \cdot 10^{-0.17 \cdot 20}} = \frac{1000}{1 + 999 \cdot 10^{-3.4}} = \frac{1000}{1.398} \approx 715$$

Thus 715 people will contract the disease after 20 days. (The calculation was done on a calculator.)

(b) "Eventually" is interpreted to mean "t very large." For t large $-.17t$ will be a large negative number and so $10^{-.17t}$ will be nearly zero (equivalently $10^{-.17t} = 1/10^{.17t}$ and $10^{.17t}$ will be very large if t is very large). Thus the denominator in N will be nearly 1 and so N itself is nearly 1000. Thus, we say that "yes, everyone of the 1000 will eventually contract the disease."

(c) We must find the t for which $N = 600$:

$$600 = \frac{1000}{1 + 999 \cdot 10^{-.17t}} \quad \text{so} \quad (600)(1 + 999 \cdot 10^{-.17t}) = 1000$$

Thus $1 + 999 \cdot 10^{-.17t} = 10/6 = 5/3$. Solving for $10^{-.17t}$, $10^{-.17t} = (2/3)999$. Therefore, $-.17t = \log_{10}((2/3) \cdot 999) \approx -3.176$ (from our calculator) and so $t = 3.176/0.17 \approx 18.68$ days.

Solved Exercises　　8. Match the graphs and functions in Fig. 7-1-8.

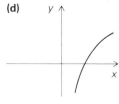

Fig. 7-1-8 Match the graphs and functions: (A) $y = 2^x$; (B) $y = \log_2 x$; (C) $y = \log_{1/2} x$; (D) $y = (\frac{1}{2})^x$.

9. Find $\log_2 4$, $\log_3 81$, and $\log_{10} 0.01$.

10. (a) Simplify: $\log_b (b^{2x}/2b)$. (b) Solve for x: $\log_2 x = \log_2 5 + 3 \log_2 3$.

Exercises　　9. Find $\log_3 3$, $\log_5 125$, and $\log_{1/2} 2$.

10. The graph of $y = \log_b x$ contains the point $(3, \frac{1}{3})$. What is b?

11. Given that $\log_7 2 \approx 0.356$, $\log_7 3 \approx 0.565$, and $\log_7 5 \approx 0.827$, find by hand:

(a) $\log_7 (7.5)$　　　(b) $\log_5 6$　　　(c) $\log_2 \sqrt{5}$　　　(d) $\log_{0.4} 0.6$

12. Solve for x:

(a) $\log_x 5 = 0$　　　　　　　　　(b) $\log_2 (x^2) = 4$

(c) $2 \log_3 x + \log_3 4 = 2$

13. Use the definition of $\log_b x$ to prove:

(a) $\log_b (x^y) = y \log_b x$　　　　(b) $\log_b x = \log_b (c) \log_c (x)$

Problems for Section 7-1

1. Simplify:

(a) $\log_2 (2^8/8^2)$　　　　　　(b) $2^{5/3}/2^{3/5}$

(c) $3^{-8/11} \cdot (1/9)^{-4/11}$　　　(d) $\log_{10} (1000)$

(e) $\log_{10} (0.001)$

2. Simplify:

(a) $2^{\log_2 4}$　　　　　　(b) $2^{\log_2 b}$

(c) $\log_2 (2^b)$　　　　　(d) $\log_2 (3^5 \cdot 4^{-6} \cdot 9^{-5/2})$

(e) $\log_b [b^2 \cdot (2b)^3 \cdot (8b)^{-2/3}]$

★3. Since $0^x = 0$ for any positive rational x, 0^0 ought to be zero. On the other hand, $b^0 = 1$ for any $b > 0$, so 0^0 ought to be 1. Are *both* choices consistent with the laws of exponents?

4. A biologist measures culture growth and gets the following data: After 1 day of growth the count is 1750 cells. After 2 days it is 3065 cells. After 4 days it is 9380 cells. Finish filling out the following table by using a table or calculator:

x = number of days of growth	1	2	4
n = number of cells	1750	3065	9380
$y = \log_{10} n$			

Verify that the data fit a curve of the form $n = Mb^x$ by examining the linear equation $y = (\log_{10} b)x + \log_{10} M$ (with respect to the y and x values in the table). Use the slope and y intercept to evaluate M and b. If the biologist counts the culture on the fifth day, predict how many cells will be found.

5. In Problem 4, suppose that you had originally known that the data would satisfy a relation of the form $n = Mb^x$. Solve for M and b without using logarithms.

6. Suppose $\log_b 10 = 2.5$. Use a $\log_{10}$ table or a calculator to find an approximate value for b.

7. Verify the formula $\log_{a^n} x = (1/n) \log_a x$. What restrictions must you make on a?

8. Solve for x: $\log_{25}(x + 1) = \log_5 x$.

★9. Show that if $b > 1$ and n is a positive integer, then

$$b^n \geq 1 + n(b - 1)$$

and

$$b^{-n} \leq \frac{1}{1 + n(b - 1)}$$

[*Hint:* Write $b^n = [1 + (b - 1)]^n$ and expand.]

10. Solve for x: $\log_x(1 - x) = 2$. Check to make sure that your answer(s) make sense.

11. Give the domain and range of the following functions:
    (a) $f(x) = \log_{10}(x^2 - 2x - 3)$
    (b) $g(x) = \log_2 [(2x + 1)/2]$
    (c) $h(x) = \log_{10}(1 - x^2)$

12. Which is larger, $\log_{11} 2$ or $\log_{10} 2$? How about $\log_{1/2} 2$ or $\log_{1/4} 2$?

13. Let $f(x) = \log_2(x - 1)$. Find a formula for the inverse function g of f. What is its domain?

14. Is the logarithm of an irrational number ever rational? If so, find an example. [*Hint:* Is 2^x ever irrational for some rational x?]

15. Give the domains and ranges of the following functions and graph them:
    (a) $y = 2^{x^2}$ (b) $y = 2^{\sqrt{x}}$ (c) $y = 2^{1/x}$

16. Write each of the following as sums of (rational) multiples of $\log_b A$, $\log_b B$, and $\log_b C$:
    (a) $\log_b(A^2 B/C)$ (b) $\log_b(\sqrt{A B^3}/C^4 B^2)$
    (c) $2 \log_b(A\sqrt{1 + B}/C^{1/3} B)$
       $- \log_b [(B + 1)/AC]$

17. Factor the following using fractional exponents. For example, $x + 2\sqrt{2xy} + 2y = (x^{1/2} + (2y)^{1/2})^2$:
    (a) $x - \sqrt{xy} - 2y$ (b) $x - y$
    (c) $\sqrt[3]{xy^2} + \sqrt[3]{yx^2} + x + y$
    (d) $x - 2\sqrt{x} - 8$ (e) $x + 2\sqrt{3x} + 3$

18. Graph and compare the following functions: $f(x) = 2 \log_2 x$; $g(x) = \log_2(x^2)$; $h(x) = 2 \log_2 |x|$. Which (if any) are the same?

19. Give the domain of the following functions. Which (if any) are the same?
    (a) $f(x) = \log_{10}\left[\dfrac{(1 - x^2)^4}{\sqrt{(x + 5)/(x^2 + 1)}}\right]$
    (b) $g(x) = 4 \log_{10}(1 - x) + 4 \log_{10}(1 + x) + \frac{1}{2} \log_{10}(x^2 + 1) - \frac{1}{2} \log_{10}(x + 5)$
    (c) $h(x) = 4 \log_{10}|1 - x| + 4 \log_{10}|1 + x| + \frac{1}{2} \log_{10}(x^2 + 1) - \frac{1}{2} \log_{10}(x + 5)$

20. Graph the exponential spiral $r = (1/1.1)^\theta$.

21. What do you see if you rotate an exponential spiral about the origin at a uniform rate? Compare with the spiral $r = \theta$.

22. A lender supplies an amount P to a borrower at an annual interest rate of r. After t years with interest compounded n times a year, the borrower will owe the lender the amount $A = P[1 + (r/n)]^{nt}$ (compound interest). Suppose $P = 100$, $r = 0.06$, $t = 2$ years. Find the amount owed for interest compounded: (a) monthly, (b) weekly, (c) daily, (d) twice daily. Draw a conclusion.

23. Color analyzers are constructed from photomultipliers and various electronic parts to give a scale reading of light intensities falling on a light probe. These scales read relative densities directly, and the scale reading S can be given by $S = k \log_{10}(I/I')$ where I is a reference intensity, I' is the new intensity, and k is a positive constant.
    (a) Show that the scale reads zero when $I = I'$.
    (b) Assume the needle is vertical on the scale when $I = I'$. Find the sign of S when $I' = 2I$ and $I' = I/2$.
    (c) In most photographic applications, the range of usable values of I' is $I/8 \leq I' \leq 8I$. What is the scale range?

24. The *opacity* of a photographic negative is the ratio I_0/I where I_0 is the reference light intensity and I the transmitted intensity (through the negative). The *density* of a negative is the quantity $D = \log_{10}(I_0/I)$. Find the density for opacities of $2, 4, 8, 10, 100, 1000$.

25. The loudness, in decibels (dB), of a sound of intensity I is $L = 10 \log_{10}(I/I_0)$ where I_0 is the threshold intensity for human hearing.
    (a) Conversations have intensity $(1,000,000)I_0$. Find the dB level.
    (b) An increase of 10 dB doubles the loudness of a particular sound. What is the effect of this increase on the intensity I?
    (c) A jet airliner on takeoff has sound intensity $10^{12} I_0$. Levels above 90 dB are considered dangerous to the ears. Is this level dangerous?

26. The Richter scale for earthquake magnitude uses the formula $R = \log_{10}(I/I_0)$, where I_0 is a minimum intensity and I is the earthquake intensity. Compare the Richter scale magnitudes of the 1906 earthquake in San Francisco, $I = 10^{8.25} I_0$, and the 1971 earthquake in Los Angeles, $I = 10^{6.7} I_0$.

27. The pH value of a substance is determined by the concentration $[H^+]$ of the hydrogen ions in the substance in moles per liter, via the formula $pH = -\log_{10}[H^+]$. The pH of distilled water is 7; acids have pH < 7; bases have pH > 7.

    (a) Tomatoes have $[H^+] = (6.3) \cdot 10^{-5}$. Are tomatoes acid?

    (b) Milk has $[H^+] = 4 \cdot 10^{-7}$. Is milk acid?

    (c) Find the hydrogen ion concentration of a skin cleanser of rated pH value 5.5.

28. Current audio industry standards require that the input RF signal strength at a tuner's antenna terminals be specified in "dBf" rather than the older "microvolts." The formula is

$$dBf = 10 \log_{10} \left[\frac{(\text{voltage})^2}{(\text{resistance})(\text{reference power})} \right]$$

The "reference power" is 10^{-15} watts, called a "femtowatt," which explains the symbol "f" in "dBf"; the voltage is measured in volts and the resistance in ohms.

    (a) Compute the dBf rating for an input signal of 5 microvolts (0.000005 volt) for a tuner of resistance 300 ohms.

    (b) How do you convert "dBf" to microvolts?

7-2 Differentiation of the Exponential and Logarithm Functions

When the number e is used as a base, the differentiation rules for the exponential and logarithm functions become particularly simple.

Since we have now defined b^x for all real x, we can attempt to differentiate with respect to x. The result is that $\exp_b$ reproduces itself up to a constant multiple when differentiated. Choosing b properly, we can make the constant equal to 1. The derivative of the corresponding logarithm function turns out to be simply $1/x$.

Goals

After studying this section, you should be able to:

Find derivatives and antiderivatives of expressions involving exponential and logarithm functions.

Express the number e as a limit.

THE DERIVATIVE OF THE EXPONENTIAL FUNCTION

Consider the function $f(x) = \exp_b(x) = b^x$ defined in the previous section. If we assume that f is differentiable at zero, we can calculate $f'(x)$ for all x, just as we did with the trigonometric functions (see Section 6-1).

The laws of exponents allow us to write

$$f(x) = b^x = b^{x-x_0} \cdot b^{x_0} = f(x - x_0) b^{x_0}$$

Differentiating by the chain rule (or shifting rule) and the constant multiple rule gives

$$f'(x) = f'(x - x_0) \cdot b^{x_0}$$

Now set $x = x_0$:

$$f'(x_0) = f'(0) b^{x_0} = f'(0) f(x_0)$$

This calculation shows that if we assume only that f is differentiable at zero, then f is differentiable at all points x_0.

In fact, $f'(0)$ does exist; the proof is rather technical and is found in the appendix to this chapter.* Thus we have the following theorem.

Theorem 3 *If $b > 0$, then $\exp_b(x) = b^x$ is differentiable and*

$$\exp_b'(x) = \exp_b'(0) \exp_b(x)$$

That is,

$$\frac{d}{dx} b^x = \exp_b'(0) b^x$$

Notice that when we differentiate an exponential function, we reproduce it, multiplied by a constant. If $b \neq 1$, then $\exp_b'(0) \neq 0$, for otherwise $\exp_b'(x)$ would be zero for all x, and $\exp_b$ would be constant (the fact that a function whose derivative is everywhere zero must be constant was proved in Section 3-3).

That the rate of growth of b^x is proportional to the value of b^x is a fact familiar to bankers and breeders. (See Section 9-1.)

Worked Example 1 Let $f(t) = 3^t$. How much faster is f increasing at $t = 5$ than at $t = 0$?

Solution By Theorem 3,

$$f'(5) = f'(0) f(5) = f'(0) \cdot 3^5$$

Thus at $t = 5$, f is increasing $3^5 = 243$ times as fast as at $t = 0$.

To use Theorem 3 effectively, we still need to find $\exp_b'(0)$ and see how it depends upon b. It would be nice to be able to adjust b so that $\exp_b'(0) = 1$, for then we would have simply $\exp_b'(x) = \exp_b(x)$. This can in fact be done.

(optional)   Let us start with the base 10 of common logarithms and try to find another base b for which $\exp_b'(0) = 1$. By definition of the logarithm,

$$b = 10^{\log_{10} b} \quad \text{(see Property 0, p. 320)}$$

Therefore, $b^x = (10^{\log_{10} b})^x = 10^{x \log_{10} b}$ and hence, $\exp_b(x) = \exp_{10}(x \log_{10} b)$.

---

*This appendix is found in the Student Guide.

(optional) Differentiate by using the chain rule:

$$\exp_b'(x) = [\exp_{10}'(x \log_{10} b)] \cdot \log_{10} b$$

Set $x = 0$: $\exp_b'(0) = \exp_{10}'(0) \cdot \log_{10} b$. If we pick b so that

$$\exp_{10}'(0) \cdot \log_{10} b = 1 \qquad (1)$$

then we will have $\exp_b'(x) = \exp_b(x)$, as desired. Solving (1) for b, we have

$$\log_{10} b = \frac{1}{\exp_{10}'(0)}$$

That is,

$$b = \exp_{10}\left[\frac{1}{\exp_{10}'(0)}\right]$$

We denote the number $\exp_{10}[1/\exp_{10}'(0)]$ by the letter e. Thus,

$$\exp_e'(x) = \exp_e(x)$$

Although we started with the arbitrary choice of 10 as a base, it is easy to show (see Solved Exercise 2) that any initial choice of base leads to the same value for e.

Since the base e is so special, we write $\exp(x)$ for $\exp_e(x) = e^x$. We shall see later that the numerical value of e is approximately 2.718.

THE NUMBER e

The number e is chosen so that

$$\exp_e'(0) = 1, \text{ that is, so that } \frac{d}{dx} e^x = e^x$$

Logarithms to the base e are called *natural logarithms*. We denote $\log_e x$ by $\ln x$. (The notation $\log x$ is generally used in calculus books for the common logarithm $\log_{10} x$.) Since $e^1 = e$, we have the important formula $\ln e = 1$.

NATURAL LOGS AND e

$\ln x$ means $\log_e x$ (natural logarithm)

$\log x$ means $\log_{10} x$ (common logarithm)

$\exp x$ means e^x

Worked Example 2 Simplify $\ln(e^5) + \ln(e^{-3})$.

Solution By the laws of logarithms (p. 320), $\ln(e^5) + \ln(e^{-3}) = \ln(e^5 \cdot e^{-3}) = \ln(e^2) = 2$.

We can now complete our differentiation formula for the general exponential function $\exp_b x$. Since $b = e^{\ln b}$, we have $b^x = e^{x \ln b}$. Using the chain rule, we find

$$\frac{d}{dx} b^x = \frac{d}{dx} e^{x \ln b} = e^{x \ln b} \frac{d}{dx}(x \ln b) = e^{x \ln b} \ln b = b^x \ln b$$

Thus the mysterious factor $\exp_b'(0)$ turns out to be just the natural logarithm of b.

Worked Example 3 Differentiate: (a) $f(x) = e^{3x}$; (b) $g(x) = 3^x$.

Solution (a) Let $u = 3x$ so $e^{3x} = e^u$ and use the chain rule:

$$\frac{d}{dx} e^u = \left(\frac{d}{du} e^u \right) \frac{du}{dx} = e^u \cdot 3 = 3e^{3x}$$

(b) $\dfrac{d}{dx} 3^x = 3^x \ln 3.$

This expression cannot be simplified further; one can find the value $\ln 3 \approx 1.0986$ in a table or with a calculator.

DIFFERENTIATION OF THE EXPONENTIAL

$$\frac{de^x}{dx} = e^x$$

$$\frac{db^x}{dx} = (\ln b) b^x$$

Solved Exercises

1. Differentiate the following functions:

   (a) xe^{3x} (b) $\exp(x^2 + 2x)$ (c) x^2

   (d) $e^{\sqrt{x}}$ (e) $e^{\sin x}$ (f) $2^{\sin x}$

2. Show that, for any base b, $\exp_b \left[\dfrac{1}{\exp_b'(0)} \right] = e.$

Exercises

1. Simplify the following expressions:

   (a) $\ln(e^{x+1}) + \ln(e^2)$ (b) $\ln(e^{\sin x}) - \ln(e^{\cos x})$

2. Differentiate the following functions:

   (a) e^{x^2+1} (b) $\sin(e^x)$ (c) $3^x - 2^{x-1}$

   (d) $e^{\cos x}$ (e) $\tan(3^{2x})$ (f) $e^{1-x^2} + x^3$

   (g) $e^{2x} - \cos(x + e^{2x})$ (h) $(e^{3x^3+x})(1 - e^x)$

3. Find the critical points of $f(x) = x^2 e^{-x}$.

THE DERIVATIVE OF THE LOGARITHM

We can differentiate the logarithm function by using the inverse function rule of Section 5-1. If $y = \ln x$, then $x = e^y$ and

$$\frac{dy}{dx} = \frac{1}{dx/dy} = \frac{1}{e^y} = \frac{1}{x}$$

Hence

$$\frac{d}{dx} \ln x = \frac{1}{x}$$

For other bases, we use the same process; setting $y = \log_b x$ and $x = b^y$:

$$\frac{d}{dx} \log_b x = \frac{1}{\dfrac{d}{dy} b^y} = \frac{1}{\ln b \cdot b^y} = \frac{1}{\ln b \cdot x}$$

That is,

$$\frac{d}{dx} \log_b x = \frac{1}{(\ln b)x}$$

The last formula may also be proved by using law 3 of logarithms (p. 320):

$$\ln x = \log_e x = \log_b x \cdot \ln b$$

so

$$\frac{d}{dx} \log_b x = \frac{d}{dx} \left(\frac{1}{\ln b} \ln x \right) = \frac{1}{\ln b} \frac{d}{dx} \ln x = \frac{1}{(\ln b)x}$$

Worked Example 4 Differentiate: (a) $\ln(3x)$ (b) $xe^x \ln x$ (c) $8 \log_3 8x$

Solution (a) Setting $u = 3x$ and using the chain rule:

$$\frac{d}{dx} \ln 3x = \frac{d}{du} (\ln u) \cdot \frac{du}{dx} = \frac{1}{3x} \cdot 3 = \frac{1}{x}$$

Alternatively, $\ln 3x = \ln 3 + \ln x$, so the derivative with respect to x is $1/x$.

(b) By the product rule:

$$\frac{d}{dx} (xe^x \ln x) = x \frac{d}{dx} (e^x \ln x) + e^x \ln x = xe^x \ln x + e^x + e^x \ln x$$

(c) From the formula $(d/dx) \log_b x = 1/[(\ln b)x]$ with $b = 3$,

$$\frac{d}{dx} 8 \log_3 8x = 8 \frac{d}{dx} \log_3 8x = 8 \left(\frac{d}{du} \log_3 u \right) \frac{du}{dx} \qquad [u = 8x]$$

$$= 8 \cdot \frac{1}{(\ln 3) \cdot u} \cdot 8 = \frac{64}{(\ln 3) 8x} = \frac{8}{(\ln 3) x}$$

Our discussion so far can be summarized as follows:

DERIVATIVE OF THE LOGARITHM

$$\frac{d}{dx}\ln x = \frac{1}{x} \qquad x > 0$$

$$\frac{d}{dx}\log_b x = \frac{1}{(\ln b)x} \qquad x > 0$$

Since the derivative of $\ln x$ is $1/x$, $\ln x$ is an antiderivative of $1/x$; that is,

$$\int \frac{1}{x}\,dx = \ln x + C \qquad x > 0$$

This integration rule fills an important gap in our earlier formula

$$\int x^n\,dx = \frac{x^{n+1}}{n+1} + C$$

from Section 1-4 which was valid only for $n \neq -1$.

DIFFERENTIATION FORMULAS FOR exp AND log

1. $\dfrac{d}{dx}e^x = e^x$ that is, $\dfrac{d}{dx}\exp x = \exp x$

2. $\dfrac{d}{dx}b^x = (\ln b)b^x$

3. $\dfrac{d}{dx}\ln x = \dfrac{1}{x} \qquad x > 0$

4. $\dfrac{d}{dx}\log_b x = \dfrac{1}{(\ln b)x} \qquad x > 0$

ANTIDIFFERENTIATION FORMULAS FOR exp AND log

1. $\displaystyle\int e^x\,dx = e^x + C$

2. $\displaystyle\int b^x\,dx = \frac{b^x}{\ln(b)} + C$

3. $\displaystyle\int \frac{1}{x}\,dx = \ln|x| + C \qquad x \neq 0$

Integration formula 3 in the preceding display is proved by dividing into the two cases $x > 0$ and $x < 0$. For $x > 0$ it is the inverse of differentiation formula 3. For $x < 0$, $(d/dx)(\ln|x|) = (d/dx)[\ln(-x)] = [1/(-x)] \cdot [-1] = 1/x$, so $\ln|x|$, $x \neq 0$, is still an antiderivative for $1/x$.

Solved Exercises

3. Differentiate:

(a) $\ln 10x$ (b) $\ln u(x)$ (c) $\ln(\sin x)$

(d) $(\sin x)\ln x$ (e) $(\ln x)/x$ (f) $\log_5 x$

4. (a) If n is any real number, prove that

$$\frac{d}{dx} x^n = nx^{n-1} \quad \text{for } x > 0$$

(b) Find $(d/dx)(x^\pi)$ (see Solved Exercise 14, p. 228).

[*Note:* Previously, we knew that the formulas $(d/dx)x^n = nx^{n-1}$ and $\int x^n\,dx = [x^{n+1}/(n+1)] + C$, $n \neq -1$, were valid only for n rational. Now we see that they are true for all real n, rational or irrational.]

5. Find the indefinite integrals:

(a) $\int e^{ax}\,dx$ (b) $\int [1/(3x+2)]\,dx$

Exercises

4. Differentiate:

(a) $\ln(2x+1)$ (b) $\ln(x^2 - 3x)$ (c) $\ln(\tan x)$

(d) $(\ln x)^3$ (e) $(x^2 - 2x)\ln(2x+1)$ (f) $e^{x+\ln x}$

(g) $[\ln(\tan 3x)]/(1 + \ln x^2)$

5. Find the following indefinite integrals:

(a) $\int (s^2 + 2/s)\,ds$ (b) $\int \ln x\,dx$ [*Hint:* Try $x\ln x$.]   (c) $\int 4e^{-2x}\,dx$

6. What is $\int [1/(3x+2)]\,dx$ for $3x + 2 < 0$? (See Solved Exercise 5.)

LOGARITHMIC DIFFERENTIATION

In order to differentiate complex expressions involving powers, it is sometimes convenient to begin by taking logarithms.

Worked Example 5 Differentiate the function $y = x^x$.

Solution We take natural logarithms,

$$\ln y = \ln(x^x) = x \ln x$$

Next we differentiate using the chain rule, remembering that y is a function of x:

$$\frac{1}{y}\frac{dy}{dx} = x \cdot \frac{1}{x} + \ln x = 1 + \ln x$$

Hence

$$\frac{dy}{dx} = y(1 + \ln x) = x^x(1 + \ln x)$$

In general, $(d/dx)\ln f(x) = f'(x)/f(x)$ is called the *logarithmic derivative* of f. The quantity $f'(x)/f(x)$ is called the *relative rate of change* of f, since it measures the rate of change of f per unit of f itself. This idea is explored in the following application.

Worked Example 6   A certain company's profits are given by $P = 5000\exp(0.3t - .001t^2)$ dollars, where t is the time in years from January 1, 1980. By what percent per year are the profits increasing on July 1, 1981?

Solution   We compute the relative rate of change of P by using logarithmic differentiation.

$$\frac{1}{P}\frac{dP}{dt} = \frac{d}{dt}(\ln P) = \frac{d}{dt}(\ln 5000 + 0.3t - 0.001t^2) = 0.3 - 0.002t$$

Substituting $t = 1.5$ corresponding to July 1, 1981, we get

$$\frac{1}{P}\frac{dP}{dt} = 0.3 - (0.002)(1.5) = 0.2970$$

Therefore on July 1, 1981, the company's profits are increasing at a rate of 29.7% per year.

Solved Exercises
6. Use logarithmic differentiation to calculate dy/dx, where $y = (2x + 3)^{3/2}/\sqrt{x^2 + 1}$.

7. Differentiate $y = x^{(x^x)}$.

8. Find the equation of the line tangent to the graph $y = xe^{2x}$ at $x = 1$.

Exercises
7. Use logarithmic differentiation to differentiate:
   (a) $y = x^{3x}$
   (b) $y = x^{\sin x}$
   (c) $y = (\sin x)^{\cos x}$
   (d) $y = (x^3 + 1)^{x^2 - 2}$
   (e) $y = (x - 2)^{2/3}(4x + 3)^{8/7}$

8. Re-do Problem 20, Section 1-3.

9. Find the equation of the tangent line to the graphs of:
   (a) $y = x^{\ln x}$ at $x = 1$
   (b) $y = \sin(\ln x)$ at $x = 1$
   (c) $y = \ln(x^2 + 1)$ at $x = 1$

10. By what percentage are the profits of the company in Worked Example 6 increasing on January 1, 1982?

THE NUMBER e AS A LIMIT; COMPOUND INTEREST

In our previous discussion, the number e was obtained in a roundabout way. Using limits, we can derive a somewhat more explicit expression for e.

Note that $\ln'(1) = 1$, so

$$1 = \lim_{\Delta x \to 0} \frac{\ln(1 + \Delta x) - \ln(1)}{\Delta x} = \lim_{\Delta x \to 0} \frac{\ln(1 + \Delta x)}{\Delta x}$$

since $\ln(1) = 0$. Now

$$e = e^{1} = \exp\left\{ \lim_{\Delta x \to 0} \left[(\ln(1 + \Delta x))/\Delta x \right] \right\}$$

Since e^{x} is continuous, we get

$$e = \lim_{\Delta x \to 0} e^{[\ln(1+\Delta x)]/\Delta x} = \lim_{\Delta x \to 0} \left[e^{\ln(1+\Delta x)} \right]^{(1/\Delta x)} = \lim_{\Delta x \to 0} (1 + \Delta x)^{1/\Delta x}$$

or

$$e = \lim_{h \to 0} (1 + h)^{1/h} \qquad [h = \Delta x]$$

We get better and better approximations for e by taking h smaller and smaller; for instance, we could let $h = \pm (1/n)$, where n is an integer becoming larger and larger. We may say:

$$e = \lim_{n \to \infty} \left(1 + \frac{1}{n} \right)^{n} = \lim_{n \to \infty} \left(1 - \frac{1}{n} \right)^{-n} \qquad \textbf{(2)}$$

Notice that the numbers $(1 + 1/n)^{n}$ and $(1 - 1/n)^{-n}$ are all rational, so e is the limit of a sequence of rational numbers. It is known that e itself is irrational.*

---

e AS A LIMIT

$$e = \lim_{h \to 0} (1 + h)^{1/h} = \lim_{n \to \infty} \left(1 + \frac{1}{n} \right)^{n} = \lim_{n \to \infty} \left(1 - \frac{1}{n} \right)^{-n}$$

Worked Example 7 Express $\ln b$ as a limit of exponential expressions by using the formula $\ln b = \exp_b'(0)$.

Solution By the definition of the derivative as a limit,

$$\ln b = \exp_b'(0) = \lim_{\Delta x \to 0} \frac{\exp_b(\Delta x) - \exp_b(0)}{\Delta x} = \lim_{\Delta x \to 0} \frac{b^{\Delta x} - 1}{\Delta x}$$

or

$$\ln b = \lim_{h \to 0} \left(\frac{b^{h} - 1}{h} \right)$$

---

*A simple proof is given in J. Marsden, Elementary Classical Analysis, W. H. Freeman, 1974, p. 27.

The limit formula (2) for e has the following generalization (see Solved Exercise 10 for a proof):

$$e^a = \lim_{n \to \infty} \left(1 + \frac{a}{n} \right)^n = \lim_{n \to \infty} \left(1 - \frac{a}{n} \right)^{-n} \tag{3}$$

This formula, which is valid for all real numbers a, has an interpretation in terms of *compound interest.*

If a bank offers $r\%$ interest on deposits, compounded n times per year, then any invested amount will grow by a factor of $1 + r/100n$ during each compounding period and hence by a factor of $(1 + r/100n)^n$ over a year. For instance, a deposit of $1000 at 6% interest will become, at the end of a year,

$$1000(1 + 6/400)^4 = \$1061.36 \text{ with quarterly compounding}$$
$$1000(1 + 6/36500)^{365} = \$1061.8314 \text{ with daily compounding}$$

and

$$1000[1 + 6/(24 \cdot 36500)]^{24 \cdot 365} = \$1061.8362 \text{ with hourly compounding}$$

Two lessons seem to come out of this calculation: the final balance is an increasing function of the number of compounding periods, but there may be an upper limit to how much interest could be earned at a given rate, even if the compounding period were to be decreased to the tiniest fraction of a second.

In fact, applying formula (3) with $a = r/100$ gives us:

$$\lim_{n \to \infty} \left(1 + \frac{r}{100n} \right)^n = e^{r/100}$$

and so $1000 invested at 6% interest can never grow in a year to more than $1000e^{.06} = \$1061.8366$, no matter how frequent the compounding. (Strictly speaking, this assertion depends on the fact that $(1 + r/100n)^n$ is really an increasing function of n. This is intuitively clear from the compound interest interpretation; proofs are outlined in Problem 18 at the end of the chapter and in Solved Exercise 10, Section 12-1.)

In general, if P_0 dollars are invested at $r\%$ interest, compounded n times a year, then the account balance after a year will be $P_0(1 + r/100n)^n$, and the limit of this as $n \to \infty$ is $P_0 e^{r/100}$. This limiting case is often referred to as *continuously compounded interest.* The actual fraction by which the funds increase with continuous compound interest is $(P_0 e^{r/100} - P_0)/P_0 = e^{r/100} - 1$.

Worked Example 8 What is the yearly percent increase on a savings account with 5% interest compounded continuously?

Solution By the formula just derived, the fraction by which funds increase is $e^{r/100} - 1$. Substituting $r = 5$ gives $e^{r/100} - 1 = e^{.05} - 1 = 0.513 = 5.13\%$.

COMPOUND INTEREST

If an initial principal P_0 is invested at r percent interest compounded n times per year,	then the balance after one year is $P_0(1 + r/100\,n)^n$
If $n \to \infty$, so that the limit of continuous compounding is reached,	then the balance after one year is $P_0\,e^{r/100}$

The period of investment need not be just a year. The amount after t years will be $P(t) = P_0(1 + r/100n)^{[nt]}$ where $[nt]$ denotes the greatest integer less than or equal to nt. The function $P(t)$ increases by a factor of $(1 + r/100n)$ whenever t is an integer multiple of $1/n$, and it is constant between those moments. In the case of continuous compounding, the balance after time t is $P_c(t) = P_0\,e^{rt/100}$, which varies smoothly with t. Notice that $P'_c(t) = (r/100)\,P_c(t)$, so that $r\%$ may be considered as the instantaneous rate of interest.

Continuous compounding of interest is an example of *exponential growth*, a topic that will be treated in detail in Section 9-1.

Solved Exercises   9. Calculate $(1 - 1/n)^{-n}$ and $(1 + 1/n)^n$ for various values of n.

10. Prove formula 3.

Exercises   11. Express $3 \ln b$ as a limit.

12. Express $3^{\sqrt{2}}$ as a limit.

13. Show that $\ln b = \lim_{n \to \infty} n(\sqrt[n]{b} - 1)$.

14. A bank offers 8% per year compounded continuously and advertises an actual yield of 8.33%. Verify.

Problems for Section 7-2

1. Simplify:
   (a) $e^{4x}[\ln(e^{3x-1}) - \ln(e^{1-x})]$
   (b) $e^{x\ln 3 + \ln 2^x}$

2. Differentiate:
   (a) $e^{x\sin x}$
   (b) x^e
   (c) $14^{x^2 - 8\sin x}$
   (d) x^{x^2}
   (e) $\ln(x^{-5} + x)$
   (f) $\sin(x^4 + 1) \cdot \log_8(14x - \sin x)$

3. Differentiate:
   (a) x^{e^x}
   (b) $(1/x)^{\tan x^2}$
   (c) $\sin(x^x)$
   (d) $\ln(x^{\sec x^2})$
   (e) $(1 + \sin x)^{5/6}(x + e^x)^{2/7}$

4. Sketch the graph of $y = e^{-x}\sin x$.

5. Find the critical points of $f(x) = \sin(xe^x)$, $-4\pi \le x \le 4\pi$.

6. Find the equation of the tangent line to the graph of
   (a) $y = \cos(\pi e^x/4)$ at $x = 0$
   (b) $y = x^2 e^{x/2}$ at $x = 2$

7. Sketch the graph of $y = xe^{-x}$; indicate on your graph the regions where y is increasing, decreasing, concave up or down.

8. Find the minimum of $y = x^x$ for x in $(0, \infty)$.

9. Differentiate:
   (a) $\log_{5/3}(\cos 2x)$
   (b) $\ln(x + \ln x)$
   (c) $3x^{\sqrt{x}}$
   (d) $3x^{x/2}$
   (e) $\ln(x^{x+1})$

10. Differentiate:
   (a) $e^x \sin(\ln x + 1)$
   (b) $6\ln(x^3 - xe^x) + e^x \ln x$
   (c) $\log_2[\sin(x^2)]$
   (d) $\sin(x^{\cos x})$
   (e) $(\sin x)^{(\cos x)^x}$

11. Express the derivatives of the following in terms of $f(x)$, $g(x)$, $f'(x)$, and $g'(x)$:
    (a) $f(x) \cdot e^x + g(x)$ (b) $e^{f(x)+x^2}$
    (c) $f(x) \cdot e^{g(x)}$ (d) $f(e^x + g(x))$
    (e) $f(x)^{g(x)}$

12. Find the following antiderivatives:
    (a) $\int [(x^2 + 1)/2x]\,dx$
    (b) $\int (e^{4x} - 2/x)dx$
    (c) $\int \log_3 x\, dx$ (See Exercise 5b; p. 330.)
    (d) $\int (s^2 + s + 1 + 1/s + 1/s^2)ds$
    (e) $\int [x/(x-1)]\,dx$ [*Hint:* Divide.]

13. Find the following antiderivatives:
    (a) $\int 3^x\,dx$ (b) $\int x^3\,dx$
    (c) $\int [x/(x+3)]\,dx$
    (d) $\int [\log_2 x + (1/x)]\,dx$ (See Exercise 5b.)
    (e) $\int [(x^2 + 2x + 2)/(x-8)]\,dx$ [*Hint:* Divide.]

14. Find the domain and range of $f(x) = \log_x 2$. Sketch a graph.

15. (a) Show that the first-order approximation to b^x, for x near zero, is $1 + x \ln b$.
    (b) Compare $2^{0.01}$ with $1 + 0.01 \ln 2$; compare $2^{0.0001}$ with $1 + 0.0001 \ln 2$. (Use a calculator or tables.)
    (c) By writing $e = (e^{1/n})^n$ and using the first-order approximation for $e^{1/n}$, obtain an approximation for e.

16. Express e^{a+1} as a limit.

17. Express $\ln(\frac{1}{2})$ as a limit.

★18. Let $a > 0$. Show that $[1 + (a/n)]^n$ is an increasing function of n by following this outline:
    (a) Suppose that $f(1) = 0$ and $f'(x)$ is positive and decreasing on $[1, \infty)$. Then show that $g(x) = xf(1 + (1/x))$ is increasing on $[1, \infty)$. [*Hint:* Compute $g'(x)$ and use the mean value theorem to show that it is positive.]
    (b) Apply the result of (a) to $f(x) = \ln(x)$.
    (c) Apply the result of (b) to
$$\frac{1}{a} \ln\left[\left(1 + \frac{a}{n}\right)^n\right]$$

(Another, purely algebraic, solution to this problem is given in Solved Exercise 10, Section 12-1.)

19. (a) What rate of interest compounded annually is equivalent to 7% compounded continuously?
    (b) How much money would you need to invest at 7% to see the difference between continuous compounding and compounding by the minute over a year?

★20. Let $r = b^\theta$ be an exponential spiral.
    (a) Show that the angle ϕ between the tangent line at any point of the spiral and the line from that point to the origin is the same for all points of the spiral. (Use the formula for the tangent line in polar coordinates given in Section 6-3).
    (b) Express ϕ in terms of b.
    (c) The tangent lines to a certain spiral make an angle of 45° with the lines to the origin. By what factor does the spiral grow after one turn about the origin?

21. A certain company's profits are given by $P = 50{,}000 \exp(0.1\,t - 0.002\,t^2 + 0.00001\,t^3)$ where t is the time in years from July 1, 1975. By what percentage are the profits growing on January 1, 1980?

22. For $0 \le t \le 1000$, the height of a redwood tree in feet, t years after being planted, is given by $h = 300(1 - \exp[-t/(1000 - t)])$. By what percent per year is the height increasing when the tree is 500 years old?

23. The amount A for principal P compounded continuously for t years at an annual interest rate of r is $A = Pe^{rt}$. Find the amount after three years for $100 principal compounded continuously at 6%.

24. One form of the Weber-Fechner law of mathematical psychology is $dS/dR = c/R$ where S = perceived sensation, R = stimulus strength. The law says, for example, that doubling or tripling the volume on a stereo system does not cause a corresponding change in perception, but rather the rate of perception goes down as the volume goes up.
    (a) Let R_0 be the threshold level for R to be perceived. Show that $S = c \cdot \ln(R/R_0)$ satisfies the Weber-Fechner law and that $S(R_0) = 0$.
    (b) The loudness L in decibels is given by $L = 10 \log_{10}(I/I_0)$ where I_0 is the least audible intensity. Find the value of the constant c in the Weber-Fechner law of loudness.

25. Carbon-14 is known to satisfy the decay law $Q = Q_0 e^{-0.0001238t}$ for the amount Q present after t years. Find the age of a bone sample in which the carbon-14 present is 70% of the original amount Q_0.

26. The pressure P in the aorta during the diastole phase—period of relaxation—can be modeled by the equation

$$\frac{dP}{dt} + \frac{C}{W}P = 0, \quad P(0) = P_0$$

The numbers C, W are positive constants.

(a) Verify that $P = P_0 e^{-Ct/W}$ is a solution.

(b) Find $\ln(P_0/P)$ after one second.

27. The pressure P in the aorta during systole can be given by

$$P = \left(P_0 + \frac{CAW^2 B}{C^2 + W^2 B^2}\right) e^{-Ct/W}$$

$$+ \frac{CAW}{C^2 + W^2 B^2}[C\sin Bt + (-WB)\cos Bt]$$

Show that $P(0) = P_0$ and $\dfrac{dP}{dt} + \dfrac{C}{W}P = CA\sin Bt$.

28. The atmospheric pressure p at x feet above sea level can be approximated by $p = 2116\,e^{-0.0000318x}$ Compute the decrease in outside pressure expected in one second by a balloon at 2000 feet which is rising at 10 ft/sec. [*Hint:* Use $dp/dt = (dp/dx)(dx/dt)$ when $dx/dt = 10$.]

29. A company truck and trailer has salvage value $y = 120{,}000e^{-0.1x}$ dollars after x years of use. (a) Find the rate of depreciation in dollars per year after five years. (b) By what percent is the value decreasing after three years?

30. Let f be a function satisfying $f'(t_0) = 0$. Show that the relative rate of change of $P = \exp f(t)$ is zero at $t = t_0$.

Review Problems for Chapter 7

1. Differentiate:
   (a) $y = x\ln(x+3)$ (b) $y = xe^{(x+2)^3}$
   (c) $y = \sin[\cos(3x+1)]$
   (d) $y = \sec_2^{-1}[(\cos 2x)+1]$
   (e) $y = \log_2(3x)$

2. Differentiate:
   (a) $x = 1/[(\ln t)^2 + 3]$
   (b) $y = e^{-x^2}/(1+x^2)$ (c) $y = x\sin^{-1}(1/x)$
   (d) $y = t^{\exp t}$ (e) $x = \cos(e^{t^2+2})$

3. Find dy/dx if:
   (a) $e^{xy}/(1+xy) = 1$ (b) $x^y + y = 3$
   (c) $\cos(xy+3x) = x$

4. Find antiderivatives for:
   (a) $f(x) = 3x^2$ (b) $f(x) = \sin 5x$
   (c) $f(x) = 1/(x+2)$
   (d) $f(x) = x^2 + \cos 5x$
   (e) $f(x) = e^{6x} + e^{-6x}$

5. Compute:
   (a) $\int e^{3x}\,dx$
   (b) $\int(\cos x + 1/3x)\,dx$
   (c) $\int(x+1)/x\,dx$
   (d) $\int(\cos 2x + \sin 6x)\,dx$
   (e) $\int dx/(1+2x^2)$

6. Find the tangent line at $(0,\ln 3)$ to the graph of the curve defined implicitly by the equation

$$e^y - 3 + \ln(x+1)\cos y = 0$$

7. The velocity of a particle moving on the line is given by

$$v(t) = 37 + 10e^{-.07t}\ \text{meters/second}$$

(a) If the particle is at $x = 0$ at $t = 0$, how far has it travelled after ten seconds? (b) How important is the term $e^{-.07t}$ in the first ten seconds of motion? In the second ten seconds?

8. Find
   (a) $\displaystyle\lim_{n\to\infty}\left(1+\frac{8}{n}\right)^n$ (b) $\displaystyle\lim_{n\to\infty}\left(1-\frac{3}{2n}\right)^{-2n}$

9. Find: (a) $\displaystyle\lim_{n\to\infty}(1+10/n)^n$; (b) $\displaystyle\lim_{n\to\infty}(1-6/n)^{2n}$

10. Find the equation of the tangent line to the graph of $y = (x+1)e^{(3x^2+4x)}$ at $(0,1)$.

11. Simplify:
    (a) $\ln(e^3) + \frac{1}{2}\ln(e^{-5})$ (b) $(3e^{-\ln 4})/(\ln e^4)$

12. Simplify
    (a) $\ln\exp(-36)$
    (b) $\exp(\ln(\exp 3 + \exp 4) + \ln(8))$

13. Sketch the graph of $y = 1/[(\ln t)^2 + 1]$.

14. Sketch the graph of $y = e^{-x}/(1+x)$.

15. Use logarithmic differentiation to find:
    (a) $(d/dx)(\ln x)^x$
    (b) $(d/dx)[(x+3)^{7/2}(x+8)^{5/3}/(x^2+1)^{6/11}]$

16. Use logarithmic differentiation to differentiate
    (a) $(\ln x)^{\exp x}$
    (b) $(3x + 2)^{1/2}(8x^2 - 6)^{3/4}(\sin x - 3)^{6/17}$

17. Show that for any $x \neq 0$ there is a number c between zero and x such that $e^x = 1 + e^c x$. Deduce that $e^x > 1 + x$.

18. We have seen that the exponential function $\exp(x)$ satisfies $\exp(x) > 0$, $\exp(0) = 1$, and $\exp'(x) = \exp(x)$. Let $f(x)$ be a function such that

    $$0 \leq f'(x) \leq f(x) \quad \text{and} \quad f(0) = 0$$

    Prove that $f(x) = 0$ for all x. [*Hint:* Consider $g(x) = f(x)/\exp(x)$.]

19. If $1000 is to double in ten years, at what rate of interest must it be invested if interest is compounded (a) continuously, (b) quarterly?

20. If a deposit of A_0 dollars is made t times and is compounded n times during each deposit interval at an interest rate of i, then

    $$A = A_0\left\{\frac{(1 + i/n)^{nt} - 1}{(1 + i/n)^n - 1}\right\}$$

    is the amount after t intervals of deposit. (Deposits occur at the *end* of each deposit interval.)
    (a) Justify the formula. [*Hint:* $x^{t-1} + \cdots + x + 1 = (x^t - 1)/(x - 1)$.]
    (b) A person deposits $400 every three months, to be compounded quarterly at 7% per annum. How much is in the bank after six years?

21. The *transmission density* of a test area in a color slide is $D = \log_{10}(I/I_0)$ where I_0 is a reference intensity and I is the intensity of light transmitted through the slide. Rewrite this equation in terms of the natural logarithm.

22. Consider two decay laws for radioactive carbon-14 $Q = Q_0 e^{-\alpha t}$, $Q = Q_0 e^{-\beta t}$, where $\alpha = 0.0001238$ and $\beta = 0.0001236$. Find the percentage error between the two exponential laws for predicting the age of a skull sample with 50% of the carbon-14 decayed. (See Problem 25, p. 335.)

23. The salvage value of a tugboat is $y = 260{,}000\, e^{-0.15x}$ dollars after x years of use. What is the expected depreciation during the fifth year?

24. Find the marginal revenue of a commodity with demand curve $p = (1 + e^{-0.05x})10^3$ dollars per unit for x units produced. (Revenue = (# of units)(price per unit) = xp; the marginal revenue is the derivative of the revenue with respect to x.)

25. A population model which takes birth and death rates into account is the *logistic model* for the population P: $dP/dt = P(a - bP)$. The constants a, b, where $a > 0$ and $b \neq 0$, are the *vital constants*.
    (a) Let $P(0) = P_0$. Check by differentiation that $P(t) = a/[b + ([a/P_0] - b)e^{-at}]$ is a solution of the logistic equation.
    (b) Show that the population size approaches a/b as t tends to ∞.

8

Basic Methods of Integration

Two new important techniques extend the range of applications of integration.

In this chapter we develop fundamental techniques of integration in a more systematic way than we did in Chapter 4. By the end of this chapter, you should be proficient at calculating a variety of integrals involving the trigonometric and exponential functions; you will also be able to use the methods of integration by substitution and integration by parts.

CONTENTS

Section 8-1 reviews the integration techniques learned in Chapter 4 and shows how these techniques can be applied to the trigonometric and exponential functions. Sections 8-2 and 8-3 complete the job of reading the differentiation formulas backwards. The new integration methods, obtained from the chain rule and the product rule, are powerful and widely used. Throughout this chapter we apply the new techniques to area and rate problems (see Section 4-3). Additional applications of integration appear in the following three chapters.

8-1 Calculating Integrals

The rules for differentiating the trigonometric and exponential functions lead to new integration formulas.

We begin this section by summarizing the basic formulas for integration learned in Chapter 4. Using the rules for differentiating the trigonometric and exponential functions, we obtain corresponding integration formulas.

Goals

After studying this section, you should be able to:

Use the sum, constant multiple, and power rules for indefinite and definite integrals.

Integrate simple combinations of trigonometric, exponential, and logarithmic functions.

Find areas under and between curves involving trigonometric and exponential functions.

POLYNOMIALS AND POWERS

Given a function $f(x)$, $\int f(x)\,dx$ denotes the general antiderivative of f, also called the indefinite integral. Thus

$$\int f(x)\,dx = F(x) + C$$

where $F'(x) = f(x)$ and C is a constant. The definite integral is obtained via the fundamental theorem of calculus by evaluating the indefinite integral between two limits (Theorem 2 of Chapter 4). Thus:

$$\int_a^b f(x)\,dx = F(x)\,\Big|_a^b = F(b) - F(a).$$

We recall the following general rules for antiderivatives (see Section 1-4).

> **SUM AND CONSTANT MULTIPLE RULES**
>
> $$\int [f(x) + g(x)]\,dx = \int f(x)\,dx + \int g(x)\,dx$$
>
> $$\int cf(x)\,dx = c \int f(x)\,dx$$

These rules follow from the corresponding differentiation rules. To check the sum rule, for instance, we must see if

$$\frac{d}{dx}\left[\int f(x)\,dx + \int g(x)\,dx\right] = f(x) + g(x)$$

But this is true by the sum rule for derivatives.

The rule for powers is given as follows:

POWER RULE

$$\int x^n\,dx = \begin{cases} \dfrac{x^{n+1}}{n+1} + C & n \neq -1 \\[2ex] \ln|x| + C & n = -1 \end{cases}$$

The power rule for integer n was introduced in Section 1-4. The power rule was extended to cover the case $n = -1$ in Section 7-2. In that section we also proved the power rule for all real numbers n, rational or irrational.

Worked Example 1   Calculate $\int (3x^{2/3} + 8/x)\,dx$.

Solution   By the sum and constant multiple rules,

$$\int \left(3x^{2/3} + \frac{8}{x}\right)dx = 3\int x^{2/3}\,dx + 8\int \frac{1}{x}\,dx$$

By the power rule, this becomes

$$3 \cdot \frac{x^{5/3}}{5/3} + 8\ln|x| + C = \frac{9}{5}x^{5/3} + 8\ln|x| + C$$

Applying the fundamental theorem to the power rule, we obtain the rule presented below. The extra conditions on a and b are imposed because the integrand must be defined and continuous on the domain of integration; otherwise the fundamental theorem does not apply. (See Solved Exercise 4.)

DEFINITE INTEGRAL OF A POWER

$$\int_a^b x^n\,dx = \frac{x^{n+1}}{n+1}\Bigg|_a^b = \frac{b^{n+1} - a^{n+1}}{n+1}$$
$$\text{for } n \text{ real}, n \neq -1$$

If $n = -2, -3, -4, \ldots$, a and b must have the same sign. If n is not an integer, a and b must be positive.

$$\int_a^b \frac{1}{x}\,dx = \ln|x|\Bigg|_a^b = \ln|b| - \ln|a|$$

a and b must have the same sign.

Worked Example 2   Evaluate $\int_0^1 (x^4 - 3\sqrt{x})\,dx$.

Solution

$$\int_0^1 (x^4 - 3\sqrt{x})\,dx = \int (x^4 - 3\sqrt{x})\,dx \,\bigg|_0^1 = \frac{x^5}{5} - 3 \cdot \frac{x^{3/2}}{3/2}\,\bigg|_0^1$$

$$= \frac{1}{5} - 2 = -9/5$$

Worked Example 3   Find the area between the x axis, the curve $y = 1/x$, and the lines $x = -e^3$ and $x = -e$.

Solution   For $-e^3 \le x \le -e$, we notice that $1/x$ is negative. Therefore, the graph of $1/x$ lies below the x axis (the graph of $y = 0$), and the area is

$$\int_{-e^3}^{-e} \left(0 - \frac{1}{x}\right)\,dx = -\ln|x|\,\bigg|_{-e^3}^{-e}$$

$$= -(\ln e - \ln e^3)$$

$$= -(1 - 3) = 2$$

See Fig. 8-1-1.

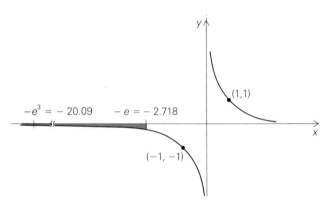

Fig. 8-1-1   Find the shaded area.

Solved Exercises*

1. Evaluate: (a) $\int [(x^3 + 8x + 3)/x]\,dx$;   (b) $\int (x^\pi + x^3)\,dx$.

2. Calculate: (a) $\int_1^2 (\sqrt{x} + 2/x)\,dx$;   (b) $\int_{1/2}^1 [(x^4 + x^6 + 1)/x^2]\,dx$.

3. Water flows into a tank at the rate of $2t + 3$ liters per minute, where t is the time measured in *hours* after noon. If the tank is empty at noon and holds 1000 liters, when will it be full?

4. We have $1/x^4 > 0$ for all x. On the other hand, we calculate $\int dx/x^4 = \int x^{-4}\,dx = (x^{-3}/-3) + C$, so

$$\int_{-1}^1 \frac{dx}{x^4} = \frac{1^{-3} - (-1)^{-3}}{-3} = \frac{1 + 1}{-3} = -\frac{2}{3}$$

How can a positive function have a negative integral?

Exercises

1. Evaluate the following integrals:

(a) $\int_{-2}^2 (x^8 + 2x^2 - 1)\,dx$ (b) $\int_{16}^{81} \sqrt[4]{s}\,ds$ (c) $\int_1^5 (1/t)\,dt$

---

*Solutions appear in Appendix A.

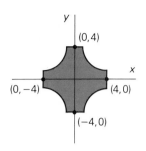

Fig. 8-1-2 Find the area of the shaded region.

2. Evaluate:

(a) $\int_3^6 (1 - y + y^2)\, dy$

(b) $\int_{-4}^{-3} (1/r^2)\, dr$

(c) $\int_{-243.8}^{243.8} (65x^{73} + 48x^{29} - 3x^{13} + 15x^5 - 2x)\, dx$

3. Check the formula $\int x\sqrt{1 + x}\, dx = \frac{2}{15}(3x - 2)(1 + x)^{3/2} + C$ and evaluate $\int_0^3 x\sqrt{1 + x}\, dx$.

4. A region containing the origin and shaped like the shaded region in Fig. 8-1-2 is cut out by the curves $y = 1/\sqrt{x}$, $y = -1/\sqrt{x}$, $y = 1/\sqrt{-x}$, and $y = -1/\sqrt{-x}$ and the lines $x = \pm 4$, $y = \pm 4$. Find the area of this region.

PROPERTIES OF THE DEFINITE INTEGRAL

We recall in the following display some general properties satisfied by the integral. These properties were proved in Chapter 4. The examples and exercises are intended to review and reinforce these properties.

PROPERTIES OF THE INTEGRAL

1. If $f(x) \le g(x)$ for all x in $[a, b]$, then

$$\int_a^b f(x)dx \le \int_a^b g(x)dx$$

2. Sum rule

$$\int_a^b [f(x) + g(x)]\, dx = \int_a^b f(x)dx + \int_a^b g(x)dx$$

3. Constant multiple rule

$$\int_a^b kf(x)dx = k\int_a^b f(x)dx, \quad k \text{ a constant}$$

4. $\displaystyle\int_a^b f(x)dx = \int_a^c f(x)dx + \int_c^b f(x)dx, \quad a < c < b$

5. Wrong way integrals

$$\int_b^a f(x)dx = -\int_a^b f(x)dx$$

Worked Example 4 Let

$$f(t) = \begin{cases} \frac{1}{2} & 0 \le t < \frac{1}{2} \\ t & \frac{1}{2} \le t \le 1 \end{cases}$$

Draw a graph of f and evaluate $\int_0^1 f(t)\, dt$.

Solution   The graph of f is drawn in Fig. 8-1-3. To evaluate the integral, we apply Property 4 in the previous display with $a = 0$, $c = \frac{1}{2}$, and $b = 1$:

$$\int_0^1 f(t)\,dt = \int_0^{1/2} f(t)\,dt + \int_{1/2}^1 f(t)\,dt = \int_0^{1/2} \tfrac{1}{2}\,dt + \int_{1/2}^1 t\,dt$$

$$= \tfrac{1}{2}t \Big|_0^{1/2} + \tfrac{1}{2}t^2 \Big|_{1/2}^1 = \tfrac{1}{4} + \tfrac{3}{8} = \tfrac{5}{8}$$

Fig. 8-1-3 The integral of f on $[0, 1]$ is the sum of its integrals on $[0, \frac{1}{2}]$ and $[\frac{1}{2}, 1]$.

Let us recall that the alternative form of the fundamental theorem of calculus states that if f is continuous, then

$$\frac{d}{dx} \int_a^x f(t)\,dt = f(x)$$

(see Theorem 3 of Chapter 4, p. 199).

Worked Example 5   Find $(d/dt) \int_0^{t^2} \sqrt{1 + 2s^3}\,ds$.

Solution   We write $g(t) = \int_0^{t^2} \sqrt{1 + 2s^3}\,ds$ as $f(t^2)$, where $f(u) = \int_0^u \sqrt{1 + 2s^3}\,ds$. By the fundamental theorem (alternative version), $f'(u) = \sqrt{1 + 2u^3}$; by the chain rule, $g'(t) = f'(t^2)[d(t^2)/dt] = \sqrt{1 + 2t^6} \cdot 2t$.

Solved Exercises

5. Interpret formula 4 in the display above in terms of areas.

6. Assuming without proof that $\int_0^{\pi/2} \sin^2 x\,dx = \int_0^{\pi/2} \cos^2 x\,dx$ (see Fig. 8-1-4), find $\int_0^{\pi/2} \sin^2 x\,dx$.

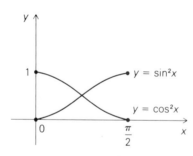

Fig. 8-1-4 The areas under the graphs of $\sin^2 x$ and $\cos^2 x$ on $[0, \pi/2]$ are equal.

7. Show that formula 4 in the display above holds even if $a < b < c$.

8. What happens to formula 1 in the display above if $b < a$?

Exercises

5. Suppose that $\int_0^2 f(t)\,dt = 5$, $\int_2^5 f(t)\,dt = 6$, and $\int_0^7 f(t)\,dt = 3$.
   (a) Find $\int_0^5 f(t)\,dt$.   (b) Find $\int_5^7 f(t)\,dt$.
   (c) Show that $f(t) < 0$ for some t in $(5, 7)$.

6. Interpret formula 1 in the display above in terms of areas.

7. Show that formulas 2 and 3 of the display above still hold if $b < a$.

8. Find $\int_1^3 [4f(s) + 3/\sqrt[3]{s}\,]\,ds$, where $\int_1^3 f(s)\,ds = 6$.

9. Find $(d/dt) \int_{t^2}^4 \sqrt{e^x + \sin 5x^2}\,dx$.

TRIGONOMETRIC FUNCTIONS

As we developed the calculus of the trigonometric and exponential functions, we obtained formulas for the antiderivatives of certain of these functions. For convenience, we summarize those formulas.

TRIGONOMETRIC FORMULAS

1. $\displaystyle\int \cos\theta \, d\theta = \sin\theta + C$ 2. $\displaystyle\int \sin\theta \, d\theta = -\cos\theta + C$

3. $\displaystyle\int \sec^2\theta \, d\theta = \tan\theta + C$ 4. $\displaystyle\int \csc^2\theta \, d\theta = -\cot\theta + C$

5. $\displaystyle\int \tan\theta \sec\theta \, d\theta = \sec\theta + C$ 6. $\displaystyle\int \cot\theta \csc\theta \, d\theta = -\csc\theta + C$

INVERSE TRIGONOMETRIC FORMULAS

1. $\displaystyle\int \frac{dx}{\sqrt{1-x^2}} = \sin^{-1}x + C, \ -1 < x < 1$

2. $\displaystyle\int \frac{-dx}{\sqrt{1-x^2}} = \cos^{-1}x + C, \ -1 < x < 1$

3. $\displaystyle\int \frac{dx}{1+x^2} = \tan^{-1}x + C, \ -\infty < x < \infty$

4. $\displaystyle\int \frac{-dx}{1+x^2} = \cot^{-1}x + C, \ -\infty < x < \infty$

5. $\displaystyle\int \frac{dx}{\sqrt{x^2(x^2-1)}} = \sec^{-1}x + C, \ -\infty < x < -1 \quad \text{or} \quad 1 < x < \infty$

6. $\displaystyle\int \frac{-dx}{\sqrt{x^2(x^2-1)}} = \csc^{-1}x + C, \ -\infty < x < -1 \quad \text{or} \quad 1 < x < \infty$

By combining these formulas and those in the tables on the endpapers of this book with the fundamental theorem of calculus, we can compute many definite integrals.

Worked Example 6   Evaluate $\int_0^\pi (x^4 + 2x + \sin x)\, dx$.

Solution   We begin by calculating the indefinite integral, using the sum and constant multiple rules, the power rule, and the fact that the antiderivative of $\sin x$ is $-\cos x + C$.

$$\int (x^4 + 2x + \sin x)\, dx = \int x^4\, dx + 2 \int x\, dx + \int \sin x\, dx$$
$$= x^5/5 + x^2 - \cos x + C$$

The fundamental theorem then gives

$$\int_0^\pi (x^4 + 2x + \sin x)\, dx = \int (x^4 + 2x + \sin x)\, dx \Big|_0^\pi$$

$$= \frac{x^5}{5} + x^2 - \cos x \Big|_0^\pi = \frac{\pi^5}{5} + \pi^2 - \cos \pi - (0 + 0 - \cos 0)$$

$$= \frac{\pi^5}{5} + \pi^2 + 1 + 1 = 2 + \pi^2 + \frac{\pi^5}{5} \approx 73.07$$

Worked Example 7   Find the area between the graphs of $\cos x$ and $\sin x$ on $[0, \pi/4]$.

Solution   Since $0 \leq \sin x \leq \cos x$ for x in $[0, \pi/4]$ (see Fig. 8-1-5), the formula for the area between two graphs (see Section 4-3) gives

$$\int_0^{\pi/4} (\cos x - \sin x)\, dx = \sin x + \cos x \Big|_0^{\pi/4} = \frac{1}{\sqrt{2}} + \frac{1}{\sqrt{2}} - 1$$
$$= \sqrt{2} - 1$$

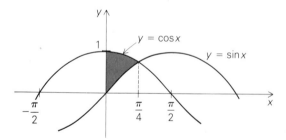

Fig. 8-1-5   Find the area of the shaded region.

Solved Exercises

9. Find $\int_{-1/2}^{1/2} dy/\sqrt{1 - y^2}$.

10. Show that the area under the graph of $f(x) = 1/(1 + x^2)$ on $[a, b]$ is less than π, no matter what the values of a and b may be.

11. Find $\int_0^{\pi/6} \cos 3x\, dx$.

Exercises

10. Evaluate the following definite integrals:

(a) $\int_{-\pi}^{\pi} \cos x\, dx$ (b) $\int_0^{\pi/4} \sec^2 x\, dx$

(c) $\int_0^{\pi/2} \sin 5x\, dx$ (d) $\int_{\sqrt{2}}^{2} du/u\sqrt{u^2 - 1}$

(e) $\int_0^{\pi/12} (3 \sin 4x + 4 \cos 3x)\, dx$

11. Find:
    (a) $\int \cos 2x \, dx$ (b) $\int (\cos^2 x - \sin^2 x) \, dx$
    (c) $\int (\cos^2 x + \sin^2 x) \, dx$
    (d) $\int \cos^2 x \, dx$ (Use parts (b) and (c).)
    (e) $\int_0^{\pi/2} \cos^2 x \, dx$ and $\int_0^{\pi/2} \sin^2 x \, dx$ (Compare with Solved Exercise 6 above.)

12. (a) Check the integral $\int (1/x\sqrt{x-1}) \, dx = 2 \tan^{-1} \sqrt{x-1} + C$.
    (b) Evaluate $\int_2^4 (1/x\sqrt{x-1}) \, dx$.

13. (a) Show that $\int \sin t \cos t \, dt = \frac{1}{2} \sin^2 t + C$.
    (b) Using the identity $\sin 2t = 2 \sin t \cos t$, show that
    $\int \sin t \cos t \, dt = -\frac{1}{4} \cos 2t + C$.
    (c) Use each of parts (a) and (b) to compute $\int_{\pi/6}^{\pi/4} \sin t \cos t \, dt$. Compare your answers.

14. Find the area of the shaded region in Fig. 8-1-6.

Fig. 8-1-6 Find the shaded area.

EXPONENTIAL FUNCTIONS

The following display summarizes the antidifferentiation formulas obtained in Chapter 7.

> ### EXPONENTIAL AND LOGARITHM
>
> $$\int e^x \, dx = e^x + C$$
>
> $$\int b^x \, dx = \frac{b^x}{\ln b} + C$$
>
> $$\int \frac{1}{x} \, dx = \ln |x| + C$$

Worked Example 8   Find $\int_{-1}^{1} 2^x \, dx$.

Solution
$$\int_{-1}^{1} 2^x \, dx = \int 2^x \, dx \,\Big|_{-1}^{1} = \frac{2^x}{\ln 2} \,\Big|_{-1}^{1}$$
$$= \frac{2}{\ln 2} - \frac{2^{-1}}{\ln 2} = \frac{3}{2 \ln 2} \approx 2.164$$

Worked Example 9   Find the area under the graph of $y = e^{2x}$ between $x = 0$ and $x = 1$.

Solution   Since $e^{2x} > 0$, the area is $\int_0^1 e^{2x} \, dx$. We begin by finding $\int e^{2x} \, dx$. Guessing e^{2x}, we find by the chain rule that $(d/dx)e^{2x} = 2e^{2x}$, so $\int e^{2x} \, dx = \frac{1}{2} e^{2x} + C$. Now we have

$$\int_0^1 e^{2x} \, dx = \frac{1}{2} e^{2x} \,\Big|_0^1 = \frac{1}{2}(e^2 - e^0) = \frac{1}{2}(e^2 - 1) \approx 3.1945$$

Solved Exercises

12. Evaluate $\int_0^1 (3e^x + 2\sqrt{x})\, dx$.

13. (a) Differentiate $x \ln x$.　(b) Find $\int \ln x\, dx$.　(c) Find $\int_2^5 \ln x\, dx$.

14. Find $\int_0^1 2^{2y}\, dy$.

Exercises

15. Evaluate:

(a) $\int_1^2 (e^{3x} + x^{2/3})\, dx$

(b) $\int_{-1}^1 e^{-4x}\, dx$

(c) $\int_0^1 \dfrac{3}{x^2 + 1}\, dx$

(d) $\int_{3/2}^2 \dfrac{dx}{20x}$

(e) $\int_3^4 e^{-2x}\, dx$

16. Find the area under the graph of each of the following functions on the stated interval:

(a) $x^3 + 1/(x^2 + 1)$; on $[0, 2]$

(b) $1/(x^2 + 1)$; on $[0, 2]$

(c) $(x^2 + 2)/\sqrt{x}$; on $[1, 4]$

(d) $\sin x - \cos 2x$; on $[\pi/4, \pi/2]$

17. Let $P(t)$ denote the population of rabbits in a certain colony at time t. Suppose that $P(0) = 100$ and that P is increasing at a rate of $20e^{3t}$ rabbits per year at time t. How many rabbits are there after 50 years?

Problems for Section 8-1

1. Evaluate the following definite integrals:

(a) $\int_0^a (6x^2 + 3x + 2)\, dx$

(b) $\int_0^\pi (3 \sin\theta + 4 \cos\theta)\, d\theta$

(c) $\int_{-2}^2 (x^{16} + x^9)\, dx$

(d) $\int_1^3 (1/x^2 + 1/x^3)\, dx$　(e) $\int_1^{81} \sqrt[4]{s}\, ds$

2. Evaluate:

(a) $\int_0^{\pi/2} (\sin^2 x + x^2)\, dx$ (See Solved Exercise 6.)

(b) $\int_1^2 (1/x^4)(1 + 2x + 3x^2 + 4x^3)\, dx$

(c) $\int_0^1 ds/(1 + s^2)$

(d) $\int_0^{\sqrt{2}/2} (4 - 4s^2)^{-1/2}\, ds$

(e) $\int_0^{\pi/4} (e^x - 3/\cos^2 x)\, dx$

3. Let $f(x) = \sin x$,

$$g(x) = \begin{cases} 1 & -\pi \le x \le 2 \\ 2 & 2 < x \le \pi \end{cases}$$

and

$$h(x) = \frac{1}{x^2}$$

Find:

(a) $\int_{-\pi/2}^{\pi/2} f(x)g(x)\, dx$　(b) $\int_1^3 g(x)h(x)\, dx$

(c) $\int_{\pi/2}^x f(t)g(t)\, dt$, for x in $(0, \pi]$. Draw a graph of this function of x.

4. Find:

(a) $\int_0^{\pi/2} (1 + \sin u)^2\, du$

(b) $\int_1^2 (x + 1/x)^2\, dx$

5. Using the identity $(1/t) - 1/(t + 1) = 1/[t(t + 1)]$, find $\int_1^e dt/[t(t + 1)]$.

6. (a) Show that $\int xe^{x^2}\, dx = \frac{1}{2}e^{x^2} + C$.

(b) Evaluate $\int_1^e (2xe^{x^2} + 3 \ln x)\, dx$. (See Solved Exercise 13.)

7. (a) Verify the formula $\int \sqrt{x^2 - 1}/x\, dx = \sqrt{x^2 - 1} - \sec^{-1} x + C$.

(b) Evaluate $\int_1^{3/2} \sqrt{x^2 - 1}/x\, dx$.

8. Let

$$f(t) = \begin{cases} 2 & -1 \le t < 0 \\ t & 0 \le t \le 2 \\ -1 & 2 < t \le 3 \end{cases}$$

Compute $\int_{-1}^3 f(t)\, dt$.

9. Let

$$h(x) = \begin{cases} x & 0 \le x < \frac{3}{4} \\ \frac{3}{4} & \frac{3}{4} \le x < 1 \end{cases}$$

Compute $\int_0^1 h(x)\, dx$.

10. Show that Formula 4 in the display on p. 343 holds for a, b, c in any order (see Solved Exercise 7).

11. Find the area of the shaded region in Fig. 8-1-7.

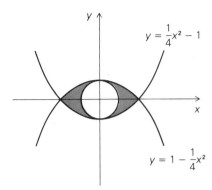

Fig. 8-1-7 Find the area of the "retina."

12. Illustrate in terms of areas the fact that

$$\int_0^{n\pi} \sin x \, dx$$

$$= \begin{cases} 2 & \text{if } n \text{ is an odd positive integer} \\ 0 & \text{if } n \text{ is an even positive integer} \end{cases}$$

13. Find the area of the shaded region in Fig. 8-1-8.

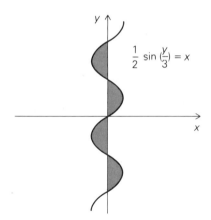

Fig. 8-1-8 Find the area of the shaded region.

14. Find the area of the shaded "flower" in Fig. 8-1-9.

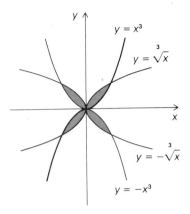

Fig. 8-1-9 Find the shaded area.

15. Each unit in a four-plex rents for $230/month. The owner will trade the property in five years. He wants to know the *capital value* of the property over a five-year period for continuous interest of 8.25%, that is, the amount he could borrow *now* at 8.25% continuous interest, to be paid back by the rents over the next five years. This amount A is given by $A = \int_0^T Re^{-kt} \, dt$, where $R =$ annual rents, $k =$ annual continuous interest rate, $T =$ period in years.
    (a) Verify that $A = (R/k)(1 - e^{-kT})$
    (b) Find A for the four-plex problem.

16. The *strain energy* V_e for a simply-supported uniform beam with a load P at its center is

$$V_e = \frac{1}{EI} \int_0^{\ell/2} \left(\frac{Px}{2} \right)^2 dx$$

    The *flexural rigidity* EI and the *bar length* ℓ are constants, $EI \neq 0$ and $\ell > 0$. Find V_e.

17. The sales of a clothing company t days after January 1 are expected to grow continuously according to the function $S(t) = 260e^{(0.1)t}$ dollars per day.
    (a) Set up a definite integral which gives the accumulated sales on $0 \le t \le 10$.
    (b) Find the accumulated sales for the first ten days.
    (c) How many days must pass before sales exceed $900 per day?

18. A particle starts at the origin with velocity $v(t) = 7 + 4t^3 + 6 \sin(\pi t)$ cm/sec. after t seconds. Find the distance traveled in 200 seconds.

19. The divorce rate $D(t)$, t years after 1920, is approximately $D(t) = 10^5 \cdot e^{25t/10^3}$ divorces per year.

   (a) Explain why $\int_{t_1}^{t_2} D(t)\,dt$ is the total number of divorces in the time interval $t_1 \leq t \leq t_2$.

   (b) How many divorces occurred in 1970?

20. A manufacturer determines by curve-fitting methods that the marginal revenue $R'(t) = 1000e^{t/2}$ and the marginal cost $C'(t) = 1000 - 2t$, t days after January 1. The revenue and cost are in dollars.

   (a) Suppose $R(0) = 0$, $C(0) = 0$. Find, by means of integration, formulas for $R(t)$ and $C(t)$.

   (b) The total profit is $P = R - C$. Find the total profit for the first seven days.

21. A rich uncle makes an endowment to his brother's firstborn son of $10,000, due on the child's twenty-first birthday. How much money should be put into a 9% continuous interest account to secure the endowment? [*Hint:* Use the formula $P = P_0 e^{kt}$, solving for P_0.]

22. The probability P that a capacitor manufactured by an electronics company will last between three years and five years in normal use is given approximately by $P = \int_3^5 (22.05)t^{-3}\,dt$.

   (a) Find the probability P.

   (b) Verify that $\int_3^7 (22.05)t^{-3}\,dt = 1$, which says that all capacitors have expected life between three and seven years.

8-2 Integration by Substitution

Integrating the chain rule leads to the method of substitution.

Effective integration requires more practice and skill than differentiation because it is less mechanical and relies more upon your ability to recognize certain combinations of functions. This section contains the first and most important of the integration methods.

Goals

After studying this section, you should be able to:

Write down the general formula for substitution in an integral.

Choose new variables for substitution, and use the method to evaluate indefinite and definite integrals.

THE METHOD OF SUBSTITUTION

The method of integration by substitution is based on the chain rule for differentiation. If F and g are differentiable functions, the chain rule tells us that $(F \circ g)'(x) = F'(g(x))g'(x)$; that is, $F(g(x))$ is an antiderivative of $F'(g(x))g'(x)$. In indefinite integral notation, we have

$$\int F'(g(x))\,g'(x)\,dx = F(g(x)) + C$$

To use this formula, it is necessary to recognize that a given integrand has the form $F'(g(x))g'(x)$. As with the chain for differentiation, it is convenient to introduce an intermediate variable $u = g(x)$; then the preceding formula becomes

$$\int F'(u)\,\frac{du}{dx}\,dx = F(u) + C$$

If we write $f(u) = F'(u)$, so that $\int f(u) \, du = F(u) + C$, we obtain the formula

$$\int f(u) \frac{du}{dx} \, dx = \int f(u) \, du \tag{1}$$

This formula is easy to remember, since one may formally "cancel the dx's." We can verify formula (1) directly by observing that $\int f(u) \, du$ is an antiderivative of $f(u) \, du/dx$. In fact, by the chain rule, $(d/dx)[\int f(u) \, du] = [(d/du) \int f(u) \, du](du/dx) = f(u)(du/dx)$.

To apply the method of integration by substitution one must find in a given integrand an expression $u = g(x)$ whose derivative $du/dx = g'(x)$ also occurs in the integrand.

Worked Example 1 Find $\int 2x\sqrt{x^2 + 1} \, dx$.

Solution None of the rules in Section 8-1 applies to this integral, so we try integration by substitution. Noticing that $2x$ is the derivative of $x^2 + 1$, we are led to write $u = x^2 + 1$. Then we have

$$\int 2x\sqrt{x^2 + 1} \, dx = \int \sqrt{x^2 + 1} \cdot 2x \, dx = \int \sqrt{u} \left(\frac{du}{dx} \right) dx$$

By formula (1), the last integral equals $\int \sqrt{u} \, du = \int u^{1/2} \, du = \frac{2}{3}u^{3/2} + C$. Substituting $x^2 + 1$ for u gives

$$\int 2x\sqrt{x^2 + 1} \, dx = \frac{2}{3}(x^2 + 1)^{3/2} + C$$

Checking our answer by differentiating has educational as well as insurance value, since it will show how the chain rule produces an integrand of the form we started with:

$$\frac{d}{dx}\left[\frac{2}{3}(x^2 + 1)^{3/2} + C \right] = \frac{2}{3} \cdot \frac{3}{2}(x^2 + 1)^{1/2} \frac{d}{dx}(x^2 + 1) = [\sqrt{x^2 + 1}] \, 2x$$

as it should be.

Sometimes the derivative of the intermediate variable is hidden as it appears in the integrand. If we are clever, however, we can still use the method of substitution, as the next example shows.

Worked Example 2 Find $\int \cos^2 x \sin x \, dx$.

Solution We are tempted to make the substitution $u = \cos x$, but du/dx is then $-\sin x$ rather than $\sin x$. No matter—we can rewrite the integral as

$$\int (-\cos^2 x)(-\sin x) \, dx$$

Putting $u = \cos x$, we have

$$\int -u^2 \frac{du}{dx}\, dx = \int -u^2\, du = -\frac{u^3}{3} + C$$

so

$$\int \cos^2 x \sin x\, dx = -\tfrac{1}{3} \cos^3 x + C$$

Check this by differentiating.

Worked Example 3   Find $\int e^x/(1 + e^{2x})\, dx$.

Solution   Here we cannot just let $u = 1 + e^{2x}$, because $du/dx = 2e^{2x} \neq e^x$. We have to recognize that $e^{2x} = (e^x)^2$ and remember that the derivative of e^x is e^x. Making the substitution $u = e^x$ and $du/dx = e^x$, we have

$$\int \frac{e^x}{1 + e^{2x}}\, dx = \int \frac{1}{1 + (e^x)^2} \cdot e^x\, dx$$
$$= \int \frac{1}{1 + u^2} \cdot \frac{du}{dx} \cdot dx = \int \frac{1}{i + u^2}\, du$$
$$= \tan^{-1} u + C = \tan^{-1}(e^x) + C$$

Again you should check this by differentiation.

We may summarize the method of substitution as developed so far.

INTEGRATION BY SUBSTITUTION

To integrate a function which involves an intermediate variable u and its derivative du/dx, write the integrand in the form $f(u)\,(du/dx)$, incorporating constant factors as required in $f(u)$. Then apply the formula

$$\int f(u) \frac{du}{dx}\, dx = \int f(u)\, du$$

Finally, evaluate $\int f(u)\, du$ if you can; then substitute for u its expression in terms of x.

Sometimes, after we make a substitution, the integral $\int f(u)\, du$ will still be something we don't know how to evaluate. In that case it may be necessary to make another substitution or use a completely different method. There is an infinite choice of substitutions available in any given situation. It takes practice to learn to choose one that works.

The techniques of differentiation are quite straightforward—to differentiate a complicated function, it is necessary only to break up the function into simpler components and then apply the rules of differentiation, perhaps several times. Integration, by contrast, is more difficult. If we know $\int f(x)\, dx$ and $\int g(x)\, dx$, for example, there is no straightforward means for generating $\int f(x)g(x)\, dx$. In general, integration is a trial-and-error process that involves a certain amount of educated guessing. What is more, the antiderivatives of such innocent-looking functions as

$$\frac{1}{\sqrt{(1 - x^2)(1 - 2x^2)}} \quad \text{and} \quad \frac{1}{\sqrt{3 - \sin^2 x}}$$

simply cannot be expressed as algebraic combinations and compositions of polynomials, trigonometric functions, or exponential functions. Despite these difficulties, you can learn to integrate many functions, but the learning process is slower than for differentiation, and practice is more important than ever.

Because integration is more difficult than differentiation, people who use calculus in practice often use tables of integrals. A short table is available on the end papers, and extensive books of tables are on the market. (Two of the most popular are Burington's and the CRC tables, both of which contain lots of mathematical data in addition to the integrals.) Before you conclude that the existence of the tables makes it unnecessary to learn the techniques of integration, you should know that even to use the tables requires a knowledge of the basic techniques—this is perhaps the ultimate justification for learning them.

Solved Exercises

1. Find $\int (2x/\sqrt{x^2 + 1}) \, dx$.

2. Find $\int \sin 2x \, dx$.

3. Find $\int x^2/(x^3 + 5) \, dx$

4. Find $\int dt/(t^2 - 6t + 10)$. [*Hint:* Complete the square in the denominator.]

5. Find $\int \sin^2 2x \cos 2x \, dx$.

Exercises

1. Evaluate each of the following integrals by making the indicated substitution, and check your answer by differentiating:
   (a) $\int 2x(x^2 + 4)^{3/2} \, dx; \; u = x^2 + 4$
   (b) $\int (x + 1)(x^2 + 2x - 4)^{-4} \, dx; \; u = x^2 + 2x - 4$
   (c) $\int (\sec^2 \theta/\tan^3 \theta) \, d\theta; \; u = \tan \theta$
   (d) $\int [(2y^7 + 1)/(y^8 + 4y - 1)^2] \, dy; \; x = y^8 + 4y - 1$
   (e) $\int [x/(1 + x^4)] \, dx; \; u = x^2$

2. Evaluate each of the following integrals by the method of substitution, and check your answer by differentiating:
   (a) $\int (x + 1) \cos (x^2 + 2x) \, dx$ (b) $\int [x^3/\sqrt{x^4 + 2}] \, dx$
   (c) $\int [x^3/(1 + x^8)] \, dx$
   (d) $\int [t^{1/3}/(t^{4/3} + 1)^{3/2}] \, dt$
   (e) $\int 2r \sin r^2 \cos^3 r^2 \, dr$ (f) $\int e^{\sin x} \cos x \, dx$

DIFFERENTIAL NOTATION

Two simple substitutions are so useful that they are worth noting explicitly. We have already used them in the preceding examples.

The first is the *shifting rule*, obtained by the substitution $u = x + a$, where a is a constant. Here $du/dx = 1$.

SHIFTING RULE

To evaluate $\int f(x + a)\, dx$,   first evaluate $\int f(u)\, du$, then substitute $x + a$ for u:

$$\int f(x + a)\, dx = F(x + a) + C \quad \text{where } F(u) = \int f(u)\, du$$

The second rule is the *scaling rule*, obtained by substituting $u = bx$, where b is a constant. Here $du/dx = b$. The substitution corresponds to a change of scale on the x axis.

SCALING RULE

To evaluate $\int f(bx)\, dx$,   evaluate $\int f(u)\, du$, divide by b, and substitute bx for u:

$$\int f(bx)\, dx = \frac{1}{b} F(bx) + C \quad \text{where } F(u) = \int f(u)\, du$$

Worked Example 4   Find $\int \sec^2(x + 7)\, dx$.

Solution   Since $\int \sec^2 u\, du = \tan u + C$, the shifting rule gives

$$\int \sec^2(x + 7)\, dx = \tan(x + 7) + C$$

Differential notation makes the substitution process more mechanical. In particular, it helps keep track of the constant factors which must be distributed between the $f(u)$ and du/dx parts of the integrand. We illustrate the device with an example before explaining why it works.

Worked Example 5   Find $\int [(x^4 + 2)/(x^5 + 10x)^5]\, dx$.

Solution   We wish to substitute $u = x^5 + 10x$; note that $du/dx = 5x^4 + 10$. Pretending that du/dx is a fraction, we may "solve for dx," writing $dx = du/(5x^4 + 10)$. Now we substitute u for $x^5 + 10x$ and $du/(5x^4 + 10)$ for dx in our integral to obtain

$$\int \frac{x^4 + 2}{(x^5 + 10x)^5}\, dx = \int \frac{x^4 + 2}{u^5}\, \frac{du}{5x^4 + 10} = \int \frac{x^4 + 2}{5x^4 + 10}\, \frac{du}{u^5} = \int \frac{1}{5}\, \frac{du}{u^5}$$

Notice that the $(x^4 + 2)$'s cancel, leaving us an integral in u which we can evaluate:

$$\frac{1}{5} \int \frac{du}{u^5} = \frac{1}{5} \left(-\frac{1}{4}\, u^{-4} \right) + C = -\frac{1}{20u^4} + C$$

so

$$\int \frac{x^4 + 2}{(x^5 + 10x)^5}\, dx = -\frac{1}{20(x^5 + 10x)^4} + C$$

Although du/dx is not really a fraction, we can still justify "solving for du" when we integrate by substitution. Suppose that we are trying to integrate $\int h(x)\, dx$ and we substitute the new variable $u = g(x)$. "Solving for dx" in the equation $du/dx = g'(x)$ amounts to replacing dx by $du/g'(x)$, which yields the integral $\int [h(x)/g'(x)]\, du$. "Expressing the integrand in terms of u" means writing $h(x)/g'(x) = f(u)$ for some function f and writing $\int h(x)\, dx = \int f(u)\, du$. To see that $\int f(u)\, du$ is really an antiderivative for $h(x)$, we can differentiate *with respect to* x. By the chain rule, we have

$$\frac{d}{dx}\int f(u)\, du = \frac{d}{du}\left[\int f(u)\, du\right]\frac{du}{dx} = f(u)\frac{du}{dx} = f(u)g'(x) = \frac{h(x)}{g'(x)}\, g'(x) = h(x)$$

Worked Example 6   Find $\int (e^{1/x}/x^2)\, dx$.

Solution   Let $u = 1/x$; $du/dx = -1/x^2$ and $dx = -x^2\, du$, so $\int (1/x^2)\, e^{1/x}\, dx = \int (1/x^2)\, e^u(-x^2\, du) = -\int e^u\, du = -e^u + C$ and therefore $\int (1/x^2)\, e^{1/x}\, dx = -e^{1/x} + C$.

INTEGRATION BY SUBSTITUTION (DIFFERENTIAL NOTATION)

To integrate $\int h(x)\, dx$ by substitution:

1. Choose a new variable $u = g(x)$.

2. Differentiate to get $du/dx = g'(x)$.

3. Solve for dx and replace in the integral to give $\int [h(x)/g'(x)]\, du$.

4. Express $h(x)/g'(x)$ in terms of u; that is, $h(x)/g'(x) = f(u)$.

   (If you cannot, try another substitution or another method.)

5. Evaluate the new integral $\int f(u)\, du$ (if you can).

6. Express the result in terms of x.

7. Check by differentiating.

Solved Exercises

6. Find $\int [(x^2 + 2x)/\sqrt[3]{x^3 + 3x^2 + 1}]\, dx$.

7. What happens in the integral $\int [(x^2 + 3x)/\sqrt[3]{x^3 + 3x^2 + 1}]\, dx$ if you make the substitution $u = x^3 + 3x^2 + 1$?

8. Find $\int \cos x\, [\cos(\sin x)]\, dx$.

9. What does the integral $\int dx/(1 + x^4)$ become if you substitute $u = x^2$?

10. Find $\int [\sqrt{1 + \ln x}/x]\, dx$.

Exercises 3. Carry out each of the following integrations:

(a) $\int [x/(x^2 + 3)^2]\, dx$

(b) $\int [x^{1/2}/(x^{3/2} + 2)^2]\, dx$

(c) $\int [(t + 1)/\sqrt{t^2 + 2t + 3}\,]\, dt$

(d) $\int u \sin(u^2)\, du$

4. Carry out the indicated integrations:

(a) $\int \sin(\theta + 4)\, d\theta$

(b) $\int (5x^4 + 1)(x^5 + x)^{100}\, dx$

(c) $\int (1 + \cos s)\sqrt{s + \sin s}\, ds$

(d) $\int dx/\sqrt{1 - 4x^2}$

(e) $\int dx/(x^2 + 4)$

5. Compute $\int \sin x \cos x\, dx$ by each of the following three methods:

(a) Substitute $u = \sin x$.

(b) Substitute $u = \cos x$.

(c) Use the identity $\sin 2x = 2 \sin x \cos x$.

Show that the three answers you get are really the same.

CHANGING VARIABLES IN THE DEFINITE INTEGRAL

We have just learned how to evaluate many indefinite integrals by the method of substitution. Using the fundamental theorem of calculus, we can use this knowledge to evaluate definite integrals as well.

Worked Example 7 Find $\int_0^2 \sqrt{x + 3}\, dx$.

Solution Substitute $u = x + 3$, $du = dx$. Then

$$\int \sqrt{x + 3}\, dx = \int \sqrt{u}\, du = \tfrac{2}{3}u^{3/2} + C = \tfrac{2}{3}(x + 3)^{3/2} + C$$

By the fundamental theorem of calculus,

$$\int_0^2 \sqrt{x + 3}\, dx = \tfrac{2}{3}(x + 3)^{3/2}\Big|_0^2 = \tfrac{2}{3}(5^{3/2} - 3^{3/2}) \approx 3.99$$

To check this result we observe that, on the interval $[0, 2]$, $\sqrt{x + 3}$ lies between $\sqrt{3} \approx 1.73$ and $\sqrt{5} \approx 2.24$, so the integral must lie between $2\sqrt{3} \approx 3.46$ and $2\sqrt{5} \approx 4.47$. (This check actually enabled the authors to spot an error in their first attempted solution of this problem.)

Notice that it is necessary to express the indefinite integral in terms of x before plugging in the endpoints 0 and 2, since the endpoints refer to values of x. It is possible, however, to evaluate the definite integral directly in the u variable—*provided that we change the endpoints.* We offer an example before stating the general procedure.

Worked Example 8 Find $\int_1^4 (x\, dx)/(1 + x^4)$.

Solution Substitute $u = x^2$, $dx = du/2x$; that is, $x\, dx = du/2$. As x runs from 1 to 4, $u = x^2$ runs from 1 to 16, so we have

$$\int_1^4 \frac{x}{1+x^4}\,dx = \int_1^{16} \frac{x}{1+x^4}\frac{du}{2x} = \frac{1}{2}\int_1^{16}\frac{du}{1+x^4} = \frac{1}{2}\int_1^{16}\frac{du}{1+u^2}$$

$$= \frac{1}{2}\tan^{-1}u\bigg|_1^{16} = \frac{1}{2}(\tan^{-1}16 - \tan^{-1}1) \approx 0.361$$

In general, suppose that we have an integral of the form $\int_a^b f(g(x))g'(x)\,dx$. We know that if $F'(u) = f(u)$, then $F(g(x))$ is an antiderivative of $f(g(x))g'(x)$; by the fundamental theorem of calculus, we have

$$\int_a^b f(g(x))g'(x)\,dx = F(g(b)) - F(g(a))$$

But the right-hand side is equal to $\int_{g(a)}^{g(b)} f(u)\,du$, so we have the formula

$$\int_a^b f(g(x))g'(x)\,dx = \int_{g(a)}^{g(b)} f(u)\,du$$

Notice that $g(a)$ and $g(b)$ are the values of $u = g(x)$ when $x = a$ and b, respectively. Thus we can evaluate an integral $\int_a^b h(x)\,dx$ by writing $h(x)$ as $f(g(x))g'(x)$ and using the formula

$$\int_a^b h(x)\,dx = \int_{g(a)}^{g(b)} f(u)\,du$$

Worked Example 9   Evaluate $\int_0^{\pi/4} \cos 2\theta\,d\theta$.

Solution   Let $u = 2\theta$; $d\theta = \frac{1}{2}du$; $u = 0$ when $\theta = 0$, $u = \pi/2$ when $\theta = \pi/4$. Thus

$$\int_0^{\pi/4} \cos 2\theta\,d\theta = \frac{1}{2}\int_0^{\pi/2} \cos u\,du = \frac{1}{2}\sin u\bigg|_0^{\pi/2} = \frac{1}{2}\left(\sin\frac{\pi}{2} - \sin 0\right) = \frac{1}{2}$$

DEFINITE INTEGRAL BY SUBSTITUTION

Given an integral $\int_a^b h(x)\,dx$ and a new variable $u = g(x)$, substitute $du/g'(x)$ for dx and try to express the integrand $h(x)/g'(x)$ in terms of u. Then change the endpoints a and b to the corresponding values, $g(a)$ and $g(b)$, of u.

$$\int_a^b h(x)\,dx = \int_{g(a)}^{g(b)} f(u)\,du \quad \text{where } f(u) = \frac{h(x)}{du/dx}$$

Since $h(x) = f(g(x))g'(x)$, this can be written as

$$\int_a^b f(g(x))g'(x)\,dx = \int_{g(a)}^{g(b)} f(u)\,du$$

Solved Exercises
11. Evaluate $\int_1^5 [x/(x^4 + 10x^2 + 25)]\ dx$.

12. Find $\int_0^{\pi/4} (\cos^2\theta - \sin^2\theta)\ d\theta$.

13. Evaluate $\int_0^1 [e^x/(1 + e^x)]\ dx$.

Exercises
6. Evaluate each of the following definite integrals:

(a) $\int_{-1}^1 \sqrt{x + 2}\ dx$ (b) $\int_0^2 x\sqrt{x^2 + 1}\ dx$

(c) $\int_2^4 (x + 1)(x^2 + 2x + 1)^{5/4}\ dx$

7. Evaluate the integral $\int_0^{\pi/2} \sin x \cos x\ dx$ in six different ways, as follows:

(a) Using each of the antiderivatives obtained in Exercise 5, evaluate at zero and $\pi/2$ and subtract.

(b) Using each of the substitutions suggested in Exercise 5, make the appropriate change of limits in the definite integral as done in Worked Examples 8 and 9.

Make sure that your six answers are the same.

8. Using the result $\int_0^{\pi/2} \sin^2 x\ dx = \pi/4$ (See Exercise 11, Section 8-1), compute the following integrals:

(a) $\int_0^{\pi} \sin^2(x/2)\ dx$ (b) $\int_{\pi/2}^{\pi} \sin^2(x - \pi/2)\ dx$ (c) $\int_0^{\pi/4} \cos^2(2x)\ dx$

Problems for Section 8-2

1. Evaluate the following indefinite integrals:

(a) $\int (1/x^2) \sin(1/x)\ dx$

(b) $\int t\sqrt{t^2 + 1}\ dt$ (c) $\int t\sqrt{t + 1}\ dt$

(d) $\int \cos^3\theta\ d\theta$
[*Hint:* Use $\cos^2\theta + \sin^2\theta = 1$.]

(e) $\int \tan x\ dx$ [*Hint:* $\tan x = \sin x/\cos x$.)]

2. Integrate:

(a) $\int dx/(x \ln x)$

(b) $\int \sqrt{4 - x^2}\ dx$ [*Hint:* Let $x = 2\sin u$.]

(c) $\int \sin^2 x\ dx$ (Use $\cos 2x = 1 - 2\sin^2 x$.)

(d) $\int [(x^3 + x - 1)/(x^2 + 1)]\ dx$

(e) $\int [\cos\theta/(1 + \sin\theta)]\ d\theta$

3. Evaluate:

(a) $\int [\sin(\ln t)/t]\ dt$ (b) $\int dx/\ln(x^x)$

(c) $\int \sec^2 x(e^{\tan x} + 1)\ dx$

(d) $\int [\sqrt[3]{3 + 1/x}/x^2]\ dx$

(e) $\int [e^{2s}/(1 + e^{2s})]\ ds$

4. Evaluate the following definite integrals:

(a) $\int_1^3 [3x/(x^2 + 5)^2]\ dx$

(b) $\int_0^{\pi/6} \sin(3\theta + \pi)\ d\theta$

(c) $\int_1^2 (\sqrt{1 + \ln x}/x)\ dx$

(d) $\int_0^1 xe^{x^2}\ dx$

5. Evaluate:

(a) $\int_1^2 [(t^2 + 1)/\sqrt{t^3 + 3t + 3}]\ dt$

(b) $\int_{-\pi/2}^{\pi/2} 5\cos^2 x \sin x\ dx$

(c) $\int_0^{\pi} \sin(\theta/2 + \pi/4)\ d\theta$

(d) $\int_{\pi/4}^{\pi/2} [\csc^2 y/(\cot^2 y + 2\cot y + 1)]\ dy$

★6. Let $f(x) = \int_1^x dt/t$. Show, using substitution, and without using logarithms, that $f(a) + f(b) = f(ab)$ if $a, b > 0$. [*Hint:* Transform $\int_a^{ab} dt/t$ by a change of variables.]

7. Evaluate the following definite integrals:

(a) $\int_0^1 t\sqrt{t^2 + 1}\ dt$

(b) $\int_1^3 [(x^3 + x - 1)/(x^2 + 1)]\ dx$ (simplify first)

(c) $\int_{\pi/4}^{\pi/2} \cos^2\theta\ d\theta$ (d) $\int_2^3 dt/(t - 1)$

(e) $\int_0^{\sqrt{\pi}} x \sin(x^2)\ dx$

8. Evaluate:

(a) $\int_0^1 [e^x/(1 + e^{2x})]\ dx$

(b) $\int_1^e [2 \ln(x^x) + 1]/x^2\ dx$

(c) $\int_0^1 [x^2/(x^3 + 1)]\ dx$

(d) $\int_1^{\pi/2} [\ln(\sin x) + (x \cot x)](\sin x)^x\ dx$

(e) $\int_{\pi/4}^{\pi/2} \cot\theta\ d\theta$

9. (a) By combining the shifting and scaling rules, find a formula for $\int f(ax + b)\ dx$.

(b) Find $\int_2^3 dx/(4x^2 + 12x + 9)$. [*Hint:* Factor the denominator.]

10. Given two functions f and g, define a function h by

$$h(x) = \int_0^1 f(x - t)g(t)\, dt$$

Show that

$$h(x) = \int_{x-1}^x g(x - t)f(t)\, dt$$

11. Give another solution to Solved Exercise 12 by writing $\cos^2\theta - \sin^2\theta = (\cos\theta - \sin\theta) \times (\cos\theta + \sin\theta)$ and using the substitution $u = \cos\theta + \sin\theta$.

12. Find the area under the graph of $y = (x + 1)/(x^2 + 2x + 2)^{3/2}$ from $x = 0$ to $x = 1$.

13. The curve $x^2/a^2 + y^2/b^2 = 1$ (Fig. 8-2-1), where a and b are positive, describes an *ellipse*. Find the area of the region inside this ellipse. [*Hint:* Write half the area as an integral and then change variables in the integral so that it becomes the integral for the area inside a semicircle.]

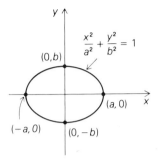

Fig. 8-2-1 Find the area inside the ellipse.

14. The curve $y = x^{1/3}$, $1 \le x \le 8$, is revolved about the y-axis to generate a surface of revolution of area s. In Chapter 11 we will prove that the area is given by $s = \int_1^2 2\pi y^3 \sqrt{1 + 9y^4}\, dy$. Evaluate this integral.

8-3 Integration by Parts

Integrating the product rule leads to the method of integration by parts.

The second of the two important new methods of integration is developed in this section. The method parallels that of substitution, with the chain rule replaced by the product rule.

Goals

After studying this section, you should be able to:

Write down the general formula for integration by parts.

Break up an integrand as a product and use the method of integration by parts to evaluate indefinite and definite integrals.

INDEFINITE INTEGRATION BY PARTS

The product rule for derivatives asserts that

$$(FG)'(x) = F'(x)G(x) + F(x)G'(x)$$

Since $F(x)G(x)$ is an antiderivative for $F'(x)G(x) + F(x)G'(x)$, we can write

$$\int [F'(x)G(x) + F(x)G'(x)]\, dx = F(x)G(x) + C$$

Applying the sum rule and transposing one term leads to the formula

$$\int F'(x)G(x)\,dx = F(x)G(x) - \int F(x)G'(x)\,dx \tag{1}$$

which is the rule for integration by parts. (There is a constant "C" implicit on both sides of this formula.) To apply formula (1) we need to break up a given integrand as a product $F'(x)G(x)$, write down the right-hand side of (1), and hope that we can integrate $F(x)G'(x)$. Practice with the method leads to an ability to see how to break up an integrand in an effective way. Integrands involving trigonometric, logarithmic, and exponential functions are often good candidates for integration by parts.

Worked Example 1   Evaluate $\int x \cos x\,dx$.

Solution   If we remember that $\cos x$ is the derivative of $\sin x$, we can write $x \cos x$ as $F'(x)G(x)$, where $F(x) = \sin x$ and $G(x) = x$. Applying formula (1), we have

$$\int x \cos x\,dx = \sin x \cdot x - \int \sin x \cdot 1\,dx = x \sin x - \int \sin x\,dx = x \sin x + \cos x + C$$

Checking by differentiation, we have

$$\frac{d}{dx}(x \sin x + \cos x) = x \cos x + \sin x - \sin x = x \cos x$$

as required.

As with integration by substitution, there is an alternative notation for integration by parts which makes the procedure at least partly mechanical. As usual, we write $f(x)$ for $F'(x)$ and $g(x)$ for $G'(x)$. Then formula (1) becomes

$$\int f(x)G(x)\,dx = F(x)G(x) - \int F(x)g(x)\,dx \tag{2}$$

We can also use differential notation. Here we write $v = F(x)$ and $u = G(x)$. Then $dv/dx = F'(x)$ and $du/dx = G'(x)$. Treating the derivatives as if they were quotients of "differentials" du, dv, and dx, we have $dv = F'(x)\,dx$ and $du = G'(x)\,dx$. Substituting these into (1) gives

$$\int u\,dv = uv - \int v\,du \tag{3}$$

See Fig. 8-3-1.

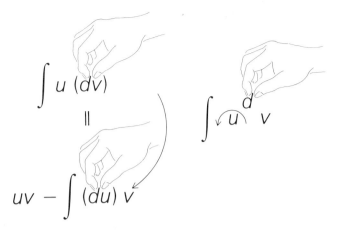

Fig. 8-3-1 You may move "d" from v to u if you switch the sign and add uv.

INTEGRATION BY PARTS

Given an integral $\int h(x)\,dx$, write $h(x)$ as a product $f(x)G(x)$, where the antiderivative $F(x)$ is known, take the derivative $g(x)$ of $G(x)$, and use the formula

$$\int f(x)G(x)\,dx = F(x)G(x) - \int F(x)g(x)\,dx$$

or write $h(x)\,dx$ as a product $u\,dv$ and use the formula

$$\int u\,dv = uv - \int v\,du$$

Worked Example 2 Find $\int x \sin x\,dx$.

Solution Let $f(x) = \sin x$, $G(x) = x$. Integrating $f(x)$ gives $F(x) = -\cos x$; also, $G'(x) = 1$, so

$$\int x \sin x\,dx = -x \cos x - \int -\cos x\,dx$$
$$= -x \cos x - (-\sin x) + C$$
$$= -x \cos x + \sin x + C$$

Worked Example 3 Find $\int x^2 \sin x\,dx$.

Solution Let $u = x^2$, $dv = \sin x\,dx$. To apply the formula for integration by parts, we need to know v. But $v = \int dv = \int \sin x\,dx = -\cos x$. (We leave out the arbitrary constant here and will put it in at the end of the problem.)

Now

$$\int x^2 \sin x \, dx = uv - \int v \, du$$

$$= -x^2 \cos x - \int -\cos x \cdot 2x \, dx$$

$$= -x^2 \cos x + 2 \int x \cos x \, dx$$

Using the result of Worked Example 1, we obtain

$$-x^2 \cos x + 2 (x \sin x + \cos x) + C = -x^2 \cos x + 2x \sin x + 2 \cos x + C$$

Check this result by differentiation; it is nice to see how much cancellation takes place.

Integration by parts is also commonly used in conjunction with e^x and ln x. The following worked example and Solved Exercise 4 illustrate the point.

Worked Example 4   Find $\int \ln x \, dx$ using integration by parts.

Solution   Here, let $u = \ln x$, $dv = 1 \, dx$. Then $du = dx/x$ and $v = \int 1 \, dx = x$. Applying the formula for integration by parts, we have

$$\int \ln x \, dx = uv - \int v \, du = (\ln x)x - \int x \frac{dx}{x}$$

$$= x \ln x - \int 1 \, dx = x \ln x - x + C$$

To remember the formula for integration by parts, it helps to keep in mind that if we call F and G the "integrals" and f and g the "derivatives," then the term outside an integral sign in formula (2) involves both integrals, while each of the other terms involves one integral and one derivative. In integration by parts, it is essential to choose as $f(x)$ a function whose antiderivative is already known. Furthermore, with a good choice of f and G, it usually should be the case that g is simpler than G, while F is not much more complicated than f, so that $\int F(x)g(x) \, dx$ is simpler than $\int f(x)G(x) \, dx$. This is only general advice, though. Experience will teach you a variety of uses for integration by parts. A last reminder—don't forget the minus sign.

Solved Exercises   1. What happens in Worked Example 2 if you choose $f(x) = x$ and $G(x) = \sin x$?

2. Find $\int \sin^{-1} x \, dx$. [*Hint:* Write $\sin^{-1} x = 1 \cdot \sin^{-1} x$.]

3. Find $\int \sin x \cos x \, dx$ by integration by parts.

4. Find $\int xe^x \, dx$.

5. Apply the technique of integration by parts twice to find $\int e^x \sin x \, dx$.

Exercises 1. Evaluate the following integrals:

(a) $\int x^2 \cos x \, dx$

(b) $\int 2t^3 \cos t^2 \, dt$ [*Hint:* Let $u = t^2$.]

(c) $\int (x^3/\sqrt{x^2+1}) \, dx$ [*Hint:* For part (c), first integrate by parts with $u = x^2$ and $v = \int (x/\sqrt{x^2+1}) \, dx$ and use substitution to evaluate v and $\int v \, du$.]

2. Evaluate:

(a) $\int [x^2/(x^2+1)^2] \, dx$ (b) $\int x^7(x^4+1)^{2/3} \, dx$

(c) $\int (1/x^3) \cos (1/x) \, dx$

3. Find $\int \sin x \cos x \, dx$ by using integration by parts with $f(x) = \cos x$, $G(x) = \sin x$. Compare your result with Solved Exercise 3.

4. Find:

(a) $\int (x+2)e^x \, dx$ (b) $\int (x^2-1)e^{2x} \, dx$ (c) $\int x \ln x \, dx$

DEFINITE INTEGRATION BY PARTS

Combining the method of integration by parts with the fundamental theorem of calculus gives us a method for calculating definite integrals.

Worked Example 5 Find $\int_{-\pi/2}^{\pi/2} x \sin x \, dx$.

Solution From Worked Example 2 we have $\int x \sin x = -x \cos x + \sin x + C$, so

$$\int_{-\pi/2}^{\pi/2} x \sin x \, dx = (-x \cos x + \sin x) \Big|_{-\pi/2}^{\pi/2}$$

$$= \left(-\frac{\pi}{2} \cos \frac{\pi}{2} + \sin \frac{\pi}{2}\right) - \left[\frac{\pi}{2} \cos \left(-\frac{\pi}{2}\right) + \sin \left(-\frac{\pi}{2}\right)\right]$$

$$= (0+1) - [0+(-1)] = 2$$

Worked Example 6 Find $\int_0^{\ln 2} e^x \ln (e^x + 1) \, dx$.

Solution Notice that e^x is the derivative of $(e^x + 1)$, so we first make the substitution $t = e^x + 1$. Then

$$\int_0^{\ln 2} e^x \ln (e^x + 1) \, dx = \int_2^3 \ln t \, dt$$

and, from Worked Example 4, $\int \ln t \, dt = t \ln t - t + C$. Therefore

$$\int_0^{\ln 2} e^x \ln (e^x + 1) \, dx = t \ln t - t \Big|_2^3 = (3 \ln 3 - 3) - (2 \ln 2 - 2)$$

$$= 3 \ln 3 - 2 \ln 2 - 1 \approx 0.9095$$

Solved Exercises   6. Find $\int_0^{\pi/2} \sin 2x \cos x \, dx$. [*Hint:* Integrate by parts twice.]

7. Find $\int_0^{\pi^2} \sin \sqrt{x} \, dx$. [*Hint:* Change variables first.]

8. Find $\int_1^e \sin(\ln x) \, dx$. [*Hint:* Change variables first.]

9. Find the area under the nth bend of $y = x \sin x$ in the first quadrant (see Fig. 8-3-2).

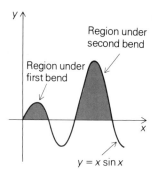

Fig. 8-3-2   What is the area under the nth bend?

Exercises   5. Calculate the following integrals:
   (a) $\int_0^{\pi/2} \sin 3x \cos 2x \, dx$ (b) $\int_1^3 \ln x^3 \, dx$
   (c) $\int_1^2 x^{1/3} (x^{2/3} + 1)^{3/2} \, dx$

6. Evaluate:
   (a) $\int_1^2 x \ln x \, dx$ (b) $\int_0^{\pi/4} (x^2 + x - 1)\cos x \, dx$
   (c) $\int_{-\pi}^{\pi} e^{2x} \sin(2x) \, dx$

7. Calculate $\int_0^{\pi/2} \sin x \cos x \, dx$ in as many different ways as you can.

INTEGRATING INVERSE FUNCTIONS

By itself, the rule for differentiating inverse functions does not lead to a rule for integrating them; however, combining this differentiation rule with the method of integration by parts does reduce the integration of f^{-1} to the integration of f. The basic idea was already used in Worked Example 4 and Solved Exercise 2; now we use it for the general case.

If f is any function, we can integrate by parts to get

$$\int f(x) \, dx = \int 1 \cdot f(x) \, dx = xf(x) - \int xf'(x) \, dx$$

Now introduce $y = f(x)$ as a new variable, with $dx = dy/f'(x)$; we get

$$\int f(x) \, dx = xy - \int x \, dy$$

If g is the inverse function to f, we have $x = g(y)$, and the preceding equation becomes

$$\int f(x)\,dx = xf(x) - \int g(y)\,dy$$

Hence

$$\int y\,dx = xy - \int x\,dy$$

Notice that the last equation looks just like the formula for integration by parts, except that here we are considering x and y as functions of one another rather than as two functions of a third variable.

We can also state our result in terms of antiderivatives. If $G(y)$ is an antiderivative for $g(y)$, then

$$F(x) = xf(x) - G(f(x))$$

is an antiderivative for f. (This can be checked, directly, by differentiation.)

Worked Example 7   Find an antiderivative for $\cos^{-1}x$.

Solution   If $f(x) = \cos^{-1}x$, then $g(y) = \cos y$ and $G(y) = \sin y$. By the last formula above, we have

$$F(x) = x\cos^{-1}x - \sin(\cos^{-1}x)$$

But $\sin(\cos^{-1}x) = \sqrt{1 - x^2}$, so

$$F(x) = x\cos^{-1}x - \sqrt{1 - x^2}$$

is an antiderivative for $\cos^{-1}x$. This may be checked by differentiation.

Worked Example 8   Find $\int \csc^{-1}\sqrt{x}\,dx$.

Solution   If $y = \csc^{-1}\sqrt{x}$, we have $\csc y = \sqrt{x}$ and $x = \csc^2 y$. Then (see Fig. 8-3-3):

$$\int \csc^{-1}\sqrt{x}\,dx = \int y\,dx = xy - \int x\,dy$$

$$= x\csc^{-1}\sqrt{x} - \int \csc^2 y\,dy$$

$$= x\csc^{-1}\sqrt{x} + \cot y + C$$

$$= x\csc^{-1}\sqrt{x} + \cot(\csc^{-1}\sqrt{x}) + C$$

$$= x\csc^{-1}\sqrt{x} + \sqrt{x - 1} + C$$

Fig. 8-3-3 $\theta = \csc^{-1}\sqrt{x}$

Finally, we have a rule for definite integrals. If f and g are inverse functions, then

$$\int_a^b f(x)\,dx = bf(b) - af(a) - \int_{f(a)}^{f(b)} g(y)\,dy$$

or

$$\int_a^b f(x)\,dx = b\beta - a\alpha - \int_\alpha^\beta g(y)\,dy$$

where $\alpha = f(a)$ and $\beta = f(b)$. We will see a geometric interpretation of this last result in Solved Exercise 12.

INTEGRATION OF INVERSE FUNCTIONS

If f and g are inverse functions with $y = f(x)$ and $x = g(y)$, then

$$\int f(x)\,dx = xf(x) - \int g(y)\,dy$$

$$\int y\,dx = xy - \int x\,dy$$

$$\int_a^b f(x)\,dx = bf(b) - af(a) - \int_{f(a)}^{f(b)} g(y)\,dy$$

Solved Exercises

10. Find $\int \sqrt{\sqrt{x}+1}\,dx$.

11. Find $\int_2^{34} f(x)\,dx$, where f is the inverse function of $g(y) = y^5 + y$.

12. Illustrate geometrically the formula for integration of inverse functions:

$$\int_a^b f(x)\,dx = bf(b) - af(a) - \int_{f(a)}^{f(b)} g(y)\,dy$$

where $0 < a < b$, $0 < f(a) < f(b)$, f is increasing on $[a,b]$, and g is the inverse function of f.

Exercises

8. Find the following indefinite integrals:

(a) $\int \sqrt{(1/y)-1}\,dy$ (b) $\int (\sqrt{x}-2)^{1/5}\,dx$

9. Compute $\int \sqrt{x}\,dx$ by the rule for inverse functions. Compare with the result given by the usual method.

10. Find the following definite integrals:

(a) $\int_0^{1/\sqrt{2}} \sin^{-1} x\,dx$ (b) $\int_0^1 \cos^{-1}(\sqrt{y})\,dy$

[*Hint:* For part (b), use $\int_0^{\pi/2} \cos^2 x\,dx = \pi/4$.]

11. Show that $\int_0^1 \sqrt{2-x^2}\,dx - \int_0^{\sqrt{2}} \sqrt{2-x^2}\,dx = (1-\pi/2)/2$.

Problems for Section 8-3 ▬▬▬▬▬▬

1. Evaluate the following indefinite integrals:
   (a) $\int x \cos(5x)\,dx$
   (b) $\int [x^5/(x^3-4)^{2/3}]\,dx$
   (c) $\int \ln(10x)\,dx$ (d) $\int x^2 \ln x\,dx$
   (e) $\int x \sin(\ln x)\,dx$

   [*Hint:* For part (b), use the method of Exercise 1(c).]

2. Evaluate:
   (a) $\int x^2 \sin x\,dx$ (b) $\int s^2 e^{3s}\,ds$
   (c) $\int \cos^{-1}(2x)\,dx$ (d) $\int \ln(9+x^2)\,dx$
   (e) $\int \tan x \ln(\cos x)\,dx$

★3. Following Exercise 1(c) and Problem 1(b), find a general formula for
   $$\int x^{2n-1}(x^n+1)^m\,dx$$
   where n and m are rational numbers with $n \neq 0$, $m \neq -1, -2$.

4. Find the following definite integrals:
   (a) $\int_0^{\pi/5} (8+5\theta)(\sin 5\theta)\,d\theta$
   (b) $\int_{1/4}^{1/2} \cos^{-1}(2x)\,dx$
   (c) $\int_1^2 x \ln x\,dx$

5. Evaluate:
   (a) $\int_0^1 x \tan^{-1} x\,dx$ (b) $\int_0^1 xe^x\,dx$
   (c) $\int_1^e (\ln x)^2\,dx$

6. (a) Prove the following *reduction formula:*
   $$\int \cos^n x\,dx = \frac{\cos^{n-1} x \sin x}{n}$$
   $$+ \frac{n-1}{n} \int \cos^{n-2} x\,dx$$

   (b) Use part (a) to show that
   $$\int \cos^2 x\,dx = \tfrac{1}{2}(\cos x \sin x + x) + C$$
   and
   $$\int \cos^4 x\,dx$$
   $$= \frac{1}{4}\left(\cos^3 x \sin x + \frac{3}{2}\cos x \sin x + \frac{3x}{2}\right) + C$$

7. Find $\int_0^{2\pi} x \sin ax$ as a function of a. What happens to this integral as a becomes larger and larger?

8. (a) By integrating by parts twice (see Solved Exercise 6), find $\int \sin ax \cos bx\,dx$, where $a^2 \neq b^2$.

   (b) Using the formula $\sin 2x = 2 \sin x \cos x$, find $\int \sin ax \cos bx\,dx$ when $a = \pm b$.

   (c) Let $g(a) = (4/\pi)\int_0^{\pi/2} \sin x \sin ax\,dx$. Find a formula for $g(a)$. (The formula will have to distinguish the cases $a^2 \neq 1$ and $a^2 = 1$.)

   ▦(d) Evaluate $g(a)$ for $a = 0.9, 0.99, 0.999, 0.9999$, and so on. Compare the results with $g(1)$. Also try $a = 1.1, 1.01, 1.001$, and so on. What do you guess is true about the function g at $a = 1$?

9. (a) Integrating by parts twice, show that
   $$\int e^{ax} \cos bx\,dx =$$
   $$e^{ax}\left(\frac{b \sin bx + a \cos bx}{a^2+b^2}\right) + C$$

   (b) Evaluate $\int_0^{\pi/10} e^{3x} \cos 5x\,dx$.

10. (a) Prove the following *reduction formula:*
   $$\int x^n e^x\,dx = x^n e^x - n \int x^{n-1} e^x\,dx$$

   (b) Evaluate $\int_0^3 x^3 e^x\,dx$

★11. (a) Suppose that $\phi'(x) > 0$ for all x in $[0,\infty)$ and $\phi(0) = 0$. Show that if $a \geq 0$, $b \geq 0$, and b is in the domain of ϕ^{-1}, then Young's inequality holds
   $$ab \leq \int_0^a \phi(x)\,dx + \int_0^b \phi^{-1}(y)\,dy$$
   where ϕ^{-1} is the inverse function to ϕ. [*Hint:* Express $\int_0^b \phi^{-1}(y)\,dy$ in terms of an integral of ϕ by using the formula for integrating an inverse function. Consider separately the cases $\phi(a) \leq b$ and $\phi(a) \geq b$. For the latter, prove the inequality $\int_{\phi^{-1}(b)}^a \phi(x)\,dx \geq$
   $\int_{\phi^{-1}(b)}^a b\,dx = b[a - \phi^{-1}(b)]$]

   (b) Prove (a) by a geometric argument based on Solved Exercise 12.

   (c) Using the result of part (a), show that if $a, b \geq 0$ and $p, q > 1$, with $1/p + 1/q = 1$, then Minkowski's inequality holds:
   $$ab \leq \frac{a^p}{p} + \frac{b^q}{q}$$

★12. If f is a function on $[0, 2\pi]$, the numbers $a_n = (1/\pi)\int_0^{2\pi} f(x) \cos nx\,dx$ and $b_n = (1/\pi)\int_0^{2\pi} f(x) \sin nx\,dx$ are called the *Fourier coefficients* of f ($n = 0, \pm1, \pm2, \dots$). Find the Fourier coefficients of:
   (a) $f(x) = 1$ (b) $f(x) = x$ (c) $f(x) = x^2$
   (d) $f(x) = \sin 2x + \sin 3x + \cos 4x$

13. The mass density of a beam is $\rho = x^2 e^{-x}$ kg/cm. The beam is 200 cm long, so its mass is $M = \int_0^{200} \rho \, dx$ kg. Find the value of M.

14. The volume of the solid formed by rotation of the plane region enclosed by $y = 0$, $y = \sin x$, $x = 0$, $x = \pi$, around the y-axis, will be shown in Chapter 10 to be given by $V = \int_0^\pi 2\pi x \sin x \, dx$. Find V.

15. The *Fourier series* analysis of the sawtooth wave requires the computation of the integral

$$b_m = \frac{\omega^2 A}{2\pi^2} \int_{-\pi/\omega}^{\pi/\omega} t \sin(m\omega t) \, dt$$

where m is an integer and ω and A are non-zero constants. Compute it.

16. The current i in an underdamped *RLC* circuit is given by

$$i = EC \left(\frac{\alpha^2}{\omega} + \omega \right) e^{-\alpha t} \sin(\omega t)$$

The constants are E = constant emf, switched on at $t = 0$, C = capacitance in farads, R = resistance in ohms, L = inductance in henrys, $\alpha = R/2L$, $\omega = (1/2L)(4L/C - R^2)^{1/2}$.

   (a) The charge Q in coulombs is given by $dQ/dt = i$, and $Q(0) = 0$. Find an integral formula for Q, using the fundamental theorem of calculus.

   (b) Determine Q by integration.

17. A critically damped *RLC* circuit with a steady emf of E volts has current $i = EC \, \alpha^2 \, t e^{-\alpha t}$, where $\alpha = R/2L$. The constants R, L, C are in ohms, henrys, and farads, respectively. The charge Q in coulombs is given by $Q(T) = \int_0^T i \, dt$. Find it explicitly, by means of integration by parts.

Review Problems for Chapter 8

1. Perform the indicated integrations:

   (a) $\int (x + \sin x) \, dx$

   (b) $\int (x + 1/\sqrt{1 - x^2}) \, dx$

   (c) $\int (e^x - x^2 - 1/x + \cos x) \, dx$

   (d) $\int (x^3 + \cos x) \, dx$

   (e) $\int (8t^4 - 5 \cos t) \, dt$

2. Integrate:

   (a) $\int (e^\theta + \theta^2) \, d\theta$

   (b) $\int [(1/\sqrt{4 - t^2}) + t^2] \, dt$

   (c) $\int (3^x - 3/x + \cos x) \, dx$

   (d) $\int (x + 2)^5 \, dx$ (e) $\int e^{3x} \, dx$

3. Find the area between the graphs of $y = -x^3 - 2x - 6$ and $y = e^x + \cos x$ from $x = 0$ to $x = \pi/2$.

4. Water is flowing in a tank at a rate of $10(t^2 + \sin t)$ gallons per minute after time t. Calculate: (a) the number of gallons stored after 30 minutes, starting at $t = 0$; (b) the average flow rate in gallons per minute over this 30-minute interval.

5. Evaluate:

   (a) $\int \cos^2 x \sin x \, dx$ (b) $\int x^2 e^{x^3} \, dx$

   (c) $\int dx/(3x + 4)$ (d) $\int x(\ln x)^2 \, dx$

   (e) $\int 3x \cos 2x \, dx$

6. Evaluate:

   (a) $\int x e^{4x} \, dx$ (b) $\int x^2 e^{4x^3} \, dx$

   (c) $\int x^2 \sin x^3 \, dx$ (d) $\int [3u/(u^2 + 2)] \, du$

   (e) $\int [\cos x] \ln(\sin x) \, dx$

7. Substitute $x = \sin u$ to evaluate

$$\int \frac{x \, dx}{\sqrt{1 - x^2}}$$

and

$$\int \frac{x^2 \, dx}{\sqrt{1 - x^2}}; \quad 0 < x < 1.$$

8. Evaluate:

   (a) $\displaystyle \int \frac{\ln x}{x} \, dx$

   (b) $\displaystyle \int_{\sqrt{3}}^{3\sqrt{3}} \frac{dx}{x^2 \sqrt{x^2 + 9}}$

   using the substitution $x = 3 \tan u$.

9. In each case, sketch the region under the graph of the given function on the given interval and find its area:

   (a) $40 - x^3$ on $[0, 3]$

   (b) $\sin x + 2x$ on $[0, 4\pi]$

   (c) $3x/\sqrt{x^2 + 9}$ on $[0, 4]$

   (d) $x \sin^{-1} x + 2$ on $[0, 1]$

   (e) $\sin x$ on $[0, \pi/4]$

   (f) $\sin 2x$ on $[0, \pi/2]$

   (g) $1/x$ on $[2, 4]$

10. Evaluate:

    (a) $\int_{a+1}^{a+2} (t/\sqrt{t-a})\, dt$ (using the substitution $x = \sqrt{t-a}$)

    (b) $\int_0^1 [\sqrt{x}/(x+1)]\, dx$ (c) $\int_0^1 x\sqrt{2x+3}\, dx$

    (d) $\int 3/(3+u^2)\, du$

11. Evaluate:

    (a) $\int_1^2 x^{-2}\cos(1/x)\, dx$

    (b) $\int_0^{\pi/2} x^2\cos(x^3)\sin(x^3)\, dx$

    (c) $\int x\tan^{-1}x\, dx$ (d) $\int e^x\tan e^x\, dx$

12. Let R_n be the region bounded by the x axis, the line $x = 1$, and the curve $y = x^n$. The area of R_n is what fraction of the area of the triangle R_1?

13. Evaluate:

    (a) $\int xe^{6x}\, dx$ (b) $\int x^2\cos x\, dx$

    (c) $\int x^2\ln x\, dx$ (d) $\int x^3\ln x\, dx$

    (e) $\int x^n\ln x\, dx$; $n \geq 0$

    (f) $\int \tan^{-1}x\, dx$ (g) $\int e^{\sqrt{x}}\, dx$

    (h) $\int [(\ln\sqrt{x})/\sqrt{x}]\, dx$

14. What happens if $\int f(x)\, dx$ is integrated by parts with $u = f(x)$, $v = x$?

15. Arthur Perverse believes that the product rule for integrals ought to be $\int f(x)g(x)\, dx = f(x) \times \int g(x)\, dx + g(x)\int f(x)\, dx$. We wish to show him that this is not a good rule.

    (a) Show that if the functions $f(x) = x^m$ and $g(x) = x^n$ satisfy Perverse's rule, then for fixed n the number m must satisfy a certain quadratic equation (assume n, $m \geq 0$).

    (b) Show that the quadratic equation of part (a) does not have any real roots for any n.

    ★(c) Are there *any* pairs of functions, f and g, which satisfy Perverse's rule? (Don't count the case where one function is zero.)

16. Evaluate:

    (a) $\int x^2 e^{2x}\, dx$ (b) $\int x\cos 3x\, dx$

    (c) $\int t\cos 2t\, dt$ (d) $\int (\ln x)^2\, dx$

    (e) $\int \sin^{-1}x\, dx$ (let $u = \sin^{-1}x$, $v = x$)

    (f) $\int e^{-x}\cos x\, dx$

17. Evaluate $\int \sin(\pi x/2)\cos(\pi x)\, dx$ by integrating by parts two different ways and comparing the results.

18. Do Problem 17 by using the product formulas for sine and cosine (see p. 273).

19. (a) Prove the following reduction formula:

    $$\int \sin^n x\, dx = -\frac{\sin^{n-1}x\cos x}{n} + \frac{n-1}{n}\int \sin^{n-2}x\, dx$$

    if $n \geq 2$, by integration by parts, with $u = \sin^{n-1}x$, $v = -\cos x$.

    (b) Evaluate $\int \sin^2 x\, dx$ by using this formula.

    (c) Evaluate $\int \sin^4 x\, dx$.

20. (a) Show that:

    $$\int x^m(\ln x)^n\, dx = \frac{x^{m+1}(\ln x)^n}{m+1} - \frac{n}{m+1}\int x^m(\ln x)^{n-1}\, dx$$

    (b) Evaluate $\int_1^2 x^2(\ln x)^2\, dx$.

21. Evaluate:

    (a) $\int 3\sin 3x\cos 3x\, dx$

    (b) $\int x\sqrt{5-x^2}\, dx$ (c) $\int x\sqrt{x+3}\, dx$

    (d) $\int xe^{x^2}\, dx$ (e) $\int x^3 e^{x^2}\, dx$

    (f) $\int x^2\sqrt{x+1}\, dx$

22. Evaluate:

    (a) $\int e^x\tan e^x\, dx$

    (b) $\int \sin 2x\cos x\, dx$

    (c) $\int x^5 e^{x^3}\, dx$

    (d) $\int dx/(x^2+2x+3)$ (Complete the square.)

23. Find the area under the graph of $f(x) = x/\sqrt{x^2+2}$ from $x = 0$ to $x = 2$.

24. Evaluate $\int \sqrt{(1+x)/(1-x)}\, dx$. [*Hint:* Multiply numerator and denominator by $\sqrt{1+x}$.]

25. Evaluate:

    (a) $\int e^x/(1+e^{2x})\, dx$

    (b) $\int [(\sqrt[3]{x^2} - x^{5/2})/\sqrt{x}]\, dx$

    (c) $\int \tan x\sec^2 x\, dx$

★26. Derive an integration formula obtained by reading the quotient rule for derivatives backwards.

★27. (a) Suppose that f is continuous on the real line and that g is a differentiable function. Let $F(x) = \int_0^{g(x)} f(t)\, dt$. What is $F'(x_0)$? [*Hint:* Use the chain rule.]

    (b) Let $F(x) = \int_1^{x^2} dt/t$. What is $F'(x)$?

★28. (a) Show that $\int_1^{x^2} dt/t = 2\int_1^x dt/t$. (Use Problem 27.)

    (b) What is the relation between $\int_1^{x^n} dt/t$ and $\int_1^x dt/t$?

★29. Find $(d/dx)\int_{g(x)}^{h(x)} f(t)\, dt$. [*Hint:* First fix b and find $(d/dx)\int_{g(x)}^b f(t)\, dt$.]

★30. Find $\int xe^{ax}\cos(bx)\, dx$.

9

Differential Equations

A function may be determined by a differential equation together with initial conditions.

Many physical phenomena (mechanical, biological, geological, etc.) can be described in terms of differential equations. Basic techniques for solving them are developed in this chapter, along with numerous applications. The use of calculus to solve differential equations is one of the most valuable tools of a scientist.

CONTENTS

The first section of this chapter concentrates on detailed methods of solution of the two differential equations $d^2x/dt^2 + \omega^2 x = 0$ and $dx/dt = \gamma x$. These are perhaps the two most widely used and basic differential equations. Following this, the second section develops a few techniques for solving some simple classes of differential equations, with a variety of applications.

This chapter relies on all of the previous chapters, and may be read at any time after Chapters 1 through 8 are completed.

9-1 Oscillations, Growth and Decay

Solutions of the equations for simple harmonic motion and population growth can be expressed in terms of trigonometric and exponential functions.

Many important laws of nature—physical, chemical, biological, economic, and sociological—can be expressed as relations between unknown functions and their derivatives. Such relations are called *differential equations*. In this section we study two of the simplest and most important equations: $(d^2x/dt^2) + \omega^2 x = 0$ describes a wide variety of oscillatory phenomena; $dx/dt = \gamma x$ describes growth or decay.

Goals

After studying this section, you should be able to:

Find the general solution of a differential equation of the form $(d^2x/dt^2) + \omega^2 x = 0$ or $dx/dt = \gamma x$.

Find the particular solution of one of these equations if the initial data are given.

THE SPRING EQUATION

A common problem in physics is to determine the motion of a particle in a given force field. Mathematically, the field is given by specifying the force F as a function of the position x. The problem is to write x as a function of the time t so that the equation

$$F = m\,\frac{d^2x}{dt^2} \tag{1}$$

is satisfied, where m is the mass of the particle. Equation (1) is called *Newton's second law of motion.**

For example, if the force is a *constant* F_0 and we rewrite equation (1) as $d^2x/dt^2 = F_0/m$, we can use our knowledge of antiderivatives to conclude that

$$\frac{dx}{dt} = \frac{F_0}{m}\,t + C_1$$

and

$$x = \frac{1}{2}\,\frac{F_0}{m}\,t^2 + C_1 t + C_2$$

where C_1 and C_2 are constants. We see that the position of a particle moving in a constant force field is a quadratic function of time (or a

---

Newton always expressed his laws of motion in words. The first one to formulate Newton's laws carefully as differential equations was the mathematical giant L. Euler around 1750. (See C. Truesdell, Essays on the History of Mechanics, Springer-Verlag, 1968.)

linear function, if the force is zero). Such a situation occurs for vertical motion under the force of gravity near the earth's surface.

If the force is not constant, then it takes more than simple antidifferentiation to find x as a function of t. For instance, if x is the downward displacement from equilibrium of a weight on a spring, then *Hooke's law* asserts that

$$F = -kx \tag{2}$$

where k is a positive constant called the *spring constant*. (See Fig. 9-1-1.) This law, discovered experimentally, is quite accurate if x is not too large. There is a minus sign in the formula for F because the force, being directed toward the equilibrium, has the opposite sign to x.

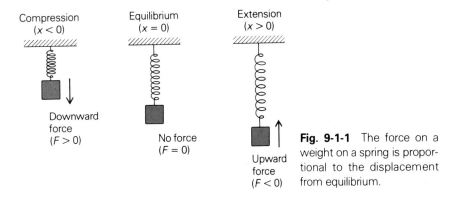

Fig. 9-1-1 The force on a weight on a spring is proportional to the displacement from equilibrium.

Substituting (2) into Newton's law (1) gives

$$-kx = m \frac{d^2 x}{dt^2} \quad \text{or} \quad \frac{d^2 x}{dt^2} = -\left(\frac{k}{m}\right) x$$

It is convenient to write the ratio k/m as ω^2, where $\omega = \sqrt{k/m}$ is a new constant. This substitution gives us the *spring equation*:

$$\frac{d^2 x}{dt^2} = -\omega^2 x \tag{3}$$

Since x is unknown, we cannot find dx/dt by antidifferentiating the right-hand side. (In particular, we do *not* get $dx/dt = -\frac{1}{2}\omega^2 x^2 + C$, since it is t rather than x which is the independent variable.) Instead, we shall use trial and error.

A good first guess, guided by the observation that weights on springs bob up and down, is

$$x = \sin t$$

Differentiating twice, we get

$$\frac{d^2 x}{dt^2} = -\sin t = -x$$

The factor ω^2 is missing, so we may be tempted to try $x = \omega^2 \sin t$. In this case, we get

$$\frac{d^2 x}{dt^2} = -\omega^2 \sin t$$

which is again $-x$. To get a *new factor* to appear upon differentiation, we must take advantage of the chain rule. If we put $x = \sin \omega t$, then

$$\frac{dx}{dt} = \cos \omega t \, \frac{d(\omega t)}{dt} = \omega \, \cos \omega t$$

and

$$\frac{d^2 x}{dt^2} = -\omega^2 \sin \omega t = -\omega^2 x$$

which is just what we wanted. Looking back at our erroneous guesses suggests that it would not hurt to put a constant factor in front, so that

$$x = B \sin \omega t$$

is also a solution for any B. Finally, we observe that $\cos \omega t$ is another solution; in fact, if A and B are any two constants, then

$$x = A \cos \omega t + B \sin \omega t \tag{4}$$

is a solution of the spring equation (3), as you may verify by differentiating (4) twice. We say that solution (4) is a *superposition* of the two solutions $\sin \omega t$ and $\cos \omega t$.

Worked Example 1   Let $x = f(t) = A \cos \omega t + B \sin \omega t$. Show that x is periodic with period $2\pi/\omega$; that is, $f(t + 2\pi/\omega) = f(t)$.

Solution   Substitute $t + 2\pi/\omega$ for t:

$$f\left(t + \frac{2\pi}{\omega}\right) = A \cos \left[\omega \left(t + \frac{2\pi}{\omega}\right)\right] + B \sin \left[\omega \left(t + \frac{2\pi}{\omega}\right)\right]$$
$$= A \cos [\omega t + 2\pi] + B \sin [\omega t + 2\pi]$$
$$= A \cos \omega t + B \sin \omega t$$
$$= f(t)$$

Here we used the fact that the sine and cosine functions are themselves periodic with period 2π.

The constants A and B are similar to the constant which arises when antiderivatives are taken. For any value of A and B we have a solution. If we assign particular values to A and B, we get a particular solution.

Worked Example 2   Find a solution of the spring equation $d^2 x/dt^2 = -\omega^2 x$ for which $x = 1$ and $dx/dt = 1$ when $t = 0$.

Solution In the solution $x = A \cos \omega t + B \sin \omega t$, we have to find A and B. When $t = 0$, $x = A$, so $A = 1$. Also, $dx/dt = \omega B \cos \omega t - \omega A \sin \omega t$; when $x = 0$, $dx/dt = \omega B$. Thus $B = 1/\omega$, and so the required solution is $x = \cos \omega t + (1/\omega) \sin \omega t$.

In general, if we are given the conditions that $x = x_0$ and $dx/dt = v_0$ when $t = 0$, then

$$x = x_0 \cos \omega t + \frac{v_0}{\omega} \sin \omega t \tag{5}$$

is the unique function of the form (4) which satisfies these conditions.

Physicists expect that the motion of a particle in a force field is completely determined once the initial values of x and dx/dt are specified. Our solution of the spring equation will meet the physicists' requirement if we can show that *every* solution of the spring equation (3) is of the form (4). Thus we wish to prove the following uniqueness theorem.

Theorem 1 *If $x = f(t)$ and $y = g(t)$ are two solutions of (3) with $f(0) = g(0)$ and $f'(0) = g'(0)$, then $f(t) = g(t)$.*

(optional)

First Proof. This is like the theorem which classifies antiderivatives, but its proof is a bit trickier. Rather than showing that the *difference* $x - y = f(t) - g(t)$ is equal to zero, it turns out to be simpler to show that the *ratio* $x/y = f(t)/g(t)$ is equal to 1. We may compute:

$$\frac{d}{dt}\left(\frac{x}{y}\right) = \frac{y(dx/dt) - x(dy/dt)}{y^2} \tag{6}$$

When $t = 0$, the numerator in (6) is zero, since $x = y$ and $dx/dt = dy/dt = 0$ there. Differentiating the numerator, we get

$$\frac{d}{dt}\left(y\frac{dx}{dt} - x\frac{dy}{dt}\right) = \frac{dy}{dt}\frac{dx}{dt} + y\frac{d^2x}{dt^2} - \frac{dx}{dt}\frac{dy}{dt} - x\frac{d^2y}{dt^2}$$

$$= y\frac{d^2x}{dt^2} - x\frac{d^2y}{dt^2}$$

$$= y(-\omega^2 x) - x(-\omega^2 y)$$

$$= 0$$

Thus the numerator in (6) is constant and as it is zero at $t = 0$, it must be identically zero. It follows that x/y is constant. But when $t = 0$, $x = y$ by assumption, so $x/y = 1$ for all t and $x = y$ for all t.

This proof is conceptually simple, but it has a technical flaw: it breaks down when the denominator $y = g(t)$ is zero. (A related proof avoiding this flaw is outlined in Problems 16 and 17 at the end of this section.)

Second Proof. There is a different way to prove Theorem 1 which is instructive. This proof is based on the fact that if $h(t)$ is a solution to (3), then the quantity $E = \{[h'(t)]^2 + [\omega \, h(t)]^2\}/2$ is constant in time. This

(optional)

expression is called the *energy* of the spring. To see that E is constant in time, we differentiate using the chain rule:

$$\frac{dE}{dt} = h'(t)h''(t) + \omega^2 h(t)h'(t) = h'(t)\{h''(t) + \omega^2 h(t)\} = 0$$

Now if f and g are solutions of (3) with $f(0) = g(0)$ and $f'(0) = g'(0)$ then $h(t) = f(t) - g(t)$ is also a solution with $h(0) = 0$ and $h'(0) = 0$. Therefore the corresponding energy $E = (1/2)\{[h'(t)]^2 + [\omega h(t)]^2\}$ is constant. But it vanishes at $t = 0$, so it is identically zero. Thus, since two non-negative numbers which add to zero must both be zero, $h'(t) = 0$ and $\omega h(t) = 0$. In particular, $h(t) = 0$ and so $f(t) = g(t)$ as required.

Any solution of the spring equation can be expressed in the form

$$x = \alpha \cos(\omega t - \theta) \tag{7}$$

where α and θ are constants. In fact, the addition formula for cosine gives

$$\alpha \cos(\omega t - \theta) = \alpha \cos \omega t \cos \theta + \alpha \sin \omega t \sin \theta$$

This will be equal to $A \cos \omega t + B \sin \omega t$ if

$$\alpha \cos \theta = A \qquad \text{and} \qquad \alpha \sin \theta = B$$

Thus α and θ must be the polar coordinates of the point whose cartesian coordinates are (A, B), and so we can always find such an α and θ with $\alpha \geq 0$. The form (7) is convenient for plotting, as shown in Fig. 9-1-2.

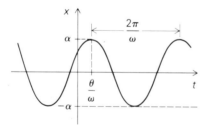

Fig. 9-1-2 The graph of $x = \alpha \cos(\omega t - \theta)$.

Worked Example 3 Sketch the graph of the solution of $(d^2x/dt^2) + 9x = 0$ satisfying $x = 1$ and $dx/dt = 6$ when $t = 0$.

Solution Using (5) with $\omega = 3$, $x_0 = 1$, and $v_0 = 6$, we have

$$x = \cos(3t) + 2\sin(3t) = \alpha \cos(3t - \theta)$$

Since $(A, B) = (1, 2)$, and (α, θ) are its polar coordinates,

$$\alpha = \sqrt{1^2 + 2^2} = \sqrt{5} \approx 2.2$$

and

$$\theta = \tan^{-1} 2 \approx 1.1 \text{ radians } (63°)$$

so $\theta/\omega \approx 0.37$. The period is $\tau = 2\pi/\omega \approx 2.1$. Thus we can plot the graph as shown in Fig. 9-1-3.

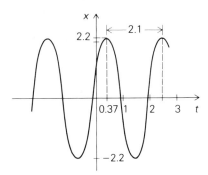

Fig. 9-1-3 The graph of $x = 2.2 \cos(3t - 1.1)$.

In Figs. 9-1-2 and 9-1-3 notice that the solution is a cosine curve which is shifted by the *phase* ω and has *amplitude* α. The number ω is called the *angular frequency*, since it is the time rate of change of the "angle" $\omega t - \theta$ at which the cosine is evaluated.

The motion described by equation (7) is called *simple harmonic motion.* It arises whenever a system is subject to a restoring force proportional to its displacement from equilibrium. Such oscillatory systems occur in physics, biology, electronics, and chemistry, so our analysis of the spring equation is of broad applicability.

SIMPLE HARMONIC MOTION

Every solution of the *spring equation* has the form $x = A \cos \omega t + B \sin \omega t$, where A and B are constants.

$$\frac{d^2 x}{dt^2} = -\omega^2 x$$

The solution can also be written $x = \alpha \cos(\omega t - \theta)$, where (α, θ) are the polar coordinates of (A, B).

If the values of x and dx/dt are specified to be x_0 and v_0 at $t = 0$, then a unique solution is determined by $x = x_0 \cos \omega t + (v_0/\omega) \sin \omega t$

Solved Exercises[*]
1. Solve for x: $d^2 x/dt^2 = -x$, $x = 0$ and $dx/dt = 1$ when $t = 0$.

2. (a) Solve for y: $(d^2 y/dx^2) + 9y = 0$, $y = 1$ and $dy/dx = -1$ when $x = 0$.

   (b) Sketch the graph of y as a function of x.

Exercises
1. Find the solution of $d^2 y/dt^2 = -4y$ for which $y = 1$ and $dy/dt = 3$ when $t = 0$.

*Solutions appear in Appendix A.

2. Let M be a weight with mass 1 gram on a spring with spring constant $\frac{3}{2}$. Let the weight be initially at an extended distance of 1 centimeter moving at a velocity of 2 centimeters per second.

   (a) How fast is M moving at $t = 3$?

   (b) What is M's acceleration at $t = 4$?

   (c) What is M's maximum displacement from the rest position? When does it occur?

   (d) Sketch a graph of the solution.

3. An observer sees a weight of 5 grams on a spring undergoing the motion $x(t) = 6.1 \cos(2t - \pi/6)$.

   (a) What is the spring constant?

   (b) What is the force acting on the weight at $t = 0$? At $t = 2$?

GROWTH AND DECAY

Many quantities, such as bank balances (see Section 7-2), population, radioactivity, and temperature, change at a rate which is proportional to the amount of the quantity present. Mathematically, this assumption can be phrased as follows: Let $f(t)$ denote the amount of the quantity present at time t; then

$$f'(t) = \gamma f(t) \tag{8}$$

where γ is a constant.

Equation (8) is a differential equation since, as an equation for the unknown function $f(t)$, it involves $f'(t)$ as well.

Worked Example 4 The temperature of a bowl of porridge decreases at a rate 0.0837 times the difference between its present temperature and room temperature (fixed at 20°C). Write down a differential equation for the temperature of the porridge. (Time is measured in minutes.)

Solution Let T be the temperature (°C) of the porridge and let $f(t) = T - 20$ be its temperature above 20°C. Then $f'(t) = dT/dt$ and so

$$f'(t) = -(0.0837)f(t)$$

Hence

$$\frac{dT}{dt} = -(0.0837)(T - 20)$$

The minus sign is used because the temperature is *decreasing*; $\gamma = -0.0837$.

We solve equation (8) by guesswork, just as we did the spring equation. The answer must be a function which reproduces itself times a constant when differentiated. Such a function is b^t, which gives $(\ln b)b^t$ when differentiated. We want $\ln b = \gamma$; that is, $e^\gamma = b$, so $b^t = e^{\gamma t}$. We are now ready to solve (8) explicitly.

Theorem 2 *For $f(0)$ given, there is one and only one solution to equation (8):*

$$f(t) = f(0)e^{\gamma t} \tag{9}$$

Proof That equation (9) solves (8) and has the given value at $t = 0$ is easy: by the chain rule

$$f'(t) = f(0) \cdot \gamma e^{\gamma t} = \gamma f(t)$$

and

$$f(0)e^{\gamma t}\big|_{t=0} = f(0) \quad \text{since } e^0 = 1$$

Next we show that this is the only solution. Suppose that $g(t)$ satisfies $g'(t) = \gamma g(t)$ and $g(0) = f(0)$. We will show that $g(t) = f(t)$, where $f(t)$ is defined by (9). To do this, consider the quotient

$$h(t) = \frac{g(t)}{e^{\gamma t}} = e^{-\gamma t} g(t)$$

and differentiate:

$$\begin{aligned} h'(t) &= -\gamma e^{-\gamma t} g(t) + e^{-\gamma t} g'(t) \\ &= -\gamma e^{-\gamma t} g(t) + \gamma e^{-\gamma t} g(t) = 0 \end{aligned}$$

Since $h'(t) = 0$, we may conclude that h is constant. We find that setting $t = 0$, this constant must be $f(0)$. Hence

$$e^{-\gamma t} g(t) = f(0)$$

Thus

$$g(t) = f(0)e^{\gamma t} = f(t)$$

as required.

Notice that the problem of zero denominators, which we encountered in the first proof of Theorem 1, does not arise here since $e^{\gamma t}$ is always positive.

Worked Example 5 Find a formula for the temperature of the bowl of porridge in Worked Example 4 if it starts at 80°C. Jane Cool refuses to eat the porridge when it is too cold—namely, if it falls below 50°C. How long does she have to come to the table?

Solution Let $f(t) = T - 20$ as in the previous solution. Then $f'(t) = -0.0837f(t)$ and $f(0) = 80 - 20 = 60$. So, by Theorem 2,

$$f(t) = 60e^{-0.0837t}$$

Hence

$$T = f(t) + 20 = 60e^{-0.0837t} + 20$$

When $T = 50$, we have

$$50 = 60e^{-0.0837t} + 20$$

$$1 = 2e^{-0.0837t}$$

$$e^{0.0837t} = 2$$

$$0.0837t = \ln 2 = 0.693$$

Thus

$$t = 8.28 \text{ minutes}$$

Jane has a little more than 8 minutes before the temperature drops to 50°C.

Note how the behavior of solution (9) depends on the sign of γ. If $\gamma > 0$, then $e^{\gamma t} \to \infty$ as $t \to \infty$ (growth); if $\gamma < 0$, then $e^{\gamma t} \to 0$ as $t \to \infty$ (decay). See Fig. 9-1-4.

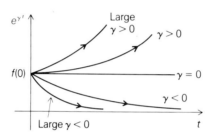

Fig. 9-1-4 Growth occurs if $\gamma > 0$, decay if $\gamma < 0$.

The situation may be summed up as follows:

NATURAL GROWTH OR DECAY

The solution of $f' = \gamma f$ is $f(t) = f(0)e^{\gamma t}$

which grows as t increases if $\gamma > 0$

and

which decays as t increases if $\gamma < 0$.

If a quantity $f(t)$ is undergoing natural growth, then by the preceding display $f(t) = f(0)e^{\gamma t}$. We notice that

$$\frac{f(t + s)}{f(t)} = \frac{f(0)e^{\gamma(t+s)}}{f(0)e^{\gamma t}} = e^{\gamma s} = \frac{f(s)}{f(0)}$$

so

$$\frac{f(t + s)}{f(t)} = \frac{f(s)}{f(0)}$$

Thus the percentage increase in f over a time interval of length s is fixed, independent of when we start. This property, characteristic of natural growth, is called *uniform growth*. It states, for example, that if you leave money in a bank (with a fixed interest rate) for a fixed period, say 6 months, then the percent increase in your deposit does not depend on the date of deposit.

We can show that if f undergoes uniform growth, then f undergoes natural growth. Indeed, take the relation

$$f(t + s) = \frac{f(t)f(s)}{f(0)}$$

and differentiate with respect to s:

$$f'(t + s) = \frac{f(t)f'(s)}{f(0)}$$

Now set $s = 0$:

$$f'(t) = \gamma f(t)$$

where $\gamma = f'(0)/f(0)$, which is the law of natural growth. Thus natural growth and uniform growth are equivalent notions.

Solved Exercises

3. The population of the planet δooμ is increasing at an instantaneous rate of 5% per year. How long until the population doubles?

4. (a) The amount of radium in a rock decreases at a rate of 0.0428% per year. How long until only half of it is left? This time interval is called the *half-life*. (b) Develop a general formula for the half-life in a decay problem.

5. A foolish king, on losing a famous bet, agrees to pay a wizard 1 cent on the first day of the month, 2 cents on the second day, 4 cents on the third, and so on, each day doubling the sum. How much is paid on the thirtieth day?

Exercises

4. (a) Solve the equation $f' = -3f$ if $f(0) = 2$. (b) Find y if $dy/dt = 8y$ and $y = 2$ when $t = 1$.

5. A certain bacterial culture undergoing natural growth doubles in size after 10 minutes. If the culture contains 100 specimens at time $t = 0$, when will the number have increased to 3000 specimens?

6. A bathtub is full of hot water at 110°F. After 10 minutes it will be 90°F. The bathroom is at 65°F. George College refuses to bathe in water below 100°F. How long can he wait to get in the tub?

THE HYPERBOLIC FUNCTIONS

Recall that $\sin t$ and $\cos t$ are solutions of the equation $(d^2x/dt^2) + x = 0$. Now we switch the sign and consider $(d^2x/dt^2) - x = 0$. (This corresponds to a negative spring constant!)

We already know one solution to this equation: $x = e^t$. Another is e^{-t}, because when we differentiate e^{-t} twice we bring down, via the chain rule, two minus signs and so recover e^{-t} again. The linear combination

$$x = A e^t + B e^{-t} \tag{10}$$

is also a solution.

If we wish to find a solution analogous to the sine function, we should pick A and B such that $x = 0$ and $dx/dt = 1$ when $t = 0$; that is,

$$0 = A + B$$
$$1 = A - B$$

so $A = \frac{1}{2}$ and $B = -\frac{1}{2}$, giving $x = (e^t - e^{-t})/2$.

Similarly, if we wish to find a solution analogous to the cosine function, we should pick A and B such that $x = 1$ and $dx/dt = 0$ when $t = 0$; that is,

$$1 = A + B$$
$$0 = A - B$$

so $A = B = \frac{1}{2}$, giving $x = (e^t + e^{-t})/2$.

This leads to the following definition.

HYPERBOLIC SINE AND COSINE

The *hyperbolic sine function*, written $\sinh t$, is defined by

$$\sinh t = \frac{(e^t - e^{-t})}{2} \tag{11}$$

The *hyperbolic cosine function*, written $\cosh t$, is defined by

$$\cosh t = \frac{(e^t + e^{-t})}{2} \tag{12}$$

(See Fig. 9-1-5.)

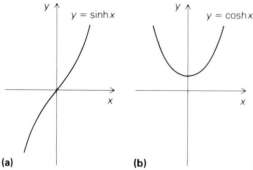

(a)   **(b)**

Fig. 9-1-5 The graph of $y = \sinh x$ and $y = \cosh x$.

The usual trigonometric functions $\sin t$ and $\cos t$ are called the *circular functions* because $(x, y) = (\cos t, \sin t)$ parametrizes the circle $x^2 + y^2 = 1$. The functions $\sinh t$ and $\cosh t$ are called *hyperbolic functions* because

$(x, y) = (\cosh t, \sinh t)$ parametrizes one branch of the hyperbola $x^2 - y^2 = 1$; that is, for any t, we have the identity

$$\cosh^2 t - \sinh^2 t = 1 \tag{13}$$

(See Fig. 9-1-6.)

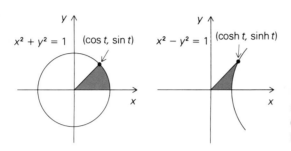

Fig. 9-1-6 The points $(\cos t, \sin t)$ lie on a circle, while $(\cosh t, \sinh t)$ lie on a hyperbola.

To prove (13), we square formulas (11) and (12):

$$\cosh^2 t = \tfrac{1}{4}(e^t + e^{-t})^2 = \tfrac{1}{4}(e^{2t} + 2 + e^{-2t})$$

and

$$\sinh^2 t = \tfrac{1}{4}(e^t - e^{-t})^2 = \tfrac{1}{4}(e^{2t} - 2 + e^{-2t})$$

Subtracting gives (13).

Worked Example 6   Prove that $e^x = \cosh x + \sinh x$.

Solution   By definition,

$$\cosh x = \frac{e^x + e^{-x}}{2} \quad \text{and} \quad \sinh x = \frac{e^x - e^{-x}}{2}$$

Adding, $\cosh x + \sinh x = e^x/2 + e^{-x}/2 + e^x/2 - e^{-x}/2 = e^x$.

The other hyperbolic functions can be introduced by analogy with the trigonometric functions:

$$\tanh x = \frac{\sinh x}{\cosh x} \qquad \coth x = \frac{\cosh x}{\sinh x}$$

$$\operatorname{sech} x = \frac{1}{\cosh x} \qquad \operatorname{csch} x = \frac{1}{\sinh x} \tag{14}$$

Various general identities can be proved exactly as we proved (13)—for instance, the addition formulas (see Solved Exercise 6):

$$\sinh(x + y) = \sinh x \cosh y + \cosh x \sinh y$$

$$\cosh(x + y) = \cosh x \cosh y + \sinh x \sinh y \tag{15}$$

Notice that in the formula for $\cosh(x + y)$ there is no minus sign. This is one of several differences in signs between the hyperbolic and circular functions. Another is in the following:

$$\frac{d}{dx}\sinh x = \cosh x \qquad (16a)$$

$$\frac{d}{dx}\cosh x = \sinh x \qquad (16b)$$

Worked Example 7   Prove formula ($16a$).

Solution   By definition $\sinh x = (e^x - e^{-x})/2$, so

$$\frac{d}{dx}\sinh x = \frac{e^x - (-1)e^{-x}}{2} = \frac{e^x + e^{-x}}{2} = \cosh x$$

We also note that

$$\sinh(-x) = -\sinh x \quad (\sinh \text{ is } odd)$$

and $\qquad\qquad\qquad\qquad\qquad\qquad\qquad\qquad\qquad$ (17)

$$\cosh(-x) = \cosh x \quad (\cosh \text{ is } even)$$

From (15), we have half-angle formulas:

$$\sinh^2 x = \frac{\cosh 2x - 1}{2} \quad \text{and} \quad \cosh^2 x = \frac{\cosh 2x + 1}{2} \qquad (18)$$

Various other analogies with the circular functions can be made (see Solved Exercise 7).

Finally, let us return to the equation $(d^2 x/dt^2) - \omega^2 x = 0$. The solution to this can be summarized as follows.

THE EQUATION $(d^2 x/dt^2) - \omega^2 x = 0$

The solution of

$$\frac{d^2 x}{dt^2} - \omega^2 x = 0 \quad (19) \qquad \text{is} \qquad x = x_0 \cosh \omega t + \frac{v_0}{\omega}\sinh \omega t \quad (20)$$

where $x = x_0$ and $dx/dt = v_0$ when $t = 0$.

That equation (20) is a solution of (19) is easy to see:

$$\frac{dx}{dt} = \omega x_0 \sinh \omega t + v_0 \cosh \omega t$$

using (16) and the chain rule. Differentiating again, we get

$$\frac{d^2x}{dt^2} = \omega^2 x_0 \cosh\omega t + \omega v_0 \sinh\omega t = \omega^2 x$$

so (19) is verified.

One may prove that (20) is the only solution of (19) just as we did for the spring equation (Problem 18).

Worked Example 8 Solve for $f(t)$: $f'' - 3f = 0$, $f(0) = 1$, $f'(0) = -2$.

Solution We use formula (20) with $\omega^2 = 3$ (so $\omega = \sqrt{3}$), $x_0 = f(0) = 1$, $v_0 = f'(0) = -2$, and with $f(t)$ in place of x. Thus

$$f(t) = \cosh\sqrt{3}\,t - \frac{2}{\sqrt{3}}\sinh\sqrt{3}\,t$$

is our solution.

The interpretation of $d^2x/dt^2 - \omega^2 x = 0$ as a spring equation with negative spring constant seems to take us away from physical reality. However, there are other, less obvious, situations which are physically realistic to which this equation applies. For example, this equation describes a freely hanging cable.* Such a cable has the same general shape as the graph of $\cosh x$.

The hyperbolic functions are also important in antidifferentiation, as we will see in the next subsection.

Solved Exercises

6. Prove the identity (15) for $\sinh(x + y)$. Use the fact that $e^x = \cosh x + \sinh x$ and $e^{-x} = \cosh x - \sinh x$.

7. Find $(d/dx)\tanh x$ and prove that $1 - \tanh^2 x = \operatorname{sech}^2 x$.

8. Find the solution of the equation $(d^2x/dt^2) - 9x = 0$, for which $x = 1$ and $dx/dt = 1$ at $t = 0$.

9. Prove that $\cosh x$ has a minimum value of 1 at $x = 0$.

Exercises

7. Prove that $\tanh^2 x + \operatorname{sech}^2 x = 1$.

8. Solve $(d^2y/dt^2) - 25y = 0$, where $y = 1$ and $dy/dt = -1$ when $t = 0$.

9. Differentiate:
    (a) $\sinh(3x + x^3)$ (b) $\cos^{-1}(\tanh x)$ (c) $3x/(\cosh x + \sinh 3x)$

THE INVERSE HYPERBOLIC FUNCTIONS

Since $(d/dx)\sinh x = \cosh x \geq 1$ (see Solved Exercise 9), $\sinh x$ is increasing. The range of $\sinh x$ is in fact $(-\infty, \infty)$ since $\sinh x \to \pm\infty$ as $x \to \pm\infty$. Thus from Theorem 1 of Chapter 5 we know that $y = \sinh x$ has an inverse function, to be denoted $\sinh^{-1} y$ by analogy with the trigonometric functions.

*A cable on a suspension bridge does not usually hang freely. It normally has a parabolic shape. (See Problem 16, Section 9-2.)

By Theorem 2 of Chapter 5,

$$\frac{d}{dy}\sinh^{-1}y = \frac{1}{(d/dx)(\sinh x)} = \frac{1}{\cosh x}$$

From $\cosh^2 x - \sinh^2 x = 1$ and the fact that $\cosh x$ is always positive, we get

$$\frac{d}{dy}\sinh^{-1}y = \frac{1}{\cosh x} = \frac{1}{\sqrt{1+\sinh^2 x}} = \frac{1}{\sqrt{1+y^2}}$$

The basic properties of $\sinh^{-1}$ are summarized in the following display.

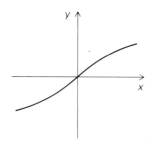

Fig. 9-1-7 The graph of $y = \sinh^{-1}x$.

INVERSE HYPERBOLIC SINE FUNCTION

1. $\sinh^{-1}x$ is the inverse function of $\sinh x$; $\sinh^{-1}x$ is defined and is increasing for all x (Fig. 9-1-7); by definition: $\sinh^{-1}x = y$ is that number such that $\sinh y = x$.

2. $(d/dx)\sinh^{-1}x = 1/\sqrt{1+x^2}$. (21)

3. $\int dx/\sqrt{1+x^2} = \sinh^{-1}x + C = \ln(x+\sqrt{1+x^2}) + C$.

The formula in part 3 of the preceding display,

$$\sinh^{-1}x = \ln(x+\sqrt{1+x^2})$$

or, using y in place of x,

$$\sinh^{-1}y = \ln(y+\sqrt{1+y^2}) \tag{22}$$

comes about as follows. We can solve $y = \sinh x = (e^x - e^{-x})/2$ explicitly for x. Multiplying through by $2e^x$ and gathering terms on the left side of the equation gives

$$2e^x y - e^{2x} + 1 = 0$$

Hence

$$(e^x)^2 - 2e^x y - 1 = 0$$

and so, by the quadratic formula,

$$e^x = \frac{2y \pm \sqrt{4y^2 + 4}}{2} = y \pm \sqrt{y^2 + 1}$$

Since e^x is positive, we must select plus and not minus, so

$$e^x = y + \sqrt{y^2 + 1} \qquad \text{and} \qquad \sinh^{-1}y = x = \ln(y+\sqrt{y^2+1})$$

Worked Example 9 Verify the formula $\int dx/\sqrt{1 + x^2} = \ln(x + \sqrt{1 + x^2}) + C$.

Solution
$$\frac{d}{dx} \ln(x + \sqrt{1 + x^2})$$

$$= \frac{1}{(x + \sqrt{1 + x^2})}\left(1 + \frac{x}{\sqrt{1 + x^2}}\right) \quad \text{(by the chain rule)}$$

$$= \left(\frac{1}{x + \sqrt{1 + x^2}}\right)\left(\frac{\sqrt{1 + x^2} + x}{\sqrt{1 + x^2}}\right)$$

$$= \frac{1}{\sqrt{1 + x^2}}$$

Thus an antiderivative for $1/\sqrt{1 + x^2}$ is $\ln(x + \sqrt{1 + x^2}) + C$.

Fig. 9-1-8 The graph of $y = \cosh^{-1}x$.

In a similar fashion we can investigate $\cosh^{-1}x$. Since $\cosh x$ is increasing on $[0, \infty)$ and has range $[1, \infty)$, $\cosh^{-1}x$ will be increasing, will be defined on $[1, \infty)$, and will have range $[0, \infty)$. Its graph can be read off from that of $\cosh x$ by the usual method of looking through the page from the other side (Fig. 9-1-8).

We find the following by the same method that we obtained (21):

$$\frac{d}{dx} \cosh^{-1}x = \frac{1}{\sqrt{x^2 - 1}} \quad (x > 1) \tag{23}$$

Worked Example 10 Find $(d/dx) \cosh^{-1}(\sqrt{x^2 + 1})$, $x \neq 0$.

Solution Let $u = \sqrt{x^2 + 1}$. Then, by the chain rule,

$$\frac{d}{dx} \cosh^{-1}(\sqrt{x^2 + 1}) = \left(\frac{d}{du} \cosh^{-1}u\right) \cdot \left(\frac{du}{dx}\right)$$

$$= \frac{1}{\sqrt{u^2 - 1}} \cdot \frac{x}{\sqrt{x^2 + 1}}$$

$$= \frac{1}{\sqrt{x^2}} \cdot \frac{x}{\sqrt{x^2 + 1}} = \frac{x}{|x|} \cdot \frac{1}{\sqrt{x^2 + 1}}$$

Therefore,

$$\frac{d}{dx} \cosh^{-1}(\sqrt{x^2 + 1}) = \begin{cases} \dfrac{1}{\sqrt{x^2 + 1}} & \text{if } x > 0 \\[4mm] \dfrac{-1}{\sqrt{x^2 + 1}} & \text{if } x < 0 \end{cases}$$

Similarly, we can consider $\tanh^{-1} x$ (see Fig. 9-1-9) and get, for $-1 < x < 1$,

$$\frac{d}{dx} \tanh^{-1} x = \frac{1}{1 - x^2} \tag{24}$$

(a) $y = \tanh x$ **(b)** $y = \tanh^{-1} x$ **Fig. 9-1-9** The graph of $y = \tanh x$ and $y = \tanh^{-1} x$.

Worked Example 11   Prove that $\tanh^{-1} x = \frac{1}{2} \ln [(1 + x)/(1 - x)]$, $-1 < x < 1$.

Solution   Let $y = \tanh^{-1} x$, so

$$x = \tanh y = \frac{\sinh y}{\cosh y} = \frac{(e^y - e^{-y})}{(e^y + e^{-y})}$$

Thus $x(e^y + e^{-y}) = e^y - e^{-y}$. Multiplying through by e^y and gathering terms on the left:

$$(x - 1)e^{2y} + x + 1 = 0$$

$$e^{2y} = \frac{1 + x}{1 - x}$$

$$2y = \ln \left(\frac{1 + x}{1 - x} \right)$$

$$y = \frac{1}{2} \ln \left(\frac{1 + x}{1 - x} \right)$$

as required.

The table on the next page is what we have found:

DIFFERENTIATION FORMULAS FOR HYPERBOLIC FUNCTIONS

Hyperbolic functions:

$$\sinh x = \frac{e^x - e^{-x}}{2} \qquad\qquad \frac{d}{dx} \sinh x = \cosh x$$

$$\cosh x = \frac{e^x + e^{-x}}{2} \qquad\qquad \frac{d}{dx} \cosh x = \sinh x$$

$$\tanh x = \frac{e^x - e^{-x}}{e^x + e^{-x}} \qquad\qquad \frac{d}{dx} \tanh x = \operatorname{sech}^2 x$$

$$\coth x = \frac{e^x + e^{-x}}{e^x - e^{-x}} \qquad\qquad \frac{d}{dx} \coth x = -\operatorname{csch}^2 x$$

$$\operatorname{sech} x = \frac{2}{e^x + e^{-x}} \qquad\qquad \frac{d}{dx} \operatorname{sech} x = -\operatorname{sech} x \tanh x$$

$$\operatorname{csch} x = \frac{2}{e^x - e^{-x}} \qquad\qquad \frac{d}{dx} \operatorname{csch} x = -\operatorname{csch} x \coth x$$

Inverse hyperbolic functions:

$$\sinh^{-1} x = \ln(x + \sqrt{x^2 + 1}) \qquad\qquad \frac{d}{dx} \sinh^{-1} x = \frac{1}{\sqrt{x^2 + 1}}$$

$$\cosh^{-1} x = \ln(x + \sqrt{x^2 - 1}) \qquad\qquad \frac{d}{dx} \cosh^{-1} x = \frac{1}{\sqrt{x^2 - 1}} \qquad |x| > 1$$

$$\tanh^{-1} x = \frac{1}{2} \ln\left(\frac{1 + x}{1 - x}\right) \qquad\qquad \frac{d}{dx} \tanh^{-1} x = \frac{1}{1 - x^2} \qquad |x| < 1$$

$$\coth^{-1} x = \frac{1}{2} \ln\left(\frac{x + 1}{x - 1}\right) \qquad\qquad \frac{d}{dx} \coth^{-1} x = \frac{1}{1 - x^2} \qquad |x| > 1$$

$$\operatorname{sech}^{-1} x = \ln\left(\frac{1 + \sqrt{1 - x^2}}{x}\right) \qquad\qquad \frac{d}{dx} \operatorname{sech}^{-1} x = \frac{-1}{x\sqrt{1 - x^2}} \qquad 0 < x < 1$$

$$\operatorname{csch}^{-1} x = \begin{cases} \ln\left(\dfrac{1 + \sqrt{1 + x^2}}{x}\right) \\[2ex] -\ln\left(\dfrac{1 + \sqrt{1 + x^2}}{-x}\right) \end{cases} \qquad \frac{d}{dx} \operatorname{csch}^{-1} x = \begin{cases} \dfrac{-1}{x\sqrt{1 + x^2}} & x > 0 \\[2ex] \dfrac{1}{x\sqrt{1 + x^2}} & x < 0 \end{cases}$$

Reading these formulas backwards, we get the corresponding antidifferentiation formulas in the following display:

ANTIDIFFERENTIATION FORMULAS FOR HYPERBOLIC FUNCTIONS

$$\int \cosh x \, dx = \sinh x + C \qquad\qquad \int \operatorname{csch}^2 x \, dx = -\coth x + C$$

$$\int \sinh x \, dx = \cosh x + C \qquad\qquad \int \operatorname{sech} x \tanh x \, dx = -\operatorname{sech} x + C$$

$$\int \operatorname{sech}^2 x \, dx = \tanh x + C \qquad\qquad \int \operatorname{csch} x \coth x \, dx = -\operatorname{csch} x + C$$

$$\int \frac{dx}{\sqrt{x^2 + 1}} = \sinh^{-1} x + C = \ln\left(x + \sqrt{x^2 + 1}\right) + C$$

$$\int \frac{dx}{\sqrt{x^2 - 1}} = \cosh^{-1} x + C = \ln\left(x + \sqrt{x^2 - 1}\right) + C \qquad |x| > 1$$

$$\int \frac{dx}{1 - x^2} = \frac{1}{2} \ln\left| \frac{1 + x}{1 - x} \right| + C \qquad |x| \neq 1$$

$$= \begin{cases} \tanh^{-1} x + C = \dfrac{1}{2} \ln\left(\dfrac{1 + x}{1 - x}\right) + C & |x| < 1 \\[3mm] \coth^{-1} x + C = \dfrac{1}{2} \ln\left(\dfrac{x + 1}{x - 1}\right) + C & |x| > 1 \end{cases}$$

$$\int \frac{dx}{x\sqrt{1 - x^2}} = -\operatorname{sech}^{-1} x + C = -\ln\left(\frac{1 + \sqrt{1 - x^2}}{x}\right) + C \qquad 0 < x < 1$$

$$\int \frac{dx}{x\sqrt{1 + x^2}} = -\operatorname{csch}^{-1} x + C = -\ln\left(\frac{1 + \sqrt{1 + x^2}}{x}\right) + C \qquad x > 0$$

Solved Exercises

10. Calculate $(d/dx)[\sinh^{-1}(3 \tanh 3x)]$.

11. Show that the antiderivative of $1/(1 - x^2)$ is $\ln[(1 + x)/(1 - x)] + C$, $|x| < 1$, by noticing that

$$\frac{1}{1 - x^2} = \frac{1}{2}\left(\frac{1}{1 - x} + \frac{1}{1 + x}\right)$$

12. Find $\sinh^{-1} 5$ numerically.

13. Find $\int dx/\sqrt{4 + x^2}$. [*Hint:* $1/\sqrt{4 + x^2} = 1/(2\sqrt{1 + (x/2)^2})$.]

Exercises

10. Differentiate:
    (a) $\sinh^{-1}(3x + \cos x)$ \qquad\qquad (b) $(x + \cosh^{-1} x)/(\sinh^{-1} x + x)$

11. Find $\tanh^{-1}(0.5)$ numerically.

12. Find $\int dx/\sqrt{3x^2 - 1}$

Problems for Section 9-1 ▬▬▬▬▬▬

1. Solve each of the following equations for x and sketch its graph:
   (a) $(d^2x/dt^2) + 4x = 0$; $x = -1$, and $dx/dt = 0$ when $t = 0$
   (b) $(dx/dt) - 3x = 0$; $x = 1$ when $t = 0$
   (c) $(d^2x/dt^2) + 25x = 0$; $x = 5$, and $dx/dt = 5$ when $t = 0$
   (d) $(d^2x/dt^2) + 25x = 0$; $x = 5$ when $t = 0$, $dx/dt = 5$ when $t = \pi/4$

2. Solve each of the following equations for $f(t)$ and sketch its graph:
   (a) $f'' = -f$, $f(0) = 5$, $f'(\pi) = 5$
   (b) $f' + 3f = 0$, $f(0) = 1$
   (c) $f' = 8f$, $f(0) = e$
   (d) $f' = 8f$, $f(1) = e$

3. A mass of 1 kg is hanging from a spring. If $x = 0$ is the equilibrium position, it is given that $x = 1$ and $dx/dt = 1$ when $t = 0$. The weight is observed to oscillate with a frequency of twice a second.
   (a) What is the spring constant?
   (b) Sketch the graph of x as a function of t, indicating the amplitude of the motion on your drawing.

4. (a) If money grows according to the formula $f' = \gamma f$, with time measured in years, we call γ "the interest rate when money is compounded continuously." Show that the *yearly* interest rate—that is, the relative increase in your money if left for a whole year—is $e^\gamma - 1$.
   (b) A credit card company advertises: "Your interest rate on the unpaid balance is 17% compounded continuously, but federal law requires us to state that your annual interest rate is 18.53%." Explain.

5. The half-life of uranium is about 0.45 billion years. If 1 gram of uranium is left undisturbed, how long will it take for 90% of it to have decayed?

6. A certain calculus textbook sells according to this formula: $S(t) = 2000 - 1000e^{-0.3t}$, where t is the time in years and $S(t)$ is the number of books sold.
   (a) Find $S'(t)$.
   (b) Find $\lim_{t \to \infty} S(t)$ and discuss.
   (c) Graph S.

7. Sketch the graph of (a) $y = \tanh 3x$; (b) $y = 3 \cosh 2x$.

8. Solve:
   (a) $y'' + 9y = 0$, $y(0) = 0$, $y'(0) = 1$
   (b) $y'' - 9y = 0$, $y(0) = 0$, $y'(0) = 1$

9. Solve:
   (a) $f'' - 81f = 0$, $f(0) = 1$, $f'(0) = -1$
   (b) $f'' + 81f = 0$, $f(0) = 1$, $f'(0) = -1$

10. Find dy/dx for:
    (a) $y = x + \sinh^{-1}(1 + x)$
    (b) $y = \cos(x + 3x \tanh^{-1} 8x)$
    (c) $[\sinh(x + y)]/xy = 1$

11. Prove that $(\cosh x + \sinh x)^n = \cosh nx + \sinh nx$.

12. It takes 300,000 years for a certain radioactive substance to decay to 30% of its original mass. What is its half-life?

13. The author of a certain calculus textbook is awake writing in the stillness of 2 AM. A sound disturbs him. He discovers that the toilet tank fills up fast at first, then slows down as the water is being shut off. Examining the insides of the tank and contemplating for a moment, he thinks that maybe during shutoff the rate of flow of water into the tank is proportional to the height left to go; that is,

    $$\frac{dx}{dt} = c(h - x)$$

    where x = height of water, h = desired height of water, and c = a constant (depending on the mechanism). Show that $x = h - Ke^{-ct}$. What is K?

    Looking at this formula for x, he says, "That explains why my tank never exactly fills up!" and goes to bed.

14. Find:
    (a) $\int \cosh 3x \, dx$
    (b) $\int dx/(4x^2 + 1)$
    (c) $\int dx/\sqrt{4x^2 + 1}$

15. Find:
    (a) $\int dx/(1 - 4x^2)$
    (b) $\int [\operatorname{csch}^2 2x + (3/x)] \, dx$
    (c) $\int dx/x\sqrt{1 - 4x^2}$

Problems 16 and 17 outline the complete proof of the following theorem using the "method of variation of constants."

Theorem 1'   Let $x = f(t)$ be a twice-differentiable function of t such that $(d^2x/dt^2) + \omega^2 x = 0$. Then $x = A \cos \omega t + B \sin \omega t$ for constants A and B.

★16. Some preliminary calculations are done first. Write

$$x = A(t) \cos \omega t + B(t) \sin \omega t \qquad (25)$$

It is possible to choose $A(t)$ and $B(t)$ in many ways, since for each t either $\sin \omega t$ or $\cos \omega t$ is nonzero. To determine $A(t)$ and $B(t)$ we add a second equation:

$$\frac{dx}{dt} = -\omega A(t) \sin \omega t + \omega B(t) \cos \omega t \qquad (26)$$

This equation is obtained by differentiating (25) *pretending* that $A(t)$ and $B(t)$ are constants. Since this is what we are trying to prove, we should be very suspicious here of circular reasoning. But push on and see what happens. By multiplying (25) by $\omega \sin \omega t$ and (26) by $\cos \omega t$ and adding, show that

$$B(t) = x \sin \omega t + \frac{dx/dt}{\omega} \cos \omega t \qquad (27)$$

Similarly, show that

$$A(t) = x \cos \omega t - \frac{dx/dt}{\omega} \sin \omega t \qquad (28)$$

★17. Use the calculations in Problem 16 to give the proof of Theorem 1, making sure to avoid circular reasoning. We are given $x = f(t)$ and ω such that $(d^2x/dt^2) + \omega^2 x = 0$. *Define* $A(t)$ and $B(t)$ by equations (27) and (28). Show that $A(t)$ and $B(t)$ are in fact constants by differentiating (27) and (28) to show that $A'(t)$ and $B'(t)$ are identically zero. Then rewrite (27) and (28) as

$$B = x(t) \sin \omega t + \frac{dx/dt}{\omega} \cos \omega t \qquad (29)$$

$$A = x(t) \cos \omega t - \frac{dx/dt}{\omega} \sin \omega t \qquad (30)$$

Use these formulas to show that

$$A \cos \omega t + B \sin \omega t = x$$

which proves the theorem.

★18. Prove that the equation $(d^2x/dt^2) - \omega^2 x = 0$, with $x = x_0$ and $dx/dt = v_0$ when $t = 0$, has a unique solution given by $x = x_0 \cosh \omega t + v_0/\omega \sinh \omega t$.

★19. (a) Suppose that $f(t)$ is given and that $y = g(t)$ satisfies $(d^2y/dt^2) + \omega^2 y = f(t)$. Show that $x = y + A \sin \omega t + B \cos \omega t$ represents the general solution of $(d^2x/dt^2) + \omega^2 x = f$; that is, x is a solution and any solution has this form. One calls y a *particular* solution and x the *general* solution.

(b) Solve $(d^2x/dt^2) + \omega^2 x = k$ if $x = 1$ and $dx/dt = -1$ when $t = 0$; k is a nonzero constant.

(c) Solve $(d^2x/dt^2) + \omega^2 x = \omega^2 t$ if $x = -1$ and $dx/dt = 3$ when $t = 0$.

20. Suppose that $x = f(t)$ satisfies the spring equation. Let $g(t) = at + b$, where a and b are constants. Show that if the composite function $f \circ g$ satisfies the spring equation, then $a = \pm 1$.

21. (a) Verify that the solution of $dy/dt = p(t)y$ is $y = y_0 \exp P(t)$, where $P(t)$ is the antiderivative of $p(t)$ with $P(0) = 0$.

(b) Solve $dy/dt = ty$; $y = 1$ when $t = 0$.

9-2 Techniques of Differential Equations

The solution of first-order, separable, and linear equations can be reduced to integration.

The previous section dealt with detailed methods for solving two particular types of differential equations, namely the spring equation and the equation of growth or decay. This section treats a few other classes of differential equations that can be solved explicitly and discusses a few properties of differential equations in general.

Goals

After studying this section, you should be able to:

Recognize and solve separable and linear first-order differential equations.

Translate simple physical problems into the language of differential equations and interpret the solution.

SEPARABLE DIFFERENTIAL EQUATIONS

A differential equation of the form

$$\frac{dy}{dx} = g(x)h(y)$$

in which the right-hand side factors into a product of a function of x and a function of y is called *separable*. The object is to find a function $y = f(x)$ that satisfies this equation. Note that we use the term separable only for *first-order* equations; that is, equations involving only the first derivative of y with respect to x.

We may formally solve the above separable equation by rewriting it in differential notation

$$\frac{dy}{h(y)} = g(x)\, dx$$

and integrating:

$$\int \frac{dy}{h(y)} = \int g(x)\, dx + C$$

(See Section 8-2 for justification of the differential notation.) If the integrations can be carried out, we obtain an expression relating x and y. If this expression can be solved for y, the problem is solved; otherwise one has an equation that implicitly defines y in terms of x. The constant of integration may be determined by giving a value y_0 to y for a given value x_0 of x; that is, by specifying *initial conditions*.

Worked Example 1   Solve $dy/dx = -3xy$, $y = 1$ when $x = 0$.

Solution   We have

$$\frac{dy}{y} = -3x \, dx$$

Integrating both sides gives

$$\ln |y| = -\frac{3x^2}{2} + C$$

and so

$$y = \pm \exp C \exp(-3x^2/2)$$

Since $y = 1$ when $x = 0$, we choose the $+$ solution and $C = 0$, to give

$$y = \exp(-3x^2/2)$$

The reader may check by using the chain rule that this function satisfies the given differential equation.

The equation of growth (or decay) $y' = \gamma y$ studied in the previous section is clearly separable, and the technique outlined above reproduces our solution $y = Ce^{\gamma x}$. The spring equation is *not* separable since it is *second order,* that is, it involves the second derivative of y with respect to x.

SEPARABLE DIFFERENTIAL EQUATIONS

To solve the equation

$$y' = g(x)h(y):$$

1. write

$$\frac{dy}{h(y)} = g(x) \, dx$$

2. integrate both sides:

$$\int \frac{dy}{h(y)} = \int g(x) \, dx + C,$$

3. solve for y if possible,
4. the constant of integration C is determined by a given value of y at a given value of x, that is, by given *initial conditions.*

Worked Example 2   Solve $dy/dx = y^2$, $y = 1$ when $x = 1$, and sketch the solution.

Solution   Here we get

$$\frac{dy}{y^2} = dx$$

and so integrating,

$$-\frac{1}{y} = x + C$$

Since $y = 1$ when $x = 1$, $C = -2$. Thus, solving for y,

$$y = \frac{1}{2 - x}$$

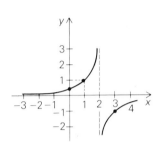

Fig. 9-2-1 The solution of $y' = y^2$, $y(1) = 1$.

The solution is sketched in Fig. 9-2-1. Notice that the solution has an asymptote at $x = 2$; thus it is undefined at $x = 2$. Extending the solution beyond $x = 2$ doesn't really make sense since that portion beyond $x = 2$ can be changed, for instance, to $1/(3 - x)$; the resulting function is still a solution, satisfying $y' = y^2$ and $y = 1$ at $x = 1$. Thus, an attempt to extend beyond $x = 2$ leads to a breakdown of continuity and uniqueness of the solution. For this reason, when dealing with differential equations, we usually confine our attention to intervals throughout which y is differentiable and satisfies the equation and initial conditions.

A Remark on Notation.   Up to now we have distinguished *variables*, which are mathematical objects that represent "quantities," and *functions*, which represent relations between quantities. Thus, when $y = f(x)$, we have written $f'(x)$ and dy/dx but *not* y', df/dx or $y(x)$. It is common in mathematical writing to use the same symbol to denote a function and its dependent variable; thus one sometimes writes $y = y(x)$ to indicate that y is a function of x and then writes "$y' = dy/dx$," "$y(3)$ is the value of y when $x = 3$" and so on. Beginning with the Solved Exercises below we will occasionally drop our scruples in distinguishing functions from variables and will use this abbreviated notation.

Solved Exercises

1. Solve $yy' = \cos 2x$, $y(0) = 1$.

2. Solve $dy/dx = x/(y + yx^2)$, $y(0) = -1$.

3. Solve $y' = x^2 y^2 + x^2 - y^2 - 1$, $y(0) = 0$.

Exercises

1. Solve $dy/dx = \cos x$, $y(0) = 1$.

2. Solve $dy/dx = y \cos x$, $y(0) = 1$.

3. Solve $dy/dx = 2xy - 2y + 2x - 2$, $y(1) = 0$.

4. Solve $dy/dx = (1 + y)/(1 + x)$, $y(0) = 1$.

APPLICATIONS OF SEPARABLE EQUATIONS

Several interesting physical problems can be solved by using separable differential equations. We will give a sample of some of these here.

Worked Example 3 *(Electric circuits)* We are told that the equation governing the electric circuit shown in Fig. 9-2-2 is

Fig. 9-2-2 A simple electric circuit.

$$L \frac{dI}{dt} + RI = E$$

and that in this case,

E = voltage, a constant,

R = resistance, a constant > 0

L = inductance, a constant > 0

and I = current, a function of time. Solve this equation for I with a given value I_0 at $t = 0$.

Solution Rewrite the differential equation as follows:

$$L \frac{dI}{dt} = E - RI$$

$$\frac{L}{E - RI} \frac{dI}{dt} = 1$$

The variables are separated, so we may integrate:

$$-\frac{L}{R} \ln |E - RI| = t + C$$

Thus

$$|E - RI| = \exp \left[-(t + C) R/L \right]$$

and so

$$E - RI = \pm \exp \left(-\frac{R}{L} t \right) \exp \left(-\frac{R}{L} C \right)$$

$$= A \exp \left(-\frac{Rt}{L} \right), \quad A = \pm \exp \left(-\frac{R}{L} C \right)$$

At $t = 0$, $I = I_0$, so $E - RI_0 = A$. Substituting this in the previous equation and simplifying gives

$$I = \frac{E}{R} + \left(I_0 - \frac{E}{R} \right) e^{-Rt/L}$$

Fig. 9-2-3 The current tends to the constant value E/R as $t \to \infty$.

As $t \to \infty$, $I \to \dfrac{E}{R}$, the *steady state part*. The term $\left(I_0 - \dfrac{E}{R} \right) e^{-Rt/L}$, which approaches zero as $t \to \infty$, is called the *transient part* of I. See Fig. 9-2-3.

Worked Example 4 ***(Predator-prey equations)*** Consider x predators that prey on y prey. Their numbers change as t changes. Imagine the following model (called the *Lotka-Volterra model*).

(i) the prey increase by normal population growth (studied in Section 9-1), at a rate by (b is a positive birth rate constant), but decrease at a rate proportional to the number of predators and the number of prey, that is, $-rxy$ (r is a positive death rate constant). Thus

$$\frac{dy}{dt} = by - rxy$$

(ii) the predators' population decreases at a rate proportional to their number due to natural decay (starvation) and increases at a rate proportional to the number of predators and the number of prey, that is,

$$\frac{dx}{dt} = -sx + cxy$$

for constants of starvation and consumption s and c.

If we eliminate t by writing (see Section 5-2)

$$\frac{dy}{dt} \bigg/ \frac{dx}{dt} = \frac{dy}{dx}$$

we get

$$\frac{dy}{dx} = \frac{by - rxy}{-sx + cxy}$$

Solve this equation implicitly.

Solution   The variables separate:

$$\frac{dy}{dx} = \frac{(b - rx)y}{x(-s + cy)}$$

$$\left(\frac{cy - s}{y} \right) dy = \left(\frac{b - rx}{x} \right) dx$$

$$cy - s \ln|y| = b \ln|x| - rx + C$$

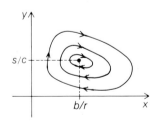

Fig. 9-2-4 Solutions of the predator-prey equation.

for a constant C. This is an implicit form for the parametric curves followed by the predator-prey population. One can show that these curves are closed curves "centered" at $(b/r, s/c)$, as shown in Fig. 9-2-4.* Variants of this model are important in ecology for predicting and studying cyclic variations in populations. (For example, this simple model already shows that if one uses an insecticide on a predator insect species, it may lead to a dramatic increase in the population of its prey, followed by an increase in the predators and so on, in cyclic fashion. Similar remarks hold for foxes and rabbits, etc.)

---

*For a proof due to Volterra, see G. F. Simmons, Differential Equations, *McGraw-Hill* (1972) p. 286. There is also a good deal of information, including many references, in Chapter 9 of Elementary Differential Equations and Boundary Value Problems *by W. Boyce and R. DiPrima, Third Edition, Wiley (1977).*

Solved Exercises

4. *Orthogonal trajectories:* Consider the family of parabolas $y = kx^2$ for various constants k. (a) Find a differential equation satisfied by this family that does not involve k by differentiating and eliminating k. (b) Write a differential equation for a family of curves orthogonal to each of the parabolas $y = kx^2$ and solve it. Sketch.

5. *Falling object in a resisting medium:* The downward force acting on a body of mass m falling in air is mg, where g is the gravitational constant. The force of air resistance is γv, where γ is a constant of proportionality and v is the downward speed. If a body is released from rest, find its velocity as a function of time t. (Assume it is released from a great enough height so that it does not hit the ground before time t.)

Exercises

5. (a) Find a differential equation satisfied by the family of hyperbolas $xy = k$ for various constants k.

   (b) Find a differential equation satisfied by the orthogonal trajectories to the hyperbolas $xy = k$, solve it, and sketch the resulting family of curves.

6. *Capacitor equation:* The equation $R(dQ/dt) + Q/C = E$ describes the charge Q on a capacitor, where R, C, E are constants. (See Fig. 9-2-5.) (a) Find Q as a function of time if $Q = 0$ at $t = 0$. (b) How long does it take for Q to attain 99% of its limiting charge?

Fig. 9-2-5 A circuit with a charging capacitor.

7. *Falling object with drag resistance:* Redo Solved Exercise 5 assuming the resistance is proportional to the square of the velocity.

DIRECTION FIELDS AND FLOW LINES

Separable differential equations are a special case of the equation

$$\frac{dy}{dx} = F(x, y)$$

where F is a function depending on both x and y. For example,

$$\frac{dy}{dx} = -x^2 y + y^3 + 3 \sin y + 1$$

is a differential equation that is not separable. There is little hope of solving such equations explicitly, except in rather special cases, such as the separable case. In general, one has to resort to numerical or other approximate methods in such cases. To do so, it is useful to have a geometric picture of what is going on.

The given data $dy/dx = F(x, y)$ tell us the slope of the solution $y = f(x)$ that we seek at each point. We can therefore imagine drawing small lines in the xy plane, with slope $F(x, y)$ at the point (x, y), as in Fig. 9-2-6.

The problem of finding a solution to the differential equation is precisely the problem of threading our way through this *direction field* with a curve which is tangent to the direction field. See Fig. 9-2-7.

Fig. 9-2-6 A plot of a direction field.

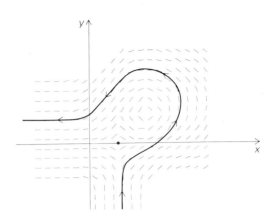

Fig. 9-2-7 A solution threads its way through the direction field.

We saw in Worked Example 4 that differential equations may be given in parametric form

$$\frac{dx}{dt} = g(x, y)$$

$$\frac{dy}{dt} = h(x, y)$$

(So $dy/dx = h(x, y)/g(x, y)$.) Here we seek a parametric curve $(x(t), y(t))$ solving these two equations. From our discussion of parametric curves in Section 5-2 we see that the pair $(g(x, y), h(x, y))$ gives the velocity of the solution curve passing through (x, y). In this formulation, we can interpret Fig. 9-2-6 as a *velocity field*. If one thinks of the motion of a fluid, one can phrase the problem of finding solutions to the above pair of differential equations as follows: given the velocity field of a fluid, find the paths that fluid particles follow. For this reason, a solution curve is often called a *flow line*.

Solved Exercises

6. Sketch the direction field for the equation $dy/dx = -x/y$ and solve the equations.

7. *Euler method:* The simplest numerical technique for solving differential equations replaces the actual solution curve by a polygonal line and follows the direction field by moving a short distance along a straight line. For $dy/dx = F(x, y)$ we start at (x_0, y_0) and break up the interval $[x_0, x_0 + a]$ into n steps x_0, $x_1 = x_0 + \dfrac{a}{n}$, $x_2 = x_0 + \dfrac{2a}{n}$, ..., $x_n = x_0 + a$. Now we recursively define

$$y_1 = F(x_0, y_0)\frac{a}{n} + y_0$$

$$y_2 = F(x_1, y_1)\frac{a}{n} + y_1$$

$$\vdots \qquad\qquad \vdots$$

$$y_n = F(x_{n-1}, y_{n-1})\frac{a}{n} + y_{n-1}$$

that is,

$$y_i - y_{i-1} = \left[\frac{dy}{dx}(x_{i-1}, y_{i-1}) \right] (x_i - x_{i-1}), \quad i = 1, 2, \ldots, n$$

to produce the desired curve shown in Fig. 9-2-8.*

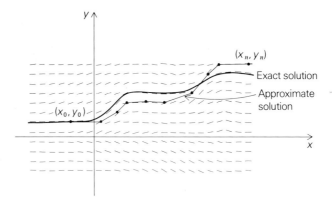

Fig. 9-2-8 The Euler method for numerically solving differential equations.

Solve the equation $dy/dx = x + \cos y$, $y(0) = 0$ from $x = 0$ to $x = \pi/4$ using a ten-step Euler method; that is, find $y(\pi/4)$ approximately.

Exercises

8. Sketch the direction field for the equation $y' = -y/x$. Solve the equation and show that the solutions are consistent with your direction field.

9. Use a ten-step Euler method to find y approximately at $x = 1$ if $dy/dx = y - x^2$ and $y(0) = 1$.

LINEAR FIRST-ORDER EQUATIONS

We have seen that separable equations can be solved directly by integration. There are a few other classes that can be solved by reducing them to integration after a suitable transformation. We shall treat a case in point now.†

We are concerned with equations that are linear in the unknown function y:

$$\frac{dy}{dx} = P(x)y + Q(x) \tag{1}$$

for given functions P and Q of x. If Q is absent, the equation (1) becomes

$$\frac{dy}{dx} = P(x)y \tag{2}$$

---

*The Euler or related methods are particularly easy to use with a programmable calculator. In practice, the Euler method is not the most accurate or efficient. Usually the Runge-Kutta or predictor-corrector method is more accurate. For details and comparative error analyses, a book such as C. W. Gear, Numerical Initial Value Problems in Ordinary Differential Equations, Prentice-Hall (1971) should be consulted.
†Another class of differential equations called exact equations is discussed in Chapter 18.

which is separable:

$$\frac{1}{y}\, dy = P(x)\, dx$$

$$\ln |y| = \int P(x)\, dx + C$$

$$|y| = \exp(C) \exp \left(\int P(x)\, dx \right)$$

Choosing $C = 0$ and $y > 0$ gives the particular solution

$$y = \exp \left(\int P(x)\, dx \right) \tag{3}$$

Now we use the solution (3) of equation (2) to help us simplify equation (1). If y solves (1), we divide it by the solution (3) obtaining a new function

$$w = y \exp \left(-\int P(x)\, dx \right) \tag{4}$$

and see what equation it satisfies. By the product and chain rules we get

$$\frac{dw}{dx} = (dy/dx) \exp \left(-\int P(x)\, dx \right) - y\, P(x) \exp \left(\int -P(x)\, dx \right)$$

$$= [P(x)y + Q(x)] \exp \left(-\int P(x)\, dx \right) - P(x)y \exp \left(\int -P(x)\, dx \right)$$

The terms involving y cancel, leaving

$$\frac{dw}{dx} = Q(x) \exp \left(-\int P(x)\, dx \right)$$

But this equation has the solution

$$w = \int Q(x) \left[\exp \left(-\int P(x)\, dx \right) \right] dx + C \tag{5}$$

Combining (4) and (5) gives the general solution y of (1). The situation is summarized in the following display.

LINEAR FIRST-ORDER EQUATIONS

The general solution of

$$\frac{dy}{dx} = P(x)y + Q(x) \quad \text{is} \quad y = \exp \left(\int P(x)\, dx \right) \left\{ \int \left[Q(x) \exp \left(-\int P(x)\, dx \right) dx \right] + C \right\} \tag{6}$$

where C is a constant

One may verify by direct substitution that the expression (6) for y in this display solves equation (1). Instead of memorizing the formula for the solution, it may be easier to remember the method as summarized in the following display.

METHOD FOR SOLVING $dy/dx = P(x)y + Q(x)$

1. Calculate $\int P(x)\,dx$, dropping the integration constant.
2. Transpose $P(x)y$ to the left side: $dy/dx - P(x)y = Q(x)$.
3. Multiply the equation by $\exp(-\int P(x)\,dx)$ (called the *integrating factor*).
4. Use the identity

$$\left[\exp\left(-\int P(x)\,dx\right)\right][(dy/dx) - P(x)y] = (d/dx)\left[y \cdot \exp\left(-\int P(x)\,dx\right)\right]$$

to rewrite the equation as

$$(d/dx)\left[y \cdot \exp\left(-\int P(x)\,dx\right)\right] = Q(x)\exp\left(-\int P(x)\,dx\right).$$

Then integrate both sides, keeping the constant of integration.
5. Solve the resulting equation for y.

Worked Example 5   Solve $dy/dx = xy + x$.

Solution   We follow the five-step procedure in the above display.

1. $P(x) = x$, so $\displaystyle\int P(x)\,dx = \frac{x^2}{2}$

2. $\dfrac{dy}{dx} - xy = x$

3. $\exp\left(-\dfrac{x^2}{2}\right)\left\{\dfrac{dy}{dx} - xy\right\} = \exp\left(-\dfrac{x^2}{2}\right)x$

4. $\dfrac{d}{dx}\left\{y\exp\left(-\dfrac{x^2}{2}\right)\right\} = x\exp\left(-\dfrac{x^2}{2}\right)$

$y\exp\left(-\dfrac{x^2}{2}\right) = \displaystyle\int x\exp\left(-\dfrac{x^2}{2}\right)dx = -\exp\left(-\dfrac{x^2}{2}\right) + C$

5. $y = C\exp\left(\dfrac{x^2}{2}\right) - 1$

Worked Example 6   (Electric circuits)   In Worked Example 3, replace E by the sinusoidal voltage

$$E = E_0 \sin \omega t$$

with L, R, E_0 constants and solve the resulting equation.

Solution The equation is

$$\frac{dI}{dt} = -\frac{RI}{L} + \frac{E_0}{L}\sin\omega t$$

We follow the five-step procedure with x replaced by t and y by I:

1. $P(t) = -\dfrac{R}{L}$, a constant, so $\int P(t)\,dt = -tR/L$

2. $\dfrac{dL}{dt} + \dfrac{RI}{L} = \dfrac{E_0}{L}\sin\omega t$

3. $\exp\left\{\dfrac{tR}{L}\right\}\left\{\dfrac{dI}{dt} + \dfrac{RI}{L}\right\} = \dfrac{E_0}{L}\exp\left(\dfrac{tR}{L}\right)\sin\omega t$

4. $\dfrac{d}{dt}\left\{\exp\left(\dfrac{tR}{L}\right)I\right\} = \dfrac{E_0}{L}\exp\left(\dfrac{tR}{L}\right)\sin\omega t$

$$\exp\left(\frac{tR}{L}\right)I = \frac{E_0}{L}\int\exp\left(\frac{tR}{L}\right)\sin\omega t\,dt$$

This integral may be evaluated by the method of Solved Exercise 5, Section 8-3, namely integration by parts twice. One gets

$$\exp\left(\frac{tR}{L}\right)I = \frac{E_0}{L}\left\{\frac{e^{tR/L}}{(R/L)^2 + \omega^2}\left(\frac{R}{L}\sin\omega t - \omega\cos\omega t\right)\right\} + C$$

5. Solving for I,

$$I = \frac{E_0}{L}\frac{1}{(R/L)^2 + \omega^2}\left(\frac{R}{L}\sin\omega t - \omega\cos\omega t\right) + Ce^{-tR/L}$$

The constant C is determined by the value of I at $t = 0$. This expression for I contains an oscillatory part, oscillating with the same frequency ω as the driving voltage (but with a phase shift; see Exercise 13) and a transient part $Ce^{-tR/L}$ which decays to zero as $t \to \infty$. See Fig. 9-2-9.

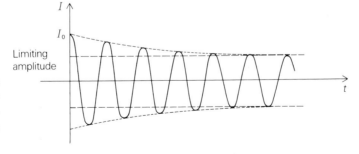

Fig. 9-2-9 The nature of the solution of a sinusoidally forced electric circuit containing a resistor and an inductor.

Solved Exercises   Solve the equations in Solved Exercises 8 and 9 with the stated initial conditions.

8. $y' = e^{-x} - y$, $y(0) = 1$

9. $y' = \cos^2 x - (\tan x)y$, $y(0) = 1$

10. *Pollution:* A small lake contains 4×10^7 liters of pure water at $t = 0$. A polluted stream carries 0.67 liter of pollutant and 10 liters of water into the lake per second. (Assume this mixes with the lake water instantly, for simplicity.) Meanwhile, 10.67 liters per second of the lake flow out in a drainage stream. Find the amount of pollutant in the lake as a function of time. What is the limiting value?

Exercises

10. Solve $y' = \dfrac{y}{x} + x$, $y(1) = 1$.

11. Solve $xy' = e^x - y$, $y(1) = 0$.

12. Solve $y' = y + \cos 5x$, $y(0) = 0$.

13. In Worked Example 6, use the method of Worked Example 3 in Section 9-1 to determine the amplitude and phase of the oscillatory part

$$\frac{E_0}{L} \cdot \frac{1}{(R/L)^2 + \omega^2} \left\{ \frac{R}{L} \sin \omega t - \omega \cos \omega t \right\}$$

14. In Solved Exercise 10, show that the lake will reach 90% of its limiting pollution value within 3.33 months.

Problems for Section 9-2 ■■■■■■■■■■■■■■■■■■■■■■

1. Solve the following differential equations (express your answer implicitly if necessary):

   (a) $y(dy/dx) = x$, $y(0) = 1$

   (b) $\dfrac{1}{y}(dy/dx) = \dfrac{1}{x}$, $y(1) = -2$

   (c) $dy/dx = xe^{-y}/(x^2 + 1)y$, $y(0) = 1$

   (d) $dy/dx = 3xy - x$

   (e) $dy/dx = \sin 2x - y \cos x$

2. Solve the following differential equations (express your answer implicitly if necessary):

   (a) $e^y (dy/dx) = 1 + e^{2y} - xe^{2y} - x$, $y(0) = 1$

   (b) $x(dy/dx) = \sqrt{1 - y^2}$, $y(1) = 0$.

   (c) $dy/dx = y^3 \sin x/(1 + 8y^4)$, $y(0) = 1$

   (d) $dy/dx = x^3 y - x^3$

   (e) $x(dy/dx) = y + x \ln x$

3. Suppose $y = f(x)$ solves the equation $dy/dx = e^x y^2 + 4xy^5$, $f(0) = 1$. Calculate $f'''(0)$.

4. Suppose $y'' + 3(y')^3 + 8e^x y^2 = 5 \cos x$, $y'(0) = 1$, and $y''(0) = 2$. Calculate $y'''(0)$.

5. (a) Find a differential equation satisfied by the family of ellipses, $x^2 + 4y^2 = k$, k a constant.

   (b) Find a differential equation satisfied by the orthogonal trajectories to this family of ellipses, solve it and sketch.

6. *Logistic law of population growth:*[*] If a population can support only P_0 members, the rate of growth of the population may be given by $dP/dt = kP(P_0 - P)$. This modification of the law of growth $dP/dt = \alpha P$ discussed in Section 9-1 is called the logistic law, or the Verhulst law. Solve this equation and show that P tends to P_0 as $t \to \infty$. Hint: In solving the equation you may wish to use the identity

$$\frac{1}{P(P_0 - P)} = \frac{1}{P_0}\left(\frac{1}{P} + \frac{1}{P_0 - P}\right)$$

[*]*See M. Braun,* Differential Equations and their Applications, *Second Edition, Springer (1978) for a detailed discussion.*

7. *Chemical reaction rates:* Chemical reactions often proceed at a rate proportional to the concentrations of each reagent. For example, consider a reaction of the type $2A + B \rightarrow C$ in which two molecules of A and one of B combine to produce one molecule of C. Concentrations are measured in moles per liter, where a mole is a definite number (6×10^{23}) of molecules. Let the concentrations of A, B, and C at time t be a, b, c, and suppose that $c = 0$ at $t = 0$. Since no molecules are destroyed, $b_0 - b = (a_0 - a)/2$, where a_0 and b_0 are the values of a and b at $t = 0$. The rate of change of a is given by $da/dt = ka^2 b$ for a constant k. Solve this equation. [*Hint:* You will need to make up an identity like the one used in Problem 6.]

8. (a) Sketch the direction field for the equation $dy/dx = 2y/x$.

   (b) Solve this equation.

9. Sketch the direction field for the equation $dy/dx = -3y + x$ and solve the equation.

10. Assuming $P(x)$ and $Q(x)$ are continuous functions of x, prove that the problem $y' = P(x)y + Q(x)$, $y(0) = y_0$ has *exactly* one solution.

11. Express the solution of the equation $y' = xy + 1$, $y(0) = 1$ in terms of an integral.

12. The equation $R(dI/dt) + (I/C) = dE/dt$ describes the current I in a circuit containing a resistor with resistance R (a constant) and a capacitor with capacitance C (a constant) as shown in Fig. 9-2-10. If $E = E_0 \sin \omega t$ and $I = 0$ at $t = 0$, find I as a function of time. Discuss your solution.

Fig. 9-2-10 A resistor and a capacitor in an electric circuit.

13. *Bernoulli's equation:* This equation has the form $dy/dx = P(x)y + Q(x)y^n$, $n = 2, 3, 4, \cdots$.

   (a) Show that the equation satisfied by $w = y^{1-n}$ is linear in w.

   (b) Use (a) to solve the equation $x(dy/dx) = x^4 y^3 - y$.

14. *Riccati equation:* This equation is $dy/dx = P(x) + Q(x)y + R(x)y^2$.

   (a) Let $y_1(x)$ be a known solution (found by inspection). Show that the general solution is $y(x) = y_1(x) + w(x)$ where w satisfies the Bernoulli equation $dw/dx = [Q(x) + 2R(x)y_1(x)]w + R(x)w^2$. (See Problem 13.)

   (b) Use (a) to solve the Riccati equation $y' = \dfrac{y}{x} + x^3 y^2 - x^5$, taking $y_1(x) = x$.

15. *Mixing problem:* A lake contains 5×10^8 liters of water, into which is dissolved 10^4 kilograms of salt at $t = 0$. Water flows into the lake at a rate of 100 liters per second and contains 1% salt; water flows out of the lake at the same rate. Find the amount of salt in the lake as a function of time. When is 90% of the limiting amount of salt reached?

16. *Hanging cable:* Consider a freely hanging cable which weighs m kilograms per meter and is subject to a tension T_0. See Fig. 9-2-11. It can be shown* that the position of the cable satisfies

$$\frac{d^2y}{dx^2} = \frac{mg}{T_0}\sqrt{1 + \left(\frac{dy}{dx}\right)^2}$$

Fig. 9-2-11 A cable hanging under its own weight.

By introducing $w = dy/dx$, show that w satisfies a separable equation. Solve for w and hence y. Express your answer in terms of the hyperbolic cosine function.

★17. Consider a family of curves defined by a separable equation $dy/dx = g(x)h(y)$. Express the family of orthogonal trajectories implicitly in terms of integrals.

18. An object of mass 10 kg is dropped from a balloon. The force of air resistance is $0.07v$ where v is the velocity. What is the object's velocity as a function of time? How far has the object traveled before it is within 10% of its terminal velocity?

19. If $y' = x + \tan y$, $y(0) = 0$, find $y(1)$ approximately using a ten-step Euler procedure.

20. Find an approximate solution for $y(1)$ if $y' = x\sqrt{1 + y^4}$ and $y(0) = 0$ using a fifteen-step Euler method.

21. Redo Solved Exercise 7 using a twenty-step Euler method, compare the answers, and discuss.

---

*See, for instance, T. M. Creese and R. M. Haralick, *Differential Equations for Engineers*, *McGraw-Hill (1978) pp. 71–75.*

Review Problems for Chapter 9

1. Solve the differential equations:
   (a) $dy/dt = 3y$, $y(0) = 1$
   (b) $d^2y/dt^2 + 3y = 0$, $y(0) = 0$, $y'(0) = 1$
   (c) $dy/dt = 3y + 1$, $y(0) = 1$
   (d) $dy/dt = y$, $y(0) = 1$

2. Solve the following equations:
   (a) $dy/dt = (\cos t)y + \cos t$, $y(0) = 0$
   (b) $d^2y/dt^2 + 9y = 0$, $y(0) = 2$, $y'(0) = 0$
   (c) $dy/dt = t^3 y^2$, $y(0) = 1$
   (d) $dy/dt + 10y = 0$, $y(1) = 1$

3. Sketch the solution of $(d^2x/dt^2) + 9x = 0$, $x(0) = 1$, $x'(0) = 0$.

4. Solve $(d^2y/dx^2) + (dy/dx) = x$, $y(0) = 0$, $y'(0) = 1$. [*Hint:* What equation does $w = dy/dx$ satisfy?]

5. If an object cools from 100°C to 80°C in an environment of 18°C in 8 minutes, how long will it take to cool to 50°C?

6. Solve for f:
   (a) $f' = 4f$; $f(0) = 1$
   (b) $f'' = 4f$; $f(0) = 1$, $f'(0) = 1$
   (c) $f'' = -4f$; $f(0) = 1$, $f'(0) = 1$

7. Solve for x:
   (a) $dx/dt = -4x$; $x = 1$ when $t = 0$
   (b) $(d^2x/dt^2) + x = 0$; $x = 1$ when $t = 0$, $x = 0$ when $t = \pi/4$
   (c) $(d^2x/dt^2) + 6x = 0$; $x = 1$ when $t = 0$, $x = 6$ when $t = 1$

8. (a) The oil consumption rate satisfies the equation $C(t) = C_0 e^{rt}$ where C_0 is the consumption rate at $t = 0$ (number of barrels per year) and r is a constant. If the consumption rate is $C_0 = 2.5 \times 10^{10}$ barrels per year in 1976 and $r = 0.06$, how long will it take before 2×10^{12} barrels (the total world's supply) are used up?

   (b) As the fuel is almost used up, the prices will probably skyrocket and other sources of energy will be turned to. Let $S(t)$ be the supply left at time t. Assume that $dS/dt = -\alpha S$, where α is a constant (the panic factor). Find $S(t)$.

9. Solve $y'' + 3yy' = 0$ for $y(x)$ if $y(0) = 1$, $y'(0) = 0$. $\left[Hint: yy' = \left(\dfrac{y^2}{2} \right)'. \right]$

10. (a) The population of the United States in 1900 was about 76 million; in 1910, about 92 million. Assume that the population growth is uniform, so $f(t) = e^{\gamma t} f(1900)$, t in years after 1900. (i) Show that $\gamma \approx 0.0191$. (ii) What would have been a reasonable prediction for the population in 1960? In 1970? (iii) At this rate, how long does it take for the population to double?

    (b) The actual U.S. population in 1960 was about 179 million and in 1970 about 203 million. (i) By what fraction did the "growth rate factor" ("γ") change between 1900–1910 and 1960–1970? (ii) Compare the percentage increase in population from 1900 to 1910 with the percentage increase from 1960 to 1970.

11. (a) A bank advertises "5% interest on savings—but you earn more because it is compounded continuously." The formula for computing the amount $M(t)$ of money in an account t days after $M(0)$ dollars was deposited (and left untouched) is
    $$M(t) = M(0)e^{.05t/365}$$
    What is the percentage increase on an amount $M(0)$ left untouched for 1 year?

    (b) A bank wants to compute its interest by the method in part (a). But it wants to give only \$5 interest on each \$100 that is left untouched for 1 year. How must it change the formula for that to occur?

12. Solve for $g(t)$: $3(d^2/dt^2)g(t) = -7g(t)$, $g(0) = 1$, $g'(0) = -2$. Find the amplitude and phase of $g(t)$. Sketch.

13. Solve for $z = f(t)$: $(d^2z/dt^2) + 5z = 0$, $f(0) = -3$, $f'(0) = 4$. Find the amplitude and phase of $f(t)$. Sketch.

14. Sketch the graph of y as a function of x if $-(d^2y/dx^2) = 2y$; $y = \frac{1}{2}$, and $dy/dx = \frac{1}{2}$ when $x = 0$.

15. "Suppose the pharaohs had built nuclear energy plants. They might have elected to store the resulting radioactive wastes inside the pyramids they built. Not a bad solution, considering how well the pyramids have lasted. But plutonium-239 stored in the oldest of them—some 4600 years ago—would today still exhibit 88 percent of its initial radioactivity." (From G. Hardin, "The Fallibility Factor," *Skeptic* 14 (1976): 12.)

    (a) What is the half-life of plutonium?

    (b) How long will it take for plutonium stored today to have only 1% of its present radioactivity? How long for $\frac{1}{1000}$?

16. A 3-foot metal rod is suspended horizontally from a spring, as shown in Fig. 9-R-1. The rod bobs up and down around the equilibrium point, 5 feet from the ground, with an amplitude of 1 foot and a frequency of two bobs per second. What is the maximum length of its shadow? How fast is its shadow changing length when the rod passes the middle of its bob?

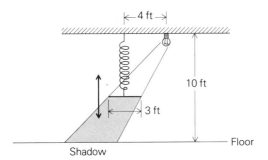

Fig. 9-R-1 Study the movement of the shadow of the bobbing rod.

17. Suppose your car radiator holds four gallons of fluid two thirds of which is water and one third is old antifreeze. The mixture begins flowing out at a rate of $\frac{1}{2}$ gallon per minute while fresh water is added at the same rate. How long does it take for the mixture to be 95% fresh water? Is it faster to wait until the radiator has drained before adding fresh water?

18. *Simple Harmonic Motion with Damping:* Consider the equation $x'' + 2\beta x' + \omega^2 x = 0$ where $0 < \beta < \omega$. (a) Show that $y = e^{\beta t}x$ satisfies a harmonic oscillator equation. (b) Show that the solution is of the form $x = e^{-\beta t}(A\cos\omega_1 t + B\sin\omega_1 t)$ where $\omega_1 = \sqrt{\omega^2 - \beta^2}$, and A and B are constants. (c) Solve $x'' + 2x' + 4x = 0$; $x(0) = 1$, $x'(0) = 0$, and sketch.

19. *Forced oscillations:*

   (a) Show that a solution of $x'' + 2\beta x' + \omega^2 x = f_0 \cos\omega_0 t$ is
   $$x_1(t) = \frac{f_0}{(\omega^2 - \omega_0^2)^2 + 4\omega_0^2\beta^2} \times$$
   $$[2\omega_0\beta\sin\omega_0 t + (\omega^2 - \omega_0^2)\cos\omega_0 t]$$

   (b) Show that the general solution is $x(t) = x_1(t) + x_0(t)$ where x_0 is the solution found in Problem 18.

   (c) *Resonance:* Show that the "amplitude" $f_0/[(\omega^2 - \omega_0^2)^2 + 4\omega_0^2\beta^2]$ of the solution is largest when ω_0 is near ω (the natural frequency) for β (the friction constant) small, by maximizing the amplitude for fixed f_0, ω, β and variable ω_0. (This is the phenomenon responsible for the Tacoma bridge disaster . . . somewhat simplified of course.)

20. *Simple pendulum:* The equation for a simple pendulum (see Fig. 9-R-2) is
    $$\frac{d^2\theta}{dt^2} = -\frac{g}{L}\sin\theta$$

Fig. 9-R-2 A simple pendulum.

   (a) Let $w(t) = d\theta/dt$. Show that
   $$\frac{d}{dt}\left(\frac{w^2}{2}\right) = \frac{g}{L}\frac{d}{dt}\cos\theta$$
   and so
   $$\frac{w^2}{2} = \frac{g}{L}(\cos\theta - \cos\theta_0)$$
   where $w = 0$ when $\theta = \theta_0$ (the maximum value of θ).

   (b) Conclude that θ is implicitly determined by
   $$\int_{\theta_0}^{\theta} \frac{d\theta}{\sqrt{\cos\theta - \cos\theta_0}} = \sqrt{\frac{2g}{L}}\, t$$

   (c) Show that the period of oscillation is
   $$T = \sqrt{\frac{2L}{g}}\int_0^{\theta_0}\frac{d\theta}{\sqrt{\cos\theta - \cos\theta_0}}$$
   $$= \sqrt{\frac{2L}{g}}\int_0^{\pi/2}\frac{d\phi}{\sqrt{1 - k^2\sin^2\phi}}$$
   where $k = \sin(\theta_0/2)$. [*Hint:* Write $\cos\theta = 1 - 2\sin^2(\theta/2)$.] The last integral is called an *elliptic integral of the first kind,* and cannot be evaluated explicitly.

21. (a) Sketch the direction field for the equation $y' = -9x/y$. Solve the equation exactly.

   (b) Find the orthogonal trajectories for the solutions in (a).

22. Solve $y' = ay + b$ for constants a and b by (a) introducing $w = y + b/a$ and a differential equation for w. (b) treating it as a separable equation and (c) treating it as a linear equation. Are your answers the same?

23. Test the accuracy of the Euler method by using a ten-step Euler method on the problem of finding $y(1)$ if $y' = y$ and $y(0) = 1$. Compare your answer with the exact solution and with a twenty-step Euler method.

24. Solve $y' = \csc y$ approximately for $0 \le x \le 1$ with $y(0) = 1$ using a ten-step Euler method.

25. Solve for $y(2)$ if $y' = y^2$ and $y(0) = 1$, numerically, using a twenty-step Euler method. Do you detect some numerical trouble? What do you think is going wrong?

10

Applications of Integration

Physical and geometric quantities can be expressed as integrals of infinitesimal pieces.

Our applications of integration in Chapter 4 were limited to area and rate problems. In this chapter, we will see how to use integrals to set up problems involving volumes, averages, centers of mass, work, energy, and power. The techniques developed in Chapter 8 make it possible to solve many of these problems completely.

CONTENTS

In Section 10-1, we show how to use integration to calculate the volume of some special objects in three-dimensional space. (In Chapter 17, we will use multiple integration to find the volumes of more general objects.) In Section 10-2, we study the average value of a function on an interval, and the center of mass, which represents the "average position" of a point in an object. Finally, Section 10-3 gives applications of integration to physical problems involving energy. In particular, the problem of computing the energy received from the sun on a given day at a given location is worked out in detail.

10-1 Volumes and Cavalieri's Principle

By slicing up a solid region, we find a formula for its volume.

In this section, we will apply integral calculus to the problem of finding the volume of regions in space. We begin with a general method called Cavalieri's principle, which we then apply to a variety of special cases. Further methods for calculating volumes will be discussed when we study multiple integration in Chapter 17.

Goals

After studying this section, you should be able to:

Use Cavalieri's principle to express the volume of a solid as an integral.

Compute the volumes of solids of revolution.

CAVALIERI'S PRINCIPLE AND THE SLICE METHOD

Cavalieri's principle is a method for computing the volume of solid regions. As we did with areas in Section 4-3, we will assume (as did Cavalieri) that the volume of any reasonably smooth solid region is a quantity defined in advance. Our task is to calculate it.

Cavalieri's principle relates the volume of a solid to the areas of certain plane sections of the solid. If S is a solid (a set of points in three-dimensional space) and P is a plane, the *plane section of S by P* consists of those points belonging to both S and P; we consider it as a region in the plane P. To visualize a plane section, we may think of S as a piece of cheese being cut by a knife which moves in the plane P. The exposed end of either of the resulting pieces has the shape of the section of S by P (see Fig. 10-1-1). By a *family of parallel planes*, we mean all the planes parallel to a given one. We now state Cavalieri's principle.

Fig. 10-1-1 Illustrating the section of S by P.

CAVALIERI'S PRINCIPLE

Given two solids, S_1 and S_2, suppose that for every plane P in a certain family of parallel planes the area of the sections of S_1 and S_2 are equal.

Then the volumes of S_1 and S_2 are equal.

S_1

S_2

Fig. 10-1-2 The solids S_1 and S_2 are sliced by the same family of parallel planes.

We will see how Cavalieri used his principle in a moment, but first let us examine how he justified it. Imagine the solids S_1 and S_2 as having been cut into thin slices by some of the planes in the given family (see Fig. 10-1-2). If we look at a pair of corresponding slices from the two solids (Fig. 10-1-3), we see that each slice is approximately a right cylinder with base equal to the cross-section of the solid by the plane and height equal to the thickness of a slice. Since the volume of such a cylinder is equal to the area of the base times the height (this can be taken as a basic principle for the volume of solids, analogous to the basic principle for the area of a rectangle), and since the two cross-sections are assumed to have the same area, we conclude that the two slices have approximately the same volume. (We say approximately because each slice is not exactly a right cylinder.) Adding up the volumes of the slices for S_1 and S_2, we conclude that the solids have approximately the same volume.

S_1 (sliced) S_2 (sliced)

Slice from S_1 Slice from S_2

Fig. 10-1-3 Corresponding slices of S_1 and S_2 have approximately the same volume.

Cavalieri went a step farther and considered each solid as being sliced into infinitely many slices of "infinitesimal" thickness. In this way, the corresponding slices of the two solids became *exactly* equal in volume, so the total solids had the same volume. This argument was considered in Cavalieri's day to be a proof (although a somewhat controversial one). Fortunately, it is possible to provide a modern proof as well; in a moment we discuss some of the problems involved.

Worked Example 1   Using Cavalieri's principle, find the volume of the oblique circular cylinder shown in Fig. 10-1-4.

r h

Fig. 10-1-4 Find the volume of the oblique cylinder.

Solution   We compare the oblique cylinder with a right cylinder with the same base and height (Fig. 10-1-5). If we cut both figures by a plane parallel to their bases, we get in each case a circular region with radius r, so both plane sections have the same area (equal to πr^2). By Cavalieri's principle, both solids have the same volume. But we know that the volume of the right cylinder is $\pi r^2 h$, so the volume of the oblique cylinder must also be $\pi r^2 h$.

Fig. 10-1-5 The oblique and right cylinders have equal cross-sectional areas, hence equal volumes.

If you take a stack of records in the form of a right cylinder, you can slide the records along one another so that the pile becomes approximately an oblique cylinder. The volume of the records does not change during this sliding operation. This experiment follows the pattern of the preceding demonstration of Cavalieri's principle.

Fig. 10-1-6 The plane P_x is at distance x from P_0.

By combining Cavalieri's principle with geometric tricks, it is possible to find the volume of many figures (see Problem 15 or any good high school text on solid geometry). Nevertheless Cavalieri's principle does not give a *systematic* procedure for computing volumes. With the aid of integral calculus, we will obtain a formula for the volume of a given solid in terms of the areas of its sections by a family of parallel planes. This formula can be combined with the techniques of integration to enable us to compute volumes quite efficiently.

To describe the formula, we begin with a family of parallel planes in space. We can label each plane in the family by its distance from a reference plane P_0. We denote by P_x the plane at distance x from P_0, with planes on opposite sides of P_0 corresponding to x's of opposite signs. (See Fig. 10-1-6.) Given a solid S, we denote by $A(x)$ the area of the plane section of S by P_x.

Worked Example 2   Let S be a ball of radius r (that is, all points having distance $\leq r$ from a fixed point) and let P_x be a family of parallel planes such that P_0 passes through the center of S. Find a formula for $A(x)$.

Solution   The plane section of S by P_x is a circular region if $-r < x < r$, a single point if $x = \pm r$, and is empty if $x < -r$ or $x > r$. The radius of the circular region is $\sqrt{r^2 - x^2}$ (see Fig. 10-1-7), and so $A(x) = \pi(\sqrt{r^2 - x^2})^2 = \pi(r^2 - x^2)$ for $-r < x < r$ and $A(x) = 0$ otherwise.

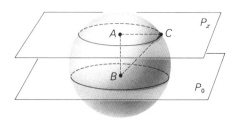

Fig. 10-1-7 $|AB| = x$ and $|AC| = r$, so $|BC| = \sqrt{r^2 - x^2}$.

Fig. 10-1-8 A "piecewise cylindrical" solid.

Cavalieri's principle says that if two surfaces give rise to the same function $A(x)$, they have the same volume; in other words, the function $A(x)$ determines the volume of S. To get a formula for the volume, we will look first at the case where S looks like a pile of n cylinders, as in Fig. 10-1-8.

If the ith cylinder C_i lies between the planes $P_{x_{i-1}}$ and P_{x_i} and has cross-sectional area k_i, then the function $A(x)$ is piecewise constant on the interval $[x_0, x_n]$; in fact, $A(x) = k_i$ for x in (x_{i-1}, x_i). The volume of C_i is the product of its base area k_i by its height $\Delta x_i = x_i - x_{i-1}$, so the volume of the total figure is $\sum_{i=1}^{n} k_i \Delta x_i$. But this is just the integral $\int_{x_0}^{x_n} A(x)\, dx$ of the piecewise constant function $A(x)$. We conclude that if S is a "piecewise cylindrical" solid between the planes P_a and P_b, then

$$\text{Volume } S = \int_a^b A(x)\, dx$$

If S is a reasonably "smooth" solid region, we expect that it can be squeezed arbitrarily closely between piecewise cylindrical regions on the inside and outside. Specifically, for every positive number ε, there should be a piecewise cylindrical region S_i inside S and another such region S_o outside S such that volume S_o − volume $S_i < \varepsilon$. If $A_i(x)$ and $A_o(x)$ are the corresponding piecewise constant area functions, we then have $A_i(x) \leq A(x) \leq A_o(x)$, so

$$\text{Volume } S_i = \int_a^b A_i(x)\, dx \leq \int_a^b A(x)\, dx \leq \int_a^b A_o(x)\, dx = \text{volume } S_o$$

But we also have

$$\text{Volume } S_i \leq \text{volume } S \leq \text{volume } S_o$$

Thus the numbers volume S and $\int_a^b A(x)\, dx$ are both in the interval [volume S_i, volume S_o], which has length less than ε. It follows that the difference between volume S and $\int_a^b A(x)\, dx$ is less than any positive number ε; the only way this can be so is if the two numbers are equal. Thus we have the following principle.

THE SLICE METHOD (CAVALIERI'S PRINCIPLE)

Let S be a solid and P_x be a family of parallel planes such that:

1. S lies between P_a and P_b.
2. The area of the section of S by P_x is an integrable function $A(x)$.

Then the volume of S is equal to

$$\int_a^b A(x)\, dx$$

Worked Example 3   Using the result of Worked Example 2, find the volume of a ball of radius *r*.

Solution   Let P_x be a family of parallel planes, with P_0 passing through the center of the ball *S*. Then *S* lies between P_{-r} and P_r, and we have $A(x) = \pi(r^2 - x^2)$. By the principle in the preceding display,

$$\text{Volume } S = \int_{-r}^{r} \pi(r^2 - x^2)\, dx = \pi\left(r^2 x - \frac{x^3}{3}\right)\Bigg|_{-r}^{r}$$

$$= \pi\left[\left(r^3 - \frac{r^3}{3}\right) - \left(-r^3 + \frac{r^3}{3}\right)\right] = \tfrac{4}{3}\pi r^3$$

This is the usual formula for the volume of a sphere.

We called the slice-method formula—volume $S = \int_a^b A(x)\, dx$—a "principle" rather than a theorem because we assumed without proof that we could always find inside and outside regions S_i and S_o whose volumes were arbitrarily close together. In fact, to prove that this is so would require a precise definition of a "sufficiently smooth" region, which is beyond the scope of this course. One way out of the difficulty is to take the slice method as a *definition* of the volume.* In any case, you can rest assured that the principle will give the correct result whenever the integral $\int_a^b A(x)\, dx$ is calculable.

Solved Exercises†   1. Find the volume of the conical solid in Fig. 10-1-9. (The base is a circle.)

Fig. 10-1-9 Find the volume of this oblique circular cone.

2. A ball of radius *r* is cut into three pieces by parallel planes at a distance of $r/3$ on each side of the center. Find the volume of each piece.

Exercises   1. Find the volume of the tent in Fig. 10-1-10. The plane section at height *x* above the base is a square of side $\frac{1}{6}(x - 6)^2 - \frac{1}{6}$. The height of the tent is 5 feet.

Fig. 10-1-10 Find the volume of this tent.

2. The conical solid in Fig. 10-1-9 is to be cut by horizontal planes into four pieces of equal volume. Where should the cuts be made? [*Hint:* What is the volume of the portion of the cone above the plane P_x?]

3. A cylindrical hole of radius $\frac{1}{2}$ is drilled through the center of a ball of radius 1. What is the volume of the resulting solid? [*Hint:* The integral does *not* go from -1 to 1.]

---

*In this case, it would be necessary to show that we get the same volume if we slice the surface in different directions. A result called "Fubini's theorem" can be used to show that this is the case. (See Section 17-2.)

†*Solutions appear in Appendix A.*

SOLIDS OF REVOLUTION

One way to construct a solid is to take a plane region R, as shown in Fig. 10-1-11, and revolve it around the x axis so that it sweeps out a solid region S. Such solids are common in woodworking shops in the form of lathe-tooled table legs, in pottery studios in the form of wheel-thrown pots, and in nature in the form of unicellular organisms.* They are called *solids of revolution* and are said to have *axial symmetry*.

Fig. 10-1-11 S is the solid of revolution obtained by revolving the plane region R about the x axis.

Suppose that region R is bounded by the lines $x = a$, $x = b$, and $y = 0$, and the graph of the function $y = f(x)$. To compute the volume of S by the slice method, we use the family of planes perpendicular to the x axis, with P_0 passing through the origin. The plane section of S by P_x is a circular disk of radius $f(x)$ (see Fig. 10-1-12), so its area $A(x)$ is $\pi[f(x)]^2$. By the basic formula of the slice method, the volume of S is

$$\int_a^b A(x)\,dx = \int_a^b \pi[f(x)]^2\,dx = \pi\int_a^b [f(x)]^2\,dx$$

Fig. 10-1-12 The volume of a solid of revolution obtained by the slice method.

VOLUME OF A SOLID OF REVOLUTION: DISK METHOD

The volume of the solid of revolution obtained by revolving the region under the graph of a (nonnegative) function $f(x)$ on $[a, b]$ about the x axis is

$$\pi\int_a^b [f(x)]^2\,dx$$

---

*See D'Arcy Thompson, On Growth and Form, *abridged edition, Cambridge University Press, 1969.*

Worked Example 4 The region under the graph of x^2 on $[0, 1]$ is revolved about the x axis. Sketch the resulting solid and find its volume.

Solution The solid, which is shaped something like a trumpet, is sketched in Fig. 10-1-13. Its volume is

$$\pi \int_0^1 (x^2)^2 \, dx = \pi \int_0^1 x^4 \, dx = \left. \frac{\pi x^5}{5} \right|_0^1 = \frac{\pi}{5}$$

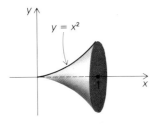

Fig. 10-1-13 The volume of this solid of revolution is $\pi \int_0^1 (x^2)^2 \, dx$.

Another way to obtain a solid S is to revolve the region R under the graph of a nonnegative function $f(x)$ on $[a, b]$ about the y axis as shown in Fig. 10-1-14. We assume $0 \le a < b$.

Fig. 10-1-14 The solid S is obtained by revolving the plane region R about the y axis.

If we break the region R into thin vertical strips in the usual way, the result of rotating such a strip is not a flat "slice" but rather a cylindrical "shell" (Fig. 10-1-15).

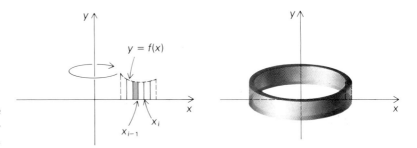

Fig. 10-1-15 The volume of a solid of revolution obtained by the shell method.

What is the volume of such a shell? Suppose for a moment that f has the constant value k_i on the interval (x_{i-1}, x_i). Then the shell is the "difference" of two cylinders of height k_i, one with radius x_i and one with radius x_{i-1}. The volume of the shell is, therefore, $\pi x_i^2 k_i - \pi x_{i-1}^2 k_i = \pi k_i (x_i^2 - x_{i-1}^2)$; we may observe that this last expression is $\int_{x_{i-1}}^{x_i} 2\pi k_i x \, dx$.

If f is piecewise constant on $[a, b]$, with adapted partition $(x_0, \dots, x_n)$ and $f(x) = k_i$ on (x_{i-1}, x_i), then the volume of the collection of n shells is

$$\sum_{i=1}^n \int_{x_{i-1}}^{x_i} 2\pi k_i x \, dx$$

But $k_i = f(x)$ on (x_{i-1}, x_i), so this is

$$\sum_{i=1}^{n} \int_{x_{i-1}}^{x_i} 2\pi x f(x) \, dx$$

which is simply $\int_a^b 2\pi x f(x) \, dx$.

We now have a formula,

$$\text{Volume} = 2\pi \int_a^b x f(x) \, dx$$

which is valid whenever $f(x)$ is piecewise constant on $[a, b]$; we claim that the same formula is valid for general f. To see this, we may squeeze f between piecewise constant functions above and below and use the same argument we used for the slice method.

Another argument uses the differential idea. If we rotate a strip of width dx and height $f(x)$ located at the point x, the result is a cylindrical shell of radius x, height $f(x)$, and thickness dx. The volume of this shell is equal to its area $2\pi x f(x)$ (unroll the shell to get a flat rectangle) times its thickness dx—that is, $2\pi x f(x) dx$. The total volume of the solid is obtained by integrating the volumes of the infinitesimal shells—that is, $\int_a^b 2\pi x f(x) \, dx$.

VOLUME OF A SOLID OF REVOLUTION: SHELL METHOD

The volume of the solid of revolution obtained by revolving about the y axis the region under the graph of a (nonnegative) function $f(x)$ on $[a, b]$ $(0 \le a < b)$ is

$$2\pi \int_a^b x f(x) \, dx$$

Worked Example 5   The region under the graph of x^2 on $[0, 1]$ is revolved about the y axis. Sketch the resulting solid and find its volume.

Solution   The solid, in the shape of a bowl, is sketched in Fig. 10-1-16. Its volume is

$$2\pi \int_0^1 x \cdot x^2 \, dx = 2\pi \int_0^1 x^3 \, dx = 2\pi \frac{x^4}{4} \bigg|_0^1 = \frac{\pi}{2}$$

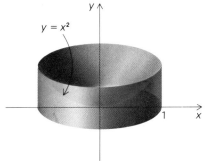

Fig. 10-1-16 Find the volume of the "bowl-like" solid.

Solved Exercises   3. The region under the graph of $\sqrt{x}$ on $[0, 1]$ is revolved around the x axis. Sketch the resulting solid and find its volume. Relate the result to Worked Example 5.

4. The region between the graphs of $\sin x$ and x on $[0, \pi/2]$ is revolved about the x axis. Sketch the resulting solid and find its volume.

5. The region under the graph of $\sin x$ on $[0, \pi]$ is revolved around the y axis. Sketch the resulting solid and find its volume.

6. The circular region with radius 1 and center $(4, 0)$ is revolved around the y axis. Sketch the resulting solid and find its volume.

Exercises   4. Sketch and find the volume of the solid obtained by revolving each of the following regions about the x axis: (a) the region under the graph of $3x + 1$ on $[0, 2]$; (b) the region under the graph of $\cos x + 1$ on $[0, 2\pi]$.

5. The region under the graph of $\sqrt{x}$ on $[0, 1]$ is revolved around the y axis. Sketch the resulting solid and find its volume. Relate the result to Worked Example 4.

6. Sketch and find the volume of the solid obtained by revolving each of the following regions about the y axis: (a) the region under the graph of e^x on $[1, 3]$; (b) the region under the graph of $2x^3 + 5x + 1$ on $[0, 1]$.

Problems for Section 10-1

1. A spherical shell of radius r and thickness h is, by definition, the region between two concentric spheres of radius $r - h/2$ and $r + h/2$.

    (a) Find a formula for the volume $V(r, h)$ of a spherical shell of radius r and thickness h.

    (b) For fixed r, what is the value of $(d/dh) V(r, h)$ when $h = 0$? Interpret your result in terms of the surface area of the sphere.

2. Find the volume of the solid shown in Fig. 10-1-17.

Fig. 10-1-17 Each plane section is a circle of radius 1.

3. The tent in Exercise 1 is to be cut into two pieces of equal volume by a plane parallel to the base. Where should the cut be made?

    (a) Express your answer as the root of a fifth degree polynomial.

    (b) Find an approximate solution using the method of bisection.

★4. A right circular cone of base radius r and height 14 is to be cut into three equal pieces by parallel planes which are parallel to the base. Where should the cuts be made?

5. Find the volume of the solid obtained by revolving each of the following regions about the axis indicated and sketch the region:

    (a) the region under the graph of $x(x - 1)^2$ on $[1, 2]$ (x axis)

    (b) the circular region with center $(a, 0)$ and radius r $(0 < r < a)$ (x axis)

    (c) the region between the graphs of $\sin x$ and x on $[0, \pi/2]$ (y axis)

6. Find the volume of the solid obtained by revolving each of the following regions about the axis indicated and sketch the region:

   (a) the circular region with center $(a, 0)$ and radius r $(0 < r < a)$ (y axis)

   (b) the triangular region with vertices $(1, 1)$, $(2, 2)$, and $(3, 1)$ (y axis)

   (c) the triangular region with vertices $(1, 1)$, $(2, 2)$, and $(3, 1)$ (x axis)

7. Let the functions f and g satisfy $0 \leq f(x) \leq g(x)$ for x in $[a, b]$, where $0 \leq a < b$. Find the volume of the region obtained by revolving the region between the graphs of $f(x)$ and $g(x)$ on $[a, b]$: (a) around the x axis (b) around the y axis.

★8. Let $f(x)$ and $g(x)$ be inverse functions with $f(a) = \alpha$, $f(b) = \beta$, $0 < a < b$, $0 < \alpha < \beta$. Show that

$$2\pi \int_{\alpha}^{\beta} y g(y)\, dy$$

$$= b\pi\beta^2 - a\pi\alpha^2 - \pi \int_{a}^{b} [f(x)]^2\, dx$$

Interpret this statement geometrically. (See Solved Exercise 3 and Exercise 5.)

9. Find the volume of the solids obtained by revolving each of the following regions about *each* axis and sketch the region:

   (a) the region under the graph of $2 - (x - 1)^2$ on $[0, 2]$

   (b) the region under the graph of $\cos 2x$ on $[0, \pi/4]$

   (c) the region between the graphs of $\sqrt{3 - x^2}$ and $5 + x$ on $[0, 1]$

10. Find the volume of the solids obtained by revolving each of the following regions about *each* axis and sketch the region:

    (a) the square region with vertices $(4, 6)$, $(5, 6)$, $(5, 7)$, and $(4, 7)$

    (b) the region in part (a) moved 2 units upward

    (c) the region in part (a) rotated by 45° around its center

    (d) the region under the graph of $\sqrt{4 - 4x^2}$ on $[0, 1]$.

11. The base of a solid S is the disk of radius 1 and center $(0, 0)$. Each section of S by a plane perpendicular to the x axis is an equilateral triangle. Find the volume of S.

★12. A wedge is cut in a tree of radius 0.5 meter by making two cuts to the tree's center, one horizontal and another at an angle of 15° to the first. Find the volume of the wedge. [*Hint:* Make up a formula for its volume using an infinitesimal argument.]

★13. Find the formula for the volume of a doughnut with outside radius R and a hole of radius r.

14. A vase with axial symmetry has the cross-section shown in Fig. 10-1-18 when it is cut by a plane through its axis of symmetry. Find the volume of the vase to the nearest cubic centimeter.

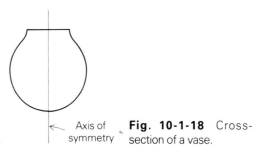

← Axis of symmetry   **Fig. 10-1-18** Cross-section of a vase.

15. Using Cavalieri's principle, without integration, find a relation between the volumes of:

    (a) a hemisphere of radius 1;

    (b) a right circular cone of base radius 1 and height 1

    (c) a right circular cylinder of base radius 1 and height 1

    [*Hint:* Consider two of the solids side by side as a single solid. The sums of two volumes will equal the third.]

10-2 Average Values and Center of Mass

The center of mass of a region is the average position of a point in the region.

The average value of a function on an interval is expressed in terms of an integral. We begin by motivating the formula for the average in terms of the formula for the average of a list of numbers. Following this, we take up the closely related topics of the mean value theorem for integrals and centers of mass.

Goals

After studying this section, you should be able to:

Find the average value of a function on an interval.

State the mean value theorem for integrals.

Set up and solve center of mass problems.

AVERAGE VALUES AND THE MEAN VALUE THEOREM

The average, or mean, of a list $a_1, \ldots, a_n$ of n numbers is defined simply as $(1/n) \sum_{i=1}^{n} a_i$.

If a grain dealer buys wheat from n farmers, buying b_i bushels from the ith farmer at the price of p_i dollars per bushel, the average price is determined not by taking the simple average of the p_i's but rather by the "weighted average":

$$p_{\text{average}} = \frac{\sum_{i=1}^{n} p_i b_i}{\sum_{i=1}^{n} b_i} = \frac{\text{total dollars}}{\text{total bushels}}$$

If a cyclist changes his speed intermittently, traveling at v_1 miles per hour from t_0 to t_1, v_2 miles per hour from t_1 to t_2, and so on up to time t_n, then the average speed for the trip is

$$v_{\text{average}} = \frac{\sum_{i=1}^{n} v_i (t_i - t_{i-1})}{\sum_{i=1}^{n} (t_i - t_{i-1})} = \frac{\text{total miles}}{\text{total hours}}$$

If, in either of the last two examples, the b_i's or $(t_i - t_{i-1})$'s are all equal, then the average value is simply the usual average of the p_i's or the v_i's. In general, we are led to the following definition:

Definition If f is a piecewise constant function of t on $[a,b]$, with adapted partition $(t_0, t_1, \ldots, t_n)$ and $f(t) = k_i$ on (t_{i-1}, t_i), then the *average value* of f on the interval $[a,b]$ is defined to be

$$\overline{f(t)}_{[a,b]} = \frac{\sum\limits_{i=1}^{n} k_i \Delta t_i}{\sum\limits_{i=1}^{n} \Delta t_i} \tag{1}$$

How can we define the average value of a function which is not piecewise constant? For instance, it is common to talk of the average temperature at a place on earth, although the temperature is not piecewise constant as a function of time. We may rewrite (1) as

$$\overline{f(t)}_{[a,b]} = \frac{\int_a^b f(t)\, dt}{b - a} \tag{2}$$

and this leads us to adopt formula (2) as the definition of the average value for any integrable function f, not just a piecewise constant function.

AVERAGE VALUE

If the function f is integrable on $[a,b]$, the average value $\overline{f(t)}_{[a,b]}$ of f on $[a,b]$ is defined by the formula

$$\overline{f(t)}_{[a,b]} = \frac{1}{b-a} \int_a^b f(t)\, dt$$

Worked Example 1 Find the average value of $f(x) = x^2$ on $[0,2]$.

Solution By definition (with x instead of t), we have

$$\overline{x^2}_{[0,2]} = \frac{1}{2-0} \int_0^2 x^2\, dx = \frac{1}{2} \cdot \frac{1}{3} x^3 \Big|_0^2 = \frac{4}{3}$$

Worked Example 2 Show that if $v = f(t)$ is the velocity of a moving object, then the definition of $\bar{v}_{[a,b]}$ agrees with the usual notion of average velocity.

Solution By the definition,

$$\bar{v}_{[a,b]} = \frac{1}{b-a} \int_a^b v\, dt$$

But $\int_a^b v \, dt$ is the distance traveled between $t = a$ and $t = b$ (see Solved Exercise 4, Section 4-2), so $\bar{v}_{[a,b]}$ = distance traveled/time of travel, which is the usual definition of average velocity.

An important property of average values is given in the following proposition.

Proposition   *If $m \le f(t) \le M$ for all t in $[a, b]$, then*

$$m \le \overline{f(t)}_{[a,b]} \le M$$

Proof   The integrals $\int_a^b m \, dt$ and $\int_a^b M \, dt$ are lower and upper sums for f on $[a, b]$, so $m(b - a) \le \int_a^b f(t) \, dt \le M(b - a)$. Dividing by $(b - a)$ gives the desired result.

As a corollary to this proposition, we obtain a mean value theorem for integrals.

Theorem 1   Mean Value Theorem for Integrals.   *Let f be continuous on $[a, b]$. Then there is a point t_0 in $[a, b]$ such that*

$$\overline{f(t)}_{[a,b]} = \frac{1}{b - a} \int_a^b f(t) \, dt$$

In other words, the average value of a continuous function on an interval is always attained somewhere on the interval.

Proof   By the extreme value theorem (p. 154), $f(t)$ attains a minimum value m and a maximum value M on $[a, b]$. Then $m \le f(t) \le M$ for t in $[a, b]$, so $\overline{f(t)}_{[a,b]}$ lies between m and M, by the preceding proposition. By the intermediate value theorem, first version (p. 112), applied to the interval between the points where $f(t) = m$ and $f(t) = M$, we conclude that there is a t_0 in this interval (and thus in $[a, b]$), such that $f(t_0) = \overline{f(t)}_{[a,b]}$.

Fig. 10-2-1 The point t_0 is such that the area of the rectangle equals the area under the graph.

The geometric content of the mean value theorem for nonnegative functions is illustrated in Fig. 10-2-1: the area under the graph of f and the rectangular area are equal. Physically, if f represents the instantaneous height of wavy water in a thin fish tank, the average value of f is the height of the water when it settles.

Solved Exercises

1. Find the average value of $\sqrt{1 - x^2}$ on $[-1, 1]$.

2. Find $\overline{x^2 \sin(x^3)}_{[0,\pi]}$.

3. Prove the mean value theorem for integrals by using the fundamental theorem of calculus and the mean value theorem for derivatives.

Exercises   1. Calculate each of the following average values:

(a) $\overline{1/(1 + t^2)}_{[-1, 1]}$ (b) $\overline{x^3}_{[0, 2]}$

(c) $\overline{\sin^{-1} x}_{[-1/2, 0]}$ (d) $\overline{\sqrt{1 - t^2}}_{[0, 1]}$

(e) $\overline{\sin x \cos 2x}_{[0, \pi/2]}$ (f) $\overline{z^3 + z^2 + 1}_{[1, 2]}$

2. Show that if f is integrable on $[a, c]$, and $a < b < c$, then

$$\overline{f(t)}_{[a, c]} = \left(\frac{b - a}{c - a} \right) \overline{f(t)}_{[a, b]} + \left(\frac{c - b}{c - a} \right) \overline{f(t)}_{[b, c]}$$

3. (a) Find $\overline{t^2 + 3t + 2}_{[0, x]}$ as a function of x. (b) Evaluate this function of x for $x = 0.1, 0.01, 0.00001$. Try to explain what is happening.

CENTER OF MASS: POINT MASSES

Center of mass   Plate

Fig. 10-2-2 The plate balances when supported at its center of mass.

l_1 l_2

m_1 m_2

Fig. 10-2-3 The support is at the center of mass when $m_1 l_1 = m_2 l_2$.

x_1 $\bar{x}$ x_2

m_1 m_2

Fig. 10-2-4 The center of mass is at $\bar{x}$ if $m_1(\bar{x} - x_1) = m_2(x_2 - \bar{x})$.

An important problem in mechanics, one which was considered early on by Archimedes, is to locate the point on which a plate of some given irregular shape will balance (Fig. 10-2-2). This point is called the center of mass, or center of gravity, of the plate. The center of mass can also be defined for solid objects, and its applications range from theoretical physics to the problem of arranging wet towels in a washing machine for the spin cycle.

To give a mathematical definition of the center of mass, we begin with the ideal case of two point masses, m_1 and m_2, attached to a light rod whose mass we neglect. (Think of a see-saw.) If we support the rod (see Fig. 10-2-3) at a point which is at distance l_1 from m_1 and distance l_2 from m_2, we find that the rod tilts toward m_1 (or m_2) if $m_1 l_1 > m_2 l_2$ (or $m_1 l_1 < m_2 l_2$). It balances only if the condition

$$m_1 l_1 = m_2 l_2 \tag{3}$$

is satisfied. One can derive this balance condition from basic physical principles,* or one may accept it as an experimental fact; we will not try to prove it here, but rather derive its consequences.

Suppose that the rod lies along the x axis, with m_1 at x_1 and m_2 at x_2. Let $\bar{x}$ be the position of the center of mass. Comparing Figs. 10-2-3 and 10-2-4, we see that $l_1 = \bar{x} - x_1$ and $l_2 = x_2 - \bar{x}$, so formula (3) may be rewritten as $m_1(\bar{x} - x_1) = m_2(x_2 - \bar{x})$. Solving for $\bar{x}$ gives the explicit formula

$$\bar{x} = \frac{m_1 x_1 + m_2 x_2}{m_1 + m_2} \tag{4}$$

We may observe that the position of the center of mass is just the *weighted average* of the positions of the individual masses. This suggests the following generalization:

*See, for instance, R. P. Feynman, Lectures on Physics, Addison-Wesley, 1964.

If n masses, $m_1, m_2, \ldots, m_n$, are placed at the points $x_1, x_2, \ldots, x_n$, respectively, their center of mass is located at

$$\bar{x} = \frac{\displaystyle\sum_{i=1}^{n} m_i x_i}{\displaystyle\sum_{i=1}^{n} m_i} \qquad (5)$$

We may accept formula (5), as we did (3), as a physical fact, or we may derive it (see Worked Example 3) from formula (3) and the following principle, which must be accepted as a general physical fact.

CONSOLIDATION PRINCIPLE

If a body B is divided into two parts, B_1 and B_2, with masses M_1 and M_2 then the center of mass of the body B is located as if B consisted of two point masses: M_1, located at the center of mass of B_1, and M_2, located at the center of mass of B_2.

Worked Example 3   Using formula (4) and the consolidation principle, derive formula (5) for the case of three masses.

Solution   We consider the body B consisting of m_1, m_2, and m_3 as divided into B_1, consisting of m_1 and m_2, and B_2, consisting of m_3 alone. (See Fig. 10-2-5.)

Fig. 10-2-5 Center of mass of three points by the consolidation principle.

By formula (4) we know that X_1, the center of mass of B_1, is given by

$$X_1 = \frac{m_1 x_1 + m_2 x_2}{m_1 + m_2}$$

The mass M_1 of B_1 is $m_1 + m_2$. The body B_2 has center of mass at $X_2 = x_3$ and mass $M_2 = m_3$. Applying (4) once again to the point masses M_1 at X_1 and M_2 at X_2 gives the center of mass $\bar{x}$ of B by the consolidation principle:

$$\bar{x} = \frac{M_1 X_1 + M_2 X_2}{M_1 + M_2} = \frac{(m_1 + m_2)\left(\dfrac{m_1 x_1 + m_2 x_2}{m_1 + m_2}\right) + m_3 x_3}{(m_1 + m_2) + m_3}$$

$$= \frac{m_1 x_1 + m_2 x_2 + m_3 x_3}{m_1 + m_2 + m_3}$$

which is exactly formula (5) for $n = 3$.

Worked Example 4 Masses of 10, 20, and 25 grams are located at $x_1 = 0$, $x_2 = 5$, and $x_3 = 12$ centimeters, respectively. Locate the center of mass.

Solution Using formula (5), we have

$$\bar{x} = \frac{10(0) + 20(5) + 25(12)}{10 + 20 + 25} = \frac{400}{55} = \frac{80}{11} \approx 7.27 \text{ centimeters}$$

Now let us study masses in the plane. Suppose that the masses $m_1, m_2, \ldots, m_n$ are located at the points $(x_1, y_1), \ldots, (x_n, y_n)$. We imagine the masses as being attached to a weightless card, and we seek a point $(\bar{x}, \bar{y})$ on the card where it will balance. (See Fig. 10-2-6.)

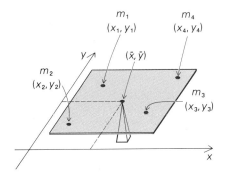

Fig. 10-2-6 The card balances at the center of mass.

To locate the center of mass $(\bar{x}, \bar{y})$, we use the following observation: If the card balances on the *point* $(\bar{x}, \bar{y})$, it will certainly balance along any *line* through $(\bar{x}, \bar{y})$. Take, for instance, a line parallel to the y axis (Fig. 10-2-7). The balance along this line will not be affected if we move each mass parallel to the line so that m_1, m_2, m_3, and m_4 are lined up parallel to the x axis (Fig. 10-2-8).

Fig. 10-2-7 If the card balances at a point, it balances along any line through that point.

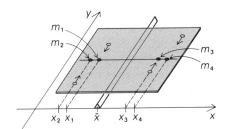

Fig. 10-2-8 Moving the masses parallel to a line does not affect the balance along this line.

Now we can apply the balance equation (5) for masses in a line to conclude that the x component $\bar{x}$ of the center of mass is equal to the weighted average

$$\bar{x} = \frac{\sum_{i=1}^{n} m_i x_i}{\sum_{i=1}^{n} m_i}$$

of the x components of the point masses.

Repeating the construction for a balance line parallel to the x axis (we urge you to draw versions of Figs. 10-2-7 and 10-2-8 for this case), and applying formula (5) to the masses as lined up parallel to the y axis, we conclude that

$$\bar{y} = \frac{\sum_{i=1}^{n} m_i y_i}{\sum_{i=1}^{n} m_i}$$

These two equations completely determine the position of the center of mass.

CENTER OF MASS IN THE PLANE

If n masses, $m_1, m_2, \ldots, m_n$, are placed at the points $(x_1, y_1), (x_2, y_2), \ldots, (x_n, y_n)$, respectively, then their center of mass is located at $(\bar{x}, \bar{y})$, where

$$\bar{x} = \frac{\sum_{i=1}^{n} m_i x_i}{\sum_{i=1}^{n} m_i} \quad \text{and} \quad \bar{y} = \frac{\sum_{i=1}^{n} m_i y_i}{\sum_{i=1}^{n} m_i} \tag{6}$$

Worked Example 5 Masses of 10, 15, and 30 grams are located at $(0, 1)$, $(1, 1)$, and $(1, 0)$. Find their center of mass.

Solution Applying equation (6), with $m_1 = 10$, $m_2 = 15$, $m_3 = 30$, $x_1 = 0$, $x_2 = 1$, $x_3 = 1$, $y_1 = 1$, $y_2 = 1$, and $y_3 = 0$, we have

$$\bar{x} = \frac{10 \cdot 0 + 15 \cdot 1 + 30 \cdot 1}{10 + 15 + 30} = \frac{9}{11}$$

and

$$\bar{y} = \frac{10 \cdot 1 + 15 \cdot 1 + 30 \cdot 0}{10 + 15 + 30} = \frac{5}{11}$$

so the center of mass is located at $(\frac{9}{11}, \frac{5}{11})$.

Solved Exercises

4. Suppose that n equal masses are located at the points $1, 2, 3, \ldots, n$ on a line. Where is their center of mass?

5. Particles of mass 1, 2, 3, and 4 are located at successive vertices of a unit square. How far from the center of the square is the center of mass?

Exercises

4. Masses of 1, 3, 5, and 7 units are located at the points 7, 3, 5, and 1, respectively, on the x axis. Where is the center of mass?

5. For each integer i from 1 to 100, a point of mass i is located at the point $x = i$. Where is the center of mass?

6. If, in formula (5), we have $a \leq x_i \leq b$ for all x_i, show that $a \leq \bar{x} \leq b$ as well. Interpret this statement geometrically.

7. (a) Equal masses are placed at the vertices of an equilateral triangle whose base is the segment from $(0, 0)$ to $(1, 0)$. Where is the center of mass? (b) The mass at $(0, 0)$ is doubled. Where is the center of mass now?

8. Verify the consolidation principle for the situation in which four masses in the plane are divided into two groups of two masses each.

CENTER OF MASS: PLATES

A flat plate is said to be of *uniform density* if there is a constant ρ such that the mass of any piece of the plate is equal to ρ times the area of the piece. The number ρ is called the *density* of the plate. We represent a plate of uniform density by a region R in the plane; we will see that the value of ρ is unimportant as far as the center of mass is concerned.

A line l is called an *axis of symmetry* for the region R, if, when the plane is flipped 180° around l (or, equivalently, reflected across l), the region R is taken into itself. For example, a square has four different axes of symmetry, a nonsquare rectangle two, and a circle infinitely many. (See Fig. 10-2-9.)

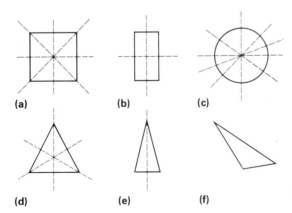

(a)   **(b)**   **(c)**

(d)   **(e)**   **(f)**

Fig. 10-2-9 The axes of symmetry of various geometric figures.

If flipping an object across the axis of symmetry l takes the region R into itself, it also takes the center of mass of R into itself. Since the only points left fixed by such a flip are on the line l, we have the following result.

SYMMETRY PRINCIPLE

If l is an axis of symmetry for the plate R of uniform density,   then the center of mass of R lies on l.

If a plate admits of more than one axis of symmetry, then the center of mass must lie on all the axes. In this case, we can conclude that the center of mass lies at the point of intersection of the axes of symmetry. Looking at parts (a) through (d) of Fig. 10-2-9, we see that in each case the center of mass is located at the "geometric center" of the figure. In case (e), we know only that the center of mass is on the altitude; in case (f), we know nothing at all.

Next suppose that R is the region under the graph of a piecewise constant function f on $[a, b]$ with $f(x) \geq 0$ on $[a, b]$; see Fig. 10-2-10. If $(x_0, x_1, \ldots, x_n)$ is an adapted partition for f (see p. 181), then R is composed of n rectangles $R_1, \ldots, R_n$. The area of R_i is $k_i(x_i - x_{i-1}) = k_i \Delta x_i$, where k_i = the value of f on (x_{i-1}, x_i), so the mass of R_i is $\rho k_i \Delta x_i = m_i$. By the symmetry principle, the center of mass of R_i is located at $(\bar{x}_i, \bar{y}_i)$, where $\bar{x}_i = \frac{1}{2}(x_{i-1} + x_i)$ and $y_i = \frac{1}{2}k_i$.

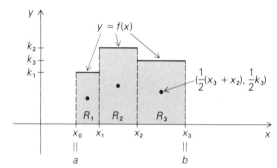

Fig. 10-2-10 The center of mass of the shaded region is obtained by the consolidation principle.

Now we use the consolidation principle, extended to a decomposition into n pieces, to conclude that the center of mass of R is the same as the center of mass of masses $m_1, \ldots, m_n$ placed at the points $(\bar{x}_1, \bar{y}_1), \ldots, (\bar{x}_n, \bar{y}_n)$. By formula (6), we have, first of all,

$$\bar{x} = \frac{\displaystyle\sum_{i=1}^{n} m_i \bar{x}_i}{\displaystyle\sum_{i=1}^{n} m_i} = \frac{\displaystyle\sum_{i=1}^{n} \rho k_i \Delta x_i \left[\frac{1}{2}(x_{i-1} + x_i)\right]}{\displaystyle\sum_{i=1}^{n} \rho k_i \Delta x_i}$$

We wish to rewrite the numerator and denominator as integrals, so that we can eventually treat the case where f is not piecewise constant. The denominator is easy to handle. Factoring out ρ gives $\rho \sum_{i=1}^{n} k_i \Delta x_i$, which we recognize as $\rho \int_a^b f(x)\, dx$, the total mass of the plate. The numerator of $\bar{x}$ equals

$$\frac{1}{2}\rho \sum_{i=1}^{n} k_i(x_i - x_{i-1})(x_i + x_{i-1}) = \frac{1}{2}\rho \sum_{i=1}^{n} k_i(x_i^2 - x_{i-1}^2)$$

We wish to interpret $k_i(x_i^2 - x_{i-1}^2)$ as an integral of something over $[x_{i-1}, x_i]$. In fact, it is $\int_{x_{i-1}}^{x_i} 2k_i x \, dx$, which we can also write as $\int_{x_{i-1}}^{x_i} 2xf(x) \, dx$, since $f(x) = k_i$ on (x_{i-1}, x_i). Now the numerator of $\bar{x}$ becomes

$$\frac{1}{2}\rho \sum_{i=1}^{n} \int_{x_{i-1}}^{x_i} 2xf(x) \, dx = \frac{1}{2}\rho \int_a^b 2xf(x) \, dx = \rho \int_a^b xf(x) \, dx$$

and we have

$$\bar{x} = \frac{\rho \displaystyle\int_a^b xf(x) \, dx}{\rho \displaystyle\int_a^b f(x) \, dx}$$

The ρ's cancel, giving

$$\bar{x} = \frac{\displaystyle\int_a^b xf(x) \, dx}{\displaystyle\int_a^b f(x) \, dx}$$

To find the y coordinates of the center of mass, we use the second half of formula (6):

$$\bar{y} = \frac{\displaystyle\sum_{i=1}^{n} m_i \bar{y}_i}{\displaystyle\sum_{i=1}^{n} m_i} = \frac{\displaystyle\sum_{i=1}^{n} \rho k_i \Delta x_i (\tfrac{1}{2}k_i)}{\displaystyle\sum_{i=1}^{n} \rho k_i \Delta x_i}$$

The denominator is the total mass $\rho \int_a^b f(x) \, dx$, as before. The numerator is $\frac{1}{2}\rho \sum_{i=1}^{n} k_i^2 \Delta x_i$, and we recognize $\sum_{i=1}^{n} k_i^2 \Delta x_i$ as the integral $\int_a^b [f(x)]^2 \, dx$ of the piecewise constant function $[f(x)]^2$. Thus:

$$\bar{y} = \frac{\dfrac{1}{2}\rho \displaystyle\int_a^b [f(x)]^2 \, dx}{\rho \displaystyle\int_a^b f(x) \, dx} = \frac{\dfrac{1}{2}\displaystyle\int_a^b [f(x)]^2 \, dx}{\displaystyle\int_a^b f(x) \, dx}$$

We have derived the formulas for $\bar{x}$ and $\bar{y}$ for the case in which $f(x)$ is piecewise constant; however, they make sense as long as $f(x)$, $xf(x)$, and $[f(x)]^2$ are integrable on $[a, b]$. We may take for granted that the same formulas still work, or we may take the formulas as a definition of the center of mass. (Also see Solved Exercise 8.)

CENTER OF MASS OF THE REGION UNDER A GRAPH

The center of mass of a plate of uniform density represented by the region under the graph of a (nonnegative) function $f(x)$ on $[a, b]$ is located at $(\bar{x}, \bar{y})$, where

$$\bar{x} = \frac{\int_a^b x f(x)\, dx}{\int_a^b f(x)\, dx} \qquad \text{and} \qquad \bar{y} = \frac{\frac{1}{2} \int_a^b [f(x)]^2\, dx}{\int_a^b f(x)\, dx} \tag{7}$$

Since the center of mass depends only upon the region in the plane, and not upon the density ρ, we usually refer to $(\bar{x}, \bar{y})$ simply as the *center of mass of the region.*

Worked Example 6   Find the center of mass of the region under the graph of x^2 from 0 to 1.

Solution   By formulas (7), with $f(x) = x^2$, $a = 0$, and $b = 1$,

$$\bar{x} = \frac{\int_0^1 x^3\, dx}{\int_0^1 x^2\, dx} = \frac{\frac{1}{4}}{\frac{1}{3}} = \frac{3}{4} \qquad \bar{y} = \frac{\frac{1}{2} \int_0^1 x^4\, dx}{\int_0^1 x^2\, dx} = \frac{\frac{1}{10}}{\frac{1}{3}} = \frac{3}{10}$$

so the center of mass is located at $(\frac{3}{4}, \frac{3}{10})$. (See Fig. 10-2-11.) You can verify this result experimentally by cutting a figure out of stiff cardboard and seeing where it balances.

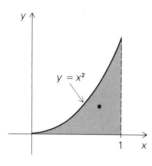

Fig. 10-2-11   The center of mass of the shaded region is located at $(\frac{3}{4}, \frac{3}{10})$.

By using the consolidation principle, we can calculate the center of mass of a region which is *not* under a graph by breaking it into simpler regions, as we did for areas in Section 4-3. (See Solved Exercise 7.)

Solved Exercises

6. Find the center of mass of a semicircular region of radius 1.

7. Find the center of mass of the region consisting of a disk of radius 1 centered at the origin and the region under the graph of $\sin x$ on $(2\pi, 3\pi)$.

8. "Derive" formulas (7) by an argument using infinitesimals.

Exercises   9. Find the center of mass of the region under the graph of $\sqrt{1-x^2}$ on $[0,1]$.

10. (a) Find the center of mass of the triangle with vertices at $(0,0)$, $(0,2)$, and $(4,0)$. (b) Find the center of mass of the triangle with vertices at $(1,0)$, $(4,0)$, and $(2,3)$.

11. (a) Find the center of mass of the region under the graph of $4/x^2$ on $[1,3]$. (b) Verify your result experimentally by balancing a sheet of cardboard.

12. Find the center of mass of the region between the graphs of $-x^4$ and x^2 on $[-1,1]$.

Problems for Section 10-2

1. Find the following average values:
   (a) $\overline{\sin^2 x}_{[0,\pi]}$
   (b) $\overline{x^3 + \sqrt{1/x}}_{[1,3]}$
   (c) $\overline{[(x^3 + x - 2)/(x^2 + 1)]}_{[-1,1]}$
   (d) $\overline{(x^2 + x - 1)\sin x}_{[0,\pi/4]}$
   (e) $\overline{\sin^{-1} x}_{[0,1]}$
   (f) $\overline{\ln x}_{[1,e]}$

2. What was the average temperature in Moose Jaw on June 13, 1857? (See Fig. 10-2-12).

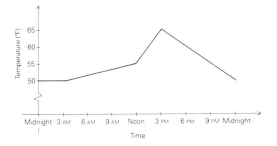

Fig. 10-2-12 Temperature in Moose Jaw on June 13, 1857.

★3. Let f be defined on the real line and let $a(x) = \overline{f(t)}_{[0,x]}$.
   (a) Derive the formula
   $$a'(x) = (1/x)[f(x) - a(x)]$$
   (b) Interpret the formula in the cases $f(x) = a(x)$, $f(x) < a(x)$, and $f(x) > a(x)$.
   (c) When baseball players strike out, it lowers their batting average more at the beginning of the season than at the end. Explain why.

4. Show that if $\overline{f'(x)}_{[a,b]} = 0$ then $f(b) = f(a)$.

5. Let a mass m_i be placed at position x_i on a line ($i = 1, \ldots, n$). Show that the function $f(x) = \Sigma_{i=1}^{n} m_i(x - x_i)^2$ is minimized when x is the center of mass of the n particles.

6. Equal masses are placed at the points (x_1, y_1), (x_2, y_2), and (x_3, y_3). Show that their center of mass is at the intersection point of the medians of the triangle at whose vertices the masses are located.

7. From a disk of radius 5, a circular hole with radius 2 and center 1 unit from the center of the disk is cut out. Sketch and find the center of mass of the resulting figure.

8. Masses of 2, 3, 4, and 5 kilograms are placed at the points $(1,2)$, $(1,4)$, $(3,5)$, and $(2,6)$, respectively. Where should a mass of 1 kilogram be placed so that the configuration of five masses has its center of mass at the origin?

9. A mass m_i is at position $x_i = f_i(t)$ at time t. Show that if the force on m_i is $F_i(t)$, and $F_1(t) + F_2(t) = 0$, then the center of mass of m_1 and m_2 moves with constant velocity.

10. Find the center of mass of each of the following plane regions:
   (a) the region under the graph of $1 + x^2 + x^4$ on $[-1,1]$
   (b) the region under the graph of $\sqrt{1 - x^2/a^2}$ on $[-a,a]$ (a semiellipse)
   (c) the region between the graphs of $\sin x$ and $\cos x$ on $[0, \pi/4]$

★11. Suppose that $f(x) \leq g(x)$ for all x in $[a, b]$. Show that the center of mass of the region between the graphs of f and g on $[a, b]$ is located at $(\bar{x}, \bar{y})$, where

$$\bar{x} = \frac{\int_a^b x\,[g(x) - f(x)]\,dx}{\int_a^b [g(x) - f(x)]\,dx}$$

$$\bar{y} = \frac{\frac{1}{2}\int_a^b [g(x) + f(x)]\,[g(x) - f(x)]\,dx}{\int_a^b [g(x) - f(x)]\,dx}$$

12. Find the center of mass of the triangular region with vertices (x_1, y_1), (x_2, y_2), and (x_3, y_3). (For convenience, you may assume that $x_1 \leq x_2 \leq x_3$, $y_1 \leq y_3$, and $y_2 \leq y_3$.) Compare with Problem 6.

★13. Suppose f' exists and is continuous on $[a, b]$. Prove the mean value theorem for derivatives from the mean value theorem for integrals.

10-3 Energy

Energy is the integral of power over time, and work is the integral of force over distance.

We began our treatment of the integral by discussing solar energy. We now conclude our study of the physical applications of integration by returning to some problems involving energy.

Goals

After studying this section, you should be able to:

Express energy and work as integrals.

Compute the energy and work required to perform simple physical processes.

ENERGY, POWER, AND WORK

Energy appears in various forms and can often be converted from one form into another. For instance, a solar cell converts the energy in light into electrical energy; a fusion reactor, in changing deuterium to helium, transforms nuclear energy into heat energy. Despite the variety of forms in which energy may appear, there is a common unit of measure for all these forms. In the MKS (meter-kilogram-second) system, it is the *joule*, which equals 1 kilogram meter meter per second per second.

Energy is an "extensive" quantity. The longer a generator runs, the more electrical energy it produces; the longer a light bulb burns, the more energy it consumes. The rate (with respect to time) at which some form of energy is produced or consumed is called the *power* output or input of the energy conversion device. Power is an *instantaneous* or "intensive" quantity. By the fundamental theorem of calculus, we can compute the total energy transformed between times a and b by integrating the power from a to b.

POWER AND ENERGY

$$P = \frac{dE}{dt}$$

Power is the rate of change of energy with respect to time.

$$E = \int_a^b P \, dt$$

The total energy over a time period is the integral of power with respect to time.

A common unit of measurement for power is the *watt*, which equals 1 joule per second. One *horsepower* is equal to 746 watts. The kilowatt-hour is a unit of energy equal to the energy obtained by using 1000 watts for 1 hour (3600 seconds)—that is, 3,600,000 joules.

Worked Example 1 The power output (in watts) of a 60-cycle generator varies with time (measured in seconds) according to the formula $P = P_0 \sin^2(120\pi t)$, where P_0 is the maximum power output. (a) What is the total energy output during an hour? (b) What is the average power output during an hour?

Solution (a) The energy output, in joules, is

$$E = \int_0^{3600} P_0 \sin^2(120\pi t) \, dt$$

Using the formula $\sin^2\theta = (1 - \cos 2\theta)/2$, we find

$$E = \frac{1}{2} P_0 \int_0^{3600} (1 - \cos 240\pi t) \, dt$$

$$= \frac{1}{2} P_0 \left[t - \frac{1}{240\pi} \sin 240\pi t \right]_0^{3600}$$

$$= \frac{1}{2} P_0 [3600 - 0 - (0 - 0)]$$

$$= 1800 P_0$$

(b) The average power output is (see Section 10-2) the energy output divided by the time, or $1800\,P_0/3600 = \frac{1}{2} P_0$, in this case, half the maximum power output.

In the beginning of Chapter 4, we said that if light with intensity I was incident upon a solar power unit from time $t = a$ to $t = b$, then the total energy stored was $\int_a^b kI\, dt$. We can now interpret this statement by observing that kI is the *power* output of the solar energy converter.

A more visible form of energy is mechanical energy—the energy stored in the movement of a massive object (*kinetic energy*) or the energy stored in an object by virtue of its position (*potential energy*). The latter is illustrated by the energy we can extract from water stored at the top of a hydroelectric power plant.

We accept the following principles from physics:

1. The kinetic energy of a mass m moving with velocity v is $\frac{1}{2}mv^2$.

2. The (gravitational) potential energy of a mass m at a height h is mgh (here g is the gravitational acceleration; $g = 9.8$ meters per second per second $= 32$ feet per second per second).

The total force on a moving object is equal to the product of the mass m and the acceleration $dv/dt = d^2x/dt^2$. (See p. 372.) The unit of force is the *newton* and is 1 kilogram meter per second per second. If the force depends upon the position of the object—that is, $F = f(x)$—we may calculate the variation of the kinetic energy $K = \frac{1}{2}mv^2$ with position. We have

$$\frac{dK}{dx} = \frac{dK/dt}{dx/dt} = \frac{(d/dt)(\frac{1}{2}mv^2)}{v} = \frac{mv\, dv/dt}{v} = m\frac{dv}{dt} = F$$

Applying the fundamental theorem of calculus, we find that the change ΔK of kinetic energy as the particle moves from a to b is $\int_a^b F\, dx$. Often we can divide the total force on an object into parts arising from identifiable sources (gravity, friction, fluid pressure). We are led to define the work W done by a particular force F on a moving object (even if there are other forces present) as $W = \int_a^b F\, dx$. If the total force F is a sum $F_1 + \cdots + F_n$, then we have

$$\Delta K = \int_a^b (F_1 + \cdots + F_n)\, dK = \int_a^b F_1\, dx + \cdots + \int_a^b F_n\, dx$$

Thus the total change in kinetic energy is equal to the sum of the works done by the individual forces.

Note that if the force F is constant, then the work done is simply the product of F with the *displacement* $\Delta x = b - a$. Accordingly, 1 joule equals 1 newton-meter.

Worked Example 2   The acceleration of gravity near the earth is $g = 9.8$ meters per second per second. How much work does a weight lifter do in raising a 50-kilogram barbell to a height of 2 meters?

Solution   We let x denote the height of the barbell above the ground. Before and after the lifts the barbell is stationary, so the net change in kinetic energy is zero. The work done by the weight lifter must be the *negative* of the work done by gravity. Since the pull of gravity is downward, its force is -9.8 m/sec$^2 \times 50$ kg $= -490$ kg m/sec$^2 = -490$ newtons; $\Delta x = 2$ m, so the work done by gravity is -980 kg m^2/sec$^2 = -980$ joules. Thus the work done by the weight lifter is 980 joules. (If the lift takes s seconds, the average power output is $980/s$ watts.)

Worked Example 3   A pump is to empty the conical tank of water shown in Fig. 10-3-1. How much energy (in joules) is required for the job? (A cubic meter of water has mass 10^3 kilograms.)

Fig. 10-3-1 To calculate the energy needed to empty the tank, we add up the energy needed to remove slabs of thickness dx.

Solution   Consider a layer of thickness dx at depth x, as shown in Fig. 10-3-1. By similar triangles, the radius is $r = \frac{3}{10}(10 - x)$, so the volume of the layer is $\pi \cdot \frac{9}{100}(10 - x)^2\, dx$ and its mass is $10^3 \cdot \pi \cdot \frac{9}{100}(10 - x)^2\, dx = 90\pi(10 - x)^2\, dx$. To lift this layer x meters to the top of the tank, we must do $90\pi(10 - x)^2\, dx \cdot g \cdot x$ joules of work, where $g = 9.8$ meters per second per second is the acceleration due to gravity (see Worked Example 2). Thus the total work done in emptying the tank is

$$90g\pi \int_0^{10} (10 - x)^2 x\, dx = 90g\pi \left[100\,\frac{x^2}{2} - 20\,\frac{x^3}{3} + \frac{x^4}{4} \right] \Bigg|_0^{10}$$

$$= 90g\pi(10^4)\left[\frac{1}{2} - \frac{2}{3} + \frac{1}{4} \right]$$

$$= (90)(9.8)(\pi)(10^4)\left(\frac{1}{12} \right)$$

$$\approx 2.3 \times 10^6 \text{ joules}$$

FORCE AND WORK

The work done by a force on a moving object is the integral of the force with respect to position:

$$W = \int_a^b F\, dx$$

Solved Exercises

1. Show that the *power* exerted by a force F on a moving object is Fv, where v is the velocity of the object.

2. The pump which is emptying the conical tank in Worked Example 3 has a power output of 10^5 joules per hour (that is, 360,000,000 watts). What is the water level at the end of 6 minutes of pumping? How fast is the water level dropping at this time?

Exercises

1. The power output of a solar cell is $25 \sin(\pi t/12)$ watts, where t is the time in *hours* after 6 AM. (a) How many joules of energy are produced between 6 AM and 6 PM? (b) How would you account for this variation of power with time of day?

2. A particle with mass 1000 grams has position $x = 3t^2 + 4$ meters at time t seconds. (a) What is the kinetic energy at time t? (b) What is the rate at which power is being supplied to the object at time $t = 10$?

3. A particle of mass 20 grams is at rest at $t = 0$, and power is applied at the rate of 10 joules per second. (a) What is the energy at time t? (b) If all the energy is kinetic energy, what is the velocity at time t? (c) How far has the particle moved at the end of t seconds? (d) What is the force on the particle at time t?

4. The gravitational force on an object at a distance r from the center of the earth is k/r^2, where k is a constant. How much work is required to move the object:

   (a) From $r = 1$ to $r = 10$?   (b) From $r = 1$ to $r = 1000$?

   (c) From $r = 1$ to $r = 10,000$?   (d) From $r = 1$ to "$r = \infty$"?

(optional) **INTEGRATING SUNSHINE**

We will now apply the theory and practice of integration to compute the total amount of sunshine received during a day, as a function of latitude and time of year. If we have a horizontal square meter of surface, then the rate at which solar energy is received by this surface—that is, the *intensity* of the solar radiation—is proportional to the sine of the angle A of elevation of the sun above the horizon.* Thus the intensity is highest when the sun is directly overhead ($A = \pi/2$) and reduces to zero at sunrise and sunset.

The total energy received on day T must therefore be equal to a constant (which can be determined only by experiment, and which we will ignore) times the integral $E = \int_{t_0(T)}^{t_1(T)} \sin A \, dt$, where t is the time of day (measured in hours from noon) and $t_0(T)$ and $t_1(T)$ are the times of sunrise and sunset on day T. (When the sun is below the horizon, although $\sin A$ is negative, the solar intensity is simply zero.)

We presented a formula for $\sin A$ (formula (4), p. 303), to be derived in the appendix to Chapter 15 and used it to determine the time of sunset (formula (6), p. 304). The time of sunrise is the negative of the time of sunset, so we have†

$$E = \int_{-s}^{s} \sin A \, dt$$

where

$$\sin A = \cos l \sqrt{1 - \sin^2\alpha \cos^2\left(\frac{2\pi T}{365}\right)} \cos\left(\frac{2\pi t}{24}\right)$$
$$+ \sin l \sin\alpha \cos\left(\frac{2\pi T}{365}\right)$$

*We will justify this assertion in the appendix to Chapter 15. We also note that, strictly speaking, it applies only if we neglect absorption by the atmosphere or assume that our surface is at the top of the atmosphere.

†All these calculations assume that there is a sunrise and sunset. In the polar regions during the summer, the calculations must be altered (see Problem 6 below and Solved Exercise 8, Section 6-3).

(optional)

and

$$S = \frac{24}{2\pi} \cos^{-1}\left[-\tan l \frac{\sin \alpha \cos(2\pi T/365)}{\sqrt{1 - \sin^2\alpha \, \cos^2(2\pi T/365)}} \right]$$

Here $\alpha \approx 23.5°$ is the inclination of the earth's axis from the perpendicular to the plane of the earth's orbit; l is the latitude of the point where the sunshine is being measured.

The integration will be simpler than you may expect. First of all, we simplify notation by writing k for the expression $\sin \alpha \, \cos(2\pi T/365)$, which appears so often. Then we have

$$E = \int_{-S}^{S} \left[\cos l \sqrt{1 - k^2} \cos\left(\frac{2\pi t}{24}\right) + (\sin l)k \right] dt$$

$$= \cos l \sqrt{1 - k^2} \int_{-S}^{S} \cos\left(\frac{2\pi t}{24}\right) dt + (\sin l)k \int_{-S}^{S} dt$$

Integration gives

$$E = \cos l \sqrt{1 - k^2} \left(\frac{24}{2\pi} \sin \frac{2\pi t}{24} \Big|_{-S}^{S} \right) + 2Sk \sin l$$

$$= \cos l \sqrt{1 - k^2} \left(\frac{24}{2\pi} \right)\left(\sin \frac{2\pi S}{24} - \frac{\sin 2\pi(-S)}{24} \right) + 2Sk \sin l$$

$$= \frac{24}{\pi} \cos l \sqrt{1 - k^2} \sin \frac{2\pi S}{24} + 2Sk \sin l$$

The expression $\sin(2\pi S/24)$ can be simplified. Since $\cos(2\pi S/24) = -(\tan l)(k/\sqrt{1 - k^2})$,

$$\sin \frac{2\pi S}{24} = \sqrt{1 - \cos^2 \frac{2\pi S}{24}} = \sqrt{1 - \frac{(\tan^2 l)k^2}{1 - k^2}} = \sqrt{\frac{1 - k^2 - (\tan^2 l)k^2}{1 - k^2}}$$

$$= \sqrt{\frac{1 - (1 + \tan^2 l)k^2}{1 - k^2}} = \sqrt{\frac{1 - (\sec^2 l)k^2}{1 - k^2}}$$

and so we get, finally,

$$E = \frac{24}{\pi} \cos l \sqrt{1 - (\sec^2 l)k^2} + \frac{24}{\pi} k \sin l \cos^{-1}\left[-\frac{(\tan l)k}{\sqrt{1 - k^2}} \right]$$

Since both k and $\sqrt{1 - k^2}$ appear, we can do even better by writing $k = \sin D$ (the number D is important in astronomy—it is called the *declination*), and we get

$$E = \frac{24}{\pi} [\cos l \sqrt{1 - \sec^2 l \sin^2 D} + \sin l \sin D \cos^{-1}(-\tan l \tan D)]$$

Since we have already ignored a constant factor in E, we will also ignore the factor $24/\pi$. Incorporating $\cos l$ into the square root, we obtain as our final result

$$E = \sqrt{\cos^2 l - \sin^2 D} + \sin l \sin D \cos^{-1}(-\tan l \tan D)$$

where $\sin D = \sin \alpha \cos(2\pi T/365)$.

(optional)

Plotting E as a function of l for various values of T leads to graphs like those on p. 152 and in Fig. 10-3-2.

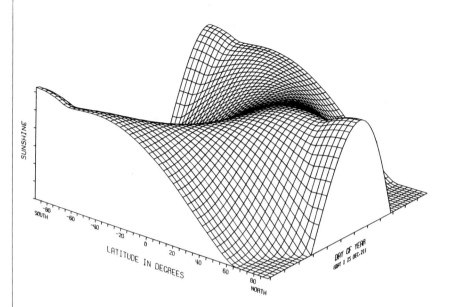

SUNSHINE

LATITUDE IN DEGREES

-80 SOUTH   -60   -40   -20   0   20   40   60   80 NORTH

DAY OF YEAR
(DAY 1 IS DEC. 21)

Fig. 10-3-2 Computer-generated graph of the daily sunshine intensity on the earth as a function of day of the year and latitude.

Worked Example 4 When does the equator receive the most solar energy? The least?

Solution At the equator, $l = 0$, so we have

$$E = \sqrt{1 - \sin^2 D} = \sqrt{1 - \sin^2\alpha \, \cos^2\left(\frac{2\pi T}{365}\right)}$$

We see by inspection that E is largest when $\cos^2(2\pi T/365) = 0$—that is, when $T/365 = \frac{1}{4}$ or $\frac{3}{4}$; that is, on the first days of fall and spring: on these days $E = 1$. We note that E is smallest on the first days of summer and winter, when $\cos^2(2\pi T/365) = 1$ and we have $E = \sqrt{1 - \sin^2\alpha} = \cos\alpha = \cos 23.5° = 0.917$, or about 92% of the maximum value.

Using this example we can standardize units in which E can be measured. One unit of E is the total energy received on a square meter at the equator on the first day of spring. All other energies may be expressed in terms of this unit.

Solved Exercises ★3. Compare the solar energy received on June 21 at the Arctic Circle ($l = 90° - \alpha$) with that received at the equator.

★4. (a) Express the total solar energy received over a whole year at latitude l by using summation notation. (b) Write down an integral which is approximately equal to this sum. Can you evaluate it?

Exercises ★5. What would the inclination of the earth need to be in order for E on June 21 to have the same value at the equator as at the latitude $90° - \alpha$?

★6. Simplify the integral in the solution of Solved Exercise 4(b) for the cases $l = 0$ (equator) and $l = 90° - \alpha$ (Arctic Circle). In each case, one of the two terms in the integrand can be integrated explicitly: find the integral of this term.

Problems for Section 10-3

1. An electric motor is operating at $15 + 2\sin(t\pi/24)$ watts, where t is the time in hours measured from midnight. How much energy is consumed in one day's operation?

2. Compute the work done by the given force acting over the given interval:
   (a) $F = k/x^2$; $1 \le x \le 6$ (k a constant)
   (b) $F = \sin^3 x \cos^2 x$; $0 \le x \le 2$ [*Hint:* Write $\sin^3 x = \sin x(1 - \cos^2 x)$.]
   (c) $F = 1/(4 + x^2)$; $0 \le x \le 1$

3. How much power must be applied to raise an object of mass 1000 grams at a rate of 10 meters per second in the earth's gravitational field?

4. (a) The power output of an electric generator is $25 \cos^2(120\pi t)$ joules per second. How much energy is produced in 1 hour? (b) The output of the generator in part (a) is converted, with 80% efficiency, into the horizontal motion of a 250-gram object. How fast is the object moving at the end of 1 minute?

5. (a) How much energy is required to pump all the water out of the swimming pool in Fig. 10-3-3? (b) Suppose that a mass equal to that of the water in the pool were moving with energy equal to the result of part (a). What would its velocity be? (c) Do parts (a) and (b) assuming that the pool is filled with a liquid three times as dense as water.

★6. (a) Find the total solar energy received at a latitude in the polar region on a day on which the sun never sets. (b) Find the total solar energy received over the entire year at a latitude in the polar region (your answer will contain a constant that depends on the units of measure chosen).

★7. How do you think the climate of the earth would be affected if the inclination α were to become: (a) 10°? (b) 40°? (In each case, discuss whether the North Pole receives more or less energy during the year than the equator.)

8. A force $F(x) = 3x \sin(\pi x/2)$ newtons acts on a particle between positions $x = 0$ and $x = 2$. What is the increase in kinetic energy of the particle between these positions?

Fig. 10-3-3 How much energy is required to empty this pool of water?

Review Problems for Chapter 10

1. Find the volume of the solid obtained by rotating the region under the graph of $y = \sin x$, $0 \leq x \leq \pi$ about (a) the y axis and (b) the x axis.

2. Find the volume of the solid obtained by rotating the region under the graph of $y = e^x$, $0 \leq x \leq \ln 2$ about (a) the y axis, (b) the x axis.

3. Let the region under the graph of a positive function $f(x)$, $a \leq x \leq b$, be revolved about the x axis to form a solid S. Suppose this solid has a mass density of $\rho(x)$ grams per cubic centimeter at a distance x along the x axis. (a) Find a formula for the mass of S. (b) If $f(x) = x^2$, $a = 0$ and $b = 1$, and $\rho(x) = (1 + x^4)$, find the mass of S.

4. Find the average value of each function on the stated interval:
   (a) $1 + t^3$, $0 \leq t \leq 1$
   (b) $t \sin (t^2)$, $\pi \leq t \leq 3\pi/2$
   (c) xe^x, $0 \leq x \leq 1$
   (d) $1/(1 + x^2)$, $1 \leq x \leq 3$

★5. (a) Prove that $1/\sqrt{2} \leq \int_0^1 dt/\sqrt{t^3 + 1} \leq 1$.
   (b) Prove that $\int_0^1 dt/\sqrt{t^3 + 1} = \sin \theta$ for some θ, $\pi/4 \leq \theta \leq \pi/2$.

6. Let μ be the average value of f on $[a, b]$. Then the average value of $[f(x) - \mu]^2$ on $[a, b]$ is called the *variance* of f on $[a, b]$, and the square root of the variance is called the *standard deviation* of f on $[a, b]$ and is denoted σ. Find the average value, variance, and standard deviation of each of the following functions on the interval specified:
   (a) x^2 on $[0, 1]$ (b) $3 + x^2$ on $[0, 1]$
   (c) xe^x on $[0, 1]$ (d) $\sin 2x$ on $[0, 4\pi]$
   (e) $f(x) = \begin{cases} 2 \text{ on } [0, 1] \\ 3 \text{ on } (1, 2] \\ 1 \text{ on } (2, 3] \\ 5 \text{ on } (3, 4] \end{cases}$ (let $[a, b] = [0, 4]$).

★7. See Problem 6 for definitions.
   (a) Suppose that $f(t)$ is piecewise constant on $[a, b]$, with adapted partition $(t_0, \ldots, t_n)$ and $f(t) = k_i$ on (t_{i-1}, t_i). Find a formula for the standard deviation of f on $[a, b]$.
   (b) Simplify your formula in part (a) for the case when all the Δt_i's are equal.
   (c) Show that if the standard deviation of a piecewise constant function is zero, then the function has the same value on all the intervals of the partition.
   (d) Give a definition for the standard deviation of a list $a_1, \ldots, a_n$ of numbers.
   (e) What can you say about a list of numbers if its standard deviation is zero?

★8. (a) Prove, by analogy with Theorem 1 (in Section 10-2), the *second mean value theorem:* If f and g are continuous on $[a, b]$ and $g(x) \geq 0$, for x in $[a, b]$, then there is a point t_0 in $[a, b]$ such that
   $$\int_a^b f(t)g(t)\, dt = f(t_0) \int_a^b g(t)\, dt$$
   (b) Show that Theorem 1 is a special case of the result in part (a).
   (c) Show that the conclusion of part (a) is false without the assumption that $g(t) \geq 0$.

★9. Show that if f is an increasing continuous function on $[a, b]$, the mean value theorem for integrals implies the conclusions of the intermediate value theorem.

10. Over a time period $0 \leq t \leq 6$ (t measured in minutes), an engine is consuming power at a rate of $20 + 5te^{-t}$ watts. What is (a) the total energy consumed? (b) the average power used?

11. Water is being pumped from a deep, irregularly shaped well at a constant rate of $3\frac{1}{2}$ cubic meters per hour. At a certain instant, it is observed that the water level is dropping at a rate of 1.2 meters per hour. What is the cross-sectional area of the well at that depth?

12. Find the center of mass of the region under the graph of $y = \ln(1 + x)$ between $x = 0$ and $x = 1$.

13. The engine in Fig. 10-R-1 is using energy at a rate of 300 joules per second to lift the weight of 600 kilograms. If the engine operates at 60% efficiency, at what speed (meters per second) can it raise the weight?

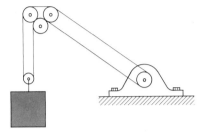

Fig. 10-R-1 The engine for Problem 13.

14. (a) Develop a formula for the work required to empty a tank which is a solid of revolution about a vertical axis of symmetry. (b) How much work is required to empty the tank shown in Fig. 10-R-2?

Circular rim

40 m

10 m

A parabola

Fig. 10-R-2 How much energy is needed to empty the tank?

★15. Show that in the context of Problem 14(a), the work needed to empty the tank equals Mgh, where M is the total mass of water in the tank and h is the distance of the center of mass of the tank below the top of the tank.

16. A force $F(x) = 30 \sin(\pi x/4)$ newtons acts on a particle between positions $x = 2$ and $x = 4$. What is the increase in kinetic energy (in joules) between these positions?

17. The pressure (force per unit area) at a depth h below the surface of a body of water is given by

$$p = \rho g h = 9800h$$

measured in newtons per square meter. (This formula derives from the fact that the force needed to support a column of water of cross-sectional area A is (volume) × (density) × (g) = $Ah\rho g$, so the force per unit area is $\rho g h$ and $\rho = 10^3$ kilograms per cubic meter, $g = 9.8$ meters per second per second).

(a) For the dam shown in Fig. 10-R-3(a) show that the total force exerted on it by the water is $F = \frac{1}{2}\int_a^b \rho g\,[f(x)]^2\,dx$. [*Hint:* First calculate the force exerted on a vertical rectangular slab.]

(b) Make up a geometric theorem relating F to the volume of a certain solid.

(c) Find the total force exerted on the dam whose face is shown in Fig. 10-R-3(b).

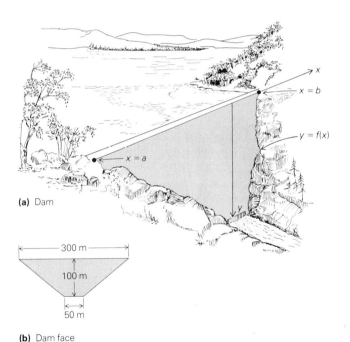

(a) Dam

300 m

100 m

50 m

(b) Dam face

Fig. 10-R-3 Calculate the force on a dam.

★18. Consider the equation for E on p. 437. For $D = \pi/6$, compute dE/dl at $l = \pi/4$. Is your answer consistent with the graph in Fig. 10-3-2 (look in the plane of constant T)?

★19. Determine whether a square meter at the equator or at the North Pole receives more solar energy: (a) during the month of February (b) during the month of April (c) during the entire year.

11

Further Development and Applications of Integration

Some simple geometric problems demand advanced methods of integration.

Beyond the basic methods of integration associated with reversing the differentiation rules, there are special methods for integrands of particular forms. Using these methods, we can compute many lengths and areas by expressing these quantities as integrals.

CONTENTS

The first section of this chapter contains some new methods of integration involving substitutions and algebraic transformations which reduce integration problems to ones we know how to handle. The following two sections are devoted to geometric applications of the theory and techniques of integration. Many of these applications lead to integrals of the type treated in the first section.

11-1 Trigonometric and Algebraic Transformations

Despite the existence of tables, some techniques of integration are indispensable.

In this section, we describe several techniques in which trigonometric and algebraic identities are used to simplify integrals. In a typical problem, several techniques must be applied; the solution of such problems proceeds by trial and error, guided by experience.

Goals

After studying this section, you should be able to:

Evaluate integrals involving powers and products of sines and cosines.

Use trigonometric substitutions to evaluate integrals involving square roots of quadratic functions.

Evaluate integrals of rational functions.

TRIGONOMETRIC INTEGRALS

We begin by considering integrals of the form

$$\int \sin^m x \cos^n x \, dx$$

where m and n are integers.

The case $n = 1$ is easy, for if we let $u = \sin x$, we find

$$\int \sin^m x \cos x \, dx = \int u^m \, du = \frac{u^{m+1}}{m+1} + C = \frac{\sin^{m+1}(x)}{m+1} + C$$

(or $\ln|\sin x| + C$, if $m = -1$). The case $m = 1$ is similar:

$$\int \sin x \cos^n x \, dx = -\frac{\cos^{n+1}(x)}{n+1} + C$$

(or $-\ln|\cos x| + C$, if $n = -1$).

If either m or n is odd, we can use the identity $\sin^2 x + \cos^2 x = 1$ to reduce the integral to the previous case.

Worked Example 1   Evaluate $\int \sin^2 x \cos^3 x \, dx$.

Solution
$$\int \sin^2 x \cos^3 x \, dx = \int \sin^2 x \cos^2 x \cos x \, dx$$
$$= \int (\sin^2 x)(1 - \sin^2 x) \cos x \, dx$$

which can be integrated by the substitution $u = \sin x$. We get

$$\int u^2(1 - u^2) \, du = \frac{u^3}{3} - \frac{u^5}{5} + C = \frac{1}{3} \sin^3 x - \frac{1}{5} \sin^5 x + C$$

If $m = 2k$ and $n = 2l$ are both even, we can use the half-angle formulas, $\sin^2 x = (1 - \cos 2x)/2$ and $\cos^2 x = (1 + \cos 2x)/2$, to write

$$\int \sin^{2k} x \cos^{2l} x \, dx = \int \left(\frac{1 - \cos 2x}{2}\right)^k \left(\frac{1 + \cos 2x}{2}\right)^l dx$$

$$= \frac{1}{2} \int \left(\frac{1 - \cos y}{2}\right)^k \left(\frac{1 + \cos y}{2}\right)^l dy$$

where $y = 2x$. Multiplying this out, we are faced with a sum of integrals of the form $\int \cos^m y \, dy$, with m ranging from zero to $k + l$. The integrals for odd m can be handled by the previous method; to those with even m we apply the half-angle formula once again. The whole process is repeated as often as necessary until everything is integrated.

Worked Example 2   Evaluate $\int_0^{2\pi} \sin^4 x \cos^2 x \, dx$.

Solution   Substitute $\sin^2 x = (1 - \cos 2x)/2$ and $\cos^2 x = (1 + \cos 2x)/2$ to get

$$\int \sin^4 x \cos^2 x \, dx = \int \frac{(1 - \cos 2x)^2}{4} \frac{(1 + \cos 2x)}{2} dx$$

$$= \tfrac{1}{8} \int (1 - 2\cos 2x + \cos^2 2x)(1 + \cos 2x) \, dx$$

$$= \tfrac{1}{8} \int (1 - \cos 2x - \cos^2 2x + \cos^3 2x) \, dx$$

$$= \tfrac{1}{16} \int (1 - \cos y - \cos^2 y + \cos^3 y) \, dy$$

where $y = 2x$. But $\int \cos^2 y \, dy = \int [(1 + \cos 2y)/2] \, dy = y/2 + (\sin 2y)/4 + C$ and $\int \cos^3 y \, dy = \int (1 - \sin^2 y) \cos y \, dy = \sin y - (\sin^3 y)/3 + C$. Thus

$$\int \sin^4 x \cos^2 x \, dx = \frac{1}{16}\left(y - \sin y - \frac{y}{2} - \frac{\sin 2y}{4} + \sin y - \frac{\sin^3 y}{3}\right) + C$$

$$= \frac{1}{16}\left(\frac{y}{2} - \frac{\sin 2y}{4} - \frac{\sin^3 y}{3}\right) + C$$

$$= \frac{1}{16}\left(x - \frac{\sin 4x}{4} - \frac{\sin^3 2x}{3}\right) + C$$

and so

$$\int_0^{2\pi} \sin^4 x \cos^2 x \, dx = \frac{\pi}{8}$$

TRIGONOMETRIC INTEGRALS

To evaluate $\int \sin^m x \cos^n x \, dx$:

1. If m is odd, $m = 2k + 1$, write

    $$\int \sin^m x \cos^n x \, dx = \int \sin^{2k} x \cos^n x \sin x \, dx$$
    $$= \int (1 - \cos^2 x)^k \cos^n x \sin x \, dx$$

    and integrate by substituting $u = \cos x$.

2. If n is odd, $n = 2l + 1$, write

    $$\int \sin^m x \cos^n x \, dx = \int \sin^m x \cos^{2l} x \cos x \, dx$$
    $$= \int \sin^m x (1 - \sin^2 x)^l \cos x \, dx$$

    and integrate by substituting $u = \sin x$.

3. (a) If $m = 2k$ and $n = 2l$ are even, write

    $$\int \sin^{2k} x \cos^{2l} x \, dx = \int \left(\frac{1 - \cos 2x}{2} \right)^k \left(\frac{1 + \cos 2x}{2} \right)^l dx$$

    Substitute $y = 2x$. Expand and apply step 2 to the odd powers of $\cos y$.

    (b) Apply step 3(a) to the even powers of $\cos y$ and continue until the integration is completed.

Solved Exercises*

1. Evaluate $\int (\sin^2 x + \sin^3 x \cos^2 x) \, dx$.

2. Evaluate $\int \sin^2 x \cos^2 x \, dx$.

Exercises

1. Evaluate $\int (\cos 2x - \cos^2 x) \, dx$.

2. Evaluate $\int \cos 2x \sin x \, dx$.

3. Evaluate $\int \sin^3 x \cos^2 x \, dx$.

4. Evaluate $\int (\sin^2 \theta / \cos^2 \theta) \, d\theta$.

MORE TRIGONOMETRIC INTEGRALS

Certain other trigonometric integrals yield to the use of the addition formulas:

$$\sin(x + y) = \sin x \cos y + \cos x \sin y \tag{1a}$$
$$\cos(x + y) = \cos x \cos y - \sin x \sin y \tag{1b}$$

and the product formulas:

$$\sin x \cos y = \tfrac{1}{2} [\sin(x + y) + \sin(x - y)] \tag{2a}$$
$$\sin x \sin y = \tfrac{1}{2} [\cos(x - y) - \cos(x + y)] \tag{2b}$$
$$\cos x \cos y = \tfrac{1}{2} [\cos(x + y) + \cos(x - y)] \tag{2c}$$

---

*Solutions appear in Appendix A.

Worked Example 3   Evaluate $\int \sin ax \sin bx \, dx$, where a and b are constants.

Solution   If we use identity ($2b$), we get

$$\int \sin ax \sin bx \, dx = \frac{1}{2} \int [\cos(a-b)x - \cos(a+b)x] \, dx$$

$$= \begin{cases} \dfrac{1}{2} \dfrac{\sin(a-b)x}{a-b} - \dfrac{1}{2} \dfrac{\sin(a+b)x}{a+b} + C & \text{if } a \neq \pm b \\[2ex] \dfrac{x}{2} - \dfrac{1}{4a} \sin 2ax + C & \text{if } a = b \\[2ex] \dfrac{1}{4a} \sin 2ax - \dfrac{x}{2} + C & \text{if } a = -b \end{cases}$$

The difference between the case $a \neq \pm b$ and the other two should be noted. The first case is "pure oscillation" in that it consists of two sine terms. The others contain the nonoscillating linear term $x/2$, called a *secular term*. This example is related to the phenomenon of *resonance:* when an oscillating system is subjected to a sinusoidally varying force, the oscillation will build up indefinitely if the force has the same frequency as the oscillator. (See Review Problem 19, Chapter 9.)

Solved Exercises

3. Evaluate $\int \sin x \cos 2x \, dx$.

4. Evaluate $\int \cos 3x \cos 5x \, dx$.

Exercises

5. Evaluate $\int \sin 4x \sin 2x \, dx$.

6. Evaluate $\int_0^{2\pi} \sin 5x \sin 2x \, dx$.

7. Show that $\int \sin^6 x \, dx = \frac{1}{192}(60x - 48\sin 2x + 4\sin^3 2x + 9\sin 4x) + C$.

8. Evaluate $\int \tan^3 x \sec^3 x \, dx$. [*Hint:* Convert to sines and cosines.]

TRIGONOMETRIC SUBSTITUTIONS

Many integrals containing factors of the form $\sqrt{a^2 \pm x^2}$, $\sqrt{x^2 - a^2}$, or $a^2 + x^2$ can be evaluated or simplified by means of trigonometric substitutions. In order to remember what to substitute it is useful to draw the appropriate right-angle triangle. See the following display.

TRIGONOMETRIC SUBSTITUTIONS

1. If $\sqrt{a^2 - x^2}$ occurs, try $x = a \sin \theta$; then $dx = a \cos \theta \, d\theta$ and $\sqrt{a^2 - x^2} = a \cos \theta$:

2. If $\sqrt{x^2 - a^2}$ occurs, try $x = a \sec \theta$; then $dx = a \tan \theta \sec \theta \, d\theta$ and $\sqrt{x^2 - a^2} = a \tan \theta$:

3. If $\sqrt{a^2 + x^2}$ or $a^2 + x^2$ occurs, try $x = a \tan \theta$; then $dx = a \sec^2 \theta \, d\theta$ and $\sqrt{a^2 + x^2} = a \sec \theta$ (one can also use $x = a \sinh \theta$; then $\sqrt{a^2 + x^2} = a \cosh \theta$):

Worked Example 4   Evaluate $\int (\sqrt{9 - x^2}/x^2) \, dx$.

Solution   Let $x = 3 \sin \theta$, so $\sqrt{9 - x^2} = 3 \cos \theta$. Thus $dx = 3 \cos \theta \, d\theta$ and

$$\int \frac{\sqrt{9 - x^2}}{x^2} \, dx = \int \frac{3 \cos \theta}{9 \sin^2 \theta} 3 \cos \theta \, d\theta$$

$$= \int \frac{\cos^2 \theta}{\sin^2 \theta} \, d\theta = \int \frac{1 - \sin^2 \theta}{\sin^2 \theta} \, d\theta$$

$$= \int (\csc^2 \theta - 1) \, d\theta = -\cot \theta - \theta + C$$

$$= -\frac{\sqrt{9 - x^2}}{x} - \sin^{-1}\left(\frac{x}{3}\right) + C$$

In the last line, we used the first figure in the preceding display to get

$$\cot \theta = \frac{\sqrt{a^2 - x^2}}{x} \quad \text{where} \quad a = 3$$

You may notice that whereas all our previous substitutions have been of the form $u = f(x)$, we use here the form $x = g(u)$. Since f and g are inverse functions, this procedure is sometimes called *inverse substitution*. (Observe that in writing the result in terms of x, we are required to use the form $u = f(x)$ in the end; hence the presence of the inverse trigonometric function.) In practice you need not be concerned whether the substitution you make is "inverse" or not. The notation, if you use it correctly, keeps track of things automatically.

Worked Example 5   Evaluate $\int dx/\sqrt{4x^2 - 1}$.

Solution   Let $x = \frac{1}{2} \sec \theta$, so $dx = \frac{1}{2} \tan \theta \sec \theta \, d\theta$ and $\sqrt{4x^2 - 1} = \tan \theta$. Thus

$$\int \frac{dx}{\sqrt{4x^2 - 1}} = \frac{1}{2} \int \frac{\tan \theta \sec \theta}{\tan \theta} \, d\theta = \frac{1}{2} \int \sec \theta \, d\theta$$

Here is a trick* for evaluating $\int \sec \theta$:

$$\int \sec \theta \, d\theta = \int \sec \theta \, \frac{\sec \theta + \tan \theta}{\sec \theta + \tan \theta} \, d\theta$$

$$= \int \frac{\sec^2 \theta + \sec \theta \tan \theta}{\sec \theta + \tan \theta} \, d\theta$$

$$= \ln |\sec \theta + \tan \theta| + C$$

Thus

$$\int \frac{dx}{\sqrt{4x^2 - 1}} = \frac{1}{2} \ln |2x + \sqrt{4x^2 - 1}| + C$$

Fig. 11-1-1 Geometry of the substitution $x = \frac{1}{2} \sec \theta$.

See Fig. 11-1-1.

This integral can also be evaluated by means of the formula $\int du/\sqrt{u^2 - 1} = \cosh^{-1} u + C$. See Solved Exercise 6(b).

These examples show that trigonometric substitutions work quite well in the presence of algebraic integrands involving square roots, such as $\int (x/\sqrt{1 - x^2}) \, dx$ or $\int [1/(x^2\sqrt{x^2 - 4})] \, dx$. You should also keep in mind the direct integration formulas involving inverse trigonometric and hyperbolic functions.

Solved Exercises

5. Evaluate $\int [1/(1 + x^2)] \, dx$ (a) as $\tan^{-1} x$ and (b) by the substitution $x = \tan u$. Compare your answers.

6. Evaluate:

   (a) $\displaystyle\int \frac{x}{\sqrt{4 - x^2}} \, dx$ (b) $\displaystyle\int \frac{1}{\sqrt{x^2 - 4}} \, dx$

7. Evaluate $\int (x^2/\sqrt{4 - x^2}) \, dx$.

Exercises

9. Evaluate $\int (\sqrt{x^2 - 4}/x) \, dx$.

10. Evaluate $\int (x^3/\sqrt{4 - x^2}) \, dx$.

11. Evaluate $\int [x^2/(1 + x^2)^{3/2}] \, dx$.

12. Evaluate $\int [s/\sqrt{4 + s^2}] \, ds$.

---

*The same trick shows that
$$\int \operatorname{csch} \theta \, d\theta = -\ln |\operatorname{csch} \theta + \coth \theta| + C$$

COMPLETING THE SQUARE

The identity

$$ax^2 + bx + c = a\left(x + \frac{b}{2a}\right)^2 - \frac{b^2}{4a} + c$$

called completing the square can be useful in reducing integrals involving $ax^2 + bx + c$ to those treated in the preceding section.

Worked Example 6   Evaluate

$$\int \frac{dx}{\sqrt{10 + 4x - x^2}}$$

Solution   Completing the square, with $a = -1$, $b = 4$, and $c = 10$, we get

$$10 + 4x - x^2 = -(x - 2)^2 + 14$$

Hence

$$\int \frac{dx}{\sqrt{10 + 4x - x^2}} = \int \frac{dx}{\sqrt{14 - (x - 2)^2}} = \int \frac{du}{\sqrt{14 - u^2}}$$

where $u = x - 2$. This integral is $\sin^{-1}(u/\sqrt{14}) + C$, so our final answer is

$$\sin^{-1}\left(\frac{x - 2}{\sqrt{14}}\right) + C$$

COMPLETING THE SQUARE

If an integral involves $ax^2 + bx + c$, write

$$ax^2 + bx + c = a\left(x + \frac{b}{2a}\right)^2 + \left(\frac{-b^2}{4a} + c\right)$$

Let $u = x + (b/2a)$ and then use a trigonometric substitution or some other method to evaluate the integral.

Solved Exercises

8. Evaluate $\int dx/(x^2 + x + 1)$.

9. Evaluate $\int dx/\sqrt{x^2 + x + 1}$.

Exercises

13. Evaluate $\int dx/\sqrt{4x^2 + x + 1}$.

14. Evaluate $\int dx/\sqrt{5 - 4x - x^2}$.

15. Evaluate $\int x\,dx/\sqrt{3x^2 + x - 1}$.

PARTIAL FRACTIONS

By the method of partial fractions, one can evaluate *any* integral of the form $\int [P(x)/Q(x)]\,dx$, where P and Q are polynomials. Here we see clearly the need for doing integrals by hand, for tables cannot include all integrals of this type (since there are infinitely many).* As an example of the method, consider the problem of integrating $(x^3 + 2)/(1 - x^2)$. In this form the term looks unwieldy, but we can use algebraic manipulation to rewrite it in a more tractable form:

$$\frac{x^3 + 2}{1 - x^2} = -x + \frac{x + 2}{1 - x^2} = -x + \frac{x}{1 - x^2} + \left(\frac{1}{1 - x} + \frac{1}{1 + x}\right)$$

The first equality is obtained by division and the second by noting that

$$\frac{1}{1 - x^2} = \frac{1}{(1 - x)(1 + x)} = \frac{1}{2}\left(\frac{1}{1 - x} + \frac{1}{1 + x}\right)$$

Thus

$$\int \frac{x^3 + 2}{1 - x^2}\,dx = -\frac{x^2}{2} - \frac{1}{2}\ln|1 - x^2| - \ln|1 - x| + \ln|1 + x| + C$$

$$= -\frac{x^2}{2} - \frac{1}{2}\ln|1 - x^2| + \ln\left|\frac{1 + x}{1 - x}\right| + C$$

The second term, $x/(1 - x^2)$, was integrated by substitution. We could also have integrated $2/(1 - x^2)$ directly, obtaining $2\tanh^{-1}x$, but the answers are the same.

It turns out that algebraic trickery of this sort can always be made to work. A systematic procedure is supplied in the following display.

*Computer systems such as MACSYMA, capable of algebraic manipulations, can perform integrations of this type automatically (by following the procedures below).

PARTIAL FRACTIONS

To integrate $P(x)/Q(x)$, where P and Q are polynomials containing no common factor:

1. If the degree of P is larger than the degree of Q, divide Q into P by long division, obtaining a polynomial plus $R(x)/Q(x)$, where the degree of R is less than that of Q. Thus we need only investigate the case where the degree of P is less than that of Q.

2. Factor the denominator Q into linear and quadratic factors—that is, factors of the form $(x - a)$ and $ax^2 + bx + c$. (Factor the quadratic expressions if possible.)

3. If $(x - a)^m$ occurs in the factorization of Q, write down a sum of the form

$$\frac{a_1}{(x - a)} + \frac{a_2}{(x - a)^2} + \cdots + \frac{a_m}{(x - a)^m}$$

where $a_1, a_2, \ldots$ are constants. Do so for each factor of this form (using constants $b_1, b_2, \ldots, c_1, c_2, \ldots$, and so on) and add the expressions you get. The constants $a_1, a_2, \ldots, b_1, b_2, \ldots$, and so on will be determined in step 5.

4. If $(ax^2 + bx + c)^p$ occurs in the factorization of Q, write down a sum of the form

$$\frac{A_1 x + B_1}{ax^2 + bx + c} + \frac{A_2 x + B_2}{(ax^2 + bx + c)^2} + \cdots + \frac{A_p x + B_p}{(ax^2 + bx + c)^p}$$

Do so for each factor of this form and add the expressions you get. The constants $A_1, A_2, \ldots, B_1, B_2, \ldots$ are determined in step 5. Add this expression to the one obtained in step 3.

5. Equate the expression obtained in steps 3 and 4 to $P(x)/Q(x)$. Multiply through by $Q(x)$ to obtain an equation between two polynomials. Comparing coefficients of these polynomials, determine equations for the constants $a_1, a_2, \ldots, A_1, A_2, \ldots, B_1, B_2, \ldots$ and solve these equations. Sometimes the constants can be determined by substitution of convenient values of x in the equality.

6. Integrate the expression obtained in step 5 by using

$$\int \frac{dx}{(x - a)^r} = -\left[\frac{1}{(r - 1)(x - a)^{r-1}}\right] + C \quad r > 1 \quad \text{and} \quad \int \frac{dx}{x - a} = \ln|x - a| + C$$

The terms with a quadratic denominator may be integrated by a manipulation which makes the derivative of the denominator appear in the numerator together with completing the square (see Worked Example 8).

We omit the proof that this procedure always works. It may be found in books on algebra.*

Worked Example 7 Integrate

$$\int \frac{dx}{x^3 - x^2}$$

---

*See, for instance, H. B. Fine, College Algebra, *Dover, 1961, p. 241.*

Solution The denominator $x^3 - x^2$ is of degree 3, and the numerator is of degree zero. Thus we proceed to step 2 and factor:

$$x^3 - x^2 = x^2(x - 1)$$

Here $x = x - 0$ occurs to the power 2, so by step 3 we write down

$$\frac{a_1}{x} + \frac{a_2}{x^2}$$

We also write down $b_1/(x - 1)$ for the second factor. Then we add:

$$\frac{a_1}{x} + \frac{a_2}{x^2} + \frac{b_1}{x - 1}$$

Since there are no quadratic factors, we omit step 4. By step 5 we equate the preceding expression to $1/(x^3 - x^2)$:

$$\frac{a_1}{x} + \frac{a_2}{x^2} + \frac{b_1}{x - 1} = \frac{1}{x^2(x - 1)}$$

Then we multiply by $x^2(x - 1)$:

$$a_1 x(x - 1) + a_2(x - 1) + b_1 x^2 = 1$$

Setting $x = 0$, we get $a_2 = -1$. Setting $x = 1$, we get $b_1 = 1$. Comparing the coefficients of x^2 on both sides of the equation gives $a_1 + b_1 = 0$, so $a_1 = -b_1 = -1$. Thus $a_2 = -1$, $a_1 = -1$, and $b_1 = 1$. (We can check by substitution into the preceding equation: the left side is $(-1)x(x - 1) - (x - 1) + x^2$, which is just 1.)

Thus

$$\frac{1}{x^3 - x^2} = -\frac{1}{x^2} - \frac{1}{x} + \frac{1}{x - 1}$$

and so

$$\int \frac{dx}{x^3 - x^2} = \frac{1}{x} - \ln|x| + \ln|x - 1| + C = \frac{1}{x} + \ln\left|\frac{x - 1}{x}\right| + C$$

Worked Example 8 Integrate

$$\int \frac{x^3}{(x - 1)(x^2 + 2x + 2)^2}\, dx$$

Solution For the factor $x - 1$ we write

$$\frac{a_1}{x - 1}$$

and for $(x^2 + 2x + 2)^2$ (which does not factor further since $x^2 + 2x + 2$ does not have real roots) we write

$$\frac{A_1 x + B_1}{x^2 + 2x + 2} + \frac{A_2 x + B_2}{(x^2 + 2x + 2)^2}$$

We then set

$$\frac{a_1}{x - 1} + \frac{A_1 x + B_1}{x^2 + 2x + 2} + \frac{A_2 x + B_2}{(x^2 + 2x + 2)^2} = \frac{x^3}{(x - 1)(x^2 + 2x + 2)^2}$$

and multiply by $(x - 1)(x^2 + 2x + 2)^2$:

$$a_1(x^2 + 2x + 2)^2 + (A_1 x + B_1)(x - 1)(x^2 + 2x + 2) + (A_2 x + B_2)(x - 1) = x^3$$

Setting $x = 1$ gives $a_1(25) = 1$ or $a_1 = \frac{1}{25}$. Expanding the left-hand side, we get:

$$\frac{1}{25}(x^4 + 4x^3 + 8x^2 + 8x + 4) + A_1 x^4 + (A_1 + B_1)x^3 + B_1 x^2$$
$$- 2A_1 x - 2B_1 + A_2 x^2 + (B_2 - A_2)x - B_2 = x^3$$

Comparing coefficients:

$$x^4: \quad \frac{1}{25} + A_1 = 0 \tag{3}$$
$$x^3: \quad \frac{4}{25} + (A_1 + B_1) = 1 \tag{4}$$
$$x^2: \quad \frac{8}{25} + B_1 + A_2 = 0 \tag{5}$$
$$x: \quad \frac{8}{25} - 2A_1 + (B_2 - A_2) = 0 \tag{6}$$
$$1: \quad \frac{4}{25} - 2B_1 - B_2 = 0 \tag{7}$$

Thus

$$A_1 = -\tfrac{1}{25} \quad \text{(from (3))}$$
$$B_1 = \tfrac{22}{25} \quad \text{(from (4))}$$
$$A_2 = -\tfrac{30}{25} \quad \text{(from (5))}$$
$$B_2 = -\tfrac{40}{25} \quad \text{(from (6))}$$

At this stage you should always check the algebra by substitution into the last equation—in this case, equation (7). Algebraic errors are easy to make in this process.

So we have

$$\frac{x^3}{(x - 1)(x^2 + 2x + 2)^2} = \frac{1}{25}\left[\frac{1}{x - 1} + \frac{-x + 22}{x^2 + 2x + 2} + \frac{-30x - 40}{(x^2 + 2x + 2)^2} \right]$$

We compute the integrals of the first two terms as follows:

$$\int \frac{1}{x-1}\, dx = \ln|x-1| + C$$

$$\int \frac{-x+22}{x^2+2x+2}\, dx = \int \frac{-x-1+23}{x^2+2x+2}\, dx$$

$$= -\frac{1}{2}\int \frac{2x+2}{x^2+2x+2}\, dx + 23\int \frac{dx}{x^2+2x+2}$$

$$= -\frac{1}{2}\ln|x^2+2x+2| + 23\int \frac{dx}{(x+1)^2+1}$$

$$= -\frac{1}{2}\ln|x^2+2x+2| + 23\tan^{-1}(x+1) + C$$

Finally, for the last term, we arrange the numerator to make the derivative of the denominator appear:

$$\int \frac{-30x-40}{(x^2+2x+2)^2}\, dx = \int \frac{-15(2x+2)-10}{(x^2+2x+2)^2}\, dx$$

$$= 15\cdot\frac{1}{(x^2+2x+2)} - 10\int \frac{1}{[(x+1)^2+1]^2}\, dx$$

Let $x + 1 = \tan\theta$, so $dx = \sec^2\theta\, d\theta$ and $(x+1)^2 + 1 = \sec^2\theta$. Then

$$\int \frac{1}{[(x+1)^2+1]^2}\, dx = \int \frac{\sec^2\theta\, d\theta}{\sec^4\theta}$$

$$= \int \cos^2\theta\, d\theta$$

$$= \int \frac{1+\cos 2\theta}{2}\, d\theta$$

$$= \frac{\theta}{2} + \frac{\sin 2\theta}{4} + C$$

$$= \frac{1}{2}\tan^{-1}(x+1) + \frac{1}{2}\sin\theta\cos\theta + C$$

$$= \frac{1}{2}\tan^{-1}(x+1) + \frac{1}{2}\cdot\frac{x+1}{(x+1)^2+1} + C$$

Fig. 11-1-2 Geometry of the substitution $x + 1 = \tan\theta$.

(See Fig. 11-1-2.) Adding all this, we find

$$\int \frac{x^3}{(x-1)(x^2+2x+2)}\, dx$$

$$= \frac{1}{25}\left[\ln|x-1| - \frac{1}{2}\ln(x^2+2x+2) + 23\tan^{-1}(x+1) + 15\frac{1}{x^2+2x+2} - 5\tan^{-1}(x+1)\right.$$

$$\left. - 5\frac{x+1}{x^2+2x+2}\right] + C = \frac{1}{25}\left[\ln\left(\frac{|x-1|}{\sqrt{x^2+2x+2}}\right) + 18\tan^{-1}(x+1) + \frac{10-5x}{x^2+2x+2}\right] + C$$

Solved Exercises

10. Integrate $\int [(x^3 + 2x + 1)/(x - 1)^5] \, dx$. [*Hint:* Do not use partial fractions!]

11. Integrate $\int [1/(x^3 - 1)] \, dx$

12. Integrate $\int [x^2/(x^2 - 2)^2] \, dx$.

Exercises

16. Integrate $\int [x/(x^4 + 2x^2 - 3)] \, dx$.

17. Integrate $\int x^2/[(x - 2)(x^2 + 2x + 2)] \, dx$.

18. Integrate $\int 1/[(x - 2)^2(x^2 + 1)^2] \, dx$.

19. Integrate $\int [(x^4 + 2x^3 + 3)/(x - 4)^6] \, dx$.

Problems for Section 11-1

1. Evaluate the following integrals:

(a) $\int \dfrac{x}{x^2 + 3} \, dx$

(b) $\int \dfrac{x}{\sqrt{x^2 - 1}} \, dx$

(c) $\int \dfrac{x}{\sqrt{x + 1}} \, dx$

(d) $\int x^3 \ln x \, dx$

(e) $\int \sqrt{1 + \sin x} \cdot \cos x \, dx$

(f) $\int (e^x + 1)^3 e^x \, dx$

2. Integrate:

(a) $\int \dfrac{x}{x^2 + 1} \, dx$

(b) $\int \sin 2\theta \cos 5\theta \, d\theta$

(c) $\int \dfrac{x}{(x^2 + 1)^2} \, dx$

(d) $\int \cot \theta \, d\theta$

(e) $\int (e^{\sqrt{x}}/\sqrt{x}) \, dx$

(f) $\int_2^3 [x/(x^2 + 1)] \, dx$

3. Evaluate the definite integrals:

(a) $\int_0^1 \dfrac{x^4}{(x^2 + 1)^2} \, dx$

(b) $\int_2^4 \dfrac{x^3 + 1}{x^3 - 1} \, dx$

(c) $\int_3^4 \dfrac{dx}{(x - 2)(x^2 + 3x + 1)}$

(d) $\int_0^{\pi/4} \sin^2 x \cos 2x \, dx$

(e) $\int_1^2 [(\ln 3x + 5)^3/x] \, dx$

(f) $\int_0^{\pi/2} \sin x \, e^{\cos x} \, dx$

4. Integrate:

(a) $\int_e^{e^4} \dfrac{\ln t^2}{t^2} \, dt$

(b) $\int_0^1 \sinh^2 x \, dx$

(c) $\int_0^{\pi/4} \dfrac{\sin \theta}{\sqrt{1 - \cos^2 \theta}} \, d\theta$

(d) $\int_0^{2\pi} \dfrac{\sin \theta}{1 + \cos \theta + \cos^2 \theta} \, d\theta$

(e) $\int \sec^2 100x \, dx$

(f) $\int dt/(3 + t^2)$

5. Substitute $u = \tan (\theta/2)$ to prove that

$$\int \frac{d\theta}{2 + \cos \theta} = \frac{2}{\sqrt{3}} \tan^{-1}\left(\frac{\tan (\theta/2)}{\sqrt{3}}\right)$$

6. Find the average value of $\sin^2 x \cos^2 x$ on the interval $[0, 2\pi]$.

7. Find the average value of $\cos^n x$ on the interval $[0, 2\pi]$ for $n = 0, 1, 2, 3, 4, 5, 6$.

8. Find the volume of the solid obtained by revolving the region under the graph $y = \sin^2 x$ on $[0, 2\pi]$ about the x axis.

9. Find the volume of the solid obtained by revolving the region under the graph $y = 1/[(1 - x)(1 - 2x)]$ on $[5,6]$ about the y axis.

10. Find the center of mass of the region under the graph of $1/(x^2 + 4)$ on $[1, 3]$.

11. Find the center of mass of the region under the graph of $1/\sqrt{x^2 + 2x + 2}$ on $[0, 1]$.

12. A plating company wishes to prepare the bill for a silver plate job of 200 parts. Each part has the shape of the region bounded by $y = \sqrt{x^2 - 9/x^2}$, $y = 0$, $x = 5$.

   (a) Find the area enclosed.

   (b) Assume all units are centimeters. Only one side of the part is to receive silver plate. The customer was charged $25 for 1460 cm^2 previously. How much should the 200 parts cost?

13. The *average power* P for a resistance R and associated current i of period T is

$$P = \frac{1}{T} \int_0^T Ri^2 \, dt$$

That is, P is the average value of the instantaneous power Ri^2 on $[0, T]$. Compute the power for $R = 2.5$, $i = 10 \sin(377t)$, $T = 2\pi/377$.

14. The current I in a certain RLC circuit is given by $I(t) = Me^{-\alpha t}[\sin^2(\omega t) + 2\cos(2\omega t)]$. Find the charge Q in coulombs, given by $Q(t) = Q_0 + \int_0^t I(s)\,ds$.

15. The *root mean square current* and *voltage* are

$$I_{\text{rms}} = \left(\frac{1}{T} \int_0^T i^2 \, dt \right)^{1/2}$$

$$E_{\text{rms}} = \left(\frac{1}{T} \int_0^T e^2 \, dt \right)^{1/2}$$

where $i(t)$ and $e(t)$ are the current through and voltage across a pure resistance R. (The current flowing through R is assumed to be periodic with period T.) Compute these numbers for $e(t) = 3 + (1.5)\cos(100t)$ volts, $i(t) = 1 + 2\sin(100t - \pi/6)$ amperes, which corresponds to period $T = 2\pi/(100)$.

16. The *average power* $P = (1/T)\int_0^T Ri^2\,dt$ for periodic waveshapes in general does not obey a superposition principle. Two voltage sources e_1 and e_2 may individually supply 5 watts (when the other is dead), but when both sources are present the power can be zero (not 10). Investigate the validity of $\int_0^T R(i_1 + i_2)^2\,dt = \int_0^T Ri_1^2\,dt + \int_0^T Ri_2^2\,dt$ when $i_1 = I_1\cos(m\omega t + \phi_1)$, $i_2 = I_2\cos(n\omega t + \phi_2)$, $m \neq n$ (m, n positive integers), $T = 2\pi/\omega$, and R, I_1, I_2, ω, ϕ_1, ϕ_2 are constants.

17. A charged particle is constrained by magnetic fields to move along a straight line, oscillating back and forth from the origin with higher and higher amplitude.

    Let $S(t)$ be the directed distance from the origin, and assume that $[S(t)]^2 S'(t) = t\sin t + \sin^2 t \cos^2 t$, for concreteness.

    (a) Show that

    $$[S(t)]^3 = 3\int_0^t (x\sin x + \sin^2 x \cos^2 x)dx$$

    (b) Find $S(t)$.

    (c) Find all zeros of $S'(t)$ for $t > 1$. Which zeros correspond to times of maximum excursion from the origin?

18. A chemical reaction problem produces the following integration problem:

$$\int \frac{dx}{(80 - x)(60 - x)} = k\int dt, \quad k = \text{constant}$$

In this formula, $x(t)$ is the number of kilograms of reaction product present after t minutes, starting with 80 kg and 60 kg of substances, respectively, obeying the *law of mass action*.

    (a) Integrate to get a logarithmic formula involving x and t.

    (b) Convert the answer to an exponential formula for x.

    (c) How much reaction product is present after 15 minutes, assuming $x = 20$ when $t = 10$?

19. Integrate:

    (a) $\displaystyle\int \frac{2x^2 - x + 2}{x^5 + 2x^3 + x}\,dx$

    (b) $\displaystyle\int_{\pi/6}^{\pi/2} \frac{\cos x\,dx}{\sin x + \sin^3 x}$

    (c) $\displaystyle\int_0^1 \frac{dx}{8x^3 + 1}$

    (d) $\displaystyle\int \frac{(\sec^2 x + 1)\sec^2 x\,dx}{1 + \tan^3 x}$

20. Partial fractions appear in electrical engineering as a convenient means of analyzing and describing circuit responses to applied voltages. By means of the *Laplace transform*, circuit responses are associated with rational functions. Partial fraction methods are used to decompose these rational functions into elementary quotients, which are recognizable to engineers as arising from standard kinds of circuit responses. For example, from

$$\frac{s + 1}{(s + 2)(s^2 + 1)(s^2 + 4)} = \frac{A}{s + 2} + \frac{Bs + C}{s^2 + 1} + \frac{Ds + 2E}{s^2 + 4}$$

an engineer can easily see that this rational function represents the response

$$Ae^{-2t} + B\cos t + C\sin t + D\cos 2t + E\sin 2t$$

Find the constants A, B, C, D, E.

11-2 Arc Length and Surface Area

Integration can be used to find the length of graphs in the plane and the area of surfaces of revolution.

In Sections 4-3 and 10-1 we developed formulas for areas under and between graphs and for volumes of solids of revolution. We continue this line of development by obtaining formulas for lengths of graphs and for areas of surfaces of revolution.

Goals

After studying this section, you should be able to:

Express the length of a graph as an integral.

Express the area of a surface of revolution as an integral.

LENGTH OF CURVES

The length of a piece of curve in the plane is sometimes called the *arc length* of the curve. As we did with areas and volumes, we assume that the length exists and will try to express it as an integral. For now, we confine our attention to curves which are graphs of functions; general curves are considered in the next section.

As our first principle for arc length, we assume that the length of a straight line segment is equal to the distance between its endpoints. Thus, if $f(x) = mx + q$ on $[a, b]$, the endpoints (see Fig. 11-2-1) of the graph are $(a, ma + q)$ and $(b, mb + q)$, and the distance between them is

$$\sqrt{(a-b)^2 + [(ma+q)-(mb+q)]^2} = \sqrt{(a-b)^2 + m^2(a-b)^2}$$
$$= (b-a)\sqrt{1+m^2}$$

(Since $a < b$, the square root of $(a-b)^2$ is $b-a$.)

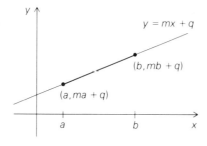

Fig. 11-2-1 The length of the dark segment is $(b-a)\sqrt{1+m^2}$.

Our strategy, as in Chapter 10, will be to interpret the arc length for a simple curve as an integral and then use the same formula for general curves. In the case of the straight line segment, $f(x) = mx + q$, whose length between $x = a$ and $x = b$ is $(b-a)\sqrt{1+m^2}$, we can interpret m as the derivative $f'(x)$, so that

$$\text{Length} = (b - a)\sqrt{1 + m^2} = \int_a^b \sqrt{1 + [f'(x)]^2}\, dx$$

Since the formula for the length is an integral of f', rather than of f, it is natural to look next at the functions for which f' is piecewise constant. If f' is constant on an interval, f is linear on that interval; thus the functions with which we will be dealing are the *piecewise linear* (also called *ramp*, or *polygonal*) functions.

To obtain a piecewise linear function, we choose a partition $(x_0, x_1, ..., x_n)$ of the interval $[a, b]$ and specify the values $(y_0, y_1, ..., y_n)$ of the function f at these points. For each $i = 1, 2, ..., n$, we then connect the point (x_{i-1}, y_{i-1}) to the point (x_i, y_i) by a straight line segment. Since the equation of the line through (x_{i-1}, y_{i-1}) and (x_i, y_i) is

$$y = y_{i-1} + \left(\frac{y_i - y_{i-1}}{x_i - x_{i-1}}\right)(x - x_{i-1})$$

the function f is given by

$$f(x) = y_{i-1} + \left(\frac{y_i - y_{i-1}}{x_i - x_{i-1}}\right)(x - x_{i-1}) \qquad \text{for } x_{i-1} \leq x \leq x_i$$

(See Fig. 11-2-2.)

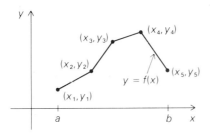

Fig. 11-2-2 The graph of a piecewise linear function.

The function $f(x)$ is differentiable on each of the intervals (x_{i-1}, x_i), where its derivative is constant and equal to the slope $(y_i - y_{i-1})/(x_i - x_{i-1})$. Thus the function $\sqrt{1 + [f'(x)]^2}$ is piecewise constant on $[a, b]$, with value

$$k_i = \sqrt{1 + \left(\frac{y_i - y_{i-1}}{x_i - x_{i-1}}\right)^2}$$

on (x_{i-1}, x_i).* We have

$$\int_a^b \sqrt{1 + [f'(x)]^2}\, dx = \sum_{i=1}^n k_i \Delta x_i$$

$$= \sum_{i=1}^n \sqrt{1 + \left(\frac{y_i - y_{i-1}}{x_i - x_{i-1}}\right)^2}(x_i - x_{i-1})$$

$$= \sum_{i=1}^n \sqrt{(x_i - x_{i-1})^2 + (y_i - y_{i-1})^2}$$

---

*Actually, $\sqrt{1 + f'(x)^2}$ is not defined at the points $x_0, x_1, ..., x_n$, but this does not matter when we take its integral, since the integral is not affected by changing the value of the integrand at isolated points.

Note that the ith term in this sum, $\sqrt{(x_i - x_{i-1})^2 + (y_i - y_{i-1})^2}$, is just the length of the segment of the graph of f between (x_{i-1}, y_{i-1}) and (x_i, y_i).

Now we invoke a second principle for arc length: if n curves are placed end to end, the length of the total curve is the sum of the lengths of the pieces. Using this principle, we see that the preceding sum is just the length of the graph of f on $[a, b]$. So we have now shown, for piecewise linear functions, that the length of the graph of f on $[a, b]$ equals the integral $\int_a^b \sqrt{1 + [f'(x)]^2}\, dx$.

Worked Example 1 Let the graph of f consist of straight line segments joining $(1, 0)$ to $(2, 1)$ to $(3, 3)$ to $(4, 1)$. Verify that the length of the graph, as computed directly, is given by the formula $\int_a^b \sqrt{1 + [f'(x)]^2}\, dx$.

Solution The graph is sketched in Fig. 11-2-3. The length is $d_1 + d_2 + d_3 = \sqrt{1 + 1} + \sqrt{1 + 2^2} + \sqrt{1 + (-2)^2} = \sqrt{2} + 2\sqrt{5}$.

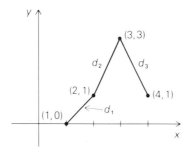

Fig. 11-2-3 The length of this graph is $d_1 + d_2 + d_3$.

On the other hand,

$$f'(x) = \begin{cases} 1 & \text{on } (1, 2) \\ 2 & \text{on } (2, 3) \\ -2 & \text{on } (3, 4) \end{cases}$$

(and is not defined at $x = 1, 2, 3, 4$). Thus, by definition of the integral of a piecewise constant function (see p. 182),

$$\int_1^4 \sqrt{1 + [f'(x)]^2}\, dx = (\sqrt{1 + 1^2}) \cdot 1 + (\sqrt{1 + 2^2}) \cdot 1 + [\sqrt{1 + (-2)^2}] \cdot 1$$
$$= \sqrt{2} + 2\sqrt{5}$$

which agrees with the preceding answer.

Justifying the passage from piecewise linear functions to general functions is more complicated than in the case of area, since we cannot squeeze a general curve between polygons as far as length is concerned. Nevertheless, the following formula is still correct.

LENGTH OF CURVES

Suppose that the function f is continuous on $[a, b]$ and that the derivative f' exists and is continuous everywhere, except possibly at finitely many points, on $[a, b]$.

Then the length of the graph of f on $[a, b]$ is:

$$L = \int_a^b \sqrt{1 + [f'(x)]^2}\, dx \tag{1}$$

We can give an "infinitesimal" argument as follows. The curve may be thought of as being composed of infinitely many infinitesimally short segments (see Fig. 11-2-4). By the theorem of Pythagoras, the length ds of each segment is equal to $\sqrt{dx^2 + dy^2}$. But $dy/dx = f'(x)$, so $dy = f'(x)\,dx$ and $ds = \sqrt{dx^2 + [f'(x)]^2\,dx^2} = \sqrt{1 + [f'(x)]^2}\,dx$. To get the total length, we take $\int_a^b ds = \int_a^b \sqrt{1 + [f'(x)]^2}\,dx$.

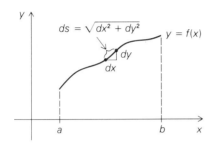

Fig. 11-2-4 An "infinitesimal segment" of the graph of f.

Let us check that formula (1) gives the right result for the length of an arc of a circle.

Worked Example 2 Use integration to find the length of the graph of $f(x) = \sqrt{1 - x^2}$ on $[0, b]$, where $0 < b < 1$. Then find the length geometrically and compare the results.

Solution By formula (1), the length is

$$\int_0^b \sqrt{1 + [f'(x)]^2}\, dx$$

where $f(x) = \sqrt{1 - x^2}$. We have

$$f'(x) = \frac{-x}{\sqrt{1 - x^2}}, \quad f'(x)^2 = \frac{x^2}{1 - x^2}, \quad 1 + [f'(x)]^2 = \frac{1}{1 - x^2}$$

Hence

$$L = \int_0^b \frac{dx}{\sqrt{1 - x^2}} = \sin^{-1}(b) - \sin^{-1}(0) = \sin^{-1}(b)$$

Examining Fig. 11-2-5, we see that $\sin^{-1}(b)$ is equal to θ, the angle intercepted by the arc whose length we are computing. By the definition of radian measure, the length of the arc is equal to the angle $\theta = \sin^{-1}(b)$, which agrees with our calculation by means of the integral.

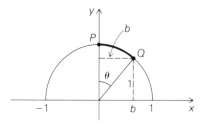

Fig. 11-2-5 The length of the arc PQ is θ, which equals $\sin^{-1} b$.

Because of the square root in formula (1), the integral obtained, even for quite simple f, is often difficult or even impossible to evaluate by elementary means. Of course, we can always approximate the result numerically. The following example shows how a simple-looking function can lead to a complicated integral for arc length.

Worked Example 3   Find the length of the parabola $y = x^2$ from $x = 0$ to $x = 1$.

Solution   We substitute $f(x) = x^2$ and $f'(x) = 2x$ into formula (1):

$$L = \int_0^1 \sqrt{1 + (2x)^2}\ dx$$

$$= 2 \int_0^1 \sqrt{(\tfrac{1}{2})^2 + x^2}\ dx$$

Now substitute $x = \tfrac{1}{2} \tan \theta$ and $\sqrt{(\tfrac{1}{2})^2 + x^2} = \tfrac{1}{2} \sec \theta$:

$$\int \sqrt{(\tfrac{1}{2})^2 + x^2}\ dx = \int (\tfrac{1}{2} \sec \theta)(\tfrac{1}{2} \sec^2 \theta\ d\theta)$$
$$= \tfrac{1}{4} \int \sec^3 \theta\ d\theta$$

We evaluate the integral of $\sec^3 \theta$ as follows:

$$\int \sec^3 \theta\ d\theta = \int \sec \theta \sec^2 \theta\ d\theta$$
$$= \int \sec \theta\ (\tan^2 \theta + 1)\ d\theta$$
$$= \int \sec \theta \tan^2 \theta\ d\theta + \int \sec \theta\ d\theta$$
$$= \int (\sec \theta \tan \theta) \tan \theta\ d\theta + \ln|\sec \theta + \tan \theta|$$

(See p. 449.) Now integrate by parts:

$$\int (\sec \theta \tan \theta) \tan \theta\ d\theta = \int \frac{d}{d\theta} (\sec \theta) \tan \theta\ d\theta$$
$$= \sec \theta \tan \theta - \int \sec \theta \sec^2 \theta\ d\theta$$
$$= \sec \theta \tan \theta - \int \sec^3 \theta\ d\theta$$

Substituting this formula into the last expression for $\int \sec^3 \theta\ d\theta$ gives

$$\int \sec^3 \theta\ d\theta = \sec \theta \tan \theta - \int \sec^3 \theta\ d\theta + \ln|\sec \theta + \tan \theta|$$

So*

$$\int \sec^3 \theta \, d\theta = \tfrac{1}{2}(\sec \theta \tan \theta + \ln |\sec \theta + \tan \theta|) + C$$

Since $2x = \tan \theta$ and $\sec \theta = 2 \cdot \sqrt{(\tfrac{1}{2})^2 + x^2} = \sqrt{1 + 4x^2}$, the integral $\int \sec^3 \theta \, d\theta$ in terms of x is

$$x\sqrt{1 + 4x^2} + \tfrac{1}{2} \ln |2x + \sqrt{1 + 4x^2}| + C$$

Substitution into the formula for L gives

$$L = \tfrac{1}{2}(x\sqrt{1 + 4x^2} + \tfrac{1}{2} \ln |2x + \sqrt{1 + 4x^2}|) \Big|_0^1$$

$$= \tfrac{1}{2}[\sqrt{5} + \tfrac{1}{2}\ln(2 + \sqrt{5})] \approx 1.479$$

Solved Exercises

1. Express the length of the graph of x^n on $[a, b]$ as an integral. For which values of n can you evaluate the integral?

2. Express the length of the graph of $f(x) = \sqrt{1 - k^2 x^2}$ (an ellipse) on $[0, b]$ as an integral.

Exercises

1. Find the length of the graph of $f(x) = (x - 1)^{3/2} + 2$ on $[1, 2]$.

2. Find the length of the graph of $f(x) = x^4/8 + (1/4x^2)$ on $[1, 3]$.

3. Express the length of the graph of $f(x) = \sin x$ on $[0, 2\pi]$ as an integral. (Do not evaluate.)

4. Find the length of the graph of $y = [x^3 + (3/x)]/6$ for $1 \le x \le 3$.

5. Find the length of the graph of $y = \sqrt{x}(4x - 3)/6$ for $1 \le x \le 9$.

THE AREA OF A SURFACE OF REVOLUTION

If we revolve the region R under the graph of $f(x)$ (assumed nonnegative) on $[a, b]$, about the x axis, we obtain a solid of revolution S. In Section 10-1, we saw how to express the volume of such a solid as an integral. Suppose now that, instead of revolving the region, we revolve the graph G itself. We obtain a curved surface Σ, called a *surface of revolution*, which forms part of the boundary of S. (The remainder of the boundary consists of the disks at the ends of the solid, which have radii $f(a)$ and $f(b)$; see Fig. 11-2-6.) Our goal is to obtain a formula for the *area* of the surface Σ.

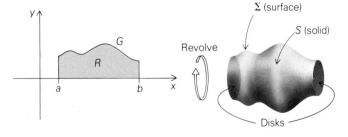

Fig. 11-2-6 The boundary of the solid of revolution S consists of the surface of revolution Σ obtained by revolving the graph G together with two disks.

*We can also write

$$\int \sec^3 \theta \, d\theta = \int \frac{\cos \theta}{\cos^4 \theta} \, d\theta = \int \frac{\cos \theta}{(1 - \sin^2 \theta)^2} \, d\theta = \int \frac{du}{(1 - u^2)^2} \qquad (u = \sin \theta)$$

The last integral may now be evaluated by partial fractions. This method is slightly longer.

If $f(x) = k$, a constant, the surface is a cylinder of radius k and height $b - a$. Unrolling the cylinder, we obtain a rectangle with dimensions $2\pi k$ and $b - a$ (see Fig. 11-2-7), whose area is $2\pi k(b - a)$, so we can say that the area of the cylinder is $2\pi k(b - a)$.

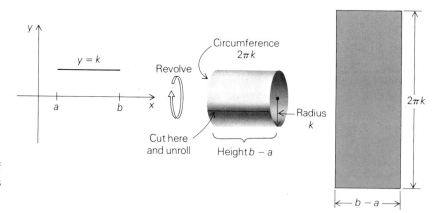

Fig. 11-2-7 The area of the shaded cylinder is $2\pi k\,(b - a)$.

Fig. 11-2-8 A frustum of a cone.

Next we look at the case where $f(x) = mx + q$, a linear function. The surface of revolution, as shown in Fig. 11-2-8, is a frustum of a cone—that is, the surface obtained from a right circular cone by cutting it with two planes perpendicular to the axis. To find the area of this surface, we may slit the frustum along a line and unroll it into the plane, as in Fig. 11-2-9, obtaining a circular sector of radius r and angle θ with a concentric sector of radius $r - s$ removed. By the definition of radian measure, we have $\theta r = 2\pi r_2$ and $\theta(r - s) = 2\pi r_1$, so $\theta s = 2\pi(r_2 - r_1)$, or $\theta = 2\pi[(r_2 - r_1)/s]$; from this we find $r = r_2 s/(r_2 - r_1)$. The area of the figure is

$$\frac{\theta}{2\pi}\left[\pi r^2 - \pi(r - s)^2\right] = \frac{\theta}{2}\left[r^2 - (r^2 - 2rs + s^2)\right] = \frac{\theta}{2}\,(2rs - s^2)$$

$$= \theta s\left(r - \frac{s}{2}\right) = 2\pi(r_2 - r_1)\left(\frac{r_2 s}{r_2 - r_1} - \frac{s}{2}\right)$$

$$= 2\pi s\left(r_2 - \frac{r_2 - r_1}{2}\right) = \pi s(r_1 + r_2)$$

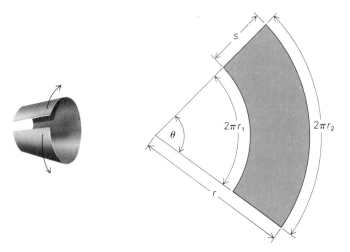

Fig. 11-2-9 The area of the frustum, found by unrolling, is $\pi s(r_1 + r_2)$.

(Notice that the proof breaks down in the case $r_1 = r_2 = k$, a cylinder, since r is then "infinite." Nevertheless, the resulting formula $2\pi ks$ for the area is still correct.)

We now wish to express $\pi s(r_1 + r_2)$ as an integral involving the function $f(x) = mx + q$. We have $r_1 = ma + q$, $r_2 = mb + q$, and $s = \sqrt{1 + m^2}\,(b - a)$ (see Fig. 11-2-1), so the area is

$$\pi\sqrt{1 + m^2}\,(b - a)[m(b + a) + 2q]$$
$$= \pi\sqrt{1 + m^2}\,[m(b^2 - a^2) + 2q(b - a)]$$
$$= 2\pi\sqrt{1 + m^2}\left[m\frac{b^2 - a^2}{2} + q(b - a)\right]$$
$$= 2\pi\sqrt{1 + m^2}\left(m\frac{x^2}{2} + qx\right)\bigg|_a^b$$

Since $m(x^2/2) + qx$ is the antiderivative of $mx + q$, we have

$$2\pi\sqrt{1 + m^2}\left(m\frac{x^2}{2} + qx\right)\bigg|_a^b = 2\pi\int_a^b \sqrt{1 + m^2}\,(mx + q)\,dx$$
$$= 2\pi\int_a^b [\sqrt{1 + f'(x)^2}\,]\,f(x)\,dx$$

so we have succeeded in expressing the surface area as an integral.

Now we are ready to work with general surfaces. If $f(x)$ is *piecewise linear* on $[a, b]$, the surface obtained is a "conoid," produced by pasting together a finite sequence of frustums of cones, as in Fig. 11-2-10.

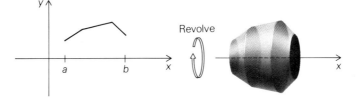

Fig. 11-2-10 The surface obtained by revolving the graph of a piecewise linear function is a "conoid" consisting of several frustums pasted together.

The area of the conoid is the sum of the areas of the component frustums. Since the area of each frustum is given by the integral of $2\pi\sqrt{1 + f'(x)^2}\,f(x)$ over the appropriate interval, the additivity of the integral implies that the area of the conoid is given by the same formula:

$$A = 2\pi\int_a^b f(x)\sqrt{1 + f'(x)^2}\,dx \tag{2}$$

Worked Example 4   The polygon joining the points $(2, 0)$, $(4, 4)$, $(7, 5)$, and $(8, 3)$ is revolved about the x axis. Find the area of the resulting surface of revolution.

Solution   The function f whose graph is the given polygon is

$$f(x) = \begin{cases} 2 & (x - 2) & 2 \le x \le 4 \\ \frac{1}{3} & (x - 4) + 4 & 4 \le x \le 7 \\ -2 & (x - 7) + 5 & 7 \le x \le 8 \end{cases}$$

Then we have

$$f'(x) = \begin{cases} 2 & 2 < x < 4 \\ \frac{1}{3} & 4 < x < 7 \\ -2 & 7 < x < 8 \end{cases}$$

Thus

$$A = 2\pi \int_2^8 f(x)\sqrt{1 + f'(x)^2}\, dx$$

$$= 2\pi \left[\int_2^4 f(x)\sqrt{1 + f'(x)^2}\, dx + \int_4^7 f(x)\sqrt{1 + f'(x)^2}\, dx + \int_7^8 f(x)\sqrt{1 + f'(x)^2}\, dx \right]$$

$$= 2\pi \left\{ \int_2^4 [2(x-2)]\sqrt{1 + 4}\, dx + \int_4^7 [\tfrac{1}{3}(x-4) + 4]\sqrt{1 + \tfrac{1}{9}}\, dx \right.$$

$$\left. + \int_7^8 [-2(x-7) + 5]\sqrt{1 + 4}\, dx \right\}$$

Using $\int (x-a)\, dx = \frac{1}{2}(x-a)^2 + C$, we find

$$A = 2\pi \left\{ \sqrt{5}\, [(x-2)^2] \Big|_2^4 + \frac{\sqrt{10}}{3} \left[\frac{1}{6}(x-4)^2 + 4(x-4) \right] \Big|_4^7 + \sqrt{5}\, [-(x-7)^2 + 5(x-7)] \Big|_7^8 \right\}$$

$$= 2\pi \left(4\sqrt{5} + \frac{9}{2}\sqrt{10} + 4\sqrt{5} \right) \approx 201.8$$

We wish to assert, as we did for arc length, that formula (2) is true for general functions f. To do this rigorously, we would need a precise definition of surface area, which is rather complicated to give (much more complicated, even, than for arc length). There are still three possible approaches. One is to take formula (2) as a *definition* of the area of the surface. A second is to consider a smoothly curved surface as being better and better approximated by conoids with more and more segments. Finally, we can give an infinitesimal argument. Namely, we may think of a smooth surface of revolution as being composed of infinitely many infinitesimal frustums, as in Fig. 11-2-11. The area of each frustum is equal to its circumference $2\pi f(x)$ times its width $ds = \sqrt{dx^2 + dy^2}$, so the total area is

$$\int_a^b 2\pi f(x)\sqrt{dx^2 + dy^2}$$

which, since $dy = f'(x)\, dx$, equals the expression in formula (2).

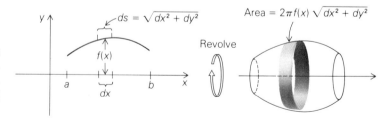

Fig. 11-2-11 The surface of revolution may be considered as composed of infinitely many infinitesimal frustums.

Worked Example 5 Find the area of the surface (a sphere of radius r) obtained by revolving the graph of $y = \sqrt{r^2 - x^2}$ on $[-r, r]$ about the x axis.

Solution As in Worked Example 2, we have $\sqrt{1 + f'(x)^2} = r/\sqrt{r^2 - x^2}$, so the area is

$$\int_{-r}^{r} 2\pi\sqrt{r^2 - x^2}\,\frac{r}{\sqrt{r^2 - x^2}}\,dx = 2\pi\int_{-r}^{r} r\,dx = 2\pi r \cdot 2r = 4\pi r^2$$

which is the usual value for the area of a sphere.

Next we find an area which can be determined only with the aid of calculus.

Worked Example 6 Find the area of the surface obtained by revolving the graph of x^3 on $[0, 1]$ about the x axis.

Solution We find that $f'(x) = 3x^2$ and $\sqrt{1 + f'(x)^2} = \sqrt{1 + 9x^4}$, so

$$A = 2\pi\int_0^1 \sqrt{1 + 9x^4}\,x^3\,dx = \frac{\pi}{2}\int_0^1 \sqrt{1 + 9u}\,du \qquad (u = x^4,\ du = 4x^3\,dx)$$

$$= \frac{\pi}{18}\int_0^9 (1 + v)^{1/2}\,dv \qquad (u = \tfrac{1}{9}v,\ du = \tfrac{1}{9}dv)$$

$$= \frac{\pi}{18}\left[\frac{2}{3}(1 + v)^{3/2}\right]\bigg|_0^9 = \frac{\pi}{27}(10^{3/2} - 1) \approx 3.56$$

AREA OF A SURFACE OF REVOLUTION

The area of the surface obtained by revolving the graph of $f(x)$ (≤ 0) on $[a, b]$ about the x axis is

$$A = 2\pi\int_a^b f(x)\sqrt{1 + f'(x)^2}\,dx = 2\pi\int_a^b y\sqrt{1 + \left(\frac{dy}{dx}\right)^2}\,dx$$

As with arc length the factor $\sqrt{1 + f'(x)^2}$ in the integrand usually makes it impossible to evaluate the surface area integrals by any way other than numerical methods (see Worked Example 5, Section 12-4).

Solved Exercises 3. Give a geometric interpretation of the fact that the integrand in Worked Example 5 is equal to $r\,dx$.

4. (a) Find a formula for the area obtained by revolving the graph of $f(x)$ on $[a, b]$ about the y axis (assume $0 \leq a \leq b$).
   (b) Calculate this area for $f(x) = x^2$ on $[1, 2]$.

Exercises   Find the area of the following surfaces of revolution.

6. The graph of $\sqrt{x+1}$ on $[0, 2]$ revolved about the x axis.

7. The graph of $x^{1/3}$ on $[1, 3]$ revolved about the y axis (see Solved Exercise 4).

8. The graph of $y = [x^3 + (3/x)]/6$, $1 \leq x \leq 3$ about the x-axis.

9. The graph of $y = \sqrt{x}\,(4x - 3)/6$, $1 \leq x \leq 9$ about the x-axis.

Problems for Section 11-2

1. Find the length of the graph of $a(x + b)^{3/2} + c$ on $[0, 1]$, where a, b, and c are constants. What is the effect of changing the value of c?

2. Find the length of the graph of $f(x) = (x^4 - 12x + 3)/6x$ on $[2, 4]$.

3. Express the length of the graph of $f(x) = x \cos x$ on $[0, 1]$ as an integral. (Do not evaluate.)

★4. Prove that, for any positive integer n, the graph of $f(x) = x^{n+1}$ on $[0, 1]$ is longer than the graph of $g(x) = x^n$ on $[0, 1]$.

5. Prove that the length of the graph of $f(x) = \cos(\sqrt{3}\,x)$ on $[0, 2\pi]$ is less than or equal to 4π.

6. Suppose that $f(x) \geq g(x)$ for all x in $[a, b]$. Does this imply that the length of the graph of f on $[a, b]$ is greater than or equal to that for g? Justify your answer by a proof or an example.

★7. Show that the length of the graph of $\sin x$ on $[0.1, 1]$ is less than the length of the graph of $1 + x^4$ on $[0.1, 1]$.

8. Express the length of the graph of $f(x) = 2x^3$ on $[-1, 2]$ as an integral. Evaluate numerically to within 1.0 by finding upper and lower sums. Compare your results with a string-and-ruler measurement.

9. Find the length of the graph of $y = x^2$ on $[0, b]$.

10. Find the length, accurate to within 1 centimeter, of the curve in Fig. 11-2-12.

Fig. 11-2-12 Find the length of this curve.

11. Find the area of the surface obtained by revolving the graph of $y = \cos x$ on $[-\pi/2, \pi/2]$ about the x axis.

★12. Write an integral representing the area of the surface obtained by revolving the graph of $1/(1 + x^2)$ about the x axis. Do not evaluate the integral, but show that it is less than $2\pi\sqrt{2}$, no matter how long an interval is taken.

13. Find the area of the surface obtained by revolving the graph of e^x on $[0, 1]$ about the x axis.

14. For each of the following functions and intervals, express as an integral: (i) the length of the curve; (ii) the area of the surface obtained by revolving the curve about the x axis. (Do not evaluate the integrals.)

  (a) $\tan x + 2x$ on $[0, \pi/2]$

  (b) $x^3 + 2x - 1$ on $[1, 3]$

  (c) $1/x + x$ on $[1, 2]$

★15. Suppose that the function f on $[a, b]$ has an inverse function g defined on $[\alpha, \beta]$. Assume $0 < a < b$ and $0 < \alpha < \beta$.

  (a) Find a formula, in terms of f, for the area of the surface obtained by revolving the graph of g on $[\alpha, \beta]$ about the x axis.

  (b) Show that this formula is consistent with the one in Solved Exercise 4(a) for the area of the surface obtained by revolving the graph of f on $[a,b]$ about the y axis.

16. Find the area, accurate to within 10 square centimeters, of the surface obtained by revolving the curve in Fig. 11-2-13 around the x axis.

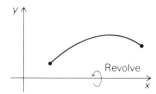

Fig. 11-2-13 Find the area of the surface obtained by revolving this curve.

17. Use upper and lower sums to find the area, accurate to within 1 unit, of the surface obtained by revolving the graph of x^4 on $[0, 1]$ about the x axis.

18. The *power* (units are horsepower, or watts) developed by a jet engine of thrust $F(t)$ traveling a path with velocity ds/dt is $P(t) = F(t)ds/dt$. The *average power* on $[t_1, t_2]$ is

$$\frac{1}{t_2 - t_1} \int_{t_1}^{t_2} P(t)\, dt$$

(a) Suppose $F(t) = 3000$ lb for $t_1 \leq t \leq t_2$, during which time the average speed of the aircraft is 600 mph. Find the average horsepower developed by the engine. [*Note:* 1 hp = 550 ft-lb = 746 watts, 600 mph = 880 ft/sec.]

(b) Suppose the aircraft follows another path for which the instantaneous speed ds/dt is not known, but the *average speed* $[s(t_2) - s(t_1)]/(t_2 - t_1)$ is known. If you assume the thrust $F(t)$ is constant, then can you determine the horsepower?

★ ▦ 19. Craftsman Cabinet Company was preparing a bid on a job that required epoxy coating of several tank interiors. The tanks were constructed from steel cylinders C ft in circumference and height H ft, with spherical steel caps welded to each end (see Fig. 11-2-14). Specifications required a $\frac{1}{8}''$ coating. The 20-year-old estimator quickly figured the cylindrical part as HC square feet. For the spherical cap he stretched a tape measure over the cap to obtain S ft.

Fig. 11-2-14 A cross section of a tank requiring an epoxy coating on its interior.

(a) Find an equation relating the surface area of the steel cap and the tape measurements S and C. [*Hint:* Revolve $y = \sqrt{R^2 - x^2}$ about the y-axis, $0 \leq x \leq C/2\pi$.]

(b) Find an equation relating the surface area of the tank, S, C, and H.

(c) Determine the cost for six tanks with H = 16 ft, C = 37.7 ft, S = 13.2 ft, given the coating costs $2.10 per square foot.

11-3 Parametric Curves and Polar Coordinates

Arc lengths and areas may be found by integral calculus for curves which are not graphs of functions.

We begin this section with a study of the differential calculus of parametric curves, a topic which was introduced in Section 5-2. The arc length of a parametric curve is then expressed as an integral, and this is applied to the problem of determining the arc length of a curve given in polar coordinates. Finally, the area of a region expressed in polar coordinates is expressed as an integral.

Goals

After studying this section, you should be able to:

Compute the arc length of a curve given in parametric form.

Compute the arc length of a curve given in polar coordinates.

Compute the area of a region given in polar coordinates.

PARAMETRIC CURVES AND THEIR TANGENTS

We recall (from Section 5-2) that a *parametric curve* in the xy-plane is specified by a pair of functions: $x = f(t)$, $y = g(t)$. The variable t, called the *parameter* of the curve, may be thought of as time; the pair $(f(t), g(t))$ then describes the path in the plane of a moving point.

Worked Example 1 Describe the motion of the point (x, y) if $x = \cos t$ and $y = \sin t$, for t in $[0, 2\pi]$.

Solution At $t = 0$, the point is at $(1, 0)$. Since $\cos^2 t + \sin^2 t = 1$, the point (x, y) satisfies $x^2 + y^2 = 1$, so it moves on the unit circle. As t increases from zero, $x = \cos t$ decreases and $y = \sin t$ increases, so the point moves in a counterclockwise direction. Finally, since $(\cos(2\pi), \sin(2\pi)) = (0, 1)$, the point makes a full rotation after 2π units of time (see Fig. 11-3-1).

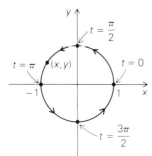

Fig. 11-3-1 The point $(\cos t, \sin t)$ moves in a circle.

Worked Example 2 Describe the motion of the point (t, t^3) for t in $(-\infty, \infty)$.

Solution   We have $x^3 = t^3 = y$, so the point is on the curve $y = x^3$. At t increases, so does x, and the point moves from left to right (see Fig. 11-3-2).

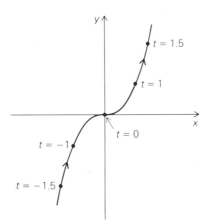

Fig. 11-3-2   The motion of the point (t, t^3).

Worked Example 2 illustrates a general fact: Any curve $y = f(x)$ which is the graph of a function can be described parametrically: we set $x = t$ and $y = f(t)$. One reason for the importance of parametric equations is that they can also describe curves which are not the graphs of functions, like the circle in Worked Example 1.

The equations

$$x = at + b, \quad y = ct + d$$

describe a straight line. To show this, we *eliminate the parameter t* in the following way. If $a \neq 0$, we solve the first equation for t, obtaining $t = (x - b)/a$. Substituting into the second equation gives $y = c[(x - b)/a] + d$, or $y = (c/a)x + (ad - bc)/a$, which is a straight line with slope c/a. If $a = 0$, we have $x = b$ and $y = ct + d$. If $c \neq 0$, then y takes all values as t varies and b is fixed, so we have the vertical line $x = b$ (which is not the graph of a function). If $c = 0$ as well as $a = 0$, then $x = b$ and $y = d$, so the graph is a "stationary" point (b, d).

Similarly, we can see that

$$x = r\cos t + x_0, \quad y = r\sin t + y_0$$

describes a circle by writing

$$\frac{x - x_0}{r} = \cos t, \quad \frac{y - y_0}{r} = \sin t$$

Therefore,

$$\left(\frac{x - x_0}{r}\right)^2 + \left(\frac{y - y_0}{r}\right)^2 = \cos^2 t + \sin^2 t = 1$$

or $(x - x_0)^2 + (y - y_0)^2 = r^2$, which is the equation of a circle with radius r and center (x_0, y_0). As t varies from zero to 2π, the point (x, y) moves once around the circle, as in Worked Example 1 (see p. 26).

PARAMETRIC EQUATIONS OF LINES AND CIRCLES

Straight line

$x = at + b$

$-\infty < t < \infty$

$y = ct + d$

(*a* and *c* not both zero) through (b, d), slope $= c/a$

Circle

$x = r \cos t + x_0$

$0 \le t < 2\pi$

$y = r \sin t + y_0$

$r > 0$, r = radius

(x_0, y_0) = center

The same geometric curve can often be represented parametrically in more than one way. For example, the line $x = at + b$, $y = ct + d$ can also be represented by

$$x = t, \qquad y = \frac{ct}{a} + \frac{ad - bc}{a}$$

or by

$$x = t^3, \qquad y = \frac{ct^3}{a} + \frac{ad - bc}{a}$$

(If we used t^2, we would get only half of the line since $t^2 \ge 0$ for all t.) Compare Worked Example 11 in Section 5-2.

We saw on p. 243 that the tangent line to the curve $(x, y) = (f(t), g(t))$ at the point $(f(t_0), g(t_0))$ has slope

$$\frac{dy}{dx} = \frac{dy/dt}{dx/dt} = \frac{g'(t_0)}{f'(t_0)}$$

If $f'(t_0) = 0$ and $g'(t_0) \ne 0$, the tangent line is vertical; if $f'(t_0)$ and $g'(t_0)$ are both zero, the tangent line is not defined (see Solved Exercise 3(c) below). Since the tangent line passes through $(f(t_0), g(t_0))$, we may write its equation in point-slope form:

$$y = \frac{g'(t_0)}{f'(t_0)} [x - f(t_0)] + g(t_0) \tag{1}$$

Worked Example 3 Find the equation of the tangent line when $t = 1$ for the curve $x = t^4 + 2\sqrt{t}$, $y = \sin t\pi$.

Solution When $t = 1$, we have $x = 3$ and $y = \sin \pi = 0$. Furthermore, $dx/dt = 4t^3 + 1/\sqrt{t}$, which equals 5 when $t = 1$; $dy/dt = \pi \cos t\pi$, which equals $-\pi$ when $t = 1$. Thus the equation of the tangent line is, by formula (1),

$$y = -\frac{\pi}{5}(x - 3) + 0 \quad \text{or} \quad y = -\frac{\pi}{5}x + \frac{3\pi}{5}$$

If a curve is given parametrically, it is natural to express its tangent line parametrically as well. To do this, we transform equation (1) to the form

$$\frac{y - g(t_0)}{g'(t_0)} = \frac{x - f(t_0)}{f'(t_0)}$$

We can set both sides of this equation equal to t, obtaining

$$x = tf'(t_0) + f(t_0), \qquad y = tg'(t_0) + g(t_0) \tag{2}$$

Reviewing the discussion on p. 471, we see that this is indeed the parametric equation for a line with slope $g'(t_0)/f'(t_0)$ if $f'(t_0) \neq 0$. If $f'(t_0) = 0$ but $g'(t_0) \neq 0$, equations (2) describe a vertical line. If $f'(t_0)$ and $g'(t_0)$ are both zero, equations (2) describe a stationary point.

It is convenient to make one more transformation of (2), so that the tangent line passes through (x_0, y_0) at the same time t_0 as the curve, rather than at $t = 0$. Substituting $t - t_0$ for t, we obtain the formulas

$$x = f'(t_0)(t - t_0) + f(t_0), \qquad y = g'(t_0)(t - t_0) + g(t_0) \tag{3}$$

Fig. 11-3-3 If the forces constraining a particle to the curve $(f(t), g(t))$ are removed at t_0, then the particle will follow the tangent line at t_0.

Notice that the functions in formulas (3) which define the tangent line to a curve are exactly the *linear approximations* (see p. 75) to the functions defining the curve itself. If we think of $(x, y) = (f(t), g(t))$ as the position of a moving particle, then the tangent line at t_0 is the path which the particle would follow if, at time t_0, all constraining forces were suddenly removed and the particle were allowed to move freely in a straight line. (See Fig. 11-3-3.)

Worked Example 4 A child is whirling an object on a string, letting out string at a constant rate, so that the object follows the path $x = (1 + t) \cos t$, $y = (1 + t) \cdot \sin t$.

(a) Sketch the path for $0 \leq t \leq 4\pi$.

(b) At $t = 4\pi$ the string breaks, so that the object follows its tangent line. Where is the object at $t = 5\pi$?

Solution (a) By plotting some points and thinking of (x, y) as moving in an ever-enlarging circle, we obtain the sketch in Fig. 11-3-4.

(b) We differentiate:

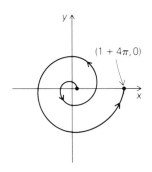

Fig. 11-3-4 The curve $((1 + t)\cos t, (1 + t)\sin t)$ for t in $[0, 4\pi]$.

$$f'(t) = \frac{dx}{dt} = (1 + t)(-\sin t) + \cos t$$

$$g'(t) = \frac{dy}{dt} = (1 + t) \cos t + \sin t$$

When $t_0 = 4\pi$, we have

$$f(t_0) = (1 + 4\pi) \cos 4\pi = 1 + 4\pi$$

$$g(t_0) = (1 + 4\pi) \sin 4\pi = 0$$

$$f'(t_0) = (1 + 4\pi) \cdot 0 + 1 = 1$$

$$g'(t_0) = (1 + 4\pi) \cdot 1 + 0 = 1 + 4\pi$$

By formulas (3), the equations of the tangent line are

$$x = t - 4\pi + (1 + 4\pi), \qquad y = (1 + 4\pi)(t - 4\pi) + 0$$

When $t = 5\pi$, the object, which is now following the tangent line, is at $x = 1 + 5\pi \approx 16.71$, $y = (1 + 4\pi)\pi \approx 42.62$.

TANGENTS TO PARAMETRIC CURVES

Let $x = f(t)$ and $y = g(t)$ be the parametric equations of a curve C. If f and g are differentiable at t_0, and $f'(t_0)$ and $g'(t_0)$ are not both zero, then the tangent line to C at t_0 is defined by the parametric equations:

$$x = f'(t_0)(t - t_0) + f(t_0), \qquad y - g'(t_0)(t - t_0) + g(t_0)$$

If $f'(t_0) \neq 0$, this line has slope $g'(t_0)/f'(t_0)$ and its equation can be written as

$$y = \frac{g'(t_0)}{f'(t_0)} [x - f(t_0)] + g(t_0)$$

If $f'(t_0) = 0$ and $g'(t_0) \neq 0$, the line is vertical; its equation is

$$x = f(t_0)$$

Solved Exercises

1. Consider the curve $x = t^3 - t$, $y = t^2$.

   (a) Plot the points corresponding to $t = -2, -1, -\frac{1}{2}, 0, \frac{1}{2}, 1, 2$.

   (b) Using these points, together with the behavior of the functions t^2 and $t^3 - t$, sketch the entire curve.

   (c) Find the slope of the tangent line at the points corresponding to $t = 1$ and $t = -1$.

   (d) Eliminate the parameter t to obtain an equation in x and y for the curve.

2. Describe the ellipse $4x^2 + 9y^2 = 1$ in parametric form.

3. (a) Sketch the curve $x = t^3$, $y = t^2$.

   (b) Find the equation of the tangent line at $t = 1$.

   (c) What happens at $t = 0$?

4. Consider the curve $x = \cos 3t$, $y = \sin t$. Find the points where the tangent is horizontal and those where it is vertical. Use this information to sketch the curve.

Exercises

1. For each of the following pairs of parametric equations, sketch the curve and find an equation in x and y by eliminating the parameter:

   (a) $x = 4t - 1$, $y = t + 2$ (b) $x = 2t + 1$, $y = t^2$

   (c) $x = \tan t$, $y = \sec t$ (d) $x = 4\sin\theta$, $y = \cos\theta - 3$

2. Find the equation of the tangent line to each of the following curves at the given point:

   (a) $x = \frac{1}{2}t^2 + t$, $y = t^{2/3}$; $t_0 = 1$

   (b) $x = 1/t$, $y = \sqrt{t + 1}$; $t_0 = 2$

   (c) $x = \cos^2(t/2)$, $y = \frac{1}{2}\sin t$; $t_0 = \pi/2$

3. Represent each of the curves in Exercise 2 in the form $y = h(x)$, and compute the slope of the tangent line by differentiating h. Compare with the results in Exercise 2.

4. Let $(x, y) = (\cos 3t, \sin 2t)$. Find the points at which the tangent line is horizontal or vertical, and use this information to aid in sketching the curve.

ARC LENGTH AND SPEED

What is the length of the curve given by $(x, y) = (f(t), g(t))$ for $a \le t \le b$? To get a formula in terms of f and g, we begin by considering the case in which the point $(f(t), g(t))$ moves along the graph of a function $y = h(x)$; that is, $g(t) = h(f(t))$.

If $f(a) = \alpha$ and $f(b) = \beta$, the length of the curve is $\int_\alpha^\beta \sqrt{1 + h'(x)^2}\, dx$, by formula (1) on p. 461. If we change variables from x to t in this integral, we have $dx = f'(t)\, dt$, so the length is

$$\int_a^b \sqrt{1 + [h'(f(t))]^2}\, f'(t)\, dt$$

To eliminate the function h from this formula, we may apply the chain rule to $g(t) = h(f(t))$: $g'(t) = h'(f(t)) \cdot f'(t)$. Solving for $h'(f(t))$ and substituting in the integral gives

$$\int_a^b \sqrt{1 + \left[\frac{g'(t)}{f'(t)}\right]^2}\, f'(t)\, dt$$

or

$$L = \int_a^b \sqrt{f'(t)^2 + g'(t)^2}\, dt \tag{4}$$

Formula (4) involves only the information contained in the parametrization. Since we can break up any reasonably behaved parametric curve into segments, each of which is the graph of a function or a vertical line (for which we see that (4) gives the correct length, since $f'(t) \equiv 0$), we conclude that (4) ought to be valid for any parametric curve.

LENGTH OF A PARAMETRIC CURVE

Suppose that a parametric curve C is given by continuous functions $x = f(t)$, $y = g(t)$, for $a \le t \le b$, and that $f'(t)$ and $g'(t)$ exist and are continuous, except possibly for finitely many points. Then the length of C is given by

$$L = \int_a^b \sqrt{f'(t)^2 + g'(t)^2}\, dt = \int_a^b \sqrt{\left(\frac{dx}{dt}\right)^2 + \left(\frac{dy}{dt}\right)^2}\, dt$$

Worked Example 5 Find the length of the circle of radius 2 given by $x = 2 \cos t + 3$, $y = 2 \sin t + 4$, $0 \le t \le 2\pi$.

Solution We find $f'(t) = dx/dt = -2 \sin t$ and $g'(t) = dy/dt = 2 \cos t$, so

$$L = \int_0^{2\pi} \sqrt{4 \sin^2 t + 4 \cos^2 t}\, dt$$

$$= \int_0^{2\pi} 2\sqrt{\sin^2 t + \cos^2 t}\, dt = \int_0^{2\pi} 2\, dt = 4\pi$$

(which equals 2π times the radius).

Given a point moving according to $x = f(t)$, $y = g(t)$, the integral

$$D(t) = \int_a^t \sqrt{f'(s)^2 + g'(s)^2}\, ds$$

is the distance (along the curve) traveled by the point between time a and time t. The derivative $D'(t)$ should then represent the speed of the point along the curve. By the fundamental theorem of calculus (alternative version), we have

$$D'(t) = \sqrt{f'(t)^2 + g'(t)^2}$$

SPEED

Let a point move according to the equations $x = f(t)$, $y = g(t)$. Then the speed of the point at time t is

$$\boxed{\sqrt{f'(t)^2 + g'(t)^2} = \sqrt{\left(\frac{dx}{dt}\right)^2 + \left(\frac{dy}{dt}\right)^2}}$$

Worked Example 6 A particle moves around the elliptical track $4x^2 + y^2 = 4$ according to the equations $x = \cos t$, $y = 2 \sin t$. When is the speed greatest? Where is it least?

Solution The speed is

$$\sqrt{\left[\frac{d(\cos t)}{dt}\right]^2 + \left[\frac{d(2 \sin t)}{dt}\right]^2} = \sqrt{\sin^2 t + 4 \cos^2 t} = \sqrt{1 + 3 \cos^2 t}$$

Without any further calculus, we observe that the speed is greatest when $\cos t = \pm 1$; that is, $t = 0$, π, 2π, and so forth. The speed is least when $\cos t = 0$; that is, $t = \pi/2$, $3\pi/2$, $5\pi/2$, and so on.

Suppose that an object is moving along the curve $x = f(t)$, $y = g(t)$ and that at time t_0 the constraining forces are removed and the particle continues along the tangent line

$$x = f'(t_0)(t - t_0) + f(t_0) \quad y = g'(t_0)(t - t_0) + g(t_0)$$

At time $t_0 + \Delta t$, the particle is at $(f'(t_0)\Delta t + f(t_0),\ g'(t_0)\Delta t + g(t_0))$, which is at distance $\sqrt{f'(t_0)^2 + g'(t_0)^2}\ \Delta t$ from $(f(t_0), g(t_0))$. Thus the distance traveled in time Δt after the force is removed is equal to Δt times the speed at t_0, so we have another justification of our formula for the speed.

Solved Exercises

5. Find the length of the curve $x = \frac{1}{2}\sin 2t$, $y = 3 + \cos^2 t$ on $[0, \pi]$.

6. Show that if $x = f(t)$, $y = g(t)$ is any curve with $(f(0), g(0)) = (0,0)$ and $(f(1), g(1)) = (0, a)$, then the length of the curve in $[0, 1]$ is at least equal to a. What can you say if the length is exactly equal to a?

Exercises

5. Find the length of $x = t^2$, $y = t^3$ on $[0, 1]$.

6. Find the length of $x = t^8$, $y = t^4$ on $[1, 3]$.

7. Express the length of $x = t\sin t$, $y = t\cos t$ on $[0, 4\pi]$ as an integral.

8. Show that if $x = a\cos t + b$ and $y = a\sin t + d$:
   (a) the speed is constant;
   (b) the length of the curve on $[t_0, t_1]$ is equal to the speed times $(t_1 - t_0)$.

LENGTH IN POLAR COORDINATES

The formula $L = \int_a^b \sqrt{(dx/dt)^2 + (dy/dt)^2}\ dt$ can be applied to a curve $r = f(\theta)$ in polar coordinates if we take the parameter to be θ in place of t. We write:

$$x = r\cos\theta = f(\theta)\cos\theta \quad \text{and} \quad y = r\sin\theta = f(\theta)\sin\theta$$

Then the length is (see Fig. 11-3-5)

$$\int_\alpha^\beta \sqrt{[f'(\theta)\cos\theta - f(\theta)\sin\theta]^2 + [f'(\theta)\sin\theta + f(\theta)\cos\theta]^2}\ d\theta$$

which simplifies to

$$\int_\alpha^\beta \sqrt{f'(\theta)^2 + f(\theta)^2}\ d\theta$$

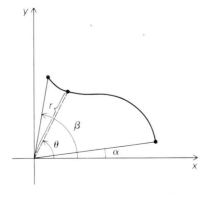

Fig. 11-3-5 The length of the curve is $\int_\alpha^\beta \sqrt{(dr/d\theta)^2 + r^2}\ d\theta$.

ARC LENGTH IN POLAR COORDINATES

The length of the curve $r = f(\theta)$, $\alpha \le \theta \le \beta$, is given by

$$L = \int_{\alpha}^{\beta} \sqrt{f'(\theta)^2 + f(\theta)^2}\, d\theta = \int_{\alpha}^{\beta} \sqrt{\left(\frac{dr}{d\theta}\right)^2 + r^2}\, d\theta$$

One can obtain the same formula by an infinitesimal argument, following Fig. 11-3-6. By Pythagoras' theorem, $ds^2 = dr^2 + r^2 d\theta^2$, or $ds = \sqrt{dr^2 + r^2 d\theta^2}$. If we use $dr = f'(\theta)\, d\theta$, this becomes

$$ds = \sqrt{\left(\frac{dr}{d\theta}\right)^2 d\theta^2 + r^2 d\theta^2} = \sqrt{\left(\frac{dr}{d\theta}\right)^2 + r^2}\, d\theta$$

so

$$L = \int_{\alpha}^{\beta} ds = \int_{\alpha}^{\beta} \sqrt{\left(\frac{dr}{d\theta}\right)^2 + r^2}\, d\theta$$

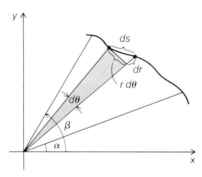

Fig. 11-3-6 The infinitesimal element of arc length ds equals $\sqrt{dr^2 + r^2 d\theta^2}$.

Worked Example 7 Find the length of the curve $r = 1 - \cos\theta$, $0 \le \theta \le 2\pi$.

Solution We find $dr/d\theta = \sin\theta$, so

$$L = \int_{0}^{2\pi} \sqrt{\sin^2\theta + (1 - \cos\theta)^2}\, d\theta = \int_{0}^{2\pi} \sqrt{2 - 2\cos\theta}\, d\theta$$

$$= \sqrt{2} \int_{0}^{2\pi} \sqrt{1 - \cos\theta}\, d\theta$$

$$= \sqrt{2} \int_{0}^{2\pi} \sqrt{2\sin^2\frac{\theta}{2}}\, d\theta = 2\int_{0}^{2\pi} \sin\frac{\theta}{2}\, d\theta$$

$$= 4\int_{0}^{\pi} \sin u\, du \qquad \left(u = \frac{\theta}{2}\right)$$

$$= 4(-\cos\theta)\Big|_{0}^{\pi} = 8$$

Solved Exercise 7. Find the length of the cardioid $r = 1 + \cos\theta$ $(0 \le \theta \le 2\pi)$.

9. Sketch and find the length of each of the following curves:

(a) $r = 3(1 + \sin\theta)$; $0 \le \theta \le 2\pi$

(b) $r = 1/(\cos\theta + \sin\theta)$; $0 \le \theta \le \pi/2$

(c) $r = 4\theta^2$; $0 \le \theta \le 3$

AREA IN POLAR COORDINATES

The curve expressed in polar coordinates by the equation $r = f(\theta)$, together with the rays $\theta = \alpha$ and $\theta = \beta$, encloses a region of the type shown (shaded) in Fig. 11-3-7. We call this the region *inside* the graph of f on $[\alpha, \beta]$.

We wish to find a formula for the area of such a region as an integral involving the function f. As usual, we begin with the simplest case, in which f is a constant function $f(\theta) = k$. The region inside the curve $r = k$ on $[\alpha, \beta]$ is then a circular sector with radius k and angle $\beta - \alpha$ (see Fig. 11-3-8). The area is $(\beta - \alpha)/2\pi$ times the area πk^2 of a circle of radius k, or $\frac{1}{2}k^2(\beta - \alpha)$. We can express this as the integral $\int_\alpha^\beta \frac{1}{2}f(\theta)^2\,d\theta$.

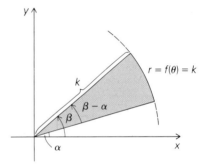

Fig. 11-3-7 The region inside the graph $r = f(\theta)$ on $[\alpha, \beta]$ is shaded.

Fig. 11-3-8 The area of the sector is $\frac{1}{2}k^2(\beta - \alpha)$.

If f is piecewise constant, with $f(\theta) = k_i$ on (θ_{i-1}, θ_i), then the region inside the graph of f is of the type shown in Fig. 11-3-9. Its area is equal to the sum of the areas of the individual sectors, or

$$\sum_{i=1}^n \frac{1}{2}k_i^2\Delta\theta_i$$

which is equal to the integral $\int_\alpha^\beta \frac{1}{2}f(\theta)^2\,d\theta$.

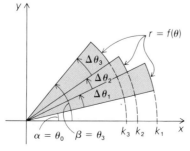

Fig. 11-3-9 The area of the shaded region is $\sum\frac{1}{2}k_i^2\Delta\theta_i$.

As usual, we conclude that the same area formula holds whenever f^2 is an integrable function.

AREA IN POLAR COORDINATES

The area of the region enclosed by the curve $r = f(\theta)$ and the rays $\theta = \alpha$ and $\theta = \beta$ is given by

$$A = \tfrac{1}{2}\int_\alpha^\beta f(\theta)^2\, d\theta = \tfrac{1}{2}\int_\alpha^\beta r^2\, d\theta$$

Worked Example 8 Find the area enclosed by one leaf of the four-leafed rose $r = \cos 2\theta$ (see p. 277).

Solution The leaf at the right is enclosed by the arc $r = \cos 2\theta$ and the rays $\theta = -\pi/4$ and $\theta = \pi/4$ (Fig. 11-3-10). Notice that the rays do not actually appear in the boundary of the figure, since the radius $r = \cos(\pm\pi/2)$ is zero there. The area is $\tfrac{1}{2}\int_{-\pi/4}^{\pi/4} r^2\, d\theta = \tfrac{1}{2}\int_{-\pi/4}^{\pi/4} \cos^2 2\theta\, d\theta$. By the half-angle formula this is

$$\frac{1}{2}\int_{-\pi/4}^{\pi/4}\frac{1+\cos 4\theta}{2}\, d\theta = \frac{1}{4}\left(\theta + \frac{\sin 4\theta}{4}\right)\Bigg|_{-\pi/4}^{\pi/4} = \frac{\pi}{8}$$

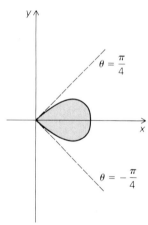

Fig. 11-3-10 One leaf of the four-leafed rose $r = \cos 2\theta$.

Solved Exercises

8. Use an "infinitesimal" argument to obtain the formula for area in polar coordinates.

9. Find the area enclosed by the cardioid $r = 1 + \cos\theta$ (see p. 278).

10. Develop a formula for the area between two curves in polar coordinates.

Exercises

10. Sketch and find the area of the region enclosed by the graph $r = f(\theta)$ and the rays $\theta = \alpha$ and $\theta = \beta$ for:

(a) $f(\theta) = \theta$; $\alpha = 0$, $\beta = 3\pi/2$

(b) $f(\theta) = \theta\cos(\theta^3)$; $\alpha = 0$, $\beta = \pi/4$

(c) $f(\theta) = \theta + \sin 4\theta$; $\alpha = \pi/4$, $\beta = \pi$

11. Sketch and find the area of the region bounded by the curve:

(a) $r = 3 \sin \theta; \ 0 \le \theta \le \pi$

(b) $r = 2(1 + \sin \theta); \ 0 \le \theta \le 2\pi$

(c) $r = 4 + \sin \theta; \ 0 \le \theta \le 2\pi$

12. Check the arc length and area formulas in polar coordinates for a circle.

Problems for Section 11-3

1. Sketch each of the following parametric curves, find an equation in x and y by eliminating the parameter, and find the points where the tangent line is horizontal or vertical.

(a) $x = t^2; \ y = \cos t$ (What happens at $t = 0$?)

(b) $x = \pi/2 - s; \ y = 2 \sin 2s$

(c) $x = \cos 2t; \ y = \sin t$

2. Sketch and find the length of the curve (t^2, t^4) on $0 \le t \le 1$.

3. Find two different parametric representations for each of the following curves:

(a) $2x^2 + y^2 = 1$ (b) $4xy = 1$

(c) $y = 3x - 2$ (d) $y = x^3 + 1$

(e) $3x^2 - y^2 = 1$ (f) $y = \cos(2x)$

(g) $y^2 = x + x^2$

4. Suppose that the distance from the origin to $(x, y) = (f(t), g(t))$ attains its maximum value at $t = t_0$. Show that the tangent line at t_0 is perpendicular to the line from the origin to $(f(t_0), g(t_0))$.

★5. (a) Find a parametric curve $x = f(t), \ y = g(t)$ passing through the points $(1, 1), (2, 2), (4, 2),$ $(5, 1), (3, 0),$ and $(1, 1)$ such that the functions f and g are both piecewise linear and the curve is a polygon whose vertices are the given points in the given order.

(b) Compute the length of this curve by formula (4) and then by elementary geometry. Compare the results.

(c) What is the area of the surface obtained by revolving the given curve about the y axis?

★6. At each point (x_0, y_0) of the parabola $y = x^2$, the tangent line is drawn and a point is marked on this line at a distance of 1 unit from (x_0, y_0) to the right of (x_0, y_0).

(a) Describe the collection of points thus obtained as a parametrized curve.

(b) Describe the collection of points thus obtained in terms of a relation between x and y.

7. An object moves from left to right along the curve $y = x^{3/2}$ at constant speed. If the point is at $(0, 0)$ at noon and at $(1, 1)$ at 1:00 PM, where is it at 1:30 PM?

8. Consider the parametrized curve $x = 2 \cos \theta$, $y = \theta - \sin \theta$.

(a) Find the equation of the tangent line at $\theta = \pi/2$.

(b) Sketch the curve.

(c) Express the length of the curve on $[0, \pi]$ as an integral.

9. Sketch and find the length (as an integral) of the graph of $r = f(\theta), \ \alpha \le \theta \le \beta$. Then find the area of the region bounded by this graph and the rays $\theta = \alpha$ and $\theta = \beta$. (The answer may be partly in the form of an integral).

(a) $r = \tan(\theta/2); \ -\pi/2 \le \theta \le \pi/2$

(b) $r = \theta + \sin \theta^2; \ -\pi/4 \le \theta \le 3\pi/4$

(c) $r = \sec \theta + 2; \ 0 \le \theta \le \pi/4$

10. Find the length of and areas bounded by the following curves between the rays indicated. Express the areas as numbers but leave the lengths as integrals.

(a) $r = \theta(1 + \cos \theta); \ \theta = 0, \ \theta = \pi/2$

(b) $r = 1/\theta; \ \theta = 1, \ \theta = \pi$

(c) $f(\theta) = \sqrt{1 + 2 \sin 2\theta}; \ \theta = 0, \ \theta = \pi$

(d) $f(\theta) = \theta^2 - (\pi/2)\,\theta + 4; \ \theta = 0, \ \theta = \pi/2$

11. Sketch and find the area of each of the regions between each of the following pairs of curves $(0 \le \theta \le 2\pi)$. Then find the length of the curves which bound the regions.

(a) $r = \cos \theta$ (b) $r = 3$
    $r = \sqrt{3} \sin \theta$ $r = 2(1 + \cos\theta)$

(c) $r = 2 \cos \theta$
    $r = 1 + \cos \theta$

12. The curve $r = e^{\theta}$ is called a logarithmic spiral.

(a) Sketch this curve. What happens for large θ? Small θ?

(b) Show that the tangent lines at all points of the logarithmic spiral make the same angle with the ray from the origin to the point.

(c) Find the length of the loop of the logarithmic spiral for θ in $[2n\pi, 2(n + 1)\pi]$.

13. Show that if

$$\frac{dx}{dt}\frac{d^2x}{dt^2} = -\frac{dy}{dt}\frac{d^2y}{dt^2}$$

then the speed of the curve $x = f(t)$, $y = g(t)$ is constant.

★14. If $x = t$ and $y = g(t)$, show that the points where the speed is maximized are points of inflection of $y = g(x)$.

★15. (a) Looking at a map of the United States, estimate the length of the coastline of Maine.

(b) Estimate the same length by looking at a map of Maine.

(c) Suppose that you used detailed local maps to compute the length of the coastline of Maine. How would the result compare with that obtained in part (b)?

(d) What is the "true" length of the coastline of Maine?

(e) What length for the coastline can you find given in an atlas or almanac?

16. A particle travels a path in space with speed $s(t) = \sin^2(\pi t) + \tan^4(\pi t)\sec^2(\pi t)$. Find the distance $\int_0^{10} s(t)dt$ traveled in the first ten seconds.

17. A Renault loaded with skiers climbs a hill to a ski resort, constantly changing gears due to variations in the incline. Assume, for simplicity, that the motion of the auto is planar: $x = x(t)$, $y = y(t)$, $0 \le t \le T$. Let $s(t)$ be the distance traveled along the road at time t (Fig. 11-3-11).

(a) The value $s(10)$ is the difference in the odometer readings from $t = 0$ to $t = 10$. Explain.

(b) The value $s'(t)$ is the speedometer reading at time t. Explain.

(c) The value $y'(t)$ is the rate of change in altitude, while $x'(t)$ is the rate of horizontal approach to the resort. Explain.

(d) What is the average rate of vertical ascent? What is the average speed for the trip? Respond in the language of calculus.

Fig. 11-3-11 A car on its way to a ski cabin.

18. The position (x, y) of a bulge in a bicycle tire as it rolls down the street can be parametrized by the angle θ shown in Fig. 11-3-12. Let the radius of the tire be a. It can be verified that $x = a\theta - a\sin\theta$, $y = a - a\cos\theta$, $0 \le \theta \le 2\pi$, by methods of plane trigonometry. (This is called a *cycloid*.)

Fig. 11-3-12 Investigate how a bulge on a tire moves.

(a) Find the distance traveled by the bulge for $0 \le \theta \le 2\pi$, using the identity $1 - \cos\theta = 2\sin^2(\theta/2)$. This distance is greater than $2\pi a$ (distance the tire rolls).

★(b) Draw a figure for one arch of the cycloid, and superimpose the circle of radius a with center at $(\pi a, a)$, together with the line segment $0 \le x \le 2\pi a$. Show that the three enclosed areas are each πa^2.

★19. On a movie set, an auto races down a street. A follow-spot lights the action from 20 meters away, keeping a constant distance from the auto in order to maintain the same reflected light intensity for the camera. The follow-spot location (x, y) is the *pursuit curve*

$$x = t - 20\operatorname{sech}\left(\frac{t}{20}\right), \quad y = 20\operatorname{sech}\left(\frac{t}{20}\right)$$

Geometrically, this is a *tractrix*. Graph it.

20. A child walks with speed k from the center of a merry-go-round to its edge, while the equipment rotates counterclockwise with constant angular speed ω. The motion of the child relative to the ground is $x = kt\cos\omega t$, $y = kt\sin\omega t$.

(a) Find the *velocities* $\dot{x} = dx/dt$, $\dot{y} = dy/dt$.

(b) Determine the *speed*.

(c) The child experiences a *Coriolis* force opposite to the direction of rotation, tangent to the edge of the merry-go-round. The magnitude of this force is the mass m of the child times the factor

$$\sqrt{\ddot{x}(0)^2 + \ddot{y}(0)^2}$$

where $\ddot{x} = d^2x/dt^2$. Find this force.

21. An elliptical orbit is parameterized by $x = a\cos\theta$, $y = b\sin\theta$, $0 \le \theta \le 2\pi$. This parameterization is 2π-periodic. In chapter 18 we shall show that for any T-periodic parameterization of a continuously differentiable closed curve $x = x(t)$, $y = y(t)$, which is *simple* (never crosses itself),

$$\text{area enclosed} = \int_0^T [x(t)\dot{y}(t) - \dot{x}(t)y(t)]\, dt$$

(a) Use this formula to verify that the area enclosed by an ellipse of semiaxes a and b is πab.

(b) Apply the formula to the case of a curve $x(t) = r\cos t$, $y(t) = r\sin t$, where $r = r(t)$, showing that area enclosed $= \int_0^T r^2\, dt$.

Review Problems for Chapter 11

1. Integrate the following:
 (a) $\int dx/(x^2 + 4x + 5)$　(b) $\int \sec^6\theta\, d\theta$
 (c) $\int \sin\sqrt{x}\, dx$　(d) $\int dx/(1 + \cos ax)$
 (e) $\int dx/(1 - \cos ax)^2$

2. Integrate:
 (a) $\int dx/(1 - x^4)$
 (b) $\int [(\sin^2 x)/(\cos x)]\, dx$
 (c) $\int \ln[(x+a)/(x-a)]\, dx$
 (d) $\int (\tan^{-1}x)/(1 + x^2)\, dx$
 (e) $\int dx/(x^4 + 1)$

3. Evaluate:
 (a) $\int [x/(x^3 - 9)]\, dx$　(b) $\int (x+5)\ln x\, dx$
 (c) $\int e^{\sqrt{x}}\, dx$　(d) $\int x^3\sqrt{1 - x^2}\, dx$
 (e) $\int dx/(1 + e^x)$

4. Evaluate the following integrals:
 (a) $\int x/(x-3)^8\, dx$
 (b) $\int [x/(x^2 - 1)]^3\, dx$
 (c) $\int \sqrt{x^2 + 2x + 3}\, dx$
 (d) $\int \sin 3x\cos 2x\, dx$
 ★(e) $\int dx/\sqrt{ax^2 + bx + c}$; $a \ne 0$.
 (Find a general formula for the integral; there will be several cases, depending upon the sign of $b^2 - 4ac$.)

5. Use the substitution $u = \tan(x/2)$ to show that $\int \sec x\, dx = \ln|\tan(x/2 + \pi/4)| + C$. Reconcile this answer with the formula $\ln|\sec x + \tan x| + C$ obtained on p. 449.

6. Recall (see Problem 12, Section 8-3) that if f is a function on $[0, 2\pi]$, then the numbers

$$a_m = \frac{1}{\pi}\int_0^{2\pi} f(x)\cos mx\, dx$$
and
$$b_m = \frac{1}{\pi}\int_0^{2\pi} f(x)\sin mx\, dx$$
$(m = 0, 1, 2, \ldots)$

are called the Fourier coefficients of f. Find all the Fourier coefficients of each of the following functions:
 (a) $\sin 2x$　(b) $\sin 5x$
 (c) $\cos 3x$　(d) $\cos 8x$
 (e) $3\cos 4x$
 (f) $2\cos 8x + \sin 7x + \cos 9x$
 (g) $\sin^2 x$　(h) $\cos^3 x$

★7. Let $f(x) = x^n$, $0 < a \le x \le b$. For which rational values of n can you evaluate the integral occurring in the formula for:
 (a) The area under the graph of f?
 (b) The length of the graph of f?
 (c) The volume of the surface obtained by revolving the region under the graph of f about the x axis? The y axis?
 (d) The area of the surface of revolution obtained by revolving the graph of f about the x axis? The y axis?
 Evaluate these integrals.

★8. Same as Problem 7, but with $f(x) = 1 + x^n$.

★9. Same as Problem 7, but with $f(x) = (1 + x^2)^n$.

10. For each of the following pairs of parametric equations, sketch the curve and find an equation in x and y by eliminating the parameter.
 (a) $x = 3\cos t$; $y = \sin t$
 (b) $x = 2t + 5$; $y = t^3$
 (c) $x = 3t$; $y = 2t + 1$
 (d) $x = t$; $y = t$　(e) $x = 0$; $y = t^4$

11. Find the equation of the tangent line to the parametric curve $x = 3 \cos t$, $y = \sin t$ at $t = \pi/4$. Sketch.

12. Find the arc length of $x = t^2$, $y = 2t^4$ from $t = 0$ to $t = 2$.

13. Sketch the curve $x^2 = y^3 - y^2$ by converting to a suitable parametric form.

14. Find the arc length (as an integral if necessary) for each of the following functions in polar coordinates:

    (a) $r = \theta^2$; $0 \leq \theta \leq \pi/2$

    (b) $r = \frac{1}{2} + \cos 2\theta$; $0 \leq \theta \leq \pi$

    (c) $r = 1/\cos\theta$; $0 \leq \theta \leq \pi/4$

15. Find the area enclosed by each of the curves and bounding rays in Problem 14.

16. Sketch and find the lengths of:

    (a) $r = 3 \cos^4 (\theta/4)$; $0 \leq \theta \leq \pi$

    (b) $r = \frac{1}{2} \sin^2 (\theta/2)$; $\pi/4 \leq \theta \leq 3\pi/4$

    (c) $r = 2 |\cos\theta|$; $0 \leq \theta \leq 2\pi$

★17. (a) Find the formula for the area of the surface obtained by revolving the graph of $r = f(\theta)$ about the x axis, $\alpha \leq \theta \leq \beta$.

    (b) Find the area of the surface obtained by revolving $r = \cos 2\theta$, $-\pi/4 \leq \theta \leq \pi/4$ about the x axis (express as an integral if necessary).

★18. Consider the integral

$$\int \frac{dx}{\sqrt{(1 - x^2)(1 - k^2 x^2)}}$$

    (a) Show that, for $k = 0$ and $k = 1$, this integral can be evaluated in terms of trigonometric and exponential functions and their inverses.

    (b) Show that, for any k, the integral may be transformed to one of the form

$$\int \frac{d\theta}{\sqrt{1 - k^2 \sin^2\theta}}$$

    (This integrand occurs in the sunshine formula—see Solved Exercise 4, p. 438.)

    (c) Show that the integral $\int \sqrt{1 - k^2 \sin^2\theta} \, d\theta$ (which also occurs in the sunshine formula) arises when one tries to find the arc length of an ellipse $x^2/a^2 + y^2/b^2 = 1$. Express k in terms of a and b.

    Because of the result of part (c), the integrals in parts (a), (b), and (c) are called *elliptic integrals*.

★19. Consider the parametric curve given by

$$x = \cos mt \quad y = \sin nt$$

    where m and n are integers. Such a curve is called a *Lissajous figure* (see Solved Exercise 4, Section 11-3).

    (a) Plot this curve for $m = 1$ and $n = 1, 2, 3, 4$.

    (b) Describe the general behavior of the curve if $m = 1$, for any value of n. Does it matter whether n is even or odd?

    (c) Plot the curve for $m = 2$ and $n = 1, 2, 3, 4, 5$.

    (d) Plot the curve for $m = 3$ and $n = 4, 5, 6$.

★20. (Lissajous figures continued). The path $x = x(t)$, $y = y(t)$ of movement of the *tri-suspension pendulum* of Fig. 11-R-1 produces a Lissajous figure of the general form $x = A_1 \cos(\omega_1 t + \theta_1)$, $y = A_2 \sin(\omega_2 t + \theta_2)$.

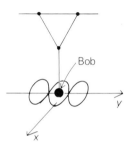

Fig. 11-R-1 The bob on this pendulum traces out a Lissajous figure.

    (a) Draw the Lissajous figures for $\omega_1 = \omega_2 = 1$, $A_1 = A_2$, for some sample values of θ_1, θ_2. The figures should come out to be straight lines, circles, ellipses.

    (b) When $\omega_1 = 1$, $\omega_2 = 3$, $A_1 = A_2 = 1$, the bob retraces its path, but has two self-intersections. Verify this using the results of Problem 19. Conjecture what happens when ω_2/ω_1 is the ratio of integers.

    (c) When $\omega_2/\omega_1 = \pi$, $A_1 = A_2 = 1$, the bob does not retrace its path, and has infinitely many self-intersections. Verify this, graphically. Conjecture what happens when ω_2/ω_1 is irrational (not the quotient of integers).

    (d) On an oscilloscope, the behavior in (a) can be reproduced visually. The idea is to apply, simultaneously, sinusoidal alternating voltages to the horizontal and vertical deflecting plates of the cathode-ray tube. Availability permitting, experiment with an oscilloscope and report on your findings.

21. The solution of the *logistic equation* of population biology, $dN/dt = (k_1 N - k_2)N$, $N(0) = N_0$, requires the evaluation of the definite integral

$$\int_{N_0}^{N(t)} \frac{du}{(k_1 u - k_2)u}$$

   (a) Evaluate by means of partial fraction methods and compare your answer with Problem 6, Section 9-2.

   (b) The integral is just the time t. Solve for $N(t)$ in terms of t, using exponentials.

22. The charge Q in coulombs for an *RC* circuit with sinusoidal switching satisfies the equation

$$\frac{dQ}{dt} + \frac{1}{0.04} Q = 100 \sin(\pi/2 - 5t), \quad Q(0) = 0$$

The solution is

$$Q(t) = 100 e^{-25t} \int_0^t e^{25x} \cos 5x \, dx$$

   (a) Find Q explicitly by means of integration by parts.

   (b) Verify that $Q(1.01) = 0.548$ coulomb. [*Hint:* Be sure to use radians throughout the calculation.]

23. Kepler's second law of planetary motion says that *the radial segment drawn from the sun to a planet sweeps out equal areas in equal times.* Locate the origin $(0, 0)$ at the sun and introduce polar coordinates (γ, θ) for the planet location. Assume the *angular momentum* of the planet (of mass m) about the sun is constant: $mr^2 \dot{\theta} = mk$, $k = $ constant. And $\dot{\theta} = d\theta/dt$. Establish Kepler's second law by showing $\int_t^{t+h} r^2 \dot{\theta} \, dt$ is the same for all times t; thus the area swept out is the same for all time intervals of length h.

24. An elliptical satellite circuits the earth in a circular orbit. The angle ϕ between its major axis and the radial direction of the earth oscillates between $+\phi_m$ and $-\phi_m$ (*librations of the earth satellite*). It is assumed that $0 < \phi_m < \pi/2$, so that the satellite does not tumble end-over-end. The time T for one complete cycle of this oscillation is given by

$$T = \frac{4}{\pi} \int_0^{\phi_m} \frac{d\phi}{\sqrt{\cos 2\phi - \cos 2\phi_m}}$$

Change variables in the integral via the formulas

$$\sin \phi = \sin \phi_m \sin \beta$$
$$\cos 2\phi = 1 - 2 \sin^2 \phi$$
$$\cos 2\phi_m = 1 - 2 \sin^2 \phi_m$$

to obtain the *elliptic integral representation*

$$T = \frac{4}{\pi\sqrt{2}} \int_0^{\pi/2} \frac{d\beta}{\sqrt{1 - k^2 \sin^2 \beta}}$$

for the period of libration T, where $k^2 = \sin^2 \phi_m$.

25. Applying Hertz' theory of impact, one can obtain the equation

$$\frac{1}{2} \rho c_0 \Omega \alpha' = k(\alpha_1^{3/2} - \alpha^{3/2})$$

for the indentation α at time t. The physical problem is that of impact of an infinite bar by a short round-headed bar, making maximum indentation α_1. The symbols ρ, c_0, Ω, k are constants. The equation is solved by an initial integration to get

$$t = \frac{\rho c_0 \Omega}{2k\sqrt{\alpha_1}} \int_0^{\alpha/\alpha_1} \frac{du}{1 - u^{3/2}}$$

   (a) Evaluate the integral by making the substitution $v = \sqrt{u}$, followed by the method of partial fractions.

   (b) Substitute $s = (4tk\sqrt{\alpha_1})/(3\rho c_0 \Omega)$ to obtain

$$s = \frac{2\pi}{\sqrt{3}} + \ln \left| \frac{1 + y + y^2}{(1 - y)^2} \right|$$
$$- 4\sqrt{3} \tan^{-1} \left(\frac{2y + 1}{\sqrt{3}} \right), \quad y = \sqrt{\alpha/\alpha_1}$$

12

Limits and L'Hôpital's Rule

Limits are used in both the theory and the applications of calculus

Our treatment of limits up to this point has been rather casual. Now, having learned some differential and integral calculus, you should be prepared to appreciate a more detailed study of limits.

In Sections 12-1 and 12-2, we develop the basic notions of limit theory. The next two sections, devoted to the "practical" side of limits, can be read independently of the earlier sections (and of each other). In Section 12-3, we present two topics connecting limits with formal differentiation and integration. First we learn how to use differential calculus to compute many limits; then we use limits to evaluate integrals over unbounded intervals and integrals of unbounded functions . Section 12-4 is devoted to numerical techniques for solving equations and evaluating integrals; in each case, the desired number is the limit of a sequence whose terms can be calculated explicitly.

---

*This appendix is found in the Student Guide.

12-1 Limits of Sequences

A convergent sequence is an infinite list of numbers which are better and better approximations to a fixed number, called the limit of the sequence.

We begin this section with some examples of limits of sequences which should already be familiar to you. These are followed by a precise definition of limits and a list of the basic laws of limits. Finally, we state a theorem which makes it possible to show, without finding limits explicitly, that certain sequences converge. In Section 12-2 we take up the related notion of limits of functions, but sequences will appear again in connection with numerical methods in Section 12-4. Furthermore, all of Chapter 13 on infinite series is based on the notion of the limit of a sequence.

Goals

After studying this section, you should be able to:

Verify the convergence or divergence of simple sequences by using the definition of limit.

Find limits by using the laws of limits.

Establish the convergence of certain sequences by showing that they are monotonic and bounded.

SOME SEQUENCES AND THEIR LIMITS

A *sequence* is just an "infinite list" of numbers: $a_1, a_2, a_3, \ldots$, with one a_n for each natural number n. Roughly speaking, a number l is called the *limit* of this sequence if a_n gets and stays arbitrarily close to l as n gets larger.

Perhaps the most familiar example of a sequence with a limit is that of an infinite decimal expansion. Consider, for instance, the equation

$$\tfrac{1}{3} = 0.333 \ldots \tag{1}$$

in which the dots on the right-hand side are often taken to stand for "infinitely many 3's." We can interpret equation (1) without recourse to any metaphysical notion of infinity: the finite decimals 0.3, 0.33, 0.333, and so on are approximations to $\tfrac{1}{3}$, and we can make the approximation as good as we wish by taking enough 3s. Our sequence $a_1, a_2, \ldots$ is defined in this case by $a_n = 0.33\ldots3$, with n 3s (here the three dots stand for only finitely many 3s). In other words,

$$a_n = \frac{3}{10} + \frac{3}{100} + \cdots + \frac{3}{10^n} \tag{2}$$

We can estimate the difference between a_n and $\tfrac{1}{3}$ by using some algebra. Multiplying equation (2) by 10 gives

$$10a_n = 3 + \frac{3}{10} + \cdots + \frac{3}{10^{n-1}} \tag{3}$$

and subtracting (2) from (3) gives

$$9a_n = 3 - \frac{3}{10^n}$$

$$a_n = \frac{1}{3} - \frac{1}{3}\left(\frac{1}{10^n}\right)$$

Finally,

$$\frac{1}{3} - a_n = \frac{1}{3}\left(\frac{1}{10^n}\right) \qquad\qquad (4)$$

As n is taken larger and larger, the denominator 10^n becomes larger and larger, and so the difference $\frac{1}{3} - a_n$ becomes smaller and smaller. In fact, if n is chosen large enough, we can make $\frac{1}{3} - a_n$ as small as we please. (See Fig. 12-1-1.)

Fig. 12-1-1 The decimal approximations to $\frac{1}{3}$ form a sequence converging to $\frac{1}{3}$.

Worked Example 1   How large must n be for the error $\frac{1}{3} - a_n$ to be less than 1 part in 1 millionth?

Solution   By (4), we must have

$$\frac{1}{3}\left(\frac{1}{10^n}\right) < 10^{-6}$$

or $10^{-n} < 3 \cdot 10^{-6}$. It suffices to have $n \geq 6$, so the finite decimal 0.333333 approximates $\frac{1}{3}$ to within 1 part in a millionth. So do the longer decimals 0.3333333, 0.33333333, and so on.

There is nothing special about the number 10^{-6} in Worked Example 1. Given *any* positive number ε ("epsilon"), we will always be able to make $\frac{1}{3} - a_n = \frac{1}{3}(1/10^n)$ less than ε by letting n be sufficiently large. We express this fact by saying that $\frac{1}{3}$ is the *limit* of the numbers

$$a_n = \frac{3}{10} + \frac{3}{100} + \cdots + \frac{3}{10^n}$$

as n becomes arbitrarily large, or

$$\lim_{n\to\infty}\left(\frac{3}{10} + \frac{3}{100} + \cdots + \frac{3}{10^n}\right) = \frac{1}{3}$$

The symbol ∞ ("infinity") does not denote a number but merely indicates that n is to become arbitrarily large.

We may think of a sequence $a_1, a_2, a_3, \ldots$ as a function whose domain consists of the natural numbers $1, 2, 3, \ldots$. (Occasionally, we allow the domain to start at zero or some other integer.) Thus we may represent a sequence graphically in two ways—either by plotting the points $a_1, a_2, \ldots$ on a number line or by plotting the pairs (n, a_n) in the plane.

Worked Example 2   Write out the first six terms of the sequence $a_n = n/(n + 1)$, $n = 1, 2, 3, \ldots$. Represent this sequence graphically in two ways. Find the value of $\lim\limits_{n \to \infty} [n/(n + 1)]$.

Solution   We obtain the terms a_1 through a_6 by substituting $n = 1, 2, 3, \ldots, 6$ into the formula for a_n, giving $\frac{1}{2}, \frac{2}{3}, \frac{3}{4}, \frac{4}{5}, \frac{5}{6}, \frac{6}{7}$. These values are plotted in Fig. 12-1-2. As n gets larger, the fraction $n/(n + 1)$ gets larger and larger but never exceeds 1; we may guess that the limit is equal to 1.

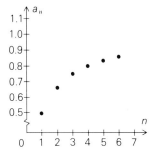

Fig. 12-1-2 The sequence $a_n = n/(n + 1)$ represented graphically in two different ways.

To verify this guess, we look at the difference $1 - n/(n + 1)$. We have

$$1 - \frac{n}{n + 1} = \frac{n + 1 - n}{n + 1} = \frac{1}{n + 1}$$

which does indeed become arbitrarily small as n gets larger, so

$$\lim_{n \to \infty} n/[(n + 1)] = 1$$

Solved Exercises*   1. (a) Do as in Worked Example 2 for the sequence $a_n = (-1)^n/n$.
   (b) Do as in Worked Example 2 for the sequence $a_n = (-1)^n n/(n + 1)$.

2. Using numerical calculations, guess the value of $\lim\limits_{n \to \infty} \sqrt[n]{n}$.

Exercises   1. Do as in Worked Example 2 for these sequences:

   (a) $a_n = 1 + \dfrac{1}{2} + \dfrac{1}{4} + \cdots + \dfrac{1}{2^n}$

   (b) $a_n = \sin(n\pi/2)$

2. How large must n be for the difference $1 - a_n$ in Worked Example 2 to become less than 1 part in 1 millionth?

3. Using numerical calculations, guess the limit as $n \to \infty$ of:
   (a) $\sqrt[n]{n/2}$ (b) $\sqrt[n]{n(n + 1)/4}$

Solutions appear in Appendix A.

(optional) THE DEFINITION OF LIMIT

We will state the precise definition of limit straightaway and then make comments.

Definition Let $a_1, a_2, \ldots$ be a sequence and l a real number. We say that l is the *limit* of the sequence $a_1, a_2, \ldots$ if, for any positive number ε, there is a positive integer N such that $|a_n - l| < \varepsilon$ whenever $n > N$. We write $\lim_{n \to \infty} a_n = l$.

Fig. 12-1-3 The relationship between a_n, ℓ, and ϵ in the definition of limit.

It is useful to think of the number ε in this definition as a *tolerance*, or allowable error. The definition specifies that if l is to be the limit of the sequence a_n, then, given any tolerance, all the terms of the sequence beyond a certain point should be within that tolerance of l. Of course, as the tolerance is made smaller, it will usually be necessary to go farther out in the sequence to bring the terms within tolerance of the limit. (See Fig. 12-1-3.)

Worked Example 3 Suppose that, for the price of n dollars, a jeweler will produce a bar of gold weighing $a_n = 100 + (\sin n)/n^2$ grams. Find the values of a_n for $n = 1, 2, 3, 4, 10, 11, 100, 102$. What minimum price N must a customer pay to guarantee that the bar weighs within ε of 100 grams?

Solution The required values of a_n are tabulated here:

n	1	2	3	4	10	11	100	102
a_n	100.84	100.22	100.02	99.95	99.995	99.991	99.99995	100.00009

Notice that a higher value of n does not necessarily give a_n closer to 100. To estimate a minimum price, we may observe that $|a_n - 100| = |(\sin n)/n^2|$ is at most $1/n^2$, since $|\sin n| \le 1$. Thus, to make $|a_n - 100| < \varepsilon$, it suffices to make $1/n^2 < \varepsilon$, or $n > 1/\sqrt{\varepsilon}$. A customer who pays more than $N = 1/\sqrt{\varepsilon}$ dollars can be sure that $|a_n - 100| < \varepsilon$. (It is possible that a lower payment might accidentally buy enough accuracy, but we cannot be sure of this without knowing the precise behavior of the numbers $\sin n$.)

A sequence which has a limit is said to be *convergent*, or to *converge* to its limit. If a sequence has no limit, it is said to be *divergent*. For instance, the sequences in Worked Examples 2 and 4 and Solved Exercise 1(a) are convergent while the sequence in Solved Exercise 1(b) is divergent.

There are a few simple facts about limits:

$$\lim_{n \to \infty} c = c \quad (a_n = c \text{ a constant})$$

and

$$\lim_{n \to \infty} \frac{1}{n} = 0 \quad \left(a_n = \frac{1}{n}\right)$$

Together with the limit theorems to be described, these facts can be used to evaluate a large number of limits fairly readily.

Worked Example 4   Prove that $\lim\limits_{n \to \infty} 1/n = 0$.

Solution   To show that the definition is satisfied we must show that for any $\varepsilon > 0$ there
(optional)   is a number N such that $|1/n - 0| < \varepsilon$ if $n > N$. If we choose $N \geq 1/\varepsilon$ we
get, for $n > N$,

$$\left| \frac{1}{n} - 0 \right| = \frac{1}{n} < \frac{1}{N} \leq \varepsilon$$

Thus the assertion is proved.

Calculator Discussion

Limits of sequences can sometimes be visualized on a calculator. Consider
the sequence obtained by taking successive square roots of a given positive num-
ber a; $a_0 = a$, $a_1 = \sqrt{a}$, $a_2 = \sqrt{\sqrt{a}}$, $a_3 = \sqrt{\sqrt{\sqrt{a}}}$, and so forth. (See
Fig. 12-1-4.) For instance, if we enter $a = 5.2$, we get

$$a_0 = 5.2$$
$$a_1 = \sqrt{5.2} = 2.2803508$$
$$a_2 = \sqrt{2.2803508} = 1.5100830$$
$$a_3 = \sqrt{1.5100830} = 1.2288544$$

and so on. After pressing the $\sqrt{}$ repeatedly you will see the numbers getting
closer and closer to 1 until round-off error causes the number 1 to appear and
then stay forever. This sequence has 1 as a limit. (Of course, the calculation
does not *prove* this fact, but does suggest it.)

Press $\sqrt{}$
key
$$\overrightarrow{a_{n+1} = \sqrt{a_n}}$$

Fig. 12-1-4   For a recursively defined sequence
$a_{n+1} = f(a_n)$, the next member in the sequence is obtained by depressing the "f"
key. Here $f = \sqrt{}$

In the calculator discussion observe that the sequence is defined *recursively*—
that is, each member of the sequence is obtained from the previous one by some
specific process. The sequence $1, 2, 4, 8, 16, 32, \ldots$ is another example; each term
is twice the previous one: $a_{n+1} = 2a_n$.

Worked Example 5   Write down the first few terms of the sequence starting at $a_0 = 0$ and defined
recursively by

$$a_{n+1} = \tfrac{1}{2}(1 + a_n)$$

Solution   We get $a_1 = \tfrac{1}{2}(1 + a_0) = \tfrac{1}{2}$, $a_2 = \tfrac{1}{2}(1 + a_1) = \tfrac{1}{2}(1 + \tfrac{1}{2}) = \tfrac{3}{4}$, $a_3 = \tfrac{1}{2}(1 + \tfrac{3}{4}) = \tfrac{7}{8}$,
and so on, so the sequence is $0, 1/2, 3/4, 7/8, 15/16, 31/32, \ldots$.

The following limit theorem is not quite so obvious as Worked Example 4.

(optional)

Theorem 1 *Let r be a number such that $|r| < 1$. Then*

$$\lim_{n \to \infty} r^n = 0$$

Proof Since $|r| < 1$, we get $1/|r| > 1$, and so we may write $1/|r| = 1 + a$, where $a > 0$. By the binomial theorem,

$$\frac{1}{|r|^n} = (1 + a)^n = 1 + na + \binom{n}{2} a^2 + \cdots + a^n > 1 + na$$

so $|r|^n < 1/(1 + na)$. If we wish to make $|r^n - 0| = |r|^n < \varepsilon$ for $n > N$, it suffices to choose N so that $1/(1 + Na) \le \varepsilon$; i.e., $N \ge [(1/\varepsilon) - 1]/a$. Then $n > N$ implies $|r|^n < 1/(1 + na) < 1/(1 + Na) < \varepsilon$.

Worked Example 6 Find $\lim_{n \to \infty} (3^n/4^n)$.

Solution Let $r = \frac{3}{4}$ and apply Theorem 1 to get $\lim_{n \to \infty} 3^n/4^n = \lim_{n \to \infty} r^n = 0$.

If $|r| \ge 1$, then $|r^n| = |r|^n \ge 1$ and so r^n will not converge to zero. For instance, if $r = 1$, then $r^n = 1$ is the constant sequence. If $r = 2$, the sequence 2^n is getting larger without bound as n increases.

Solved Exercises ★3. Prove that $\lim_{n \to \infty} a_n = 1$, where a_n is the sequence in Worked Example 5.

★4. Prove that if $a_n = c$ for all n (c is a constant), then $\lim_{n \to \infty} a_n = c$.

Exercises ★4. Write the first five terms of the indicated sequences.
  (a) $a_n = (n - 1)/(n + 1)$; $n = 0, 1, 2, \ldots$
  (b) $a_n = (2n + 1)/n^2$; $n = 1, 2, \ldots$ (c) $b_n = \sqrt{n} + 1$; $n = 0, 1, \ldots$
  (d) $c_n = (1/n!) \sin(n\pi/2)$; $n = 0, 1, 2, \ldots$
  (e) $g_n = (-1)^n n(n + 1)$; $n = 1, 2, \ldots$ (f) $l_n = (4/3) - (1/n)$; $n = 1, 2, \ldots$

★5. Write the first five terms of the sequences defined recursively as indicated.
  (a) $a_{n+1} = \frac{2}{3} a_n$; $a_0 = 1$ (b) $a_{n+1} = \sqrt{1 + a_n}$; $a_0 = 0$
  (c) $a_{n+1} = 9 - (a_n)^2$; $a_0 = 1$ (d) $a_{n+1} = -a_n/(n + 1)$; $a_0 = 1$

★6. (a) Prove that $\lim_{n \to \infty} [3/(2n + 1)] = 0$.
  (b) Prove that $\lim_{n \to \infty} (1 + r^n) = 1$ if $|r| < 1$. [*Hint:* Use Theorem 1 and the definition of $\lim_{n \to \infty} a_n = l$.]

PROPERTIES OF LIMITS

The following limit theorem makes it possible to evaluate many limits without using the definition.

Theorem 2 *Suppose that the sequences $a_1, a_2, \ldots$ and $b_1, b_2, \ldots$ are convergent, $\lim_{n \to \infty} a_n = l$ and $\lim_{n \to \infty} b_n = m$. Then*

1. $\lim_{n \to \infty} (a_n + b_n) = \lim_{n \to \infty} a_n + \lim_{n \to \infty} b_n$

2. $\lim_{n \to \infty} (ca_n) = c \lim_{n \to \infty} a_n$

3. $\lim_{n \to \infty} (a_n b_n) = \lim_{n \to \infty} a_n \cdot \lim_{n \to \infty} b_n$

4. *If* $\lim_{n \to \infty} b_n \neq 0$ *and* $b_n \neq 0$ *for all n, then*

$$\lim_{n \to \infty} \frac{a_n}{b_n} = \frac{\lim_{n \to \infty} a_n}{\lim_{n \to \infty} b_n}$$

5. *If f is continuous at l, then*

$$\lim_{n \to \infty} f(a_n) = f(\lim_{n \to \infty} a_n)$$

A proof of Theorem 2 is given in the appendix to this chapter.

Three other useful rules are as follows (see Worked Example 4, Solved Exercise 4, and Theorem 1 for proofs):

6. $\lim_{n \to \infty} c = c$

7. $\lim_{n \to \infty} (1/n) = 0$

8. If $|r| < 1$, then $\lim_{n \to \infty} r^n = 0$

Worked Example 7   Find $\lim_{n \to \infty} (3 + n)/(2n + 1)$.

Solution $\lim_{n \to \infty} \dfrac{3 + n}{2n + 1} = \lim_{n \to \infty} \dfrac{3/n + 1}{2 + 1/n} = \dfrac{3 \lim_{n \to \infty} 1/n + \lim_{n \to \infty} 1}{\lim_{n \to \infty} 2 + \lim_{n \to \infty} 1/n} = \dfrac{3 \cdot 0 + 1}{2 + 0} = \dfrac{1}{2}$

This solution used parts 1, 2, and 4 of Theorem 2, together with the facts that $\lim_{n \to \infty} 1/n = 0$ (rule 7) and $\lim_{n \to \infty} c = c$ (rule 6).

Worked Example 8   Find $\lim_{n \to \infty} \sin [\pi n/(2n + 1)]$.

Solution   Since $\sin x$ is a differentiable and hence continuous function, we can use rule 5 in Theorem 2 to get

$$\lim_{n \to \infty} \sin \left(\frac{\pi n}{2n + 1} \right) = \sin \left[\lim_{n \to \infty} \left(\frac{\pi n}{2n + 1} \right) \right]$$

$$= \sin \left[\lim_{n \to \infty} \left(\frac{\pi}{2 + 1/n} \right) \right]$$

$$= \sin \left(\frac{\pi}{2} \right) = 1$$

Solved Exercises

5. Find $\lim_{n \to \infty} [(8n^2 + 2)/(3n^2 + n)]$.

6. Find $\lim_{n \to \infty} (\pi + (\tfrac{2}{3})^n)^3$.

7.★(a) If $a_n \to 0$ and $|b_n| \leq |a_n|$, show that $b_n \to 0$. (This is sometimes called the *comparison test*.)

   (b) Find: (i) $\lim_{n \to \infty} (\sin n/n)$; (ii) $\lim_{n \to \infty} [(1 + n)/2n]^n$.

8. Let $a_n = 1$ if n is even and -1 if n is odd. Does $\lim_{n \to \infty} a_n$ exist?

Exercises

7. Find the following limits.

(a) $\lim_{n \to \infty} [(8 - 2n)/5n]$

(b) $\lim_{n \to \infty} [1 - (2 + n)/(3n + 1)]$

(c) $\lim_{n \to \infty} \sqrt[3]{(8n^3 - 2n + 1)/(n^3 + 1)}$

8. Does $\lim_{n \to \infty} \{[(1 + n) \cos n] / n\}$ exist?

9. Suppose that $\lim_{n \to \infty} a_n = a$ and that $a > 0$. Prove that there is a positive integer N such that $a_n > 0$ for all $n > N$.

10. Find the limit or prove that the limit does not exist:

(a) $\lim_{n \to \infty} \left(\dfrac{1}{2} + (-1)^n \left(\dfrac{1}{2} - \dfrac{1}{n} \right) \right)$

(b) $\lim_{n \to \infty} \left(\dfrac{(-1)^n \sin (n \pi / 2)}{n} \right) \left(\dfrac{n^2 + 1}{n + 1} \right)$

(c) $\lim_{n \to \infty} \left(\dfrac{3n}{4n + 1} + \dfrac{(-1)^n \sin n}{n + 1} \right)$

(optional) **MONOTONIC SEQUENCES**

Up to now, we have demonstrated the convergence of sequences by finding their limits explicitly. It is sometimes useful, particularly in connection with infinite series (to be studied in Chapter 13), to be able to show that a limit exists without finding the limit itself.

A sequence may be given recursively, for instance, in which case applying Theorem 2 becomes awkward. For example, let a_n be defined as follows:

$$a_0 = 0 \qquad a_1 = \sqrt{3} \qquad a_2 = \sqrt{3 + a_1} = \sqrt{3 + \sqrt{3}}$$

$$a_3 = \sqrt{3 + a_2} = \sqrt{3 + \sqrt{3 + \sqrt{3}}} \ldots$$

And, in general,

$$a_n = \sqrt{3 + a_{n-1}}$$

If we attempt to write out a_n "explicitly" we quickly find ourselves in a notational nightmare. However, numerical computation suggests that the sequence may be convergent:

$a_1 = 1.73205 \qquad a_2 = 2.17533 \qquad a_3 = 2.27493$

$a_4 = 2.29672 \qquad a_5 = 2.30146 \qquad a_6 = 2.30249$

$a_7 = 2.30271 \qquad a_8 = 2.30276 \qquad a_9 = 2.30277$

$a_{10} = 2.30278 \qquad a_{11} = 2.30278 \qquad a_{12} = 2.30278 \ldots$

The sequence appears to be converging to a number $l \approx 2.30278 \ldots$, but the numerical evidence only *suggests* that the sequence converges. How can we really demonstrate convergence?* Fortunately, there is a simple and very useful test for this. To state it, we introduce some terminology.

Once we know that this sequence converges, we can find its limit algebraically (see Solved Exercise 9).

(optional)

Definition

1. A sequence $a_1, a_2, \ldots$ is called *increasing* (*decreasing*) if $a_n \leq a_{n+1}$ ($a_n \geq a_{n+1}$) for each n; a sequence which is either increasing or decreasing is called *monotonic*.

2. A sequence $a_1, a_2, \ldots$ is *bounded* if there is a constant M such that $|a_n| \leq M$ for every n; that is, if all terms of the sequence lie in some finite interval $[-M, M]$. A sequence, such as $a_n = n$, which is not wholly contained in some finite interval is called an *unbounded* sequence.

If a_n is increasing, it will be bounded if and only if there is a constant M such that $a_n \leq M$ for all n, because all a_n's then lie between a_0 and M—that is, in the interval $[a_0, M]$. In these circumstances, it is natural to speak of the sequence a_n as being *bounded above*. (See Fig. 12-1-5.) Similarly, a decreasing sequence is bounded if and only if it is bounded below.

(a)

(b)

Fig. 12-1-5 An increasing sequence (a) and a sequence bounded above by M (b).

Worked Example 9 Let $a_n = n/(n + 1)$. Show that a_n is increasing and is bounded above by M if M is any number ≥ 1.

Solution The sequence $\frac{1}{2}, \frac{2}{3}, \frac{3}{4}, \frac{4}{5}, \ldots$ appears to be increasing and bounded above by 1. To prove that it is increasing, we must show that $a_n \leq a_{n+1}$—that is, that

$$\frac{n}{n+1} \leq \frac{n+1}{(n+1)+1} \qquad \text{or} \qquad n(n+2) \leq (n+1)$$

or

$$n^2 + 2n \leq n^2 + 2n + 1 \qquad \text{or} \qquad 0 \leq 1$$

Reversing steps shows that $a_n \leq a_{n+1}$. Since $n < n + 1$, we have $a_n = n/(n + 1) < 1$, so $a_n < M$ if $M \geq 1$.

Theorem 3 *If a_n is an increasing (decreasing) sequence which is bounded above (below), then a_n converges to some number a as $n \to \infty$.*

This theorem, proved in the appendix to this chapter, expresses a simple idea. If the sequence is increasing, the numbers a_n keep getting larger, but they can never go past M. What else could they do but converge?

For example, consider

$$a_1 = 0.3, \quad a_2 = 0.33, \quad a_3 = 0.333$$

(optional) and so forth. These a_n's are increasing (in fact, strictly increasing) and are bounded above by 0.4, so we know from Theorem 3 that they must converge. In fact, Theorem 3 shows that any infinite decimal expansion converges and so represents a real number.

Solved Exercises ★ 9. Let a_n be defined by $a_0 = 0$, $a_1 = \sqrt{3}$, $a_2 = \sqrt{3 + \sqrt{3}}, ..., a_n = \sqrt{3 + a_{n-1}}$. Prove by induction that a_n is increasing and bounded above by 5. Apply Theorem 3 to prove that a_n converges to a limit l. Find l by using rule 5 of Theorem 2.

★10. Prove that $a_n = (1 + 1/n)^n$ is increasing and bounded above as follows:*

(a) If $0 \le a < b$, prove that

$$\frac{b^{n+1} - a^{n+1}}{b - a} < (n + 1)b^n$$

That is, prove that $b^n[(n + 1)a - nb] < a^{n+1}$.

(b) Let $a = 1 + [1/(n + 1)]$ and $b = 1 + (1/n)$ and deduce that a_n is increasing.

(c) Let $a = 1$ and $b = 1 + (1/2n)$ and deduce that $(1 + 1/2n)^{2n} < 4$.

(d) Use parts (b) and (c) to show that $a_n < 4$.

Conclude that a_n converges to some number (the number is e—see p. 332).

Exercises ★11. Give definitions for a sequence of numbers a_n to be (a) strictly decreasing; (b) bounded below.

★12. Show that the following sequences are increasing (or decreasing) and bounded above (or below):

(a) $a_n = 2n/(n + 3)$ (b) $a_n = n/(n^2 + 1)$

(c) $b_n = n \sin(1/n)$

★13. Let $B > 0$ and $a_0 = 1$; $a_{n+1} = \frac{1}{2}(a_n + B/a_n)$. Show that $a_n \to \sqrt{B}$.

★14. Let $a_{n+1} = 3 - (1/a_n)$; $a_0 = 1$. Prove that this sequence is increasing and bounded above. What is $\lim_{n \to \infty} a_n$?

THE INTEGRAL AS A LIMIT OF SUMS

Let f be defined on an interval $[a, b]$. In our definition of the integral $\int_a^b f(t)\, dt$ in Section 4-1, we considered partitions $(t_0, t_1, ..., t_n)$ of $[a, b]$ and formed:

Lower sums:

$$\sum_{i=1}^n k_i(t_i - t_{i-1}) \quad \text{where } k_i < f(t) \text{ for all } t \text{ in } (t_{i-1}, t_i)$$

Upper sums:

$$\sum_{i=1}^n l_i(t_i - t_{i-1}) \quad \text{where } f(t) < l_i \text{ for all } t \text{ in } (t_{i-1}, t_i)$$

The integral of f was defined as that number S, if it exists, for which every $s < S$ is a lower sum and every $s > S$ is an upper sum.

---

*See R. F. Johnsonbaugh, "*Another Proof of an Estimate for e,*" American Mathematical Monthly, *November 1974, p. 1011.*

(optional)   We also mentioned approximating sums

$$S_n = \sum_{i=1}^{n} f(c_i)(t_i - t_{i-1}) \quad \text{where } c_i \text{ is in } [t_{i-1}, t_i]$$

called *Riemann sums* and said that the integral could be considered as a limit of Riemann sums. Here is a precise version of that statement.

Theorem 4   *Let f be a bounded function on* $[a, b]$:

1. *Assume that f is integrable and that the maximum of the numbers* $\Delta t_i = t_i - t_{i-1}$ *goes to zero as* $n \to \infty$. *Then for any choice of* c_i *in* $[t_{i-1}, t_i]$,

$$\lim_{n \to \infty} S_n = \int_a^b f(t)\, dt \tag{5}$$

2. *Suppose that for every choice of* c_i *and* t_i *with the maximum of* Δt_i *tending to zero as* $n \to \infty$, $\lim_{n \to \infty} S_n = S$. *Then f is integrable with integral S.*

This theorem shows that formula (5) may be taken as an alternative definition of the integral. Its proof is given in the appendix (see the Student Guide).

Solved Exercises ★11. Let $S_n = \Sigma_{i=1}^{n} (1 + i/n)(1/n)$. Prove that $S_n \to \frac{3}{2}$ as $n \to \infty$ (a) directly and (b) using Riemann sums.

★12. Use Theorem 4 to demonstrate that

$$\int_a^b [f(x) + g(x)]\, dx = \int_a^b f(x)\, dx + \int_a^b g(x)\, dx$$

Exercises ★15. (a) Prove that

$$\lim_{n \to \infty} \left(\frac{1}{n+1} + \frac{1}{n+2} + \cdots + \frac{1}{n+n} \right) = \int_0^1 \frac{dx}{1+x} = \ln 2$$

(b) Evaluate the sum for $n = 10$ and compare with $\ln 2$.

★16. Use Theorem 4 to demonstrate the following:

(a) $\int_a^b cf(x)\, dx = c \int_a^b f(x)\, dx$

(b) If $f(x) \le g(x)$ for all x in $[a, b]$, then $\int_a^b f(x)\, dx \le \int_a^b g(x)\, dx$.

Problems for Section 12-1

1. Write down the first six terms of the following sequences:

(a) $k_n = n^2 - 2\sqrt{n}$; $n = 0, 1, 2, \ldots$

(b) $a_n = (-1)^{n+1} [(n-1)/n!]$; $n = 0, 1, 2, \ldots$ $(0! = 1)$

(c) $b_n = nb_{n-1}/(1 + n)$; $b_0 = \frac{1}{7}$

2. Write down the first eight terms of the following sequences:

(a) $c_{n+1} = -c_n / [2n(4n + 1)]$; $c_1 = 2$

(b) $a_{n+1} = [1/(n + 1)] \Sigma_{i=0}^{n} a_i$; $a_0 = \frac{1}{2}$

(c) $k_n = \sqrt{3n^2 + 2n}$; $n = 1, 2, 3, \ldots$

3. Show that

(a) $\lim_{n \to \infty} \{ [(-1)^n \cdot 2] /(n + 1) \} = 0$

(b) $\lim_{n \to \infty} \{ [\cos(\pi \sqrt{n})] / n^2 \} = 0$

(c) $\lim_{n \to \infty} \{ [1 + (\frac{3}{4})^n] / [8 + (\frac{2}{3})^n] \} = \frac{1}{8}$

4. Find the following limits if they exist. (If any do not exist, say "the limit does not exist.")

   (a) $\lim\limits_{n \to \infty} [(n - 3n^2)/(n^2 + 1)]$

   (b) $\lim\limits_{n \to \infty} [(\sin n)^2/(n + 2)]$

   (c) $\lim\limits_{n \to \infty} \left[\dfrac{3n^2 - 2n + 1}{n(n + 1)} - \dfrac{n(n + 2)}{(n + 1)(n + 3)} \right]^2$

   (d) $\lim\limits_{n \to \infty} \tan \left[\dfrac{(-1)^n \pi}{\frac{1}{2} - \frac{1}{n}} + (-1)^{3n} \dfrac{\pi}{2} \right]$

5. Find the following limits if they exist:

   (a) $\lim\limits_{n \to \infty} \dfrac{n^3 + 3n^2 + 1}{n^4 + 8n^2 + 2}$

   (b) $\lim\limits_{n \to \infty} \dfrac{(1 + n) \cos (n + 1)}{n^2 + 1}$

   (c) $\lim\limits_{n \to \infty} \dfrac{2 + 1/n}{(n^2 - 2)/(n^2 + 1)}$

   (d) $\lim\limits_{n \to \infty} \left[\dfrac{3n^2 + 2n}{(n + 1)(n + 2)} \right] \left[2 - \left(\dfrac{7}{8} \right)^n \right]$

6. Find:

   (a) $\lim\limits_{n \to \infty} \dfrac{n + (\frac{3}{4})^n}{n^2 + 2}$

   (b) $\lim\limits_{n \to \infty} \left[\dfrac{3b + (\frac{1}{2})^{2n}}{n^2 - 1} \right]^3$; b a constant

   (c) $\lim\limits_{n \to \infty} \left[\dfrac{(-1)^n}{3} + (-1)^{2n+1} \left(\dfrac{1}{3} + \dfrac{2}{n} + \dfrac{1}{n^2} \right) \right]$

   (d) $\lim\limits_{n \to \infty} \left\{ \dfrac{(-1)^n}{2} + n \left[\dfrac{1 + (-1)^{n+1}(3n + 1)}{6n^2 - 5n + 2} \right] \right\}$

★7. (a) Define, in a way you think is reasonable, what $\lim\limits_{n \to \infty} a_n = \infty$ means.

   (b) Prove, using your definition in part (a), that $\lim\limits_{n \to \infty} [(1 + n^2)/(1 + 8n)] = \infty$.

★8. Verify Theorem 3 for the following sequences:

   (a) $a_n = \dfrac{1}{2n} - \dfrac{1}{n + 1}$

   (b) $a_n = \dfrac{n^2 - (n + 1)(n + 2)}{n!}$

★9. Let $B > 0$ and let $a_0 = 1$; $a_{n+1} = \frac{1}{2}(a_n + B/a_n^2)$. Prove that a_n converges to $\sqrt[3]{B}$.

★10. Let $a_{n+1} = \frac{1}{2} a_n + \sqrt{a_n}$; $a_0 = 1$. Prove that a_n is increasing and bounded above. What is $\lim\limits_{n \to \infty} a_n$?

11. If a radioactive substance has a half-life of T, so that half of it decays after time T, write a sequence a_n showing the fraction remaining after time nT. What is $\lim\limits_{n \to \infty} a_n$?

★12. Give another proof that $\lim\limits_{n \to \infty} r^n = 0$ if $0 < r < 1$ as follows. Show that r^n decreases and is bounded below by zero. If the limit (guaranteed to exist by Theorem 3) is l, show that $rl = l$ and conclude that $l = 0$.

★13. Suppose that $a_n, b_n,$ and c_n, $n = 1, 2, 3, ...,$ are sequences of numbers such that for each n, $a_n < b_n < c_n$.

   (a) If $\lim\limits_{n \to \infty} a_n = L$ and if $\lim\limits_{n \to \infty} b_n$ exists, show that $\lim\limits_{n \to \infty} b_n \geq L$. [*Hint:* Suppose not!]

   (b) If $\lim\limits_{n \to \infty} a_n = L = \lim\limits_{n \to \infty} c_n$, prove that $\lim\limits_{n \to \infty} b_n = L$.

14. Imported products are transported through wholesalers $W_1, ..., W_n$. Wholesaler W_k charges x_k percent for his services. Find the price P_n charged for a \$200 import which went through $W_1, ..., W_n$.

15. Enter the display value 1.0000000 on your calculator and repeatedly press the "sin" key. This process generates display numbers $a_1 = 1.0000000$, $a_2 = 0.84147$, $a_3 = 0.74562$, ...

   (a) Write a formula for a_n, using function notation.

   (b) Conjecture the value of $\lim\limits_{n \to \infty} a_n$. Explain with a graph.

16. Display the number 2 on your calculator. Repeatedly press the "x^2" key. You should get the numbers 2, 4, 16, 256, 65536, ... Express the display value a_n after n repetitions by a formula.

17. Let $f(x) = 1 + 1/x$. Equipped with a calculator with reciprocal function, complete the following:

   (a) Write out $f(f(f(f(f(f(2))))))$ as a division problem, and calculate the value. We abbreviate this as $f^{(6)}(2)$, meaning to display the value 2, press the "$1/x$" key and add 1, successively six times.

   (b) Experiment to determine $\lim\limits_{n \to \infty} [1/f^{(n)}(2)]$ to five decimal places.

★18. A portion of an insect population in a limited environment dies each year. The next year's population is born from eggs fertilized by the deceased adult insects the previous year. The number present initially is N_0, and N_k is the number present in generation k. The number N_D dying from the kth generation and the number N_B of emerging young satisfy $N_D = A + a N_k$, $N_B = B + b N_{k-1}$, where A, B are constants and a is the *death rate*, b the *birth rate*.

(a) Assume $N_B = N_D$ (the number of births and deaths are equal). Rearrange this equation to obtain the *difference equation*

$N_k = (b/a) N_{k-1} + \overline{N}$

where $\overline{N} = (B - A)/a$.

(b) Compute the sequence $N_0, N_1, N_2, N_3, ...$ of generation numbers from part (a), showing that

$$N_k = \left[\frac{1 - (b/a)^k}{1 - (b/a)}\right] \overline{N} + (b/a)^k N_0$$

(c) In practice, $b < 0$, $a > 0$. Show that for $|b| < |a|$, $\lim_{k \to \infty} N_k = \overline{N}/[1 - (b/a)]$, and interpret biologically.

(d) Assume $b/a = -1$ (i.e., $|b| = |a|$). Verify that the population size oscillates about the optimum size $\overline{N}/2$. Explain why $\lim_{k \to \infty} N_k$ does not exist and interpret biologically.

★19. Write down Riemann sums for the given functions. Sketch.

(a) $f(x) = x/(x + 1)$ for $1 \le x \le 6$ with $t_0 = 1$, $t_1 = 2$, $t_2 = 3$, $t_3 = 4$, $t_4 = 5$, $t_5 = 6$; $c_i = i$ on $[t_{i-1}, t_i] = [i, i + 1]$.

(b) $f(x) = x + \sin[(\pi/2)x]$, $0 \le x \le 6$ with $t_i = i$, $i = 0,1,2,3,4,5,6$. Find Riemann sums S_6 with

$c_i = i$ on $[i - 1, i]$

and

$c_i = i - \frac{1}{2}$ on $[i - 1, i]$

★20. Write each of the following integrals as a limit.

(a) $\int_1^3 [1/(x^2 + 1)]\, dx$; partition $[1,3]$ into n equal parts and use a suitable choice of c_i.

(b) $\int_0^\pi (\cos \frac{1}{2}x + x)\, dx$; partition $[0, \pi]$ into n equal parts and use a suitable choice of c_i.

★21. Let

$$S_n = \sum_{i=1}^n \left(\frac{i}{n} + \frac{i^2}{n^2}\right)\frac{1}{n}$$

Prove that $S_n \to \frac{5}{6}$ as $n \to \infty$ by using Riemann sums.

★22. Expressing the following sums as Riemann sums, show that:

(a) $\lim_{n\to\infty} \sum_{i=1}^n \left[\sqrt{\frac{i}{n}} - \left(\frac{i}{n}\right)^{3/2}\right]\frac{1}{n} = \frac{4}{15}$

(b) $\lim_{n\to\infty} \sum_{i=1}^n \frac{3n}{(2n + i)^2} = \frac{1}{2}$

★23. Write down a Riemann sum for $f(x) = x^3 + 2$ on $[-2, 3]$ with $t_i = -2 + (i/2)$; $i = 0, 1, 2, ..., 10$.

★24. Write $\int_{-\pi/4}^{\pi/4} (1 + \tan x)\, dx$ as a limit. (Partition $[-\pi/4, \pi/4]$ into $2n$ equal parts and choose c_i appropriately.)

★25. Use Theorem 4 to prove that $\int_a^c f(x)\, dx = \int_a^b f(x)\, dx + \int_b^c f(x)\, dx$.

26. A gold mine goes into production at $t = 0$ producing ore at a rate of 800,000 tons per year. Productivity is to decline via the rate function $(20,000)e^{-t}$ tons per year per year, $0 \le t \le 10$, t in years.

(a) Explain why the production rate $p(t)$ is given by $p(t) = 780 + 20e^{-t}$ kilo-tons.

(b) The amount of ore extracted in time Δt_k years, t_k years in the future, is $p(t_k)\Delta t_k$. Explain.

(c) An approximate value for the total ore extracted in ten years is $\sum_{k=1}^n p(t_k)\Delta t_k$. Explain the meaning of t_k, Δt_k, n to make this valid, and write the true value as a definite integral.

★27. P dollars is deposited in an account each day for a year. The account earns interest at a rate r (e.g., $r = .05$ means 5%) compounded continuously. Use Riemann sums to show that the amount in the account at the end of the year is approximately $(365P/r)(e^r - 1)$.

28. There is a simple method which may be used to determine areas of irregular regions R in the plane. The idea is to cut the region to scale out of cardboard and weigh it. Then divide this weight by the weight of a square of the same cardboard with unit side.

(a) Discuss the failings of this method.

(b) Devise a way to find average temperature by this method, assuming the temperature is recorded in a roll-paper graph.

29. The barometric pressure is recorded on a roll-paper graph, continuously, by means of heat imprint on the heat-sensitive roll-paper. Explain how to determine the *average barometric pressure*, by the following methods. Pay close attention to the units (millibars, hours).

(a) Average the pressures between 8 A.M. and 6 P.M., at 30-minute intervals.

(b) Make use of a *planimeter*, a precision instrument which measures the area of any plane region surrounded by a closed curve.

(optional) # 12-2 Limits of Functions

There are many kinds of limits, but they all obey similar laws.

In Section 12-1, we studied limits of the form $\lim\limits_{n\to\infty} a_n$. We begin this section by replacing the integer variable n with a real variable x tending to $+\infty$ or $-\infty$ and considering the behavior of a function $f(x)$. We then study limits of the form $\lim\limits_{x\to x_0} f(x)$, which have been used in differential calculus.

Goals

After studying this section, you should be able to:

State the definition of $\lim\limits_{x\to x_0} f(x)$.

Evaluate limits of the form

$$\lim_{x\to\pm\infty} f(x), \quad \lim_{x\to x_0} f(x), \quad \lim_{x\to x_0\pm} f(x)$$

Find horizontal and vertical asymptotes.

HORIZONTAL ASYMPTOTES AND $\lim\limits_{x\to\pm\infty} f(x)$

Let $y = f(t)$ be the length at time t of a spring with a bobbing mass on the end. If there are no frictional forces acting, the motion is sinusoidal, given by an equation of the form $f(t) = y_0 + a \cos \omega t$ (see Section 9-1). In reality, a spring does not go on bobbing forever; frictional forces cause what is called *damping*, and the actual motion is described by the function

$$y = f(t) = y_0 + ae^{-bt} \cos \omega t \tag{1}$$

where b is another positive constant. A graph of this function is sketched in Fig. 12-2-1.*

Fig. 12-2-1 The motion of a damped spring has the form $y = f(t) = y_0 + ae^{-bt} \cos \omega t$.

As time passes, we observe that the length becomes and remains arbitrarily near to the equilibrium length y_0. (Even though $y = y_0$ for $t = \pi/2\omega$, this is not the same thing because near these times the length does not *remain* near y_0.) We express this mathematical property of the function f by writing $\lim\limits_{t\to\infty} f(t) = y_0$. The limiting behavior appears graphically as the fact that the graph of f comes and remains closer to the line $y = y_0$ as we look farther to the right.

The precise definition is analogous to that for sequences. As is usual in our general definitions, we denote the independent variable by x rather than t.

---

*If you have not studied Chapter 9, you may ignore the formula for $f(t)$ and simply examine the graph.

(optional)

Definition Let f be a function whose domain contains an interval of the form (a, ∞), l a real number. We say that *l is the limit of $f(x)$ as x approaches* ∞, if, for every positive number ε, there is a number $A > a$ such that $|f(x) - l| < \varepsilon$ whenever $x > A$. We write $\lim_{x \to \infty} f(x) = l$. There is an analogous definition for $\lim_{x \to -\infty} f(x) = l$. When $\lim_{x \to \infty} f(x) = l$ or $\lim_{x \to -\infty} f(x) = l$, the line $y = l$ is called a *horizontal asymptote* of the graph $y = f(x)$.

We illustrate this definition in Figs. 12-2-2 and 12-2-3 by shading the region consisting of those points (x, y) for which $x \leq A$ or for which $x > A$ and $|f(x) - l| < \varepsilon$. If $\lim_{x \to \infty} f(x) = l$, we should be able to "catch" the graph of f in this region by choosing A large enough—that is, by sliding the point A sufficiently far to the right.

Fig. 12-2-2 When $\lim_{x \to \infty} f(x) = \ell$, we can catch the graph in the shaded region by sliding the region sufficiently far to the right. This is true no matter how small ε may be.

(a)   (b)

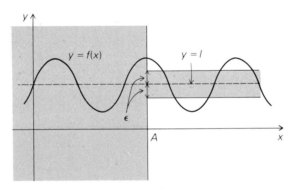

Fig. 12-2-3 When it is not true that $\lim_{x \to \infty} f(x) = l$, then for some ε we can never catch that graph of f in the shaded region, no matter how far to the right we slide the region.

Worked Example 1 Prove that $\lim_{x \to \infty} [x^2/(1 + x^2)] = 1$.

Solution Given $\varepsilon > 0$, we must choose A such that $|x^2/(1 + x^2) - 1| < \varepsilon$ for $x > A$. We have

$$\left| \frac{x^2}{1 + x^2} - 1 \right| = \left| \frac{x^2 - 1 - x^2}{1 + x^2} \right| = \frac{1}{|1 + x^2|} < \frac{1}{x^2}$$

To make this less than ε, we observe that $1/x^2 < \varepsilon$ whenever $x > 1/\sqrt{\varepsilon}$, so we may choose $A = 1/\sqrt{\varepsilon}$. (See Fig. 12-2-4.)

(optional)

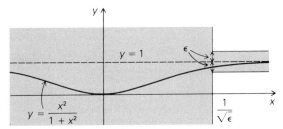

Fig. 12-2-4 Illustrating the fact that

$$\lim_{x\to\infty} \frac{x^2}{1+x^2} = 1$$

The job of finding limits of functions is made easier by the use of algebraic rules. The rules (and their proofs) are analogous to those for sequences. We simply state them below.

LIMITS OF FUNCTIONS AS $x \to \infty$

Constant function rule:

$$\lim_{x\to\infty} c = c$$

$1/x$ rule:

$$\lim_{x\to\infty} \frac{1}{x} = 0$$

Sum rule:

$$\lim_{x\to\infty} [f(x) + g(x)] = \lim_{x\to\infty} f(x) + \lim_{x\to\infty} g(x)$$

Product rule:

$$\lim_{x\to\infty} [f(x)g(x)] = \lim_{x\to\infty} f(x) \lim_{x\to\infty} g(x)$$

Quotient rule:

If $\lim_{x\to\infty} g(x) \neq 0$, then

$$\lim_{x\to\infty} \left[\frac{f(x)}{g(x)} \right] = \frac{\lim_{x\to\infty} f(x)}{\lim_{x\to\infty} g(x)}$$

Replacement rule:

If for some real number A the functions $f(x)$ and $g(x)$ agree for all $x > A$, then

$$\lim_{x\to\infty} f(x) = \lim_{x\to\infty} g(x)$$

Continuous function rule:

If h is continuous at $\lim_{x\to\infty} f(x)$, then

$$\lim_{x\to\infty} h(f(x)) = h(\lim_{x\to\infty} f(x))$$

All these rules remain true if we replace ∞ by $-\infty$ (and "$> A$" by "$< A$" in the replacement rule).

Worked Example 2 Find

$$\lim_{x\to\infty} \left(\frac{1}{x} + \frac{3}{x^2} + 5 \right)$$

Solution   We have

(optional)

$$\lim_{x\to\infty} \left(\frac{1}{x} + \frac{3}{x^2} + 5 \right) = \lim_{x\to\infty} \frac{1}{x} + 3\left(\lim_{x\to\infty} \frac{1}{x} \right)^2 + \lim_{x\to\infty} 5 = 0 + 3 \cdot 0^2 + 5 = 5$$

Worked Example 3   Find

$$\lim_{x\to\infty} \frac{8x+2}{3x-1}$$

Solution   We cannot simply apply the quotient rule, since the limits of the numerator and denominator do not exist. Instead we use a trick: if $x \neq 0$, we can multiply numerator and denominator by $1/x$ to obtain

$$\frac{8x+2}{3x-1} = \frac{8 + (2/x)}{3 - (1/x)} \quad \text{for } x \neq 0$$

By the replacement rule (with $A = 0$), we have

$$\lim_{x\to\infty} \frac{8x+2}{3x-1} = \lim_{x\to\infty} \frac{8 + (2/x)}{3 - (1/x)} = \frac{8+0}{3-0} = \frac{8}{3}$$

(The values of $(8x + 2)/(3x - 1)$ for $x = 10^2, 10^4, 10^6, 10^8$ are $2.682\ldots, 2.66682\ldots,$ $2.6666682\ldots, 2.666666682\ldots$)

Solved Exercises ★1. Find $\lim\limits_{x\to\infty} (\sqrt{x^2+1} - x)$. [*Hint:* Multiply numerator and denominator by $\sqrt{x^2+1} + x$.] Interpret the result geometrically in terms of right triangles.

★2. Find the horizontal asymptotes of $f(x) = x/\sqrt{x^2+1}$. Sketch.

★3. Prove that, for $k < 0$, $\lim\limits_{x\to\infty} e^{kx} = 0$.

Exercises ★1. Find the following limits:

(a) $\lim\limits_{x\to\infty} [(10x^2 - 2)/(15x^2 - 3)]$ (b) $\lim\limits_{x\to\infty} [(-4x + 3)/(x + 2)]$

(c) $\lim\limits_{x\to\infty} \{[x + 2 + (1/x)] / [2x + 3 + (2/x)]\}$

(d) $\lim\limits_{x\to\infty} [\sqrt{x^2 + a^2} - x]$ (interpret your answer geometrically)

★2. Show that $\lim\limits_{x\to\infty} [1/\ln x] = 0$.

★3. Find the horizontal asymptotes of the graph of $\sqrt{x^2+1} - (x+1)$. Sketch.

THE LIMIT OF A FUNCTION AT A POINT

Having looked at limits of sequences and of functions at infinity, we now turn our attention to limits of the form $\lim\limits_{x\to x_0} f(x)$. In Section 1-2, we presented a definition of such limits which was adapted to our definition of continuity. We now give another definition which is more like the ones already given in this chapter. (A proof that the two definitions are equivalent will be given in the appendix to this chapter; see the Student Guide.)

Recall that the statement $\lim\limits_{x\to x_0} f(x) = l$ is taken to mean, roughly speaking, that $f(x)$ comes and remains arbitrarily close to l as x comes closer to x_0. Thus we start with a positive "tolerance" ε and try to make $|f(x) - l|$ less than ε by requiring x to be close to x_0. The closeness of x to x_0 is to be measured by another positive number—mathematical tradition dictates the use of the Greek letter δ (lowercase delta) for this number. Here, then, is the famous "ε-δ" definition of a limit—it was first stated in this form by Karl Weierstrass around 1850.

(optional)

> **Definition** Let f be a function defined at all points of an open interval containing x_0, except perhaps at x_0 itself, and let l be a real number. We say that *l is the limit of $f(x)$ as x approaches x_0* if, for every positive number ε, there is a positive number δ such that $|f(x) - l| < \varepsilon$ whenever $|x - x_0| < \delta$ and $x \neq x_0$. We write $\lim\limits_{x \to x_0} f(x) = l$.

We can illustrate this definition by figures like the ones we used for limits at infinity. (See Fig. 12-2-5.) This time we shade the region consisting of those (x, y) for which

1. $|x - x_0| > \delta$ (region I in Fig. 12-2-5(b))

2. $x = x_0$ (the vertical line II in Fig. 12-2-5(b))

3. $x \neq x_0$, $|x - x_0| < \delta$, and $|f(x) - l| < \varepsilon$ (region III in Fig. 12-2-5(b))

Fig. 12-2-5 When $\lim\limits_{x \to x_0} f(x) = \ell$ we can, for any $\epsilon > 0$, catch the graph of f in the shaded region by making δ small enough. The value of f at x_0 is irrelevant, since the line $x = x_0$ is always "shaded."

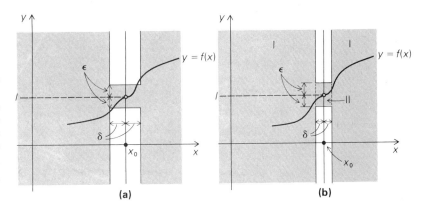

(a)   (b)

If $\lim\limits_{x \to x_0} f(x) = l$, then we can catch the graph of f in the shaded region by making δ small enough—that is, by making the unshaded strips sufficiently narrow.

Notice the statement $x \neq x_0$ in the definition. This means that the limit depends only upon the values of $f(x)$ for x *near* x_0, and not $f(x_0)$ itself. (In fact, $f(x_0)$ might not even be defined.) We will give one example of the use of the "ε-δ" definition.

Worked Example 4 Prove that $\lim\limits_{x \to 2} (x^2 + 3x) = 10$.

Solution Here $f(x) = x^2 + 3x$, $x_0 = 2$, and $l = 10$. Given $\varepsilon > 0$ we must find $\delta > 0$ such that $|f(x) - l| < \varepsilon$ if $|x - x_0| < \delta$.

A useful general rule is to write down $f(x) - l$ first and express it in terms of $x - x_0$ as much as possible, by writing $x = (x - x_0) + x_0$. In our case we replace x by $(x - 2) + 2$:

$$\begin{aligned}
f(x) - l &= x^2 + 3x - 10 \\
&= (x - 2 + 2)^2 + 3(x - 2 + 2) - 10 \\
&= (x - 2)^2 + 4(x - 2) + 4 + 3(x - 2) + 6 - 10 \\
&= (x - 2)^2 + 7(x - 2)
\end{aligned}$$

Now we use the properties of the absolute value to note that

$$|f(x) - l| \leq |x - 2|^2 + 7|x - 2|$$

(optional)   If this is to be less than ε, we should choose δ so that $\delta^2 + 7\delta \le \varepsilon$. We may require at the outset that $\delta \le 1$. Then $\delta^2 \le \delta$, so $\delta^2 + 7\delta \le 8\delta$. Hence we might try picking δ so that $\delta \le 1$ and $\delta \le \varepsilon/8$.

With this choice of δ, we must verify that $|f(x) - l| < \varepsilon$ whenever $|x - x_0| < \delta$. In our case $|x - x_0| < \delta$ means $|x - 2| < \delta$, so, for such an x,

$$
\begin{aligned}
|f(x) - l| &\le |x - 2|^2 + 7|x - 2| \\
&< \delta^2 + 7\delta \\
&\le \delta + 7\delta \\
&= 8\delta \\
&\le \varepsilon
\end{aligned}
$$

and so $|f(x) - l| < \varepsilon$.

In practice, one seldom uses the definition of limits. It is more efficient to use the properties of limits given by the following theorem, whose proof is in the appendix to this chapter.

Theorem 5   *All the limit rules in the display on p. 503 are true if we replace* $\lim\limits_{x \to \infty}$ *by* $\lim\limits_{x \to x_0}$, *with the following alterations. The "$1/x$ rule" becomes the "x rule":* $\lim\limits_{x \to x_0} x = x_0$; *in the replacement rule, the phrase "all $x > A$ for some A" becomes "all $x \ne x_0$ in some interval about x_0."*

At this point, we suggest that you review the first part of Section 1-2 to see how the limit rules are used and that you write out explicitly the limit rules described in Theorem 5. Here are a few more examples.

Worked Example 5   Using the fact that $\lim\limits_{\theta \to 0} (1 - \cos \theta)/\theta = 0$, find $\lim\limits_{\theta \to 0} \tan [(1 - \cos \theta)/\theta]$.

Solution   The continuous function rule says that $\lim\limits_{x \to x_0} g(f(x)) = g(\lim\limits_{x \to x_0} f(x))$ if g is continuous at $\lim\limits_{x \to x_0} f(x)$. We let $f(\theta) = (1 - \cos \theta)/\theta$, and $g(\theta) = \tan \theta$ so that $g(f(\theta)) = \tan [(1 - \cos \theta)/\cos \theta]$. Hence the required limit is

$$
\lim_{\theta \to 0} g(f(\theta)) = g\left(\lim_{\theta \to 0} \frac{1 - \cos \theta}{\theta} \right) = \tan 0 = 0
$$

since g is continuous at $\theta = 0$.

Worked Example 6   Find $\lim\limits_{x \to 2} [(x^2 - 5x + 6)/(x - 2)]$.

Solution   Since the denominator vanishes at $x = 2$, we cannot plug in this value. The numerator factors, however, and for any $x \ne 2$ our function is

$$
\frac{x^2 - 5x + 6}{x - 2} = \frac{(x - 3)(x - 2)}{x - 2} = x - 3
$$

Thus, by the replacement rule,

$$
\lim_{x \to 2} \frac{x^2 - 5x + 6}{x - 2} = \lim_{x \to 2} (x - 3) = 2 - 3 = -1
$$

Worked Example 7   Find $\lim\limits_{x\to 1}[(x-1)/(\sqrt{x}-1)]$.

Solution   Again we cannot plug in $x=1$. However, we can rationalize the denominator
(optional)   by multiplying numerator and denominator by $\sqrt{x}+1$. Thus (if $x\neq 1$):

$$\frac{x-1}{\sqrt{x}-1}=\frac{(x-1)(\sqrt{x}+1)}{(\sqrt{x}-1)(\sqrt{x}+1)}=\frac{(x-1)(\sqrt{x}+1)}{x-1}=\sqrt{x}+1$$

As x approaches 1, this approaches 2, so $\lim\limits_{x\to 1}(x-1)/(\sqrt{x}-1)=2$.

Solved Exercises   ★4.   Verify by the ε-δ definition that:

     (a) $\lim\limits_{x\to a}x^2=a^2$ (b) $\lim\limits_{x\to 3}(x^3+2x^2+2)=47$

   ★5.   (a)   Suppose that $\lim\limits_{x\to x_0}f(x)=0$ and that $|g(x)|\le|f(x)|$ for all $x\neq x_0$. Prove
            that $\lim\limits_{x\to x_0}g(x)=0$.

        (b)   Using part (a), show that $\lim\limits_{x\to 0}[x\sin(1/x)]=0$.

   ★6.   Evaluate $\lim\limits_{x\to 0}\{[(3+x)^2-9]/x\}$.

Exercises   ★4.   Verify by the ε-δ definition that $\lim\limits_{x\to 3}(x^2-2x+4)=7$.

   ★5.   Find the following limits:

     (a) $\lim\limits_{x\to 3}(x^2-2x+2)$ (b) $\lim\limits_{x\to -2}[(x^2-4)/(x^2+4)]$

     (c) $\lim\limits_{x\to 2}[(x^2-4)/(x^2-5x+6)]$ (d) $\lim\limits_{x\to 27}[(\sqrt[3]{x}-3)/(x-27)]$

   ★6.   Using the fact that $\lim\limits_{\theta\to 0}[(\sin\theta)/\theta]=1$, find $\lim\limits_{\theta\to 0}\cos[(\pi\sin\theta)/(4\theta)]$.

INFINITE LIMITS, ONE-SIDED LIMITS, AND VERTICAL ASYMPTOTES

Consider the limits $\lim\limits_{x\to 0}\sin(1/x)$ and $\lim\limits_{x\to 0}1/x^2$. Neither limit exists, but the functions $\sin(1/x)$ and $1/x^2$ behave quite differently as $x\to 0$. (See Fig. 12-2-6.) In the first case, for x near zero, say, $0<|x|<\delta$, the quantity $1/x$ ranges over all numbers with absolute value greater than $1/\delta$, and $\sin(1/x)$ oscillates back and forth infinitely often. It takes each value between -1 and 1 infinitely often but remains close to no particular number. In the case of $1/x^2$, the value of the function is again near no particular number, but there is a definite "trend" to be seen: as x comes nearer to zero, $1/x^2$ becomes a larger positive number; we may say that $\lim\limits_{x\to 0}1/x^2=\infty$. Here is a precise definition.

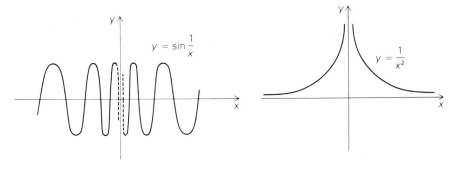

Fig. 12-2-6 $\lim\limits_{x\to 0}f(x)$ does not exist for either of these functions.

(optional)

Definition Let f be a function defined in an interval about x_0, except possibly at x_0 itself. We say that $f(x)$ *approaches* ∞ $[-\infty]$ as x *approaches* x_0 if, given any real number B, there is a positive number δ such that, for all x satisfying $|x - x_0| < \delta$ and $x \neq x_0$, we have $f(x) > B$ $[f(x) < B]$. We write $\lim_{x \to x_0} f(x) = \infty$ $[\lim_{x \to x_0} f(x) = -\infty]$.

Remarks

1. In the preceding definition, we usually think of δ as being small, while B is large positive if the limit is ∞ and large negative if the limit is $-\infty$.

2. If $\lim_{x \to x_0} f(x)$ is equal to $\pm\infty$, we still may say that "$\lim_{x \to x_0} f(x)$ does not exist," since it is not equal to any *number*.

3. One can define the statements $\lim_{x \to \infty} f(x) = -\infty$, $\lim_{n \to \infty} a_n = \infty$, and so forth in an analogous way. We leave the details to you.

The following theorem provides a useful technique for detecting "infinite limits."

Theorem 6 Let f be defined in an open interval about x_0, except possibly at x_0 itself. Then $\lim_{x \to x_0} f(x) = \infty$ $[-\infty]$ if and only if:

1. *For all* $x \neq x_0$ *in some interval about* x_0, $f(x)$ *is positive [negative]*.

2. $\lim_{x \to x_0} 1/f(x) = 0$.

The proof of Theorem 6 is given in the appendix to this chapter, but the basic idea is very simple: $f(x)$ is very large if and only if $1/f(x)$ is very small. Of course, a similar result is true for limits of the form $\lim_{x \to \infty} f(x)$ and $\lim_{n \to \infty} a_n$. (See Worked Example 8(b).)

Worked Example 8 Find the following limits:

(a) $\lim_{x \to 1} [1/(x - 1)^2]$ (b) $\lim_{x \to \infty} e^x$ (c) $\lim_{n \to \infty} [(1 - n^2)/n^{3/2}]$

Solution (a) We note that $1/(x - 1)^2$ is positive for all $x \neq 1$. We look at the reciprocal: $\lim_{x \to 1} (x - 1)^2 = 0$; by Theorem 6, $\lim_{x \to 1} 1/(x - 1)^2 = \infty$.

(b) We note that e^x is always positive. We have $\lim_{x \to \infty} 1/e^x = \lim_{x \to \infty} e^{-x} = 0$, by Solved Exercise 3, so $\lim_{x \to \infty} e^x = \infty$, by Theorem 6.

(c) For $n > 1$, $(1 - n^2)/n^{3/2}$ is negative. Now we have

$$\lim_{n \to \infty} \frac{n^{3/2}}{1 - n^2} = \lim_{n \to \infty} \frac{1}{n^{-3/2} - n^{1/2}} = \lim_{n \to \infty} \frac{1}{n^{1/2}} \frac{1}{1/n^2 - 1} = \lim_{n \to \infty} \frac{1}{n^{1/2}} \lim_{n \to \infty} \frac{1}{1/n^2 - 1}$$

$$= 0(-1) = 0$$

so $\lim_{n \to \infty} (1 - n^2)/n^{3/2} = -\infty$, by Theorem 6.

If we look at the function $f(x) = 1/(x - 1)$ near $x_0 = 1$ we find that $\lim_{x \to 1} 1/f(x) = 0$, but $f(x)$ has different signs on opposite sides of 1, so $\lim_{x \to 1} 1/(x - 1)$ is neither ∞ nor $-\infty$. This example suggests the introduction of the notion of a "one-sided limit." Here is the definition.

(optional)

> **Definition** Let f be defined for all x in an interval of the form (x_0, b) $[(a, x_0)]$. We say that $f(x)$ *approaches l as x approaches x_0 from the right* [left] if, for any positive number ε, there is a positive number δ such that, for all x such that $x_0 < x < x_0 + \delta$ $[x_0 - \delta < x < x_0]$, we have $|f(x) - l| < \varepsilon$. We write $\lim_{x \to x_0+} f(x) = l$ $[\lim_{x \to x_0-} f(x) = l]$.

In the definition of a one-sided limit, only the values of $f(x)$ for x on one side of x_0 are taken into account. Precise definitions of statements like $\lim_{x \to x_0+} f(x) = \infty$ are left to you. We remark that Theorem 6 extends to one-sided limits.

Worked Example 9 Find $\lim_{x \to 1+} [1/(1 - x)]$ and $\lim_{x \to 1-} [1/(1 - x)]$.

Solution For $x > 1$, we find $1/(1 - x)$ is negative, and we have $\lim_{x \to 1} (1 - x) = 0$, so $\lim_{x \to 1+} [1/(1 - x)] = -\infty$. Similarly, $\lim_{x \to 1-} [1/(1 - x)] = +\infty$.

If a one-sided limit of $f(x)$ at x_0 is equal to ∞ or $-\infty$, then the graph of f lies closer and closer to the line $x = x_0$; we call this line a *vertical asymptote* of the graph.

Worked Example 10 Find the vertical asymptotes of the graph of $f(x) = 1/[(x - 1)(x - 2)^2]$. Sketch.

Solution Vertical asymptotes occur where $\lim_{x \to x_0} [1/f(x)] = 0$; in this case, they occur at $x_0 = 1$ and $x_0 = 2$. We observe that $f(x)$ is negative on $(-\infty, 1)$, positive on $(1, 2)$, and positive on $(2, \infty)$. Thus we have $\lim_{x \to 1-} f(x) = -\infty$, $\lim_{x \to 1+} f(x) = \infty$, $\lim_{x \to 2-} f(x) = \infty$, and $\lim_{x \to 2+} f(x) = \infty$. The graph of f is sketched in Fig. 12-2-7.

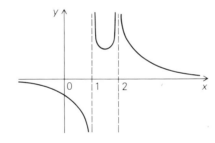

Fig. 12-2-7 The graph

$$y = \frac{1}{(x - 1)(x - 2)^2}$$

has the lines $x = 1$ and $x = 2$ as vertical asymptotes.

Solved Exercises ★7. Find the vertical asymptotes of $f(x) = 1/(x^2 - 5x + 6)$. Sketch.

★8. Let $f(x) = |x|$.
  (a) Find $f'(x)$ and sketch its graph.
  (b) Find $\lim_{x \to 0-} f'(x)$ and $\lim_{x \to 0+} f'(x)$.
  (c) Does $\lim_{x \to 0} f'(x)$ exist?

Exercises ★7. Find the horizontal and vertical asymptotes of the following functions. Sketch.
  (a) $f(x) = 1/(2x + 3)$ (b) $f(x) = 1/(x^2 - 1)$

★8. Show that $\tan x$ and $\sec x$ have the same vertical asymptotes. What are they? What about $\cot x$ and $\csc x$? Do these functions have horizontal asymptotes?

★9. Find the following limits:

  (a) $\lim_{x \to 3+} \dfrac{e^x}{x - 1}$ (b) $\lim_{x \to 1-} \dfrac{\ln 2x}{x - 1}$ (c) $\lim_{x \to 0-} \dfrac{(x - 1)(x - 2)}{x(x + 1)(x + 2)}$

Problems for Section 12-2 ▰▰▰▰▰▰▰▰▰▰▰▰▰▰▰▰▰▰

1. Find the following limits, if they exist. (Include limits of the form ∞ and $-\infty$.)

   (a) $\lim\limits_{x \to 1} \dfrac{3 + 4x}{4 + 5x}$

   (b) $\lim\limits_{x \to 1} \dfrac{x^3 - 1}{x - 1}$

   (c) $\lim\limits_{x \to 1} \dfrac{x^3 - 1}{x^2 - 1}$

   (d) $\lim\limits_{x \to 2} \dfrac{x - 2}{x^2 + 3x + 2}$

   (e) $\lim\limits_{x \to 2} \dfrac{x - 2}{x^2 - 3x + 2}$

2. Find the following limits:

   (a) $\lim\limits_{x \to -3} \dfrac{x^2 + 2x - 3}{x^2 + x - 6}$

   (b) $\lim\limits_{x \to 1} \dfrac{x^n - 1}{x - 1}$

   (c) $\lim\limits_{x \to -1} \dfrac{x^{2n+1} + 1}{x + 1}$

   (d) $\lim\limits_{x \to 2} \dfrac{x - 2}{\sqrt{x} - \sqrt{2}}$

   (e) $\lim\limits_{x \to 2+} \dfrac{x^2 - 4}{(x - 2)^2}$

3. Find:

   (a) $\lim\limits_{x \to 2-} \dfrac{x^2 - 4}{(x - 2)^2}$

   (b) $\lim\limits_{x \to \infty} \dfrac{3x^2 + 2x + 4}{5x^2 + x + 7}$

   (c) $\lim\limits_{x \to -\infty} \ln x^2$

   (d) $\lim\limits_{n \to \infty} \dfrac{(-1)^n n^2}{n + 1}$

   (e) $\lim\limits_{n \to \infty} \dfrac{n^2}{2n + 5}$

4. Find:

   (a) $\lim\limits_{x \to 2} \dfrac{x^2 + 3x + 6}{9x - 1}$

   (b) $\lim\limits_{x \to \infty} \sin(1/x)$

   (c) $\lim\limits_{x \to \infty} \dfrac{x^2 + xe^{-x}}{6x^2 + 2}$

   (d) $\lim\limits_{x \to 1+} \dfrac{e^x - 1}{x - 1}$

   (e) $\lim\limits_{x \to \infty} \sin\left(\dfrac{\pi x^2 + 4}{6x^2 + 9}\right)$

5. Find the horizontal and vertical asymptotes of:

   (a) $y = \dfrac{x}{x^2 - 1}$

   (b) $y = \dfrac{(x + 1)(x - 1)}{(x - 2)x(x + 2)}$

   (c) $\dfrac{e^x + 2x}{e^x - 2x}$

6. Sketch the graph of $y = x^2/(x^2 - 1)$.

★7. Show that if $f(x)$ and $g(x)$ are polynomials such that $\lim\limits_{x \to \infty} f(x)/g(x) = l$, then $\lim\limits_{x \to -\infty} f(x)/g(x)$ is equal to l as well. What happens if $l = \infty$ or $-\infty$?

★8. How close to 3 does x have to be to ensure that $|x^3 - 2x - 21| < \frac{1}{1000}$?

★9. Prove that $\lim\limits_{x \to 3} (x^3 + 2x) = 33$ by using the ε-δ definition.

★10. Prove that $\lim\limits_{x \to \infty} 3x/(x^2 + 2) = 0$ directly by using the definition of $\lim\limits_{x \to \infty} f(x)$.

★11. (a) Give a precise definition of this statement: $\lim\limits_{x \to \infty} f(x) = -\infty$.

   (b) Draw figures like Figs. 12-2-2, 12-2-3, and 12-2-5 to illustrate your definition.

★12. Draw figures like Figs. 12-2-2, 12-2-3, and 12-2-5 to illustrate the definition of these statements:

   (a) $\lim\limits_{x \to x_0 +} f(x) = l$

   (b) $\lim\limits_{x \to x_0 +} f(x) = \infty$

   [*Hint:* The shaded region should include all points with $x \le x_0$.]

13. (a) Graph $y = f(x)$, where
$$f(x) = \begin{cases} |x|/x & x \neq 0 \\ 0 & x = 0 \end{cases}$$
   Does $\lim\limits_{x \to 0} f(x)$ exist?

   (b) Graph $y = g(x)$, where
$$g(x) = \begin{cases} x + 1 & x < 0 \\ 2x - 1 & x \ge 0 \end{cases}$$
   Does $\lim\limits_{x \to 0} g(x)$ exist?

   (c) Let $f(x)$ be as in part (a) and $g(x)$ as in part (b). Graph $y = f(x) + g(x)$. Does $\lim\limits_{x \to 0} [f(x) + g(x)]$ exist? Conclude that the limit of a sum can exist even though the limits of the summands do not.

14. The number $N(t)$ of individuals in a population at time t is given by
$$N(t) = N_0 \dfrac{e^{3t}}{(3/2) + e^{3t}}$$
   Find the value of $\lim\limits_{t \to \infty} N(t)$ and discuss its biological meaning.

15. The current in a certain RLC circuit is given by $I(t) = [(1/3) \sin t + \cos t] e^{-t/2} + 4$ amperes. The value of $\lim\limits_{t \to \infty} I(t)$ is called the *steady-state current*; it represents the current present after a long period of time. Find it.

16. The temperature $T(x, t)$ at time t at position x of a rod located along $0 \leq x \leq l$ on the x-axis is given approximately by $T(x, t) = B_1 e^{-\mu_1 t} \cdot \sin \lambda_1 x + B_2 e^{-\mu_2 t} \sin \lambda_2 x + B_3 e^{-\mu_3 t} \sin \lambda_3 x$ where $\mu_1, \mu_2, \mu_3, \lambda_1, \lambda_2, \lambda_3$ are all positive. Show that $\lim_{t \to \infty} T(x, t) = 0$ for each fixed location x along the rod. The model applies to a rod without heat sources, with the heat allowed to radiate from the right end of the rod; zero limit means all heat eventually radiates out the right end.

17. A psychologist doing some manipulations with testing theory wishes to replace the reliability factor

$$R = \frac{nr}{1 + (n - 1)r} \quad (Spearman\text{-}Brown\ formula)$$

by unity, because someone told her that she could do this for large extension factors n. She formally replaces n by $1/x$, then sets $x = 0$, to obtain 1. What has she done, in the language of limits?

12-3 L'Hôpital's Rule and Improper Integrals

Differentiation can be used to evaluate limits, and limits to evaluate integrals.

In this section we turn to the practical side of limits, treating two topics in which the theory of limits is connected to the computation of derivatives and integrals. If you have not read Section 12-2, you should review the informal discussion of infinite limits and limits at infinity on pp. 136 and 137, since the material of this section depends upon those notions.

Goals

After studying this section, you should be able to:

Use L'Hôpital's rule to evaluate indeterminate forms of the type $\frac{0}{0}$ or $\frac{\infty}{\infty}$.

Evaluate integrals of unbounded functions and integrals over infinite intervals.

L'HÔPITAL'S RULE

L'Hôpital's rule* deals with limits of the form $\lim_{x \to x_0} [f(x)/g(x)]$, where $\lim_{x \to x_0} f(x)$ and $\lim_{x \to x_0} g(x)$ are both equal to zero or infinity, so that the quotient rule cannot be applied. (One can also replace x_0 by ∞, x_0+, or x_0-.) Such limits are called *indeterminate forms*.

Our first object is to calculate $\lim_{x \to x_0} [f(x)/g(x)]$ if $f(x_0) = 0$ and $g(x_0) = 0$. Substituting $x = x_0$ gives us $\frac{0}{0}$, so we say that we are dealing with an *indeterminate form of type* $\frac{0}{0}$. Such forms occurred when we considered the derivative as a limit of difference quotients; in Section 1-2 we used the limit rules to evaluate some simple derivatives. Now we can work the other way around, using our ability to calculate derivatives

---

In 1696, Guillaume F. A. L'Hôpital published in Paris the first calculus textbook: Analyse des Infiniment Petits (Analysis of the infinitely small). Included was a proof of what is now referred to as L'Hôpital's rule; the idea, however, probably came from J. Bernoulli. This rule was the subject of some work by A. Cauchy, who clarified its proof in his Cours d'Analyse (Course in analysis) in 1823. The foundations were in debate until almost 1900. See, for instance, the very readable article, "The Law of the Mean and the Limits $\frac{0}{0}$, $\frac{\infty}{\infty}$," by W. F. Osgood, Annals of Mathematics, Volume 12 (1898–1899), pp. 65–78.

in order to evaluate quite complicated limits: L'Hôpital's rule provides the means for doing this.

The following theorem is the simplest version of L'Hôpital's rule.

Theorem 7 *Let f and g be differentiable in an open interval containing x_0; assume that $f(x_0) = g(x_0) = 0$. If $g'(x_0) \neq 0$, then*

$$\lim_{x \to x_0} \frac{f(x)}{g(x)} = \frac{f'(x_0)}{g'(x_0)}$$

Proof Since $f(x_0) = 0 = g(x_0)$, we can write

$$\frac{f(x)}{g(x)} = \frac{f(x) - f(x_0)}{g(x) - g(x_0)} = \frac{[f(x) - f(x_0)]/(x - x_0)}{[g(x) - g(x_0)]/(x - x_0)}$$

As x tends to x_0, the numerator tends to $f'(x_0)$ and the denominator to $g'(x_0) \neq 0$, so the result follows from the quotient rule for limits.

Worked Example 1 Find $\lim\limits_{x \to 0} [(\cos x - 1)/\sin x]$.

Solution We apply L'Hôpital's rule with $f(x) = \cos x - 1$ and $g(x) = \sin x$. We have $f(0) = 0$, $g(0) = 0$, and $g'(0) = 1 \neq 0$, so

$$\lim_{x \to 0} \frac{\cos x - 1}{\sin x} = \frac{f'(0)}{g'(0)} = \frac{-\sin(0)}{\cos(0)} = 0$$

This procedure does not cover all $\frac{0}{0}$ limits. For example, suppose we wish to find

$$\lim_{x \to 0} \frac{\sin x - x}{x^3}$$

If we differentiate the numerator and denominator, we will then get $(\cos x - 1)/3x^2$, which becomes $\frac{0}{0}$ when we set $x = 0$. This suggests that we use L'Hôpital's rule again, but to do so we need to know that $\lim\limits_{x \to x_0} [f(x)/g(x)]$ is equal to $\lim\limits_{x \to x_0} [f'(x)/g'(x)]$, even when $f'(x_0)/g'(x_0)$ is again indeterminate. The following strengthened version of Theorem 7 is the result we need. The proof is given in the appendix to this chapter.

Theorem 7' *Let f and g be differentiable on an open interval containing x_0, except perhaps at x_0 itself. Assume that $g(x) \neq 0$ and $g'(x) \neq 0$ for x in an interval about x_0, $x \neq x_0$, that f and g are continuous at x_0 with $f(x_0) = g(x_0) = 0$, and that $\lim\limits_{x \to x_0} f'(x)/g'(x) = l$. Then*

$$\lim_{x \to x_0} \frac{f(x)}{g(x)} = l$$

Worked Example 2   Calculate $\lim_{x \to 0} [(\sin x - x)/x^3]$.

Solution   By Theorem 7′, $\lim_{x \to 0} [(\sin x - x)/x^3] = \lim_{x \to 0} [(\cos x - 1)/3x^2]$ if the latter limit can be shown to exist. However, we can write, again by Theorem 7′,

$$\lim_{x \to 0} \frac{\cos x - 1}{3x^2} = \lim_{x \to 0} \frac{-\sin x}{6x}$$

Applying the procedure once more, we find

$$\lim_{x \to 0} \frac{-\sin x}{6x} = \lim_{x \to 0} \frac{-\cos x}{6} = -\frac{1}{6}$$

Thus

$$\lim_{x \to 0} \frac{\sin x - x}{x^3} = -\frac{1}{6}$$

Similar statements hold for one-sided limits, limits as $x \to \infty$, or if we have indeterminates of the form $\frac{\infty}{\infty}$. To investigate the form $\frac{\infty}{\infty}$, it is tempting to write $f(x)/g(x) = [1/g(x)]/[1/f(x)]$, which is in the $\frac{0}{0}$ form; however, this gives

$$\lim_{x \to x_0} \frac{f(x)}{g(x)} = \lim_{x \to x_0} \frac{-g'(x)/[g(x)]^2}{-f'(x)/[f(x)]^2}$$

which is no easier to handle.

A similar trick does work for the $\frac{0}{0}$ form in case $x \to +\infty$; we set $t = 1/x$, so that $x = 1/t$ and $t \to 0+$ as $x \to +\infty$. Then

$$\lim_{x \to +\infty} \frac{f'(x)}{g'(x)} = \lim_{t \to 0+} \frac{f'(1/t)}{g'(1/t)}$$

$$= \lim_{t \to 0+} \frac{-t^2 f'(1/t)}{-t^2 g'(1/t)}$$

$$= \lim_{t \to 0+} \frac{(d/dt) f(1/t)}{(d/dt) g(1/t)} \qquad \text{(by the chain rule)}$$

$$= \lim_{t \to 0+} \frac{f(1/t)}{g(1/t)} \qquad \text{(by Theorem 7′)}$$

$$= \lim_{x \to +\infty} \frac{f(x)}{g(x)}$$

The proof of L'Hôpital's rule for the $\frac{\infty}{\infty}$ form (either as $x \to x_0$ or $x \to \infty$) is a little harder than Theorem 7′ although the idea is the same; see the appendix to this chapter for the proof.

Worked Example 3 Find $\lim\limits_{x \to \infty} (\ln x)/x^p$, where $p > 0$.

Solution This is of the form $\frac{\infty}{\infty}$. Differentiating the numerator and denominator, we find

$$\lim_{x \to \infty} \frac{\ln x}{x^p} = \lim_{x \to \infty} \frac{1/x}{px^{p-1}} = \lim_{x \to \infty} \frac{1}{px^p} = 0$$

since $p > 0$. Thus, although $\ln x$ goes to infinity as $x \to \infty$, it does so more slowly than any positive power of x.

Certain expressions which do not appear to be in the form $f(x)/g(x)$ can be put in that form with some manipulation.

Worked Example 4 Find $\lim\limits_{x \to 0+} x \ln x$.

Solution This might be said to be in the form $0 \cdot (-\infty)$, since $\lim\limits_{x \to 0+} x = 0$ while $\lim\limits_{x \to 0+} \ln x = -\infty$. We can write $x \ln x$ as $(\ln x)/(1/x)$, which is now in form $\frac{\infty}{\infty}$. Thus

$$\lim_{x \to 0+} x \ln x = \lim_{x \to 0+} \frac{\ln x}{1/x} = \lim_{x \to 0+} \frac{1/x}{-1/x^2} = \lim_{x \to 0+} (-x) = 0$$

Worked Example 5 Find $\lim\limits_{x \to 0+} x^x$.

Solution We could call this an indeterminate form of the type 0^0. (It is "indeterminate" because zero to any power is zero, while any number to the zeroth power is 1.) To obtain a form to which L'Hôpital's rule is applicable, we write x^x as $\exp(x \ln x)$. By Worked Example 4, we have $\lim\limits_{x \to 0+} x \ln x = 0$. Since $g(x) = \exp(x)$ is continuous, the continuous function rule applies, giving $\lim\limits_{x \to 0+} \exp(x \ln x) = \exp(\lim\limits_{x \to 0+} x \ln x) = e^0 = 1$, so $\lim\limits_{x \to 0+} x^x = 1$. Numerically, $0.01^{0.1} = 0.79$, $0.001^{0.001} = 0.993$, and $0.00001^{0.00001} = 0.99988$.

The use of L'Hôpital's rule is summarized in the following display.

L'HÔPITAL'S RULE

To find $\lim\limits_{x \to x_0} [f(x)/g(x)]$, where $\lim\limits_{x \to x_0} f(x)$ and $\lim\limits_{x \to x_0} g(x)$ are both zero or both infinity, differentiate the numerator and denominator and take the limit of the new fraction; repeat the process as many times as necessary.

If $\lim\limits_{x \to x_0} f(x) = \lim\limits_{x \to x_0} g(x) = 0$ (or both are $\pm\infty$), then

$$\lim_{x \to x_0} \frac{f(x)}{g(x)} = \lim_{x \to x_0} \frac{f'(x)}{g'(x)}$$

(x_0 may be replaced by $\pm\infty$ or $x_0\pm$).

Solved Exercises

1. Find $\lim\limits_{x \to 0} \; [(\sin x - x)/(\tan x - x)]$.

2. Is it true that $\lim\limits_{x \to x_0} \; [f(x)/g(x)] = f'(x_0)/g'(x_0)$ if we just assume that $g(x_0) = 0$ and $g'(x_0) \neq 0$ without assuming $f(x_0) = 0$? Prove or give a counterexample.

3. Find $\lim\limits_{x \to 1} \; x^{1/(x-1)}$.

Exercises

1. Evaluate the following limits, using L'Hôpital's rule where it is applicable.

(a) $\lim\limits_{x \to 0} \dfrac{\sqrt{1 + x^2} - 1}{\sin 2x}$

(b) $\lim\limits_{x \to 1} \dfrac{\sqrt{x^2 - 1}}{\cos (\pi x/2)}$

(c) $\lim\limits_{x \to 0} \dfrac{1 - \cos x}{x^2}$

(d) $\lim\limits_{x \to 1} \dfrac{1 - x^2}{1 + x^2}$

2. Find $\lim\limits_{x \to 0} \; x^p \ln x$, where p is positive.

3. Show that as $x \to \infty$, $x^n/e^x \to 0$ for any integer n; that is, e^x goes to infinity faster than any power of x.

INTEGRALS OVER UNBOUNDED INTERVALS

The definite integral $\int_a^b f(x) \, dx$ of a function f which is nonnegative on the interval $[a, b]$ equals the area of the region under the graph of f between a and b. If we let b go to infinity, the region becomes unbounded, as in Fig. 12-3-1. One's first inclination upon seeing such unbounded regions is to assert that their areas are "infinite"; however, examples suggest otherwise.

Fig. 12-3-1 The region under the graph of f on $[a, \infty)$ is unbounded.

Worked Example 6   Find $\int_1^b (1/x^4) \, dx$. What happens as b goes to infinity?

Solution   We have

$$\int_1^b \frac{dx}{x^4} = \int_1^b x^{-4} \, dx = \left. \frac{x^{-3}}{-3} \right|_1^b = \frac{1/b^3 - 1}{-3} = \frac{1 - 1/b^3}{3}$$

As b becomes larger and larger, this integral always remains less than $\frac{1}{3}$; furthermore, we have

$$\lim_{b \to \infty} \int_1^b \frac{dx}{x^4} = \lim_{b \to \infty} \frac{1 - 1/b^3}{3} = \frac{1}{3}$$

Worked Example 6 suggests that $\frac{1}{3}$ is the area of the unbounded region consisting of those points (x, y) such that $1 \leq x$ and $0 \leq y \leq$

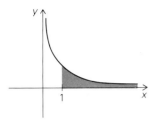

Fig. 12-3-2 The region under the graph of $1/x^4$ on $[1, \infty)$ has finite area. It is $\frac{1}{3} = \int_1^\infty (dx/x^4)$.

$1/x^4$. (See Fig. 12-3-2.) In accordance with our notation for finite intervals, we denote this area by $\int_1^\infty dx/x^4$. Guided by this example, we *define* integrals over unbounded intervals as limits of integrals over finite intervals.

Definition Suppose that, for a fixed, f is integrable on $[a, b]$ for all $b > a$. If the limit $\lim\limits_{b \to \infty} \int_a^b f(x)\, dx$ exists, we say that the *improper integral* $\int_a^\infty f(x)\, dx$ *is convergent*, and we define $\int_a^\infty f(x)\, dx = \lim\limits_{b \to \infty} \int_a^b f(x)\, dx$.

Similarly, if, for fixed b, f is integrable on $[a, b]$ for all $a < b$, we define $\int_{-\infty}^b f(x)\, dx = \lim\limits_{a \to -\infty} \int_a^b f(x)\, dx$, if the limit exists.

Finally, if f is integrable on $[a, b]$ for all $a < b$, we define $\int_{-\infty}^\infty f(x)\, dx = \int_{-\infty}^0 f(x)\, dx + \int_0^\infty f(x)\, dx$, if the improper integrals on the right-hand side are both convergent.

If an improper integral is not convergent, it is called *divergent*.

Worked Example 7 For which values of the exponent r is $\int_1^\infty x^r\, dx$ convergent?

Solution We have

$$\lim_{b \to \infty} \int_1^b x^r\, dx = \lim_{b \to \infty} \left. \frac{x^{r+1}}{r+1} \right|_1^b = \lim_{b \to \infty} \frac{b^{r+1} - 1}{r+1}$$

If $r + 1 > 0$ (that is, $r > -1$), the limit $\lim\limits_{b \to \infty} b^{r+1}$ does not exist and the integral is divergent. If $r + 1 < 0$ (that is, $r < -1$), we have $\lim\limits_{b \to \infty} b^{r+1} = 0$ and the integral is convergent—its value is $-1/(r + 1)$. Finally, if $r = -1$ we have $\int_1^b x^{-1}\, dx = \ln b$, which does not converge as $b \to \infty$. We conclude that $\int_1^\infty x^r\, dx$ is convergent just for $r < -1$.

The following test for the convergence of improper integrals is useful because it does not require us to evaluate the finite integrals explicitly.

Theorem 8 Comparison Test: *Suppose f and g are functions such that*

    (i) $|f(x)| \leq g(x)$ *for all $x \geq a$*

and   (ii) $\int_b^a f(x)dx$ *and* $\int_a^b g(x)dx$ *exist for every $b > a$*

Then   (1) *If $\int_a^\infty g(x)$ is convergent, so is $\int_a^\infty f(x)\, dx$*

and   (2) *If $\int_a^\infty f(x)\, dx$ is divergent, so is $\int_a^\infty g(x)\, dx$*

A similar result is true for integrals of the type $\int_{-\infty}^a$ and $\int_{-\infty}^\infty$. The proof is given in the appendix to this chapter, but we explain the basic idea here. If $f(x)$ and $g(x)$ are both positive functions (Fig. 12-3-3(a)), then the region under the graph of f is contained in the region under the graph of g, so the integral $\int_a^b f(x)\, dx$ increases and remains bounded as $b \to \infty$. We expect, therefore, that they should converge to some limit. In the general case (Fig. 12-3-3(b)), the sums of the plus areas and the minus areas are both bounded by $\int_a^\infty g(x)\, dx$, and the cancellations can only help the integral to converge.

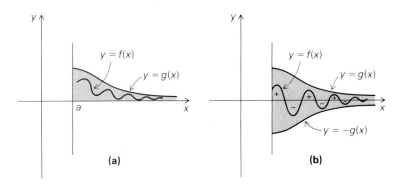

Fig. 12-3-3 Illustrating the comparison test.

Worked Example 8   Show that $\int_0^\infty dx/\sqrt{1 + x^8}$ is convergent by comparison with $1/x^4$.

Solution   We have $1/\sqrt{1 + x^8} < 1/\sqrt{x^8} = 1/x^4$, so it is tempting to compare with $\int_0^\infty dx/x^4$. Unfortunately, the latter integral is not defined because $1/x^4$ is unbounded near zero. However, we can break the integral into two parts:

$$\int_0^\infty \frac{dx}{\sqrt{1 + x^8}} = \int_0^1 \frac{dx}{\sqrt{1 + x^8}} + \int_1^\infty \frac{dx}{\sqrt{1 + x^8}}$$

The first integral on the right-hand side exists because $1/\sqrt{1 + x^8}$ is continuous on $[0, 1]$. The second integral is convergent by Theorem 8, taking $g(x) = 1/x^4$ and $f(x) = 1/\sqrt{1 + x^8}$. Thus $\int_0^\infty dx/\sqrt{1 + x^8}$ is convergent.

Solved Exercises   4. Show that $\int_0^\infty [\sin x/(1 + x)^2]\ dx$ converges (without attempting to evaluate the integral).

5. Find
$$\int_{-\infty}^\infty \frac{dx}{1 + x^2}$$

6. Show that
$$\int_1^\infty \frac{dx}{\sqrt{1 + x^2}}$$
is divergent.

Exercises   4. Which of the following integrals are convergent?

(a) $\displaystyle\int_1^\infty \frac{1}{x^2}\left(1 - \frac{1}{x}\right) dx$

(b) $\displaystyle\int_1^\infty \frac{1}{(4x - 3)^{1/3}}\ dx$

(c) $\displaystyle\int_0^\infty \left[\cos x + \frac{1}{(x + 1)^2}\right] dx$

(d) $\displaystyle\int_{-\infty}^{-2} \left(\frac{1}{x^{6/5}} - \frac{1}{x^{4/3}}\right) dx$

5. Evaluate the following improper integrals:

(a) $\displaystyle\int_0^\infty e^{-x}\ dx$

(b) $\displaystyle\int_2^\infty \frac{dx}{x^2 - 1}$

(c) $\displaystyle\int_0^\infty \frac{du}{u^{1/2} + u^{3/2}}$

[*Hint:* Substitute $x = u^{1/2}$.]

INTEGRALS OF UNBOUNDED FUNCTIONS

If the graph of f has a vertical asymptote at one endpoint of the interval $[a, b]$, then the integral $\int_a^b f(x)\, dx$ is not defined in the usual sense, since the function f is not bounded on the interval $[a, b]$ (see Solved Exercise 11 on p. 187). As with integrals of the form $\int_a^\infty f(x)\, dx$, we are dealing with areas of unbounded regions in the plane—this time the unboundedness is in the vertical direction rather than the horizontal. Following our earlier procedure, we can define the integrals of unbounded functions as limits, which are again called improper integrals.

Definition Suppose that the graph of f has $x_0 = b$ as a vertical asymptote and that, for a fixed, f is integrable on $[a, q]$ for all q in $[a, b)$. If the limit $\lim\limits_{q \to b-} \int_a^q f(x)\, dx$ exists, we can say that the *improper integral* $\int_a^b f(x)\, dx$ *is convergent* and we define $\int_a^b f(x)\, dx$ to be $\lim\limits_{q \to b-} \int_a^q f(x)\, dx$.

Similarly, if $x = a$ is a vertical asymptote, we define $\int_a^b f(x)\, dx$ to be $\lim\limits_{p \to a+} \int_p^b f(x)\, dx$, if the limit exists. (See Fig. 12-3-4.)

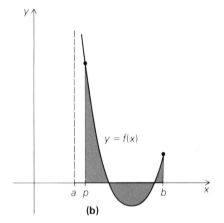

Fig. 12-3-4 Improper integrals defined by (a) the limit $\lim\limits_{q \to b-} \int_a^q f(x)\, dx$ and (b) the limit $\lim\limits_{p \to a+} \int_p^b f(x)\, dx$.

If both $x = a$ and $x = b$ are vertical asymptotes, or if there are vertical asymptotes in the interior (a, b), we may break up $[a, b]$ into subintervals such that the integral of f on each subinterval is of the type considered in the preceding definition. If each part is convergent, we may add the results to get $\int_a^b f(x)\, dx$. The comparison test (Theorem 8) is valid for all these improper integrals.

Worked Example 9 For which values of r is $\int_0^1 x^r\, dx$ convergent?

Solution If $r \geq 0$, x^r is continuous on $[0, 1]$ and the integral exists in the ordinary sense. If $r < 0$, we have $\lim\limits_{x \to 0+} x^r = \infty$, so we must take a limit. We have

$$\lim_{p \to 0+} \int_p^1 x^r\, dx = \lim_{p \to 0+} \left. \frac{x^{r+1}}{r+1} \right|_p^1 = \frac{1}{r+1}\left(1 - \lim_{p \to 0+} p^{r+1}\right)$$

If $r + 1 > 0$ (that is, $r > -1$), we have $\lim\limits_{p \to 0+} p^{r+1} = 0$, so the integral is convergent and equals $1/(r + 1)$. If $r + 1 < 0$ (that is, $r < -1$), $\lim\limits_{p \to 0+} p^{r+1} = \infty$, so the integral is divergent. Finally, if $r + 1 = 0$, we have $\lim\limits_{p \to 0+} \int_p^1 x^r \, dx = \lim\limits_{p \to 0+} (0 - \ln p) = \infty$. Thus the integral $\int_0^1 x^r \, dx$ converges just for $r > -1$. (Compare with Worked Example 7.)

Improper integrals arise in connection with arc length problems for graphs with vertical tangents.

Worked Example 10   Find the length of the curve $y = \sqrt{1 - x^2}$ for x in $[-1, 1]$. Interpret your result geometrically.

Solution   By formula (1), Section 11-2, the arc length is

$$\int_{-1}^{1} \sqrt{1 + (dy/dx)^2} \, dx = \int_{-1}^{1} \sqrt{1 + (-x/\sqrt{1 - x^2})^2} \, dx$$
$$= \int_{-1}^{1} \frac{dx}{\sqrt{1 - x^2}}$$

The integral is improper at both ends, since

$$\lim_{x \to -1+} \frac{1}{\sqrt{1 - x^2}} = \lim_{x \to 1-} \frac{1}{\sqrt{1 - x^2}} = \infty$$

We break it up as

$$\int_{-1}^{0} \frac{dx}{\sqrt{1 - x^2}} + \int_{0}^{1} \frac{dx}{\sqrt{1 - x^2}}$$
$$= \lim_{p \to -1+} \int_{p}^{0} \frac{dx}{\sqrt{1 - x^2}} + \lim_{q \to 1-} \int_{0}^{q} \frac{dx}{\sqrt{1 - x^2}}$$
$$= \lim_{p \to -1+} (\sin^{-1} 0 - \sin^{-1} p) + \lim_{q \to 1-} (\sin^{-1} q - \sin^{-1} 0)$$
$$= 0 - \left(-\frac{\pi}{2}\right) + \frac{\pi}{2} - 0 = \pi$$

(see Fig. 6-2-5, p. 289.)

Geometrically, the curve whose arc length we just found is a semicircle of radius 1, so we recover the fact that the circumference of a circle of radius 1 is 2π.

Worked Example 11   Luke Skyrunner has just been knocked out in his spaceship by his archenemy, Captain Tralfamadore. The evil captain has set the controls to send the spaceship into the sun! His perverted mind insists on a slow death, so he sets the controls so the ship makes a constant angle of 30° with the sun (Fig. 12-3-5, p. 520). What path will Luke's ship follow? How long does Luke have to wake up if he is 10 million miles from the sun and his ship keeps a constant velocity of a million miles per hour?

Fig. 12-3-5 Luke Sky-runner's ill-fated ship.

Solution To answer these questions, we phrase the problem mathematically. We use polar coordinates to describe a curve $(r(t), \theta(t))$ such that the radius makes a constant angle α with the tangent ($\alpha = 30°$ in the problem). See Fig. 12-3-6(*a*). From this figure we can see that

$$\Delta r \approx \frac{r\,\Delta\theta}{\tan\alpha} \qquad \text{so} \qquad \frac{dr}{d\theta} = \frac{r}{\tan\alpha} \tag{1}$$

We can derive this rigorously, but also more laboriously, by calculating the slope of the tangent line in polar coordinates and setting it equal to $\tan(\theta + \alpha)$ as in Fig. 12-3-6(*b*). This approach gives

$$\frac{\tan\theta\,dr/d\theta + r}{dr/d\theta - r\tan\theta} = \tan(\theta + \alpha) = \frac{\tan\theta + \tan\alpha}{1 - \tan\theta\tan\alpha}$$

so that again

$$\frac{dr}{d\theta} = \frac{r}{\tan\alpha}$$

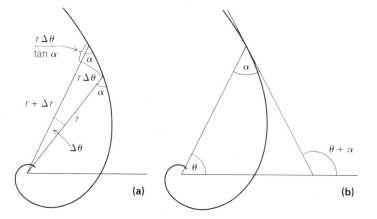

Fig. 12-3-6 The geometry of Luke's path.

The solution of (1) is

$$r(\theta) = r(0)e^{\theta/\tan\alpha} \tag{2}$$

For this solution to be valid we must regard θ as a continuous variable ranging from $-\infty$ to ∞, not as being between zero and 2π. As $\theta \to \infty$, $r(\theta) \to \infty$ and as $\theta \to -\infty$, $r(\theta) \to 0$ so the curve spirals outward as

θ increases and inward as θ decreases (if $0 < \alpha < \pi/2$). This answers the first question: Luke follows the spiral given by (2) where $\theta = 0$ is chosen as the starting point.

From p. 478 the distance Luke has to travel is the arc length of equation (2) from $\theta = 0$ to $\theta = -\infty$—namely, the improper integral

$$\int_{-\infty}^{0} \sqrt{\left(\frac{dr}{d\theta}\right)^2 + r^2}\, d\theta = \int_{-\infty}^{0} \sqrt{\frac{r^2}{\tan^2 \alpha} + r^2}\, d\theta$$

$$= \int_{-\infty}^{0} r(0)e^{\theta/\tan \alpha}\, \frac{1}{\sin \alpha}\, d\theta$$

$$= \frac{r(0)}{\cos \alpha}\, e^{\theta/\tan \alpha}\, \bigg|_{-\infty}^{0} = \frac{r(0)}{\cos \alpha}$$

With velocity $= 10^6$, $r(0) = 10^7$, and $\cos \alpha = \cos 30° = \sqrt{3}/2$, the time to go this distance is

$$\text{Time} = \frac{\text{distance}}{\text{velocity}}$$

$$= \frac{10^7}{\sqrt{3}/2} \times \frac{1}{10^6} = \frac{20}{\sqrt{3}} \approx 11.547 \text{ hours}$$

Thus Luke has less than 11.547 hours to wake up!

The logarithmic spiral turns up in another interesting situation. Place four love bugs at the corners of a square (Fig. 12-3-7). Each bug, being in love, walks directly toward the bug in front of it, at constant top bug speed. The result is that the bugs all spiral in to the center of the square following logarithmic spirals. The time required for the bugs to reach the center can be calculated as in Worked Example 11.

Fig. 12-3-7 These love bugs follow logarithmic spirals.

Solved Exercises

7. Find $\int_0^1 \ln x\, dx$.

8. Show that the improper integral $\int_0^\infty (e^{-x}/\sqrt{x})\, dx$ is convergent.

Exercises

6. Which of the following integrals are improper? For the improper integrals determine their values if they exist.

(a) $\displaystyle\int_1^{10} dx/x$

(b) $\displaystyle\int_0^1 dx/x^2$

(c) $\displaystyle\int_0^1 dx/\sqrt{1-x}$

(d) $\displaystyle\int_0^1 dx/(1-x^2)$

(e) $\displaystyle\int_{-1}^1 dx/(1+x^2)$

(f) $\displaystyle\int_{-\infty}^\infty dx/(1+x^2)$

7. Show that $\displaystyle\lim_{A \to 0^+} (\int_{-3}^{-A} dx/x + \int_A^2 dx/x)$ exists and determine its value.

8. Find α in Fig. 12-3-7 (α is defined in Worked Example 11). Find the time required for the bugs to reach the center in terms of their velocity and their initial distance from the center.

Problems for Section 12-3 ▰▰▰▰▰▰▰▰▰▰▰▰▰▰▰▰▰▰▰▰▰▰▰▰▰▰▰

1. Evaluate the following limits by using L'Hôpital's rule where appropriate.

(a) $\lim\limits_{x \to \infty} \dfrac{x}{x^2 + 1}$

(b) $\lim\limits_{x \to 0} \dfrac{x + \sin 2x}{2x + \sin 3x}$

(c) $\lim\limits_{x \to 5+} \dfrac{\sqrt{x^2 - 25}}{x - 5}$

(d) $\lim\limits_{x \to 5+} \dfrac{\sqrt{x^2 - 25}}{x + 5}$

2. Evaluate:

(a) $\lim\limits_{x \to \infty} \dfrac{(2x + 1)^3}{x^3 + 2}$

(b) $\lim\limits_{x \to -1} \dfrac{x^2 + 2x + 1}{x^2 - 1}$

(c) $\lim\limits_{x \to 1-} x^{1/(1 - x^2)}$

(d) $\lim\limits_{x \to 0} \dfrac{\cos x - 1 - x^2/2}{x^4}$

3. Evaluate:

(a) $\lim\limits_{x \to 1} \dfrac{\ln x}{e^x - 1}$

(b) $\lim\limits_{x \to \pi} \dfrac{1 + \cos x}{x - \pi}$

(c) $\lim\limits_{x \to \pi/2} (x - \pi/2)\tan(x)$

(d) $\lim\limits_{x \to 0} \dfrac{\sin x - x - (1/6)x^3}{x^5}$

4. (a) The limit

$$\lim_{x \to 0}\left(\frac{1}{x\sin x} - \frac{1}{x^2}\right)$$

is an indeterminate form of the type $\infty - \infty$. Evaluate this limit by first converting it to the form $\frac{0}{0}$.

(b) Evaluate $1/(x\sin x) - 1/x^2$ for $x = 10^{-1}$, 10^{-3}, 10^{-6}, and 10^{-8} by using a calculator. What do you learn about calculators from this computation? Could you rewrite $1/(x\sin x) - 1/x^2$ in such a way as to obtain more accurate results?

★5. Graph the function $f(x) = x^x$, $x > 0$.

6. Discuss the convergence of the following integrals:

(a) $\displaystyle\int_0^1 \dfrac{1}{\sqrt[3]{t^2}}\,dt$

(b) $\displaystyle\int_1^2 \dfrac{1}{\sqrt{t - 1}}\,dt$

(c) $\displaystyle\int_1^\infty \dfrac{1}{t^3}\,dt$

(d) $\displaystyle\int_2^\infty \dfrac{1}{t^2 - 1}\,dt$

7. Discuss the convergence of the following integrals:

(a) $\displaystyle\int_{-1}^\infty \dfrac{\tan^{-1}(x)}{(2 + x)^3}\,dx$

(b) $\displaystyle\int_{-\infty}^\infty \dfrac{\cos(x^2 + 1)}{x^2}\,dx$

(c) $\displaystyle\int_{-\infty}^\infty \dfrac{x}{(x^2 + 1)^{3/2}}\,dx$

(d) $\displaystyle\int_0^\infty \dfrac{1}{1 + x^2}\,dx$

8. Which of these are convergent?

(a) $\displaystyle\int_1^\infty \dfrac{1}{(4x - 3)^{1/3}}\,dx$

(b) $\displaystyle\int_0^\infty \left[\cos x + \dfrac{1}{(x + 2)^2}\right]dx$

(c) $\displaystyle\int_{-\infty}^2 \left(\dfrac{1}{x^{5/3}} - \dfrac{1}{x^{4/3}}\right)dx$

(d) $\displaystyle\int_{-4}^{10} \left[\dfrac{1}{(x + 4)^{2/3}} + \dfrac{1}{(x - 10)^{2/3}}\right]dx$

9. Discuss the following "calculations":

(a) $\displaystyle\int_{-1}^1 \dfrac{dx}{x^2} = -\dfrac{1}{x}\Big|_{-1}^1 = -1 + (-1) = -2$

(b) $\displaystyle\int_{\pi/2}^{5\pi/2} \dfrac{\cos x}{(1 + \sin x)^3}\,dx$

$= -\dfrac{1}{2}\cdot\dfrac{1}{(1 + \sin x)^2}\Big|_{\pi/2}^{5\pi/2} = 0$

10. Evaluate the following improper integrals, if they converge:

(a) $\displaystyle\int_2^\infty dx/x^2$

(b) $\displaystyle\int_2^\infty dx/(x\ln x)$

(c) $\displaystyle\int_0^\infty e^{-x}\ln x\,dx$

11. Find the area under the graph of $f(x) = (3x + 5)/(x^3 - 1)$ from $x = 2$ to $x = \infty$.

12. Consider the spirals defined in polar coordinates by the parametric equations $\theta = t$, $r = t^{-k}$. For which values of k does the spiral have finite arc length for $\pi/2 \le t < \infty$? (Use the comparison test.)

13. In Worked Example 11, suppose that Luke's airhoses melt down when he is 10^6 miles from the sun. Now how long does he have to wake up?

14. You can simulate the logarithmic spiral yourself as follows: Stand in an open field containing a lone tree and lock your neck muscles so that your head is pointed at a fixed angle α to your body. Walk forward in such a way that you are always looking at the tree. Prove that you will walk along a logarithmic spiral.

★15. Consider the surface of revolution obtained by revolving the graph of $f(x) = 1/x$ on the interval $[1, \infty)$ about the x axis.

   (a) Show that the area of this surface is infinite.

   (b) Show that the volume of the solid of revolution bounded by this surface is finite.

   (c) The results of parts (a) and (b) suggest that one could fill the solid with a finite amount of paint, but it would take an infinite amount of paint to paint the surface. Explain this paradox.

   Next consider the surface of revolution obtained by revolving the curve $y = 1/x^r$ for x in $[1, \infty)$ about the x axis.

   (d) For which values of r does this surface have finite area?

   (e) For which values of r does the solid surrounded by this surface have finite volume? Compute the volume for these values of r.

16. The curve $y = e^{-x}$ is rotated about the x-axis to form a solid of revolution. Find the volume obtained by discarding the portion on $-\infty < x \leq 10$ (after slicing the solid at $x = 10$).

17. Determine the lateral surface area of the surface of revolution obtained by revolving $y = e^{-x}$, $0 \leq x < \infty$, about the x-axis.

18. The probability P that a phonograph needle will last in excess of 150 hours is $P = \int_{150}^{\infty} (\frac{1}{100})e^{-t/100} \, dt$. Find the value of P.

19. The probability p that the score on a reading comprehension test is no greater than the value a is

$$p = \int_{-\infty}^{a} \frac{1}{\sqrt{2\pi\sigma^2}} e^{-(\tau - \mu)^2/2\sigma^2} \, d\tau$$

   (a) Let $x = (\tau - \mu)/\sigma$ and $x_1 = (a - \mu)/\sigma$. Show that

$$p = \int_{-\infty}^{x_1} \frac{1}{\sqrt{2\pi}} e^{-x^2/2} \, dx$$

   (b) Justify $\int_{-\infty}^{\infty} e^{-x^2/2} \, dx < \infty$. [*Hint:* Use the comparison test in Theorem 8.]

20. Pearson and Lee studied the inheritance of physical characteristics in families in 1903. One law that resulted from these studies is

$$P = \int_{-\infty}^{(\tau - \mu)/\sigma} \frac{1}{\sqrt{2\pi}} e^{-x^2/2} \, dx$$

for the probability P that a mother's height in inches is not greater than τ inches. The estimated values of μ and σ are $\mu = 62.484$ inches, $\sigma^2 = 5.7140$ inches.

   (a) Determine the value of P by appeal to integral tables for

$$\int_{-\infty}^{u} \frac{1}{\sqrt{2\pi}} e^{-x^2/2} \, dx$$

   using $\tau = 63$ inches. Look in a mathematical table under *probability functions*, or *normal distribution*.

   (b) According to the study, how many mothers out of 100 are likely to have height not exceeding 63 inches?

21. A measure of the average IQ score among 112 kindergarten children using the Stanford Revision of the Binet-Simon Intelligence Scale is the *expected value* $\mu = \int_{-\infty}^{\infty} x f(x) \, dx$, where $f(x)$ is the *probability distribution* of observed frequencies of IQ scores. The distribution is estimated to be

$$f(x) = \frac{1}{(16.23761)\sqrt{2\pi}} e^{-(x - 104.5)^2/527.32}$$

   (a) Perform the change of variables $u = (x - 104.5)/(16.23761)$ in the integral. <u>Use $\int_{-\infty}^{\infty} u \, e^{-u^2/2} \, du = 0$, $\int_{-\infty}^{\infty} e^{-u^2/2} \, du = \sqrt{2\pi}$</u> to find μ, the IQ score considered to be "average."

   (b) The observed frequency of IQ scores between 119 and 110 was 23 scores. The theoretical value is $(112)\left[\int_{-\infty}^{120} f(x) \, dx - \int_{-\infty}^{110} f(x) \, dx\right]$. Apply the change of variables in part (a) to get the preceding equal to

$$112\left[\int_{-\infty}^{0.95457} e^{-u^2/2} \, du - \int_{-\infty}^{0.33872} e^{-u^2/2} \, du\right] / \sqrt{2\pi}$$

   (c) Use a table for the *normal distribution* or *gaussian distribution* in a mathematics table to evaluate the theoretical frequency of part (b). This is the expected number of IQ scores between 110 and 119. What is the percentage error committed by the theoretical prediction?

12-4 Newton's Method and Simpson's Rule

Solutions of equations and integrals of functions can be found as the limits of sequences.

If the solution to a mathematical problem cannot be found exactly, it is often possible to find an approximate solution by numerical methods. Such methods are becoming increasingly important as electronic calculators and computers proliferate.

We give in this section two examples of numerical methods in which the solution of a problem is expressed as the limit of a sequence of numbers which can be calculated in a mechanical way. If you have not read Section 12-1, it would be useful to look at pp. 488 to 490 before beginning this section.

Goals

After studying this section, you should be able to:

Solve equations by Newton's method.

Evaluate integrals by Simpson's rule.

NEWTON'S METHOD

Many questions in mathematics and physics lead to the problem of solving an equation of the form

$$f(x) = 0 \tag{1}$$

where f is some function. The solutions of (1) are called the *roots* or *zeros* of f. If f is a polynomial of degree at most 4, one can find the roots of f by substituting the coefficients of f into a general formula (see pp. 13 and 140). On the other hand, if f is a polynomial of degree 5 or greater, or a transcendental function, there may be no explicit formula for the roots of f, and one may have to search for the solution numerically.

The method we are about to describe, devised by Newton, is an example of an *iterative* method. In such a method, we start with a guess x_0 for a root of the equation. The method will then yield a new number, x_1, which is a better guess. Repeating the method over and over gives a sequence $x_1, x_2, \ldots$. If the method is a good one, the sequence $x_1, x_2, \ldots$ will converge to a limit $\bar{x}$ with the property that $f(\bar{x}) = 0$. By going further and further out in the sequence, we get more and more accurate estimates of the root $\bar{x}$.

With Newton's method, the procedure for passing from each x_n to the next uses the linear approximations to the function f. Let x_0 be the first guess. We seek to correct this guess by an amount Δx so that $f(x_0 + \Delta x) = 0$. Solving this equation for Δx is no easier than solving the original equation (1), so we manufacture an easier problem by replacing f by its first-order approximation at x_0; that is, we replace $f(x_0 + \Delta x)$ by $f(x_0) + f'(x_0)\Delta x$. If $f'(x_0)$ is not equal to zero, we can solve the equation $f(x_0) + f'(x_0)\Delta x = 0$ to get $\Delta x = -f(x_0)/f'(x_0)$, so that our new guess is $x_1 = x_0 + \Delta x = x_0 - f(x_0)/f'(x_0)$. Geometrically, we have found x_1 by following the tangent line to the graph of f at $(x_0, f(x_0))$ until it meets the x axis; the point where it meets is $(x_1, 0)$ (see Fig. 12-4-1).

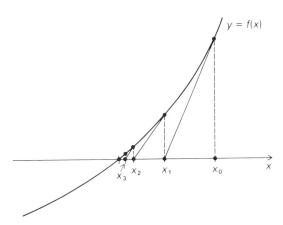

Fig. 12-4-1 The geometry of Newton's method.

Now we find x_2 by repeating the procedure with x_1 in place of x_0; that is,

$$x_2 = x_1 - \frac{f(x_1)}{f'(x_1)}$$

In general, once we have found x_n we define x_{n+1} by

$$x_{n+1} = x_n - \frac{f(x_n)}{f'(x_n)} \qquad (2)$$

Let us see how the method works in a case where we know the answer in advance. (This iteration procedure is particularly easy to use on a programmable calculator.)

Worked Example 1   Use Newton's method to find the first few approximations to a solution of the equation $x^2 = 4$, taking $x_0 = 1$.

Solution   To put the equation $x^2 = 4$ in the form $f(x) = 0$, we let $f(x) = x^2 - 4$. Then $f'(x) = 2x$, and the general iteration rule (2) becomes $x_{n+1} = x_n - (x_n^2 - 4)/2x_n$, which may be simplified to $x_{n+1} = \frac{1}{2}(x_n + 4/x_n)$. Applying this formula repeatedly, with $x_0 = 1$, we get (to the limits of our calculator's accuracy)

$$x_1 = 2.5$$
$$x_2 = 2.05$$
$$x_3 = 2.000609756$$
$$x_4 = 2.000000093$$
$$x_5 = 2$$
$$x_6 = 2$$
$$\vdots$$

and so on forever. The number 2 is precisely the root of our equation $x^2 = 4$.

Now here is an example where we do not know the root in advance.

Worked Example 2   Use Newton's method to find a positive number x such that $\sin x = x/2$.

Solution   With $f(x) = \sin x - x/2$, the iteration formula becomes

$$x_{n+1} = x_n - \frac{\sin x_n - x_n/2}{\cos x_n - 1/2} = \frac{2(x_n \cos x_n - \sin x_n)}{2 \cos x_n - 1}$$

Taking $x_0 = 0$ as our first guess, we get $x_1 = 0$, $x_2 = 0$, and so forth, since zero is already a root of our equation. To find a positive root, we try another guess, say $x_0 = 1$. We get

$$x_1 = -7.47274064 \qquad x_2 = 14.47852089$$
$$x_3 = 6.935114649 \qquad x_4 = 16.63566898$$

This does not seem to be converging to anything. To see what is going on, we draw a sketch (Fig. 12-4-2). The many bumps in the graph cause our sequence to oscillate wildly. Looking at the graph, we see that a better first guess might be $x_0 = 3$. Now we get

$$x_1 = 2.087995413 \qquad x_2 = 1.912229258 \qquad x_3 = 1.895652628$$
$$x_4 = 1.895494282 \qquad x_5 = 1.895494267 \qquad x_6 = 1.895494267$$
$$x_7 = 1.895494267 \qquad x_8 = 1.895494267$$

We conclude that our root is somewhere near 1.895494267. Substituting this value for x in $\sin x - x/2$ gives 1.0×10^{-11}. There may be further doubt about the last figure, due to internal round-off errors in the calculator; we are probably safe to announce our result as 1.89549427.

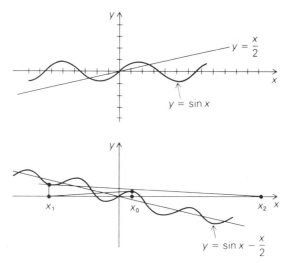

Fig. 12-4-2   Newton's method goes awry.

Worked Example 2 illustrates several important features of Newton's method. First of all, it is important to start with an initial guess which is reasonably close to a root—graphing is a help in making such a guess. Second, we notice that once we get near a root, then convergence becomes very rapid—in fact, the number of correct decimal places is approximately *doubled* with each iteration. Finally, we notice that the process for passing from x_n to x_{n+1} is the same for each value of n; this feature makes Newton's method particularly attractive for use with a programmable calculator or a computer. Human intelligence still comes into play in the choice of the first guess, however.

NEWTON'S METHOD

To find a root of the equation $f(x) = 0$, where f is a differentiable function such that $f'x$ is continuous, start with a guess x_0 which is reasonably near to a root. Then produce the sequence $x_0, x_1, x_2, \ldots$ by the iterative procedure:

$$x_{n+1} = x_n - \frac{f(x_n)}{f'(x_n)}$$

If $\lim_{n \to \infty} x_n = \bar{x}$, then $f(\bar{x}) = 0$.

Solved Exercises

1. Justify the last sentence in the preceding display. (Assume $f'(x_n) \neq 0$.)

2. Use Newton's method to locate a root of $x^5 - x^4 - x + 2 = 0$. Compare what happens with various starting values of x_0 and attempt to explain the phenomenon.

Exercises

1. Use Newton's method to locate a zero for $f(x) = x^4 - 2x^3 - 1$. Use $x_0 = 2, 3,$ and -1 as starting values and compare the results.

2. Use Newton's method to find the following numbers:
   (a) $\sqrt{2}$ (b) $\sqrt[3]{2}$
   (c) Three different solutions of $\tan x = x$.
   (d) All real roots of $x^3 - x + \frac{1}{10}$.

NUMERICAL INTEGRATION

If f is integrable on $[a, b]$, we know that the number $\int_a^b f(x)\,dx$ exists—but it is not always easy to compute its value by finding an antiderivative for f. It might also happen that the functions we wish to integrate are presented to us numerically as a table of values; for example, we can imagine being given power readings from an energy cell at selected times and being asked to integrate to find the energy stored. In either case, it is necessary to use a method of numerical integration to find an approximate value for the integral.

In using a numerical method, it is important to be able to estimate one's errors so that the final answer can be said with confidence to be correct to so many significant figures. These errors include errors in the method, round-off errors, and cumulative errors in addition and multiplication. The task of keeping careful track of possible errors is a complicated and fascinating one, of which we can give only some simple examples.*

---

*For a further discussion of error analysis in numerical integration, see, for example, P. J. Davis, Interpolation and Approximation, *Wiley, 1963.*

The simplest method of numerical integration is based upon the fact that the integral is a limit of Riemann sums (see p. 187 or 498). Suppose that we are given $f(x)$ on $[a,b]$, and divide up $[a,b]$ into subintervals $a = x_0 < x_1 < \cdots < x_n = b$. (We might be given the data $f(a)$, $f(x_1)$, ..., $f(x_{n-1})$, $f(b)$ directly.) Then $\int_a^b f(x)\,dx$ is approximated by $\Sigma_{i=1}^n f(c_i)\,\Delta x_i$, where c_i lies in $[x_{i-1}, x_i]$. Usually, the points x_i are taken to be equally spaced, so $\Delta x_i = (b-a)/n$ and $x_i = a + i(b-a)/n$. Choosing $c_i = x_i$ or x_{i+1} gives the method in the following display.

RIEMANN SUMS

To calculate an approximation to $\int_a^b f(x)\,dx$, let $x_i = a + i(b-a)/n$ and form the sum

$$\frac{b-a}{n}[f(x_0) + f(x_1) + \cdots + f(x_{n-1})] \tag{3a}$$

or

$$\frac{b-a}{n}[f(x_1) + f(x_2) + \cdots + f(x_n)] \tag{3b}$$

Unfortunately, this method is inefficient and inaccurate. Many points x_i are needed to get an accurate estimate of the integral. For this reason we will seek alternatives to the method of Riemann sums.

Worked Example 3 Let $f(x) = \cos x$. Evaluate $\int_0^{\pi/2} \cos x\,dx$ by the method of Riemann sums, taking 10 equally spaced points: $x_0 = 0$, $x_1 = \pi/20$, $x_2 = 2\pi/20$, ..., $x_{10} = 10\pi/20 = \pi/2$, and $c_i = x_i$. Compare the answer with the actual value.

Solution Method 1 (formula 3a) gives

$$\int_0^{\pi/2} \cos x\,dx \approx \frac{\pi}{20}\left(1 + \cos\frac{\pi}{20} + \cos\frac{2\pi}{20} + \cdots + \cos\frac{9\pi}{20}\right)$$

$$= \frac{\pi}{20}(1 + 0.9877 + 0.95106 + \cdots + 0.15643)$$

$$= \frac{\pi}{20}(6.8531) = 1.07648$$

The actual value is $\sin(\pi/2) - \sin(0) = 1$. Thus our estimate is about 7.6% off.

To get a better method, we might try to estimate the area in each interval $[x_{i-1}, x_i]$ more accurately. The second method does this by replacing the rectangular approximation by a trapezoidal one. See Fig. 12-4-3. We join the points $(x_i, f(x_i))$ by straight line segments to obtain a set of approximating trapezoids. The area of the trapezoid between x_{i-1} and x_i is

$$A_i = \tfrac{1}{2}[f(x_{i-1}) + f(x_i)]\,\Delta x_i$$

since the area of a trapezoid is its average height times its width.

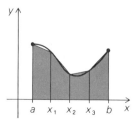

(a) Riemann sums method

(b) Trapezoidal method

Fig. 12-4-3 Comparing two methods of numerical integration.

The approximation to $\int_a^b f(x)\,dx$ given by the trapezoidal rule is $\sum_{i=1}^n \frac{1}{2}[f(x_{i-1}) + f(x_i)]\,\Delta x_i$. This becomes simpler if the points x_i are equally spaced. Then $\Delta x_i = (b-a)/n$, $x_i = a + i(b-a)/n$, and the sum is

$$\left(\frac{b-a}{n}\right) \sum_{i=1}^n \frac{1}{2}[f(x_{i-1}) + f(x_i)]$$

which can be rewritten as

$$\frac{b-a}{2n}[f(x_0) + 2f(x_1) + \cdots + 2f(x_{n-1}) + f(x_n)]$$

since every term occurs twice except those from the endpoints. Although we used areas to obtain this formula, we may apply it even if $f(x)$ takes negative values.

TRAPEZOIDAL RULE

To calculate an approximation to $\int_a^b f(x)\,dx$, let $x_i = a + i(b-a)/n$ and form the sum

$$\frac{b-a}{2n}[f(x_0) + 2f(x_1) + \cdots + 2f(x_{n-1}) + f(x_n)] \qquad \textbf{(4)}$$

Formula (4) turns out to be more accurate than the method of Riemann sums. Notice that (4) is the average of the Riemann sums (3a) and (3b). Using results of Section 13-3, it is possible to show that the error in the *method* (apart from other possible round-off or cumulative errors) is $\leq [(b-a)/12]\,M_2(\Delta x)^2$, where M_2 is the maximum of $|f''(x)|$ on $[a,b]$. Of course, if we are given only numerical data, we have no way of estimating M_2, but if a formula for f is given, M_2 can be determined. Note, however, that the error depends on $(\Delta x)^2$, so if we divide $[a,b]$ into k times as many divisions, the error goes down by $1/k^2$. The error in the Riemann sums method, on the other hand, is $\leq (b-a)M_1(\Delta x)$, where M_1 is the maximum of $|f'(x)|$ on $[a,b]$. Here Δx occurs only to the first power. Thus even if we do not know how large M_1 and M_2 are, if n is taken large enough the error in the trapezoidal rule will eventually be much smaller than that in the Riemann sums method.

Worked Example 4 Repeat Worked Example 3 by using the trapezoidal rule. Compare the answer with the true value.

Solution Now formula (4) becomes

$$\frac{\pi/2}{2\cdot 10}\left(\cos 0 + 2\cos\frac{\pi}{20} + \cdots + 2\cos\frac{9\pi}{20} + \cos\frac{\pi}{2}\right)$$

$$= \frac{\pi}{40}[1 + 2(0.9877 + 0.95106 + \cdots) + 0]$$

$$= 0.997943$$

The answer is correct to within about 0.2%, much better than the accuracy in Worked Example 3.

Worked Example 5 Use the trapezoidal rule with $n = 10$ to estimate numerically the area of the surface obtained by revolving the graph of $y = x/(1 + x^2)$ about the x axis, $0 \leq x \leq 1$.

Solution The area is given by the formula on p. 467:

$$A = 2\pi \int_0^1 \left(\frac{x}{1 + x^2}\right) \sqrt{1 + \left[\frac{d}{dx}\left(\frac{x}{1 + x^2}\right)\right]^2} \, dx$$

$$= 2\pi \int_0^1 \frac{x\sqrt{(1 + x^2)^4 + (1 - x^2)^2}}{(1 + x^2)^3} \, dx$$

There is little hope of carrying out this integration, so a numerical approach seems appropriate. We use the trapezoidal rule with

x_i	0	0.1	0.2	0.3	0.4	0.5	0.6	0.7	0.8	0.9	1.0
$y_i = f(x_i)$	0	0.13797	0.25713	0.34668	0.40650	0.44369	0.46684	0.48204	0.49215	0.49807	0.50000

where $f(x) = x\sqrt{(1 + x^2)^4 + (1 - x^2)^2} / (1 + x^2)^3$. Inserting these data in the formula

$$\int_a^b f(x) \, dx \approx \left(\frac{b - a}{2n}\right) [f(x_0) + 2f(x_1) + \cdots + 2f(x_{n-1}) + f(x_n)]$$

with $x_i = a + [i(b - a)/n]$, $a = 0$, and $b = 1$, gives

$$\int_0^1 \frac{x\sqrt{(1 + x^2)^4 + (1 - x^2)^2}}{(1 + x^2)^3} \, dx \approx 0.37811$$

so the area is $A \approx (2\pi)(0.37811) = 2.3757$. Of course, we can't be sure how many decimal places in this result are correct without an error analysis (see Problem 11).

There is a yet more powerful method of numerical integration called Simpson's rule,* which is based on approximating the graph by parabolas rather than straight lines. To determine a parabola we need to specify three points through which it passes; we will choose the adjacent points

$$(x_{i-1}, f(x_{i-1})) \quad (x_i, f(x_i)) \quad (x_{i+1}, f(x_{i+1}))$$

It is easily proved (see Problem 8) that the integral from x_{i-1} to x_{i+1} of the quadratic function whose graph passes through these three points is

$$A_i = \frac{\Delta x}{3} [f(x_{i-1}) + 4f(x_i) + f(x_{i+1})]$$

---

*It was discussed by Thomas Simpson in his book, Mathematical Dissertations on Physical and Analytical Subjects (1743).

where $\Delta x = x_i - x_{i-1} = x_{i+1} - x_i$ (equally spaced points). See Fig. 12-4-4.

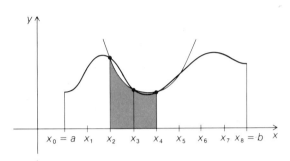

Fig. 12-4-4 Illustrating Simpson's rule.

If we do this for every set of three adjacent points, starting at the left endpoint a—that is, for $\{x_0, x_1, x_2\}$, then $\{x_2, x_3, x_4\}$, then $\{x_4, x_5, x_6\}$, and so on—we will get an approximate formula for the area. In order for the points to fill the interval exactly, n should be even, say $n = 2m$.

As in the trapezoidal rule, the contributions from endpoints a and b are counted only once, as are those from the center points of triples $\{x_{i-1}, x_i, x_{i+1}\}$ (that is, x_i for i odd) while the others are counted twice. Thus we are led to Simpson's rule.

SIMPSON'S RULE

To calculate an approximation to $\int_a^b f(x)\,dx$, let $n = 2m$ be even and $x_i = a + i(b-a)/n$. Form the sum

$$\frac{b-a}{3n}\left[f(x_0) + 4f(x_1) + 2f(x_2) + 4f(x_3) + 2f(x_4) + \cdots + 2f(x_{n-2}) + 4f(x_{n-1}) + f(x_n)\right] \tag{5}$$

This method is very accurate; the error does not exceed $[(b-a)/180]\,M_4(\Delta x)^4$, where M_4 is the maximum of the fourth derivative of $f(x)$ on $[a,b]$. As Δx is taken smaller and smaller, this error gets small much faster than in the other two methods. It is remarkable that juggling the coefficients to give formula (5) in place of (3) or (4) can increase the accuracy so much.

Worked Example 6   Repeat Worked Example 3 by using Simpson's rule. Compare the answer with the true value.

Solution   Using a calculator, we can evaluate formula (5) by

$$\frac{\pi/2}{3 \cdot 10}\left(\cos 0 + 4\cos\frac{\pi}{20} + 2\cos\frac{2\pi}{20} + 4\cos\frac{3\pi}{20} + 2\cos\frac{4\pi}{20} + \right.$$
$$\left. 4\cos\frac{5\pi}{20} + 2\cos\frac{6\pi}{20} + 4\cos\frac{7\pi}{20} + 2\cos\frac{8\pi}{20} + 4\cos\frac{9\pi}{20} + \cos\frac{\pi}{2}\right)$$
$$= \frac{\pi}{60}(1 + 3.95075 + \cdots + 0) = \frac{\pi}{60} \cdot 19.09866$$
$$= 1.0000034$$

The answer is accurate to four parts in a million.

Solved Exercises 3. Suppose that you are given the following table of data:

$$f(0) = 0.846 \qquad f(0.4) = 1.121 \qquad f(0.8) = 2.321$$
$$f(0.1) = 0.928 \qquad f(0.5) = 1.221 \qquad f(0.9) = 3.101$$
$$f(0.2) = 0.882 \qquad f(0.6) = 1.661 \qquad f(1.0) = 3.010$$
$$f(0.3) = 0.953 \qquad f(0.7) = 2.101$$

Evaluate $\int_0^1 f(x)\, dx$ by Simpson's rule.

4. How small must we take Δx in the trapezoidal rule to evaluate $\int_2^4 e^{-x^2}\, dx$ to within 10^{-6}? For Simpson's rule?

Exercises 3. Use the indicated numerical method(s) to approximate the following integrals.

(a) $\int_{-1}^1 (x^2 + 1)\, dx$. Use Riemann sums with $n = 10$ (that is, divide $[-1, 1]$ into 10 subintervals of equal length). Compare with the actual value.

(b) $\int_0^{\pi/2} (x + \sin x)\, dx$. Use Riemann sums and the trapezoidal rule with $n = 8$. Compare these two approximate values with the actual value.

(c) $\int_1^3 [(\sin \pi x/2)/(x^2 + 2x - 1)]\, dx$. Use the trapezoidal rule and Simpson's rule with $n = 12$.

(d) $\int_0^2 (1/\sqrt{x^3 + 1})\, dx$. Use Simpson's Rule with $n = 20$.

4. Suppose that you are given the following table of data:

$$f(0) = 1.384 \qquad f(0.4) = 0.915 \qquad f(0.8) = 0.935$$
$$f(0.1) = 1.179 \qquad f(0.5) = 0.768 \qquad f(0.9) = 1.262$$
$$f(0.2) = 0.973 \qquad f(0.6) = 0.511 \qquad f(1.0) = 1.425$$
$$f(0.3) = 1.000 \qquad f(0.7) = 0.693$$

Numerically evaluate $\int_0^1 (x + f(x))\, dx$ by Simpson's rule.

5. Numerically evaluate $\int_0^1 2f(x)\, dx$ by the trapezoidal rule, where $f(x)$ is the function in Exercise 4 above.

Problems for Section 12-4

1. (a) Use Newton's method to find a solution of $x^3 - 8x^2 + 2x + 1 = 0$.

(b) Use division and the quadratic formula to find the other two roots.*

2. Use Newton's method to locate roots of $f(x) = x^5 + x^2 - 3$ with starting values $x_0 = 0$, $x_0 = 2$.

3. Use Newton's method to locate a root of $\tan x = x$.

4. Use Simpson's rule with $n = 10$ to find an approximate value for $\int_0^1 (x/\sqrt{x^3 + 2})\, dx$.

5. How large must n be taken in the trapezoidal rule to guarantee an accuracy of 10^{-5} in the evaluation of the integral in Problem 4? Answer the same question for Simpson's rule.

*For a computer, this method is preferable to using the formula for the roots of a cubic!

6. Suppose that you are given the following table of data:

$f(0.0) = 2.037$ $f(1.3) = 0.819$

$f(0.2) = 1.980$ $f(1.4) = 1.026$

$f(0.4) = 1.843$ $f(1.5) = 0.799$

$f(0.6) = 1.372$ $f(1.6) = 0.662$

$f(0.8) = 1.196$ $f(1.7) = 0.538$

$f(1.0) = 0.977$ $f(1.8) = 0.555$

$f(1.2) = 0.685$

Numerically evaluate $\int_0^{1.8} f(x)\,dx$ by using Simpson's rule. [*Hint:* Watch out for the spacing of the points.]

7. Numerically evaluate $\int_0^{1.8} f(x)\,dx$ by using the trapezoidal rule, where $f(x)$ is the function in Problem 6 above.

8. Evaluate $\int_{x_1}^{x_2} (ax^2 + bx + c)\,dx$. Verify that Simpson's rule gives the exact answer. What happens if you use the trapezoidal rule? Discuss.

★9. (*Another numerical integration method.*)

(a) Let $(x_1, y_1), (x_2, y_2), \ldots, (x_n, y_n)$ be n points in the plane such that all the x_i's are different. Show that the polynomial of degree no more than $n - 1$ whose graph passes through the given points is

$$P(x) = y_1 L_1(x) + y_2 L_2(x) + \cdots + y_n L_n(x)$$

where $L_i(x) = A_i(x)/A'(x_i)$

$$A(x) = (x - x_1)(x - x_2) \cdots (x - x_n)$$

$$A_i(x) = A(x)/(x - x_i), \quad i = 1, 2, \ldots, n$$

(Here P is called the *Lagrange interpolation polynomial.*)

(b) Suppose that you are given the following data for an unknown function $f(x)$:

$f(0) = 0.01$

$f(0.1) = 0.12$

$f(0.2) = 0.82$

$f(0.3) = 1.18$

$f(0.4) = 0.91$

Estimate the value of $f(0.16)$ by using the Lagrange interpolation formula.

(c) Estimate $\int_0^{0.4} f(x)\,dx$ (1) by using the trapezoidal rule, (2) by using Simpson's rule, and (3) by integrating the Lagrange interpolation polynomial.

★10. Estimate $\int_0^{\pi/2} \cos x\,dx$ by using a Lagrange interpolation polynomial with $n = 4$. (See Problem 9.) Compare your result with those obtained by the trapezoidal and Simpson's rules.

★11. How many digits in the approximate value $A \approx 2.3757$ in Worked Example 5 can be justified by an error analysis?

12. The equation $\tan x = \alpha x$ appears in heat conduction problems to determine values $\lambda_1, \lambda_2, \lambda_3, \ldots$ that appear in the expression for the temperature distribution. The numbers $\lambda_1, \lambda_2, \ldots$ are the positive solutions of $\tan x = \alpha x$, listed in increasing order. Find the numbers $\lambda_1, \lambda_2, \lambda_3$ for $\alpha = 2, 3, 5$, by Newton's method. Display your answers in a table.

13. The *efficiency* of Newton's method for the location of a root of $f(\bar{x}) = 0$ is measured by the inequality $|x - [f(x)/f'(x)] - \bar{x}| \le K |x - \bar{x}|^2$, which is valid on $a \le x \le b$ provided $K = M/2m$, with $|f'(x)| \ge m$ and $|f''(x)| \le M$ on $[a, b]$. For this reason, the method is called a *second-order method*.

Apply this result to $f(x) = x^2 - 2$, verifying that $m = 2$, $M = 2$ on $[1, 2]$ assuming that the initial guess x_0 satisfies $|x_0 - \sqrt{2}| < 1/2$. Justify that the Newton iterates $x_0, x_1, \ldots$, satisfy $|x_0 - \sqrt{2}| < \frac{1}{2}$, $|x_1 - \sqrt{2}| < 1/2^3$, $|x_2 - \sqrt{2}| < 1/2^7$, $\ldots$. Finally, show that $x_6 = \sqrt{2}$ with 20-digit accuracy.

★14. Experiment with Newton's method for evaluation of the root $1/e$ of the equation $e^{-ex} = 1/e$. Is there any way to speed up the convergence of the iterates to the solution?

15. A tank 15 m by 60 m is filled to a depth of 3.2 m above the bottom. The time T it takes to empty half the tank through an orifice 0.5 m wide by 0.2 m high placed 0.1 m from the bottom is given by

$$\frac{(2.2)T}{10^6} = \frac{1}{\sqrt{19.6}} \int_{130}^{190} \frac{dx}{(x + 20)^{3/2} - x^{3/2}}$$

Compute T from Simpson's rule with $n = 6$, showing $T = 376$ seconds.

16. Estimate to five-place accuracy with $n = 4$ the value of $\int_1^3 e^{\sqrt{x}}\,dx$, using Simpson's rule. [*Hint:* Check your answer using $x = u^2$, $dx = 2\,u\,du$.]

17. A metropolitan sports and special events complex is circular in shape with an irregular roof that appears from a distance to be almost hemispherical (Fig. 12-4-5).

Fig. 12-4-5 The profile of the roof of a sports complex.

A summer storm severely damaged the roof, requiring a roof replacement to go out for bid. Responding contractors were supplied with plans of the complex from which to determine an estimate. Estimators had to find the roof profile $y = f(x)$, $0 \le x \le H$, which generates the roof by revolution about the x-axis (x and y in feet, x vertical, y horizontal).

(a) Find the square footage of the roof via a surface area formula. This number determines the amount of roofing material required.

(This problem continues at the top of the next column.)

17. (b) To check against construction errors, a tape measure is tossed over the roof and the measurement recorded. Give a formula for this measurement using the arc length formula.

(c) Suppose the curve f is not given explicitly in the plans, but instead $f(0)$, $f(4)$, $f(8)$, $f(12)$, ..., $f(H)$ are given (complex center-to-ceiling distances every four feet). Discuss how to use this information to numerically evaluate the integrals in (a), (b) above, using Worked Example 5 as a guide.

(d) Find an expression which approximates the surface area of the roof by assuming it is a *conoid* produced by a piecewise linear function constructed from the numbers $f(0)$, $f(4)$, $f(8)$, ..., $f(H)$.

18. *Gaussian quadrature* is an approximation method for integration on $[-1, 1]$ based on interpolation. The formula is $\int_{-1}^{1} f(x)\, dx = f(1/\sqrt{3}) + f(-1/\sqrt{3}) + R$ where the remainder R satisfies $|R| \le M/135$, M being the largest value of $f^{(4)}(x)$ on $-1 \le x \le 1$.

(a) The remainder R is zero for cubic polynomials. Check it for x^3, $x^3 - 1$, $x^3 + x + 1$.

(b) Find $\int_{-1}^{1} [x^2/(1 + x^4)]\, dx$ to four places.

(c) What is R for $\int_{-1}^{1} x^6\, dx$? Why is it so large?

Review Problems for Chapter 12

1. Find each of the following limits:

    (a) $\lim\limits_{n \to \infty} [n/(n + 2)]$ (b) $\lim\limits_{x \to \infty} [x/(x + 2)]$

    (c) $\lim\limits_{x \to 0} [(x + 3)/(3x + 8)]$

    (d) $\lim\limits_{x \to 0} [(\sin 5x)/x]$

2. Find:

    (a) $\lim\limits_{x \to \infty} x^2 e^{-x}$ (b) $\lim\limits_{x \to 0} x \cot x$

    (c) $\lim\limits_{x \to 2} \{ [\sin(x - 2) - x + 2]/(x - 2)^3 \}$

    (d) $\lim\limits_{x \to 0} \{ (x^4 + 8x)/(3x^4 + 2) \}$

3. Find each of the following limits:

    (a) $\lim\limits_{n \to \infty} \tan [3n/(n + 8)]$

    (b) $\lim\limits_{x \to 0} [\tan(x + 3) - \tan 3]/x$

    (c) $\lim\limits_{x \to 0} x^3 (\ln x)^2$ (d) $\lim\limits_{x \to \infty} [(\ln x)^2/x]$

4. Find:

    (a) $\lim\limits_{n \to \infty} (n^2 + 3n + 1)e^{-n}$

    (b) $\lim\limits_{n \to \infty} [(n - 3)/n]^{-2n}$

    (c) $\lim\limits_{x \to 1} [(x^x - 1)/(x - 1)]$

    (d) $\lim\limits_{x \to 1} [(e^{2x} - 1)/(2x - 1)]$

★5. Using ε's and δ's, or ε's and N's, show that:

    (a) $\lim\limits_{x \to 2} (x^2 - 8x + 8) = -4$

    (b) $\lim\limits_{n \to \infty} [(n^2 + 2n)/(3n^2 + 1)] = \frac{1}{3}$

6. Find $\lim\limits_{n \to \infty} a_n$, where $a_0 = 2$, $a_{n+1} = \frac{3}{4} a_n$.

★7. Suppose that $a_0 = 1$, $a_{n+1} = 1 + 1/(1 + a_n)$. Show that a_n converges and find the limit.

8. Let $f(x) = x^{1/\sin(x - 1)}$. How should $f(1)$ be defined in order to make f continuous?

9. Find the following:

(a) $\displaystyle\lim_{x\to 0}\frac{1-\cos x}{3^x-2^x}$

(b) $\displaystyle\lim_{x\to 0^+}(1+\sin 2x)^{1/x}$

(c) $\displaystyle\lim_{x\to 0^+}\left(\frac{1}{\sin x}-\frac{1}{x}\right)$

(d) $\displaystyle\lim_{x\to 0}\sin(\sqrt{x^2+2}-x)$

10. Find the vertical asymptotes for the graph of $y=1/(x^2-3x-10)^2$. Sketch.

11. Find the horizontal and vertical asymptotes of the following functions and sketch:

(a) $f(x)=(x-1)/(x^2+1)$

(b) $f(x)=(2x+3)/(3x+5)$

12. Let $f(x)=\cos x$ for $x\ge 0$ and $f(x)=1$ for $x<0$. Decide whether or not f is continuous or differentiable or both.

13. (a) Show that

$$f''(x_0)=\lim_{h\to 0}\frac{f(x_0+h)-2f(x_0)+f(x_0-h)}{h^2}$$

if f'' is continuous at x_0. [*Hint:* Use L'Hôpital's rule.]

★ (b) Find a similar formula for $f'''(x_0)$.

14. Show that

$$f''(x_0)=\lim_{\Delta x\to 0}\frac{f(x_0+2\Delta x)+f(x_0)-2f(x_0+\Delta x)}{(\Delta x)^2}$$

if f'' is continuous at x_0.

★15. Let

$$S_n=\sum_{i=1}^{2n}\left[2-\cos\left(\frac{i}{n}\pi\right)\right]\frac{1}{n}$$

Prove that $\displaystyle\lim_{n\to\infty}S_n=4$.

★16. Write $\int_0^\pi dx/\sqrt{1-\tfrac{1}{2}\sin^2 x}$ as a limit.

17. Find the following limits:

(a) $\displaystyle\lim_{x\to 0}(\tan^2 x)/x^2$

(b) $\displaystyle\lim_{n\to\infty}\sum_{i=1}^{n}(\ln n-\ln i)/n$

(c) $\displaystyle\lim_{x\to\infty}(x^3+8x+9)/(4x^3-9x^2+10)$

(d) $\displaystyle\lim_{x\to 0}\int_x^1 dt/\sqrt{t}$

18. Find:

(a) $\displaystyle\lim_{x\to 0}(24\cos x-24+12x^2-x^4)/x^5$

(b) $\displaystyle\lim_{x\to 0}(\sqrt{x^2+9}-3)/\sin x$

(c) $\displaystyle\lim_{x\to\infty}\int_0^x dx/(x^2+x+1)$

(d) $\displaystyle\lim_{x\to 0}(\sqrt[3]{x^3+27}-3)/x$

19. Which of the following integrals are convergent? Evaluate when possible.

(a) $\displaystyle\int_1^\infty (1/x^2)\,dx$

(b) $\displaystyle\int_1^\infty dx/\sqrt{x^2+8x+12}$

(c) $\displaystyle\int_2^\infty dx/\ln x$ [*Hint:* Prove that $\ln x\le x$ for $x\ge 2$.]

(d) $\displaystyle\int_{-\infty}^\infty [\sin x/(x^2+3)]\,dx$

(e) $\displaystyle\int_1^2 [1/\sqrt{x-1}]\,dx$

20. Evaluate when convergent:

(a) $\displaystyle\int_{-1}^0 [(x+1)/\sqrt{1-x^2}]\,dx$

(b) $\displaystyle\int_0^1 dx/(1-x)^{2/5}$

(c) $\displaystyle\int_1^\infty x^2 e^{-x}\,dx$ 　(d) $\displaystyle\int_0^1 x\ln x\,dx$

(e) $\displaystyle\int_0^\infty (x+2)e^{-(x^2+4x)}\,dx$

21. Find:

(a) $\displaystyle\lim_{x\to 0+}(\cot x)/\ln x$

(b) $\displaystyle\lim_{x\to 1}\left(\frac{1}{\ln x}-\frac{1}{x-1}-\frac{1}{2}\right)$

(c) $\displaystyle\lim_{x\to 0+}x^{\sin x}$

(d) $\displaystyle\lim_{x\to\infty}(\sin e^{-x})^{1/\sqrt{x}}$

(e) $\displaystyle\lim_{n\to\infty}(1+8/n)^n$

22. Find a function on $[0,1]$ which is integrable (as an improper integral) but whose square is not.

23. Show that $\int_0^\infty [(\sin x)/(1+x)]\,dx$ is convergent. [*Hint:* Integrate by parts.]

24. Use Newton's method to locate the roots of $x^3-3x^2+8=0$.

▦25. (a) Evaluate $(2/\sqrt{\pi}) \int_0^1 e^{-t^2}\, dt$ by using Simpson's rule with 10 subdivisions.

(b) Give an upper bound for the error in part (a). (See Solved Exercise 4 of Section 12-4.)

(c) What does Simpson's rule with 10 subdivisions give for $(2/\sqrt{\pi}) \int_0^{10} e^{-t^2}\, dt$?

(d) The function $(2/\sqrt{\pi}) \int_0^x e^{-t^2}\, dt$ is denoted $\mathrm{erf}(x)$ and is called the error function. Its values are tabulated. (For example: *Handbook of Mathematical Functions*, National Bureau of Standards, Applied Mathematics Series 55, June 1964, pp. 310–311.) Compare your results with the tabulated results. *Note:* $\lim_{x\to\infty} \mathrm{erf}(x) = 1$, and $\mathrm{erf}(10)$ is so close to 1 that it probably won't be listed in the tables. Explain your result in part (c).

★26. Evaluate $\lim_{x\to\pi^2} [(\cos\sqrt{x} + 1)/(x - \pi^2)]$ by recognizing the limit to be a derivative.

★27. Evaluate:

$$\lim_{x\to\pi^2} \left[\frac{\sin\sqrt{x}}{(\sqrt{x} - \pi)(\sqrt{x} + \pi)} + \tan\sqrt{x} \right]$$

★28. Prove the chain rule, $(f \circ g)'(x_0) = f'(g(x_0)) \cdot g'(x_0)$, as follows:

(a) Let $y = g(x)$ and $z = f(y)$, and write

$$\Delta y = g'(x_0)\Delta x + \rho(x)$$

Show that $\lim_{\Delta x\to 0} \dfrac{\rho(x)}{\Delta x} = 0$

Also write

$$\Delta z = f'(y_0)\Delta y + \sigma(y); \quad y_0 = g(x_0)$$

and show that

$$\lim_{\Delta y\to 0} \frac{\sigma(y)}{\Delta y} = 0$$

(b) Show that

$$\Delta z = f'(y_0)g'(x_0)\Delta x + f'(y_0)\rho(x)$$
$$+ \sigma(g(x))$$

(c) Note that $\sigma(g(x)) = 0$ if $\Delta y = 0$. Thus show that

$$\frac{\sigma(g(x))}{\Delta x} = \begin{cases} \dfrac{\sigma(g(x))}{\Delta y}\dfrac{\Delta y}{\Delta x} & \text{if } \Delta y \neq 0 \\[2ex] 0 & \text{if } \Delta y = 0 \end{cases} \to 0$$

as $\Delta x \to 0$.

(d) Use parts (b) and (c) above to show that $\lim_{\Delta x\to 0} \Delta z/\Delta x = f'(y_0)g'(x_0)$. (This proof avoids the problem of division by zero mentioned on p. 235.)

★29. Limits can sometimes be evaluated by geometric techniques. An important instance occurs when the curve $y = f(x)$ is trapped between the two intersecting lines through (a, L) with slopes m and $-m$, $0 < |x - a| \leq h$. The geometry says $\lim_{x\to a} f(x) = L$, because approaches on $y = f(x)$ from the left or right are forced into a vertex, and therefore to the geometric point (a, L).

(a) The equations of the lines are $y = L + m(x - a)$, $y = L - m(x - a)$. Draw these on a figure and insert a representative graph for f which stays between the lines.

(b) Show that the algebraic condition that f stay between the two straight lines is

$$\left| \frac{f(x) - L}{x - a} \right| \leq m, \; 0 < |x - a| \leq h$$

This is called a *Lipschitz condition*.

(c) Argue that a Lipschitz condition implies $\lim_{x\to a} f(x) = L$, by appeal to the definition of limit.

★30. Another geometric technique for evaluation of limits is obtained by requiring that $y = f(x)$ be trapped on $0 \leq |x - a| \leq h$ between two power curves

$$y = L + m(x - \alpha)^\alpha \qquad y = L - m(x - a)^\alpha$$

where $\alpha > 0$, $m > 0$. The resulting algebraic condition is called a *Hölder condition*:

$$\frac{|f(x) - L|}{|x - a|^\alpha} \leq m, \; 0 < |x - a| \leq h$$

(a) Verify that the described geometry leads to the Hölder condition.

(b) Argue geometrically that, in the presence of a Hölder condition, $\lim_{x\to a} f(x) = L$.

(c) Prove the contention in (b) by appeal to the definition of limit.

13

Infinite Series

Infinite sums can be used to represent numbers and functions.

The decimal expansion $\frac{1}{3} = 0.3333\ldots$ is a representation of $\frac{1}{3}$ as a sum $\frac{3}{10} + \frac{3}{100} + \frac{3}{1000} + \frac{3}{10,000} + \cdots$ of infinitely many rational numbers. In this chapter, we will see how to represent many numbers by infinite sums of "simpler" numbers. Moreover, we will be able to represent whole functions of x by infinite sums whose terms are constant multiples of powers of x. For example, we will see that

$$\ln 2 = 1 - \frac{1}{2} + \frac{1}{3} - \frac{1}{4} + \cdots$$

and

$$\sin x = x - \frac{x^3}{1 \cdot 2 \cdot 3} + \frac{x^5}{1 \cdot 2 \cdot 3 \cdot 4 \cdot 5} - \cdots$$

CONTENTS

In Section 13-1, the sum of an infinite series is defined as the limit, if it exists, of a sequence of finite sums. When the limit exists, the series is said to be *convergent;* otherwise it is *divergent.* We find the sum of a geometric series and present several tests for determining whether a given series converges or not. Further tests are presented in Section 13-2.

Section 13-3 is devoted to sums whose terms are of the form ax^i. Such a sum, if convergent, defines a function of x. We show how to compute with such series as if they were polynomials, as well as how to obtain the series representation of a given function.

*This is found in Appendix B.

13-1 The Sum of an Infinite Series

The sum of infinitely many numbers may be finite.

An *infinite series* is a sequence of numbers whose terms are to be added up. If the resulting sum is finite, the series is said to be *convergent*. In this section, we define the notion of convergence, give the simplest examples, and present some basic tests.

Goals

After studying this section, you should be able to:

Identify and sum geometric series.

Explain the divergence of the harmonic series.

Determine whether certain series converge or diverge by comparing them with geometric or harmonic series.

PARTIAL SUMS, CONVERGENCE, AND DIVERGENCE

Our first example (in Section 12-1) of the limit of a sequence was an expression for the number $\frac{1}{3}$:

$$\frac{1}{3} = \lim_{n \to \infty} \left(\frac{3}{10} + \frac{3}{100} + \cdots + \frac{3}{10^n} \right)$$

This expression suggests that we consider $\frac{1}{3}$ as the sum

$$\frac{3}{10} + \frac{3}{100} + \cdots + \frac{3}{10^n} + \cdots$$

of infinitely many terms. Of course, not every sum of infinitely many terms gives rise to a number (consider $1 + 1 + 1 + \cdots$), so we must be precise about what we mean by adding together infinitely many numbers. Following the idea used in the theory of improper integrals (in Section 12-3), we will define the sum of an infinite series by taking finite sums and then passing to the limit as the sum includes more and more terms.

Definition Let a_1, a_2, ... be a sequence of numbers. The number $S_n = a_1 + a_2 + \cdots + a_n = \Sigma_{i=1}^{n} a_i$ is called the *nth partial sum* of the a_i's. If the sequence S_1, S_2, ... of partial sums converges to a limit S as $n \to \infty$, we say that *the series* $a_1 + a_2 + \cdots = \Sigma_{i=1}^{\infty} a_i$ *converges*, and we write

$$\sum_{i=1}^{\infty} a_i = S$$

Hence

$$\sum_{i=1}^{\infty} a_i = \lim_{n \to \infty} \sum_{i=1}^{n} a_i$$

The limit of the partial sums is called the *sum* of the series. If the series $\Sigma_{i=1}^{\infty} a_i$ does not converge, we say that it *diverges*. In this case, the series has no sum.

CONVERGENCE OF SERIES

A series $\displaystyle\sum_{i=1}^{\infty} a_i$ converges if $\qquad$ $\displaystyle\lim_{n \to \infty} \sum_{i=1}^{n} a_i$ exists (and is finite).

A series $\displaystyle\sum_{i=1}^{\infty} a_i$ diverges if $\qquad$ $\displaystyle\lim_{n \to \infty} \sum_{i=1}^{n} a_i$ does not exist (or is infinite).

Do not confuse a *sequence* with a *series*. A sequence is simply an infinite list of numbers (separated by commas): $a_1, a_2, a_3, \ldots$. A series is an infinite list of numbers (separated by plus signs) which are meant to be added together: $a_1 + a_2 + a_3 + \cdots$. Of course, the terms in an infinite series may themselves be considered as a sequence, but the most important sequence associated with the series $a_1 + a_2 + \cdots$ is its sequence of partial sums: $S_1, S_2, S_3, \ldots$—that is, the sequence

$$a_1, a_1 + a_2, a_1 + a_2 + a_3, \cdots$$

A finite sum $a_1 + a_2 + \cdots + a_n$ may be considered as an infinite series if we follow it by infinitely many zeros: $a_1 + a_2 + \cdots + a_n = a_1 + a_2 + \cdots + a_n + 0 + 0 + \cdots$. Such a series is clearly convergent, since its partial sums after the nth are all the same, but it is not terribly interesting.

Perhaps the simplest example of a "truly infinite" series is the *geometric series*

$$a + ar + ar^2 + \cdots$$

in which the ratio between each two successive terms is the same. To write a geometric series in summation notation, it is convenient to allow the index i to start at zero, so that $a_0 = a$, $a_1 = ar$, $a_2 = ar^2$, and so on. The general term is then $a_i = ar^i$, and the series is compactly expressed as $\Sigma_{i=0}^{\infty} ar^i$. In our notation $\Sigma_{i=1}^{\infty} a_i$ for a general series, the index i will start at 1, but in special examples we are free to start it wherever we wish. Also, we may replace the index i by any other letter: $\Sigma_{i=1}^{\infty} a_i = \Sigma_{j=1}^{\infty} a_j = \Sigma_{n=1}^{\infty} a_n$, and so forth.

To find the sum of the geometric series, we must first evaluate the partial sums $S_n = \Sigma_{i=0}^{n} ar^i$. We write

$$S_n = a + ar + ar^2 + \cdots + ar^n$$
$$rS_n = \qquad ar + ar^2 + \cdots + ar^n + ar^{n+1}$$

Subtracting the second equation from the first and solving for S_n, we find

$$S_n = \frac{a(1 - r^{n+1})}{1 - r} \tag{1}$$

The sum of the entire series is the limit

$$\lim_{n \to \infty} S_n = \left(\frac{a}{1 - r} \right) \left(1 - \lim_{n \to \infty} r^{n+1} \right)$$

If $|r| < 1$, then $\lim_{n \to \infty} r^{n+1} = 0$ (see Theorem 1 of Chapter 12), so $\lim_{n \to \infty} S_n = a/(1 - r)$ and hence $\sum_{i=0}^{\infty} ar^i$ is convergent. If $|r| > 1$, or $r = -1$, the limit $\lim_{n \to \infty} r^{n+1}$ does not exist, so $\sum_{i=0}^{\infty} ar^i$ diverges (as long as $a \neq 0$). Finally, if $r = 1$, formula (1) makes no sense. Going back to the definition $S_n = \sum_{i=0}^{n} ar^i$ and substituting $r = 1$, we find that $S_n = a(n + 1)$, so the series is divergent in this case as well. See Fig. 13-1-1.

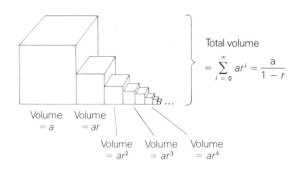

Total volume

$$= \sum_{i=0}^{\infty} ar^i = \frac{a}{1 - r}$$

Volume = a Volume = ar Volume = ar^2 Volume = ar^3 Volume = ar^4

Fig. 13-1-1 The sum of a geometric series represented as the total volume of a sequence of blocks, each r times smaller in volume than its predecessor.

GEOMETRIC SERIES

If $|r| < 1$ and a is any number, then $a + ar + ar^2 + \cdots = \sum_{i=0}^{\infty} ar^i$ converges and the sum is $a/(1 - r)$.

If $|r| \geq 1$ and $a \neq 0$, then $\sum_{i=0}^{\infty} ar^i$ diverges.

Worked Example 1 Sum the series $1 + \frac{1}{3} + \frac{1}{9} + \frac{1}{27} + \frac{1}{81} + \cdots$.

Solution This is a geometric series with $r = \frac{1}{3}$ and $a = 1$. (Note that a is the first term and r is the ratio of any term to the preceding one.) Thus

$$1 + \frac{1}{3} + \frac{1}{9} + \cdots = \sum_{i=0}^{\infty} \left(\frac{1}{3} \right)^i = \frac{1}{1 - 1/3} = \frac{3}{2}$$

Two useful general rules for summing series are presented in the following display.

ALGEBRAIC RULES

Sum rule

If $\sum\limits_{i=1}^{\infty} a_i$ and $\sum\limits_{i=1}^{\infty} b_i$ converge,   then $\sum\limits_{i=1}^{\infty} (a_i + b_i)$ converges and

$$\sum_{i=1}^{\infty} (a_i + b_i) = \sum_{i=1}^{\infty} a_i + \sum_{i=1}^{\infty} b_i.$$

Constant multiple rule

If $\sum\limits_{i=1}^{\infty} a_i$ converges and c is any real number,   then $\sum\limits_{i=1}^{\infty} ca_i$ converges and

$$\sum_{i=1}^{\infty} ca_i = c \sum_{i=1}^{\infty} a_i.$$

To prove the validity of these rules, one simply notes that the identities $\Sigma_{i=1}^{n} (a_i + b_i) = \Sigma_{i=1}^{n} a_i + \Sigma_{i=1}^{n} b_i$ and $\Sigma_{i=1}^{n} ca_i = c\, \Sigma_{i=1}^{n} a_i$ are satisfied by the partial sums. Taking limits as $n \to \infty$ and applying the sum and constant multiple rules for limits of sequences (see p. 493), we obtain the rules in the preceding display.

Worked Example 2 Sum the series $\Sigma_{i=0}^{\infty} (3^i - 2^i)/6^i$.

Solution We may write the ith term as

$$\frac{3^i}{6^i} - \frac{2^i}{6^i} = \left(\frac{1}{2}\right)^i - \left(\frac{1}{3}\right)^i = \left(\frac{1}{2}\right)^i + (-1)\left(\frac{1}{3}\right)^c$$

Since the series $\Sigma_{i=0}^{\infty} (1/2)^i$ and $\Sigma_{i=0}^{\infty} (1/3)^i$ are convergent, with sums 2 and $\frac{3}{2}$ respectively, the algebraic rules imply that

$$\sum_{i=0}^{\infty} \frac{3^i - 2^i}{6^i} = \sum_{i=0}^{\infty} \left[\left(\frac{1}{2}\right)^i + (-1)\left(\frac{1}{3}\right)^i \right]$$
$$= \sum_{i=0}^{\infty} \left(\frac{1}{2}\right)^i + (-1) \sum_{i=0}^{\infty} \left(\frac{1}{3}\right)^i$$
$$= 2 - \frac{3}{2} = \frac{1}{2}$$

Finally, we remark that the sum rule implies that *we may change (or remove—that is, change to zero) finitely many terms of a series without affecting its convergence.* In fact, changing finitely many terms of the series $\Sigma_{i=0}^{\infty} a_i$ is equivalent to adding to it a series whose terms are all zero beyond a certain point. As we already remarked, such a finite series is always convergent, so adding it to a convergent series produces a convergent result. Of course, the sum of the new series is *not* the same as that of the old one.

Solved Exercises*

1. Zeno's paradox concerns a race between Achilles and a tortoise. The tortoise begins with a head start of 10 meters, and Achilles ought to overtake it. After a certain elapsed time from the start, Achilles reaches the point A where the tortoise started, but the tortoise has moved ahead to point B (Fig. 13-1-2). After a certain further interval of time, Achilles reaches point B, but the tortoise has moved ahead to a point C, and so on, forever. Zeno concludes from this argument that Achilles can never pass the tortoise. **Where is the fallacy?**

Fig. 13-1-2   Will the runner overtake the tortoise?

2. Sum the series:

   (a) $\displaystyle\sum_{n=0}^{\infty} 1/6^{n/2}$

   (b) $\displaystyle\sum_{i=1}^{\infty} 1/5^{i}$

3. Show that the series $1\frac{1}{2} + 3\frac{3}{4} + 7\frac{7}{8} + 15\frac{15}{16} + \cdots$ diverges. [*Hint:* Write it as the difference of a divergent and a convergent series.]

Exercises

1. Sum the following series:

   (a) $\displaystyle\sum_{j=1}^{\infty} \frac{1}{13^{j}}$

   (b) $\displaystyle\sum_{n=1}^{\infty} \frac{2^{n} + 3^{n}}{6^{n}}$

   (c) $\displaystyle\sum_{k=1}^{\infty} \frac{3^{2k} + 1}{27^{k}}$

   (d) $\displaystyle\sum_{n=5}^{\infty} \frac{2^{n+1}}{3^{n-2}}$

2. Give an example to show that $\sum_{i=1}^{\infty} (a_i + b_i)$ may converge while both $\sum_{i=1}^{\infty} a_i$ and $\sum_{i=1}^{\infty} b_i$ diverge.

A NECESSARY CONDITION FOR CONVERGENCE

The terms of a series $\sum_{i=0}^{\infty} a_i$ can be reconstructed from the partial sums $S_n = \sum_{i=0}^{n} a_i$; a_{i+1} is just the difference $S_{i+1} - S_i$. (If the S_i's are points on a number line, then a_i represents the "jump" from one partial sum to the next.) If the series is convergent, the sequence of partial sums approaches a limit S, and it follows that the differences $a_i = S_i - S_{i-1}$ must go to zero as $i \to \infty$. In fact,

$$\lim_{i \to \infty} a_i = \lim_{i \to \infty} (S_i - S_{i-1}) = \lim_{i \to \infty} S_i - \lim_{i \to \infty} S_{i-1} = S - S = 0$$

Hence for a series $\sum_{i=0}^{\infty} a_i$ to be convergent, it is necessary that $\lim_{i \to \infty} a_i$ be equal to zero.

Solutions appear in Appendix A.

Worked Example 3   Test for convergence: $\Sigma_{i=1}^{\infty} i/(1 + i)$.

Solution   Here

$$a_i = \frac{i}{1 + i} = \frac{1}{1/i + 1} \to 1$$

as $i \to \infty$.* Since a_i does not tend to zero, the series must diverge.

To show that the condition "$\lim_{i \to \infty} a_i = 0$" is not *sufficient* for convergence, we will consider the series

$$1 + \frac{1}{2} + \frac{1}{3} + \frac{1}{4} + \cdots = \sum_{i=1}^{\infty} \frac{1}{i}$$

which is called the *harmonic series*. Its *i*th term, $1/i$, approaches zero as $i \to \infty$, but we can show that the series *does not converge*. To do this, we notice a pattern:

$$1$$

$$\frac{1}{2}$$

$$\frac{1}{3} + \frac{1}{4} \qquad\qquad > \quad \frac{1}{4} + \frac{1}{4} = \frac{1}{2}$$

$$\frac{1}{5} + \frac{1}{6} + \frac{1}{7} + \frac{1}{8} \qquad > \quad \frac{1}{8} + \frac{1}{8} + \frac{1}{8} + \frac{1}{8} = \frac{1}{2}$$

$$\frac{1}{9} + \cdots + \frac{1}{16} \qquad > \quad \frac{1}{16} + \cdots + \frac{1}{16} = \frac{1}{2}$$

$$\frac{1}{17} + \cdots + \frac{1}{32} \qquad > \quad \frac{1}{32} + \cdots + \frac{1}{32} = \frac{1}{2}$$

$$\vdots \qquad\qquad\qquad \vdots \quad \vdots \qquad\qquad \vdots$$

and so on. Thus the partial sum S_4 is greater than $1 + \frac{1}{2} + \frac{1}{2} = 1 + \frac{2}{2}$, $S_8 > 1 + \frac{1}{2} + \frac{1}{2} + \frac{1}{2} = 1 + \frac{3}{2}$ and, in general $S_{2^n} > 1 + n/2$, which becomes arbitrarily large as n becomes large. Therefore the harmonic series diverges.

Worked Example 4   Show that the series $\frac{1}{2} + \frac{1}{4} + \frac{1}{6} + \frac{1}{8} + \cdots$ diverges.

Solution   This series is $\Sigma_{i=1}^{\infty} 1/2i$. If it converged, so would the series $\Sigma_{i=1}^{\infty} 2 \cdot (1/2i)$, by the constant multiple rule. But $\Sigma_{i=1}^{\infty} 2 \cdot 1/2i = \Sigma_{i=1}^{\infty} 1/i$, which we have shown to diverge.

THE $\lim_{i \to \infty} a_i$ TEST	
If $\Sigma_{i=1}^{\infty} a_i$ converges,	then $\lim_{i \to \infty} a_i = 0$.
If $\lim_{i \to \infty} a_i \neq 0$,	then $\Sigma_{i=1}^{\infty}$ diverges.
If $\lim_{i \to \infty} a_i = 0$,	then $\Sigma_{i=1}^{\infty} a_i$ may converge (for example, geometric series) or diverge (for example, harmonic series).

Recall that the notation "$a_i \to l$ as $i \to \infty$" means the same as "$\lim_{i \to \infty} a_i = l$"

Solved Exercises

4. Test for convergence: $\sum_{i=1}^{\infty} (-1)^i (i/\sqrt{1+i})$.

5. Test for convergence: $\sum_{i=1}^{\infty} 1/(1+i)$.

Exercises

3. Test for convergence:

(a) $\displaystyle\sum_{i=1}^{\infty} i/\sqrt{i+1}$

(b) $\displaystyle\sum_{i=1}^{\infty} (\sqrt{i}+1)/(\sqrt{i}+8)$

(c) $\displaystyle\sum_{i=1}^{\infty} 3/(5+5i)$ [*Hint:* Factor out $\frac{3}{5}$.]

(d) $1 + \frac{1}{2} + \underbrace{\frac{1}{4} + \frac{1}{4}}_{2} + \underbrace{\frac{1}{8} + \frac{1}{8} + \frac{1}{8} + \frac{1}{8}}_{4} + \underbrace{\frac{1}{16} + \cdots}_{8}$

(e) $1 + \underbrace{\frac{1}{4} + \frac{1}{4}}_{2} + \underbrace{\frac{1}{16} + \frac{1}{16} + \frac{1}{16} + \frac{1}{16}}_{4} + \underbrace{\frac{1}{64} + \frac{1}{64} + \cdots}_{8}$

4. Show that the series $\sum_{j=1}^{\infty} (1 - 2^{-j})/j$ diverges.

COMPARISON TESTS

Series often arise which, unlike the geometric series, cannot be summed explicitly. It is still useful to be able to prove that such a series converges, for then we can approximate the sum to any desired accuracy by adding up sufficiently many terms of the series.

One way to tell whether a series converges or diverges is to compare it with a series which we already know to converge or diverge. As a fringe benefit of such a "comparison test," we sometimes get an estimate of the difference between the nth partial sum and the exact sum. Thus if we desire to find the sum with a given accuracy, we know how many terms to take.

(optional)   We will work our way toward a practical comparison test by starting with the following observation.*

Theorem 1 *Suppose that $a_i \geq 0$ and that there is a number T such that $S_n \leq T$ for all $n = 1, 2, \ldots$, where $S_n = \sum_{i=1}^{n} a_i$ is the nth partial sum. Then $\sum_{i=1}^{\infty} a_i$ converges and $\sum_{i=1}^{\infty} a_i \leq T$.*

Proof (See Fig. 13-1-3.) Since $a_i \geq 0$, $S_{n+1} = S_n + a_{n+1} \geq S_n$, so S_n is an increasing sequence. By assumption, this sequence is bounded above by T, so by Theorem 3 of Section 12-1 it converges. That the sum is less than or equal to T follows from this fact about limits: If $S_n \leq T$ and S_n converges to S, then $S \leq T$. (See Problem 13(a), Section 12-1.)

Fig. 13-1-3 The partial sums of the series $\sum_{i=1}^{\infty} a_i$ are bounded above by T.

You may wish to omit the optional material in small type throughout this section. Consult your instructor.

(optional)

Let us also observe that if $a_i \geq 0$ and $S = \Sigma_{i=1}^{\infty} a_i$ is convergent, then $S_n \leq S$. Indeed, as in the proof just given, S_n is an *increasing* sequence converging to S.

Normally Theorem 1 is not easy to use in practice because it may be hard to find a bound T. A useful stepping-stone to more practical tests is the concept of absolute convergence.

Definition Let $\Sigma_{i=1}^{\infty} a_i$ be any series. If the series $\Sigma_{i=1}^{\infty} |a_i|$ of absolute values converges, the series $\Sigma_{i=1}^{\infty} a_i$ is said to be *absolutely convergent*.

The basic fact about absolute convergence is:

Theorem 2 *An absolutely convergent series is convergent; moreover,* $|\Sigma_{i=1}^{\infty} a_i| \leq \Sigma_{i=1}^{\infty} |a_i|$.

(optional)

Proof Let the series be $\Sigma_{i=1}^{\infty} a_i$. We define two new series, $\Sigma_{i=1}^{\infty} b_i$ and $\Sigma_{i=1}^{\infty} c_i$, by the formulas

$$b_i = \left\{ \begin{array}{ll} |a_i| & \text{if } a_i \geq 0 \\ 0 & \text{if } a_i < 0 \end{array} \right\} = \left\{ \begin{array}{ll} a_i & \text{if } a_i \geq 0 \\ 0 & \text{if } a_i < 0 \end{array} \right\}$$

$$c_i = \left\{ \begin{array}{ll} |a_i| & \text{if } a_i \leq 0 \\ 0 & \text{if } a_i > 0 \end{array} \right\} = \left\{ \begin{array}{ll} -a_i & \text{if } a_i \leq 0 \\ 0 & \text{if } a_i > 0 \end{array} \right\}$$

These are the "positive and negative parts" of the series $\Sigma_{i=1}^{\infty} a_i$. It is easy to check that $a_i = b_i - c_i$. The series $\Sigma_{i=1}^{\infty} b_i$, $\Sigma_{i=1}^{\infty} c_i$ are both convergent; in fact since $b_i \leq |a_i|$, we have $\Sigma_{i=1}^{n} b_i \leq \Sigma_{i=1}^{n} |a_i| \leq \Sigma_{i=1}^{\infty} |a_i|$, which is finite since we assumed the series $\Sigma_{i=1}^{\infty} a_i$ to be absolutely convergent. Since $b_i \geq 0$ for all i, $\Sigma_{i=1}^{\infty} b_i$ is convergent by Theorem 1. The same argument proves that $\Sigma_{i=1}^{\infty} c_i$ is convergent. The sum and constant multiple rules now apply to give the convergence of $\Sigma_{i=1}^{\infty} a_i = \Sigma_{i=1}^{\infty} b_i - \Sigma_{i=1}^{\infty} c_i$.

Finally, we note that, by the triangle inequality,

$$\left| \sum_{i=1}^{n} a_i \right| \leq \sum_{i=1}^{n} |a_i| \leq \sum_{i=1}^{\infty} |a_i|$$

Since this is true for all n, and

$$\left| \sum_{i=1}^{\infty} a_i \right| = \left| \lim_{n \to \infty} \sum_{i=1}^{n} a_i \right| = \lim_{n \to \infty} \left| \sum_{i=1}^{n} a_i \right| \quad \text{(the absolute value function is continuous),}$$

it follows that $|\Sigma_{i=1}^{\infty} a_i| \leq \Sigma_{i=1}^{\infty} |a_i|$. (Here we used the fact that if $b_n \leq M$ for all n and b_n converges to b, then $b \leq M$; compare Problem 13(a), p. 499.)

Theorems 1 and 2 can now be used to give us our first practical test.

Theorem 3 *Comparison Test: If $|a_i| \leq b_i$ for all i, and $\Sigma_{i=1}^{\infty} b_i$ is convergent, then $\Sigma_{i=1}^{\infty} a_i$ is (absolutely) convergent.*

(optional)

Proof Let $S_n = \Sigma_{i=1}^{n} |a_i|$ and $T_n = \Sigma_{i=1}^{n} b_i$. Since $b_i \geq |a_i| \geq 0$ for all i, the sequence T_n is increasing. By assumption, it converges to $T = \Sigma_{i=1}^{\infty} b_i$, so $T_n \leq T$ for all n. Since $|a_i| \leq b_i$, we have $S_n \leq T_n$, so $S_n \leq T$ as well. Thus, by Theorem 1, $\Sigma_{i=1}^{\infty} |a_i|$ is convergent; that is, $\Sigma_{i=1}^{\infty} a_i$ is absolutely convergent. By Theorem 2, $\Sigma_{i=1}^{\infty} a_i$ is convergent.

Corollary *If $a_i \geq b_i \geq 0$ and if $\Sigma_{i=1}^{\infty} b_i$ diverges, then so does $\Sigma_{i=1}^{\infty} a_i$.*

Proof If $\Sigma_{i=1}^{\infty} a_i$ converged, so would $\Sigma_{i=1}^{\infty} b_i$, by Theorem 3.

The following worked example shows how the comparison test can be used to prove the convergence of a series. In Solved Exercise 6, we will show how to estimate the difference between a partial sum and the exact sum.

Worked Example 5 Prove that $\Sigma_{i=1}^{\infty} (-1)^i / (i3^{i+1})$ converges.

Solution We can compare the series with $\Sigma_{i=1}^{\infty} 1/3^i$. Let $a_i = (-1)^i/(i3^{i+1})$ and $b_i = 1/3^i$. Since $i3^{i+1} = (3i) \cdot 3^i > 3^i$, we have

$$|a_i| = \frac{1}{i3^{i+1}} < \frac{1}{3^i} = b_i$$

Therefore, since $\Sigma_{i=1}^{\infty} b_i$ converges (it is a geometric series), so does $\Sigma_{i=1}^{\infty} a_i$, by Theorem 3.

Sometimes a series "resembles" another one, so that we may expect both of them either to converge or diverge. For example, if we look at $\Sigma_{i=1}^{\infty} 2/(4 + i)$, we may observe that, for large i, the 4 in the denominator becomes negligible with respect to the i, so that a_i "behaves" like $2/i$. To make this observation useful, we have the following theorem.

Theorem 4 *Let $\Sigma_{i=1}^{\infty} b_i$ be a series with $b_i > 0$ for all i, and let $\Sigma_{i=1}^{\infty} a_i$ be another series.*

1. *If $\lim\limits_{i \to \infty} |a_i|/b_i < \infty$ and $\Sigma_{i=1}^{\infty} b_i$ is convergent, then $\Sigma_{i=1}^{\infty} a_i$ is absolutely convergent.*

2. *If $\lim\limits_{i \to \infty} a_i/b_i$ is strictly positive or ∞, $a_i \geq 0$ for all i, and $\Sigma_{i=1}^{\infty} b_i$ is divergent, then $\Sigma_{i=1}^{\infty} a_i$ is divergent.*

The proof, outlined in Problem 8, uses a comparison of $\Sigma_{i=1}^{\infty} a_i$ with a constant times $\Sigma_{i=1}^{\infty} b_i$.

Worked Example 6   Prove that $\Sigma_{i=1}^{\infty} 2/(4 + i)$ diverges.

Solution   Let $a_i = 2/(4 + i)$, $b_i = 1/i$. Then

$$\lim_{i \to \infty} \frac{a_i}{b_i} = \lim_{i \to \infty} \frac{2/(4 + i)}{1/i} = \lim_{i \to \infty} \frac{2i}{4 + i} = \lim_{i \to \infty} \frac{2}{(4/i) + 1} = \frac{2}{0 + 1} = 2$$

Now we may apply Theorem 4(2). Since $2 > 0$, and $\Sigma_{i=1}^{\infty} 1/i$ is divergent, it follows that $\Sigma_{i=1}^{\infty} 2/(4 + i)$ is divergent as well.

COMPARISON TESTS

Let $\Sigma_{i=1}^{\infty} a_i$ and $\Sigma_{i=1}^{\infty} b_i$ be series, with $b_i > 0$ for all i.

If (1) $\lvert a_i \rvert \leq b_i$ for all i, or if	If (1) $a_i \geq b_i$ for all i, or if
$$\lim_{i \to \infty} \frac{\lvert a_i \rvert}{b_i} < \infty$$	$$\lim_{i \to \infty} \frac{a_i}{b_i} > 0$$
and	and
(2) $\Sigma_{i=1}^{\infty} b_i$ is convergent,	(2) $\Sigma_{i=1}^{\infty} b_i$ is divergent,
then $\Sigma_{i=1}^{\infty} a_i$ is convergent.	then $\Sigma_{i=1}^{\infty} a_i$ is divergent.

At the moment we have only the geometric and harmonic series to use as a basis for comparison. We will learn some others in the next section.

Solved Exercises   6. Find the partial sum $\Sigma_{i=1}^{3} (-1)^i/(i3^{i+1})$ (see Worked Example 5) and estimate the difference between this partial sum and the sum of the entire series.

7. Test for convergence: $\Sigma_{i=1}^{\infty} 1/(2^i - i)$.

Exercises

5. Test the following series for convergence:

(a) $\displaystyle\sum_{n=1}^{\infty} \frac{3}{4^n + 2}$

(b) $\displaystyle\sum_{n=1}^{\infty} \left(\frac{-4}{2^n + 3}\right)^n$

(c) $\displaystyle\sum_{i=1}^{\infty} \frac{1}{2^i + 3^i}$

(d) $\displaystyle\sum_{i=1}^{\infty} \frac{(\frac{1}{2})^i}{i + 6}$

(e) $\displaystyle\sum_{i=1}^{\infty} \frac{1}{3i + 1/i}$

6. Estimate the error in approximating the given series $\Sigma_{i=1}^{\infty} \, a_i$ by the specified partial sum S_n, $(S_n = \Sigma_{i=1}^{n} \, a_i)$.

(a) $\displaystyle\sum_{i=1}^{\infty} \frac{3}{3^i + 2}$ by S_5

(b) $\displaystyle\sum_{i=1}^{\infty} \left(\frac{1}{2i + 3}\right)^{i+2}$ by S_6

(c) $\displaystyle\sum_{n=1}^{\infty} \frac{n+1}{1 + 4^n \cdot n}$ by S_5

Problems for Section 13-1

1. Test for convergence:

(a) $\displaystyle\sum_{i=1}^{\infty} \frac{2}{2i + 1}$

(b) $\displaystyle\sum_{n=1}^{\infty} \frac{4^n + 5^n}{2^n 3^n}$

(c) $\displaystyle\sum_{i=1}^{\infty} \left(\frac{1}{i+2}\right)^i$

(d) $\dfrac{1}{3} + \dfrac{1}{5} + \dfrac{1}{9} + \dfrac{1}{17} + \dfrac{1}{33} + \dfrac{1}{65} +$

$\cdots + \dfrac{1}{2^n + 1} + \cdots$

(e) $\displaystyle\sum_{i=1}^{\infty} \frac{3i}{2^i}$

2. Test for convergence:

(a) $\displaystyle\sum_{i=1}^{\infty} \frac{1}{\sqrt{i + 2}}$

(b) $\displaystyle\sum_{j=1}^{\infty} \frac{\sin j}{2^j}$

(c) $\displaystyle\sum_{i=1}^{\infty} \frac{(-2)^i}{3^i + 1}$

(d) $\displaystyle\sum_{i=1}^{\infty} \frac{1 + (-1)^i}{8i + 2^{i+1}}$

(e) $\displaystyle\sum_{n=1}^{\infty} \frac{\sqrt{3 + n}}{4^n}$

3. Test for convergence:

(a) $\displaystyle\sum_{i=1}^{\infty} \left(\frac{1}{i} + \frac{2}{i^2} + \frac{3}{i^3}\right)$

(b) $\displaystyle\sum_{i=1}^{\infty} \frac{3}{1 + 3^i}$

(c) $\displaystyle\sum_{n=1}^{\infty} e^{-n}$

(d) $\displaystyle\sum_{i=1}^{\infty} \frac{1}{\sqrt{i + 1}}$

(e) $\displaystyle\sum_{j=1}^{\infty} \frac{2^j}{j}$

(f) $\displaystyle\sum_{i=2}^{\infty} \frac{1}{\ln i}$

4. Sum the following series if they converge.

(a) $\displaystyle\sum_{i=0}^{\infty} \frac{2^{3i+4}}{3^{2i+5}}$

(b) $\displaystyle\sum_{j=-3}^{\infty} \left(\frac{1}{3}\right)^j$

(c) $\displaystyle\sum_{k=1}^{\infty} \left(\frac{4}{5}\right)^k$

(d) $\displaystyle\sum_{l=0}^{\infty} \frac{4^{4l+2}}{5^{3l+80}}$

(e) $\displaystyle\sum_{i=4}^{\infty} 5\left(\frac{1}{3}\right)^{i+1/2}$

(f) $\displaystyle\sum_{i=1}^{\infty} \left[\left(\frac{1}{2}\right)^i + \left(\frac{1}{3}\right)^{2i} + \left(\frac{1}{4}\right)^{3i+1}\right]$

5. If $0 \le a_n \le ar^n$, $r < 1$, show that the error in approximating $\Sigma_{i=1}^{\infty} \, a_i$ by $\Sigma_{i=1}^{n} \, a_i$ is less than or equal to $ar^{n+1}/(1 - r)$.

6. Find the sum of each of the following series with an error of no more than 0.01.

(a) $\displaystyle\sum_{j=1}^{\infty} \frac{1}{j4^j}$

(b) $\displaystyle\sum_{k=0}^{\infty} \frac{k}{2^k}$ [*Hint:* Compare with $\Sigma_{k=0}^{\infty} \left(\frac{2}{3}\right)^k$.]

(c) $\displaystyle\sum_{n=1}^{\infty} \frac{2^n - 1}{5^n + 1}$ (d) $\displaystyle\sum_{p=1}^{\infty} \frac{(-1)^p}{2^p + p}$

★7. A rubber ball is released from a height h. Each time it strikes the floor, it rebounds with two-thirds of its previous velocity.

(a) How far does the ball rise on each bounce? (Use the fact that the height y of the ball at time t from the beginning of each bounce is of the form $y = vt - \frac{1}{2}gt^2$ during the bounce. The constant g is the acceleration of gravity.)

(b) How long does each bounce take?

(c) Show that the ball stops bouncing after a finite time has passed.

(d) How far has the ball traveled when it stops bouncing?

(e) How would the results differ if this experiment were done on the moon?

★8. (a) Prove Theorem 4(1) as follows: Assume $\lim_{i\to\infty} |a_i|/b_i = L$, pick $M > L$, and show that, for some N, $|a_i|/b_i \le M$ if $i \ge N$. Deduce that $\Sigma_{i=N}^{\infty} |a_i|$ is convergent by the comparison test and hence conclude that $\Sigma_{i=1}^{\infty} |a_i|$ is convergent.

(b) Use similar ideas to prove Theorem 4(2).

9. A *telescoping series*, like a geometric series, can be summed. A series $\Sigma_{n=1}^{\infty} a_n$ is telescoping if its nth term a_n can be expressed as $a_n = b_{n+1} - b_n$ for some sequence $\{b_n\}_{n=1}^{\infty}$.

(a) Verify that $a_1 + a_2 + a_3 + \cdots + a_n = b_{n+1} - b_1$; therefore the series converges exactly when $\lim_{n\to\infty} b_{n+1}$ exists, and $\Sigma_{n=1}^{\infty} a_n = \lim_{n\to\infty} b_{n+1} - b_1$.

(b) Use partial fraction methods to write $1/[n(n+1)]$ as $b_{n+1} - b_n$ for some sequence $\{b_n\}$. Then evaluate $\Sigma_{n=1}^{\infty} 1/[n(n+1)]$.

10. An experiment is performed, during which time successive excursions of a deflected plate are recorded. Initially, the plate has amplitude b_0. The plate then deflects downward to form a "dish" of depth b_1, then a "dome" of height b_2, and so on. (See Fig. 13-1-4.) The a's and b's are related by $a_1 = b_0 - b_1$, $a_2 = b_1 - b_2$, $a_3 = b_2 - b_3$ .... The value a_n measures the amplitude "lost" at the nth oscillation (due to friction, say).

Fig. 13-1-4 The deflecting plate in Problem 10.

(a) Find $\Sigma_{n=1}^{\infty} a_n$. Explain why $b_0 - \Sigma_{n=1}^{\infty} a_n$ is the "average height" of the oscillating plate after a large number of oscillations.

(b) Suppose the "dishes" and "domes" decay to zero, that is, $\lim_{n\to\infty} b_{n+1} = 0$. Show that $\Sigma_{n=1}^{\infty} a_n = b_0$, and explain why this is physically obvious.

11. The *average absorption coefficient* $\bar{a}$ for sound at a particular frequency colliding with the walls and furniture of a listening room is given by

$$\bar{a} = \frac{s_1 a_1 + s_2 a_2 + s_3 a_3 + \cdots + s_{n+1} a_{n+1}}{s}$$

The term $s_i a_i$ is the absorption coefficient a_i of a particular wall, floor, or ceiling area, times its area s_i. The last term $s_{n+1} a_{n+1}$ accounts for furniture absorption. The value s is the total area of the walls, ceiling, floor.

(a) Put $A_k = a_1 + a_2 + \cdots + a_k$. Verify that $\Sigma_{k=1}^{n+1} a_k s_k = \Sigma_{k=1}^{n-1} A_k(s_k - s_{k+1}) + a_n s_n + a_{n+1} s_{n+1}$. (This is called *partial summation*.)

(b) Assume that the wall, floor, and ceiling areas s_i are all equal to a fixed value s^*, $1 \le i \le n$. Verify that $\bar{a}$ is the ratio

$$\bar{a} = \frac{\left(\displaystyle\sum_{i=1}^{n} a_i\right) s^* + a_{n+1} s_{n+1}}{ns^*}$$

12. The joining of the transcontinental railroads occurred as follows. The East and West crews were setting track 12 miles apart, the East crew working at 5 miles per hour, the West crew working at 7 miles per hour. The official with the Golden Spike traveled feverishly by carriage back and forth between the crews until the rails joined. His speed was 20 miles per hour, and he started from the East.

    (a) Assume the carriage reversed direction with no waiting time, at each encounter with an East or West crew. Let t_k be the carriage transit time for trip k. Verify that $t_{2n+2} = r^{n+1} \cdot (12/13)$, $t_{2n+1} = r^n \cdot (12/27)$ where $r = (13/27) \cdot (15/25)$, $n = 0, 1, 2, \ldots$.

    (b) Since the crews meet in one hour, the total time for the carriage travel is one hour, i.e.,
    $$\lim_{n \to \infty} (t_1 + t_2 + t_3 + t_4 + \cdots + t_n) = 1.$$
    Verify this formula by direct computation of the sequential limit, using the summation formula
    $$1 + r + r^2 + \cdots + r^n = \frac{r^{n+1} - 1}{r - 1} \quad (r \neq 1).$$

13. The celebrated example due to Karl Weierstrass of a *nowhere differentiable continuous function* $f(x)$ in $-\infty < x < \infty$ is given by
    $$f(x) = \sum_{n=0}^{\infty} \left(\frac{3}{4}\right)^n \phi(4^n x)$$
    where $\phi(x + 2) = \phi(x)$, and $\phi(x)$ on $0 \leq x \leq 2$ is the "triangle" through $(0, 0)$, $(1, 1)$, $(2, 0)$. By construction, $0 \leq \phi(4^n x) \leq 1$. Verify by means of the comparison test that the series converges for any value of x. [See *Counterexamples in Analysis* by B. R. Gelbaum and J. M. H. Olmsted, Holden-Day (1964), p. 38 for the proof that f is nowhere differentiable.]

★14. There are planar paths which are unlikely candidates for the paths of particles because they have infinite arc length. One such is $y = x \sin(1/x)$, $0 \leq x \leq 1/\pi$.

    The idea is based on the formula $(2/\pi)(\frac{1}{2} + \frac{1}{3} + \frac{1}{4} + \frac{1}{5} + \cdots) = \infty$. The procedure is to verify the inequality
    $$\int_{1/(n+1)\pi}^{1/n\pi} \sqrt{1 + (y')^2}\, dx > \frac{2}{(n+1)\pi}$$
    $(n = 1, 2, 3, \ldots)$,
    sum both sides and conclude that the arc length of $y = x \sin(1/x)$, $1/n\pi \leq x \leq 1/\pi$, is as large as we like.

    (a) Verify that $\int_{1/n\pi}^{1/\pi} ds$ is the sum of $\int_{1/2\pi}^{1/\pi} ds$, $\int_{1/3\pi}^{1/2\pi} ds$, $\int_{1/4\pi}^{1/3\pi} ds$, ..., $\int_{1/n\pi}^{1/(n-1)\pi} ds$, by using properties of integrals.

    (b) On the interval $1/[(n + 1)\pi] \leq x \leq 1/n\pi$, construct a triangle with vertices at $(1/[(n + 1)\pi], 0)$, $(2/[(2n + 1)\pi], (-1)^n)$, $(1/n\pi, 0)$. Show that the triangle lies between $y = x \sin(1/x)$ and $y = 0$.

    (c) Use the Pythagorean theorem to find the arc length of the triangular path.

    (d) Verify the cited inequality.

13-2 Further Tests for Convergence and Divergence

Calculus provides some new convergence tests.

In this section we present three more methods for determining whether a series converges.

Goals

After studying this section, you should be able to:

Recognize alternating series.

Apply the integral test and ratio test.

Determine whether certain series converge or diverge by comparing them with p series.

ALTERNATING SERIES

We saw in Section 13-1 that the harmonic series

$$1 + \tfrac{1}{2} + \tfrac{1}{3} + \tfrac{1}{4} + \cdots$$

is divergent even though $\lim\limits_{i \to \infty} 1/i = 0$. If we put a minus sign in front of every other term to obtain the series

$$1 - \tfrac{1}{2} + \tfrac{1}{3} - \tfrac{1}{4} + \cdots$$

we might hope that the alternating positive and negative terms "neutralize" one another and cause the series to converge.

Definition   A series $\Sigma_{i=1}^{\infty} a_i$ is called *alternating* if the terms a_i are alternately positive and negative and if the absolute values $|a_i|$ are decreasing to zero; that is, if:

1. $a_1 > 0$, $a_2 < 0$, $a_3 > 0$, $a_4 < 0$, and so on (or $a_1 < 0$, $a_2 > 0$, and so on)

2. $|a_1| \ge |a_2| \ge |a_3| \ge \cdots$

3. $\lim\limits_{i \to \infty} |a_i| = 0$

Conditions 1, 2, and 3 are often easy to verify. See Fig. 13-2-1.

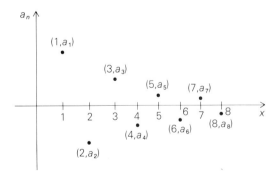

Fig. 13-2-1 Alternating series have terms that are alternately above and below the x − axis, getting successively closer to it and, in the limit, approaching it.

Worked Example 1   Is the series $1 - \frac{1}{2} + \frac{1}{3} - \frac{1}{4} \cdots$ alternating?

Solution   The terms alternate in sign, $+ - + - \cdots$, so condition 1 holds. Since the ith term $a_i = (-1)^{i+1} (1/i)$ has absolute value $1/i$, and $1/i > 1/(i + 1)$, the terms are decreasing in absolute value, so condition 2 holds. Finally, since $\lim_{i \to \infty} |a_i| = \lim_{i \to \infty} 1/i = 0$, condition 3 holds. Thus the series is alternating.

The following theorem is so general that it may surprise you.

Theorem 5   *Every alternating series converges.*

(optional)

Proof   Using condition 1, we may write the series as $b_1 - c_1 + b_2 - c_2 + b_3 - c_3 + \cdots$, where $b_i \geq 0$ and $c_i \geq 0$, by choosing $b_1 = a_1$, $c_1 = -a_2$, $b_2 = a_3$, $c_2 = -a_4$, $\cdots$. Then $b_1 \geq c_1 \geq b_2 \geq c_2 \geq b_3 \geq c_3 \geq \cdots$, by condition 2, and $\lim_{i \to \infty} b_i = 0$ and $\lim_{i \to \infty} c_i = 0$ by condition 3. Let

$$T_n = b_1 - c_1 + \cdots + b_{n-1} - c_{n-1} + b_n$$

be the $(2n - 1)$st partial sum and

$$U_n = b_1 - c_1 + \cdots + b_{n-1} - c_{n-1} + b_n - c_n$$

be the $2n$th partial sum. Notice that since $b_i - c_i \geq 0$ and $b_n > 0$, we have $T_n > 0$ and $U_n > 0$. Note also that $T_n \geq U_n$. Furthermore,

$$T_{n+1} = T_n + b_{n+1} - c_n \leq T_n \qquad \text{since } b_{n+1} - c_n \leq 0$$

and

$$U_{n+1} = U_n + b_{n+1} - c_{n+1} \geq U_n \qquad \text{since } b_{n+1} - c_{n+1} \geq 0$$

Thus, for each n,

$$U_1 \leq U_2 \leq \cdots \leq U_{n-1} \leq U_n \leq T_n \leq T_{n-1} \leq \cdots \leq T_2 \leq T_1$$

The sequence U_1, U_2, ... is therefore increasing and bounded above by T_1, so (by Theorem 3 of Section 12-1) it converges to a limit U. Similarly, the decreasing sequence T_1, T_2, ... converges to a limit T. But we have $T_n - U_n = c_n$, so

$$T - U = \lim_{n \to \infty} (T_n - U_n) = \lim_{n \to \infty} c_n$$

which is zero by condition 3 in the definition of an alternating series, so $T = U$. Since every partial sum is either a U_n or a T_n, the partial sums converge to the common value $T = U$, and so the series is convergent.

We note that in the preceding proof the limit $T = U$ lies between U_n and T_n, so that $0 \leq T_n - T \leq T_n - U_n = c_n$. Also, $0 \leq U - U_n \leq T_{n+1} - U_n = b_{n+1}$. Thus we have also proved:

Corollary *In an alternating series, the error made in approximating the sum by a partial sum is no greater than the first term omitted.*

Worked Example 2 Find $1 - \frac{1}{2} + \frac{1}{3} - \frac{1}{4} + \frac{1}{5} - \cdots$ with an error of no more than 0.04.

Solution To make the error at most $0.04 = \frac{1}{25}$, we must add up all the terms through $\frac{1}{24}$. Using a calculator, we find

$$S_{24} = 1 - \tfrac{1}{2} + \tfrac{1}{3} - \cdots - \tfrac{1}{24} = 0.6727$$

(Since the sum lies between S_{24} and $S_{25} = 0.7127$, an even better estimate is the midpoint $\frac{1}{2}(S_{24} + S_{25}) = 0.6927$, which can differ from the sum by at most 0.02.)

ALTERNATING SERIES

If $\Sigma_{i=1}^{\infty} a_i$ is a series such that the a_i alternate in sign, are decreasing in absolute value, and tend to zero then it converges.

The error made in approximating the sum by $S_n = \Sigma_{i=1}^{n} a_i$ is not greater than $|a_{n+1}|$.

Solved Exercises

1. Give an example of a convergent series which is not absolutely convergent.

2. Test for convergence: $\Sigma_{i=1}^{\infty} (-1)^i/(1+i)^2$.

3. Test for convergence: $\frac{2}{2} - \frac{1}{2} + \frac{2}{3} - \frac{1}{3} + \frac{2}{4} - \frac{1}{4} + \frac{2}{5} - \frac{1}{5} + \frac{2}{6} - \frac{1}{6} + \cdots$.

Exercises

1. Test for convergence:

   (a) $\displaystyle\sum_{n=1}^{\infty} \frac{1}{\sqrt{n}}$

   (b) $\displaystyle\sum_{n=1}^{\infty} \frac{(-1)^n}{\sqrt{n}}$

   (c) $\displaystyle\sum_{k=1}^{\infty} \frac{k}{k+1}$

   (d) $\displaystyle\sum_{k=1}^{\infty} \frac{(-1)^k k}{k+1}$

2. Find $1 - \frac{1}{3} + \frac{1}{5} - \frac{1}{7} + \cdots$ with an error of at most 0.02.

3. Test for convergence: $\frac{1}{2} + \frac{1}{2} - \frac{1}{4} - \frac{1}{4} + \frac{1}{6} + \frac{1}{6} - \frac{1}{8} - \frac{1}{8} + \cdots$.

THE INTEGRAL TEST AND p SERIES

We now establish a connection between infinite series and improper integrals of the form $\int_a^\infty f(x)\, dx$, which will lead to a useful test for convergence.

Given a series $\Sigma_{i=1}^\infty a_i$, we define a piecewise constant function $g(x)$ on $[1, \infty)$ by the formulas:

$$g(x) = a_1 \quad (1 \le x < 2)$$
$$g(x) = a_2 \quad (2 \le x < 3)$$
$$\vdots$$
$$g(x) = a_i \quad (i \le x < i + 1)$$
$$\vdots$$

Since $\int_i^{i+1} g(x) = a_i$, the partial sum $\Sigma_{i=1}^n a_i$ is equal to $\int_1^{n+1} g(x)\, dx$, and the sum $\Sigma_{i=1}^\infty a_i = \lim_{n \to \infty} \Sigma_{i=1}^n a_i$ exists if and only if the integral $\int_1^\infty g(x)\, dx = \lim_{b \to \infty} \int_1^b g(x)\, dx$ does.

By itself, this relation between series and integrals is not very useful. But suppose now, as is often the case, that the formula which defines the term a_i as a function of i makes sense when i is a real number, not just an integer. In other words, suppose that there is a function $f(x)$, defined for $x \in [1, \infty)$, such that $f(i) = a_i$ when $i = 1, 2, 3, \dots$. Suppose, further, that f satisfies these conditions:

1. $f(x) > 0$ for all x in $[1, \infty)$.

2. $f(x)$ is decreasing on $[1, \infty)$.

(For example, if $a_i = 1/i$, the harmonic series, we may take $f(x) = 1/x$.)

We may now compare $f(x)$ with the piecewise constant function $g(x)$. For $x \in [i, i + 1)$, $x \ge i$, and so

$$0 \le f(x) \le f(i) = a_i = g(x)$$

Hence

$$0 \le f(x) \le g(x)$$

(See Fig. 13-2-2.)

Fig. 13-2-2   The area under the graph of f is less than the shaded area, so

$$0 \le \int_1^{n+1} f(x)\, dx \le \sum_{i=1}^n a_i$$

It follows that, for any n,

$$0 \le \int_1^{n+1} f(x)\, dx \le \int_1^{n+1} g(x)\, dx = \sum_{i=1}^n a_i \tag{1}$$

We conclude that if the series $\Sigma_{i=1}^{\infty} a_i$ converges, then the integrals $\int_1^{n+1} f(x)\, dx$ are bounded above by the sum $\Sigma_{i=1}^{\infty} a_i$, so that the indefinite integral $\int_1^{\infty} f(x)\, dx$ converges (see p. 546 and the appendix to Chapter 12). In other words, if the integral $\int_1^{\infty} f(x)\, dx$ *diverges*, then so does the series $\Sigma_{i=1}^{\infty} a_i$.

Worked Example 3   Show that

$$1 + \frac{1}{2} + \cdots + \frac{1}{n} \geq \ln(n + 1)$$

and so obtain a new proof that the harmonic series diverges.

Solution   We take our function $f(x)$ to be $1/x$. Then, from formula (1) above, we get

$$1 + \frac{1}{2} + \cdots + \frac{1}{n} = \sum_{i=1}^{n} \frac{1}{i} \geq \int_1^{n+1} \frac{1}{x}\, dx = \ln(n + 1)$$

Since $\lim_{n \to \infty} \ln(n + 1) = \infty$, the integral $\int_1^{\infty} (1/x)\, dx$ diverges; hence the series $\Sigma_{i=1}^{\infty} 1/i$ diverges, too.

We would like to turn around the preceding argument to show that if $\int_1^{\infty} f(x)\, dx$ converges, then $\Sigma_{i=1}^{\infty} a_i$ converges as well. Inequality (1) will not suffice for this demonstration; we must do a little trick first. Cutting off the first rectangle in Fig. 13-2-2 and moving the remaining rectangles by 1 unit to the left (see Fig. 13-2-3), we obtain the graph of a new piecewise constant function $h(x)$ on $[1, \infty)$ defined by

$$h(x) = a_{i+1} \quad (i \leq x < i + 1)$$

Now we have $\int_i^{i+1} h(x)\, dx = a_{i+1}$, so $\Sigma_{i=2}^{n} a_i = \int_1^{n} h(x)\, dx$.

Fig. 13-2-3 The area under the graph of f is greater than the shaded area, so

$$0 \leq \sum_{i=2}^{n} a_i \leq \int_1^{n} f(x)\, dx$$

If $x \in [i, i + 1)$, then $x < i + 1$, and so

$$f(x) \geq f(i + 1) = a_{i+1} = h(x) \geq 0$$

Hence

$$f(x) \geq h(x) \geq 0$$

(See Fig. 13-2-3.) Thus

$$\int_1^{n} f(x)\, dx \geq \int_1^{n} h(x)\, dx = \sum_{i=2}^{n} a_i \geq 0 \qquad\qquad (2)$$

If the integral $\int_1^\infty f(x)\,dx$ converges, then the partial sums $\Sigma_{i=1}^n\,a_i = a_1 + \Sigma_{i=2}^n\,a_i$ are bounded above by $a_1 + \int_1^\infty f(x)\,dx$, and the series $\Sigma_{i=1}^\infty\,a_i$ is convergent by Theorem 1.

Worked Example 4 Show that $1 + \frac{1}{4} + \frac{1}{9} + \frac{1}{16} + \cdots$ converges.

Solution This series is $\Sigma_{i=1}^\infty\,1/i^2$. We let $f(x) = 1/x^2$; then

$$\int_1^\infty \frac{1}{x^2}\,dx = \lim_{b\to\infty}\int_1^b \frac{1}{x^2}\,dx = \lim_{b\to\infty}\left(1 - \frac{1}{b}\right) = 1$$

The indefinite integral converges, so the series does too.

INTEGRAL TEST

To test the convergence of a series $\Sigma_{n=1}^\infty\,a_i$ of positive decreasing terms, find a positive, decreasing function $f(x)$ on $[1,\infty)$ such that $f(i) = a_i$. Then:

$$\text{If } \int_1^\infty f(x)\,dx \text{ converges, so does } \sum_{i=1}^\infty a_i.$$

$$\text{If } \int_1^\infty f(x)\,dx \text{ diverges, so does } \sum_{i=1}^\infty a_i.$$

Worked Examples 3 and 4 are special cases of a rule called the *p*-series test, which arises from the integral test with $f(x) = 1/x^p$. We recall that $\int_1^\infty x^n\,dx$ converges if $n < -1$ and diverges if $n \geq -1$ (see Worked Example 7, Section 12-3). Thus we arrive at the test in the following display.

---

p SERIES

$$\text{If } p \leq 1, \text{ then } \sum_{i=1}^\infty \frac{1}{i^p} \text{ diverges.}$$

$$\text{If } p > 1, \text{ then } \sum_{i=1}^\infty \frac{1}{i^p} \text{ converges.}$$

The *p* series are often useful in conjunction with the comparison test.

Worked Example 5 Test for convergence:

$$\sum_{i=1}^\infty \frac{1}{1 + i^2}$$

Solution   We compare the given series with the convergent p series $\Sigma_{i=1}^{\infty} 1/i^2$. Let $a_i = 1/(1 + i^2)$ and $b_i = 1/i^2$. Then $0 < a_i < b_i$ and $\Sigma_{i=1}^{\infty} b_i$ converges, so $\Sigma_{i=1}^{\infty} a_i$ does too.

Solved Exercises

4. Show that $\Sigma_{m=2}^{\infty} 1/(m\sqrt{\ln m})$ diverges but $\Sigma_{m=2}^{\infty} 1/[m(\ln m)^2]$ converges.

5. Test for convergence:

   (a) $\displaystyle\sum_{j=1}^{\infty} \frac{j^2 + 2j}{j^4 - 3j^2 + 10}$ (b) $\displaystyle\sum_{n=1}^{\infty} \frac{3n + \sqrt{n}}{2n^{3/2} + 2}$

   [*Hint:* Use Theorem 4.]

6. (a) Show that $\Sigma_{n=1}^{N} 1/n^p$ approximates $\Sigma_{n=1}^{\infty} 1/n^p$ with error not exceeding $1/[(p - 1)N^{p-1}]$.

   (b) It is known that $\Sigma_{n=1}^{\infty} 1/n^2 = \pi^2/6$. Use this equation to calculate $\pi^2/6$ with error less than 0.05.*

Exercises

4. Test each of the following series for convergence:

   (a) $\displaystyle\sum_{n=1}^{\infty} \frac{\sin n}{n^{3/2}}$ (b) $\displaystyle\sum_{n=1}^{\infty} \frac{n}{n^3 + 4}$ (c) $\displaystyle\sum_{n=1}^{\infty} \frac{n}{n^2 + 4}$

5. For which values of p does $\Sigma_{n=2}^{\infty} 1/n(\ln n)^p$ converge?

6. (a) Let $f(x)$ be a positive, decreasing function on $[1, \infty)$ such that $\int_1^{\infty} f(x)\,dx$ converges. Show that

$$\left| \sum_{n=1}^{\infty} f(n) - \sum_{n=1}^{N} f(n) \right| \le \int_N^{\infty} f(x)\,dx$$

   (See Solved Exercise 6.)

   (b) Estimate $\Sigma_{n=1}^{\infty} (1 + n^2)/(1 + n^8)$ to within 0.02. (Use the comparison test *and* the integral test.)

THE RATIO TEST

The ratio test provides a general way to compare a series with a geometric series, but it formulates the hypotheses in a way which is particularly convenient, since no explicit comparison is needed. Here is the test:

Theorem 6   *Ratio Test: Let $\Sigma_{i=1}^{\infty} a_i$ be a series.*

1. *If $\lim\limits_{i \to \infty} |a_i/a_{i-1}| < 1$, then the series converges (absolutely).*

2. *If $\lim\limits_{i \to \infty} |a_i/a_{i-1}| > 1$, then the series diverges.*

3. *If $\lim\limits_{i \to \infty} |a_i/a_{i-1}| = 1$, the test is inconclusive.*

For a proof using only elementary calculus, see Y. Matsuoka, "An Elementary Proof of the Formula $\Sigma_{k=1}^{\infty} 1/k^2 = \pi^2/6$," American Mathematical Monthly 68(1961): 485–487 (reprinted in T. M. Apostol (ed.), Selected Papers on Calculus, Dickenson, 1969, p. 372). The formula may also be proved using Fourier series; see for instance J. Marsden, Elementary Classical Analysis, W. H. Freeman and Company (1974), ch. 10.

(optional)

Proof By definition of limit, $|a_i/a_{i-1}|$ will be close to its limit l for i large. Let $r = (l+1)/2$ be the midpoint between l and 1, so that $l < r < 1$. Thus there is an N such that

$$\left| \frac{a_i}{a_{i-1}} \right| < r \quad \text{if } i > N$$

We will show this implies that the given series converges.

We have $|a_{N+1}/a_N| < r$ so $|a_{N+1}| < |a_N|r$, $|a_{N+2}/a_{N+1}| < r$ so $|a_{N+2}| < |a_{N+1}|r < |a_N|r^2$ and, in general, $|a_{N+j}| < |a_N|r^j$. But $\sum_{j=1}^{\infty} |a_N| r^j = |a_N| \sum_{j=1}^{\infty} r^j$ is a convergent geometric series since $r < 1$. Hence, by the comparison test, $\sum_{j=1}^{\infty} |a_{N+j}|$ converges. Since we have omitted only $|a_1|, |a_2|, \ldots, |a_N|$, the series $\sum_{j=1}^{\infty} |a_j|$ converges as well and part 1 is proved.

For part 2 we find, as in part 1, that $|a_{N+j}| > |a_N|r^j$, where $r > 1$. As $j \to \infty$, $r^j \to \infty$, so $|a_{N+j}| \to \infty$. Thus the series cannot converge, since its terms do not converge to zero.

Finally, to prove part 3 we will exhibit a convergent series for which the limit of $|a_i/a_{i-1}|$ is 1 and a divergent series for which the limit is also 1. For the first case, consider $\sum_{i=1}^{\infty} 1/i^2$, which converges by the p-series test. Here $a_i = 1/i^2$ and

$$|a_i/a_{i-1}| = \frac{1/i^2}{1/(i-1)^2} = \left(\frac{i-1}{i} \right)^2 \to 1 \quad \text{as } i \to \infty$$

For the second case, we may consider the divergent harmonic series $\sum_{i=1}^{\infty} 1/i$ and note that $|a_i/a_{i-1}| = (i-1)/i \to 1$ as $i \to \infty$.

Worked Example 6 Test for convergence: $2 + 2^2/2^8 + 2^3/3^8 + 2^4/4^8 + \cdots$.

Solution We have $a_i = 2^i/i^8$. The ratio a_i/a_{i-1} is

$$\frac{2^i}{i^8} \cdot \frac{(i-1)^8}{2^{i-1}} = 2 \cdot \left(\frac{i-1}{i} \right)^8$$

so

$$\lim_{i \to \infty} \frac{a_i}{a_{i-1}} = 2 \left[\lim_{i \to \infty} \left(\frac{i-1}{i} \right) \right]^8 = 2 \cdot 1^8 = 2$$

which is greater than 1, so the series diverges.

RATIO TEST

To test for convergence of a series $\sum_{i=1}^{\infty} a_i$, write $|a_i/a_{i-1}|$ and take the limit as $i \to \infty$. If the limit exists and is

< 1, the series converges

> 1, the series diverges

$= 1$, you may draw no conclusion

If $\lim_{i \to \infty} |a_i/a_{i-1}| = \infty$, the series diverges.

The tests we have covered enable us to deal with many of the series which commonly arise. Of course, if the series is geometric, it may be summed. Otherwise, either the ratio test, comparison with a p series, the integral test, or the alternating series test will usually work.*

Solved Exercises   7. Test for convergence:

(a) $\sum_{n=1}^{\infty} \frac{1}{n!}$, where $n! = n(n-1) \cdots 3 \cdot 2 \cdot 1$

(b) $\sum_{j=1}^{\infty} \frac{b^j}{j!}$, b any constant

8. (a) Show that if $|a_n/a_{n-1}| < r < 1$ for $n > N$, then the error made in approximating $\sum_{n=1}^{\infty} a_n$ by $\sum_{n=1}^{N} a_n$ is no greater than $|a_N| r/(1-r)$.

(b) What is the error made in approximating $\sum_{n=1}^{\infty} 1/n!$ by $\sum_{n=1}^{4} 1/n!$?

★ 9. (a) (Root test) If $\lim_{n\to\infty} |a_n|^{1/n} < 1$, show that $\sum_{n=1}^{\infty} a_n$ converges absolutely.

(b) Does $\sum_{n=1}^{\infty} 1/n^n$ converge?

Exercises   7. Test for convergence:

(a) $\sum_{n=1}^{\infty} \frac{2\sqrt{n}}{3^n}$

(b) $\sum_{n=1}^{\infty} \frac{3^n}{2\sqrt{n}}$

(c) $\sum_{i=1}^{\infty} \frac{i^3 \cdot 3^i}{i!}$

(d) $\sum_{n=1}^{\infty} \frac{2n^2 + n!}{n^5 + (3n)!}$

8. Estimate $\sum_{n=1}^{\infty} 1/n!$:

(a) To within 0.05.   (b) To within 0.005.

(c) How many terms would you need to calculate to get an accuracy of five decimal places?

★ 9. Complete the root test of Solved Exercise 9 by showing:

(a) If $\lim_{n\to\infty} |a_n|^{1/n} > 1$, then $\sum_{n=1}^{\infty} a_n$ diverges.

(b) If $\lim_{n\to\infty} |a_n|^{1/n} = 1$, the test is inconclusive. (You may use the fact that $\lim_{n\to\infty} n^{1/n} = 1$.)

Problems for Section 13-2

1. Test for convergence:

(a) $\sum_{i=1}^{\infty} \frac{1}{i^4}$

(b) $\sum_{n=1}^{\infty} \frac{(-1)^n(n+1)}{2n+1}$

(c) $\sum_{k=2}^{\infty} \frac{\cos k\pi}{\ln k}$

(d) $\sum_{j=1}^{\infty} \frac{(j+1)^{100}}{j!}$

2. Test for convergence:

(a) $\sum_{m=1}^{\infty} \frac{(-1)^m(m+1)}{m^2+1}$

(b) $\sum_{j=3}^{\infty} \frac{j^2 + \cos j}{j^4 + \sin j}$

(c) $\sum_{n=1}^{\infty} \frac{\sqrt{n} + n + n^{3/2}}{\sqrt{n} + n + n^{5/2} + n^3}$

(d) $\sum_{j=1}^{\infty} (-1)^j \sin\left(\frac{\pi}{4j}\right)$

*More refined tests may be found in Chapter 3 of K. Knopp, Theory and Applications of Infinite Series, Hafner, 1951.

3. Test for convergence and absolute convergence:

 (a) $1 - \frac{1}{2} + \frac{2}{3} - \frac{3}{4} + \frac{4}{5} - \cdots$

 (b) $1 - \frac{1}{2} + \frac{1}{4} - \frac{1}{8} + \frac{1}{16} - \cdots$

 (c) $\sum_{i=1}^{\infty} (-1)^i [i/(i^2 + 1)]$

 (d) $\sum_{i=1}^{\infty} a_i$, where $a_i = 1/(2^i)$ if i is even and $a_i = 1/i$ if i is odd.

 (e) $\sum_{n=1}^{\infty} (-1)^n \ln [(n + 1)/n]$

4. Estimate the sum of the given series with an error of no more than that specified.

 (a) $\sum_{i=1}^{\infty} \frac{(-1)^i}{3^i + 1}; 0.01$ (b) $\sum_{n=1}^{\infty} \frac{(-1)^n n}{4n^3 + 1}; 0.005$

 (c) $\sum_{n=1}^{\infty} \left(\frac{(-1)^n}{2n} + \frac{1}{5^n} \right); 0.02$

5. Estimate each of the following sums to within 0.05 (see Solved Exercises 6 and 8):

 (a) $\sum_{n=1}^{\infty} \frac{1 + n}{2^n}$ (b) $\sum_{n=1}^{\infty} \frac{\sin n}{n^4}$

 (c) $\sum_{n=0}^{\infty} \frac{\pi^{2n+1}}{(2n + 1)!}$ (d) $\sum_{n=0}^{\infty} \frac{(\pi/2)^{2n+1}}{(2n + 1)!}$

6. For which values of p and q is the series $\sum_{n=2}^{\infty} 1/[n^p (\ln n)^q]$ convergent?

★7. (a) Let $f(x)$ be positive and decreasing on $[1, \infty)$, and suppose that $f(i) = a_i$ for $i = 1, 2, 3, \dots$. Show that

$$S - \tfrac{1}{2} f(n) \le \sum_{i=1}^{\infty} a_i \le S + \tfrac{1}{2} f(n)$$

where $S = \sum_{i=1}^{n} f(i) + \frac{1}{2} \int_n^{n+1} f(x)\, dx + \int_{n+1}^{\infty} f(x)\, dx$. [*Hint:* Look at the proof of the integral test and show that $\int_{n+1}^{\infty} f(x)\, dx \le \sum_{i=n+1}^{\infty} a_i \le \int_n^{\infty} f(x)\, dx$.]

 (b) Estimate $\sum_{n=1}^{\infty} 1/n^4$ to within 0.0001. How many terms did you use? How much work do you save by using the method of part (a) instead of that of Solved Exercise 6?

★8. (a) Show that

$$\sum_{n=1}^{\infty} \frac{(-1)^{n+1}(4n - 1)}{2n(2n - 1)}$$

is an alternating series. (Use differential calculus to prove that a certain function is decreasing.)

 (b) Test for convergence: $1 + \frac{1}{2} - \frac{1}{3} - \frac{1}{4} + \frac{1}{5} + \frac{1}{6} - \cdots$. [*Hint:* Group the terms by twos.]

★9. For which values of p does $\sum_{i=1}^{\infty} [\sin(1/i)]^p$ converge?

10. Using Fourier analysis, it is possible to show that

$$\frac{\pi^4}{96} = 1 + \frac{1}{3^4} + \frac{1}{5^4} + \frac{1}{7^4} + \cdots$$

 (a) Show directly that the series on the right is convergent, by means of the integral test.

 (b) Determine how many terms are needed to compute $\pi^4/96$, accurate to 20 digits.

11. A bar of length L is loaded by a weight W at its midpoint. At $t = 0$ the load is removed. The deflection $y(t)$ at the midpoint, measured from the straight profile $y = 0$, is given by

$$y(t) = \frac{2WL^2}{\pi^4 EI} \left[\cos r + \frac{\cos (9r)}{3^4} + \frac{\cos (25r)}{5^4} + \cdots \right]$$

where

$$r = \left(\frac{\pi^2}{L^2} \sqrt{\frac{EIg}{\gamma \Omega}} \right) t$$

The numbers EI, g, γ, Ω, L are positive constants.

 (a) Show by substitution that the bracketed terms are the first three terms of the infinite series

$$\sum_{n=0}^{\infty} \frac{\cos [(2n + 1)^2 r]}{(2n + 1)^4}$$

 (b) Make accurate graphs of the first three partial sums

$$S_1(r) = \cos (r)$$

$$S_2(r) = \cos (r) + \frac{1}{3^4} \cos (9r)$$

$$S_3(r) = \cos (r) + \frac{1}{3^4} \cos (9r) + \frac{1}{5^4} \cos (25r)$$

Up to a magnification factor, these graphs approximate the motion of the midpoint of the bar.

 (c) Using the integral test and the comparison test, show the series converges.

13-3 Power Series and Taylor's Formula

Many functions can be expressed as "polynomials with infinitely many terms."

A series of the form $\sum_{i=0}^{\infty} a_i(x - x_0)^i$, where the a_i's and x_0 are constants and x is a variable, is called a *power series* (since we are summing the powers of $(x - x_0)$). In the first part of this section, we show how a power series may be considered as a function of x, defined on a certain interval. In the second part of the section, we begin with an arbitrary function and show how to find the power series which represents it (if there is such a series).

Goals

After studying this section, you should be able to:

Find the interval of convergence for a given power series.

Apply the operations of algebra and calculus to power series.

Find the first few terms of a Taylor series by using Taylor's formula or some manipulations.

Estimate the error in Taylor series approximations.

Reproduce from memory the Taylor series of the sine, cosine, exponential, and logarithmic functions.

POWER SERIES

We first consider power series in which $x_0 = 0$; that is, those of the form

$$f(x) = a_0 + a_1 x + a_2 x^2 + a_3 x^3 + \cdots = \sum_{i=0}^{\infty} a_i x^i$$

where the a_i are given constants. The domain of f can be taken to consist of those x for which the series converges.

If there is an integer N such that $a_i = 0$ for all $i > N$, then the power series is equal to a finite sum, $\sum_{i=0}^{N} a_i x^i$, which is just a polynomial of degree N. In general, we may think of a power series as a polynomial of "infinite degree"; we will see that, as long as they converge, power series may be manipulated (added, subtracted, multiplied, divided, differentiated) just like ordinary polynomials.

The simplest power series, after the polynomials, is the geometric series

$$f(x) = 1 + x + x^2 + \cdots$$

which converges when $|x| < 1$; the sum is $1/(1 - x)$. Thus we have written $1/(1 - x)$ as a power series:

$$\frac{1}{1 - x} = 1 + x + x^2 + x^3 + \cdots \quad \text{if } |x| < 1$$

Facts about general power series are often proved by comparison with the geometric series. Here is an example.

Theorem 7 *Suppose that $\sum_{i=0}^{\infty} a_i x^i$ converges for some particular value of x, say $x = x_0$. Then:*

1. *There is an integer N such that $\sqrt[i]{|a_i|} < 1/|x_0|$ for all $i \geq N$.*

2. *If $|y| < |x_0|$, then $\sum_{i=0}^{\infty} a_i y^i$ converges absolutely.*

(optional)

Proof For part 1, suppose that $\sqrt[i]{|a_i|} \geq 1/|x_0|$ for arbitrarily large values of i. Then for these values of i we have $|a_i| \geq 1/|x_0|^i$, and $|a_i x_0^i| \geq 1$. But then we could not have $a_i x_0^i \to 0$, as is required for convergence.

For part 2, let $r = |y|/|x_0|$, so that $|r| < 1$. By part 1, $|a_i y^i| = |a_i||x_0|^i r^i < r^i$ for all $i \geq N$. By the comparison test, the series $\sum_{i=N}^{\infty} a_i y^i$ converges absolutely; it follows that the entire series converges absolutely as well.

Let I be the set of x such that $\sum_{i=0}^{\infty} a_i x^i$ converges. It follows from Theorem 7 that I must be an interval of the form $[-R, R]$, $(-R, R)$, $[-R, R)$, or $(-R, R]$, where R could be zero or (in the case $(-R, R)$) ∞. (Strictly speaking, this result requires the completeness axiom. See the appendix to Chapter 2.) We call R the *radius of convergence* of the power series $\sum_{i=0}^{\infty} a_i x^i$.

The radius of convergence of a series can often be determined with the aid of the following theorem.

Theorem 8 *Let $\sum_{i=0}^{\infty} a_i x^i$ be a power series. Suppose that $\lim_{i \to \infty} |a_i/a_{i-1}| = r$ exists. Then $R = 1/r$ is the radius of convergence of the series. (If $r = 0$, then $R = \infty$; and if $r = \infty$, then $R = 0$.)*

(optional)

Proof Consider the ratio

$$\left| \frac{a_i x^i}{a_{i-1} x^{i-1}} \right| = |x| \left| \frac{a_i}{a_{i-1}} \right|$$

of successive terms. By hypothesis, as $i \to \infty$, this ratio converges to $|x| \cdot r$, which is less than 1 if $|x| < 1/r$ and greater than 1 if $|x| > 1/r$. By the ratio test, the series converges absolutely in the first case and diverges in the second. Worked Example 1 and Solved Exercise 1 show that the series may converge for either, both, or neither of the values $x = \pm R$.

Worked Example 1 For which x does $\sum_{i=0}^{\infty} [i/(i + 1)] x^i$ converge?

Solution Here $a_i = i/(i + 1)$. Then

$$\frac{a_i}{a_{i-1}} = \frac{i/(i+1)}{(i-1)/i} = \frac{i^2}{(i+1)(i-1)}$$

$$= \frac{1}{(1 + 1/i)(1 - 1/i)} \to 1 \quad \text{as } i \to \infty$$

Hence $r = 1$. Thus the series converges if $|x| < 1$ and diverges if $|x| > 1$. If $x = 1$, then $\lim_{i \to \infty} [i/(i + 1)] x^i = 1$, so the series diverges at $x = 1$ since the terms do not go to zero. If $x = -1$, $\lim_{i \to \infty} [i/(i + 1)] x^i = \lim_{i \to \infty} [i/(i + 1)](-1)^i$ does not exist, so again the series diverges.

Worked Example 2   Determine the radius of convergence of $\Sigma_{k=0}^{\infty} [k^5/(k + 1)!] x^k$.

Solution   To use Theorem 8, we look at

$$r = \lim_{k \to \infty} \left| \frac{a_k}{a_{k-1}} \right|$$

Here $a_k = k^5/(k + 1)!$, so

$$r = \lim_{k \to \infty} \left| \frac{k^5}{(k + 1)!} \cdot \frac{k!}{(k - 1)^5} \right|$$
$$= \lim_{k \to \infty} \left(\frac{k}{k - 1} \right)^5 \cdot \lim_{k \to \infty} \frac{1}{k + 1} = 1 \cdot 0 = 0$$

Thus $r = 0$, so $R = \infty$ and the radius of convergence is infinite (that is, the series converges for all x).

Sometimes we consider series of the form $\Sigma_{i=0}^{\infty} a_i (x - x_0)^i$. The theory of these series, also called power series, is essentially the same as for those we have been considering, because $\Sigma_{i=0}^{\infty} a_i (x - x_0)^i$ may be written as $\Sigma_{i=0}^{\infty} a_i w^i$, where $w = x - x_0$, and the results may be used with w in place of x. See Fig. 13-3-1.

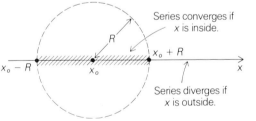

Series converges if x is inside.

R

$x_0 + R$

$x_0 - R$ x_0 x

Series diverges if x is outside.

Fig. 13-3-1 The radius of convergence of the power series $\sum_{i=0}^{\infty} a_i(x - x_0)^i$.

CONVERGENCE OF A POWER SERIES

If $\Sigma_{i=0}^{\infty} a_i (x - x_0)^i$ is a power series, there is a number R (possibly ∞), called the *radius of convergence*, such that the series converges absolutely when $|x - x_0| < R$ and diverges when $|x - x_0| > R$. Convergence or divergence at $x_0 \pm R$ depends upon the particular series.

If $\lim_{i \to \infty} |a_i/a_{i-1}| = r$ exists, then $R = 1/r$ is equal to the radius of convergence.

Worked Example 3 For which x does the series $\Sigma_{n=0}^{\infty} (4^n/\sqrt{2n+5})(x+5)^n$ converge?

Solution This series is of the form $\Sigma_{i=0}^{\infty} a_i(x-x_0)^i$, with $a_i = 4^i/\sqrt{2i+5}$ and $x_0 = -5$. We have

$$r = \lim_{i\to\infty} \frac{a_i}{a_{i-1}} = \lim_{i\to\infty} \frac{4^i}{\sqrt{2i+5}} \cdot \frac{\sqrt{2(i-1)+5}}{4^{i-1}}$$

$$= \lim_{i\to\infty} 4\sqrt{\frac{2i+3}{2i+5}} = 4$$

so the radius of convergence is $\frac{1}{4}$. Thus the series converges for $|x+5| < \frac{1}{4}$ and diverges for $|x+5| > \frac{1}{4}$. When $x = -5\frac{1}{4}$, the series becomes $\Sigma_{i=0}^{\infty} (-1)^i/\sqrt{2i+5}$, which converges because it is alternating. When $x = -4\frac{3}{4}$, the series is $\Sigma_{i=0}^{\infty} 1/\sqrt{2i+5}$, which diverges by comparison with $\Sigma_{i=1}^{\infty} 1/\sqrt{3i}$ (or by the integral test). Thus our power series converges when $-5\frac{1}{4} \le x < -4\frac{3}{4}$.

Solved Exercises
1. (a) For which x does $\Sigma_{i=1}^{\infty} x^i/i$ converge? (b) For which x does $\Sigma_{i=0}^{\infty} x^i/i^2$ converge?

2. For which x does $\Sigma_{i=0}^{\infty} x^i/i!$ converge?

3.★(a) If $\lim_{i\to\infty} |a_i|^{1/i} = \rho$, show that $\Sigma_{i=0}^{\infty} a_i x^i$ converges for $|x| < 1/\rho$. (See Solved Exercise 9, Section 13-2.)
   (b) For which x does $\Sigma_{i=1}^{\infty} x^i/(2+1/i)^i$ converge?

Exercises
1. For which x do the following series converge?

(a) $\sum_{i=0}^{\infty} \frac{2}{i+1} x^i$ (b) $\sum_{i=0}^{\infty} (2i+1)x^i$

(c) $\sum_{n=1}^{\infty} \frac{3}{n^2} x^n$ (d) $\sum_{n=1}^{\infty} \frac{(-1)^n 2^n}{n(n+1)} x^n$

(e) $\sum_{i=1}^{\infty} \frac{5i+1}{i}(x-1)^i$ (f) $\sum_{r=0}^{\infty} \frac{r!}{3^{2r}}(x+2)^r$

(g) $\sum_{n=2}^{\infty} \frac{1}{n!\sin(\pi/n)} x^n$ (h) $\sum_{i=14}^{\infty} \frac{i(i+3)}{i^3-4i+7} x^i$

(i) $\sum_{n=1}^{\infty} (-1)^n n^n x^n$ (j) $\sum_{s=1}^{\infty} \left(\frac{2^s+1}{8s^7}\right)^{3/2} x^s$

2. Find a power series which converges just when $-1 < x \le 1$.

DIFFERENTIATION AND INTEGRATION OF POWER SERIES

Let $f(x) = \Sigma_{i=0}^{\infty} a_i x^i$, defined where the series converges. By analogy with ordinary polynomials, we might guess that

$$f'(x) = \sum_{i=1}^{\infty} ia_i x^{i-1}$$

and that

$$\int f(x)\, dx = \sum_{i=0}^{\infty} \frac{a_i x^{i+1}}{i+1} + C$$

The following theorem asserts that this is true.

Theorem 9 *Let* $f(x) = \sum_{i=0}^{\infty} a_i x^i$ *be a power series with radius of convergence R. Then the series*

$$g(x) = \sum_{i=1}^{\infty} i a_i x^{i-1}$$

and

$$h(x) = \sum_{i=0}^{\infty} \frac{a_i}{i+1} x^{i+1}$$

have the same radius of convergence R as $f(x)$. *Moreover, for* $|x| < R$,

$$f'(x) = g(x)$$

and

$$\int f(x)\, dx = h(x) + C$$

A similar result holds for series of the form $\sum_{i=0}^{\infty} a_i (x - x_0)^i$.

The proof of this theorem is contained in the (moderately difficult) problems 21 through 25 at the end of the section.

Worked Example 4 Let $f(x) = \sum_{i=0}^{\infty} [i/(i+1)]\, x^i$. Find a series expression for $f'(x)$. Where is it valid?

Solution By Worked Example 1, $f(x)$ converges for $|x| < 1$. Thus $f'(x)$ also converges if $|x| < 1$ and we may differentiate term by term:

$$f'(x) = \sum_{i=0}^{\infty} \frac{i^2}{i+1} x^{i-1} \qquad |x| < 1$$

Notice that $f'(x)$ is again a power series, so it too can be differentiated. Since this can be repeated, we conclude that f can be differentiated as many times as we please. We say that f is *infinitely differentiable*.

DIFFERENTIATION AND INTEGRATION OF POWER SERIES

To differentiate or integrate a power series within its radius of convergence R, you may differentiate or integrate it term by term:

$$\text{if } |x - x_0| < R, \qquad \frac{d}{dx} \sum_{i=0}^{\infty} a_i (x - x_0)^i = \sum_{i=1}^{\infty} i a_i (x - x_0)^{i-1}$$

$$\int \left[\sum_{i=0}^{\infty} a_i (x - x_0)^i \right] dx = \sum_{i=0}^{\infty} \frac{a_i}{i+1} (x - x_0)^{i+1} + C$$

Worked Example 5 Write down power series for $x/(1 + x^2)$ and $\ln(1 + x^2)$. Where do they converge?

Solution First, we expand $1/(1 + x^2)$ as a geometric series $1/(1 - r) = 1 + r + r^2 + \cdots$, with r replaced by $-x^2$, obtaining $1 - x^2 + x^4 - \cdots$. Multiplying by x gives $x/(1 + x^2) = x - x^3 + x^5 - \cdots$, which converges for $|x| < 1$ by Theorem 8. It diverges for $x = \pm 1$.

Now we observe that $(d/dx) \ln(1 + x^2) = 2x/(1 + x^2)$, so

$$\ln(1 + x^2) = 2 \int \frac{x}{1 + x^2} dx = 2 \int (x - x^3 + x^5 - \cdots) \, dx$$

$$= 2 \left(\frac{x^2}{2} - \frac{x^4}{4} + \frac{x^6}{6} - \cdots \right) = x^2 - \frac{x^4}{2} + \frac{x^6}{3} - \frac{x^8}{4} + \cdots$$

(The integration constant was dropped because $\ln(1 + 0^2) = 0$.) This series converges for $|x| < 1$, and also for $x = \pm 1$, because it is alternating.

Solved Exercises 4. (a) Write down a power series representing the integral of $1/(1 - x)$ for $|x| < 1$.

(b) Write a power series for $\ln x = \int dx/x$ in powers of $1 - x$. Where is it valid?

5. If $f(x) = \sum_{n=0}^{\infty} x^n/n!$, show that $f'(x) = f(x)$. Conclude that $f(x) = e^x$. (See Theorem 2, p. 379.)*

Exercises 3. Let $f(x) = \sum_{i=1}^{\infty} (i + 1)x^i$.

(a) Find the radius of convergence of this series.

(b) Find the series for $\int_0^x f(t) \, dt$.

(c) Use the result of part (b) to sum the series $f(x)$.

(d) Sum the series $\frac{2}{2} + \frac{3}{4} + \frac{4}{8} + \frac{5}{16} + \cdots$.

4. Let $f(x) = x - x^3/3! + x^5/5! - \cdots$. Show that f is defined and is differentiable for all x. Show that $f''(x) + f(x) = 0$. Use the uniqueness of solutions of this equation* to show that $f(x) = \sin x$.

5. By differentiating the result of Exercise 4, find a series representation for $\cos x$.

*This exercise depends upon Section 9-1 and should be skipped if you have not yet studied differential equations.

6. Write down power series representations for the following functions:
   (a) e^{-x^2} and its derivative.
   (b) $\tan^{-1} x$ and its derivative. [*Hint:* Do the derivative first.]
   (c) The second derivative of $1/(1 - x)$.

7. Why can't $x^{1/3}$ be represented in the form of a series $\Sigma_{i=0}^{\infty} a_i x^i$, convergent near $x = 0$?

ALGEBRAIC OPERATIONS ON POWER SERIES

The operations of addition and multiplication by a constant may be performed term by term on power series, just as on polynomials. The following theorem is a consequence of the algebraic rules on p. 543.

Theorem 10 *Let $f(x) = \Sigma_{i=0}^{\infty} a_i x^i$ and $g(x) = \Sigma_{i=0}^{\infty} b_i x^i$ be power series with radii of convergence R and S, respectively. Then:*

1. *$f(x) + g(x) = \Sigma_{i=0}^{\infty} (a_i + b_i) x^i$, with radius of convergence at least as large as the smaller of R and S.*

2. *If c is any real number, then $cf(x) = \Sigma_{i=0}^{\infty} ca^i x^i$, with radius of convergence R (unless $c = 0$, in which case the radius of convergence is ∞).*

As usual, similar results hold for series in powers of $(x - x_0)$.

Worked Example 6 Write down power series of the form $\Sigma_{i=0}^{\infty} a_i x^i$ for $2/(3 - x)$, $5/(4 - x)$, and $(23 - 7x)/[(3 - x)(4 - x)]$. What are their radii of convergence?

Solution We may write

$$\frac{2}{3 - x} = \frac{2}{3} \left(\frac{1}{1 - x/3} \right) = \frac{2}{3} \sum_{i=0}^{\infty} \left(\frac{x}{3} \right)^i = \sum_{i=0}^{\infty} \frac{2}{3^{i+1}} x^i$$

The ratio of successive coefficients is $(1/3^{i+1})/(1/3^i) = 1/3$, so the radius of convergence is 3.
   Similarly,

$$\frac{5}{4 - x} = \sum_{i=0}^{\infty} \frac{5}{4^{i+1}} x^i$$

with radius of convergence 4. Finally, since $2/(3 - x) + 5/(4 - x) = (23 - 7x)/[(3 - x)(4 - x)]$ (see the discussion of partial fractions in Section 11-1), we have

$$\frac{23 - 7x}{(3 - x)(4 - x)} = \sum_{i=0}^{\infty} \left(\frac{2}{3^{i+1}} + \frac{5}{4^{i+1}} \right) x^i$$

By Theorem 10, the radius of convergence of this series is at least 3. In fact, a limit computation shows that the ratio of successive coefficients approaches $\frac{1}{3}$, so the radius of convergence is exactly 3.

Multiplication of series follows the same pattern as multiplication of polynomials. We state the following without proof.

Theorem 11 *Let* $f(x) = \Sigma_{i=0}^{\infty} a_i x^i$ *and* $g(x) = \Sigma_{i=0}^{\infty} b_i x^i$ *be power series with radii of convergence R and S, respectively. Then*

$$f(x)g(x) = \sum_{i=0}^{\infty} c_i x^i \qquad where \qquad c_i = \sum_{j=0}^{i} a_j b_{i-j}$$

The radius of convergence of the new series is at least as large as the smaller of R and S.

In practice, we do not use the formula in Theorem 11 but merely multiply the series for f and g term by term; in the product, we collect the terms involving each power of x.

Worked Example 7 Write down the terms through x^4 in the series for $e^x/(1-x)$. (See Solved Exercise 5.)

Solution We have $e^x = 1 + x + x^2/2 + x^3/6 + x^4/24 + \cdots$ (from Solved Exercise 5) and $1/(1 - x) = 1 + x + x^2 + x^3 + x^4 + \cdots$. We multiply terms in the first series by terms in the second series, in all possible ways.

	1	x	$\dfrac{x^2}{2}$	$\dfrac{x^3}{6}$	$\dfrac{x^4}{24}$	$\cdots$
1	1	x	$\dfrac{x^2}{2}$	$\dfrac{x^3}{6}$	$\dfrac{x^4}{24}$	$\cdots$
x	x	x^2	$\dfrac{x^3}{2}$	$\dfrac{x^4}{6}$	$\cdots$	
x^2	x^2	x^3	$\dfrac{x^4}{2}$	$\cdots$		
x^3	x^3	x^4	$\cdots$			
x^4	x^4	$\cdots$				

(Since we want the product series only through x^4, we may neglect the terms in higher powers of x.) Reading along diagonals from lower left to upper right, we collect the powers of x to get

$$\frac{e^x}{1-x} = 1 + (x + x) + \left(x^2 + x^2 + \frac{x^2}{2}\right) + \left(x^3 + x^3 + \frac{x^3}{2} + \frac{x^3}{6}\right)$$

$$+ \left(x^4 + x^4 + \frac{x^4}{2} + \frac{x^4}{6} + \frac{x^4}{24}\right) + \cdots$$

$$= 1 + 2x + \frac{5}{2}x^2 + \frac{8}{3}x^3 + \frac{65}{24}x^4 + \cdots$$

To find the series for $f(x)/g(x)$, one may first find the series for $1/g(x)$ and then multiply. The procedure for finding the series for a reciprocal is best given in the form of an example. The precise theorem is rather complicated to state.

Worked Example 8 Write down the terms through x^4 in the power series for $1/e^x$.

Solution We may solve this in three ways.

1. Long division:

$$
\begin{array}{r}
1 - x + \dfrac{x^2}{2} - \dfrac{x^3}{6} + \dfrac{x^4}{24}\cdots \\[2mm]
\end{array}
$$

$$
1 + x + \frac{x^2}{2} + \frac{x^3}{6} + \frac{x^4}{24}\ \Big|\ 1
$$

$$1 + x + \frac{x^2}{2} + \frac{x^3}{6} + \frac{x^4}{24}\cdots$$

$$-x - \frac{x^2}{2} - \frac{x^3}{6} - \frac{x^4}{24}\cdots$$

$$-x - x^2 - \frac{x^3}{2} - \frac{x^4}{6}\cdots$$

$$\frac{x^2}{2} + \frac{x^3}{3} + \frac{x^4}{8}\cdots$$

$$\frac{x^2}{2} + \frac{x^3}{2} + \frac{x^4}{4}\cdots$$

$$-\frac{x^3}{6} - \frac{x^4}{8}\cdots$$

$$-\frac{x^3}{6} - \frac{x^4}{6}\cdots$$

$$\frac{x^4}{24}\cdots$$

$$\frac{x^4}{24}$$

2. Undetermined coefficients. In this method, we write $1/e^x = b_0 + b_1 x + b_2 x^2 + b_3 x^3 + b_4 x^4 + \cdots$, where the b_i's are to be determined from the relation $e^x(1/e^x) = 1$. We have

$$e^x\left(\frac{1}{e^x}\right) = \left(1 + x + \frac{x^2}{2} + \frac{x^3}{6} + \frac{x^4}{24} + \cdots\right) \times$$

$$(b_0 + b_1 x + b_2 x^2 + b_3 x^3 + b_4 x^4 + \cdots)$$

Multiplying out the right-hand side gives

$$b_0 + (b_0 + b_1)x + (\tfrac{1}{2}b_0 + b_1 + b_2)x^2 + (\tfrac{1}{6}b_0 + \tfrac{1}{2}b_1 + b_2 + b_3)x^3$$
$$+ (\tfrac{1}{24}b_0 + \tfrac{1}{6}b_1 + \tfrac{1}{2}b_2 + b_3 + b_4)x^4 + \cdots$$

Setting this equal to $1 = 1 + 0 \cdot x + 0 \cdot x^2 + 0 \cdot x^3 + 0 \cdot x^4 + \cdots$ and comparing the coefficients of like powers of x, we get a sequence of equations for the b_i's:

$$b_0 = 1$$

$$b_0 + b_1 = 0$$

$$\tfrac{1}{2}b_0 + b_1 + b_2 = 0$$

$$\tfrac{1}{6}b_0 + \tfrac{1}{2}b_1 + b_2 + b_3 = 0$$

$$\tfrac{1}{24}b_0 + \tfrac{1}{6}b_1 + \tfrac{1}{2}b_2 + b_3 + b_4 = 0$$

Solving these equations in succession, we find $b_0 = 1$, $b_1 = -1$, $b_2 = \tfrac{1}{2}$, $b_3 = -\tfrac{1}{6}$, and $b_4 = \tfrac{1}{24}$, so that

$$\frac{1}{e^x} = 1 - x + \frac{x^2}{2} - \frac{x^3}{6} + \frac{x^4}{24} \cdots$$

3. The easy way: Since $1/e^x$ equals e^{-x}, substituting $-x$ for x in the series for e^x gives the terms we have already calculated.

ALGEBRAIC OPERATIONS ON POWER SERIES

Let $f(x) = \Sigma_{i=0}^{\infty} a_i x^i$, with radius of convergence R.

Let $g(x) = \Sigma_{i=0}^{\infty} b_i x^i$, with radius of convergence S.

If T is the smaller of R and S, then

$$f(x) + g(x) = \sum_{i=0}^{\infty} (a_i + b_i)x^i \qquad \text{for } |x| < T$$

$$cf(x) = \sum_{i=0}^{\infty} (ca_i)x^i \qquad \text{for } |x| < R$$

$$f(x)g(x) = \sum_{i=0}^{\infty} \left(\sum_{j=0}^{i} a_j b_{i-j} \right) x^i \qquad \text{for } |x| < T$$

If $b_o \neq 0$, then $\dfrac{f(x)}{g(x)} = \displaystyle\sum_{i=0}^{\infty} c_i x^i$ for x near zero, where the c_i's may be determined by long division or the method of undetermined coefficients. The radius of convergence of f/g cannot be predicted, but it is positive as long as $R > 0$ and $S > 0$.

Solved Exercises

6. Using the result of Exercise 4, write down the terms through x^6 in a power series expansion of $\sin^2 x$.

7. Find series $f(x)$ and $g(x)$ such that the series $f(x) + g(x)$ is not identically zero but has a larger radius of convergence than either $f(x)$ or $g(x)$.

Exercises

8. (a) By dividing the series for $\sin x$ by that for $\cos x$, find the terms through x^5 in the series for $\tan x$.

(b) Find the terms through x^4 in the series for $\sec^2 x = (d/dx)\tan x$.

(c) Using the result of part (b), find the terms through x^4 in the series for $1/\sec^2 x$.

9. Find the series for $1/[(1 - x)(2 - x)]$ by writing

$$\frac{1}{(1 - x)(2 - x)} = \frac{A}{1 - x} + \frac{B}{2 - x}$$

and adding the resulting geometric series.

10. Find series $f(x)$ and $g(x)$, each of them having radius of convergence 2, such that $f(x) + g(x)$ has radius of convergence 3.

TAYLOR'S FORMULA

Up to now, to construct a power series representation for a function, we have had to use manipulations with geometric series or differential equations. We now show how to find the coefficients of the power series for any function which admits of a series representation.

Suppose that $f(x) = \Sigma_{i=0}^{\infty} a_i(x - x_0)^i$ for $|x - x_0| < R$. We know that f is infinitely differentiable. Let us relate the coefficients a_i to the derivatives of f at $x = x_0$:

$$f(x) = a_0 + a_1(x - x_0) + a_2(x - x_0)^2 + a_3(x - x_0)^3 + \cdots \quad \text{so} \quad f(x_0) = a_0$$

$$f'(x) = a_1 + 2a_2(x - x_0) + 3a_3(x - x_0)^2 + 4a_4(x - x_0)^3 + \cdots \quad \text{so} \quad f'(x_0) = a_1$$

$$f''(x) = 2a_2 + 3 \cdot 2a_3(x - x_0) + 4 \cdot 3a_4(x - x_0)^2 + \cdots \quad \text{so} \quad f''(x_0) = 2a_2$$

$$f'''(x) = 3 \cdot 2a_3 + 4 \cdot 3 \cdot 2a_4(x - x_0) + \cdots \quad \text{so} \quad f'''(x_0) = 3 \cdot 2a_3$$

$$f''''(x) = 4 \cdot 3 \cdot 2a_4 + \cdots \quad \text{so} \quad f''''(x_0) = 4 \cdot 3 \cdot 2a_4$$

Solving for the a_i's, we have $a_0 = f(x_0)$, $a_1 = f'(x_0)$, $a_2 = f''(x_0)/2$, $a_3 = f'''(x_0)/2 \cdot 3$, and, in general, $a_i = f^{(i)}(x_0)/i!$. Here $f^{(i)}$ denotes the ith derivative of f, and we recall that $i! = i \cdot (i - 1) \cdots 3 \cdot 2 \cdot 1$ read "i factorial." (We use the conventions that $f^{(0)} = f$ and $0! = 1$.)

This argument shows that if a function $f(x)$ can be written as a power series in $(x - x_0)$, then this series *must* be of the form

$$\sum_{i=0}^{\infty} \frac{f^{(i)}(x_0)}{i!} (x - x_0)$$

For any f, this series is called the *Taylor series* of f about the point $x = x_0$. (This formula is responsible for the presence of factorials in so many important power series.)

The point x_0 is often chosen to be zero, in which case the series becomes

$$\sum_{i=0}^{\infty} \frac{f^{(i)}(0)}{i!} x^i$$

and is called the *Maclaurin series* of f.*

TAYLOR AND MACLAURIN SERIES

If f is infinitely differentiable on some interval containing x_0,

the series

$$\sum_{i=0}^{\infty} \frac{f^{(i)}(x_0)}{i!} (x - x_0)^i$$

is called the *Taylor series* of f at x_0.

When $x_0 = 0$,

the series has the simpler form

$$\sum_{i=0}^{\infty} \frac{f^{(i)}(0)}{i!} x^i$$

and is called the *Maclaurin series* of f.

Worked Example 9   Write down the Maclaurin series for $\sin x$.

Solution   We have:

$$f(x) = \sin x \qquad\qquad f(0) = 0$$
$$f'(x) = \cos x \qquad\qquad f'(0) = 1$$
$$f''(x) = -\sin x \qquad\qquad f''(0) = 0$$
$$f^{(3)}(x) = -\cos x \qquad\qquad f^{(3)}(0) = -1$$
$$f^{(4)}(x) = \sin x \qquad\qquad f^{(4)}(0) = 0$$

and the pattern repeats over and over from here on. Hence the Maclaurin series is

$$\frac{f(0)}{0!} x^0 + \frac{f'(0)}{1!} x + \frac{f^{(2)}(0)}{2!} x^2 + \cdots = x - \frac{x^3}{3!} + \frac{x^5}{5!} - \frac{x^7}{7!} + \cdots$$

Given any infinitely differentiable function f, we can write down its Taylor (or Maclaurin) series and ask whether it converges to f. To

---

Brook Taylor (1685–1731) and Colin Maclaurin (1698–1746) participated in the development of calculus following Newton and Leibniz. According to the Guinness Book of World Records, *Maclaurin has the distinction of being the youngest full professor of all time at age 19 in 1717. He was recommended by Newton. Another mathematician-physicist, Lord Kelvin, holds the record for the youngest and fastest graduation from college—between October 1834 and November 14, 1834, at age 10.*

answer this question, we may proceed as follows. Using the fundamental theorem of calculus, write

$$f(x) = f(x_0) + \int_{x_0}^{x} f'(t)\, dt \tag{1}$$

Now integrate this expression by parts by using $u = f'(t)$ and $v = x - t$. The result is:

$$\int_{x_0}^{x} f'(t)\, dt = -\int_{x_0}^{x} u\, dv = -\left(uv \,\Big|_{x_0}^{x} - \int_{x_0}^{x} v\, du \right)$$

$$= f'(x_0)(x - x_0) + \int_{x_0}^{x} (x - t) f''(t)\, dt$$

Thus we have proved the identity

$$f(x) = f(x_0) + f'(x_0)(x - x_0) + \int_{x_0}^{x} (x - t) f''(t)\, dt \tag{2}$$

Note that the first two terms on the right side of formula (2) equal the first two terms in the Taylor series of f. If we integrate by parts again with

$$u = f''(t) \qquad v = \frac{(x - t)^2}{2}$$

we get

$$\int_{x_0}^{x} (x - t) f''(t)\, dt = -\int_{x_0}^{x} u\, dv = -uv \,\Big|_{x_0}^{x} + \int_{x_0}^{x} v\, du$$

$$= \frac{f''(x_0)}{2}(x - x_0)^2 + \int_{x_0}^{x} \frac{(x - t)^2}{2} f'''(t)\, dt$$

so, substituting into (2),

$$f(x) = f(x_0) + f'(x_0)(x - x_0) + \frac{f''(x_0)}{2}(x - x_0)^2 + \int_{x_0}^{x} \frac{(x - t)^2}{2} f'''(t)\, dt \tag{3}$$

If we repeat the procedure n times, we obtain the formula

$$f(x) = f(x_0) + f'(x_0)(x - x_0) + \frac{f''(x_0)}{2}(x - x_0)^2 + \cdots$$

$$+ \frac{f^{(n)}(x_0)}{n!}(x - x_0)^n + \int_{x_0}^{x} \frac{(x - t)^n}{n!} f^{(n+1)}(t)\, dt \tag{4}$$

which is called *Taylor's formula with remainder in integral form*. The expression

$$R_n(x) = \int_{x_0}^{x} \frac{(x - t)^n}{n!} f^{(n+1)}(t)\, dt \tag{5}$$

is called the *remainder,* so formula (4) may be rewritten in the form

$$f(x) = \sum_{i=0}^{n} \frac{f^{(i)}(x_0)}{i!}(x - x_0)^i + R_n(x) \tag{6}$$

By the second mean value theorem of integral calculus (Problem 8, p. 440), we can write

$$R_n(x) = f^{(n+1)}(c)\left[\int_{x_0}^{x} \frac{(x-t)^n}{n!}\,dt\right] = f^{(n+1)}(c)\frac{(x-x_0)^{n+1}}{(n+1)!} \tag{7}$$

for some point c between x_0 and x. Substituting (7) into (6), we have

$$f(x) = \sum_{i=0}^{n} \frac{f^{(i)}(x_0)}{i!}(x - x_0)^i + \frac{f^{(n+1)}(c)}{(n+1)!}(x-x_0)^{n+1} \tag{8}$$

Formula (8), which is called *Taylor's formula with remainder in derivative form,* reduces to the usual mean value theorem when we take $n = 0$; that is,

$$f(x) = f(x_0) + f'(c)(x - x_0)$$

for some c between x_0 and x.

If $R_n(x) \to 0$ as $n \to \infty$, then formula (6) tells us that the Taylor series of f will converge to f.

The following theorem summarizes our discussion of Taylor series.

Theorem 12
1. If $f(x) = \sum_{i=0}^{\infty} a_i(x - x_0)^i$ is a convergent power series on an open interval I centered at x_0, then f is infinitely differentiable and $a_i = f^{(i)}(x_0)/i!$, so

$$f(x) = \sum_{i=0}^{\infty} \frac{f^{(i)}(x_0)}{i!}(x - x_0)^i$$

2. If f is infinitely differentiable on an open interval I centered at x_0 and if $R_n(x) \to 0$ as $n \to \infty$ for $x \in I$, where $R_n(x)$ is defined by formula (5), then the Taylor series of f converges on I and equals f:

$$f(x) = \sum_{i=0}^{\infty} \frac{f^{(i)}(x_0)}{i!}(x - x_0)^i$$

The following is a useful sufficient condition for f to equal its Taylor series.

Corollary   *Suppose that f is infinitely differentiable on the interval $I = (x_0 - R, x_0 + R)$ (or $(-\infty, \infty)$) and there is a constant M such that*

$$|f^{(n)}(x)| \le M^n \quad \text{for any } n = 0, 1, 2, \ldots \text{ and } x \text{ in } I$$

Then the Taylor series of f converges and equals f. That is, for any x in I,

$$f(x) = \sum_{i=0}^{\infty} \frac{f^{(i)}(x_0)}{i!} (x - x_0)^i$$

Proof We must show that $R_n(x) \to 0$. By formula (7),

$$|R_n(x)| = \left| f^{(n+1)}(c) \frac{(x - t)^{n+1}}{(n + 1)!} \right| \leq \frac{M^{n+1}|x - t|^{n+1}}{(n + 1)!}$$

For any number b, however, $b^n/n! \to 0$, since $\Sigma_{i=0}^{\infty} b^i/i!$ converges by Solved Exercise 7, Section 13-2. Choosing $b = M|x - t|$, we conclude that $R_n(x) \to 0$, so part 2 of Theorem 12 applies.

Worked Example 10   Prove that:

(a) $\sin x = x - \dfrac{x^3}{3!} + \dfrac{x^5}{5!} - \dfrac{x^7}{7!} + \cdots$ for x in $(-\infty, \infty)$

(b) $1 = \dfrac{\pi}{2} - \dfrac{\pi^3}{2^3 \cdot 3!} + \dfrac{\pi^5}{2^5 \cdot 5!} - \dfrac{\pi^7}{2^7 \cdot 7!} + \cdots$

Solution   We verify (a) by applying the conditions of the corollary. Since $f'(x) = \cos x$, $f''(x) = -\sin x$, ..., we see that f is infinitely differentiable.

Notice that $f^{(n)}(x)$ is $\pm\cos x$ or $\pm\sin x$, so $|f^{(n)}(x)| \leq 1$. Thus we can choose $M = 1$. Hence $\sin x$ equals its Maclaurin series, which was shown in Worked Example 9 to be $x - x^3/3! + x^5/5! - \cdots$. For part (b), let $x = \pi/2$ in part (a).

(optional)

Some discussion of the limitations of Theorem 12(2) is in order. Consider, for example, the function $f(x) = 1/(1 + x^2)$, whose Maclaurin series is $1 - x^2 + x^4 - x^6 + \cdots$. Even though the function f is infinitely differentiable on the whole real line, its Maclaurin series converges only for $|x| < 1$. If we wish to represent $f(x)$ for x near 1 by a series, we must use a Taylor series with $x_0 = 1$ (see Solved Exercise 8).

Another important example is the function $g(x) = e^{-1/x^2}$, $g(0) = 0$. This function is infinitely differentiable, but all of its derivatives at $x = 0$ are equal to zero (see Problem 10). Thus the Maclaurin series of g is $\Sigma_{i=0}^{\infty} 0 \cdot x^i$, which converges for all x, but not to the function g. There also exist infinitely differentiable functions with Taylor series having radius of convergence zero.* In each of these examples, the hypothesis that $R_n(x) \to 0$ as $n \to \infty$ fails, so Theorem 12(2) is not contradicted. It simply does not apply.

A function f defined on an open interval (a, b) is said to be *analytic* if, for every x_0 in (a, b), the Taylor series for f at x_0 has positive radius of convergence and converges to f.

*See B. R. Gelbaum and J. M. H. Olmsted, Counterexamples in Analysis, *Holden-Day*, 1964, p. 68.*

TAYLOR SERIES

To prove that a function $f(x)$ equals its Taylor series

$$\sum_{i=0}^{\infty} \frac{f^{(i)}(x_0)}{i!}(x - x_0)^i \quad \text{on } I$$

it is sufficient to show:

1. That f is infinitely differentiable on I.

2. That the derivatives of f grow no faster than the powers of a constant M; that is, for x in I,

$$|f^{(n)}(x)| \le M^n \quad n = 0, 1, 2, 3, \ldots$$

The following display contains the most basic series expansions. They are worth memorizing.

SOME IMPORTANT TAYLOR AND MACLAURIN SERIES

Geometric:

$$\frac{1}{1 - x} = 1 + x + x^2 + \cdots \qquad = \sum_{i=0}^{\infty} x^i \qquad R = 1$$

Binomial:

$$(1 + x)^\alpha = 1 + \alpha x + \frac{\alpha(\alpha - 1)}{2!}x^2 + \cdots \qquad = \sum_{i=0}^{\infty} \frac{\alpha(\alpha - 1)\cdots(\alpha - i + 1)}{i!}x^i \qquad R = 1$$

Sine:

$$\sin x = x - \frac{x^3}{3!} + \frac{x^5}{5!} - \cdots \qquad = \sum_{i=0}^{\infty} \frac{(-1)^i x^{2i+1}}{(2i + 1)!} \qquad R = \infty$$

Cosine:

$$\cos x = 1 - \frac{x^2}{2!} + \frac{x^4}{4!} - \cdots \qquad = \sum_{i=0}^{\infty} (-1)^i \frac{x^{2i}}{(2i)!} \qquad R = \infty$$

Exponential:

$$e^x = 1 + x + \frac{x^2}{2!} + \frac{x^3}{3!} + \cdots \qquad = \sum_{i=0}^{\infty} \frac{x^i}{i!} \qquad R = \infty$$

Logarithm:

$$\ln x = (x - 1) - \frac{(x - 1)^2}{2} + \frac{(x - 1)^3}{3} - \cdots = \sum_{i=1}^{\infty} (-1)^{i+1} \frac{(x - 1)^i}{i} \qquad R = 1$$

or

$$\ln(1 + x) = x - \frac{x^2}{2} + \frac{x^3}{3} - \cdots \qquad = \sum_{i=1}^{\infty} (-1)^{i+1} \frac{x^i}{i} \qquad R = 1$$

The only formula in the display which has not yet been justified is the binomial series. It may be proved by applying the corollary to Theorem 12 and evaluating the derivatives of $f(x) = (1 + x)^\alpha$ at $x = 0$. (The details are omitted.) If $\alpha = n$ is a positive integer, the series terminates and we get the formula

$$(1 + x)^n = 1 + \binom{n}{1} x + \binom{n}{2} x^2 + \cdots + \binom{n}{n} x^n$$

where

$$\binom{n}{k} = \frac{n(n - 1) \cdots (n - k + 1)}{k!}$$

is the number of ways of choosing k objects from a collection of n objects.

Solved Exercises

8. Find the terms through x^3 in the Taylor series for $1/(1 + x^2)$ at $x_0 = 1$.

9. (a) Expand the function $f(x) = 1/(1 + x^2)$ in a Maclaurin series.

   (b) Use part (a) to find $f''''''(0)$ and $f'''''''(0)$ without calculating derivatives of f directly.

   (c) Integrate the series in part (a) to prove that

$$\tan^{-1} x = x - \frac{x^3}{3} + \frac{x^5}{5} - \frac{x^7}{7} + \cdots \quad \text{for } |x| < 1$$

   (d) Justify the formula of Euler:

$$\frac{\pi}{4} = 1 - \frac{1}{3} + \frac{1}{5} - \frac{1}{7} + \cdots$$

10. Expand $\sqrt{1 + x^2}$ about $x_0 = 0$.

Exercises

11. Write down the Maclaurin series for $f(x) = (1 + x^2)^2$ in two ways:

    (a) By multiplying out the polynomial.

    (b) By taking successive derivatives and evaluating them at $x = 0$ (without multiplying out).

12. Write down the Taylor series for $\ln x$ at $x_0 = 2$.

13. (a) Write out the Maclaurin series for $1/\sqrt{1 + x^2}$. (Use the binomial series.)

    (b) What is $(d^{20}/dx^{20})(1/\sqrt{1 + x^2})|_{x=0}$?

TAYLOR POLYNOMIALS AND APPROXIMATIONS

Taylor's formula with remainder,

$$f(x) = \sum_{i=0}^{n} \frac{f^{(i)}(x_0)}{i!} (x - x_0)^i + R_n(x)$$

can be used to obtain approximations to $f(x)$; we can estimate the accuracy of these approximations by measuring the error $R_n(x)$ using the formula

$$R_n(x) = \frac{f^{(n+1)}(c)}{(n + 1)!} (x - x_0)^{n+1}$$

(for some c between x and x_0) and estimating $f^{(n+1)}$ on the interval between x and x_0. The partial sum of the Taylor series,

$$\sum_{i=0}^{n} \frac{f^{(i)}(x_0)}{i!} (x - x_0)^i$$

is a polynomial of degree n in x called the nth *Taylor (or Maclaurin) polynomial for f at x_0*, or the nth-order approximation to f at x_0. The first Taylor polynomial,

$$f(x_0) + f'(x_0)(x - x_0)$$

is just the *linear approximation* to $f(x)$ at x_0; the formula $R_1(x) = [f''(c)/2](x - x_0)^2$ shows that we can estimate the error in the first-order approximation in terms of the size of the second derivative f'' on the interval between x and x_0.

The essence of Taylor's theorem is that, for many functions, we can improve upon the linear approximation by using Taylor polynomials of higher order.

Worked Example 11 Sketch the graph of $\sin x$ along with the graphs of its Maclaurin polynomials of degree 1, 2, and 3. Evaluate the polynomials at $x = 0.02$, 0.2, and 2, and compare with the exact value of $\sin x$.

Solution The Maclaurin polynomials of order 1, 2, and 3 are x, $x + 0x^2$, and $x - x^3/6$. They are sketched in Fig. 13-3-2. Evaluating at $x = 0.02$, 0.2, 2, and 20 gives the results shown in the table:

x	$x - x^3/6$	$\sin x$
0.02	0.0199986667	0.0199986667
0.2	0.1986666	0.1986693
2	0.666666	0.909
20	-1313	0.912

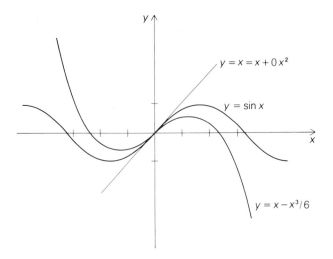

Fig. 13-3-2 The first- and third-order approximations to $\sin x$.

The first Maclaurin polynomials through degree 71 for sin x are shown in Fig. 13-3-3.* Notice that as n increases, the interval on which the nth Taylor polynomial is a good approximation to sin x becomes larger and larger; if we go beyond this interval, however, the polynomials of higher degree "blow up" more quickly than the lower ones.

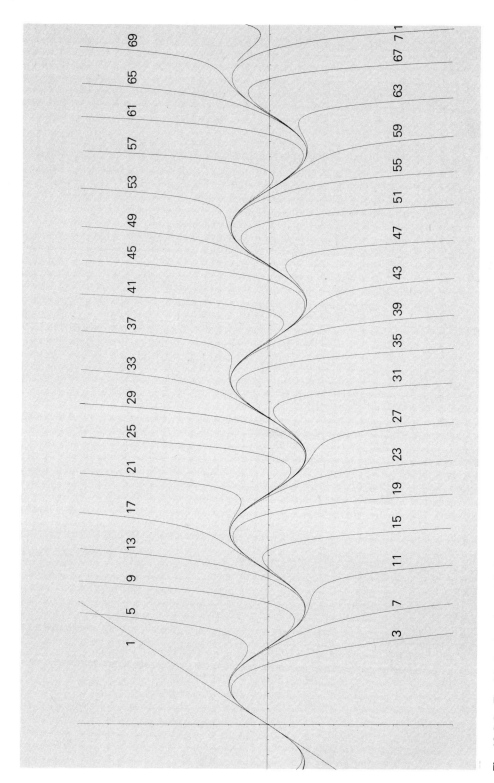

Fig. 13-3-3 The Maclaurin polynomials for sin x through order 71. (The graphs to the left of the y axis are obtained by rotating the figure through 180°.)

*We thank H. Ferguson for providing us with this computer-generated figure.

The following example shows how errors may be estimated.

Worked Example 12 Write down the Taylor polynomials of degrees 1 and 2 for $\sqrt[3]{x}$ at $x_0 = 27$. Use these polynomials to approximate $\sqrt[3]{28}$, and estimate the error in the second-order approximation by using the formula for $R_2(x)$.

Solution Let $f(x) = x^{1/3}$, $x_0 = 27$, $x = 28$. Then $f'(x) = \frac{1}{3}x^{-2/3}$, $f''(x) = -\frac{2}{9}x^{-5/3}$, and $f'''(x) = \frac{10}{27}x^{-8/3}$. Thus $f(27) = 3$, $f'(27) = \frac{1}{27}$, and $f''(27) = -(2/3^7)$, so the Taylor polynomials of degree 1 and 2 are respectively

$$3 + \frac{1}{27}(x-27) \quad \text{and} \quad 3 + \frac{1}{27}(x-27) - \frac{1}{3^7}(x-27)^2$$

Evaluating these at $x = 28$ gives 3.0370... and 3.0365798... for the first- and second-order approximations. The error in the second-order approximation is at most $1/3!$ times the largest value of $(10/27)x^{-8/3}$ on $[27, 28]$, which is

$$\frac{1}{6}\frac{10}{27}\frac{1}{3^8} = \frac{5}{3^{12}} \le 0.00001$$

(Actually, $\sqrt[3]{28} = 3.0365889....$)

Solved Exercises 11. By integrating a series for e^{-x^2}, calculate $\int_0^1 e^{-x^2}\,dx$ to within 0.001.

12. Calculate $\sin(\pi/4 + 0.06)$ to within 0.0001 by using the Taylor series about $x_0 = \pi/4$. How many terms would have been necessary if you had used the Maclaurin series?

Exercises 14. Calculate $\ln(1.1)$ to within 0.001 by using a power series.

15. Sketch the graphs of the Maclaurin polynomials through degree 4 for $\cos x$.

16. Calculate $e^{\ln 2 + 0.02}$ to within 0.0001 by using the Taylor series about $x_0 = \ln 2$. How many terms would have been necessary if you had used the Maclaurin series?

Problems for Section 13-3

1. Find the radius of convergence of each of the following series:

(a) $1 + x + \dfrac{x^2}{2!} + \dfrac{x^3}{3!} + \cdots$

(b) $1 + \dfrac{x}{2} + \dfrac{2!}{4!}x^2 + \dfrac{3!}{6!}x^3 + \cdots$

(c) $\dfrac{\sin 1}{2}x + \dfrac{\sin 2}{4}x^2 + \dfrac{\sin 3}{8}x^3 + \cdots$

(d) $1 + \dfrac{x}{2} + \dfrac{x^2}{3} + \dfrac{x^3}{4} + \cdots$

2. Find the radius of convergence R of the series $\Sigma_{n=0}^{\infty} a_n x^n$ for each of the following choices of a_n. Discuss convergence at $\pm R$.

(a) $a_n = 1/(n+1)^n$ (b) $a_n = (-1)^n/(n+1)$

(c) $a_n = (n^2 + n^3)/(1 + n)^5$

(d) $a_n = n$ (e) $a_n = \tan^{-1}(n)$

3. Express the following functions as Maclaurin series:

(a) $\dfrac{1}{1-x}$ (b) $\dfrac{1}{1+x}$

(c) $\dfrac{1}{1-x} - \dfrac{1}{1+x}$ (d) $\dfrac{1}{2}\left(\dfrac{1}{1-x} + \dfrac{1}{1+x}\right)$

(e) $\dfrac{1}{1-x^2}$

4. Let $f(x) = a_0 + a_1 x + a_2 x^2 + \cdots$. Find a_0, a_1, a_2, and a_3 for each of the following functions:

(a) $\sec x$ (b) $\sqrt{1-x^2}$

(c) $(d/dx)\sqrt{1-x^2}$ (d) e^{1+x}

5. Find Maclaurin expansions through the term in x^5 for each of the following functions:

(a) $(1 - \cos x)/x^2$ (b) $\sec x$

(c) $1 + \sin x$ (d) $\sqrt[3]{(1-x)/(1+x)}$

(e) $(d^2/dx^2)(1/\sqrt{1+x^2})$

6. Find the Taylor polynomial of degree 4 for $\ln x$ at:

(a) $x_0 = 1$ (b) $x_0 = e$ (c) $x_0 = 2$

7. (a) Let

$$f(x) = \begin{cases} (\sin x)/x & x \neq 0 \\ 1 & x = 0 \end{cases}$$

Find $f'(0), f''(0)$, and $f'''(0)$.

(b) Find the Maclaurin expansion for $(\sin x)/x$.

8. (a) Can we use the binomial expansion of $\sqrt{1+x}$ to obtain a convergent series for $\sqrt{2}$. Why or why not?

(b) Writing $2 = \frac{9}{4} \cdot \frac{8}{9}$, we have $\sqrt{2} = \frac{3}{2}\sqrt{\frac{8}{9}}$. Use this equation, together with the binomial expansion, to obtain an approximation to $\sqrt{2}$ correct to two decimal places.

(c) Use the method of part (b) to obtain an approximation to $\sqrt{3}$ correct to two decimal places.

9. By using Maclaurin expansion for $1/(1 + x)$, approximate $\int_0^{1/2} dx/(1 + x)$ to within 0.01.

★10. Let

$$g(x) = \begin{cases} e^{-1/x^2} & x \neq 0 \\ 0 & x = 0 \end{cases}$$

(a) Show that $g'(0) = 0$. [Hint: Use L'Hôpital's rule to evaluate $\lim_{x \to 0} (1/x)\, e^{-1/x^2}$.]

(b) Show that $g''(0) = 0$.

(c) Show that g is infinitely differentiable at zero and that $g^{(n)}(0) = 0$ for all n.

★11. Show that $1 - \frac{1}{2} + \frac{1}{3} - \frac{1}{4} \cdots = \ln 2$ by integrating the Maclaurin series for $1/(1 + x)$ and evaluating at $x = 1$. Follow the pattern of Solved Exercise 9(d) to justify your answer.

12. Use the power series for $\ln(1 + x)$ to calculate $\ln 2\frac{1}{2}$, correct to within 0.1. [Hint: $2\frac{1}{2} = \frac{3}{2} \cdot \frac{5}{3}$.]

13. Find a power series expansion for $\int_1^x \ln t\, dt$. Compare this with the expansion for $x \ln x$. What is your conclusion?

14. Find the first four nonvanishing terms in the power series expansion for:

(a) $\ln(1 + e^x)$ (b) e^{x^2+x}

(c) $\sin(e^x)$ (d) $e^x \cos x$

★15. (a) Write down the Maclaurin series for $1/\sqrt{1-x^2}$ and $\sin^{-1} x$. Where do they converge?

(b) Find the terms through x^3 in the series for $\sin^{-1}(\sin x)$ by substituting the series for $\sin x$ in the series for $\sin^{-1} x$; that is, if $\sin^{-1} x = a_0 + a_1 x + a_2 x^2 + \cdots$, then

$$\sin^{-1}(\sin x) = a_0 + a_1\left(x - \frac{x^3}{3!} + \frac{x^5}{5!} - \cdots\right)$$
$$+ a_2\left(x - \frac{x^3}{3!} + \frac{x^5}{5!} - \cdots\right)^2 + \cdots$$

(c) Use the substitution method of part (b) to *obtain* the first five terms of the series for $\sin^{-1} x$ by using the relation $\sin^{-1}(\sin x) = x$ and *solving* for a_0 through a_5.

(d) Find the terms through x^5 of the Maclaurin series for the inverse function $g(x)$ of $f(x) = x^3 + x$. (Use the relation $g(f(x)) = x$ and solve for the coefficients in the series for g.)

16. Using the Taylor series for $\sin x$ and $\cos x$, find the terms through x^6 in the series for $(\sin x)^2 + (\cos x)^2$.

17. Continue the work of Worked Example 12 by finding the third, fourth, fifth (and so on) order approximations to $\sqrt[3]{28}$. Stop when the round-off errors on your calculator become greater than the remainder term of the series.

18. (a) Use the second-order approximation at x_0 to derive the approximation

$$\int_{x_0-R}^{x_0+R} f(x)\, dx \approx 2Rf(x_0) + \frac{2f''(x_0)}{3!} R^3$$

Find an estimate for the error.

(b) Using the formula in part (a), find an approximate value for $\int_{-1/2}^{1/2} dx/\sqrt{1+x^2}$. Compare the answer with that obtained from Simpson's rule with $n = 4$ (p. 531).

19. (a) Find a power series expansion for a function $f(x)$ such that $f(0) = 0$ and $f'(x) - f(x) = x$. (Write $f(x) = a_0 + a_1 x + a_2 x^2 + \cdots$ and solve for the a_i's one after another.)

    (b) Find a formula for the function whose series you found in part (a).

★20. Using Taylor's formula, prove the following inequalities:

    (a) $e^x - 1 \geq x$ for $x \geq 0$.

    (b) $6x - x^3 + x^5/20 \geq 6 \sin x \geq 6x - x^3$ for $x \geq 0$.

    (c) $x^2 - x^4/12 \leq 2 - 2 \cos x \leq x^2$ for $x \geq 0$.

Problems 21 through 25 contain the proof of Theorem 9 on the differentiation and integration of power series. For simplicity, we stick to the case $x_0 = 0$.

★21. Prove that the series $f(x) = \Sigma_{i=0}^{\infty} a_i x^i$, $g(x) = \Sigma_{i=1}^{\infty} i a_i x^{i-1}$, and $h(x) = \Sigma_{i=0}^{\infty} [a_i/(i+1)] x^{i+1}$ all have the same radius of convergence. [*Hint:* Use Theorem 7, the comparison test, and the fact that $\Sigma_{i=0}^{\infty} i r^i$ converges absolutely if $|r| < 1$.]

★22. Prove that if $0 < R_1 < R$, where R is the radius of convergence of $f(x) = \Sigma_{i=0}^{\infty} a_i x^i$, then given any $\varepsilon > 0$, there is a number M such that, for all N greater than M, the difference $|f(x) - \Sigma_{i=0}^{N} a_i x^i|$ is less than ε for all x in $[-R_1, R_1]$. [*Hint:* Compare $\Sigma_{i=N+1}^{\infty} a_i x^i$ with a geometric series, using Theorem 7(1).]

★23. Prove that if $|x_0| < R$, where R is the radius of convergence of $f(x) = \Sigma_{i=0}^{\infty} a_i x^i$, then f is continuous at x_0. [*Hint:* Use Problem 22, together with the fact that the polynomial $\Sigma_{i=0}^{N} a_i x^i$ is continuous. Given $\varepsilon > 0$ write $f(x) - f(x_0)$ as a sum of terms, each of which is less than $\varepsilon/3$, by choosing N large enough and $|x - x_0|$ less than some δ.]

★24. Prove that if $|x| < R$, where R is the radius of convergence of $f(x) = \Sigma_{i=0}^{\infty} a_i x^i$, then the integral $\int_0^x f(t) \, dt$ (which exists by Problem 23) is equal to $\Sigma_{i=0}^{\infty} [a_i/(i+1)] x^{i+1}$. [*Hint:* Use the result of Problem 22 to show that the difference $|\int_0^x f(t) \, dt - \Sigma_{i=0}^{\infty} a_i x^i|$ is less than any positive number ε.]

★25. Prove that if $f(x) = \Sigma_{i=0}^{\infty} a_i x^i$ and $g(x) = \Sigma_{i=1}^{\infty} i a_i x^{i-1}$ have radius of convergence R, then $f'(x) = g(x)$ on $(-R, R)$. [*Hint:* Apply the result of Problem 24 to $\int_0^x g(t) \, dt$; then use the alternative version of the fundamental theorem of calculus.]

26. On the average, airplanes depart at the rate of one plane every 30 seconds from Los Angeles International Airport during a portion of the day. Assume the probability $p_k(x)$ that k planes depart in x minutes can be approximated by $p_k(x) = (2x)^k e^{-2x}/k!$.

    (a) Calculate $\Sigma_{k=0}^{\infty} p_k(x)$.

    (b) Interpret the result in (a).

    (c) Prove the identity $\Sigma_{k=0}^{\infty} k(ax)^k/k! = ax \, e^{ax}$.

    (d) The expected number of planes departing in x minutes is given by $n(x) = \Sigma_{k=0}^{\infty} k \, p_k(x)$. Check that $n(\frac{1}{2}) = 1$; that is, one plane leaves every 30 seconds.

27. An automobile travels on a straight highway. At noon it is 20 miles from the next town, traveling at 50 mph, with its acceleration kept between 20 mph/hr and -10 mph/hr. Use the formula

$$x(t) = x(0) + x'(0)t + \int_0^t (t - s)x''(s) \, ds$$

    to estimate the auto's distance from the town 15 minutes later.

28. A numerical analyst is about to compute $\sin(36°)$, when the batteries in his hand calculator give out. He quickly grabs a backup unit, only to find it is made for statistics and does not have a "sin" key. Unperturbed, he enters 3.1415926, divides by 5, enters the result into the memory, called "x" hereafter. Then he computes $x(1 - x^2/6)$ and uses it for the value of $\sin(36°)$.

    (a) What was his answer?

    (b) How good was it?

    (c) Explain what he did in the language of Taylor series expansions.

    (d) Describe a similar method for computing $\tan(10°)$.

★29. (a) The *Gaussian distribution* is the classical "bell-shaped curve" encountered in the distribution of grades and IQ scores. Tables give the values of $\int_{-\infty}^{x} (1/\sqrt{2\pi}) \, e^{-t^2/2} \, dt$, the integral of the Gaussian distribution. Discuss the evaluation of this integral by replacement of the exponential term by $\Sigma_{n=0}^{\infty} [(-t^2/2)^n]/n!$, and subsequent term-by-term integration.

    (b) Mathematical tables for the *normal probability distribution* give values of

$$\frac{1}{\sqrt{2\pi}} \int_{-\infty}^{x} e^{-t^2/2} \, dt, \quad 0 \leq x \leq 4$$

    Armed with the power series $e^{-t^2/2} = \Sigma_{n=0}^{\infty} [(-t^2/2)^n]/n!$ and theorems on integration of power series, give an argument which explains why the tables consider only $0 \leq x \leq 4$.

★30. *Planck's radiation law* gives the radiation density ψ as $\psi = 8\,(h/c^3)\int_0^\infty \nu^3\,(e^{h\nu/KT} - 1)^{-1}\,d\nu$. The symbols are: ν = frequency (sec^{-1}), T = temperature (deg. abs.), $h = 6.554 \times 10^{-27}$ erg sec, $c = 2.998 \times 10^{10}$ cm/sec, $K = 1.372 \times 10^{-16}$ erg/deg.

(a) Expand the integrand in a series, integrate term-by-term, and show that the integral formula reduces to the *Stefan-Boltzmann law* $\psi = aT^4$.

(b) Compute a and specify its units.

31. The maximum deflection $\bar{y}$ of a cantilever beam of length L inches and weight W lb/in. subjected to a horizontal compressive force P lb at the free end is $\bar{y} = (WEI/P^2)[1 - (\theta^2/2) - \sec\theta + \theta\tan\theta]$, where $\theta = (\sqrt{P/EI})L$ and EI = flexural rigidity.

(a) Apply the theory of series to obtain

$$\bar{y} = \frac{WEI}{P^2}\left(\frac{\theta^4}{8} + \frac{7\theta^6}{144} + \frac{113\theta^8}{5760} + \cdots\right)$$

(b) Suppose the beam is wrought-iron, $2'' \times 4'' \times 12'$, with $E = 15 \times 10^6$ lb/in.^{2} weighing 490 lb/ft^3 and $P = 1500$ lb. The 2-inch side is horizontal. Compute $I = \int_{-2}^{2} 2y^2\,dy$, then evaluate $\bar{y}$ by both formulas. Compare the accuracy.

Review Problems for Chapter 13

1. Test each of the following series for convergence:

(a) $\sum_{n=1}^{\infty} 5^{-n}$

(b) $\sum_{n=1}^{\infty} 4^n/(2n+1)!$

(c) $\sum_{n=1}^{\infty} 2^{n^2}/n!$

(d) $\sum_{i=1}^{\infty} i/(i^3 + 8)$

(e) $\sum_{j=0}^{\infty} (-1)^j j/(j^2 + 8)$

2. Test for convergence:

(a) $\sum_{k=1}^{\infty} k/3^k$

(b) $\sum_{n=1}^{\infty} 2n/(n+3)$

(c) $\sum_{n=1}^{\infty} 2n/(n^2 + 3)$

(d) $\sum_{n=1}^{\infty} (-1)^n n/3^n$

(e) $\sum_{n=1}^{\infty} (-1)^{2n}/n$

3. Tell whether each of the following statements is true or false. Justify your answer:

(a) If $a_n \to 0$, then $\sum_{n=1}^{\infty} a_n$ converges.

(b) Every geometric series $\sum_{i=1}^{\infty} r^i$ converges.

(c) Convergence or divergence of any series may be determined by the ratio test.

(d) $\sum_{i=1}^{\infty} 1/2^i = 1$.

(e) $e^{2x} = 1 + 2x + x^2 + x^3/3 + \cdots$.

(f) If a series converges, it must also converge absolutely.

(g) The error made in approximating a convergent series by a partial sum is no greater than the first term omitted.

(h) $\cos x = \sum_{k=0}^{\infty} (-1)^k x^{2k}/(2k)!$.

(i) If $\sum_{j=1}^{\infty} a_j$ and $\sum_{k=0}^{\infty} b_k$ are convergent, then $\sum_{j=1}^{\infty} a_j + \sum_{k=0}^{\infty} b_k = b_0 + \sum_{i=1}^{\infty} (a_i + b_i)$.

4. For the following, prove or give a counterexample:

(a) The convergence of $\sum_{n=1}^{\infty} a_n$ implies the convergence of $\sum_{n=1}^{\infty} (a_n + a_{n+1})$.

(b) The convergence of $\sum_{n=1}^{\infty} (a_n + a_{n+1})$ implies the convergence of $\sum_{n=1}^{\infty} a_n$.

(c) The convergence of $\sum_{n=1}^{\infty} (|a_n| + |b_n|)$ implies the convergence of $\sum_{n=1}^{\infty} |a_n|$.

★(d) The convergence of $\sum_{n=1}^{\infty} a_n^2$ and $\sum_{n=1}^{\infty} b_n^2$ implies absolute convergence of $\sum_{n=1}^{\infty} a_n b_n$.

5. Test for convergence (using the comparison test):

(a) $\sum_{n=1}^{\infty} ne^{-n^2}$

(b) $\sum_{n=1}^{\infty} \dfrac{\sqrt{n}}{n^2 - \sin^2 99n}$

(c) $\sum_{n=2}^{\infty} \dfrac{1}{(\ln n)^{\ln n}}$

(d) $\sum_{n=1}^{\infty} \left(\dfrac{1}{n} - \dfrac{1}{\sqrt{n}}\right)$

(e) $\sum_{n=1}^{\infty} \dfrac{n}{(n+1)!}$

6. Find Maclaurin series for:

(a) $f(x) = (d/dx)(\sin x - x)$

(b) $g(k) = (d^2/dk^2)(\cos k^2)$

(c) $f(x) = \int_0^x [(e^t - 1)/t]\,dt$

(d) $g(y) = \int_0^y \sin t^2\,dt$

7. Let

$$\frac{z}{e^z - 1} = 1 + B_1 z + \frac{B_2}{2!}z^2 + \frac{B_3}{3!}z^3 + \cdots$$

Determine the numbers B_1, B_2, B_3. (The B_i are known as the *Bernoulli numbers*.)

8. Determine how many terms are needed to compute the sum of $1 + r + r^2 + \cdots$ with error less than 0.01 when (a) $r = 0.5$ and (b) $r = 0.09$.

9. Find the sum of the given series:

   (a) $\displaystyle\sum_{n=1}^{\infty} \frac{1}{n(n+1)}$ [*Hint:* Use partial fractions.]

   (b) $\displaystyle\sum_{n=1}^{\infty} \frac{n}{(n+1)!}$ [*Hint:* Write the numerator as $n + 1 - 1$.]

   (c) $\displaystyle\sum_{n=1}^{\infty} \frac{1+n}{2^n}$ [*Hint:* Differentiate a certain power series.]

10. Show in two ways that $\sum_{n=1}^{\infty} na^n = a/(1-a)^2$ for $|a| < 1$ as follows.

    (a) Consider
    $$s_n = a + 2a^2 + 3a^3 + \cdots + na^n$$
    $$as_n = \qquad a^2 + 2a^3 + \cdots + (n-1)a^n + na^{n+1}$$
    and subtract.

    (b) Differentiate $\sum_{n=0}^{\infty} a^n = 1/(1-a)$ with respect to a, and then add your answer to $\sum_{n=0}^{\infty} a^n = 1/(1-a)$.

11. In each of the following, evaluate the indicated derivative:

    (a) $f^{(12)}(0)$, where $f(x) = x/(1 + x^2)$

    (b) $f^{(10)}(0)$, where $f(x) = x^6 e^{x+1}$

12. Find the radius of convergence:

    (a) $1 - \dfrac{x^2}{2!} + \dfrac{x^4}{3!} - \dfrac{x^6}{4!} + \cdots$

    (b) $1 + 3x + 5x^2 + 7x^3 + \cdots$

    (c) $\displaystyle\sum_{n=0}^{\infty} \frac{x^n}{(3n)!}$ (d) $\displaystyle\sum_{n=0}^{\infty} \frac{(x-1/2)^n}{(n+1)!}$

    (e) $\displaystyle\sum_{n=0}^{\infty} \frac{(-1)^n}{2^n} x^n$ (f) $\displaystyle\sum_{n=0}^{\infty} \frac{n^n}{n!} x^n$

★13. (a) Show that if the radius of convergence of $\sum_{n=1}^{\infty} a_n x^n$ is R, then the radius of convergence $\sum_{n=1}^{\infty} a_n x^{2n}$ is $\sqrt{R}$.

    (b) Find the radius of convergence of the series $\sum_{n=0}^{\infty} (\pi/4)^n x^{2n}$.

14. Find $\lim_{x \to 0} (1/x^2)(1 - \cos x)$:

    (a) By using L'Hôpital's rule (Section 12-4).

    (b) By finding the power series expansion for $(1/x^2)(1 - \cos x)$.

    Which way is easier?

★15. Let $f(x) = \sum_{i=0}^{\infty} a_i x^i$ and $g(x) = f(x)/(1-x)$.

    (a) By multiplying the power series for $f(x)$ and $1/(1-x)$, show that $g(x) = \sum_{i=0}^{\infty} b_i x^i$, where $b_i = a_0 + \cdots + a_i$ is the ith partial sum of the series $\sum_{i=0}^{\infty} a_i$.

    (b) Suppose that the radius of convergence of $f(x)$ is greater than 1 and that $f(1) \neq 0$. Show that $\lim_{i \to \infty} b_i$ exists and is not equal to zero. What does this tell you about the radius of convergence of $g(x)$?

    (c) Let $e^x/(1-x) = \sum_{i=0}^{\infty} b_i x^i$. What is $\lim_{i \to \infty} b_i$?

16. (a) Use a power series for $\sqrt{1+x}$ to calculate $\sqrt{\frac{5}{4}}$ correct to 0.01.

    (b) Use the result of part (a) to calculate $\sqrt{5}$. How accurate is your answer?

17. Sum each of the following series to within 0.05:

    (a) $1 - \frac{1}{4} + \frac{1}{16} - \frac{1}{32} + \cdots$

    (b) $1 - \frac{2}{4} + \frac{3}{16} - \frac{4}{32} + \cdots$

    ★ (c) $1 + \frac{2}{4} + \frac{4}{8} + \frac{9}{16} + \frac{16}{32} + \frac{25}{64} + \cdots$

★18. (a) Find the second-order approximation at $T = 0$ to the day-length function S (see p. 304) for latitude 38° (or your own latitude).

    (b) How many minutes earlier (compared with $T = 0$) does the sun set when $T = 1, 2, 10, 30$?

    (c) Compare the results in part (b) with those obtained from the exact formula and with listings in your local newspaper.

    (d) For how many days before and after June 21 is the second-order approximation correct to within 1 minute? Within 5 minutes?

19. Verify that $\sum_{n=0}^{\infty} x^2(1 + x^2)^{-n}$ is a convergent geometric series for $x \neq 0$ with sum $1 + x^2$. It also converges to 0 when $x = 0$. (This shows that the sum of an infinite series of continuous terms need not be continuous. Compare with Theorem 9.)

20. A beam of length L ft, supported at its ends, carries a concentrated load of P lb at its center. The maximum deflection D of the beam from equilibrium is

$$D = \frac{2L^3 P}{EI\pi^4} \sum_{n=1}^{\infty} \frac{|\sin(n\pi/2)|}{n^4}$$

(a) Use the formula $\sum_{n=1}^{\infty} 1/n^4 = \pi^4/90$ to show that

$$\sum_{k=1}^{\infty} \frac{1}{(2k)^4} = \left(\frac{1}{2^4}\right)\left(\frac{\pi^4}{90}\right)$$

[*Hint:* Factor out 2^{-4}.]

(b) Show that

$$\sum_{k=0}^{\infty} \frac{1}{(2k+1)^4} = \left(\frac{15}{16}\right)\left(\frac{\pi^4}{90}\right)$$

hence $D = (1/48)(L^3 P/EI)$. [*Hint:* A series is the sum of its even and odd terms.]

(c) Use the first two terms in the series for D to obtain a simpler formula for D. Show that this result differs at most by 0.23% from the theoretical value.

21. The deflection $y(x, t)$ of a string from its straight profile at time t, measured vertically at location x along the string, $0 \le x \le L$, is

$$y(x, t) = \sum_{n=1}^{\infty} A_n \sin\left(\frac{n\pi x}{L}\right) \cos\left(\frac{n\pi ct}{L}\right)$$

(a) Explain what this equation means in terms of limits of partial sums for x, t fixed.

(b) Initially (at $t = 0$), the deflection of the string is

$$\sum_{n=1}^{\infty} A_n \sin\left(\frac{n\pi x}{L}\right)$$

Find the deflection value as an infinite series at the midpoint $x = L/2$.

22. In the study of saturation of a two-phase motor servo, an engineer starts with a transfer function equation $V(s)/E(s) = K/(1 + s\tau)$, then goes to the first-order approximation $V(s)/E(s) = K(1 - s\tau)$, from which he obtains an approximate equation for the saturation dividing line.

(a) Show that $1/(1 + s\tau) = \sum_{n=0}^{\infty} (-s\tau)^n$, by appeal to the theory of geometric series. Which values of $s\tau$ are allowed?

(b) Discuss the replacement of $1/(1 + s\tau)$ by $1 - s\tau$; include an error estimate in terms of the value of $s\tau$.

★23. Find the area bounded by the curves $xy - \sin x = 0$, $x = 1$, $x = 2$, $y = 0$. Make use of the Taylor expansion of $\sin x$.

★24. In highway engineering, a *transitional spiral* is defined to be a curve whose curvature varies directly as the arc length. Assume this curve starts at $(0, 0)$ as the continuation of a road coincident with the negative x-axis. Then the parametric equations of the spiral are

$$x = k\int_0^{\phi} \frac{\cos\theta}{\sqrt{\theta}} d\theta, \qquad y = k\int_0^{\phi} \frac{\sin\theta}{\sqrt{\theta}} d\theta$$

(a) By means of infinite series methods, find the ratio x/y for $\phi = \pi/4$.

(b) Try to graph the transitional spiral for $k = 1$, using accurate graphs of $(\cos\theta)/\sqrt{\theta}$, $(\sin\theta)/\sqrt{\theta}$ and the area interpretation of the integral.

★25. The free vibrations of an elastic circular membrane can be described by infinite series, the terms of which involve trigonometric functions and *Bessel functions*. The latter functions are named after F. W. Bessel (1784–1846), who inaugurated modern practical astronomy at Königsberg Observatory. The series

$$\sum_{k=0}^{\infty} \frac{(-1)^k}{k!(n+k)!}(x/2)^{n+2k}$$

is called the *Bessel Function* $J_n(x)$; n is an integer ≥ 0.

(a) Establish convergence by the ratio test.

(b) The frequencies of oscillation of the circular membrane are essentially solutions of the equation $J_n(x) = 0$, $x > 0$. Examine the equation $J_0(x) = 0$, and see if you can explain why $J_0(2.404) = 0$ is possible.

26. A wire of length L inches and weight w lb/in., clamped at its lower end at a small angle $\tan^{-1} P_0$ to the vertical, deflects $y(x)$ inches due to bending. The displacement $y(L)$ at the upper end is given by

$$y(L) = \frac{2P_0}{3L^{1/3}} \frac{\int_0^{L^{3/2}} u(az)dz}{u(aL^{3/2})}$$

where $a = \frac{2}{3}\sqrt{W/EI}$, and

$$u(az) = \frac{az}{2}^{-1/3} \sum_{k=0}^{\infty} (-1)^k \frac{a^{2k}z^{2k}}{2^{2k} k!\, \Gamma(k + \frac{2}{3})}$$

The values of the *gamma function* Γ may be found in a mathematical table. The function u is the *Bessel function* of order $-\frac{1}{3}$.

(a) Find the smallest positive root of $u(az) = 0$ by using the first four terms of the series.

(b) Evaluate $y(L)$ approximately by using the first four terms of the series.

14

Vectors

The vector representation of a number pair or triple is an arrow in the plane or in space.

Many interesting mathematical and physical quantities depend upon more than one independent variable. The length z of the hypotenuse of a right triangle, for instance, depends upon the lengths x and y of the two other sides; the dependence is given by the Pythagorean formula $z = \sqrt{x^2 + y^2}$. Similarly, the growth rate of a plant may depend upon the amounts of sunlight, water, and fertilizer it receives; in this case, the dependence relation must be determined experimentally.

The calculus of functions of a single variable, which we have been studying since the beginning of this book, does not suffice for the description of functions which depend upon several variables—what we require is the *calculus of functions of several variables.* In our last five chapters, we present an introduction to this general calculus.

The present and following chapter set out the algebraic and geometric preliminaries for the calculus of several variables. In this respect, they are somewhat like Chapter R at the front of the book, but not so elementary. Chapter 16 is then devoted to the differential calculus, and Chapters 17 and 18 to the integral calculus, of functions of several variables.

CONTENTS

In Section 14-1, vectors are defined as classes of directed line segments with the same magnitude and direction. Vector addition and scalar multiplication are introduced, as well as the plotting of points in three-dimensional space.

Sections 14-2 and 14-3 are devoted to the operations of dot product and cross-product, with geometric applications. Determinants are presented from a geometric viewpoint.

14-1 Vectors in the Plane and in Space

Number pairs and triples have an algebraic structure.

An ordered pair (x, y) of real numbers has been considered up to now as a point in the plane—that is, as a *geometric* object. We begin this section by giving the number pairs an *algebraic* structure. Next we introduce the notion of a vector. Finally, we discuss the representation of points in space by triples (x, y, z) of real numbers, and we extend the vector concept to three dimensions. In Section 14-2, we apply the algebra of vectors to the solution of geometric problems.

Goals

After studying this section, you should be able to:

Add vectors and multiply them by scalars.

Plot and identify points in three-dimensional cartesian coordinates.

THE ALGEBRA OF ORDERED PAIRS

The authors of this book, and probably many of its readers, were brought up mathematically on the precept that "you cannot add apples to oranges." If we have x apples and y oranges, the number $x + y$ represents the number of pieces of fruit but confounds the apples with the oranges. By using the ordered pair (x, y) instead of the sum $x + y$, we can keep track of both our apples and our oranges without losing any information in the process. Furthermore, if someone adds to our fruit basket X apples and Y oranges, which we might denote by (X, Y), our total accumulation is now $(x + X, y + Y)$—that is, $x + X$ apples and $y + Y$ oranges. This addition of items kept in separate categories is useful in many contexts.

Definition   If (x_1, y_1) and (x_2, y_2) are ordered pairs of real numbers, the ordered pair $(x_1 + x_2, y_1 + y_2)$ is called their *sum* and is denoted by $(x_1, y_1) + (x_2, y_2)$.

The operation just defined is sometimes called *componentwise addition*, since each component of the sum is obtained by adding the corresponding components of the summands.

Worked Example 1   (a) Calculate $(-3, 2) + (4, 6)$.

(b) Calculate $(1, 4) + (1, 4) + (1, 4)$.

(c) Given the pairs (a, b) and (c, d), find (x, y) such that $(a, b) + (x, y) = (c, d)$.

Solution   (a) $(-3, 2) + (4, 6) = (-3 + 4, 2 + 6) = (1, 8)$

(b) We have not yet defined the sum of three ordered pairs, so we take the problem to mean $[(1, 4) + (1, 4)] + (1, 4)$, which is $(2, 8) + (1, 4) = (3, 12)$. Notice that this is $(1 + 1 + 1, 4 + 4 + 4)$, or $(3 \cdot 1, 3 \cdot 4)$.

(c) The equation $(a, b) + (x, y) = (c, d)$ means $(a + x, b + y) = (c, d)$.

Since two ordered pairs are equal only when their corresponding components are equal, the last equation is equivalent to the two numerical equations

$$a + x = c \quad \text{and} \quad b + y = d$$

Solving these equations for x and y gives $x = c - a$ and $y = d - b$, or $(x, y) = (c - a, d - b)$.

Following Worked Example 1(b), we may observe that the sum $(x, y) + (x, y) + \cdots + (x, y)$, with n terms, is equal to (nx, ny). Thinking of the sum as "n times (x, y)," we denote it by $n(x, y)$, so we have the equation

$$n(x, y) = (nx, ny)$$

Now we may note that the right-hand side of this equation makes sense when n is any real number, not just a positive integer, so we take this as a definition.

Definition If (x, y) is an ordered pair and r is a real number, the ordered pair (rx, ry) is called the *product* of r and (x, y) and is denoted by $r(x, y)$.

To distinguish ordinary numbers from ordered pairs, we sometimes call numbers *scalars*. The operation just defined is called *scalar multiplication*. The scalar is always written to the left of the ordered pair. Notice that we have not defined the product of two ordered pairs—we will do so in Section 14-2.

Worked Example 2 (a) Calculate $4(2, -3) + 4(3, 5)$.

(b) Calculate $4[(2, -3) + (3, 5)]$.

Solution (a) $4(2, -3) + 4(3, 5) = (8, -12) + (12, 20) = (20, 8)$.

(b) $4[(2, -3) + (3, 5)] = 4(5, 2) = (20, 8)$.

It is sometimes a useful shorthand to denote an ordered pair by a single letter such as $P = (x, y)$. This makes the algebra of ordered pairs look more like the algebra of ordinary numbers.

Worked Example 3 Show that if a is a number and P_1 and P_2 are ordered pairs, then $a(P_1 + P_2) = aP_1 + aP_2$.

Solution We introduce components for P_1 and P_2, writing $P_1 = (x_1, y_1)$ and $P_2 = (x_2, y_2)$. Then

$$P_1 + P_2 = (x_1 + x_2, y_1 + y_2)$$
$$a(P_1 + P_2) = (a(x_1 + x_2), a(y_1 + y_2)) = (ax_1 + ax_2, ay_1 + ay_2)$$
$$= (ax_1, ay_1) + (ax_2, ay_2) = a(x_1, y_1) + a(x_2, y_2)$$
$$= aP_1 + aP_2$$

as required.

All the usual algebraic identities which make sense for numbers and ordered pairs are true, and they can be used in computations with ordered pairs (see Exercise 3).

ALGEBRA OF ORDERED PAIRS

Addition: $(x_1, y_1) + (x_2, y_2) = (x_1 + x_2, y_1 + y_2)$

Scalar multiplication: $r(x, y) = (rx, ry)$

Solved Exercises*

1. (a) Find real numbers a_1, a_2, a_3 such that $a_1(3, 1) + a_2(6, 2) + a_3(-1, 1) = (5, 6)$.

   (b) Is the solution in part (a) unique?

   (c) Can you find a solution in which a_1, a_2, and a_3 are integers?

2. Show that every ordered pair is a sum of scalar multiples of $(1, 0)$ and $(0, 1)$.

3. Interpret the chemical equation $2NH_2 + H_2 = 2NH_3$ as a relation in the algebra of ordered pairs.

4. If P and Q are points in the plane, the point R such that $Q + R = P$ is called $P - Q$. Find the coordinates of $P - Q$ in terms of the coordinates of P and Q.

Exercises

1. Complete the following computations:

   (a) $(1, 2) + (3, 7) =$ (b) $(-2, 6) - 6(2, -10) =$

   (c) $3[(1, 1) - 2(3, 0)] =$

2. Solve for the unknown quantities in the following expressions if possible. If it is not possible, explain why.

   (a) $(1, 2) + (0, y) = (1, 3)$ (b) $r(7, 3) = (14, 6)$

   (c) $a(2, -1) = (6, -\pi)$ (d) $(7, 2) + (x, y) = (3, 10)$

   (e) $2(1, b) + (b, 4) = (3, 4)$ (f) $(x, 2) + (-3)(x, y) = (-2x, 1)$

   (g) $0(3, a) = (3, a)$

3. Throughout the following, P, Q, and R are to denote ordered pairs; 0 is the pair $(0, 0)$; if $P = (x_1, y_1)$, then $-P = (-x_1, -y_1)$; and a and b are numbers. Show the following. (Part (g) has already been done as Worked Example 3.)

   (a) $P + 0 = P$ (b) $P + (-P) = 0$

   (c) $(P + Q) + R = P + (Q + R)$ (d) $P + Q = Q + P$

   (e) $a(bP) = (ab)P$ (f) $1 \cdot P = P$

   (g) $a(P + Q) = aP + aQ$ (h) $(a + b)P = aP + bP$

4. (a) Write the chemical equation $kSO_3 + lS_2 = mSO_2$ as an equation in ordered pairs (see Solved Exercise 3).

   (b) Write the equation in part (a) as a pair of simultaneous equations in k, l, and m.

   (c) Solve the equation in part (b) with positive integer values of k, l, and m (see Solved Exercise 1).

---

*Solutions appear in Appendix A.

THE VECTOR REPRESENTATION

The algebraic operations on ordered pairs may be visualized easily if we represent the pairs by directed line segments rather than by points.

Definition The directed line segment $\overrightarrow{PQ}$ is said to *represent* the ordered pair (x, y) if $Q - P$ has coordinates (x, y).

If $P = (x_1, y_1)$ and $Q = (x_2, y_2)$, then $Q - P = (x_2 - x_1, y_2 - y_1)$; that is, $x = x_2 - x_1$ and $y = y_2 - y_1$. (See Worked Example 1(c) and Solved Exercise 4 above.) The relationship between P, Q and x, y is shown in Fig. 14-1-1.

Every directed line segment represents a uniquely determined ordered pair. A given ordered pair has many representations, though, as the following proposition shows.

Proposition 1

1. *The directed line segments $\overrightarrow{PQ}$ and $\overrightarrow{P'Q'}$ represent the same ordered pair if and only if they are equal in length, are parallel, and point in the same direction—that is, if and only if $PQQ'P'$ is a parallelogram.*

2. *Given any point P and any ordered pair (x, y), there is a uniquely determined point Q such that $\overrightarrow{PQ}$ represents (x, y).*

(optional)

Fig. 14-1-2 The quadrilateral $PQQ'P'$ is a parallelogram if and only if PRQ and $P'R'Q'$ are congruent.

Proof For part 1, let $P = (x_1, y_1)$, $Q = (x_2, y_2)$, $P' = (x_1', y_1')$, $Q' = (x_2', y_2')$. Constructing the points $R = (x_2, y_1)$ and $R' = (x_2', y_1')$, we obtain right triangles PRQ and $P'R'Q'$ as in Fig. 14-1-2. It is evident from the figure (and can be proved by elementary geometry) that the quadrilateral $PQQ'P'$ is a parallelogram if and only if the right triangles PRQ and $P'R'Q'$ are congruent—that is, if and only if length PR = length $P'R'$ and length QR = length $Q'R'$. But these lengths are just the x and y components* of $Q - P$ and $Q' - P'$, so $PQQ'P'$ is a parallelogram if and only if $Q - P = Q' - P'$—that is, if and only if $\overrightarrow{PQ}$ and $\overrightarrow{P'Q'}$ represent the same ordered pair.

For part 2, given P and (x, y), the point Q for which $\overrightarrow{PQ}$ represents (x, y) must satisfy $Q - P = (x, y)$—that is, $Q = P + (x, y)$.

The representation of ordered pairs by directed line segments leads to a simple description of addition.

Proposition 2 *If the ordered pair (x_1, y_1) is represented by $\overrightarrow{PQ}$ and the ordered pair (x_2, y_2) is represented by $\overrightarrow{QR}$, then the ordered pair $(x_1, y_1) + (x_2, y_2)$ is represented by $\overrightarrow{PR}$.*

(optional)

Proof We have $(x_1, y_1) = Q - P$ and $(x_2, y_2) = R - Q$. Adding these equations gives $(x_1, y_1) + (x_2, y_2) = Q - P + R - Q = R - P$, so $\overrightarrow{PR}$ represents $(x_1, y_1) + (x_2, y_2)$.

---

*The lengths are actually the absolute values of the components, so that this argument must be modified if the components of $Q - P$ and $Q' - P'$ are not positive.

In other words, if representatives of (x_1, y_1) and (x_2, y_2) are chosen such that the "head" of the first equals the "tail" of the second, then $(x_1, y_1) + (x_2, y_2)$ is represented by the directed line segment pointing from the tail of the first to the head of the second (see Fig. 14-1-3).

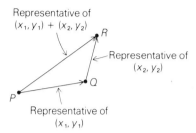

Representative of $(x_1, y_1) + (x_2, y_2)$

Representative of (x_2, y_2)

Representative of (x_1, y_1)

Fig. 14-1-3 If the ordered pairs (x_1, y_1) and (x_2, y_2) are represented by two sides of a triangle, then their sum is represented by the third side.

Worked Example 4 If $\overrightarrow{PQ}$ represents the ordered pair (x, y), describe geometrically the points R_1 and R_2 such that:

(a) $\overrightarrow{PR}_1$ represents $2(x, y)$.   (b) $\overrightarrow{PR}_2$ represents $-(x, y)$.

Solution (a) Since $2(x, y) = (x, y) + (x, y)$, we may put another "copy" of $\overrightarrow{PQ}$ with its tail at Q. The tip R_1 of this segment lies on the line through P and Q, twice as far from P as Q is. (See Fig. 14-1-4.)

(b) Since $(x, y) + [-(x, y)] = (0, 0)$, and the ordered pair $(0, 0)$ is represented by $\overrightarrow{PP}$, one representative of $-(x, y)$ must be $\overrightarrow{QP}$. To get a representative $\overrightarrow{PR}_2$ whose tail is at P, we must choose R_2 along the line through P and Q such that P is halfway between Q and R_2. (See Fig. 14-1-4.)

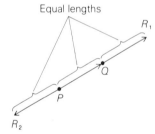

Equal lengths

R_1

Q

P

R_2

Fig. 14-1-4 If $\overrightarrow{PQ}$ represents (x, y), then $\overrightarrow{PR}_1$ represents $2(x, y)$ and $\overrightarrow{PR}_2$ represents $-(x, y)$.

The general rule for scalar multiplication, following the pattern of Worked Example 4, is given by the following proposition, whose proof is left to the reader (see Problem 3).

Proposition 3 *If the ordered pair (x, y) is represented by $\overrightarrow{PQ}$, then the ordered pair $r(x, y)$ is represented by $\overrightarrow{PR}$, where*

1. *R lies on the line through P and Q.*

2. *The distance from P to R is $|r|$ times the distance from P to Q.*

3. *R is on the same side of P as Q if $r > 0$, and P is between R and Q if $r < 0$.*

(See Fig. 14-1-5.)

R

Q

P

Fig. 14-1-5 If (x, y) represents $\overrightarrow{PQ}$, then $\overrightarrow{PR}$ represents $r(x, y)$ if $\overrightarrow{PR}$ is r times as long as $\overrightarrow{PQ}$. (In this figure, r is between 2 and 3.)

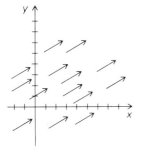

(a) The vector corresponding to (2, 1)

(b) The vector corresponding to (0, 2)

Fig. 14-1-6 Pictures of two vectors.

We wish to think of a vector as a directed line segment whose length and direction are specified, but whose location is left undetermined. The precise definition of a vector goes as follows.

Definition A *vector* in the plane is a *collection of directed line segments* consisting of all the directed line segments which represent a given ordered pair of real numbers. We denote vectors by boldface letters (**u**, **v**, **w**, and so on).

How can we draw a picture of a vector? To comply in full with the definition, we must draw *all* the directed line segments which represent a given pair. Since, by Proposition 1(2), there is one representation $\overrightarrow{PQ}$ for each choice of P, we would fill the plane with black lines and the resulting picture would be useless. We must be content, therefore, with drawing *several* of the directed line segments representing a given ordered pair, as in Fig. 14-1-6.

In practice, it becomes rather tedious to draw so many directed line segments, so we draw just one segment—with the understanding that the segment can be translated to any other location in the plane to suit our needs. Thus we usually have pictures like Fig. 14-1-7.

Note that there is a one-to-one correspondence between ordered pairs and vectors. If a vector **v** consists of all the directed line segments representing the ordered pair (x, y), we will sometimes say that one of these line segments is a representative of the vector **v**. In fact, we will often go even further and identify the vector **v** with the ordered pair (x, y). Thus we may speak of "the vector $(2, 3)$." (See Fig. 14-1-8.)

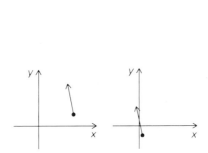

Fig. 14-1-7 Two pictures of the same vector.

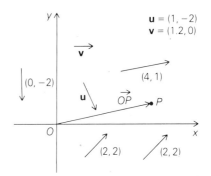

Fig. 14-1-8 Labeling vectors by names or by ordered pairs. We identify the point P and the vector $\overrightarrow{OP}$.

If $\overrightarrow{PQ}$ is a directed line segment, we will refer to the corresponding vector which it represents as "the vector from P to Q," ((the vector joining P to Q," or even "the vector $\overrightarrow{PQ}$." Since ordered pairs are also identified with points in the plane, one may carry all these identifications to their logical conclusion and identify the point P whose coordinates are (x, y) with the vector (x, y). In other words: We identify the point P with the vector from the origin to P.

We define the algebraic operations on vectors in terms of the operations on ordered pairs.

Definition

1. If $\mathbf{u}_1$ and $\mathbf{u}_2$ are the vectors corresponding to (x_1, y_1) and (x_2, y_2), then $\mathbf{u}_1 + \mathbf{u}_2$ is defined as the vector corresponding to $(x_1, y_1) + (x_2, y_2) = (x_1 + x_2, y_1 + y_2)$.

2. If $\mathbf{u}$ is the vector corresponding to (x, y) and r is a real number, then $r\mathbf{u}$ is defined as the vector corresponding to $r(x, y) = (rx, ry)$.

Worked Example 5   In Fig. 14-1-9, which vector is (a) $\mathbf{u} + \mathbf{v}$? (b) $3\mathbf{u}$? (c) $-\mathbf{v}$?

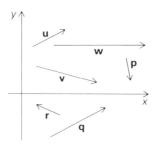

Fig. 14-1-9   Find $\mathbf{u} + \mathbf{v}$, $3\mathbf{u}$, and $-\mathbf{v}$.

Solution   (a) To construct $\mathbf{u} + \mathbf{v}$, according to Proposition 2 we must represent $\mathbf{u}$ and $\mathbf{v}$ by directed line segments so that the head of the first coincides with the tail of the second. (See Fig. 14-1-10.) Now we fill in the third side of the triangle to obtain $\mathbf{u} + \mathbf{v}$ (see Fig. 14-1-11). Comparing Fig. 14-1-11 with Fig. 14-1-9, we find that $\mathbf{u} + \mathbf{v} = \mathbf{w}$.

(b) $3\mathbf{u} = \mathbf{q}$.

(c) $-\mathbf{v} = (-1)\mathbf{v} = \mathbf{r}$ (see Fig. 14-1-12).

Fig. 14-1-10   To find $\mathbf{u} + \mathbf{v}$, place the head of one vector at the tail of the other.

Fig. 14-1-11   Completion of the triangle gives $\mathbf{u} + \mathbf{v}$.

Fig. 14-1-12   To find $3\mathbf{u}$, draw a vector in the same direction as $\mathbf{u}$, three times as long; $-\mathbf{v}$ is a vector having the same length as $\mathbf{v}$, pointing in the opposite direction.

Worked Example 6   Draw a vector diagram to illustrate the identity $\mathbf{u} + \mathbf{v} = \mathbf{v} + \mathbf{u}$.

Solution   See Fig. 14-1-13. This figure illustrates another geometric interpretation of vector addition. To add $\mathbf{u}$ and $\mathbf{v}$, we choose representatives $\overrightarrow{PQ}$ and $\overrightarrow{PR}$ having their tails at the same point P. If we complete the figure to a parallelogram $PQSR$, then the diagonal $\overrightarrow{PS}$ represents $\mathbf{u} + \mathbf{v}$. For this reason, physical quantities which combine by vector addition are sometimes said to "obey the parallelogram law."

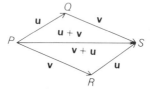

Fig. 14-1-13   Illustrating the identity $\mathbf{u} + \mathbf{v} = \mathbf{v} + \mathbf{u}$ and the parallelogram law of addition.

Definition   If **u** is the vector corresponding to the ordered pair (x, y), then x and y are called the *components* of **u**.

Worked Example 7   Find the components of the vector from $(3, 5)$ to $(4, 7)$.

Solution   Let $P = (3, 5)$ and $Q = (4, 7)$. Then the directed line segment $\overrightarrow{PQ}$ represents the ordered pair $Q - P = (1, 2)$ (see p. 593), so the vector **u** joining P to Q corresponds to the ordered pair $(1, 2)$; its components are 1 and 2.

Solved Exercises   5. Add the vector **v** from $(-1, 0)$ to $(2, -3)$ and the vector **w** from $(2, 0)$ to $(1, 1)$.

6. Multiply the vector **v** in Solved Exercise 5 by 8. If this vector is represented by the directed line segment from $(5, 6)$ to Q, what is Q?

7. If **v** and **w** are vectors, give a geometric description of $\mathbf{v} - \mathbf{w}$.

Exercises   5. Let $\mathbf{u} = (1, 2)$ and $\mathbf{v} = (2, 1)$. Draw each of the following vectors:

   (a) $\mathbf{u} + \mathbf{v}$ (b) $\mathbf{u} - \mathbf{v}$ (c) $2\mathbf{u}$ (d) $-4\mathbf{v}$ (e) $2\mathbf{u} - 4\mathbf{v}$

6. Let $P = (1, 2)$; $Q = (3, 3)$; $R = (4, 1)$.

   (a) Draw (on the same diagram) these vectors: **v** joining P to Q; **w** joining Q to R; **u** joining R to P.

   (b) What are the components of **v**, **w**, and **u**?

   (c) What is $\mathbf{v} + \mathbf{w} + \mathbf{u}$?

7. (a) Draw the vector $\mathbf{v}_1$ joining $(1, 0)$ to $(1, 1)$.

   (b) What are the components of $\mathbf{v}_1$?

   (c) Draw $\mathbf{v}_2$ joining $(1, 0)$ to $(1, \frac{5}{2})$ and write the components of $\mathbf{v}_2$.

   (d) Do the same for $\mathbf{v}_3$ joining $(1, 0)$ to $(1, -2)$.

   (e) What are the components of an arbitrary point on the vertical (that is, parallel to the y axis) line through $(1, 0)$?

   (f) What are the components of the vector **v** joining $(1, 0)$ to such a point?

8. (a) Draw the vector **v** joining $(-1, 1)$ to $(1, 1)$.

   (b) What are the components of **v**?

   (c) Sketch the vectors $\mathbf{v}_t = (-1, 1) + t\mathbf{v}$ when $t = 0, \frac{1}{4}, \frac{1}{2}, \frac{3}{4}$, and 1.

   (d) Describe, geometrically, the set of vectors $\mathbf{v}_t = (-1, 1) + t\mathbf{v}$, where t takes on all values between 0 and 1. (Assume that all the vectors have their tails at the origin.)

POINTS AND VECTORS IN SPACE

The plane is two-dimensional, but space is three-dimensional—that is, it requires *three* numbers to specify the position of a point in space. For instance, the location of a bird is specified not only by the two coordinates of the point on the ground directly below it, but also by its height. Accordingly, our algebraic model for space will be the set of *triples* (x, y, z) rather than pairs of real numbers.

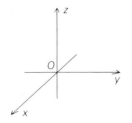

Fig. 14-1-14   Coordinate axes in space.

If one starts with abstract "space" as studied in elementary solid geometry, the first step in the introduction of coordinates is the choice of an origin O and three directed lines, each perpendicular to the other two, called the x, y, and z axes. We will usually draw figures in space with the axes oriented as in Fig. 14-1-14.

Think of the x axis as pointing toward you, out of the paper. Notice that if you wrap the fingers of your right hand around the z axis, with your fingers curling in the usual (counterclockwise) direction of rotation in the xy plane, then your thumb points toward the positive z axis. For this reason, we say that the choice of axes obeys the *right-hand rule*.

Worked Example 8   Which sets of axes in Fig. 14-1-15 obey the right-hand rule? Think of all horizontal and vertical arrows as being in the plane of the paper, while slanted arrows point out toward you.

Fig. 14-1-15   Which axes obey the right-hand rule?

Solution   Only (a) and (d) obey the right-hand rule.

Given a point P in space, we may now drop a perpendicular from P to each of the axes. By measuring the (directed) distance from the origin to the foot of each of these perpendiculars, we obtain numbers (x, y, z) which we call the *coordinates* of the point (see Fig. 14-1-16).

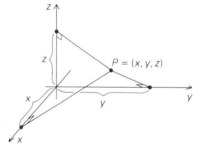

Fig. 14-1-16   We obtain the coordinates of the point P by dropping perpendiculars to the x, y, and z axes.

You may have some difficulty in visualizing the lines through P in Fig. 14-1-16 as being perpendicular to the axes. It may help to draw in some additional lines which are *parallel* to the axes. In drawing them, we will use the following useful drawing convention: Lines which are parallel to one another in space are drawn parallel. This simple convention, sometimes called the rule of "parallel projection," does not conform to ordinary rules of perspective (think of railroad tracks "converging at infinity"), but it is reasonably accurate if your distance from an object is great compared to the size of the object.

Now look at Fig. 14-1-17. Observe that the point Q, obtained by dropping a perpendicular from P to the xy plane, has coordinates $(x, y, 0)$. Similarly, the points R and S, obtained by dropping perpendiculars from P to the yz and xz planes, have coordinates $(0, y, z)$ and $(x, 0, z)$, respectively.

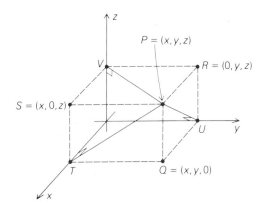

Fig. 14-1-17 Lines which are parallel in space are drawn parallel.

Worked Example 9   What are the coordinates of the points T, U, and V in Fig. 14-1-17?

Solution   The coordinates of T, U, and V are $(x, 0, 0)$, $(0, y, 0)$, and $(0, 0, z)$.

As in the plane, we can find the unique point which has a given ordered triple (x, y, z) as its coordinates. To do so, we begin by finding the point $Q = (x, y, 0)$ in the xy plane by the method of Section R-2; drawing a line through Q and parallel to the z axis, we can locate the point P at a (directed) distance of z units from Q along this line. The process just described can be carried out graphically, as in the following example.

Worked Example 10   Plot the point $(1, 2, -3)$.

Solution   We begin by plotting $(1, 2, 0)$ in the xy plane (Fig. 14-1-18a). Then we draw the line through this point parallel to the z axis and measure 3 units downward (Fig. 14-1-18b).

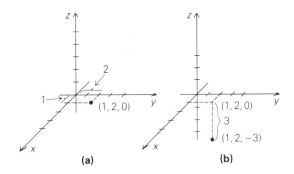

Fig. 14-1-18 Plotting the point $(1, 2, -3)$.

Warning If you are given a picture consisting simply of three axes (with units of measure) and a point, it is not possible to determine the coordinates of the point from these data alone, since some information must be lost in making a two-dimensional picture of three-dimensional space. (See Solved Exercise 8 below.)

Addition and scalar multiplication are defined coordinatewise for ordered triples just as for pairs.

Definition

1. If (x_1, y_1, z_1) and (x_2, y_2, z_2) are ordered triples of real numbers, the ordered triple $(x_1 + x_2, y_1 + y_2, z_1 + z_2)$ is called their *sum* and is denoted by $(x_1, y_1, z_1) + (x_2, y_2, z_2)$.

2. If (x, y, z) is an ordered triple and r is a real number, the triple (rx, ry, rz) is called the *product* of r and (x, y, z) and is denoted by $r(x, y, z)$.

Worked Example 11   Find $(3, 2, -2) + (-1, -2, -1)$ and $(-6)(2, -1, 1)$.

Solution   We have $(3, 2, -2) + (-1, -2, -1) = (2, 0, -3)$ and $(-6)(2, -1, 1) = (-12, 6, -6)$.

The representation of ordered triples by vectors in space proceeds just as for ordered pairs. You may (in fact, you should) reread pages 590 through 597, substituting everywhere "ordered triple" for "ordered pair" and "space" for "the plane." We will rewrite only the definition.

Definition   A *vector* in space is a *collection of directed line segments* consisting of all the directed line segments $\overrightarrow{PQ}$ which represent a given ordered triple (x, y, z) in the sense that $Q - P = (x, y, z)$. We denote vectors by boldface letters (**u**, **v**, **w**, and so on).

Worked Example 12   Let **v** be the vector $(3, 2, -2)$ and let **w** be the vector from $(2, 1, 3)$ to $(-1, 0, -1)$. Find **v** + **w**. Illustrate with a sketch.

Solution   Since **w** is the vector $(-1, 0, -1) - (2, 1, 3) = (-3, -1, -4)$, we find that **v** + **w** is the vector $(3, 2, -2) + (-3, -1, -4) = (0, 1, -6)$, as illustrated in Fig. 14-1-19.

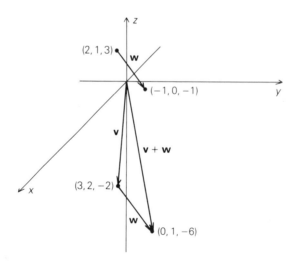

Fig. 14-1-19 Adding **v** = $(3, 2, -2)$ to **w**, the vector from $(2, 1, 3)$ to $(-1, 0, -1)$.

To describe vectors in space, it is convenient to introduce three special vectors along the x, y, and z axes.

$$\mathbf{i} = (1, 0, 0), \quad \mathbf{j} = (0, 1, 0), \quad \mathbf{k} = (0, 0, 1)$$

These *standard basis vectors* are illustrated in Fig. 14-1-20. In the plane one has, analogously,

$$\mathbf{i} = (1, 0) \quad \text{and} \quad \mathbf{j} = (0, 1)$$

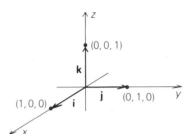

Fig. 14-1-20 The standard basis vectors.

Now, let $\mathbf{v}$ be any vector, and let (a, b, c) be the corresponding ordered triple. Then

$$\mathbf{v} = a\mathbf{i} + b\mathbf{j} + c\mathbf{k}$$

since the right-hand side is

$$a(1, 0, 0) + b(0, 1, 0) + c(0, 0, 1) = (a, 0, 0) + (0, b, 0) + (0, 0, c) = (a, b, c)$$

Thus we can express every vector as a sum of scalar multiples of $\mathbf{i}$, $\mathbf{j}$, and $\mathbf{k}$.

Worked Example 13 Express the vector $\mathbf{v}$ joining $(2, 0, 1)$ to $(\frac{3}{2}, \pi, -1)$ by using the standard basis.

Solution The vector $\mathbf{v}$ corresponds to the ordered pair $(\frac{3}{2}, \pi, -1) - (2, 0, 1) = (-\frac{1}{2}, \pi, -2)$, so $\mathbf{v} = -\frac{1}{2}\mathbf{i} + \pi\mathbf{j} - 2\mathbf{k}$.

Solved Exercises

8. Find at least *two* possible sets of coordinates for the point P in Fig. 14-1-21.

Fig. 14-1-21 Find at least two possible sets of coordinates for P.

9. Sketch $-2\mathbf{v}$, where $\mathbf{v}$ is $(-1, 1, 2)$.

10. Show that $\mathbf{v} - \frac{1}{3}\mathbf{w}$ and $3\mathbf{v} - \mathbf{w}$ are parallel.

11. Express the vector $(e, \pi, -\sqrt{3})$ by using the standard basis.

Exercises
9. Plot the following points:

(a) $(1, 0, 0)$ (b) $(0, 2, 4)$ (c) $(3, -1, 5)$ (d) $(2, -1, \frac{1}{2})$

10. What restrictions must be made on x, y, z so that the triple (x, y, z) will represent a point on the y axis? On the z axis? In the xy plane? In the xz plane?

11. Complete the following calculations and illustrate graphically:

(a) $(2, 1, 1) + (3, 0, -1) =$ (b) $(1, 0, 2) + (-2)(\frac{1}{2}, 0, 1) =$

12. Sketch $\mathbf{v}$, $2\mathbf{v}$, and $-\mathbf{v}$, where $\mathbf{v}$ is $(1, -1, -1)$.

13. What are the components of the vector from $(1, 0, 0)$ to (a) $(2, -1, 1)$? (b) $(3, -2, 2)$?

14. Express each of the following vectors in terms of the standard basis:

(a) $(-1, 2, 3)$ (b) $(0, 2, 2)$ (c) $(1, 1, 1) - (0, 1, 2)$

PHYSICAL APPLICATIONS OF VECTORS*

A simple example of a physical quantity represented by a vector is that of a displacement. Suppose that, on a part of the earth's surface small enough to be considered flat, we introduce coordinates so that the x axis points east, the y axis points north, and the unit of length is the kilometer. If we are at a point P and wish to get to a point Q, the *displacement vector* $\mathbf{u}$ joining P to Q tells us the direction and distance we have to travel. If x and y are the components of this vector, the displacement of Q from P is "x kilometers east, y kilometers north."

Worked Example 14   Suppose that two navigators, who cannot see one another but can communicate by radio, wish to determine the relative position of their ships. Explain how they can do this if they can each determine their displacement vector to the same lighthouse.

Solution   Let P_1 and P_2 be the positions of the ships and Q be the position of the lighthouse. The displacement of the lighthouse from the ith ship is the vector $\mathbf{u}_i$ joining P_i to Q. The displacement of the second ship from the first is the vector $\mathbf{v}$ joining P_1 to P_2. By Proposition 2, we have $\mathbf{v} + \mathbf{u}_2 = \mathbf{u}_1$ (Fig. 14-1-22), so $\mathbf{v} = \mathbf{u}_1 - \mathbf{u}_2$. That is, the displacement from one ship to the other is the difference of the displacements from the ships to the lighthouse.

Fig. 14-1-22 Vector methods can be used to locate objects.

Historical note. Many scientists resisted the use of vectors in favor of the more complicated theory of quaternions until around 1900. The book which popularized vector methods was Vector Analysis, by E. B. Wilson (reprinted by Dover in 1960), based on lectures of J. W. Gibbs at Yale in 1899–1900. Wilson was reluctant to take Gibbs' course since he had just completed a full-year course in quaternions at Harvard under J. M. Pierce, a champion of quaternionic methods, but was forced by a dean to add the course to his program. (For more details, see M. J. Crowe, A History of Vector Analysis, University of Notre Dame Press, 1967.)

Fig. 14-1-23 If an object moves northeast at 10 kilometers per hour, its velocity vector has components $(10/\sqrt{2}, 10/\sqrt{2})$.

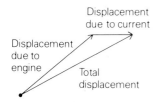

Fig. 14-1-24 The total displacement is the sum of the displacements due to the engine and the current.

We can also represent the velocity of a moving object as a vector. For the moment, we will consider only objects moving at uniform speed along straight lines—the general case is discussed in Section 15-3. Suppose, for example, that a boat is steaming across a lake at 10 kilometers per hour in the northeast direction. After 1 hour of travel, the displacement from the original position will be a vector whose components are $(10/\sqrt{2}, 10/\sqrt{2}) \approx (7.07, 7.07)$ (see Fig. 14-1-23).

We say that the vector whose components are $(10/\sqrt{2}, 10/\sqrt{2})$ is the *velocity vector* of the boat. In general, if an object is moving uniformly along a straight line, *its velocity vector is the displacement vector from the position at any moment to the position 1 unit of time later.*

If a current appears on the lake, moving due eastward at 2 kilometers per hour, and the boat continues to point in the same direction with its engine running at the same rate, its displacement after 1 hour will have components $(10/\sqrt{2} + 2, 10/\sqrt{2})$. (See Fig. 14-1-24.) The new velocity vector, therefore, has components $(10/\sqrt{2} + 2, 10/\sqrt{2})$. We note that this is the sum of the original velocity vector $(10/\sqrt{2}, 10/\sqrt{2})$ of the boat and the velocity vector $(2, 0)$ of the current.

Similarly, consider a seagull which flies in calm air with velocity vector **v**. If a wind comes up with velocity **w** and the seagull continues flying "the same way," its actual velocity will be **v** + **w**. One can "see" the direction of the vector **v** because it points along the "axis" of the seagull; by comparing the direction of actual motion with the direction of **v**, you can get an idea of the wind direction (see Fig. 14-1-25).

Fig. 14-1-25 The velocity **w** of the wind can be estimated by comparing the "wingflap" velocity **v** with the actual velocity **v** + **w**.

Another example comes from medicine. An electrocardiograph detects the flow of electricity in the heart. Both the magnitude and the direction of the net flow are of importance. This information can be summarized at every instant by means of a vector called the *cardiac vector*. The motion of this vector (see Fig. 14-1-26) gives physicians useful information about the heart's function.*

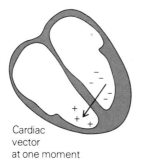

Cardiac vector at one moment

Tip of cardiac vector moving in space

Fig. 14-1-26 The magnitude and direction of electrical flow in the heart are indicated by the cardiac vector.

---

*See M. J. Goldman, Principles of Clinical Electrocardiography, *8th edition, Lange, 1973, Chapters 14 and 19.*

12. A bird is flying in a straight line with velocity vector $10\mathbf{i} + 6\mathbf{j} + \mathbf{k}$ (in kilometers per hour). Suppose that (x, y) are coordinates on the ground and z is the height above the ground.

    (a) If the bird is at position $(1, 2, 3)$ at a certain moment, where is it 1 minute later?

    (b) How many seconds does it take the bird to climb 10 meters?

13. Physical forces have magnitude and direction and may thus be represented by vectors. If several forces act at once on an object, the resultant force is represented by the sum of the individual force vectors. Suppose that forces $\mathbf{i} + \mathbf{k}$ and $\mathbf{j} + \mathbf{k}$ are acting on a body. What third force must we impose to counteract these two—that is, to make the total force equal to zero?

15. A ship at position $(1, 0)$ on a nautical chart (with north in the positive y direction) sights a rock at position $(2, 4)$. What is the vector joining the ship to the rock? What angle does this vector make with due north? This is called the *bearing* of the rock from the ship.

16. Suppose that the ship in Exercise 15 is pointing due north and traveling at a speed of 4 knots relative to the water. There is a current flowing due east at 1 knot. (The units on the chart are nautical miles; 1 knot = 1 nautical mile per hour.)

    (a) If there were no current, what vector $\mathbf{u}$ would represent the velocity of the ship relative to the sea bottom?

    (b) If the ship were just drifting with the current, what vector $\mathbf{v}$ would represent its velocity relative to the sea bottom?

    (c) What vector $\mathbf{w}$ represents the total velocity of the ship?

    (d) Where would the ship be after 1 hour?

    (e) Should the captain change course?

    (f) What if the rock were an iceberg?

17. Suppose that a cardiac vector is $\cos t\,\mathbf{i} + \sin t\,\mathbf{j} + \mathbf{k}$ at time t.

    (a) Draw the cardiac vector for $t = 0, \pi/4, \pi/2, 3\pi/4, \pi, 5\pi/4, 3\pi/2, 7\pi/4, 2\pi$.

    (b) Describe the motion of the tip of the cardiac vector in space if the tail is fixed at the origin.

Problems for Section 14-1

1. Complete the following computations:
    (a) $(1, 3, 5) + 4(-1, -3, -5) =$
    (b) $(6, 0, 5) + (5, 0, 6) =$
    (c) $(0, 0, 0) + (0, 0, 0) =$

2. If possible, find numbers a and b such that:
    (a) $a(1, 0) + b(0, 1) = (3, 7)$
    (b) $6(1, 0) + b(0, 1) = (6, 2)$
    (c) $3(1, 0) + b(0, 1) = (4, 1)$
    (d) $a(1, 1) + b(1, -1) = (3, 5)$
    (e) $(a, 1) - (2, b) = (0, 0)$
    (f) $(3a, b) + (b, a) = (1, 1)$
    (g) $(3a, a) + (3b, b) = (2, 3)$
    (h) $a(3a, 1) + a(1, -1) = (1, 0)$

★3. Prove Proposition 3.

4. Let $\mathbf{u} = 2\mathbf{i} + 3\mathbf{j}$. Sketch the vectors $\mathbf{u}$, $2\mathbf{u}$, and $-3\mathbf{u}$ on the same set of axes.

5. Let $\mathbf{v} = 3\mathbf{i} + 4\mathbf{j} + 5\mathbf{k}$ and $\mathbf{w} = \mathbf{i} - \mathbf{j} + \mathbf{k}$. Compute:
    (a) $\mathbf{v} + \mathbf{w}$ (b) $3\mathbf{v}$ (c) $-2\mathbf{w}$ (d) $6\mathbf{v} + 8\mathbf{w}$
    (e) The vector $\mathbf{u}$ from the tip of $\mathbf{w}$ to the tip of $\mathbf{v}$. (Assume that the tails of $\mathbf{w}$ and $\mathbf{v}$ are at the same point.)

6. Express each of the following vectors in terms of the standard basis vectors $\mathbf{i}$, $\mathbf{j}$, and $\mathbf{k}$:
    (a) $(7, 2, 3)$ (b) $3(1, 1, 2) - 6(3, 1, 7)$
    (c) The vector from $(3, 0, 5)$ to $(2, 7, 6)$

7. Describe geometrically the set of all ordered pairs $m(0, 1) + n(1, 1)$, where m and n are integers. (A sketch will do.)

8. Write the chemical equation $CO + H_2O = H_2 + CO_2$ (see Solved Exercise 3) as an equation in ordered triples, and illustrate it by a vector diagram in space.

9. Let $\mathbf{v} = \mathbf{i} + \mathbf{j}$ and $\mathbf{w} = -\mathbf{i} + \mathbf{j}$. Find numbers a and b such that $a\mathbf{v} + b\mathbf{w} =$
    (a) $\mathbf{i}$ (b) $\mathbf{j}$ (c) $3\mathbf{i} + 7\mathbf{j}$

10. Plot on one set of axes the eight points of the form (a, b, c), where a, b, and c are each equal to 1 or -1. Of what geometric figure are these the vertices?

11. Let $P_t = (1, 0, 0) + t(2, 1, 1)$, where t is a real number.
    (a) Compute the coordinates of P_t for $t = -1$, 0, 1, and 2.
    (b) Sketch these four points on the same set of axes.
    (c) Try to describe geometrically the set of all the P_t.

12. The z coordinate of the point P in Fig. 14-1-27 is 3. What are the x and y coordinates?

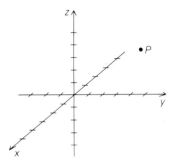

Fig. 14-1-27 Let $P = (x, y, 3)$. What are x and y?

★13. We would like to say that two vectors are parallel if they are scalar multiples of one another. That is, $\mathbf{v}$ is parallel to $\mathbf{w}$ if there is a number r with $\mathbf{v} = r\mathbf{w}$.
    (a) Explain why this definition is not precisely what we want. Is $(0, 0)$ parallel to $(1, 1)$? Is $(1, 1)$ parallel to $(0, 0)$? A better definition might be: $\mathbf{v}$ and $\mathbf{w}$ are parallel if there are numbers r and s, not both zero, such that $r\mathbf{v} + s\mathbf{w} = 0$.
    (b) Let $\mathbf{v}$ and $\mathbf{w}$ be vectors in the plane given by $\mathbf{v} = (a, b)$ and $\mathbf{w} = (c, d)$. Show that $\mathbf{v}$ is parallel to $\mathbf{w}$ if and only if $ad = bc$. [*Hint:* For one implication you might use three cases: $b \neq 0$, $d \neq 0$, and $b = d = 0$.]
    (c) Suppose that $\mathbf{v}$ and $\mathbf{w}$ are vectors in the plane which are not parallel. Show that for any vector $\mathbf{u}$ in the plane there are numbers x and y such that $x\mathbf{v} + y\mathbf{w} = \mathbf{u}$.

★14. Let $P = (a, b)$ and $Q = (c, d)$ be points in the plane. (You may assume that $0 < a < c$ and $b > d > 0$ to make the picture unambiguous.) Compute the area of the parallelogram with vertices at O, P, Q, and $P + Q$. Comment on the relationship between this and Problem 13(c).

15. A 1-kilogram mass located at the origin is suspended by ropes attached to the points $(1, 1, 1)$ and $(-1, -1, 1)$. If the force of gravity is pointing in the direction of the vector $-\mathbf{k}$, what is the vector describing the force along each rope? [*Hint:* Use the symmetry of the problem. A 1-kilogram mass weighs 9.8 newtons.]

16. An airplane is located at position $(3, 4, 5)$ at noon and traveling with velocity $400\mathbf{i} + 500\mathbf{j} - \mathbf{k}$ kilometers per hour. The pilot spots an airport at position $(23, 29, 0)$.

    (a) At what time will the plane pass directly over the airport? (Assume that the earth is flat and that the vector $\mathbf{k}$ points straight up.)

    (b) How high above the airport will the plane be when it passes?

17. The wind velocity $\mathbf{v}_1$ is 40 mi/hr from east to west while an airplane travels with air speed $\mathbf{v}_2$ of 100 mi/hr due north. The speed of the airplane relative to the earth is the vector sum $\mathbf{v}_1 + \mathbf{v}_2$.

    (a) Find $\mathbf{v}_1 + \mathbf{v}_2$.

    (b) Draw a figure to scale.

18. Classify the following as *scalar quantities or vector quantities*. Explain the classification by citing examples.

    (a) mass   (b) displacement
    (c) electric field   (d) length
    (e) velocity   (f) time
    (g) temperature   (h) acceleration
    (i) power   (j) electric charge
    (k) force   (l) position

19. A force of 50 lb is 50° above horizontal, pointing to the right. Determine its horizontal and vertical components. Display all results in a figure.

20. Two persons pull horizontally on ropes attached to a post, the angle between the ropes being 60°. A pulls with a force of 150 lb, while B pulls with a force of 110 lb.

    (a) The resultant force is the vector sum of the two forces. Draw a figure to scale which graphically represents the three forces.

    (b) Using trigonometry, determine formulas for the vector components of the two forces. Perform the algebraic addition, and find the angle the resultant force makes with A.

14-2 The Geometry of Vectors and the Dot Product

Algebraic operations on vectors can be used to solve geometric problems.

In this section we apply the algebra of vectors to the description of lines in planes in space and to the solution of other geometric problems. The operation of *dot product,* which incorporates the notions of both length and angle, is indispensible in the applications of vectors. Another useful product, the *cross-product,* is introduced in the next section.

Goals

After studying this section, you should be able to:

Calculate lengths, angles, and dot products of vectors.

Use vector methods to prove geometric theorems.

Find equations for lines and planes.

VECTOR GEOMETRY AS AN EXTENSION OF ANALYTIC GEOMETRY

The invention of analytic geometry made it possible to solve geometric problems in the plane or space by reducing them to algebraic problems involving number pairs or triples. Vector methods also convert geometric problems to algebraic ones; moreover, the vector calculations are often simpler than those from analytic geometry, since we do not need to write down all the components.

Worked Example 1   Use vector methods to prove that the diagonals of a parallelogram bisect each other.

Solution   Let *PQRS* be the parallelogram, **w** the vector joining *P* to *Q*, and **v** the vector joining *P* to *S* (see Fig. 14-2-1). Since *PQRS* is a parallelogram, **w** is also the vector joining *S* to *R*, so **v** + **w** is the vector joining *P* to *R*. The vector joining *P* to the midpoint M_1 of the diagonal *PR* is thus $\frac{1}{2}$(**v** + **w**). On the other hand, the vector joining *Q* to *S* is **v** − **w**, so the vector joining *Q* to the midpoint M_2 of the diagonal *QS* is $\frac{1}{2}$(**v** − **w**).

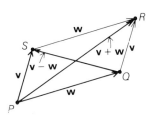

To show that the diagonals bisect each other, it suffices to show that the midpoints M_1 and M_2 are the same. The vector joining *P* to M_2 is the sum

Fig. 14-2-1 The diagonals of a parallelogram bisect each other.

$$\mathbf{w} + \tfrac{1}{2}(\mathbf{v} - \mathbf{w}) = \mathbf{w} + \tfrac{1}{2}\mathbf{v} - \tfrac{1}{2}\mathbf{w} = \mathbf{w} - \tfrac{1}{2}\mathbf{w} + \tfrac{1}{2}\mathbf{v}$$
$$= \tfrac{1}{2}\mathbf{w} + \tfrac{1}{2}\mathbf{v} = \tfrac{1}{2}(\mathbf{v} + \mathbf{w})$$

which is the same as the vector joining *P* to M_1. It follows that M_1 and M_2 are the same point.

Solved Exercises

1. Consider the cube in space with vertices at $(0, 0, 0)$, $(1, 0, 0)$, $(1, 1, 0)$, $(0, 1, 0)$, $(0, 0, 1)$, $(1, 0, 1)$, $(1, 1, 1)$, $(0, 1, 1)$. Use vector methods to locate the point one-third of the way from the origin to the middle of the face $(0, 1, 0)$, $(0, 1, 1)$, $(1, 1, 1)$, $(1, 1, 0)$.

2. Prove that the figure obtained by joining the midpoints of successive sides of any quadrilateral is a parallelogram.

Exercises

1. Find the point in the plane two-thirds of the way from the origin to the midpoint of the line segment between $(1, 1)$ and $(2, -2)$.

2. Show that the line segment joining the midpoints of two sides of a triangle is parallel to and half the length of the third side.

3. Prove that the medians of a triangle intersect in a point two-thirds of the way along any median from a vertex to the midpoint of the opposite side.

4. Prove the proposition in Worked Example 1 by using analytic geometry. Is this easier or harder than the vector method?

PARAMETRIC EQUATIONS OF LINES

In Section R-2 we discussed the equations of lines in the plane. These equations can be conveniently described in terms of vectors, and this description is equally applicable whether the line is in the plane or in space. We will now find the equation in parametric form (see Section 11-3 for a discussion of parametric curves).

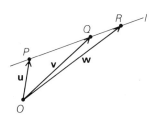

Suppose that we wish to find the equation of the line *l* passing through the two points *P* and *Q*. Let *O* be the origin and let **u** and **v** be the vectors from *O* to *P* and *O* to *Q*, as in Fig. 14-2-2. Let *R* be an arbitrary point on *l* and let **w** be the vector from *O* to *R*. Then the fact that *R* is on *l* may be expressed by saying that the vector **w** − **u** from *P* to *R* is a multiple of the vector **v** − **u** from *P* to *Q*—that is, **w** − **u** = *t*(**v** − **u**) for some number *t*. Solving for **w**, **w** = **u** + *t*(**v** − **u**) =

Fig. 14-2-2 Since *P*, *Q*, and *R* lie on a line, the vector **w** − **u** is a multiple of **v** − **u**.

$(1 - t)\mathbf{u} + t\mathbf{v}$, and the point (that is, ordered triple) R equals $(1 - t)P + tQ$. This is the parametric equation for the line l: for each real number t, we obtain the point $R = (1 - t)P + tQ$ on the line l.

PARAMETRIC EQUATION OF A LINE: POINT-POINT FORM

The equation of the line l through the points $P = (x_1, y_1, z_1)$ and $Q = (x_2, y_2, z_2)$ is

$$R = (1 - t)P + tQ$$

In coordinate form, one has the three equations

$$x = (1 - t)x_1 + tx_2$$
$$y = (1 - t)y_1 + ty_2$$
$$z = (1 - t)z_1 + tz_2$$

where $R = (x, y, z)$ is the typical point of l and the parameter t takes on all real values.

Worked Example 2 Find the equation of the line through $(2, 1, -3)$ and $(6, -1, -5)$.

Solution We have $P = (2, 1, -3)$ and $Q = (6, -1, -5)$, so

$$\begin{aligned}
(x, y, z) = R = (1 - t)P + tQ &= (1 - t)(2, 1, -3) + t(6, -1, -5) \\
&= (2 - 2t, 1 - t, -3 + 3t) + (6t, -t, -5t) \\
&= (2 + 4t, 1 - 2t, -3 - 2t)
\end{aligned}$$

or, since corresponding components of equal vectors are equal,

$$x = 2 + 4t, \qquad y = 1 - 2t, \qquad z = -3 - 2t$$

Fig. 14-2-3 If the line through P and R has the direction of the vector $\mathbf{d}$, then the vector from P to R is a multiple of $\mathbf{d}$.

We can also ask for the equation of the line which passes through a given point P in the direction of a given vector $\mathbf{d}$. By reasoning like that above (see Problem 12), one comes to the following conclusions (see Fig. 14-2-3).

PARAMETRIC EQUATION OF A LINE: POINT-DIRECTION FORM

The equation of the line through the point $P = (x_0, y_0, z_0)$ and pointing in the direction of the vector $\mathbf{d} = a\mathbf{i} + b\mathbf{j} + c\mathbf{k}$ is

$$R = P + t\mathbf{d}$$

In coordinate form, the equations are

$$x = x_0 + at$$
$$y = y_0 + bt$$
$$z = z_0 + ct$$

where $R = (x, y, z)$ is the typical point on l and the parameter t takes on all real values.

We note that the equation $R = P + t\mathbf{d}$ works in the plane as well as in space.

Worked Example 3 Find the equation of the line through the point $(1, -6)$ in the direction of $5\mathbf{i} - \pi\mathbf{j}$.

Solution   Here $P = (1, -6)$ and $\mathbf{d} = 5\mathbf{i} - \pi\mathbf{j}$, so the line is

$$R = (1, -6) + (5t, -\pi t) = (5t + 1, -6 - \pi t)$$

or

$$x = 1 + 5t, \quad y = -6 - \pi t$$

Worked Example 4   In what direction does the line $x = -3t + 2$, $y = -2(t - 1)$, $z = 8t + 2$ point?

Solution   Using the preceding display, we construct the direction $\mathbf{d} = a\mathbf{i} + b\mathbf{j} + c\mathbf{k}$ from the coefficients: $a = -3$, $b = -2$, $c = 8$. Thus the line points in the direction of $\mathbf{d} = -3\mathbf{i} - 2\mathbf{j} + 8\mathbf{k}$.

Solved Exercises   3.   Do the lines $R_1 = (t, -6t + 1, 2t - 8)$ and $R_2 = (3t + 1, 2t, 0)$ intersect?

   4.   Find the "equation" of the line segment between $(1, 1, 1)$ and $(2, 1, 2)$.

Exercises   5.   Write parametric equations for the following lines:
   (a)   The line through $(1, 1, 0)$ and $(0, 0, 1)$.
   (b)   The line through $(2, 0, 0)$ and $(0, 1, 0)$.
   (c)   The line through $(0, 0, 0)$ and $(1, 1, 1)$.

   6.   Write parametric equations for the following lines:
   (a)   The line through $(1, 1, 0)$ in the direction $-\mathbf{i} - \mathbf{j} + \mathbf{k}$.
   (b)   The line through $(0, 1, 0)$ in the direction $\mathbf{j}$.

   7.   At what point does the line through $(0, 1, 2)$ with direction $\mathbf{i} + \mathbf{j} + \mathbf{k}$ cross the xy plane?

DISTANCE BETWEEN POINTS AND LENGTH OF VECTORS

Since all the line segments representing given vector $\mathbf{v} = a\mathbf{i} + b\mathbf{j} + c\mathbf{k}$ have the same length, we may define the length of $\mathbf{v}$ to be the length of any of these segments. To calculate the length of $\mathbf{v}$, it is convenient to use the segment $\overrightarrow{OP}$, where $P = (a, b, c)$, so that the length of $\mathbf{v}$ is just the distance from $(0, 0, 0)$ to (a, b, c).

Two applications of the Pythagorean theorem enable us to calculate this distance. (See Fig. 14-2-4.) Let $Q = (a, b, 0)$ and $R = (a, 0, 0)$. Then $|OR| = |a|$ and $|RQ| = |b|$, so $|OQ| = \sqrt{a^2 + b^2}$. Now $|QP| = |c|$, so applying Pythagoras again, this time to the right triangle OQP, we obtain $|OP| = \sqrt{a^2 + b^2 + c^2}$. We denote the length of a vector $\mathbf{v}$ by $\|\mathbf{v}\|$.

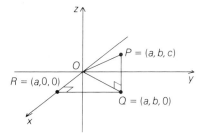

Fig. 14-2-4
$|OP| = \sqrt{a^2 + b^2 + c^2}$

LENGTH OF A VECTOR

The length $\|\mathbf{v}\|$ of a vector $\mathbf{v}$ is the square root of the sum of the squares of the components of $\mathbf{v}$:

$$\|a\mathbf{i} + b\mathbf{j} + c\mathbf{k}\| = \sqrt{a^2 + b^2 + c^2}$$

Worked Example 5   Find the length of $\mathbf{v} = 2\mathbf{i} - 6\mathbf{j} + 7\mathbf{k}$.

Solution $\|\mathbf{v}\| = \sqrt{2^2 + (-6)^2 + 7^2} = \sqrt{4 + 36 + 49} = \sqrt{89} \approx 9.434$

Fig. 14-2-5 The law of cosines applied to vectors.

Some basic properties of length may be deduced from the law of cosines (see Section 6-1 for its proof). In terms of vectors, the law of cosines states (see Fig. 14-2-5) that

$$\|\mathbf{w} - \mathbf{v}\|^2 = \|\mathbf{v}\|^2 + \|\mathbf{w}\|^2 - 2\|\mathbf{v}\|\,\|\mathbf{w}\| \cos \theta$$

where θ is the angle between the vectors $\mathbf{v}$ and $\mathbf{w}$, $0 \le \theta \le \pi$. In particular, since $\cos \theta \le 1$, we get

$$\begin{aligned}
\|\mathbf{w} - \mathbf{v}\|^2 &= \|\mathbf{w}\|^2 + \|\mathbf{v}\|^2 - 2\|\mathbf{v}\|\,\|\mathbf{w}\| \cos \theta \\
&\ge \|\mathbf{w}\|^2 + \|\mathbf{v}\|^2 - 2\|\mathbf{v}\|\,\|\mathbf{w}\| \\
&= (\|\mathbf{w}\| - \|\mathbf{v}\|)^2
\end{aligned}$$

Taking square roots and remembering that $\sqrt{x^2} = |x|$, we get

$$\|\mathbf{w} - \mathbf{v}\| \ge |\,\|\mathbf{w}\| - \|\mathbf{v}\|\,|$$

Hence

$$-(\|\mathbf{w} - \mathbf{v}\|) \le \|\mathbf{w}\| - \|\mathbf{v}\| \le \|\mathbf{w} - \mathbf{v}\|$$

In particular, from the right-hand inequality we get

$$\|\mathbf{w}\| \le \|\mathbf{w} - \mathbf{v}\| + \|\mathbf{v}\|$$

That is, the length of one side of a triangle is less than or equal to the sum of the lengths of the other sides. If we write $\mathbf{u} = \mathbf{w} - \mathbf{v}$, then $\mathbf{w} = \mathbf{u} + \mathbf{v}$, and the inequality above takes the useful form

$$\|\mathbf{u} + \mathbf{v}\| \le \|\mathbf{u}\| + \|\mathbf{v}\|$$

which is called the *triangle inequality*.

The relation between length and scalar multiplication is given by

$$\|r\mathbf{v}\| = |r|\,\|\mathbf{v}\|$$

since if $\mathbf{v} = a\mathbf{i} + b\mathbf{j} + c\mathbf{k}$, then

$$\|r\mathbf{v}\| = \sqrt{(ra)^2 + (rb)^2 + (rc)^2} = \sqrt{r^2}\,\sqrt{a^2 + b^2 + c^2} = |r|\,\|\mathbf{v}\|$$

PROPERTIES OF LENGTH

If **u**, **v**, and **w** are any vectors and r is any number:

1. $\|\mathbf{u}\| \geq 0$

2. $\|\mathbf{u}\| = 0$ if and only if $\mathbf{u} = 0$

3. $\|r\mathbf{u}\| = |r|\,\|\mathbf{u}\|$

4. $\|\mathbf{u} + \mathbf{v}\| \leq \|\mathbf{u}\| + \|\mathbf{v}\|$

4′. $\|\mathbf{w} - \mathbf{v}\| \geq |\,\|\mathbf{w}\| - \|\mathbf{v}\|\,|$ $\Big\}$ triangle inequality

Worked Example 6 Verify the triangle inequality (4) for $\mathbf{u} = \mathbf{i} + \mathbf{j}$ and $\mathbf{v} = 2\mathbf{i} + \mathbf{j} + \mathbf{k}$.

Solution We have $\mathbf{u} + \mathbf{v} = 3\mathbf{i} + 2\mathbf{j} + \mathbf{k}$, so $\|\mathbf{u} + \mathbf{v}\| = \sqrt{9 + 4 + 1} = \sqrt{14}$. On the other hand, $\|\mathbf{u}\| = \sqrt{2}$ and $\|\mathbf{v}\| = \sqrt{6}$, so the triangle inequality asserts that $\sqrt{14} \leq \sqrt{2} + \sqrt{6}$. This is indeed true, since $\sqrt{14} \approx 3.74$, while $\sqrt{2} + \sqrt{6} \approx 1.41 + 2.45 = 3.86$.

Worked Example 7 Suppose that an object is moving uniformly along a straight line. What physical quantity is represented by the length of its velocity vector?

Solution Let **v** be the velocity vector. Then the displacement vector from its position P at any time to its position Q t units of time later is $t\mathbf{v}$. The distance between P and Q is then $|t|\|\mathbf{v}\|$, so the length $\|\mathbf{v}\|$ of the velocity vector represents the ratio of distance traveled to elapsed time—it is called the *speed*.

A vector **u** is called a *unit vector* if its length is equal to 1. If **v** is any nonzero vector, $\|\mathbf{v}\| \neq 0$ and we can obtain a unit vector pointing in the direction of **v** by taking $\mathbf{u} = (1/\|\mathbf{v}\|)\mathbf{v}$. In fact,

$$\|\mathbf{u}\| = \left\|\frac{1}{\|\mathbf{v}\|}\mathbf{v}\right\| = \frac{1}{\|\mathbf{v}\|}\|\mathbf{v}\| = 1$$

We call **u** the *normalization* of **v**.

Worked Example 8 Normalize $\mathbf{v} = 2\mathbf{i} + 3\mathbf{j} - \tfrac{1}{2}\mathbf{k}$.

Solution We have $\|\mathbf{v}\| = \sqrt{2^2 + 3^2 + \tfrac{1}{4}} = \tfrac{1}{2}\sqrt{53}$, so the normalization of $\|\mathbf{v}\|$ is

$$\mathbf{u} = \frac{1}{\|\mathbf{v}\|}\mathbf{v} = \frac{4}{\sqrt{53}}\mathbf{i} + \frac{6}{\sqrt{53}}\mathbf{j} - \frac{1}{\sqrt{53}}\mathbf{k}$$

Finally, we can use the formula for the length of vectors to obtain a formula for the distance between any two points in space. If $P_1 = (x_1, y_1, z_1)$ and $P_2 = (x_2, y_2, z_2)$, then the distance between P_1 and P_2 is the length of the vector from P_2 to P_1; that is,

$$|P_2 P_1| = \|(x_1 - x_2)\mathbf{i} + (y_1 - y_2)\mathbf{j} + (z_1 - z_2)\mathbf{k}\|$$
$$= \sqrt{(x_1 - x_2)^2 + (y_1 - y_2)^2 + (z_1 - z_2)^2}$$

DISTANCE FORMULA

If P_1 has coordinates (x_1, y_1, z_1) and P_2 has coordinates (x_2, y_2, z_2),

then the distance between P_1 and P_2 is

$$\sqrt{(x_1 - x_2)^2 + (y_1 - y_2)^2 + (z_1 - z_2)^2}$$

Worked Example 9 Find the distance between $(2, 1, 0)$ and $(3, -2, 6)$.

Solution The distance is

$$\sqrt{(2 - 3)^2 + [1 - (-2)]^2 + (0 - 6)^2} = \sqrt{1^2 + 3^2 + (-6)^2}$$
$$= \sqrt{1 + 9 + 36} = \sqrt{46}$$

Solved Exercises

5. Find the values of c for which $\|\mathbf{i} + \mathbf{j} + c\mathbf{k}\| = 4$.

6. Prove that $\|\mathbf{u} - \mathbf{v}\| \le \|\mathbf{u} - \mathbf{w}\| + \|\mathbf{w} - \mathbf{v}\|$ for any vectors $\mathbf{u}$, $\mathbf{v}$, $\mathbf{w}$. Illustrate with a figure in which $\mathbf{u}$, $\mathbf{v}$, and $\mathbf{w}$ are all drawn with the same base point, that is, the same "tail."

7. Find unit vectors $\mathbf{u}$, $\mathbf{v}$, and $\mathbf{w}$ in the plane such that $\mathbf{u} + \mathbf{v} = \mathbf{w}$.

8. Let $P_t = t(1, 1, 1)$:

   (a) What is the distance from P_t to $(3, 0, 0)$?

   (b) For what value of t is that distance shortest?

   (c) What is that shortest distance?

Exercises

8. Compute the length of each of the following vectors:

   (a) $\mathbf{i} + \mathbf{j} + \mathbf{k}$ (b) $2\mathbf{i} + \mathbf{j}$

   (c) $\mathbf{i} + \mathbf{k}$ (d) $3\mathbf{i} + 4\mathbf{j}$

   (e) $2\mathbf{i} + 2\mathbf{k}$

9. Normalize each of the vectors in Exercise 8.

10. (a) Verify the triangle inequality, $\|\mathbf{u} + \mathbf{v}\| \le \|\mathbf{u}\| + \|\mathbf{v}\|$, for the vectors in parts (a) and (c) of Exercise 8.

    (b) When does equality hold in the triangle inequality? (You might try using the law of cosines as was done in the text.) Test your conclusion on the vectors in parts (c) and (e) of Exercise 8.

11. Show that $(0, 0, 1)$, $(0, 1, 0)$, and $(1, 0, 0)$ are the vertices of an equilateral triangle. How long is each side?

12. A boat whose top speed in still water is 12 knots points north and steams at full power. If there is an eastward current of 5 knots, what is the speed of the boat?

THE DOT PRODUCT

To introduce the dot product, we will calculate the angle θ between two vectors in terms of the components of the vectors.

If $\mathbf{v}_1$ and $\mathbf{v}_2$ are two vectors, we have seen (Fig. 14-2-5) that

$$\|\mathbf{v}_2 - \mathbf{v}_1\|^2 = \|\mathbf{v}_1\|^2 + \|\mathbf{v}_2\|^2 - 2\|\mathbf{v}_1\| \|\mathbf{v}_2\| \cos \theta$$

where θ is the angle between $\mathbf{v}_1$ and $\mathbf{v}_2$; therefore

$$2\|\mathbf{v}_1\| \|\mathbf{v}_2\| \cos\theta = \|\mathbf{v}_1\|^2 + \|\mathbf{v}_2\|^2 - \|\mathbf{v}_2 - \mathbf{v}_1\|^2 \tag{1}$$

If $\mathbf{v}_1 = a_1\mathbf{i} + b_1\mathbf{j} + c_1\mathbf{k}$ and $\mathbf{v}_2 = a_2\mathbf{i} + b_2\mathbf{j} + c_2\mathbf{k}$, then the right-hand side of equation (1) is

$$(a_1^2 + b_1^2 + c_1^2) + (a_2^2 + b_2^2 + c_2^2) - [(a_1 - a_2)^2 + (b_1 - b_2)^2 + (c_1 - c_2)^2]$$
$$= 2(a_1 a_2 + b_1 b_2 + c_1 c_2)$$

Thus we have proved that

$$\|\mathbf{v}_1\| \|\mathbf{v}_2\| \cos\theta = a_1 a_2 + b_1 b_2 + c_1 c_2 \tag{2}$$

This very convenient formula enables us to compute $\cos\theta$ and hence θ. The quantity on the right-hand side deserves a special name.

Definition If $\mathbf{v}_1 = a_1\mathbf{i} + b_1\mathbf{j} + c_1\mathbf{k}$ and $\mathbf{v}_2 = a_2\mathbf{i} + b_2\mathbf{j} + c_2\mathbf{k}$ are two vectors, the quantity $a_1 a_2 + b_1 b_2 + c_1 c_2$ is called their *dot product* and is denoted by $\mathbf{v}_1 \cdot \mathbf{v}_2$.

The dot product in the plane is defined analogously; just think of c_1 and c_2 as being zero. Notice that the dot product of two vectors is a number, not a vector. The dot product is sometimes called the *scalar product* (do not confuse this with scalar *multiplication*) or the *inner product*.

Worked Example 10 If $\mathbf{v}_1 = 3\mathbf{i} + \mathbf{j} - 2\mathbf{k}$ and $\mathbf{v}_2 = \mathbf{i} - \mathbf{j} + \mathbf{k}$, calculate $\mathbf{v}_1 \cdot \mathbf{v}_2$.

Solution $\mathbf{v}_1 \cdot \mathbf{v}_2 = 3 \cdot 1 + 1 \cdot (-1) + (-2) \cdot 1 = 3 - 1 - 2 = 0$

Combining formula (2) with the definition of dot product gives

$$\|\mathbf{v}_1\| \|\mathbf{v}_2\| \cos\theta = \mathbf{v}_1 \cdot \mathbf{v}_2 \tag{3}$$

where θ is the angle between $\mathbf{v}_1$ and $\mathbf{v}_2$.
We may solve (3) for $\cos\theta$ to obtain the formula

$$\cos\theta = \frac{\mathbf{v}_1 \cdot \mathbf{v}_2}{\|\mathbf{v}_1\| \|\mathbf{v}_2\|}; \quad \text{i.e., } \theta = \cos^{-1}\left(\frac{\mathbf{v}_1 \cdot \mathbf{v}_2}{\|\mathbf{v}_1\| \|\mathbf{v}_2\|}\right)$$

Worked Example 11 Find the angle between the vectors $\mathbf{i} + \mathbf{j} + \mathbf{k}$ and $\mathbf{i} + \mathbf{j} - \mathbf{k}$.

Solution Let $\mathbf{v}_1 = \mathbf{i} + \mathbf{j} + \mathbf{k}$ and $\mathbf{v}_2 = \mathbf{i} + \mathbf{j} - \mathbf{k}$. Then $\|\mathbf{v}_1\| = \sqrt{3}$, $\|\mathbf{v}_2\| = \sqrt{3}$, and $\mathbf{v}_1 \cdot \mathbf{v}_2 = 1 \cdot 1 + 1 \cdot 1 - 1 \cdot 1 = 1$. Hence $\cos\theta = \frac{1}{3}$, so $\theta = \cos^{-1}(\frac{1}{3}) \approx 1.23$ radians (70°32′).

Worked Example 12 Derive the *Schwartz inequality* $|\mathbf{v}_1 \cdot \mathbf{v}_2| \le \|\mathbf{v}_1\| \|\mathbf{v}_2\|$.

Solution This inequality follows from formula (3); since $|\cos\theta| \le 1$,

$$|\mathbf{v}_1 \cdot \mathbf{v}_2| = \|\mathbf{v}_1\| \|\mathbf{v}_2\||\cos\theta| \le \|\mathbf{v}_1\| \|\mathbf{v}_2\|$$

Since two nonzero vectors are perpendicular when $\theta = \pi/2$—that is, when $\cos\theta = 0$—we have an algebraic test for perpendicularity: The vectors $\mathbf{v}_1$ and $\mathbf{v}_2$ are perpendicular when $\mathbf{v}_1 \cdot \mathbf{v}_2 = 0$. We adopt the convention that the zero vector is perpendicular to every vector. The synonyms "orthogonal" or "normal" are also used for "perpendicular."

Worked Example 13 Find a unit vector in the plane which is orthogonal to $\mathbf{v} = \mathbf{i} - 3\mathbf{j}$.

Solution If $\mathbf{w} = a\mathbf{i} + b\mathbf{j}$ is perpendicular to $\mathbf{i} - 3\mathbf{j}$, we must have $0 = \mathbf{v} \cdot \mathbf{w} = a - 3b$; that is, $a = 3b$. If $\mathbf{w}$ is to be a unit vector, we must also have $1 = a^2 + b^2 = (3b)^2 + b^2 = 10b^2$, so $b = \pm 1/\sqrt{10}$ and $a = \pm 3/\sqrt{10}$. Thus there are two possible solutions: $(\pm 1/\sqrt{10})(3\mathbf{i} + \mathbf{j})$. (See Fig. 14-2-6.)

Fig. 14-2-6 There are two unit vectors orthogonal to $\mathbf{i} - 3\mathbf{j}$.

The dot product has some useful algebraic properties. They are summarized in the display below; proofs are left to the reader (see Problem 13).

THE DOT PRODUCT

Algebraic definition:

$$(a_1\mathbf{i} + b_1\mathbf{j} + c_1\mathbf{k}) \cdot (a_2\mathbf{i} + b_2\mathbf{j} + c_2\mathbf{k}) = a_1 a_2 + b_1 b_2 + c_1 c_2$$

Geometric interpretation:

$$\mathbf{v}_1 \cdot \mathbf{v}_2 = \|\mathbf{v}_1\| \, \|\mathbf{v}_2\| \cos \theta$$

where $\theta \in [0, \pi]$ is the angle between $\mathbf{v}_1$ and $\mathbf{v}_2$. In particular,

$$\|\mathbf{v}\|^2 = \mathbf{v} \cdot \mathbf{v}.$$

Properties:

1. $\mathbf{u} \cdot \mathbf{u} \geq 0$ for any vector $\mathbf{u}$
2. $\mathbf{u} \cdot \mathbf{u} = 0$ only if $\mathbf{u} = \mathbf{0}$
3. $\mathbf{u} \cdot \mathbf{v} = \mathbf{v} \cdot \mathbf{u}$
4. $(\mathbf{u} + \mathbf{v}) \cdot \mathbf{w} = \mathbf{u} \cdot \mathbf{w} + \mathbf{v} \cdot \mathbf{w}$
5. $(a\mathbf{u}) \cdot \mathbf{v} = a(\mathbf{u} \cdot \mathbf{v})$

Worked Example 14 Let $\mathbf{u}$ and $\mathbf{v}$ be vectors in the plane; assume that $\mathbf{u}$ is nonzero.
(a) Show that $\mathbf{w} = \mathbf{v} - (\mathbf{v} \cdot \mathbf{u}/\mathbf{u} \cdot \mathbf{u})\mathbf{u}$ is orthogonal to $\mathbf{u}$.
(b) Sketch the vectors $\mathbf{u}$, $\mathbf{v}$, $\mathbf{w}$, and $\mathbf{v} - \mathbf{w}$. The vector $\mathbf{v} - \mathbf{w}$ is called the *orthogonal projection* of $\mathbf{v}$ on $\mathbf{u}$. Why?

Solution (a) We compute:

$$\mathbf{u} \cdot \mathbf{w} = \mathbf{u} \cdot \left(\mathbf{v} - \frac{\mathbf{v} \cdot \mathbf{u}}{\mathbf{u} \cdot \mathbf{u}} \mathbf{u} \right)$$

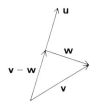

Fig. 14-2-7 The vector $\mathbf{v} - \mathbf{w} = (\mathbf{v} \cdot \mathbf{u}/\mathbf{u} \cdot \mathbf{u})\mathbf{u}$ is the orthogonal projection of $\mathbf{v}$ on $\mathbf{u}$.

By the algebraic properties of the dot product, this is equal to

$$\mathbf{u} \cdot \mathbf{v} - \frac{\mathbf{v} \cdot \mathbf{u}}{\mathbf{u} \cdot \mathbf{u}} \mathbf{u} \cdot \mathbf{u} = \mathbf{u} \cdot \mathbf{v} - \mathbf{v} \cdot \mathbf{u} = 0$$

so $\mathbf{u} \cdot \mathbf{w} = 0$ and $\mathbf{w}$ is orthogonal to $\mathbf{u}$.

(b) We note that $\mathbf{v} - \mathbf{w} = (\mathbf{v} \cdot \mathbf{u}/\mathbf{u} \cdot \mathbf{u})\mathbf{u}$ is a multiple of $\mathbf{u}$. Thus the configuration of vectors must be as in Fig. 14-2-7. The vector $\mathbf{v} - \mathbf{w}$ is called the *orthogonal projection* of $\mathbf{v}$ on $\mathbf{u}$ because it is obtained by dropping a perpendicular from the "tip" of $\mathbf{v}$ to the line determined by $\mathbf{u}$. (The base points of $\mathbf{u}$ and $\mathbf{v}$ must be the same for this construction.)

Solved Exercises

9. Prove that $\|\mathbf{v}\|^2 = \mathbf{v} \cdot \mathbf{v}$.

10. Find the angle between $3\mathbf{i} + \mathbf{j} - 2\mathbf{k}$ and $\mathbf{i} - \mathbf{j} + \mathbf{k}$.

11. Find the orthogonal projection of $\mathbf{i} + \mathbf{j}$ on $\mathbf{i} - 2\mathbf{j}$.

Exercises

13. Compute the following dot products:
   (a) $(\mathbf{i} + \mathbf{j} + \mathbf{k}) \cdot (\mathbf{i} + \mathbf{j} + 2\mathbf{k})$ (b) $(\mathbf{i} + \mathbf{j} + \mathbf{k}) \cdot (\mathbf{i} + \mathbf{k})$
   (c) $\mathbf{i} \cdot \mathbf{j}$ (d) $(3\mathbf{i} + 4\mathbf{j}) \cdot (3\mathbf{j} + 4\mathbf{k})$

14. Find the angle between each of the pairs of vectors in Exercise 13.

15. Use the formula $(\mathbf{i} + \mathbf{j} + \mathbf{k}) \cdot \mathbf{i} = 1$ to find the angle between the diagonal of a cube and one of its edges. Sketch.

16. (a) Show that if $\|\mathbf{u}\| = \|\mathbf{v}\|$, then $\mathbf{u} + \mathbf{v}$ and $\mathbf{u} - \mathbf{v}$ are perpendicular.
   (b) Use the result of part (a) to prove that any triangle inscribed in a circle, with one side of the triangle as a diameter, is a right triangle.

17. Use vector methods to prove that a triangle is isosceles if and only if its base angles are equal.

18. Show that the length of the orthogonal projection of $\mathbf{v}$ on $\mathbf{u}$ is equal to $\|\mathbf{v}\||\cos\theta|$, where θ is the angle between $\mathbf{v}$ and $\mathbf{u}$.

EQUATION OF A PLANE

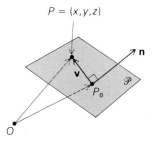

Fig. 14-2-8 The plane $\mathscr{P}$ is perpendicular to the vector $\mathbf{n}$.

The dot product makes it simple to determine the equation of a plane. Suppose that a plane $\mathscr{P}$ passes through a point $P_0 = (x_0, y_0, z_0)$ and is perpendicular to a vector $\mathbf{n} = A\mathbf{i} + B\mathbf{j} + C\mathbf{k}$. See Fig. 14-2-8.

Let $P = (x, y, z)$ be a point on $\mathscr{P}$. Then $\mathbf{n}$ must be perpendicular to the vector $\mathbf{v}$ from P_0 to P; that is, $\mathbf{n} \cdot \mathbf{v} = 0$, or

$$(A\mathbf{i} + B\mathbf{j} + C\mathbf{k}) \cdot [(x - x_0)\mathbf{i} + (y - y_0)\mathbf{j} + (z - z_0)\mathbf{k}] = 0$$

Hence

$$A(x - x_0) + B(y - y_0) + C(z - z_0) = 0$$

We call $\mathbf{n}$ the *normal vector* of the plane.

If we let $D = -(Ax_0 + By_0 + Cz_0)$, the equation of a plane becomes

$$Ax + By + Cz + D = 0$$

Worked Example 15 Find the equation of the plane through $(1,1,1)$ with normal vector $2\mathbf{i} + \mathbf{j} - 2\mathbf{k}$.

Solution Here $A\mathbf{i} + B\mathbf{j} + C\mathbf{k} = 2\mathbf{i} + \mathbf{j} - 2\mathbf{k}$, and $(x_0, y_0, z_0) = (1,1,1)$, so we get

$$A(x-1) + B(y-1) + C(z-1) = 0$$

Hence

$$2(x-1) + (y-1) - 2(z-1) = 0$$
$$2x + y - 2z = 1$$

Worked Example 16 Find a unit normal vector to the plane $3x + y - z = 10$. Sketch the plane.

Solution A normal vector is obtained by making a vector out of the coefficients of x, y, and z; that is, $(3, 1, -1)$. Normalizing, we get

$$(3, 1, -1)/\sqrt{9 + 1 + 1} = (3/\sqrt{11}, 1/\sqrt{11}, -1/\sqrt{11})$$

We may sketch the plane by noting where it meets the coordinate axes. For example, setting $y = z = 0$, we see that $(\frac{10}{3}, 0, 0)$ lies on the plane (see Fig. 14-2-9).

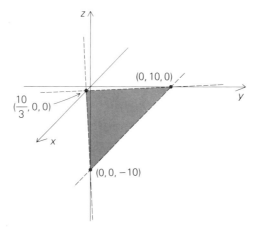

Fig. 14-2-9 The plane $3x + y - z = 10$.

EQUATION OF A PLANE IN SPACE

The equation of the plane through (x_0, y_0, z_0) with normal vector $A\mathbf{i} + B\mathbf{j} + C\mathbf{k}$ is

$$A(x - x_0) + B(y - y_0) + C(z - z_0) = 0$$

or

$$Ax + By + Cz + D = 0$$

Solved Exercises

12. Where does the line through the origin in the direction of $\mathbf{i} + \mathbf{j} + 2\mathbf{k}$ meet the plane $x + y + 2z = 5$? Use your answer to find the distance from the origin to this plane. Sketch.

13. Find the equation of the plane containing the points $(1, 1, 1)$, $(2, 0, 0)$, and $(1, 1, 0)$.

Exercises

19. Give the equation for each of the following planes:
   (a) The plane through the origin orthogonal to $\mathbf{i} + \mathbf{j} + \mathbf{k}$.
   (b) The plane through $(1, 0, 0)$ orthogonal to $\mathbf{i} + \mathbf{j} + \mathbf{k}$.
   (c) The plane through the origin orthogonal to $\mathbf{i}$.
   (d) The plane containing the points $(1, 0, 0)$, $(0, 2, 0)$, and $(0, 0, 3)$.
   (e) The plane containing (a, b, c) with normal vector $a\mathbf{i} + b\mathbf{j} + c\mathbf{k}$.

20. Find a unit normal vector to each of the following planes:
   (a) The plane given by $2x + 3y + z = 0$.
   (b) The plane through the origin containing the points $(1, 1, 1)$ and $(1, 1, -1)$.
   (c) The plane containing the line $(1 + t, 1 - t, t)$ and the point $(1, 1, 1)$.

21. Show that the locus of points equidistant from two given points is a plane and give an equation for that plane in terms of the coordinates of the two given points.

Problems for Section 14-2

1. Show that the diagonals of a rhombus are perpendicular to each other.

2. Let $\mathbf{u}$, $\mathbf{v}$, and $\mathbf{w}$ be the vectors from an origin O to the vertices of a triangle. Show that the vector $\frac{1}{3}(\mathbf{u} + \mathbf{v} + \mathbf{w})$ lies on all three medians of the triangle. What theorem from elementary geometry does this prove?

3. Describe geometrically the following sets of points:
   (a) All (x, y, z) such that $z = 0$.
   (b) All (x, y, z) such that $y = z = 0$.
   (c) All (x, y, z) such that $x^2 + y^2 + z^2 = 1$.
   (d) All (x, y, z) such that $x^2 + y^2 = 1$.

4. Find the equation of:
   (a) The plane containing $(0, 1, 0)$, $(1, 0, 0)$, $(0, 0, 1)$.
   (b) The line through $(1, 2, 1)$ and $(-1, 1, 0)$.
   (c) The line through $(1, 1, 1)$ and orthogonal to the plane in part (a).
   (d) The plane through $(0, 0, 0)$ orthogonal to the line in part (b).

5. Compute the following:
   (a) $(3\mathbf{i} + 2\mathbf{j} - \mathbf{k}) \cdot (\mathbf{j} - \mathbf{k})$
   (b) $[(3\mathbf{i} - \mathbf{j} - \mathbf{k}) - (\mathbf{i} + \mathbf{j})] \cdot \mathbf{j}$
   (c) The distance between $(1, 0, 2)$ and $(3, 2, 4)$.
   (d) The length of $(\mathbf{i} - \mathbf{j} - \mathbf{k}) + (2\mathbf{j} - \mathbf{k} + \mathbf{i})$.

6. Find the following:
   (a) A unit normal to the plane $x - 2y + z = 0$.
   (b) A vector orthogonal to $\mathbf{i} - \mathbf{j} + \mathbf{k}$ and to $\mathbf{i} + \mathbf{j} + \mathbf{k}$.
   (c) The angle between $2\mathbf{i} + \mathbf{j} + \mathbf{k}$ and $\mathbf{k} - \mathbf{i}$.
   (d) A vector in space making an angle of $45°$ with $\mathbf{i}$ and $60°$ with $\mathbf{j}$.

7. Let P_1 and P_2 be points in the plane. Give an equation of the form $ax + by = c$ for the perpendicular bisector of the line segment between P_1 and P_2.

★8. Derive the point-point form equation of a line obtained in Section R-2 from the parametric form obtained in this section. Comment on the case in which $x_1 = x_2$.

9. A ship starting at $(0,0)$ proceeds at a speed of 10 knots directly toward a buoy located at $(3, 4)$. (The chart is measured in nautical miles; a knot equals 1 nautical mile per hour.)

    (a) What is the ship's point of closest approach to a rock located at $(2, 2)$?

    (b) After how long does it reach this point?

    (c) How far is this point from the rock?

10. Use vector methods to show that if three parallel lines in the plane cut off equal segments on one transversal, then they do so on any transversal.

11. Let:

    $L_1 =$ the line $(2, 1, 1) + t(1, 1, 1)$

    $L_2 =$ the line $(1 + 7t, 7t - 2, 2 + 7t)$

    $L_3 =$ the line $(1, 0, 8) + t(1, 1, 9)$

    $L_4 =$ the line through the points $(-1, 0, 1)$ and $(1, 2, 19)$

    (a) Determine whether each of the following pairs of lines is parallel or intersects. If the lines intersect, find the point of intersection.

    L_1 and L_2 L_2 and L_3

    L_1 and L_3 L_2 and L_4

    L_1 and L_4 L_3 and L_4

    (b) For each pair of lines in part (a) which lie in a plane (that is, are not skew), find an equation for that plane.

    (c) For each of the lines L_1 to L_4, find the point of closest approach to the origin and an equation for the plane perpendicular to that line through that point.

★12. Derive the point-direction form for the parametric equation of a line from the point-point form. [*Hint:* If a line through P is to have direction **d**, what other point must lie on the line?]

13. Establish the following properties of the dot product:

    (a) $\mathbf{u} \cdot \mathbf{u} \geq 0$ for any vector **u**.

    (b) If $\mathbf{u} \cdot \mathbf{u} = 0$, then $\mathbf{u} = 0$.

    (c) $\mathbf{u} \cdot \mathbf{v} = \mathbf{v} \cdot \mathbf{u}$ for any vector **u** and **v**.

    (d) $(a\mathbf{u} + b\mathbf{v}) \cdot \mathbf{w} = a(\mathbf{u} \cdot \mathbf{w}) + b(\mathbf{v} \cdot \mathbf{w})$ for any numbers a and b and any vectors **u**, **v**, **w**.

★14. Let the equation $Ax + By + Cz + D = 0$ determine a plane and let $P = (a, b, c)$ be a point not on that plane.

    (a) Show that the line $R = P + t(A, B, C)$ is perpendicular to the given plane.

    (b) Find the point Q at which the line intersects the plane.

    (c) Show that the distance from P to Q is less than the distance from P to any other point in that plane.

    (d) Conclude that the distance from P to the given plane is

    $$\frac{|aA + bB + cC + D|}{\sqrt{A^2 + B^2 + C^2}}$$

    (e) Show that the distance from a point P to the plane through P_0 with unit normal vector **n** is given by $|\mathbf{v} \cdot \mathbf{n}|$, where **v** is the vector from P_0 to P.

15. Suppose that $\mathbf{e}_1$ and $\mathbf{e}_2$ are perpendicular unit vectors in the plane, and let **v** be an arbitrary vector. Show that $\mathbf{v} = (\mathbf{v} \cdot \mathbf{e}_1)\mathbf{e}_1 + (\mathbf{v} \cdot \mathbf{e}_2)\mathbf{e}_2$. The numbers $\mathbf{v} \cdot \mathbf{e}_1$ and $\mathbf{v} \cdot \mathbf{e}_2$ are called the *components* of **v** in the directions of $\mathbf{e}_1$ and $\mathbf{e}_2$. This expression of **v** as a sum of vectors pointing in the directions of $\mathbf{e}_1$ and $\mathbf{e}_2$ is called the *orthogonal decomposition* of **v** relative to $\mathbf{e}_1$ and $\mathbf{e}_2$.

16. Consider the vectors $\mathbf{e}_1 = (1/\sqrt{2})(\mathbf{i} + \mathbf{j})$ and $\mathbf{e}_2 = (1/\sqrt{2})(\mathbf{i} - \mathbf{j})$ in the plane. Check that $\mathbf{e}_1$ and $\mathbf{e}_2$ are unit vectors perpendicular to each other and express each of the following vectors in the form $\mathbf{v} = a_1\mathbf{e}_1 + a_2\mathbf{e}_2$ (that is, as a *linear combination* of $\mathbf{e}_1$ and $\mathbf{e}_2$):

    (a) $\mathbf{v} = \mathbf{i}$ (b) $\mathbf{v} = \mathbf{j}$

    (c) $\mathbf{v} = 2\mathbf{i} + \mathbf{j}$ (d) $\mathbf{v} = -2\mathbf{i} - \mathbf{j}$

17. Suppose that a force **F** (for example, gravity) is acting vertically downward on an object sitting on a plane which is inclined at an angle of 45° to the horizontal. Express this force as a sum of a force acting parallel to the plane and one acting perpendicular to it.

18. Suppose that an object moving in direction $\mathbf{i} + \mathbf{j}$ is acted on by a force given by the vector $2\mathbf{i} + \mathbf{j}$. Express this force as the sum of a force in the direction of motion and a force perpendicular to the direction of motion.

★19. Suppose that $R = P_0 + t(a, b, c)$ is the line through P_0 in the direction $\mathbf{d} = a\mathbf{i} + b\mathbf{j} + c\mathbf{k}$. Let $\mathbf{u} = \mathbf{d}/\|\mathbf{d}\| = (\mu, \lambda, \nu)$. Let:

α = angle from $\mathbf{i}$ to $\mathbf{d}$

β = angle from $\mathbf{j}$ to $\mathbf{d}$

γ = angle from $\mathbf{k}$ to $\mathbf{d}$

These are called the *direction angles* of the line. The numbers $\cos\alpha$, $\cos\beta$, and $\cos\gamma$ are called its *direction cosines* (see Fig. 14-2-10).

(a) Show that $P_0 + s(\mu, \lambda, \nu)$ gives the same line. (What values of s and t correspond to the same points on the line?)

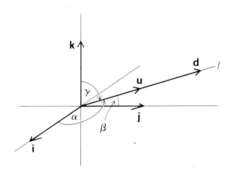

Fig. 14-2-10 The direction angles of the line *l* are α, β, and γ.

(b) Show that $\cos\alpha = \mathbf{i} \cdot \mathbf{u} = \mu$; $\cos\beta = \mathbf{j} \cdot \mathbf{u} = \lambda$; $\cos\gamma = \mathbf{k} \cdot \mathbf{u} = \nu$.

(c) Show that $\cos^2\alpha + \cos^2\beta + \cos^2\gamma = 1$.

(d) Determine the direction angles and cosines of each of the lines in Problem 11.

(e) Which lines through the origin have direction angles $\alpha = \beta = \gamma$?

★20. Let P_1 and P_2 be points in the plane with *polar* coordinates (r_1, θ_1) and (r_2, θ_2), respectively, and let $\mathbf{u}_1$ be the vector from O to P_1 and $\mathbf{u}_2$ the vector from O to P_2. Show that

$$\mathbf{u}_1 \cdot \mathbf{u}_2 = r_1 r_2 \cos(\theta_1 - \theta_2)$$

[*Hint:* Use a trigonometric identity.]

21. A force of 6 newtons makes an angle of $\pi/4$ radians with the x-axis, pointing to the right. The force acts against the movement of an object along the straight line connecting $(1, 2)$ to $(5, 4)$.

(a) Find a formula for the vector force $\mathbf{F}$.

(b) Find the angle θ between the displacement direction $\mathbf{D} = (5 - 1)\mathbf{i} + (4 - 2)\mathbf{j}$ and the force direction $\mathbf{F}$.

(c) The *work* done is $\mathbf{F} \cdot \mathbf{D}$, or equivalently, $\|\mathbf{F}\| \, \|\mathbf{D}\| \cos\theta$. Compute the work from both formulas and compare.

22. A construction worker is checking the architect's plans for some sheet metal construction. One diagram contains a triangle with sides 12.5, 16.7, 20.9, but no angles have been included. The worker gets out a calculator to check that $(12.5)^2 + (16.7)^2 \approx (20.9)^2$, then marks the angle opposite the long side as 90°.

(a) Explain from the law of cosines the reason why the worker's actions are essentially correct.

(b) The angle is not exactly 90°, from the data given. What percentage error is present?

23. A fluid flows across a plane surface with uniform vector velocity $\mathbf{v}$. Let $\mathbf{n}$ be a unit normal to the plane surface. Show that $\mathbf{v} \cdot \mathbf{n}$ is the volume of fluid that passes through a unit area of the plane in unit time.

24. The potential V produced by charges q_1 and q_2 of opposite sign placed at distances x_1 and x_2 from the origin along the x-axis is given by

$$V = \frac{q_1}{4\pi\varepsilon_0\|\mathbf{r} - \mathbf{r}_1\|} + \frac{q_2}{4\pi\varepsilon_0\|\mathbf{r} - \mathbf{r}_2\|}$$

In this formula, $\mathbf{r}$ is the vector from the origin to the point (x, y, z). The vectors $\mathbf{r}_1$ and $\mathbf{r}_2$ are vectors from the origin to the respective charges q_1 and q_2.

(a) Express the formula for V entirely in terms of the scalar quantities x, y, z, x_1, x_2, q_1, q_2, ε_0.

(b) Show that the locus of points (x, y, z) for which $V = 0$ is a sphere of radius $\sqrt{x_1 x_2}$ with center on the x-axis.

★25. (a) Look to the side as you walk in the rain. Now stop walking. How does the (apparent) direction of the falling rain change? (The rain may be falling at an angle because of wind.)

(b) Explain the observation in (a) in terms of vectors.

(c) Suppose the rain is falling vertically and that you know your walking speed. How could you measure the speed at which the rain is falling?

(d) Do part (c) if the rain is falling at an angle.

14-3 The Cross-Product and Determinants

The cross-product of two vectors in space is a new vector that is perpendicular to the first two.

We introduce in this section a new multiplication operation for vectors in space—the *cross-product*. A useful formula for the cross-product involves the notion of a *determinant*. The treatment of matrices and determinants which concludes this section will also be of use in Chapters 16 and 17.

Goals

After studying this section, you should be able to:

Calculate cross-products of vectors.

Use the cross-product to solve geometric problems.

Evaluate 2×2 and 3×3 determinants.

THE CROSS-PRODUCT

The behavior of rapidly spinning objects does not agree with most people's intuition of what should happen. Take a spinning top or gyroscope, anchored at one point, and give its axis a push (see Fig. 14-3-1).* The axis will

Anchor point

Fig. 14-3-1 What happens when you push on the axis of a gyroscope?

move, but not in the direction of your push. Instead, it will move in a direction *perpendicular* to the way you pushed. In terms of vectors, the effect can be described in the following way. Referring to Fig. 14-3-2, let

$\mathbf{v}_1$ = vector from anchor point to the point where the force is applied (assume the spin is clockwise around $\mathbf{v}_1$ as indicated in the figure)

$\mathbf{v}_2$ = force vector of the push

$\mathbf{v}$ = vector whose direction is the direction of motion of a reference point on gyroscope's axis and whose length is the speed of such a motion

Fig. 14-3-2 The vector **v** giving the motion of the axis is perpendicular to the force vector **v**₂.

This experiment is best done in an orbiting earth satellite, where the effects of gravity do not confuse the issue. Indeed, the Skylab astronauts had great fun carrying out such experiments. (See Henry S. Cooper, Jr., A House in Space, Holt, Rinehart, and Winston, 1976.) Fortunately, earthbound experiments still display the general principles involved.

It turns out that the vector **v** is proportional (with proportionality constant depending on the units being used and the mass and spin rate of the gyroscope) to a vector **w** with the following properties:

Fig. 14-3-3 *Right-hand rule:* Place the palm of your hand so that your fingers curl from **v**₁ in the direction of **v**₂ through the angle θ. Then your thumb points in the direction of **v**₁ × **v**₂.

1. $\|\mathbf{w}\| = \|\mathbf{v}_1\|\,\|\mathbf{v}_2\|\sin\theta$, where θ is the angle between $\mathbf{v}_1$ and $\mathbf{v}_2$.

2. If $\theta \neq 0$ (so that $\mathbf{w} \neq 0$), **w** is perpendicular to both $\mathbf{v}_1$ and $\mathbf{v}_2$, and the triple $(\mathbf{v}_1, \mathbf{v}_2, \mathbf{w})$ of vectors obeys the right-hand rule (see Fig. 14-3-3).

Note that condition 1 says the rate of movement of the axis is proportional to the product of the magnitude of the force and the distance from the point of application of force to the anchor point; furthermore, the rate is greatest when the force is perpendicular to the axis.

Conditions 1 and 2 determine **w** uniquely in terms of $\mathbf{v}_1$ and $\mathbf{v}_2$; **w** is called the *cross-product* (or *vector product*) of $\mathbf{v}_1$ and $\mathbf{v}_2$ and is denoted by $\mathbf{v}_1 \times \mathbf{v}_2$.

We will now determine some properties of the cross-product operation. Our ultimate goal is to find a formula for the components of $\mathbf{v}_1 \times \mathbf{v}_2$ in terms of the components of $\mathbf{v}_1$ and $\mathbf{v}_2$.

Worked Example 1   Show that $\|\mathbf{v}_1 \times \mathbf{v}_2\|$ is equal to the area of the parallelogram spanned by $\mathbf{v}_1$ and $\mathbf{v}_2$.

Solution   See Fig. 14-3-4. Drop the perpendicular *PS* as shown in the figure. Then

$$A = |OR||PS| = |OR||OP|\sin\theta = \|\mathbf{v}_1\|\,\|\mathbf{v}_2\|\sin\theta$$

Fig. 14-3-4   Calculating the area of *ORQP.*

Worked Example 2   Find all the cross-products between the standard basis vectors **i**, **j**, and **k**.

Solution   We observe first that $\mathbf{i} \times \mathbf{i} = \mathbf{j} \times \mathbf{j} = \mathbf{k} \times \mathbf{k} = 0$, because the angle between any vector and itself is zero, and $\sin 0 = 0$. Next we observe that $\mathbf{i} \times \mathbf{j}$ must be a multiple of **k**, since it is perpendicular to **i** and **j**. On the other hand, $\|\mathbf{i} \times \mathbf{j}\| = \|\mathbf{i}\|\,\|\mathbf{j}\|\sin 90° = 1$, so $\mathbf{i} \times \mathbf{j}$ must be **k** or $-\mathbf{k}$. The right-hand rule then shows that $\mathbf{i} \times \mathbf{j} = \mathbf{k}$ (see Fig. 14-3-5). Next, $\mathbf{j} \times \mathbf{i}$ must be **k** or $-\mathbf{k}$; this time the right-hand rule gives $-\mathbf{k}$ as the answer. Similarly, $\mathbf{j} \times \mathbf{k} = \mathbf{i}$, $\mathbf{k} \times \mathbf{i} = \mathbf{j}$, $\mathbf{k} \times \mathbf{j} = -\mathbf{i}$, and $\mathbf{i} \times \mathbf{k} = -\mathbf{j}$.

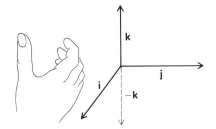

Fig. 14-3-5   The right-hand rule requires that **i** × **j** equal **k,** not $-\mathbf{k}$.

Fig. 14-3-6 As we go around the circle, the cross-product of any two consecutive vectors equals the third vector.

A good way to remember these products is to write $\mathbf{i}$, $\mathbf{j}$, and $\mathbf{k}$ in a circle, as in Fig. 14-3-6.

We will now obtain a formula for the cross-product

$$(a_1\mathbf{i} + a_2\mathbf{j} + a_3\mathbf{k}) \times (b_1\mathbf{i} + b_2\mathbf{j} + b_3\mathbf{k})$$

To do so, we *assume* at first (and prove below) that the cross-product obeys the following algebraic rules:

$$c(\mathbf{v}_1 \times \mathbf{v}_2) = c\mathbf{v}_1 \times \mathbf{v}_2 = \mathbf{v}_1 \times c\mathbf{v}_2 \tag{1}$$

$$\mathbf{v}_1 \times (\mathbf{v}_2 + \mathbf{v}_3) = \mathbf{v}_1 \times \mathbf{v}_2 + \mathbf{v}_1 \times \mathbf{v}_3 \tag{2}$$

$$(\mathbf{v}_1 + \mathbf{v}_2) \times \mathbf{v}_3 = \mathbf{v}_1 \times \mathbf{v}_3 + \mathbf{v}_2 \times \mathbf{v}_3 \tag{3}$$

Using rules (1), (2), and (3) and the result of Worked Example 2, we find

$$(a_1\mathbf{i} + b_1\mathbf{j} + c_1\mathbf{k}) \times (a_2\mathbf{i} + b_2\mathbf{j} + c_2\mathbf{k})$$
$$= a_1\mathbf{i} \times (a_2\mathbf{i} + b_2\mathbf{j} + c_2\mathbf{k}) + b_1\mathbf{j} \times (a_2\mathbf{i} + b_2\mathbf{j} + c_2\mathbf{k}) + c_1\mathbf{k} \times (a_2\mathbf{i} + b_2\mathbf{j} + c_2\mathbf{k})$$
$$= a_1 b_2\mathbf{k} + a_1 c_2(-\mathbf{j}) + b_1 a_2(-\mathbf{k}) + b_1 c_2\mathbf{i} + c_1 a_2\mathbf{j} + c_1 b_2(-\mathbf{i})$$

Collecting terms, we have

$$(a_1\mathbf{i} + b_1\mathbf{j} + c_1\mathbf{k}) \times (a_2\mathbf{i} + b_2\mathbf{j} + c_2\mathbf{k})$$
$$= (b_1 c_2 - c_1 b_2)\mathbf{i} + (c_1 a_2 - a_1 c_2)\mathbf{j} + (a_1 b_2 - b_1 a_2)\mathbf{k} \tag{4}$$

Now we will show directly that the right-hand side of (4), which we will denote by $\mathbf{u}$, satisfies conditions 1 and 2 in the definition of the cross-product of $\mathbf{v}_1 \times \mathbf{v}_2$, where $\mathbf{v}_1 = a_1\mathbf{i} + b_1\mathbf{j} + c_1\mathbf{k}$ and $\mathbf{v}_2 = a_2\mathbf{i} + b_2\mathbf{j} + c_2\mathbf{k}$. This argument will *not* use rules (1), (2), and (3).

First, we consider the length of $\mathbf{u}$. Its length squared is

$$(b_1 c_2 - c_1 b_2)^2 + (c_1 a_2 - a_1 c_2)^2 + (a_1 b_2 - b_1 a_2)^2$$
$$= b_1^2 c_2^2 - 2b_1 c_1 b_2 c_2 + c_1^2 b_2^2 + c_1^2 a_2^2 - 2a_1 c_1 a_2 c_2 + a_1^2 c_2^2 + a_1^2 b_2^2 - 2a_1 b_1 a_2 b_2 + b_1^2 a_2^2$$

Now we compute the square of $\|\mathbf{v}_1\| \, \|\mathbf{v}_2\| \sin\theta$ and show that we get the same result:

$$\|\mathbf{v}_1\|^2 \|\mathbf{v}_2\|^2 \sin^2\theta = \|\mathbf{v}_1\|^2 \|\mathbf{v}_2\|^2 (1 - \cos^2\theta)$$
$$= \|\mathbf{v}_1\|^2 \|\mathbf{v}_2\|^2 - (\|\mathbf{v}_1\| \, \|\mathbf{v}_2\| \cos\theta)^2$$
$$= \|\mathbf{v}_1\|^2 \|\mathbf{v}_2\|^2 - (\mathbf{v}_1 \cdot \mathbf{v}_2)^2$$
$$= (a_1^2 + b_1^2 + c_1^2)(a_2^2 + b_2^2 + c_2^2) - (a_1 a_2 + b_1 b_2 + c_1 c_2)^2$$

which, when it is multiplied out and terms are collected, is the same as $\|\mathbf{u}\|^2$, so $\|\mathbf{u}\| = \|\mathbf{v}_1\| \, \|\mathbf{v}_2\| \sin\theta$.

Next we check that $\mathbf{u}$ is perpendicular to $\mathbf{v}_1$ and $\mathbf{v}_2$. We have

$$\begin{aligned}
\mathbf{u} \cdot \mathbf{v}_1 &= (b_1 c_2 - c_1 b_2)a_1 + (c_1 a_2 - a_1 c_2)b_1 + (a_1 b_2 - b_1 a_2)c_1 \\
&= b_1 c_2 a_1 - c_1 b_2 a_1 + c_1 a_2 b_1 - a_1 c_2 b_1 + a_1 b_2 c_1 - b_1 a_2 c_1 \\
&= 0
\end{aligned}$$

since the terms cancel in pairs. Similarly, $\mathbf{u} \cdot \mathbf{v}_2 = 0$, so $\mathbf{u}$ is perpendicular to $\mathbf{v}_1$ and $\mathbf{v}_2$.

To check the right-hand rule would require a precise mathematical definition of that rule, which we will not attempt to give. Instead, we will merely remark that the rule is satisfied for all products of standard basis vectors.

We have now shown that the vector $\mathbf{u}$ on the right-hand side of (4) satisfies the conditions in the definition of $\mathbf{v}_1 \times \mathbf{v}_2$, so it must *be* $\mathbf{v}_1 \times \mathbf{v}_2$. Note that we used rules (1), (2), and (3) only to guess the formula for $\mathbf{v}_1 \times \mathbf{v}_2$, not to prove it. Now one can go back to prove these and other rules on the basis of formula (4). (See Solved Exercise 3 and Exercise 2.)

THE CROSS-PRODUCT

Geometric definition: $\mathbf{v}_1 \times \mathbf{v}_2$ is the vector such that:

1. $\|\mathbf{v}_1 \times \mathbf{v}_2\| = \|\mathbf{v}_1\| \, \|\mathbf{v}_2\| \sin\theta$ is the area of the parallelogram spanned by $\mathbf{v}_1$ and $\mathbf{v}_2$ (θ is the angle between $\mathbf{v}_1$ and $\mathbf{v}_2$).

2. $\mathbf{v}_1 \times \mathbf{v}_2$ is perpendicular to $\mathbf{v}_1$ and $\mathbf{v}_2$, and the triple $(\mathbf{v}_1, \mathbf{v}_2, \mathbf{v}_1 \times \mathbf{v}_2)$ obeys the right-hand rule.

Component formula:

$$\begin{aligned}
&(a_1 \mathbf{i} + b_1 \mathbf{j} + c_1 \mathbf{k}) \times (a_2 \mathbf{i} + b_2 \mathbf{j} + c_2 \mathbf{k}) \\
&= (b_1 c_2 - c_1 b_2)\mathbf{i} + (c_1 a_2 - a_1 c_2)\mathbf{j} + (a_1 b_2 - b_1 a_2)\mathbf{k}
\end{aligned}$$

Algebraic rules:

1. $\mathbf{v}_1 \times \mathbf{v}_2 = 0$ if and only if $\mathbf{v}_1$ and $\mathbf{v}_2$ are parallel or $\mathbf{v}_1$ or $\mathbf{v}_2$ is zero.

2. $\mathbf{v}_1 \times \mathbf{v}_2 = -\mathbf{v}_2 \times \mathbf{v}_1$.

3. $(\mathbf{v}_1 + \mathbf{v}_2) \times \mathbf{v}_3 = \mathbf{v}_1 \times \mathbf{v}_3 + \mathbf{v}_2 \times \mathbf{v}_3$.

4. $(a\mathbf{v}_1) \times \mathbf{v}_2 = a(\mathbf{v}_1 \times \mathbf{v}_2)$.

Multiplication table:

		Second Factor		
	$\times$	$\mathbf{i}$	$\mathbf{j}$	$\mathbf{k}$
First	$\mathbf{i}$	0	$\mathbf{k}$	$-\mathbf{j}$
Factor	$\mathbf{j}$	$-\mathbf{k}$	0	$\mathbf{i}$
	$\mathbf{k}$	$\mathbf{j}$	$-\mathbf{i}$	0

Worked Example 3   Find the area of the parallelogram spanned by the vectors $\mathbf{v}_1 = \mathbf{i} + 2\mathbf{j} + 3\mathbf{k}$ and $\mathbf{v}_2 = -\mathbf{i} - \mathbf{k}$.

Solution   We calculate the cross-product of $\mathbf{v}_1$ and $\mathbf{v}_2$ by applying the component formula, with $a_1 = 1$, $b_1 = 2$, $c_1 = 3$, $a_2 = -1$, $b_2 = 0$, $c_2 = -1$.

$$\mathbf{v}_1 \times \mathbf{v}_2 = [(2)(-1) - (3)(0)]\mathbf{i} + [(3)(-1) - (1)(-1)]\mathbf{j}$$
$$+ [(1)(0) - (2)(-1)]\mathbf{k}$$
$$= -2\mathbf{i} - 2\mathbf{j} + 2\mathbf{k}$$

Thus the area is

$$\|\mathbf{v}_1 \times \mathbf{v}_2\| = \sqrt{(-2)^2 + (-2)^2 + (2)^2} = 2\sqrt{3}$$

It is often easier to use the algebraic rules and the multiplication table directly, rather than using the component formula.

Worked Example 4   Find a unit vector which is orthogonal to the vectors $\mathbf{i} + \mathbf{j}$ and $\mathbf{j} + \mathbf{k}$.

Solution   A vector perpendicular to both $\mathbf{i} + \mathbf{j}$ and $\mathbf{j} + \mathbf{k}$ is the vector

$$(\mathbf{i} + \mathbf{j}) \times (\mathbf{j} + \mathbf{k}) = \mathbf{i} \times \mathbf{j} + \mathbf{i} \times \mathbf{k} + \mathbf{j} \times \mathbf{j} + \mathbf{j} \times \mathbf{k}$$
$$= \mathbf{k} - \mathbf{j} + 0 + \mathbf{i}$$
$$= \mathbf{i} - \mathbf{j} + \mathbf{k}$$

Since $\|\mathbf{i} - \mathbf{j} + \mathbf{k}\| = \sqrt{3}$, the vector $(1/\sqrt{3})(\mathbf{i} - \mathbf{j} + \mathbf{k})$ is a unit vector perpendicular to $\mathbf{i} + \mathbf{j}$ and $\mathbf{j} + \mathbf{k}$.

Solved Exercises

1. Find $\mathbf{i} \times (\mathbf{i} \times \mathbf{j})$ and $(\mathbf{i} \times \mathbf{i}) \times \mathbf{j}$. Are they equal?

2. Use the cross-product to find the equation of the plane containing the points $(1, 1, 1)$, $(2, 0, 0)$, and $(1, 1, 0)$. (Compare Solved Exercise 13 of Section 14-2.)

3. Prove formula (2) by using the component formula (4).

Exercises

1. Compute the following cross-products:
   (a) $(\mathbf{i} + 2\mathbf{j} + 3\mathbf{k}) \times (\mathbf{i} + 3\mathbf{k})$ (b) $(\mathbf{i} + \mathbf{j} + \mathbf{k}) \times (\mathbf{i} + \mathbf{k})$
   (c) $(\mathbf{i} + \mathbf{k}) \times (\mathbf{i} + \mathbf{j} + \mathbf{k})$

2. Prove formulas (1) and (3) by using the component formula (4).

3. Find the equation of the plane through the points $(1, 2, 0)$, $(0, 1, -2)$, and $(4, 0, 1)$.

4. Find the area of the triangle whose vertices are $(0, 1, 2)$, $(3, 4, 5)$, and $(-1, -1, 0)$. [*Hint:* The triangle is half a parallelogram.]

2×2 DETERMINANTS AND MATRICES

We will approach determinants from the point of view of geometry and vectors. If $\mathbf{v}_1 = a_1\mathbf{i} + b_1\mathbf{j}$ and $\mathbf{v}_2 = a_2\mathbf{i} + b_2\mathbf{j}$ are vectors in the plane, to compute the area of the parallelogram spanned by $\mathbf{v}_1$ and $\mathbf{v}_2$ we may consider them as vectors in space (with $c_1 = c_2 = 0$) and take the cross-product $\mathbf{v}_1 \times \mathbf{v}_2 = (a_1 b_2 - b_1 a_2)\mathbf{k}$. The area is then $\|\mathbf{v}_1 \times \mathbf{v}_2\| = |a_1 b_2 - b_1 a_2|$, but the sign of $a_1 b_2 - b_1 a_2$ also gives us some information: it is positive if the sense of (shortest) rotation from $\mathbf{v}_1$ to $\mathbf{v}_2$ is counterclockwise and negative if the sense of rotation is clockwise (see Fig. 14-3-7). We may say that the sign of $a_1 b_2 - b_1 a_2$ determines the *orientation* of the ordered pair of vectors $(\mathbf{v}_1, \mathbf{v}_2)$.

(a) $a_1 b_2 - b_1 a_2 > 0$ **(b)** $a_1 b_2 - b_1 a_2 < 0$

Fig. 14-3-7 The sign of $a_1 b_2 - b_1 a_2$ determines the orientation of the ordered pair $(a_1\mathbf{i} + b_1\mathbf{j}, a_2\mathbf{i} + b_2\mathbf{j})$.

The combination $a_1 b_2 - b_1 a_2$ of the four numbers a_1, b_1, a_2, and b_2 is denoted by

$$\begin{vmatrix} a_1 & b_1 \\ a_2 & b_2 \end{vmatrix}$$

and is called the *determinant* of these four numbers.

2×2 DETERMINANTS

If a, b, c, d are any four numbers, we write

$$\begin{vmatrix} a & b \\ c & d \end{vmatrix} = ad - bc$$

The absolute value of

$$\begin{vmatrix} a & b \\ c & d \end{vmatrix}$$

equals

the area of the parallelogram spanned by $\mathbf{v} = a\mathbf{i} + b\mathbf{j}$ and $\mathbf{w} = c\mathbf{i} + d\mathbf{j}$.

The sign of

$$\begin{vmatrix} a & b \\ c & d \end{vmatrix}$$

tells

the orientation of the pair $(\mathbf{v}, \mathbf{w})$.

Worked Example 5   Show that

$$\begin{vmatrix} 1 & 2 \\ 3 & 4 \end{vmatrix} = \begin{vmatrix} 1 & 2 \\ 4 & 6 \end{vmatrix} = \begin{vmatrix} 1 & 4 \\ 2 & 6 \end{vmatrix}$$

What is their common value?

Solution $\begin{vmatrix} 1 & 2 \\ 3 & 4 \end{vmatrix} = (1)(4) - (2)(3) = -2; \quad \begin{vmatrix} 1 & 2 \\ 4 & 6 \end{vmatrix} = (1)(6) - (2)(4) = -2;$

$\begin{vmatrix} 1 & 4 \\ 2 & 6 \end{vmatrix} = (1)(6) - (4)(2) = -2.$ Their common value is -2.

Worked Example 6   Express $(a_1\mathbf{i} + b_1\mathbf{j} + c_1\mathbf{k}) \times (a_1\mathbf{i} + b_2\mathbf{j} + c_2\mathbf{k})$ in terms of 2×2 determinants.

Solution   All the components on the right-hand side of (4) are determinants. We have:

$(a_1\mathbf{i} + b_1\mathbf{j} + c_1\mathbf{k}) \times (a_2\mathbf{i} + b_2\mathbf{j} + c_2\mathbf{k})$

$= \begin{vmatrix} b_1 & c_1 \\ b_2 & c_2 \end{vmatrix}\mathbf{i} + \begin{vmatrix} c_1 & a_1 \\ c_2 & a_2 \end{vmatrix}\mathbf{j} + \begin{vmatrix} a_1 & b_1 \\ a_2 & b_2 \end{vmatrix}\mathbf{k}$

The middle term can also be written

$-\begin{vmatrix} a_1 & c_1 \\ a_2 & c_2 \end{vmatrix}\mathbf{j}$

(See Solved Exercise 5.)

Sometimes we wish to refer to the array

$a \quad b$

$c \quad d$

of numbers without taking the combination $ad - bc$. In this case, we use the notation

$\begin{bmatrix} a & b \\ c & d \end{bmatrix}$

and refer to this object as a (2×2) *matrix*. Two matrices are considered equal only when all their corresponding entries are equal; thus, in contrast to the equalities in Worked Example 5, the matrices

$\begin{bmatrix} 1 & 2 \\ 3 & 4 \end{bmatrix} \quad \begin{bmatrix} 1 & 2 \\ 4 & 6 \end{bmatrix} \quad \begin{bmatrix} 1 & 4 \\ 2 & 6 \end{bmatrix}$

are all different. The determinant

$\begin{vmatrix} a & b \\ c & d \end{vmatrix}$

is a *single* number obtained by combining the *four* numbers in the matrix

$\begin{bmatrix} a & b \\ c & d \end{bmatrix}$

Geometrically, we may think of a matrix as representing a parallelogram (that is, a geometric figure), while the determinant represents only the area of the parallelogram (that is, a number).

Solved Exercises

4. Find the determinant of

$$\begin{bmatrix} -\pi & 3 \\ \pi/2 & 6 \end{bmatrix}$$

5. Prove that

$$\begin{vmatrix} a & b \\ c & d \end{vmatrix} = - \begin{vmatrix} b & a \\ d & c \end{vmatrix}$$

(the two colums are interchanged).

Exercises

5. Evaluate each of the following determinants:

(a) $\begin{vmatrix} 1 & 1 \\ -1 & 1 \end{vmatrix}$

(b) $\begin{vmatrix} 1 & 2 \\ -1 & 0 \end{vmatrix}$

(c) $\begin{vmatrix} 6 & 5 \\ 12 & 10 \end{vmatrix}$

(d) $\begin{vmatrix} 0 & 0 \\ 3 & 17 \end{vmatrix}$

(e) $\begin{vmatrix} 1-x & -1 \\ 1 & -1-x \end{vmatrix}$

6. Show that if

$$\begin{vmatrix} a & b \\ c & d \end{vmatrix} \neq 0$$

then the solutions to the equations

$ax + by = e$

$cx + dy = f$

are given by the formulas

$$x = \frac{\begin{vmatrix} e & b \\ f & d \end{vmatrix}}{\begin{vmatrix} a & b \\ c & d \end{vmatrix}} \qquad y = \frac{\begin{vmatrix} a & e \\ c & f \end{vmatrix}}{\begin{vmatrix} a & b \\ c & d \end{vmatrix}}$$

[*Hint:* Just substitute the expressions for x and y in the equations.] This result is called *Cramer's rule.*

7. Use Cramer's rule (Exercise 6) to solve the equations

$4x + 3y = 2$

$2x - 6y = 1$

8. If

$$\begin{bmatrix} a & b \\ c & d \end{bmatrix}$$

is a matrix, its *transpose* is defined to be the matrix

$$\begin{bmatrix} a & c \\ b & d \end{bmatrix}$$

obtained by reflection across the main (upper left to lower right) diagonal. Show that the determinant of a matrix is equal to the determinant of its transpose:

$$\begin{vmatrix} a & b \\ c & d \end{vmatrix} = \begin{vmatrix} a & c \\ b & d \end{vmatrix}$$

3×3 DETERMINANTS AND MATRICES

A 3×3 matrix consists of nine numbers in a square array, such as

$$\begin{bmatrix} 1 & 4 & 6 \\ 0 & 2 & 9 \\ 3 & 1 & -5 \end{bmatrix}$$

To define the determinant

$$D = \begin{vmatrix} a_1 & b_1 & c_1 \\ a_2 & b_2 & c_2 \\ a_3 & b_3 & c_3 \end{vmatrix}$$

of a 3×3 matrix

$$\begin{bmatrix} a_1 & b_1 & c_1 \\ a_2 & b_2 & c_2 \\ a_3 & b_3 & c_3 \end{bmatrix}$$

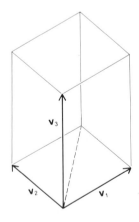

Fig. 14-3-8 The absolute value of a determinant equals the volume of the parallelepiped spanned by the rows.

we proceed by analogy with the 2×2 case. The rows of the matrix give us three vectors in space:

$$\mathbf{v}_1 = a_1\mathbf{i} + b_1\mathbf{j} + c_1\mathbf{k}$$
$$\mathbf{v}_2 = a_2\mathbf{i} + b_2\mathbf{j} + c_2\mathbf{k}$$
$$\mathbf{v}_3 = a_3\mathbf{i} + b_3\mathbf{j} + c_3\mathbf{k}$$

Since the determinant of a 2×2 matrix represents an area, we should define the determinant in such a way that its absolute value is the *volume* of the parallelepiped spanned by $\mathbf{v}_1$, $\mathbf{v}_2$, and $\mathbf{v}_3$. (See Fig. 14-3-8.)

To compute this volume in terms of the nine entries in the matrix, we drop a perpendicular PQ from the tip of $\mathbf{v}_3$ to the plane spanned by $\mathbf{v}_1$ and $\mathbf{v}_2$. It is a theorem of elementary solid geometry that the volume of the parallelepiped is equal to the length of PQ times the area of the parallelogram spanned by $\mathbf{v}_1$ and $\mathbf{v}_2$ (see Fig. 14-3-9).

Looking at the right triangle OPQ, we find that the length of $\overrightarrow{PQ}$ is $\|\mathbf{v}_3\| |\cos\theta|$, where θ is the angle between $\mathbf{v}_3$ and $\overrightarrow{PQ}$. On the other hand, PQ is perpendicular to $\mathbf{v}_1$ and $\mathbf{v}_2$, so it is parallel to $\mathbf{v}_1 \times \mathbf{v}_2$, so θ is also the angle between $\mathbf{v}_3$ and $\mathbf{v}_1 \times \mathbf{v}_2$. Now we have

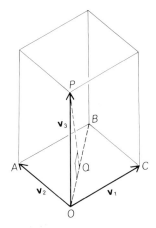

Volume = (area of base)(height)

$$= \|\mathbf{v}_1 \times \mathbf{v}_2\| \|\mathbf{v}_3\| |\cos\theta|$$

$$= \|\|\mathbf{v}_1 \times \mathbf{v}_2\| \|\mathbf{v}_3\| \cos\theta\|$$

$$= |(\mathbf{v}_1 \times \mathbf{v}_2) \cdot \mathbf{v}_3|$$

Fig. 14-3-9 The volume of the parallelepiped is its height $|PQ|$ times the area of its base.

Since the volume is to be the absolute value of the determinant, we define the determinant to be the expression inside the bars:

$$\begin{vmatrix} a_1 & b_1 & c_1 \\ a_2 & b_2 & c_2 \\ a_3 & b_3 & c_3 \end{vmatrix} = [(a_1\mathbf{i} + b_1\mathbf{j} + c_1\mathbf{k}) \times (a_2\mathbf{i} + b_2\mathbf{j} + c_2\mathbf{k})] \cdot (a_3\mathbf{i} + b_3\mathbf{j} + c_3\mathbf{k})$$

By using the component formula for the cross-product, in the form derived in Worked Example 6, we find

$$\begin{vmatrix} a_1 & b_1 & c_1 \\ a_2 & b_2 & c_2 \\ a_3 & b_3 & c_3 \end{vmatrix} = \begin{vmatrix} b_1 & c_1 \\ b_2 & c_2 \end{vmatrix} a_3 + \begin{vmatrix} c_1 & a_1 \\ c_2 & a_2 \end{vmatrix} b_3 + \begin{vmatrix} a_1 & b_1 \\ a_2 & b_2 \end{vmatrix} c_3$$

or

$$\begin{vmatrix} a_1 & b_1 & c_1 \\ a_2 & b_2 & c_2 \\ a_3 & b_3 & c_3 \end{vmatrix} = \begin{vmatrix} b_1 & c_1 \\ b_2 & c_2 \end{vmatrix} a_3 - \begin{vmatrix} a_1 & c_1 \\ a_2 & c_2 \end{vmatrix} b_3 + \begin{vmatrix} a_1 & b_1 \\ a_2 & b_2 \end{vmatrix} c_3 \tag{5}$$

We take equation (5) as the definition of the determinant of a 3×3 matrix; the volume of the parallelepiped spanned by three vectors is thus equal to the absolute value of the determinant of the matrix whose rows are the components of the vectors. (The sign of the determinant is interpreted geometrically in Solved Exercise 7.)

Worked Example 7 Evaluate the determinant

$$\begin{vmatrix} 0 & 0 & 4 \\ 2 & -1 & 6 \\ 3 & 1 & 2 \end{vmatrix}$$

Solution By equation (5),

$$\begin{vmatrix} 0 & 0 & 4 \\ 2 & -1 & 6 \\ 3 & 1 & 2 \end{vmatrix} = \begin{vmatrix} 0 & 4 \\ -1 & 6 \end{vmatrix} (3) - \begin{vmatrix} 0 & 4 \\ 2 & 6 \end{vmatrix} (1) + \begin{vmatrix} 0 & 0 \\ 2 & -1 \end{vmatrix} (2)$$

$$= (4)(3) - (-8)(1) + (0)(2)$$

$$= 12 + 8 = 20$$

We can express the cross-product of two vectors as a single 3×3 determinant. In fact, comparing equation (5) with the result of Worked Example 6, we may write

$$(a_1 \mathbf{i} + b_1 \mathbf{j} + c_1 \mathbf{k}) \times (a_2 \mathbf{i} + b_2 \mathbf{j} + c_2 \mathbf{k}) = \begin{vmatrix} a_1 & b_1 & c_1 \\ a_2 & b_2 & c_2 \\ \mathbf{i} & \mathbf{j} & \mathbf{k} \end{vmatrix}$$

Worked Example 8 Write $(\mathbf{i} + \mathbf{k}) \times (\mathbf{j} - 2\mathbf{k})$ as a determinant.

Solution

$$(\mathbf{i} + \mathbf{k}) \times (\mathbf{j} - 2\mathbf{k}) = \begin{vmatrix} 1 & 0 & 1 \\ 0 & 1 & -2 \\ \mathbf{i} & \mathbf{j} & \mathbf{k} \end{vmatrix}$$

Formula (5) is worth memorizing. To do so, notice that the ith entry in the third row is multiplied by the determinant obtained by crossing out the ith column and third row of the original matrix. Such a 2×2 determinant is called a *minor*, and formula (5) is called the *expansion by minors of the third row*.

It turns out that a determinant can be evaluated by expanding in minors of any row or column. To do this, multiply each entry in a given row or column by the 2×2 determinant obtained by crossing out the row and column of the given entry. Signs are assigned to the products according to the checkerboard pattern:

$$\begin{vmatrix} + & - & + \\ - & + & - \\ + & - & + \end{vmatrix}$$

(Remember the plus sign in the upper left-hand corner, and you can always reconstruct this pattern.)

Worked Example 9 Evaluate the determinant

$$\begin{vmatrix} 0 & 0 & 4 \\ 2 & -1 & 6 \\ 3 & 1 & 2 \end{vmatrix}$$

of Worked Example 7 by expanding in minors of the first row.

Solution
$$\begin{vmatrix} 0 & 0 & 4 \\ 2 & -1 & 6 \\ 3 & 1 & 2 \end{vmatrix} = \begin{vmatrix} -1 & 6 \\ 1 & 2 \end{vmatrix}(0) - \begin{vmatrix} 2 & 6 \\ 3 & 2 \end{vmatrix}(0) + \begin{vmatrix} 2 & -1 \\ 3 & 1 \end{vmatrix}4$$
$$= 0 - 0 + (5)(4) = 20$$

Since we do not need to evaluate the minors of the zero entries, the expansion by minors of the first row results in a simpler calculation (for this particular matrix) than the expansion by minors of the third row.

To prove that the expansions by minors of rows or columns all give the same result, it is sufficient to compare these expansions for the "general" 3×3 determinant. (See Solved Exercise 8 and Problem 24.)

3×3 DETERMINANTS

$$\begin{vmatrix} a_1 & b_1 & c_1 \\ a_2 & b_2 & c_2 \\ a_3 & b_3 & c_3 \end{vmatrix} = \begin{vmatrix} b_1 & c_1 \\ b_2 & c_2 \end{vmatrix}a_3 - \begin{vmatrix} a_1 & c_1 \\ a_2 & c_2 \end{vmatrix}b_3 + \begin{vmatrix} a_1 & b_1 \\ a_2 & b_2 \end{vmatrix}c_3$$

The determinant can be expanded by minors of any row or column.

If $\mathbf{v}_1 = a_1\mathbf{i} + b_1\mathbf{j} + c_1\mathbf{k}$

$\mathbf{v}_2 = a_2\mathbf{i} + b_2\mathbf{j} + c_2\mathbf{k}$

$\mathbf{v}_3 = a_3\mathbf{i} + b_3\mathbf{j} + c_3\mathbf{k}$

then the determinant equals

$(\mathbf{v}_1 \times \mathbf{v}_2) \cdot \mathbf{v}_3$, which is also called the *triple product* of $\mathbf{v}_1$, $\mathbf{v}_2$, and $\mathbf{v}_3$.

The absolute value of the determinant equals

the volume of the parallelepiped spanned by $\mathbf{v}_1$, $\mathbf{v}_2$, and $\mathbf{v}_3$.

The sign of the determinant tells

whether the triple $(\mathbf{v}_1, \mathbf{v}_2, \mathbf{v}_3)$ is right- or left-handed.

Solved Exercises

6. Find the volume of the parallelepiped spanned by the following vectors: $\mathbf{i} + 3\mathbf{k}$, $2\mathbf{i} + \mathbf{j} - 2\mathbf{k}$, and $5\mathbf{i} + 4\mathbf{k}$.

7. Let $\mathbf{v}_1 = a_1\mathbf{i} + b_1\mathbf{j} + c_1\mathbf{k}$, $\mathbf{v}_2 = a_2\mathbf{i} + b_2\mathbf{j} + c_2\mathbf{k}$, and $\mathbf{v}_3 = a_3\mathbf{i} + b_3\mathbf{j} + c_3\mathbf{k}$. Show that the triple $(\mathbf{v}_1, \mathbf{v}_2, \mathbf{v}_3)$ is right-handed (left-handed) if the determinant

$$\begin{vmatrix} a_1 & b_1 & c_1 \\ a_2 & b_2 & c_2 \\ a_3 & b_3 & c_3 \end{vmatrix}$$

is positive (negative). What does it mean if the determinant is zero?

8. Show that expanding the determinant

$$\begin{vmatrix} a_1 & b_1 & c_1 \\ a_2 & b_2 & c_2 \\ a_3 & b_3 & c_3 \end{vmatrix}$$

by minors of the second column gives the correct result.

Exercises

9. Evaluate the following determinants:

(a) $\begin{vmatrix} -1 & 0 & 1 \\ 2 & 1 & 3 \\ 0 & 1 & 2 \end{vmatrix}$

(b) $\begin{vmatrix} 2 & 1 & 0 \\ 0 & 2 & 1 \\ 1 & 0 & 2 \end{vmatrix}$

(c) $\begin{vmatrix} a & d & e \\ 0 & b & f \\ 0 & 0 & c \end{vmatrix}$

(d) $\begin{vmatrix} -1 & 0 & 1 \\ 2 & 1 & 3 \\ 3 & 1 & 2 \end{vmatrix}$

10. Find the volume of the parallelepiped spanned by the vectors $\mathbf{i} + \mathbf{j} + \mathbf{k}$, $\mathbf{i} - \mathbf{j} + \mathbf{k}$, and $3\mathbf{k}$.

11. Suppose that the determinant

$$D = \begin{vmatrix} a_1 & b_1 & c_1 \\ a_2 & b_2 & c_2 \\ a_3 & b_3 & c_3 \end{vmatrix}$$

is unequal to zero. Show that the solution of the equations

$$a_1 x + b_1 y + c_1 z = d_1$$
$$a_2 x + b_2 y + c_2 z = d_2$$
$$a_3 x + b_3 y + c_3 z = d_3$$

is given by the formulas

$$x = \frac{\begin{vmatrix} d_1 & b_1 & c_1 \\ d_2 & b_2 & c_2 \\ d_3 & b_3 & c_3 \end{vmatrix}}{D} \qquad y = \frac{\begin{vmatrix} a_1 & d_1 & c_1 \\ a_2 & d_2 & c_2 \\ a_3 & d_3 & c_3 \end{vmatrix}}{D} \qquad z = \frac{\begin{vmatrix} a_1 & b_1 & d_1 \\ a_2 & b_2 & d_2 \\ a_3 & b_3 & d_3 \end{vmatrix}}{D}$$

This result is called Cramer's rule for 3×3 systems.* (See Exercise 6.)

12. Use Cramer's rule (Exercise 11) to solve

$$-x + y \qquad = 14$$
$$2x + y + z = 8$$
$$x + y + 5z = -1$$

13. Check that

$$\begin{vmatrix} a & d & g \\ b & e & h \\ c & f & i \end{vmatrix} = \begin{vmatrix} a & b & c \\ d & e & f \\ g & h & i \end{vmatrix}$$

That is, check that the determinant of the transpose of a 3×3 matrix is equal to the determinant of the original matrix.

---

*Gabriel Cramer (1704–1752) published this rule in his book, Introduction à l'analyse des lignes courbes algébriques (1750). However, it was probably known to Maclaurin in 1729. For systems of n equations in n unknowns, there is a generalization of this rule, but it is inefficient to compute with when n is large.

Problems for Section 14-3

1. Compute the following cross-products:
   (a) $(\mathbf{i} - \mathbf{j} + \mathbf{k}) \times (\mathbf{j} - \mathbf{k})$
   (b) $\mathbf{i} \times (\mathbf{j} - \mathbf{k})$
   (c) $(\mathbf{i} + \mathbf{j}) \times [(\mathbf{k} - \mathbf{j}) + (3\mathbf{j} - 2\mathbf{i} + \mathbf{k})]$

2. Compute the following cross-products:
   (a) $(a\mathbf{i} + \mathbf{j} - \mathbf{k}) \times \mathbf{i}$
   (b) $[(3\mathbf{i} + 2\mathbf{j}) \times 3\mathbf{j}] \times (2\mathbf{i} - \mathbf{j} + \mathbf{k})$
   (c) $(\mathbf{i} \times \mathbf{j}) \times (\mathbf{i} + \mathbf{j} + \mathbf{k})$

3. Compute the following determinants:

   (a) $\begin{vmatrix} 1 & 2 \\ 3 & 4 \end{vmatrix}$
   (b) $\begin{vmatrix} 1 & 2 \\ 2 & 4 \end{vmatrix}$

   (c) $\begin{vmatrix} a & b \\ 0 & c \end{vmatrix}$
   (d) $\begin{vmatrix} 1 & 1 & 2 \\ 2 & -1 & 1 \\ 1 & 0 & 0 \end{vmatrix}$

   (e) $\begin{vmatrix} 1 & 2 & 3 \\ -1 & -1 & 2 \\ 0 & 1 & -1 \end{vmatrix}$

4. Compute the following determinants:

   (a) $\begin{vmatrix} 4 & 3 \\ 1 & 7 \end{vmatrix}$
   (b) $\begin{vmatrix} a & b \\ -b & a \end{vmatrix}$

   (c) $\begin{vmatrix} 1 & 0 & 1 \\ 0 & 1 & 0 \\ 1 & 0 & 1 \end{vmatrix}$
   (d) $\begin{vmatrix} 1 & 1 & 0 \\ 0 & 1 & 1 \\ 0 & 1 & 1 \end{vmatrix}$

   (e) $\begin{vmatrix} 2 & -1 & 0 \\ 4 & 3 & 2 \\ 3 & 0 & 1 \end{vmatrix}$

5. By using the cross-product of the vectors $\cos\theta\mathbf{i} + \sin\theta\mathbf{j}$ and $\cos\psi\mathbf{i} + \sin\psi\mathbf{j}$, show that $\sin(\theta - \psi) = \sin\theta\cos\psi - \cos\theta\sin\psi$

6. Show that the plane through the points $A = (a_1, a_2, a_3)$, $B = (b_1, b_2, b_3)$, and $C = (c_1, c_2, c_3)$ consists of the points $P = (x, y, z)$ given by

   $$\begin{vmatrix} a_1 - x & a_2 - y & a_3 - z \\ b_1 - x & b_2 - y & b_3 - z \\ c_1 - x & c_2 - y & c_3 - z \end{vmatrix} = 0$$

   [*Hint:* Try writing the determinant as a triple product.]

7. Find a unit vector perpendicular to $\mathbf{i} + \mathbf{j}$ and to $\mathbf{i} - \mathbf{j} - \frac{1}{2}\mathbf{k}$ and with a positive $\mathbf{k}$ coordinate.

8. Find the volume of the parallelepiped spanned by $\mathbf{i} - \mathbf{j}$, $\mathbf{j} - \mathbf{k}$, and $\mathbf{k} + \mathbf{i}$.

9. Find the volume of the parallelepiped with one vertex at $(1, 1, 2)$ and three adjacent vertices at $(2, 0, 2)$, $(3, 1, 3)$, and $(2, 2, -3)$.

10. Find the equation of the plane passing through the points $(0, 0, 0)$, $(2, 0, -1)$, and $(0, 4, -3)$.

11. Prove the following:
    (a) *The Jacobi identity:*

    $(\mathbf{u} \times \mathbf{v}) \times \mathbf{w} + (\mathbf{v} \times \mathbf{w}) \times \mathbf{u} + (\mathbf{w} \times \mathbf{u}) \times \mathbf{v} = \mathbf{0}$
    [*Hint:* Express $\mathbf{u}$, $\mathbf{v}$, and $\mathbf{w}$ in terms of $\mathbf{i}$, $\mathbf{j}$, and $\mathbf{k}$.]

    (b) $(\mathbf{u} \times \mathbf{v}) \cdot (\mathbf{a} \times \mathbf{b}) = \begin{vmatrix} \mathbf{u} \cdot \mathbf{a} & \mathbf{u} \cdot \mathbf{b} \\ \mathbf{v} \cdot \mathbf{a} & \mathbf{v} \cdot \mathbf{b} \end{vmatrix}$.

12. Let $\mathscr{L}$ be a line through a point P_0 in direction $\mathbf{d}$. Show that the distance from a point P to $\mathscr{L}$ is given by $\|(\overrightarrow{P_0 P}) \times \mathbf{d}\| / \|\mathbf{d}\|$.

13. Let a line in the plane be given by the equation $ax + by = c$. Show that the distance from a point $P = (x, y)$ to this line is given by

    $$\frac{|ax + by - c|}{\sqrt{a^2 + b^2}}$$

14. Show that if the two rows of a 2×2 matrix are interchanged, the determinant changes sign.

15. Show that if any two rows of a 3×3 matrix are interchanged, the determinant changes sign. [*Hint:* Expand by minors.]

16. Show that if any two columns of a matrix are interchanged, the determinant changes sign. [*Hint:* Do it directly, or use Problems 14 and 15 and Exercises 8 and 13.]

17. (a) Using the result of Problem 15, prove the identity

    $$(\mathbf{v}_1 \times \mathbf{v}_2) \cdot \mathbf{v}_3 = (\mathbf{v}_3 \times \mathbf{v}_1) \cdot \mathbf{v}_2$$

    (b) Show that if $(\mathbf{v}_1, \mathbf{v}_2, \mathbf{v}_3)$ is a right-handed triple, then so is $(\mathbf{v}_2, \mathbf{v}_3, \mathbf{v}_1)$.

18. Show that if two rows or columns of a matrix are equal, then the determinant of the matrix is zero.

19. Show that adding a multiple of the first row of a matrix to the second row leaves the determinant unchanged; that is,

    $$\begin{vmatrix} a_1 & b_1 & c_1 \\ a_2 + \lambda a_1 & b_2 + \lambda b_1 & c_2 + \lambda c_1 \\ a_3 & b_3 & c_3 \end{vmatrix} = \begin{vmatrix} a_1 & b_1 & c_1 \\ a_2 & b_2 & c_2 \\ a_3 & b_3 & c_3 \end{vmatrix}$$

20. Show that adding a multiple of any row (column) of a matrix to another row (column) leaves the determinant unchanged. [*Hint:* Use the result of Problem 19.]

21. (a) Justify the steps in the following computation:

$$\begin{vmatrix} 1 & 2 & 3 \\ 4 & 5 & 6 \\ 7 & 8 & 10 \end{vmatrix} = \begin{vmatrix} 1 & 2 & 3 \\ 0 & -3 & -6 \\ 7 & 8 & 10 \end{vmatrix}$$

$$= \begin{vmatrix} 1 & 2 & 3 \\ 0 & -3 & -6 \\ 0 & -6 & -11 \end{vmatrix} = \begin{vmatrix} -3 & -6 \\ -6 & -11 \end{vmatrix} = 33 - 36 = -3$$

[*Hint:* See Problem 20.]

(b) Follow the technique of part (a) to evaluate the determinant

$$\begin{vmatrix} 1 & 2 & 5 \\ 2 & 3 & 6 \\ -1 & 2 & 1 \end{vmatrix}$$

(Add -2 times the first row to the second row, then add the first row to the third row.)

(c) Use the technique in part (a) to evaluate the 3×3 determinants in Problems 3 and 4.

22. Use Cramer's rule (see Exercises 6 and 11) to solve the systems:

   (a) $2x + 3y = 5$
   $3x - 2y = 9$

   (b) $x + y + z = 3$
   $x - y + z = 4$
   $x + y - z = 5$

23. Show that $\|\mathbf{u} \times \mathbf{v}\|^2 = \|\mathbf{u}\|^2 \|\mathbf{v}\|^2 - (\mathbf{u} \cdot \mathbf{v})^2$ in two ways:

   (a) By using the component formulas for the cross and dot products.

   (b) By using the geometric definitions of the cross and dot products, together with a trigonometric identity.

24. Check that expanding the determinant

$$\begin{vmatrix} a_1 & b_1 & c_1 \\ a_2 & b_2 & c_2 \\ a_3 & b_3 & c_3 \end{vmatrix}$$

by (a) minors of the second row and (b) minors of the third column gives the correct result.

25. Two media of indices of refraction n_1 and n_2 are separated by a plane surface perpendicular to the unit vector $\mathbf{N}$. Let $\mathbf{a}$ and $\mathbf{b}$ be unit vectors along the incident and refracted rays, respectively, their directions being those of the light rays. Show that $n_1(\mathbf{N} \times \mathbf{a}) = n_2(\mathbf{N} \times \mathbf{b})$. (You may use *Snell's law* $\sin \theta_1 / \sin \theta_2 = n_2/n_1$, where θ_1 and θ_2 are the angles of incidence and refraction, respectively.)

26. The angular velocity $\mathbf{w}$ of rotation of a rigid body has direction equal to the axis of rotation and magnitude equal to the scalar angular velocity. The sense of $\mathbf{w}$ is determined by the right-hand rule.

   (a) Let $\mathbf{r}$ be a vector from the axis to a point P on the rigid body. Show that the quantity $\mathbf{v} = \mathbf{w} \times \mathbf{r}$ is the velocity vector of P.

   (b) Interpret the result for the rotation of a carousel about its axis, with P a point on the circumference.

27. In mechanics, *the moment M of a force* $\mathbf{F}$ *about a point* 0 is defined to be the magnitude of $\mathbf{F}$ times the perpendicular distance from 0 to the line of action of $\mathbf{F}$. The *vector moment* $\mathbf{M}$ is the vector of magnitude M whose direction is perpendicular to the plane of 0 and $\mathbf{F}$, determined by the right-hand rule. Show that $\mathbf{M} = \mathbf{R} \times \mathbf{F}$, where $\mathbf{R}$ is any vector from 0 to the line of action of $\mathbf{F}$.

Review Problems for Chapter 14

1. Carry out the following calculations:

   (a) $(1, 2, 3) + 2(-1, -2, 7)$
   (b) $2[(-1, 0, 1) + (6, 0, 2)] - (0, 0, 1)$
   (c) $(8\mathbf{i} + 3\mathbf{j} - \mathbf{k}) - 6(\mathbf{i} - \mathbf{j} - \mathbf{k})$
   (d) $(8\mathbf{i} + 3\mathbf{j} - \mathbf{k}) \times (\mathbf{i} - \mathbf{j} - \mathbf{k})$
   (e) $(8\mathbf{i} + 3\mathbf{j} - \mathbf{k}) \cdot (\mathbf{i} - \mathbf{j} - \mathbf{k})$

2. Find a unit vector which is:

   (a) Orthogonal to the plane $x - 6y + z = 12$.
   (b) Parallel to the line
   $$x = 3t + 1, \quad y = 16t - 2, \quad z = -(t + 2)$$
   (c) Orthogonal to $\mathbf{i} + 2\mathbf{j} - \mathbf{k}$ and $\mathbf{k}$.
   (d) Parallel to both the planes
   $$8x + y + z = 1 \quad \text{and} \quad x - y - z = 0$$
   (e) At an angle of $30°$ to $\mathbf{i}$ and makes equal angles with $\mathbf{j}$ and $\mathbf{k}$.

3. Write an equation, or set of equations, to describe each of the following sets:

   (a) The x axis in space.

   (b) The sphere of radius 2 with center at the origin.

   (c) The line through $(1, 1, 2)$ and $(2, 2, 3)$.

   (d) The plane through $(1, 1, 2)$, $(2, 2, 3)$, and $(0, 0, 0)$.

   (e) The line perpendicular to the plane in part (d) and passing through $(0, 0, 3)$.

4. Suppose that $\mathbf{v}$ and $\mathbf{w}$ each are parallel to the (x, y) plane. What can you say about $\mathbf{v} \times \mathbf{w}$?

5. Suppose that $\mathbf{u}$, $\mathbf{v}$, and $\mathbf{w}$ are three vectors. Explain how to find the angle between $\mathbf{w}$ and the plane determined by $\mathbf{u}$ and $\mathbf{v}$.

6. Describe the set of all lines through the origin in space which make an angle of $\pi/3$ with the x axis.

7. Find the intersection of the medians of the triangle with vertices at $(0, 0)$, $(1, \frac{1}{2})$, and $(2, 0)$.

★8. (a) Suppose that $\mathbf{v} \cdot \mathbf{w} = 0$ for all vectors $\mathbf{w}$. Show that $\mathbf{v} = \mathbf{0}$. [*Note:* This is not the same thing as showing that $\mathbf{0} \cdot \mathbf{w} = 0$.]

   (b) Suppose that $\mathbf{u} \cdot \mathbf{w} = \mathbf{v} \cdot \mathbf{w}$ for all vectors $\mathbf{w}$. Show that $\mathbf{u} = \mathbf{v}$.

   (c) Suppose that $\mathbf{v} \cdot \mathbf{i} = \mathbf{v} \cdot \mathbf{j} = \mathbf{v} \cdot \mathbf{k} = 0$. Show that $\mathbf{v} = \mathbf{0}$.

   (d) Suppose that $\mathbf{u} \cdot \mathbf{i} = \mathbf{v} \cdot \mathbf{i}$; $\mathbf{u} \cdot \mathbf{j} = \mathbf{v} \cdot \mathbf{j}$; and $\mathbf{u} \cdot \mathbf{k} = \mathbf{v} \cdot \mathbf{k}$. Show that $\mathbf{u} = \mathbf{v}$.

9. Find the equation of the plane through $(1, 2, -1)$ which is parallel to both $\mathbf{i} - \mathbf{j} + 2\mathbf{k}$ and $\mathbf{i} - 3\mathbf{k}$.

10. (a) Solve by using determinants:

    $$x - y + 2z = 4$$
    $$3x + y + z = 1$$
    $$4x - y - z = 2$$

    (b) Use Problem 20, p. 633, to show that

    $$\begin{vmatrix} 66 & 628 & 246 \\ 88 & 435 & 24 \\ 2 & -1 & 1 \end{vmatrix} = \begin{vmatrix} 68 & 627 & 247 \\ 86 & 436 & 23 \\ 2 & -1 & 1 \end{vmatrix}$$

    (c) Evaluate $\begin{vmatrix} 6 & 2 & -3 \\ 2 & 2 & 3 \\ 4 & 8 & -1 \end{vmatrix}$

    (d) Find the area of the parallelogram spanned by $3\mathbf{i} - 2\mathbf{j} + \mathbf{k}$ and $8\mathbf{i} - \mathbf{k}$.

    (e) Find the volume of the parallelepiped spanned by $\mathbf{i} - \mathbf{j} - \mathbf{k}$, $2\mathbf{i} + \mathbf{j} - 5\mathbf{k}$, and $\frac{8}{3}\mathbf{i} - \mathbf{j} + \frac{1}{2}\mathbf{k}$.

11. Let PQR be a triangle in the plane. For each side of the triangle, construct the vector perpendicular to that side, pointing into the triangle, and having the same length as the side. Prove that the sum of the three vectors is zero.

12. Let A, B, C, D be four points in space. The solid figure bounded by the triangles $\triangle_1 = BCD$, $\triangle_2 = ACD$, $\triangle_3 = ABD$, and $\triangle_4 = ABC$ is called a *tetrahedron*. The triangle $\triangle_i$ is called the *ith face of the tetrahedron*. For each i, there is a unique vector $\mathbf{v}_i$ defined as follows: $\mathbf{v}_i$ is perpendicular to the face $\triangle_i$ and points *into* the tetrahedron; the length of $\mathbf{v}_i$ is equal to the area of $\triangle_i$.

    (a) Prove that for any tetrahedron $ABCD$, the sum $\mathbf{v}_1 + \mathbf{v}_2 + \mathbf{v}_3 + \mathbf{v}_4$ is zero. [*Hint:* Use algebraic properties of the cross-product.]

   ★(b) Try to generalize the result of part (a) to more complicated polyhedra. (A polyhedron is a solid which is bounded by planar figures.) In other words, show that the sum of the inward normals, with lengths equal to areas of the sides, is zero. You may want to do some numerical calculations if you can't prove anything. You may want to restrict yourself to a special class of figures (distorted cubes, decapitated tetrahedra, figures with five vertices, and so forth).

    (c) There is a physical interpretation to the results in parts (a) and (b). If the polyhedron is immersed in a fluid under *constant* pressure p, then the force acting on the ith face is $p\mathbf{v}_i$. Interpret the result $\Sigma \mathbf{v}_i = 0$ in this context. Does this contradict the fact that water pressure tends to buoy up an immersed object? [*Note:* There is a version of all this material for smooth surfaces. It is related to a result called the *divergence theorem* and involves partial differentiation and surface integrals. See Chapter 18.]

13. Let $P = (1, 2)$ and $Q = (2, 1)$. Describe geometrically and sketch the set of points in the plane of the form $rP + sQ$, where r and s are:

    (a) Positive integers   (b) Integers
    (c) Positive real numbers   (d) Real numbers

14. Repeat Problem 13 with $P = (1, 2)$ and $Q = (2, 4)$.

★15. Repeat Problem 13 with $P = (1, 2)$ and $Q = (\pi, 2\pi)$. (You may have to guess parts (a) and (b).)

★16. A *regular tetrahedron* is a solid bounded by four equilateral triangles. Use vector methods to find the angle between the planes containing two of the faces.

17. An airplane flying in a straight line at 500 miles per hour for 12 minutes moves 35 miles north and 93.65 miles east. How much does its altitude change? Can you determine whether the airplane is climbing or descending? (Ignore the curvature of the earth.)

18. Consider the set of all points P in space such that the vector from O to P has length 2 and makes an angle of 45° with $\mathbf{i} + \mathbf{j}$.

    (a) What kind of geometric object is this set?

    (b) Describe this set by a pair of equations in x, y, and z.

★19. There are two unit vectors such that if they were drawn on the axes of Fig. 14-R-1, their heads and tails would appear to be at the same point (that is, they would be viewed head on). Approximately what are these vectors? [*Hint:* Suppose that when you tried to plot a point P, the resulting dot on the paper fell right where the axes cross.]

Fig. 14-R-1 Which unit vectors would be drawn as the dot at the origin?

20. Use Problem 20, p. 633, to show that

$$\begin{vmatrix} n & n+1 & n+2 \\ n+3 & n+4 & n+5 \\ n+6 & n+7 & n+8 \end{vmatrix}$$

has the same value no matter what n is. What is this value?

★21. Suppose that the three vectors $a_i\mathbf{i} + b_i\mathbf{j} + c_i\mathbf{k}$ for $i = 1$, 2, and 3 are unit vectors, each orthogonal to the other two. Find the value of

$$\begin{vmatrix} a_1 & b_1 & c_1 \\ a_2 & b_2 & c_2 \\ a_3 & b_3 & c_3 \end{vmatrix}$$

22. A sphere of radius 10 cm with center at $(0, 0, 0)$ rotates about the z-axis with angular velocity 4 radians/sec such that the rotation looks clockwise from the positive z-axis.

    (a) Find the rotation vector $\mathbf{w}$ (see Section 14-3, Problem 26).

    (b) Find the velocity $\mathbf{v} = \mathbf{w} \times \mathbf{r}$ when $\mathbf{r} = 5\sqrt{2} (\mathbf{i} - \mathbf{j})$ is on the "equator."

23. Let a triangle have adjacent sides $\mathbf{a}$ and $\mathbf{b}$.

    (a) Show that $\mathbf{c} = \mathbf{b} - \mathbf{a}$ is the third side.

    (b) Show that $\mathbf{c} \times \mathbf{a} = \mathbf{c} \times \mathbf{b}$.

    (c) Derive the law of sines.

24. The work w done in moving an object from $(0, 0)$ to $(7, 2)$ subject to a force $\mathbf{F}$ is $w = \mathbf{F} \cdot \mathbf{r}$ where $\mathbf{r}$ is the vector with head at $(7, 2)$ and tail at $(0, 0)$. The units are feet and pounds.

    (a) Suppose the force $\mathbf{F} = 10 \cos \theta \, \mathbf{i} + 10 \sin \theta \, \mathbf{j}$. Find w in terms of θ.

    (b) Suppose the force $\mathbf{F}$ has magnitude 6 lb and makes an angle of $\pi/6$ radian with the horizontal, pointing right. Find w in ft-lbs.

25. The volume of a *tetrahedron* with concurrent edges $\mathbf{a}$, $\mathbf{b}$, $\mathbf{c}$ is given by $V = (1/6) \, \mathbf{a} \cdot (\mathbf{b} \times \mathbf{c})$.

    (a) Express the volume as a determinant.

    (b) Evaluate V when $\mathbf{a} = \mathbf{r} + \mathbf{j} + \mathbf{k}$, $\mathbf{b} = \mathbf{i} - \mathbf{j} + \mathbf{k}$, $\mathbf{c} = \mathbf{i} + \mathbf{j}$.

26. A tetrahedron sits in xyz-coordinates with one vertex at $(0, 0, 0)$, and the three edges concurrent at $(0, 0, 0)$ are coincident with the vectors $\mathbf{a}$, $\mathbf{b}$, $\mathbf{c}$.

    (a) Draw a figure.

    (b) Label the heads of the vectors as $\mathbf{a}$, $\mathbf{b}$, $\mathbf{c}$. Find the center of mass of each of the four triangular faces of the tetrahedron.

    (c) Find a formula for the center of mass of the tetrahedron.

27. Let $\mathbf{r}_1, \ldots, \mathbf{r}_n$ be vectors from 0 to the masses $m_1, \ldots, m_n$. The *center of mass* is the vector

$$\mathbf{c} = \left(\sum_{i=1}^{n} m_i \mathbf{r}_i \right) \Big/ \left(\sum_{i=1}^{n} m_i \right)$$

    (a) Show that for any vector $\mathbf{r}$,

$$\sum_{i=1}^{n} m_i \|\mathbf{r} - \mathbf{r}_i\|^2 = \sum_{i=1}^{n} m_i \|\mathbf{r}_i - \mathbf{c}\|^2 + m\|\mathbf{r} - \mathbf{c}\|^2$$

    where $m = \sum_{i=1}^{n} m_i$ is the total mass of the system. [*Hint:* $\|\mathbf{a}\|^2 = \mathbf{a} \cdot \mathbf{a}$, for any vector $\mathbf{a}$.]

    (b) The sum in (a) is a *weighted moment* (mass times square of the lever arm). The formula says that the weighted moment can be computed by considering the head of $\mathbf{c}$ to be the origin. Explain this in the case when there are three masses along a line, and $\mathbf{r}$ has its head on this line.

28. A pair of dipoles are located at a distance r from each other. The magnetic potential energy P is given by $P = -\mathbf{m}_1 \cdot \mathbf{B}_2$ *(dipole-dipole interaction potential)* where the first dipole has moment $\mathbf{m}_1$ in the external gradient field $\mathbf{B}_2$ of the second dipole. In MKS units,

$$\mathbf{B}_2 = \mu_0 \frac{-\mathbf{m}_2 + 3(\mathbf{m}_2 \cdot \mathbf{l}_r)\mathbf{l}_r}{4\pi r^3}$$

where $\mathbf{l}_r$ is a unit vector, and μ_0 is a scalar constant.

(a) Show that

$$P = \mu_0 \frac{\mathbf{m}_1 \cdot \mathbf{m}_2 - 3(\mathbf{m}_2 \cdot \mathbf{l}_r)(\mathbf{m}_1 \cdot \mathbf{l}_r)}{4\pi r^3}$$

(b) Find P when $\mathbf{m}_1$ and $\mathbf{m}_2$ are perpendicular.

15

Surfaces and Curves in Space

Some three-dimensional geometry is a prerequisite for understanding the differential calculus of functions of two variables.

One-variable calculus builds on a knowledge of graphs of functions of one variable and analytic geometry in the plane. Likewise, two-variable calculus builds on graphs of functions of two variables and analytic geometry in space.

CONTENTS

Sections 15-1 and 15-2 are parallel to Sections R-3 and R-4; the latter should be reviewed at this time. Section 15-1 is the most important prerequisite for proceeding to Chapter 16. Space curves may be regarded as vector functions of one variable; since only one-variable calculus is needed for their description, we present this material in Section 15-3, before the calculus of several variables. As an application of spatial geometry and vectors, the derivation of the sunshine formula (used in Sections 6-3 and 10-3) is given in the appendix.

---

*This is found in Appendix B.

15-1 Functions of Several Variables, Graphs, and Level Surfaces

Functions of two and three variables have various graphic representations.

The graph of a function $z = f(x, y)$ of two variables is a surface in (x, y, z) space. The method of level sets allows us to represent $z = f(x, y)$ by a family of curves in the xy plane, as well as to represent a function $w = f(x, y, z)$ by a family of surfaces in space.

Goals

After studying this section, you should be able to:

Evaluate functions of two and three variables.

Sketch graphs of simple functions of two variables.

Sketch level curves and level surfaces.

FUNCTIONS OF SEVERAL VARIABLES

On the daily weather map of North America (see Fig. 15-1-1) are shown the temperatures, at a fixed time, of various locations throughout the continent. If we let x be the longitude and y the latitude of a point, the temperature T at that point may be thought of as a function of the *pair* (x, y). Weather maps often contain curves through points with the same temperature. These curves, called *isotherms*, help us to visualize the temperature function; for instance, they help us to locate a hot spot in the southwestern U.S. and a cold spot in Canada.

Fig. 15-1-1 *Isotherms* are lines of constant temperature (in degrees Celsius).

Functions of two variables arise in many other contexts as well. For instance, in topography the height h of the land depends on the two coordinates that give the location. The reaction rate σ of two chemicals A and B depends on their concentrations a and b. The altitude α of the sun in the sky on June 21 depends on the latitude l and the number of hours t after midnight.

Many quantities depend on more than two variables. For instance, the temperature can be regarded as a function of the time t as well as of x and y to give a function of three variables. (Try to imagine visualizing this function by watching the isotherms move and wiggle as the day progresses.) The rate of a reaction involving 10 chemicals is a function of 10 variables.

In this book we limit our attention to functions of two and three variables. Readers who have mastered this material can construct for themselves, or find in a more advanced work,* the generalizations of the concepts presented here to functions of four and more variables.

The mathematical development of functions of several variables begins with some definitions.

Definition A *function of two variables* is a rule which assigns a number $f(x, y)$ to each point (x, y) of a domain in the xy plane.

Worked Example 1 Describe the domain of $f(x, y) = x/(x^2 + y^2)$. Evaluate $f(1, 0)$ and $f(1, 1)$.

Solution As given, this function is defined as long as $x^2 + y^2 \neq 0$; that is, as long as $(x, y) \neq (0, 0)$. We have

$$f(1, 0) = \frac{1}{1^2 + 0^2} = 1 \quad \text{and} \quad f(1, 1) = \frac{1}{1^2 + 1^2} = \frac{1}{2}$$

Definition The *graph* of a function $f(x, y)$ of two variables consists of all points (x, y, z) in space such that (x, y) is in the domain of the function and

$$z = f(x, y)$$

Some particularly simple graphs can be drawn on the basis of our work in earlier chapters.

Worked Example 2 Sketch the graph of $f(x, y) = x - y + 2$.

Solution We recognize $z = x - y + 2$ (that is, $x - y - z + 2 = 0$) as the equation of a plane. Its normal is $(1, -1, -1)$ and it meets the axes at $(-2, 0, 0)$, $(0, 2, 0)$, $(0, 0, 2)$. From this information we sketch its graph in Fig. 15-1-2 at the top of the next page.

---

*See, for example, J. Marsden and A. Tromba, Vector Calculus, *Second Edition*, W. H. Freeman and Co., 1980.

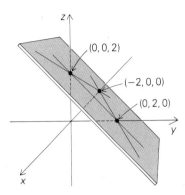

Fig. 15-1-2 The graph of $z = x - y + 2$ is a plane.

Definition Let f be a function of two variables and let c be a constant. The set of all (x, y) in the plane such that $f(x, y) = c$ is called a *level curve* of f (with value c).

Isotherms are just the level curves of a temperature function.

Worked Example 3 Sketch the level curves with values -1, 0, 1 for $f(x, y) = x - y + 2$.

Solution The level curve with value -1 is obtained by setting $f(x, y) = -1$; that is,

$$x - y + 2 = -1 \qquad \text{that is,} \qquad x - y + 3 = 0$$

which is a straight line in the plane (see Fig. 15-1-3). The level curve with value zero is the line

$$x - y + 2 = 0$$

and the curve with value 1 is the line

$$x - y + 2 = 1 \qquad \text{that is,} \qquad x - y + 1 = 0$$

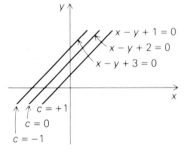

Fig. 15-1-3 Three level curves of the function $f(x,y) = x - y + 2$.

Definition A *function of three variables* is a rule which assigns a number $f(x, y, z)$ to each point (x, y, z) of a domain in (x, y, z) space.

The graph of a function $w = f(x, y, z)$ of three variables would have to lie in four-dimensional space, so we cannot visualize it. But the concept of level curve has a natural extension.

Definition Let f be a function of three variables and let c be a constant. The set of all points (x, y, z) in space such that $f(x, y, z) = c$ is called a *level surface* of f (with value c).

Worked Example 4 Let $f(x, y, z) = x - y + z + 2$. Sketch the level surfaces with values 1, 2, 3.

Solution In each case we set $f(x, y, z) = c$:

$$c = 1: \quad x - y + z + 2 = 1 \quad \text{(that is, } x - y + z + 1 = 0\text{)}$$

$$c = 2: \quad x - y + z + 2 = 2 \quad \text{(that is, } x - y + z = 0\text{)}$$

$$c = 3: \quad x - y + z + 2 = 3 \quad \text{(that is, } x - y + z - 1 = 0\text{)}$$

These surfaces are parallel planes and are sketched in Fig. 15-1-4.

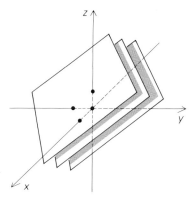

Fig. 15-1-4 Three level surfaces of the function $f(x, y, z) = x - y + z + 2$.

Solved Exercises*

1. Sketch the graph of $f(x, y) = 3x$.

2. Sketch the level surface of $f(x, y, z) = x^2 + y^2 + z^2 - 8$ with value 1.

3. How is the intersection of the plane $z = c$ with the graph of $f(x, y)$ related to the level curves of f? Sketch.

Exercises

1. Describe the domain of each of the following functions and compute the values at the indicated points:
   (a) $f(x, y) = y/x$; $(1, 0)$, $(1, 1)$
   (b) $f(x, y) = (x + y)/(x - y)$; $(1, -1)$, $(1, 0.9)$
   (c) $f(x, y, z) = (2x + y - z)/(x^2 + y^2 + z^2 - 1)$; $(1, 1, 1)$, $(0, 0, 2)$

2. Sketch the graphs of the following functions:
   (a) $f(x, y) = 1 - x - y$ $\qquad$ (b) $f(x, y) = -1 - x - y$

3. Sketch the level curves $f(x, y) = 1$ and $f(x, y) = -1$ for each of the functions in Exercise 2.

SURFACES IN SPACE

Plotting surfaces in space is usually more difficult than plotting curves in the plane. It is rare that plotting a few points on a surface will give us enough information to sketch the surface. Instead we often plot several *curves* on the surface and then interpolate between the curves. This technique, called the *method of sections*, is useful for plotting all surfaces in space, whether they be graphs of functions of two variables or level surfaces of functions of three variables. The idea behind the method of sections is to obtain a picture of the surface in space by looking at its slices by planes parallel to one of the coordinate planes. For instance, for a graph $z = f(x, y)$ the section $z = c$ is illustrated in Fig. 15-1-5.

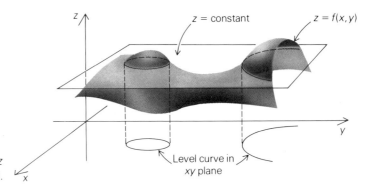

Fig. 15-1-5 The section $z = c$ of the graph $z = f(x, y)$.

Worked Example 5 Sketch the surface in the space given by $z = -y^2$.

Solution Since x is missing, all sections $x = $ constant look the same: they are copies of the parabola $z = -y^2$. Thus we draw the parabola $z = -y^2$ in the yz plane and extend it parallel to the x axis as shown in Fig. 15-1-6. The surface is called a *parabolic cylinder*.

Fig. 15-1-6 The graph $z = -y^2$ is a parabolic cylinder.

Worked Example 6 Sketch the graph of $f(x, y) = x^2 + y^2$ (this graph is called a *paraboloid of revolution*).

Solution If we set $z = $ constant, we get $x^2 + y^2 = c$, a circle. Taking $c = 1^2$, 2^2, 3^2, 4^2 we get circles of radius 1, 2, 3, and 4. These are placed on the planes $z = 1^2 = 1$, $z = 2^2 = 4$, $z = 3^2 = 9$, and $z = 4^2 = 16$ to give the graph shown in Fig. 15-1-7.

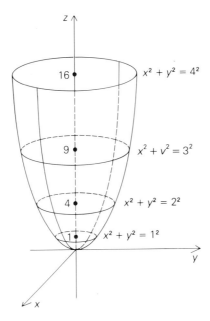

Fig. 15-1-7 The sections of the graph $z = x^2 + y^2$ by planes $z = c$ are circles.

If we set $x = 0$, we obtain the parabola $z = y^2$; if we set $y = 0$, we obtain the parabola $z = x^2$. The graph is symmetric about the z axis since z depends only on $r = \sqrt{x^2 + y^2}$.

In the next section we will use our knowledge of conic sections to plot the graphs of more complicated quadratic functions.

PLOTTING SURFACES: METHOD OF SECTIONS

1. See if any variables x, y, or z are missing from the equation. If so, the surface is a "cylinder" parallel to the axis of the missing variable and with cross-section the curve in the other variables (see Worked Example 5).

2. If the surface is a graph

$z = f(x, y)$

find the level curves $f(x, y) = c$ for various convenient values of c and draw these curves on the planes $z = c$. Smoothly join these curves with a surface in space. Draw the curves obtained by setting $x = 0$ and $y = 0$ or other convenient values to help reinforce the picture.

3. If the surface has the form $F(x, y, z) = c$, then either:
   (a) Solve for one of the variables and use step 2 if it is convenient to do so.
   (b) Set x equal to various constant values to obtain curves in y and z; draw these curves on the corresponding $x = $ constant planes. Repeat with $y = $ constant or $z = $ constant or both. Fill in the curves obtained with a surface.

4. Note any symmetries of the graph.

The computer can help us graph surfaces that may be tedious or impossible to plot by hand. The computer draws the graph either by drawing sections perpendicular to the x and y axes or by sections perpendicular to the z axis—that is, level curves lifted to the graph. When this is done for the function $z = x^2 + y^2$ (Worked Example 6), Figs. 15-1-8(a) and (b) result. (The pointed tips appear because a rectangular domain has been chosen for the function.)*

The computer-generated graph in Fig. 15-1-9 on the next page shows the function

$$z = (x^2 + 3y^2)e^{1-(x^2+y^2)}$$

Fig. 15-1-10 shows the level curves of this function in the xy plane, viewed first from an angle and then from above. Study these pictures to help develop your powers of three-dimensional visualization; attempt to reconstruct the graph in your mind by looking at the level curves.

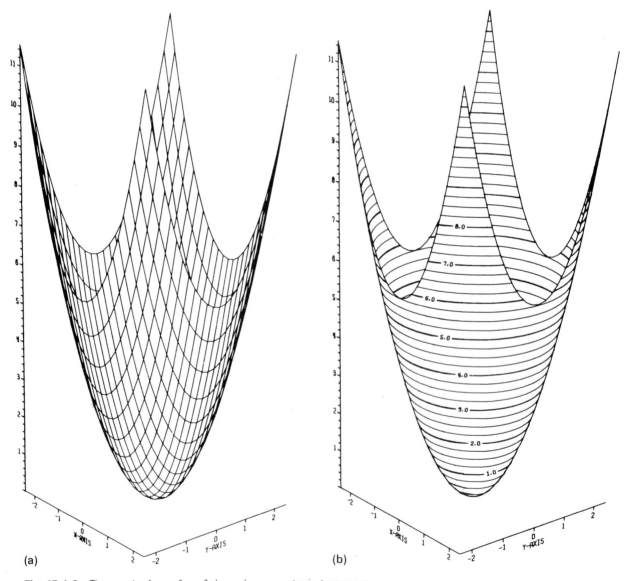

(a)   (b)

Fig. 15-1-8 The graph of $z = x^2 + y^2$ drawn by computer in two ways.

The authors are indebted to Jerry Kazdan for preparing most of the computer-generated graphs in this book.

Fig. 15-1-9 Computer-generated graph of $z = (x^2 + 3y^2)e^{1-(x^2+y^2)}$.

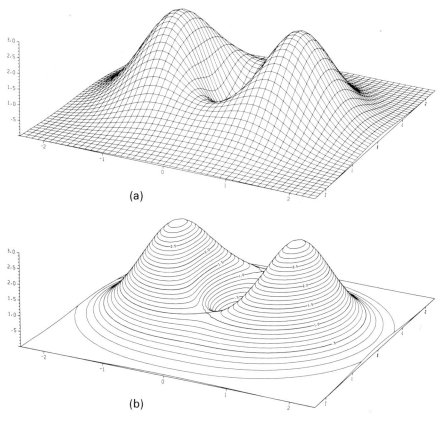

(a)

(b)

Fig. 15-1-10 Level curves for the function $z = (x^2 + 3y^2)e^{1-(x^2+y^2)}$.

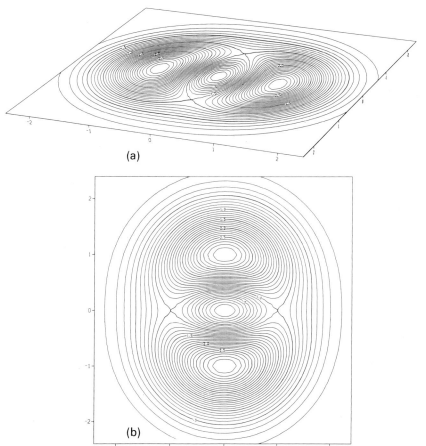

(a)

(b)

An example of a surface in space which is not given by an explicit formula is the hull of a sailing ship. Fig. 15-1-11 shows how such a surface may be displayed by the method of sections. Curves are drawn showing what the cross-sections look like in each of three directions. This procedure is called "taking the lines of the ship."

Fig. 15-1-11 Lines of the Cup Defender MAGIC, of New York. (Illustration of the ship "Magic" is reproduced from *The History of American Sailing Ships* by Howard I. Chapelle, with the permission of W.W. Norton & Company, Inc. Copyright 1935 by W. W. Norton & Company, Inc. Copyright renewed 1963 by Howard I. Chapelle.)

Solved Exercises

4. Sketch the surface $x^2 + y^2 = 25$ in (x, y, z) space.

5. Sketch the surface $z = x^2 + y^2 - 4x - 6y + 13$. [*Hint:* Complete the square.]

Exercises

4. Sketch the surface in space defined by each of the following equations:
   (a) $z^2 + x^2 = 4$ (b) $z = |y|$
   (c) $z = 3x^2 + 3y^2 - 6x + 12y + 15$

5. Sketch each of these surfaces:
   (a) $z = \sqrt{x^2 + y^2}$
   (b) $z = \max(|x|, |y|)$ [*Note:* $\max(|x|, |y|)$ is the maximum of $|x|$ and $|y|$.]
   (c) $x^2 + y^2 + z^2 = 4$

Problems for Section 15-1

1. Describe the domain of each of the following functions and evaluate each function at the indicated points:
   (a) $f(x, y) = (x + y)/(x^2 + y^2 - 1)$; $(1, 1)$, $(-1, 1)$
   (b) $f(x, y) = 2xy/(x^2 + y^2)$; $(1, 1)$, $(-1, 1)$
   (c) $f(x, y) = 1/(x^2 - 4y^2 - 1)$; $(1, 0.5)$, $(0.1, 0.5)$
   (d) $f(x, y) = (2x - \sin y)/(1 + \cos x)$; $(0, \pi/4)$, $(\pi/4, \pi)$

2. For each of the functions of parts (a), (b), and (c) of Problem 1, sketch the level curves for $c = -2, -1, 1, 2$. Describe the general level curve $f(x, y) = c$ for the function.

3. Sketch the graph (in space) of:
   (a) $z = (x - 1)^2 + y^2$
   (b) $x = -8z^2 + z$
   (c) $z = x^2 + y^2 - 2x + 8$

4. Sketch the graph (in space) of:

    (a) $z = x - y$

    (b) $z = \sin x$ (the "washboard")

    (c) $z = 1/(1 + y^2)$

★5. Describe the behavior of the level curve $f(x, y) = c$ as c varies for each of these functions:

    (a) $f(x, y) = x^2 + y^2 + 1$

    (b) $f(x, y) = 1 - x^2 - y^2$

    (c) $f(x, y) = x^2 + xy$

    (d) $f(x, y) = x^3 - x$

6. Let $f(x, y) = e^{-1/(x^2+y^2)}$.

    (a) Sketch the level curve $f(x, y) = c$ for $c = 0.001$, $c = 0.01$, $c = 0.5$, and $c = 0.9$.

    (b) What happens if c is less than zero or greater than 1?

    (c) Sketch the cross-section of the graph in the vertical plane $y = 0$ (that is, the intersection of the graph with the xz plane).

    (d) Argue that this cross-section looks the same in any vertical plane through the origin.

    (e) Describe the graph in words and try to sketch it.

7. Draw the level curves $f(x, y) = c$—first in the xy plane and then lifted to the graph—for:

    (a) $f(x, y) = x^2 + 2y^2$; $c = 0, 1, 2$

    (b) $f(x, y) = x^2 - y^2$; $c = -1, 0, 1$

    (c) $f(x, y) = x - y^2$; $c = -2, 0, 2$

8. Sketch the surface $4x^2 + y^2 + 9z^2 = 1$ by taking several sections parallel to each of the coordinate axes.

▦9. The formula

$$z = \frac{2x}{(y - y^{-1})^2 + (2x)^2}$$

appears in the study of steady state motions of a mechanical system with viscous damping subjected to a harmonic external force. The *average power input* by the external force is proportional to the variable z (with proportionality constant $k > 0$). The variable y is the ratio of input frequency to natural frequency. The variable x measures the viscous damping constant.

    (a) Plot z versus y for $x = 0.2, 0.5, 2.0$ on the same axes. Use the range of values $0 \le y \le 2.0$.

    (b) The average power input is a maximum when $y = 1$, that is, when the input and natural frequencies are the same. Verify this both graphically and algebraically.

10. The potential difference E between electrolyte solutions separated by a membrane is given by

$$E = \frac{RT}{F} \frac{x - y}{x + y} \ln |z|.$$

[The symbols R, T, F are the universal gas constant, absolute temperature and Faraday unit, respectively—these are constants. The symbols x and y measure the mobilities of Na^+ and Cl^-, respectively. The symbol z is c_1/c_2, where c_1 and c_2 are the mean salt (NaCl) concentrations on each side of the membrane.] Assume hereafter that $RT/F = 25$.

    (a) Write the level surface $E = -12$ as $z = f(x, y)$.

    (b) In practice, $y = 3x/2$. Plot E versus z in this case.

15-2 Three-Dimensional Analytic Geometry

Quadric surfaces are defined by quadratic equations in x, y, and z.

In the first part of this section, we discuss a special class of surfaces called *quadrics*, which are defined by quadratic equations in x, y, and z. To prepare for this discussion, you should study Section R-4 on conics. The second part of the section is devoted to the cylindrical and spherical coordinate systems. These are three-dimensional analogs of the polar coordinate system in the plane.

Goals

After studying this section, you should be able to:

Sketch the graph of simple quadric surfaces.

Plot points and simple graphs in cylindrical and spherical coordinates.

QUADRIC SURFACES

The methods of Section 15-1, together with our knowledge of conics, enable us to graph a number of interesting surfaces defined by quadratic equations.

Worked Example 1   Sketch the graph of $z = f(x, y) = x^2 - y^2$ (a *hyperbolic paraboloid*).

Solution   To visualize this surface, we first draw the level curves; we graph the equations $x^2 - y^2 = c$. Consider $c = 0, \pm 1, \pm 4$. For $c = 0$ we have $y^2 = x^2$ (that is, $y = \pm x$), so this level set consists of two straight lines through the origin. For $c = 1$ the level curve is $x^2 - y^2 = 1$, which is a hyperbola that passes vertically through the x axis at the points $(\pm 1, 0)$ (see Fig. 15-2-1). Similarly, for $c = 4$ the level curve is $x^2/4 - y^2/4 = 1$, the hyperbola passing vertically through the x axis at $(\pm 2, 0)$. For $c = -1$ we obtain the curve $x^2 - y^2 = -1$, that is, the hyperbola passing horizontally through the y axis at $(0, \pm 1)$. And for $c = -4$ the hyperbola through $(0, \pm 2)$ is obtained. These level curves are shown in Fig. 15-2-1. To aid us in visualizing the graph of f, we will also compute two sections. First, set $x = 0$ to obtain $z = -y^2$, a parabola opening downward. Second, setting $y = 0$ gives the parabola $z = x^2$ opening upward.

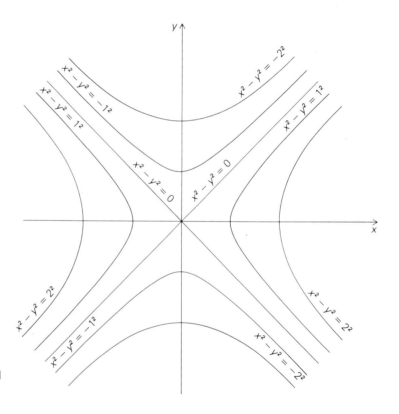

Fig. 15-2-1 Some level curves of $f(x) = x^2 - y^2$.

The graph may now be visualized by lifting the level curves to the appropriate heights and smoothing out the resulting surface. Their placement is aided by the use of the parabolic sections. This procedure generates the saddle-shaped surface indicated in Fig. 15-2-2. The graph is unchanged under reflection in the yz plane and in the xz plane. When accurately plotted by a computer, this graph has the appearance of Fig. 15-2-3; the level curves are shown in Fig. 15-2-4. (The graph has been rotated by 90° about the z axis.)

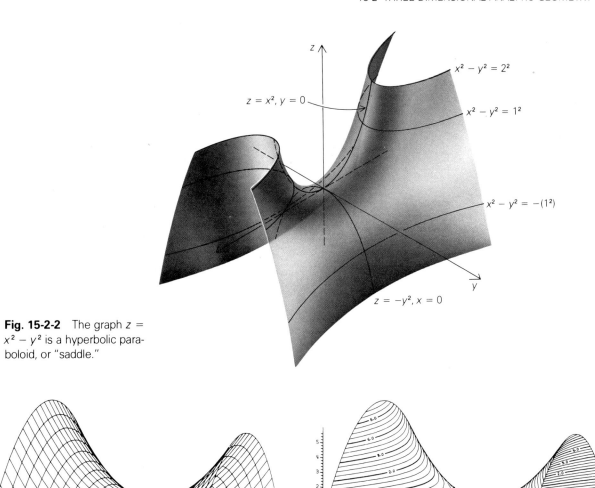

Fig. 15-2-2 The graph $z = x^2 - y^2$ is a hyperbolic paraboloid, or "saddle."

(a)

(b)

Fig. 15-2-3 Computer-generated graph of $z = x^2 - y^2$.

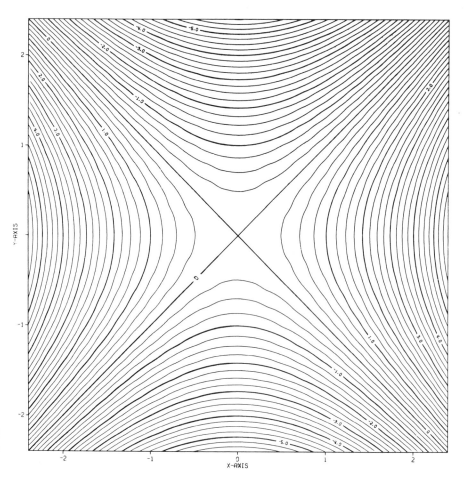

Fig. 15-2-4 Level curves of $z = x^2 - y^2$ drawn by computer.

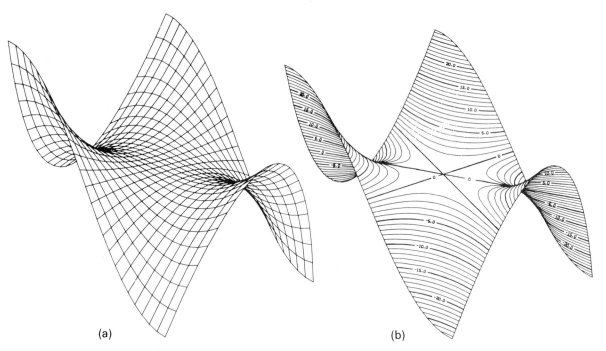

(a)   (b)

Fig. 15-2-5 Graph of the monkey saddle: $z = x^3 - 3xy^2$.

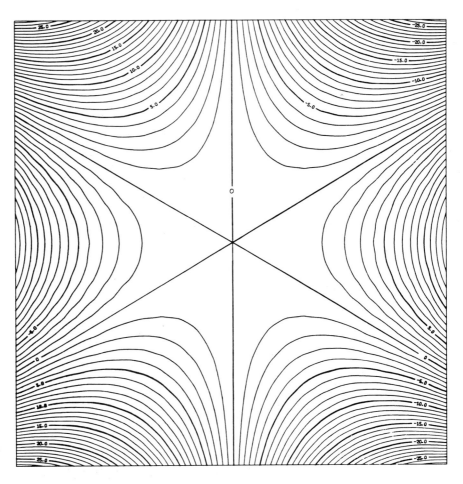

Fig. 15-2-6 Contour lines for the monkey saddle.

The origin is called a *saddle point* for the function $z = x^2 - y^2$ because of the saddlelike appearance of the graph. We will return to the study of saddle points in Chapter 16, but it is worth noting another kind of saddle here. Figure 15-2-5 shows the graph of $z = x^3 - 3xy^2$, again plotted by a computer using sections and contours. The origin now is called a *monkey saddle*, since there are two places for the legs and one for the tail. Figure 15-2-6 shows the contour lines in the plane. Figure 15-2-7 shows the four-legged, or *dog saddle:* $z = 4x^3y - 4xy^3$.

A *quadric surface* is a three-dimensional figure defined by a quadratic equation in three variables:

$$ax^2 + by^2 + cz^2 + dxy + exz + fyz + gx + hy + kz + m = 0$$

The quadric surfaces are the three-dimensional versions of the conic sections, studied in Section R-4, which were defined by quadratic equations in two variables.

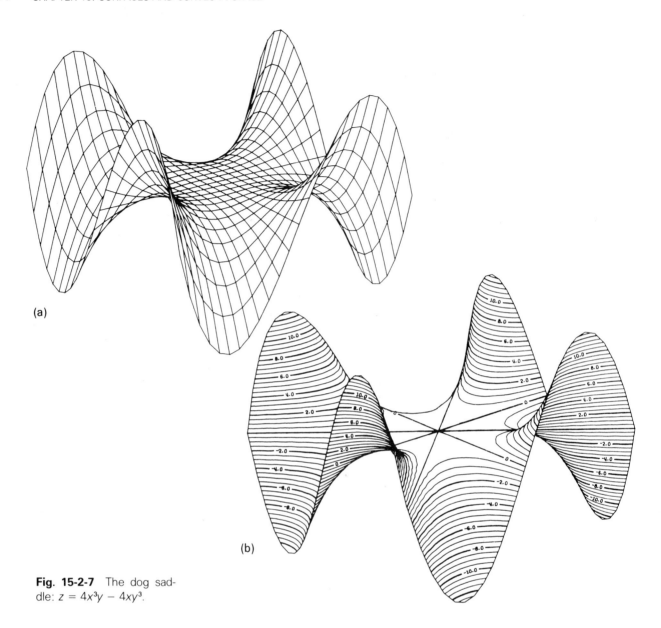

(a)

(b)

Fig. 15-2-7 The dog saddle: $z = 4x^3y - 4xy^3$.

Worked Example 2 Particular conic sections can degenerate to points or lines. Similarly, some quadric surfaces can degenerate to points, lines, or planes. Match the sample equations to the appropriate description.

(a) $x^2 + 3y^2 + z^2 = 0$ (1) No points at all

(b) $z^2 = 0$ (2) A single point

(c) $x^2 + y^2 = 0$ (3) A line

(d) $x^2 + y^2 + z^2 + 1 = 0$ (4) One plane

(e) $x^2 - y^2 = 0$ (5) Two planes

Solution Equation (a) matches (2) since only $(0, 0, 0)$ satisfies the equation; (b) matches (4) since this is the plane $z = 0$; (c) matches (3) since this is the z axis, where $x = 0$ and $y = 0$; (d) matches (1) since a nonnegative number added to 1 can never be zero; (e) matches (5) since the equation $x^2 - y^2 = 0$ is equivalent to the two equations $x + y = 0$ or $x - y = 0$, which define two planes.

If one variable is missing from the equation, we only have to find a curve in one plane and then extend it parallel to the axis of the missing variable. This procedure produces a generalized *cylinder,* either *elliptic, parabolic,* or *hyperbolic.*

Worked Example 3   Sketch the surface $z = y^2 + 1$.

Solution   The intersection of this surface with a plane $x = $ constant is a parabola of the form $z = y^2 + 1$. The surface, a *parabolic cylinder,* is sketched in Fig. 15-2-8. (See also Fig. 15-1-6.)

Fig. 15-2-8   The surface $z = y^2 + 1$ is a parabolic cylinder.

The surface defined by an equation of the form $x^2/a^2 + y^2/b^2 - z^2/c^2 = -1$ is called a *hyperboloid of two sheets.*

Worked Example 4   Sketch the surface $x^2 + 4y^2 - z^2 = -4$.

Solution   The section by the plane $z = c$ has the equation $x^2 + 4y^2 = c^2 - 4$. This is an ellipse when $|c| > 2$, a point when $c = \pm 2$, and is empty when $|c| < 2$. The section with the xz plane is the hyperbola $x^2 - z^2 = -4$, and the section with the yz plane is the hyperbola $4y^2 - z^2 = -4$. The surface is symmetric with respect to each of the coordinate planes. A sketch is given in Fig. 15-2-9.

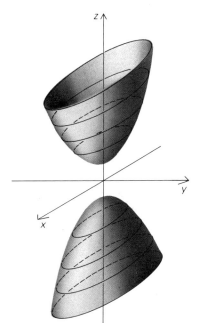

Fig. 15-2-9   The surface $x^2 + 4y^2 - z^2 = -4$ is a hyperboloid of two sheets (shown with some of its sections by planes of the form $z = $ constant).

Solved Exercises

1. The surface defined by an equation of the form $x^2/a^2 + y^2/b^2 + z^2/c^2 = 1$ is called an *ellipsoid*. Sketch the surface $x^2/9 + y^2/16 + z^2 = 1$.

2. The surface defined by an equation of the form $x^2/a^2 + y^2/b^2 - z^2/c^2 = 1$ is called a *hyperboloid of one sheet*. Sketch the surface $x^2 + y^2 - z^2 = 4$.

Exercises

1. Sketch the surfaces in three-dimensional space defined by each of the following equations, and describe each in words.
   (a) $y^2 + z^2 = 1$ (b) $9x^2 + 4z^2 = 36$
   (c) $x^2 + y^2 + z^2/4 = 1$ (d) $4x^2 + y^2 = 2$
   (e) $x^2 + y^2 = 0$ (f) $z^2 - 8y^2 = 0$

2. Sketch the graph of $x = 2z^2 - y^2$.

3. Sketch the graph of $16z^2 = 4x^2 + y^2 + 16$.

4. This exercise concerns the *elliptic paraboloid:*
   (a) Sketch the graph of $z = 2x^2 + y^2$.
   (b) Sketch the surface given by $x = -3y^2 - 2z^2$.
   (c) Consider the equation $z = ax^2 + by^2$, where a and b are both positive or both negative. Describe the horizontal cross-sections where $z =$ constant. Describe the sections obtained in the planes $x = 0$ and $y = 0$. What is the section obtained in the vertical plane $x = c$? (The special case in which $a = b$ is a *paraboloid of revolution* as in Worked Example 6, Section 15-1.)

CYLINDRICAL COORDINATES

In Sections 6-1, 6-2, and 11-3, we saw the usefulness of polar coordinates in the plane. In space there are two coordinate systems analogous to polar coordinates. The first is the cylindrical coordinate system.

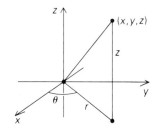

Fig. 15-2-10 The cylindrical coordinates of the point (x, y, z).

Definition The *cylindrical coordinates* of a point (x, y, z) in space are the numbers (r, θ, z), where r and θ are the polar coordinates of (x, y); that is,

$$x = r \cos \theta \qquad y = r \sin \theta \qquad z = z$$

See Fig. 15-2-10.

As with polar coordinates, we can solve for r and θ in terms of x and y.

CYLINDRICAL COORDINATES

If the cartesian coordinates of a point in space are (x, y, z),

then the cylindrical coordinates of the point are (r, θ, z), where

$$x = r \cos \theta \qquad y = r \sin \theta \qquad z = z$$

or

$$r = \sqrt{x^2 + y^2}$$

$$\theta = \begin{cases} \tan^{-1}(y/x) & \text{if } x \geq 0 \\ \tan^{-1}(y/x) + \pi & \text{if } x < 0 \end{cases}$$

Worked Example 5 (a) Find the cylindrical coordinates of $(6, 6, 8)$. Plot.
(b) If a point has cylindrical coordinates $(3, -\pi/6, -4)$, what are its cartesian coordinates? Plot.

Solution (a) Here $r = \sqrt{6^2 + 6^2} = 6\sqrt{2}$ and $\theta = \tan^{-1}\left(\frac{6}{6}\right) = \tan^{-1}(1) = \pi/4$. Thus the cylindrical coordinates are $(6\sqrt{2}, \pi/4, 8)$. See Fig. 15-2-11(a).
(b) We have $x = r \cos\theta = 3 \cos(-\pi/6) = 3\sqrt{3}/2$, and $y = r \sin\theta = 3 \sin(-\pi/6) = -3/2$. Thus the cartesian coordinates are $(3\sqrt{3}/2, -3/2, -4)$. See Fig. 15-2-11(b).

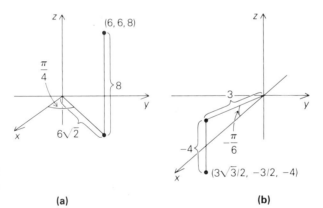

Fig. 15-2-11 Comparing the cylindrical and cartesian coordinates of two points.

Many surfaces are easier to describe in cylindrical than in cartesian coordinates, just as many curves are easier to work with using polar rather than cartesian coordinates.

Worked Example 6 Plot the surface described in cylindrical coordinates by $r = 3$.

Solution Note that r is the distance from the given point to the z axis. Therefore the points with $r = 3$ lie on a cylinder of radius 3 centered on the z axis. See Fig. 15-2-12 at the top of the next page.

Fig. 15-2-12 The cylinder has a very simple equation in cylindrical coordinates.

Solved Exercises

3. (a) Let a point have cartesian coordinates $(2, -3, 6)$. Find its cylindrical coordinates and plot.

   (b) Let a point have cylindrical coordinates $(2, 3\pi/4, 1)$. Find its cartesian coordinates and plot.

4. Sketch the surface given in cylindrical coordinates by $r = \cos 2\theta$.

5. Describe the geometric meaning of replacing (r, θ, z) by $(r, \theta + \pi, -z)$.

6. Show that the surface $r = f(z)$ is a surface of revolution.

Exercises

5. Convert from cartesian to cylindrical coordinates and plot:

   (a) $(1, -1, 0)$ (b) $(\sqrt{2}, 1, 1)$ (c) $(3, -2, 1)$
   (d) $(0, 6, -2)$ (e) $(6, 0, -2)$

6. Convert from cylindrical to cartesian coordinates and plot:

   (a) $(1, \pi/2, 0)$ (b) $(3, 45°, 8)$ (c) $(-1, \pi/6, 4)$
   (d) $(2, 0, 1)$ (e) $(0, \pi/18, 6)$

7. Sketch the surface described in cylindrical coordinates by $r = 1 + 2\cos\theta$.

8. Describe the geometric meaning of replacing:

   (a) (r, θ, z) by $(r, \theta, -z)$ (b) (r, θ, z) by $(2r, \theta, z)$
   (c) (r, θ, z) by $(2r, \theta, -z)$

9. Describe the surfaces $r = $ constant, $\theta = $ constant, and $z = $ constant in cylindrical coordinates.

SPHERICAL COORDINATES

Cylindrical coordinates are best adapted for problems which have cylindrical symmetry—that is, a symmetry about the z axis. Similarly, for problems with spherical symmetry—that is, symmetry with respect to rotations about the origin in space—the spherical coordinate system is useful.

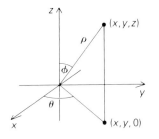

Fig. 15-2-13 Spherical coordinates.

Definition The spherical coordinates of a point (x, y, z) in space are the numbers (ρ, θ, ϕ) defined as follows:

ρ = distance from (x, y, z) to the origin

θ = cylindrical coordinate θ [angle from the positive x axis to the point (x, y)]

ϕ = the angle (in $[0, \pi]$) from the positive z axis to the line from origin to (x, y, z)

See Fig. 15-2-13.

To express the cartesian coordinates in terms of spherical coordinates, we first observe that the cylindrical coordinate $r = \sqrt{x^2 + y^2}$ is equal to $\rho \sin \phi$ and that $z = \rho \cos \phi$ (see Fig. 15-2-13). Therefore

$$x = r \cos \theta = \rho \sin \phi \cos \theta$$

$$y = r \sin \theta = \rho \sin \phi \sin \theta$$

$$z = \rho \cos \phi$$

We may also solve for ρ, θ, and ϕ. The results are given in the following display.

SPHERICAL COORDINATES

If the cartesian coordinates of a point in space are (x, y, z),

then the spherical coordinates of the point are (ρ, θ, ϕ), where

$$x = \rho \sin \phi \cos \theta$$

$$y = \rho \sin \phi \sin \theta$$

$$z = \rho \cos \phi$$

or

$$\rho = \sqrt{x^2 + y^2 + z^2}$$

$$\theta = \begin{cases} \tan^{-1}(y/x) & \text{if } x \geq 0 \\ \tan^{-1}(y/x) + \pi & \text{if } x < 0 \end{cases}$$

$$\phi = \cos^{-1} \frac{z}{\sqrt{x^2 + y^2 + z^2}}$$

Notice that the spherical coordinates θ and ϕ are similar to the geographic coordinates of longitude and latitude if we take the earth's axis to be the z axis. There are differences, though: the geographical longitude is $|\theta|$ and is called east or west longitude according to whether θ is positive or negative; the geographical latitude is $|\pi/2 - \phi|$ and is called north or south latitude according to whether $\pi/2 - \phi$ is positive or negative.

Worked Example 7 (a) Find the spherical coordinates of $(1, -1, 1)$ and plot.
(b) Find the cartesian coordinates of $(3, \pi/6, \pi/4)$ and plot.

Solution (a) $\rho = \sqrt{x^2 + y^2 + z^2} = \sqrt{1^2 + (-1)^2 + 1^2} = \sqrt{3}$

$$\theta = \tan^{-1}\left(\frac{y}{x}\right) = \tan^{-1}\left(\frac{-1}{1}\right) = -\frac{\pi}{4}$$

$$\phi = \cos^{-1}\left(\frac{z}{\rho}\right) = \cos^{-1}\left(\frac{1}{\sqrt{3}}\right) \approx 0.955 \approx 54.74°$$

See Fig. 15-2-14(a).

(b) $x = \rho \sin \phi \cos \theta = 3 \sin\left(\dfrac{\pi}{4}\right) \cos\left(\dfrac{\pi}{6}\right) = 3 \left(\dfrac{1}{\sqrt{2}}\right) \dfrac{\sqrt{3}}{2} = \dfrac{3\sqrt{3}}{2\sqrt{2}}$

$y = \rho \sin \phi \sin \theta = 3 \sin\left(\dfrac{\pi}{4}\right) \sin\left(\dfrac{\pi}{6}\right) = 3 \left(\dfrac{1}{\sqrt{2}}\right)\left(\dfrac{1}{2}\right) = \dfrac{3}{2\sqrt{2}}$

$z = \rho \cos \phi = 3 \cos\left(\dfrac{\pi}{4}\right) = \dfrac{3\sqrt{2}}{2}$

See Fig. 15-2-14(b).

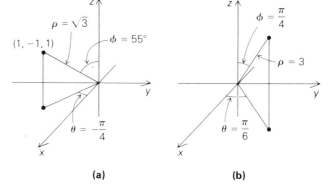

Fig. 15-2-14 Finding the spherical coordinates of the point $(1, -1, 1)$ and the cartesian coordinates of $(3, \dfrac{\pi}{6}, \dfrac{\pi}{4})$.

Worked Example 8 Find the equation in spherical coordinates of $x^2 + y^2 - z^2 = 4$ (a hyperboloid of revolution—see Solved Exercise 2).

Solution To take advantage of the relationship $x^2 + y^2 + z^2 = \rho^2$, write

$$x^2 + y^2 - z^2 = (x^2 + y^2 + z^2) - 2z^2 = \rho^2 - 2\rho^2 \cos^2 \phi$$

since $z = \rho \cos \phi$. Also, we can note

$$\rho^2 - 2\rho^2 \cos^2 \phi = \rho^2 (1 - 2 \cos^2 \phi) = -\rho^2 \cos 2\phi$$

Thus the surface is

$$\rho^2 \cos 2\phi + 4 = 0$$

Solved Exercises

7. (a) Let a point have cartesian coordinates $(2, -3, 6)$. Find its spherical coordinates and plot.

  (b) Let a point have spherical coordinates $(1, -\pi/2, \pi/4)$. Find its cartesian coordinates and plot.

8. Describe the surface given in spherical coordinates by $\rho = 3$.

9. Describe the geometric meaning of replacing (ρ, θ, ϕ) by $(\rho, \theta + \pi, \phi)$.

10. Show that the surface $\rho = f(\phi)$ is a surface of revolution.

Exercises

10. Convert from cartesian to spherical coordinates and plot:
  (a) $(0, 1, 1)$ (b) $(1, 0, 1)$ (c) $(-2, 1, -3)$
  (d) $(1, 2, 3)$ (e) $(-3, -2, -4)$

11. Convert from spherical to cartesian coordinates and plot:
  (a) $(1, \pi/6, \pi/3)$ (b) $(2, -\pi/6, \pi/3)$ (c) $(3, 2\pi, 0)$
  (d) $(3, \pi/3, \pi)$ (e) $(8, -\pi/3, \pi)$

12. Describe the surface given in spherical coordinates by $\theta = \pi/4$.

13. Express the surface $xz = 1$ in spherical coordinates.

14. Describe the geometric meaning of replacing (ρ, θ, ϕ) by $(2\rho, \theta, \phi)$.

Problems for Section 15-2

1. Sketch the graph of:
  (a) $x^2 - z^2 = 1$
  (b) $z^2 + 4y^2 = x^2 + 4$
  (c) $x^2/4 + y^2 + z^2/4 = 1$

2. Sketch:
  (a) $8x^2 + 3z^2 = 0$ (b) $x^2 + 9y^2 - z^2 = 1$
  (c) $x^2 = 4z^2 + 9$ (d) $x^2 + y^2 + 4z^2 = 1$

3. This problem concerns the *elliptic cone*. Consider the equation $x^2 + y^2 - z^2 = 0$.
  (a) What are the horizontal cross-sections for $z = \pm 1, \pm 2, \pm 3$?
  (b) What are the vertical cross-sections for $x = 0$ or $y = 0$? (Sketch and describe.)
  (c) Show that this surface really is a cone by showing that any straight line through the origin making a 45° angle with the z axis lies in the surface.
  (d) Sketch this surface and compare the result to Exercise 5(a), Section 15-1.

  A more general form of the above equation is $x^2/a^2 + y^2/b^2 - z^2/c^2 = 0$.
  (e) Describe the horizontal cross-section $z =$ constant.
  (f) Describe the vertical cross-section for $x = 0$ and $y = 0$.
  (g) Show that this surface has the property that if it contains the point (x_0, y_0, z_0), then it contains the whole line through $(0, 0, 0)$ and (x_0, y_0, z_0).

4. This problem concerns the *hyperbolic paraboloid*. (A surface of this kind was studied in Worked Example 1.) A standard form for the equation is $z = ax^2 - by^2$, with a and b both positive or both negative.
  (a) Sketch the graph of $z = x^2 - 2y^2$.
  (b) Show that $z = xy$ also determines a hyperbolic paraboloid. Sketch some of its level curves.

5. Convert each of the following points from cartesian to cylindrical and spherical coordinates and plot:
  (a) $(0, 3, 4)$ (b) $(-\sqrt{2}, 1, 0)$
  (c) $(0, 0, 0)$ (d) $(-1, 0, 1)$
  (e) $(-2\sqrt{3}, -2, 3)$

6. The following points are given in cylindrical coordinates. Convert to cartesian and spherical coordinates:
  (a) $(1, \pi/4, 1)$ (b) $(3, \pi/6, -4)$
  (c) $(0, \pi/4, 1)$ (d) $(2, -\pi/2, 1)$
  (e) $(-2, -\pi/2, 1)$

7. The following points are given in spherical coordinates. Convert to cartesian and cylindrical coordinates and plot:
  (a) $(1, \pi/2, \pi)$ (b) $(2, -\pi/2, \pi/6)$
  (c) $(0, \pi/8, \pi/35)$ (d) $(2, -\pi/2, -\pi)$
  (e) $(-1, \pi, \pi/6)$

8. Sketch the surface given in cylindrical coordinates by $r = 1 + \cos \theta$.

9. Sketch the surface given in spherical coordinates by ϕ = constant.

10. Describe the geometric meaning of replacing (r, θ, z) by $(2r, \theta + \pi, z)$ in cylindrical coordinates.

11. Describe the geometric meaning of replacing (ρ, θ, ϕ) by $(\rho, \theta, \phi + \pi/2)$ in spherical coordinates.

12. Express the surface $z = x^2 - y^2$ (a hyperbolic paraboloid—see Worked Example 1) in (a) cylindrical and (b) spherical coordinates.

13. Show that in spherical coordinates:
    (a) ρ is the length of $x\mathbf{i} + y\mathbf{j} + z\mathbf{k}$.
    (b) $\phi = \cos^{-1}(\mathbf{v} \cdot \mathbf{k}/\|\mathbf{v}\|)$, where $\mathbf{v} = x\mathbf{i} + y\mathbf{j} + z\mathbf{k}$.
    (c) $\theta = \cos^{-1}(\mathbf{u} \cdot \mathbf{i}/\|\mathbf{u}\|)$, where $\mathbf{u} = x\mathbf{i} + y\mathbf{j}$.

★14. Describe the surface given in spherical coordinates by $\rho = \cos 2\theta$.

★15. The quadric surfaces may be shifted and rotated in space just as the conic sections may be shifted in the plane. These transformations will produce more complicated cases of the general quadratic equation in three variables. Complete squares to bring the following to one of the standard forms (shifted) and sketch the resulting surfaces:
    (a) $4x^2 + y^2 + 4z^2 + 8x - 4y - 8z + 8 = 0$
    (b) $2x^2 + 3y^2 - 4z^2 + 4x + 9y - 8z + 10 = 0$

16. A circular membrane lies over the region $x^2 + y^2 \leq a^2$ in space. The maximum deflection z of the membrane is b. Assume that (x, y, z) is a point on the deflected membrane. Show that the corresponding point (r, θ, z) in cylindrical coordinates belong to the set P described by $0 \leq r \leq a$, $0 \leq \beta \leq 2\pi$, $|z| \leq b$.

17. A tank in the shape of a right circular cylinder of radius 10 ft and height 16 ft is half filled and lying on its side. Describe the air space inside the tank by cylindrical coordinates.

18. A vibrometer is to be designed which withstands the heating effects of its spherical enclosure of diameter d, which is buried to a depth $d/3$ in the earth, the upper portion being heated by the sun. Heat conduction analysis requires a description of the buried portion of the enclosure, in spherical coordinates. Find it.

19. An oil filter cartridge is a right porous cylinder in which the liquid inside diffuses from the axis to the outer curved surface. Describe the cartridge in cylindrical coordinates, given that the diameter of the filter is 4.5″, the height is 5.6″ and the center of the cartridge is drilled through from the top to admit a $\frac{5}{8}$″ diameter bolt.

15-3 The Calculus of Curves in Space

Tangents, velocity, and arc length of curves in space can be computed by vector methods.

We continue our study of geometry in space by considering curves in space. We can consider tangents to these curves by using calculus, since only the calculus of functions of one variable and a knowledge of vectors are required. (To determine tangent planes to surfaces, we will need the calculus of functions of several variables.)

Before proceeding, you should review the portion of Section 11-3 dealing with parametric curves in the plane.

Goals

After studying this section, you should be able to:

Sketch simple parametric curves in space.

Differentiate vector functions of one variable.

Find tangent lines and arc lengths for parametric curves in space.

PARAMETRIC CURVES

Recall that a parametric curve in the plane consists of a pair of functions $(x, y) = (f(t), g(t))$. As t ranges through some interval (on which f and g are defined), the point (x, y) traces out a curve in the plane; see Fig. 15-3-1.

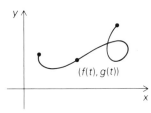

Fig. 15-3-1 A parametric curve in the plane.

Worked Example 1 What curve is traced out by $(\sin t,\ 2 \cos t)$, $0 \le t \le 2\pi$?

Solution Since $x = \sin t$ and $y/2 = \cos t$, (x, y) satisfies $x^2 + y^2/4 = 1$, so the curve traced out is an ellipse. As t goes from zero to 2π, the moving point goes once around the ellipse, starting and ending at P (Fig. 15-3-2).

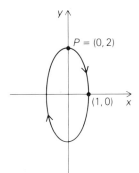

Fig. 15-3-2 The ellipse traced out by $(\sin t, 2 \cos t)$.

Definition A *parametric curve in space* consists of three functions $(x, y, z) = (f(t), g(t), h(t))$ defined for t in some interval on which f, g, and h are defined.

The curve we "see" is the path traced out by the point (x, y, z) as t varies, just as for curves in the plane.

Worked Example 2 Sketch the parametric curve $(x, y, z) = (3t + 2, 8t - 1, t)$.

Solution If we write $P = (x, y, z)$, then

$$P = (2, -1, 0) + t(3, 8, 1)$$

which is a straight line through $(2, -1, 0)$ in direction $(3, 8, 1)$ (see Section 14-2). To sketch it, we pick the points obtained by setting $t = 0$ and $t = 1$, that is, $(2, -1, 0)$ and $(5, 7, 1)$; see Fig. 15-3-3 at the top of the next page.

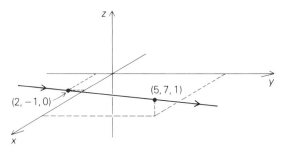

Fig. 15-3-3 The parametric curve $(3t + 2, 8t - 1, t)$ is a line.

Worked Example 3 Sketch the curve given by $x = \cos t$, $y = \sin t$, $z = t$; $-\infty < t < \infty$.

Solution As t varies, the point (x, y) traces out a circle in the plane. Thus (x, y, z) is a path which circles around the z axis, but at value t its height above the xy plane is $z = t$. Thus we get the helix shown in Fig. 15-3-4. (To be exact, it is called a *right circular helix*, since it lies on the right circular cylinder $x^2 + y^2 = 1$.)

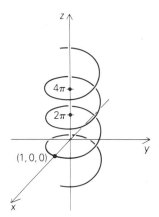

Fig. 15-3-4 The curve $(\cos t, \sin t, t)$ is a helix.

In Fig. 15-3-4 the z axis has been drawn with a different scale than the x and y axes so that more coils of the helix can be shown. It is often useful to do something like this when displaying sketches of curves or graphs. You should be careful, however, not to give a false impression—label the axes to show the scale when necessary.

Worked Example 4 Sketch the curve $(t, 2t, \cos t)$.

Solution If we ignore z temporarily, we note that $(t, 2t)$ describes the line $y = 2x$ in the xy plane. As t varies, $(t, 2t)$ moves along this line. Thus $(t, 2t, \cos t)$ moves along a curve over this line with the z component oscillating like $\cos t$. Thus we get the curve shown in Fig. 15-3-5.

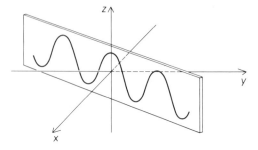

Fig. 15-3-5 The curve $(t, 2t, \cos t)$ lies in the plane $y = 2x$.

Solved Exercises   1. Sketch the curve $(\cos t, 2 \sin t, 2t)$.

2. Sketch the curve $x = 3t^2 + 2$, $y = t^3 - 8$, $z = 4t^3 + 3$.

Exercises   1. Sketch the following curves:

(a) $x = 2t - 1$; $y = t + 2$; $z = t$; $-\infty < t < \infty$.

(b) $x = -t$; $y = t$; $z = t^2$; $0 \leq t \leq 3$.

(c) $x = -t$; $y = 2t$; $z = 1/t$; $1 \leq t \leq 3$.

(d) $x = \cos t$; $y = \sin t$; $z = t/2\pi$; $-2\pi \leq t \leq 2\pi$.

2. (a) Sketch the following curves. On each curve, indicate the points obtained when $t = 0, \frac{1}{4}, \frac{1}{2}, \frac{3}{4}, 1$.

(i) $x = -t$; $y = 2t$; $z = 3t$; $0 \leq t \leq 1$.

(ii) $x = -t^2$; $y = 2t^2$; $z = 3t^2$; $-1 \leq t \leq 1$.

(iii) $x = -\sin(\pi t/2)$; $y = 2\sin(\pi t/2)$; $z = 3\sin(\pi t/2)$; $0 \leq t \leq 1$.

(b) Show that the point set in space covered by each of these curves is the same. Discuss differences between the curves thought of as functions of t. (How fast do you move along the curve as t changes? How many times is each point covered?)

DIFFERENTIATION OF VECTOR FUNCTIONS

In doing calculus with parametric curves, it is useful to identify the point $P = (x, y, z) = (f(t), g(t), h(t))$ with the vector $\mathbf{r} = x\mathbf{i} + y\mathbf{j} + z\mathbf{k} = f(t)\mathbf{i} + g(t)\mathbf{j} + h(t)\mathbf{k}$. This vector is a *function* of t, according to the following definition.

Definition   A *vector function* of one variable is a rule σ which associates a vector $\mathbf{r} = \sigma(t)$ in space (or the plane) to each real number t in some domain.

If σ is a vector function and t is in its domain, we can express $\sigma(t)$ in terms of the standard basis vectors $\mathbf{i}$, $\mathbf{j}$, and $\mathbf{k}$. The coefficients will themselves depend upon t, so we may write

$$\sigma(t) = f(t)\mathbf{i} + g(t)\mathbf{j} + h(t)\mathbf{k}$$

where f, g, and h are scalar (real-valued) functions with the same domain as σ. Notice that the functions $f(t)$, $g(t)$, $h(t)$ define a parametric curve such that the displacement vector from the origin to $(f(t), g(t), h(t))$ is just $\sigma(t)$. The functions f, g, and h are called the *component functions* of the vector function $\sigma(t)$. To summarize, we may say that parametric curves, vector functions, and triples of scalar functions are mathematically equivalent objects; we simply visualize them differently. For instance, the wind velocity at a fixed place on earth, or the cardiac vector (see p. 603), may be visualized as a vector depending on time.

Worked Example 5 Let **u**, **v**, and **w** be three vectors such that **v** and **w** are perpendicular and have the same length r, and let $\boldsymbol{\sigma}(t) = \mathbf{u} + \mathbf{v}\cos t + \mathbf{w}\sin t$.*

(a) Describe the motion of the tip of $\boldsymbol{\sigma}(t)$ if the tail of $\boldsymbol{\sigma}(t)$ is fixed at the origin. (That is, describe the parametric curve corresponding to $\boldsymbol{\sigma}(t)$.)

(b) Find the component functions of $\boldsymbol{\sigma}(t)$ if $\mathbf{u} = 2\mathbf{i} + \mathbf{j}$, $\mathbf{v} = \mathbf{j} - \mathbf{k}$, and $\mathbf{w} = \mathbf{j} + \mathbf{k}$.

Solution (a) We observe first that the vector $\mathbf{v}\cos t + \mathbf{w}\sin t$ always lies in the plane spanned by **v** and **w** and that the square of its length is

$$(\mathbf{v}\cos t + \mathbf{w}\sin t) \cdot (\mathbf{v}\cos t + \mathbf{w}\sin t)$$
$$= \mathbf{v}\cdot\mathbf{v}\cos^2 t + 2\mathbf{v}\cdot\mathbf{w}\sin t\cos t + \mathbf{w}\cdot\mathbf{w}\sin^2 t$$
$$= r^2\cos^2 t + r^2\sin^2 t$$
$$= r^2(\cos^2 t + \sin^2 t) = r^2$$

so the tip of the vector $\mathbf{v}\cos t + \mathbf{w}\sin t$ moves in a circle of radius r if its tail is fixed. Adding **u** to $\mathbf{v}\cos t + \mathbf{w}\sin t$ to get $\boldsymbol{\sigma}(t)$, we find that the tip of $\boldsymbol{\sigma}(t)$ moves in a circle of radius r whose center is at the tip of **u**. (See Fig. 15-3-6.)

Fig. 15-3-6 The tip of $\mathbf{u} + \mathbf{v}\cos t + \mathbf{w}\sin t$ moves in a circle of radius r with center at the tip of **u** and in a plane parallel to that spanned by **v** and **w**.

(b) We have

$$\boldsymbol{\sigma}(t) = \mathbf{u} + \mathbf{v}\cos t + \mathbf{w}\sin t$$
$$= 2\mathbf{i} + \mathbf{j} + (\mathbf{j} - \mathbf{k})\cos t + (\mathbf{j} + \mathbf{k})\sin t$$
$$= 2\mathbf{i} + (1 + \cos t + \sin t)\mathbf{j} + (-\cos t + \sin t)\mathbf{k}$$

so the component functions are 2, $1 + \cos t + \sin t$, and $-\cos t + \sin t$.

We wish now to define the rate of change, or *derivative*, of a vector function $\boldsymbol{\sigma}(t)$ with respect to t. If $\boldsymbol{\sigma}(t)$ is the displacement from a fixed origin to a moving point, this derivative will represent the velocity of the point. To see how the derivative should be defined, we examine the case of uniform rectilinear motion.

---

*Formulas involving vector functions sometimes look neater if scalars are written to the right of vectors. Any expression of the form $\mathbf{v}f(t)$ is to be interpreted as $f(t)\mathbf{v}$.

Worked Example 6 Let $\sigma(t) = \mathbf{u} + t\mathbf{v}$, so that $\sigma(t)$ is the displacement from the origin to a point moving uniformly with velocity vector $\mathbf{v}$. (See pp. 602–604.) Let $\mathbf{u} = a\mathbf{i} + b\mathbf{j} + c\mathbf{k}$ and $\mathbf{v} = l\mathbf{i} + m\mathbf{j} + n\mathbf{k}$.
(a) Find the component functions of $\sigma(t)$.
(b) Show that the components of the velocity vector are obtained by differentiating the component functions of $\sigma(t)$.

Solution (a) We have $\sigma(t) = \mathbf{u} + t\mathbf{v} = a\mathbf{i} + b\mathbf{j} + c\mathbf{k} + t(l\mathbf{i} + m\mathbf{j} + n\mathbf{k}) = (a + lt)\mathbf{i} + (b + mt)\mathbf{j} + (c + nt)\mathbf{k}$, so the component functions are $a + lt$, $b + mt$, and $c + nt$.
(b) The derivatives of the component functions of $\sigma(t)$ are the constant functions l, m, and n; these are precisely the components of the velocity vector $\mathbf{v}$.

Worked Example 7 Let $\sigma(t) = f(t)\mathbf{i} + g(t)\mathbf{j}$ be a vector function in the plane. Show that the tangent line at time t_0 to the parametric curve corresponding to $\sigma(t)$ (with the tail of $\sigma(t)$ fixed at zero) has the direction of the vector $f'(t_0)\mathbf{i} + g'(t_0)\mathbf{j}$.

Solution Recall from Section 11-3 that if $(f(t), g(t))$ is a parametrized curve in the plane, then the slope of its tangent line at $(f(t_0), g(t_0))$ is $g'(t_0)/f'(t_0)$. A line in the direction of $f'(t_0)\mathbf{i} + g'(t_0)\mathbf{j}$ has slope $g'(t_0)/f'(t_0)$, so it is in the same direction as the tangent line.

Guided by Worked Examples 6 and 7, we make the following definition.

Definition   Let $\sigma(t) = f(t)\mathbf{i} + g(t)\mathbf{j} + h(t)\mathbf{k}$ be a vector function. If the coordinate functions f, g, and h are all differentiable at t_0, then we say that σ is *differentiable* at t_0, and we define the *derivative* $\sigma'(t_0)$ to be the vector $f'(t_0)\mathbf{i} + g'(t_0)\mathbf{j} + h'(t_0)\mathbf{k}$.

The derivative of σ is a function of the number at which the derivative is taken. Thus $\sigma'(t)$ is a new vector function, and we may consider the second derivative $\sigma''(t)$, as well as higher derivatives.

We will sometimes use Leibniz notation for derivatives of vector functions: if $\mathbf{r} = \sigma(t)$, we will write $d\mathbf{r}/dt$ for $\sigma'(t)$ and $d^2\mathbf{r}/dt^2$ for $\sigma''(t)$.

Worked Example 8 Let $\sigma(t)$ be the vector function of Worked Example 5, with $\mathbf{u}$, $\mathbf{v}$, and $\mathbf{w}$ as in part (b) of that example. Find $\sigma'(t)$ and $\sigma''(t)$.

Solution In terms of components,

$$\sigma(t) = 2\mathbf{i} + (1 + \cos t + \sin t)\mathbf{j} + (-\cos t + \sin t)\mathbf{k}$$

Differentiating the components, we have

$$\sigma'(t) = (-\sin t + \cos t)\mathbf{j} + (\sin t + \cos t)\mathbf{k}$$

and

$$\sigma''(t) = (-\cos t - \sin t)\mathbf{j} + (\cos t - \sin t)\mathbf{k}$$

The differentiation of vector functions is facilitated by algebraic rules which follow from the corresponding rules for scalar functions. We list the rules in the following box, leaving their proof to the reader. (See Solved Exercise 3 below.)

DIFFERENTIATION RULES FOR VECTOR FUNCTIONS

To differentiate a vector function

$$\boldsymbol{\sigma}(t) = f(t)\mathbf{i} + g(t)\mathbf{j} + h(t)\mathbf{k},$$

differentiate it component by component:

$$\boldsymbol{\sigma}'(t) = f'(t)\mathbf{i} + g'(t)\mathbf{j} + h'(t)\mathbf{k}$$

Let $\boldsymbol{\sigma}(t)$, $\boldsymbol{\sigma}_1(t)$, and $\boldsymbol{\sigma}_2(t)$ be vector functions and let $p(t)$ and $q(t)$ be scalar functions.

Sum Rule: $\dfrac{d}{dt}[\boldsymbol{\sigma}_1(t) + \boldsymbol{\sigma}_2(t)] = \boldsymbol{\sigma}'_1(t) + \boldsymbol{\sigma}'_2(t)$

Scalar Multiplication Rule: $\dfrac{d}{dt}[p(t)\boldsymbol{\sigma}(t)] = p'(t)\boldsymbol{\sigma}(t) + p(t)\boldsymbol{\sigma}'(t)$

Dot Product Rule: $\dfrac{d}{dt}[\boldsymbol{\sigma}_1(t)\cdot\boldsymbol{\sigma}_2(t)] = \boldsymbol{\sigma}'_1(t)\cdot\boldsymbol{\sigma}_2(t) + \boldsymbol{\sigma}_1(t)\cdot\boldsymbol{\sigma}'_2(t)$

Cross-Product Rule: $\dfrac{d}{dt}[\boldsymbol{\sigma}_1(t) \times \boldsymbol{\sigma}_2(t)] = \boldsymbol{\sigma}'_1(t) \times \boldsymbol{\sigma}_2(t) + \boldsymbol{\sigma}_1(t) \times \boldsymbol{\sigma}'_2(t)$

Chain Rule: $\dfrac{d}{dt}[\boldsymbol{\sigma}(q(t))] = q'(t)\boldsymbol{\sigma}'(q(t))$

Worked Example 9 Show that if $\boldsymbol{\sigma}(t)$ is a vector function such that $\|\boldsymbol{\sigma}(t)\|$ is constant, then $\boldsymbol{\sigma}'(t)$ is perpendicular to $\boldsymbol{\sigma}(t)$ for all t.

Solution If $\|\boldsymbol{\sigma}(t)\|$ is constant, so is its square $\|\boldsymbol{\sigma}(t)\|^2 = \boldsymbol{\sigma}(t)\cdot\boldsymbol{\sigma}(t)$. The derivative of this constant is zero, so by the dot product rule we have

$$0 = \frac{d}{dt}[\boldsymbol{\sigma}(t)\cdot\boldsymbol{\sigma}(t)] = \boldsymbol{\sigma}'(t)\cdot\boldsymbol{\sigma}(t) + \boldsymbol{\sigma}(t)\cdot\boldsymbol{\sigma}'(t)$$

$$= 2\boldsymbol{\sigma}(t)\cdot\boldsymbol{\sigma}'(t)$$

So $\boldsymbol{\sigma}(t)\cdot\boldsymbol{\sigma}'(t) = 0$; that is, $\boldsymbol{\sigma}'(t)$ is perpendicular to $\boldsymbol{\sigma}(t)$.

Solved Exercises 3. Prove the dot product rule.

4. If $\boldsymbol{\sigma}(t) = f(t)\mathbf{i} + g(t)\mathbf{j} + h(t)\mathbf{k}$ is a vector function, we may define $\lim_{t\to t_0} \boldsymbol{\sigma}(t)$ componentwise; that is,

$$\lim_{t\to t_0} \boldsymbol{\sigma}(t) = [\lim_{t\to t_0} f(t)]\mathbf{i} + [\lim_{t\to t_0} g(t)]\mathbf{j} + [\lim_{t\to t_0} h(t)]\mathbf{k}$$

if the three limits on the right-hand side all exist. Using this definition, show that

$$\lim_{\Delta t \to 0} \frac{1}{\Delta t} [\boldsymbol{\sigma}(t_0 + \Delta t) - \boldsymbol{\sigma}(t_0)] = \boldsymbol{\sigma}'(t_0)$$

5. Suppose that $\boldsymbol{\sigma}(t)$ is a vector function such that $\boldsymbol{\sigma}'(t) = -\boldsymbol{\sigma}(t)$. Show that $\boldsymbol{\sigma}(t) = e^{-t}\boldsymbol{\sigma}(0)$. What is the behavior of $\boldsymbol{\sigma}(t)$ as $t \to \infty$?

Exercises   3. Let

$$\boldsymbol{\sigma}_1(t) = e^t\mathbf{i} + (\sin t)\mathbf{j} + t^3\mathbf{k}$$

$$\boldsymbol{\sigma}_2(t) = e^{-t}\mathbf{i} + (\csc t)\mathbf{j} - 2t^3\mathbf{k}$$

Find each of the following derivatives in two different ways:

(a) $\dfrac{d}{dt} [\boldsymbol{\sigma}_1(t) + \boldsymbol{\sigma}_2(t)]$ (b) $\dfrac{d}{dt} [\boldsymbol{\sigma}_1(t) \cdot \boldsymbol{\sigma}_2(t)]$

(c) $\dfrac{d}{dt} [\boldsymbol{\sigma}_1(t) \times \boldsymbol{\sigma}_2(t)]$ (d) $\dfrac{d}{dt} e^t\boldsymbol{\sigma}_1(t)$

(e) $\dfrac{d}{dt} [\boldsymbol{\sigma}_1(t^2)]$

4. (a) Give the "natural" domain for this vector function:

$$\boldsymbol{\sigma}(t) = \frac{1}{t}\mathbf{i} + \frac{1}{t-1}\mathbf{j} + \frac{1}{t-2}\mathbf{k}$$

(b) Find $\boldsymbol{\sigma}'$ and $\boldsymbol{\sigma}''$.

5. Prove (for vector functions): (a) The sum rule; (b) The scalar multiplication rule; (c) The cross-product rule; (d) The chain rule.

VELOCITY, SPEED, AND ACCELERATION OF SPACE CURVES

Definition   Let $(f(t), g(t), h(t))$ be a parametric curve. If f, g, and h are differentiable at t_0, the vector $f'(t_0)\mathbf{i} + g'(t_0)\mathbf{j} + h'(t_0)\mathbf{k}$ is called the *velocity vector* of the curve at t_0.

Fig. 15-3-7 The velocity vector of a parametric curve is the derivative of the vector $\boldsymbol{\sigma}(t)$ from the origin to the curve.

Notice that if $\boldsymbol{\sigma}(t)$ is the vector function corresponding to $(f(t), g(t), h(t))$, then the velocity vector at t_0 is just $\boldsymbol{\sigma}'(t_0)$ (see Fig. 15-3-7). We often write $\mathbf{v}$ for the velocity vector—that is, $\mathbf{v} = \boldsymbol{\sigma}'(t)$. In Leibniz notation, if $\mathbf{r} = \boldsymbol{\sigma}(t)$ we have $\mathbf{v} = d\mathbf{r}/dt$.

Several other quantities of interest may be defined in terms of the velocity vector. If $\mathbf{v} = \boldsymbol{\sigma}'(t_0)$ is the velocity vector of a curve at t_0, then the length $v = \|\mathbf{v}\| = \|\boldsymbol{\sigma}'(t_0)\|$ is called the *speed* along the curve at t_0, and the line through $\boldsymbol{\sigma}(t_0)$ in the direction of $\boldsymbol{\sigma}'(t_0)$ (assuming $\boldsymbol{\sigma}'(t_0) \neq 0$) is called the *tangent line* to the curve (see Worked Example 7).

For a curve describing uniform rectilinear motion, the velocity vector is constant (see Worked Example 6). In general, the velocity vector is a vector function $\mathbf{v} = \boldsymbol{\sigma}'(t)$ which depends on t. The derivative $\mathbf{a} = d\mathbf{v}/dt = \boldsymbol{\sigma}''(t)$ is called the *acceleration vector* of the curve. Notice that if the curve is $(f(t), g(t), h(t))$, then the acceleration vector is

$$\mathbf{a} = f''(t)\mathbf{i} + g''(t)\mathbf{j} + h''(t)\mathbf{k}$$

The terms *velocity, speed,* and *acceleration* come from physics, where parametric curves represent the motion of particles.

Worked Example 10   A particle moves in a helical path along the curve $(\cos t, \sin t, t)$. Find its velocity and acceleration vectors.

Solution   Differentiating the components, we have $\mathbf{v} = -\sin t\,\mathbf{i} + \cos t\,\mathbf{j} + \mathbf{k}$, and $\mathbf{a} = d\mathbf{v}/dt = -\cos t\,\mathbf{i} - \sin t\,\mathbf{j}$. Notice that the acceleration vector points directly from $(\cos t, \sin t, t)$ to the z axis and is perpendicular to the axis as well as to the velocity vector (see Fig. 15-3-8).

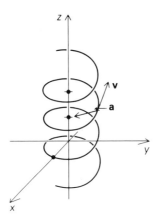

Fig. 15-3-8 The velocity and acceleration of a particle moving on a helix.

Solved Exercises   6. Find the speed of the particle in Worked Example 10.

7. A particle moves in such a way that its acceleration is constantly equal to $-\mathbf{k}$. If the position when $t = 0$ is $(0, 0, 1)$ and the velocity at $t = 0$ is $\mathbf{i} + \mathbf{j}$, when and where does the particle fall below the plane $z = 0$? Describe the path traveled by the particle.

Exercises   6. Show that if the acceleration of an object is always perpendicular to the velocity, then the speed of the object is constant. [*Hint:* See Worked Example 9.]

7. For each of the curves in Exercise 1 in this section, find the velocity vector and an expression for the line tangent to the curve at $t = 1$.

8. For each curve in Exercise 2, find the velocity vector $\mathbf{v}$ and the speed v as functions of t. Compute $\int_a^b v\,dt$, where $[a, b]$ is the defining interval for t in each case. What should this number represent? Explain the difference between the result for part (ii) and that for (i) and (iii).

9. Describe and sketch the curves specified by the following data (see Solved Exercise 7):
   (a) $\boldsymbol{\sigma}'(t) = (1, 0, 1)$; $\boldsymbol{\sigma}(0) = (0, 0, 0)$
   (b) $\boldsymbol{\sigma}'(t) = (-1, 1, 1)$; $\boldsymbol{\sigma}(0) = (1, 2, 3)$
   (c) $\boldsymbol{\sigma}'(t) = (-1, 1, 1)$; $\boldsymbol{\sigma}(0) = (0, 0, 0)$

ARC LENGTH

In Section 11-3 we found the formula

$$L = \int_a^b \sqrt{f'(t)^2 + g'(t)^2}\, dt$$

for the length of a parametric curve in the plane. We use the same formula, with one term added, for curves in space.

Definition Let $(x, y, z) = (f(t), g(t), h(t))$ be a parametric curve in space. The *length* of the curve, for t in the interval $[a, b]$, is defined to be

$$L = \int_a^b \sqrt{(f'(t))^2 + (g'(t))^2 + (h'(t))^2}\, dt$$

or

$$L = \int_a^b \sqrt{\left(\frac{dx}{dt}\right)^2 + \left(\frac{dy}{dt}\right)^2 + \left(\frac{dz}{dt}\right)^2}\, dt$$

Worked Example 11 Find the length of the helix $(\cos t, \sin t, t)$ for $0 \le t \le \pi$.

Solution Here $f'(t) = -\sin t$, $g'(t) = \cos t$, and $h'(t) = 1$, so the integrand in the arc length formula is $\sqrt{\sin^2 t + \cos^2 t + 1} = \sqrt{2}$, a constant. Thus the length is simply

$$L = \int_0^\pi \sqrt{2}\, dt = \pi\sqrt{2}$$

Notice that the integrand in the arc length formula is precisely the speed $\|\boldsymbol{\sigma}'(t)\|$ of a particle moving along the parametric curve. Thus the arc length, which can be written as $L = \int_a^b \|\boldsymbol{\sigma}'(t)\|\, dt$, is the integral of speed with respect to time and represents the total distance traveled by the particle between time a and time b.

Solved Exercises*
8. Find the arc length of $(\cos t, \sin t, t^2)$, $0 \le t \le \pi$.

9. Find the arc length of (e^t, t, e^t), $0 \le t \le 1$. [*Hint:* Use the substitution $u = \sqrt{1 + 2e^{2t}}$ to evaluate the integral.]

Exercises Find the arc length of the given curve on the specified interval.

10. $(1, 3t^2, t^3)$; $0 \le t \le 1$

11. $(\sin 3t, \cos 3t, 2t^{3/2})$; $0 \le t \le 1$

12. $(t, t\sin t, t\cos t)$; $0 \le t \le \pi$

---

*Several of these exercises make use of the formula

$$\int \sqrt{x^2 + a^2}\, dx = \tfrac{1}{2}[x\sqrt{x^2 + a^2} + a^2 \ln(x + \sqrt{x^2 + a^2})] + C$$

(optional) **NEWTON'S SECOND LAW**

If a particle of mass m moves in space, the total force $\mathbf{F}$ acting on it at any time is a vector which is related to the acceleration by *Newton's second law* (see Section 6-3):

$$\mathbf{F} = m\mathbf{a}$$

Many interesting problems involve determining the path of a particle, given its mass, initial position, and velocity and given a force law. If the path is described by a vector function $\mathbf{r} = \boldsymbol{\sigma}(t)$, Newton's law becomes a differential equation for $\boldsymbol{\sigma}(t)$, and techniques of differential equations can be used to solve it (as we solved the spring equation in Section 9-1). For example, a planet moving around the sun (considered to be located at the origin) obeys to a high degree of precision the law

$$m\frac{d^2\mathbf{r}}{dt^2} = -\frac{GmM}{\|\mathbf{r}\|^3}\mathbf{r} = -\frac{GmM}{r^3}\mathbf{r}$$

where $\mathbf{r}$ is the vector pointing from the sun to the planet at time t, M is the mass of the sun, m that of the planet, $r = \|\mathbf{r}\|$, and G is the gravitational constant ($G = 6.67 \times 10^{-11}$ newton meter^2 per kilogram^2). The law of force given by $\mathbf{F} = -GmM\mathbf{r}/r^3$ is called *Newton's law of gravitation*. We will not investigate the general solution of these equations in this book but will content ourselves with the following special case.

Worked Example 12   A particle of mass m is moving in the xy plane at constant speed v in a circular path of radius r_0. Find the acceleration.

Solution   Let $\mathbf{r}$ be the vector from the center of the circle to the particle at time t. Motion of the type described is given by

$$\mathbf{r} = r_0 \cos\left(\frac{tv}{r_0}\right)\mathbf{i} + r_0 \sin\left(\frac{tv}{r_0}\right)\mathbf{j}$$

Differentiating twice, we see that

$$\mathbf{a} = \frac{d^2\mathbf{r}}{dt^2} = -\frac{v^2}{r_0}\cos\left(\frac{tv}{r_0}\right)\mathbf{i} - \frac{v^2}{r_0}\sin\left(\frac{tv}{r_0}\right)\mathbf{j} = -\frac{v^2}{r_0^2}\mathbf{r}$$

In uniform circular motion the acceleration vector points in a direction opposite to $\mathbf{r}$—that is, it is directed toward the center of the circle (see Fig. 15-3-9). This acceleration, multiplied by the mass of the particle, is called the *centripetal force*. Note that even though the speed is constant, the *direction* of the velocity vector is continually changing, which is why there is an acceleration. By Newton's law, some force must cause the acceleration which keeps the particle moving in its circular path. In whirling a rock at the end of a string, you must constantly be pulling on the string. If you stop that force by releasing the string, the rock will fly off in a straight line tangent to the circle. The force needed to keep a planet or satellite bound into an elliptical or circular orbit is supplied by gravity. The force needed to keep a car going around a curve must be supplied by the friction of the tires against the road or by direct pressure if the road is banked (see Problem 17).

Suppose that a satellite is moving with a speed v around a central body with mass M in a circular orbit of radius r_0. Then, by Worked Example 12 and Newton's law,

Fig. 15-3-9 The acceleration vector of a particle in uniform circular motion points to the center of the circle.

(optional)

$$-\frac{v^2}{r_0^2}\mathbf{r} = -\frac{GM}{r_0^3}\mathbf{r}$$

The lengths of the vectors on both sides of this equation must be equal. Hence

$$v^2 = \frac{GM}{r_0} \tag{1}$$

If T is the period of one revolution, then $2\pi r_0/T = v$ (distance/time = speed); substituting this value for v in equation (1) and solving for T^2, we obtain the rule:

$$T^2 = r_0^3 \frac{(2\pi)^2}{GM}$$

The square of the period is proportional to the cube of the radius. This law is one of the famous three which were discovered empirically by Kepler before Newton's laws were formulated; it enables one to compute the period of a satellite given the radius of its orbit or to determine the radius of the orbit if the period is prescribed.

Solved Exercises ★10. Let $\mathbf{r} = \boldsymbol{\sigma}(t)$ be the vector from a fixed point to the position of an object, $\mathbf{v}$ its velocity, and $\mathbf{a}$ its acceleration. Suppose that $\mathbf{F}$ is the force acting at time t. Prove that $(d/dt)(m\mathbf{r} \times \mathbf{v}) = \boldsymbol{\sigma} \times \mathbf{F}$, (that is, "rate of change of *angular momentum = torque*"). What can you conclude if $\mathbf{F}$ is parallel to $\mathbf{r}$? Is this the case in planetary motion?

★11. Continue the investigations in Solved Exercise 10 to prove that a planet moving about the sun does so in a fixed plane. (This is another of Kepler's laws.)

Exercises ★13. For a falling body near the surface of the earth, the force of gravity can be approximated very well as a constant downward force with magnitude $F = GmM/R^2$, where G is the gravitational constant, M the mass of the earth, m the mass of the body, and R the radius of the earth.

$M = 5.98 \times 10^{24}$ kilograms

$R = 6.37 \times 10^6$ meters

$G = 6.67 \times 10^{-11}$ newton meter^2 per kilogram^2

(a) Show that this approximation means that any body falling freely (neglecting air resistance) near the surface of the earth experiences a constant acceleration of $g = 9.8$ meters per second per second. Note that this acceleration is independent of m. Any two bodies fall at the same rate.

(b) Show that the flight path of a projectile or a baseball is a parabola (see Solved Exercise 7).

★14. Suppose we want to have a satellite in circular orbit about the earth in such a way that it stays fixed in the sky over one point on the equator. What should be the radius of such an orbit?

ELEMENTARY DIFFERENTIAL GEOMETRY OF SPACE CURVES

Differential geometry is the branch of mathematics in which calculus is used to study the geometry of curves, surfaces, and more complicated objects. When we studied the arc length of curves, we were already doing differential geometry—now we will go further and introduce the important idea of curvature.

The *curvature* of a curve in the plane or in space is a measure of the rate at which the direction of motion along the curve is changing. A curve with curvature *(optional)* zero is just a straight line. We can define the curvature as the rate of change of the velocity vector if the length of this vector happens to be 1; otherwise the change in length of the velocity vector confuses the issue. We make, therefore, the following definitions.

Definition Let $\mathbf{r} = \sigma(t)$ be a parametric curve.

1. The curve is called *regular* if $\mathbf{v} = \sigma'(t)$ is not equal to zero for any t.

2. If the curve is regular, the vector $\mathbf{T} = \mathbf{v}/\|\mathbf{v}\| = \sigma'(t)/\|\sigma'(t)\|$ is called the *unit tangent vector* to the curve.

3. If the length of $\sigma'(t)$ is constant and equal to 1 (in which case $\mathbf{T} = \mathbf{v}$), the curve is said to be *parametrized by arc length*.

Worked Example 13 Suppose that the curve $\mathbf{r} = \sigma(t)$ is parametrized by arc length. Show that the length of the curve between $t = a$ and $t = b$ is simply $b - a$.

Solution The integrand in the arc length formula on p. 671 is constant and equal to 1 if the curve is parametrized by arc length. Thus

$$L = \int_a^b 1\, dt = b - a$$

If a curve $\mathbf{r} = \sigma(t)$, as it is presented to us, is regular but is not parametrized by arc length, we can introduce a new independent variable so that the new curve *is* parametrized by arc length. In fact, we can choose a value a in the domain of the curve and define $s = p(t)$ to be the arc length $\int_a^t \|\sigma'(u)\|\,du$ of the curve between a and t. We have $ds/dt = \|\sigma'(t)\| > 0$ since the curve is regular, so the inverse function $t = q(s)$ exists (see Section 5-1). Now look at the new curve $\mathbf{r} = \sigma_1(s) = \sigma(q(s))$, which goes through the same points in space as the original curve but at a different speed. In fact, the new speed is

$$\|\sigma_1'(s)\| = \|q'(s)\sigma'(q(s))\| \quad \text{(chain rule)}$$
$$= q'(s)\|\sigma'(q(s))\|$$
$$= \frac{1}{p'(q(s))}\|\sigma'(q(s))\|$$

But by the fundamental theorem of calculus and the definition of p, $p'(t) = \|\sigma'(t)\|$, so $\|\sigma_1'(s)\| = 1$, and so the new curve is parametrized by arc length.

The arc length parametrization is mostly useful for theoretical purposes, since the integral in its definition is usually impossible to evaluate. Still, the *existence* of this parametrization makes the definitions which follow much simpler.

Worked Example 14 Find the arc length parametrization for the helix $(\cos t, \sin t, t)$.

Solution We have $\|\mathbf{v}\| = \sqrt{\sin^2 t + \cos^2 t + 1} = \sqrt{2}$. Taking $a = 0$, we have $s = p(t) = \int_0^t \sqrt{2}\, du = \sqrt{2}\,t$, so $s = \sqrt{2}\,t$, and $t = s/\sqrt{2}$; the curve in arc length parametrization is therefore $(\cos(s/\sqrt{2}), \sin(s/\sqrt{2}), s/\sqrt{2})$.

Whenever a curve is parametrized by arc length, we will denote this parameter by s. Notice that in this case $\mathbf{T} = \mathbf{v} = d\mathbf{r}/ds$.

(optional)

> *Definition* Let $\mathbf{T}$ be the unit tangent vector of a curve parametrized by arc length. The scalar $k = \|d\mathbf{T}/ds\|$ is called the *curvature* of the curve. If $k \neq 0$, the unit vector $\mathbf{N} = (d\mathbf{T}/ds)/\|d\mathbf{T}/ds\|$ is called the *principal normal vector* to the curve.

Worked Example 15 Show that the principal normal vector is perpendicular to the unit tangent vector.

Solution Since $\mathbf{T}$ has constant length, we know by Worked Example 9 that $d\mathbf{T}/ds$ is perpendicular to $\mathbf{T}$. Since $\mathbf{N}$ has the same direction as $d\mathbf{T}/ds$, it is perpendicular to $\mathbf{T}$ as well.

Worked Example 16 Compute the curvature and principal normal vector of the helix in Worked Example 14.

Solution We have $\mathbf{T} = -(1/\sqrt{2})\sin(s/\sqrt{2})\mathbf{i} + (1/\sqrt{2})\cos(s/\sqrt{2})\mathbf{j} + (1/\sqrt{2})\mathbf{k}$, so $d\mathbf{T}/ds = -\frac{1}{2}\cos(s/\sqrt{2})\mathbf{i} - \frac{1}{2}\sin(s/\sqrt{2})\mathbf{j}$; the curvature is

$$\sqrt{\frac{1}{4}\cos^2\left(\frac{s}{\sqrt{2}}\right) + \frac{1}{4}\sin^2\left(\frac{s}{\sqrt{2}}\right)} = \sqrt{\frac{1}{4}} = \frac{1}{2}$$

and the principal normal vector is $-\cos(s/\sqrt{2})\mathbf{i} - \sin(s/\sqrt{2})\mathbf{j}$.

If a curve is not parametrized by arc length, it is possible to compute its curvature and principal normal vector directly by the following formulas:

$$k = \frac{\|\mathbf{v} \times \mathbf{v}'\|}{\|\mathbf{v}\|^3} \tag{2}$$

$$\mathbf{N}(t) = \frac{(\mathbf{v}\cdot\mathbf{v})\mathbf{v}' - (\mathbf{v}'\cdot\mathbf{v})\mathbf{v}}{\|(\mathbf{v}\cdot\mathbf{v})\mathbf{v}' - (\mathbf{v}'\cdot\mathbf{v})\mathbf{v}\|} \tag{3}$$

These formulas are derived in Solved Exercise 12 and Exercise 16.

Worked Example 17 Find the curvature of the exponential spiral $(e^{-t}\cos t, e^{-t}\sin t, 0)$ (Fig. 15-3-10). What happens as $t \to \infty$?

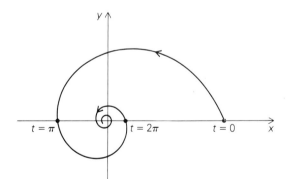

Fig. 15-3-10 Graph of the exponential spiral in the (x, y) plane.

Solution We have

$$\mathbf{v} = (-e^{-t}\cos t - e^{-t}\sin t)\mathbf{i} + (-e^{-t}\sin t + e^{-t}\cos t)\mathbf{j}$$

and

$$\mathbf{v}' = (e^{-t}\cos t + e^{-t}\sin t + e^{-t}\sin t - e^{-t}\cos t)\mathbf{i}$$
$$+ (e^{-t}\sin t - e^{-t}\cos t - e^{-t}\cos t - e^{-t}\sin t)\mathbf{j}$$
$$= 2e^{-t}(\sin t\,\mathbf{i} - \cos t\,\mathbf{j})$$

(optional) | Then

$$\mathbf{v} \times \mathbf{v}' = 2e^{-2t} \begin{vmatrix} -\cos t - \sin t & -\sin t + \cos t \\ \sin t & -\cos t \end{vmatrix} \mathbf{k}$$

$$= 2e^{-2t}(\cos^2 t + \cos t \sin t + \sin^2 t - \cos t \sin t)\mathbf{k}$$

$$= 2e^{-2t}\mathbf{k}$$

so

$$\|\mathbf{v} \times \mathbf{v}'\| = 2e^{-2t}$$

Since $\|\mathbf{v}\| = \{[e^{-t}(\cos t + \sin t)]^2 + [e^{-t}(\cos t - \sin t)]^2\}^{1/2}$,

$$\|\mathbf{v}\|^3 = e^{-3t}[(\cos t + \sin t)^2 + (\cos t - \sin t)^2]^{3/2}$$

$$= e^{-3t}2^{3/2}$$

By formula (2) above,

$$k = \frac{2e^{-2t}}{e^{-3t}2^{3/2}} = \frac{e^t}{\sqrt{2}}$$

As $t \to \infty$, the curvature goes to infinity as the spiral wraps more and more tightly about the origin.

Solved Exercises ★12. Prove formula (2) above.

★13. Show that if the curvature of a curve is identically zero, then the curve is a straight line.

Exercises ★15. Show that a circle of radius r has constant curvature $1/r$.

★16. Derive formula (3) above by using the methods of Solved Exercise 12.

★17. Compute the curvature vector and principal normal vector for the helix $(r \cos wt, r \sin wt, ct)$ in terms of r, w, and c.

★18. Find the curvature of the ellipse $x^2 + 2y^2 = 1$, $z = 0$. (Choose a suitable parametrization.)

Problems for Section 15-3

1. Sketch each of the following curves:
   (a) $(4 \cos t, 2 \sin t, t)$; $0 \le t \le 2\pi$
   (b) $(t, 1/t, t)$; $1 \le t \le 3$

2. Sketch the curves:
   (a) $(t, -t, t^2)$; $0 \le t \le 2$
   (b) $(\cosh t, \sinh t, t)$; $-1 \le t \le 1$

3. Let θ and ϕ be fixed angles, and consider these curves:
   (a) $x = \sin \phi \cos t$
      $y = \sin \phi \sin t$ $\quad 0 \le t \le 2\pi$
      $z = \cos \phi$
   (b) $x = \sin t \cos \theta$
      $y = \sin t \sin \theta$ $\quad 0 \le t \le 2\pi$
      $z = \cos t$

   Show that each is a circle lying in the sphere of radius 1 centered at the origin. Find the center and radius of each circle. Sketch the curves for $\phi = 45°$ and for $\theta = 45°$.

4. Suppose that $0 \le \theta \le 2\pi$ and $0 \le \phi \le 2\pi$, and let

$$\sigma(\theta, \phi) = ((2 + \cos\phi)\cos\theta, (2 + \cos\phi)\sin\theta, \sin\phi)$$

(Note that this is a vector function of two variables.)

(a) Describe each of the following curves:

   (i) $\sigma(\theta, 0)$; $0 \le \theta \le 2\pi$

   (ii) $\sigma(\theta, \pi)$; $0 \le \theta \le 2\pi$

   (iii) $\sigma(\theta, \pi/2)$; $0 \le \theta \le 2\pi$

   (iv) $\sigma(0, \phi)$; $0 \le \phi \le 2\pi$

   (v) $\sigma(\pi/2, \phi)$; $0 \le \phi \le 2\pi$

   (vi) $\sigma(\pi/4, \phi)$; $0 \le \phi \le 2\pi$

(b) Show that $\sigma(\theta, \phi)$ lies on the circle of radius $2 + \cos\phi$ parallel to the xy plane and centered at $(0, 0, \sin\phi)$.

(c) Show that $\sigma(\theta, \phi)$ lies on the doughnut-shaped surface (a *torus*) shown in Fig. 15-3-11.

(d) Describe and sketch the curve

$$((2 + \cos t)\cos t, (2 + \cos t)\sin t, \sin t)$$

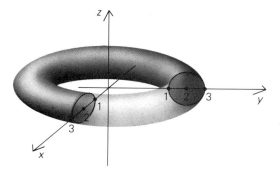

Fig. 15-3-11 The points $\sigma(\theta, \phi)$ (Problem 4) lie on this surface.

5. Compute the velocity and acceleration vectors for the curves in Problems 1 and 2.

6. For each of the following curves, determine the velocity and acceleration vectors and the equation for the tangent line at the specified value of t.

(a) $(6t, 3t^2, t^3)$; $t = 0$

(b) $(\sin 3t, \cos 3t, 2t^{3/2})$; $t = 1$

(c) $(\cos^2 t, 3t - t^3, t)$; $t = 0$

(d) $(t \sin t, t \cos t, \sqrt{3}\, t)$; $t = 0$

(e) $(\sqrt{2}\, t, e^t, e^{-t})$; $t = 0$

7. Suppose that a curve has the velocity vector $\mathbf{v} = \sigma'(t) = (\sin t, -\cos t, d)$, where d is a constant. Describe the curve. Sketch the curve if you know that $\sigma(0) = \mathbf{i}$. What if $d = 0$?

8. Suppose that a curve $\sigma(t)$ has the velocity vector $\sigma'(t) = (a, b, \sin t)$, where a and b are constants. Sketch the curve if $a = -1$, $b = 2$, and $\sigma(0) = (0, 0, 1)$.

★9. A satellite is in a circular orbit 500 miles above the surface of the earth. What is the period of the orbit?

10. Suppose that a particle follows a path $(e^t, e^{-t}, \cos t)$ until it flies off on a tangent at $t = 1$. Where is it at $t = 2$?

★11. Let $\mathbf{r} = \sigma(t)$ be a parametric curve.

(a) Suppose there is a unit vector $\mathbf{u}$ (constant) such that $\sigma(t) \cdot \mathbf{u} = 0$ for all values of t. What can you say about the curve $\sigma(t)$?

(b) What can you say if $\sigma(t) \cdot \mathbf{u} = c$ for some constant c? [*Hint:* Differentiate.]

(c) What cay you say if $\sigma(t) \cdot \mathbf{u} = b\|\sigma(t)\|$ for some constant b with $0 < b < 1$?

12. Show that, at a local maximum or minimum of $\|\sigma(t)\|$, $\sigma'(t)$ is perpendicular to $\sigma(t)$.

13. (a) Let $\sigma_1(t)$ and $\sigma_2(t)$ satisfy the differential equation $\sigma''(t) = -\sigma(t)$. Show that for any constants A_1 and A_2, $A_1\sigma_1(t) + A_2\sigma_2(t)$ satisfies the equation as well.

(b) Find as many solutions of $\sigma''(t) = -\sigma(t)$ as you can.

14. Suppose that $\sigma(t)$ satisfies the equation $\sigma''(t) + \omega^2\sigma(t) = \mathbf{0}$. Describe and sketch the curve if $\sigma(0) = (0, 0, 1)$ and $\sigma'(0) = (0, \omega, \omega)$.

15. Suppose that $P_0 = (x_0, y_0, 0)$ is a point on the unit circle in the xy plane. Describe the set of points lying directly above or below P_0 on the right circular helix of Worked Example 3. What is the vertical distance between coils of the helix?

★16. Sketch and discuss the curve given by

$$x = r \cos \omega t$$

$$y = r \sin \omega t$$

$$z = ct$$

where r, ω, and c are positive constants and $-\infty < t < \infty$.

(a) What path is traced out by (x, y) in the plane?

(b) The curve in space lies on what cylinder?

(c) It is a right-circular helix. Sketch it.

(d) What is the vertical distance between coils?

(e) For what t_0 does the curve trace out one coil of the helix as t goes through the interval $0 \le t \le t_0$?

★17. (a) Suppose that a car is going around a circular curve of radius r at speed v. Then it will exert an outward horizontal force on the roadway due to the centripetal acceleration and a vertical force due to gravity. At what angle θ should the roadway be banked so that the total force tends to press the car directly into (perpendicular to) the roadway? (See Fig. 15-3-12.) How does the bank angle depend on r and on the speed v?

(b) Discuss how you might treat the design problem in part (a) for a curve that is not part of a circle. Design an elliptical racetrack with major axis 800 meters, minor axis 500 meters, and speed 160 kilometers per hour.

Fig. 15-3-12 For what value of θ does the total force press directly into the roadway?

★18. A particle with charge q moving with velocity vector **v** through a magnetic field is acted on by the force $\mathbf{F} = (q/c)\,\mathbf{v} \times \mathbf{B}$, where c is the speed of light and **B** is a vector describing the magnitude and direction of the magnetic field. Suppose that:

(1) The particle has mass m and is following a path $\mathbf{r} = \boldsymbol{\sigma}(t) = x\mathbf{i} + y\mathbf{j} + z\mathbf{k}$.

(2) $\boldsymbol{\sigma}(0) = \mathbf{i}$; $\boldsymbol{\sigma}'(0) = a\mathbf{j} + c\mathbf{k}$.

(3) The magnetic field is constant and uniform given by a vector $\mathbf{B} = b\mathbf{k}$.

(a) Use the equation $\mathbf{F} = (q/c)\,\mathbf{v} \times \mathbf{B} - mg\mathbf{k}$ to write differential equations relating the components of $\mathbf{a} = \boldsymbol{\sigma}''(t)$ and $\mathbf{v} = \boldsymbol{\sigma}'(t)$.

(b) Solve these equations to obtain the components of $\boldsymbol{\sigma}(t)$. [*Hint:* Integrate the equations for d^2x/dt^2 once, use item (2) in the list above to determine the constant of integration, and substitute the resulting expression for dx/dt into the equation for d^2y/dt^2 to get an equation similar to the spring equation solved in Section 9-1.]

(c) Show that the path is a right circular helix. What are the radius and axis of the cylinder on which it lies? (The dimensions of the helix followed by a particle in a magnetic field in a bubble chamber are used to measure the charge to mass ratio of the particle.)

19. Compute the arc length for the following curve between the indicated limits:

$$\left(t + 1, \frac{2\sqrt{2}}{3} t^{3/2} + 7, \frac{1}{2} t^2 \right) \quad \text{for } 1 \le t \le 2$$

20. Compute the arc length of the curve

$$(t, t, t^2) \quad \text{for } 1 \le t \le 2$$

★21. Let **T** and **N** be the unit tangent and principal normal vectors to a space curve $\mathbf{r} = \boldsymbol{\sigma}(t)$. Define a third unit vector perpendicular to them by $\mathbf{B} = \mathbf{T} \times \mathbf{N}$. This is called the *binormal* vector. Together, **T**, **N**, and **B** form a right-handed system of mutually orthogonal unit vectors which may be thought of as moving along the curve (see Fig. 15-3-13).

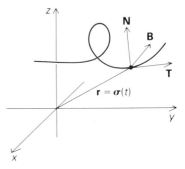

Fig. 15-3-13 The vectors **T**, **N**, and **B** form a "moving basis" along the curve.

(a) Show that

$$\mathbf{B} = (\mathbf{v} \times \mathbf{a})/\|\mathbf{v} \times \mathbf{a}\| = (\mathbf{v} \times \mathbf{a})/(k\|\mathbf{v}\|^3)$$

where **a** is the acceleration vector.

(b) Show that $(d\mathbf{B}/dt)\cdot\mathbf{B} = 0$. [*Hint:* $\|\mathbf{B}\|^2 = 1$ is constant.]

(c) Show that $(d\mathbf{B}/dt)\cdot T = 0$. [*Hint:* Take derivatives in $\mathbf{B}\cdot\mathbf{T} = 0$.]

(d) Show that $d\mathbf{B}/dt$ is a scalar multiple of **N**.

(e) Using part (d) we can define a scalar-valued function τ called *torsion* by $d\mathbf{B}/dt = -\tau\|\mathbf{v}\|\mathbf{N}$. Show that

$$\tau = \frac{[\boldsymbol{\sigma}'(t) \times \boldsymbol{\sigma}''(t)] \cdot \boldsymbol{\sigma}'''(t)}{\|\boldsymbol{\sigma}'(t) \times \boldsymbol{\sigma}''(t)\|^2}$$

★22. (a) Show that if the curve lies in a plane, then the torsion τ is identically zero. [*Hint:* The vector function $\sigma(t)$ must satisfy an equation like $\sigma(t) \cdot \mathbf{n} = 0$. By taking successive derivatives show that σ', σ'', and σ''' all lie in the same plane through the origin. What does this do to the triple product in Problem 21(e)?]

   (b) Show that $\mathbf{B}$ is then constant and is a normal vector to the plane in which the curve lies.

★23. If the torsion is not zero, it gives a measure of how fast the curve is tending to twist out of the plane. Compute the binormal vector and the torsion for the helix of Worked Example 14.

★24. Using the results of Problems 21 and 22, prove the following *Frenet formulas* for a curve parametrized by arc length:

$$\frac{d\mathbf{T}}{ds} = k\mathbf{N}$$

$$\frac{d\mathbf{N}}{ds} = -k\mathbf{T} + \tau\mathbf{B}$$

$$\frac{d\mathbf{B}}{ds} = -\tau\mathbf{N}$$

[*Hint:* To get the last formula from the previous ones, note that $\mathbf{N} \cdot \mathbf{N}$, $\mathbf{N} \cdot \mathbf{B}$, and $\mathbf{N} \cdot \mathbf{T}$ are constant. Take derivatives and use earlier formulas to get $(d\mathbf{N}/dt) \cdot \mathbf{B}$ and $(d\mathbf{N}/dt) \cdot \mathbf{T}$.]

Review Problems for Chapter 15

1. Sketch the curves or surfaces described by the following equations:
   (a) $z = x + y$ (b) $z = -(x^2 + y^2)$
   (c) $\sigma(t) = 3\sin t\mathbf{i} + t\mathbf{j} + \cos t\mathbf{k}$
   (d) $z^2 = x^2 - 4y^2$

2. Sketch the following curves or surfaces in space:
   (a) $z^2 = -(x^2 + y^2/4)$
   (b) $\sigma(t) = (\sin t, t + 1, 2t - 1)$
   (c) $z^2 = -x^2 - 3y^2 + 2$
   (d) $x^2 + 2xz + z^2 = 0$

3. This problem concerns the *elliptic hyperboloid of one sheet*. An example of this type was studied in Section 15-2, Solved Exercise 2. A standard form for the equation is

$$\frac{x^2}{a^2} + \frac{y^2}{b^2} - \frac{z^2}{c^2} = 1 \quad (a, b, \text{ and } c \text{ positive})$$

   (a) What are the horizontal cross-sections obtained by holding z constant?
   (b) What are the vertical cross-sections obtained by holding either x or y constant?
   (c) Sketch the surface defined by

$$\frac{x^2}{4} + \frac{y^2}{1} - \frac{z^2}{1} = 1$$

   In a yz plane, sketch the cross-section curves obtained from this surface when x is held constantly equal to 0, 1, 2, and 3 ($x = 2$ is especially interesting).

4. Sketch and describe each of the following surfaces:
   (a) $x^2 + 4y^2 + z^2 = 1$ (b) $x^2 + 4y^2 + z^2 = 0$
   (c) $x^2 + 4y^2 - z^2 = 1$ (d) $x^2 + 4y^2 - z^2 = 0$

5. Sketch and classify the following surfaces:
   (a) $x^2 + 4y^2 - z = 1$
   (b) $x^2 + 4y^2 - z = 0$
   (c) $x^2 + 4y^2 + z = 1$
   (d) $x^2 + 4y^2 + z = 0$

6. Find the equation of the line tangent to each of the following curves at the indicated point:
   (a) $(t^3 + 1, e^{-t}, \cos(\pi t/2))$; $t = 1$
   (b) $(t^2 - 1, \cos t^2, t^4)$; $t = \sqrt{\pi}$

7. Find the velocity and acceleration vectors for the following curves:
   (a) $\sigma_1(t) = e^t\mathbf{i} + \sin t\mathbf{j} + \cos t\mathbf{k}$
   (b) $\sigma_2(t) = \dfrac{t^2}{1 + t^2}\mathbf{i} + t\mathbf{j} + \mathbf{k}$
   (c) $\sigma(t) = \sigma_1(t) + \sigma_2(t)$, where σ_1 and σ_2 are given in parts (a) and (b).

8. Find the arc length of $\sigma(t) = t\mathbf{i} + \ln t\mathbf{j} + 2\sqrt{2t}\,\mathbf{k}$; $1 \le t \le 2$.

9. Write in parametric form the curve described by the equations $x - 1 = 2y + 1 = 3z + 2$.

★10. (a) Write in parametric form the curve which is the intersection of the surfaces
$$x^2 + y^2 + z^2 = 3 \quad \text{and} \quad y = 1$$
   (b) Find the equation of the line tangent to this curve at $(1, 1, 1)$.
   (c) Write an integral expression for the arc length of this curve. What is the value of this integral?

★11. Describe the level curves $f(x, y) = c$ for each of the following functions. In particular, discuss any special values of c at which the behavior of the level curves changes suddenly. Sketch the curves for $c = -1, 0,$ and 1.

(a) $f(x, y) = x + 2y$ (b) $f(x, y) = x^2 - y^2$

(c) $f(x, y) = y^2 - x^2$ (d) $f(x, y) = x^2 + y^2$

(e) $f(x, y) = xy$ (f) $f(x, y) = y - 2x^2$

12. Describe the level surfaces $f(x, y, z) = c$ for each of the following functions. Sketch for $c = 1$ and $c = 5$.

(a) $f(x, y, z) = x - y - z$

(b) $f(x, y, z) = x^2 + y^2 + z^2 + 1$

(c) $f(x, y, z) = x^2 + 2y^2 + 3z^2$

13. (a) Describe the level surfaces of the function $f(x, y, z) = x^2 + y^2 - z^2$. In particular, discuss the surface $f(x, y, z) = c$ when c is positive, when c is negative, and when c is zero.

(b) Several level surfaces of f are sketched in Fig. 15-R-1. Find the value of c associated with each.

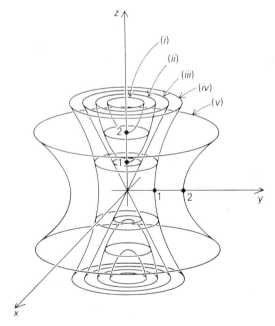

Fig. 15-R-1   Level surfaces of $x^2 + y^2 - z^2$.

(c) Describe how the appearance of the level surfaces changes if we consider instead the function $g(x, y, z) = x^2 + 2y^2 - z^2$.

14. Let $f(x, y) = y/x$.

(a) Sketch the level curves $f(x, y) = c$ for $c = -10, -1, 0, 1, 2,$ and 10.

(b) Describe the intersection of the graph of f with the vertical planes $x = 1$, $x = -1$, $x = 2$, $y = 1$, $y = 2$, $y = -1$.

(c) Describe the intersection of the graph of f with the cylinder of radius R (that is, $r = R$ in cylindrical coordinates).

(d) Sketch the graph of f.

15. The contour lines on a topographical map are the level curves of the function giving height above sea level as a function of position. Figure 15-R-2 is a portion of the U.S. Geological Survey map of Yosemite Valley. There is a heavy contour line for every 200 feet of elevation and a lighter line at each 40-foot interval between these.

(a) What does it mean in terms of the terrain when these contour lines are far apart?

(b) What if they are close together?

(c) What does it mean when several contour lines seem to merge together for a distance? Is elevation really a function of position at such points? (Look at the west face of Half Dome. Does this seem like a good direction from which to climb it?)

(d) Sketch a cross-section of terrain along a north-south line through the top of Half Dome.

(e) A hiker is not likely to follow the straight north-south path in part (d) and is probably more interested in the behavior of the terrain along the trail he or she will follow. The 3-mile route from the Merced River (altitude approximately 6100 feet) to the top of Half Dome (approximately 8842 feet) along the Sunrise and Half Dome Trails has been emphasized in Fig. 15-R-2. Show how a cross-section of the terrain along this trail behaves by plotting altitude above sea level as a function of miles along the trail. (A piece of string or flexible wire may be of aid in measuring distances along the trail.)

Fig. 15-R-2 Yosemite Valley (portion). (*U.S. Department of Interior Geological Survey.*)

16. Fill in the blanks and plot:

	Rectangular coordinates	Cylindrical coordinates	Spherical coordinates
(a)	$(1, -1, 1)$		
(b)			$(3, -\pi/6, \pi/4)$
(c)		$(5, \pi/12, 4)$	
(d)			$(10, \pi/4, \pi/2)$
(e)		$(8, 3\pi/2, 2)$	
(f)	$(1, 0, 3)$		

17. A surface is described in cylindrical coordinates by $3r^2 = z^2 + 1$. Convert to rectangular coordinates and plot.

18. Show that a surface described in spherical coordinates by $f(\rho, \phi) = 0$ is a surface of revolution.

19. A particle is constrained to move around the unit circle in the xy plane according to the formula $(x, y, z) = (\cos(t^2), \sin(t^2), 0)$, $t \geq 0$.

    (a) At what point on the circle should the particle be released to hit a target at $(2, 0, 0)$? (Be careful about which direction the particle is moving around the circle.)

    (b) At what time t should the release take place? (Use the smallest $t > 0$ which will work.)

    (c) What are the velocity vector and speed of the particle as functions of t?

    (d) What are the velocity and speed at the time of release?

    (e) At what time is the target hit?

★20. Let n be a positive integer and consider the curve

$$x = \cos t \cos (4nt)$$

$$y = \cos t \sin (4nt) \qquad -\frac{\pi}{2} \leq t \leq \frac{\pi}{2}$$

$$z = \sin t$$

    (a) Show that the path traced out lies on the surface of the sphere of radius 1 centered at the origin.

    (b) How many times does the curve wind around the z axis?

    (c) Where does the curve cross the xy plane?

    (d) Sketch the curve when $n = 1$ and when $n = 2$.

★21. Let $\mathbf{u}_1$ and $\mathbf{u}_2$ be unit vectors, and define curve $\boldsymbol{\sigma}(t)$ by

$$\boldsymbol{\sigma}(t) = \frac{\mathbf{u}_1 \cos t + \mathbf{u}_2 \sin t}{|\mathbf{u}_1 \cos t + \mathbf{u}_2 \sin t|} \qquad 0 \leq t \leq \frac{\pi}{2}$$

    (a) Find $\boldsymbol{\sigma}(0)$ and $\boldsymbol{\sigma}(\pi/2)$.

    (b) On what surfaces does $\boldsymbol{\sigma}(t)$ lie for all t?

    (c) Find, by geometry, the arc length of the curve $\boldsymbol{\sigma}(t)$ for $0 \leq t \leq \pi/2$.

    (d) Compute the arc length of the curve $\boldsymbol{\sigma}(t)$ for $0 \leq t \leq \pi/2$ by expressing it as an integral.

    (e) Find a curve $\boldsymbol{\sigma}_1(t)$ which traverses the same path as $\boldsymbol{\sigma}(t)$ for $0 \leq t \leq \pi/2$ and such that the speed $\|\boldsymbol{\sigma}_1'(t)\|$ is constant.

★22. Let $\mathbf{r} = \boldsymbol{\sigma}(t)$ be a curve in space and $\mathbf{N}$ be its principal normal vector. Consider the "parallel curve" $\mathbf{r} = \boldsymbol{\mu}(t) = \boldsymbol{\sigma}(t) + \mathbf{N}(t)$, where $\boldsymbol{\sigma}(t)$ is the displacement vector to $P(t)$ from a fixed origin.

    (a) Under what conditions does $\boldsymbol{\mu}(t)$ have zero velocity for some t_0?

    (b) Find the parametric equation of the parallel curve to the ellipse $(\frac{1}{4} \cos t, 4 \sin t, 0)$.

★23. (a) Show that the hyperboloid $x^2 + y^2 - z^2 = 4$ is a ruled surface by finding two straight lines lying in the surface through each point. [*Hint:* Let (x_0, y_0, z_0) lie on the surface; write the equation of the line in the form $x = x_0 + at$, $y = y_0 + bt$, $z = z_0 + t$; write out $x^2 + y^2 - z^2 = 4$ using $x_0^2 + y_0^2 - z_0^2 = 4$ to obtain two equations for a and b representing a line and a circle in the (a, b) plane. Show that these equations have two solutions by showing that the distance from the origin to the line is less than the radius of the circle.]

    (b) Is the hyperboloid $x^2 + y^2 - z^2 = -4$ a ruled surface? Explain.

    (c) Generalize the results of parts (a) and (b).

★24. Find a formula for the curvature of the graph $y = f(x)$ in terms of f and its derivatives.

25. Describe by means of cylindrical coordinates a solenoid consisting of a copper rod of radius 5 cm and length 15 cm wound on the outside with copper wire to a thickness of 1.2 cm. Give separate descriptions of the rod and the winding.

26. A gasoline storage tank has two spherical cap ends of arclength 56.55 ft. The cylindrical part of the tank has length 16 ft and circumference 75.40 ft. Let $(0, 0, 0)$ be the geometric center of the tank. See Fig. 15-R-3.

75.40 ft

56.55 ft

16 ft

Fig. 15-R-3 The gasoline storage tank for Problem 26.

    (a) Describe the cylindrical part of the tank via cylindrical coordinates.

    (b) Describe the end caps with spherical coordinates. [Set up spherical coordinates using the centers of the cap ends as the origin.]

16

Partial Differentiation

A function of several variables can be differentiated with respect to one variable at a time.

The rate of change of a function of several variables is not just a single function, since the independent variables may vary in many ways. All the rates of change, for a function of *n* variables, are described by *n* functions called its *partial derivatives*. These derivatives may be assembled into a single vector called the *gradient,* which is useful in geometric and physical applications. To deal with several functions of several variables at once, we use a matrix to arrange all the rates of change in a convenient form.

CONTENTS

In Section 16-1, partial derivatives are defined and methods of calculation are presented along with simple applications. Most interesting applications involve the chain rule, the basic form of which is introduced in Section 16-2; the chain rule in full generality appears in Section 16-4. The theory of maxima and minima for functions of two variables, analogous to that for one variable given in Chapters 2 and 3, is presented in Section 16-3.

16-1 Partial Derivatives

The partial derivatives of a function of several variables are its ordinary derivatives with respect to each variable separately.

We begin this section with a definition of the partial derivatives of a function of several variables. After the motivation, definition, and some computational practice, we will learn the geometric meaning of partial derivatives. In the case of two variables, they are the "slopes" of a plane tangent to the graph of the function.

Goals

After studying this section, you should be able to:

Compute first and second partial derivatives.

Find the equation of the tangent plane to the graph of a function of two variables.

Find linear approximations to functions of two and three variables.

THE DEFINITION OF PARTIAL DERIVATIVES

(a)

(b)

Fig. 16-1-1 Moving water in a narrow tank shown at two different instants of time.

Fig. 16-1-2 The motion of the water is depicted by a graph in (t, y, z) space; sections by planes of the form $t = t_0$ show the configuration of the water at various instants of time.

Two of the most important problems in the historical development of partial differentiation concerned *wave motion* and *heat conduction*. We will use wave motion to illustrate the notion of partial differentiation before we give the precise definition.

Consider water in motion in a narrow tank, as illustrated in Fig. 16-1-1. We will assume that the motion of the water is gentle enough so that, at any instant of time, the height z of the water above the bottom of the tank is a function of the position y measured along the long direction of the tank; this means that there are no "breaking waves" and that the height of water is constant along the short direction of the tank. Since the water is in motion, the height z depends on the time as well as on y, so we may write $z = f(t, y)$; the domain of the function f consists of all pairs (t, y) such that t lies in an interval of time relevant for the experiment, and $a \leq y \leq b$, where a and b mark the ends of the tank.

We can graph the entire function f as a surface in (t, y, z) space lying over the strip $a \leq y \leq b$ (see Fig. 16-1-2); the section of this surface

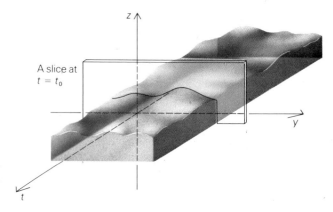

A slice at $t = t_0$

by a plane of the form $t = t_0$ is a curve which shows the configuration of the water at the moment t_0 (such as each of the "snapshots" in Fig. 16-1-1). This curve is the graph of a function of *one* variable, $z = g(y)$, where g is defined by $g(y) = f(t_0, y)$. If we take the derivative of the function g at a point y_0 in (a, b), we get a number $g'(y_0)$ which represents the slope of the water's surface at the time t_0 and at the location y_0. (See Fig. 16-1-3.) It could be observed as the slope of a small stick floating on the water at that time and position.

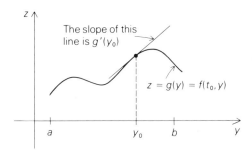

Fig. 16-1-3 The derivative $g'(y_0)$ of the function $g(y) = f(t_0, y)$ represents the slope of the water's surface at time t_0 and position y_0.

This number, $g'(y_0)$, is obtained from the function f by:

1. Fixing t at the value t_0

2. Differentiating the resulting function of y

3. Setting y equal to y_0

The number $g'(y_0)$ is called the *partial derivative of f with respect to y at the point* (t_0, y_0). For now, we will denote this partial derivative by $f_y(t_0, y_0)$; the subscript y indicates the variable with respect to which we are differentiating.

Worked Example 1   Suppose that $f(t, y) = e^{-y} \cos t + \sin(y + t^2)$. What is $f_y(2, 3)$?

Solution   To find the partial derivative with respect to y, we first set t equal to $t_0 = 2$, obtaining the function $g(y) = f(2, y) = e^{-y} \cos 2 + \sin(y + 4)$. Next we differentiate this function of y:

$$g'(y) = -e^{-y} \cos 2 + \cos(y + 4)$$

Finally, we set y equal to $y_0 = 3$ to obtain

$$f_y(2, 3) = g'(3) = -e^{-3} \cos 2 + \cos 7 \approx 0.7746$$

We can also define the partial derivative of f with respect to t at (t_0, y_0); it is obtained by

1. Fixing y at the value y_0

2. Differentiating the resulting function of t

3. Setting t equal to t_0

The result is denoted $f_t(t_0, y_0)$. In the first step, we obtain the function $h(t) = f(t, y_0)$, which represents the vertical motion of the water's surface observed at the fixed position y_0. The derivative with respect to t is, therefore, the *vertical velocity* of the surface at the position y_0. It could be observed as the vertical velocity of a cork floating on the water at that position. Finally, setting t equal to t_0 merely involves observing the velocity at the specific time t_0.

Worked Example 2   Find $f_t(2, 3)$ for the function in Worked Example 1.

Solution   Setting $y = y_0 = 3$ gives $h(t) = f(t, 3) = e^{-3} \cos t + \sin(3 + t^2)$. Then $h'(t) = -e^{-3} \sin t + \cos(3 + t^2)2t$, and so

$$f_t(2, 3) = h'(2) = -e^{-3} \sin 2 + 4 \cos 7 \approx 2.970$$

Now we come to the general definition of the partial derivatives of functions of two variables. We revert to the usual x, y notation for our variables, although one can use any names desired.

Definition   Let $f(x, y)$ be a function of two variables and let (x_0, y_0) be a point of its domain.
   The derivative with respect to x of the function $f(x, y_0)$ of x, evaluated at $x = x_0$, is called the *partial derivative of f with respect to x at (x_0, y_0)* and is denoted by $f_x(x_0, y_0)$.
   The derivative with respect to y of the function $f(x_0, y)$ of y, evaluated at $y = y_0$, is called the *partial derivative of f with respect to y at (x_0, y_0)* and is denoted by $f_y(x_0, y_0)$.

If f is a function of three variables, we define its partial derivatives by holding fixed two of the variables at a time.

Definition   Let $f(x, y, z)$ be a function of three variables and let (x_0, y_0, z_0) be a point of its domain.
   The derivative with respect to x of the function $f(x, y_0, z_0)$ of x, evaluated at $x = x_0$, is called the *partial derivative of f with respect to x at (x_0, y_0, z_0)* and is denoted by $f_x(x_0, y_0, z_0)$. Partial derivatives with respect to y and z are defined similarly.

Worked Example 3   Let $f(x, y, z) = \sin(xy/z)$. Find $f_z(1, 2, 3)$.

Solution   We first fix $x_0 = 1$ and $y_0 = 2$, obtaining the function $\sin(2/z)$ of z. Differentiating with respect to z gives $\cos(2/z)(-2/z^2)$, and evaluating at $z = 3$ gives $f_z(1, 2, 3) = -\frac{2}{9} \cos(\frac{2}{3})$.

Solved Exercises*   1. In defining the partial derivative f_t in the wave motion example, we used a function $h(t) = f(t, y_0)$. How is the graph of this function obtained from the graph of f?

2. Find $f_y(1, 2, 3)$ for the function in Worked Example 3.

Exercises   1. Let $f(x, y) = 3x^2 + 2\sin(x/y^2) + y^3(1 - e^x)$. Find:
   (a) $f_x(2, 3)$ (b) $f_y(1, 1)$
   (c) $f_x(0, 1)$ (d) $f_y(-1, -1)$

---

*Solutions appear in Appendix A.

2. Let $g(t, u, v) = \ln(t + u + v) - \tan(tuv)$. Find:

(a) $g_t(0, 0, 1)$ (b) $g_u(1, 2, 3)$

(c) $g_v(-1, 3, 5)$ (d) $g_t(1, 0, 0)$

(e) $g_u(2, 3, 1)$ (f) $g_v(-1, 5, 3)$

PARTIAL DERIVATIVES AS NEW FUNCTIONS

In defining partial derivatives, we computed them for one point at a time. Just as with ordinary derivatives, though, the partial derivatives may be considered as functions of the point where they are evaluated.

Worked Example 4   Let $f(x, y) = \sqrt{x^2 + y^2}$. Find a general formula for $f_x(x_0, y_0)$.

Solution   To compute f_x, we hold y fixed at y_0, obtaining the function $f(x, y_0) = \sqrt{x^2 + y_0^2}$. Now we differentiate with respect to x:

$$\frac{d}{dx}\sqrt{x^2 + y_0^2} = \frac{d}{dx}(x^2 + y_0^2)^{1/2} = \frac{1}{2}(x^2 + y_0^2)^{-1/2} \cdot 2x = \frac{x}{\sqrt{x^2 + y_0^2}}$$

Evaluating at x_0 gives the required general formula

$$f_x(x_0, y_0) = \frac{x_0}{\sqrt{x_0^2 + y_0^2}}$$

Since the result of this calculation is true for all values of (x_0, y_0) (except $(0, 0)$, where the partial derivative does not exist), we can replace (x_0, y_0) by (x, y) and think of the result as a new function:

$$f_x(x, y) = \frac{x}{\sqrt{x^2 + y^2}}$$

PARTIAL DERIVATIVES AS FUNCTIONS

If $f(x, y)$ is a function of two variables, f_x is the function whose value at (x, y) is the partial derivative of f with respect to x at (x, y).

The function f_y is defined similarly, as are f_x, f_y, and f_z for functions of three variables.

In practice, when we calculate the partial derivative functions, we do not use the symbols x_0, y_0, z_0. We merely *think of two of the variables as being constant while differentiating with respect to the third.* If you find this a little difficult to do at first, rely on our earlier procedures.

Worked Example 5   Let $f(x, y, z) = \sin(xy/z)$. Find $f_z(x, y, z)$ and then substitute to find $f_z(1, 2, 3)$. Compare with Worked Example 3.

Solution We differentiate $\sin(xy/z)$ with respect to z, thinking of x and y as constants; the result is $f_z(x, y, z) = \cos(xy/z)(-xy/z^2) = -(xy/z^2)\cos(xy/z)$. Substituting $(1, 2, 3)$ for (x, y, z) gives

$$f_z(1, 2, 3) = -\frac{1 \cdot 2}{3^2}\cos\left(\frac{1 \cdot 2}{3}\right) = -\frac{2}{9}\cos\left(\frac{2}{3}\right)$$

as in Worked Example 3.

If $y = f(x)$ is a function of one variable, the Leibniz notation for the derivative $f'(x)$ is dy/dx or $df(x)/dx$. There is a similar notation for several variables, which enables us to dispense with "f's and g's" and work directly with the formulas defining the functions.

If $z = f(x, y)$, the Leibniz notation for the partial derivative $f_x(x, y)$ is $\partial z/\partial x$, which is to be read "partial z, partial x" or, occasionally, "del z, del x." The other partial derivatives are denoted in an analogous way.*

Worked Example 6 Find $(\partial/\partial y)(1/\sqrt{x^2 + y^2})$.

Solution This is $f_y(x, y)$, where $z = f(x, y) = 1/\sqrt{x^2 + y^2}$. We differentiate with respect to y, thinking of x as a constant:

$$\frac{\partial}{\partial y}(x^2 + y^2)^{-1/2} = -\frac{1}{2}(x^2 + y^2)^{-3/2}(2y) = -\frac{y}{(x^2 + y^2)^{3/2}}$$

PARTIAL DIFFERENTIATION

If f is a function of several variables, to calculate the partial derivative with respect to a certain variable: treat the remaining variables as constants and differentiate as usual by using the rules of one-variable calculus.

Since the partial derivatives are themselves functions, we can take their partial derivatives to obtain higher derivatives. For a function of two variables, there are four ways to take a second derivative. If $z = f(x, y)$, we may compute

$$f_{xx}(x, y) = \frac{\partial}{\partial x}\left(\frac{\partial z}{\partial x}\right) = \frac{\partial^2 z}{\partial x^2} \qquad f_{yy}(x, y) = \frac{\partial}{\partial y}\left(\frac{\partial z}{\partial y}\right) = \frac{\partial^2 z}{\partial y^2}$$

$$f_{xy}(x, y) = \frac{\partial}{\partial y}\left(\frac{\partial z}{\partial x}\right) = \frac{\partial^2 z}{\partial y\,\partial x} \qquad f_{yx}(x, y) = \frac{\partial}{\partial x}\left(\frac{\partial z}{\partial y}\right) = \frac{\partial^2 z}{\partial x\,\partial y}$$

Worked Example 7 Compute the second partial derivatives of $z = xy^2 + ye^{-x} + \sin(x - y)$.

Solution We compute the first partials:

$$\frac{\partial z}{\partial x} = y^2 - ye^{-x} + \cos(x - y)$$

The symbol ∂ seems to have first been used by Clairaut and Euler around 1740 to avoid confusion with d. The notation $D_x f$ or $D_1 f$ for f_x and $D_y f$ or $D_2 f$ for f_y is also used.

and

$$\frac{\partial z}{\partial y} = 2xy + e^{-x} - \cos(x - y)$$

Now we differentiate again:

$$\frac{\partial^2 z}{\partial x^2} = ye^{-x} - \sin(x - y) \qquad \frac{\partial^2 z}{\partial y^2} = 2x - \sin(x - y)$$

$$\frac{\partial^2 z}{\partial y\,\partial x} = \frac{\partial}{\partial y}\left(\frac{\partial z}{\partial x}\right) = 2y - e^{-x} + \sin(x - y)$$

$$\frac{\partial^2 z}{\partial x\,\partial y} = \frac{\partial}{\partial x}\left(\frac{\partial z}{\partial y}\right) = 2y - e^{-x} + \sin(x - y)$$

Worked Example 8   Let $f(x, y, z) = e^{xy} + z \cos x$. Find f_{zx} and f_{xz}.

Solution   We find $f_x(x, y, z) = ye^{xy} - z \sin x$; $f_z(x, y, z) = \cos x$; $f_{zx}(x, y, z) = (\partial/\partial x)(\cos x) = -\sin x$; $f_{xz}(x, y, z) = (\partial/\partial z)(ye^{xy} - z \sin x) = -\sin x$.

In the preceding examples, note that the mixed partials taken in different orders, like $\partial^2 z/\partial x\,\partial y$ and $\partial^2 z/\partial y\,\partial x$, or f_{xz} and f_{zx}, are equal. This is no accident.

Theorem 1   *If $u = f(x, y)$ has continuous second partial derivatives, then the mixed partials are equal; that is,*

$$\frac{\partial^2 u}{\partial x\,\partial y} = \frac{\partial^2 u}{\partial y\,\partial x} \quad \text{or} \quad f_{yx} = f_{xy}$$

L. Euler discovered this result around 1734 in connection with problems in hydrodynamics. We give a proof later in this section, using limits and continuity for functions of several variables.

Solved Exercises
3. If $f(x, y)$ is a function of two variables, express f_x as a limit.
4. If $h = rs^2 \sin(r^2 + s^2)$, find $\partial h/\partial s$.
5. The temperature (in degrees Celsius) near Dawson Creek at noon on April 15, 1901 is given by $T = -(0.0003)x^2y + (0.9307)y$, where x and y are the latitude and longitude (in degrees). At what rate is the temperature changing if we proceed directly north? (The latitude and longitude of Dawson Creek are $x = 55.7°$ and $y = 120.2°$.)
6. If $u = y \cos(xz) + x \sin(yz)$, calculate $\partial^2 u/\partial x\,\partial z$ and $\partial^2 u/\partial z\,\partial x$.

Exercises
3. Let $z = (\sin x)e^{-xy}$.
   (a) Find $\partial z/\partial y$.
   (b) Compute $\partial z/\partial x$ and $\partial z/\partial y$ at the following points: $(0, 0)$, $(0, \pi/2)$, $(\pi/2, 0)$, and $(\pi/2, \pi/2)$.
4. Let $u = e^{-xyz}(xy + xz + yz)$. Find $\partial u/\partial x$ and $\partial u/\partial z$.

5. If $f(x,y,z)$ is a function of three variables, express f_z as a limit.

6. In the situation of Solved Exercise 5, how fast is the temperature changing if we proceed directly west? (The longitude y is increasing as we go west.)

7. Compute $\partial^2 u/\partial x^2$, $\partial^2 u/\partial y\,\partial x$, $\partial^2 u/\partial y^2$, and $\partial^2 u/\partial x\,\partial y$ for each of the following functions. Check that Theorem 1 works for each.

   (a) $u = 2xy/(x^2 + y^2)^2$ (b) $u = \cos(xy^2)$

   (c) $u = e^{-xy^2} + y^3 x^4$ (d) $u = 1/(\cos^2 x + e^{-y})$

LINEAR APPROXIMATIONS AND TANGENT PLANES

In the calculus of functions of one variable, the simplest functions are the linear functions $l(x) = mx + b$. The derivative of such a function is the constant m, which represents the slope of the graph or the rate of change of y with respect to x. If $f(x)$ is any differentiable function, its tangent line at x_0 is the graph of the *linear approximation* $y = f(x_0) + f'(x_0)(x - x_0)$.

To extend these ideas to functions of two variables, we begin by looking at linear functions of the form $z = l(x,y) = ax + by + c$, the graphs of which are the planes which contain no line parallel to the z axis. Such a plane has two "slopes," the numbers a and b, which determine the direction of its normal vector $-a\mathbf{i} - b\mathbf{j} + \mathbf{k}$ (see p. 615). These slopes can be recovered from the function l as the partial derivatives $l_x = a$ and $l_y = b$. By analogy with the situation for one variable, we define the *linear approximation* at (x_0,y_0), for a general function f of two variables, to be the linear function

$$l(x,y) = f(x_0,y_0) + f_x(x_0,y_0)(x - x_0) + f_y(x_0,y_0)(y - y_0)$$

which is of the form $ax + by + c$ with

$$a = f_x(x_0,y_0), \qquad b = f_y(x_0,y_0)$$

and

$$c = f(x_0,y_0) - x_0 f_x(x_0,y_0) - y_0 f_y(x_0,y_0)$$

This is the unique linear function which has at (x_0,y_0) the same value *and* the same partial derivatives as f. The graph

$$z = f(x_0,y_0) + f_x(x_0,y_0)(x - x_0) + f_y(x_0,y_0)(y - y_0)$$

of the linear approximation is a plane through $(x_0,y_0,f(x_0,y_0))$, with normal vector $-f_x(x_0,y_0)\mathbf{i} - f_y(x_0,y_0)\mathbf{j} + \mathbf{k}$; it is called the *tangent plane* at (x_0,y_0) to the graph of f (see Fig. 16-1-4). In the next section we present a geometric description of the tangent plane.

Fig. 16-1-4 The tangent plane at (x_0, y_0) to the graph $z = f(x, y)$ has the equation
$z = f(x_0, y_0) + f_x(x_0, y_0)(x - x_0) + f_y(x_0, y_0)(y - y_0)$
The normal to the plane is
$\mathbf{n} = -f_x(x_0, y_0)\mathbf{i} - f_y(x_0, y_0)\mathbf{j} + \mathbf{k}$

Worked Example 9   Find the equation of the plane tangent to the hemisphere $z = \sqrt{1 - x^2 - y^2}$ at a point (x_0, y_0). Interpret your result geometrically.

Solution   With $f(x, y) = \sqrt{1 - x^2 - y^2}$, we have $f_x(x, y) = -x/\sqrt{1 - x^2 - y^2}$ and $f_y(x, y) = -y/\sqrt{1 - x^2 - y^2}$. The equation of the tangent plane at (x_0, y_0, z_0) is

$$z = \sqrt{1 - x_0^2 - y_0^2} - \frac{x_0}{\sqrt{1 - x_0^2 - y_0^2}}(x - x_0) - \frac{y_0}{\sqrt{1 - x_0^2 - y_0^2}}(y - y_0)$$

or

$$z = z_0 - \frac{x_0}{z_0}(x - x_0) - \frac{y_0}{z_0}(y - y_0)$$

A normal vector is

$$\frac{x_0}{z_0}\mathbf{i} + \frac{y_0}{z_0}\mathbf{j} + \mathbf{k}$$

Multiplying by z_0, we find that another normal vector is $x_0\mathbf{i} + y_0\mathbf{j} + z_0\mathbf{k}$. Thus we have recovered the geometric result that the tangent plane at a point P of a sphere is perpendicular to the vector from the center of the sphere to P.

The linear approximation may be defined as well for a function of three variables. We include its definition in the following display.

LINEAR APPROXIMATION AND TANGENT PLANE

The linear approximation at (x_0, y_0) to $f(x, y)$ is the linear function:

$$l(x, y) = f(x_0, y_0) + f_x(x_0, y_0)(x - x_0) + f_y(x_0, y_0)(y - y_0)$$

The graph $z = l(x, y)$ is called the *tangent plane* to the graph of f at (x_0, y_0). It has normal vector $-f_x(x_0, y_0)\mathbf{i} - f_y(x_0, y_0)\mathbf{j} + \mathbf{k}$.

The linear approximation at (x_0, y_0, z_0) to $f(x, y, z)$ is the linear function:

$$l(x, y, z) = f(x_0, y_0, z_0) + f_x(x_0, y_0, z_0)(x - x_0) + f_y(x_0, y_0, z_0)(y - y_0) + f_z(x_0, y_0, z_0)(z - z_0)$$

Just as in one-variable calculus, we use the linear approximation for approximate numerical computations. Suppose that the number $z = f(x, y)$ depends on both x and y and we want to know how much z changes as x and y are changed a little. The partial derivative $f_x(x_0, y_0)$ gives the rate of change of z with respect to x at (x_0, y_0). Thus the change in z which results from a change Δx in x should be about

$$f_x(x_0, y_0) \Delta x$$

Similarly, the change in z caused by the shift in y should be about

$$f_y(x_0, y_0) \Delta y$$

Thus the total change in z should be approximately

$$\Delta z \approx f_x(x_0, y_0) \Delta x + f_y(x_0, y_0) \Delta y$$

Notice that the change in z is obtained by simply *adding* the changes due to Δx and Δy. If we write $\Delta x = x - x_0$ and $\Delta y = y - y_0$, then the right-hand side is the linear approximation to $f(x, y) - f(x_0, y_0)$ at (x_0, y_0).

Worked Example 10 Calculate an approximate value for $(0.99e^{0.02})^8$. Compare with the value from a calculator.

Solution Let $z = f(x, y) = (xe^y)^8$ and let $x_0 = 1$ and $y_0 = 0$, so $f(1, 0) = 1$. We get

$$\frac{\partial z}{\partial x} = 8x^7 e^{8y} \quad \text{which is 8 at } (1, 0)$$

and

$$\frac{\partial z}{\partial y} = 8x^8 e^{8y} \quad \text{which is 8 at } (1, 0)$$

Thus if we let $x = 0.99$ and $y = 0.02$ so that $x - x_0 = -0.01$ and $y - y_0 = 0.02$, the linear approximation is

$$1 + 8(-0.01) + 8(0.02) = 1.08$$

The value for $(0.99e^{0.02})^8$ obtained on our calculator is 1.082850933.

Solved Exercises 7. Find the equation of the plane tangent to the graph of $f(x, y) = (x^2 + y^2)/xy$ at $(x_0, y_0) = (1, 2)$.

8. Find a formula for the unit normal vector to the graph of a function $f(x, y)$ at a point (x_0, y_0).

9. Find an approximate value for $\sin(0.01) \cdot \cos(0.99\pi)$.

Exercises 8. Find equations for the planes tangent to the following surfaces at the indicated points:

(a) $z = x^3 + y^3 - 6xy$; $(1, 2, -3)$

(b) $z = (\cos x)(\cos y)$; $(0, \pi/2, 0)$

(c) $z = (\cos x)(\sin y)$; $(0, \pi/2, 1)$

(d) $xyz = 1$; $(1, 1, 1)$

9. Find an approximate value for each of the following:

(a) $(0.99)^3 + (2.01)^3 - 6(0.99)(2.01)$

(b) $(0.98) \sin \left(\dfrac{0.99}{1.03} \right)$ 　　　　　　 (c) $(0.98)(0.99)(1.03)$

10. In the setup of Solved Exercise 5, give an estimate for the temperature, near Dawson Creek, at $x = 55°$ and $y = 120°$. At Dawson Creek, is the temperature increasing or decreasing as you proceed south? As you proceed east?

(optional) **LIMITS AND CONTINUITY**

The theoretical aspects of functions of several variables require the concepts of limit and continuity. We define these just as we did for functions of one variable, with distances on the line replaced by distances in the plane or in space. The reader who wishes to study this material thoroughly should review Sections 12-1 and 12-2.

In the plane or space, the analog of an open interval is a disk or a ball.

Definition   The *disk* [*ball*] about (x_0, y_0) $[(x_0, y_0, z_0)]$ with radius r consists of points (x, y) $[(x, y, z)]$ such that the distance from (x, y) to (x_0, y_0) is less than r [the distance from (x, y, z) to (x_0, y_0, z_0) is less than r].

Let us write $d((x, y), (x_0, y_0)) = \sqrt{(x - x_0)^2 + (y - y_0)^2}$ for the distance between (x, y) and (x_0, y_0), with a similar notation $d((x, y, z), (x_0, y_0, z_0))$ in space. Thus the disk about (x_0, y_0) with radius r consists of all (x, y) such that $d((x, y), (x_0, y_0)) < r$. See Fig. 16-1-5.

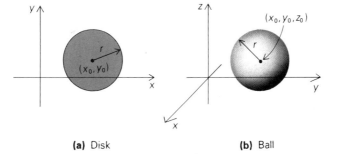

Fig. 16-1-5 The disk and ball.

(a) Disk 　　　　　　 **(b)** Ball

The limit concept now can be defined by the same ε-δ technique as in one variable.

Definition Suppose that f is defined on a region including a disk about (x_0, y_0), except possibly at (x_0, y_0) itself. We write

$$\lim_{(x, y) \to (x_0, y_0)} f(x, y) = l$$

if, for every $\varepsilon > 0$, there is a $\delta > 0$ such that $|f(x, y) - l| < \varepsilon$ whenever $0 < d((x, y), (x_0, y_0)) < \delta$. A similar definition is made for functions of three variables.

(optional)

The similarity between this definition and the one in Chapter 12 should be evident. The rules for limits, including rules for sums, products, and quotients, may be proved as in the appendix to Chapter 12.

Worked Example 11 Find $\lim\limits_{(x,y)\to(0,0)} (xy^2 + x^2y^4)/(2xy^2 + x^3y^5)$.

Solution Dividing numerator and denominator by xy^2 gives, by the replacement rule,

$$\lim_{(x,y)\to(0,0)} \frac{xy^2 + x^2y^4}{2xy^2 + x^3y^5} = \lim_{(x,y)\to(0,0)} \frac{1 + xy^2}{2 + x^2y^3} = \frac{1}{2}$$

using the limit rules for products, quotients, and the "obvious" limits:

$$\lim_{(x,y)\to(0,0)} x = 0 \qquad \lim_{(x,y)\to(0,0)} y = 0$$

We can base the continuity concept on that of limits (compare p. 108).

Definition Let f be defined in a disk about (x_0, y_0). Then we say f is *continuous* at (x_0, y_0) if

$$\lim_{(x,y)\to(x_0,y_0)} f(x,y) = f(x_0, y_0).$$

There is a similar definition for functions of three variables.

The complete development of the theory linking continuity and differentiability is not given in this text. It proceeds as in one-variable calculus, but there are added complications and subtleties.

Using these ideas of limit and continuity we can now give the proof of Theorem 1. It uses the mean value theorem for functions of one variable (p. 165).

Theorem 1 *If f has continuous second partial derivatives, then the mixed partials are equal; that is, if $z = f(x,y)$, then*

$$\frac{\partial^2 z}{\partial x \, \partial y} = \frac{\partial^2 z}{\partial y \, \partial x}$$

Proof Consider the expression

$$f(x_0 + \Delta x, y_0 + \Delta y) - f(x_0 + \Delta x, y_0) - f(x_0, y_0 + \Delta y) + f(x_0, y_0) \qquad \textbf{(1)}$$

We fix y_0 and Δy and introduce the function

$$g(x) = f(x, y_0 + \Delta y) - f(x, y_0)$$

so that formula (1) equals $g(x_0 + \Delta x) - g(x_0)$. By the mean value theorem for functions of one variable, this equals $g'(\bar{x})\Delta x$ for some $\bar{x}$ between x_0 and $x_0 + \Delta x$. Hence (1) equals

$$\left[\frac{\partial z}{\partial x}(\bar{x}, y_0 + \Delta y) - \frac{\partial z}{\partial x}(\bar{x}, y_0) \right] \Delta x$$

Applying the mean value theorem again we get for (1)

$$\frac{\partial^2 z}{\partial y \, \partial x}(\bar{x}, \bar{y})\Delta x \, \Delta y$$

(optional)

Since $\partial^2 z / \partial y\, \partial x$ is continuous and $(\bar{x}, \bar{y}) \rightarrow (x_0, y_0)$ as $(\Delta x, \Delta y) \rightarrow (0, 0)$, it follows that

$$\frac{\partial^2 z}{\partial y\, \partial x}(x_0, y_0) =$$

$$\lim_{(\Delta x, \Delta y) \rightarrow (0,0)} \frac{[f(x_0 + \Delta x, y_0 + \Delta y) - f(x_0 + \Delta x, y_0) - f(x_0, y_0 + \Delta y) + f(x_0, y_0)]}{\Delta x \, \Delta y} \quad (2)$$

The right-hand side of formula (2) is symmetric in x and y, so that in this derivation we can reverse the roles of x and y. In other words, in the same manner we prove that $\partial^2 z / \partial x\, \partial y$ is given by the same limit, and so we obtain the desired result.

One additional result to illustrate the ideas involved in continuity is the following.

Theorem 2 *Suppose that f is continuous at (x_0, y_0) and $f(x_0, y_0) > 0$. Then there is a disk about (x_0, y_0) on which $f(x, y) > 0$.*

Proof Choose $\varepsilon = f(x_0, y_0)/2$. By the definition of continuity and limit, there is a $\delta > 0$ such that

$$|f(x, y) - f(x_0, y_0)| < \varepsilon$$

if $d((x, y), (x_0, y_0)) < \delta$. However,

$$|f(x, y) - f(x_0, y_0)| < \frac{f(x_0, y_0)}{2}$$

implies

$$f(x_0, y_0) - f(x, y) < \frac{f(x_0, y_0)}{2}$$

Hence

$$f(x, y) > \frac{f(x_0, y_0)}{2} > 0$$

Thus a disk which works is the disk of radius δ about (x_0, y_0).

Solved Exercises ★10. Prove the "obvious" limit, $\displaystyle\lim_{(x,y)\rightarrow(x_0,y_0)} x = x_0$, using ε's and δ's.

★11. Show that

$$\lim_{(x,y)\rightarrow(0,0)} \frac{\partial}{\partial x} \sqrt{x^2 + y^2}$$

does not exist. A rigorous argument is not necessary. [*Hint:* Look at the limits along the x and y axes.]

Exercises ★11. Find:

(a) $\displaystyle\lim_{(x,y)\rightarrow(0,0)} \frac{x + y^2}{2x + y}$

(b) $\displaystyle\lim_{(x,y)\rightarrow(1,1)} \frac{x + y - 2}{(x + y)^2 - 4}$

(c) $\displaystyle\lim_{(x,y)\rightarrow(0,0)} \sin(xy)$

(d) $\displaystyle\lim_{(x,y)\rightarrow(0,0)} \frac{1}{1 + \ln(1/xy)}$

★12. Prove, using ε's and δ's, that $\lim\limits_{(x,y)\to(x_0,y_0)} (x + y) = x_0 + y_0$.

★13. Show that

$$\lim_{(x,y,z)\to(0,0,0)} \frac{\partial}{\partial z}(x^2 + y^2 + z^2)^{2/3}$$

does not exist.

Problems for Section 16-1

1. Compute f_x and f_y (and f_z, if applicable) for each of the following functions and evaluate them at the indicated points:
   (a) $f(x,y) = xy$; (1,1)
   (b) $f(x,y) = x/y$; (1,1)
   (c) $f(x,y) = \tan^{-1}(x - 3y^2)$; (1,0)
   (d) $f(x,y,z) = xyz$; (1,1,1)
   (e) $f(x,y,z) = \sqrt{x^2 + y^2 + z^2}$; (3,0,4)

2. Compute f_x and f_y (and f_z, if applicable) for each of the following functions and evaluate them at the indicated points.
   (a) $f(x,y) = \sqrt{x^2 + y^2}$; (1,-1)
   (b) $f(x,y) = e^{xy}\sin(x + y)$; (0,0)
   (c) $f(x,y) = \ln(x^2 + y^2 + 1)$; (0,0)
   (d) $f(x,y,z) = \cos(xy^2) + e^{3xyz}$; ($\pi$,1,1)
   (e) $f(x,y,z) = x^{yz}$; (1,1,0)

3. Suppose that $z = f(x,y) = x^2 + y^2$.
   (a) Find $\dfrac{\partial z}{\partial y}\Big|_{(1,1)}$
   (b) Describe the curve obtained by intersecting the graph of f with the plane $x = 1$.
   (c) Find a tangent vector to this curve at the point $(1,1,f(1,1))$.

4. Repeat Problem 3 for $z = f(x,y) = e^{xy}$.

5. Find the equation of the plane tangent to the graph of $z = f(x,y) = x^2 + 2y^3 + 1$ at each of the following points:
   (a) (1,1,4)   (b) (-1,-1,0)   (c) (0,0,1)

6. Find the equation of the tangent plane of the graph of f at the point $(x_0,y_0,f(x_0,y_0))$ for:
   (a) $f(x,y) = x - y + 2$; $(x_0,y_0) = (1,1)$
   (b) $f(x,y) = x^2 + 4y^2$; $(x_0,y_0) = (2,-1)$
   (c) $f(x,y) = xy$; $(x_0,y_0) = (1,1)$

7. For each of the functions in Problem 6, find a unit normal vector to the graph at $(x_0,y_0,f(x_0,y_0))$.

8. Let $f(x,y) = (1 - x^2 - y^2)^{1/2}$ for (x,y) such that $x^2 + y^2 < 1$. Show that the plane tangent to the graph of f at $(x_0,y_0,f(x_0,y_0))$ is orthogonal to the vector $(x_0,y_0,f(x_0,y_0))$. Interpret this geometrically.

9. Find the partial derivatives $\partial u/\partial x$ and $\partial u/\partial y$ for each of the following functions:
   (a) $u = 3x^2 + 2y^2$
   (b) $u = \sin(x^2 - 3xy)$
   (c) $u = (2x^2 + 7x^2y)/3xy$
   (d) $u = x^2y^3e^{2xy}$

10. If three resistors R_1, R_2, and R_3 are connected in parallel, the total electrical resistance is determined by the equation

$$\frac{1}{R} = \frac{1}{R_1} + \frac{1}{R_2} + \frac{1}{R_3}$$

   (a) What is $\partial R/\partial R_1$?
   (b) Suppose that R_1, R_2, and R_3 are variable resistors set at 100, 200, and 300 ohms, respectively. How fast is R changing with respect to R_1?
   (c) If, in part (b), R_1 is increased by 1 ohm, R_2 is decreased by 2 ohms, and R_3 is increased by 4 ohms, use the linear approximation to calculate the change in R. Compare with a direct calculation (on a calculator).

11. Consider the topographical map of Yosemite Valley in Fig. 15-R-2. Let r represent the east-west coordinate on the map, increasing from west to east. Let s be the north-south coordinate, increasing as you go north. (East is the positive r direction, north the positive s direction.) Let h be the elevation above sea level.
   (a) Explain how $\partial h/\partial r$ and $\partial h/\partial s$ are related to the distances between contour lines and their directions.
   (b) At the center of the letter o in Half Dome, what is the sign of $\partial h/\partial r$? Of $\partial h/\partial s$?

12. Find first-order approximations for each of the following quantities:
    (a) $(1.01)^2 [1 - \sqrt{1.98}]$
    (b) $\tan [(\pi + 0.01)/3.97]$
    (c) $\sqrt{(4.01)^2 + (3.98)^2 + (2.02)^2}$

13. Let $f(a, v)$ be the length of a side of a cube whose surface area is a and whose volume is v. Find the first-order approximation to $f(6 + \Delta a, 1 + \Delta v)$.

14. Find the partial derivatives $\partial^2 u/\partial x^2$, $\partial^2 u/\partial x \, \partial y$, $\partial^2 u/\partial y \, \partial x$, $\partial^2 u/\partial y^2$ for each of the functions in Problem 9.

15. Let $f(x, y, z) = x^2 y + xy^2 + yz^2$. Find f_{xy}, f_{yz}, f_{zx}, and f_{xyz}.

16. Let $z = x^4 y^3 - x^8 + y^4$.
    (a) Compute $\partial^3 z/\partial y \, \partial x \, \partial x$, $\partial^3 z/\partial x \, \partial y \, \partial x$, and $\partial^3 z/\partial x \, \partial x \, \partial y$.
    (b) Compute $\partial^3 z/\partial x \, \partial y \, \partial y$, $\partial^3 z/\partial y \, \partial x \, \partial y$, and $\partial^3 z/\partial y \, \partial y \, \partial x$.

17. A function $u = f(x, y)$ with continuous second partial derivatives satisfying Laplace's equation

    $$\frac{\partial^2 u}{\partial x^2} + \frac{\partial^2 u}{\partial y^2} = 0$$

    is called a *harmonic function*. Show that $u(x, y) = x^3 - 3xy^2$ is harmonic.

18. Which of the following functions satisfy Laplace's equation? (See Problem 17.)
    (a) $f(x, y) = x^2 - y^2$ (b) $f(x, y) = x^2 + y^2$
    (c) $f(x, y) = xy$ (d) $f(x, y) = x^3 - 3xy^2$
    (e) $f(x, y) = y^3 + 3x^2 y$
    (f) $f(x, y) = \sin x \cosh y$
    (g) $f(x, y) = e^x \sin y$

19. Let f and g be differentiable functions of one variable. Set $\varphi = f(x - t) + g(x + t)$.
    (a) Prove that φ satisfies the *wave equation*: $\partial^2 \varphi/\partial t^2 = \partial^2 \varphi/\partial x^2$.
    (b) Sketch the graph of φ against t and x if $f(x) = x^2$ and $g(x) = 0$.

20. (a) Show that the function $g(x, t) = 2 + e^{-t} \sin x$ satisfies the *heat equation*: $g_t = g_{xx}$. (Here $g(x, t)$ represents the temperature in a rod at position x and time t.)
    (b) Sketch the graph of g for $t \geq 0$. [*Hint:* Look at sections by the planes $t = 0$, $t = 1$, and $t = 2$.]
    (c) What happens to $g(x, t)$ as $t \to \infty$? Interpret this limit in terms of the behavior of heat in a rod.

21. Let $f(x, y) = x^2 + y^2$ and suppose that (x, y) lies along the curve $(x(t), y(t)) = (\cos t, e^t)$.
    (a) Let $g(t) = f(x(t), y(t))$ and compute directly $g'(t_0)$.
    (b) Show that this is the same as

    $$f_x(x(t_0), y(t_0)) \, x'(t_0) + f_y(x(t_0), y(t_0)) \, y'(t_0)$$

22. Let $f(x, y, z) = x^2 + 2y - z$ and suppose that (x, y, z) moves along the parametric curve $(1, t, t^2)$.
    (a) Let $g(t) = f(1, t, t^2)$ and compute $g'(t)$.
    (b) Show that this is equal to

    $$f_x \frac{dx}{dt} + f_y \frac{dy}{dt} + f_z \frac{dz}{dt}$$

★23. Let

    $$f(x, y) = \begin{cases} \dfrac{xy(x^2 - y^2)}{x^2 + y^2} & (x, y) \neq (0, 0) \\ 0 & (x, y) = (0, 0) \end{cases}$$

    (a) If $(x, y) \neq (0, 0)$, compute f_x and f_y.
    (b) What is the value of $f(x, 0)$ and $f(0, y)$?
    (c) Show that $f_x(0, 0) = 0 = f_y(0, 0)$.

★24. Consider the function f in Problem 23.
    (a) Show that $f_x(0, y) = -y$ when $y \neq 0$.
    (b) What is $f_y(x, 0)$ when $x \neq 0$?
    (c) Show that $f_{yx}(0, 0) = 1$ and $f_{xy}(0, 0) = -1$. [*Hint:* Express them as limits.]
    (d) What went wrong? Why are the mixed partials not equal?

25. Chicago Skate Company produces three roller skates. The cost in dollars for producing x, y and z units of each, respectively, is $c(x, y, z) = 3000 + 27x + 36y + 47z$.
    (a) The value of $\partial c/\partial x$ is the change in cost due to a one unit increase in production of the least expensive skate, the levels of production of the higher-priced units being held fixed. Find it.
    (b) Find $\partial c/\partial z$, and interpret.

26. The productivity z per employee per week of a company depends on the size x of the labor force and the amount y of investment capital in millions of dollars. A typical formula is $z(x, y) = 60xy - x^2 - 4y^2$.
    (a) The value of $\partial z/\partial x$ at $x = 5$, $y = 3$ is the marginal productivity of labor per worker at a labor force of 5 people and investment level of 3 million dollars. Find it.
    (b) Find $\partial z/\partial y$ at $x = 5$, $y = 3$, and interpret.

16-2 The Chain Rule

The derivative of a composite function with several intermediate variables is a sum of products.

In Chapter 5 we developed the chain rule for functions of one variable: If y is a function of x and z is a function of y, then z also may be regarded as a function of x, and

$$\frac{dz}{dx} = \frac{dz}{dy} \cdot \frac{dy}{dx}$$

For functions of several variables the chain rule is more complicated. Here we consider the case where z is a function of x and y, and x and y are functions of t; we can then regard z as a function of t. In this case the chain rule states that

$$\frac{dz}{dt} = \frac{\partial z}{\partial x}\frac{dx}{dt} + \frac{\partial z}{\partial y}\frac{dy}{dt}$$

Goals

After studying this section, you should be able to:

Differentiate a function of one variable which is given in terms of two or three intermediate variables.

Calculate gradients and directional derivatives.

Find the equation of the tangent plane to a level surface of a function of three variables.

STATEMENT OF THE CHAIN RULE

The chain rule applies when quantities in which we are interested depend in a known way upon other quantities which in turn depend upon a third set of quantities.

Suppose, for example, that the temperature T on the surface of a certain pond is a function $f(x, y)$ of the position coordinates (x, y). If a duck swims on the pond according to the parametric equations $x = g(t)$, $y = h(t)$, it will feel the water temperature varying with time according to the function $T = F(t) = f(g(t), h(t))$. The rate at which this temperature changes with respect to time is the derivative dT/dt. By analogy with the chain rule in one variable, we may expect this derivative to depend upon the direction and magnitude of the duck's velocity, as given by the derivatives dx/dt and dy/dt, as well as upon the partial derivatives $\partial T/\partial x$ and $\partial T/\partial y$ of temperature with respect to position. The correct formula relating all these derivatives is given in the following display.

THE CHAIN RULE

To find dz/dt, when $z = f(x, y)$ has continuous partial derivatives and $x = g(t)$ and $y = h(t)$ are differentiable,

multiply the partial derivatives of z with respect to each of the intermediate variables x and y by the derivative of that intermediate variable with respect to t and add the products.

If $F(t) = f(g(t), h(t))$,

$$F'(t) = f_x(g(t), h(t))g'(t) + f_y(g(t), h(t))h'(t)$$

In Leibniz notation,

$$\frac{dz}{dt} = \frac{\partial z}{\partial x}\frac{dx}{dt} + \frac{\partial z}{\partial y}\frac{dy}{dt}$$

For three intermediate variables, if u depends on x, y, and z, and x, y, and z depend on t,

$$\frac{du}{dt} = \frac{\partial u}{\partial x}\frac{dx}{dt} + \frac{\partial u}{\partial y}\frac{dy}{dt} + \frac{\partial u}{\partial z}\frac{dz}{dt}$$

Worked Example 1 Suppose that a duck is swimming in a circle, $x = \cos t$, $y = \sin t$, while the water temperature is given by the formula $T = x^2 e^y - xy^3$. Find dT/dt in two ways: (a) by the chain rule; (b) by expressing T directly in terms of t and differentiating.

Solution (a) $\partial T/\partial x = 2xe^y - y^3$; $\partial T/\partial y = x^2 e^y - 3xy^2$; $dx/dt = -\sin t$; $dy/dt = \cos t$. By the chain rule, $dT/dt = (\partial T/\partial x)(dx/dt) + (\partial T/\partial y)(dy/dt)$, so

$$\frac{dT}{dt} = (2xe^y - y^3)(-\sin t) + (x^2 e^y - 3xy^2)\cos t$$

$$= (2\cos t\, e^{\sin t} - \sin^3 t)(-\sin t) + (\cos^2 t\, e^{\sin t} - 3\cos t \sin^2 t)\cos t$$

$$= -2\cos t \sin t\, e^{\sin t} + \sin^4 t + \cos^3 t\, e^{\sin t} - 3\cos^2 t \sin^2 t$$

(b) Substituting for x and y in the formula for T, we get $T = \cos^2 t\, e^{\sin t} - \cos t \sin^3 t$, and differentiating this gives

$$\frac{dT}{dt} = 2\cos t(-\sin t)e^{\sin t} + \cos^2 t\, e^{\sin t} \cos t + \sin t \sin^3 t - \cos t\, 3\sin^2 t \cos t$$

$$= -2\cos t \sin t\, e^{\sin t} + \cos^3 t\, e^{\sin t} + \sin^4 t - 3\cos^2 t \sin^2 t$$

which is the same as the answer in part (a).

An intuitive argument for the chain rule is based on the linear approximation of Section 16-1. If the position of the duck changes from (x, y) to $(x + \Delta x, y + \Delta y)$, the temperature change ΔT is approximately $(\partial T/\partial x) \Delta x + (\partial T/\partial y) \Delta y$. On the other hand, the linear approximation for functions of one variable gives $\Delta x \approx (dx/dt) \Delta t$ and $\Delta y \approx (dy/dt)\Delta t$. Putting together these two approximations gives

$$\Delta T \approx \frac{\partial T}{\partial x}\frac{dx}{dt}\Delta t + \frac{\partial T}{\partial y}\frac{dy}{dt}\Delta t$$

Hence

$$\frac{\Delta T}{\Delta t} \approx \frac{\partial T}{\partial x}\frac{dx}{dt} + \frac{\partial T}{\partial y}\frac{dy}{dt} \tag{1}$$

As $\Delta t \to 0$, the approximations become more and more accurate and the ratio $\Delta T/\Delta t$ approaches dT/dt, so the approximation formula (1) becomes the chain rule.

In trying to make a proof out of this intuitive argument, one discovers that more than differentiability is required: the partial derivatives of T should be continuous. Everyone should master the statement of the following theorem; the proof is outlined in Problem 19.

Theorem 3   *Let $z = f(x,y)$ be a real-valued function of x and y such that the partial derivatives $\partial z/\partial x$ and $\partial z/\partial y$ are continuous. Let $x = g(t)$ and $y = h(t)$ be differentiable functions with continuous derivatives. Then*

$$\frac{dz}{dt} = \frac{\partial z}{\partial x}\frac{dx}{dt} + \frac{\partial z}{\partial y}\frac{dy}{dt}$$

where $\partial z/\partial x$ and $\partial z/\partial y$ are evaluated at $(x,y) = (g(t),\ h(t))$ and dx/dt and dy/dt are evaluated at t.

The chain rule for the case of two intermediate variables has an important geometric interpretation involving the tangent plane. Recall that the tangent plane at (x_0, y_0) to the graph $z = f(x,y)$ was defined as the graph of the linear approximation $z = f(x_0,y_0) + f_x(x_0,y_0)(x - x_0) + f_y(x_0,y_0)(y - y_0)$ to f at (x_0,y_0). For this definition to be consistent with the definition of the tangent line to a curve, we would like the following statement to be true.

Proposition   *If $(x, y, z) = (g(t),\ h(t),\ k(t))$ is any curve on the surface $z = f(x,y)$, with $(g(t_0),\ h(t_0)) = (x_0,y_0)$, then the tangent line to the curve at t_0 lies in the tangent plane to the surface at (x_0,y_0). (As above, all derivatives are assumed to be continuous.)*

Proof   We will show that this proposition is, in fact, equivalent to the chain rule. The tangent line lies in the tangent plane if and only if the velocity vector $\mathbf{v} = g'(t_0)\mathbf{i} + h'(t_0)\mathbf{j} + k'(t_0)\mathbf{k}$ is perpendicular to the normal $\mathbf{n} = -f_x(x_0,y_0)\mathbf{i} - f_y(x_0,y_0)\mathbf{j} + \mathbf{k}$ to the tangent plane. Expressing this orthogonality in terms of the dot product gives

$$0 = \mathbf{v}\cdot\mathbf{n} = -f_x(x_0,y_0)g'(t_0) - f_y(x_0,y_0)h'(t_0) + k'(t_0)$$

or

$$k'(t_0) = f_x(x_0,y_0)g'(t_0) + f_y(x_0,y_0)h'(t_0) \tag{2}$$

On the other hand, the fact that the curve $(x,y,z) = (g(t),\ h(t),\ k(t))$ lies on the surface $z = f(x,y)$ is expressed by the equation

$$k(t) = f(g(t), h(t))$$

so that

$$k'(t_0) = \frac{d}{dt}f(g(t), h(t))\bigg|_{t=t_0} \tag{3}$$

The equality of the right-hand sides of formulas (2) and (3) is precisely the chain rule.

You may think of this proposition as the "geometric statement" of the chain rule. It is illustrated in Fig. 16-2-1.

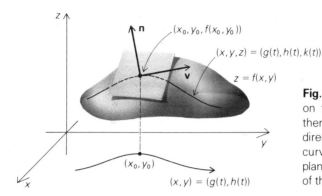

Fig. 16-2-1 If a curve lies on the surface $z = f(x, y)$, then the tangent line (with direction vector **v**) to the curve lies in the tangent plane (with normal vector **n**) of the surface.

Worked Example 2 Show that for any curve $\sigma(t)$ in the upper hemisphere $z = \sqrt{1 - x^2 - y^2}$, the velocity vector $\sigma'(t)$ is perpendicular to $\sigma(t)$.

Solution Let $(x, y, z) = \sigma(t)$. By the preceding proposition, $\sigma'(t)$ is perpendicular to the normal vector to the hemisphere at (x, y, z). But, by Worked Example 9 of the previous section, this normal vector is just $x\mathbf{i} + y\mathbf{j} + z\mathbf{k} = \sigma(t)$.

Solved Exercises

1. Verify the chain rule for $u = xe^{yz}$ and $(x, y, z) = (e^t, t, \sin t)$.

2. What do you get if you apply the chain rule to the case $z = xy$, where x and y are arbitrary functions of t?

3. A bug is swimming along the surface of the water whose motion is described by the function studied in Worked Examples 1 and 2 of Section 16-1. At $t = 2$, the bug is at the position $y = 3$ and its horizontal velocity dy/dt is equal to 5. What is its vertical velocity dz/dt at that moment?

Exercises

1. Show that applying the chain rule to $z = x/y$ (x and y arbitrary functions of t) gives the quotient rule for functions of one variable.

2. Use the chain rule to find a formula for $(d/dt)\,(f(t)g(t)h(t))$.

3. Verify the chain rule for $u = x/y + y/z + z/x$, $x = e^t$, $y = e^{t^2}$, $z = e^{t^3}$.

4. Show that the tangent plane at each point (x_0, y_0, z_0) of the cone $z = \sqrt{x^2 + y^2}$ ($(x, y) \neq (0, 0)$) contains the line through (x_0, y_0, z_0) and the origin.

THE GRADIENT OF A FUNCTION

The right-hand side of the chain rule

$$\frac{du}{dt} = \frac{\partial u}{\partial x}\frac{dx}{dt} + \frac{\partial u}{\partial y}\frac{dy}{dt} + \frac{\partial u}{\partial z}\frac{dz}{dt}$$

has the appearance of a dot product—in fact it is the dot product of the vectors

$$\mathbf{v}_1 = \frac{\partial u}{\partial x}\mathbf{i} + \frac{\partial u}{\partial y}\mathbf{j} + \frac{\partial u}{\partial z}\mathbf{k} \qquad\qquad \textbf{(4)}$$

and

$$\mathbf{v}_2 = \frac{dx}{dt}\mathbf{i} + \frac{dy}{dt}\mathbf{j} + \frac{dz}{dt}\mathbf{k} \qquad\qquad \textbf{(5)}$$

We recognize the vector $\mathbf{v}_2$ as the *velocity vector* of a parametric curve; if $\boldsymbol{\sigma}(t)$ is the vector representation of the curve, $\mathbf{v}_2(t)$ is just $\boldsymbol{\sigma}'(t)$.

The vector $\mathbf{v}_1$ is something new: it depends upon the function $u = f(x, y, z)$ and contains in vector form all three partial derivatives of f.

Worked Example 3 Find the vector $\mathbf{v}_1$ if $u = f(x, y, z) = xy - z^2$.

Solution Substituting the partial derivatives of f into formula (4), we find that the vector $\mathbf{v}_1$ is $y\mathbf{i} + x\mathbf{j} - 2z\mathbf{k}$.

Notice that the components of the vector $\mathbf{v}_1$ in Worked Example 3 are functions of the point (x, y, z) in space; in other words, this vector is a function of the point in space where the partial derivatives are evaluated. The following terminology and notation will prove convenient.

Definition A rule which assigns a vector in space to each point of some domain in space (or a vector in the plane to each point in the plane) is called a *vector field*.

Definition Let $f(x, y, z)$ be a function whose partial derivatives exist in some domain D in space. The vector field whose value at (x, y, z) is the vector

$$f_x(x, y, z)\mathbf{i} + f_y(x, y, z)\mathbf{j} + f_z(x, y, z)\mathbf{k}$$

defined for (x, y, z) in D, is called the *gradient vector field*, or simply the *gradient*, of f. It is denoted by ∇f. The value $\nabla f(x_0, y_0, z_0)$ of the gradient vector field at a specific point $P = (x_0, y_0, z_0)$ is called the *gradient of f at P*.

If f is defined in a domain in the plane, the gradient vector field is $\nabla f(x, y) = f_x(x, y)\mathbf{i} + f_y(x, y)\mathbf{j}$.

We may sketch a vector field $\Phi(x, y)$ in the plane by choosing several values for (x, y), evaluating $\Phi(x, y)$ at each point, and drawing the vector $\Phi(x, y)$ with its tail at the point (x, y). The same thing may be done for vector fields in space.

Worked Example 4   Sketch the gradient vector field of the function $f(x, y) = x^2/10 + y^2/6$.

Solution   The partial derivatives are $f_x(x, y) = x/5$ and $f_y(x, y) = y/3$. Evaluating these for various values of x and y and plotting, we obtain the sketch in Fig. 16-2-2.

Remark   In sketching a vector field $\Phi(x, y)$, we sometimes find that the vectors are so long that they overlap one another, making the drawing confusing. In this case, it is better to sketch $\varepsilon\Phi(x, y)$, where ε is a small positive number. (See Solved Exercise 5.)

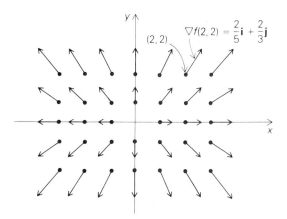

Fig. 16-2-2 The gradient vector field ∇f, where $f(x, y) = (x^2/10) + (y^2/6)$.

THE GRADIENT

If $u = f(x, y, z)$ is a function of three variables, its gradient vector field ∇f is defined by

$$\nabla f(x, y, z)$$
$$= f_x(x, y, z)\mathbf{i} + f_y(x, y, z)\mathbf{j} + f_z(x, y, z)\mathbf{k}$$
$$= \frac{\partial u}{\partial x}\mathbf{i} + \frac{\partial u}{\partial y}\mathbf{j} + \frac{\partial u}{\partial z}\mathbf{k}$$

If $z = f(x, y)$ is a function of two variables, its gradient vector field ∇f is defined by

$$\nabla f(x, y) = f_x(x, y)\mathbf{i} + f_y(x, y)\mathbf{j}$$
$$= \frac{\partial z}{\partial x}\mathbf{i} + \frac{\partial z}{\partial y}\mathbf{j}$$

Now we may state the chain rule in terms of gradients.

THE CHAIN RULE FOR FUNCTIONS AND CURVES

Let f be a function of two [three] variables, $\boldsymbol{\sigma}(t)$ a parametric curve in the plane [in space], and $h(t) = f(\boldsymbol{\sigma}(t))$ the composite function.

Then $h'(t) = \nabla f(\boldsymbol{\sigma}(t)) \cdot \boldsymbol{\sigma}'(t)$; that is, $\dfrac{d}{dt} f(\boldsymbol{\sigma}(t)) = \nabla f(\boldsymbol{\sigma}(t)) \cdot \boldsymbol{\sigma}'(t)$

In this form, the chain rule looks more like it did for functions of one variable:

$$\frac{d}{dt}(f(g(t))) = f'(g(t))g'(t)$$

Worked Example 5 Using the result of Worked Example 3, find $(d/dt)(\sin t \cos t - e^{2t})$.

Solution We set $u = f(x, y, z) = xy - z^2$ and $(x, y, z) = (\sin t, \cos t, e^t)$. The gradient vector of f is $y\mathbf{i} + x\mathbf{j} - 2z\mathbf{k}$ by Worked Example 3; the velocity vector is $\cos t\, \mathbf{i} - \sin t\, \mathbf{j} + e^t \mathbf{k}$. By the chain rule,

$$\frac{du}{dt} = \nabla f \cdot \boldsymbol{\sigma}' = (y\mathbf{i} + x\mathbf{j} - 2z\mathbf{k}) \cdot (\cos t\, \mathbf{i} - \sin t\, \mathbf{j} + e^t \mathbf{k})$$

$$= y \cos t - x \sin t - 2ze^t = \cos^2 t - \sin^2 t - 2e^{2t}$$

Solved Exercises

4. Let $\mathbf{r} = x\mathbf{i} + y\mathbf{j} + z\mathbf{k}$ and $r = \|\mathbf{r}\| = \sqrt{x^2 + y^2 + z^2}$. Show that

$$\nabla\left(\frac{1}{r}\right) = -\frac{\mathbf{r}}{r^3} \quad r \neq 0$$

What is $\|\nabla(1/r)\|$?

5. (a) Sketch the vector field $\boldsymbol{\Phi}(x, y) = 3y\mathbf{i} - 3x\mathbf{j}$ by sketching $\frac{1}{6}\boldsymbol{\Phi}(x, y)$ instead.

   (b) Using the law of equality of mixed partial derivatives, show that this vector field $\boldsymbol{\Phi}(x, y)$ is *not* the gradient vector field of any function.

6. Suppose that f takes the value 2 at all points of a curve $\boldsymbol{\sigma}(t)$ with velocity vector $\mathbf{v}(t)$. What can you say about $\nabla f(\boldsymbol{\sigma}(t))$ and $\boldsymbol{\sigma}'(t)$?

Exercises

5. Compute the gradient ∇f for each of the following functions:
   (a) $f(x, y, z) = \sqrt{x^2 + y^2 + z^2}$
   (b) $f(x, y, z) = xy + yz + xz$
   (c) $f(x, y) = \ln(\sqrt{x^2 + y^2}); (x, y) \neq (0, 0)$

6. For each of the following, find ∇f and the velocity vector $\boldsymbol{\sigma}'(t)$; evaluate $(d/dt)(f(\boldsymbol{\sigma}(t)))$ at $t = 1$ in two ways.
   (a) $f(x, y, z) = xz + yz + xy$; $\boldsymbol{\sigma}(t) = (e^t, \cos t, \sin t)$
   (b) $f(x, y, z) = e^{xyz}$; $\boldsymbol{\sigma}(t) = (6t, 3t^2, t^3)$
   (c) $f(x, y, z) = \sqrt{x^2 + y^2 + z^2}$; $\boldsymbol{\sigma}(t) = (\sin t, \cos t, t)$

7. Let f and g be functions of three variables. Show that
   $\nabla(fg) = f\nabla g + g\nabla f$.

DIRECTIONAL DERIVATIVES

Let $u = f(x, y, z)$ be a function (with continuous partial derivatives) and $(x, y, z) = \boldsymbol{\sigma}(t)$ a parametrized curve in space. The derivative with respect to t of the composed function $f(\boldsymbol{\sigma}(t))$ may be thought of as "the derivative of f *along the curve* $\boldsymbol{\sigma}(t)$." According to the chain rule, the value of this derivative at $t = t_0$ is the dot product of the gradient $\nabla f(\boldsymbol{\sigma}(t))$ and the velocity vector $\boldsymbol{\sigma}'(t_0)$; that is, $\nabla f(\boldsymbol{\sigma}(t)) \cdot \boldsymbol{\sigma}'(t)$. We may write this dot product as

$$\|\nabla f(\boldsymbol{\sigma}(t_0))\| \, \|\boldsymbol{\sigma}'(t_0)\| \cos \theta$$

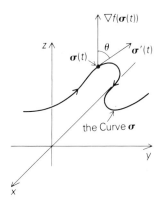

where θ is the angle between the gradient vector and the velocity vector (Fig. 16-2-3). If we fix the function f and differentiate it along various curves through a given point $\mathbf{r}$ (here, as usual, we identify a point with the vector from the origin to the point), the derivative will be proportional to the *speed* $\|\boldsymbol{\sigma}'(t_0)\|$ and to the cosine of the angle between the gradient and velocity vectors. To describe how the derivative of f varies as we change the direction of the curve along which it is differentiated, we fix $\mathbf{r}$ and choose $\boldsymbol{\sigma}(t) = \mathbf{r} + t\mathbf{n}$ for $\mathbf{n}$ a unit vector. Note that since $\mathbf{n}$ is a unit vector, the speed of the curve $\boldsymbol{\sigma}(t)$ is 1, so 1 unit of time corresponds to unit distance along the curve. We can use these curves to measure the rate of change of f. We make the following definition.

Fig. 16-2-3 The derivative of f along the curve $\boldsymbol{\sigma}(t)$ is

$$\frac{d}{dt} f(\boldsymbol{\sigma}(t)) = \nabla f(\boldsymbol{\sigma}(t)) \cdot \boldsymbol{\sigma}'(t)$$
$$= \|\nabla f(\boldsymbol{\sigma}(t))\| \, \|\boldsymbol{\sigma}'(t)\| \cos \theta$$

Definition Let $f(x, y, z)$ be a function of three variables, $\mathbf{r}$ a point in its domain, $\mathbf{n}$ a unit vector. Define the parametric curve $\boldsymbol{\sigma}(t)$ by $\boldsymbol{\sigma}(t) = \mathbf{r} + t\mathbf{n}$. The derivative $(d/dt) f(\boldsymbol{\sigma}(t))|_{t=0}$ is called the *directional derivative* of f at $\mathbf{r}$ in the direction of $\mathbf{n}$.

Since $\boldsymbol{\sigma}'(t) = \mathbf{n}$ and $\|\mathbf{n}\| = 1$, we see that if f has continuous partial derivatives, the directional derivative at $\mathbf{r}$ in the direction of $\mathbf{n}$ is

$$\nabla f(\mathbf{r}) \cdot \mathbf{n} = \|\nabla f(\mathbf{r})\| \cos \theta$$

Notice that the directional derivatives in the directions of $\mathbf{i}$, $\mathbf{j}$, and $\mathbf{k}$ are just the partial derivatives. For instance, choosing $\mathbf{n} = \mathbf{i}$, $\nabla f \cdot \mathbf{i} = (f_x \mathbf{i} + f_y \mathbf{j} + f_z \mathbf{k}) \cdot \mathbf{i} = f_x$. Similarly, $\nabla f \cdot \mathbf{j} = f_y$ and $\nabla f \cdot \mathbf{k} = f_z$.

As we let $\mathbf{n}$ vary, the directional derivative takes its maximum value when $\cos \theta = 1$, that is, when $\mathbf{n}$ points in the direction of $\nabla f(\mathbf{r})$. The maximum value of the directional derivative is just the length $\|\nabla f(\mathbf{r})\|$. We state the last result as a theorem.

Theorem 4 *Assume that f has continuous partial derivatives and that some directional derivative of f at $\mathbf{r}$ is not equal to zero. Then the direction of the gradient $\nabla f(\mathbf{r})$ is that of the unit vector for which the directional derivative is greatest; the length of $\nabla f(\mathbf{r})$ is the maximum value of the directional derivatives at $\mathbf{r}$.*

GRADIENTS AND DIRECTIONAL DERIVATIVES

The *directional derivative* at $\mathbf{r}$ in the direction of a unit vector $\mathbf{n}$ is the rate of change of f along the straight line through $\mathbf{r}$ in direction $\mathbf{n}$; $\boldsymbol{\sigma}(t) = \mathbf{r} + t\mathbf{n}$.

The directional derivative at $\mathbf{r}$ in the direction $\mathbf{n}$ equals $\nabla f(\mathbf{r}) \cdot \mathbf{n}$ (dot product). It is greatest (for fixed $\mathbf{r}$) when $\mathbf{n}$ points in the same direction as $\nabla f(\mathbf{r})$ and least when $\mathbf{n}$ points in the same direction as $-\nabla f(\mathbf{r})$.

If one wishes to move from $\mathbf{r} = (x, y, z)$ in a direction in which f is increasing most quickly, one should move in the direction $\nabla f(\mathbf{r})$. Likewise, $-\nabla f(\mathbf{r})$ is the direction in which f is decreasing at the fastest rate.

Worked Example 6 Let $u = f(x, y, z) = (\sin xy)e^{-z^2}$. In what direction from $(1, \pi, 0)$ should one proceed to increase f most rapidly?

Solution We compute the gradient at $(1, \pi, 0)$:

$$\nabla f = \frac{\partial u}{\partial x}\mathbf{i} + \frac{\partial u}{\partial y}\mathbf{j} + \frac{\partial u}{\partial z}\mathbf{k}$$
$$= y\cos(xy)e^{-z^2}\mathbf{i} + x\cos(xy)e^{-z^2}\mathbf{j} + (-2z\sin xy)e^{-z^2}\mathbf{k}$$

At $(1, \pi, 0)$ this becomes

$$\pi\cos(\pi)\mathbf{i} + \cos(\pi)\mathbf{j} = -\pi\mathbf{i} - \mathbf{j}$$

Thus one should proceed in the direction of the unit vector

$$(1/\sqrt{1 + \pi^2})(-\pi\mathbf{i} - \mathbf{j})$$

Directional derivatives are also defined for functions of two variables. In this case, we have a geometric interpretation of the directional derivatives of $f(x, y)$ in terms of the graph $z = f(x, y)$. Given a point (x_0, y_0) in the plane and a unit vector $\mathbf{n} = a\mathbf{i} + b\mathbf{j}$, we can intersect the graph with the plane $\mathscr{P}$ in space which lies above the line through (x_0, y_0) with direction $\mathbf{n}$. (See Fig. 16-2-4.)

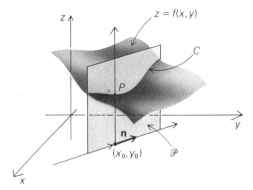

Fig. 16-2-4 The slope at the point P of the curve C in the plane $\mathscr{P}$ is the directional derivative at (x_0, y_0) of f in the direction $\mathbf{n}$.

The result is a curve C which may be parametrized by the formula $(x, y, z) = (x_0 + at, y_0 + bt, f(x_0 + at, y_0 + bt))$. The tangent vector to this curve at $P = (x_0, y_0, f(x_0, y_0))$ is

$$\mathbf{v} = a\mathbf{i} + b\mathbf{j} + \frac{d}{dt} f(x_0 + at, y_0 + bt) \bigg|_{t=0} \mathbf{k}$$

$$= a\mathbf{i} + b\mathbf{j} + [af_x(x_0, y_0) + bf_y(x_0, y_0)] \mathbf{k}$$

The slope of C in the plane $\mathscr{P}$ is the ratio of the vertical component $af_x(x_0, y_0) + bf_y(x_0, y_0)$ of $\mathbf{v}$ to the length $\sqrt{a^2 + b^2}$ of the horizontal component. But $\sqrt{a^2 + b^2} = 1$, since $\mathbf{n} = a\mathbf{i} + b\mathbf{j}$ is a unit vector. Hence the slope of C in the plane $\mathscr{P}$ is just $af_x(x_0, y_0) + bf_y(x_0, y_0) = \mathbf{n} \cdot \nabla f(x_0, y_0)$, which is precisely the directional derivative of f at (x_0, y_0) in the direction of $\mathbf{n}$.

If we let the vector $\mathbf{n}$ rotate in the xy plane, then the plane $\mathscr{P}$ will rotate about the vertical line through (x_0, y_0) and the curve C will change. The slopes at P of all these curves are determined by the two numbers $f_x(x_0, y_0)$ and $f_y(x_0, y_0)$, and the tangent lines to all these curves lie in the tangent plane to $z = f(x, y)$ at P.

Solved Exercises

7. What rate of change does $\nabla f(x, y, z) \cdot (-\mathbf{j})$ represent?

8. Let $\mathbf{r} = x\mathbf{i} + y\mathbf{j} + z\mathbf{k}$ and $r = \|\mathbf{r}\|$. Compute ∇r. In what direction is r increasing the fastest? Interpret your answer geometrically.

9. Let $f(x, y) = x^2 - y^2$. In what direction from $(0, 1)$ should one proceed in order to increase f the fastest? Illustrate your answer with a sketch.

10. Captain Astro is drifting in space near the sunny side of Mercury and notices that the hull of her ship is beginning to melt. The temperature in her vicinity is given by $T = e^{-x} + e^{-2y} + e^{3z}$. If she is at $(1, 1, 1)$, in what direction should she proceed in order to *cool* fastest?

Exercises

8. Compute the directional derivatives of the following functions at the indicated points in the given directions.
    (a) $f(x, y) = x + 2x^2 - 3xy$; $(x_0, y_0) = (1, 1)$; $\mathbf{n} = (\frac{3}{5}, \frac{4}{5})$
    (b) $f(x, y) = \log(\sqrt{x^2 + y^2})$; $(x_0, y_0) = (1, 0)$; $\mathbf{n} = (2\sqrt{5}/5, \sqrt{5}/5)$
    (c) $f(x, y, z) = xyz$; $(x_0, y_0, z_0) = (1, 1, 1)$; $\mathbf{n} = (1/\sqrt{2})\mathbf{i} - (1/\sqrt{2})\mathbf{k}$
    (d) $f(x, y, z) = e^x + yz$; $(x_0, y_0, z_0) = (1, 1, 1)$; $\mathbf{n} = (1/\sqrt{3})(\mathbf{i} - \mathbf{j} + \mathbf{k})$

9. Suppose that a mountain has the shape of an elliptic paraboloid $z = c - ax^2 - by^2$, where a, b, and c are positive constants, x and y are the east-west and north-south map coordinates, and z is the altitude above sea level (x, y, z are all measured in meters). At the point $(1, 1)$, in what direction is the altitude increasing most rapidly? If a marble were released at $(1, 1)$, in what direction would it begin to roll?

10. An engineer wishes to build a railroad up the mountain of Exercise 9. Straight up the mountain is much too steep for the power of the engines. At the point $(1, 1)$, in what directions may the track be laid so that it will be climbing with a 3% grade—that is, an angle whose tangent is 0.03. (There are two possibilities.) Make a sketch of the situation indicating a few level curves for altitude, the gradient vector at $(1, 1)$, and the two possible directions for a 3% grade at $(1, 1)$.

11. A function $f(x, y)$ has at the point $(1, 3)$ directional derivatives of $+2$ in the direction toward $(2, 3)$ and -2 in the direction toward $(1, 4)$. Determine the gradient vector at $(1, 3)$ and compute the directional derivative in the direction toward $(3, 6)$.

GRADIENTS, LEVEL SURFACES, AND LEVEL CURVES

Recall that the tangent plane to a graph $z = f(x, y)$ was defined as the graph of the linear approximation to f. We found (p. 702) that the tangent plane at a point could also be characterized as the plane containing the tangent *lines* to all curves on the surface through the given point. For a general surface, we take this as a definition:

Definition Let S be a surface in space, $\mathbf{r}_0$ a point of S. If there is a plane which contains the tangent lines at $\mathbf{r}_0$ to all curves through $\mathbf{r}_0$ which lie in S, then this plane is called *the tangent plane to S at* $\mathbf{r}_0$. A normal to the tangent plane is sometimes said to be *perpendicular to S.*

The next theorem tells us how to find the tangent plane to a level surface.

Theorem 5 *Let* $\mathbf{r}_0$ *lie on the surface* S *defined by* $f(x, y, z) = c$, *and suppose that* $\nabla f(\mathbf{r}_0) \neq 0$. *Then the plane through* $\mathbf{r}_0$ *with normal vector* $\nabla f(\mathbf{r}_0)$ *is the tangent plane to* S *at* $\mathbf{r}_0$; *that is, if* $\mathbf{r} = \boldsymbol{\sigma}(t)$ *is any curve in* S, *with* $\boldsymbol{\sigma}(t_0) - \mathbf{r}_0$, *then* $\nabla f(\mathbf{r}_0)$ *is perpendicular to the velocity vector* $\boldsymbol{\sigma}'(t_0)$. *See Fig. 16-2-5.*

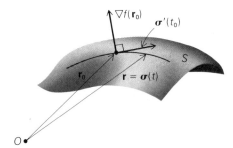

Fig. 16-2-5 The gradient of f at $\mathbf{r}_0$ is perpendicular to the tangent vector of any curve in the level surface.

Proof Since $\boldsymbol{\sigma}(t)$ lies in S, $f(\boldsymbol{\sigma}(t)) = c$. Hence

$$\frac{d}{dt}f(\boldsymbol{\sigma}(t)) = 0$$

By the chain rule for gradients, this gives

$$\nabla f(\boldsymbol{\sigma}(t)) \cdot \boldsymbol{\sigma}'(t) = 0$$

Setting $t = t_0$, we have $\nabla f(\mathbf{r}_0) \cdot \boldsymbol{\sigma}'(t_0) = 0$ as required. (We required $\nabla f(\mathbf{r}_0) \neq 0$ so there would be a well-defined plane orthogonal to $\nabla f(\mathbf{r}_0)$.)

Worked Example 7 Let $u = f(x, y, z) = x^2 + y^2 - z^2$. Find $\nabla f(0, 0, 1)$. Plot this on the level surface $f(x, y, z) = -1$.

Solution We have

$$\nabla f = \frac{\partial u}{\partial x}\mathbf{i} + \frac{\partial u}{\partial y}\mathbf{j} + \frac{\partial u}{\partial z}\mathbf{k} = 2x\mathbf{i} + 2y\mathbf{j} - 2z\mathbf{k}$$

At $(0, 0, 1)$, $\nabla f(0, 0, 1) = -2\mathbf{k}.$

The level surface $x^2 + y^2 - z^2 = -1$ is a hyperboloid of two sheets (see p. 655). If we plot $\nabla f(0, 0, 1)$ on it (Fig. 16-2-6), we see that it is indeed perpendicular to the surface.

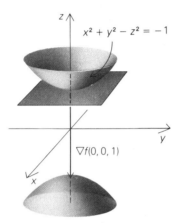

Fig. 16-2-6 $\nabla f(0, 0, 1)$ is perpendicular to the surface.

A similar result is true for functions of two variables: the tangent line to a level *curve* of a function $f(x, y)$ is perpendicular to the gradient of f at each point. Combining this fact with Theorem 4, we see that the direction in which the function f is increasing or decreasing most rapidly is perpendicular to the level curves of f. This fact is illustrated by the wisdom of the mountain goat: to get down most directly from the top of a hill, you should always proceed in a direction which is perpendicular to the level contours. (See Fig. 16-2-7.)

(a) Steepest descent of a hill

(b) Contour map of hill 2037 feet high

Fig. 16-2-7 The curve of steepest descent is perpendicular to the level curves.

Theorem 5 actually enables us to compute the tangent plane to the level surface S at $\mathbf{r}_0$. Indeed, $\nabla f(\mathbf{r}_0)$ will be a normal to this plane, which passes through $\mathbf{r}_0$. Therefore its equation can be read off immediately (see p. 616).

Worked Example 8   Compute the equation of the plane tangent to the surface $3xy + z^2 = 4$ at $(1, 1, 1)$.

Solution   Here $f(x, y, z) = 3xy + z^2$ and $\nabla f = (3y, 3x, 2z)$, which at $(1, 1, 1)$ is the vector $3\mathbf{i} + 3\mathbf{j} + 2\mathbf{k}$. Thus the tangent plane is

$$3(x - 1) + 3(y - 1) + 2(z - 1) = 0 \qquad \text{or} \qquad 3x + 3y + 2z = 8$$

GRADIENTS, LEVEL SURFACES, AND LEVEL CURVES

The normal to the tangent plane at $\mathbf{r}_0 = (x_0, y_0, z_0)$ of the level surface $f(x, y, z) = c$ is

$$\nabla f(\mathbf{r}_0)$$

The equation of the plane is

$$f_x(x_0, y_0, z_0)(x - x_0) + f_y(x_0, y_0, z_0)(y - y_0) + f_z(x_0, y_0, z_0)(z - z_0) = 0$$

The equation of the tangent line at (x_0, y_0) to the curve $f(x, y) = c$ is

$$f_x(x_0, y_0)(x - x_0) + f_y(x_0, y_0)(y - y_0) = 0$$

Solved Exercises

11. Find a unit normal to the surface $\sin(xy) = e^z$ at $(1, \pi/2, 0)$.

12. Let $z = g(x, y)$. The graph of g may be defined as the level surface $f(x, y, z) = 0$, where $f(x, y, z) = z - g(x, y)$. Compute the gradient of f and verify that it is perpendicular to the tangent plane of the graph $z = g(x, y)$ as defined in Section 16-1.

13. The gravitational force exerted on a mass m at (x, y, z) by a mass M at the origin is, by Newton's law of gravitation,

$$\mathbf{F} = -\frac{GMm}{r^3}\mathbf{r} \quad \text{where} \quad \mathbf{r} = x\mathbf{i} + y\mathbf{j} + z\mathbf{k}$$

Write $\mathbf{F}$ as the negative gradient of a function V (called the gravitational potential) and verify that $\mathbf{F}$ is orthogonal to the level surfaces of V. [*Hint:* Study Solved Exercise 4.]

Exercises

12. (a) Find a unit normal to the ellipsoid

$$x^2 + 2y^2 + 3z^2 = 10$$

at each of the points $(\sqrt{10}, 0, 0)$, $(-\sqrt{10}, 0, 0)$, $(1, 0, \sqrt{3})$, and $(-1, 0, -\sqrt{3})$.

(b) Do the vectors you have found point to the inside or outside of the ellipsoid?

(c) Give equations for the tangent planes to the surface at the two points of the surface with $x_0 = y_0 = 1$.

13. Find the equation for the tangent line to each curve at the indicated point.

    (a) $x^2 + 2y^2 = 3$; $(1, 1)$ (b) $xy = 17$; $(x_0, 17/x_0)$

14. Find the equation for the tangent plane to each surface at the indicated point.

    (a) $x^2 + 2y^2 + 3z^2 = 10$; $(1, \sqrt{3}, 1)$

    (b) $xyz^2 = 1$; $(1, 1, 1)$

IMPLICIT DIFFERENTIATION AND RELATED RATES REVISITED

Like Molière's character Jourdain,* without knowing it we have been using functions of several variables and their partial derivatives ever since we studied implicit differentiation and related rates in Section 5-2.

Worked Example 9   Suppose that y is defined implicitly in terms of x by the following formula: $x^3 + 8x \sin y = 0$. Find dy/dx by implicit differentiation.

Solution   Considering y as a function of x and differentiating with respect to x (using the chain rule for functions of one variable), we get

$$3x^2 + 8 \sin y + 8x \cos y \, \frac{dy}{dx} = 0$$

so

$$\frac{dy}{dx} = -\frac{3x^2 + 8 \sin y}{8x \cos y}$$

We can easily interpret this example in terms of functions of two variables. In writing the equation $x^3 + 8x \sin y = 0$, we are saying that the graph $y = f(x)$ is contained in the level curve $F(x,y) = 0$, where $F(x,y) = x^3 + 8x \sin y$. The numerator and denominator in the fraction for dy/dx are just the partial derivatives $F_x(x,y)$ and $F_y(x,y)$.

Let us consider the general case: let $F(x,y)$ be a function of two variables, and $y = f(x)$ be a function whose graph is contained in the level curve $F(x,y) = 0$; in other words,

$$F(x, f(x)) = 0$$

We may consider $(x, f(x))$ as a parametric curve in the plane, where the parameter is x. Since $F(x, f(x)) = 0$, the chain rule gives

$$0 = \frac{d}{dx} [F(x, f(x))] = F_x(x, f(x)) \frac{dx}{dx} + F_y(x, f(x)) \frac{dy}{dx}$$

Since $dx/dx = 1$, we may solve this for dy/dx to get

$$\frac{dy}{dx} = -\frac{F_x(x, f(x))}{F_y(x, f(x))}$$

---

*"Good heavens! For more than forty years I have been speaking prose without knowing it" (Molière, Le Bourgeois Gentilhomme (1670)).

Or, if we put $z = F(x, y)$,

$$\frac{dy}{dx} = -\frac{\partial z/\partial x}{\partial z/\partial y} \tag{6}$$

Notice that it is *incorrect* to "cancel the ∂z's," because the minus sign would be left.

Worked Example 10   Find dy/dx in Worked Example 9 by using formula (6).

Solution   With $z = x^3 + 8x \sin y$, we have $\partial z/\partial x = 3x^2 + 8 \sin y$ and $\partial z/\partial y = 8x \cos y$, so

$$\frac{dy}{dx} = -\frac{\partial z/\partial x}{\partial z/\partial y} = -\frac{3x^2 + 8 \sin y}{8x \cos y}$$

In related rate problems, we usually have a parametric curve $(x, y) = (g(t), h(t))$ which lies in a level curve $F(x, y) = 0$. Differentiating with respect to t by the chain rule, we get

$$0 = F_x(x, y) \frac{dx}{dt} + F_y(x, y) \frac{dy}{dt} \tag{7}$$

which is a relation between the rates dx/dt and dy/dt. Such relations were obtained in Section 5-2 by one-variable calculus.

Worked Example 11   Suppose that $x = g(t)$ and $y = h(t)$ satisfy the relation $x^2 - y^2 = xy$. Find a relation between dx/dt and dy/dt:
(a) By one-variable calculus.
(b) By formula (7).

Solution   (a) Differentiating the relation $x^2 - y^2 = xy$ with respect to t by one-variable calculus, we obtain $2x (dx/dt) - 2y (dy/dt) = y (dx/dt) + x (dy/dt)$, or $(2x - y)(dx/dt) - (2y + x)(dy/dt) = 0$.
(b) To apply formula (7), we set $F(x, y) = x^2 - y^2 - xy$. Then $F_x(x, y) = 2x - y$ and $F_y(x, y) = -(2y + x)$, so formula (7) gives $(2x - y)(dx/dt) - (2y + x)(dy/dt) = 0$ as well.

Solved Exercises

14. What happens in formula (6) when the denominator $\partial z/\partial y$ becomes zero? (Assume that $\partial z/\partial x \neq 0$.) Illustrate with reference to the example $x - y^3 = 0$.

15. Suppose that $x = g(t)$ and $y = h(t)$ satisfy the relation $x^y = 2$. Find a relation between dx/dt and dy/dt.

Exercises

15. In each of the following, suppose that y is defined implicitly as a function of x by the given equation; express dy/dx in terms of y and x.
(a) $x^2 + 2y^2 = 3$ (b) $x^2 - y^2 = 7$ (c) $x/y = 10$

16. (a) Show that the curve $x^2 - y^2 = c$, for *any* value of c, satisfies the differential equation $dy/dx = x/y$.

(b) Draw in a few of the curves $x^2 - y^2 = c$, say for $c = \pm 1$. At several points (x, y) along each of these curves, draw a short segment of slope x/y; check that these segments appear to be tangent to the curve. What happens when $y = 0$? What happens when $c = 0$?

17. In each of the following, assume that x and y are functions of t which satisfy the given relation, and find a relation between dx/dt and dy/dt.

    (a) $x \ln y = 1$ (b) $\sin(xy) + \cos(xy) = 1$

    (c) $x^4 + y^4 = 1$ (d) $x^2 + 3y^2 = 2$

Problems for Section 16-2

1. Verify the chain rule for each of the following functions and curves:

    (a) $f(x, y) = (x^2 + y^2) \ln(\sqrt{x^2 + y^2})$;
        $\sigma(t) = (e^t, e^{-t})$

    (b) $f(x, y, z) = x + y^2 + z^3$;
        $\sigma(t) = (\cos t, \sin t, t)$

    (c) $f(x, y) = xe^{x^2+y^2}$; $\sigma(t) = (t, -t)$

2. Suppose that the temperature at the point (x, y, z) in space is $T(x, y, z) = x^2 + y^2 + z^2$. Let a particle follow the right circular helix $\sigma(t) = (\cos t, \sin t, t)$ and let $T(t)$ be its temperature at time t.

    (a) What is $T'(t)$?

    (b) Find an approximate value for the temperature at $t = (\pi/2) + 0.01$.

3. Compute the gradient of each of the following functions:

    (a) $f(x, y) = (x^2 + y^2) \ln\sqrt{x^2 + y^2}$

    (b) $f(x, y, z) = x + y^2 + z^3$

    (c) $f(x, y) = xe^{x^2+y^2}$

4. Suppose that f and g are real-valued functions with continuous partial derivatives. Show that:

    (a) $\nabla f = \mathbf{0}$ if f is constant.

    (b) $\nabla(f + g) = \nabla f + \nabla g$.

    (c) $\nabla(cf) = c\nabla f$ if c is a number.

    (d) $\nabla(fg) = f\nabla g + g\nabla f$.

    (e) $\nabla(f/g) = (g\nabla f - f\nabla g)/g^2$ at points where $g \neq 0$.

5. (a) Using the notation of Solved Exercise 4, show that $\nabla r = \mathbf{r}/r$.

    (b) Use part (a) and Problem 4 to show that $\nabla(1/r) = -\mathbf{r}/r^3$ $(r \neq 0)$. (Compare Solved Exercise 4.)

    (c) Find $\nabla(1/r^2)$ $(r \neq 0)$.

6. Compute the directional derivative of each function at the given point (x_0, y_0) in the given direction $\mathbf{n}$.

    (a) $f(x, y) = x^2 + y^2 - 3xy^3$; $(x_0, y_0) = (1, 2)$; $\mathbf{n} = (\frac{1}{2}, \sqrt{3}/2)$

    (b) $f(x, y) = e^x \cos y$; $(x_0, y_0) = (0, \pi/4)$; $\mathbf{n} = (\mathbf{i} + 3\mathbf{j})/\sqrt{10}$

    (c) $f(x, y) = 17x^y$; $(x_0, y_0) = (1, 1)$; $\mathbf{n} = (\mathbf{i} + \mathbf{j})/\sqrt{2}$

7. Compute the directional derivative of each function at the given point (x_0, y_0, z_0) in the given direction $\mathbf{n}$.

    (a) $f(x, y, z) = x^2 - 2xy + 3z^2$; $(x_0, y_0, z_0) = (1, 1, 2)$; $\mathbf{n} = (\mathbf{i} + \mathbf{j} - \mathbf{k})/\sqrt{3}$

    (b) $f(x, y, z) = e^{-(x^2+y^2+z^2)}$; $(x_0, y_0, z_0) = (1, 10, 100)$; $\mathbf{n} = (1, -1, -1)/\sqrt{3}$

    (c) $f(x, y, z) = \sin(xyz)$; $(x_0, y_0, z_0) = (1, 1, \pi/4)$; $\mathbf{n} = (1/\sqrt{2}, 0, -1/\sqrt{2})$

8. Find the plane tangent to each of the following surfaces at the indicated point.

    (a) $x^2 + 2y^2 + 3xz = 10$; $(1, 2, \frac{1}{3})$

    (b) $y^2 - x^2 = 3$; $(1, 2, 8)$

    (c) $xyz = 1$; $(1, 1, 1)$

    (d) $xy/z = 1$; $(1, 1, 1)$

9. (a) In what direction is the directional derivative of $f(x, y) = (x^2 - y^2)/(x^2 + y^2)$ at $(1, 1)$ equal to zero?

    (b) How about at an arbitrary point (x_0, y_0) in the first quadrant?

    (c) Describe the level curves of f. In particular, discuss them in terms of the result of part (b).

10. Suppose that a particle is ejected from the surface $x^2 + y^2 - z^2 = -1$ at the point $(1, 1, \sqrt{3})$ in a direction normal to the surface at time $t = 0$ with a speed of 10 units per second. When and where does it cross the xy plane?

11. In what direction is each of the following increasing fastest at $(1, 1)$?

    (a) $f(x, y) = x^2 + 2y^2$

    (b) $g(x, y) = x^2 - 2y^2$

    (c) $h(x, y) = e^x \sin y$

12. Captain Astro is once again in trouble near the sunny side of Mercury. She is at location $(1, 1, 1)$, and the temperature of the ship's hull when she is at location (x, y, z) will be given by

    $$T(x, y, z) = e^{-x^2 - 2y^2 - 3z^2}$$

    where x, y, and z are measured in meters.

    (a) In what direction should she proceed in order to decrease the temperature most rapidly?

    (b) If the ship travels at e^8 meters per second, how fast will the temperature decrease if she proceeds in that direction?

    (c) Unfortunately, the metal of the hull will crack if cooled at a rate greater than $\sqrt{14}\, e^2$ degrees per second. Describe the set of possible directions in which she may proceed to bring the temperature down at that rate.

13. Suppose that $f(x, y)$ has continuous partial derivatives. At $(1, 1)$ the directional derivative in the direction toward $(2, 4)$ is 2 and in the direction toward $(2, 2)$ it is 3. Find the gradient of f at $(1, 1)$ and the directional derivative there in the direction toward $(2, 3)$.

14. Find the equation of the lines normal to the given surface at the given point:

    (a) $e^{-(x^2 + y^2 + z^2)} = e^{-3}$; $(1, 1, 1)$

    (b) $2x^2 + 3y^2 + z^2 = 9$; $(1, 1, 2)$

    (c) $x/yz = 1$; $(1, 1, 1)$

15. (a) Use the chain rule to find $(d/dx)(x^x)$ by using the function $f(y, z) = y^z$.

    (b) Calculate $(d/dx)(x^x)$ by using one-variable calculus.

    (c) Which way do you prefer?

16. The height h of the Hawaiian volcano Mauna Loa is (roughly) described by the function $h(x, y) = 2.59 - 0.00024y^2 - 0.0065x^2$, where h is the height above sea level in miles and x and y measure east-west and north-south distances in miles from the top of the mountain. At $(x, y) = (-2, -4)$:

    (a) How fast is the height increasing in the direction $(1, 1)$ (that is, northeastward)? Express your answer in miles of height per mile of horizontal distance traveled.

    (b) In what direction is the steepest upward path?

    (c) In what direction is the steepest downward path?

    (d) In what direction(s) is the path level?

    (e) If you proceed south, are you ascending or descending? At what rate?

    (f) If you move northwest, are you ascending or descending? At what rate?

    (g) In what direction(s) may you proceed in order to be climbing with a grade of 3%?

★17. Let V be a function defined on a domain in space. The force associated with V is $\mathbf{F} = \Phi(x, y, z) = -\nabla V(x, y, z)$; we call V the *potential* of Φ (see Solved Exercise 13). Let a point with mass m move on a parametric curve $\sigma(t)$ and satisfy Newton's second law (see p. 672), $m\mathbf{a} = \mathbf{F}$, where $\mathbf{a}$ is the acceleration of the curve. Use the chain rule to prove the law of conservation of energy:

    $$E = \tfrac{1}{2} m \|\sigma'(t)\|^2 + V[\sigma(t)]$$

    is constant, where $\sigma(t)$ is the position vector of the curve.

★18. (Continuation of Problem 17.) The level surfaces of a potential function V are called *equipotential surfaces*.

    (a) What is the relation between the force vector and the equipotential surfaces?

    (b) Explain why "sea level" is approximately an equipotential surface for the earth's gravitational field. What spoils the approximation?

★19. Prove the chain rule by filling in the details in the following argument. Let $z = f(g(t), h(t))$, where g and h are differentiable and f has continuous partial derivatives.

    (a) Show that

    $$\frac{\Delta z}{\Delta t} = \frac{1}{\Delta t}\{[f(g(t + \Delta t), h(t + \Delta t)) - f(g(t), h(t + \Delta t))] + [f(g(t), h(t + \Delta t)) - f(g(t), h(t))]\}$$

    (b) Apply the mean value theorem for functions of one variable to each of the expressions in square brackets.

    (c) Take the limit as $\Delta t \to 0$. (You will use the continuity of partial derivatives at this point.)

20. The possible time T in minutes of a scuba dive is given by $T = 32\,V/(x + 32)$ where V is the volume of air at 15 psi (pounds per square inch) which is compressed into the air tanks and x is the depth of the scuba dive in feet.

    (a) How long can a 27-ft dive last when $V = 65$?

    (b) Find $\partial T/\partial x$ and $\partial T/\partial V$ when $x = 27$, $V = 65$. Interpret.

21. The productivity z of a company is given by $z(x,y) = 100\,xy - 2x^2 - 6y^2$ where $x \cdot 10^3$ people work for the company, and the capital investment of the company is y million dollars.

    (a) Find the marginal productivity of labor $\partial z/\partial x$. This number is the expected change in production for an increase of 1000 staff with fixed capital investment.

    (b) Find $\partial z/\partial y$ when $x = 5$ and $y = 3$. Interpret.

22. In electrostatistics, the force $\mathbf{P}$ of attraction between two particles of opposite charge is $\mathbf{P} = k\,(\mathbf{r}/|\mathbf{r}|^3)$ (*Coulomb's law*), where k is a constant and $\mathbf{r} = x\mathbf{i} + y\mathbf{j} + z\mathbf{k}$. Show that $\mathbf{P}$ is the gradient of $f = -k/|\mathbf{r}|$.

23. The potential V due to two infinite, parallel filaments of charge of linear densities λ and $-\lambda$ is given by $V = (\lambda/2\pi\varepsilon_0)\,\ln(r_2/r_1)$ where $(r_2/r_1)^2 = [(x + x_0)^2 + y^2]\,/\,[(x - x_0)^2 + y^2]$. We think of the filaments as being in the z-direction, passing through the xy-plane at $(-y_0, 0)$ and $(x_0, 0)$.

    (a) Find grad $V(x,y)$, using the chain rule.

    (b) Verify the *flux law* $\partial^2 V/\partial x^2 + \partial^2 V/\partial y^2 = 0$.

24. The *ideal gas law* $PV = nRT$ involves a constant R, the number n of moles of the gas, the volume V, the Kelvin temperature T, and the pressure P.

    (a) Show that each of n, P, T, V is a function of the remaining variables, and determine explicitly the defining equations.

    (b) The quantity $\partial P/\partial T$ is a rate of change. Discuss this in detail, and illustrate with an example which identifies the variables held constant.

    (c) Calculate $\partial V/\partial T$, $\partial T/\partial P$, $\partial P/\partial V$ and show their product equals -1.

25. The specific volume V, pressure P, temperature T of a Van der Waals gas are related by $P = [RT/(V - \beta)] - \alpha/V^2$ where α, β, R are considered to be constants.

    (a) Explain why any two of V, P, T, can be considered independent variables which determine the third variable.

    (b) Find $\partial T/\partial P$, $\partial P/\partial V$, $\partial V/\partial T$. Identify which variables are constant, and interpret each partial derivative physically.

    (c) Verify that $(\partial T/\partial P)\,(\partial P/\partial V)\,(\partial V/\partial T) = -1$ (not $+1$!).

16-3 Maxima and Minima

First and second derivative tests are developed for locating maximum and minimum points for functions of two variables.

In studying maxima and minima for functions of one variable, we found that the basic tests involved the vanishing of the first derivative and the sign of the second derivative. In this section we develop related tests, involving first and second partial derivatives, for locating maxima and minima of functions of two variables. (Similar tests for functions of three variables are outlined in Problems 20 and 21.) In addition, we introduce the method of Lagrange multipliers for finding extrema in the presence of side conditions.

Goals

After studying this section, you should be able to:

Locate the critical points of a function of two variables.

Determine whether a critical point of a function of two variables is a maximum, minimum, or saddle point.

Find extreme points of a function of two variables on a level curve of a second function.

NECESSARY CONDITIONS: THE FIRST DERIVATIVE TEST

The definition of local maxima for functions of two variables is similar to that in the one-variable case, except that we use disks instead of intervals. Recall that the disk of radius r about (x_0, y_0) consists of all (x, y) such that the distance $\sqrt{(x - x_0)^2 + (y - y_0)^2}$ is less than r.

Definition Let $f(x, y)$ be a function of two variables. We say that (x_0, y_0) is a *local minimum point* for f if there is a disk (of positive radius) about (x_0, y_0) such that $f(x, y) \geq f(x_0, y_0)$ for all (x, y) in the disk.

Similarly, if $f(x, y) \leq f(x_0, y_0)$ for all (x, y) in some disk (of positive radius) about (x_0, y_0), we call (x_0, y_0) a *local maximum point* for f.

A point which is either a local maximum or minimum point is called a *local extremum*.

We may also define global maximum and minimum points to be those at which a function attains the greatest and least values for all points in the domain.

Worked Example 1 Refer to Fig. 16-3-1, a computer-drawn graph of $z = 2(x^2 + y^2)e^{-x^2-y^2}$. Where are the maximum and minimum points?

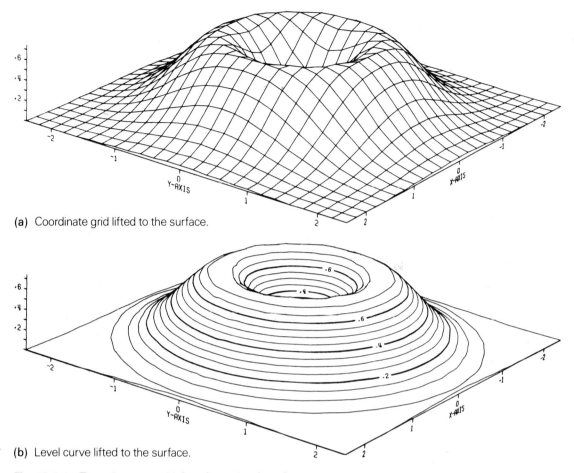

(a) Coordinate grid lifted to the surface.

(b) Level curve lifted to the surface.

Fig. 16-3-1 The volcano: $z = 2(x^2 + y^2) \exp(-x^2 - y^2)$.

Solution There is a local (in fact, global) minimum at the volcano's center $(0, 0)$, where $z = 0$. There are maximum points all around the crater's rim (the circle $x^2 + y^2 = 1$).

The following theorem is the analog in two variables of the first derivative test (see Theorem 1, p. 153).

Theorem 6 *Suppose that (x_0, y_0) is a local extremum of f and that the partial derivatives of f exist at (x_0, y_0). Then $f_x(x_0, y_0) = f_y(x_0, y_0) = 0$.*

Proof We consider the case of a local minimum; the proof for a local maximum is essentially the same.

By assumption, there is a disk of radius r about (x_0, y_0) on which $f(x, y) \geq f(x_0, y_0)$. In particular, if $|x - x_0| < r$, then $f(x, y_0) \geq f(x_0, y_0)$, so the function $g(x) = f(x, y_0)$ has a local minimum at x_0. By the first derivative test of one-variable calculus, $g'(x_0) = 0$. But $g'(x_0)$ is just $f_x(x_0, y_0)$. Similarly, the function $h(y) = f(x_0, y)$ has a local minimum at y_0, so $f_y(x_0, y_0) = 0$.

Theorem 6 has a simple geometric interpretation: at a local extremum of f, the tangent plane to the graph $z = f(x, y)$ is horizontal (that is, parallel to the xy plane).

Points at which f_x and f_y *both* vanish are called *critical points* of f. As in one-variable calculus, finding critical points is only the first step in finding local extrema. A critical point could be a local maximum, local minimum, or neither. After looking at some examples, we will present the second derivative test for functions of two variables.

Worked Example 2   Let $z = x^2 - y^2$. Show that $(0, 0)$ is a critical point. Is it a local extremum?

Solution   The partial derivatives $\partial z / \partial x = 2x$ and $\partial z / \partial y = -2y$ vanish at $(0, 0)$, so the origin is a critical point. It is neither a local maximum nor minimum since $f(x, y) = x^2 - y^2$ is zero at $(0, 0)$ and can be either positive (on the x axis) or negative (on the y axis) arbitrarily near the origin. This is also clear from examination of the graph (see Fig. 15-2-2), which has a saddle point at $(0, 0)$.

If we know in advance that a function has a minimum point, and that the partial derivatives exist there, then we can use the first derivative test to locate the point.

Worked Example 3   Find the minimum distance from the origin to a point on the plane $x + 3y - z = 6$.

Solution   The distance in question is $d = \sqrt{x^2 + y^2 + z^2}$, where $z = x + 3y - 6$. It is simpler to minimize $d^2 = x^2 + y^2 + (x + 3y - 6)^2$. By Theorem 6, we must have

$$\frac{\partial(d^2)}{\partial x} = 0 \quad \text{(that is, } 2x + 2(x + 3y - 6) = 0\text{)}$$

and

$$\frac{\partial(d^2)}{\partial y} = 0 \quad \text{(that is, } 2y + 6(x + 3y - 6) = 0\text{)}$$

Solving these equations gives $y = \frac{18}{11}$, $x = \frac{6}{11}$. Thus $z = x + 3y - 6 = -\frac{6}{11}$ and so the minimum distance is $d = \sqrt{x^2 + y^2 + z^2} = 6\sqrt{11}/11$. (Geometric intuition tells us that d^2 has a minimum—any plane has a point on it closest to the origin, so the point we found must be that one. See Fig. 16-3-2.)

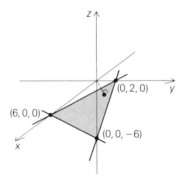

Fig. 16-3-2 The point nearest to the origin on the plane $z = x + 3y - 6$ is $(6/11, 18/11, -6/11)$.

Solved Exercises   1. Find the minimum distance from $(1, 2, 0)$ to the cone $z^2 = x^2 + y^2$.

2. A rectangular box, open at the top, is to hold 256 cubic centimeters

of wheat germ. Find the dimensions for which the surface area (bottom and four sides) is minimized.

3. Verify that the critical points of the function in Worked Example 1 occur at $(0, 0)$ and on the circle $x^2 + y^2 = 1$.

Exercises

1. Find the critical points of each of the following functions. Decide, if you can, whether each of the critical points is a local maximum, minimum, or neither.

   (a) $f(x, y) = x^2 + 2y^2$ 　　　　　(b) $f(x, y) = x^2 - 2y^2$

   (c) $f(x, y) = x^2 + y^2 + 6x - 4y + 13$ [*Hint:* Complete the square.]

   (d) $f(x, y) = \exp(-x^2 - 7y^2 + 3)$ 　　(e) $f(x, y) = \exp(x^2 + 2y^2)$

2. Suppose that the material for the bottom of the box in Solved Exercise 2 costs b cents per square centimeter, while that for the sides costs s cents per square centimeter. Find the dimensions which minimize the cost of the material.

3. Find the distance from the plane $x + 2y + 3z - 10 = 0$:

   (a) To the origin. 　　　　　　　　(b) To the point $(1, 1, 1)$.

4. A computer-generated graph of $z = (\sin \pi r)/\pi r$, $r = \sqrt{x^2 + y^2}$, is shown in Fig. 16-3-3. Show, by calculation, that all critical points of the function lie on circles whose radius satisfies the equation $\pi r = \tan(\pi r)$. Which points are maxima? Minima? What symmetries does the graph have?

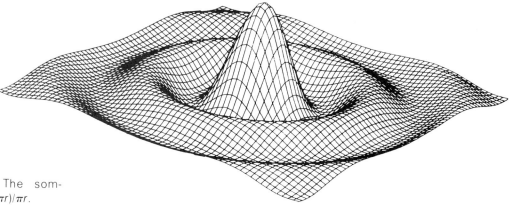

Fig. 16-3-3 The sombrero: $z = (\sin \pi r)/\pi r$.

SUFFICIENT CONDITIONS: THE SECOND DERIVATIVE TEST

Captain Astro is being held captive by Tralfamadorians who are studying human intelligence. She is in a room where a loudspeaker emits a piercing noise. There are two knobs on the wall, whose positions, x and y, seem to affect the loudness of the noise. The knobs are initially at $x = 0$ and $y = 0$ and, when the first knob is turned, the noise gets even louder for $x < 0$ and for $x > 0$. So the captain leaves $x = 0$ and turns the second knob both ways, but, alas, the noise again gets louder. Finally, she sees the formula $f(x, y) = x^2 + 3xy + y^2 + 16$ printed on the wall.

Worked Example 4 What to do?

Solution First of all, she notices that $f(x, 0) = x^2 + 16$ and $f(0, y) = y^2 + 16$, so the function f, like the loudness of the noise, increases if either x or y is moved away from zero. But look! If we put $y = -x$, then the "$3xy$" term becomes negative. In fact, $f(x, -x) = x^2 - 3x^2 + x^2 + 16 = -x^2 + 16$. Captain Astro rushes to the dials and turns them both at once, in opposite directions. (Why?) The noise subsides (and the Tralfamadorians cheer).

The function $f(x, y) = x^2 + 3xy + y^2 + 16$ has a critical point at $(0, 0)$, and the functions $g(x) = f(x, 0)$ and $h(y) = f(0, y)$ both have zero as a local minimum point. But $(0, 0)$ is not a local minimum point for f, because $f(x, -x) = -x^2 + 16$ is less than $f(0, 0) = 16$ for arbitrarily small x.

This example shows that to tell whether a critical point (x_0, y_0) of a function $f(x, y)$ is a local extremum, we must look at the behavior of f along lines passing through (x_0, y_0) in all directions, not just those parallel to the axes.

The following theorem enables us to determine the nature of the critical point $(0, 0)$ for any function of the form $Ax^2 + 2Bxy + Cy^2$.

Theorem 7 *Let* $g(x, y) = Ax^2 + 2Bxy + Cy^2$, *where* A, B, *and* C *are constants.*

1. *If* $AC - B^2 > 0$, *and* $A > 0$, $[A < 0]$, *then for all* $(x, y) \neq (0, 0)$, $g(x, y)$ *has a minimum [maximum] at* $0, 0$.

2. *If* $AC - B^2 < 0$, *then* $g(x, y)$ *takes both positive and negative values for* (x, y) *near* $(0, 0)$, *so* $(0, 0)$ *is not a local extremum for* g.

Proof

1. If $AC - B^2 > 0$, then A cannot be zero (why?), so we may write

$$g(x, y) = A\left(x^2 + \frac{2B}{A}xy + \frac{C}{A}y^2\right)$$
$$= A\left(x^2 + \frac{2B}{A}xy + \frac{B^2y^2}{A^2} + \frac{C}{A}y^2 - \frac{B^2y^2}{A^2}\right)$$
$$= A\left(x + \frac{B}{A}y\right)^2 + \frac{1}{A}(AC - B^2)y^2 \qquad \textbf{(1)}$$

Both terms on the right-hand side of (1) have the same sign as A, and they are both zero only when $x + (B/A)y = 0$ and $y = 0$—that is, when $(x, y) = (0, 0)$. Thus $(0, 0)$ is a minimum point for g if $A > 0$ and a maximum point if $A < 0$.

2. If $AC - B^2 < 0$ and $A \neq 0$, then formula (1) still applies, but now the terms on the right-hand side have opposite signs. By suitable choices of x and y (see Problem 22), we can make either term zero and the other nonzero. If $A = 0$, then $g(x, y) = y(2Bx + Cy)$, and we can again achieve both signs.

In case 2 of Theorem 7, $(0,0)$ is called a *saddle point* for $g(x,y)$. (See Problems 22 and 23 for a further discussion of this case and the case $AC - B^2 = 0$.)

Worked Example 5   Apply Theorem 7 to Captain Astro's problem.

Solution   The noise function is $f(x,y) = x^2 + 3xy + y^2 + 16$. We may write this as $g(x,y) + 16$, where $g(x,y) = x^2 + 3xy + y^2$ has the form used in Theorem 7, with $A = 1$, $B = \frac{3}{2}$, and $C = 1$. Since $AC - B^2 = 1 - \frac{9}{4}$ is negative, there exist choices of x and y making $g(x,y)$ negative, so the noise level can be reduced. (Equation (1) gives $g(x,y) = (x + \frac{3}{2}y)^2 - \frac{9}{4}y^2$, so moving along the line $x = -\frac{3}{2}y$ would be a good choice.)

Worked Example 6   Determine whether $(0,0)$ is a maximum point, a minimum point, or neither, of $g(x,y) = 3x^2 - 5xy + 3y^2$.

Solution   Here $A = 3$, $B = -\frac{5}{2}$, and $C = 3$, so that $A = 3 > 0$ and $AC - B^2 = 9 - \frac{25}{4} > 0$. Thus $(0,0)$ is a minimum point by Theorem 7, part 1.

We are ready to go on to the second derivative test for general functions of two variables. Note that for the function $g(x,y)$, the constants A, B, and C entering into Theorem 7 can be recovered from g by the formulas

$$A = \frac{1}{2}\frac{\partial^2 g}{\partial x^2} \qquad B = \frac{1}{2}\frac{\partial^2 g}{\partial x\, \partial y} \qquad C = \frac{1}{2}\frac{\partial^2 g}{\partial y^2}$$

so that the signs of A and $AC - B^2$ are the same as those of $\partial^2 g/\partial x^2$ and $(\partial^2 g/\partial x^2)(\partial^2 g/\partial y^2) - (\partial^2 g/\partial x\, \partial y)^2$.

Theorem 8   *Let $f(x,y)$ have continuous second partial derivatives, and suppose that (x_0, y_0) is a critical point for f. Let $D(x_0, y_0) = f_{xx}(x_0, y_0)f_{yy}(x_0, y_0) - [f_{xy}(x_0, y_0)]^2$.*

1. *If $D(x_0, y_0) > 0$, then (x_0, y_0) is a local minimum or maximum point for f according to whether $f_{xx}(x_0, y_0)$ is positive or negative.*

2. *If $D(x_0, y_0) < 0$, then (x_0, y_0) is not a local extremum for f. (In this case, (x_0, y_0) is called a **saddle point**.)*

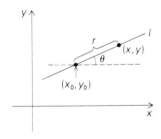

Fig. 16-3-4   The point (x, y) is $(x_0 + r\cos\theta, y_0 + r\sin\theta)$.

Proof   We look at f along straight lines through (x_0, y_0). Specifically, for each fixed θ in $[0, 2\pi]$, we will consider the function $h(r) = f(x_0 + r\cos\theta, y_0 + r\sin\theta)$, which describes the behavior of f along the line through (x_0, y_0) in the direction of $\cos\theta\mathbf{i} + \sin\theta\mathbf{j}$. (See Fig. 16-3-4.)

For each θ, $h(r)$ is a function of one variable with a critical point at $r = 0$. To analyze the behavior of $h(r)$ near $r = 0$ by using the second derivative test for functions of one variable, we differentiate $h(r)$ by using the chain rule of Section 16-2. Let $x = x_0 + r\cos\theta$ and $y = y_0 + r\sin\theta$; then

$$h'(r) = f_x(x,y)\frac{dx}{dr} + f_y(x,y)\frac{dy}{dr} = f_x(x,y)\cos\theta + f_y(x,y)\sin\theta$$

We differentiate again, applying the chain rule to f_x and f_y:

$$h''(r) = f_{xx}(x, y) \cos \theta \frac{dx}{dr} + f_{xy}(x, y) \cos \theta \frac{dy}{dr}$$

$$+ f_{yx}(x, y) \sin \theta \frac{dx}{dr} + f_{yy}(x, y) \sin \theta \frac{dy}{dr}$$

Since $f_{xy} = f_{yx}$ by Theorem 1, this becomes

$$h''(r) = f_{xx}(x, y) \cos^2 \theta + 2f_{xy}(x, y) \cos \theta \sin \theta + f_{yy}(x, y) \sin^2 \theta$$

or

$$h''(r) = f_{xx}(x_0 + r \cos \theta, y_0 + r \sin \theta) \cos^2 \theta$$
$$+ 2f_{xy}(x_0 + r \cos \theta, y_0 + r \sin \theta) \cos \theta \sin \theta$$
$$+ f_{yy}(x_0 + r \cos \theta, y_0 + r \sin \theta) \sin^2 \theta \tag{2}$$

Setting $r = 0$, we get

$$h''(0) = f_{xx}(x_0, y_0) \cos^2 \theta + 2f_{xy}(x_0, y_0) \cos \theta \sin \theta + f_{yy}(x_0, y_0) \sin^2 \theta$$

which has the form $Ax^2 + 2Bxy + Cy^2$, with $x = \cos \theta$, $y = \sin \theta$, $A = f_{xx}(x_0, y_0)$ $B = f_{xy}(x_0, y_0)$, $C = f_{yy}(x_0, y_0)$, and $AC - B^2 = D(x_0, y_0)$.

Now suppose that $D(x_0, y_0) > 0$ and $f_{xx}(x_0, y_0) > 0$. By Theorem 7, $h''(0) > 0$, so h has a local minimum at $r = 0$. Since this is true for all values of θ, f has a local minimum along each line through (x_0, y_0). It is thus plausible that f has a local minimum at (x_0, y_0), so we will end the proof at this point. (Actually, more work is needed; for further details, see Problems 24 and 25.)

If $D(x_0, y_0) > 0$ and $f_{xx}(x_0, y_0) < 0$, then f has a local maximum along every line through (x_0, y_0); if $D(x_0, y_0) < 0$, then f has a local minimum along some lines through (x_0, y_0) and a local maximum along others.

SECOND DERIVATIVE TEST

Assume that $f_x(x_0, y_0) = 0$ and $f_y(x_0, y_0) = 0$.

Let $A = f_{xx}(x_0, y_0)$, $B = f_{xy}(x_0, y_0)$, and $C = f_{yy}(x_0, y_0)$.

If:	Then:
$A > 0$, $AC - B^2 > 0$	(x_0, y_0) is a local minimum;
$A < 0$, $AC - B^2 > 0$	(x_0, y_0) is a local maximum;
$AC - B^2 < 0$	(x_0, y_0) is a saddle point;
$AC - B^2 = 0$	the test fails.

Worked Example 7 Find the maxima, minima, and saddle points of $z = (x^2 - y^2)e^{(-x^2 - y^2)/2}$.

Solution First we locate the critical points by setting $\partial z/\partial x = 0$ and $\partial z/\partial y = 0$. Here

$$\frac{\partial z}{\partial x} = [2x - x(x^2 - y^2)]\, e^{(-x^2 - y^2)/2}$$

and

$$\frac{\partial z}{\partial y} = [-2y - y(x^2 - y^2)]\, e^{(-x^2 - y^2)/2}$$

so the critical points are the solutions of

$$x\,[2 - (x^2 - y^2)] = 0, \qquad y\,[-2 - (x^2 - y^2)] = 0$$

This has solutions $(0,0)$, $(\pm\sqrt{2}, 0)$, and $(0, \pm\sqrt{2})$.
  The second derivatives are

$$\frac{\partial^2 z}{\partial x^2} = [2 - 5x^2 + x^2(x^2 - y^2) + y^2]\, e^{(-x^2 - y^2)/2}$$

$$\frac{\partial^2 z}{\partial x\,\partial y} = xy(x^2 - y^2) e^{(-x^2 - y^2)/2}$$

$$\frac{\partial^2 z}{\partial y^2} = [5y^2 - 2 + y^2(x^2 - y^2) - x^2]\, e^{(-x^2 - y^2)/2}$$

Using the second derivative test results in the following data:

Point	A	B	C	$AC - B^2$	Type
$(0,0)$	2	0	-2	-4	saddle
$(\sqrt{2}, 0)$	$-4/e$	0	$-4/e$	$16/e^2$	maximum
$(-\sqrt{2}, 0)$	$-4/e$	0	$-4/e$	$16/e^2$	maximum
$(0, \sqrt{2})$	$4/e$	0	$4/e$	$16/e^2$	minimum
$(0, -\sqrt{2})$	$4/e$	0	$4/e$	$16/e^2$	minimum

Compare this table with the computer-generated graph in Fig. 16-3-5.

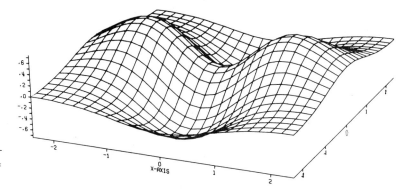

Fig. 16-3-5 Computer-generated graph of $z = (x^2 - y^2)e^{(-x^2 - y^2)/2}$.

Solved Exercises 4. Let $f(x, y) = 3 + 2x^2 - xy + y^2$. Is $(0, 0)$ a maximum or minimum?

5. Let $z = (x^2 + y^2)\cos(x + 2y)$. Show that $(0, 0)$ is a critical point. Is it an extremum?

6. Test for extrema: $z = \ln(x^2 + y^2 + 1)$.

Exercises 5. For each of the following, decide whether $(0, 0)$ is a maximum, minimum, or saddle point:

(a) $f(x, y) = x^2 + xy + y^2$ (b) $f(x, y) = x^2 - xy + y^2 + 1$

(c) $f(x, y) = y^2 - x^2 + 3xy$ (d) $f(x, y) = x^2 + y^2 + 3x - 2y + 1$

(e) $f(x, y) = y^2$

6. Apply the second derivative test to the critical point in Solved Exercise 2.

7. Find and classify the critical points of each of the following functions:

(a) $f(x, y) = x^2 - y^2 + xy$ (b) $f(x, y) = x^2 + y^2 - xy$

(c) $f(x, y) = e^{1 + x^2 - y^2}$

(d) $f(x, y) = x^2 - 3xy + 5x - 2y + 6y^2 + 8$

(e) $f(x, y) = 3x^2 + 2xy + 2y^2 - 3x + 2y + 10$

(f) $f(x, y) = \sin(x^2 + y^2)$ [Consider only the critical point $(0, 0)$.]

8. Show that $z = (x^3 - 3x)/(1 + y^2)$ has exactly one local maximum and one local minimum. What symmetries does the graph have? (It is computer-drawn in Fig. 16-3-6.)

Fig. 16-3-6 Computer-drawn graph of $z = (x^3 - 3x)/(1 + y^2)$.

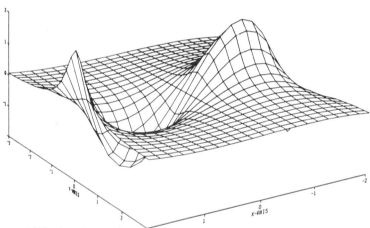

(a) Coordinate grid lifted to the graph.

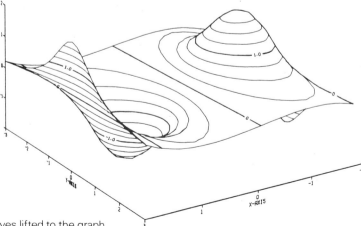

(b) Level curves lifted to the graph.

CONSTRAINED EXTREMA AND LAGRANGE MULTIPLIERS

When we studied maxima and minima for functions of one variable (Chapter 3), finding critical points was only the first step. In many problems, the function to be maximized or minimized was restricted to an interval, and the extreme points of the function on the interval could be either:

1. Critical points of f in the open interval (a, b)
2. Singular points of f
3. Boundary points of $[a, b]$ (the endpoints a and b)

Suppose now that D is a region in the plane. We would like to find extreme points (x_0, y_0) of f in D. Essentially the same three things can happen as above:

1. Suppose that (x_0, y_0) is an extremum lying inside D and that the partial derivatives of f exist at (x_0, y_0). Then our earlier analysis applies and (x_0, y_0) must be a critical point (point P_1 in Fig. 16-3-7).

2. (x_0, y_0) might be a singular point of f. (For example, $f(x, y) = \sqrt{x^2 + y^2}$ has a minimum at $(0,0)$. The graph is a cone which comes to a point at $(0,0)$, and f is not differentiable there.)

3. (x_0, y_0) might lie on the boundary of D (P_2 in Fig. 16-3-7).

This gives us a guide of where to look for extreme values:

Fig. 16-3-7 If the interior point P_1 is an extreme point of f on D, then the partial derivatives of f at P_1, if they exist, must be zero. If the boundary point P_2 is an extreme point, the partial derivatives there might not be zero.

EXTREME VALUES ON A REGION

To find extreme values for a function f on a region D,

1. Analyze the behavior of f at critical points inside D.
2. Examine singular points.
3. Find extreme values of f as a function restricted to the boundary of D and compare these values to nearby values inside D.
4. Compare the values of f at all these points.

The third step is hardest to handle. For an interval there was no difficulty, since the boundary consisted of two points and we had only to compute the value of f there. For a region in the plane, the boundary is a curve, and we must find the extrema of f when the "competition" is restricted to points on this curve.

If we can parametrize the boundary curve, say by $\sigma(t)$ for t in $[a, b]$, then the restriction of f to the boundary becomes a function of one variable, $f(\sigma(t))$, to which the methods of one-variable calculus apply.

Worked Example 8 Find the extreme values of $z = f(x, y) = x^2 + 2y^2$ on the disk D consisting of points (x, y) satisfying $x^2 + y^2 \le 1$.

Solution 1. At a critical point, $\partial z/\partial x = 2x = 0$ and $\partial z/\partial y = 4y = 0$. So the only critical point is $(0,0)$. It is clearly a minimum point for f; we may also verify this by the second derivative test:

$$\frac{\partial^2 z}{\partial x^2} = 2 > 0$$

and

$$\left(\frac{\partial^2 z}{\partial x^2}\right)\left(\frac{\partial^2 z}{\partial y^2}\right) - \left(\frac{\partial^2 z}{\partial x \, \partial y}\right)^2 = 2 \cdot 4 - 0 > 0$$

This result confirms that $(0, 0)$ is a local minimum.

2. There are no singular points.

3. The boundary of D is just the unit circle, which we may parametrize by $(\cos t, \sin t)$. Along the boundary,

$$h(t) = f(\cos t, \sin t) = \cos^2 t + 2 \sin^2 t = 1 + \sin^2 t$$

Since $h'(t) = 2 \sin t \cos t$, $h'(t) = 0$ at $t = 0, \pi/2, \pi$, and $3\pi/2$ (2π gives the same point as zero). We see that h takes a minimum of 1 at $t = 0$ or π and a maximum of 2 at $t = \pi/2$ or $3\pi/2$. So, along the boundary, f has a minimum of 1 at $(1, 0)$ and $(-1, 0)$ and a maximum of 2 at $(0, 1)$ and $(0, -1)$. As we move from these points into the disk, x^2 and y^2 get smaller, so f decreases and $(\pm 1, 0)$ are not even local minimum points for f on D; $(0, \pm 1)$ *are* maximum points.

4. Combining these results, we see that f has a minimum value of zero at $(0, 0)$ and a maximum of 2 at $(0, \pm 1)$ in D. (See Fig. 16-3-8.)

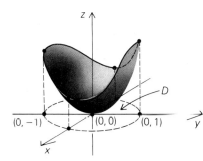

Fig. 16-3-8 The function $f(x, y) = x^2 + y^2$ on the disk D has a minimum point at $(0, 0)$ and maximum points at $(0, 1)$ and $(0, -1)$.

Often we are not so fortunate as to be able to find a parametrization for the curve C on which we are searching for extrema. Instead the curve C may be given as a level curve of a function $g(x, y)$. In this case, by merely assuming that a parametrization exists we can derive a first derivative test for local maxima and minima. The following result leads to the *method of Lagrange multipliers*.

Theorem 9   *Let f and g be functions of two variables with continuous partial derivatives. Suppose that the function f, when restricted to the level curve C defined by $g(x, y) = c$, has a local extremum at the point (x_0, y_0) and that $\nabla g(x_0, y_0) \neq 0$. Then there is a number λ such that*

$$\nabla f(x_0, y_0) = \lambda \nabla g(x_0, y_0)$$

If $\lambda \neq 0$, this formula says that the level curve of f and g through (x_0, y_0) have the same tangent line at (x_0, y_0).

Proof Choose a parametrization $(x, y) = \sigma(t)$ for C near (x_0, y_0), with $\sigma(0) = (x_0, y_0)$ and $\sigma'(0) \neq 0$.* Since f has a local extremum at (x_0, y_0), the function $h(t) = f(\sigma(t))$ has a local extremum at $t = 0$, so $h'(0) = 0$. According to the chain rule (p. 706), $h'(0) = \nabla f(x_0, y_0) \cdot \sigma'(0)$, so $\nabla f(x_0, y_0)$ must be perpendicular to $\sigma'(0)$. But we already know that the gradient $\nabla g(x_0, y_0)$ is perpendicular to the tangent vector $\sigma'(0)$ to the level curve C (p. 710). In the plane, any two vectors perpendicular to a given nonzero vector must be parallel, so $\nabla f(x_0, y_0) = \lambda \nabla g(x_0, y_0)$ for some number λ. If $\lambda \neq 0$, the tangent line to the level curve of f through (x_0, y_0), which is perpendicular to $\nabla f(x_0, y_0)$, must also be perpendicular to $\nabla g(x_0, y_0)$. But the tangent line to C is also perpendicular to $\nabla g(x_0, y_0)$, so the level curves of f and g through (x_0, y_0) must have the same tangent line.

There is another geometric way of seeing the result in Theorem 9. If the level curves of f and g had different tangent lines at (x_0, y_0), then the level curves would cross one another. It would follow that the level curve C of g would intersect level curves of f for both higher and lower values of f, so the point (x_0, y_0) would not be an extremum (see Fig. 16-3-9).

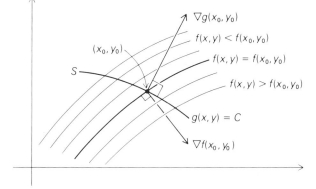

Fig. 16-3-9 If $\nabla g(x_0, y_0)$ and $\nabla f(x_0, y_0)$ are not parallel, the level curve $g(x, y) = C$ cuts all nearby level curves of f.

The technique of applying Theorem 9 depends upon the nature of each problem. Sometimes it is easiest to use the geometric condition of tangency directly. A more systematic approach is to look for a point (x_0, y_0) on C and a constant λ, called a *Lagrange multiplier*, such that $\nabla f(x_0, y_0) = \lambda \nabla g(x_0, y_0)$. This means we wish to solve the three simultaneous equations

$$
\begin{aligned}
f_x(x, y) &= \lambda g_x(x, y) \\
f_y(x, y) &= \lambda g_y(x, y) \\
g(x, y) &= c
\end{aligned}
\tag{3}
$$

for the three unknown quantities x, y, and λ. Another way of looking at equations (3) is that we seek the critical points of the auxiliary function $k(x, y, \lambda) = f(x, y) - \lambda [g(x, y) - c]$. By a critical point of a function

The implicit function theorem guarantees that such a parametrization exists; see J. Marsden, Elementary Classical Analysis, W. H. Freeman, 1974, p. 241.

of three variables, we mean a point where all three of its partial derivatives vanish (see Problem 20). Here

$$k_x = f_x - \lambda g_x$$
$$k_y = f_y - \lambda g_y$$
$$k_\lambda = c - g$$

and setting these equal to zero produces equations (3). We call this attack on the problem the *method of Lagrange multipliers*.

METHOD OF LAGRANGE MULTIPLIERS

To find the extreme points of $f(x, y)$ subject to the constraint $g(x, y) = c$, seek points (x, y) and numbers λ such that $\nabla f(x, y) = \lambda \nabla g(x, y)$. Equivalently: Look among the critical points of the auxiliary function

$$k(x, y, \lambda) = f(x, y) - \lambda [g(x, y) - c]$$

That is, solve system (3).

Worked Example 9   Find the extreme values of $f(x, y) = x^2 - y^2$ along the circle S of radius 1 centered at the origin.

Solution   The circle S is the level curve $g(x, y) = x^2 + y^2 = 1$. So we want x, y, and λ such that

$$f_x(x, y) = \lambda g_x(x, y)$$
$$f_y(x, y) = \lambda g_y(x, y)$$
$$g(x, y) = 1$$

That is,

$$2x = \lambda 2x$$
$$2y = -\lambda 2y$$
$$x^2 + y^2 = 1$$

From the first equation, either $x = 0$ or $\lambda = 1$. If $x = 0$, then from the third equation $y = \pm 1$ and then from the second, $\lambda = -1$. If $\lambda = 1$, then $y = 0$ and $x = \pm 1$. So the eligible points are $(x, y) = (0, \pm 1)$ with $\lambda = -1$ and $(x, y) = (\pm 1, 0)$ with $\lambda = 1$. We must now check them to see if they really are extrema and, if so, what kind.

$$f(0, 1) = f(0, -1) = -1$$
$$f(1, 0) = f(-1, 0) = 1$$

Since x^2 and y^2 are each between 0 and 1 along C, it is now clear that $+1$ and -1 are the maximum and minimum values.

The multiplier λ was introduced as an artificial device enabling us to find maxima and minima, but sometimes it represents something meaningful.

Worked Example 10   Suppose that the output of a manufacturing firm is a quantity Q of product which is a function $f(K, L)$ of the amount K of capital equipment or investment and the amount L of labor used. If the price of labor is p, the price of capital is q, and the firm can spend no more than B dollars, how do you find the amount of capital and labor to maximize the output Q?

Solution   It is useful to think about the problem before applying our machinery. We would expect that if the amount of capital or labor is increased, then the output Q should also increase; that is,

$$\frac{\partial Q}{\partial K} \geq 0 \quad \text{and} \quad \frac{\partial Q}{\partial L} \geq 0$$

We also expect that as more and more labor is added to a given amount of capital equipment, we get less and less additional output for our effort; that is,

$$\frac{\partial^2 Q}{\partial L^2} < 0$$

Similarly,

$$\frac{\partial^2 Q}{\partial K^2} < 0$$

It is reasonable to expect the level curves of output (called *isoquants*) $Q(K, L) = c$ to look something like the curves sketched in Fig. 16-3-10, with $c_1 < c_2 < c_3$.

Fig. 16-3-10 What is the largest value of Q in the shaded triangle?

We can interpret the convexity of the isoquants as follows: As you move to the right along a given isoquant, it takes more and more capital to replace a unit of labor and still produce the same output. The budget constraint means that we must stay inside the triangle bounded by the axes and the line $pL + qK = B$. Geometrically, it is clear that we produce the most by spending all our money in such a way as to pick the isoquant which just touches, but does not cross, the budget line.

Since the maximum point lies on the boundary of our domain, to find it we apply the method of Lagrange multipliers. To maximize $Q = f(K, L)$ subject to the constraint $pL + qK = B$, we look for critical points of the auxiliary function,

$$h(K, L, \lambda) = f(K, L) - \lambda(pL + qK - B)$$

So we want

$$\frac{\partial Q}{\partial K} = \lambda q, \qquad \frac{\partial Q}{\partial L} = \lambda p, \qquad pL + qK = B$$

These are the conditions we must meet in order to maximize output. (You are asked to work out a specific case in Solved Exercise 7.)

In this example, λ does represent something interesting and useful. Let $k = qK$ and $l = pL$, so that k is the dollar value of the capital used and l is the dollar value of the labor used. Then the first two equations become

$$\frac{\partial Q}{\partial k} = \frac{1}{q} \frac{\partial Q}{\partial K} = \lambda = \frac{1}{p} \frac{\partial Q}{\partial L} = \frac{\partial Q}{\partial l}$$

Thus at the optimum production point the marginal change in output per dollar's worth of additional capital investment is equal to the marginal change of output per dollar's worth of additional labor, and λ is this common value. At the optimum point, the exchange of a dollar's worth of capital for a dollar's worth of labor does not change the output. Away from the optimum point the marginal outputs are different, and one exchange or the other will increase the output.*

Solved Exercises

7. Carry out the analysis of Worked Example 10 for the production function $Q(K, L) = AK^\alpha L^{1-\alpha}$, where A and α are positive constants and $\alpha < 1$. This *Cobb-Douglas production function* is sometimes used as a simple model for the national economy. Then Q is the output of the entire economy for a given input of capital and labor.

8. Find the point(s) furthest from and closest to the origin on the curve $x^6 + y^6 = 1$.

Exercises

9. Find the extrema of f subject to the stated constraints.
   (a) $f(x, y) = x + y$; $x^2 + y^2 = 1$
   (b) $f(x, y) = x - y$; $x^2 - y^2 = 2$
   (c) $f(x, y) = 3x + 2y$; $2x^2 + 3y^2 \le 3$
   (d) $f(x, y) = xy$; $2x + 3y \le 10$, $0 \le x$, $0 \le y$

---

*More of this type of mathematical analysis in economics can be found in Microeconomic Theory, *by James Henderson and Richard Quandt, McGraw-Hill, 1958. This reference discusses a second derivative test for the Lagrange multiplier method. (See also J. Marsden and A. Tromba, *Vector Calculus, *Second Edition, *W. H. Freeman and Co., 1980.)*

10. Let $f(x, y) = x^2 + xy + y^2$.

    (a) Find the maximum and minimum points and values of f along the circle $x^2 + y^2 = 1$.

    (b) What is happening at the points $(\pm 1, 0)$ and $(0, \pm 1)$?

    (c) Find extreme points and values for f in the disk D consisting of all (x, y) such that $x^2 + y^2 \le 1$.

11. A firm uses wool and cotton fiber to produce cloth. The amount of cloth produced is given by $Q(x, y) = xy - x - y + 1$, where x is the number of pounds of wool, y is the number of pounds of cotton, and $x > 1$ and $y > 1$. If wool costs p dollars per pound, cotton costs q dollars per pound, and the firm can spend B dollars on material, what should the mix of cotton and wool be to produce the most cloth?

Problems for Section 16-3

1. Find the critical points of each of the following functions and classify them as local maxima, minima, or neither:

   (a) $x^2 - y^2 + xy - 7$

   (b) $x^2 + y^2 + 3xy + 10$

   (c) $f(x, y) = \ln(2 + \sin xy)$

   (d) $f(x, y) = x^2 + y^2 - 6x - 14y + 100$

2. Find and classify the extreme points and values (if any) of the functions given:

   (a) $f(x, y) = y^2 - x^2$

   (b) $f(x, y) = 2x^2 - 2xy + y^2 - 2x + 1$

   (c) $f(x, y) = x^2 + xy^2 + y^4$

   (d) $f(x, y) = (x^2 + y^2)e^{x^2 - y^2}$

3. Analyze the behavior of $z = x^5y + xy^5 + xy$ at its critical points.

4. Minimize the distance from the plane $x - y + 2z = 3$ to the origin.

5. Analyze the behavior of the following functions at the indicated points:

   (a) $f(x, y) = x^2 - y^2 + 3xy$; $(0, 0)$

   (b) $f(x, y) = x^2 + y^2 + Cxy$; $(0, 0)$

   Determine what happens for various values of C. At what values of C does the behavior change qualitatively?

6. Analyze the critical point at $(0, 0)$ for the function $f(x, y) = x^2 + y^3$. Make a sketch.

7. Minimize the distance from $(0, 0, 0)$ to each of the following surfaces. [*Hint:* Write the square of the distance as a function of x and y.]

   (a) $z = \sqrt{x^2 - 1}$ (b) $z = 6xy + 7$

   (c) $z = 1/xy$

8. Find the local maxima and minima for $z = (x^2 + 3y^2)e^{1-x^2-y^2}$. (See Fig. 15-1-10.)

METHOD OF LEAST SQUARES (PROBLEMS 9–14)

It often happens that the theory behind an experiment indicates that the data should lie along a straight line of the form $y = mx + b$. The actual results, of course, will never match exactly with theory, so we are faced with the problem of finding the straight line which best fits some experimental data $(x_1, y_1), \ldots, (x_n, y_n)$ as in Fig. 16-3-11. For the straight line $y = mx + b$, each point will deviate vertically from the line by an amount $d_i = y_i - (mx_i + b)$. We would like to choose m and b in such a way as to make the total effect of these deviations as small as possible. Since some deviations are negative and some are positive, however, a better measure of the total error might be the sum of the squares of these deviations. So we are led to the problem of finding m and b to minimize the function

$$s = f(m, b) = d_1^2 + d_2^2 + \cdots + d_n^2$$
$$= \sum_{i=1}^{n} (y_i - mx_i - b)^2$$

where $x_1, \ldots, x_n$ and $y_1, \ldots, y_n$ are given data.

Fig. 16-3-11 The method of least squares finds a straight line which "best" approximates a set of data.

9. For each set of three data points, plot the points, write down the function $f(m, b)$, find m and b to give the best straight-line fit according to the method of least squares, and plot the straight line.

   (a) $(x_1, y_1) = (1, 1)$
   $(x_2, y_2) = (2, 3)$
   $(x_3, y_3) = (4, 3)$

   (b) $(x_1, y_1) = (0, 0)$
   $(x_2, y_2) = (1, 2)$
   $(x_3, y_3) = (2, 3)$

10. Show that if only two data points (x_1, y_1) and (x_2, y_2) are given, then this method produces the line through (x_1, y_1) and (x_2, y_2).

11. Show that the equations for a critical point, $\partial s/\partial b = 0$ and $\partial s/\partial m = 0$, are equivalent to

$$m(\Sigma x_i) + nb = (\Sigma y_i)$$
and $m(\Sigma x_i^2) + b(\Sigma x_i) = \Sigma x_i y_i$

   where all the sums run from $i = 1$ to $i = n$.

12. If $y = mx + b$ is the best-fitting straight line to the data points $(x_1, y_1), \ldots, (x_n, y_n)$ according to the least squares method, show that

$$\sum_{i=1}^{n} (y_i - mx_i - b) = 0$$

   That is, show that the positive and negative deviations cancel (see Problem 11).

★13. Use the second derivative test to show that the critical point of f is actually a minimum.

14. Use the method of least squares to find the straight line that best fits the points $(0, 1)$, $(1, 3)$, $(2, 2)$, $(3, 4)$, and $(4, 5)$. Plot your points and line.

15. Locate any maxima, minima, or saddle points of $f(x, y) = \ln(ax^2 + by^2 + 1)$, $a, b > 0$.

16. Find the extreme points and values of each function subject to the stated constraints:

   (a) $f(x, y) = xy$; $x + y = 1$
   (b) $f(x, y) = \cos^2 x + \cos^2 y$; $x + y = \pi/4$
   (c) $f(x, y) = x - 3y$; $x^2 + y^2 = 1$

★17. Locate extreme points and values for the function

$$f(x, y) = x^2 + y^2 - x - y + 1$$

   in the disk $x^2 + y^2 \le 1$.

18. A rectangular mirror with area A square feet is to have trim along the edges. If the trim along the horizontal edges costs p cents per foot and that for the vertical edges costs q cents per foot, find the dimensions which will minimize the total cost.

★19. The Baraboo, Wisconsin, plant of International Widget Co. uses aluminum, iron, and magnesium to produce high-quality widgets. The quantity of widgets which may be produced using x tons of aluminum, y tons of iron, and z tons of magnesium is $Q(x, y, z) = xyz$. The cost of raw materials is aluminum, \$6 per ton; iron, \$4 per ton; and magnesium, \$8 per ton. How many tons each of aluminum, iron, and magnesium should be used to manufacture 1000 widgets at the lowest possible cost? [*Hint:* You want an extreme value for what function? Subject to what constraint?]

20. (a) Show that if (x_0, y_0, z_0) is a local minimum or maximum point of $w = f(x, y, z)$, then $\partial w/\partial x$, $\partial w/\partial y$, and $\partial w/\partial z$ are all zero at (x_0, y_0, z_0).

   (b) Find the critical points of the function $\sin(x^2 + y^2 + z^2)$.

   (c) Find the point in space which minimizes the sum of the squares of the distances from $(0, 0, 0)$, $(1, 0, 0)$, $(0, 1, 0)$, and $(0, 0, 1)$.

★21. Suppose that (x_0, y_0, z_0) is a critical point for the restriction of the function $f(x, y, z)$ to the surface $g(x, y, z) = c$. The method of Lagrange multipliers tells us that in this case the partial derivatives with respect to x, y, z, and λ of the function of *four* variables

$$k(x, y, z, \lambda) = f(x, y, z) - \lambda[g(x, y, z) - c]$$

   are equal to zero.

   (a) Interpret this fact as a statement about the gradient vectors of f and g at (x_0, y_0, z_0).

   (b) Find the maxima and minima of xyz on the sphere $x^2 + y^2 + z^2 = 1$.

   (c) Rework Solved Exercise 2 (p. 720) by minimizing a function of x, y, and h subject to the constraint $xyh = 256$.

★22. Complete the proof of Theorem 7(2) by following these steps:

   (a) If $A \ne 0$, show that

$$g(x, y) = A\left[\left(x + \frac{B}{A}y\right) - ey\right]\left[\left(x + \frac{B}{A}y\right) + ey\right]$$

   for some number e. What is e?

   (b) Show that the set where $g(x, y) = 0$ consists of two intersecting lines. What are their equations?

   (c) Show that $g(x, y)$ is positive on two of the regions cut out by the lines in part (b) and negative on the other two.

   (d) If $A = 0$, $g(x, y) = 2Bxy + Cy^2$. B must be nonzero. (Why?) Write $g(x, y)$ as a product of linear functions and repeat parts (b) and (c).

23. Discuss the function $Ax^2 + 2Bxy + Cy^2$ in the case where $AC - B^2 = 0$.

    (a) If $A \neq 0$, use formula (1) in the proof of Theorem 7.

    (b) Sketch a graph of the function $x^2 + 2xy + y^2$.

    (c) What happens if $A = 0$?

★24. Let $f(x, y) = 3x^4 - 4x^2y + y^2$.

    (a) Show that $f(x, y)$ has a critical point at the origin.

    (b) Show that for all values of θ, the function $h(r) = f(r \cos \theta, r \sin \theta)$ has a local minimum at $r = 0$.

    (c) Show that, nevertheless, the origin is not a local minimum point for f. [*Hint:* Consider the curve $(x, y) = (t, 2t^2)$.]

    (d) Find the set of (x, y) for which $f(x, y) = 0$.

    (e) Sketch the regions in the plane where $f(x, y)$ is positive and negative.

    (f) Discuss why parts (b) and (c) do not contradict one another.

★25. Complete the proof of Theorem 8 by following this outline:

    (a) We begin with the case $D(x_0, y_0) > 0$ and $f_{xx}(x_0, y_0) > 0$. Using Theorem 2 (p. 697), show that there is a number $\varepsilon > 0$ such that, whenever (x, y) lies in the disk of radius ε about (x_0, y_0),

    $$D(x, y) = f_{xx}(x, y) f_{yy}(x, y) - f_{xy}(x, y)^2$$

    and $f_{xx}(x, y)$ are both positive.

    (b) Show that the function $h(r)$ is concave upward on the interval $(-\varepsilon, \varepsilon)$ for any choice of θ.

    (c) Conclude that $f(x, y) \geq f(x_0, y_0)$ for all (x, y) in the disk of radius ε about (x_0, y_0).

    (d) Complete the case $D(x_0, y_0) > 0$ and $f_{xx}(x_0, y_0) < 0$.

    (e) Complete the case $D(x_0, y_0) < 0$ by showing that f takes values near (x_0, y_0) which are greater and less than $f(x_0, y_0)$.

26. Drug reactions can be measured by functions of the form $R(u, t) = u^2(c - u)t^2 e^{-t}$, $0 \leq u \leq c$, $t \geq 0$. The symbols u and t are drug units and time in hours, respectively. Find the dosage u and time t at which R is a maximum.

27. The work w done in a compressor with $k + 1$ compression cylinders is given by

    $$w = c_1 y + c_2, \quad c_1 \neq 0$$

    where

    $$y = \sum_{i=0}^{k} T_i (p_i / p_{i+1})^{(n-1)/n}$$

    The symbols T_i and p_i stand for temperature and pressure in cylinder i, $1 \leq i \leq k + 1$; the pressures $p_1, \ldots, p_k$ are the independent variables while $p_0, p_{k+1}, T_0, \ldots, T_k$ and $n > 1$ are given.

    (a) Find the relations corresponding to $w = $ minimum, using critical point theory for several variables.

    (b) Find p_1, p_2, p_3 explicitly for the case $k = 3$.

28. *Planck's law* gives the relationship of the energy E emitted by a blackbody to the wavelength λ and temperature T:

    $$E = \frac{2\pi k^5 T^5}{h^4 c^3} \frac{x^5}{e^x - 1} \quad \text{where} \quad x = hc/\lambda kT$$

    The constants are $h = 6.6256 \times 10^{-34}$ joule seconds (Planck's constant), $k = 1.3805 \times 10^{-23}$ joule kilograms^{-1} (Boltzmann's constant), $c = 2.9979 \times 10^8$ meter second^{-1} (velocity of light). The plot of E versus λ for fixed T is called a *Planck curve*.

    (a) The maximum along each Planck curve is obtained by setting $\partial E / \partial \lambda = 0$ and solving for $\lambda_{\max}$. The relationship so derived is called *Wien's displacement law*. Show this law is just $\lambda_{\max} = hc/kTx_0$ where $5 - x_0 - 5e^{-x_0} = 0$.

    (b) Clearly x_0 is close to 5. By examining the sign of $f(x) \equiv 5 - x - 5e^{-x}$, use your calculator to complete the expansion $x_0 = 4.965 \ldots$ to a full six digits.

    (c) Improve upon the displacement law $\lambda_{\max} = 0.00288/T$ by giving a slightly better constant. The peak for the earth ($288°$ K) is about 10 micro-meters, the peak for the sun ($6000°$ K) about 0.48 micrometers, so the maximum occurs in the infared and visible range, respectively.

29. A sphere of mass m, radius a and uniform density has *potential u* and *gravitational force* **F**, at a distance r from the center $(0, 0, 0)$, given by

$$u = \frac{3m}{2a} - \frac{mr^2}{2a^3}, \quad \mathbf{F} = -\frac{m}{a^3}\mathbf{r} \quad (r \le a)$$

$$u = \frac{m}{r}, \quad \mathbf{F} = -\frac{m}{r^3}\mathbf{r} \quad (r > a)$$

Here, $r = \|\mathbf{r}\|$, $\mathbf{r} = x\mathbf{i} + y\mathbf{j} + z\mathbf{k}$.

(a) Verify that $\mathbf{F} = \text{grad } u$ on the inside and outside of the sphere.

(b) Check that u satisfies Poisson's equation: $\partial^2 u/\partial x^2 + \partial^2 u/\partial y^2 + \partial^2 u/\partial z^2 = \text{constant}$ inside the sphere.

(c) Show u satisfies Laplace's equation: $\partial^2 u/\partial x^2 + \partial^2 u/\partial y^2 + \partial^2 u/\partial z^2 = 0$ outside the sphere.

30. The *pressure gradient* **G** is a vector quantity that points from regions of high pressure to regions of low pressure, normal to the lines of constant pressure (*isobars*).

(a) In an xy-coordinate system,

$$\mathbf{G} = -\frac{\partial P}{\partial x}\mathbf{i} - \frac{\partial P}{\partial y}\mathbf{j}$$

Write a formula for the magnitude of the pressure gradient.

(b) If the horizontal pressure gradient provided the only horizontal force acting on the air, the wind would blow directly across the isobars in the direction of **G**, and for a given air mass, with acceleration proportional to the magnitude of **G**. Explain.

(c) *Buys-Ballot's law* states: "If in the Northern Hemisphere, you stand with your back to the wind, the high pressure is on your right and the low pressure on your left." Draw a figure and introduce xy-coordinates so that **G** points in the proper direction.

(d) State and graphically illustrate Buys-Ballot's law for the Southern Hemisphere, in which the orientation of high and low pressure is reversed.

31. Cascade Container Company produces a cardboard shipping crate at three different plants in amounts x, y, z, respectively, producing an annual revenue of $R(x, y, z) = 8xyz^2 - 200,000(x + y + z)$. The company is to produce 100,000 units annually. How should production be handled to maximize the revenue?

32. The temperature T on the spherical surface $x^2 + y^2 + z^2 = 1$ satisfies the equation $T(x, y, z) = xz + yz$. Find all the hot spots.

33. A pipeline of length ℓ is to be constructed from one pipe of length ℓ_1 and diameter D_1 connected to another pipe of length ℓ_2, diameter D_2. The finished pipe must deliver Q liters per second at pressure loss h. The expense is reduced to a minimum by minimization of the cost $C = \ell_1(a + bD_1) + \ell_2(a + bD_2)$ ($a, b = $ constants) subject to the conditions

$$\ell_1 + \ell_2 = \ell$$

$$h = kQ^m \left\{ \frac{\ell_1}{D_1^{m_1}} + \frac{\ell_2}{D_2^{m_2}} \right\}$$

where k, m, m_1, m_2 are constants. Show that the pipe must be the same diameter throughout its entire length ℓ, in order to achieve the minimum cost.

(optional) # 16-4 Matrix Multiplication and the Chain Rule

The derivative matrix of a composite function is the product of two matrices.

The chain rule of Section 5-2 enabled us to differentiate a single function which depended on a single independent variable through a single intermediate variable. In Section 16-2, this result was extended to the case of two or three intermediate variables. When we allow the number of functions and independent variables to grow to two or three, the chain rule may be expressed in terms of matrix multiplication.

Goals

After studying this section, you should be able to:

Multiply matrices.

Differentiate functions of several variables which are given in terms of several intermediate variables.

DERIVATIVE MATRICES

An *m* × *n matrix* is a rectangular array of *mn* numbers, called the *entries* of the matrix, arranged in *m* rows and *n* columns. The entry in the *i*th row and *j*th column is called the (i, j) entry. (See Fig. 16-4-1.)

Fig. 16-4-1 The (i, j) entry of this matrix is a_{ij}.

In the chain rule of Section 16-2, we represented both the derivative of one function of three variables and the derivatives of three functions of one variable by vectors of the form $a\mathbf{i} + b\mathbf{j} + c\mathbf{k}$. We may distinguish these vectors by writing the gradient $(\partial u / \partial x)\,\mathbf{i} + (\partial u / \partial y)\,\mathbf{j} + (\partial u / \partial z)\,\mathbf{k}$ as a 1 × 3 matrix, or *row vector:*

$$\left[\frac{\partial u}{\partial x}, \frac{\partial u}{\partial y}, \frac{\partial u}{\partial z} \right] = \frac{\partial u}{\partial(x, y, z)} \tag{1}$$

and the velocity vector $(dx/dt)\,\mathbf{i} + (dy/dt)\,\mathbf{j} + (dz/dt)\,\mathbf{k}$ as a 3 × 1 matrix, or *column vector:*

$$\begin{bmatrix} \dfrac{dx}{dt} \\[2mm] \dfrac{dy}{dt} \\[2mm] \dfrac{dz}{dt} \end{bmatrix} = \frac{\partial(x, y, z)}{\partial t} \tag{2}$$

(If there are only two intermediate variables, our row and column vectors will be 1 × 2 and 2 × 1 matrices.)

To express the chain rule in this new notation, we define a product between row and column vectors of the same length.

(optional)

Definition Let

$$A = [a_1, a_2, ..., a_n] \quad \text{and} \quad B = \begin{bmatrix} b_1 \\ b_2 \\ \vdots \\ b_m \end{bmatrix}$$

be a row vector and a column vector, respectively. If $m = n$, we define the product AB to be the number $a_1 b_1 + a_2 b_2 + \cdots a_n b_n = \Sigma_{i=1}^n a_i b_i$. (If $m \neq n$, the product AB is *not defined*.)

In terms of this definition, the chain rule for three intermediate variables becomes

$$\frac{du}{dt} = \begin{bmatrix} \dfrac{\partial u}{\partial x} & \dfrac{\partial u}{\partial y} & \dfrac{\partial u}{\partial z} \end{bmatrix} \begin{bmatrix} \dfrac{dx}{dt} \\ \dfrac{dy}{dt} \\ \dfrac{dz}{dt} \end{bmatrix} = \frac{\partial u}{\partial(x, y, z)} \frac{\partial(x, y, z)}{\partial t} \tag{3}$$

which looks very much like the chain rule of one-variable calculus.

The product of row and column vectors has many other applications. For example, every linear function $f(x, y, z) = ax + by + cz + d$ can be written as

$$f(x, y, z) = \begin{bmatrix} a & b & c \end{bmatrix} \begin{bmatrix} x \\ y \\ z \end{bmatrix} + d$$

Your bill at the fruit market can be expressed as PQ, where $P = [p_1 \quad p_2 \quad \cdots \quad p_n]$ is the *price vector* whose ith entry is the price of the ith fruit in dollars per kilogram, and

$$Q = \begin{bmatrix} q_1 \\ \vdots \\ q_n \end{bmatrix}$$

is the *quantity vector* whose ith entry is the number of kilograms of the ith fruit purchased.

Worked Example 1 Find AB if

$$A = [1 \quad 2 \quad 3 \quad 4] \quad \text{and} \quad B = \begin{bmatrix} -1 \\ 1 \\ -1 \\ 1 \end{bmatrix}$$

Solution $AB = (1)(-1) + (2)(1) + (3)(-1) + (4)(1) = -1 + 2 - 3 + 4 = 2$.

(optional)   Having described the derivative of m functions of one variable by an $m \times 1$ matrix, and the derivative of one function of n variables by a $1 \times n$ matrix, it is natural for us to describe the derivative of m functions of n variables by an $m \times n$ matrix. For example, if $x = f(u, v)$, $y = g(u, v)$, and $z = h(u, v)$, we may put all six partial derivatives into a 3×2 matrix:

$$
\begin{bmatrix}
\dfrac{\partial x}{\partial u} & \dfrac{\partial x}{\partial v} \\[2ex]
\dfrac{\partial y}{\partial u} & \dfrac{\partial y}{\partial v} \\[2ex]
\dfrac{\partial z}{\partial u} & \dfrac{\partial z}{\partial v}
\end{bmatrix}
= \frac{\partial(x, y, z)}{\partial(u, v)}
$$

The rows of this matrix are the gradient vectors of f, g, and h. The columns are the "partial velocity vectors" with respect to u and v of the vector-valued function $\mathbf{r}(u, v) = (f(u, v), g(u, v), h(u, v))$.

In general, we may define the *derivative matrix* of m functions of n variables. (You may assume that m and n are 1, 2, or 3 if you wish.)

Definition Let $u_1 = f_1(x_1, x_2, \ldots, x_n)$, $u_2 = f_2(x_1, x_2, \ldots, x_n)$, $\ldots$, $u_m = f_m(x_1, x_2, \ldots, x_n)$ be m functions of the n variables $x_1, \ldots, x_n$. The *derivative matrix* of the u_i's with respect to the x_j's is the $m \times n$ matrix:

$$
\frac{\partial(u_1, \ldots, u_m)}{\partial(x_1, \ldots, x_n)} =
\begin{bmatrix}
\dfrac{\partial u_1}{\partial x_1} & \dfrac{\partial u_1}{\partial x_2} & \cdots & \dfrac{\partial u_1}{\partial x_n} \\[2ex]
\dfrac{\partial u_2}{\partial x_1} & \dfrac{\partial u_2}{\partial x_2} & \cdots & \dfrac{\partial u_2}{\partial x_n} \\[2ex]
\vdots & \vdots & & \vdots \\[2ex]
\dfrac{\partial u_m}{\partial x_1} & \dfrac{\partial u_m}{\partial x_2} & & \dfrac{\partial u_m}{\partial x_n}
\end{bmatrix}
$$

whose (i, j) entry is the partial derivative $\partial u_i / \partial x_j$.

The entries of the derivative matrix are functions of $(x_1, \ldots, x_n)$. If we fix values $(x_1^0, \ldots, x_n^0)$ for the independent variables, then the derivative matrix becomes a matrix of numbers and is denoted by

$$
\frac{\partial(u_1, \ldots, u_m)}{\partial(x_1, \ldots, x_n)} \bigg|_{(x_1^0, \ldots, x_n^0)}
$$

Worked Example 2   Let $u = x^2 + y^2$, $v = x^2 - y^2$, and $w = xy$. Find $\partial(u, v, w)/\partial(x, y)$ and evaluate

$$
\frac{\partial(u, v, w)}{\partial(x, y)} \bigg|_{(-2, 3)}
$$

Solution   Applying the definition, with $m = 3$, $n = 2$, $u_1 = u$, $u_2 = v$, $u_3 = w$, $x_1 = x$, and $x_2 = y$, we get

$$
\frac{\partial(u, v, w)}{\partial(x, y)} =
\begin{bmatrix}
2x & 2y \\
2x & -2y \\
y & x
\end{bmatrix}
$$

(optional) | Substituting $x = -2$ and $y = 3$, we get

$$\left. \frac{\partial(u, v, w)}{\partial(x, y)} \right|_{(-2, 3)} = \begin{bmatrix} -4 & 6 \\ -4 & -6 \\ 3 & -2 \end{bmatrix}$$

Notice that the derivative matrix of one function $u = f(t)$ of one variable is a 1×1 matrix $\partial(u)/\partial(t)$ whose single entry is just the ordinary derivative du/dt. Thus the chain rule (3) can be rewritten as

$$\frac{\partial(u)}{\partial(t)} = \frac{\partial(u)}{\partial(x, y, z)} \frac{\partial(x, y, z)}{\partial(t)} \tag{4}$$

In the remainder of this section, we will show how to multiply matrices of all sizes and, thereby, to generalize the chain rule (4) to several independent and dependent variables. Solved Exercises 2 and 3 below will play an important role in what follows.

Solved Exercises ★1. Let B be the $m \times 1$ column vector

$$\begin{bmatrix} \dfrac{1}{m} \\ \dfrac{1}{m} \\ \vdots \\ \dfrac{1}{m} \end{bmatrix}$$

If $A = [a_1 \cdots a_m]$ is any row vector, what is AB?

★2. Suppose that $u = ax + by + cz + d$ and $v = ex + fy + gz + h$, where $a, b, ..., h$ are constants.

(a) Express u and v by using products of row and column vectors.

(b) Find the derivative matrix $\partial(u, v)/\partial(x, y, z)$.

★3. Suppose that x, y, and z in Solved Exercise 2 are linear functions of t: $x = mt + n$, $y = pt + q$, and $z = rt + s$.

(a) Express u and v in terms of t and find the derivative matrix $\partial(u, v)/\partial(t)$.

(b) Express the elements of $\partial(u, v)/\partial(t)$ as products of row and column vectors.

★4. Let $x = r \cos \theta$ and $y = r \sin \theta$ be the functions which convert polar to cartesian coordinates. Find the derivative matrix $\partial(x, y)/\partial(r, \theta)$ and evaluate at $(5, \pi/6)$.

Exercises ★1. Find the following products if they are defined:

(a) $[1 \quad 2 \quad 3] \begin{bmatrix} 4 \\ 5 \\ 6 \end{bmatrix}$

(b) $[\frac{1}{4} \quad \frac{1}{2} \quad \frac{1}{4}] \begin{bmatrix} 1 \\ 2 \\ 1 \end{bmatrix}$

(c) $[2 \quad 6] \begin{bmatrix} 3 \\ 4 \\ 5 \end{bmatrix}$

(d) $[1 \quad 2 \quad 3 \quad 4 \quad 5] \begin{bmatrix} 1 \\ 2 \\ 3 \\ 4 \\ 5 \end{bmatrix}$

★2. Express the polar coordinates r and θ in terms of the cartesian coordinates x and y, and find the derivative matrix $\partial(r, \theta)/\partial(x, y)$.

★3. Find the following derivative matrices and evaluate at the given points:

    (a) $\partial(x, y)/\partial(u, v)$; $x = u \sin v$, $y = e^{uv}$; evaluate at $(0, 1)$.

    (b) $\partial(x, y, z)/\partial(r, \theta, \phi)$; $x = r \sin \phi \cos \theta$, $y = r \sin \phi \sin \theta$, $z = r \cos \theta$; evaluate at $(2, \pi/3, \pi/4)$.

    (c) $\partial(u, v)/\partial(x, y, z)$; $u = xyz$, $v = x + y + z$; evaluate at $(3, 3, 3)$.

(optional)   **MATRIX MULTIPLICATION**

Guided by Solved Exercises 2 and 3, we define the product AB of two matrices, A and B, by multiplying the rows of A by the columns of B.

Definition   Let A be an $m \times n$ matrix and B a $p \times k$ matrix. If $n = p$, we define the product AB to be the $m \times k$ matrix whose (i, j) element is the product of the ith row of A and the jth column of B. (If $n \neq p$, the product is not defined.)

MATRIX MULTIPLICATION

Let A and B be two matrices and assume that the number of columns of A equals the number of rows of B. To form $C = AB$, take the product of the first row of A and first column of B and let it be the $(1, 1)$ entry of C. Then take the product of the first row and second column of B and let it be the $(1, 2)$ entry of C. Repeat. In general, the product of the ith row of A and jth column of B is the (i, j) entry of C.

Worked Example 3   Let

$$A = \begin{bmatrix} 1 & -1 \\ 2 & 0 \\ 8 & -3 \end{bmatrix} \quad \text{and} \quad B = \begin{bmatrix} 2 & 0 \\ 1 & -1 \end{bmatrix}$$

Find AB.

Solution

$$\begin{bmatrix} 1 & -1 \\ 2 & 0 \\ 8 & -3 \end{bmatrix} \begin{bmatrix} 2 & 0 \\ 1 & -1 \end{bmatrix} = \begin{bmatrix} 1 \cdot 2 + (-1) \cdot 1 & \\ & \end{bmatrix} = \begin{bmatrix} 1 & \\ & \end{bmatrix}$$

(going across the first row of A and down the first column of B). Moving to the second column of B:

$$\begin{bmatrix} 1 & -1 \\ 2 & 0 \\ 8 & -3 \end{bmatrix} \begin{bmatrix} 2 & 0 \\ 1 & -1 \end{bmatrix} = \begin{bmatrix} 1 & 1 \cdot 0 + (-1)(-1) \\ & \end{bmatrix}$$

$$= \begin{bmatrix} 1 & 1 \\ & \end{bmatrix}$$

742 CHAPTER 16: PARTIAL DIFFERENTIATION

(optional) Moving to the second and third rows of A, we fill in the remaining entries:

$$\begin{bmatrix} 1 & -1 \\ 2 & 0 \\ 8 & -3 \end{bmatrix} \begin{bmatrix} 2 & 0 \\ 1 & -1 \end{bmatrix} = \begin{bmatrix} 1 & 1 \\ 4 & 0 \\ 13 & 3 \end{bmatrix}$$

Worked Example 4 Find BA if A and B are as in Worked Example 3.

Solution BA is not defined since the number of columns of B is not equal to the number of rows of A.

Worked Example 5 Find:

$$\begin{bmatrix} 1 & 1 \\ 2 & 0 \end{bmatrix}\begin{bmatrix} 1 & 2 \\ 0 & 1 \end{bmatrix} \quad \text{and} \quad \begin{bmatrix} 1 & 2 \\ 0 & 1 \end{bmatrix}\begin{bmatrix} 1 & 1 \\ 2 & 0 \end{bmatrix}$$

Solution $$\begin{bmatrix} 1 & 1 \\ 2 & 0 \end{bmatrix}\begin{bmatrix} 1 & 2 \\ 0 & 1 \end{bmatrix} = \begin{bmatrix} 1 & 3 \\ 2 & 4 \end{bmatrix} \quad \text{and} \quad \begin{bmatrix} 1 & 2 \\ 0 & 1 \end{bmatrix}\begin{bmatrix} 1 & 1 \\ 2 & 0 \end{bmatrix} = \begin{bmatrix} 5 & 1 \\ 2 & 0 \end{bmatrix}$$

Worked Example 5 shows that even if AB and BA are defined, they may not be equal. In other words, matrix multiplication is *not commutative*.

Solved Exercises ★5. Let

$$I = \begin{bmatrix} 1 & 0 \\ 0 & 1 \end{bmatrix} \quad \text{and} \quad A = \begin{bmatrix} 1 & 2 \\ 0 & 1 \end{bmatrix}$$

Find a matrix B such that $AB = I$.

★6. For 2×2 matrices A and B, verify that $|AB| = |A| |B|$, where $|A|$ denotes the determinant of A.

★7. Interpret the result of Solved Exercise 3(b) in terms of matrix multiplication.

Exercises ★4. Compute the following matrix products:

(a) $\begin{bmatrix} 1 & 2 \\ 0 & 1 \end{bmatrix}\begin{bmatrix} 2 & 3 \\ 4 & 5 \end{bmatrix}$

(b) $\begin{bmatrix} 1 & 0 & 1 \\ 0 & 2 & 1 \end{bmatrix}\begin{bmatrix} 2 & 1 \\ 3 & 0 \\ 4 & 5 \end{bmatrix}$

(c) $\begin{bmatrix} 1 & 0 \\ 0 & 0 \end{bmatrix}\begin{bmatrix} 0 & 0 \\ 0 & 1 \end{bmatrix}$

(d) $\begin{bmatrix} 1 & 0 & 0 \\ 0 & 1 & 0 \\ 0 & 0 & 1 \end{bmatrix}\begin{bmatrix} a & b & c \\ d & e & f \\ g & h & i \end{bmatrix}$

(e) $\left(\begin{bmatrix} 1 & 0 \\ 2 & 3 \end{bmatrix}\begin{bmatrix} 2 & 4 \\ 1 & -1 \end{bmatrix}\right)\begin{bmatrix} 1 & 1 \\ 0 & 1 \end{bmatrix}$

(f) $\begin{bmatrix} 1 & 0 \\ 2 & 3 \end{bmatrix}\left(\begin{bmatrix} 2 & 4 \\ 1 & -1 \end{bmatrix}\begin{bmatrix} 1 & 1 \\ 0 & 1 \end{bmatrix}\right)$

★5. In Solved Exercise 5, show that we also have

$$BA = \begin{bmatrix} 1 & 0 \\ 0 & 1 \end{bmatrix}$$

Show that the solution of the equation

$$A\begin{bmatrix} x \\ y \end{bmatrix} = \begin{bmatrix} e \\ f \end{bmatrix} \quad \text{is} \quad \begin{bmatrix} x \\ y \end{bmatrix} = B\begin{bmatrix} e \\ f \end{bmatrix}$$

★6. Let A be the derivative matrix $\partial(x,y)/\partial(r,\theta)$ of Solved Exercise 4. Let B be the derivative matrix $\partial(r,\theta)/\partial(x,y)$ of Exercise 2, with its entries expressed in terms of r and θ. Find AB and BA.

(optional)   **THE GENERAL CHAIN RULE**

The general chain rule should now come as no surprise.

Theorem 10 *Let $u_1 = f_1(x_1, \ldots, x_n), \ldots, u_m = f_m(x_1, \ldots, x_n)$ be m functions of n variables, with continuous partial derivatives. Let $x_1 = g_1(t_1, \ldots, t_k)$, $\ldots, x_n = g_n(t_1, \ldots, t_k)$ be n functions of k variables, with continuous partial derivatives.*

Consider the u_i's as functions of the t_j's by

$$u_i = f_i(g_1(t_1, \ldots, t_k), \ldots, g_n(t_1, \ldots, t_k))$$

Then

$$\frac{\partial(u_1, \ldots, u_m)}{\partial(t_1, \ldots, t_k)} = \frac{\partial(u_1, \ldots, u_m)}{\partial(x_1, \ldots, x_n)} \frac{\partial(x_1, \ldots, x_n)}{\partial(t_1, \ldots, t_k)}$$

Hence

$$\frac{\partial u_i}{\partial t_j} = \frac{\partial u_i}{\partial x_1}\frac{\partial x_1}{\partial t_j} + \frac{\partial u_i}{\partial x_2}\frac{\partial x_2}{\partial t_j} + \cdots + \frac{\partial u_i}{\partial x_n}\frac{\partial x_n}{\partial t_j}$$

(Note that there are as many terms in the sum as there are intermediate variables.)

Proof We will carry out the proof for the "typical" case $m = 2$, $n = 3$, $k = 2$. We must prove that

$$\begin{bmatrix} \dfrac{\partial u_1}{\partial t_1} & \dfrac{\partial u_1}{\partial t_2} \\[2.5ex] \dfrac{\partial u_2}{\partial t_1} & \dfrac{\partial u_2}{\partial t_2} \end{bmatrix} = \begin{bmatrix} \dfrac{\partial u_1}{\partial x_1} & \dfrac{\partial u_1}{\partial x_2} & \dfrac{\partial u_1}{\partial x_3} \\[2.5ex] \dfrac{\partial u_2}{\partial x_1} & \dfrac{\partial u_2}{\partial x_2} & \dfrac{\partial u_2}{\partial x_3} \end{bmatrix} \begin{bmatrix} \dfrac{\partial x_1}{\partial t_1} & \dfrac{\partial x_1}{\partial t_2} \\[2.5ex] \dfrac{\partial x_2}{\partial t_1} & \dfrac{\partial x_2}{\partial t_2} \\[2.5ex] \dfrac{\partial x_3}{\partial t_1} & \dfrac{\partial x_3}{\partial t_3} \end{bmatrix}$$

This matrix equation represents four ordinary equations. We will prove a typical one:

$$\frac{\partial u_2}{\partial t_1} = \frac{\partial u_2}{\partial x_1}\frac{\partial x_1}{\partial t_1} + \frac{\partial u_2}{\partial x_2}\frac{\partial x_2}{\partial t_1} + \frac{\partial u_2}{\partial x_3}\frac{\partial x_3}{\partial t_1} \tag{5}$$

In taking the partial derivatives with respect to t_1, we hold t_2 fixed and take ordinary derivatives with respect to t_1. With this understood, we may rewrite (5) as

$$\frac{du_2}{dt_1} = \frac{\partial u_2}{\partial x_1}\frac{dx_1}{dt_1} + \frac{\partial u_2}{\partial x_2}\frac{dx_2}{dt_1} + \frac{\partial u_2}{\partial x_3}\frac{dx_3}{dt_1} \tag{6}$$

We are now in the situation of Section 16-2—we have the independent variable t_1, the dependent variable u_2, and the intermediate variables (x_1, x_2, x_3). But the chain rule for this case is just formula (6), so (6) is true and, hence, (5) is proved.

Worked Example 6
(optional)

Let (x, y) be cartesian coordinates in the plane and let (r, θ) be polar coordinates. If $z = f(x, y)$ is a function on the plane, express the partial derivatives $\partial z / \partial r$ and $\partial z / \partial \theta$ in terms of $\partial z / \partial x$ and $\partial z / \partial y$.

Solution By Theorem 10,

$$\begin{bmatrix} \dfrac{\partial z}{\partial r} & \dfrac{\partial z}{\partial \theta} \end{bmatrix} = \begin{bmatrix} \dfrac{\partial z}{\partial x} & \dfrac{\partial z}{\partial y} \end{bmatrix} \begin{bmatrix} \dfrac{\partial x}{\partial r} & \dfrac{\partial x}{\partial \theta} \\[2mm] \dfrac{\partial y}{\partial r} & \dfrac{\partial y}{\partial \theta} \end{bmatrix} = \begin{bmatrix} \dfrac{\partial z}{\partial x} & \dfrac{\partial z}{\partial y} \end{bmatrix} \begin{bmatrix} \cos\theta & -r\sin\theta \\ \sin\theta & r\cos\theta \end{bmatrix}$$

Multiplying out, we have

$$\frac{\partial z}{\partial r} = \frac{\partial z}{\partial x}\cos\theta + \frac{\partial z}{\partial y}\sin\theta$$

$$\frac{\partial z}{\partial \theta} = r\left[-\frac{\partial z}{\partial x}\sin\theta + \frac{\partial z}{\partial y}\cos\theta \right]$$

Worked Example 7 Verify the chain rule for $\partial p / \partial x$, where

$$p = f(u, v, w) = u^2 + v^2 - w$$

$$u = x^2 y$$

$$v = y^2$$

and

$$w = e^{-xz}$$

Solution $\qquad f(u, v, w) = (x^2 y)^2 + y^4 - e^{-xz} = x^4 y^2 + y^4 - e^{-xz}$

Thus

$$\frac{\partial p}{\partial x} = 4x^3 y^2 + ze^{-xz}$$

On the other hand,

$$\frac{\partial p}{\partial u}\frac{\partial u}{\partial x} + \frac{\partial p}{\partial v}\frac{\partial v}{\partial x} + \frac{\partial p}{\partial w}\frac{\partial w}{\partial x} = 2u(2xy) + 2v \cdot 0 + ze^{-xz}$$

$$= (2x^2 y)(2xy) + ze^{-xz}$$

which is the same.

Solved Exercises ★8. Express $\partial^2 z / \partial r^2$ in cartesian coordinates.

★9. Suppose that $(t, s) = (f(x, y), g(x, y))$, $x = u - 2v$, and $y = u + 3v$. Express the derivative matrix $\partial(t, s) / \partial(u, v)$ in terms of $\partial(t, s) / \partial(x, y)$.

Exercises ★7. Compute $\partial z / \partial x$ and $\partial z / \partial y$ by the chain rule and by direct substitution for each of the following:

(a) $z = u^2 + v^2$; $u = 2x + 7$, $v = 3x + y + 7$

(b) $z = u^2 + 3uv - v^2$; $u = \sin x$, $v = -\cos x + \cos y$

(c) $z = \sin u \cos v$; $u = 3x^2 - 2y$, $v = x - 3y$

★8. Suppose that a function is given in terms of rectangular coordinates by $u = f(x, y, z)$. If

$$x = r \cos \theta \sin \phi$$

$$y = r \sin \theta \sin \phi$$

$$z = r \cos \phi$$

express $\partial u / \partial r$, $\partial u / \partial \theta$, and $\partial u / \partial \phi$ in terms of $\partial u / \partial x$, $\partial u / \partial y$, and $\partial u / \partial z$.

★9. Suppose that x, y, z are as in Exercise 8 and $u = x^2 + y^2 + z^2$. Find $\partial u / \partial r$, $\partial u / \partial \theta$, and $\partial u / \partial \phi$.

Problems for Section 16-4

1. Compute the following matrix products or explain why they are not defined:

(a) $\begin{bmatrix} 1 & 0 \\ 0 & 0 \end{bmatrix} \begin{bmatrix} a & b \\ c & d \end{bmatrix}$ (b) $\begin{bmatrix} 0 & 0 \\ 0 & 1 \end{bmatrix} \begin{bmatrix} a \\ b \end{bmatrix}$

(c) $\begin{bmatrix} 0 & 1 \\ 2 & 3 \end{bmatrix} \begin{bmatrix} 1 \\ 2 \\ 3 \end{bmatrix}$

2. Multiply these matrices if the product is defined:

(a) $\begin{bmatrix} 1 & 2 \\ 3 & 4 \\ 5 & 6 \end{bmatrix} \begin{bmatrix} 0 & 0 & 0 \\ 3 & 2 & 1 \end{bmatrix}$ (b) $\begin{bmatrix} 1 & 0 \\ 0 & 1 \end{bmatrix} \begin{bmatrix} a & b \\ c & d \end{bmatrix}$

(c) $\begin{bmatrix} 0 & 1 \\ 1 & 0 \end{bmatrix} \begin{bmatrix} a & b \\ c & d \end{bmatrix}$ (d) $\begin{bmatrix} 1 \\ 2 \\ 3 \end{bmatrix} \begin{bmatrix} 4 \\ 5 \\ 6 \end{bmatrix}$

3. Use the method of Solved Exercise 5 to find a matrix B such that

$$B \begin{bmatrix} 1 & 2 \\ 2 & 5 \end{bmatrix} = \begin{bmatrix} 1 & 0 \\ 0 & 1 \end{bmatrix}$$

Using the result of Exercise 5, solve each of the following systems of equations:

(a) $x + 2y = 1$ (b) $x + 2y = 0$
     $2x + 5y = 2$ $2x + 5y = 0$

★4. Express Simpson's rule (p. 531) by using a product of row and column vectors.

5. Suppose that A and B are matrices. Show that AB and BA are both defined if and only if A and B are square matrices of the same size.

★6. Public health officials have located four persons, x_1, x_2, x_3, and x_4, known to be carrying a new strain of flu. Three persons, y_1, y_2, and y_3, report possible contact, and a first-order contact matrix A is set up whose i, jth entry is 1 if there was contact between x_i and y_j and zero otherwise. Five other people, z_1, z_2, z_3, z_4, and z_5, are questioned for possible contact with y_1, y_2, and y_3, and another first-order contact matrix is set up whose i, jth entry is 1 if y_i has contacted z_j and zero otherwise.

(a) Show that the product matrix $C = AB$ counts the number of second-order contacts. That is, the i, jth entry of C is the number of possible paths of disease communication from x_i to z_j.

(b) Write down the three matrices for the situation shown in Fig. 16-4-2. Check the conclusion of part (a).

Fig. 16-4-2 Contacts between three groups of people.

7. (a) Compute derivative matrices $\partial(x, y) / \partial(t, s)$ and $\partial(u, v) / \partial(x, y)$ if

$$x = t + s \quad y = t - s$$

$$u = x^2 + y^2 \quad v = x^2 - y^2$$

(b) Express (u, v) in terms of (t, s) and calculate $\partial(u, v) / \partial(t, s)$.

(c) Verify that the chain rule holds.

8. Do as in Problem 7 for the functions

$$x = t^2 - s^2 \quad y = ts$$

$$u = \sin(x + y) \quad v = \cos(x - y)$$

9. Verify the chain rule in each case:

   (a) $x = ts$, $y = ts$; $u = x$, $v = -y$

   (b) $x = t^2 + s^2$, $y = t^2 - s^2$, $z = 2ts$; $u = xy$, $v = xz$, $w = xz$

10. If $(f_1(x_1, \ldots, x_n), \ldots, f_n(x_1, \ldots, x_n)) = (u_1, \ldots, u_n)$ are n functions of n variables, then the (square) matrix of partial derivatives is called the *jacobian matrix*. Its determinant is called the *jacobian determinant* and is denoted by

$$\left| \frac{\partial(u_1, \ldots, u_n)}{\partial(x_1, \ldots, x_n)} \right|$$

   (a) Suppose that $n = 2$. Show that

$$\left| \frac{\partial(u_1, u_2)}{\partial(x, y)} \right|_{(a,b)}$$

   is the area of the parallelogram spanned by

$$\left(\frac{\partial u_1}{\partial x} \bigg|_{(a,b)}, \frac{\partial u_2}{\partial x} \bigg|_{(a,b)} \right)$$

   and

$$\left(\frac{\partial u_1}{\partial y} \bigg|_{(a,b)}, \frac{\partial u_2}{\partial y} \bigg|_{(a,b)} \right)$$

   (b) Suppose that $n = 3$. Show that

$$\left| \frac{\partial(u_1, u_2, u_3)}{\partial(x_1, x_2, x_3)} \right|_{(a,b,c)}$$

   is the volume of the parallelepiped spanned by the vectors

$$\left(\frac{\partial u_1}{\partial x_i} \bigg|_{(a,b,c)}, \frac{\partial u_2}{\partial x_i} \bigg|_{(a,b,c)}, \frac{\partial u_3}{\partial x_i} \bigg|_{(a,b,c)} \right)$$

   for $i = 1, 2, 3$.

11. Compute the following jacobian determinants (see Problem 10):

   (a) $(x, y) = (r \cos \theta, r \sin \theta)$. Find

$$\left| \frac{\partial(x, y)}{\partial(r, \theta)} \right|$$

   (b) Let $(x, y, z) = (r \cos \theta, r \sin \theta, z)$. Find

$$\left| \frac{\partial(x, y, z)}{\partial(r, \theta, z)} \right|$$

   (c) Let $(x, y, z) = (r \cos \theta \sin \phi, r \sin \theta \sin \phi, r \cos \phi)$. Find

$$\left| \frac{\partial(x, y, z)}{\partial(r, \theta, \phi)} \right|$$

12. Compute the jacobian determinants (see Problem 10) of the following functions at the indicated points:

   (a) $(x, y) = (t^2 + s^2, t^2 - s^2)$; $(t, s) = (1, 2)$

   (b) $(u, v) = (x + y, xy)$; $(x, y) = (5, -3)$

   (c) Compute u, v in terms of t and s from parts (a) and (b) at $(t, s) = (1, 2)$.

13. Prove the following equations (notation from Problem 10) in light of the chain rule and the multiplicative property of determinants found in Solved Exercise 6.

   (a) $\left| \dfrac{\partial(u, v)}{\partial(x, y)} \right| \left| \dfrac{\partial(x, y)}{\partial(t, s)} \right| = \left| \dfrac{\partial(u, v)}{\partial(t, s)} \right|$

   (b) $\left| \dfrac{\partial(x, y)}{\partial(t, s)} \right| \left| \dfrac{\partial(t, s)}{\partial(x, y)} \right| = 1$

★14. Suppose that f is a differentiable function of one variable and that a function $u = g(x, y)$ is defined by

$$u = g(x, y) = xyf\left(\frac{x + y}{xy} \right)$$

   Show that u satisfies a (partial differential) equation of the form

$$x^2 \frac{\partial u}{\partial x} - y^2 \frac{\partial u}{\partial y} = G(x, y)u$$

   and find the function $G(x, y)$.

15. The coordinates u, φ, θ are defined by $x = au \sin \varphi \cos \theta$, $y = bu \sin \varphi \sin \theta$, $z = cu \cos \varphi$ for $u \geq 0$, $0 \leq \varphi \leq \pi$, $0 \leq \theta \leq 2\pi$.

   (a) Show that the surfaces $u = $ constant are the *ellipsoids*

$$\left(\frac{x}{au} \right)^2 + \left(\frac{y}{bu} \right)^2 + \left(\frac{z}{cu} \right)^2 = 1$$

   (b) Show that the surfaces $\varphi = $ constant are *elliptical cones*.

   (c) Show the surface $\theta = $ constant is a *plane*.

   (d) Volume calculations involve the determinant of $\partial(x, y, z)/\partial(u, \varphi, \theta)$. Show it equals $abcu^2 \sin \varphi$.

16. Let v_1, v_2, v_3 be the components of a vector field **v**, u a scalar function, a, b, ρ constants. Express in vector-matrix notation the following equations of elasticity.

$$\rho(\partial^2 v_1/\partial t^2) = (a + b)(\partial u/\partial x)$$
$$+ b(\partial^2 v_1/\partial x^2 + \partial^2 v_1/\partial y^2 + \partial^2 v_1/\partial z^2)$$

$$\rho(\partial^2 v_2/\partial t^2) = (a + b)(\partial u/\partial y)$$
$$+ b(\partial^2 v_2/\partial x^2 + \partial^2 v_2/\partial y^2 + \partial^2 v_2/\partial z^2)$$

$$\rho(\partial^2 v_3/\partial t^2) = (a + b)(\partial u/\partial z)$$
$$+ b(\partial^2 v_3/\partial x^2 + \partial^2 v_3/\partial y^2 + \partial^2 v_3/\partial z^2)$$

17. A rotation in two dimensions is given by

$$\begin{bmatrix} x' \\ y' \end{bmatrix} = \begin{bmatrix} \cos\theta & \sin\theta \\ -\sin\theta & \cos\theta \end{bmatrix} \begin{bmatrix} x \\ y \end{bmatrix}$$

where θ is the angle of rotation of

$$\begin{bmatrix} x \\ y \end{bmatrix} \quad \text{into} \quad \begin{bmatrix} x' \\ y' \end{bmatrix}$$

Show by means of matrix multiplication that a rotation of θ_1 followed by a rotation of θ_2 is the same as a rotation of θ_2 followed by a rotation of θ_1.

18. The matrix equation

$$\begin{bmatrix} x' \\ y' \\ 1 \end{bmatrix} = \begin{bmatrix} \cos\theta & \sin\theta & 0 \\ -\sin\theta & \cos\theta & 0 \\ 0 & 0 & 1 \end{bmatrix} \begin{bmatrix} x \\ y \\ 1 \end{bmatrix}$$

can be viewed as the *rotation* of a plane optical ray through the angle θ. Similarly, the equation

$$\begin{bmatrix} x' \\ y' \\ 1 \end{bmatrix} = \begin{bmatrix} 1 & 0 & a \\ 0 & 1 & b \\ 0 & 0 & 1 \end{bmatrix} \begin{bmatrix} x \\ y \\ 1 \end{bmatrix}$$

can be viewed as the *translation* of a plane optical ray.

(a) Use matrix multiplication to find the transition equation for a rotation followed by a translation.

(b) Is the rotation of an optical ray followed by a translation the same as a translation followed by a rotation? What does this mean about matrix multiplication?

Review Problems for Chapter 16

1. Calculate all first partial derivatives for each of the following:
   (a) $u = f(x, y, z) = \ln[1 + e^{-x}\cos(xy)]$
   (b) $u = g(x, y) = \sin(\pi x)/(1 + y^2)$
   (c) $u = h(x, y, z) = \cos(e^{-x^2 - 2y^2})$

2. Calculate all first partial derivatives for each of the following:
   (a) $u = f(x, z) = x/\sqrt{1 + \exp(\cos(xz^2))}$
   (b) $u = g(x, y, z) = xz + e^z(\int_0^x t^2 e^t \, dt)$
   (c) $u = h(x, y, z) = \dfrac{zx + y^2 + yz}{\sqrt[3]{1 + x^3 + y^3 + z^2 y}}$

3. Calculate the gradient at the origin for each function in Problem 1.

4. Calculate the gradient at a general point for each function in Problem 2.

5. Check that $\partial^2 u/\partial x\,\partial y = \partial^2 u/\partial y\,\partial x$ for each function in Problem 1.

6. Check that $\partial^2 u/\partial x\,\partial z = \partial^2 u/\partial z\,\partial x$ for the functions in Problem 2.

7. At time $t = 0$ a particle is ejected from the surface $x^2 + 2y^2 + 3z^2 = 6$ at the point $(1, 1, 1)$ in a direction normal to the surface at a speed of 10 units per second. At what time does it cross the sphere $x^2 + y^2 + z^2 = 103$?

8. Verify the chain rule for the function $f(x, y, z) = \ln(1 + x^2 + 2z^2)/(1 + y^2)$ and the curve $\sigma(t) = (t, 1 - t^2, \cos t)$.

9. For each of the following, find the maximum and minimum values attained by the function f along the curve $\sigma(t)$:
   (a) $f(x, y) = xy$; $\sigma(t) = (\cos t, \sin t)$; $0 \le t \le 2\pi$
   (b) $f(x, y) = x^2 + y^2$; $\sigma(t) = (\cos t, 2\sin t)$; $0 \le t \le 2\pi$

★10. (a) A function $u = f(x_1, \dots, x_m)$ is called *homogeneous of degree n* if
$$f(tx_1, \dots, tx_m) = t^n f(x_1, \dots, x_m)$$
Show that such a function satisfies the differential equation
$$x_1 \frac{\partial u}{\partial x_1} + x_2 \frac{\partial u}{\partial x_2} + \cdots + x_m \frac{\partial u}{\partial x_m}$$
$$= nf(x_1, \dots, x_m)$$

(b) Show that each of the following functions satisfies a differential equation of the type in part (a), find n, and check directly that f is homogeneous of degree n.
   (i) $f(x, y) = x^2 + xy + y^2$
   (ii) $f(x, y, z) = x + 3y - \sqrt{xz}$; $xz > 0$
   (iii) $f(x, y, z) = xyz + x^3 - x^2 y$

11. A computer-generated graph of

$$z = \frac{\sin(\pi x)}{1 + y^2}$$

is shown in Fig. 16-R-1. Verify analytically that this function has alternating maxima and minima on the x axis with no other critical points.

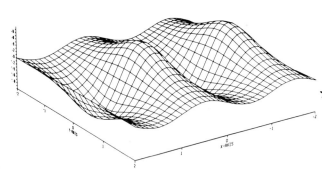

Fig. 16-R-1 Computer-generated graph of $\sin(\pi x)/(1 + y^2)$.

12. Suppose that $f(x,y) = x^2 + y$. Find the maximum and minimum values of f for (x,y) on a circle of radius 1 centered at the origin in two ways:

(a) By parametrizing the circle.

(b) By Lagrange multipliers.

13. Consider the two surfaces

S_1: $x^2 + y^2 + z^2 = f(x,y,z) = 6$

S_2: $2x^2 + 3y^2 + z^2 = g(x,y,z) = 9$

(a) Find the normal vectors and tangent planes to S_1 and S_2 at $(1,1,2)$.

(b) Find the angle between the tangent planes.

(c) Find an expression for the line tangent at $(1,1,2)$ to the curve of intersection of S_1 and S_2. [*Hint:* It lies in both tangent planes.]

★14. Find a family of curves orthogonal to the level curves of $f(x,y) = x^2 - y^2$ as follows:

(a) Find an expression for a vector normal to the level curve of f through (x_0, y_0) at (x_0, y_0).

(b) Use this expression to find a vector tangent to the level curve of f through (x_0, y_0) at (x_0, y_0).

(c) Find a function g which has these vectors as its gradient.

(d) Explain why the level curves of g should intersect those of f orthogonally.

(e) Draw a few of the level curves of f and g to illustrate this result.

15. (a) Suppose that $\mathbf{F}(x,y) = P(x,y)\mathbf{i} + Q(x,y)\mathbf{j}$. Show that if there is a function $f(x,y)$ with continuous second partial derivatives such that $\mathbf{F} = \nabla f$, then $P_y = Q_x$.

(b) Suppose that $\mathbf{F}(x,y,z) = P(x,y,z)\mathbf{i} + Q(x,y,z)\mathbf{j} + R(x,y,z)\mathbf{k}$. Show that if there is a function $f(x,y,z)$ with continuous second partial derivatives such that $\mathbf{F} = \nabla f$, then

$$P_y = Q_x \qquad P_z = R_x \qquad Q_z = R_y$$

(c) Let $\mathbf{F} = 3xy\mathbf{i} - ye^x\,\mathbf{j}$. Is there an f such that $\mathbf{F} = \nabla f$?

★16. (Continuation of Problem 15). Suppose that P and Q have continuous partial derivatives everywhere in the xy plane and that $P_y = Q_x$. Follow the steps below to prove that there is a function f such that $\nabla f = P\mathbf{i} + Q\mathbf{j}$; that is, $f_x = P$ and $f_y = Q$.

(a) Let $g(x,y)$ be an antiderivative of Q with respect to y; that is, $g_y = Q$. Establish that $P - g_x$ is a function of x alone by showing that $(P - g_x)_y = 0$.

(b) If $P - g_x = 0$, then we may simply take $g = f$. Otherwise let $h(x)$ be an antiderivative of $P - g_x$; that is, $h'(x) = P - g_x$. Show that $f(x,y) = g(x,y) + h(x)$ satisfies $\nabla f = P\mathbf{i} + Q\mathbf{j}$.

★17. For each of the following vector fields $P\mathbf{i} + Q\mathbf{j}$, find a function f such that $f_x = P$ and $f_y = Q$ or show that no such function exists. (See Problems 15 and 16.)

(a) $(x^2 y^2 + 2x)e^{xy^2}\mathbf{i} + 2x^3 ye^{xy^2}\mathbf{j}$

(b) $(x^2 y^2 + 2x)e^{xy^3}\mathbf{i} + 2x^3 y\,e^{xy^3}\mathbf{j}$

(c) $\dfrac{2x}{1 + x^2 + y^2}\mathbf{i} + \dfrac{2y}{1 + x^2 + y^2}\mathbf{j}$

(d) $\dfrac{2y}{1 + x^2 + y^2}\mathbf{i} + \dfrac{2x}{1 + x^2 + y^2}\mathbf{j}$

18. Find and describe the critical points of $f(x,y) = y\sin(\pi x)$ (see Fig. 16-R-2).

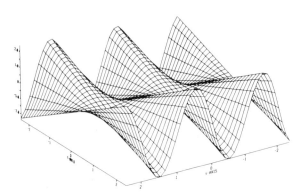

Fig. 16-R-2 Computer-generated graph of $z = y\sin(\pi x)$.

19. Compute $\partial z/\partial x$ and $\partial z/\partial y$ if

$$z = \frac{u^2 + v^2}{u^2 - v^2} \qquad u = e^{-x-y} \qquad v = e^{xy}$$

20. (a) Suppose that $u = f(x,y)$ and $v = g(x,y)$ have continuous partial derivatives which satisfy the *Cauchy-Riemann equations*:

$$\frac{\partial u}{\partial x} = \frac{\partial v}{\partial y} \quad \text{and} \quad \frac{\partial u}{\partial y} = -\frac{\partial v}{\partial x}$$

Show that the level curves of u are perpendicular to the level curves of v.

(b) Confirm this result for the functions $u = x^2 - y^2$ and $v = 2xy$. Sketch some of the level curves of these functions (all on the same set of axes).

★21. (*The gradient and laplacian in polar coordinates.*) Let r and θ be polar coordinates in the plane and let f be a given function of (x,y). Write

$$u = f(x,y) = f(r\cos\theta, r\sin\theta)$$

Let $\mathbf{i}_r = \cos\theta\,\mathbf{i} + \sin\theta\,\mathbf{j}$ and $\mathbf{i}_\theta = -\sin\theta\,\mathbf{i} + \cos\theta\,\mathbf{j}$.

(a) Show that when based at $\mathbf{v} = x\mathbf{i} + y\mathbf{j}$ the vectors $\mathbf{i}_r$ and $\mathbf{i}_\theta$ are orthogonal unit vectors in the directions of increasing r and θ, respectively.

(b) Show that

$$\nabla f = \left(\cos\theta\,\frac{\partial u}{\partial r} - \frac{\sin\theta}{r}\,\frac{\partial u}{\partial \theta}\right)\mathbf{i}$$

$$+ \left(\sin\theta\,\frac{\partial u}{\partial r} + \frac{\cos\theta}{r}\,\frac{\partial u}{\partial \theta}\right)\mathbf{j}$$

$$= \frac{\partial u}{\partial r}\,\mathbf{i}_r + \frac{1}{r}\,\frac{\partial u}{\partial \theta}\,\mathbf{i}_\theta$$

(c) Show that

$$\frac{\partial^2 u}{\partial x^2} + \frac{\partial^2 u}{\partial y^2} = \frac{1}{r}\,\frac{\partial}{\partial r}\left(r\,\frac{\partial u}{\partial r}\right) + \frac{1}{r^2}\,\frac{\partial^2 u}{\partial \theta^2}$$

22. Prove that

$$z = \frac{3x^4 - 4x^3 - 12x^2 + 18}{12(1 + 4y^2)}$$

has one local maximum, one local minimum, and one saddle point. (The computer-generated graph is shown in Fig. 16-R-3.)

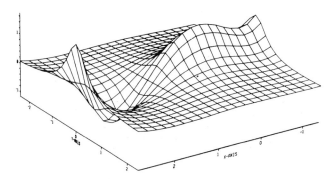

Fig. 16-R-3 Computer-generated graph of
$$z = \frac{3x^4 - 4x^3 - 12x^2 + 18}{12(1 + 4y^2)}$$

23. Find the maxima, minima, and saddles of $z = (2 + \cos\pi x)(\sin\pi y)$, which is graphed in Fig. 16-R-4.

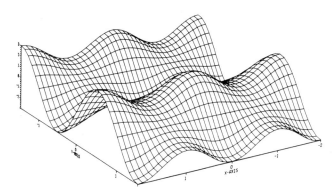

Fig. 16-R-4 Computer-generated graph of $z = (2 + \cos\pi x)(\sin\pi y)$.

24. Suppose that $z = f(x,y)$ is defined, has continuous second partial derivatives, and is harmonic:

$$\frac{\partial^2 z}{\partial x^2} + \frac{\partial^2 z}{\partial y^2} = 0$$

Assume that $(\partial^2 z/\partial x^2)(x_0, y_0) \neq 0$. Prove that f cannot have a local maximum or minimum at (x_0, y_0).

★25. (a) Figure 16-R-5 shows the function $z = (x^2 - y^2)/(x^2 + y^2)$. Show that z has different limits if we come in along the x or y axis.

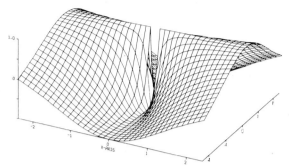

Fig. 16-R-5 Computer-generated graph of $z = (x^2 - y^2)/(x^2 + y^2)$.

(b) Figure 16-R-6 shows the function $z = 2xy^2/(x^2 + y^4)$. Show that if we approach the origin on any straight line, z approaches zero, but z has different limits when $(0, 0)$ is approached along the parabolas $x = \pm y^2$.

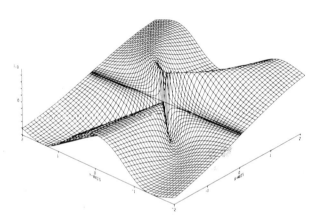

Fig. 16-R-6 Computer-generated graph of $z = 2xy^2/(x^2 + y^4)$.

★26. Let $f(x, y) = y^3/(x^2 + y^2)$.

(a) Compute $f_x, f_y, f_x(0,0)$, and $f_y(0,0)$.

(b) Show that, for any θ, the directional derivative $(d/dr) f(r \cos \theta, r \sin \theta)\big|_{\theta=0}$ exists.

(c) Show that the directional derivatives are not all given by dotting the direction vector with the gradient vector (see Fig. 16-R-7). Why does this not contradict the result on p. 708?

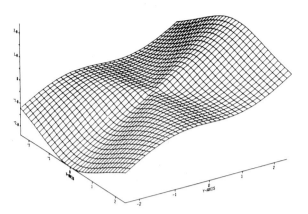

Fig. 16-R-7 Computer-generated graph of $z = y^3/(x^2 + y^2)$.

★27. Do the same things as in Problem 26 for $z = (x^3 - 3xy^2)/(x^2 + y^2)$, which is graphed in Fig. 16-R-8.

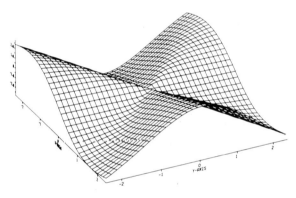

Fig. 16-R-8 Computer-generated graph of $z = (x^3 - 3xy^2)/(x^2 + y^2)$.

★28. Let $z = xy(x^2 - y^2)/(x^2 + y^2)$. Show that $\partial z/\partial x$ and $\partial z/\partial y$ exist and are zero at $(0, 0)$ (use limits!). Show that $\partial^2 z/\partial x\, \partial y$ and $\partial^2 z/\partial y\, \partial x$ exist at the origin but are unequal. Why does this not contradict Theorem 1? (See Fig. 16-R-9.)

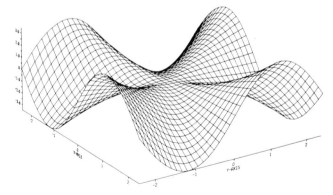

Fig. 16-R-9 Computer-generated graph of $z = [xy(x^2 - y^2)]/(x^2 + y^2)$.

29. A water main consists of two sections of pipe of fixed lengths ℓ_1, ℓ_2 carrying fixed amounts Q_1 and Q_2 liters per second. For a given total loss of head h, the (variable) diameters D_1, D_2 of the pipe will result in a minimum cost if

    $$C = \ell_1(a + bD_1) + \ell_2(a + bD_2) = \text{minimum}$$

    subject to the condition

    $$h = \frac{c\,\ell_1 Q_1^2}{D_1^5} + \frac{c\,\ell_2 Q_2^2}{D_2^5}$$

    Find the ratio D_2/D_1 by the method of Lagrange multipliers.

★30. Ammonia, NH_3, is to be produced at fixed temperature T and pressure p. The pressures of N_2, H_2, NH_3 are labeled as u, v, w, and are known to satisfy $u + v + w = p$, $w^2 = cuv^3$ (c = positive constant). Due to the nature of the reaction $N_2 + 3H_2 \rightleftharpoons 2NH_3$, the maximum ammonia production occurs for w = maximum. Find the maximum pressure $w_{\max}$.

31. The capacitance per unit length of a parallel pair of wires of radii R and axis-to-axis separation D is given by

    $$C = \frac{\pi \varepsilon_0}{\ln\left(\dfrac{D + \sqrt{D^2 - 4R^2}}{2R}\right)}$$

    The capacitance between a wire and a plane parallel to it is

    $$C^* = \frac{2\pi\varepsilon_0}{\ln\left(\dfrac{h + \sqrt{h^2 - R^2}}{R}\right)}$$

    where h = distance from the wire to the plane.

    (a) Find the expected change in capacitance for two parallel wires, separated 2 cm, 0.40 cm in radius, due to a radius increase of 0.01 cm.

    (b) A wire of 0.57 cm radius has its central axis at a uniform distance of 3 cm from a conducting plane. Due to heating, the wire increases 0.02 cm in radius, but due to bowing of the wire it can be assumed that the axis of the wire was raised to 3.15 cm above the plane. What is the expected change in capacitance?

32. The *potential temperature* θ is defined in terms of temperature T and pressure p by

    $$\theta = T\left(\frac{1000}{p}\right)^{0.286}$$

    The temperature and pressure may be thought of as functions of position (x, y, z) in the atmosphere and also of time t.

    (a) Find formulas for $\partial\theta/\partial x$, $\partial\theta/\partial y$, $\partial\theta/\partial z$, $\partial\theta/\partial t$ in terms of partial derivatives of T and p.

    (b) The condition $\partial\theta/\partial z < 0$ is labeled as *unstable atmosphere*, for it corresponds to large vertical excursions of air parcels from a single upward or downward impetus. Meteorologists use the formula (assumed hereafter)

    $$\partial\theta/\partial z = \frac{\theta}{T}\left(\frac{\partial T}{\partial z} + \frac{g}{C_p}\right)$$

    where $g = 32.2$, C_p = constant > 0. How does the temperature change in the upward direction for an unstable atmosphere?

33. The traffic speed $u(x, t)$ along a highway at position x and time t is given by $u = k - k\tanh(kx - u(u - k)t)$.

    (a) Find u_x, u_t and verify

    $$ku_t + u(u - k)u_x = 0.$$

    (b) Show that *shocks* arise (that is, $u_t = \infty$) when

    $$t = k\{u(2k - u)(2u - k)\}^{-1}$$

    and

    $$kx = \tanh^{-1}\left(\frac{k - u}{k}\right) + u(u - k)t$$

34. An irrigation canal in Arizona has concrete sides and bottom with trapezoidal cross section of area $A = y(x + y\tan\theta)$ and wetted perimeter $P = x + 2y/\cos\theta$ where x = bottom width, y = water depth, θ = side inclination, measured from vertical. The best design for fixed inclination θ is found by solving P = minimum subject to the condition A = constant. Show that $y^2 = A\cos\theta/(2 - \sin\theta)$, while x is determined from the first relation.

35. The friction in an open air aqueduct is proportional to the wetted perimeter of the cross section. Show that the best form of a rectangular cross section is one with the width x equal to twice the depth y, by solving the problem perimeter = $2y + x$ = minimum, area = xy = constant.

17

Multiple Integration

Functions can be integrated over regions in the plane and in space.

This chapter is an introduction to the integration of real-valued functions of two and three variables. As in the previous chapter, our intent is to introduce the ideas and methods of calculation, leaving a detailed examination of the theoretical points for a later course.

Double and triple integrals enable us to "sum" the values of real-valued functions of two or three variables; we evaluate them by integration with respect to one variable at a time, using the methods of one-variable calculus.

CONTENTS

Double and triple integrals are introduced in Sections 17-1 and 17-2. Section 18-1 on line integrals does not depend on these sections, however, so it can be read independently. The reader who wishes to skip triple integrals should proceed directly to pp. 785–786 after Section 17-1 to study double integrals in polar coordinates and then carry on with Sections 18-1 and the first three subsections of Section 18-2.

17-1 Double Integrals

The double integral of a nonnegative function over a region in the plane is equal to the volume under its graph.

The definite integral $\int_a^b f(x)\,dx$, defined in Chapter 4, represents a "sum" of the values of f at the (infinitely many) points of the interval $[a, b]$. To "sum" the values of a function $f(x, y)$ over the points of a region D in the plane, we will define the *double integral* $\int\int_D f(x, y)\,dx\,dy$. We recommend a rapid review of Section 4-1 as preparation for the present section.

Goals

After studying this section, you should be able to:

Express double integrals over rectangles and more general regions as iterated integrals, and evaluate these integrals.

Find the average value of a function on a region in the plane.

Express the surface area of a graph as a double integral.

THE DOUBLE INTEGRAL OVER A RECTANGLE

(a) The closed rectangle
$a \le x \le b, c \le y \le d$

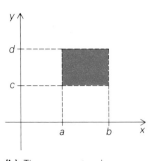

(b) The open rectangle
$a < x < b, c < y < d$

Fig. 17-1-1 Examples of closed and open rectangles.

To illustrate the process of summation which is represented by the double integral, we may use an example connected with solar energy. The intensity of solar radiation is a "local" quantity, which may be measured at any point on the earth's surface. Since the solar intensity is really a rate of power input (see Section 10-3), we can measure it in units of watts per square meter.

If the solar intensity is uniform over a region, the total power received is equal to the intensity times the area of the region. In practice, the intensity is a function of position (in particular, it is a function of latitude—see p. 152 or 436). To find the total power over a region requires a process of *double integration* analogous to the process by which we integrate instantaneous power input with respect to time to obtain the total energy received during a time interval.

Our development of double integrals will be similar to that of definite integrals in Chapter 4. We will give the formal definition of double integrals in this subsection. However, this definition is not directly relevant to working specific examples. What is relevant is the statement of Theorem 1 below and the associated Worked Examples. Thus the less theoretically inclined student may wish to just skim the material up to that point.

The sets in the plane which will play the role of "intervals" in double integration are the rectangles (see Fig. 17-1-1). A *closed rectangle D* consists of all x and y such that $a \le x \le b$ and $c \le y \le d$; it is denoted by $[a, b] \times [c, d]$. The *interior* of D consists of all x and y such that $a < x < b$ and $c < y < d$; it is called an *open rectangle* and is denoted by $(a, b) \times (c, d)$. The *area* of D is the product $(b - a)(d - c)$.

We say that a function $g(x, y)$ defined on $[a, b] \times [c, d]$ is *piecewise constant* if there are partitions $a = t_0 < t_1 < t_2 < \cdots < t_n = b$ of $[a, b]$ and $c = s_0 < s_1 < s_2 < \cdots s_m = d$ of $[c, d]$ such that, in each of the mn open rectangles $R_{ij} = (t_{i-1}, t_i) \times (s_{j-1}, s_j)$, $g(x, y)$ has a constant value k_{ij}. The graph of a piecewise constant function is shown in Fig. 17-1-2.

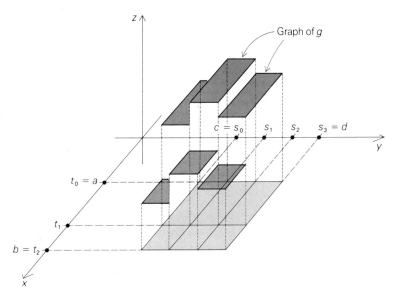

Fig. 17-1-2 g is piecewise constant since it is constant on each subrectangle.

By analogy with our definition for piecewise constant functions of one variable (p. 182), we define:

$$\iint_D g(x, y)\, dx\, dy = \sum_{i=1, j=1}^{n,m} (k_{ij})(\text{area } R_{ij}) = \sum_{i=1, j=1}^{n,m} (k_{ij})(\Delta t_i)(\Delta s_j)$$

where $\Delta t_i = t_i - t_{i-1}$ and $\Delta s_j = s_j - s_{j-1}$. The summation symbol means that we sum over all i and j with i ranging from 1 to n and j from 1 to m; there are nm terms in the sum corresponding to the nm rectangles R_{ij}.

If $g(x, y) \geq 0$, the integral of g is exactly the volume under its graph. Indeed, the height of the box over the rectangle R_{ij} is k_{ij}, so its volume is $k_{ij} \times \text{area } (R_{ij}) = k_{ij}\Delta t_i \Delta s_j$; the integral of g is the sum of these, the total volume.

Worked Example 1   Let g take the values on the rectangles as shown in Fig. 17-1-3. Calculate the integral of g.

Fig. 17-1-3
Find $\iint_D g(x, y)\, dx\, dy$ if g
takes the values shown.

Solution   The integral of g is the sum of the values of g times the areas of the rectangles:

$$\iint_D g(x, y)\, dx\, dy = -8 \times 2 + 2 \times 3 + 6 \times 2 + 3 \times 3 + 4 \times 2 - 1 \times 3 = 16$$

We proceed now to define the integral of a function f, defined on a closed rectangle D, in the same manner as we did in Section 4-1.

Definition The integral over D of a piecewise constant function g such that $g(x, y) \le f(x, y)$ on D is called a *lower sum* for f on D. The integral over D of a piecewise constant function h such that $h(x, y) \ge f(x, y)$ on D is called an *upper sum* for f on D.

The sets of lower and upper sums form intervals which may or may not have a common endpoint, as in Fig. 4-1-11.

Definition We say that f is *integrable* on D if there is a number S_0 such that every $S < S_0$ is a lower sum for f on D and every $S > S_0$ is an upper sum. The number S_0 is called the *integral* of f over D and is denoted by

$$\iint_D f(x, y)\, dx\, dy$$

The basic properties of the double integral are similar to those of the ordinary integral:

1. Every continuous function is integrable.

2. If a rectangle D is divided by a line segment into two rectangles D_1 and D_2, and if $f(x, y)$ is integrable on D_1 and D_2, then f is integrable on D and

$$\iint_D f(x, y)\, dx\, dy = \iint_{D_1} f(x, y)\, dx\, dy + \iint_{D_2} f(x, y)\, dx\, dy$$

3. If f_1 and f_2 are integrable on D and if $f_1 \le f_2$ on D, then

$$\iint_D f_1(x, y)\, dx\, dy \le \iint_D f_2(x, y)\, dx\, dy$$

4. If $f(x, y) = k$ on D,

$$\iint_D f(x, y)\, dx\, dy = (k)(\text{area of } D)$$

5. $$\iint_D [(f_1(x, y) + f_2(x, y)]\, dx\, dy$$
    $$= \iint_D f_1(x, y)\, dx\, dy + \iint_D f_2(x, y)\, dx\, dy$$

6. $$\iint_D cf(x, y)\, dx\, dy = c \iint_D f(x, y)\, dx\, dy$$

We omit the proofs of these results since they are similar to the one-variable case.

We observed earlier that if $g(x, y) \geq 0$ and g is piecewise constant, then the integral of g is the volume under its graph. If $f(x, y) \geq 0$ is any integrable function, then the volume under the graph of f lies between the volumes under the graphs of piecewise constant functions $g \leq f$ and $h \geq f$; that is, between lower and upper sums. Since the integral has exactly this property, we conclude as for functions of one variable that *the integral of f over D equals the volume under the graph of f if $f \geq 0$.* See Fig. 17-1-4.

Fig. 17-1-4 The volume of R, the region under the graph of f, equals $\int\int_D f(x, y)dx\, dy.$

For the moment, we can evaluate integrals only approximately or by appealing to geometric formulas for volumes of special solids. In the next subsection we show how the fundamental theorem of calculus can be brought into play.

Solved Exercises*

1. Let D be the rectangle $0 \leq x \leq 2$, $1 \leq y \leq 3$, and let $f(x, y) = x^2 y$. Choose a piecewise constant function $h(x, y) \geq f(x, y)$ to show that $\int\int_D f(x, y)\, dx\, dy \leq 25$.

2. Suppose that a rectangular plate D has mass density $\rho(x, y)$ grams per square centimeter. Argue that $\int\int_D \rho(x, y)\, dx\, dy$ is the mass of the plate.

Exercises

1. (a) Let D be the rectangle determined by $-1 \leq x \leq 1$ and $2 \leq y \leq 4$. Let $f(x, y) = x(1 + y)$. Find a piecewise constant function $g(x, y) \leq f(x, y)$ to show that $\int\int_D f(x, y)\, dx\, dy \geq -9$.

   (b) Sketch the graph of $f(x, y)$ over D. Use symmetry to argue that the value of the integral in part (a) must in fact be zero.

2. Let D be the rectangle $[0, 3] \times [0, 3]$. Estimate $\int\int_D f(x, y)\, dx\, dy$ to within ± 2, where $f(x, y) = 1 + y \sin \pi x$ on the shaded portion in Fig. 17-1-5 and $f(x, y) = 0$ on the remainder.

3. Let D be the rectangle $1 \leq x \leq 4$, $0 \leq y \leq 2$. Suppose that D is a plate with mass density $\rho(x, y) = 2xy^2 + \cos \pi y + 1$ (grams per square centimeter). Find piecewise constant functions to show that the mass m of the plate D satisfies 30 grams $\leq m \leq$ 62 grams.

Fig. 17-1-5 Estimate $\int\int_D [1 + y \sin (\pi x)]\, dx\, dy.$

Solutions appear in Appendix A.

ITERATED INTEGRALS

Before establishing the fundamental result which will enable us to use one-variable techniques to evaluate double integrals, we must explain some notation. The *iterated integral*

$$\int_c^d \left[\int_a^b f(x, y)\, dx \right] dy$$

is evaluated, like most parenthesized expressions, from the inside out. One first holds y fixed and evaluates the integral $\int_a^b f(x, y)\, dx$ with respect to x; the result is a function of y which is then integrated from c to d.

The expression $\int_a^b [\int_c^d f(x, y)\, dy]\, dx$ is defined similarly; this time the integral with respect to y is evaluated first. Iterated integrals are often written without parentheses as $\int_c^d \int_a^b f(x, y)\, dx\, dy$ and $\int_a^b \int_c^d f(x, y)\, dy\, dx$.

Worked Example 2   Evaluate $\int_0^2 \int_1^3 x^2 y\, dy\, dx$.

Solution

$$\int_0^2 \int_1^3 x^2 y\, dy\, dx = \int_0^2 \left(\int_1^3 x^2 y\, dy \right) dx$$

$$= \int_0^2 \left(\left. \frac{x^2 y^2}{2} \right|_{y=1}^3 \right) dx$$

$$= \int_0^2 x^2 \left(\frac{9}{2} - \frac{1}{2} \right) dx$$

$$= 4 \int_0^2 x^2\, dx = 4 \left. \frac{x^3}{3} \right|_0^2 = \frac{4}{3} 8 = \frac{32}{3}$$

Notice that the second step in this calculation is essentially the inverse of a partial differentiation.

Theorem 1*   *Assume that $f(x, y)$ is integrable on the rectangle $D = [a, b] \times [c, d]$. Then any iterated integral which exists is equal to the double integral; that is,*

$$\iint_D f(x, y)\, dx\, dy = \int_c^d \left[\int_a^b f(x, y)\, dx \right] dy$$

$$= \int_a^b \left[\int_c^d f(x, y)\, dy \right] dx$$

(optional)

Proof   We first show that the theorem is true for piecewise constant functions. Let g be a piecewise constant function, with $g(x, y) = k_{ij}$ on $(t_{i-1}, t_i) \times (s_{j-1}, s_j)$, so that

$$\iint_D g(x, y)\, dx\, dy = \sum_{i=1, j=1}^{n,m} k_{ij} \Delta t_i \Delta s_j$$

---

*This result is sometimes called Fubini's theorem.

(optional)

If the summands $k_{ij}\,\Delta t_i\,\Delta s_j$ are laid out in a rectangular array, they may be added by first adding along rows and then adding up the subtotals, as follows:

$$
\begin{array}{ccccc}
k_{11}\,\Delta t_1\,\Delta s_1 & k_{21}\,\Delta t_2\,\Delta s_1 & \cdots & k_{n1}\,\Delta t_n\,\Delta s_1 & \longrightarrow \left(\displaystyle\sum_{i=1}^{n} k_{i1}\,\Delta t_i\right)\Delta s_1 \\[2ex]
k_{12}\,\Delta t_1\,\Delta s_2 & k_{22}\,\Delta t_2\,\Delta s_2 & \cdots & k_{n2}\,\Delta t_n\,\Delta s_2 & \longrightarrow \left(\displaystyle\sum_{i=1}^{n} k_{i2}\,\Delta t_i\right)\Delta s_2 \\[2ex]
\vdots & \vdots & \vdots & \vdots & \vdots \\[2ex]
k_{1m}\,\Delta t_1\,\Delta s_m & k_{2m}\,\Delta t_2\,\Delta s_m & \cdots & k_{nm}\,\Delta t_n\,\Delta s_m & \longrightarrow \left(\displaystyle\sum_{i=1}^{n} k_{im}\,\Delta t_1\right)\Delta s_m
\end{array}
$$

$$
\sum_{j=1}^{m}\left(\sum_{i=1}^{n} k_{ij}\,\Delta t_i\right)\Delta s_j
$$

The coefficient of Δs_j in the sum over the jth row, $\sum_{i=1}^{n} k_{ij}\,\Delta t_i$, is equal to $\int_a^b g(x,y)\,dx$ for any y with $s_{i-1} < y < s_i$, since for y fixed $g(x,y)$ is a piecewise constant function of x. Thus the integral $\int_a^b g(x,y)\,dx$ is a piecewise constant function of y, and its integral with respect to y is the sum:

$$
\int_c^d \left[\int_a^b g(x,y)\,dx\right]dy = \sum_{j=1}^{m}\left(\sum_{i=1}^{n} k_{ij}\,\Delta t_i\right)\Delta s_j
$$

$$
= \iint_D g(x,y)\,dx\,dy
$$

Similarly, by summing first over columns and then over rows, we obtain

$$
\iint_D g(x,y)\,dx\,dy = \int_a^b\left[\int_c^d g(x,y)\,dy\right]dx
$$

The theorem is therefore true for piecewise constant functions.

Now let f be integrable on $D = [a,b] \times [c,d]$ and assume that the iterated integral $\int_a^b \left[\int_c^d f(x,y)\,dy\right]\,dx$ exists. Denoting this integral by S_0, we will show that every lower sum for f on D is less than or equal to S_0, while every upper sum is greater than or equal to S_0, so S_0 must be the integral of f over D.

To carry out our program, let g be any piecewise constant function such that

$$
g(x,y) \le f(x,y) \tag{1}
$$

for all (x,y) in D. Integrating equation (1) with respect to y and using property 5 of the one-variable integral (see p. 201), we obtain

$$
\int_c^d g(x,y)\,dy \le \int_c^d f(x,y)\,dy \tag{2}
$$

for all x in $[c,d]$. Integrating (2) with respect to x and applying property 5 once more gives

$$
\int_a^b\left[\int_c^d g(x,y)\,dy\right]dx \le \int_a^b\left[\int_c^d f(x,y)\,dy\right]dx \tag{3}
$$

(optional)

Since g is piecewise constant, it follows from the first part of this proof that the left-hand side of (3) is equal to the lower sum $\iint_D g(x, y)\, dx\, dy$; the right-hand side of (3) is just S_0, so we have shown that every lower sum is less than or equal to S_0. The proof that every upper sum is greater than or equal to S_0 is similar, and we are done.

Worked Example 3   Let $f(x, y) = e^{2x+y}$. Evaluate the integral of f over $D = [0, 1] \times [0, 3]$.

Solution

$$\iint_D f(x, y)\, dx\, dy = \int_0^3 \left(\int_0^1 e^{2x+y}\, dx \right) dy \quad \text{(by Theorem 1, first equality)}$$

$$= \int_0^3 \left(\frac{1}{2} e^{2x+y} \Big|_{x=0}^{1} \right) dy = \frac{1}{2} \int_0^3 (e^{2+y} - e^y)\, dy$$

$$= \frac{1}{2}(e^2 - 1) \int_0^3 e^y\, dy = \frac{(e^2 - 1)(e^3 - 1)}{2} \quad (\approx 60.9693)$$

Check that if we integrate with respect to y first, we get the same answer.

Worked Example 4   Evaluate $\int_1^3 \int_0^2 x^2 y\, dx\, dy$ and compare with Worked Example 2.

Solution

$$\int_1^3 \int_0^2 x^2 y\, dx\, dy = \int_1^3 \left(\int_0^2 x^2 y\, dx \right) dy = \int_1^3 \left(\frac{x^3 y}{3} \Big|_{x=0}^{2} \right) dy$$

$$= \int_1^3 \frac{8}{3} y\, dy = \frac{4}{3} y^2 \Big|_1^3 = \frac{4}{3}(3^2 - 1^2) = \frac{32}{3}$$

The answer is the same as that in Worked Example 2, as predicted by Theorem 1. (It is also consistent with Solved Exercise 1.)

The formulas established in Theorem 1 are closely related to Cavalieri's principle as discussed in Section 10-1. Suppose that $f(x, y) \geq 0$, so that $\iint_D f(x, y)\, dx\, dy$ represents the volume of the region R under the graph of f. If we take this volume and slice it by a plane parallel to the yz plane at a distance x from the origin, we get a two-dimensional region with area $A(x) = \int_c^d f(x, y)\, dy$ (see Fig. 17-1-6).

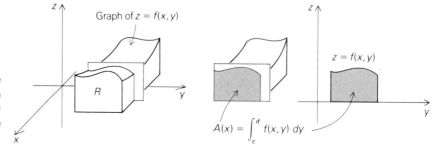

Fig. 17-1-6 The area of the cross-section is the area under the graph of $z = f(x, y)$ from $y = c$ to $y = d$ (where x is fixed).

By Cavalieri's principle, the total volume is the integral of the area function $A(x)$. Thus

$$\iint_D f(x, y)\, dx\, dy = \text{volume of } R$$

$$= \int_a^b A(x)\, dx$$

$$= \int_a^b \left[\int_c^d f(x, y)\, dy \right] dx$$

This is exactly the second equality in Theorem 1. In the same way, if we use planes parallel to the xz plane, we get

$$\iint_D f(x, y)\, dx\, dy = \int_c^d \left[\int_a^b f(x, y)\, dx \right] dy$$

We see that Cavalieri's principle gives a geometric "proof" of Theorem 1; the analytically inclined reader may prefer, however, to take our proof of Theorem 1 as a justification for Cavalieri's principle.

DOUBLE AND ITERATED INTEGRALS

To evaluate the double integral of $f(x, y)$ over a rectangle $D = [a, b] \times [c, d]$, evaluate the iterated integral of $f(x, y)$: either variable may be integrated first:

$$\iint_D f(x, y)\, dx\, dy = \int_c^d \left[\int_a^b f(x, y)\, dx \right] dy = \int_a^b \left[\int_c^d f(x, y)\, dy \right] dx$$

Solved Exercises

3. Compute $\iint_D \sin(x + y)\, dx\, dy$, where $D = [0, \pi] \times [0, 2\pi]$.

4. If D is a plate defined by $1 \le x \le 2$, $0 \le y \le 1$, and the mass density is $\rho(x, y) = ye^{xy}$ grams per square centimeter, find the mass of the plate (see Solved Exercise 2).

5. Find the volume under the graph of $f(x, y) = x^2 + y^2$ between the planes $x = 0$, $x = 3$, $y = -1$, $y = 1$.

Exercises

4. Compute $\iint_D f(x, y)\, dx\, dy$ for:
   (a) $f(x, y) = (x + 2y)^2$; $\quad D = [-1, 2] \times [0, 2]$
   (b) $f(x, y) = y^3 \cos^2 x$; $\quad D = [-\pi/2, \pi] \times [1, 2]$
   (c) $f(x, y) = x^2 + 2xy - y\sqrt{x}$; $\quad D = [0, 1] \times [-2, 2]$

5. Sketch and find the volume under the graph of f between the planes $x = a$, $x = b$, $y = c$, and $y = d$ for each of the following:
   (a) $f(x, y) = x^3 + y^2 + 2$; $\quad a = -1, b = 1, c = 1, d = 3$
   (b) $f(x, y) = 2x + 3y^2 + 2$; $\quad a = 0, b = 3, c = -2, d = 1$

6. Find the mass of the plate described in Exercise 3.

THE DOUBLE INTEGRAL OVER GENERAL REGIONS

Applying the methods just obtained to the problem of integrating solar power, we could (at least in principle) find the total power received by the state of Colorado, which is a rectangle in longitude-latitude coordinates. The problem for Utah is also tractable, since that state is composed of two rectangles, but what happens if we are interested in Michigan or Florida?

We need to define the integral $\iint_D f(x, y)\, dx\, dy$ on regions D more general than rectangles. To do so, we will use a trick to reduce the problem to one of integrating over rectangles.

Definitions   A region D in the plane is said to be *bounded* if it is contained in some rectangle.

Let D be a bounded region in the plane and let f be a function whose domain contains D. Let f^* be the function defined by

$$f^*(x, y) = \begin{cases} f(x, y) & \text{if } (x, y) \in D \\ 0 & \text{if } (x, y) \notin D \end{cases}$$

(See Fig. 17-1-7.) Let D^* be any rectangle containing D. If f^* is integrable on D^*, then we say that f is *integrable* on D, and we define $\iint_D f(x, y)\, dx\, dy$ to be $\iint_{D^*} f^*(x, y)\, dx\, dy$.

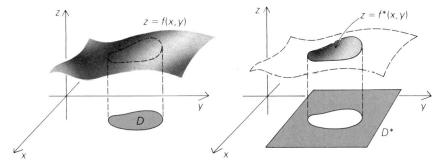

Fig. 17-1-7 Given f and D, we construct f^* by setting $f^*(x, y)$ equal to zero outside D.

We must verify that the definition does not depend upon the choice of the surrounding rectangle D^*. But that is simple to check, since the function f^* is zero on the region between two such rectangles.

We may evaluate integrals over general regions by using iterated integrals. Before describing the general procedure, we give an example to illustrate the above definition. The most important method, however, is given in the display on p. 765 and in Worked Example 7.

Worked Example 5   Find $\iint_D (x^2 + y^2)\, dx\, dy$, where D is the triangle defined by $x \geq 0$, $y \geq 0$, $x + y \leq 1$.

Solution   The triangle is shaded in Fig. 17-1-8. We will take D^* to be the rectangle $[0, 1] \times [0, 1]$. Then the function $f^*(x, y)$ is defined by

$$f^*(x, y) = \begin{cases} x^2 + y^2 & \text{if } x + y \leq 1 \\ 0 & \text{if } x + y > 1 \end{cases}$$

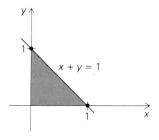

Fig. 17-1-8 Integrate $x^2 + y^2$ over the shaded triangle.

To compute $\iint_D f(x, y)\, dx\, dy = \iint_{D^*} f^*(x, y)\, dx\, dy$, we take the iterated integral

$$\int_0^1 \left[\int_0^1 f^*(x, y)\, dx \right] dy$$

If we fix a value of y, then the inside integral becomes $\int_0^1 g(x)\, dx$, where

$$g(x) = \begin{cases} x^2 + y^2 & \text{if } x \le 1 - y \\ 0 & \text{if } x > 1 - y \end{cases}$$

(See Fig. 17-1-9.)

Fig. 17-1-9 The graph of $g(x) = f^*(x, y)$ for a fixed y.

Since the integral of g from $1 - y$ to 1 is zero, we have

$$\int_0^1 g(x)\, dx = \int_0^{1-y} (x^2 + y^2)\, dx$$

which we may evaluate as

$$\left. \frac{x^3}{3} + y^2 x \right|_{x=0}^{1-y} = \frac{(1-y)^3}{3} + y^2(1 - y)$$

Now we must integrate this function of y from 0 to 1:

$$\begin{aligned} \iint_D f(x, y)\, dy &= \int_0^1 \left[\frac{(1-y)^3}{3} + y^2(1-y) \right] dy \\ &= \left. \left(-\frac{(1-y)^4}{12} + \frac{y^3}{3} - \frac{y^4}{4} \right) \right|_0^1 \\ &= \left(\frac{1}{3} - \frac{1}{4} \right) - \left(-\frac{1}{12} \right) = \frac{1}{6} \end{aligned}$$

Notice that when we perform the integration over x, the limit of integration $1 - y$ is a function of y.

Fig. 17-1-10 A region D is of type 1 if it is the region between the graphs of two functions, $y = \phi_1(x)$ and $y = \phi_2(x)$.

To describe the general procedure for evaluating integrals over plane regions, we first define two simple types of regions, which we will call *elementary regions*. Complicated regions can often be broken into elementary ones.

Suppose that we are given two continuous real-valued functions ϕ_1 and ϕ_2 on $[a, b]$ which satisfy $\phi_1(t) \le \phi_2(t)$ for all t in $[a, b]$. Let D be the set of all points (x, y) such that

$$x \text{ is in } [a, b] \quad \text{and} \quad \phi_1(x) \le y \le \phi_2(x)$$

This region D is said to be of *type 1*. (See Fig. 17-1-10.) The curves and straight line segments that enclose the region constitute the *boundary* of D.

We say that a region D is of *type 2* if there are continuous functions ψ_1 and ψ_2 on $[c, d]$ such that D is the set of points (x, y) satisfying

$$y \text{ is in } [c, d] \quad \text{and} \quad \psi_1(y) \le x \le \psi_2(y)$$

where $\psi_1(t) \le \psi_2(t)$ for t in $[c, d]$. (See Fig. 17-1-11.) Again the boundary of the region consists of the curves and line segments enclosing the region.

Fig. 17-1-11 A region D is of type 2 if it is the region between the graphs of $x = \psi_1(y)$ and $x = \psi_2(y)$.

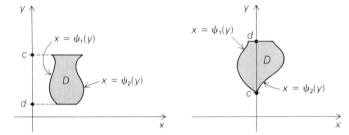

The following example shows that a given region may be of types 1 and 2 at the same time.

Worked Example 6 Show that the region D defined by $x^2 + y^2 \le 1$ (the unit disk) is a region of types 1 and 2.

Solution Descriptions of the disk, showing that it is of both types, are given in Fig. 17-1-12.

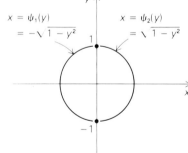

Fig. 17-1-12 The unit disk as type 1 region and type 2 region.

(a) Type 1 region

(b) Type 2 region

We will use without proof* the following fact: If f is continuous on an elementary region D, then f is integrable on D. This fact, combined with Theorem 1, enables us to evaluate integrals over elementary regions by iterated integration.

If $D^* = [a, b] \times [c, d]$ is a rectangle containing D, then by Theorem 1

$$\iint_D f(x, y)\, dx\, dy = \iint_{D^*} f^*(x, y)\, dx\, dy = \int_a^b \int_c^d f^*(x, y)\, dy\, dx$$
$$= \int_c^d \int_a^b f^*(x, y)\, dx\, dy$$

where f^* equals f in D and is zero outside D. Assume that D is a region of type 1 determined by functions ϕ_1 and ϕ_2 on $[a, b]$. Consider the iterated integral

$$\int_a^b \int_c^d f^*(x, y)\, dy\, dx$$

and, in particular, the inner integral $\int_c^d f^*(x, y)\, dy$ for some fixed x (Fig. 17-1-13). By definition, $f^*(x, y) = 0$ if $y < \phi_1(x)$ or $y > \phi_2(x)$, so

$$\int_c^d f^*(x, y)\, dy = \int_{\phi_1(x)}^{\phi_2(x)} f^*(x, y)\, dy = \int_{\phi_1(x)}^{\phi_2(x)} f(x, y)\, dy$$

A similar construction works for type 2 regions.

Fig. 17-1-13 The integral of f^* over D^* equals that of f over D.

DOUBLE INTEGRALS

If D is a region of type 1 (Fig. 17-1-10),

$$\iint_D f(x, y)\, dx\, dy = \int_a^b \left[\int_{\phi_1(x)}^{\phi_2(x)} f(x, y)\, dy \right] dx \tag{4}$$

If D is of type 2 (Fig. 17-1-11),

$$\iint_D f(x, y)\, dx\, dy = \int_c^d \left[\int_{\psi_1(y)}^{\psi_2(y)} f(x, y)\, dx \right] dy \tag{5}$$

If D is of both types, either (4) or (5) is applicable.

*For a proof, see virtually any advanced calculus text, such as J. Marsden, Elementary Classical Analysis, W. H. Freeman, 1974, Chapter 8.

Worked Example 7 Find $\iint_D (x + y)\, dx\, dy$, where D is the shaded region in Fig. 17-1-14.

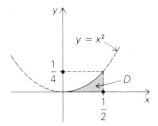

Fig. 17-1-14 Find $\iint_D (x + y)\, dx\, dy.$

Solution D is a region of type 1, with $[a, b] = [0, \tfrac{1}{2}]$, $\varphi_1(x) = 0$, and $\varphi_2(x) = x^2$. By formula (4) in the preceding display,

$$\iint_D (x + y)\, dx\, dy = \int_0^{1/2} \int_0^{x^2} (x + y)\, dy\, dx = \int_0^{1/2} \left[\left(xy + \frac{y^2}{2} \right) \Big|_{y=0}^{x^2} \right] dx$$

$$= \int_0^{1/2} \left(x^3 + \frac{x^4}{2} \right) dx = \left(\frac{x^4}{4} + \frac{x^5}{10} \right) \Big|_0^{1/2} = \frac{1}{64} + \frac{1}{320} = \frac{3}{160}$$

D is also a region of type 2, with $\psi_1(y) = \sqrt{y}$ and $\psi_2(y) = \tfrac{1}{2}$. We leave it to you to verify that the double integral calculated by formula (5) is also $\tfrac{3}{160}$.

General regions can often be broken into elementary regions, and double integrals over these regions can be computed a piece at a time.

Worked Example 8 Find $\iint_R x^2\, dx\, dy$, where R is the shaded region in Fig. 17-1-15.

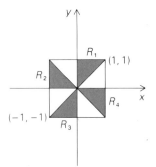

Fig. 17-1-15 Find $\iint_R x^2 dx\, dy$, where R is the shaded region.

Solution Each of the regions R_1, R_2, R_3, R_4 is of types 1 and 2, so we may integrate over each one separately and sum the results. Using formula (4), we get

$$\iint_{R_1} x^2\, dx\, dy = \int_0^1 \int_x^1 x^2\, dy\, dx = \int_0^1 (x^2 y) \Big|_{y=x}^1 dx$$

$$= \int_0^1 (x^2 - x^3)\, dx = \frac{x^3}{3} - \frac{x^4}{4} \Big|_0^1$$

$$= \tfrac{1}{3} - \tfrac{1}{4} = \tfrac{1}{12}$$

Similarly, we find

$$\iint_{R_2} x^2\, dx\, dy = \int_{-1}^{0}\int_{0}^{-x} x^2\, dy\, dx = \tfrac{1}{4}$$

$$\iint_{R_3} x^2\, dx\, dy = \int_{-1}^{0}\int_{-1}^{x} x^2\, dy\, dx = \tfrac{1}{12}$$

$$\iint_{R_4} x^2\, dx\, dy = \int_{0}^{1}\int_{-x}^{0} x^2\, dy\, dx = \tfrac{1}{4}$$

so

$$\iint_{R} x^2\, dx\, dy = \tfrac{1}{12} + \tfrac{1}{4} + \tfrac{1}{12} + \tfrac{1}{4} = \tfrac{2}{3}$$

Check that formula (5) gives the same answers.

Solved Exercises

6. Evaluate $\int_0^1 \int_{x^3}^{x^2} xy\, dy\, dx$. Sketch the region for the corresponding double integral.

7. Write $\int_0^1 \int_0^{\sqrt{1-x^2}} \sqrt{1-y^2}\, dy\, dx$ as an integral over a region. Sketch the region and show that it is of types 1 and 2. Reverse the order of integration and evaluate.

8. Compute the volume of the solid in space bounded by the planes $x = 0$ and $y = 0$, the plane $3x + 4y = 10$, and the graph of $z = x^2 + y^2$.

Exercises

7. Evaluate the following integrals. Sketch and identify the corresponding regions by type.
   (a) $\int_0^\pi \int_{\sin x}^{3\sin x} x(1 + y)\, dy\, dx$
   (b) $\int_0^1 \int_{x-1}^{x\cos 2\pi x} (x^2 + xy + 1)\, dy\, dx$
   (c) $\int_{-1}^1 \int_{y^{2/3}}^{(2-y)^2} (y\sqrt{x} + y^3 - 2y)\, dx\, dy$

8. Interchange the order of integration, sketch, and evaluate: $\int_0^1 \int_x^1 xy\, dy\, dx$.

9. Find the volume of the region bounded by the plane $x = 0$, the surface $x = -4y^2 + 3$, and the surface $z = x^3 y$.

APPLICATIONS OF THE DOUBLE INTEGRAL

We have observed that if $f(x, y) \geq 0$ on D, then $\iint_D f(x, y)\, dx\, dy$ represents the volume of the region R defined by (x, y) in D, $0 \leq z \leq f(x, y)$. An "infinitesimal argument" for this result goes as follows. Consider R to be made of "infinitesimal rectangular prisms" with base dx by dy and height $f(x, y)$ (see Fig. 17-1-16 at the top of the next page). The total volume is obtained by integrating (that is, "summing") the volumes of these cylinders.

If $f(x, y)$ is identically equal to 1, the volume of the region under the graph is just the area of D, so we have the result that *the area of D is equal to* $\iint_D dx\, dy$. Applying the method of iterated integration to evaluate this integral just leads us back to the formulas of Section 4-3 for the area of the region between two graphs. We can also use the method of upper and lower sums to estimate the area of a region (see Problem 23).

Fig. 17-1-16 The region under the graph of *f* on *D* may be thought of as being composed of infinitesimal rectangular prisms.

By reasoning similar to that for one-variable calculus (see Section 10-2), we are led to the following definition: If *f* is an integrable function on *D*, the ratio of the integral to the area of *D*,

$$\frac{\iint_D f(x, y)\, dx\, dy}{\iint_D dx\, dy}$$

is called the *average value* of *f* on *D* (see Solved Exercise 9).

Double integration also allows us to find the center of mass of a plate with variable density: applying the consolidation principle (see Section 10-2) to the infinitesimal rectangles, one can derive the formulas

$$\bar{x} = \frac{\iint_D x\, \rho(x, y)\, dx\, dy}{\iint_D \rho(x, y)\, dx\, dy} \quad \text{and} \quad \bar{y} = \frac{\iint_D y\, \rho(x, y)\, dx\, dy}{\iint_D \rho(x, y)\, dx\, dy}$$

for the coordinates of the center of mass. (If *D* represents a plate with variable density $\rho(x, y)$, we can imagine breaking *D* into infinitesimal elements with mass $\rho(x, y)\, dx\, dy$; the total mass is thus $\iint_D \rho(x, y)\, dx\, dy$, as we stated in Solved Exercise 2.)

Worked Example 9 Find the center of mass of the rectangle $[0, 1] \times [0, 1]$ if the density is e^{x+y}.

Solution First we compute the mass:

$$\iint_D e^{x+y}\, dx\, dy = \int_0^1 \int_0^1 e^{x+y}\, dx\, dy = \int_0^1 \left(e^{x+y} \Big|_{x=0}^1 \right) dy$$

$$= \int_0^1 (e^{1+y} - e^y)\, dy$$

$$= e^{1+y} - e^y \Big|_{y=0}^1 = e^2 - e - (e - 1) = e^2 - 2e + 1$$

The numerator in the formula for $\bar{x}$ is

$$\int_0^1 \int_0^1 xe^{x+y}\, dx\, dy = \int_0^1 (xe^{x+y} - e^{x+y}) \Big|_{x=0}^{1} dy$$

$$= \int_0^1 [e^{1+y} - e^{1+y} - (0e^y - e^y)]\, dy$$

$$= \int_0^1 e^y\, dy = e^y \Big|_{y=0}^{1} = e - 1$$

So

$$\bar{x} = \frac{e-1}{e^2 - 2e + 1} = \frac{e-1}{(e-1)^2} = \frac{1}{e-1} \approx 0.582$$

The roles of x and y may be interchanged in all these calculations, so $\bar{y} = 1/(e - 1) \approx 0.582$ as well.

In Section 11-2 we used ordinary integration to determine the area of surfaces of revolution. Using the double integral, we can find the area of general curved surfaces. We confine ourselves here to an infinitesimal argument—the rigorous theory of surface area is quite subtle.*

To find the area of the graph $z = f(x, y)$ of a function f over the plane region D, we divide D into "infinitesimal rectangles" of the form $[x, x + dx] \times [y, y + dy]$. The image of this infinitesimal rectangle on the graph of f is approximately an "infinitesimal parallelogram" with vertices at

$$P_1 = (x, y, f(x, y))$$

$$P_2 = (x + dx, y, f(x + dx, y)) \approx (x + dx, y, f(x, y) + f_x(x, y)dx)$$

$$P_3 = (x, y + dy, f(x, y + dy)) \approx (x, y + dy, f(x, y) + f_y(x, y)dy)$$

$$P_4 = (x + dx, y + dy, f(x + dx, y + dy))$$
$$\approx (x + dx, y + dy, f(x, y) + f_x(x, y)dx + f_y(x, y)dy)$$

(See Fig. 17-1-17 at the top of the next page.)

We compute the area dA of this parallelogram by taking the length of the cross-product of the vectors from P_1 to P_2 and from P_1 to P_3 (see p. 621). The vectors in question are $dx\, \mathbf{i} + f_x(x, y)\, dx\, \mathbf{k}$ and $dy\, \mathbf{j} + f_y(x, y)dy\, \mathbf{k}$; the cross-product is

$$\begin{vmatrix} dx & 0 & f_x(x,y)dx \\ 0 & dy & f_y(x,y)dy \\ \mathbf{i} & \mathbf{j} & \mathbf{k} \end{vmatrix} = -f_x(x,y)dx\, dy\, \mathbf{i} - f_y(x,y)dx\, dy\, \mathbf{j} + dx\, dy\, \mathbf{k}$$

and the length of this vector is $dA = \sqrt{1 + f_x(x, y)^2 + f_y(x, y)^2}\, dx\, dy$.

*See T. Radó, Length and Area, *American Mathematical Society Colloquium Publications, Volume 30, 1958.*

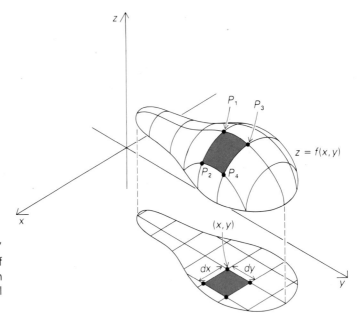

Fig. 17-1-17 The "image" on the surface $z = f(x, y)$ of an infinitesimal rectangle in the plane is the infinitesimal parallelogram $P_1 P_2 P_4 P_3$.

To get the area of the surface, we "sum" the areas of the infinitesimal parallelograms by integrating over D:

$$\text{Area} = \iint_D dA = \iint_D \sqrt{1 + f_x(x, y)^2 + f_y(x, y)^2} \, dx \, dy$$

$$= \iint_D \sqrt{1 + \left(\frac{\partial z}{\partial x}\right)^2 + \left(\frac{\partial z}{\partial y}\right)^2} \, dx \, dy$$

Note the similarity of this expression with the formula for the arc length of a graph on p. 461. As with arc length, the presence of the square root makes the evaluation of surface area integrals difficult or even impossible in all but a few accidentally simple cases.

Worked Example 10   Find the surface area of the part of the sphere $x^2 + y^2 + z^2 = 1$ lying above the ellipse $x^2 + (y^2/a^2) \leq 1$; $0 < a \leq 1$.

Solution   The region described by $x^2 + (y^2/a^2) \leq 1$ is type 1 with $\phi_1(x) = -a\sqrt{1 - x^2}$ and $\phi_2(x) = a\sqrt{1 - x^2}$; $-1 \leq x \leq 1$. The upper hemisphere may be described by the equation $z = f(x, y) = \sqrt{1 - x^2 - y^2}$. The partial derivatives of f are $\partial z/\partial x = -x/\sqrt{1 - x^2 - y^2}$ and $\partial z/\partial y = -y/\sqrt{1 - x^2 - y^2}$, so the area integrand is

$$\sqrt{1 + \frac{x^2}{1 - x^2 - y^2} + \frac{y^2}{1 - x^2 - y^2}} = \frac{1}{\sqrt{1 - x^2 - y^2}}$$

and the area is

$$A = \iint_D \frac{dx\,dy}{\sqrt{1-x^2-y^2}} = \int_{-1}^{1}\left(\int_{-a\sqrt{1-x^2}}^{a\sqrt{1-x^2}} \frac{dy}{\sqrt{1-x^2-y^2}}\right) dx$$

$$= \int_{-1}^{1}\left(\sin^{-1}\frac{y}{\sqrt{1-x^2}}\,\bigg|_{-a\sqrt{1-x^2}}^{a\sqrt{1-x^2}}\right) dx$$

$$= 2\int_{-1}^{1}\sin^{-1}a\,dx = 4\sin^{-1}a$$

(See Fig. 17-1-18.) As a check on our answer, note that if $a = 1$, we get $4\sin^{-1}1 = 4\cdot\pi/2 = 2\pi$, the correct formula for the area of a hemisphere of radius 1.

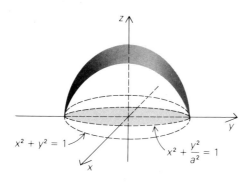

Fig. 17-1-18 The area of the hemisphere above the ellipse $x^2 + y^2/a^2 \leq 1$ is $4\sin^{-1}a$.

Solved Exercises

9. Find the average value of $f(x,y) = x\sin^2(xy)$ on $D = [0,\pi] \times [0,\pi]$.

10. The equation in (x,y,z) space of the surface obtained by revolving the graph $y = f(x)$ about the x axis is $y^2 + z^2 = f(x)^2$. Express as a double integral the area of the part of this surface lying between the planes $x = a$ and $x = b$. Carry out the integration over y. Do you recognize the remaining integral over x?

Exercises

10. Find the average value of each of the following functions on the given region:

   (a) $f(x,y) = y\sin xy$; $D = [0,\pi] \times [0,\pi]$

   (b) $f(x,y) = x^2 + y^2$; $D =$ the ring between the circles $x^2 + y^2 = \frac{1}{2}$ and $x^2 + y^2 = 1$

11. Find the center of mass of the bounded region between $y = x^2$ and $y = x$ if the density is $x + y$.

12. Find the area of the graph of $f(x,y) = \frac{2}{3}(x^{3/2} + y^{3/2})$ over the domain $D = [0,1] \times [0,1]$.

Problems for Section 17-1 ▮▮▮▮▮▮▮▮▮▮▮▮▮▮▮▮▮▮▮▮▮▮▮

1. Compute $\iint_D f(x, y)\, dx\, dy$ for:
   (a) $f(x, y) = xy^3 e^{x^2 y^2}$, where $D = [1, 3] \times [1, 2]$.
   (b) $f(x, y) = xy + x/(y + 1)$, where $D = [1, 4] \times [1, 2]$.
   (c) $f(x, y) = y^5 \sin x\; e^{y^3 \cos x}$; $D = [0, 1] \times [-1, 0]$.

2. Compute the following. Sketch and identify the region by type.
   (a) $\int_2^4 \int_{y^2-1}^{y^3} 3\, dx\, dy$
   (b) $\int_0^1 \int_{x^2}^x (x + y)^2\, dy\, dx$
   (c) $\int_0^1 \int_0^{3y} e^{x+y}\, dx\, dy$

3. Find $\iint_D f(x, y)\, dx\, dy$ for:
   (a) $f(x, y) = x - y$; D the triangle with vertices $(0, 0)$, $(1, 0)$, and $(2, 1)$.
   (b) $f(x, y) = x^3 y + \cos x$; D the triangle defined by $0 \le x \le \pi/2$, $0 \le y \le x$.
   (c) $f(x, y) = (x^2 + 2xy^2 + 2)$; D the region bounded by the graph of $y = -x^2 + x$, the x axis, and the lines $x = 0$ and $x = 2$.

4. Compute the volume beneath the graph of $f(x, y) = 1 + \sin(\pi y/2) + x$ on the parallelogram in the xy plane with vertices $(0, 0)$, $(1, 2)$, $(2, 0)$, $(3, 2)$. Sketch.

5. Compute the volume beneath the graph of $f(x, y) = 4x^2 + 3y^2 + 27$ on the disk of radius 2 centered at $(0, 1)$. Sketch.

6. Compute the volume beneath the graph of $f(x, y) = (\cos y)e^{1-\cos 2x} + xy$ on the region bounded by the line $y = 2x$, the x axis, and the line $x = \pi/2$.

7. Compute the volume between the graphs of the functions
   $$f(x, y) = 2x + 1 \quad \text{and} \quad g(x, y) = -x - 3y - 6$$
   on the region bounded by the y axis and the curve $x = 4 - y^2$. Sketch.

8. Sketch the regions involved in each of the following integrations and evaluate:
   (a) $\displaystyle\int_0^2 \int_{-3(\sqrt{4-x^2})/2}^{3(\sqrt{4-x^2})/2} \left(\frac{5}{\sqrt{2 + x}} + y^3 \right) dy\, dx$
   (b) $\int_0^1 \int_0^{x^2} (x^2 + xy - y^2)\, dy\, dx$

9. In each of the following, interchange the order of integration, sketch, and evaluate:
   (a) $\int_0^{\pi/2} \int_0^{\cos \theta} \cos \theta\, dr\, d\theta$
   (b) $\int_0^1 \int_{1-y}^1 (x + y)^2\, dx\, dy$

10. Find the center of mass of the region in Worked Example 7 if the density is $143x$.

11. A sculptured gold plate D is defined by $0 \le x \le 2\pi$ and $0 \le y \le \pi$ (centimeters) and has mass density $\rho(x, y) = y^2 \sin^2 4x + 2$ (grams per square centimeter). If gold sells for \$7 per gram, how much is the gold in the plate worth?

12. Find the volume under the graph of $f(x, y)$ between the planes $x = a$, $x = b$, $y = c$, and $y = d$, where:
    (a) $f(x, y) = x \sin y + 3$; $a = 0$, $b = 2$, $c = \pi$, $d = 3\pi$.
    (b) $f(x, y) = xy^5 + 2x^4 y + 6$; $a = 0$, $b = 1$, $c = -1$, $d = 1$.

13. Find the volume under the graph of $f(x, y)$ between the planes $x = a$, $x = b$, $y = c$, and $y = d$, where:
    (a) $f(x, y) = 2x^4 + 3y^{2/3}$; $a = -1$, $b = 1$, $c = 0$, $d = 2$.
    (b) $f(x, y) = xy\sqrt{x^2 + 3y^2}$; $a = 1$, $b = 2$, $c = 1$, $d = 2$.

14. Find the average value of each of the following functions over the indicated region:
    (a) $f(x, y) = e^{x+y}$; $D =$ the triangle with vertices at $(0, 0)$, $(0, 1)$, and $(1, 0)$.
    (b) $f(x, y) = 1/(x + y)$; $D = [e, e^2] \times [e, e^2]$.

15. Find the average value of $x^2 + y^2$ over each of the following regions:
    (a) The square $[0, 1] \times [0, 1]$.
    (b) The square $[a, a + 1] \times [0, 1]$, where $a > 0$.
    (c) The square $[0, a] \times [0, a]$, where $a > 0$.
    (d) The set of (x, y) such that $x^2 + y^2 < a^2$.

★16. (a) Relate the integrand in the surface area formula to the angle between $\mathbf{k}$ and the normal to the surface $z = f(x, y)$.
     (b) Express the ratio of the area of the graph of f over D to the area of D as the average value of some geometrically defined quantity.

★17. Express as a double integral the volume enclosed by the surface of revolution in Solved Exercise 10. Carry out the integration over y and show that the resulting integral over x is a formula in Section 10-1.

18. Show that if a plate D has constant density, then the average values of x and y on D are the coordinates of the center of mass.

19. Find the center of mass of the region bounded by $y = x^3$ and $y = \sqrt[3]{x}$ if the density is $(x - y)^2$. Try to minimize your work by exploiting some symmetry in the problem.

20. Find the center of mass of the disk $(x - 1)^2 + y^2 \leq 1$ if the density is: (a) x^2; (b) y^2.

21. Find the area cut out of the cylinder $x^2 + z^2 = 1$ by the cylinder $x^2 + y^2 = 1$.

22. Find the area of the part of the cone $z^2 = x^2 + y^2$ lying in the region of space defined by $x \geq 0$, $y \geq 0$, $z \leq 1$.

23. Let D be the region defined by $x^2 + y^2 \leq 1$.

    (a) Estimate $\iint_D dx\ dy$ (the area of D) within 0.1 by taking a rectangular grid in the plane and counting the number of rectangles: (i) contained in D (lower sum); (ii) intersecting D (upper sum).

    (b) Compute $\iint_D dx\ dy$ exactly by using an iterated integral.

24. (a) Prove that the area on a sphere of radius r cut out by a cone of angle ϕ is $2\pi r^2 (1 - \cos \phi)$ (Fig. 17-1-19).

Fig. 17-1-19 The area of the cap is $2\pi r^2 (1 - \cos \phi)$.

    (b) A sphere of radius 1 sits with its center on the surface of a sphere of radius $r > 1$. Show that the area of surface on the second sphere cut out by the first sphere is π. Does something about this result surprise you?

★25. Let n right circular cylinders of radius r intersect such that their axes lie in a plane, meeting at one point with equal angles. Find the volume of their intersection.

26. Which states in the United States are regions of type 1? Type 2? (Take x = longitude, y = latitude.)

27. A uniform rectangular steel plate of sides a and b rotates about its center of gravity with constant angular velocity ω.

    (a) Kinetic energy equals $\frac{1}{2}$ (mass) (velocity)^2. Argue that the kinetic energy of an element of mass $\rho dx dy$ (ρ = constant) is $\rho(\omega^2/2)(x^2 + y^2)\ dx\ dy$, provided the origin $(0, 0)$ is placed at the center of gravity of the plate.

    (b) Justify the formula for kinetic energy:

$$\text{K.E.} = \iint_{\text{plate}} \rho \frac{\omega^2}{2} (x^2 + y^2)\ dx\ dy$$

    (c) Evaluate the integral, assuming the plate is described by $-a/2 \leq x \leq a/2$, $-b/2 \leq y \leq b/2$.

28. A membrane in the xy-plane is subjected to a disturbance, resulting in a z-deflection given by the double series

$$z = \sum_{m=1}^{\infty} \sum_{n=1}^{\infty} B_{mn} \sin (m\pi x/a) \sin (n\pi y/b) \cos (\alpha_{mn} t)$$

    The coefficients B_{mn} are given by

$$B_{mn} = \frac{4}{ab} \int_0^b \int_0^a f(x, y) \sin (m\pi x/a) \sin (n\pi y/b)\, dx\, dy$$

    where $f(x, y) = h(ax - x^2)(by - y^2)$, with h, a, b = constants. Integrate to obtain

$$B_{mn} = \frac{16a^2 b^2 h}{\pi^6 m^3 n^3} (1 - \cos (m\pi))(1 - \cos (n\pi))$$

17-2 Triple Integrals

Integrals over regions in three-dimensional physical space require the triple integral.

The basic ideas developed in Section 17-1 for double integrals can be readily extended to triple integrals. As with double integrals, one of the most powerful methods for evaluating these integrals is reduction to iterated integrals. A second important method, which we discuss at the end of the section, is the method of changing variables. For functions of one variable this is one of our key integration methods. For double and triple integrals this method is particularly useful in connection with polar, cylindrical, or spherical coordinates.

Goals

After studying this section, you should be able to:

Express triple integrals over boxes and general regions as iterated integrals, and evaluate these integrals.

Use triple integration to find volumes, average values, and centers of mass.

Simplify double and triple integrals in problems with symmetry by converting them to polar, cylindrical, or spherical coordinates.

THE TRIPLE INTEGRAL OVER A BOX

If the temperature inside an oven is not uniform, determining the average temperature involves "summing" the values of the temperature function at all points in the solid region enclosed by the oven walls. Such a sum is expressed mathematically as a triple integral.

We formalize the ideas in a manner analogous to double integrals. Suppose that W is a box (that is, rectangular parallelepiped) in space bounded by the planes $x = a$, $x = b$, $y = c$, $y = d$, and $z = p$, $z = q$, as in Fig. 17-2-1. We denote this box by $[a, b] \times [c, d] \times [p, q]$. Let $f(x, y, z)$ be a function defined for (x, y, z) in W—that is, for

$$a \le x \le b \qquad c \le y \le d \qquad p \le z \le q$$

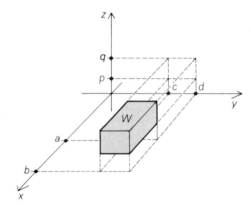

Fig. 17-2-1 The box $W = [a, b] \times [c, d] \times [p, q]$ consists of points (x, y, z) satisfying $a \le x \le b$, $c \le y \le d$, and $p \le z \le q$.

In order to define the *triple integral*

$$\iiint_W f(x, y, z)\, dx\, dy\, dz$$

we first define the concept of a piecewise constant function.

Definition A function $g(x, y, z)$ defined on $[a, b] \times [c, d] \times [p, q]$ is called *piecewise constant* if there are partitions

$$a = t_0 < t_1 < \cdots < t_n = b \text{ of } [a, b]$$

$$c = s_0 < s_1 < \cdots < s_m = d \text{ of } [c, d]$$

$$p = r_0 < r_1 < \cdots < r_l = q \text{ of } [p, q]$$

such that $g(x, y, z)$ has the constant value k_{ijk} for (x, y, z) in the open box

$$W_{ijk} = (t_{i-1}, t_i) \times (s_{j-1}, s_j) \times (r_{k-1}, r_k)$$

We cannot draw the graphs of functions of three variables; however, we can indicate the value k_{ijk} associated with each box. See Fig. 17-2-2.

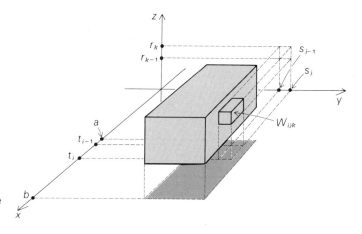

Fig. 17-2-2 *g* has the value k_{ijk} on the small box W_{ijk}.

The integral of g is defined as a sum of *nml* terms:

$$\iiint_W g(x, y, z) \, dx \, dy \, dz = \sum_{i=1, j=1, k=1}^{n, m, l} k_{ijk} (\text{volume } W_{ijk})$$

$$= \sum_{i=1, j=1, k=1}^{n, m, l} k_{ijk} (\Delta t_i)(\Delta s_i)(\Delta r_i)$$

If f is any function on W, its lower [upper] sums are defined, as before, as the integrals of piecewise constant functions g [h] such that $g(x, y, z) \leq f(x, y, z)$ [$h(x, y, z) \geq f(x, y, z)$] for all points (x, y, z) in W.

Definition We say that f is *integrable* on W if there is a number S_0 such that every $S < S_0$ is a lower sum for f on W and every $S > S_0$ is an upper sum. The number S_0 is called the *integral* of f over W and is denoted by

$$\iiint_W f(x, y, z) \, dx \, dy \, dz$$

At this point you should look back at the basic properties of double integrals listed on p. 756. Similar properties hold for triple integrals. Furthermore, there is a similar reduction to iterated integrals, which we state formally.

Theorem 2 *Assume that $f(x, y, z)$ is integrable on the box $W = [a, b] \times [c, d] \times [p, q]$. Then any iterated integral which exists is equal to the triple integral; that is,*

$$\iiint_W f(x, y, z)\, dx\, dy\, dz = \int_p^q \int_c^d \int_a^b f(x, y, z)\, dx\, dy\, dz$$

$$= \int_p^q \int_a^b \int_c^d f(x, y, z)\, dy\, dx\, dz$$

$$= \int_a^b \int_p^q \int_c^d f(x, y, z)\, dy\, dz\, dx$$

and so on. (There are six possible orders altogether.)

The proof of Theorem 2 is just like that of Theorem 1 in Section 17-1, so we omit it.

Worked Example 1 Let W be the box $[0, 1] \times [-\frac{1}{2}, 0] \times [0, \frac{1}{3}]$. Evaluate

$$\iiint_W (x + 2y + 3z)^2\, dx\, dy\, dz$$

Solution According to Theorem 2, this integral may be evaluated as the iterated integral

$$\int_0^{1/3} \int_{-1/2}^0 \int_0^1 (x + 2y + 3z)^2\, dx\, dy\, dz$$

$$= \int_0^{1/3} \int_{-1/2}^0 \left[\frac{(x + 2y + 3z)^3}{3} \bigg|_{x=0}^{1} \right] dy\, dz$$

$$= \int_0^{1/3} \int_{-1/2}^0 \tfrac{1}{3}\left[(1 + 2y + 3z)^3 - (2y + 3z)^3 \right] dy\, dz$$

$$= \int_0^{1/3} \frac{1}{24}\left[(1 + 2y + 3z)^4 - (2y + 3z)^4 \right] \bigg|_{y=-1/2}^{0} dz$$

$$= \int_0^{1/3} \frac{1}{24}\left[(3z + 1)^4 - 2(3z)^4 + (3z - 1)^4 \right] dz$$

$$= \frac{1}{24 \cdot 15}\left[(3z + 1)^5 - 2(3z)^5 + (3z - 1)^5 \right] \bigg|_{z=0}^{1/3}$$

$$= \frac{1}{24 \cdot 15}(2^5 - 2) = \frac{1}{12}$$

Triple integrals afford even more opportunities for error than double integrals. Fortunately, you have five ways to check any answer.

TRIPLE AND ITERATED INTEGRALS

To evaluate the triple integral of $f(x, y, z)$ over a box W: $W = [a, b] \times [c, d] \times [p, q]$,

evaluate the iterated integral of $f(x, y, z)$: the variables may be integrated in any of six orders:

$$\iiint_W f(x, y, z)\, dx\, dy\, dz = \int_p^q \int_c^d \int_a^b f(x, y, z)\, dx\, dy\, dz$$

$$= \int_c^d \int_a^b \int_p^q f(x, y, z)\, dz\, dx\, dy$$

and so forth.

Solved Exercises

1. Verify that we get the same answer in Worked Example 1 if the integration is done in one of the other orders—say, y first, then z, and then x.

2. If the mass density (kilograms per cubic meter) of a box W is given by $\rho(x, y, z)$, what does the integral of ρ over W represent?

3. Evaluate the integral of e^{x+y+z} over the box $[0, 1] \times [0, 1] \times [0, 1]$.

Exercises

1. Evaluate $\iiint_W (2x + 3y + z)\, dx\, dy\, dz$, where $W = [1, 2] \times [-1, 1] \times [0, 1]$, in at least two ways.

2. The cube $[1, 2] \times [1, 2] \times [1, 2]$ has mass density $\rho(x, y, z) = (1 + x)e^z y$. Find the mass of the box.

3. Write out the property for triple integrals corresponding to property 2 of double integrals on p. 756.

THE TRIPLE INTEGRAL OVER GENERAL REGIONS

As in the two-variable case, we define the integral of a function f over a bounded region W by defining a new function f^*, equal to f on W and zero outside W, and then setting

$$\iiint_W f(x, y, z)\, dx\, dy\, dz = \iiint_{W^*} f^*(x, y, z)\, dx\, dy\, dz$$

where W^* is any box containing the region W.

As before, we restrict our attention to particularly simple regions. A three-dimensional region W will be said to be of *type I* if there is an elementary region D in the xy plane and a pair of continuous functions, $\gamma_1(x, y)$ and $\gamma_2(x, y)$ defined on D, such that W consists of those triples (x, y, z) for which $(x, y) \in D$ and $\gamma_1(x, y) \leq z \leq \gamma_2(x, y)$. The region D may itself be of type 1 or type 2, so there are two possible descriptions of a type I region:

$$a \le x \le b, \quad \phi_1(x) \le y \le \phi_2(x)$$
$$\gamma_1(x, y) \le z \le \gamma_2(x, y) \quad \text{(if } D \text{ is of type 1)} \tag{1}$$

or

$$c \le y \le d, \quad \psi_1(y) \le x \le \psi_2(y)$$
$$\gamma_1(x, y) \le z \le \gamma_2(x, y) \quad \text{(if } D \text{ is of type 2)} \tag{2}$$

Figure 17-2-3 shows two regions of type I that are described by conditions (1) and (2), respectively.

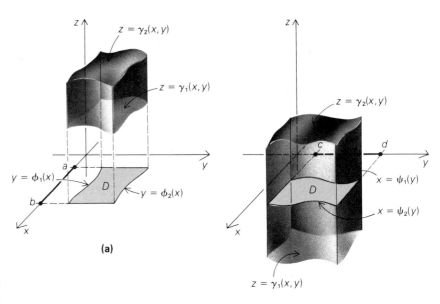

(a)

Fig. 17-2-3 A region of type I lies between two graphs $z = \gamma_1(x, y)$ and $z = \gamma_2(x, y)$.

(b)

A region W is of *type II* if it can be expressed in form (1) or (2) with the roles of x and z interchanged, and W is of *type III* if it can be expressed in form (1) or (2) with y and z interchanged. See Fig. 17-2-4.

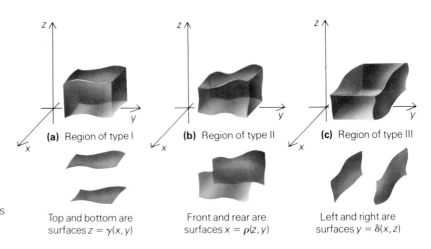

(a) Region of type I

(b) Region of type II

(c) Region of type III

Fig. 17-2-4 The three types of regions in space.

Top and bottom are surfaces $z = \gamma(x, y)$

Front and rear are surfaces $x = \rho(z, y)$

Left and right are surfaces $y = \delta(x, z)$

Notice that a given region may be of two or even three types at once. (See Fig. 17-2-5.) As with regions in the plane, we call a region of type I, II, or III in space an *elementary region.*

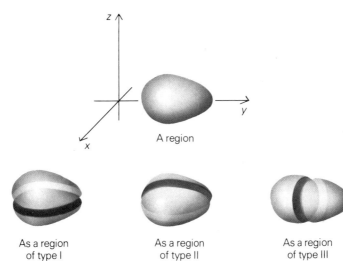

A region

As a region
of type I

As a region
of type II

As a region
of type III

Fig. 17-2-5 Regions in space can be of more than one type. This one is of all three types.

Worked Example 2   Show that the unit ball $x^2 + y^2 + z^2 \le 1$ is a region of all three types.

Solution   As a type I region, we can express it as

$$-1 \le x \le 1, \qquad \sqrt{1 - x^2} \le y \le \sqrt{1 - x^2}$$
$$-\sqrt{1 - x^2 - y^2} \le z \le \sqrt{1 - x^2 - y^2}$$

In doing this we first write the top and bottom hemispheres as $z = \sqrt{1 - x^2 - y^2}$ and $z = -\sqrt{1 - x^2 + y^2}$, where x and y vary over the unit disk (that is, $-\sqrt{1 - x^2} \le y \le \sqrt{1 - x^2}$ and x varies between -1 and 1). (See Fig. 17-2-6.) We write the region as a type II or III region in a similar manner by interchanging the roles of x, y, and z in the defining inequalities.

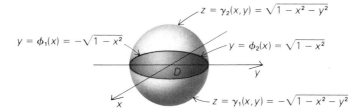

$z = \gamma_2(x, y) = \sqrt{1 - x^2 - y^2}$

$y = \phi_1(x) = -\sqrt{1 - x^2}$

$y = \phi_2(x) = \sqrt{1 - x^2}$

$z = \gamma_1(x, y) = -\sqrt{1 - x^2 - y^2}$

Fig. 17-2-6 The unit ball described as a region of type I.

As with integrals in the plane, any function of three variables which is continuous over an elementary region is integrable on that region. An argument like that on p. 765 shows that a triple integral over an elementary region can be rewritten as an iterated integral in which the limits of integration are functions. The formulas for such iterated integrals are given in the following display.

TRIPLE INTEGRALS

Suppose that W is type I. Then either

$$\iiint_W f(x, y, z)\, dx\, dy\, dz = \int_a^b \int_{\phi_1(x)}^{\phi_2(x)} \int_{\gamma_1(x,y)}^{\gamma_2(x,y)} f(x, y, z)\, dz\, dy\, dx \tag{3}$$

(see Fig. 17-2-3a) or

$$\iiint_W f(x, y, z)\, dx\, dy\, dz = \int_c^d \int_{\psi_1(y)}^{\psi_2(y)} \int_{\gamma_1(x,y)}^{\gamma_2(x,y)} f(x, y, z)\, dz\, dx\, dy \tag{4}$$

(see Fig. 17-2-3b).

For regions of type II or III interchange x and z or y and z (see Solved Exercise 4).

Another way to write formula (3) is

$$\iiint_W f(x, y, z)\, dx\, dy\, dz = \iint_D \left[\int_{\gamma_1(x,y)}^{\gamma_2(x,y)} f(x, y, z)\, dz \right] dy\, dx$$

and for formula (4),

$$\iiint_W f(x, y, z)\, dx\, dy\, dz = \iint_D \left[\int_{\gamma_1(x,y)}^{\gamma_2(x,y)} f(x, y, z)\, dz \right] dx\, dy$$

Notice that the triple integral $\iiint_W dx\, dy\, dz$ is simply the volume of W.

Worked Example 3 Verify the formula for the volume of a ball of radius 1: $\iiint_W dx\, dy\, dz = \frac{4}{3}\pi$, where W is the set of (x, y, z) with $x^2 + y^2 + z^2 \le 1$.

Solution As explained in Worked Example 2, the ball is a region of type I. By formula (3), the integral is

$$\int_{-1}^1 \int_{-\sqrt{1-x^2}}^{\sqrt{1-x^2}} \int_{-\sqrt{1-x^2-y^2}}^{\sqrt{1-x^2-y^2}} dz\, dy\, dx$$

Holding y and x fixed and integrating with respect to z yields

$$\int_{-1}^1 \int_{-\sqrt{1-x^2}}^{\sqrt{1-x^2}} \left[z \Big|_{-\sqrt{1-x^2-y^2}}^{\sqrt{1-x^2-y^2}} \right] dy\, dx$$

$$= 2 \int_{-1}^1 \left[\int_{-\sqrt{1-x^2}}^{\sqrt{1-x^2}} (1 - x^2 - y^2)^{1/2}\, dy \right] dx$$

Since x is fixed in the integral over y, this integral can be expressed as $\int_{-a}^a (a^2 - y^2)^{1/2}\, dy$, where $a = (1 - x^2)^{1/2}$. This integral represents the area of a semicircular region of radius a, so that

$$\int_{-a}^{a} (a^2 - y^2)^{1/2}\, dy = \frac{a^2}{2}\,\pi$$

(We could also have evaluated this integral by using a trigonometric substitution.) Thus

$$\int_{-\sqrt{1-x^2}}^{\sqrt{1-x^2}} (1 - x^2 - y^2)\, dy = \frac{1 - x^2}{2}\,\pi$$

and so

$$2 \int_{-1}^{1} \int_{-\sqrt{1-x^2}}^{\sqrt{1-x^2}} (1 - x^2 - y^2)^{1/2}\, dy\, dx = 2 \int_{-1}^{1} \pi\, \frac{1 - x^2}{2}\, dx$$

$$= \int_{-1}^{1} \pi(1 - x^2)\, dx = \pi\left(x - \frac{x^3}{3}\right)\Bigg|_{x=-1}^{1} = \frac{4}{3}\,\pi$$

Worked Example 4 Let W be the region bounded by the planes $x = 0$, $y = 0$, $z = 2$ and the surface $z = x^2 + y^2$. Compute $\int\int\int_W x\, dx\, dy\, dz$ and sketch the region.

Solution The region W is sketched in Fig. 17-2-7. We may write this as a region of type I with $\gamma_1(x, y) = x^2 + y^2$, $\gamma_2(x, y) = 2$, $\phi_1(x) = 0$, $\phi_2(x) = \sqrt{2 - x^2}$, $a = 0$, and $b = \sqrt{2}$. By formula (3),

$$\int\int\int_W x\, dx\, dy\, dz = \int_0^{\sqrt{2}} \left[\int_0^{\sqrt{2-x^2}} \left(\int_{x^2+y^2}^{2} x\, dz \right) dy \right] dx$$

$$= \int_0^{\sqrt{2}} \int_0^{\sqrt{2-x^2}} x(2 - x^2 - y^2)\, dy\, dx$$

$$= \int_0^{\sqrt{2}} x \left[(2 - x^2)^{3/2} - \frac{(2 - x^2)^{3/2}}{3} \right] dx$$

$$= \int_0^{\sqrt{2}} \frac{2x}{3} (2 - x^2)^{3/2}\, dx = \frac{-2(2 - x^2)^{5/2}}{15} \Bigg|_0^{\sqrt{2}}$$

$$= 2 \cdot \frac{2^{5/2}}{15} = \frac{8\sqrt{2}}{15}$$

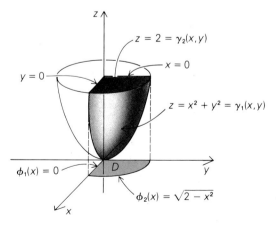

Fig. 17-2-7 W is the region below the plane $z = 2$, above the paraboloid $z = x^2 + y^2$, and on the positive sides of the planes $x = 0$, $y = 0$.

Solved Exercises   4. Write general formulas analogous to (3) and (4) for the triple integral over a region of type II.

5. Do Worked Example 4 by writing W as a region of type II.

6. Evaluate $\int_0^1 \int_0^x \int_{x^2+y^2}^1 dz\,dy\,dx$. Sketch the region W of integration and interpret your answer.

Exercises   4. Write general formulas analogous to (3) and (4) for the triple integral over a region of type III.

5. Do Worked Example 3 by writing W as a region of type III.

6. Evaluate $\int_0^2 \int_0^x \int_0^{x+y} dz\,dy\,dx$. Sketch the region of integration.

7. Find the volume of the region bounded by $z = x^2 + y^2$ and $z = 10 - x^2 - 2y^2$. Sketch.

APPLICATIONS OF TRIPLE INTEGRALS

Some of the applications in Section 17-1 carry over directly from double to triple integrals. We can compute the volume, mass, and center of mass of a region with variable density $\rho(x, y, z)$ by the formulas

$$\left.\begin{array}{ll} \text{Volume} & = \iiint_W dx\,dy\,dz \\[2em] \text{Mass} & = \iiint_W \rho(x, y, z)\,dx\,dy\,dz \end{array}\right\} \tag{5}$$

Center of mass $= (\bar{x}, \bar{y}, \bar{z})$, where

$$\left.\begin{array}{l} \bar{x} = \dfrac{\iiint_W x\rho(x, y, z)\,dx\,dy\,dz}{\text{mass}} \\[2em] \bar{y} = \dfrac{\iiint_W y\rho(x, y, z)\,dx\,dy\,dz}{\text{mass}} \\[2em] \bar{z} = \dfrac{\iiint_W z\rho(x, y, z)\,dx\,dy\,dz}{\text{mass}} \end{array}\right\} \tag{6}$$

The average value of a function f on a region W is defined by

$$\frac{\iiint_W f(x, y, z)\,dx\,dy\,dz}{\iiint_W dx\,dy\,dz} \tag{7}$$

Worked Example 5 Find the center of mass of the hemispherical region W defined by $x^2 + y^2 + z^2 \le 1$, $z \ge 0$. (Assume that the density is constant.)

Solution By symmetry, the center of mass must lie on the z axis, so $\bar{x} = \bar{y} = 0$. To find $\bar{z}$, we must compute, by formula (6), $I = \iiint_W z\, dx\, dy\, dz$. The hemisphere is of types I, II, and III; we will consider it to be of type III. Then the integral I becomes

$$I = \int_0^1 \int_{-\sqrt{1-z^2}}^{\sqrt{1-z^2}} \int_{-\sqrt{1-y^2-z^2}}^{\sqrt{1-y^2-z^2}} z\, dx\, dy\, dz$$

Since z is a constant for the x and y integrations, we can bring it out to obtain

$$I = \int_0^1 z \left(\int_{-\sqrt{1-z^2}}^{\sqrt{1-z^2}} \int_{-\sqrt{1-y^2-z^2}}^{\sqrt{1-y^2-z^2}} dx\, dy \right) dz$$

Instead of calculating the inner two integrals explicitly, we observe that they are simply the double integral $\iint_D dx\, dy$ over the disk $x^2 + y^2 \le 1 - z^2$, considered as a type II region. The area of this disk is $\pi(1 - z^2)$, so

$$I = \pi \int_0^1 z(1 - z^2)\, dz = \pi \int_0^1 (z - z^3)\, dz = \pi \left[\frac{z^2}{2} - \frac{z^4}{4} \right]_0^1 = \frac{\pi}{4}$$

The volume of the hemisphere is $\frac{2}{3}\pi$, by Worked Example 3, so $\bar{z} = (\pi/4)/\frac{2}{3}\pi = \frac{3}{8}$.

The formulas for volume and surface area derived in Section 17-1, when extended from two to three variables, will represent "hypervolume" and "hyperarea" of objects in four-dimensional space. Geometric objects in spaces of four or more dimensions are of interest both from a purely mathematical standpoint and for applications (as in statistical mechanics[*]), but a discussion of such objects here would take us too far afield.

An interesting physical application of triple integration is the determination of the gravitational fields of solid objects. Solved Exercise 13, p. 712, showed that the gravitational force field $\mathbf{F}(x, y, z)$ is the negative of the gradient of a function $V(x, y, z)$ called the *gravitational potential*. If there is a point of mass m at (x, y, z), then the gravitational potential at (x_1, y_1, z_1) due to this mass is $Gm\left[(x - x_1)^2 + (y - y_1)^2 + (z - z_1)^2\right]^{-1/2}$, where G is the universal gravitational constant (again, see Solved Exercise 13, p. 712). If our attracting object is an extended domain W with density $\rho(x, y, z)$, we may think of it as made of infinitesimal box-shaped masses $dm = \rho(x, y, z)\, dx\, dy\, dz$ located at (x, y, z). The total gravitational potential for W is then obtained by "summing" the potentials from the infinitesimal masses—that is, as a triple integral:

$$V(x_1, y_1, z_1) = G \iiint_W \frac{\rho(x, y, z)\, dx\, dy\, dz}{\sqrt{(x - x_1)^2 + (y - y_1)^2 + (z - z_1)^2}} \tag{8}$$

(See Fig. 17-2-8 at the top of the next page.)

[*]*See, for example, A. Sommerfeld,* Thermodynamics and Statistical Mechanics, *Academic Press, 1964, Chapter 4.*

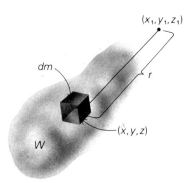

Fig. 17-2-8 The gravitational potential at (x_1, y_1, z_1) arising from the mass $dm = \rho(x, y, z)\, dx\, dy\, dz$ at (x, y, z) is $[G\rho(x, y, z)\, dx\, dy\, dz]/r$.

The evaluation of the integral for the gravitational potential is usually quite difficult. The few examples which can be carried out completely require the use of cylindrical or spherical coordinate systems, which we introduce in the next subsection.

Historical note. Newton withheld publication of his gravitational theories for quite some time, until he could prove that a spherical planet has the same gravitational field that it would have if its mass were all concentrated at the planet's center. Using multiple integrals and spherical coordinates, we can solve Newton's problem quite easily (see Solved Exercise 11 and Problem 14); Newton's published solution used only euclidean geometry.

Solved Exercises

7. The temperature at points in the cube $W = [-1, 1] \times [-1, 1] \times [-1, 1]$ is proportional to the square of the distance from the origin.

   (a) What is the average temperature?

   (b) At which points of the cube is the temperature equal to the average temperature?

8. Let W be a region of constant density and total mass M. Show that the gravitational potential is given by

$$V(x_1, y_1, z_1) = \overline{\left(\frac{1}{r}\right)} GM$$

   where $\overline{(1/r)}$ is the average over W of

$$f(x, y, z) = \frac{1}{\sqrt{(x - x_1)^2 + (y - y_1)^2 + (z - z_1)^2}}$$

Exercises

8. Evaluate the integral in Worked Example 5 by considering the hemisphere as a region of type I.

9. Redo Solved Exercise 7 for the cube $W = [-c, c] \times [-c, c] \times [-c, c]$. [*Hint:* Guess the answer to part (b) first.]

10. (a) Find the center of mass of the box $[0, \frac{1}{2}] \times [0, 1] \times [0, 2]$, assuming the density to be uniform.

    (b) Same problem as part (a), but with a mass density $\rho(x, y, z) = x^2 + 3y^2 + z + 1$.

INTEGRALS IN POLAR, CYLINDRICAL, AND SPHERICAL COORDINATES

The evaluation of many multiple integrals can be greatly simplified by the use of polar, cylindrical, or spherical coordinates. Our arguments here will be intuitive; proofs are omitted.

We deal first with polar coordinates. Recall that a double integral

$$\iint_D f(x, y)\, dx\, dy$$

may be thought of as a "sum of" the values of f over infinitesimal rectangles with area $(dx) \cdot (dy)$. As in Section 11-3, however, we can also describe a region in polar coordinates and can use infinitesimal rectangles appropriate to those coordinates. The infinitesimal area now is $r\, dr\, d\theta$ as is evident from Fig. 17-2-9. If $u = f(x, y)$, then u may be expressed in terms of r and θ by the formula $u = f(r \cos \theta, r \sin \theta)$.

Fig. 17-2-9 The area of the infinitesimal shaded region is $r\, dr\, d\theta$.

Therefore we obtain:

INTEGRALS IN POLAR COORDINATES

$$\iint_D f(x, y)\, dx\, dy = \iint_{D'} f(r \cos \theta, r \sin \theta)\, r\, dr\, d\theta \tag{9}$$

where D' is the region corresponding to D in the variables r and θ.

Worked Example 6 Evaluate $\iint_{D_a} e^{-(x^2 + y^2)}\, dx\, dy$, where D_a is the disk $x^2 + y^2 \le a^2$.

Solution The presence of $r^2 = x^2 + y^2$ in the integrand and the disk suggests a change to polar coordinates. The disk is described by $0 \le r \le a$, $0 \le \theta \le 2\pi$, so by formula (9) we get

$$\iint_{D_a} e^{-(x^2 + y^2)}\, dx\, dy = \int_0^{2\pi} \int_0^a e^{-r^2}\, r\, dr\, d\theta$$

$$= \int_0^{2\pi} \left(-\tfrac{1}{2} e^{-r^2} \right) \Bigg|_0^a d\theta$$

$$= \int_0^{2\pi} -\tfrac{1}{2}(e^{-a^2} - 1)\, d\theta$$

$$= \pi(1 - e^{-a^2})$$

There is no direct way to evaluate this integral in (x, y) coordinates!

There is a remarkable application of the result of Worked Example 6 to single-variable calculus: we will evaluate the gaussian integral $\int_{-\infty}^{\infty} e^{-x^2}\,dx$, which is of basic importance in probability theory and quantum mechanics. There is no known way to evaluate this integral directly. If we bring in two-variable calculus, however, the solution is surprisingly simple. Letting a go to ∞ in the formula $\iint_{D_a} e^{-(x^2+y^2)}\,dx\,dy = \pi(1 - e^{-a^2})$, we find that the limit $L = \lim_{a\to\infty} \iint_{D_a} e^{-(x^2+y^2)}\,dx\,dy$ exists and equals π. By analogy with the definition of improper integrals on the line, we may consider L as the (improper) integral of $e^{-(x^2+y^2)}$ over the entire plane, since the disks D_a exhaust the plane as $a \to \infty$. On the other hand, the rectangles $R_a = [-a, a] \times [-a, a]$ exhaust the plane, too, so we must have

$$\lim_{a\to\infty} \iint_{R_a} e^{-(x^2+y^2)}\,dx\,dy = \pi$$

as well. (See Solved Exercise 12 for the details of the proof.) But

$$
\begin{aligned}
\iint_{R_a} e^{-(x^2+y^2)}\,dx\,dy &= \int_{-a}^{a}\int_{-a}^{a} e^{-(x^2+y^2)}\,dx\,dy \\
&= \int_{-a}^{a}\int_{-a}^{a} e^{-x^2}e^{-y^2}\,dx\,dy \\
&= \left(\int_{-a}^{a} e^{-x^2}\,dx\right)\left(\int_{-a}^{a} e^{-y^2}\,dy\right) \\
&= I_a^2
\end{aligned}
$$

where $I_a = \int_{-a}^{a} e^{-x^2}\,dx$. Thus

$$
\begin{aligned}
\int_{-\infty}^{\infty} e^{-x^2}\,dx &= \lim_{a\to\infty} I_a = \sqrt{\lim_{a\to\infty} I_a^2} = \sqrt{\lim_{a\to\infty}\iint_{R_a} e^{-(x^2+y^2)}\,dx\,dy} \\
&= \sqrt{\pi}
\end{aligned}
$$

THE GAUSSIAN INTEGRAL

$$\int_{-\infty}^{\infty} e^{-x^2}\,dx = \sqrt{\pi}$$

We will now evaluate triple integrals in cylindrical and spherical coordinates. At this point you should review the basic features of these coordinates as discussed in Section 15-2.

Cylindrical coordinates consist of polar coordinates in the xy plane, together with the z coordinate. Therefore the infinitesimal "volume element" has volume $r\,dr\,d\theta\,dz$. See Fig. 17-2-10. Hence we conclude the following:

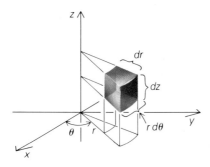

Fig. 17-2-10 The infinites-imal shaded region has vol-ume $r\,dr\,d\theta\,dz$.

INTEGRALS IN CYLINDRICAL COORDINATES

$$\iiint_W f(x, y, z)\,dx\,dy\,dz = \iiint_{W'} f(r\cos\theta, r\sin\theta, z)\,r\,dr\,d\theta\,dz \qquad \textbf{(10)}$$

where W' is the region in r, θ, z coordinates corresponding to W.

Worked Example 7 Evaluate $\iiint_W (z^2 x^2 + z^2 y^2)\,dx\,dy\,dz$, where W is the cylindrical region $x^2 + y^2 \le 1$, $-1 \le z \le 1$.

Solution By formula (10) we have

$$\begin{aligned}
\iiint_W (z^2 x^2 + z^2 y^2)\,dx\,dy\,dz &= \int_{-1}^{1} \int_0^{2\pi} \int_0^1 (z^2 r^2)\,r\,dr\,d\theta\,dz \\
&= \int_{-1}^{1} \int_0^{2\pi} z^2 \frac{r^4}{4}\bigg|_0^1 \,d\theta\,dz \\
&= \int_{-1}^{1} \frac{2\pi}{4} z^2\,dz = \frac{\pi}{3}
\end{aligned}$$

Finally, we turn to spherical coordinates. The volume in space corresponding to infinitesimal changes $d\rho$, $d\theta$, and $d\phi$ is shown in Fig. 17-2-11.

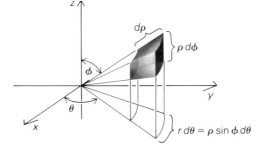

Fig. 17-2-11 The infinites-imal shaded region has vol-ume $\rho^2 \sin\phi\,d\rho\,d\theta\,d\phi$.

The sides of this "box" have lengths $d\rho$, $r\,d\theta = \rho\sin\phi\,d\theta$, and $\rho\,d\phi$ as shown. Therefore its volume is $\rho^2 \sin\phi\,d\rho\,d\theta\,d\phi$. Hence we get the following:

INTEGRALS IN SPHERICAL COORDINATES

$$\iiint_W f(x, y, z)\, dx\, dy\, dz = \iiint_W f(\rho \sin\phi \cos\theta, \rho \sin\phi \sin\theta, \rho \cos\phi)\, \rho^2 \sin\phi\, d\rho\, d\theta\, d\phi \qquad \text{(11)}$$

where W is the region in ρ, θ, ϕ space corresponding to W.

Worked Example 8   Find the volume of the ball $x^2 + y^2 + z^2 \leq R^2$ by using spherical coordinates.

Solution   The ball is described in spherical coordinates by $0 \leq \theta \leq 2\pi$, $0 \leq \phi \leq \pi$, and $0 \leq \rho \leq R$. Therefore, by formula (11),

$$\iiint_D dx\, dy\, dz = \int_0^\pi \int_0^{2\pi} \int_0^R \rho^2 \sin\phi\, d\rho\, d\theta\, d\phi = \frac{R^3}{3} \int_0^\pi \int_0^{2\pi} \sin\phi\, d\theta\, d\phi$$

$$= \frac{2\pi R^3}{3} \int_0^\pi \sin\phi\, d\phi = \frac{2\pi R^3}{3} \{-[\cos(\pi) - \cos(0)]\} = \frac{4\pi R^3}{3}$$

which is the familiar formula for the volume of a ball. Compare the effort involved with Worked Example 3.

Solved Exercises

9. Evaluate $\iint_D \ln(x^2 + y^2)\, dx\, dy$, where D is the region in the first quadrant lying between the circles $x^2 + y^2 = 1$ and $x^2 + y^2 = 4$.

10. Evaluate $\iiint_W \exp[(x^2 + y^2 + z^2)^{3/2}]\, dx\, dy\, dz$, where W is the ball $x^2 + y^2 + z^2 \leq 1$.

11. Let W be the region between the concentric spheres $\rho = \rho_1$ and $\rho = \rho_2$. Find the gravitational potential function $V(x_1, y_1, z_1)$ given by formula (8). Assume that the density is constant.

★12. Refer to Worked Example 6. Prove, without appeal to improper integrals, that

$$\lim_{a \to \infty} \iint_{D_a} e^{-(x^2+y^2)}\, dx\, dy = \lim_{a \to \infty} \iint_{R_a} e^{-(x^2+y^2)}\, dx\, dy$$

Exercises

11. Evaluate $\iint_D (x^2 + y^2)^{3/2}\, dx\, dy$, where D is the disk $x^2 + y^2 \leq 4$.

12. Evaluate $\iiint_W (dx\, dy\, dz)/\sqrt{1 + x^2 + y^2 + z^2}$, where W is the ball $x^2 + y^2 + z^2 \leq 1$.

13. Find the volume of the region bounded by
$$x^2 + y^2 + z^2 = 1 \quad \text{and} \quad x^2 + y^2 = \tfrac{1}{4}$$

14. Find the mass of the solid bounded by the cylinder $x^2 + y^2 = 2x$ and the cone $z^2 = x^2 + y^2$ if the density is $\rho = \sqrt{x^2 + y^2}$.

Problems for Section 17-2

1. Evaluate the following integrals:

   (a) $\iiint_W x^2\, dx\, dy\, dz$; $W = [0, 1] \times [-1, 1] \times [0, 1]$.

   (b) $\iiint_W x^2 \cos z\, dx\, dy\, dz$; W is the region bounded by $z = 0$, $z = \pi$, $y = 0$, $y = 1$, $x = 0$, and $x + y = 1$.

   (c) $\iint_D (x^2 + y^2)^{5/2}\, dx\, dy$; D is the disk $x^2 + y^2 \le 1$.

2. Evaluate the following integrals:

   (a) $\iiint_W (x^2 + y^2 + z^2)^{5/2}\, dx\, dy\, dz$; W is the ball $x^2 + y^2 + z^2 \le 1$.

   (b) $\iiint_R (x^2 + y^2 + z^2)\, dx\, dy\, dz$; R is the region bounded by $x + y + z = a$ (where $a > 0$), $x = 0$, $y = 0$, and $z = 0$.

   (c) $\iiint_W z \sin(x^2 + y^2)\, dx\, dy\, dz$; W is the region bounded by the planes $x = 0$, $y = 0$, $z = 0$, $z = 1$, and the cylinder $x^2 + y^2 = 1$, with $x \ge 0$, $y \ge 0$.

3. Find the volume of the following regions:

   (a) The solid bounded by $x^2 + 2y^2 = 2$, $z = 0$, and $x + y + 2z = 2$.

   (b) The solid bounded by $x^2 + y^2 + z^2 = 1$ and $z^2 \ge x^2 + y^2$.

   (c) The region common to the intersecting cylinders $x^2 + y^2 \le a^2$ and $x^2 + z^2 \le a^2$.

4. Find the volumes of the following regions:

   (a) The solid bounded by the circular cylinder $r = 2a \cos \theta$, the cone $z = r$, and the plane $z = 0$.

   (b) The solid bounded by the cone $z = \sqrt{x^2 + y^2}$ and the paraboloid $z = x^2 + y^2$.

   (c) The region bounded by $y = x^2$, $y = x + 2$, $4z = x^2 + y^2$, and $z = x + 3$.

5. (a) Sketch the region for the integral

$$\int_0^1 \int_0^x \int_0^y f(x, y, z)\, dz\, dy\, dx$$

   (b) Write the integral with the integration order $dx\, dy\, dz$.

6. (a) Show that the triple integral of a product over a box is the product of three ordinary integrals; that is, if $D = [a, b] \times [c, d] \times [p, q]$, then

$$\iiint_D f(x)g(y)h(z)\, dx\, dy\, dz$$
$$= \int_a^b f(x)\, dx \int_c^d g(y)\, dy \int_p^q h(z)\, dz$$

   (b) Use the result of part (a) to do Solved Exercise 3.

7. If $f(x, y, z) = F(x, y)$ for some function F—that is, if $f(x, y, z)$ does not depend on z—what is the triple integral of f over a box W?

8. Evaluate

$$\iiint_S \frac{dx\, dy\, dz}{(x^2 + y^2 + z^2)^{3/2}}$$

   where S is the solid bounded by the spheres $x^2 + y^2 + z^2 = a^2$ and $x^2 + y^2 + z^2 = b^2$, where $a > b > 0$.

9. Show that the formula using triple integrals for the volume under the graph of a function $f(x, y)$, on an elementary region D in the plane, reduces to the double integral of f over D.

★10. Prove that

$$\int_0^x \left[\int_0^t F(u)\, du \right] dt = \int_0^x (x - u) F(u)\, du$$

   [*Hint:* Interchange the order of integration.]

11. Find the center of mass of the following solid S with constant density ρ:

   (a) S is bounded by $x + y + z = 2$, $x = 0$, $y = 0$, $z = 0$.

   (b) S is bounded by the parabolic cylinder $z = 4 - x^2$ and the planes $x = 0$, $y = 0$, $y = 6$, $z = 0$.

12. The *moment of inertia* about the x axis of a solid S with density ρ is defined by

$$I_x = \iiint_S \rho(y^2 + z^2)\, dx\, dy\, dz$$

   Similarly,

$$I_y = \iiint_S \rho(x^2 + z^2)\, dx\, dy\, dz$$

$$I_z = \iiint_S \rho(x^2 + y^2)\, dx\, dy\, dz$$

   For each of the following solids, set up an integral for I_z; assume that the density is a constant.

   (a) The solid above the xy plane, bounded by the paraboloid $z = x^2 + y^2$ and the cylinder $x^2 + y^2 = a^2$.

   (b) The solid bounded above by the plane $z = a$ and below by the cone described in spherical coordinates by $\phi = k$ where k is a constant $0 < k < \pi/2$.

★13. Find the gravitational potential in Solved Exercise 11 when $\rho_1 < R < \rho_2$. [*Hint:* Break the integral over ρ into two parts.]

★14. Use an infinitesimal argument and Solved Exercise 11 to show that the gravitational potential of a spherically symmetric body whose density can be a function of the radius is given, exterior to the body, by $V(x_1, y_1, z_1) = GM/R$, where M is the mass of the body and $R = \sqrt{x_1^2 + y_1^2 + z_1^2}$ is the distance to the center of the body.

★15. The *general change of variables formula in two dimensions* reads

$$\iint_D f(x, y)\, dx\, dy$$
$$= \iint_{D^*} h(u, v) \left| \frac{\partial(x, y)}{\partial(u, v)} \right| du\, dv$$

where $h(u, v) = f(x(u, v),\ y(u, v))$ and $|\partial(x, y)/\partial(u, v)|$ is the determinant

$$\begin{vmatrix} \dfrac{\partial x}{\partial u} & \dfrac{\partial x}{\partial v} \\[2ex] \dfrac{\partial y}{\partial u} & \dfrac{\partial y}{\partial v} \end{vmatrix}$$

Here $x(u, v)$ and $y(u, v)$ are the functions relating the variables (u, v) to the variables (x, y).

   (a) Show that this formula is plausible by using the geometric interpretation of derivatives and determinants.

   (b) Show that the formula reduces to our earlier one for polar coordinates.

   (c) Write down the general three-dimensional formula and show that it reduces to our earlier ones for cylindrical and spherical coordinates.

★16. (a) By using the change of variables formula in Problem 15 and $u = x + y$, $y = uv$, show that

$$\int_{x=0}^{1} \int_{y=0}^{1-x} e^{y/(x+y)}\, dy\, dx = \frac{e - 1}{2}$$

Also graph the region in the xy plane and uv plane.

   (b) Let D be the region bounded by $x + y = 1$, $x = 0$, $y = 0$. Show that

$$\iint_D \cos\left(\frac{x - y}{x + y}\right) dx\, dy = \frac{\sin 1}{2}$$

and graph D on an xy plane and a uv plane, with $u = x - y$ and $v = x + y$.

17. The *flexural rigidity EI* of a uniform beam is the product of its Young's modulus of elasticity E and the moment of inertia I of the cross section of the beam at x with respect to a horizontal line ℓ passing through the center of gravity of this cross section. Here

$$I = \iint_R D^2(x, y)\, dx\, dy,$$

where $D(x, y) =$ the distance from (x, y) to ℓ, $R =$ the cross section of the beam being considered.

   (a) Assume the cross section R is the rectangle $-1 \le x \le 1$, $-2 \le y \le 2$ and ℓ is the x-axis. Find I.

   (b) Assume the cross section R is a circle of radius 4, and ℓ is the x-axis. Find I, using polar coordinates.

18. A body W in xyz-coordinates is *symmetrical with respect to a plane* if for every particle on one side of the plane there is a particle of equal mass located at its mirror image through the plane.

   (a) Discuss the planes of symmetry for the body of an automobile.

   (b) Let the plane of symmetry be the xy-plane, and denote by $W+$ and $W-$ the portions of W above and below the plane, respectively. By assumption, the mass density $\rho(x, y, z)$ satisfies $\rho(x, y, -z) = \rho(x, y, z)$. Justify these steps:

$$\bar{z} \cdot \iiint_W \rho(x, y, z)\, dx\, dy\, dz$$
$$= \iiint_W z\rho(x, y, z)\, dx\, dy\, dz$$
$$= \iiint_{W+} z\rho(x, y, z)\, dx\, dy\, dz$$
$$+ \iiint_{W-} z\rho(x, y, z)\, dx\, dy\, dz$$
$$= \iiint_{W+} z\rho(x, y, z)\, dx\, dy\, dz$$
$$+ \iiint_{W+} -w\rho(u, v, -w)\, du\, dv\, dw$$
$$= 0$$

   (c) Explain why part (b) proves that if a body is symmetrical with respect to a plane, then its center of mass lies in that plane.

   (d) Derive this law of mechanics:
   "If a body is symmetrical in two planes, then its center of mass lies on their line of intersection."

19. If a body is composed of two or more parts whose centers of mass are known, then the center of mass of the composite body can be computed by regarding the component parts as single particles located at their respective centers of mass. Apply this consolidation principle below.

    (a) Find the center of mass of an aluminum block of density ρ (= constant), of base 4 cm $\times$ 6 cm, height 10 cm, with a hole drilled

through. The cylinder removed is 2 cm in diameter and 6 cm long with its axis of symmetry 8 cm above the base and symmetrically placed.

    (b) Repeat for a solid formed by pouring epoxy into a hemispherical form, of radius 20 cm, which contains a balloon of diameter 8 cm placed at the center of the circular base.

Review Problems for Chapter 17

1. Evaluate the following integrals:

    (a) $\displaystyle\int_2^3 \int_4^8 [x^3 + \sin(x + y)]\, dx\, dy$

    (b) $\displaystyle\iint_D [x^3 + \sin(x + y)]\, dx\, dy$;
    D = rectangle $[1, 2] \times [-3, 2]$

    (c) $\displaystyle\int_1^2 \int_2^3 \int_3^4 (x + y + z)\, dx\, dy\, dz$

2. Evaluate:

    (a) $\displaystyle\iiint_W [e^x + (y + z)^5]\, dx\, dy\, dz$;
    W = cube $[0, 1] \times [0, 1] \times [0, 1]$.

    (b) $\displaystyle\iint_D (x^3 + y^2 x)\, dx\, dy$; D = region under the graph of $y = x^2$ from $x = 0$ to $x = 2$.

    (c) $\displaystyle\iiint_W (x^2 + y^2 + z^2)\, dx\, dy\, dz$; W = solid hemisphere $x^2 + y^2 + z^2 \le 1$, $z \ge 0$.

3. Find the volume of each of the following solids:

    (a) The region bounded by the planes $x = 0$, $x = 1$, $y = 0$, $y = 2$, $z = 0$, and the paraboloid $z = x^2 + y^2$.

    (b) The cone defined by $(x - \frac{1}{2}z)^2 + y^2 \le 4z^2$, $0 \le z \le 3$.

    (c) The ellipsoid $x^2/a^2 + y^2/b^2 + z^2/c^2 \le 1$.

    (d) The intersection of a ball B of radius 1 with a ball of radius $\frac{1}{2}$ whose center is on the boundary of B.

4. Sketch each of the following solids and find its volume:

    (a) The spherical sector $x^2 + y^2 + z^2 \le 1$, $z \ge 0$, $x^2 + y^2 \le az^2$.

    (b) The region between the graphs of $f(x, y) = \cos^2(y + x)$ and $g(x, y) = -\sin^2(y + x)$ on the domain $D = [0, 1] \times [0, 1]$.

    (c) The "partially eaten ice cream cone" $x^2 + y^2 \le \frac{1}{5}z^2$, $0 \le z$, $x^2 + y^2 + (z - 5)^2 \ge 5$.

    (d) The region below the plane $z + y = 1$ and inside the cylinder $x^2 + y^2 \le 1$, $0 \le z \le 1$.

★5. Justify the formula

$$\iint_D f(x, y)\, dx\, dy = \int_a^b \left(\int_{\phi_1(x)}^{\phi_2(x)} f(x, y)\, dy \right) dx$$

for an integral over a region of type 1 by means of Cavalieri's principle.

6. The *tetrahedron* defined by $x \ge 0$, $y \ge 0$, $z \ge 0$, $x + y + z \le 1$ is to be sliced into n segments of equal volume by planes parallel to the plane $x + y + z = 1$. Where should the slices be made?

7. (a) Evaluate $\iint_D \sec(x^2 + y^2)\, dx\, dy$, where D is the region defined by $x^2 + y^2 \le 1$.

    (b) Evaluate $\iiint_W (x^2 + 8yz)\, dx\, dy\, dz$, where W is the region bounded by the surfaces $z = x^2 - y^2$, $z = 0$, $y = \pm 1$, and $x = 0, 4$.

8. Sketch and find the volume under the graph of f between the planes $x = a$, $x = b$, $y = c$, and $y = d$ for each of the following:

    (a) $f(x, y) = x^2 + \sin(2y) + 1$; $a = -3$, $b = 1$, $c = 0$, $d = \pi$.

    (b) $f(x, y) = 10 - x^2 - y^2$; $a = -2$, $b = 2$, $c = -1$, $d = 1$.

9. Find the average value of $f(x, y)$ on $D = [a, b] \times [c, d]$ for the functions and regions given in Problem 8.

10. Show that the volume obtained by cutting the rectangular cylinder $a \le x \le b$, $c \le y \le d$ by the planes $z = 0$ and $z = px + qy + r$ is equal to the area of the base times the average of the heights of the four vertical edges. Assume that $px + qy + r \ge 0$ for (x, y) in the rectangle $[a, b] \times [c, d]$. (See Fig. 17-R-1.)

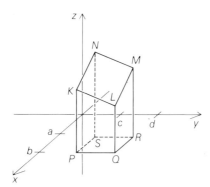

Fig. 17-R-1   The volume of the solid *PQRS KLMN* is equal to the area of the base *PQRS* times the average of the lengths of *PK, QL, RM,* and *SN*.

11. (a) Interpret the result in Problem 10 as a fact about the average value of a linear function on a rectangle.

    ★(b) What is the average value of a linear function on a parallelogram?

★12. Write $x = r \cos \theta$, $y = r \sin \theta$, and

$$dx = \frac{\partial x}{\partial r} dr + \frac{\partial x}{\partial \theta} d\theta \qquad dy = \frac{\partial y}{\partial r} dr + \frac{\partial y}{\partial \theta} d\theta$$

as is suggested by the chain rule. If you formally multiply $dx \cdot dy$, do you get the right area element $r \, dr \, d\theta$? What should the products $dr \, dr$, $d\theta \, dr$, and $d\theta \, d\theta$ be?

13. Use cylindrical coordinates to find the center of mass of the "dish": $x^2 + y^2 \le 1$, $x^2 + y^2 + (z - 2)^2 \le 25/4$, $z \le 0$.

14. Find the surface area of the graph of each of the following functions over the unit disk $x^2 + y^2 \le 1$:

    (a) $f(x, y) = xy$ (b) $f(x, y) = x^2 + y^2$

    (c) $f(x, y) = x^2 - y^2$

    [*Hint:* Use polar coordinates.]

15. Show that the area of the surface of the sphere $x^2 + y^2 + z^2 = 1$ lying above the rectangle $[-a, a] \times [-a, a]$ $(2a^2 < 1)$ is

$$A = 2 \int_{-a}^{a} \sin^{-1}\left(\frac{a}{\sqrt{1 - x^2}}\right) dx$$

16. As is well known, the density of a typical planet is not constant throughout the planet. Assume that planet I.K.U. has a mass density (in grams per cubic centimeter)

$$\rho(x, y, z) = \begin{cases} \dfrac{3 \times 10^4}{r} & r \ge 10^4 \text{ cm,} \\[2mm] 3 & r \le 10^4 \text{ cm} \end{cases}$$

where $r = \sqrt{x^2 + y^2 + z^2}$

and a radius of 5×10^8 centimeters. Find a formula for the gravitational attraction outside I.K.U.

★17. If the world were two-dimensional,* the laws of physics would predict that the gravitational potential of a mass point is proportional to the logarithm of the distance from the point. Using polar coordinates, find the gravitational potential of a disk of constant density.

18. Evaluate:

    (a) $\int_0^\pi \int_0^{\pi/2} \int_0^{\sin \phi} \rho^2 \sin \phi \, d\rho \, d\phi \, d\theta$

    (b) $\int_0^1 \int_0^{\pi/2} \int_0^{\cos \theta} r \, z \, dr \, d\theta \, dz$

    (c) $\displaystyle\iint_D \frac{dx \, dy}{x^2 + y^2}$, where D is the region in the first quadrant bounded by the circles $x^2 + y^2 = 1$ and $x^2 + y^2 = 2$

19. Derive the following three laws (see Problem 18, Section 17-2):

    (a) If a body is symmetrical about an axis, then its center of mass lies on that axis.

    (b) If a body is symmetrical in three planes with a common point, then that point is its center of mass.

    (c) If a body has spherical symmetry about a point (that is, if the density depends only on the distance from that point), then that point is its center of mass.

---

*See Flatland by Edwin A. Abbot (Barnes and Noble, 1963) for an amusing description of such a world.

★20. There is an interesting application of double integrals to the problem of differentiation under the integral sign.*

Theorem   Let $D = [a, b] \times [c, d]$ and let $g(x, y)$ be continuous, with $g_x(x, y)$ continuous on D. Then

$$\frac{d}{dx} \int_c^d g(x, y)\, dy = \int_c^d g_x(x, y)\, dy$$

$a < x < b$

      The idea of the proof is this: By the fundamental theorem of calculus,

$$\int_c^d g(x, y)\, dy$$
$$= \int_c^d \left[\int_a^x g_x(s, y)\, ds + g(a, y) \right] dy$$

Interchanging the order of integration, this equals

$$\int_a^x \left[\int_c^d g_x(s, y)\, dy \right] ds + \int_c^d g(a, y)\, dy$$

---

*See "Fubini Implies Leibniz Implies $F_{yx} = F_{xy}$" by R. T. Seeley, American Mathematical Monthly 68(1961):56–57.

We can show that $\int_c^d g_x(s, y)\, dy$ is a continuous function of s, so we can use the fundamental theorem of calculus (alternative version) to get

$$\frac{d}{dx} \int_c^d g(x, y)\, dy$$
$$= \frac{d}{dx} \int_a^x \left[\int_c^d g_x(s, y)\, dy \right] ds$$
$$+ \frac{d}{dx} \left[\int_c^d g(a, y)\, dy \right]$$
$$= \int_c^d g_x(x, y)\, dy + 0$$

as asserted.

(a) Verify by direct integration that

$$\frac{d}{dx} \int_0^{\pi/2} \sin(xy)\, dy = \int_0^{\pi/2} \frac{\partial}{\partial x} \sin(xy)\, dy$$

(b) Show that $F(k) = \int_0^{\pi/2} dx / \sqrt{1 - k \cos^2 x}$, for $0 \le k < 1$, is an increasing function of k. How fast is it increasing at $k = 0$?

★21. Use the discussion of the theorem in Problem 20 to show that interchanging the order of integration allows one to prove that the mixed partials of a function are equal.

18

Vector Analysis

The fundamental theorem of calculus is extended from the line to the plane and to space.

The major theorems of vector analysis relate double and triple integrals to integrals over curves and surfaces. These results have their origins in problems of fluid flow and electromagnetic theory. Thus, it is not surprising that they are terribly important for physics as well as for their own mathematical beauty. Enough applications are given so the student can appreciate the physical meaning of the various terms occurring in the theorems.

CONTENTS

Green's and Stokes' theorems involve the concept of a line integral. This concept is developed in Section 18-1 and applied to force problems. Stokes' and Gauss' theorems involve the concept of a surface integral. This concept is defined in Section 18-2 through the motivation of fluid flow. These ideas and double and triple integrals are then brought together in the theorems of Green, Stokes, and Gauss. The student who has not studied triple integrals can omit Gauss' theorem.

18-1 Line Integrals

Vector fields may be integrated along curves to produce real numbers.

The integration of vector fields along curves is of fundamental importance in physics. We will use the concept of *work* to motivate the material in this section. In the next section we establish Green's and Stokes' theorems, which relate line integrals, partial derivatives, and double integrals.

Before beginning this section, you should review the material on parametric curves in Section 15-3.

Goals

After studying this section, you should be able to:

Evaluate the line integral of a given vector field along a given path.

Find line integrals of conservative fields by using antiderivatives.

THE WORK DONE BY A FORCE ALONG A PATH

To introduce the idea of line integration, we resume the discussion of work and energy begun on p. 434. Using vector methods, we can now discuss motions in three dimensions instead of just one.

The motion of an object is described by a parametric curve—that is, by a vector function $\mathbf{r} = \sigma(t)$. By differentiating this function, we obtain (see Section 15-3):

$$\mathbf{v} = \sigma'(t) = \text{velocity at time } t$$

$$v = \|\mathbf{v}\| = \|\sigma'(t)\| = \text{speed at time } t$$

$$\mathbf{a} = \frac{d\mathbf{v}}{dt} = \sigma''(t) = \text{acceleration at time } t$$

If the mass of the object is m, the kinetic energy K is defined by

$$K = \tfrac{1}{2}mv^2 = \tfrac{1}{2}m\|\sigma'(t)\|^2$$

According to Newton's second law (see p. 672)

$$\mathbf{F} = m\mathbf{a} = m\frac{d^2\mathbf{r}}{dt^2} = m\sigma''(t)$$

where $\mathbf{F}$ is the total force acting on the object.

To investigate the relationship between force and kinetic energy, we differentiate the formula

$$K = mv^2 = \tfrac{1}{2}m\mathbf{v} \cdot \mathbf{v}$$

obtaining

$$\frac{dK}{dt} = \frac{1}{2} m\left(\frac{d\mathbf{v}}{dt}\cdot\mathbf{v} + \mathbf{v}\cdot\frac{d\mathbf{v}}{dt}\right) = m\frac{d\mathbf{v}}{dt}\cdot\mathbf{v} = m\mathbf{a}\cdot\mathbf{v}$$
$$= \mathbf{F}\cdot\mathbf{v}$$

The total change in kinetic energy from time t_1 to time t_2 is therefore the integral of dK/dt:

$$\Delta K = \int_{t_1}^{t_2}\frac{dK}{dt}\,dt = \int_{t_1}^{t_2}\mathbf{F}\cdot\mathbf{v}\,dt = \int_{t_1}^{t_2}\mathbf{F}\cdot\frac{d\mathbf{r}}{dt}\,dt$$

The integral

$$\int_{t_1}^{t_2}\mathbf{F}\cdot\frac{d\mathbf{r}}{dt}\,dt \tag{1}$$

denoted W is called the *work* done by the force $\mathbf{F}$ along the path $\mathbf{r} = \boldsymbol{\sigma}(t)$.

It is often possible to express the total force on an object as a sum of forces which are due to identifiable sources (such as gravity, friction, and fluid pressure). If $\mathbf{F}$ represents a force of a particular type, then the integral (1) is still called the work done by this particular force.

Let us suppose now that the force $\mathbf{F}$ at time t depends only on the position $\mathbf{r} = \boldsymbol{\sigma}(t)$. That is, we assume that there is a *vector field* $\boldsymbol{\Phi}(\mathbf{r})$ such that $\mathbf{F} = \boldsymbol{\Phi}(\boldsymbol{\sigma}(t))$. (Examples of such *position-dependent* forces are those caused by gravitational and electrostatic attraction; frictional and magnetic forces are *velocity-dependent*.) Then we may write integral (1) as

$$W = \int_{t_1}^{t_2}\boldsymbol{\Phi}(\boldsymbol{\sigma}(t))\cdot\boldsymbol{\sigma}'(t)\,dt \tag{2}$$

In the one-dimensional case discussed in Section 10-3, we were able to transform integral (2) into an integral over the position variable x; this calculation showed that the work of a given force, as a body moved from position x_1 to position x_2, depends only upon x_1 and x_2 and not upon the details of the motion.

To a certain extent, the same situation remains true for motion in space: the work done by a force field as a particle moves along a path does not depend upon how the particle moves along the path; if different paths are taken between the same endpoints, however, the work may be different.

Worked Example 1 Find the work done by the force field $\boldsymbol{\Phi}(x,y,z) = y\mathbf{i} - x\mathbf{j} + \mathbf{k}$ as a particle is moved from $(1,0,0)$ to $(1,0,1)$ along each of the following paths:
(a) $(x,y,z) = (\cos t, \sin t, t/2\pi)$; $0 \le t \le 2\pi$
(b) $(x,y,z) = (\cos t^3, \sin t^3, t^3/2\pi)$; $0 \le t \le \sqrt[3]{2\pi}$
(c) $(x,y,z) = (\cos t, -\sin t, t/2\pi)$; $0 \le t \le 2\pi$

Solution   The (helical) paths are sketched in Fig. 18-1-1.

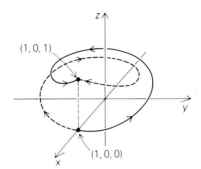

Fig. 18-1-1 Paths (a) and (b) follow the solid line; path (c) follows the dashed line.

(a) Applying integral (2), with $t_1 = 0$, $t_2 = 2\pi$, $\boldsymbol{\sigma}(t) = \cos t\,\mathbf{i} + \sin t\,\mathbf{j} + (t/2\pi)\mathbf{k}$, and $\boldsymbol{\sigma}'(t) = -\sin t\,\mathbf{i} + \cos t\,\mathbf{j} + (1/2\pi)\mathbf{k}$, we have

$$W_a = \int_0^{2\pi} (\sin t\,\mathbf{i} - \cos t\,\mathbf{j} + \mathbf{k})\cdot\left(-\sin t\,\mathbf{i} + \cos t\,\mathbf{j} + \frac{1}{2\pi}\mathbf{k}\right)dt$$

$$= \int_0^{2\pi}\left(-\sin^2 t - \cos^2 t + \frac{1}{2\pi}\right)dt$$

$$= 2\pi\left(-1 + \frac{1}{2\pi}\right) = -2\pi + 1 \approx -5.28$$

(b) This time,

$$\boldsymbol{\sigma}'(t) = -(\sin t^3)(3t^2)\mathbf{i} + (\cos t^3)(3t^2)\mathbf{j} + \frac{3t^2}{2\pi}\mathbf{k}$$

$$= 3t^2\left[-\sin t^3\,\mathbf{i} + \cos t^3\,\mathbf{j} + \frac{1}{2\pi}\mathbf{k}\right]$$

and

$$W_b = \int_0^{\sqrt[3]{2\pi}} (\sin t^3\,\mathbf{i} - \cos t^3\,\mathbf{j} + \mathbf{k})\cdot(3t^2)\left(-\sin t^3\,\mathbf{i} + \cos t^3\,\mathbf{j} + \frac{1}{2\pi}\mathbf{k}\right)dt$$

$$= \int_0^{\sqrt[3]{2\pi}}\left(-\sin^2 t^3 - \cos^2 t^3 + \frac{1}{2\pi}\right)3t^2\,dt$$

$$= \int_0^{\sqrt[3]{2\pi}}\left(-1 + \frac{1}{2\pi}\right)3t^2\,dt = \left(-1 + \frac{1}{2\pi}\right)t^3\bigg|_{t=0}^{\sqrt[3]{2\pi}}$$

$$= \left(-1 + \frac{1}{2\pi}\right)(2\pi) = 1 - 2\pi$$

just as in part (a).

(c) Here

$$\boldsymbol{\sigma}'(t) = -\sin t\,\mathbf{i} - \cos t\,\mathbf{j} + \frac{1}{2\pi}\mathbf{k}$$

so

$$W_c = \int_0^{2\pi} (-\sin t\,\mathbf{i} - \cos t\,\mathbf{j} + \mathbf{k}) \cdot \left(-\sin t\,\mathbf{i} - \cos t\,\mathbf{j} + \frac{1}{2\pi}\,\mathbf{k} \right) dt$$

$$= \int_0^{2\pi} \left(\sin^2 t + \cos^2 t + \frac{1}{2\pi} \right) dt$$

$$= 2\pi \left(1 + \frac{1}{2\pi} \right) = 2\pi + 1 \approx 7.28$$

In the case of path (c), the motion is "with the force," so the work is positive; for paths (a) and (b), the motion is "against the force" and the work is negative.

The equality of the work along paths (a) and (b) in Worked Example 1 is no accident: it is a consequence of the fact that the two paths are simply two different parametrizations of the same curve in space. In the next subsection, we will prove that the work along a path is independent of the parametrization; later on, we will try to characterize those force fields for which the work is independent of the path itself.

Solved Exercises*

1. Suppose that in the force field of Worked Example 1, you pick up a unit mass at $(1, 0, 0)$, carry it along path (a), and leave it at $(1, 0, 1)$. How much work have you done? [*Hint:* Compare Worked Example 2, p. 434.]

2. Consider the gravitational force field defined (for $(x, y, z) \neq (0, 0, 0)$) by

$$\mathbf{\Phi}(x, y, z) = -\frac{1}{(x^2 + y^2 + z^2)^{3/2}} (x\mathbf{i} + y\mathbf{j} + z\mathbf{k})$$

Show that the work done by the gravitational force as a particle moves from (x_1, y_1, z_1) to (x_2, y_2, z_2) depends only on the radii $R_1 = \sqrt{x_1^2 + y_1^2 + z_1^2}$ and $R_2 = \sqrt{x_2^2 + y_2^2 + z_2^2}$.

Exercises

1. Calculate the work done by the force field $\mathbf{\Phi}(x, y, z) = x\mathbf{i} + y\mathbf{j}$ when a particle is moved along the path $(3t^2, t, 1)$, $0 \leq t \leq 1$.

2. (a) Show that if a particle is moved along the closed curve $(\cos t, \sin t, 0)$, $0 \leq t \leq 2\pi$, then the force field in Worked Example 1 does a nonzero amount of work on the particle. How much is the work?

   (b) Show that if a particle is moved along any closed curve (that is, $\boldsymbol{\sigma}(t_1) = \boldsymbol{\sigma}(t_2)$), then the work done on it by the gravitational field in Solved Exercise 2 is zero.

3. Let $\mathbf{\Phi}(x, y) = [1/(x^2 + y^2)] (-y\mathbf{i} + x\mathbf{j})$ be a force field in the plane (minus the origin). Compute the work done by this force along each of the following paths:

   (a) $(\cos t, \sin t)$; $\quad 0 \leq t \leq \pi$ \qquad (b) $(\cos t, -\sin t)$; $\quad 0 \leq t \leq \pi$

   (c) $(\cos t, \sin t)$; $\quad 0 \leq t \leq 2\pi$

Solutions appear in Appendix A.

LINE INTEGRALS

To define the line integral, we use formula (2) abstracted from its physical interpretation.

Definition   Let $\boldsymbol{\Phi}$ be a vector field defined in some region of space, and let $\mathbf{r} = \boldsymbol{\sigma}(t)$ be a parametric curve in that region defined for t in $[t_1, t_2]$. The integral

$$\int_{t_1}^{t_2} \boldsymbol{\Phi}(\boldsymbol{\sigma}(t)) \cdot \boldsymbol{\sigma}'(t) \, dt \tag{3}$$

is called the *line integral* of the vector field $\boldsymbol{\Phi}$ along this curve.

The work done by a force field on a moving particle is therefore the line integral of the force along the path traveled by the particle.

Worked Example 2   Find the line integral of the vector field $e^y \mathbf{i} + e^x \mathbf{j} + e^z \mathbf{k}$ along the curve $(0, t, t^2)$, t in $[0, \ln 2]$.

Solution   The velocity vector $\boldsymbol{\sigma}'(t)$ is $\mathbf{j} + 2t\mathbf{k}$, so the line integral is

$$\int_0^{\ln 2} (e^t \mathbf{i} + e^0 \mathbf{j} + e^{t^2} \mathbf{k}) \cdot (\mathbf{j} + 2t\mathbf{k}) \, dt = \int_0^{\ln 2} (1 + e^{t^2} 2t) \, dt$$

$$= (t + e^{t^2}) \bigg|_0^{\ln 2} = \ln 2 + e^{(\ln 2)^2} - 1 = \ln 2 + 2^{\ln 2} - 1 \approx 1.31$$

The next theorem shows that the line integral does not depend upon how the path of integration is parametrized.

Theorem 1   *Given $\boldsymbol{\Phi}$ and $\boldsymbol{\sigma}$ as in the preceding definition, let $t = f(u)$ be a differentiable function defined on the interval $[u_1, u_2]$ such that $f(u_1) = t_1$ and $f(u_2) = t_2$. Let $\boldsymbol{\sigma}_1(u)$ be the parametric curve defined by $\boldsymbol{\sigma}_1(u) = \boldsymbol{\sigma}(f(u))$. Then*

$$\int_{u_1}^{u_2} \boldsymbol{\Phi}(\boldsymbol{\sigma}_1(u)) \cdot \boldsymbol{\sigma}_1'(u) \, du = \int_{t_1}^{t_2} \boldsymbol{\Phi}(\boldsymbol{\sigma}(t)) \cdot \boldsymbol{\sigma}'(t) \, dt$$

Proof   The basic idea was illustrated in Worked Example 1. We apply the chain rule to $\boldsymbol{\sigma}_1(u) = \boldsymbol{\sigma}(f(u))$, obtaining $\boldsymbol{\sigma}_1'(u) = \boldsymbol{\sigma}'(f(u))f'(u)$. Hence

$$\int_{u_1}^{u_2} \boldsymbol{\Phi}(\boldsymbol{\sigma}_1(u)) \cdot \boldsymbol{\sigma}_1'(u) \, du = \int_{u_1}^{u_2} \boldsymbol{\Phi}(\boldsymbol{\sigma}(f(u)) \cdot \boldsymbol{\sigma}'(f(u))f'(u) \, du$$

We next change variables from u to $t = f(u)$. Since $dt = f'(u) \, du$, the integral becomes $\int_{t_1}^{t_2} \boldsymbol{\Phi}(\boldsymbol{\sigma}(t)) \cdot \boldsymbol{\sigma}'(t) \, dt$ as required.

A *geometric curve* C is the collection of points $\boldsymbol{\sigma}(t)$ associated to some parametric curve. Theorem 1 suggests that the line integral of a vector field along a geometric curve is well defined. This is actually true, with these reservations:

1. One must state the direction in which the parametrization is to go—that is, a start and a finish for C. See Fig. 18-1-2.

2. The parametrization chosen for C should traverse C *exactly once.*

To explain reservation 1, we observe that if the direction of parametrization is reversed, then the line integral changes sign. (See Worked Example 3(a) below.) To explain reservation 2, consider the case in which C is *closed*—that is, $\boldsymbol{\sigma}(t_1) = \boldsymbol{\sigma}(t_2)$. If a parametrization is chosen which traverses C twice, then the line integral doubles its value. (See Fig. 18-1-3 and Worked Example 3(b).)

Fig. 18-1-2 A geometric curve must be parametrized in a specified direction.

Fig. 18-1-3 If a parametrization is chosen which traverses C twice, the line integral doubles its value.

Worked Example 3   (a) Let C be the line segment joining $(0, 0, 0)$ to $(1, 0, 0)$ and let $\boldsymbol{\sigma}_1(t) = (t, 0, 0)$, $0 \le t \le 1$. Find the line integral of $\boldsymbol{\Phi}(x, y, z) = \mathbf{i}$ along this curve. If C is parametrized by $\boldsymbol{\sigma}_2(t) = (1 - t, 0, 0)$, $0 \le t \le 1$, find the line integral.

(b) Let C be the circle described by $x^2 + y^2 = 1$, $z = 0$. Let $\boldsymbol{\sigma}_1(t) = (\cos t, \sin t, 0)$, $0 \le t \le 2\pi$. Find the integral of $\boldsymbol{\Phi}(x, y, z) = -y\mathbf{i} + x\mathbf{j}$ along this curve. If C is parametrized by $\boldsymbol{\sigma}_2(t) = (\cos t, \sin t, 0)$, $0 \le t \le 4\pi$, find the line integral.

Solution   (a) Here $\boldsymbol{\sigma}_1'(t) = \mathbf{i}$, $t_1 = 0$, $t_2 = 1$, and $\boldsymbol{\Phi}(\boldsymbol{\sigma}_1(t)) = \mathbf{i}$, so integral (3) gives

$$\int_{t_1}^{t_2} \boldsymbol{\Phi}(\boldsymbol{\sigma}_1(t)) \cdot \boldsymbol{\sigma}_1'(t)\, dt = \int_0^1 \mathbf{i} \cdot \mathbf{i}\, dt = \int_0^1 dt = 1$$

For $\boldsymbol{\sigma}_2$, we similarly have $t_1 = 0$, $t_2 = 1$, $\boldsymbol{\sigma}_2'(t) = -\mathbf{i}$, and $\boldsymbol{\Phi}(\boldsymbol{\sigma}_2(t)) = \mathbf{i}$, so

$$\int_{t_1}^{t_2} \boldsymbol{\Phi}(\boldsymbol{\sigma}_2(t)) \cdot \boldsymbol{\sigma}_2'(t)\, dt = \int_0^1 -\mathbf{i} \cdot \mathbf{i}\, dt = -\int_0^1 dt = -1$$

Here the geometric curve C is the same, but the two parametrizations, $\boldsymbol{\sigma}_1$ and $\boldsymbol{\sigma}_2$, traverse C in opposite directions. See Fig. 18-1-4 at the top of the next page.

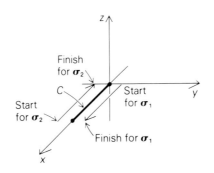

Fig. 18-1-4 σ_1 and σ_2 traverse C in opposite directions.

(b) The line integral for σ_1 is obtained from integral (3) by substituting $t_1 = 0$, $t_2 = 2\pi$, $\sigma_1(t) = (\cos t, \sin t, 0)$, $\Phi(x, y, z) = -y\mathbf{i} + x\mathbf{j}$, $\Phi(\sigma_1(t)) = -\sin t\,\mathbf{i} + \cos t\,\mathbf{j}$, $\sigma_1(t) = \cos t\,\mathbf{i} + \sin t\,\mathbf{j}$, and $\sigma_1'(t) = -\sin t\,\mathbf{i} + \cos t\,\mathbf{j}$, as follows:

$$\int_{t_1}^{t_2} \Phi(\sigma_1(t)) \cdot \sigma_1'(t)\, dt = \int_0^{2\pi} (\sin^2 t + \cos^2 t)\, dt = 2\pi$$

If we use $\sigma_2(t)$, the only change is that t_2 is changed to 4π, so we get

$$\int_{t_1}^{t_2} \Phi(\sigma_2(t)) \cdot \sigma_2'(t)\, dt = 4\pi$$

This is double our first answer, since σ_2 traversed the circle C twice. See Fig. 18-1-5.

Fig. 18-1-5 σ_1 goes around C once while σ_2 goes around twice.

If C is a closed curve, the choice of starting point does not affect the line integral, but the direction is still important. For instance, in Worked Example 3(b) we can choose any θ and parametrize C by the equation $\sigma(t) = (\cos(t + \theta), \sin(t + \theta), 0)$, $0 \le t \le 2\pi$; this will start and finish at $(\cos\theta, \sin\theta, 0)$. If we go backwards along C, using $\sigma(t) = (\cos(-t), \sin(-t), 0)$, we will get the negative of our earlier answer.

A useful notation for the line integral is suggested by the Leibniz notation. Let us write $\mathbf{r} = \sigma(t)$ so that $\sigma'(t) = d\mathbf{r}/dt$ and integral (3) becomes $\int_{t_1}^{t_2} \Phi(\mathbf{r}) \cdot (d\mathbf{r}/dt)dt$. It is tempting to change variables to $\mathbf{r}$ and write $\int_{\mathbf{r}_1}^{\mathbf{r}_2} \Phi(\mathbf{r}) \cdot d\mathbf{r}$, where $\mathbf{r}_1 = \sigma(t_1)$ and $\mathbf{r}_2 = \sigma(t_2)$, but this notation does not display the curve traced out by $\mathbf{r} = \sigma(t)$, only its endpoints. If we use the letter C to denote the path of integration, however, we can define

$$\int_C \Phi(\mathbf{r}) \cdot d\mathbf{r} = \int_{t_1}^{t_2} \Phi(\sigma(t)) \cdot \sigma'(t)\, dt \qquad \textbf{(4)}$$

where $\sigma(t)$ is any parametrization of C (subject to the reservations discussed above).

The notation $d\mathbf{r}$ for $\boldsymbol{\sigma}'(t)\ dt$ in formula (4) can be justified on an intuitive basis by reference to Fig. 18-1-6.

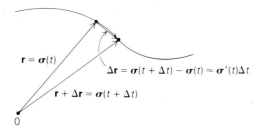

$\mathbf{r} = \boldsymbol{\sigma}(t)$

$\Delta\mathbf{r} = \boldsymbol{\sigma}(t + \Delta t) - \boldsymbol{\sigma}(t) \approx \boldsymbol{\sigma}'(t)\Delta t$

$\mathbf{r} + \Delta\mathbf{r} = \boldsymbol{\sigma}(t + \Delta t)$

0

Fig. 18-1-6 $\Delta\mathbf{r} \approx \boldsymbol{\sigma}'(t)\ \Delta t.$

Worked Example 4   Let C be the straight line segment joining $(2, 1, 3)$ to $(-4, 6, 8)$. Find

$$\int_C \boldsymbol{\Phi}(\mathbf{r}) \cdot d\mathbf{r}$$

where $\boldsymbol{\Phi}(x, y, z) = x\mathbf{i} - y\mathbf{j} + xy\,\mathbf{k}$.

Solution   We may choose any parametrization of C; the simplest is probably

$$\begin{aligned}
\boldsymbol{\sigma}(t) &= (1 - t)(2, 1, 3) + t(-4, 6, 8)\\
&= (2 - 6t, 1 + 5t, 3 + 5t)\quad 0 \le t \le 1
\end{aligned}$$

As t varies from 0 to 1, $\boldsymbol{\sigma}(t)$ moves along C from $(2, 1, 3)$ to $(-4, 6, 8)$. By formula (4), we get

$$\begin{aligned}
\int_C &\boldsymbol{\Phi}(\mathbf{r}) \cdot d\mathbf{r}\\
&= \int_0^1 [(2 - 6t)\mathbf{i} - (1 + 5t)\mathbf{j} + (2 - 6t)(1 + 5t)\mathbf{k})] \cdot (-6\mathbf{i} + 5\mathbf{j} + 5\mathbf{k})\,dt\\
&= \int_0^1 (-7 + 31t - 150t^2)\ dt = -\tfrac{83}{2}
\end{aligned}$$

The formulas

$$\int_a^b f(x)\ dx + \int_b^c f(x)\ dx = \int_a^c f(x)\ dx \tag{5}$$

and

$$\int_a^b f(x)\ dx = -\int_b^a f(x)\ dx \tag{6}$$

for ordinary integrals have their counterparts in line integration. If we choose a point on a curve C, it divides C into two curves, C_1 and C_2 (see Fig. 18-1-7(a)). We write $C = C_1 + C_2$. Then (4) and (5) give

(a)

(b)

Fig. 18-1-7   "Algebraic" operations on curves.

$$\int_{C_1 + C_2} \boldsymbol{\Phi}(\mathbf{r}) \cdot d\mathbf{r} = \int_{C_1} \boldsymbol{\Phi}(\mathbf{r}) \cdot d\mathbf{r} + \int_{C_2} \boldsymbol{\Phi}(\mathbf{r}) \cdot d\mathbf{r} \tag{7}$$

Let $-C$ be the curve C traversed in the reverse direction (see Fig. 18-1-7(b)). Then (4) and (6) give

$$\int_{-C} \boldsymbol{\Phi}(\mathbf{r}) \cdot d\mathbf{r} = -\int_{C} \boldsymbol{\Phi}(\mathbf{r}) \cdot d\mathbf{r} \tag{8}$$

Addition formula (7) suggests a way to define line integrals over curves with "corners"—that is, continuous curves on which the tangent vector is undefined at certain points. If C is such a curve, we write $C = C_1 + C_2 + C_3 + \cdots + C_n$ by dividing it at the corner points, and we *define* $\int_C \boldsymbol{\Phi}(\mathbf{r}) \cdot d\mathbf{r}$ to be

$$\sum_{i=1}^{n} \int_{C_i} \boldsymbol{\Phi}(\mathbf{r}) \cdot dr$$

See Fig. 18-1-8.

Fig. 18-1-8
If $C = C_1 + \ldots + C_n$, then
$\int_C \boldsymbol{\Phi}(\mathbf{r}) \cdot d\mathbf{r} = \sum_{i=1}^{n} \int_{C_i} \boldsymbol{\Phi}(\mathbf{r}) \cdot d\mathbf{r}$

Worked Example 5   Let C be the perimeter of the unit square $[0, 1] \times [0, 1]$ in the plane, traversed in the counterclockwise direction (see Fig. 18-1-9). Evaluate the line integral $\int_C \boldsymbol{\Phi}(\mathbf{r}) \cdot d\mathbf{r}$, where $\boldsymbol{\Phi}(x, y) = x^2 \mathbf{i} + xy \mathbf{j}$.

Solution   We do this problem by integrating along each of the sides C_1, C_2, C_3, C_4 separately and adding the results (see Fig. 18-1-9). The parametrizations are

$$
\begin{aligned}
C_1: & \quad (t, 0), \quad 0 \le t \le 1; \quad \boldsymbol{\sigma}_1(t) = t\mathbf{i} \\
C_2: & \quad (1, t), \quad 0 \le t \le 1; \quad \boldsymbol{\sigma}_2(t) = \mathbf{i} + t\mathbf{j} \\
C_3: & \quad (1 - t, 1), \quad 0 \le t \le 1; \quad \boldsymbol{\sigma}_3(t) = (1 - t)\mathbf{i} + \mathbf{j} \\
C_4: & \quad (0, 1 - t), \quad 0 \le t \le 1; \quad \boldsymbol{\sigma}_4(t) = (1 - t)\mathbf{j}
\end{aligned}
$$

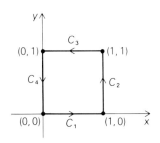

Fig. 18-1-9 The perimeter of the unit square broken into four pieces.

Thus, by (4),

$$\int_{C_1} \boldsymbol{\Phi}(\mathbf{r}) \cdot d\mathbf{r} = \int_0^1 t^2 \, dt = \tfrac{1}{3}$$

$$\int_{C_2} \boldsymbol{\Phi}(\mathbf{r}) \cdot d\mathbf{r} = \int_0^1 t \, dt = \tfrac{1}{2}$$

$$\int_{C_3} \boldsymbol{\Phi}(\mathbf{r}) \cdot d\mathbf{r} = \int_0^1 (1 - t)^2 \, (-1) \, dt = -\tfrac{1}{3}$$

$$\int_{C_4} \boldsymbol{\Phi}(\mathbf{r}) \cdot d\mathbf{r} = \int_0^1 0 \cdot dt = 0$$

Adding, we get

$$\int_C \mathbf{\Phi}(\mathbf{r}) \cdot d\mathbf{r} = \tfrac{1}{3} + \tfrac{1}{2} - \tfrac{1}{3} + 0 = \tfrac{1}{2}$$

Another notation for the line integral arises if we write our vectors in components. Suppose that $\mathbf{\Phi}(x, y, z) = a(x, y, z)\mathbf{i} + b(x, y, z)\mathbf{j} + c(x, y, z)\mathbf{k}$.

The expression for the derivative,

$$\frac{d\mathbf{r}}{dt} = \frac{dx}{dt}\mathbf{i} + \frac{dy}{dt}\mathbf{j} + \frac{dz}{dt}\mathbf{k}$$

can be written formally as

$$d\mathbf{r} = dx\,\mathbf{i} + dy\,\mathbf{j} + dz\,\mathbf{k}$$

and so

$$\int_C \mathbf{\Phi}(\mathbf{r}) \cdot d\mathbf{r}$$

$$= \int_C [a(x, y, z)\mathbf{i} + b(x, y, z)\mathbf{j} + c(x, y, z)\mathbf{k}] \cdot (dx\,\mathbf{i} + dy\,\mathbf{j} + dz\,\mathbf{k})$$

$$= \int_C a(x, y, z)\,dx + b(x, y, z)\,dy + c(x, y, z)\,dz \tag{9}$$

The expression inside the last integral is called a *differential form*. To evaluate the integral, we choose a parametrization of C. Then x, y, and z become functions of t; dx, dy, dz are expressed in terms of t and dt; and the integrand becomes an ordinary integrand in t which may be integrated over the parametrization interval.

Worked Example 6   Evaluate $\int_C \cos z\,dx + e^x\,dy + e^y\,dz$, where C is parametrized by $(x, y, z) = (1, t, e^t)$, $0 \le t \le 2$.

Solution   We compute $dx = 0$, $dy = dt$, $dz = e^t\,dt$, and so

$$\int_C \cos z\,dx + e^x\,dy + e^y\,dz = \int_0^2 (0 + e^1 + e^{2t})\,dt$$

$$= [et + \tfrac{1}{2}e^{2t}]_0^2 = 2e + \tfrac{1}{2}e^4 - \tfrac{1}{2}$$

LINE INTEGRALS

The line integral of a vector function (x, y, z)
along a curve $\boldsymbol{\sigma}(t)$, $t_1 \le t \le t_2$, is

$$\int_{t_1}^{t_2} \boldsymbol{\Phi}(\boldsymbol{\sigma}(t)) \cdot \boldsymbol{\sigma}'(t)\, dt$$

If $\boldsymbol{\Phi}(x, y, z) = a(x, y, z)\mathbf{i} + b(x, y, z)\mathbf{j} + c(x, y, z)\mathbf{k}$, and on the curve $\boldsymbol{\sigma}$, (x, y, z) are written as functions of t,

$$\int_{t_1}^{t_2} \boldsymbol{\Phi}(\boldsymbol{\sigma}(t)) \cdot \boldsymbol{\sigma}'(t)\, dt$$

$$= \int_{t_1}^{t_2} a(x(t), y(t), z(t)) \frac{dx}{dt}\, dt + \int_{t_1}^{t_2} b(x(t), y(t), z(t)) \frac{dy}{dt}\, dt + \int_{t_1}^{t_2} c(x(t), y(t), z(t)) \frac{dz}{dt}\, dt$$

which is also written

$$\int_C a\, dx + b\, dy + c\, dz \quad \text{or} \quad \int_C \boldsymbol{\Phi}(\mathbf{r}) \cdot d\mathbf{r} \quad \text{or simply} \quad \int_C \boldsymbol{\Phi}$$

Solved Exercises

3. Find $\int_C \sin \pi x\, dy - \cos \pi y\, dz$, where C is the triangle whose vertices are $(1, 0, 0)$, $(0, 1, 0)$, and $(0, 0, 1)$ in that order.

4. Suppose that $\boldsymbol{\Phi}(\mathbf{r})$ is orthogonal to $\boldsymbol{\sigma}'(t)$ at each point of the curve $\boldsymbol{\sigma}(t)$. What can you say about the line integral $\int_C \boldsymbol{\Phi}(\mathbf{r}) \cdot d\mathbf{r}$?

5. Let C be parametrized by

$$x = \cos^3 \theta \qquad y = \sin^3 \theta \qquad z = \theta \qquad 0 \le \theta \le \frac{7\pi}{2}$$

Evaluate the integral $\int_C \sin z\, dx + \cos z\, dy - (xy)^{1/3}\, dz$.

Exercises

4. Let $\boldsymbol{\sigma}(t) = (\sin t, \cos t, t)$, $0 \le t \le 2\pi$, and let $\boldsymbol{\Phi}(x, y, z) = x\mathbf{i} + y\mathbf{j} + z\mathbf{k}$. Evaluate the line integral of $\boldsymbol{\Phi}$ along this path.

5. Evaluate $\int_C x^2\, dx + xy\, dy + dz$, where C is parametrized by $\boldsymbol{\sigma}(t) = (t, t^2, 1)$, $0 \le t \le 1$.

6. Evaluate $\int_C \boldsymbol{\Phi}(\mathbf{r}) \cdot d\mathbf{r}$, where $\boldsymbol{\Phi}(x, y, z) = \sin z\mathbf{i} + \cos \sqrt{y}\mathbf{j} + x^3\mathbf{k}$ and C is the line segment from $(1, 0, 0)$ to $(0, 0, 3)$.

CONSERVATIVE FORCES AND PATH-INDEPENDENT LINE INTEGRALS

A parametric curve $\boldsymbol{\sigma}(t)$ defined on $[t_1, t_2]$ is called *closed* if its endpoints coincide—that is, if $\boldsymbol{\sigma}(t_1) = \boldsymbol{\sigma}(t_2)$. A vector field $\boldsymbol{\Phi}(x, y, z)$ is called *conservative* if the line integral of $\boldsymbol{\Phi}$ along every closed curve is equal to zero. (A conservative force field is thus one in which energy is conserved if a particle goes around a closed path.)

Theorem 2 *A vector field* $\Phi(x, y, z)$ *defined in all of space is conservative if and only if there is a function* $f(x, y, z)$ *such that, for every parametric curve* $\sigma(t)$, $t_1 \leq t \leq t_2$, *the line integral of* Φ *along the curve equals* $f(\sigma(t_2)) - f(\sigma(t_1))$.

Notice the resemblance to the fundamental theorem of calculus. By analogy with one-variable calculus, we will call f an *antiderivative* for the conservative vector field Φ.

(optional)

Proof of Theorem 2 It is clear that if a function f with the desired property exists, then Φ is conservative; in fact, if $\sigma(t)$ is a closed curve, then the line integral of Φ around the curve is $f(\sigma(t_2)) - f(\sigma(t_1)) = 0$, since $\sigma(t_2) = \sigma(t_1)$.

Given a conservative field Φ, we can construct the function f by "integrating Φ" as follows. Fix a point (x_0, y_0, z_0) and set $f(x_0, y_0, z_0) = 0$. For every point (x, y, z), choose a parametric curve $\mu(t)$ on $[0, 1]$ with $\mu(0) = (x_0, y_0, z_0)$ and $\mu(1) = (x, y, z)$—for example, $\mu(t) = (1 - t)(x_0, y_0, z_0) + t(x, y, z)$. Define $f(x, y, z)$ to be the line integral of Φ along this curve.

We must now show that the function f we have constructed meets the condition of the theorem—that is, that it is really an antiderivative for Φ. So let $\sigma(t)$ be a parametric curve defined on $[t_1, t_2]$ and let $\sigma(t_1) = (x_1, y_1, z_1)$ and $\sigma(t_2) = (x_2, y_2, z_2)$. For $i = 1$ or 2, let $\mu_i(t)$ be the chosen parametric curve from (x_0, y_0, z_0) to (x_i, y_i, z_i) and let C_i be the geometric curve traced out by $\mu_i(t)$ for $0 \leq t \leq 1$. Let C be the curve traced out by $\sigma(t)$ for $t_1 \leq t \leq t_2$. We need to show that

$$\int_C \Phi(\mathbf{r}) \cdot d\mathbf{r} = f(\sigma(t_2)) - f(\sigma(t_1))$$

That is, we need to show that

$$\int_C \Phi(\mathbf{r}) \cdot d\mathbf{r} = \int_{C_2} \Phi(\mathbf{r}) \cdot d\mathbf{r} - \int_{C_1} \Phi(\mathbf{r}) \cdot d\mathbf{r} \tag{10}$$

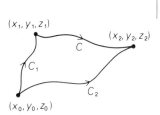

Fig. 18-1-10 The curves C_1, C, and C_2 together form a closed curve if we traverse C_2 backwards.

To prove equation (10), we will manufacture a closed curve C' from C_1, C_2, and C and use the fact that $\int_{C'} \Phi(\mathbf{r}) \cdot d\mathbf{r} = 0$ since Φ is conservative. A closed curve is obtained by traversing C_1, then C, then C_2 in the reverse direction (see Fig. 18-1-10), so we let $C' = C_1 + C + (-C_2)$. By formulas (7) and (8) above,

$$\int_{C_1} \Phi(\mathbf{r}) \, d\mathbf{r} + \int_C \Phi(\mathbf{r}) \, d\mathbf{r} - \int_{C_2} \Phi(\mathbf{r}) \, d\mathbf{r} = \int_{C'} \Phi(\mathbf{r}) \, d\mathbf{r} \tag{11}$$

But the right-hand side of (11) is zero, since C' is closed and Φ is conservative. Transposing terms in (11) gives (10). (Geometry related to this proof is discussed in Problem 13.)

Fig. 18-1-11 If we choose two points P_1 and P_2 on a closed curve, then we can write $C = C_1 - C_2$, where C_1 and C_2 have the same endpoints.

Notice that if Φ is conservative, the line integral depends only on the endpoints and not on the particular path chosen. We say that the line integral is *path-independent*. The converse is also true. Indeed, suppose that Φ has path-independent line integrals. If C is a closed curve, write $C = C_1 - C_2$, where C_1 and C_2 are determined as in Fig. 18-1-11. Then $\int_C \Phi(\mathbf{r}) \cdot d\mathbf{r} = \int_{C_1} \Phi(\mathbf{r}) \cdot d\mathbf{r} - \int_{C_2} \Phi(\mathbf{r}) \cdot d\mathbf{r} = 0$, since C_1 and C_2 have the same endpoints. Thus Φ is conservative.

Worked Example 7 If f_1 and f_2 are both antiderivatives for the conservative field Φ, then show that $f_1 - f_2$ is constant.

Solution Choose a path C from the origin to (x, y, z). Since f_1 and f_2 are both antiderivatives,

$$\int_C \Phi(\mathbf{r}) \cdot d\mathbf{r} = f_1(x, y, z) - f_1(0, 0, 0)$$
$$= f_2(x, y, z) - f_2(0, 0, 0)$$

Thus $f_1(x, y, z) - f_2(x, y, z) = f_1(0, 0, 0) - f_2(0, 0, 0)$, which is independent of (x, y, z).

Theorem 2 is not yet useful for doing computations. The next result will remedy the situation. (See Section 16-2 if you need to review gradients.)

Theorem 3

1. *If f is a function with continuous partial derivatives, then ∇f is conservative and f is an antiderivative for ∇f.*

2. *If Φ is conservative and f is an antiderivative for Φ with continuous partial derivatives, then $\Phi = \nabla f$.*

(optional)

Proof To prove part 1, let C be a curve from (x_1, y_1, z_1) to (x_2, y_2, z_2). We will show that $\int_C \nabla f(\mathbf{r}) \cdot d\mathbf{r} = f(x_2, y_2, z_2) - f(x_1, y_1, z_1)$, so that f is an antiderivative for ∇f and f is conservative. Choose a parametrization $\sigma(t)$ for C such that $\sigma(t_i) = (x_i, y_i, z_i)$ for $i = 1$ or 2. Now

$$\int_C \nabla f(\mathbf{r}) \cdot d\mathbf{r} = \int_{t_1}^{t_2} \nabla f(\sigma(t)) \cdot \sigma'(t) \, dt$$

By the chain rule for gradients and curves (Section 16-2), $\nabla f(\sigma(t)) \cdot \sigma'(t) = (d/dt) [f(\sigma(t))]$, so

$$\int_C \nabla f(\mathbf{r}) \cdot d\mathbf{r} = \int_{t_1}^{t_2} \frac{d}{dt} [f(\sigma(t))] \, dt = f(\sigma(t_2)) - f(\sigma(t_1)) = f(x_2, y_2, z_2) - f(x_1, y_1, z_1)$$

In the next-to-last equality, we used the one-variable fundamental theorem of calculus (Section 4-2).

To prove part 2, let f be any antiderivative for the conservative vector field Φ. Since f is also an antiderivative for ∇f, we conclude that for any curve C

$$\int_C \Phi(\mathbf{r}) \cdot d\mathbf{r} = \int_C \nabla f(\mathbf{r}) \cdot d\mathbf{r}$$

since both integrals are equal to the difference between the values of f at the endpoints. Subtracting, we conclude that

$$\int_C [\Phi(\mathbf{r}) - \nabla f(\mathbf{r})] \cdot d\mathbf{r} = 0$$

for any path C. But this implies that $\Phi(\mathbf{r}) - \nabla f(\mathbf{r})$ is identically zero (if not, its integral over a suitably chosen short straight path would be nonzero), so $\Phi = \nabla f$.

Worked Example 8   Evaluate $\int_C y\,dx + x\,dy$ if C is parametrized by $(t^9, \sin^9(\pi t/2))$, $0 \le t \le 1$.

Solution   The vector field we are integrating is $\boldsymbol{\Phi} = y\mathbf{i} + x\mathbf{j}$. We recognize this (by guesswork) as the gradient of $f(x, y) = xy$ since

$$f_x = y \quad \text{and} \quad f_y = x$$

Therefore, by Theorem 3(1),

$$\int_C \boldsymbol{\Phi}(\mathbf{r}) \cdot d\mathbf{r} = f(\boldsymbol{\sigma}(1)) - f(\boldsymbol{\sigma}(0)) = f(1, 1) - f(0, 0) = 1$$

We summarize Theorems 2 and 3 in the following display.

CONSERVATIVE FIELDS

A vector field $\boldsymbol{\Phi}$ is called conservative when   the line integral of $\boldsymbol{\Phi}$ around any closed curve is zero (which is the same as the line integrals of $\boldsymbol{\Phi}$ being path-independent).

A vector field $\boldsymbol{\Phi}$ is conservative if and only if $\boldsymbol{\Phi} = \nabla f$ for some f.

$$\int_C \nabla f(\mathbf{r}) \cdot d\mathbf{r} = f(P_2) - f(P_1) \tag{12}$$

where P_1 and P_2 are the endpoints of C.

In differential notation, formula (12) reads

$$\int_C f_x\,dx + f_y\,dy + f_z\,dz = f(P_2) - f(P_1) \tag{13}$$

(Omit the term $f_z\,dz$ in the plane.) We call differential forms of the type $f_x\,dx + f_y\,dy + f_z\,dz$ *exact differential forms*. Thus another way of phrasing Theorem 3 is that $\boldsymbol{\Phi}$ is conservative if and only if $\boldsymbol{\Phi}(\mathbf{r}) \cdot d\mathbf{r}$ is exact.

In the following section we obtain another important criterion for determining when a vector field is conservative.

Solved Exercises

6. If $f(x, y, z) = x^3 y^2$, evaluate $\int_C f_x\,dx + f_y\,dy$, where C is parametrized by $(e^{\sqrt{t}}, \sin^3(\pi t/2), \ln(1 + t))$, $0 \le t \le 1$.

7. If $a\,dx + b\,dy$ is an exact differential in the plane, and a and b have continuous partial derivatives, prove that $a_y = b_x$. (Assume that these partials are continuous.)

8. Show that $x^2\,dx + xy\,dy$ is not exact by (a) showing that its integral around a closed path is not zero and (b) using Solved Exercise 7.

Exercises   7. Let $f(x, y, z) = xze^y - z^3/(1 + y^2)$. Evaluate the line integral

$$\int_C \frac{\partial f}{\partial x} \, dx + \frac{\partial f}{\partial y} \, dy + \frac{\partial f}{\partial z} \, dz, \text{ where } C \text{ is parametrized by:}$$

(a) $\sigma(t) = \left(\sqrt{t} \, \sin \frac{\pi}{4} (1 + t), t^2 - 1, \frac{2 - t^2}{2 + t^2} \right); \quad 0 \le t \le 1$

(b) $\sigma(t) = (\cos^4 t, e^{\sin(\pi t/2)}, 2 - t); \quad 0 \le t \le 1$

(c) $\sigma(t) = \left(\cos \frac{\pi}{2} t, \sin \frac{\pi}{2} t, t \right); \quad -1 \le t \le 1$

8. If $f(x, y, z) = x^3 - y^3 + \sin(\pi yz/2)$ evaluate $\int_C \nabla f(\mathbf{r}) \cdot d\mathbf{r}$ along the curve

$$\sigma(t) = \left(\frac{t^3}{1 + t^2}, \sin \frac{\pi}{4} t^5, t^2 + 2 \right); \quad 0 \le t \le 1$$

9. Which of the following are exact? [*Hint:* Use Solved Exercise 7 and some guesswork.]

(a) $x^2 y \, dx + (x^3 y/3) \, dy$ (b) $x^2 y \, dx + (x^3/3) \, dy$

(c) $x^2 \, dx + (x^3 y/3) \, dy$ (d) $xy^2 \, dx + (y^3/3) \, dy$

Problems for Section 18-1

1. Let $\Phi(x, y, z) = x\mathbf{i} - y\mathbf{j} + z\mathbf{k}$. Evaluate the integral of Φ along each of the following paths:

(a) $\sigma(t) = (t, t, t); \quad 0 \le t \le 1$

(b) $\sigma(t) = (\cos t, \sin t, 0); \quad 0 \le t \le \pi/2$

(c) $\sigma(t) = (\cos t, \sin t, 0); \quad 0 \le t \le \pi/2$

2. Evaluate the line integral of the given vector field Φ along the given path $\sigma(t)$:

(a) $\Phi(x, y, z) = \sin z \mathbf{i} + \cos \sqrt{y} \, \mathbf{j} + x^3 \mathbf{k}$
   $\sigma(t) = (\sin t, t^2, t); \quad 0 \le t \le 2\pi$

(b) $\Phi(x, y, z) = xz\mathbf{i} + xy\mathbf{j} + yz\mathbf{k}$
   $\sigma(t) = (\cos t, \sec t, \tan t); \quad -\pi/4 \le t \le \pi/4$

(c) $\Phi(x, y, z) = [1/(z^2 + 1)]\mathbf{i} + x(1 + y^2)\mathbf{j} + e^y \mathbf{k}$
   $\sigma(t) = ((1 + t^2)^2, 1, t); \quad 0 \le t \le 1$

(d) $\Phi(x, y, z) = (x^2 + x)\mathbf{i} + [(x - y)/(x + y)]\mathbf{j} + (z - z^3)\mathbf{k}$
   $\sigma(t) = 3t\mathbf{i} + (t - 1)\mathbf{j} + t^2\mathbf{k}; \quad 0 \le t \le 1$

3. Let $\Phi(x, y, z) = x^2\mathbf{i} - xy\mathbf{j} + \mathbf{k}$. Evaluate the line integral of Φ along each of the following curves:

(a) The straight line joining $(0, 0, 0)$ to $(1, 1, 1)$.

(b) The circle of radius 1, center at the origin and lying in the yz plane, traversed counterclockwise as viewed from the positive x-axis.

(c) The parabola $z = x^2$ between $(-1, 0, 1)$ and $(1, 0, 1)$.

4. Calculate the work done in moving a mass m from a distance r_1 out to a distance r_2 in the gravitational field of a mass M. The force is $\mathbf{F} = -(GMm/r^3)\mathbf{r}$, where $\mathbf{r} = x\mathbf{i} + y\mathbf{j} + z\mathbf{k}$ and the mass M is located at the origin. Do this by:

(a) Using Solved Exercise 2.

(b) Using formula (12) and the fact that $\mathbf{F}$ is a gradient (see Solved Exercise 13, Section 16-2).

5. Compute the work done in moving a particle along the given path with the given force:

(a) $\sigma(t) = (\cos t, \sin t, t); \, 0 \le t \le 2\pi$
   $\Phi(x, y, z) = x\mathbf{i} + y\mathbf{j} + z\mathbf{k}$

(b) $\sigma(t) = (\cos t, \sin t, 0); \quad 0 \le t \le 2\pi$
   $\Phi(x, y, z) = x\mathbf{i} + y\mathbf{j} + z\mathbf{k}$

(c) Same as part (b) but with $0 \le t \le \pi$.

(d) $\sigma(t) = (a \cos t, 0, b \sin t); \quad 0 \le t \le 2\pi$
   $\Phi(\mathbf{r}) = -\mathbf{r}/\|\mathbf{r}\|^3$

6. Let $\Phi(x, y, z) = (z^3 + 2xy)\mathbf{i} + x^2\mathbf{j} + 3xz^2\mathbf{k}$. Show that the integral of Φ around the circumference of the unit square $[0, 1] \times [0, 1]$ is zero by:

(a) Evaluating directly.

(b) Demonstrating that Φ is the gradient of some function f.

7. Let $f(x, y, z) =$

$e^x \cos(yz + x^8) + [\sin(yz)] \ln(1 + x^2)$

Show that

$$\int_{C_1} \nabla f(\mathbf{r}) \cdot d\mathbf{r} = \int_{C_2} \nabla f(\mathbf{r}) \cdot d\mathbf{r}$$

where C_1 is the straight line joining $(0, 1, 0)$ to $(1, 1, 0)$ and C_2 is parametrized by $(\sin t, \cos 4t, \sin 4t)$; $0 \le t \le \pi/2$.

8. The mass of the earth is approximately 6×10^{27} grams; the mass of the sun is 330,000 times as much. The gravitational constant (in units of grams, seconds, and centimeters) is 6.7×10^{-8}. The distance of the earth from the sun is about 1.5×10^{12} centimeters. Compute, approximately, the work necessary to increase the distance of the earth from the sun by 1 centimeter.

9. The line integral of a scalar function f along a parametric curve $\boldsymbol{\sigma}(t)$, $t_1 \le t \le t_2$, is defined by

$$\int_{t_1}^{t_2} f(\boldsymbol{\sigma}(t)) \|\boldsymbol{\sigma}'(t)\| \, dt$$

Note that if $f = 1$, this is just the arc length of the curve. Evaluate the line integrals of the following functions along the indicated curves:

(a) $f(x, y, z) = x + y + yz$ and $\boldsymbol{\sigma}(t) = (\sin t, \cos t, t)$; $0 \le t \le 2\pi$

(b) $f(x, y, z) = x + \cos^2 z$ and $\boldsymbol{\sigma}$ as in part (a)

(c) $f(x, y, z) = x \cos z$, $\boldsymbol{\sigma}(t) = t\mathbf{i} + t^2\mathbf{j}$; $0 \le t \le 1$

(d) $f(x, y, z) = \exp \sqrt{z}$ and $\boldsymbol{\sigma}(t) = (1, 2, t^2)$; $0 \le t \le 1$

(e) $f(x, y, z) = yz$ and $\boldsymbol{\sigma}(t) = (t, 3t, 2t)$; $1 \le t \le 3$

(f) $f(x, y, z) = (x + y)/(y + z)$ and $\boldsymbol{\sigma}(t) = (t, \frac{2}{3}t^{3/2}, t)$; $1 \le t \le 2$

★10. Show that the value of the line integral of a scalar function over a parametric curve, defined in Problem 9, is unchanged if the curve is reparametrized. [*Hint:* See Theorem 1.]

11. Suppose that the kinetic energy of a particle which moves in a circular path increases after it makes one circuit. Can the force field governing the particle's motion be conservative?

12. In the earth's gravitational field, show that if a mass is taken on a journey, however tortuous, the net amount of work done against the gravitational field is zero, provided the mass ends up in the same spot it began, with the same speed.

★13. *Two-color problem.* Several intersecting circles are drawn in the plane. Show that the resulting "map" can be colored with two colors in such a way that adjacent regions have different colors (as in Fig. 18-1-12). [*Hint:* First show that every closed curve crosses the union of the circles an even number of times. Then divide the regions into two classes according to whether an arc from the region to a fixed point crosses the circles an even or odd number of times. Compare the proof of Theorem 2.]

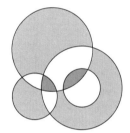

Fig. 18-1-12 Adjacent regions have opposite colors.

18-2 Green's, Stokes' and Gauss' Theorems

A line integral along a closed curve can be converted to a double integral over a region.

Green's theorem relates the line integral of a vector field (or differential form) around a closed curve to the double integral of a certain function over the region bounded by the curve. Among the many applications of this theorem is a criterion for deciding which vector fields in the plane are gradients. Stokes' and Gauss' theorems generalize Green's theorem in two different ways.

Goals

After studying this section, you should be able to:

Use Green's theorem to convert line integrals over closed curves in the plane to double integrals and vice versa.

Determine whether a given vector field in the plane is conservative (and whether a given differential form is exact).

Use Stokes' theorem to relate line integrals over closed curves in space to surface integrals and Gauss' theorem to relate integrals over closed surfaces to volume integrals.

State the physical significance of line and surface integrals.

GREEN'S THEOREM*

Consider a region of type 1 in the plane, as shown in Fig. 18-2-1. The *boundary* of this region is defined to be the closed curve C which goes once around the region in the counterclockwise direction. If we start at the upper left-hand corner, the boundary curve first traverses the vertical line l_1 (call this C_1) then goes along the graph of ϕ_1 (call this C_2), then up l_2 (call this C_3), and, finally, backwards along the graph of ϕ_2 (call this C_4).

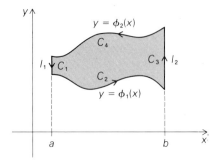

Fig. 18-2-1 The boundary of this type 1 region consists of the graphs $y = \phi_1(x)$ and $y = \phi_2(x)$ and segments of the lines l_1 and l_2.

George Green (1793–1841), an English mathematician and physicist, was one of the early investigators of electricity and magnetism. His work on potentials led to what is commonly called Green's theorem (although it may really be due to Cauchy). This theorem, as generalized to three dimensions by Lord Kelvin, Stokes, Gauss, and Ostrogradsky, was crucial to later developments in electromagnetic, gravitational, and other physical theories. Some of these applications are discussed later in this section. For more history, see "The History of Stokes' Theorem" by V. J. Katz, Mathematics Magazine, 52 (1979) 146–156.

The following lemma is a preliminary form of Green's theorem.

Lemma *Let D be a type 1 region as above, with C its bounding curve. Let $P = f(x, y)$ have continuous partial derivatives in D and on C. Then*

$$\int_C P \, dx = - \int\int_D \frac{\partial P}{\partial y} \, dx \, dy \tag{1}$$

(optional)

Proof Since the double integral may be evaluated as an iterated integral (Theorem 1, Section 17-1), we have

$$\int\int_D \frac{\partial P}{\partial y} \, dx \, dy = \int_a^b \int_{\phi_1(x)}^{\phi_2(x)} f_y(x, y) \, dy \, dx$$

$$= \int_a^b [f(x, \phi_2(x)) - f(x, \phi_1(x))] \, dx$$

The latter equality uses the fundamental theorem of calculus.

To compute the line integral, we parametrize each of the segments of C:

C_1: $(x, y) = (a, \phi_2(a) - t)$; t in $[0, \phi_2(a) - \phi_1(a)]$; $dx = 0$, $dy = -dt$

C_2: $(x, y) = (t, \phi_1(t))$; t in $[a, b]$; $dx = dt$, $dy = \phi_1'(t) \, dt$

C_3: $(x, y) = (b, \phi_1(b) + t)$; t in $[0, \phi_2(b) - \phi_1(b)]$; $dx = 0$, $dy = dt$

C_4: $(x, y) = (-t, \phi_2(-t))$; t in $[-b, -a]$; $dx = -dt$, $dy = -\phi_2'(-t) \, dt$

Now $\int_C P \, dx$ is the sum of the line integrals over the four C_i's. The integrals over C_1 and C_3 are zero, since $dx = 0$ on those curves (x is constant). The integrals over C_2 and C_4 are given by

$$\int_{C_2} P \, dx = \int_a^b f(t, \phi_1(t)) \, dt$$

and

$$\int_{C_4} P \, dx = \int_{-b}^{-a} f(-t, \phi_2(-t))(-dt)$$

$$= \int_b^a f(t, \phi_2(t)) \, dt \quad \text{(substituting } -t \text{ for } t)$$

$$= - \int_a^b f(t, \phi_2(t)) \, dt$$

Thus

$$\int_C P \, dx = \int_a^b f(t, \phi_1(t)) \, dt - \int_a^b f(t, \phi_2(t)) \, dt$$

$$= - \int_a^b [f(t, \phi_2(t)) \, dt - f(t, \phi_1(t))] \, dt$$

$$= - \int\int_D \frac{\partial P}{\partial y} \, dx \, dy$$

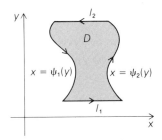

Fig. 18-2-2 The boundary of this type 2 region consists of the two curves, $x = \psi_1(y)$ and $x = \psi_2(y)$, and segments of the lines, l_1 and l_2.

In exactly the same way, we can prove that if D is a type 2 region with boundary curve C traversed counterclockwise (Fig. 18-2-2), then for $Q = g(x, y)$ with continuous partial derivatives,

$$\int_C Q \, dy = \iint_D \frac{\partial Q}{\partial x} \, dx \, dy \qquad \qquad \textbf{(2)}$$

If we have a region which is of both types 1 and 2, then equations (1) and (2) are both valid. Adding them yields Green's theorem.

Theorem 4 (Green's theorem) *If D is a region of types 1 and 2 with boundary curve C traversed counterclockwise, and if $P = f(x, y)$ and $Q = g(x, y)$ have continuous partial derivatives in D and on C, then*

$$\int_C P \, dx + Q \, dy = \iint_D \left(\frac{\partial Q}{\partial x} - \frac{\partial P}{\partial y} \right) dx \, dy$$

Worked Example 1 Verify Green's theorem for $P = x$ and $Q = xy$, where D is the unit disk $x^2 + y^2 \le 1$.

Solution We can evaluate both sides in Green's theorem directly. The boundary C of D is the unit circle parametrized by $x = \cos t$, $y = \sin t$, $0 \le t \le 2\pi$, so

$$\int_C P \, dx + Q \, dy = \int_0^{2\pi} [(\cos t)(-\sin t) + \cos t \sin t \cos t] \, dt$$
$$= \left[\frac{\cos^2 t}{2} \right]_0^{2\pi} + \left[-\frac{\cos^3 t}{3} \right]_0^{2\pi} = 0$$

On the other hand,

$$\iint_D \left(\frac{\partial Q}{\partial x} - \frac{\partial P}{\partial y} \right) dx \, dy = \iint_D y \, dx \, dy$$

which is zero also, since the contributions from the upper and lower half-circles cancel one another. Thus Green's theorem is verified in this case.

Green's theorem applies as well to many regions other than just those of types 1 and 2. Often one can show this by dividing up the region, as in the following example.

Worked Example 2 Show that Green's theorem is valid for the region D shown in Fig. 18-2-3.

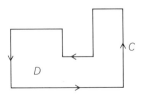

Fig. 18-2-3 Is Green's theorem valid for D?

Solution   Figure 18-2-4 shows how to divide up D into three regions, D_1, D_2, D_3, each of which is of types 1 and 2. Let C_1, C_2, C_3 be the boundary curves of these regions. Then

$$\iint_{D_i} \left(\frac{\partial Q}{\partial x} - \frac{\partial P}{\partial y} \right) dx\, dy = \int_{C_i} P\, dx + Q\, dy$$

Fig. 18-2-4 Breaking a region up into smaller regions, each of which is both type 1 and type 2.

The double integral over the D_i's adds up to the double integral over D, so

$$\iint_{D} \left(\frac{\partial Q}{\partial x} - \frac{\partial P}{\partial y} \right) dx\, dy = \int_{C_1} P\, dx + Q\, dy + \int_{C_2} P\, dx + Q\, dy$$
$$+ \int_{C_3} P\, dx + Q\, dy$$

However, the dotted portions of the boundaries shown in Fig. 18-2-4 are traversed twice in opposite directions; these cancel in the line integrals. Thus we are left with

$$\iint_{D} \left(\frac{\partial Q}{\partial x} - \frac{\partial P}{\partial y} \right) dx\, dy = \int_{C} P\, dx + Q\, dy$$

and so Green's theorem is valid.

GREEN'S THEOREM

If D is a region and C is the boundary of D oriented as in Fig. 18-2-5, then

$$\int_{C} P\, dx + Q\, dy = \iint_{D} \left(\frac{\partial Q}{\partial x} - \frac{\partial P}{\partial y} \right) dx\, dy$$

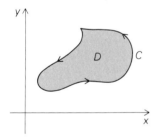

Fig. 18-2-5 The boundary C of a region D must be oriented counterclockwise in Green's theorem.

We can use Green's theorem to obtain a formula for the *area* of a region bounded by a curve C.

Corollary   *If C is a curve that bounds a region (to which Green's theorem applies), then the area of the region D bounded by C is*

$$A = \frac{1}{2} \int_{C} x\, dy - y\, dx \tag{3}$$

Proof Let $P = -y$, $Q = x$; then by Green's theorem we have

$$\frac{1}{2} \int_C x\, dy - y\, dx = \frac{1}{2} \iint_D \left(\frac{\partial x}{\partial x} - \frac{\partial(-y)}{\partial y} \right) dx\, dy = \iint_D dx\, dy$$

which is the area of D.

Worked Example 3 Verify formula (3) in the case where D is the disk $x^2 + y^2 \le r^2$.

Solution The area is πr^2. Formula (3) with $x = r\cos t$, $y = r\sin t$, $0 \le t \le 2\pi$, gives

$$A = \tfrac{1}{2} \int_C x\, dy - y\, dx = \tfrac{1}{2} \int_0^{2\pi} (r\cos t)(r\cos t)\, dt - (r\sin t)(-r\sin t)\, dt$$

$$= \tfrac{1}{2} \int_0^{2\pi} r^2\, dt = \pi r^2$$

so formula (3) checks.

Solved Exercises

1. Let $\Phi(x, y) = y\mathbf{i} - x\mathbf{j}$ and let C be a circle of radius r. Write $\int_C \Phi(\mathbf{r}) \cdot d\mathbf{r}$ as a double integral using Green's theorem. Evaluate.

2. Under the conditions of Green's theorem, verify that

$$\int_C PQ\, dx + PQ\, dy = \iint_D \left[Q\left(\frac{\partial P}{\partial x} - \frac{\partial P}{\partial y}\right) + P\left(\frac{\partial Q}{\partial x} - \frac{\partial Q}{\partial y}\right) \right] dx\, dy$$

3. Let C be the perimeter of the square $[0,1] \times [0,1]$ oriented counterclockwise. Evaluate $\int_C (y^4 + x^3)\, dx + 2x^6\, dy$.

Exercises

1. Show how to decompose each of the regions shown in Fig. 18-2-6 into subregions, each of which is of types 1 and 2.

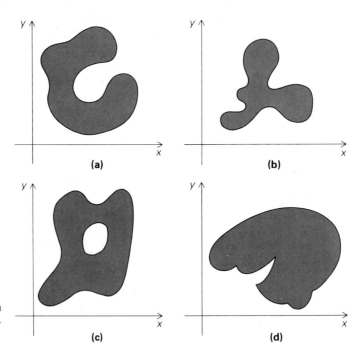

(a) (b) (c) (d)

Fig. 18-2-6 Subdivide each of these regions into subregions of types 1 and 2.

2. (a) Use the corollary to Green's theorem to compute the area bounded by the ellipse $C: x^2/a^2 + y^2/b^2 = 1$.

   (b) Let $\boldsymbol{\Phi}(x, y) = xy^2\mathbf{i} - yx^2\mathbf{j}$. Write $\int_C \boldsymbol{\Phi}(\mathbf{r}) \cdot d\mathbf{r}$ as a double integral using Green's theorem. Evaluate.

3. Use the corollary to Green's theorem to determine the area of the regions in the plane bounded by:

   (a) The triangle with vertices $(1, 0)$, $(3, 4)$, and $(5, -1)$.

   (b) The rhombus with vertices $(0, -1)$, $(3, 0)$, $(1, 2)$, and $(4, 3)$.

EXACT DIFFERENTIALS AND EXACT DIFFERENTIAL EQUATIONS

We showed in the previous section that a vector field $\boldsymbol{\Phi}(x, y) = a(x, y)\mathbf{i} + b(x, y)\mathbf{j}$ is conservative if and only if there is a function $f(x, y)$ such that $f_x = a$ and $f_y = b$. In differential notation, the differential form $P\,dx + Q\,dy$ is *exact* if its integral around every closed curve is zero, or, equivalently, if $P = \partial u/\partial x$ and $Q = \partial u/\partial y$ for some function $u = f(x, y)$.

In Solved Exercise 7, Section 18-1, we found a simple *necessary* condition for a vector field to be conservative: since the mixed partial derivatives of a function f are equal, if $f_x = a$ and $f_y = b$ then we must have

$$a_y = f_{xy} = f_{yx} = b_x$$

That is, $a_y = b_x$. In differential notation, we must have $\partial P/\partial y = \partial Q/\partial x$ if $P\,dx + Q\,dy$ is to be exact.

Using Green's theorem, we can show that the condition $\partial P/\partial y = \partial Q/\partial x$ is also *sufficient* for a differential form defined on the plane to be exact.

Theorem 5 *Let $P\,dx + Q\,dy$ be a differential form defined in the plane. If P and Q have continuous partials and $\partial P/\partial y = \partial Q/\partial x$, then $P\,dx + Q\,dy$ is exact—that is, there is a function $u = f(x, y)$ such that $P = \partial u/\partial x$ and $Q = \partial u/\partial y$. (In other words, the vector field $P\mathbf{i} + Q\mathbf{j}$ is the gradient of f.)*

(optional)

Proof Define $u = f(x, y)$ as follows. Let C be the path which consists of the horizontal segment C_1 from $(0, 0)$ to $(x, 0)$ followed by the vertical segment C_2 from $(x, 0)$ to (x, y). (See Fig. 18-2-7.) Then set $u = f(x, y) = \int_C P\,dx + Q\,dy$. It is easy to show that $\partial u/\partial y = Q$. In fact, if $P = a(x, y)$ and $Q = b(x, y)$, we have

$$f(x, y) = \int_0^x a(t, 0)\,dt + \int_0^y b(x, t)\,dt$$

Differentiating this equation with respect to y is easy; by the alternative version of the fundamental theorem of calculus, we get

$$f_y(x, y) = 0 + b(x, y)$$

or

Fig. 18-2-7 The path used to construct $u = f(x, y)$.

$$\frac{\partial u}{\partial y} = Q \qquad\qquad (4)$$

(optional)

Fig. 18-2-8 The path used to construct $\hat{u} = f(x, y)$.

It is not so easy to show that $\partial u / \partial x = P$. Define another function $\hat{u} = \hat{f}(x, y)$ by letting $\hat{C}$ be the path which consists of the vertical segment $\hat{C}_1$ from $(0, 0)$ to $(0, y)$, followed by the horizontal segment $\hat{C}_2$ from $(0, y)$ to (x, y) (see Fig. 18-2-8). Then

$$\hat{f}(x, y) = \int_0^y b(0, t) \, dt + \int_0^x a(t, y) \, dt$$

and

$$\hat{f}_x(x, y) = 0 + a(x, y)$$

Hence

$$\frac{\partial \hat{u}}{\partial x} = P \tag{5}$$

We apply Green's theorem to the rectangle D bounded by $C_1 + C_2 + (-\hat{C}_2) + (-\hat{C}_1)$ and get

$$\iint_D \left(\frac{\partial P}{\partial x} - \frac{\partial Q}{\partial y} \right) dx \, dy = \int_{C_1 + C_2 + (-\hat{C}_2) + (-\hat{C}_1)} (P \, dx + Q \, dy)$$

$$= \int_{C_1 + C_2} (P \, dx + Q \, dy) - \int_{\hat{C}_1 + \hat{C}_2} (P \, dx + Q \, dy)$$

$$= u - \hat{u}$$

But the double integral over D is zero, since $\partial P / \partial x = \partial Q / \partial y$ by assumption. Thus $u = \hat{u}$, so (5) implies that $\partial u / \partial x = P$ and we are done.

If we are willing to take a few liberties, Theorem 5 also follows directly from Theorems 2 and 3 and Green's theorem. Indeed, we need only to show that $P \, dx + Q \, dy$ is conservative—that is, $\int_C P \, dx + Q \, dy = 0$ for a closed curve C. If we assume that C bounds a region D to which Green's theorem applies, then, since $\partial Q / \partial x = \partial P / \partial y$,

$$\int_C P \, dx + Q \, dy = \iint_D \left(\frac{\partial Q}{\partial x} - \frac{\partial P}{\partial y} \right) dx \, dy = 0$$

The proof of Theorem 5, however, gives an explicit method for finding u, given P and Q. Review Problem 16 of Chapter 16 showed how to prove Theorem 5 directly without using line integrals. It also gave an alternative problem-solving technique.

Worked Example 4 Is $(2xy \cos y + x^2) \, dx + (x^2 \cos y - x^2 y \sin y) \, dy$ an exact differential? If so, find u such that $\partial u / \partial x = 2xy \cos y + x^2$ and $\partial u / \partial y = x^2 \cos y - x^2 y \sin y$.

Solution To test for exactness, we let $P = 2xy \cos y + x^2$ and $Q = x^2 \cos y - x^2 y \sin y$ and compute that

$$\frac{\partial P}{\partial y} = -2xy \sin y + 2x \cos y = \frac{\partial Q}{\partial x}$$

and so a u exists. To find $u = f(x, y)$ we use the formula from the proof:

$$f(x, y) = \int_0^x a(t, 0)\, dt + \int_0^y b(x, t)\, dt$$

where $P = a(x, y)$ and $Q = b(x, y)$. Here $a(x, y) = 2xy \cos y + x^2$ and $b = x^2 \cos y - x^2 y \sin y$, so we get

$$f(x, y) = \int_0^x t^2\, dt + \int_0^y (x^2 \cos t - x^2 t \sin t)\, dt$$

Evaluating the last integral by parts gives

$$u = f(x, y) = \frac{x^3}{3} + \int_0^y x^2 \cos t\, dt - \int_0^y x^2 \cos t\, dt + x^2 t \cos t \Big|_0^y$$

$$= \frac{x^3}{3} + x^2 y \cos y$$

We check that $\partial u / \partial x$ and $\partial u / \partial y$ are as required.

There is another method for finding $u = f(x, y)$ which does not use line integrals. (To understand it, you should study Review Problem 16 on p. 748.) We must have

$$\frac{\partial u}{\partial x} = P \quad \text{and} \quad \frac{\partial u}{\partial y} = Q$$

Hence

$$\frac{\partial u}{\partial x} = 2xy \cos y + x^2 \quad \text{and} \quad \frac{\partial u}{\partial y} = x^2 \cos y - x^2 y \sin y$$

Integrating the first equation with respect to x gives $u = x^2 y \cos y + x^3/3 + h_1(y)$; the "constant" of integration h_1 must be a function of y alone, since $\partial u / \partial x = (\partial / \partial x)(x^2 y \cos y + x^3/3)$, and so $(h_1)_x = 0$. Similarly, integrating the second equation with respect to y, $u = x^2 y \cos y + h_2(x)$. These expressions match if we choose $h_1(y) = 0$ and $h_2(x) = x^3/3$, so $u = x^3/3 + x^2 y \cos y$, as above.

EXACT DIFFERENTIALS

To determine whether there is a function $u = f(x, y)$ such that $\partial u / \partial x = P = a(x, y)$ and $\partial u / \partial y = Q = b(x, y)$, check whether

$$\frac{\partial P}{\partial y} = \frac{\partial Q}{\partial x}$$

If so, f may be constructed by the formula

$$f(x, y) = \int_0^x a(t, 0)\, dt + \int_0^y b(x, t)\, dt$$

If $\partial P / \partial y \neq \partial Q / \partial x$, no such f can exist.

In Chapter 9 we studied a number of useful classes of differential equations. A new class may be solved by the methods of this section. Namely, a differential equation of the form

$$P(x, y) + Q(x, y) \frac{dy}{dx} = 0 \qquad\qquad\qquad \textbf{(6)}$$

will be called *exact* if the corresponding differential form $P\,dx + Q\,dy$ is exact. These equations may be solved as follows: there is a function $f(x, y)$ such that $P = f_x$ and $Q = f_y$. The curves

$$f(x, y) = C \qquad\qquad\qquad \textbf{(7)}$$

for C a constant are solutions of (6) assuming that they define y as a function of x. Indeed, differentiating (7) implicitly by the chain rule, we get

$$f_x + f_y \frac{dy}{dx} = 0 \qquad \text{i.e.,} \qquad P + Q \frac{dy}{dx} = 0$$

as required. The following display summarizes the situation.

EXACT DIFFERENTIAL EQUATIONS

To test if the equation

$$P + Q \frac{dy}{dx} = 0$$

is exact,

see if $\dfrac{\partial P}{\partial y} = \dfrac{\partial Q}{\partial x}$

To solve an exact equation

$$P + Q \frac{dy}{dx} = 0$$

find $f(x, y)$ such that

$$P = f_x \quad \text{and} \quad Q = f_y$$

(See the preceding display.) The solutions are $f(x, y) = C$ for any constant C.

Worked Example 5   Find the solution of the differential equation

$$2xy \cos y + x^2 + (x^2 \cos y - x^2 y \sin y) \frac{dy}{dx} = 0$$

that passes through $(1, 0)$.

Solution   By Worked Example 4, this equation is exact and

$$f(x, y) = \frac{x^3}{3} + x^2 y \cos y$$

The solution is thus $(x^3/3) + x^2 y \cos y = C$. Since $y = 0$ when $x = 1$, $C = \frac{1}{3}$. Thus our solution is given implicitly by

$$\frac{x^3}{3} + x^2 y \cos y = \frac{1}{3}$$

Solved Exercises

4. Is $x\mathbf{i} + xy\mathbf{j}$ a gradient? If so, of what function f?

5. (a) Let $(x, y) = (e^{t-1}, \sin(\pi/t))$, $1 \le t \le 2$, be a parametrization of the curve C. Calculate $\int_C 2x \cos y \, dx - x^2 \sin y \, dy$.

   (b) Find the solution of $2x \cos y = x^2 \sin y \dfrac{dy}{dx}$ that satisfies $y(3) = 0$.

6. Let $P = -y/(x^2 + y^2)$ and $Q = x/(x^2 + y^2)$. Show that $\partial P/\partial y = \partial Q/\partial x$, but that the integral of $P \, dx + Q \, dy$ around the unit circle is not zero. Conclude that $P \, dx + Q \, dy$ is not exact. What is wrong?

7. Show that we can also choose

$$f(x, y) = \int_0^1 xa(tx, ty) \, dt + yb(tx, ty) \, dt$$

   in the display entitled "exact differentials."

Exercises

4. (a) Is $2xy\mathbf{i} + (x^2 + \cos y)\mathbf{j}$ a gradient? If so, find f.

   (b) Is $x^2 y\mathbf{i} + (\frac{1}{2}x^3 + ye^y)\mathbf{j}$ a gradient? If so, find f.

5. Let C_1 be parametrized by $(t^2 + 1, t \sin(\pi t/8))$, $0 \le t \le 4$, and C_2 by $((t^2 + 1) \cos \pi t, t)$, $0 \le t \le 4$. Determine whether $\int_{C_1} P \, dx + Q \, dy = \int_{C_2} P \, dx + Q \, dy$, where:

   (a) $P \, dx + Q \, dy = \frac{1}{2}(x^4 + xy^2) \, dx + (xy + e^y) \, dy$

   (b) $P = -y/(x^2 + y^2 + 1)$; $Q = x/(x^2 + y^2 + 1)$

6. (a) Verify that $\partial P/\partial y = \partial Q/\partial x$ and find $u = f(x, y)$ such that $P = \partial u/\partial x$ and $Q = \partial u/\partial y$ in each case:

   (i) $P = x^2 + y^3$; $Q = 3xy^2 + 1$

   (ii) $P = y \cos x$; $Q = y + \sin x$

   (b) Solve the following differential equations with the given conditions:

   (i) $x^2 + y^3 + (3xy^2 + 1) \dfrac{dy}{dx} = 0$; $y(0) = 1$

   (ii) $y \cos x + (y + \sin x) \dfrac{dy}{dx} = 0$; $y(\pi/4) = 1$

CIRCULATION AND STOKES' THEOREM

If $\boldsymbol{\Phi}$ is a vector field defined at points in the plane, we can write $\boldsymbol{\Phi} = P\mathbf{i} + Q\mathbf{j}$. The line integral of $\boldsymbol{\Phi}$ around a curve C, namely

$$\int_C \boldsymbol{\Phi} = \int_C P \, dx + Q \, dy$$

occurs on the left-hand side of the equation in the statement of Green's theorem (see Theorem 4). The terminology "the *circulation* of Φ around C" is often used for the number $\int_C \Phi$. This terminology arose historically through the application of Green's theorem to fluid mechanics. We shall now discuss this application.

Imagine a fluid moving in the plane. Each particle of the fluid (or piece of dust suspended in the fluid) has a well-defined velocity. If, at a particular time, we assign to each point (x, y) of the plane the velocity $\mathbf{V}(x, y)$ of the fluid particle moving through (x, y) at the particular time in question, we define a vector field $\mathbf{V}$ on the plane. Similarly, a fluid moving in space defines a velocity vector field in space. See Fig. 18-2-9. The integral $\int_C \mathbf{V}$ of $\mathbf{V}$ around a closed curve C represents, intuitively, the sum of the tangential components of $\mathbf{V}$ around C. Thus, if C is traversed counterclockwise, and $\int_C \mathbf{V} > 0$, there is a net counterclockwise motion of the fluid. Likewise, if $\int_C \mathbf{V} < 0$, the fluid is circulating clockwise. This intuition explains the origin of the term "circulation" and is illustrated in Fig. 18-2-10.

Path of a fluid particle

$\mathbf{V}(x, y) =$ velocity of a fluid particle

Fig. 18-2-9   The velocity field of a fluid.

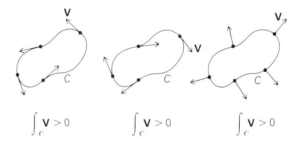

$\int_C \mathbf{V} > 0$ $\int_C \mathbf{V} > 0$ $\int_C \mathbf{V} > 0$

Fig. 18-2-10   The intuitive meaning of the possible signs of $\int_C \mathbf{V}$.

In the previous section we interpreted the line integral of a force vector field along a curve as the work done by the force on a particle traversing the curve. Notice that the single mathematical concept of a line integral is subject to different physical interpretations, depending on the type of vector field chosen.

The integrand on the right-hand side of Green's theorem, namely

$$\frac{\partial Q}{\partial x} - \frac{\partial P}{\partial y}$$

becomes an object of importance, since when integrated over the region whose boundary is C, it produces the circulation of Φ around C.

Definition   If $\boldsymbol{\Phi} = P\mathbf{i} + Q\mathbf{j}$ is a vector field in the plane,

$$\frac{\partial Q}{\partial x} - \frac{\partial P}{\partial y}$$

is called the *scalar curl* of $\boldsymbol{\Phi}$.

Worked Example 6   In the plane, the vector field

$$\mathbf{V}(x, y) = \frac{y\,\mathbf{i}}{x^2 + y^2} - \frac{x\,\mathbf{j}}{x^2 + y^2}$$

approximates the velocity field of water flowing down a drain (see Fig. 18-2-11). (a) Calculate its scalar curl. (b) Is Green's theorem valid for this field on the unit disk D, defined by $x^2 + y^2 \le 1$?

Fig. 18-2-11 The velocity field near a drain.

Solution   Here $P(x, y) = y/(x^2 + y^2)$ and $Q(x, y) = -x/(x^2 + y^2)$ so the scalar curl is

$$\begin{aligned}
\frac{\partial Q}{\partial x} - \frac{\partial P}{\partial y} &= -\frac{\partial}{\partial x}\left(\frac{x}{x^2 + y^2}\right) - \frac{\partial}{\partial y}\left(\frac{y}{x^2 + y^2}\right) \\
&= \frac{-(x^2 + y^2) + 2x^2}{(x^2 + y^2)^2} + \frac{-(x^2 + y^2) + 2y^2}{(x^2 + y^2)^2} \\
&= 0
\end{aligned}$$

The circulation of $\mathbf{V}$ about the circle $x^2 + y^2 = 1$ is

$$\int_C P\,dx + Q\,dy = \int_C y\,dx - x\,dy = -2 \cdot (\text{area of disk}) = -2\pi$$

(See Worked Example 3.) This is an apparent contradiction to Green's theorem! The resolution is the fact that the hypotheses of Green's theorem are not satisfied; $\mathbf{V}$ is not defined at $(0, 0)$.

Because fluids can move in space as well as in the plane, it is reasonable to search for a generalization of Green's theorem. There is such a theorem, called *Stokes' theorem*. We shall prove a special case of this theorem.

Consider a region D in the xy plane and a function f defined in D. The equation $z = f(x, y)$ defines a surface over D. If we refer to the discussion on p. 692, we see that a normal vector to this surface is given by

$$-f_x \mathbf{i} - f_y \mathbf{j} + \mathbf{k}$$

and so a unit normal is

$$\mathbf{n} = \frac{-f_x \mathbf{i} - f_y \mathbf{j} + \mathbf{k}}{\sqrt{f_x^2 + f_y^2 + 1}}$$

The area element on the surface is given by

$$dA = \sqrt{1 + f_x^2 + f_y^2}\; dx\, dy$$

as was shown in Section 17-1, p. 769.

Definition If $\boldsymbol{\Phi} = P\mathbf{i} + Q\mathbf{j} + R\mathbf{k}$ is a vector field in space and S is the surface $z = f(x, y)$, the *surface integral* of $\boldsymbol{\Phi}$ over S is the integral of the normal component of $\boldsymbol{\Phi}$ over S:

$$\iint_S \boldsymbol{\Phi} = \iint_S \boldsymbol{\Phi} \cdot \mathbf{n}\; dA$$

$$= \iint_D (-Pf_x - Qf_y + R)\; dx\, dy \qquad (8)$$

We shall discuss a physical interpretation of this definition in the following subsection.

Worked Example 7 Let $\boldsymbol{\Phi} = x^2 \mathbf{i} + y^2 \mathbf{j} + z\mathbf{k}$. Evaluate $\iint_S \boldsymbol{\Phi}$ where S is the graph of $z = x + y + 1$ over the rectangle $0 \le x \le 1$, $0 \le y \le 1$.

Solution By definition, with $P(x, y, z) = x^2$, $Q(x, y, z) = y^2$, $R(x, y, z) = z$, and $z = f(x, y) = x + y + 1$,

$$\iint_S \boldsymbol{\Phi} = \iint_D [-x^2 \cdot 1 - y^2 \cdot 1 + (x + y + 1)]\; dx\, dy$$

$$= \int_0^1 \int_0^1 (x + y + 1 - x^2 - y^2)\; dx\, dy$$

$$= \frac{1}{2} + \frac{1}{2} + 1 - \frac{1}{3} - \frac{1}{3} = \frac{4}{3}$$

Stokes' theorem will involve another concept defined as follows:

Definition Let $\boldsymbol{\Phi} = P\mathbf{i} + Q\mathbf{j} + R\mathbf{k}$ be a vector field in space. Its *curl* is defined by

$$\text{curl } \boldsymbol{\Phi} = (R_y - Q_z)\mathbf{i} + (P_z - R_x)\mathbf{j} + (Q_x - P_y)\mathbf{k}$$

If Φ is a vector field in the plane, but regarded as being in space with $R = 0$ and P, Q independent of z, then the curl of Φ is just the scalar curl of Φ times $\mathbf{k}$.

Formally, we can write

$$\text{curl } \Phi = \nabla \times \Phi = \begin{vmatrix} \dfrac{\partial}{\partial x} & \dfrac{\partial}{\partial y} & \dfrac{\partial}{\partial z} \\ P & Q & R \\ \mathbf{i} & \mathbf{j} & \mathbf{k} \end{vmatrix}$$

which helps one remember the formula.

Worked Example 8   Find the curl of $xy\mathbf{i} - \sin z\mathbf{j} + \mathbf{k}$.

Solution

$$\text{curl } \Phi = \nabla \times \Phi = \begin{vmatrix} \dfrac{\partial}{\partial x} & \dfrac{\partial}{\partial y} & \dfrac{\partial}{\partial z} \\ xy & -\sin z & 1 \\ \mathbf{i} & \mathbf{j} & \mathbf{k} \end{vmatrix}$$

$$= \begin{vmatrix} \dfrac{\partial}{\partial y} & \dfrac{\partial}{\partial z} \\ -\sin z & 1 \end{vmatrix} \mathbf{i} - \begin{vmatrix} \dfrac{\partial}{\partial x} & \dfrac{\partial}{\partial z} \\ xy & 1 \end{vmatrix} \mathbf{j} + \begin{vmatrix} \dfrac{\partial}{\partial x} & \dfrac{\partial}{\partial y} \\ xy & -\sin z \end{vmatrix} \mathbf{k}$$

$$= \cos z\, \mathbf{i} - x\, \mathbf{k}$$

Now we are ready to state Stokes' theorem.

Theorem 6 (Stokes' theorem)   *Let D be a region in the plane to which Green's theorem applies and S be the surface $z = f(x, y)$, where f is twice continuously differentiable. Let ∂D be the boundary of D traversed counterclockwise and ∂S the corresponding boundary of S (see Fig. 18-2-12). If Φ is a continuously differentiable vector field in space, then*

$$\int_{\partial S} \Phi(\mathbf{r}) \cdot d\mathbf{r} = \iint_S (\nabla \times \Phi) \cdot \mathbf{n}\, dA$$

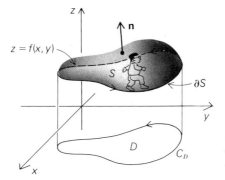

Fig. 18-2-12 As you traverse ∂S counterclockwise, the surface is on your left.

(optional)

Proof Let $\mathbf{\Phi} = P\mathbf{i} + Q\mathbf{j} + R\mathbf{k}$ so that

$$\nabla \times \mathbf{\Phi} = (R_y - Q_z)\mathbf{i} + (P_z - R_x)\mathbf{j} + (Q_x - P_y)\mathbf{k}$$

We may use formula (8) to write

$$\iint_S \operatorname{curl} \mathbf{\Phi} = \int_D \left[\left(\frac{\partial R}{\partial y} - \frac{\partial Q}{\partial z} \right) \left(-\frac{\partial z}{\partial x} \right) \right.$$
$$\left. + \left(\frac{\partial P}{\partial z} - \frac{\partial R}{\partial x} \right) \left(-\frac{\partial z}{\partial y} \right) + \left(\frac{\partial Q}{\partial x} - \frac{\partial P}{\partial y} \right) \right] dA \qquad \textbf{(9)}$$

On the other hand, if $\boldsymbol{\sigma}(t) = x(t)\mathbf{i} + y(t)\mathbf{j}$ parametrizes ∂D, then $\boldsymbol{\eta}(t) = x(t)\mathbf{i} + y(t)\mathbf{j} + f(x(t), y(t))\mathbf{k}$ is the orientation-preserving parametrization of the oriented simple closed curve ∂S discussed above. Thus

$$\int_{\partial S} \mathbf{\Phi} = \int_a^b \left(P \frac{dx}{dt} + Q \frac{dy}{dt} + R \frac{dz}{dt} \right) dt \qquad \textbf{(10)}$$

But by the Chain Rule

$$\frac{dz}{dt} = \frac{\partial z}{\partial x} \frac{dx}{dt} + \frac{\partial z}{\partial y} \frac{dy}{dt}$$

Substituting this expression into (10) we obtain

$$\int_{\partial S} \mathbf{\Phi} = \int_a^b \left[\left(P + R \frac{\partial z}{\partial x} \right) \frac{dx}{dt} + \left(Q + R \frac{\partial z}{\partial y} \right) \frac{dy}{dt} \right] dt$$
$$= \int_\sigma \left(P + R \frac{\partial z}{\partial x} \right) dx + \left(Q + R \frac{\partial z}{\partial y} \right) dy$$
$$= \int_{\partial D} \left(P + R \frac{\partial z}{\partial x} \right) dx + \left(Q + R \frac{\partial z}{\partial y} \right) dy \qquad \textbf{(11)}$$

Applying Green's Theorem to (11) yields

$$\iint_D \left[\frac{\partial (Q + R\, \partial z/\partial y)}{\partial x} - \frac{\partial (P + R\, \partial z/\partial x)}{\partial y} \right] dA$$

Now we use the chain rule, remembering that P, Q, and R are functions of x, y, and z, and z is a function of x and y, to obtain

$$\int_{\partial S} \mathbf{\Phi}$$
$$= \iint_D \left[\left(\frac{\partial Q}{\partial x} + \frac{\partial Q}{\partial z} \cdot \frac{\partial z}{\partial x} + \frac{\partial R}{\partial x} \cdot \frac{\partial z}{\partial y} + \frac{\partial R}{\partial z} \cdot \frac{\partial z}{\partial x} \cdot \frac{\partial z}{\partial y} + R \cdot \frac{\partial^2 z}{\partial x\, \partial y} \right) \right.$$
$$\left. - \left(\frac{\partial P}{\partial y} + \frac{\partial P}{\partial z} \cdot \frac{\partial z}{\partial y} + \frac{\partial R}{\partial y} \cdot \frac{\partial z}{\partial x} + \frac{\partial R}{\partial z} \cdot \frac{\partial z}{\partial y} \cdot \frac{\partial z}{\partial x} + R \frac{\partial^2 z}{\partial y\, \partial x} \right) \right] dA$$

The last two terms in each parenthesis cancel each other, and we can rearrange terms to obtain the integral (9), which completes the proof.

As with Green's theorem, Stokes' theorem is valid for a much wider class of surfaces than graphs, but for simplicity we have treated only this case.

Worked Example 9 Let $\boldsymbol{\Phi} = ye^z\mathbf{i} + xe^z\mathbf{j} + xye^z\mathbf{k}$. Show that the integral of $\boldsymbol{\Phi}$ around an oriented simple closed curve C that is the boundary of a surface S is 0. (Assume S is the graph of a function, as in Theorem 6.)

Solution By Stokes' theorem,

$$\int_C \boldsymbol{\Phi} = \int_S \nabla \times \boldsymbol{\Phi}$$

However,

$$\nabla \times \boldsymbol{\Phi} = \begin{vmatrix} \mathbf{i} & \mathbf{j} & \mathbf{k} \\ \dfrac{\partial}{\partial x} & \dfrac{\partial}{\partial y} & \dfrac{\partial}{\partial z} \\ ye^z & xe^z & xye^z \end{vmatrix} = \mathbf{0}$$

so $\int_C \boldsymbol{\Phi} = \int_S \nabla \times \boldsymbol{\Phi} = \int_S \mathbf{0} = 0$.

Solved Exercises

8. If f is a twice differentiable function in space, prove that $\nabla \times (\nabla f) = \mathbf{0}$.

9. Evaluate the integral of $\mathbf{F}(x, y, z) = x^2\mathbf{i} + y^2\mathbf{j} - z\mathbf{k}$ around the triangle with vertices $(0, 0, 0)$, $(0, 2, 0)$ and $(0, 0, 3)$ using Stokes' theorem.

Exercises

7. Calculate the curl of $\mathbf{F}(x, y, z) = e^z\mathbf{i} - \cos(xy)\mathbf{j} + z^3y\mathbf{k}$ and of $\boldsymbol{\Phi}(x, y, z) = xz\cos x\,\mathbf{i} - yz\sin x\,\mathbf{j} - xy\tan y\,\mathbf{k}$.

8. Show that Stokes' theorem reduces to $0 = 0$ for $\boldsymbol{\Phi} = \nabla f$, by evaluating each side directly.

9. Let C consist of straight lines joining $(2, 0, 0)$, $(0, 1, 0)$, and $(0, 0, 3)$. Evaluate the integral of $\boldsymbol{\Phi}(x, y, z) = xy\,\mathbf{i} + yz\,\mathbf{j} + xz\,\mathbf{k}$ around C by using Stokes' theorem.

10. Let $\boldsymbol{\Phi}$ be perpendicular to the tangent vector of the boundary ∂S of a surface S. Show that

$$\int_S \nabla \times \boldsymbol{\Phi} = 0$$

FLUX AND THE DIVERGENCE THEOREM

Let $\mathbf{V}$ be the velocity field of a fluid moving in the plane. In the previous subsection we explained why the line integral of $\mathbf{V}$ around a closed curve C is called the circulation of $\mathbf{V}$ around C. The line integral is the integral of the tangential component of $\mathbf{V}$. The integral of the normal component of $\mathbf{V}$ around C also has physical meaning.

Imagine first that **V** is constant and C is a line segment; see Fig. 18-2-13. Suppose we consider a parallelogram consisting of a unit area of fluid: the shaded area in Fig. 18-2-13. The parallelogram's base is one unit in length along C and has its other side parallel to **V**. Since its area is one, the other side has length $d = 1/\cos\theta$, where θ is the angle between **n** and **V**. It takes this parallelogram $t = d/\|V\| = 1/\cos\theta \|V\|$ units of time to cross C. Thus $l \cos\theta \|V\|$ square units of fluid cross C per unit time, where l is the length of C. Since **n** has unit length, this rate equals $l\,\mathbf{n}\cdot\mathbf{V}$.

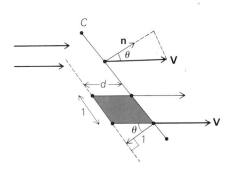

Fig. 18-2-13 The amount of fluid crossing C per unit time is the normal component of **V** times the length of C.

If we now imagine C to consist of straight line segments and **V** to be constant across each one, we are led to interpret the integral of the normal component of **V** along C, that is

$$\int_C \mathbf{V}\cdot\mathbf{n}\, ds$$

as the amount of fluid crossing C per unit of time. This integral is the *flux* of **V** across C.

Let C be parametrized by $\boldsymbol{\sigma}(t) = x(t)\mathbf{i} + y(t)\mathbf{j}$. Then a unit tangent vector is

$$\mathbf{t} = \frac{x'\mathbf{i} + y'\mathbf{j}}{\sqrt{(x')^2 + (y')^2}} \quad \text{where } x' = dx/dt \text{ and } y' = dy/dt$$

The element of length is

$$ds = \sqrt{(x')^2 + (y')^2}\, dt$$

and a unit normal is

$$\mathbf{n} = \frac{y'\mathbf{i} - x'\mathbf{j}}{\sqrt{(x')^2 + (y')^2}}$$

Fig. 18-2-14
$\mathbf{n} = \dfrac{y'\mathbf{i} - x'\mathbf{j}}{\sqrt{(x')^2 + (y')^2}}$ is the
unit outward normal.

This **n** has length one and is perpendicular to **t** as is easily checked. We chose this **n** and not its negative so that if C is a closed curve traversed counter-clockwise, **n** will be the unit *outward* normal, as shown in Fig. 18-2-14.

If $\mathbf{V} = P\mathbf{i} + Q\mathbf{j}$ then substitution of the above formulae for **n** and ds gives

$$\mathbf{V}\cdot\mathbf{n}\, ds = (Py' - Qx')dt$$

This leads to the following definition.

Definition   The *flux* of **V** across *C* is defined to be

$$\int_C \mathbf{V} \cdot \mathbf{n} \, ds = \int_C P\,dy - Q\,dx$$

If *C* is parametrized by $\boldsymbol{\sigma}(t) = x(t)\mathbf{i} + y(t)\mathbf{j}$, $a \le t \le b$, the flux equals

$$\int_a^b \left(P(x(t), y(t)) \frac{dy}{dt} - Q(x(t), y(t)) \frac{dx}{dt} \right) dt$$

as our calculations above show.

The divergence theorem relates the flux of a vector field **V** across *C* to the integral of the *divergence* of **V** over *D* defined as follows.

Definition   The *divergence* of a vector field $\mathbf{V} = P\mathbf{i} + Q\mathbf{j}$ is the function given by

$$\operatorname{div} \mathbf{V} = \nabla \cdot \mathbf{V} = \frac{\partial P}{\partial x} + \frac{\partial Q}{\partial y}$$

We can remember the formula for div **V** in the plane by writing the "dot product" of $\nabla = (\partial/\partial x)\,\mathbf{i} + (\partial/\partial y)\,\mathbf{j}$ with **V**; where instead of multiplying $\partial/\partial x$ and *P*, we let $\partial/\partial x$ *operate on P*. Similarly, our expression for $\nabla \times \boldsymbol{\Phi}$ in space on p. 825 may be regarded as the cross product of $\nabla = \partial/\partial x\,\mathbf{i} + \partial/\partial y\,\mathbf{j} + \partial/\partial z\,\mathbf{k}$ with $\boldsymbol{\Phi}$.

Theorem 7 (Gauss' divergence theorem in the plane)   *Let D be a region in the plane to which Green's theorem applies and let C be its boundary traversed in a counterclockwise direction. Then*

$$\int_C \mathbf{V} \cdot \mathbf{n} \, ds = \iint_D (\operatorname{div} \mathbf{V}) \, dx \, dy$$

Proof   The left side equals

$$\int_C P\,dy - Q\,dx$$

By Green's theorem, this equals (with *P* replaced by $-Q$ and *Q* by *P* in the statement of Green's theorem)

$$\iint_D \left(\frac{\partial P}{\partial x} - \frac{\partial(-Q)}{\partial y} \right) dx \, dy = \iint_D (\operatorname{div} \mathbf{V}) \, dx \, dy$$

Worked Example 10 Calculate the flux of $V = x \cos y \; \mathbf{i} - \sin y \; \mathbf{j}$ across the boundary of the unit square in the plane, with vertices $(0, 0)$, $(1, 0)$, $(1, 1)$, and $(0, 1)$.

Solution The divergence of V is

$$\text{div } V = \frac{\partial}{\partial x}(x \cos y) + \frac{\partial}{\partial y}(-\sin y)$$
$$= \cos y - \cos y = 0$$

so by the divergence theorem, the flux across any closed curve is zero. Thus the flux across the given boundary is zero.

Notice that directly computing the flux across C is possible, but more tedious.

A vector field in the plane is called *incompressible* or *divergence free* if div $V = 0$. This terminology arises from the divergence theorem and the example in which V is the velocity of a fluid. In fact, theorem 7 implies that the flux across closed curves is zero, that is, the net area of fluid entering and leaving the region enclosed by C is zero. If a fluid were compressible, one can imagine situations in which the fluid is being squeezed and a net amount of fluid is entering a region. The flux across C would be negative, and this occurs if div $V < 0$. Likewise, if div $V > 0$ in a region, the fluid is expanding. See Fig. 18-2-15.

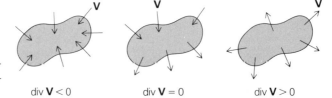

Fig. 18-2-15 A compressing, incompressible and expanding fluid.

div $V < 0$ div $V = 0$ div $V > 0$

We saw that a generalization of Green's theorem to three dimensions which relies on the idea of circulation is given by Stokes' theorem. It is natural to also seek a generalization of the divergence theorem to three dimensions.

Let V be a vector field defined in space. Reasoning as we did in the plane, we see that if V represents the velocity field of a fluid, and S is a surface, then the surface integral $\iint_S V \cdot \mathbf{n} \; dA$ is the volume of fluid crossing the surface S in the direction of the normal $\mathbf{n}$ per unit time. (The surface integral was defined and discussed in the previous subsection). Thus we call $\iint_S V \cdot \mathbf{n} \; dA$ the *flux* of V across S.

If $V = P\mathbf{i} + Q\mathbf{j} + R\mathbf{k}$ we can generalize our definition of two-dimensional divergence by setting

$$\text{div } V = \frac{\partial P}{\partial x} + \frac{\partial Q}{\partial y} + \frac{\partial R}{\partial z}$$

called the *divergence of* V.

Theorem 8 (Gauss' divergence theorem in space) *Let W be a region in space which is of types I, II and III (see Section 17-2), and let ∂W be the surface of W and $\mathbf{n}$ the outward pointing unit normal. Let $\mathbf{V}$ be a (smooth) vector field defined on W. Then*

$$\iiint_W (\text{div } \mathbf{V}) \, dx \, dy \, dz = \iint_{\partial W} (\mathbf{V} \cdot \mathbf{n}) \, dA$$

In words, the total flux across the boundary of W equals the total divergence in W. The proof of theorem 8 is similar to the proof of Green's theorem in the plane and is outlined in Review Problem 23.

Worked Example 11   Evaluate $\iint_S \mathbf{F} \cdot \mathbf{n} \, dA$ where $\mathbf{F}(x, y, z) = xy^2 \mathbf{i} + x^2 y \mathbf{j} + y \mathbf{k}$ and S is the surface of the cylinder W defined by $x^2 + y^2 \leq 1$, $-1 < z < 1$.

Solution   One can compute this integral directly, but it is easier to use the divergence theorem.

Since S is the boundary of the region, W, theorem 8 gives $\iint_S \mathbf{F} \cdot \mathbf{n} \, dA = \iiint_W \text{div } \mathbf{F}$. We have $\text{div } \mathbf{F} = (\partial/\partial x)(xy^2) + (\partial/\partial y)(x^2 y) + (\partial/\partial z)(y) = x^2 + y^2$, and so

$$\iiint_W \text{div } \mathbf{F} \, dx \, dy \, dz = \iiint_W (x^2 + y^2) \, dx \, dy \, dz$$

$$= \int_{-1}^{1} \left(\iint_{x^2 + y^2 \leq 1} (x^2 + y^2) \, dx \, dy \right) dz$$

$$= 2 \iint_{x^2 + y^2 \leq 1} (x^2 + y^2) \, dx \, dy$$

Before evaluating the double integral, we note that $\iint_S \mathbf{F} \cdot \mathbf{n} \, dA = 2 \iint_{x^2 + y^2 \leq 1} (x^2 + y^2) \, dx \, dy > 0$. This means that the net flux of $\mathbf{F}$ out of the cylinder is positive, which agrees with the fact that $0 \leq \text{div } \mathbf{F} = x^2 + y^2$ inside the cylinder.

We change variables to polar coordinates to evaluate the double integral:

$$x = r \cos \theta \qquad y = r \sin \theta \qquad 0 \leq r \leq 1 \qquad 0 \leq \theta \leq 2\pi$$

Replacing $x^2 + y^2$ by r^2 and $dx \, dy$ by $r \, dr \, d\theta$ (see p. 785) we have

$$\iint_{x^2 + y^2 \leq 1} (x^2 + y^2) \, dx \, dy = \int_0^{2\pi} \left(\int_0^1 r^3 \, dr \right) d\theta = \frac{1}{2} \pi$$

Therefore $\iint_S \mathbf{F} \cdot \mathbf{n} \, dA = \iiint_W \text{div } \mathbf{F} \, dx \, dy \, dz = \pi$.

There are numerous mathematical and physical examples of Stokes' and Gauss' theorems beyond those we have given.*

---

* *We refer the reader to the problems and to a more thorough text on the subject, such as* Vector Calculus *by J. Marsden and A. Tromba, for further information.*

Solved Exercises

10. Let $\mathbf{V} = 2x\,\mathbf{i} + 2y\,\mathbf{j} + 2z\,\mathbf{k}$ and S the unit sphere $x^2 + y^2 + z^2 = 1$. Calculate the flux of $\mathbf{V}$ across S.

11. A basic law of electrostatics is that the electric field E in space satisfies div $\mathbf{E} = \rho$ where ρ is the charge density. Show that the flux of $\mathbf{E}$ across "any" closed surface equals the total charge *inside* the surface.

Exercises

11. Calculate the divergence of
    (a) $\boldsymbol{\Phi}(x,y) = x^3\mathbf{i} - x\sin(xy)\mathbf{j}$
    (b) $\mathbf{V}(x,y,z) = e^{xy}\mathbf{i} - e^{xy}\mathbf{j} + e^{yz}\mathbf{k}$

12. Calculate the flux of $\boldsymbol{\Phi}(x,y) = x^2\mathbf{i} - y^3\mathbf{j}$ across the perimeter of the square with vertices $(-1,-1)$, $(-1,1)$, $(1,1)$, $(1,-1)$.

13. Calculate the flux of $\mathbf{V}(x,y,z) = x^3\mathbf{i} + y^3\mathbf{j} + z^3\mathbf{k}$ across the surface of the unit sphere $x^2 + y^2 + z^2 = 1$. [*Hint:* Use spherical coordinates.]

14. Suppose a vector field $\mathbf{V}$ is tangent to the boundary of a region W in space. Prove that $\iiint_W (\text{div } \mathbf{V})\, dx\, dy\, dz = 0$.

Problems for Section 18-2

1. Let C be the boundary of the rectangle $[1,2] \times [1,2]$. Evaluate $\int_C x^2 y\, dx + 3yx^2\, dy$ by using Green's theorem.

2. Check the validity of Green's theorem for the region between the curves $y = x^2$ and $y = x$ between $x = 0$ and $x = 1$, with $P = xy$ and $Q = x$.

3. Suppose that $P\mathbf{i} + Q\mathbf{j}$ is parallel to the tangent vector of a closed curve C.
   (a) Show that $Q\mathbf{i} - P\mathbf{j}$ is perpendicular to the tangent vector.
   (b) Show that $\iint_D (\partial P/\partial x + \partial Q/\partial y)\, dx\, dy = 0$, where D is the region whose boundary is C.

4. Use the corollary of Green's theorem to recover the formula

$$A = \frac{1}{2}\int_\alpha^\beta r^2\, d\theta$$

for a region in polar coordinates (see Section 11-3).

5. Sketch the proof of Green's theorem for the region shown in Fig. 18-2-16.

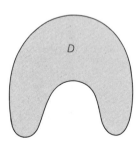

Fig. 18-2-16 Prove Green's theorem for this region.

6. Let C be the perimeter of the rectangle with sides $x = 1$, $y = 2$, $x = 3$, and $y = 3$. Evaluate:
   (a) $\int_C (2y^2 + x^5)\, dx + 3y^6\, dy$
   (b) $\int_C (xy^2 - y^3)\, dx + (-5x^2 + y^3)\, dy$
   (c) $\int_C (3x^4 + 5)\, dx + (y^5 + 3y^2 - 1)\, dy$

7. Find the work done by the force field $(3x + 4y)\mathbf{i} + (8x + 9y)\mathbf{j}$ on a particle which moves once around the ellipse $4x^2 + 9y^2 = 36$ by (a) directly evaluating the line integral and by (b) using Green's theorem.

8. Find the area between the curves $y = x^3$ and $y = \sqrt{x}$ by using Green's theorem.

9. Is $4x\cos^2(y/2)\mathbf{i} - x^2\sin y\mathbf{j}$ a gradient? If so, find f.

10. Is $[2x + e^{xy}(xy + 1)]\, dx + x^2 e^{xy}\, dy$ an exact differential? If so, find f.

★11. (a) If D is a region to which Green's theorem applies, write the identity of Solved Exercise 2 this way:

$$\iint_D \left(P\frac{\partial Q}{\partial x} - P\frac{\partial Q}{\partial y} \right) dx\, dy$$
$$= \int_C PQ\, dx + PQ\, dy$$
$$- \iint_D \left(Q\frac{\partial P}{\partial x} - Q\frac{\partial P}{\partial y} \right) dx\, dy$$

How is this like integration by parts?

This problem continues on the next page.

★11. (b) Discuss the following statement: Green's theorem is "the fundamental theorem of calculus" in the plane since it relates a double integral to an integral around the boundary, just as the fundamental theorem relates an integral over an interval to a sum over the boundary of the interval (that is, the two endpoints).

12. Let $P = y$ and $Q = x$. What is $\int_C P\,dx + Q\,dy$ if C is a closed curve?

★13. *Project:* The formula $A = \frac{1}{2}\int_C x\,dy - y\,dx$ is the basis for the operation of the *planimeter*, a mechanical device for measuring areas. Find out about planimeters from an encyclopedia or other source and relate their operation to this integral.

14. We call $\nabla^2 u = \partial^2 u/\partial x^2 + \partial^2 u/\partial y^2$ the *laplacian* of $u = f(x, y)$. Prove:

(a) $$\iint_D u\nabla^2 v\,dx\,dy = -\iint_D \nabla u \cdot \nabla v\,dx\,dy$$
$$\int_D u\frac{\partial v}{\partial x}\,dy - u\frac{\partial v}{\partial y}\,dx$$

(b) $$\iint_D (u\nabla^2 v - v\nabla^2 u)\,dx\,dy$$
$$= \int_C \left(u\frac{\partial v}{\partial x} - v\frac{\partial u}{\partial x}\right)dy$$
$$- \left(u\frac{\partial v}{\partial y} - v\frac{\partial u}{\partial y}\right)dx$$

These are called *Green's first and second identities.* [*Hint:* To prove part (b), write down part (a) again with u and v interchanged and subtract.]

(c) Use Green's first identity with $v = u$ to prove that if $\nabla^2 u = 0$ on D and $u = 0$ on C, then $\nabla u = 0$ on D, and hence $u = 0$ on D.

15. Suppose that $\partial^2 u/\partial x^2 + \partial^2 u/\partial y^2 = 0$ on D. Show that
$$\int_C \frac{\partial u}{\partial y}\,dx - \frac{\partial u}{\partial x}\,dy = 0$$

16. Show that the area enclosed by the hypocycloid $x = a\cos^3\theta$, $y = a\sin^3\theta$, $0 \le \theta \le 2\pi$ is $\frac{3}{8}\pi a^2$. (Use Green's theorem.)

17. Verify Green's theorem when D is the disk of radius r, center $(0, 0)$, and $P = xy^2$, $Q = -yx^2$.

18. Evaluate $\int_C (x^5 - 2xy^3)\,dx - 3x^2 y^2\,dy$, where C is parametrized by (t^8, t^{10}), $0 \le t \le 1$.

19. Find the area bounded by one arc of the cycloid $x = a(\theta - \sin\theta)$, $y = a(1 - \cos\theta)$, $a > 0$, $0 \le \theta \le 2\pi$.

20. Let $P = \ln(x^2 + 1) - 2xe^{-y}$ and $Q = x^2 e^{-y} - \ln(y^2 + 1)$. Determine whether
$$\int_{C_1} P\,dx + Q\,dy = \int_{C_2} P\,dx + Q\,dy$$
where C_1 and C_2 are two curves with the same endpoints.

21. Is there a function $u = f(x, y)$ such that $\partial u/\partial x = x\sqrt{x^2 y^2 + 1}$ and $\partial u/\partial y = y\sqrt{x^2 y^2 + 1}$? If so, find it.

22. Solve the following differential equations:

(a) $ye^x + e^y + (xe^y + e^x)\dfrac{dy}{dx} = 0$, $y(0) = 2$

(b) $e^y + (xe^y + 2y)\dfrac{dy}{dx} = 0$, $y(0) = 1$

(c) $3x^2 + 2xy + (x^2 + y^2)\dfrac{dy}{dx} = 0$, $y(1) = 2$

23. Determine which of the following equations are exact and solve the ones that are:

(a) $x\dfrac{dy}{dx} + y + x^2 y^2 - 1 = 0$

(b) $y - x^3 + (x + y^3)\dfrac{dy}{dx} = 0$

(c) $xy^2 + 3x^2 y + (x + y)x^2\dfrac{dy}{dx} = 0$

(d) $e^x\sin y - e^y\sin x\,\dfrac{dy}{dx} = 0$

(e) $\cos y\sin x - \sin y\cos x\,\dfrac{dy}{dx} = 0$

24. Let $\mathbf{F} = x^2\mathbf{i} + (2xy + x)\mathbf{j} + z\mathbf{k}$. Let C be the circle $x^2 + y^2 = 1$ oriented counterclockwise and S the disk $x^2 + y^2 \le 1$. Determine

(a) the flux of $\mathbf{F}$ through S.

(b) the circulation of $\mathbf{F}$ around C.

(c) Find the flux of $\nabla \times \mathbf{F}$ through S. Verify Stokes' theorem directly in this case.

25. Evaluate $\iint_{\partial W} \mathbf{F}\cdot\mathbf{n}\,dA$ where $\mathbf{F}(x, y, z) = x\mathbf{i} + y\mathbf{j} - z\mathbf{k}$ and W is the unit cube in the first octant. Perform the calculation directly and check by using the divergence theorem.

26. Evaluate the flux of $\boldsymbol{\Phi}(x, y, z) = 3xy^2\mathbf{i} + 3x^2 y\mathbf{j} + z^3\mathbf{k}$ out of the unit sphere.

27. Let **E** and **H** be time-dependent electric and magnetic fields in space. Let S be a surface with boundary C. We define

$$\int_C \mathbf{E} \cdot d\mathbf{r} = \text{voltage drop around } C$$

$$\int_S \mathbf{H} \cdot \mathbf{n}\, dA = \text{magnetic flux across } S$$

Faraday's law (see Fig. 18-2-17) states that the voltage around C equals the negative rate of change of magnetic flux through S. Assuming that you may differentiate under the integral sign,* show that Faraday's law follows from the following differential equation (one of the Maxwell equations):

$$\nabla \times \mathbf{E} = -\frac{\partial H}{\partial t}$$

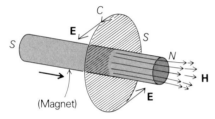

Fig. 18-2-17   Faraday's law.

28. *Ampere's law* states that if an electric current density is described by a vector field **j** and the induced magnetic field is **H**, then the circulation of **H** around the boundary C of a surface S equals the flux of **j** across S (i.e., the total current crossing S). See Fig. 18-2-18. Show that this is implied by the Maxwell equation $\nabla \times \mathbf{H} = \mathbf{j}$.

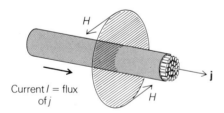

Fig. 18-2-18   Ampere's law.

29. (i) If $\boldsymbol{\Phi}$ is a twice continuously differentiable vector field in space, show that
$$\nabla \cdot (\nabla \times \boldsymbol{\Phi}) = 0.$$

(ii) Prove that $\dfrac{\partial}{\partial t}(\nabla \cdot \mathbf{H}) = 0$ from Faraday's law (Problem 27).

(iii) Prove that $\nabla \cdot \mathbf{j} = 0$ from Ampere's law (Problem 28).

(iv) Argue physically that the flux of **j** through any closed surface is zero (conservation of charge). Use this to deduce $\nabla \cdot \mathbf{j} = 0$ from Gauss' theorem.

30. Figure 18-2-19 shows some flow lines for a fluid moving in the plane. Let **V** be the velocity field. At which of the indicated points A, B, C, D is

(a) div $\mathbf{V} > 0$?   (b) div $\mathbf{V} < 0$?

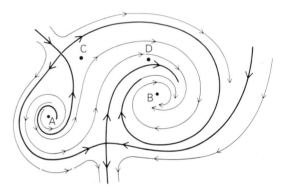

Fig. 18-2-19   The flow lines of a fluid moving in the plane.

---

*These issues are dealt with in Review Problem 20 of Chapter 17.

Review Problems For Chapter 18

1. Evaluate:

   (a) $\int_C xy\,dx + x\sin y\,dy$, where C is the straight line joining $(0,1,1)$ to $(2,2,-3)$

   (b) $\int_C x\,dy$, where C is the unit circle $x^2 + y^2 = 1$ traversed counterclockwise

   (c) $\int_C \nabla(xy^2\cos z)\cdot d\mathbf{r}$ where C is a curve in space joining $(0,1,0)$ to $(8,2,\pi)$

2. Evaluate:

   (a) $\int_C x^2\,dx + y^2\,dy$, where C is the unit circle $x^2 + y^2 = 1$, traversed counterclockwise

   (b) $\int_C \dfrac{\partial}{\partial x}(e^{xyz})dx + \dfrac{\partial}{\partial y}(e^{xyz})\,dy + \dfrac{\partial}{\partial z}(e^{xyz})dz$

   where C is a curve in space joining $(1,1,1)$ to $(0,2,0)$

   (c) $\int_C xe^y\,dx - ye^x\,dy$ where C is the straight line joining $(0,1,0)$ to $(2,1,1)$

3. Find the work done moving a particle along the path $\boldsymbol{\sigma}(t) = (t, t^2)$, $0 \le t \le 1$ subject to the force $\boldsymbol{\Phi}(x,y) = e^x\mathbf{i} - xe^{xy}\mathbf{j}$.

4. Find the work done moving a particle from $(0,0,0)$ to $(1,1,1)$ along the path $(x,y,z) = (t^3, t, t^2)$ subject to the force $\boldsymbol{\Phi}(x,y,z) = z\cos x\mathbf{j} + 8yz\mathbf{k}$.

5. (a) Evaluate $\int_C \sin x\,dx - \ln z\,dy + xy\,dz$, where C is parametrized by $(2t+1, \ln t, t^2)$, $1 \le t \le 2$.

   (b) Evaluate $\int_C xy^2\,dx + (y+z)\,dy + [\sin(e^y)]\,dz$, where C is parametrized by $(\sin 2t, \cos t, 5)$, $0 \le t \le 2\pi$.

6. Let $f(x,y) = x^2\sin y - xy^2$. Evaluate $\int_C (\partial f/\partial x)\,dx + (\partial f/\partial y)\,dy$, where C is parametrized by:

   (a) $\boldsymbol{\sigma}(t) = (\tan \pi t, \ln(t+1))$; $-\frac14 \le t \le \frac14$

   (b) $\boldsymbol{\sigma}(t) = (t\cos^2(\pi/4)\,t, e^{1+t})$; $0 \le t \le 2$

   (c) $\boldsymbol{\sigma}(t) = (t^3 - 2t^2 + 1, \sin^5(\pi t/2) - 2t^{3/2})$; $0 \le t \le 1$

7. If $f(x,y,z) = e^{x^2+y} - \ln(z^2 + 1)$, evaluate $\int_C \nabla f(\mathbf{r})\cdot d\mathbf{r}$ along the curve $\boldsymbol{\sigma}(t) = (-t^2, t^3 + 1, t\sin\frac{15}{2}\pi t)$, $0 \le t \le 1$.

8. If D is a region in the plane with boundary curve C traversed counterclockwise, express the following in terms of the area of D:

   (a) $\int_C x\,dy$ (b) $\int_C y\,dx$ (c) $\int_C x\,dx$

★9. (a) Use the corollary of Green's theorem to find a formula for the area of the triangle with vertices (x_1,y_1), (x_2,y_2), and (x_3,y_3).

   (b) Use the corollary of Green's theorem to find a formula for the area of the n-sided polygon whose consecutive vertices are (x_1,y_1), (x_2,y_2), ..., (x_n,y_n).

10. Which of the following vector fields are conservative?

   (a) $\boldsymbol{\Phi}(x,y) = x\tan y\mathbf{i} + x\sec^2 y\mathbf{j}$

   (b) $\boldsymbol{\Phi}(x,y) = \tan y\mathbf{i} + x\sec^2 y\mathbf{j}$

   (c) $\boldsymbol{\Phi}(x,y) = 3x^2y^2\mathbf{i} + 2x^3y^2\mathbf{j}$

   (d) $\boldsymbol{\Phi}(x,y) = 3x^2y^2\mathbf{i} + 2x^3y\mathbf{j}$

11. For each of the conservative vector fields $\boldsymbol{\Phi}$ in Problem 10, find f such that $\boldsymbol{\Phi} = \nabla f$.

12. (a) If $\boldsymbol{\Phi}(x,y,z) = a(x,y,z)\mathbf{i} + b(x,y,z)\mathbf{j} + c(x,y,z)\mathbf{k}$ is conservative, prove that $a_y = b_x$, $a_z = c_x$, and $b_z = c_y$.

   (b) Is $\boldsymbol{\Phi}(x,y,z) = 2xy\mathbf{i} + (3z^3 + x^2)\mathbf{j} + 9yz^2\mathbf{k}$ conservative? If so, find a function f such that $\nabla f = \boldsymbol{\Phi}$.

13. Which of the following differential forms are exact? Find a potential function for those that are.

   (a) $(e^y\sin x + xe^y\cos x)\,dx + xe^y\sin x\,dy$

   (b) $(y\cos x + \sin z)\,dx + (z\cos y + \sin x)\,dy + (x\cos z + \sin y)\,dz$
   (You may wish to study Problem 12 first.)

   (c) $xe^y\,dx + ye^x\,dy$

   (d) $\exp(x^2 + y^2)(x\,dx + y\,dy)$

14. Solve the following differential equations:

   (a) $y\cos x + 2xe^y + (\sin x + x^2e^y + 2)\dfrac{dy}{dx} = 0$,
   $y(\pi/2) = 0$

   (b) $3x^2 - 2y + e^xe^y + (e^xe^y - 2x + 4)\dfrac{dy}{dx} = 0$, $y(0) = 0$

   (c) $2xy + (x^2 + 1)\dfrac{dy}{dx} = 1$, $y(1) = 1$

15. Test each of the following equations for exactness and solve the exact ones.

   (a) $xy - (x^2 + 3y^2)\dfrac{dy}{dx} = 0$

   (b) $\sin x\sin y - xe^y + (e^y + \cos x\cos y)\dfrac{dy}{dx} = 0$

   (c) $5x^3y^4 - 2y + (3x^2y^5 + x)\dfrac{dy}{dx} = 0$

   (d) $9x^2 + y = (4y - x)\dfrac{dy}{dx} + 1$

   (e) $2xy - 2x + (x^2 + 1)\dfrac{dy}{dx} = 0$

16. *Integrating factors:* Consider a differential equation of the form $M + N\dfrac{dy}{dx} = 0$. If μ is a function (of x and y) such that $\mu M + \mu N \dfrac{dy}{dx} = 0$ is exact, we call μ an *integrating factor* for the given equation.

   (a) Show that μ is an integrating factor when
$$(N\mu_x - M\mu_y)/\mu = M_y - N_x$$

   (b) If $(M_y - N_x)/N = h$ is a function of x alone, show that
$$\mu = \exp(\textstyle\int h(x)\,dx)$$
   is an integrating factor.

   (c) Solve the equation $x\dfrac{dy}{dx} - x^5 + x^3 y - y = 0$.

   (d) Solve the linear equation $\dfrac{dy}{dx} + P(x)y + Q(x) = 0$ using the method of integrating factors. (See Section 9-2 for the answer.)

17. (a) Let $\mathbf{F} = y\mathbf{i} - x\mathbf{j} + zx^3 y^2\,\mathbf{k}$. Calculate $\nabla \times \mathbf{F}$ and $\nabla \cdot \mathbf{F}$.

   (b) Evaluate $\iint_{S_1} (\nabla \times \mathbf{F})\cdot\mathbf{n}\,dA$ where S_1 is the surface $x^2 + y^2 + z^2 = 1$, $z \le 0$.

   (c) Evaluate $\iint_{S_2} \mathbf{F}\cdot\mathbf{n}\,dA$ where S_2 is the surface of the unit cube in the first octant.

18. (a) Let $\mathbf{F} = (x^2 + y - 4)\mathbf{i} + 3xy\,\mathbf{j} + (2xz + z^2)\mathbf{k}$. Calculate the divergence and curl of $\mathbf{F}$.

   (b) Find the flux of the curl of $\mathbf{F}$ across the surface $x^2 + y^2 + z^2 = 16$, $z \ge 0$.

   (c) Find the flux of $\mathbf{F}$ across the surface of the unit cube in the first quadrant.

19. Express as a surface integral the work done by a force field $\mathbf{F}$ going around a closed curve C in space.

20. (a) Let $\mathbf{\Phi}$ be a vector field in space. Follow the proof of Theorem 5, replacing Green's theorem by Stokes' theorem to show that $\mathbf{\Phi} = \nabla f$ for some f (that is, $\mathbf{\Phi}$ is conservative) if and only if $\nabla \times \mathbf{\Phi} = 0$.

   (b) Is $\mathbf{F} = (2xyz + \sin x)\mathbf{i} + x^2 z\mathbf{j} + x^2 y\mathbf{k}$ a gradient? If so, find f.

21. In Worked Example 2, Section 18-2 we gave an example of a region to which Green's theorem as stated did not apply, but which could be cut up into smaller regions to which it did apply. In this way, Green's theorem was extended. Give an example of a similar procedure for Stokes' theorem.

22. For a region W in space with boundary ∂W, unit outward normal $\mathbf{n}$ and functions f and g defined on W and ∂W, prove *Green's identities*

   (a) $\iint_{\partial W} f(\nabla g)\cdot\mathbf{n}\,dA$
       $= \iiint_W (f\nabla^2 g + \nabla f\cdot\nabla g)\,dx\,dy\,dz$

   and

   (b) $\iint_{\partial W} (f\nabla g - g\nabla f)\cdot\mathbf{n}\,dA$
       $= \iiint_W (f\nabla^2 g - g\nabla^2 f)\,dx\,dy\,dz$

   where $\nabla^2 f = \dfrac{\partial^2 f}{\partial x^2} + \dfrac{\partial^2 f}{\partial y^2} + \dfrac{\partial^2 f}{\partial z^2}$ is the laplacian of f.

23. Fill in the details of the proof of Gauss' divergence theorem in space (Theorem 8), outlined as follows:
   It is sufficient to prove the three equalities

$$\iint_{\partial W} P\mathbf{i}\cdot\mathbf{n}\,dA = \iiint_W \frac{\partial P}{\partial x}\,dx\,dy\,dz$$

$$\iint_{\partial W} Q\mathbf{j}\cdot\mathbf{n}\,dA = \iiint_W \frac{\partial Q}{\partial y}\,dx\,dy\,dz$$

$$\iint_{\partial W} R\,\mathbf{k}\cdot\mathbf{n}\,dA = \iiint_W \frac{\partial R}{\partial z}\,dx\,dy\,dz$$

   Prove the last one by expressing W by the inequalities

$$f_2(x,y) \le z \le f_1(x,y),\ (x,y)\text{ in } D$$

   for functions f_1 and f_2 on a domain D in the xy plane. (See Fig. 18-R-1.) Prove that

$$\iiint_W \frac{\partial R}{\partial z}\,dx\,dy\,dz$$
$$= \iint_D [R(x,y,f_1(x,y)) - R(x,y,f_2(x,y))]\,dx\,dy$$

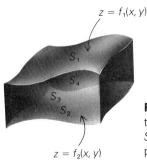

$z = f_1(x,y)$

$z = f_2(x,y)$

Fig. 18-R-1 A region W of type 1. The four sides of $\partial\Omega$ S_3, S_4, S_5, S_6 have normals perpendicular to the z-axis.

The boundary of W is a closed surface whose top S_1 is the graph of $z = f_1(x,y)$, (x,y) in D, and whose bottom S_2 is the graph of $z = f_2(x,y)$, (x,y) in D. The four other sides of ∂W (if they exist) consist of surfaces S_3, S_4, S_5, and S_6 whose normals are always perpendicular to the z-axis. By definition

This problem continues on the next page.

23. (cont'd)

$$\iint_{\partial W} R\mathbf{k}\cdot\mathbf{n}\,dA = \iint_{S_1} R\mathbf{k}\cdot\mathbf{n}_1\,dA$$

$$+ \iint_{S_2} R\mathbf{k}\cdot\mathbf{n}_2\,dA + \sum_{i=3}^{6}\iint_{S_i} R\mathbf{k}\cdot\mathbf{n}_i\,dA$$

Since on each of S_3, S_4, S_5, and S_6 the normal $\mathbf{n}_i$ is perpendicular to $\mathbf{k}$, we have $\mathbf{k}\cdot\mathbf{n}_i = 0$ along these faces, and so the integral reduces to

$$\iint_{\partial W} R\mathbf{k}\cdot\mathbf{n}\,dA = \iint_{S_1} R\mathbf{k}\cdot\mathbf{n}_1\,dA$$

$$+ \iint_{S_2} R\mathbf{k}\cdot\mathbf{n}_2\,dA$$

The surface S_2 is defined by $z = f_2(x,y)$, so

$$\mathbf{n}_2 = \frac{\dfrac{\partial f_2}{\partial x}\mathbf{i} + \dfrac{\partial f_2}{\partial y}\mathbf{j} - \mathbf{k}}{\sqrt{\left(\dfrac{\partial f_2}{\partial x}\right)^2 + \left(\dfrac{\partial f_2}{\partial y}\right)^2 + 1}}$$

Thus

$$\mathbf{n}_2\cdot\mathbf{k} = \frac{-1}{\sqrt{\left(\dfrac{\partial f_2}{\partial x}\right)^2 + \left(\dfrac{\partial f_2}{\partial y}\right)^2 + 1}}$$

and so

$$\iint_{S_2} R(\mathbf{k}\cdot\mathbf{n}_2)\,dA$$

$$= \iint_D R(x,y,f_2(x,y))\left(\frac{-1}{\sqrt{\left(\dfrac{\partial f_2}{\partial x}\right)^2 + \left(\dfrac{\partial f_2}{\partial y}\right)^2 + 1}}\right)$$

$$\times \sqrt{\left(\dfrac{\partial f_2}{\partial x}\right)^2 + \left(\dfrac{\partial f_2}{\partial y}\right)^2 + 1}\ dA$$

$$= -\iint_D R(x,y,f_2(x,y))\,dx\,dy$$

Show that

$$\iint_{S_1} R(\mathbf{k}\ \mathbf{n}_1)\,dA$$

$$= \iint_D R(x,y,f_1(x,y))\,dx\,dy$$

Conclude that

$$\iiint_W \frac{\partial R}{\partial z}\,dx\,dy\,dz = \iint_{\partial W} R(\mathbf{k}\cdot\mathbf{n})\,dA$$

24. Surface integrals apply to the study of heat flow. Let $T(x,y,z)$ be the temperature at a point (x,y,z) in W where W is some region in space and T is a function with continuous partial derivatives. Then

$$\nabla T = \frac{\partial T}{\partial x}\mathbf{i} + \frac{\partial T}{\partial y}\mathbf{j} + \frac{\partial T}{\partial z}\mathbf{k}$$

represents the temperature gradient, and heat "flows" with the vector field $-k\,\nabla T = \mathbf{F}$, where k is a positive constant. Therefore $\iint_S \mathbf{F}\cdot\mathbf{n}\,dA$ is the total rate of heat flow or flux across the surface S. ($\mathbf{n}$ is the unit outward normal.)

   Suppose a temperature function is given as $T(x,y,z) = x^2 + y^2 + z^2$, and let S be the unit sphere $x^2 + y^2 + z^2 = 1$. Find the heat flux across the surface S if $k = 1$.

25. (a) Express conservation of thermal energy by means of the statement that for any volume W in space

$$\frac{d}{dt}\iiint_W e\,dx\,dy\,dz = \iint_{\partial W}\mathbf{F}\cdot\mathbf{n}\,dA$$

where $\mathbf{F} = -k\,\nabla T$ as in Problem 24 and $e = c\rho_0 T$ where c is the specific heat (a constant) and ρ_0 is the mass density (another constant). Use the divergence theorem to show that this statement of conservation of energy is equivalent to the statement

$$\frac{\partial T}{\partial t} = \frac{k}{c\rho_0}\nabla^2 T \quad \text{(heat equation)}$$

where $\nabla^2 T = \text{div grad } T = \dfrac{\partial^2 T}{\partial x^2} + \dfrac{\partial^2 T}{\partial y^2} + \dfrac{\partial^2 T}{\partial z^2}$

is the *laplacian* of T.

★(b) Make up an integral statement of conservation of mass for fluids that is equivalent to the *continuity equation*

$$\frac{\partial \rho}{\partial t} + \text{div}(\rho\,\mathbf{V})$$

where ρ is the mass density of a fluid and $\mathbf{V}$ is the fluid's velocity field.

Contents for the Appendices

Appendix A: Solved Exercises

CHAPTER 1

Solved Exercises for Section 1-1

1. The slope of the line $y = 8x + 8$ is $m = 8$. Thus the velocity is $v = 8$ at any x_0 and in particular at $x_0 = 2$.

2. The average velocity for a time interval Δx starting at $x_0 = 1$ is given by:

$$\frac{\Delta y}{\Delta x} = \frac{f(1 + \Delta x) - f(1)}{\Delta x}$$

where $f(x) = 3x^2 + 8x$. Thus,

$$\frac{\Delta y}{\Delta x} = \frac{[3(1 + \Delta x)^2 + 8(1 + \Delta x)] - [3 + 8]}{\Delta x}$$

$$= \frac{[3(1 + 2\Delta x + (\Delta x)^2) + 8 + 8\Delta x] - 11}{\Delta x}$$

$$= \frac{6\Delta x + 3(\Delta x)^2 + 8\Delta x}{\Delta x}$$

$$= 14 + 3\Delta x$$

As Δx gets close to zero the term $3\Delta x$ gets close to zero as well, and so $\Delta y / \Delta x$ gets close to 14, so the required velocity is $v = 14$.

3. Here the function is linear so the derivative equals the slope: $f'(10) = 12$; alternatively

$$\frac{\Delta y}{\Delta x} = \frac{(12(10 + \Delta x) + 2) - (12 \cdot 10 + 2)}{\Delta x}$$

$$= \frac{12\Delta x}{\Delta x} = 12$$

4. Here

$$\frac{\Delta y}{\Delta x} = \frac{[10(1 + \Delta x)^2 + 2(1 + \Delta x)] - [10 + 2]}{\Delta x}$$

$$= \frac{10(1 + 2\Delta x + (\Delta x)^2) + 2\Delta x + 2 - 12}{\Delta x}$$

$$= \frac{20\Delta x + 10(\Delta x)^2 + 2\Delta x}{\Delta x}$$

$$= 22 + 10\Delta x$$

which gets close to 22 as Δx gets close to zero. Thus the derivative of $f(x) = 10x^2 + 2x$ at $x_0 = 1$ is $f'(1) = 22$. The value of y_0 is $f(x_0) = f(1) = 12$ and so the equation of the tangent line, $y = y_0 + f'(x_0)(x - x_0)$ becomes $y = 12 + 22(x - 1)$ or $y = 22x - 10$.

5. $f'(x) = 3$ for all x, so $f'(8) = 3$.

6. The velocity at time t is $f'(t)$, where $f(t) = 4.9t^2$. By the quadratic function rule $f'(t) = 2(4.9)t = 9.8t$; at $t = 3$, this is 29.4 meters per second.

7. $f'(x) = 2 \cdot 3x + 4 = 6x + 4$, so $f'(1) = 10$. Also, $f(1) = 9$, so the equation of the tangent line is $y = 9 + 10(x - 1)$, or $y = 10x - 1$.

8. Let $f(x) = ax^2 + bx + c$. Then $f'(x) = 2ax + b$ and the derivative of this is $f''(x) = 2a$. Hence $f''(x)$ is equal to zero when $a = 0$—that is, when $f(x)$ is a linear function $bx + c$.

9. The rate of change is the slope: $m = 3$. The equation of a straight line with this slope is $y = 3x + b$, where b is to be determined. Since $y = 2$ when $x = 0$, b must be 2; hence $y = 3x + 2$.

10. Watch the words *of* and *to*! The rise in S *of* 10 means that $\Delta S = 10$; the decline of P *to* 161 means that $\Delta P = 161 - 163 = -2$. Thus the rate of change is $-\frac{2}{10} = -\frac{1}{5}$. (That this is negative reflects the fact that the direction of price change is opposite to the direction of supply change.) We have $P = -\frac{1}{5}S + b$ for some b. Since $P = 163$ when $S = 50$, we have $163 = -\frac{1}{5} \cdot 50 + b$, or $b = 163 + 10 = 173$, so $P = -\frac{1}{5}S + 173$. When $S = 30$, this gives $P = -6 + 173 = 167$. At this point, then, pork will cost $1.67 a pound.

11. The rate of change is the derivative of $f(S) = 0.01S^2 - 0.03S + 160$ with respect to $x = S$ at $x_0 = 50$. By the quadratic function rule $f'(S) = 0.02S - 0.03$. When $S = 50$ we get $f'(50) = 1 - 0.03 = 0.97$.

12. The velocity in Solved Exercise 6 is $f'(t) = 9.8t$. The acceleration is $f''(t) = 9.8$. Thus the acceleration at $t = 3$ is 9.8 meters per second^2.

Solved Exercises for Section 1-2

1. $\lim\limits_{x \to 4} (4x - 2) = \lim\limits_{x \to 4} 4 \cdot \lim\limits_{x \to 4} x - \lim\limits_{x \to 4} 2 = 4 \cdot 4 - 2 = 14$, so
$$\lim\limits_{x \to 4} [1/(4x - 2)] = 1/\lim\limits_{x \to 4} (4x - 2) = \tfrac{1}{14}$$

   The rational function rule gives the result directly. With $f(x) = 1/(4x - 2)$, we have $\lim\limits_{x \to 4} f(x) = f(4) = 1/(4 \cdot 4 - 2) = \frac{1}{14}$.

2. Let $f(x) = (x^3 - 3x^2 + 5x - 3)/(x - 1)$. We find $f(1.1) = 2.01$, $f(1.001) = 2.000001$, $f(0.9) = 2.01$, $f(0.999) = 2.000001$. These results suggest that $\lim\limits_{x \to 1} f(x) = 2$. To verify this, we divide the numerator by $x - 1$, obtaining
$$f(x) = [(x - 1)(x^2 - 2x + 3)]/(x - 1)$$

   which, for $x \neq 1$, is equal to $x^2 - 2x + 3$. Applying the replacement rule and the polynomial rule, we have
$$\lim\limits_{x \to 1} f(x) = \lim\limits_{x \to 1} (x^2 - 2x + 3) = 1^2 - 2 \cdot 1 + 3 = 2$$

3. (a) $\lim\limits_{x \to a} (cf(x)) = (\lim\limits_{x \to a} c)(\lim\limits_{x \to a} f(x)) = c \lim\limits_{x \to a} f(x)$. The first equality uses the product rule; the second uses the constant function rule.

   (b) We write $f(x)/g(x)$ as $f(x) \cdot [1/g(x)]$. By the reciprocal rule,
$$\lim\limits_{x \to a} \frac{1}{g(x)} = \frac{1}{\lim\limits_{x \to a} g(x)}$$
   (The hypothesis $\lim\limits_{x \to a} g(x) \neq 0$ is the same in both rules.) By the product rule,
$$\lim\limits_{x \to a} \frac{f(x)}{g(x)} = \lim\limits_{x \to a} f(x) \cdot \lim\limits_{x \to a} \frac{1}{g(x)}$$
$$= \lim\limits_{x \to a} f(x) \cdot \frac{1}{\lim\limits_{x \to a} g(x)} = \frac{\lim\limits_{x \to a} f(x)}{\lim\limits_{x \to a} g(x)}$$

4. We have $t_0 = 0$ and $\Delta t = 1$. The average velocity is $[f(1) - f(0)]/1 = 4 - 0 = 4$. The instantaneous velocity at time t is $f'(t) = 6t + 1$, which is equal to 4 when $t = \frac{1}{2}$. (It is a peculiarity of quadratic functions that the average velocity on an interval is always equal to the instantaneous velocity at the *midpoint* of the interval.)

5. (a) $[f(0.5) - f(0)]/0.5 = 2.5$, taking $t_0 = 0$ and $\Delta t = 0.5$

   (b) $[f(0.1) - f(0)]/0.1 = 1.3$

   (c) $[f(0.01) - f(0)]/0.01 = 1.03$

   (d) $[f(0.001) - f(0)]/0.001 = 1.003$

   The instantaneous velocity is $f'(t) = 6t + 1$. When $t = 0$, this is equal to 1. The average velocities get close to this number as the time interval becomes short.

6. Letting $f(x) = x^3$, we have
$$f'(x_0) = \lim\limits_{\Delta x \to 0} \frac{f(x_0 + \Delta x) - f(x_0)}{\Delta x}$$
$$= \lim\limits_{\Delta x \to 0} \frac{(x_0 + \Delta x)^3 - x_0^3}{\Delta x}$$
$$= \lim\limits_{\Delta x \to 0} \frac{x_0^3 + 3x_0^2 \Delta x + 3x_0(\Delta x)^2 + (\Delta x)^3 - x_0^3}{\Delta x}$$
   (expanding the cube)
$$= \lim\limits_{\Delta x \to 0} \frac{3x_0^2 \Delta x + 3x_0(\Delta x)^2 + (\Delta x)^3}{\Delta x}$$
$$= \lim\limits_{\Delta x \to 0} (3x_0^2 + 3x_0 \Delta x + (\Delta x)^2)$$
   (by the replacement rule)
$$= 3x_0^2 \quad \text{(using the polynomial rule, setting } \Delta x = 0)$$

   The derivative of x^3 is therefore $3x^2$.

7. Let $f(x) = x^2$, so $f'(x) = 2x$. Thus the first-order approximation to $f(x_0 + \Delta x)$ is $f(x_0) + f'(x_0)\Delta x = x_0^2 + 2x_0\Delta x$.

8. Let $x_0 = 1$ and $\Delta x = 0.03$; use Solved Exercise 7. The approximate value is $1 + 2\Delta x = 1.06$. The exact value is 1.0609. If $\Delta x = 0.0003$, the approximate value is 1.0006 (very easy to compute), while the exact value is 1.00060009 (slightly harder to compute). If $\Delta x = 0.0000003$, the approximate value is 1.0000006, while the exact value is 1.00000060000009. Notice that the error gets small even faster than Δx gets small.

9. $A = f(r) = r^2$. The first-order approximation around $r_0 = 2$ is given by $f(r_0) + f'(r_0)(r - r_0) = r_0^2 + 2r_0(r - r_0) = 4 + 4(r - r_0)$. When $r - r_0 = 0.01$, this is 4.04. The required figure is shown in Fig. 1-2-9. It differs from the square of side 2.01 only by the small shaded square in the corner, whose area is $(0.01)^2 = 0.0001$.

Fig. 1-2-9 The first-order approximation to the change in area with respect to a side has error equal to the shaded area. (Diagram not to scale.)

10. $dy/dx = f'(x) = 16x + 2$.

11. $V = a^2 h$, so $dV/da = 2ah$, by the quadratic function rule, since h is constant.

12. The variable is m so we use $(d/dx)(ax^2 + bx + c) = 2ax + b$ with $a = 4$, $b = 7$, $c = 6$ and with $m = x$ to get $(d/dm)(4m^2 + 7m + 6) = 8m + 7$.

13. $dy/dx = 6x + 5$, so $d^2y/dx^2 = (d/dx)(6x + 5) = 6$.

14. $(d^2/dx^2)(x^3) = (d/dx)(d/dx)(x^3) = (d/dx)(3x^2) = 6x$. (The second equality is from Solved Exercise 6.)

Solved Exercises for Section 1-3

1. $du/dx = 6x + 5$ and $dv/dx = 4x + 5$, thus $du/dx + dv/dx = 10x + 10$. On the other hand, $u + v = 5x^2 + 10x + 9$, so $(d/dx)(u + v) = 10x + 10$, and the sum rule is verified.

2. We apply the sum rule twice:
$$\frac{d}{dx}[f(x) + g(x) + h(x)]$$
$$= \frac{d}{dx}f(x) + \frac{d}{dx}[g(x) + h(x)]$$
$$= \frac{d}{dx}f(x) + \frac{d}{dx}g(x) + \frac{d}{dx}h(x)$$

3. (a) By the sum rule, we sum the derivatives of x^5 and $8x$—that is, $5x^4$ and 8. The answer is thus $5x^4 + 8$.

   (b) By the constant multiple rule, the derivative is five times the derivative of x^3—that is, $(d/dx)(5x^3) = 5(d/dx)(x^3) = 5 \cdot 3x^2 = 15x^2$.

   (c) We differentiate each of the four terms and add:
$$\frac{d}{dx}(x^5 + 6x^2 + 8x + 2)$$
$$= \frac{d}{dx}(x^5) + 6\frac{d}{dx}(x^2) + 8\frac{d}{dx}(x) + \frac{d}{dx}(2)$$
$$= 5x^4 + 12x + 8$$

   (d) $\dfrac{d}{ds}(s^{10} + 8s^9 + 5s^8 + 2)$
$$= 10s^9 + 8 \cdot 9s^8 + 5 \cdot 8s^7 + 0$$
$$= 10s^9 + 72s^8 + 40s^7$$

4. Let $f(x) = x^2 + 2$ and $g(x) = x + 8$. Then $f'(x) = 2x$ and $g'(x) = 1$. By the product rule, the derivative of $f(x)g(x)$ is
$$f'(x)g(x) + f(x)g'(x) = 2x(x + 8) + (x^2 + 2) \cdot 1$$
$$= 2x^2 + 16x + x^2 + 2$$
$$= 3x^2 + 16x + 2$$
If we multiply out first, we find $f(x)g(x) = x^3 + 8x^2 + 2x + 16$; differentiating this directly gives $3x^2 + 8 \cdot 2x + 2 = 3x^2 + 16x + 2$, which agrees with the answer given by the product rule.

5. The velocity at time t is
$$f'(t) = 32t + (0.03) \cdot 4t^3 = 32t + 0.12t^3$$
When $t = 8$, we get $f'(8) = 32 \cdot 8 + 0.12 \cdot 8^3 = 317.44$.

6. The tangent line has the equation $y = f(1) + f'(1)(x - 1)$. Here $f'(x) = 8x^7 + 4x$, so $f'(1) = 12$. Since $f(1) = 4$, the equation is $y = 4 + 12(x - 1)$, or $y = 12x - 8$.

7. Let $f(x) = x^4$; we approximate $f(2.94)$ by the first-order approximation $f(3) + f'(3)(2.94 - 3)$. Since $f'(x) = 4x^3$, this becomes $81 + 108(-0.06) = 81 - 6.48 = 74.52$. (The exact answer is $74.71\ldots$.)

8. By the power of a function rule, with $u = x^4 + 2x^2$ and $n = 3$,

$$\frac{d}{dx}(x^4 + 2x^2)^3 = 3(x^4 + 2x^2)^2 \frac{d}{dx}(x^4 + 2x^2)$$

$$= 3(x^4 + 2x^2)^2(4x^3 + 4x)$$

If we expand the cube first, we get $(x^4 + 2x^2)^3 = x^{12} + 6x^{10} + 12x^8 + 8x^6$, so

$$\frac{d}{dx}(x^4 + 2x^2)^3 = \frac{d}{dx}(x^{12} + 6x^{10} + 12x^8 + 8x^6)$$

$$= 12x^{11} + 60x^9 + 96x^7 + 48x^5$$

To compare the two answers, we expand the first one:

$$3(x^4 + 2x^2)^2(4x^3 + 4x)$$
$$= 3(x^8 + 4x^6 + 4x^4) \cdot 4(x^3 + x)$$
$$= 12(x^{11} + 5x^9 + 8x^7 + 4x^5)$$
$$= 12x^{11} + 60x^9 + 96x^7 + 48x^5$$

which checks.

If we had used the *incorrect* formula $3[f(x)]^2$ for the derivative of $[f(x)]^3$, we would have obtained $3(x^4 + 2x^2)^2 = 3(x^8 + 4x^6 + 4x^4)$, which does not agree with the result obtained by multiplying out first.

9. First of all,

$$\frac{d}{dx}(x^2 + 1)^{27} = 27(x^2 + 1)^{26} \cdot 2x$$

and

$$\frac{d}{dx}(x^4 + 3x + 1)^8 = 8(x^4 + 3x + 1)^7(4x^3 + 3)$$

Now, by the product rule, for

$$f(x) = (x^2 + 1)^{27}(x^4 + 3x + 1)^8,$$
$$f'(x) = 27(x^2 + 1)^{26} \cdot 2x \cdot (x^4 + 3x + 1)^8 +$$
$$(x^2 + 1)^{27} \cdot 8(x^4 + 3x + 1)^7(4x^3 + 3)$$

To simplify this, we can factor out the highest powers of $x^2 + 1$ and $x^4 + 3x + 1$ to get

$$f'(x) = (x^2 + 1)^{26}(x^4 + 3x + 1)^7 \times$$
$$[27 \cdot 2x(x^4 + 3x + 1) + (x^2 + 1) \cdot 8(4x^3 + 3)]$$

We can consolidate the expression in square brackets to a single polynomial of degree 5, getting $f'(x) = 2(x^2 + 1)^{26}(x^4 + 3x + 1)^7(43x^5 + 16x^3 + 93x^2 + 27x + 12)$. [*Note:* Consult your instructor regarding the amount of simplification required.]

10. $(d^2/dx^2)(u^n) = (d/dx)(nu^{n-1} \, du/dx)$. Now we use the constant multiple and product rules to get

$$n\frac{d}{dx}\left(u^{n-1}\frac{du}{dx}\right) =$$
$$n\left[(n-1)u^{n-2}\frac{du}{dx}\frac{du}{dx} + u^{n-1}\frac{d^2u}{dx^2}\right]$$

so

$$\frac{d^2}{dx^2}(u^n) = nu^{n-2}\left[u\frac{d^2u}{dx^2} + (n-1)\left(\frac{du}{dx}\right)^2\right]$$

11. By the quotient rule, with $f(x) = x^2$ and $g(x) = x^3 + 5$,

$$\frac{d}{dx}\left(\frac{x^2}{x^3 + 5}\right) = \frac{2x(x^3 + 5) - x^2(3x^2)}{(x^3 + 5)^2}$$
$$= \frac{x}{(x^3 + 5)^2}(2x^3 + 10 - 3x^3) = \frac{x(-x^3 + 10)}{(x^3 + 5)^2}$$

12. $f'(x) = [2(3x + 1) - (2x + 1)3]/(3x + 1)^2 = -1/(3x + 1)^2$. The equation of the tangent line is

$$y = f(1) + f'(1)(x - 1)$$
$$= \tfrac{3}{4} - \tfrac{1}{16}(x - 1)$$

or

$$y = -\tfrac{1}{16}x + \tfrac{13}{16}.$$

13. Let $h(x) = 1/x$. Applying the quotient rule with $f(x) = 1$ and $g(x) = x$, we get

$$\frac{d}{dx}\left(\frac{1}{x}\right) = \frac{0 \cdot x - 1 \cdot 1}{x^2} = -\frac{1}{x^2}$$

The linear approximation to $1/x$ around $x = 1$ is therefore $1 - (x - 1) = 2 - x$. If $x = 0.98$, this gives 1.02.

14. $(d/dx)[1/(x^3 + 5x)^4] = (d/dx)(x^3 + 5x)^{-4} = -4(x^3 + 5x)^{-5}(3x^2 + 5) = -4(3x^2 + 5)/(x^3 + 5x)^5$. (Don't forget the factor $(3x^2 + 5)$.)

15.
$$\left(\frac{f}{g}\right)'(x) = \left(f \cdot \frac{1}{g}\right)'(x)$$
$$= f'(x) \cdot \frac{1}{g(x)} + f(x) \cdot \left(\frac{1}{g}\right)'(x)$$
$$= \frac{f'(x)}{g(x)} + f(x) \cdot \frac{-g'(x)}{g(x)^2}$$
$$= \frac{f'(x)g(x) - f(x)g'(x)}{g(x)^2}$$

(This calculation gives another way to reconstruct the quotient rule if you forget it—assuming, of course, that you have remembered the reciprocal rule.)

16. Let $f(x) = (x^2 + 3)(x^2 + 4)$. By the product rule, $f'(x) = 2x(x^2 + 4) + (x^2 + 3)2x = 4x^3 + 14x$. By the reciprocal rule, the derivative of $1/f(x)$ is

$$-\frac{f'(x)}{f(x)^2} = -\frac{4x^3 + 14x}{(x^2 + 3)^2(x^2 + 4)^2}$$

Solved Exercises for Section 1-4

1. We may begin by looking for an antiderivative for x^4. If we guess x^5, the derivative is $5x^4$, which is five times too big, so we make a new guess, $\frac{1}{5}x^5$, which works. An antiderivative for 5 is $5x$. Adding our two results gives $\frac{1}{5}x^5 + 5x$; differentiating $\frac{1}{5}x^5 + 5x$ gives $x^4 + 5$, so $\frac{1}{5}x^5 + 5x$ is an antiderivative for $x^4 + 5$. We may add an arbitrary constant to get the antiderivatives $\frac{1}{5}x^5 + 5x + C$.

2. Let v be the velocity. Then $dv/dt = 9.8$; since the antiderivative of 9.8 is $9.8t$, we have $v = 9.8t + C$. At $t = 0$, $v = v_0$, so $v_0 = (9.8)0 + C = C$, and so $v = 9.8t + v_0$. If x is the position, $dx/dt = v = 9.8t + v_0$. Since the antiderivative of $9.8t$ is $(9.8/2)t^2 = 4.9t^2$ and the antiderivative of v_0 is v_0t, we have $x = 4.9t^2 + v_0t + D$. At $t = 0$, $x = x_0$, so $x_0 = 4.9(0)^2 + v_0 \cdot 0 + D = D$, and so $x = 4.9t^2 + v_0t + x_0$. (The fact that the velocity and position of a falling body are linear and quadratic functions of time was discovered experimentally by Galileo. Newton later explained Galileo's empirical law by his own laws of gravitation and motion.)

3. If we try ax^n as the antiderivative of x^{-1}, we must have $nax^{n-1} = x^{-1}$. Comparing exponents, we find that n must be zero; but then we must have $0 \cdot a = 1$, which is impossible. The formula $x^{n+1}/(n + 1)$ makes no sense for $n = -1$ because the denominator is zero. [It turns out that $1/x = x^{-1}$ does have an antiderivative, but it is a logarithm function rather than a power of x. We will introduce the logarithm function in Chapter 7.]

4. We take the notation to mean $\int [1/(3x + 1)^5] \, dx$, so we are looking for an antiderivative of $1/(3x + 1)^5$. The power of a function rule suggests that we guess $a/(3x + 1)^4$. Differentiating, we have

$$\frac{d}{dx}\frac{a}{(3x + 1)^4} = \frac{-4a}{(3x + 1)^5}\frac{d}{dx}(3x + 1) = \frac{-12a}{(3x + 1)^5}$$

Comparing with $1/(3x + 1)^5$, we see that we must have $a = -\frac{1}{12}$, so

$$\int \frac{dx}{(3x + 1)^5} = \frac{-1}{12(3x + 1)^4} + C$$

5. By the definitions of velocity and acceleration, we have $v = dx/dt$ and $a = dv/dt$. It follows that

$$v = \int a \, dt \quad \text{and} \quad x = \int v \, dt$$

CHAPTER 2

Solved Exercises for Section 2-1

1. Since f is continuous at x_0, it satisfies conditions 1 and 2 in the definition of continuity. Since we are given $f(x_0) > 0$, we may substitute $c_1 = 0$ in condition 1. Then the conclusion of condition 1 is that there is an open interval I about x_0 such that $0 < f(x)$ for all x in I; that is, f is positive on I.

2. The graph of f is shown in Fig. 2-1-7. We must establish conditions 1 and 2 in the definition of continuity. First we check condition 2. Let c_2 be such that $f(x_0) = f(0) = 0 < c_2$; that is, $c_2 > 0$. We must find an open interval J about zero such that $f(x) < c_2$ for all x in J. From Fig. 2-1-7 we see that we should try $J = (-c_2, c_2)$. For $x \geq 0$ and x in J, we have $f(x) = x < c_2$. For $x < 0$ and x in J, we have $f(x) = -x$. Since $x > -c_2$, we have $-x < c_2$; that is, $f(x) < c_2$. Thus, for all x in J, $f(x) < c_2$. For condition 1 we have $c_1 < f(0) = 0$. We can take I to be any open interval about zero, even $(-\infty, \infty)$, since $c_1 < 0 \leq f(x)$ for all real numbers x. Hence f is continuous at zero.

   Notice that the interval J has to be chosen smaller and smaller as $c_2 > 0$ is nearer and nearer to zero.

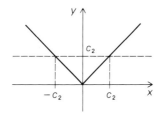

Fig. 2-1-7 The absolute value function is continuous at zero.

3. (a) The function is discontinuous at ... -2, -1, 0, 1, 2, ... (Take $c_2 = \frac{1}{2}$ in condition 2 of the definition of continuity.)

   (b) The function is continuous, even though it is not differentiable at the "corners" of the graph.

   (c) The function is continuous, even though its graph cannot be drawn without removing pencil from paper. In fact, if you take any point (x_0, y_0) *on the curve*, the part of the curve lying over some open interval about x_0 *can* be drawn without removing pencil from paper.

   (d) The function is not continuous at 1. (Use condition 2 of the definition with $c_1 = \frac{3}{2}$.)

4. This is a quotient of polynomials, and the denominator takes the value $4 \neq 0$ at $x = 1$, so the function is continuous there by part 2 of the corollary to Theorem 1.

5. The absolute value function (see Solved Exercise 2) was shown to be continuous at $x_0 = 0$. We will show in a moment that the same function fails to be differentiable at zero. Thus a function which is continuous at a point need not be differentiable there, so the converse of Theorem 1 is false. Notice that our example does not contradict Theorem 1 itself; since the absolute value function is not differentiable at zero, Theorem 1 simply has nothing to say about it.

   To show that the absolute value function $f(x) = |x|$ is not differentiable at $x_0 = 0$, we observe that the difference quotient $[f(\Delta x) - f(0)]/\Delta x = |\Delta x|/\Delta x$ is equal to -1 for all $\Delta x < 0$ and $+1$ for all $\Delta x > 0$, so it has no limit as $\Delta x \to 0$.

6. By part 2 of the corollary to Theorem 1, this rational function is continuous wherever its denominator $x^3 - 8$ is unequal to zero—that is, except where $x^3 = 8$. The only real x for which $x^3 = 8$ is $x = 2$, so $(x^4 + 1)/(x^3 - 8)$ is continuous on $(-\infty, 2)$ and $(2, \infty)$, the domain of definition.

7. The denominator vanishes at $x = \pm 1$. By the corollary to Theorem 1, f is continuous on $(-\infty, -1)$, $(-1, 1)$, and $(1, \infty)$. Since $[-\frac{1}{2}, \frac{1}{2}]$ lies inside $(-1, 1)$, f is surely continuous there.

8. We saw in Worked Example 5 that the equation has a solution in the interval $[0, 2]$. To locate the solution more precisely, we evaluate $f(1) = 1^5 - 1 = 0$. Thus $f(1) < 3 < f(2)$, so there is a root in $(1, 2)$. Now we bisect $[1, 2]$ into $[1, 1.5]$ and $[1.5, 2]$ and repeat: $f(1.5) \approx 6.09 > 3$, so there is a root in $(1, 1.5)$; $f(1.25) = 1.80 < 3$, so there is a root in $(1.25, 1.5)$; $f(1.375) \approx 3.54 > 3$, so there is a root in $(1.25, 1.375)$; thus $x_0 = 1.3$ is within 0.1 of a root. Further accuracy can be obtained by means of further bisections. (Related techniques for root-finding are suggested in the problems for this section.)

9. The function $f(x) = 1/(x - 1)$ is not continuous on the interval $[0, 2]$ since it is not defined at $x = 1$; therefore the intermediate value theorem does not apply to f and there is no contradiction.

10. Let $f(x)$ be your height in meters at x years after your birth. Assume that
    1. f is continuous
    2. $f(0) < 1$
    3. $f(b) > 1$, where b is your present age

    Then the intermediate value theorem applied to f on $[0, b]$ shows that $f(x_0) = 1$ for some x_0 in $(0, b)$.

Solved Exercises for Section 2-2

1. If $x < r_1$, then $x < r_2$ as well, so $x - r_1$ and $x - r_2$ are both negative. Thus $f(x) < 0$ if $x < r_1$. If $r_1 < x < r_2$ then $x - r_1 > 0$ and $x - r_2 < 0$ and so $f(x) < 0$. Thus f changes sign from positive to negative at r_1; the interval (a, b) in the definition may be taken to be $(-\infty, r_2)$.

2. We calculate $f'(x)$ by the quotient rule:

$$f'(x) = \frac{(x - 1) - (x + 1)}{(x - 1)^2} = -\frac{2}{(x - 1)^2}$$

   Since $f'(0) = -2$, which is negative, f is decreasing at zero by Theorem 3 (part 2), so it is getting colder.

3. $f'(x) = 6x^2 - 18x + 12 = 6(x^2 - 3x + 2) = 6(x - 1)(x - 2)$. This is positive for x in $(-\infty, 1)$ and x in $(2, \infty)$, zero for $x = 1$ and 2, and negative for x in $(1, 2)$. By Theorem 3, we conclude that f is increasing at each x in $(-\infty, 1)$ and $(2, \infty)$ and that f is decreasing at x in $(1, 2)$. The theorem does not apply at $x = 1$ and $x = 2$.

4. (a) The function $y = x^n$ changes sign at $x = 0$ whenever the positive integer n is odd.

   (b) The derivative at zero of each of these functions is zero, so Theorem 3 does not give us any information: we must use the definition of increasing and decreasing functions.

   For $f(x) = x^3$, we have $f(x) = x \cdot x^2$. Since $x^2 \geq 0$ for all x, x^3 has the same sign as x; that is, $f(x) < 0 = f(0)$ for $x < 0$ and $f(x) > 0 = f(0)$ for $x > 0$. This means that f is increasing at zero; we can take the interval (a, b) to be $(-\infty, \infty)$. A similar argument shows that $-x^3 = -x \cdot x^2$ is decreasing at zero.

   Finally, for $f(x) = x^2$, we have $f(x) \geq 0 = f(0)$ for all x. The conditions for increasing or decreasing can never be met, because we would need to have $f(x) < f(0)$ for x on one side or the other of zero. Thus x^2 is neither increasing nor decreasing at zero.

5. $f'(x) = 3x^2 - 2$ is a polynomial and so is continuous on $(-\infty, \infty)$. It is zero when $3x^2 - 2 = 0$; that is, for $x = \pm\sqrt{\frac{2}{3}}$. By the test in the display on p. 122, f is monotonic on each of the intervals $(-\infty, -\sqrt{\frac{2}{3}}]$, $[-\sqrt{\frac{2}{3}}, \sqrt{\frac{2}{3}}]$, and $[\sqrt{\frac{2}{3}}, \infty)$. Choosing points in each interval, we find $f'(-1) = 1$, $f'(0) = -2$, and $f'(1) = 1$, so f is increasing on $(-\infty, -\sqrt{\frac{2}{3}}]$ and $[\sqrt{\frac{2}{3}}, \infty)$ and decreasing on $[-\sqrt{\frac{2}{3}}, \sqrt{\frac{2}{3}}]$.

6. The graph of f is sketched in Fig. 2-2-19. We see that f is continuous on $(-\infty, \infty)$ and differentiable everywhere except at zero. The derivative is

$$f'(x) = \begin{cases} 1 & \text{for } x < 0 \\ 2 & \text{for } x > 0 \end{cases}$$

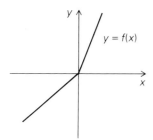

Fig. 2-2-19 The function f is monotonic on $(-\infty, \infty)$.

We can apply the corollary of Theorem 5 to the intervals $(-\infty, 0)$ and $(0, \infty)$ and conclude that f is increasing on each of these intervals. In fact, f is actually increasing on all of $(-\infty, \infty)$, although the corollary does not apply directly since $f'(0)$ does not exist.

7. Function 1 is decreasing for $x < 0$ and increasing for $x > 0$. The only functions in the right-hand column which are negative for $x < 0$ and positive for $x > 0$ are a and c. We notice, further, that the derivative of function 1 is not constant for $x < 0$ (the slope of the tangent is constantly changing), which eliminates a. Similar reasoning leads to the rest of the answers, which are: 1-c, 2-b, 3-e, 4-a, 5-d.

8. We begin by finding the critical points:

$$f'(x) = 12x^3 - 24x^2 + 12x = 12x(x^2 - 2x + 1)$$

$= 12x(x - 1)^2$; the critical points are thus 0 and 1. Since $(x - 1)^2$ is always nonnegative, the only sign change is from negative to positive at zero. Thus zero is a local minimum point, while the critical point 1 is not a turning point.

9. $(d/dx)x^n = nx^{n-1}$. This changes sign from negative to positive at zero if n is even (but $n \neq 0$); there is no sign change if n is odd. Thus zero is a local minimum point if n is even and at least 2; otherwise there is no turning point.

10. $g'(x) = -f'(x)/f(x)^2$, which is zero when $f'(x)$ is zero, negative when $f'(x)$ is positive, and positive when $f'(x)$ is negative (as long as $f(x) \neq 0$). Thus, as long as $f(x) \neq 0$, $g'(x)$ changes sign when $f'(x)$ does, but in the opposite direction. Hence we have:

1. Local minimum points of $g(x)$ = local maximum points of $f(x)$ where $f(x) \neq 0$.

2. Local maximum points of $g(x)$ = local minimum points of $f(x)$ where $f(x) \neq 0$.

Note that $g(x_0)$ is not defined if $f(x_0) = 0$, so g cannot have a turning point there, even if f does.

Solved Exercises for Section 2-3

1. $f'(x) = 3x^2 - 1$ and $f''(x) = 6x$. The critical points are zeros of $f'(x)$; that is, $x = \pm(1/\sqrt{3})$; $f''(-1/\sqrt{3}) = -(6/\sqrt{3}) < 0$ and $f''(1/\sqrt{3}) = 6/\sqrt{3} > 0$. By the second derivative test, $-(1/\sqrt{3})$ is a local maximum point and $1/\sqrt{3}$ is a local minimum point.

2. Let $g = f'$. Then $g'(x_0) = 0$ and $g''(x_0) > 0$, so x_0 is a local minimum point for $g(x) = f'(x)$, and so $f'(x)$ cannot change sign at x_0. Thus x_0 cannot be a turning point for f.

3. The function $l(x) = 0$ is the linear approximation to all three graphs at zero. For $x \neq 0$, $x^4 > 0$ and $-x^4 < 0$, so x^4 is concave upward at zero and $-x^4$ is concave downward. The function x^3 is negative for $x < 0$ and positive for $x > 0$, so no matter how small the interval (a, b) containing zero is made, neither condition 1 nor 2 in the definition on p. 130 can be satisfied.

4. $f'(x) = 9x^2 - 8$, $f''(x) = 18x$. Thus f is concave upward when $18x > 0$ (that is, when $x > 0$) and concave downward when $x < 0$. Before sketching the graph, we notice that turning points occur when $x = \pm\sqrt{\frac{8}{9}} = \pm\frac{2}{3}\sqrt{2}$. See Fig. 2-3-7.

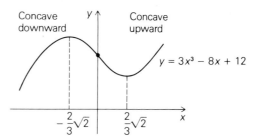

Fig. 2-3-7 The turning points and concavity of $3x^3 - 8x + 12$.

5. (a) If $f''(x_0) < 0$ (or $f''(x_0) > 0$) the linear approximation to f at x_0 is always greater (or less) than $f(x)$ for x near x_0.

   (b) Let $f(x) = 1/x$; $f'(x) = -1/x^2$; $f''(x) = 2/x^3$. We have $f''(1) = 2$, so f is concave upward at 1. Thus, for x near 1, the linear approximation $l(x) = f(1) + f'(1)(x - 1) = 1 - 1(x - 1) = 2 - x$ is always less than $1/x$. (We leave it to the reader as an exercise to prove directly from the laws of inequalities that $2 - x < 1/x$ for all $x > 0$.)

6. For each of the functions listed, $f'(0) = f''(0) = f'''(0) = 0$. We already saw (Solved Exercise 3) that zero is a local minimum point for x^4 and a local maximum point for $-x^4$. For x^5, we have $f''(x) = 20x^3$, which changes sign from negative to positive at $x = 0$, so zero is a point of inflection at which the graph crosses the tangent line from below to above. (You can see this directly: $x^5 < 0$ for $x < 0$ and $x^5 > 0$ for $x > 0$.) Similarly, zero is an inflection point for $-x^5$ at which the graph of the function crosses the tangent line from above to below.

7. Suppose that f'' changes sign from negative to positive at x_0. If $h(x) = f(x) - [f(x_0) + f'(x_0) \cdot (x - x_0)]$, then $h'(x) = f'(x) - f'(x_0)$ and $h''(x) = f''(x)$. Since $f''(x)$ changes sign from negative to positive at x_0, so does $h''(x)$; thus x_0 is a local minimum point for $h'(x)$. Since $h'(x_0) = 0$, we have $h'(x) > 0$ for all $x \neq x_0$ in some open interval (a, b) about x_0. It follows from the corollary to Theorem 5 that h is increasing on $(a, x_0]$ and on $[x_0, b)$. Since $h(x_0) = 0$, this gives $h(x) < 0$ for x in (a, x_0) and $h(x) > 0$ for x in (x_0, b); that is, h changes sign from negative to positive at x_0. The case where f'' changes sign from positive to negative at x_0 is similar.

8. We have $f'(x) = 96x^3 - 96x^2 + 18x$, so $f''(x) = 288x^2 - 192x + 18$ and $f'''(x) = 576x - 192$. To find inflection points, we begin by solving $f''(x) = 0$; the quadratic formula gives $x = (4 \pm \sqrt{7})/12$. Using our knowledge of parabolas, we can conclude that f'' changes from positive to negative at $(4 - \sqrt{7})/12$ and from negative to positive at $(4 + \sqrt{7})/12$; thus both are inflection points. One could also evaluate $f'''((4 \pm \sqrt{7})/12)$, but this would be more complicated.)

Solved Exercises for Section 2-4

1. We carry out the six-step procedure:

   1. $f(-x) = -x/(1 + x)$, which is not equal to $f(x)$ or $-f(x)$, so f is neither even or odd.

   2. We carried out this step in Worked Example 1 (see Fig. 2-4-3).

   3. $f'(x) = [1 - x - x(-1)]/(1 - x)^2 = 1/(1 - x)^2$, which is positive for all $x \neq 1$. Thus f is always increasing; there are no maxima or minima.

   4. $f''(x) = 2(1 - x)/(1 - x)^4 = 2/(1 - x)^3$. This is negative when $x > 1$ and positive when $x < 1$, so f is concave upward for $x < 1$ and downward when $x > 1$. There are no points of inflection.

   5. We note that $f(0) = 0$, $f'(0) = 1$; $f(2) = -2$, $f'(2) = 1$; $f(-1) = -\frac{1}{2}$, $f'(-1) = \frac{1}{4}$. The only solution of $f(x) = 0$ is $x = 0$.

      The information obtained in steps 1 to 5 is placed tentatively on the graph in Fig. 2-4-13.

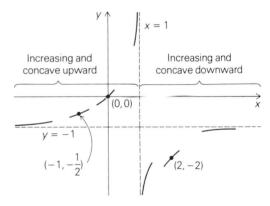

Fig. 2-4-13 The graph of $y = x/(1 - x)$ after steps 1 to 5.

6. Examining Fig. 2-4-13 we find that some adjustments are necessary. There is no way to connect the point $(2, -2)$ with the piece of curve to the right of the asymptote $x = 1$ and to keep f increasing. Similarly, to connect the piece of curve above the asymptote $y = -1$ with the point $(-1, -\frac{1}{2})$ and the piece of tangent there, it would be necessary to violate the condition that f be concave upward. The plotted points must stay where they are; we draw a smooth curve through those points with the required asymptotic behavior, resulting in a sketch like Fig. 2-4-14.

2

1. (*cont'd*)

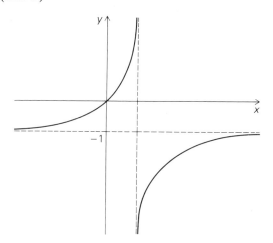

Fig. 2-4-14 The completed graph of $y = x/(1-x)$.

The graph looks like that of $g(x) = -(1/x)$, moved one unit to the right and one unit down. We may check this by simplifying:

$$g(x-1) - 1 = -\frac{1}{x-1} - 1 = \frac{-1-x+1}{x-1}$$

$$= \frac{x}{1-x} = f(x)$$

We see that plotting a good graph can help us discover an algebraic fact. Of course, if we had noticed the algebraic fact first, we could have used it in plotting the graph.

2. Again we carry out the six-step procedure:

1. $f(-x) = -x/[1 + (-x)^2] = -x/[1 + x^2] = -f(x)$; f is odd, so we need only study $f(x)$ for $x \geq 0$.

2. Since the denominator $1 + x^2$ is never zero, the function is defined everywhere; there are no vertical asymptotes. For $x \neq 0$, we have

$$f(x) = \frac{x}{1+x^2} = \frac{1}{x + (1/x)}$$

Since $1/x$ becomes small as x becomes large, $f(x)$ looks like $1/(x + 0) = 1/x$ for x large. Thus $y = 0$ is a horizontal asymptote; the graph is below it for x large negative and above it for x large positive.

3. $f'(x) = [1 + x^2 - x(2x)]/(1 + x^2)^2 = (1 - x^2)/(1 + x^2)^2$, which vanishes when $x = \pm 1$. Also, $f'(0) = 1, f'(-2) = \frac{-3}{25}, f'(2) = \frac{-3}{25}$. Thus f is decreasing on $(-\infty, -1)$ and on $(1, \infty)$ and f is increasing on $(-1, 1)$. Hence -1 is a local minimum and 1 is a local maximum. (See p. 122.)

2. (*cont'd*)
4. $f''(x)$

$$= \frac{(1 + x^2)^2(-2x) - (1 - x^2) \cdot 2(1 + x^2)2x}{(1 + x^2)^4}$$

$$= \frac{2x(x^2 - 3)}{(1 + x^2)^3}$$

This is zero when $x = 0, \sqrt{3}$, and $-\sqrt{3}$. Since the denominator of f'' is positive, we can determine the sign by evaluating the numerator. Evaluating at $-2, -1, 1$, and 2 we get $-4, 4. -4$, and 4, so f is concave downward on $(-\infty, -\sqrt{3})$ and $(0, \sqrt{3})$ and concave upward on $(-\sqrt{3}, 0)$ and $(\sqrt{3}, \infty)$; $-\sqrt{3}, 0$, and $\sqrt{3}$ are points of inflection.

5. $f(0) = 0; f'(0) = 1$
$f(1) = \frac{1}{2}; f'(1) = 0$
$f(\sqrt{3}) = \frac{1}{4}\sqrt{3}; f'(\sqrt{3}) = -\frac{1}{8}$

The only solution of $f(x) = 0$ is $x = 0$.

The information obtained in steps 1 through 5 is placed tentatively on the graph in Fig. 2-4-15. As we said in step 1, we need do this only for $x \geq 0$.

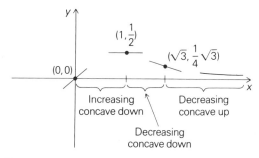

Fig. 2-4-15 The graph of $y = x/(1 + x^2)$ after steps 1 to 5.

6. From our experience in Solved Exercise 1, we have learned to draw the asymptotic part of the curve *after* plotting the specific points. No more corrections are necessary; we draw the final graph, remembering to obtain the left-hand side by reflecting the right-hand side in both axes. (You can get the same effect by rotating the graph 180°, keeping the origin fixed.) The result is shown in Fig. 2-4-16.

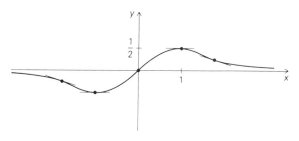

Fig. 2-4-16 The completed graph of $y = x/(1 + x^2)$.

3. The six steps are as follows:

   1. $f(-x) = -2x^3 - 8x + 1$ is not equal to $f(x)$ or $-f(x)$, so f is neither even nor odd.

   2. f is defined everywhere. We may write

      $$f(x) = 2x^3 \left(1 + \frac{4}{x^2} + \frac{1}{2x^3}\right)$$

      For x large, the factor $1 + (4/x^2) + (1/2x^3)$ is near 1, so $f(x)$ is large positive for x large positive and large negative for x large negative. There are no horizontal or vertical asymptotes.

   3. $f'(x) = 6x^2 + 8$, which vanishes nowhere and is always positive. Thus f is increasing on $(-\infty, \infty)$ and has no turning points.

   4. $f''(x) = 12x$, which is negative for $x < 0$ and positive for $x > 0$. Thus f is concave downward on $(-\infty, 0)$ and concave upward on $(0, \infty)$; zero is a point of inflection.

   5. $f(0) = 1; f'(0) = 8$
      $f(1) = 11; f'(1) = 14$
      $f(-1) = -9; f'(-1) = 14$

      The information obtained so far is plotted in Fig. 2-4-17.

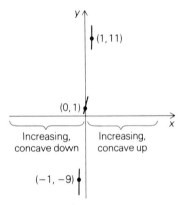

Fig. 2-4-17 The graph of $y = 2x^3 + 8x + 1$ after steps 1 to 5.

   6. A look at Fig. 2-4-17 suggests that the graph will be very long and thin. In fact, $f(2) = 33$, which is way off the graph. To get a useful picture, we may stretch the graph horizontally by changing units on the x axis so that a unit on the x axis is, say, four times as large as a unit on the y axis. We add a couple of additional points by calculating

      $f(\tfrac{1}{2}) = 5\tfrac{1}{4}$ $f'(\tfrac{1}{2}) = 9\tfrac{1}{2}$
      $f(-\tfrac{1}{2}) = -3\tfrac{1}{4}$ $f'(-\tfrac{1}{2}) = 9\tfrac{1}{2}$

      Then we draw a smooth curve as in Fig. 2-4-18.

3. (*cont'd*)

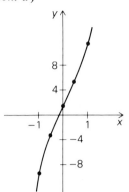

Fig. 2-4-18 The completed graph of $y = 2x^3 + 8x + 1$.

4. Again we use the six-step procedure:

   1. There are no symmetries.

   2. There are no asymptotes. As in Solved Exercise 3, $f(x)$ is large positive for x large positive and large negative for x large negative.

   3. $f'(x) = 6x^2 - 8$, which is zero when $x = \pm\sqrt{\frac{4}{3}} = \pm 2/\sqrt{3} \approx \pm 1.15$. Also, $f'(-2) = f'(2) = 16$ and $f'(0) = -8$, so f is increasing on $(-\infty, -2/\sqrt{3}\,]$ and $[2/\sqrt{3}, \infty)$ and decreasing on $[-2/\sqrt{3}, 2/\sqrt{3}\,]$. Thus $-2/\sqrt{3}$ is a local maximum point, $2/\sqrt{3}$ a local minimum point.

   4. $f''(x) = 12x$, so f is concave downward on $(-\infty, 0)$ and concave upward on $(0, \infty)$, as in Solved Exercise 3. Zero is an inflection point.

   5. $\quad f(0) = 1, f'(0) = -8$
      $f(-2/\sqrt{3}) = 1 + 32/3\sqrt{3}$
      $\quad\quad \approx 7.16; f'(-2/\sqrt{3}) = 0$
      $f(2/\sqrt{3}) = 1 - 32/3\sqrt{3}$
      $\quad\quad \approx -5.16; f'(2/\sqrt{3}) = 0$
      $f(-\tfrac{1}{2}) = 4\tfrac{3}{4}; f'(-\tfrac{1}{2}) = -6\tfrac{1}{2}$
      $f(\tfrac{1}{2}) = -2\tfrac{3}{4}; f'(\tfrac{1}{2}) = -6\tfrac{1}{2}$
      $f(-2) = 1; f'(-2) = 16$
      $f(2) = 1; f'(2) = 16$

      The data are plotted in Fig. 2-4-19. The scale is stretched by a factor of 4 in the x direction, as in Fig. 2-4-18.

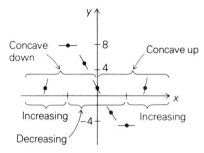

Fig. 2-4-19 The graph of $y = 2x^3 - 8x^2 + 1$ after steps 1 to 5.

4. *(cont'd)*

6. We draw the graph (Fig. 2-4-20).

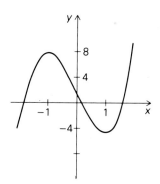

Fig. 2-4-20 The completed graph of $y = 2x^3 - 8x^2 + 1$.

5. $\frac{1}{2}(2x^3 + 3x^2 + x + 1) = x^3 + \frac{3}{2}x^2 + \frac{1}{2}x + \frac{1}{2}$. Substituting $x - \frac{1}{2}$ for x (since the coefficient of x^2 is $\frac{3}{2}$) gives

$$(x - \tfrac{1}{2})^3 + \tfrac{3}{2}(x - \tfrac{1}{2})^2 + \tfrac{1}{2}(x - \tfrac{1}{2}) + \tfrac{1}{2}$$
$$= x^3 - \tfrac{3}{2}x^2 + \tfrac{3}{4}x - \tfrac{1}{8} + \tfrac{3}{2}(x^2 - x + \tfrac{1}{4}) + \tfrac{1}{2}(x - \tfrac{1}{2}) + \tfrac{1}{2}$$
$$= x^3 - \tfrac{1}{4}x + \tfrac{1}{2}$$

Thus, after being shifted along the x and y axes, the cubic becomes $x^3 - \frac{1}{4}x$. Since $c < 0$, it is of type III.

3

CHAPTER 3

Solved Exercises for Section 3-1

1. The solutions are indicated in Fig. 3-1-5.

In case (a), there is no highest point on the graph; hence, no maximum. In case (c), one might be tempted to call 2 the maximum value, but it is not attained at any point of the interval $(-1, 1)$. Study this example well; it will be a useful test case for the general statements to be made later in this section.

	I	Graph	Maximum points	Maximum value	Minimum points	Minimum value
(a)	$(-\infty, \infty)$		None	None	0	1
(b)	$(0, \infty)$		None	None	None	None
(c)	$(-1, 1)$		None	None	0	1
(d)	$[-1, 1]$		$-1, 1$	2	0	1
(e)	$(0, 1]$		1	2	None	None
(f)	$\left[\frac{1}{2}, 1\right]$		1	2	$\frac{1}{2}$	$\frac{5}{4}$
(g)	$(-2, 1]$		None	None	0	1
(h)	$[-2, 1)$		-2	5	0	1

→ means that the graph goes off to infinity

○ means that the endpoint does not belong to the graph

Fig. 3-1-5 Solutions to Solved Exercise 1.

2. The maximum value is about 13 kilowatt-hours, attained at $x = 90$ (the North Pole); the minimum value is about 9 kilowatt-hours, attained at $x = 0$ (the equator). (There are local maximum and minimum points at about $x = 40$ and $x = 60$, but these are not what we are looking for.)

    The explanation for this unexpected result lies in the fact that although the intensity of the sun's radiation is lowest at the pole, the summer day is longest there, resulting in a larger amount of accumulated energy. At the surface of the earth, absorption of solar energy by the atmosphere is greatest near the pole, since the low angle of the sun makes the rays pass through more air. The resulting graph is more like the one in Fig. 3-1-6, which is in better correspondence with the earth's climate.

Fig. 3-1-6 The solar energy received on June 21 at the surface of the earth, assuming clear skies. (Reproduced from W.G. Kendrew, *Climatology*, Oxford University Press, 1949.)

3. Since m_I is the minimum value of f on I, there is at least one x_1 in I such that $f(x_1) = m_I$ (part 2 of the definition). Since I is contained in J, x_1 belongs to J as well, so $m_J \leq f(x_1)$, because m_J is the minimum value of f on J (part 1 of the definition). Thus $m_J \leq f(x_1) = m_I$.

    For instance, without looking at any data, we can be sure that the lowest temperature recorded at Mount Washington during the last 10 years is no lower than the lowest temperature recorded during the last 35 years. (Under what conditions would the two low temperatures be equal?)

4. There are no singular points. The other answers are given in the following table. (Note that a point counts as an endpoint only if it belongs to the interval.)

4. *(cont'd)*

	Critical points	Endpoints	Maximum or minimum points
(a)	0	None	0
(b)	None	None	None
(c)	0	None	0
(d)	0	$-1, 1$	$-1, 0, 1$
(e)	None	1	1
(f)	None	$\frac{1}{2}, 1$	$\frac{1}{2}, 1$
(g)	0	1	0
(h)	0	-2	$-2, 0$

The lists of critical points and endpoints are the same as the maximum-minimum points except in case (g), where the endpoint 1 is neither a maximum nor a minimum point.

5. Here $f'(x) = 3x^2$, so zero is the only critical point. The only endpoint of $[-1, \infty)$ is -1. A look at the graph (Fig. 3-1-7) tells us that -1 is a minimum point, while there is no maximum point. In particular, the critical point zero is neither a maximum nor a minimum point.

Fig. 3-1-7 -1 is a minimum point for $f(x) = x^3$ on $[-1, \infty)$.

6. Looking at Fig. 2-4-15, we see that -1 and 1 are the critical points and that -1 is a minimum point and 1 is a maximum point.

7. We draw a graph (Fig. 3-1-8). The derivative $f'(x)$ is equal to -1 on $(-2, 0)$ and 1 on $(0, 1)$, so there are no critical points. In this case 1 is an endpoint and zero is a singular point. Zero is also a minimum point; there is no maximum point.

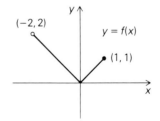

Fig. 3-1-8 Zero is the minimum point of $f(x) = |x|$ on $(-2, 1]$. There is no maximum.

8. We know, by the extreme value theorem, that the function in question must be discontinuous. Let f be defined by

$$f(x) = \begin{cases} 0 & x = 0 \\ 1 - x & 0 < x \le 1 \end{cases}$$

(See Fig. 3-1-9.) The maximum value must be at least 1, but no such value is attained for any x in $[0,1]$, so there is no maximum value.

Fig. 3-1-9 This function has no maximum on [0, 1].

9. We have $f'(x) = 4x^3 - 8x = 4x(x^2 - 2)$, so the critical points are $-\sqrt{2}, 0$, and $\sqrt{2}$. The list required by step 1 of the closed interval test is, therefore, $-4, -\sqrt{2}, 0, \sqrt{2}, 2$, since there are no points where f is not differentiable. We make a table:

x	-4	$-\sqrt{2}$	0	$\sqrt{2}$	2
$f(x)$	199	3	7	3	7

The maximum value is 199; the minimum value is 3. (The maximum *point*, which was not demanded in the problem, is -4; the minimum points are $-\sqrt{2}$ and $\sqrt{2}$.)

10. The derivative of f is given by

$$f'(x) = \begin{cases} 5 & x < 0 \\ -3 & x > 0 \end{cases}$$

and $f'(0)$ does not exist. The list $x_1, \dots, x_n$ of critical points, endpoints, and singular points is just $-1, 0, 1$; evaluating f at these points gives the table:

x	-1	0	1
$f(x)$	-5	0	-3

The maximum value is zero; the minimum value is -5.

11. Here $(d/dx)(-3x^2 + 2x + 1) = -6x + 2$, which is zero when $x = \frac{1}{3}$. The second derivative is $-6 < 0$, so $\frac{1}{3}$ is a maximum point and $\frac{4}{3}$ is the maximum value, by the turning point test. Since there are no other critical points, endpoints, or singular points, there are no minima. (In fact, $-3x^2 + 2x + 1$ becomes arbitrarily large negative as x becomes arbitrarily large.)

12. We have $f'(x) = p - (q/x^2)$. Setting $f'(x) = 0$, we get $x^2 = q/p$. If p and q have opposite signs, this equation has no solutions. Since there are no endpoints or singular points in $(0, \infty)$, there are no maximum or minimum points. If p and q have the same sign, then $f'(x)$ has the unique root $\sqrt{q/p}$ in $(0, \infty)$.

Case 1: $p > 0, q > 0$. For $x_+ > \sqrt{q/p}$, we have $x_+^2 > q/p$, and $p > q/x_+^2$, so $f'(x_+) = p - (q/x_+^2) > 0$. Similarly, for $x_- < \sqrt{q/p}$, we have $f'(x_-) < 0$. By the turning point test, $\sqrt{q/p}$ is a minimum point. (We could also have used the second derivative test: $f''(x) = 2q/x^3$, which is positive for all $x > 0$.)

The minimum value is $f(\sqrt{q/p}) = p\sqrt{q/p} + q\sqrt{p/q} = 2\sqrt{pq}$. (Notice that p and q enter symmetrically into the minimum value; can you explain why?)

Case 2: $p < 0, q < 0$. In this case, $\sqrt{q/p}$ is a maximum point and $2\sqrt{pq}$ is the maximum value.

13. We have $f'(x) = 3x^2 - 3$, which is zero when $x = \pm 1$. Moreover, $f''(x) = 6x$, which is negative for $x = -1$ and positive for $x = 1$. We now sketch a very rough graph (Fig. 3-1-10).

Our candidate for a maximum point is -1. To see if it really is one, we need to know whether the peak at $x = -1$ is higher than the loose end near $x = \frac{3}{2}$. Computing, we find $f(-1) = 7$, while $f(\frac{3}{2}) = 3\frac{7}{8}$, so -1 is indeed a maximum point, and 7 is the maximum value.

Making the same test to decide whether 1 is a minimum point, we find $f(1) = 3$, while $f(-3) = -13$, so $f(x)$ is certainly less than 3 for x near the left-hand endpoint of $(-3, \frac{3}{2})$, and 1 is not a minimum point. Since there are no other critical points, endpoints, or singular points, there is no minimum point. A better graph is now drawn (Fig. 3-1-11).

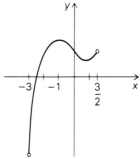

Fig. 3-1-10 A rough sketch of the graph of $x^3 - 3x + 5$ on $(-3, \frac{3}{2})$.

Fig. 3-1-11 A more accurate graph of $x^3 - 3x + 5$.

Solved Exercises for Section 3-2

1. Problem 2: We place the first source at $x = 0$ and the second source at $x = 10$ on the real line (Fig. 3-2-4). Denote by I_1 and I_2 the intensities of the two sources. Let L_1 = illumination at x from the first source and L_2 = illumination at x from the second source. Then $L = L_1 + L_2$ is the total illumination.

Intensity I_1 Intensity I_2

0   x 10

Fig. 3-2-4 Light sources of intensities I_1 and I_2 are placed at $x = 0$ and $x = 10$. The observer is at x.

The given relations are

$$I_2 = 4I_1$$

$$L_1 = \frac{kI_1}{x^2} \quad (k > 0 \text{ is a proportionality constant})$$

$$L_2 = \frac{kI_2}{(10 - x)^2}$$

We want to minimize L.

Problem 3: Call the unknown number x. We want to minimize $y = (x - a)^2 + (x - b)^2 + (x - c)^2 + (x - d)^2$.

Problem 4: x is the number of days we run the car. At this point in solving the problem, it is completely legitimate to pace around the room, asking "What should be minimized?" This is not clearly stated in the problem, so we must determine it ourselves. A reasonable objective is to minimize the total amount of money to be paid. How is this to be done? Well, as soon as the cost of running the car exceeds 50 cents per day, we should switch to the bus. So let y = the cost per day of running the car at day x. We want the first x for which $y \geq 50$. The relation between the variables is

$$y = \frac{d}{dx}\left(\frac{x^2}{100} + 10x\right)$$

$$= \frac{x}{50} + 10$$

2. Problem 2: We want to minimize

$$L = L_1 + L_2 = kI_1\left[\frac{1}{x^2} + \frac{4}{(10 - x)^2}\right]$$

Since the point is to be between the sources, we must have $0 < x < 10$. To minimize L on $(0, 10)$, we compute:

2. (*cont'd*)

$$L'(x) = kI_1\left[-\frac{2}{x^3} + \frac{4(20 - 2x)}{(10 - x)^4}\right]$$

$$= 2kI_1\left[-\frac{1}{x^3} + \frac{4}{(10 - x)^3}\right]$$

The critical points occur when

$$\frac{4}{(10 - x)^3} = \frac{1}{x^3}$$

Hence

$$\frac{10 - x}{x} = \sqrt[3]{4}$$

$$10 - x - \sqrt[3]{4} \cdot x = 0$$

$$x = \frac{10}{1 + \sqrt[3]{4}}$$

Thus there is one critical point; we use the turning point test to determine if it is a maximum or minimum point.

Let $x_- = 0.0001$. Then

$$L'(x_-) = 2kI_1\left[-\frac{1}{(0.0001)^3} + \frac{4}{(10 - 0.0001)^3}\right]$$

Without calculating this explicitly, we can see that the term $-(2/0.0001^3)$ is negative and much larger in size than the other term, so $L'(x_-) < 0$.

Let $x_+ = 9.9999$. Then

$$L'(x_+) = 2kI_1\left[-\frac{1}{(9.9999)^3} + \frac{4}{(0.0001)^3}\right]$$

Now it is the second term which dominates; since it is positive, $L'(x_+) > 0$. By the turning point test, $x_0 = 10/(1 + \sqrt[3]{4})$ must be a minimum point.

The darkest point is thus at a distance of $10/(1 + \sqrt[3]{4}) \approx 3.86$ meters from the smaller source. It is interesting to compare the distances of the darkest point from the two sources. The ratio

$$\frac{10 - [10/(1 + \sqrt[3]{4})]}{10/(1 + \sqrt[3]{4})}$$

is simply $\sqrt[3]{4}$.

Problem 3: $y = (x - a)^2 + (x - b)^2 + (x - c)^2 + (x - d)^2$. There are no restrictions on x.

$$\frac{dy}{dx} = 2(x - a) + 2(x - b) + 2(x - c) + 2(x - d)$$

$$= 2(4x - a - b - c - d)$$

which is zero only if $x = \frac{1}{4}(a + b + c + d)$.

2. (*cont'd*)

Since $d^2y/dx^2 = 8$ is positive, $\frac{1}{4}(a + b + c + d)$ is the minimum point. The number required is thus $\frac{1}{4}(a + b + c + d)$, the average, or arithmetic mean, of a, b, c, and d.

Problem 4: This is not a standard maximum-minimum problem. We minimize our expenses by selling the car at the time when the cost per day, $(x/50) + 10$, reaches the value 50. (See Fig. 3-2-5). We solve $(x/50) + 10 = 50$, getting $x = 2000$. Thus the car should be kept for 2000 days.

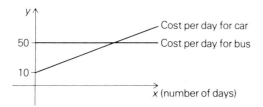

Fig. 3-2-5 When does the car become more expensive than the bus?

3. A square of side 1.

4. A circle, since it is the most symmetric figure. (The proof of this result is not so simple.*)

5. Halfway between the sources.

6. Let u be the velocity, relative to the ground, of the boat heading upstream. Its velocity relative to the water is $v = u + 5$. The cost per river mile is $10v^3$. The boat goes 1 land mile in $1/u$ hours, during which time it goes $v \cdot 1/u$ river miles. Thus the cost per land mile is

$$10v^3 \frac{v}{u} = \frac{10v^4}{u} = \frac{10(u + 5)^4}{u}$$

We want to minimize $f(u) = [10(u + 5)^4]/u$ for u in $(0, \infty)$. The critical points are given by $f'(u) = 0$. That is,

$$\frac{40(u + 5)^3 u - 10(u + 5)^4}{u^2} = 0$$

$$4u - (u + 5) = 0$$

$$u = \tfrac{5}{3}$$

Since $f'(u) = [10(u + 5)^3/u^2](3u - 5)$ changes sign from negative to positive at $u = \frac{5}{3}$, $\frac{5}{3}$ is a minimum point. Thus the most economical speed is $1\frac{2}{3}$ miles per hour relative to the land, or $6\frac{2}{3}$ miles per hour in the water.

---

*See, for instance, G. Polya, Induction and Analogy in Mathematics, *Princeton University Press, 1954*, Chapter X: or J. Marsden, Elementary Classical Analysis, *W. H. Freeman, 1974, p. 432. For the history of the problem, see M. Kline, Mathematical Thought from Ancient to Modern Times,* Oxford (1972), pp. 838–9, 576.

7. The profit is

$$
\begin{aligned}
P(x) &= R(x) - C(x) \\
&= (50x - 0.01x^2) - (800 + 30x + 0.02x^2) \\
&= 20x - 0.01x^2 - 800
\end{aligned}
$$

The limits on x are only that $0 < x < \infty$. The critical points are where $P'(x) = 0$; i.e., $20 - 0.02x = 0$, i.e., $x = 1000$. This is a maximum since $P''(x) = -0.02$ is negative; at $x = 1000$, $P(1000) = 9200$. (The endpoint $x = 0$ is not a maximum since $P(0) = -800$ is negative.) Thus the production level should be set at $x = 1000$.

Solved Exercises for Section 3-3

1. Here $a = -2$ and $b = 3$, so

$$
\begin{aligned}
[f(b) - f(a)] / (b - a) &= [27 - (-8)] / [3 - (-2)] \\
&= 35/5 = 7
\end{aligned}
$$

This should equal $f'(x_0) = 3x_0^2$, for some x_0 between -2 and 3. In fact, we must have $3x_0^2 = 7$ or $x_0 = \pm\sqrt{7/3} \approx \pm 1.527$. Each of these values is between -2 and 3, so either will do as x_0. The situation is sketched in Fig. 3-3-3.

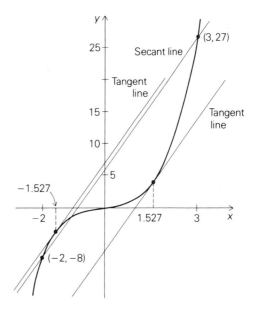

Fig. 3-3-3 Two tangent lines are parallel to a secant line of $y = x^3$.

2. If, in Corollary 1, we take $S = (0, \infty)$, then the hypothesis $f'(x) \in S$ becomes $f'(x) > 0$, and the conclusion $[f(x_2) - f(x_1)]/(x_2 - x_1) \in S$ becomes $[f(x_2) - f(x_1)]/(x_2 - x_1) > 0$, or $f(x_2) > f(x_1)$ when $x_2 > x_1$. Thus Corollary 1, in this case, is just the increasing function theorem (Theorem 5, p. 121).

3. Let $f(t)$ be the position of the train at time t; let a and b be the beginning and ending times of the trip. By Corollary 1, with $S = [40, 50]$, we have $40 \le [f(b) - f(a)]/(b - a) \le 50$. But $f(b) - f(a) = 200$, so

$$40 \le \frac{200}{b - a} \le 50$$

$$\frac{1}{5} \le \frac{1}{b - a} \le \frac{1}{4}$$

$$5 \ge b - a \ge 4$$

Hence the trip takes somewhere between 4 and 5 hours.

4. The derivative of $1/x$ is $-(1/x^2)$, so by Corollary 3, $F(x) = (1/x) + C$, where C is a constant, on any interval not containing zero. So $F(x) = (1/x) + C_1$ for $x < 0$ and $F(x) = (1/x) + C_2$ for $x > 0$. Hence C_1 and C_2 are constants, but they are not necessarily equal.

5. Since $f(0) = 0$ and $f(2) = 0$, Rolle's theorem shows that f' is zero at some x_0 in $(0, 2)$; that is, $0 < x_0 < 2$.

6. Apply the horserace theorem with $f_1(x) = f(x)$, $f_2(x) = x^2$, and $[a, b] = [0, 1]$.

CHAPTER 4

Solved Exercises for Section 4-1

1. $\sum_{j=-2}^{2} j^3 = (-2)^3 + (-1)^3 + (0)^3 + (1)^3 + (2)^3$
   $= -8 - 1 + 0 + 1 + 8 = 0$.

2. As in Worked Example 4, we may apply summation property 3 to obtain

$$\sum_{i=m}^{n} i = \sum_{i=1}^{n} i - \sum_{i=1}^{m-1} i$$

$$= \tfrac{1}{2}n(n + 1) - \tfrac{1}{2}(m - 1)(m - 1 + 1)$$

$$= \tfrac{1}{2}[n(n + 1) - (m - 1)m]$$

We may simplify this to

$$\tfrac{1}{2}[n^2 + n - m^2 + m] = \tfrac{1}{2}[n^2 - m^2 + n + m]$$

$$= \tfrac{1}{2}[(n - m)(n + m) + n + m]$$

$$= \tfrac{1}{2}(n + m)(n - m + 1)$$

Note that the sum is equal to the average of the first and last terms multiplied by the number of terms. (You can also obtain this result directly by imitating the proof of the formula for the sum of the first n integers.)

3. We can do this problem by making the substitution $i = j + 6$. As j runs from 1 to 102, i runs from 7 to 108, and we get

$$\sum_{j=1}^{102} (j + 6) = \sum_{i=7}^{108} i$$

By the result of Solved Exercise 2, this is $\tfrac{1}{2} \cdot 115 \cdot 102 = 5865$.

4. Let $a_k = \sin(k^2)$ and $b_k = 1$. Since $\sin(k^2) \le 1$ for all k, we have by property 5

$$\sum_{k=1}^{1000} \sin(k^2) \le \sum_{k=1}^{1000} 1$$

By property 4 the right-hand side equals 1000.

5. Since the function $f(t) = 9 - t^2$ is decreasing on the interval $[0, 2]$, the value of f anywhere on a subinterval is at least equal to the value of f at the upper endpoint; for example, for $1.6 \le t \le 1.7$, we have $f(t) \ge f(1.7) = 9 - (1.7)^2 = 6.11$. Evaluating $f(t)$ at all the points $0.1, 0.2, \ldots, 1.9, 2.0$, we obtain the following table:

On the interval with endpoints:

0.0	0.1	0.2	0.3	0.4	0.5	0.6
0.1	0.2	0.3	0.4	0.5	0.6	0.7

0.7	0.8	0.9	1.0	1.1	1.2	1.3
0.8	0.9	1.0	1.1	1.2	1.3	1.4

1.4	1.5	1.6	1.7	1.8	1.9
1.5	1.6	1.7	1.8	1.9	2.0

the value of $9 - t^2$ is at least:

8.99	8.96	8.91	8.84	8.75	8.64	8.51

8.36	8.19	8.00	7.79	7.56	7.31	7.04

6.75	6.44	6.11	5.76	5.39	5.00

It follows that

$$\Delta E \ge \sum_{i=1}^{20} \kappa I_i \Delta t_i = (0.1)(8.99)(0.1) + (0.1)(8.96)(0.1)$$
$$+ \cdots + (0.1)(5.39)(0.1) + (0.1)(5.00)(0.1)$$

or

$$1.513 \le \Delta E$$

You are asked to derive an upper estimate for ΔE in Exercise 5.

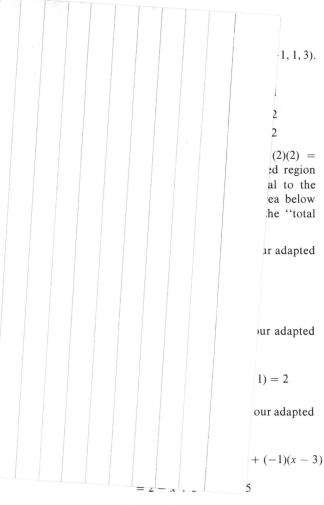

·1, 1, 3).

(2)(2) =
:d region
al to the
ea below
he "total

ır adapted

ur adapted

1) = 2

our adapted

+ (−1)(x − 3)

= z − x ·   ·   5

Summarizing, we have

$$\int_0^x f(t)\, dt = \begin{cases} 2x & \text{if } 0 < x \le 1 \\ 2 & \text{if } 1 < x \le 3 \\ -x + 5 & \text{if } 3 < x \le 4 \end{cases}$$

(b) The graph of $F(x)$ is shown in Fig. 4-1-15. Functions like F are sometimes called *piecewise linear*, *polygonal*, or *ramp* functions.

Fig. 4-1-15 The graph of $y = F(x) = \int_0^x f(t)\, dt$ is polygonal.

7. (*cont'd*)

(c) We can see from the graph that F is differentiable everywhere on $[0, 4]$ except at the points 0, 1, 3, and 4. We have:

$$F'(x) = \begin{cases} 2 & \text{if } 0 < x < 1 \\ 0 & \text{if } 1 < x < 3 \\ -1 & \text{if } 3 < x < 4 \end{cases}$$

Comparing this with the definition of $f(t)$ on p. 184, we see that F' is the same as f, except at the points where the piecewise constant function f has a jump. This example illustrates the general fact that the derivative of the integral of a function, with respect to an endpoint of integration, is more or less equal to the original function. The fundamental theorem of calculus is just this fact, made precise and extended to functions which are not necessarily piecewise constant.

8. Let f be piecewise constant on $[a, b]$. Let $g(t) = f(t)$. Then g is piecewise constant on $[a, b]$, and $g(t) \le f(t)$ for all t, so

$$S_0 = \int_a^b g(t)\, dt = \int_a^b f(t)\, dt$$

is a lower sum. (Here the integral signs refer to the integral as defined on p. 182.) Since every number less than a lower sum is again a lower sum, we conclude that every number less than S_0 is a lower sum. But we also have $g(t) \ge f(t)$ for all t, so S_0 is an upper sum as well, and so is every number greater than S_0. Thus, by the definition on p. 185, f is integrable on $[a, b]$, and its integral is S_0, which was the integral of f according to the definition on p. 182.

9. The integral exists, since $1/t$ is continuous on $[1, 2]$. To estimate it to within $\frac{1}{10}$, we try to find lower and upper sums which are within $\frac{2}{10}$ of one another. We divide the interval into three parts and use the best possible piecewise constant function, as shown in Fig. 4-1-16. For a lower sum we have

$$\int_1^2 g(t)\, dt = \frac{1}{4/3}\left(\frac{4}{3} - 1\right) + \frac{1}{5/3}\left(\frac{5}{3} - \frac{4}{3}\right)$$
$$+ \frac{1}{2}\left(2 - \frac{5}{3}\right)$$
$$= \frac{3}{4}\cdot\frac{1}{3} + \frac{3}{5}\cdot\frac{1}{3} + \frac{1}{2}\cdot\frac{1}{3}$$
$$= \frac{1}{3}\left(\frac{3}{4} + \frac{3}{5} + \frac{1}{2}\right)$$
$$= \frac{1}{3}\left(\frac{37}{20}\right) = \frac{37}{60}$$

9. (*cont'd*)

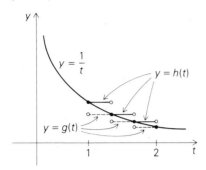

Fig. 4-1-16 Illustrating upper and lower sums for $1/t$ on $[1, 2]$.

For an upper sum we have

$$\int_1^2 h(t)\, dt = \frac{1}{1} \cdot \frac{1}{3} + \frac{1}{4/3} \cdot \frac{1}{3} + \frac{1}{5/3} \cdot \frac{1}{3}$$

$$= \frac{1}{3}\left(1 + \frac{3}{4} + \frac{3}{5}\right)$$

$$= \frac{1}{3}\left(\frac{47}{20}\right) = \frac{47}{60}$$

It follows that

$$\frac{37}{60} \le \int_1^2 \frac{1}{t}\, dt \le \frac{47}{60}$$

Since the integral lies in the interval $\left[\frac{37}{60}, \frac{47}{60}\right]$, whose length is $\frac{1}{6}$, we may take the midpoint $\frac{42}{60} = \frac{7}{10}$ as our estimate; it will differ from the true integral by no more than $\frac{1}{2} \cdot \frac{1}{6} = \frac{1}{12}$, which is less than $\frac{1}{10}$.

10. If we examine the upper and lower sums in Solved Exercise 9, we see that they differ by the "outer terms"; the difference is

$$\frac{1}{3}\left(1 - \frac{1}{2}\right) = \frac{1}{3} \cdot \frac{1}{2} = \frac{1}{6}$$

If, instead of three steps, we used n steps, the difference between the upper and lower sums would be $(1/n) \cdot (1/2)$. To estimate the integral to within $\frac{1}{100}$, we would have to make $(1/2n) \le \frac{1}{50}$, which we could do by putting $n = 25$.

Computing these upper and lower sums with a calculator, we obtain

$$0.6832 \le \int_1^2 \frac{1}{t}\, dt \le 0.7032$$

so our estimate is 0.6932.

11. Since f takes arbitrarily large values on $[0, 1]$, there are no upper sums for f, and so condition 2 in the definition of integrability cannot be satisfied; that is, f is not integrable on $[0, 1]$. On the interval $\left[\frac{1}{8}, 1\right]$, f is continuous, so it is integrable by Theorem 1.

12. Since $h(t) = t_i = a + [i(b - a)/n]$ on (t_{i-1}, t_i) (see Fig. 4-1-17), we have

$$\int_a^b h(t)\, dt = \sum_{i=1}^n k_i \Delta t_i$$

$$= \sum_{i=1}^n \left[a + \frac{i(b-a)}{n}\right]\left(\frac{b-a}{n}\right)$$

$$= \sum_{i=1}^n \left[\frac{a(b-a)}{n} + \frac{i(b-a)^2}{n^2}\right]$$

$$= \sum_{i=1}^n \frac{a(b-a)}{n} + \frac{(b-a)^2}{n^2}\sum_{i=1}^n i$$

$$= n \cdot \frac{a(b-a)}{n} + \frac{(b-a)^2}{n^2}\frac{n(n+1)}{2}$$

$$= a(b-a) + \frac{(b-a)^2(n+1)}{2n}$$

$$= ab - a^2 + \frac{(b-a)^2}{2} + \frac{(b-a)^2}{2n}$$

$$= \frac{b^2 - a^2}{2} + \frac{(b-a)^2}{2n}$$

Fig. 4-1-17 An upper sum for $f(t) = t$ on $[a, b]$.

13. We repeat the calculation of $\int_a^b t\, dt$, except that we make $g(t) = 5t_{i-1}$ for $t_{i-1} \le t < t_i$ and $h(t) = 5t_i$ for $t_{i-1} \le t < t_i$. Then all the terms in the integrals of g and h are multiplied by 5, so

$$\int_a^b g(t)\, dt = \frac{5(b^2 - a^2)}{2} - \frac{5(b-a)^2}{2n}$$

and

$$\int_a^b h(t)\, dt = \frac{5(b^2 - a^2)}{2} + \frac{5(b-a)^2}{2n}$$

13. (*cont'd*)

Since $g(t) \le 5t \le h(t)$ on $[a, b]$, we have, for all n,

$$\frac{5(b^2 - a^2)}{2} - \frac{5(b - a)^2}{2n} \le \int_a^b 5t\, dt \le$$
$$\frac{5(b^2 - a^2)}{2} + \frac{5(b - a)^2}{2n}$$

It follows that $\int_a^b 5t\, dt = [5(b^2 - a^2)]/2$.

14. (a) The region is sketched in Fig. 4-1-18.

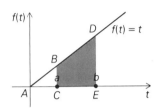

Fig. 4-1-18 The shaded area equals $\frac{1}{2}(b^2 - a^2)$.

(b) The area of the region $BCED$ is the difference between the areas of the triangles ADE and ABC; that is, $\frac{1}{2}b^2 - \frac{1}{2}a^2 = \frac{1}{2}(b^2 - a^2)$, which is equal to $\int_a^b t\, dt$.

Solved Exercises for Section 4-2

1. An antiderivative of x^4 is $\frac{1}{5}x^5$, since $(d/dx)(\frac{1}{5}x^5) = \frac{1}{5} \cdot 5x^4 = x^4$, so

$$\int_0^1 x^4\, dx = \frac{1}{5}x^5 \Big|_0^1 = \frac{1}{5}(1^5) - \frac{1}{5}(0^5) = \frac{1}{5}.$$

Notice that we use x everywhere here just as we used t before.

2. By the sum, constant multiple, and power rules for antiderivatives, an antiderivative for $t^2 + 3t$ is $(t^3/3) + (3t^2/2)$, so

$$\int_0^3 (t^2 + 3t)\, dt = \left(\frac{t^3}{3} + \frac{3t^2}{2}\right) \Big|_0^3$$
$$= \frac{3^3}{3} + \frac{3 \cdot 3^2}{2} = \frac{45}{2}$$

3. We saw in Section 4-1 that the change in energy stored is $\int_a^b \kappa I\, dt$ if light with intensity I shines from $t = a$ to $t = b$. For the two units in this exercise, the amounts of energy added are $\int_0^1 \kappa t^2\, dt$ and $\int_0^1 \kappa 5t^3\, dt$. (Here κ is the same for both units, since they are identical.) Antiderivatives for κt^2 and $5\kappa t^3$ are $\frac{1}{3}\kappa t^3$ and $\frac{5}{4}\kappa t^4$, respectively, so we have

3. (*cont'd*)

$$\int_0^1 \kappa t^2\, dt = \frac{1}{3}\kappa t^3 \Big|_0^1 = \frac{1}{3}\kappa$$

and

$$\int_0^1 \kappa 5t^3\, dt = \frac{5}{4}\kappa t^4 \Big|_0^1 = \frac{5}{4}\kappa$$

Since $\frac{5}{4} > \frac{1}{3}$, the second unit accumulates more energy. (Note that the answer is independent of the value of κ.)

4. Velocity is defined as the time derivative of position; that is, $v = dx/dt$, where $x = F(t)$ is the position at time t. The fundamental theorem gives the equation

$$F(b) - F(a) = \int_a^b F'(t)\, dt = \int_a^b f(t)\, dt$$

or

$$\Delta x = \int_a^b \frac{dx}{dt}\, dt = \int_a^b v\, dt$$

So the integral of the velocity v over the time interval $[a, b]$ is the *total displacement* of the object from time a to time b.

5. By the fundamental theorem, $R(100) - R(0) = \int_0^{100} R'(x)\, dx$. But $R(0) = 0$ and $R'(x)$, the marginal revenue, is $15 - 0.1\, x$. Thus

$$R(100) = \int_0^{100} (15 - 0.1x)\, dx$$
$$= \left(15x - \frac{0.1x^2}{2}\right) \Big|_0^{100}$$
$$= 1500 - 500 = 1000$$

6. By Corollary 1 to the mean value theorem, we must have

$$\frac{F(\frac{1}{3}) - F(0)}{\frac{1}{3} - 0} < 2$$

and

$$\frac{F(2) - F(\frac{1}{3})}{2 - \frac{1}{3}} < 1$$

Hence

$$F(\tfrac{1}{3}) - F(0) < \tfrac{2}{3}$$

and

$$F(2) - F(\tfrac{1}{3}) < \tfrac{5}{3}$$

Adding the last two equations gives the conclusion

$$F(2) - f(0) < \tfrac{7}{3}$$

This is a simple example of one of the basic ideas involved in the proof of the fundamental theorem of calculus.

7. The indefinite integral notation, $\int f(t)\, dt$, denotes the antiderivative of $f(t)$. By Theorem 3, the function $\int_a^t f(s)\, ds$ *is* an antiderivative of $f(t)$; the most general antiderivative is obtained by adding an arbitrary constant C. This example shows what is "indefinite" in an indefinite integral: it is one of the endpoints of integration.

8. $\displaystyle\int_{-2}^{3} (x^4 + 5x^2 + 2x + 1)\, dx$

$$= \left(\int (x^4 + 5x^2 + 2x + 1)\, dx \right) \Bigg|_{-2}^{3}$$

(fundamental theorem)

$$= \left(\int x^4\, dx + 5\int x^2\, dx + 2\int x\, dx + \int dx \right) \Bigg|_{-2}^{3}$$

(sum and constant multiple rules)

$$= \left(\frac{x^5}{5} + \frac{5x^3}{3} + x^2 + x \right) \Bigg|_{-2}^{3} \quad \text{(power rule)}$$

$$= \frac{1}{5}[3^5 - (-2)^5] + \frac{5}{3}[3^3 - (-2)^3] + 3^2 - (-2)^2 + 3 - (-2)$$

$$= \frac{275}{5} + \frac{175}{3} + 5 + 5 = \frac{370}{3} = 123\tfrac{1}{3}$$

9. By the fundamental theorem, we have

$$F(t) - F(a) = \int_a^t F'(s)\, ds$$

Adding $F(a)$ to both sides gives

$$F(t) = \int_a^t F'(s)\, ds + F(a)$$

Thus we recover the function F exactly by integrating its derivative, letting the endpoint vary, and adding the constant $F(a)$.

10. $\displaystyle\int_1^2 \frac{x^2 + 2x + 2}{x^4}\, dx = \int \left(\frac{1}{x^2} + \frac{2}{x^3} + \frac{2}{x^4} \right) dx \Bigg|_1^2$

(dividing and using the fundamental theorem)

$$= \left(-\frac{1}{x} - \frac{1}{x^2} - \frac{2}{3x^3} \right) \Bigg|_1^2 \quad \text{(power rule)}$$

$$= \left(-\frac{1}{2} - \frac{1}{4} - \frac{2}{3\cdot 8} \right) - \left(-1 - 1 - \frac{2}{3} \right) = \frac{11}{6}$$

11. If $f(x) \le g(x)$ on (a,b), then $(F - G)'(x) = F'(x) - G'(x) = f(x) - g(x) \le 0$ for x in (a,b). By Corollary 1, p. 166, with $S = (-\infty, 0)$, $\{[F(b) - G(b)] - [F(a) - G(a)]\}/(b - a) \le 0$. Since $b - a > 0$,

$$[F(b) - G(b)] - [F(a) - G(a)] \le 0$$

$$F(b) - F(a) \le G(b) - G(a)$$

By the fundamental theorem of calculus, the last inequality can be written

$$\int_a^b f(x)\, dx \le \int_a^b g(x)\, dx$$

as required.

12. $\int_6^2 t^3\, dt = (t^4/4)\,|_6^2 = \frac{1}{4}(16 - 1296) = -320$. Although the function $f(t) = t^3$ is positive, the integral is negative. To explain this, we remark that as t goes from 6 to 2, "dt is negative."

13. $\displaystyle\int_a^b f(x + k)\, dx = \int_{a+k}^{b+k} f(y)\, dy$

[This formula can be proven by noting that each side is $F(b + k) - F(a + k)$, where F is an antiderivative for f (assuming f is continuous).]

Solved Exercises for Section 4-3

1. (a) By the quotient rule, $(d/dx)[1/(1 + x^2)] = -2x/(1 + x^2)^2$.

   (b) The function $x/(1 + x^2)^2$ is not negative for x in $[0, 1]$, so the area under its graph is

$$\int_0^1 \frac{x}{(1 + x^2)^2}\, dx = \frac{1}{(-2)} \int_0^1 \frac{(-2)x}{(1 + x^2)^2}\, dx$$

$$= \frac{1}{(-2)} \int_0^1 \frac{d}{dx} \left(\frac{1}{1 + x^2} \right) dx$$

(by (a))

$$= \frac{1}{(-2)} \left(\frac{1}{1 + x^2} \right) \Bigg|_0^1$$

(by the fundamental theorem)

$$= \frac{1}{(-2)} \left(\frac{1}{2} - 1 \right)$$

$$= \frac{1}{4}$$

2. Place the coordinate system as shown in Fig. 4-3-19 and let the (shifted) parabola have equation $y = ax^2 + b$. Since $y = 8$ when $x = 0$, $b = 8$. Also, $y = 0$ when $x = 3$, so $0 = a3^2 + 8$, so $a = -\frac{8}{9}$. Thus the parabola is $y = -\frac{8}{9}x^2 + 8$. The area under its graph is

$$\int_{-3}^{3} \left(-\frac{8}{9}x^2 + 8 \right) dx = \left. -\frac{8}{27}x^3 + 8x \right|_{-3}^{3}$$

$$= \left(-\frac{8}{27} \cdot 27 + 8 \cdot 3 \right) - \left(\frac{8}{27} \cdot 27 - 8 \cdot 3 \right) = 32$$

Thus 32 square feet have been cut out.

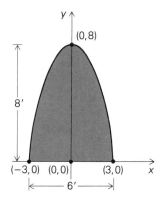

Fig. 4-3-19   Find the area of the parabolic doorway.

3. The region in question is sketched in Fig. 4-3-20. The area is

$$\int_{0}^{2} (4 - x^2)\, dx = \left. 4x - \frac{x^3}{3} \right|_{0}^{2}$$

$$= 4 \cdot 2 - \frac{2^3}{3}$$

$$= \frac{16}{3}$$

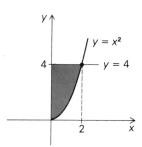

Fig. 4-3-20   What is the area of the shaded region?

4. If we plot x as a function of y, we obtain the graphs and region shown in Fig. 4-3-21. We use our general rule for the area between graphs, on p. 210, which gives

$$A = \int_{-\sqrt{2}}^{\sqrt{2}} \left(1 + \frac{1}{2}y^2 - y^2 \right) dy$$

$$= \int_{-\sqrt{2}}^{\sqrt{2}} \left(1 - \frac{1}{2}y^2 \right) dy = \left. y - \frac{1}{6}y^3 \right|_{-\sqrt{2}}^{\sqrt{2}}$$

$$= \sqrt{2} - \frac{1}{6}(\sqrt{2})^3 - \left[-\sqrt{2} - \frac{1}{6}(-\sqrt{2})^3 \right]$$

$$= 2\sqrt{2} - \frac{1}{3}(\sqrt{2})^3$$

$$= \sqrt{2}\left(2 - \frac{2}{3} \right) = \frac{4}{3}\sqrt{2} = \frac{2^{5/2}}{3}$$

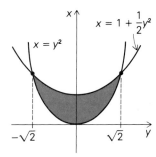

Fig. 4-3-21   The bounded region determined by $x = y^2$ and $x = 1 + \frac{1}{2}y^2$.

5. Refer to Fig. 4-3-22. We know that

$$\int_{0}^{2} (x^2 - 1)\, dx = \int_{0}^{1} (x^2 - 1)\, dx + \int_{1}^{2} (x^2 - 1)\, dx$$

and that $\int_{1}^{2} (x^2 - 1)\, dx$ represents the area of the region R_2 under the graph $x^2 - 1$ on $[1, 2]$, since $x^2 - 1 \geq 0$ for x in $[1, 2]$.

What is represented by $\int_{0}^{1} (x^2 - 1)\, dx$? Looking at the region R_1, we see that it is the region between the graphs of $y = x^2 - 1$ and $y = 0$. Since $x^2 - 1 \leq 0$ on $[0, 1]$, the area of R_1 is

$$\int_{0}^{1} [0 - (x^2 - 1)]\, dx = -\int_{0}^{1} (x^2 - 1)\, dx$$

Thus $\int_{0}^{1} (x^2 - 1)\, dx$ is the *negative* of the area of R_1. We conclude that $\int_{0}^{2} (x^2 - 1)\, dx$ is the area of R_2 *minus* the area of R_1.

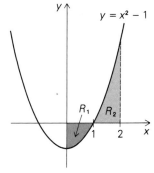

Fig. 4-3-22 $\int_{0}^{2} (x^2 - 1)\, dx$ is the difference between the areas of R_2 and R_1.

CHAPTER 5

Solved Exercises for Section 5-1

1. We set $y = mx + b$ and solve for x, obtaining $x = y/m - b/m$. Thus x is completely determined by y, and the inverse function is $g(y) = y/m - b/m$. (Notice that if $m = 0$, there is no solution for x unless $y = b$; the constant function is not invertible.)

2. The domain S consists of all x such that $x \neq -d/c$. Solving $y = (ax + b)/(cx + d)$ for x in terms of y gives:

   $y(cx + d) = ax + b$

   $(cy - a)x = -dy + b$

   If $cy - a \neq 0$, we have the unique solution

   $$x = \frac{-dy + b}{cy - a} = g(y)$$

   It appears that the inverse function g is defined for all $y \neq a/c$; we must, however, check whether or not $y \pm a/c$ whenever $y = (ax + b)/(cx + d)$. Note that

   $$cy - a = c\left(\frac{ax + b}{cx + d}\right) - a$$

   $$= \frac{cax + bc - acx - ad}{cx + d} = \frac{bc - ad}{cx + d}$$

   Thus there are two cases to consider:
   Case 1: $bc - ad \neq 0$. In this case, f is invertible on S, and T consists of all y which are unequal to a/c.
   Case 2: $bc - ad = 0$. In this case, $b = ad/c$ and

   $$y = \frac{ax + ad/c}{cx + d} = \frac{1}{c}\left(\frac{acx + ad}{cx + d}\right)$$

   $$= \frac{a}{c}\left(\frac{cx + d}{cx + d}\right) = \frac{a}{c}$$

   so f is a constant function, which is not invertible.

3. Writing $y = x^2 + 2x + 1 = (x + 1)^2$, we get $x = -1 \pm \sqrt{y}$. In this case, there is *not* a unique solution for x in terms of y; in fact, for $y > 0$ there are two solutions, while for $y < 0$ there are no solutions. Looking at the graph of f (Fig. 5-1-11), we see that the range of values of f is $[0, \infty]$ and that f becomes invertible if we restrict it to $(-\infty, -1]$ or $[-1, \infty)$. Since we want an interval containing zero, we choose $[-1, \infty)$. Now if $x \geq -1$ in the equation $x = -1 \pm \sqrt{y}$, we must choose the positive square root, so the inverse function g has domain $T = [0, \infty)$ and is defined by $g(y) = -1 + \sqrt{y}$. (See Fig. 5-1-12.) Hence $g(9) = -1 + \sqrt{9} = -1 + 3 = 2$. (Note that $f(2) = 2^2 + 2 \cdot 2 + 1 = 9$.) Also, $g(x) = -1 + \sqrt{x}$. (Again, one just substitutes. Any letter can be used to denote the variable in a function.)

3. *(cont'd)*

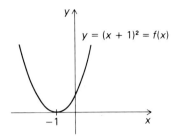

Fig. 5-1-11 This function has an inverse if we restrict x to $[-1, \infty]$.

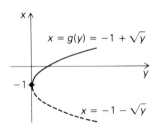

Fig. 5-1-12 The graph of the inverse (Solved Exercise 3).

4. The graphs, which we obtain by viewing the graphs in Fig. 5-1-7 from the reverse side of the page, are shown in Fig. 5-1-13.

Fig. 5-1-13 Graphs of the inverse functions (compare Fig. 5-1-7).

5. A simple test is the following: The function is invertible if each horizontal line meets the graph in at most one point. Applying this test, we find that the functions (a) and (c) are invertible while (b) is not.

6. (a) $f'(x) = 5x^4 + 1 > 0$, so by Theorem 1, f is invertible on $[-2, 2]$. The domain of the inverse is $[f(-2), f(2)]$, which is $[-34, 34]$.

   (b) Since $f'(x) > 0$ for all x in $(-\infty, \infty)$, f is increasing on $(-\infty, \infty)$. Now f takes arbitrarily large positive and negative values as x varies over $(-\infty, \infty)$; it takes all values in between by the intermediate value theorem, so the domain of f^{-1} is $(-\infty, \infty)$. There is no simple formula for $f^{-1}(y)$, the solution of $x^5 + x = y$, in terms of y, but we can calculate $f^{-1}(y)$ for any specific values of y to any desired degree of accuracy. (This is really no worse than the situation for $\sqrt{x}$. If the inverse function to $x^5 + x$ had as many applications as the square root function, we would learn about it in high school, tables would be readily available for it, calculators would calculate it at the touch of a key, and there would be a standard notation like $\sqrt[\varphi]{y}$ for the solution of $x^5 + x = y$, just as $\sqrt[5]{y}$ is the standard notation for the solution of $x^5 = y$.)*

   (c) Since $f(1) = 1^5 + 1 = 2$, $f^{-1}(2)$ must equal 1.

   (d) To calculate $f^{-1}(3)$—that is, to seek an x such that $x^5 + x = 3$—we use the method of bisection described in Solved Exercise 8, p. 113. Since $f(1) = 2 < 3$ and $f(2) = 34 > 3$, x must lie between 1 and 2. We can squeeze toward the correct answer by calculating:

   $f(1.5) = 9.09375$ so $1 < x < 1.5$
   $f(1.25) = 4.30176$ so $1 < x < 1.25$
   $f(1.1) = 2.71051$ so $1.1 < x < 1.25$
   $f(1.15) = 3.16135$ so $1.1 < x < 1.15$
   $f(1.14) = 3.06541$ so $1.1 < x < 1.14$
   $f(1.13) = 2.97244$ so $1.13 < x < 1.14$
   $f(1.135) = 3.01856$ so $1.13 < x < 1.135$

   Thus, to two decimal places, $x = 1.13$. (About 10 minutes of further experimentation gave $f(1.132997566) = 3.000000002$ and $f(1.132997565) = 2.999999991$. What does this tell you about $f^{-1}(3)$?

7. We find $f'(x) = 5x^4 - 1$, which has roots at $\pm\sqrt[4]{\frac{1}{5}}$. It is easy to check that each of these roots is a turning point, so f is invertible on $(-\infty, -\sqrt[4]{\frac{1}{5}}]$, $[-\sqrt[4]{\frac{1}{5}}, \sqrt[4]{\frac{1}{5}}]$, and $[\sqrt[4]{\frac{1}{5}}, \infty)$.

8. Here $f'(x) = nx^{n-1}$. Since n is odd, $n - 1$ is even and $f'(x) > 0$ for all $x \neq 0$. Thus f is increasing on $(-\infty, \infty)$ and so it is invertible there. Since $f(x)$ is continuous and x^n takes arbitrarily large positive and negative values, it must take on all values in between, so the range of values of f, which is the domain of f^{-1}, is $(-\infty, \infty)$. We usually denote $f^{-1}(y)$ by $y^{1/n}$ or $\sqrt[n]{y}$. We have just proved that if n is odd, every real number has a unique real nth root.

9. We find that $f'(x) = nx^{n-1}$ changes sign at zero if n is even. Thus f is invertible on $(-\infty, 0]$ and on $[0, \infty)$. Since x^n is positive for all n and takes arbitrarily large values, the range of values for f is $[0, \infty)$, whether the domain is $(-\infty, 0]$ or $[0, \infty)$. The inverse function to f on $[0, \infty)$ is usually denoted by $y^{1/n} = \sqrt[n]{y}$; the inverse function to f on $(-\infty, 0]$ is denoted by $-y^{1/n}$ or $-\sqrt[n]{y}$. Remember that these are defined only for $y \geq 0$.

10. We have $y = f(x) = (ax + b)/(cx + d)$, so

$$f'(x) = \frac{a(cx + d) - c(ax + b)}{(cx + d)^2} = \frac{ad - bc}{(cx + d)^2}$$

(Notice that this is never zero if $ad - bc \neq 0$.)

   In Solved Exercise 2, we saw that $x = g(y) = (-dy + b)/(cy - a)$, so

$$g'(y) = \frac{-d(cy - a) - c(-dy + b)}{(cy - a)^2}$$

$$= \frac{ad - bc}{(cy - a)^2}$$

As they stand, $f'(x) = (ad - bc)/(cx + d)^2$ and $g'(y) = (ad - bc)/(cy - a)^2$ do not appear to be reciprocals of one another. However, the inverse function theorem says that $g'(y) = 1/[f'(g(y))]$, so we must substitute $g(y) = (-dy + b)/(cy - a)$ for x in $f'(x)$. We obtain

$$f'(g(y)) = \frac{ad - bc}{\left[c\left(\dfrac{-dy + b}{cy - a} \right) + d \right]^2}$$

$$= \frac{ad - bc}{\left[\dfrac{-cdy + cb + cdy - ad}{cy - a} \right]^2}$$

$$= \frac{ad - bc}{\left[\dfrac{bc - ad}{cy - a} \right]^2} = \frac{(cy - a)^2}{ad - bc}$$

which *is* $1/[g'(y)]$.

11. Since f is a polynomial, it is continuous. The derivative is $f'(x) = 3x^2 + 2$. This is positive for $x > 0$ and in particular for x in $(0, 2)$. By Theorem 1, f has an inverse function $g(y)$. We notice that $g(4) = 1$, since $f(1) = 4$. From Theorem 2, with $y_0 = 4$ and $x_0 = 1$,

$$g'(4) = \frac{1}{f'(1)} = \frac{1}{3 \cdot 1^2 + 2} = \frac{1}{5}$$

12. By Worked Example 4, $dy/dx = 1/(2\sqrt{x})$. At $x = 2$ we get $dy/dx = 1/(2\sqrt{2})$. Thus the tangent line has the equation

$$y - \sqrt{2} = \frac{1}{2\sqrt{2}}(x - 2)$$

or

$$y = \frac{1}{2\sqrt{2}}x + \sqrt{2} - \frac{1}{\sqrt{2}}$$

$$= \frac{1}{2\sqrt{2}}x + \frac{1}{\sqrt{2}}$$

13. (a) $(d/dx)(10x^{1/8}) = 10 \cdot \frac{1}{8} \cdot x^{(1/8)-1} = \frac{5}{4}x^{-7/8}$
    (b) $(d/dx)(x^{3/5}) = \frac{3}{5}x^{-2/5}$
    (c) $(d/dx)[x^2(x^{1/3} + x^{4/3})]$
    $= (d/dx)(x^{7/3} + x^{10/3}) = \frac{7}{3}x^{4/3} + \frac{10}{3}x^{7/3}$

    (You could also differentiate the original expression by the product rule and combine terms later.)
    (d) We must write $1/\sqrt{x}$ as $x^{-1/2}$. The derivative is then $-\frac{1}{2}x^{-3/2}$, or $-1/[2(\sqrt{x})^3]$.

14. We have obtained the rule $(d/dx)\, x^n = nx^{n-1}$ first for positive integer n, then for integer n, then for rational n. It is natural to conjecture that if we define things properly, the rule should hold for all real n. Thus we guess $(d/dx)\, x^\pi = \pi x^{\pi - 1}$. (That this is indeed true will be seen in Chapter 7.)

15. Since $(d/dr)x^{r+1} = (r + 1)x^r$, we have

$$\int x^r \, dx = \frac{x^{r+1}}{r + 1} + C$$

The only excluded case is $r = -1$, when the denominator becomes zero. Notice that if $r = p/q$ and q is even, our formula is valid only for $x > 0$.

Solved Exercises for Section 5-2

1. We multiply 5 meters/second by -0.095 gsc/meter to obtain the rate of -0.475 gsc/second. It is negative because the pressure is decreasing as the airplane climbs. (Note that neither the value of 1500 meters for the altitude nor the pressure at that altitude enters directly in the calculation.)

2. If $u = g(t)$ and $p = h(t) = f(u) = f(g(t))$, we have $h'(t_0) = dp/dt$, $f'(u_0) = dp/du$, and $g'(t_0) = du/dt$, so the chain rule becomes $dp/dt = (dp/du)(du/dt)$. (Notice how the notation behaves as if these were ordinary fractions—one is tempted to prove the chain rule by "cancelling the du's." Although this not a proof, it is a good way to remember the chain rule.)

3. We find $(f \circ g)(x) = f(g(x)) = f(x + 1) = (x + 1)^2$. (Or $(f \circ g)(x) = f(u) = u^2 = (x + 1)^2$.) On the other hand, $(g \circ f)(x) = g(f(x)) = g(x^2) = x^2 + 1$. Note that $g \circ f \neq f \circ g$.

4. We observe that $h(x)$ can be expressed in terms of x^{12} as $(x^{12})^2 + 3(x^{12}) + 1$. If $u = g(x) = x^{12}$, then $h(x) = f(g(x))$, where $f(u) = u^2 + 3u + 1$.

5. (a) $D_f = (-\infty, \infty)$, $D_g = [0, \infty)$.
    (b) The domain of $f \circ g$ consists of all x in D_g for which $g(x) \in D_f$; that is, all x in $[0, \infty)$ for which $\sqrt{x} \in (-\infty, \infty)$. Since $\sqrt{x}$ is always in $(-\infty, \infty)$ the domain of $f \circ g$ is $[0, \infty)$. We have $(f \circ g)(x) = f(g(x)) = f(\sqrt{x}) = \sqrt{x} - 1$.

    The domain of $g \circ f$ consists of all x in D_f for which $f(x) \in D_g$; that is, all x in $(-\infty, \infty)$ for which $x - 1 \in [0, \infty)$. But $x - 1 \in [0, \infty)$ means $x - 1 \geq 0$, or $x \geq 1$, so the domain of $g \circ f$ is $[1, \infty)$. We have $(g \circ f)(x) = g(x - 1) = \sqrt{x - 1}$.
    (c) $(f \circ g)(2) = \sqrt{2} - 1$; $(g \circ f)(2) = \sqrt{2 - 1} = \sqrt{1} = 1$.
    (d) See Fig. 5-2-7.

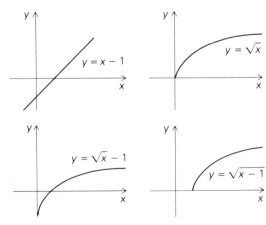

Fig. 5-2-7 Compositions of $f(x) = x - 1$ and $g(x) = \sqrt{x}$.

6. $(i \circ f)(x) = i(f(x)) = f(x)$, so $i \circ f = f$; $(f \circ i)(x) = f(i(x)) = f(x)$, so $f \circ i = f$.

7. Let $u = x^3$, so $y = (1 - u)/(1 + u)$. Then

$$\frac{dy}{du} = \frac{-(1 + u) - (1 - u)}{(1 + u)^2} = \frac{-2}{(1 + u)^2}$$

$$\frac{du}{dx} = 3x^2$$

Hence

$$\frac{dy}{dx} = \frac{dy}{du} \frac{du}{dx} = \frac{-2}{(1 + u)^2} \cdot 3x^2 = \frac{-6x^2}{(1 + x^3)^2}$$

Directly by the quotient rule:

$$\frac{dy}{dx} = \frac{d}{dx}\left(\frac{1 - x^3}{1 + x^3}\right)$$

$$= \frac{(1 + x^3)(-3x^2) - (1 - x^3)(3x^2)}{(1 + x^3)^2}$$

$$= \frac{-6x^2}{(1 + x^3)^2}$$

and so the answers agree.

8. Let $u = g(x) = x^2$, so $h(x) = f(u)$. Then $h'(x) = f'(u) \cdot g'(x) = f'(x^2) \cdot 2x$. Thus

$$h'(x) = f'(x^2) \cdot 2x$$

9. Let $f(x) = \sqrt{x}$. By the shifting rule, the derivative of $f(x + 3)$ is $f'(x + 3)$. But $f'(x) = \frac{1}{2}x^{-1/2} = 1/(2\sqrt{x})$, so

$$\frac{d}{dx}(\sqrt{x + 3}) = \frac{1}{2\sqrt{x + 3}}$$

The graphs $y = \sqrt{x}$ and $y = \sqrt{x + 3}$ are plotted in Fig. 5-2-8.

Fig. 5-2-8 Two graphs related by shifting.

10. Let $u = x^4 + 2x + 1$ and $y = u^{5/8} = (x^4 + 2x + 1)^{5/8}$; then

$$dy/dx = (dy/du)(du/dx) = \frac{5}{8}u^{-3/8} \cdot (4x^3 + 2)$$
$$= \frac{5}{8}(x^4 + 2x + 1)^{-3/8}(4x^3 + 2)$$

11. By the power of a function rule,

$$\frac{d}{dx}(x^4 + 2x + 1)^{5/8}$$

$$= \frac{5}{8}(x^4 + 2x + 1)^{-3/8} \cdot \frac{d}{dx}(x^4 + 2x + 1)$$

$$= \frac{5}{8}(x^4 + 2x + 1)^{-3/8}(4x^3 + 2)$$

That this is the same result as was obtained in Solved Exercise 10 should be no surprise. In fact, the calculation in Solved Exercise 10 was essentially the same as the derivation of the general power of a function rule.

12. We write this as $x^2 \cdot (x^2 + 1)^{-1/3}$ and use the product rule. (We could also use the form $x^2/(x^2 + 1)^{1/3}$ and use the quotient rule.)

$$\frac{d}{dx}[x^2 \cdot (x^2 + 1)^{-1/3}]$$

$$= 2x \cdot (x^2 + 1)^{-1/3} + x^2 \cdot (-\tfrac{1}{3})(x^2 + 1)^{-4/3} \cdot 2x$$

$$= 2x\left[\frac{1}{(x^2 + 1)^{1/3}} - \frac{x^2}{3(x^2 + 1)^{4/3}}\right]$$

$$= 2x\left[\frac{3(x^2 + 1) - x^2}{3(x^2 + 1)^{4/3}}\right] = \frac{2x(2x^2 + 3)}{3(x^2 + 1)^{4/3}}$$

You may wonder why this last form is preferable to the form of the derivative on the second line. The reason is that we can easily look at the factored form to determine the zeros of the derivative and the intervals on which it is positive and negative—information which is useful in graphing and maximum-minimum problems. For this reason, it is important for you to become adept at the algebra required to simplify derivatives.

13. Let A = area, p = population, t = time (days). Then $dA/dt = (dA/dp)(dp/dt) = \frac{1}{1000} \cdot 10000 = 10$ square miles per day.

14. $\frac{3}{2}$ inches per minute.

15. Refer to Fig. 5-2-9. By similar triangles, $y/(y - x) = 10/2$, so $y = 5x/4$. Then $dy/dt = (dy/dx)(dx/dt) = (\frac{5}{4})3 = 3\frac{3}{4}$ feet per second.

Fig. 5-2-9 Dog trotting proudly away from lamp post.

16. $(d/dx)(x^2 + y^2) = (d/dx)\, 3 = 0$; that is, $2x + 2y\,(dy/dx) = 0$; that is, $dy/dx = -x/y$. At $x = 0$, $y = \sqrt{3}$, $dy/dx = 0$.

17. Since we are asked to find dx/dy, we think of x as a function of y and differentiate $x^3 + y^3 = xy$ with respect to y:

$$\frac{d}{dy}(x^3 + y^3) = \frac{d}{dy}(xy)$$

$$3x^2\frac{dx}{dy} + 3y^2 = \frac{dx}{dy}\cdot y + x\cdot 1$$

so

$$\frac{dx}{dy} = \frac{x - 3y^2}{3x^2 - y}$$

18. Refer to Fig. 5-2-10. Let x denote the radius of the cylinder and h its height. Being inscribed in the sphere means that

$$R^2 = \left(\frac{h}{2}\right)^2 + x^2 \qquad (1)$$

Fig. 5-2-10 Cylinder inscribed in a sphere.

18. (*cont'd*)

We want to maximize

$$V = \pi x^2 h. \qquad (2)$$

The direct approach would be to solve equation (1) for h in terms of x and substitute the result in (2). This would lead to rather messy calculations. Instead, to find out when $dV/dx = 0$, we may differentiate both equations implicitly, regarding V and h as functions of x. We have

$$0 = \frac{1}{4}\cdot 2h\frac{dh}{dx} + 2x \qquad (3)$$

and

$$\frac{dV}{dx} = \pi\left(2xh + x^2\frac{dh}{dx}\right) \qquad (4)$$

Solving (3) for dh/dx gives $dh/dx = -4x/h$, and substituting this in (4) gives

$$\frac{dV}{dx} = \pi\left(2xh - \frac{4x^3}{h}\right) = 2\pi x\left(h - \frac{2x^2}{h}\right)$$

Thus $dV/dx = 0$ when $x = 0$ or $h - 2x^2/h = 0$. When $x = 0$, we get zero volume, so the maximum must occur when $h - 2x^2/h = 0$; that is, $h^2 = 2x^2$, so $h = \sqrt{2}\,x$; that is, the height is $\sqrt{2}$ times the radius.

To get the actual values of x and h, we may substitute into equation (1):

$$R^2 = \left(\frac{h}{2}\right)^2 + \left(\frac{h}{\sqrt{2}}\right)^2 = \frac{h^2}{4} + \frac{h^2}{2} = \frac{3}{4}h^2$$

so $h = \dfrac{2}{\sqrt{3}}R$ and $x = \dfrac{2}{\sqrt{6}}R$

These actual values tell us less about the *shape* of the optimal cylinder, though, than the earlier result $h = \sqrt{2}\,x$.

We remark that when the method of implicit differentiation is used in solving minimum-maximum problems, it is sometimes hard to verify that the point you find is actually a minimum or maximum, but one can often justify the result by a physical or geometric argument.

19. Let the radius of the balloon be denoted by r and the volume by V. Thus $V = \frac{4}{3}\pi r^3$ and

$$\frac{dV}{dt} = 4\pi r^2\frac{dr}{dt}$$

At the instant in question, $dV/dt = 50$ and $r = 10$. Thus

$$50 = 4\pi(10)^2\frac{dr}{dt}$$

19. (*cont'd*)

and so

$$\frac{dr}{dt} = \frac{50}{400\pi} = \frac{1}{8\pi} = 0.04 \text{ centimeter per second.}$$

20. Differentiating $x^2 + y^2 = 1$ gives $2x\,(dx/dt) + 2y\,(dy/dt) = 0$, so $dy/dt = -x/y\,(dx/dt)$. If $x = y = 1/\sqrt{2}$, $dy/dt = -dx/dt = -1$ centimeter per second.

21. Multiplying $x = at + b$ by c, multiplying $y = ct + d$ by a, and subtracting, we get

$$cx - ay = bc - ad$$

so $y = (c/a)x + (1/a)(ad - bc)$, which is the equation of a line with slope c/a. (If $a = 0$, x is constant and the line is vertical, unless $c = 0$ too.) Note that the slope can also be obtained as $(dy/dt)/(dx/dt)$, since $dy/dt = c$ and $dx/dt = a$.

CHAPTER 6

Solved Exercises for Section 6-1

1. $36° \rightarrow 36 \times 0.01745 = 0.6282$ radian; $160° \rightarrow 160 \times 0.01745 = 2.792$ radians; $280° \rightarrow 280 \times 0.01745 = 4.886$ radians; $300° \rightarrow 300 \times 0.01745 = 5.235$ radians, or $300 \times \pi/180 = 5\pi/3$ radians.

2. $5\pi/18 \rightarrow 5\pi/18 \times 180/\pi = 50°$; $2.6 \rightarrow 2.6 \times 57.296 = 148.97°$; $6.27 \rightarrow 6.27 \times 57.296 = 359.25°$; $0.2 \rightarrow 0.2 \times 57.296 = 11.46°$.

3. $5\pi/2 = 2\pi + \pi/2$ radians (that is, $\pi/2$ radians); $470° = 360° + 110° = 110°$.

4. $22° \rightarrow 22 \times 0.01745 = 0.3839$ radian. Since $\theta = s/r$, $s = \theta r = 0.3839 \times 10 = 3.839$ meters. (In doing calculations of this type it is essential to use radians, not degrees.)

5. Refer to Fig. 6-1-32.

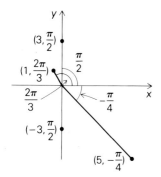

Fig. 6-1-32 Points plotted using polar coordinates.

6. Since $0 \le r \le 2$, we can range from the origin to 2 units from the origin. Our angle with the x axis varies from 0 to π, but not including π. Thus we are confined to the region in Fig. 6-1-33. The negative x axis is dashed since it is not included in the region.

Fig. 6-1-33 The region $0 \le r \le 2, 0 \le \theta < \pi$.

7. In Fig. 6-1-34, $\tan \theta = |AB|/|OA|$ and $\cot(\pi/2 - \theta) = |AB|/|OA|$ as well, so $\tan\theta = \cot(\pi/2 - \theta)$.

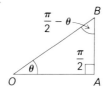

Fig. 6-1-34 θ and $\pi/2 - \theta$ are complementary angles.

8. Referring to Fig. 6-1-35 we see that if θ is switched to $-\theta$, this changes the sign of $y = \sin\theta$. Hence $\sin(-\theta) = -\sin\theta$.

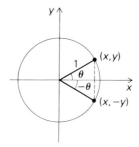

Fig. 6-1-35 $\sin\theta$ is an odd function.

9. Refer to Fig. 6-1-36. Now $|AB| = |OA|\tan 53° = 50 \tan 53° = 50(1.3270) = 66.35$ meters.

Fig. 6-1-36 Trigonometry used to find the height of a tree.

10. Refer to Fig. 6-1-37, which gives an overhead view of the mountain. To determine digging directions from the given point, choose a point P, visible from both A and B, such that $\angle APB$ is a right angle. The lengths a and b of AP and BP are measured, and then the angles θ and ϕ are determined by the formulas $\tan \theta = a/b$ and $\phi = \pi/2 - \theta$ (or from scale drawings). Using these angles, workers starting from A and B can proceed to dig in a level, straight line, and they will meet somewhere under the mountain.* In large projects the curvature of the earth or other factors may necessitate some corrections to this procedure.

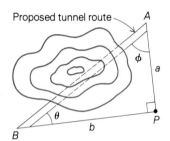

Proposed tunnel route

Fig. 6-1-37 Heron's geometry for a tunnel route.

11. $\tan (\theta + \phi) = \dfrac{\sin (\theta + \phi)}{\cos (\theta + \phi)}$

$\left(\text{from } \tan \psi = \dfrac{\sin \psi}{\cos \psi} \right)$

$= \dfrac{\sin \theta \cos \phi + \cos \theta \sin \phi}{\cos \theta \cos \phi - \sin \theta \sin \phi}$

(from the addition formulas for sin and cos)

$= \dfrac{\tan \theta + \tan \phi}{1 - \tan \theta \tan \phi}$

(divide numerator and denominator by $\cos \theta \cos \phi$).

12. By the half-angle formula:

$\sin^2 15° = \dfrac{1 - \cos 30°}{2} = \dfrac{1 - \sqrt{3}/2}{2}$

$= \dfrac{2 - \sqrt{3}}{4}$

Hence

$\sin 15° = \dfrac{\sqrt{2 - \sqrt{3}}}{2} \quad (\approx 0.258819)$

---

* *For more on this problem, see J. Goodfield, "The Tunnel of Eupalinus," Scientific American, June 1964, pp. 104–112.*

13. From the addition formulas:

$\sin (\theta + \tfrac{\pi}{2}) = \sin \theta \cos \tfrac{\pi}{2} + \cos \theta \sin \tfrac{\pi}{2}$

$= \cos \theta$

and

$\sin (\tfrac{3\pi}{2} + \theta) = \sin (\tfrac{3\pi}{2}) \cos \theta + \cos (\tfrac{3\pi}{2}) \sin \theta$

$= -\cos \theta$

14. From the law of cosines:

$x^2 = 3^2 + 2^2 - 2 \cdot 3 \cdot 2 \cos 20°$

$= 13 - 12(0.9397) = 1.7236$

$x = 1.3129$

15. Referring to Fig. 6-1-38 we see that $|BD|$ can be computed by using ΔOBD or ΔABD; we get

$|BD| = a \sin \gamma, \quad |BD| = c \sin (\pi - \alpha) = c \sin \alpha$

since $\sin (\pi - \alpha) = \sin \pi \cos \alpha - \cos \pi \sin \alpha = \sin \alpha$ ($\sin \pi = 0$, $\cos \pi = -1$). Hence

$a \sin \gamma = c \sin \alpha, \quad \dfrac{\sin \gamma}{c} = \dfrac{\sin \alpha}{a}$

Similarly,

$\dfrac{\sin \alpha}{a} = \dfrac{\sin \beta}{b}$

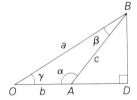

Fig. 6-1-38 Geometry for the proof of the law of sines.

16. We obtain $y = 3 \cos 5\theta$ by compressing the graph of $y = \cos \theta$ horizontally by a factor of 5 and stretching it vertically by a factor of 3 (see Fig. 6-1-39).

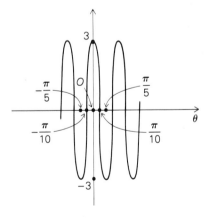

Fig. 6-1-39 The graph of $y = 3 \cos 5\theta$.

17. Recall that an inflection point is a point where the second derivative changes sign—that is, a point between different types of concavity (see Section 2-3). On the graph of $\tan \theta$, notice that the graph is concave upward on $(0, \pi/2)$ and downward on $(-\pi/2, 0)$. Hence zero is an inflection point, as are π, $-\pi$, and so forth. The general inflection point is $n\pi$, where $n = 0, \pm 1, \pm 2, \ldots$. (But $\pi/2$ is not an inflection point because $\tan \theta$ is not defined there.)

    We expect, from the graph, that $\tan \theta$ will be a differentiable function of θ except at $\pm\pi/2$, $\pm 3\pi/2, \ldots$.

18. (a) $r = \sqrt{5^2 + (-2)^2} = \sqrt{29} \approx 5.3852$, $\cos \theta = 5/\sqrt{29} \approx 0.9285$, so $\theta \approx -22° \approx -0.3805$ radian (minus since $(5, -2)$ is in the fourth quadrant).

    (b) $x = 2 \cos \pi/6 = \sqrt{3} \approx 1.7321$; $y = 2 \sin \pi/6 = 1$.

19. Symmetry in the x axis means $f(-\theta) = \cos(-2\theta) = \cos 2\theta = f(\theta)$, which is true. Symmetry in the y axis means $f(\pi - \theta) = f(\theta)$; and, indeed,

    $$f(\pi - \theta) = \cos[2(\pi - \theta)]$$
    $$= \cos 2\pi \cos 2\theta + \sin 2\pi \sin 2\theta$$
    $$= \cos 2\theta = f(\theta)$$

    Finally,

    $$f(\theta + \tfrac{\pi}{2}) = \cos 2(\theta + \tfrac{\pi}{2})$$
    $$= \cos(2\theta + \pi)$$
    $$= \cos 2\theta \cos \pi - \sin 2\theta \sin \pi$$
    $$= -\cos 2\theta = -f(\theta)$$

    Thus the graph is unchanged when we reflect in the x axis and the y axis. When we rotate by 90°, $r = f(\theta)$ reflects through the origin; that is, r changes to $-r$.

20. The graph is symmetric in the x axis and, moreover, $f(\theta + \pi/3) = -f(\theta)$. This means that we need only sketch the graph for $0 \le \theta \le \pi/3$ and obtain the rest by reflection and rotations. Thus we expect a six-leafed rose. As θ varies from 0 to $\pi/3$, 3θ varies from 0 to π and $\cos 3\theta$ decreases from 1 to 0. Hence we get the graph in Fig. 6-1-40.

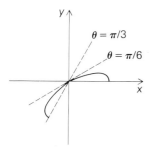

Fig. 6-1-40 Beginning the graph of $r = \cos 3\theta$.

20. *(cont'd)*
    Reflect across the axis to complete the leaf and then rotate by $\pi/3$ and reflect through the origin; see Fig. 6-1-41.

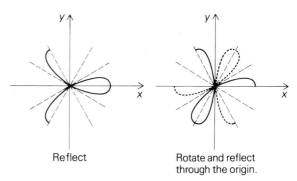

Reflect   Rotate and reflect through the origin.

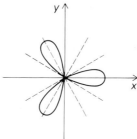

Full graph.

Fig. 6-1-41 The graphing of $r = \cos 3\theta$.

21. This means that the graph will have the same appearance if it is rotated by 90°, since replacing θ by $\theta + \pi/2$ means that we rotate through an angle $\pi/2$.

Solved Exercises for Section 6-2

1. Our calculator gave:

    $$\frac{1 - \cos \theta}{\theta} = 9.999665 \times 10^{-3} \quad \text{for } \theta = 0.02$$

    $$\frac{1 - \cos \theta}{\theta} = 5.001 \times 10^{-4} \quad \text{for } \theta = 0.001$$

    and

    $$\frac{\sin \theta}{\theta} = 0.9999333 \quad \text{for } \theta = 0.02$$

    $$\frac{\sin \theta}{\theta} = 0.9999998 \quad \text{for } \theta = 0.001$$

    Your answer may differ because of calculator inaccuracies. However, these numbers confirm that $(1 - \cos \theta)/\theta$ is small for θ small and $(\sin \theta)/\theta$ is near 1 for θ small.

2. Since $\tan \theta = \sin \theta / \cos \theta$, $\tan' \theta = (\cos \theta \sin' \theta - \sin \theta \cos' \theta)/\cos^2 \theta$; at $\theta = 0$, $\cos 0 = 1$, $\sin 0 = 0$, $\sin' 0 = 1$, $\cos' 0 = 0$, so $\tan' 0 = 1$.

3. $\lim_{\theta \to 0} (\sin a\theta)/\theta = f'(0)$, where $f(\theta) = \sin a\theta$. However, by the chain rule, $f'(\theta) = a \sin'(a\theta)$, so $f'(0) = a$. Thus the required limit is a.

4. (a) By the product rule,

$$\frac{d}{dx} \sin x \cos x = \cos x \cos x + \sin x (-\sin x)$$

$$= \cos^2 x - \sin^2 x$$

(The answer $\cos 2x$ is also correct.)

(b) By the quotient and chain rules,

$$\frac{d}{dx} \frac{\tan 3x}{1 + \sin^2 x} =$$

$$\frac{(3 \sec^2 3x)(1 + \sin^2 x) - (\tan 3x)(2 \sin x \cos x)}{(1 + \sin^2 x)^2}$$

(c) $\dfrac{d}{dx}(1 - \csc^2 5x) = -2 \csc 5x \dfrac{d}{dx}(\csc 5x)$

$$= 10 \csc^2 5x \cot 5x$$

5. By the chain rule,

$$\frac{d}{d\theta} \sin(\sqrt{3\theta^2 + 1})$$

$$= \cos(\sqrt{3\theta^2 + 1}) \frac{d}{d\theta} \sqrt{3\theta^2 + 1}$$

$$= \cos(\sqrt{3\theta^2 + 1}) \cdot \frac{1}{2} \frac{3 \cdot 2\theta}{\sqrt{3\theta^2 + 1}}$$

$$= \frac{3\theta}{\sqrt{3\theta^2 + 1}} \cos \sqrt{3\theta^2 + 1}$$

6. $\sin'(t_0) = \lim_{\Delta t \to 0} \dfrac{\sin(t_0 + \Delta t) - \sin t_0}{\Delta t}$

$$= \lim_{\Delta t \to 0} \frac{\sin t_0 \cos \Delta t + \cos t_0 \sin \Delta t - \sin t_0}{\Delta t}$$

$$= \sin t_0 \lim_{\Delta t \to 0} \frac{\cos \Delta t - 1}{\Delta t} + \cos t_0 \lim_{\Delta t \to 0} \frac{\sin \Delta t}{\Delta t}$$

$$= \sin t_0 \cdot 0 + \cos t_0 \cdot 1 = \cos t_0$$

7. If $f(x) = \sin^2 x$, $f'(x) = 2 \sin x \cos x$ and $f''(x) = 2(\cos^2 x - \sin^2 x)$. The first derivative vanishes when either $\sin x = 0$ or $\cos x = 0$, at which points f'' is positive and negative, yielding minima and maxima. Thus the minima of f are at $0, \pm\pi, \pm 2\pi, \ldots$, where $f = 0$, and the maxima are at $\pm\pi/2, \pm 3\pi/2, \ldots$, where $f = 1$.

The function $f(x)$ is concave upward when $f''(x) > 0$ (that is, $\cos^2 x > \sin^2 x$) and downward when $f''(x) < 0$ (that is, $\cos^2 x < \sin^2 x$). Also, $\cos x = \pm\sin x$ exactly if $x = \pm\pi/4, \pm\pi/4 \pm \pi, \pm\pi/4 \pm 2\pi$, and so on (see the graphs of sine and cosine). These are then inflection points separating regions of concavity and convexity. The graph is shown in Fig. 6-2-10.

7. *(cont'd)*

Fig. 6-2-10 The graph of $\sin^2 x$.

8. We differentiate by using the chain rule:

$$\frac{d}{d\theta}[-f(\cos \theta)] = -f'(\cos \theta) \frac{d}{d\theta} \cos \theta$$

$$= -\frac{1}{\cos \theta}(-\sin \theta)$$

$$= \frac{\sin \theta}{\cos \theta} = \tan \theta$$

Thus $-f(\cos \theta)$ is an antiderivative for $\tan \theta$, and so is $-f(\cos \theta) + C$ for any constant C.

9. If we guess $-\cos 4u$ as the antiderivative, we find $(d/du)(-\cos 4u) = 4 \sin 4u$, which is four times too big, so

$$\int \sin 4u \, du = -\tfrac{1}{4} \cos 4u + C$$

10. (a) Since $\sin(\pi/6) = \frac{1}{2}$, $\sin^{-1}(\frac{1}{2}) = \pi/6$. Similarly, $\sin^{-1}(\sqrt{3}/2) = -\pi/3$. Finally, $\sin^{-1}(2)$ is not defined since 2 is not in the domain $[-1, 1]$ of $\sin^{-1}$.

(b) From Fig. 6-2-11 we see that $\theta = \sin^{-1} x$ (that is, $\sin \theta = |AB|/|OB| = x$) and $\tan \theta = x/\sqrt{1 - x^2}$, so $\tan(\sin^{-1} x) = x/\sqrt{1 - x^2}$.

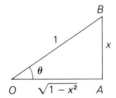

Fig. 6-2-11 $\tan(\sin^{-1} x) = x/\sqrt{1 - x^2}$.

11. By the power rule and the chain rule,

$$\frac{d}{dx}(\sin^{-1} 2x)^{3/2} = \frac{3}{2}(\sin^{-1} 2x)^{1/2} \frac{d}{dx} \sin^{-1} 2x$$

$$= \frac{3}{2}(\sin^{-1} 2x)^{1/2} \cdot 2 \cdot \frac{1}{\sqrt{1 - (2x)^2}}$$

$$= 3\left(\frac{\sin^{-1} 2x}{1 - 4x^2}\right)^{1/2}$$

12. By the chain rule,

$$\frac{d}{dx} \sin^{-1} \sqrt{1 - x^2}$$

$$= \frac{1}{\sqrt{1 - (\sqrt{1 - x^2})^2}} \cdot \frac{d}{dx} \sqrt{1 - x^2}$$

$$= \frac{1}{\sqrt{x^2}} \cdot \frac{-x}{\sqrt{1 - x^2}} = \frac{-1}{\sqrt{1 - x^2}},$$

since $0 < x < 1$.

13. We find that $\cos^{-1}(-\tfrac{1}{2}) = 2\pi/3$ since $\cos(2\pi/3) = -\tfrac{1}{2}$, as is seen from Fig. 6-2-12 or the table on p. 269. Similarly, $\tan^{-1}(1) = \pi/4$ since $\tan \pi/4 = 1$. Finally, $\csc^{-1}(2/\sqrt{3}) = \pi/3$ since $\csc(\pi/3) = 2/\sqrt{3}$.

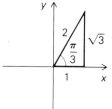

Fig. 6-2-12 Evaluating some inverse trigonometric functions.

14. No. For example, if $y = 1/\sqrt{3}$, then $\tan^{-1}(1/\sqrt{3}) = \pi/6$, while $\cot^{-1}(1/\sqrt{3}) = \pi/3$. What is true is that

$$\cot^{-1}\left(\frac{1}{y}\right) = \tan^{-1} y$$

since

$$\cot x = \frac{1}{\tan x}$$

15. (a) By the chain rule,

$$\frac{d}{dy} \sec^{-1} y^2 = \frac{1}{\sqrt{y^4(y^4 - 1)}} \cdot 2y$$

$$= \frac{2}{y\sqrt{y^4 - 1}}$$

15. (*cont'd*)

   (b) By the chain rule,

$$\frac{d}{dx} \cot^{-1}\left(\frac{x^3 + 1}{x^3 - 1}\right)$$

$$= \frac{-1}{1 + \left[\dfrac{x^3 + 1}{x^3 - 1}\right]^2} \frac{d}{dx}\left(\frac{x^3 + 1}{x^3 - 1}\right)$$

$$= \frac{-(x^3 - 1)^2}{(x^3 - 1)^2 + (x^3 + 1)^2} \times$$

$$\left[\frac{(x^3 - 1) \cdot 3x^2 - (x^3 + 1)3x^2}{(x^3 - 1)^2}\right]$$

$$= \frac{6x^2}{(x^3 - 1)^2 + (x^3 + 1)^2}$$

$$= \frac{3x^2}{x^6 + 1}$$

16. Let $f(x)$ be one of the functions $\sin x$, $\tan x$, or $\sec x$ and let $g(x)$ be the corresponding cofunction $\cos x$, $\cot x$, or $\csc x$. Then we know that

$$f\left(\frac{\pi}{2} - x\right) = g(x)$$

If we let y denote this common value, then we get

$$\frac{\pi}{2} - x = f^{-1}(y) \qquad \text{and} \qquad x = g^{-1}(y)$$

so

$$\frac{\pi}{2} - f^{-1}(y) = g^{-1}(y)$$

It follows by differentiation in y that

$$-\frac{d}{dy} f^{-1}(y) = \frac{d}{dy} g^{-1}(y)$$

Hence the derivatives of $f^{-1}(y)$ and $g^{-1}(y)$ are negatives, which is the general reason why this same phenomenon occurred three times in Table 6-2-2.

17. We know an antiderivative for $1/(1 + x^2)$; it is $\tan^{-1} x$. Now we write

$$\int \frac{x^2}{1 + x^2}\, dx = \int \frac{1 + x^2}{1 + x^2}\, dx - \int \frac{1}{1 + x^2}\, dx$$

$$= \int 1\, dx - \int \frac{1}{1 + x^2}\, dx$$

$$= x - \tan^{-1} x + C$$

6

Solved Exercises for Section 6-3

1. Let $(x, 0)$ be a point on the x axis. The distance from $(0, 1)$ is $\sqrt{1 + x^2}$ and from (p, q) it is $\sqrt{(x - p)^2 + q^2}$. The sum of the distances is $S = \sqrt{1 + x^2} + \sqrt{(x - p)^2 + q^2}$. To minimize, we find:

$$\frac{dS}{dx} = \frac{1}{2}(1 + x^2)^{-1/2}(2x)$$

$$+ \frac{1}{2}[(x - p)^2 + q^2]^{-1/2}2(x - p)$$

$$= \frac{x}{\sqrt{1 + x^2}} + \frac{x - p}{\sqrt{(x - p)^2 + q^2}}$$

Setting this equal to zero gives

$$\frac{x}{\sqrt{1 + x^2}} = \frac{p - x}{\sqrt{(x - p)^2 + q^2}}$$

Instead of solving for x, we will interpret the preceding equation geometrically. In Fig. 6-3-16, $\sin \theta_1 = x/\sqrt{1 + x^2}$ and $\sin \theta_2 = (p - x)/\sqrt{(x - p)^2 + q^2}$; our equation says that these are equal, so $\theta_1 = \theta_2$. Thus $(x, 0)$ is located at the point for which the lines from $(x, 0)$ to $(0, 1)$ and (p, q) make equal angles with the y axis. This result is sometimes called the *law of reflection*.

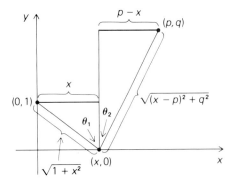

Fig. 6-3-16 The shortest path from $(0, 1)$ to (p, q) via the x axis has $\theta_1 = \theta_2$.

2. Refer to Fig. 6-3-17. The length of PQ is

$$f'(\theta) = \frac{a}{\sin \theta} + \frac{b}{\cos \theta}$$

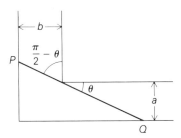

Fig. 6-3-17   The pole in the corner.

The minimum of $f(\theta)$, $0 \leq \theta \leq \pi/2$, will give the length of the longest pole which will fit around the corner. The derivative is

$$f'(\theta) = -\frac{a \cos \theta}{\sin^2 \theta} + \frac{b \sin \theta}{\cos^2 \theta}$$

which is zero when $a \cos^3 \theta = b \sin^3 \theta$; that is, when $\tan^3 \theta = a/b$; hence $\theta = \tan^{-1}(\sqrt[3]{a/b})$. Since f is large positive near 0 and $\pi/2$, and there are no other critical points, this is a global minimum. (You can also use the second derivative test.) Thus the answer is

$$\frac{a}{\sin \theta} + \frac{b}{\cos \theta}$$

where $\theta = \tan^{-1}(\sqrt[3]{a/b})$. Using $\sin(\tan^{-1}\alpha) = \alpha/\sqrt{1 + \alpha^2}$ and $\cos(\tan^{-1}\alpha) = 1/\sqrt{1 + \alpha^2}$ (Fig. 6-3-18), one can express the answer, after some simplification, as $(a^{2/3} + b^{2/3})^{3/2}$.

Fig. 6-3-18 $\phi = \tan^{-1}\alpha$.

One way to check the answer (which the authors actually used to catch an error) is to note its "dimension." The result must have the dimension of a length. Thus an answer like $a^{1/3}(a^{2/3} + b^{2/3})^{3/2}$, which has dimension of $(\text{length})^{1/3} \times$ length, cannot be correct.

3. It is convenient to replace the diagram by a more abstract one (Fig. 6-3-19). We assume that the passenger's eye is located at a point O on the line AB when he is sitting upright and that he wishes to maximize the angle $\angle POQ$ subtended by the window PQ. Denote by x the distance from O to B, which can be varied. Let the width of the window be w and the distance from AB to the window be d. Then we have

$$\angle POQ = \angle BOQ - \angle POB$$

$$\tan(\angle BOQ) = \frac{w+d}{x}$$

and

$$\tan(\angle POB) = \frac{d}{x}$$

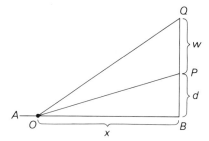

Fig. 6-3-19 Which value of x maximizes $<POQ$?

So we wish to maximize

$$f(x) = \angle POQ = \tan^{-1}\left(\frac{w+d}{x}\right) - \tan^{-1}\left(\frac{d}{x}\right)$$

Differentiating, we have

$$f'(x) = \frac{1}{1+[(w+d)/x]^2}\cdot\left(-\frac{w+d}{x^2}\right)$$
$$- \frac{1}{1+(d/x)^2}\cdot\left(-\frac{d}{x^2}\right)$$
$$= -\frac{w+d}{x^2+(w+d)^2} + \frac{d}{x^2+d^2}$$

Setting $f'(x) = 0$ yields the equation

$$(x^2+d^2)(w+d) = [x^2+(w+d)^2]d$$

or

$$wx^2 = (w+d)^2 d - d^2(w+d)$$

or

$$x^2 = (w+d)\cdot d$$

The solution, therefore, is $x = \sqrt{(w+d)d}$.

3. *(cont'd)*

For example (all distances measured in feet), if $d = 1$ and $w = 5$, we should take $x = \sqrt{6} \approx 2.45$. Thus it is probably better to take the second seat from the window, rather than the window seat.

There is a geometric interpretation for the solution of this problem. We may rewrite the solution as

$$x^2 = d^2 + wd \quad \text{or} \quad x^2 = (d+\tfrac{1}{2}w)^2 - (\tfrac{1}{2}w)^2$$

This second formula leads to the following construction (which you may be able to carry out mentally before choosing your seat). Draw a line RP through P and parallel to AB. Now construct a circle with center at the midpoint M of PQ and with radius MB. Let Z be the point where the circle intersects RP. Then $x = ZP$. (See Fig. 6-3-20.)

Fig. 6-3-20 Geometric construction for the best seat.

4. (a) Using the chain rule, we get $f'(x) = 3x\sqrt{x^2+1}$. Hence $f'(x) < 0$ (so f is decreasing) on $(-\infty, 0)$, and $f'(x) > 0$ (so f is increasing) on $(0, \infty)$.

(b) By the first derivative test, $x = 0$ is a minimum point. Note that $f(x)$ is an even function and

$$f''(x) = 3\left(\frac{x^2}{\sqrt{x^2+1}} + \sqrt{x^2+1}\right) > 0$$

so f is concave upward. Thus we can sketch its graph as in Fig. 6-3-21. There are no cusps.

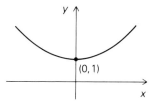

Fig. 6-3-21 The graph of $y = (x^2+1)^{3/2}$.

5. Letting $f(x) = (x + 1)^{2/3} x^2$, we have

$$f'(x) = \tfrac{2}{3}(x + 1)^{-1/3} x^2 + (x + 1)^{2/3} \cdot 2x$$
$$= [2x/3(x + 1)^{1/3}] (4x + 3)$$

For x near -1, but $x > -1$, $f'(x)$ is large positive, while for $x < -1$, $f'(x)$ is large negative. Since f is continuous at -1, this is a local minimum and a cusp.

   The other critical points are $x = 0$ and $x = -\tfrac{3}{4}$. From the first derivative test (or second derivative test if you like it better), $-\tfrac{3}{4}$ is a local maximum and zero is a local minimum. For $x > 0$, f is increasing since $f'(x) > 0$; for $x < -1$, f is decreasing since $f'(x) < 0$. Thus we can sketch the graph as in Fig. 6-3-22. (We located the inflection points at $(-33 \pm \sqrt{609})/40 \approx -0.208$ and -1.442 by setting the second derivative equal to zero.)

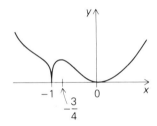

Fig. 6-3-22 The graph of $y = x^2(x + 1)^{2/3}$ has a cusp at $(-1, 0)$.

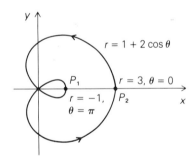

Fig. 6-3-23 The maxima and minima of $1 + 2 \cos \theta$ correspond to the points P_1 and P_2.

6. Here $f(\theta) = \cos 3\theta$, so the slope is, by formula (3), p. 302,

$$\frac{f'(\theta) \tan \theta + f(\theta)}{f'(\theta) - f(\theta) \tan \theta} = \frac{-3 \sin 3\theta \tan \theta + \cos 3\theta}{-3 \sin 3\theta - \cos 3\theta \tan \theta}$$
$$= \frac{1 - 3 \tan 3\theta \tan \theta}{-\tan \theta - 3 \tan 3\theta}$$

Hence at $\theta = \pi/3$, slope $= 1/-1.732 \approx -0.577$.

7. Here $dr/d\theta = -2 \sin \theta$, which vanishes if $\theta = 0, \pi$. Also, $d^2 r/d\theta^2 = -2 \cos \theta$, which is -2 at $\theta = 0$, $+2$ at $\theta = \pi$. Hence $r = 3$, $\theta = 0$ is a local maximum and $r = -1$, $\theta = \pi$ is a local minimum. The tangent lines are vertical there. The curve passes through $r = 0$ when $\theta = \pm 2\pi/3$. The curve is symmetric in the x axis and can thus be plotted as in Fig. 6-3-23.

8. If $l > \pi/2 - \alpha$, we are above the Arctic Circle. Equation (5) (p. 304) has a solution only when

$$-1 \le -\tan l \frac{\sin \alpha \cos (2\pi T/365)}{\sqrt{1 - \sin^2 \alpha \cos^2 (2\pi T/365)}} \le 1$$

since $\cos 2\pi S/24$ lies between -1 and 1. Thus there will be a sunset on those days T for which

$$\tan^2 l \frac{\sin^2 \alpha \cos^2 2\pi T/365}{1 - \sin^2 \alpha \cos^2 2\pi T/365} \le 1$$

or $\cos^2(2\pi T - 365) \le \cos^2 l / \sin^2 \alpha$. Thus there is no sunset until $n = (365/2\pi) \cos^{-1}(\cos l/\sin \alpha)$ days after June 21. We find n reaches its maximum value of $365/4$ when $l = \pi/2$, and decreases to 0 when $l = \pi/2 - \alpha$. At the pole ($l = \pi/2$) half the year, the quarters before and after June 21, between vernal and autumnal equinox, are without sunset. As we move away from the pole, more and more days will have a sunset until we get below the arctic circle ($l = \pi/2 - \alpha = 66.5°$ N latitude). Then every day has a sunset. (See Fig. 6-3-24.)

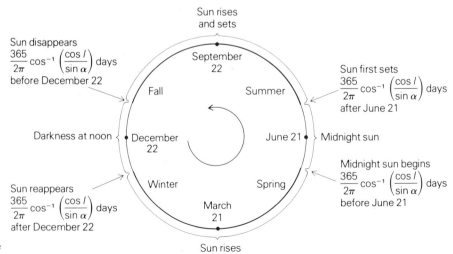

Fig. 6-3-24 The sun cycle in the north polar region.

CHAPTER 7

Solved Exercises for Section 7-1

1. We must show that $(b^{1/n})^m$ is the nth root of b^m. But $[(b^{1/n})^m]^n = (b^{1/n})^{mn} = (b^{1/n})^{nm} = [(b^{1/n})^n]^m = b^m$; this calculation used only the laws of *integer* exponents and the fact that $(b^{1/n})^n = b$.

2. $8^{-2/3} = 1/8^{2/3} = 1/(\sqrt[3]{8})^2 = 1/2^2 = 1/4$; $9^{3/2} = (\sqrt{9})^3 = 3^3 = 27$.

3. $(x^{2/3})^{5/2}/x^{1/4} = x^{2/3 \cdot 5/2 - 1/4} = x^{5/3 - 1/4} = x^{17/12}$.

4. If $0 < b < 1$, then $1/b > 1$, so if $p < q$ we have $-q < -p$, and thus $(1/b)^{-q} < (1/b)^{-p}$. But $(1/b)^{-q} = b^q$ and $(1/b)^{-p} = b^p$, so we have $b^q < b^p$ whenever $0 < b < 1$ and $p < q$.

5. It is the reflection in the y axis, since $\exp_{1/b} x = (1/b)^x = (b^{-1})^x = b^{-x} = \exp_b(-x)$.

6. $(2^{\sqrt{3}} + 2^{-\sqrt{3}})(2^{\sqrt{3}} - 2^{-\sqrt{3}}) = (2^{\sqrt{3}})^2 - (2^{-\sqrt{3}})^2 = 2^{2\sqrt{3}} - 2^{-2\sqrt{3}}$.

7. (a)–(C); (b)–(B); (c)–(D); (d)–(A).

8. (a)–(C); (b)–(A); (c)–(D); (d)–(B).

9. $\log_2 4 = 2$ since $2^2 = 4$; $\log_3 81 = 4$ since $3^4 = 81$; and $\log_{10} 0.01 = -2$ since $10^{-2} = 0.01$.

10. (a) $\log_b(b^{2x}/2b) = \log_b(b^{2x-1}/2) = \log_b(b^{2x-1}) - \log_b 2 = (2x - 1) - \log_b 2$.

    (b) $\log_2 x = \log_2 5 + 3 \log_2 3 = \log_2 5 \cdot 3^3 = \log_2 135$, so $x = \exp_2(\log_2 x) = \exp_2(\log_2 135) = 135$.

Solved Exercises for Section 7-2

1. (a) $3xe^{3x} + e^{3x}$

   (b) $(2x + 2)\exp(x^2 + 2x)$

   (c) $2x$

   (d) $e^{\sqrt{x}}(1/2\sqrt{x})$ by the chain rule.

   (e) $e^{\sin x} \cdot \cos x$

   (f) By the chain rule, $(d/dx)b^{u(x)} = b^{u(x)} \ln b \cdot du/dx$. Here $(d/dx)2^{\sin x} = 2^{\sin x} \cdot \ln 2 \cdot \cos x$.

2. We have $\exp_b'(0) = \ln b$, so

$$\exp_b\left[\frac{1}{\exp_b'(0)}\right] = \exp_b\left(\frac{1}{\ln b}\right)$$
$$= b^{1/\ln b} = (e^{\ln b})^{1/\ln b}$$
$$= e^{\ln b(1/\ln b)} = e^1 = e$$

3. (a) By the chain rule, $(d/dx) \ln 10x = [(d/du) \ln u](du/dx)$, where $u = 10x$; that is, $(d/dx) \ln 10x = 10 \cdot 1/10x = 1/x$.

   (b) $(d/dx) \ln u(x) = u'(x)/u(x)$ by the chain rule.

   (c) $\cos x/\sin x = \cot x$ (We use the preceding solution.)

   (d) Use the product rule: $(d/dx)(\sin x \ln x) = \cos x \ln x + (\sin x) \cdot 1/x$.

   (e) Use the quotient rule: $(d/dx)(\ln x/x) = (x \cdot 1/x - \ln x)/x^2 = (1 - \ln x)/x^2$.

   (f) By differentiation formula 4 on p. 329: $(d/dx) \log_5 x = 1/(\ln 5)x$.

4. (a) Write $x^n = e^{(\ln x)n}$ and differentiate using the chain rule:

$$\frac{d}{dx}x^n = \frac{d}{dx}e^{(\ln x) \cdot n} = \frac{n}{x} \cdot e^{(\ln x)n}$$
$$= \frac{n}{x}x^n = nx^{n-1}$$

   using the laws of exponents.

   (b) $(d/dx)x^\pi = \pi x^{\pi - 1}$, by part (a).

5. (a) $(d/dx)e^{ax} = ae^{ax}$, by the chain rule, so we must have $\int e^{ax} dx = (1/a)e^{ax} + C$.

   (b) $(d/dx) \ln(3x + 2) = [1/(3x + 2)] \cdot 3$, by the chain rule, so $\int 1/(3x + 2)dx = \frac{1}{3} \ln(3x + 2) + C$ (for $3x + 2 > 0$; what if $3x + 2 < 0$?).

6. $\ln y = \ln[(2x + 3)^{3/2}/(x^2 + 1)^{1/2}] = \frac{3}{2} \ln(2x + 3) - \frac{1}{2} \ln(x^2 + 1)$, so

$$\frac{1}{y}\frac{dy}{dx} = \frac{3}{2} \cdot \frac{2}{2x + 3} - \frac{1}{2} \cdot \frac{2x}{x^2 + 1}$$
$$= \frac{3}{2x + 3} - \frac{x}{x^2 + 1} = \frac{(x^2 - 3x + 3)}{(2x + 3)(x^2 + 1)}$$

and hence

$$\frac{dy}{dx} = \frac{(2x + 3)^{3/2}}{(x^2 + 1)^{1/2}} \cdot \frac{(x^2 - 3x + 3)}{(2x + 3)(x^2 + 1)}$$
$$= \frac{(x^2 - 3x + 3)(2x + 3)^{1/2}}{(x^2 + 1)^{3/2}}$$

7. We find that $\ln y = x^x \ln x$, so using the fact that $(d/dx)x^x = x^x(1 + \ln x)$ from Worked Example 5, we get

$$\frac{1}{y}\frac{dy}{dx} = x^x(1 + \ln x)\ln x + \frac{x^x}{x}$$

so

$$\frac{dy}{dx} = x^{(x^x)}[x^x(1 + \ln x)\ln x + x^{x-1}]$$

8. If $y = xe^{2x}$, then $dy/dx = 2xe^{2x} + e^{2x}$, which at $x = 1$ is $2e^2 + e^2 = 3e^2$. The tangent line has equation

$$y = y_0 + m(x - x_0)$$

Thus

$$y = e^2 + (3e^2)(x - 1)$$
$$= 3e^2 x - 2e^2$$

9. One can achieve a fair degree of accuracy before round-off errors make the operations meaningless. For example, on our calculator we obtained:

n	$\left(1 - \dfrac{1}{n}\right)^{-n} = \left(\dfrac{n}{n-1}\right)^n$	$\left(1 + \dfrac{1}{n}\right)^n = \left(\dfrac{n+1}{n}\right)^n$	
2	4.000000	2.250000	
4	3.160494	2.441406	
8	2.910285	2.565785	
16	2.808404	2.637928	
32	2.762009	2.676990	
64	2.739827	2.697345	
128	2.728977	2.707739	
256	2.723610	2.712992	
512	2.720941	2.715632	
1024	2.719610	2.716957	
2048	2.718947	2.717617	
4096	2.718611	2.717954	
8192	2.718443	2.718109	
16,384	2.718370	2.718192	
32,768	2.718367	2.718278	
65,536	2.718299	2.718299	
131,072	2.718131	2.718131	
262,144	2.718492	2.718492	Calculator
524,288	2.717782	2.717782	round-off
1,048,576	2.719209	2.719209	errors significant
16,777,216	2.736372	2.736372	below here
536,870,912	2.926309	2.926309	
2,147,483,648	1.00000	1.00000	

10. We generalize the procedure in the text as follows:

$$e^a = e^{\ln'(1/a)}$$

$$= \exp \lim_{\Delta x \to 0} \left[\frac{\ln(1/a + \Delta x) - \ln(1/a)}{\Delta x} \right]$$

$$= \lim_{\Delta x \to 0} \left[\exp \ln \left(\frac{1/a + \Delta x}{1/a} \right) \right]^{1/\Delta x}$$

$$= \lim_{\Delta x \to 0} (1 + a\Delta x)^{1/\Delta x}$$

Therefore, if we write $h = \Delta x$ or $n = \pm(1/\Delta x)$ we get

$$e^a = \lim_{h \to 0} (1 + ah)^{1/h}$$

$$= \lim_{n \to \infty} \left(1 + \frac{a}{n} \right)^n$$

$$= \lim_{n \to \infty} \left(1 - \frac{a}{n} \right)^{-n}$$

CHAPTER 8

Solved Exercises for Section 8-1

1. (a) $\int \dfrac{x^3 + 8x + 3}{x}\, dx = \int \left(x^2 + 8 + \dfrac{3}{x}\right) dx$

$$= \dfrac{x^3}{3} + 8x + 3 \ln |x| + C$$

(b) $\int (x^\pi + x^3)\, dx$

$$= \dfrac{x^{\pi+1}}{\pi + 1} + \dfrac{x^4}{4} + C$$

2. (a) $\displaystyle\int_1^2 \left(\sqrt{x} + \dfrac{2}{x}\right) dx$

$$= \left(\dfrac{x^{3/2}}{3/2} + 2 \ln |x|\right)\Bigg|_1^2$$

$$= \dfrac{2}{3} 2^{3/2} + 2 \ln 2 - \left(\dfrac{2}{3} + 0\right)$$

$$= \dfrac{4\sqrt{2} - 2}{3} + 2 \ln 2$$

(b) $\displaystyle\int_{1/2}^1 \left[\dfrac{x^4 + x^6 + 1}{x^2}\right] dx$

$$= \int_{1/2}^1 \left(x^2 + x^4 + \dfrac{1}{x^2}\right) dx$$

$$= \left(\dfrac{x^3}{3} + \dfrac{x^5}{5} - \dfrac{1}{x}\right)\Bigg|_{1/2}^1$$

$$= \left(\dfrac{1}{3} + \dfrac{1}{5} - 1\right) - \left(\dfrac{1}{3 \cdot 8} + \dfrac{1}{5 \cdot 32} - 2\right)$$

$$= \dfrac{713}{480}$$

3. First we should express everything in terms of the same unit of time. Choosing hours, we convert the rate of $2t + 3$ liters per minute to $60(2t + 3) = 120t + 180$ liters per hour. The total amount of water in the tank at time T hours past noon is the integral

$$\int_0^T (120t + 180)\, dt = \dfrac{120}{2}(T^2 - 0^2) + 180(T - 0)$$

$$= 60T^2 + 180T$$

The tank is full when $60T^2 + 180T = 1000$. Solving for T by the quadratic equation, we find $T \approx 2.849$ hours past noon, so the tank is full at 2:51 PM.

4. The integral $\int_{-1}^1 dx/x^4$ is not defined, since the function $1/x^4$ is not integrable on $[-1, 1]$. It is not defined at zero, and there is no way to define it there which would make the function bounded. The fundamental theorem of calculus requires that the function under the integral sign be integrable, so it does not and cannot apply in this situation. A positive function *cannot* have a negative integral.

5. If we consider the integral as the area under the graph, then formula 4 is just the principle of addition of areas (see Fig. 8-1-10).

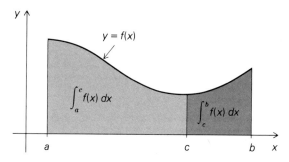

Fig. 8-1-10 The area of the entire figure is $\int_a^b f(x)\, dx = \int_a^c f(x)\, dx + \int_c^b f(x)\, dx$ which is the sum of the areas of the two subfigures.

6. Integrating the identity $\sin^2 x + \cos^2 x = 1$ from 0 to $\pi/2$ and using formula 2 gives

$$\int_0^{\pi/2} \sin^2 x\, dx + \int_0^{\pi/2} \cos^2 dx = \int_0^{\pi/2} 1\, dx$$

The two integrals on the left-hand side are equal by assumption and the right-hand side is $\pi/2$, so

$$\int_0^{\pi/2} \sin^2 x\, dx = \dfrac{\pi}{4}$$

(Note that we did this problem without ever finding $\int \sin^2 x\, dx$.)

7. If $a < b < c$, then formula 4 with c and b interchanged gives

$$\int_a^c f(x)\, dx = \int_a^b f(x)\, dx + \int_b^c f(x)\, dx$$

By our definition of wrong-way integrals, $\int_b^c f(x)\, dx = -\int_c^b f(x)\, dx$, so

$$\int_a^c f(x)\, dx = \int_a^b f(x)\, dx - \int_c^b f(x)\, dx$$

or

$$\int_a^b f(x)\, dx = \int_a^c f(x)\, dx + \int_c^b f(x)\, dx$$

which is formula 4 as originally stated.

8

8. If $b < a$,

$$\int_a^b f(x)\, dx = -\int_b^a f(x)\, dx$$

and

$$\int_a^b g(x)\, dx = -\int_b^a g(x)\, dx$$

But

$$\int_b^a f(x)\, dx \le \int_b^a g(x)\, dx$$

and so multiplying by -1,

$$-\int_b^a f(x)\, dx \ge -\int_b^a g(x)\, dx$$

Therefore

$$\int_a^b f(x)\, dx \ge \int_a^b g(x)\, dx$$

when $f(x) \le g(x)$ for all x in $[b, a]$.

9. From the table on p. 293, we find that $(d/dy)(\sin^{-1} y) = 1/\sqrt{1 - y^2}$ for $-1 < y < 1$, so we have

$$\int \frac{1}{\sqrt{1 - y^2}}\, dy = \sin^{-1} y + C$$

and so by the Fundamental Theorem

$$\int_{-1/2}^{1/2} \frac{1}{\sqrt{1 - y^2}}\, dy = \sin^{-1} y \,\Big|_{-1/2}^{1/2}$$

$$= \sin^{-1}\left(\frac{1}{2}\right) - \sin^{-1}\left(-\frac{1}{2}\right)$$

$$= \frac{\pi}{6} - \left(-\frac{\pi}{6}\right) = \frac{\pi}{3}$$

10. The area (shaded in Fig. 8-1-11) is given by the integral

$$\int_a^b \frac{dx}{1 + x^2} = \int \frac{dx}{1 + x^2} \,\Big|_a^b = \tan^{-1} x \,\Big|_a^b$$

$$= \tan^{-1} b - \tan^{-1} a$$

By the definition of the inverse tangent function (see p. 292), the value of $\tan^{-1} x$ always lies in the interval $(-\pi/2, \pi/2)$, so the difference between two values must be less than π, regardless of the length of the interval $[a, b]$.

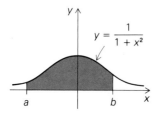

$$y = \frac{1}{1 + x^2}$$

Fig. 8-1-11   The shaded area remains less than π no matter how large the interval $[a, b]$ may be.

11. An antiderivative of $\cos 3x$ is, by guesswork, $\frac{1}{3} \sin 3x$. Thus

$$\int_0^{\pi/6} \cos 3x\, dx = \frac{1}{3} \sin 3x \,\Big|_0^{\pi/6} = \frac{1}{3} \sin \frac{\pi}{2} = \frac{1}{3}$$

12.
$$\int_0^1 (3e^x + 2\sqrt{x})\, dx = 3\int_0^1 e^x\, dx + 2\int_0^1 x^{1/2}\, dx$$

$$= 3e^x \,\Big|_0^1 + 2\left(\frac{x^{3/2}}{3/2}\right)\Big|_0^1$$

$$= 3(e^1 - e^0) + \tfrac{4}{3}(1^{3/2} - 0^{3/2})$$

$$= 3e - 3 + \tfrac{4}{3}$$

$$= 3e - \tfrac{5}{3} \approx 6.488$$

13. (a) By the product rule for derivatives,

$$\frac{d}{dx}(x \ln x) = \ln x + x \cdot 1/x = \ln x + 1.$$

(b) From (a), $\int (\ln x + 1)\, dx = x \ln x + C$. Therefore, $\int \ln x\, dx = x \ln x - x + C$.

(c)
$$\int_2^5 \ln x\, dx = (x \ln x - x) \,\Big|_2^5$$

$$= (5 \ln 5 - 5) - (2 \ln 2 - 2)$$

$$= 5 \ln 5 - 2 \ln 2 - 3$$

14. By a law of exponents, $2^{2y} = (2^2)^y = 4^y$. Thus,

$$\int_0^1 2^{2y}\, dy = \int_0^1 4^y\, dy = \frac{4^y}{\ln 4}\,\Big|_0^1 = \frac{1}{\ln 4}(4 - 1)$$

$$= \frac{3}{2 \ln 2}$$

Solved Exercises for Section 8-2

1. We substitute $u = x^2 + 1$. Then

$$\int \frac{2x\, dx}{\sqrt{x^2 + 1}} = \int \frac{1}{\sqrt{u}} \frac{du}{dx}\, dx = \int \frac{1}{\sqrt{u}}\, du$$

$$= 2u^{1/2} + C = 2\sqrt{x^2 + 1} + C$$

Checking, we have

$$\frac{d}{dx}(2\sqrt{x^2 + 1} + C) = 2 \cdot \frac{1}{2}(x^2 + 1)^{-1/2} \cdot 2x$$

$$= \frac{2x}{\sqrt{x^2 + 1}}$$

2. If there were a 2 in the integral, we could put $u = 2x$. We do so anyway, by writing $1 = \frac{1}{2}(2)$:

$$\int \sin 2x \, dx = \int \frac{1}{2}(\sin 2x)2dx$$

$$= \int \frac{1}{2}\sin u \frac{du}{dx}dx \quad = \int \frac{1}{2}\sin u \, du$$

$$= \frac{1}{2}\int \sin u \, du = -\frac{1}{2}\cos u + C$$

Thus

$$\int \sin 2x \, dx = -\frac{1}{2}\cos 2x + C$$

Check this answer by differentiating.

3. Put $u = x^3 + 5$; $du/dx = 3x^2$. Then

$$\int \frac{x^2}{x^3 + 5}dx = \int \frac{1}{3(x^3+5)}3x^2 \, dx$$

$$= \frac{1}{3}\int \frac{1}{u}\frac{du}{dx}dx$$

$$= \frac{1}{3}\int \frac{du}{u} = \frac{1}{3}\ln|u| + C$$

$$= \frac{1}{3}\ln|x^3 + 5| + C$$

4. Completing the square, we find

$$t^2 - 6t + 10 = (t^2 - 6t + 9) - 9 + 10$$
$$= (t-3)^2 + 1 \quad \text{(see Section R-1)}$$

We put $u = t - 3$; $du/dt = 1$. Then

$$\int \frac{dt}{t^2 - 6t + 10} = \int \frac{dt}{1 + (t-3)^2}$$

$$= \int \frac{1}{1+u^2}\frac{du}{dt}dt$$

$$= \int \frac{1}{1+u^2}du = \tan^{-1}u + C$$

so

$$\int \frac{dt}{t^2 - 6t + 10} = \tan^{-1}(t-3) + C$$

5. We first substitute $u = 2x$, as in Solved Exercise 2. Since $du/dx = 2$,

$$\int \sin^2 2x \cos 2x \, dx = \int \sin^2 u \cos u \frac{1}{2}\frac{du}{dx}dx$$

$$= \frac{1}{2}\int \sin^2 u \cos u \, du$$

5. *(cont'd)*

At this point, we notice that another substitution is appropriate: we put $s = \sin u$ and $ds/du = \cos u$. Then

$$\frac{1}{2}\int \sin^2 u \cos u \, du = \frac{1}{2}\int s^2 \frac{ds}{du}du$$

$$= \frac{1}{2}\int s^2 \, ds = \frac{1}{2}\frac{1}{3}s^3 + C$$

$$= \frac{s^3}{6} + C$$

Now we must put our answer in terms of x. Since $s = \sin u$ and $u = 2x$, we have

$$\int \sin^2 2x \cos 2x \, dx = \frac{s^3}{6} + C = \frac{\sin^3 u}{6} + C$$

$$= \frac{\sin^3 2x}{6} + C$$

You should check this formula by differentiating.

You may have noticed that we could have done this problem in one step by substituting $u = \sin 2x$ in the beginning. We did the problem the long way to show that you can solve an integration problem without seeing everything at once.

6. Let $u = x^3 + 3x^2 + 1$; $du/dx = 3x^2 + 6x$, so $dx = du/(3x^2 + 6x)$ and

$$\int \frac{x^2 + 2x}{\sqrt[3]{x^3 + 3x^2 + 1}}dx = \int \frac{1}{\sqrt[3]{u}}\frac{x^2+2x}{3x^2+6x}du$$

$$= \frac{1}{3}\int \frac{1}{\sqrt[3]{u}}du$$

$$= \frac{1}{3}\frac{3}{2}u^{2/3} + C$$

Thus

$$\int \frac{x^2 + 2x}{\sqrt[3]{x^3 + 3x^2 + 1}}dx = \frac{1}{2}(x^3 + 3x^2 + 1)^{2/3} + C$$

7. With the same substitution as in Solved Exercise 6, we get

$$\int \frac{x^2 + 3x}{\sqrt[3]{x^3 + 3x^2 + 1}}dx = \int \frac{1}{\sqrt[3]{u}}\frac{x^2+3x}{3x^2+6x}du$$

$$= \int \frac{1}{\sqrt[3]{u}}\frac{x+3}{3x+6}du$$

There is no simple way to express the quantity $(x+3)/(3x+6)$ in terms of u. (We would have to solve the equation $u = x^3 + 3x^2 + 1$ for x in terms of u.) We conclude that the substitution was not effective in this case.

8. Let $u = \sin x$; $du/dx = \cos x$, $dx = du/\cos x$, so

$$\int \cos x \, [\cos(\sin x)] \, dx$$

$$= \int \cos x \, [\cos(\sin x)] \, \frac{du}{\cos x}$$

$$= \int \cos u \, du = \sin u + C$$

and therefore

$$\int \cos x \, [\cos(\sin x)] \, dx = \sin(\sin x) + C$$

9. If $u = x^2$, $du/dx = 2x$ and $dx = du/2x$, so

$$\int \frac{dx}{1 + x^4} = \int \frac{1}{1 + u^2} \frac{du}{2x}$$

We must solve $u = x^2$ for x, obtaining $x = \sqrt{u}$, so

$$\int \frac{dx}{1 + x^4} = \int \frac{du}{2\sqrt{u} \, (1 + u^2)}$$

Unfortunately, we do not know how to evaluate the integral in u, so we have succeeded only in equating two unknown quantities.

10. Let $u = 1 + \ln x$; $du/dx = 1/x$, $dx = x \, du$, so

$$\int \frac{\sqrt{1 + \ln x}}{x} \, dx = \int \frac{\sqrt{1 + \ln x}}{x} \, (x \, du)$$

$$= \int u^{1/2} \, du = \frac{2}{3} u^{3/2} + C$$

and therefore

$$\int \frac{\sqrt{1 + \ln x}}{x} \, dx = \frac{2}{3} (1 + \ln x)^{3/2} + C$$

11. Seeing the denominator in terms of x^2, we try $u = x^2$, $dx = du/(2x)$; $u = 1$ when $x = 1$ and $u = 25$ when $x = 5$. Thus

$$\int_1^5 \frac{x}{x^4 + 10x^2 + 25} \, dx = \frac{1}{2} \int_1^{25} \frac{du}{u^2 + 10u + 25}$$

Now we notice that the denominator is $(u + 5)^2$, so we put $v = u + 5$, $du = dv$; $v = 6$ when $u = 1$, $v = 30$ when $u = 25$. Thus

11. *(cont'd)*

$$\frac{1}{2} \int_1^{25} \frac{du}{u^2 + 10u + 25} = \frac{1}{2} \int_6^{30} \frac{dv}{v^2}$$

$$= \frac{1}{2} \left(-\frac{1}{v} \right) \Big|_6^{30}$$

$$= -\frac{1}{60} + \frac{1}{12} = \frac{1}{15}$$

If you see the substitution $v = x^2 + 5$ right away, you can do the problem in one step instead of two.

12. It is not obvious what substitution is appropriate here, so a little trial and error is called for. If we remember from our trigonometric identities that $\cos 2\theta = \cos^2 \theta - \sin^2 \theta$, we can proceed easily:

$$\int_0^{\pi/4} (\cos^2 \theta - \sin^2 \theta) \, d\theta = \int_0^{\pi/4} \cos 2\theta \, d\theta$$

$$= \int_0^{\pi/2} \cos u \, \frac{du}{2}$$

$$(u = 2\theta)$$

$$= \frac{\sin u}{2} \Big|_0^{\pi/2}$$

$$= \frac{1 - 0}{2} = \frac{1}{2}$$

(See Problem 11 for another method.)

13. Let $u = 1 + e^x$; $du = e^x \, dx$, $dx = du/e^x$; $u = 1 + e^0 = 2$ when $x = 0$ and $u = 1 + e$ when $x = 1$. Thus

$$\int_0^1 \frac{e^x}{1 + e^x} \, dx = \int_2^{1+e} \frac{1}{u} \, du = \ln u \Big|_2^{1+e}$$

$$= \ln(1 + e) - \ln 2 = \ln \left(\frac{1 + e}{2} \right)$$

Solved Exercises for Section 8-3

1. $F(x) = x^2/2$ and $g(x) = \cos x$. We obtain

$$\int x \sin x \, dx = \frac{x^2}{2} \sin x - \frac{1}{2} \int x^2 \cos x \, dx$$

The new integral on the right is more complicated than the one we started with, so this choice of f and G is not suitable.

2. Put $u = \sin^{-1} x$, $dv = 1 \, dx = dx$. Then $v = x$, $du = (1/\sqrt{1 - x^2}) \, dx$, and

$$\int \sin^{-1} x \, dx = uv - \int v \, du$$

$$= x \sin^{-1} x - \int \frac{x \, dx}{\sqrt{1 - x^2}}$$

2. (*cont'd*)

The integral on the right is perfectly set up for substitution. We put $t = 1 - x^2$, $dx = dt/-2x$, and

$$\int \frac{x\,dx}{\sqrt{1-x^2}} = \int \frac{x}{\sqrt{t}}\,\frac{dt}{-2x} = -\frac{1}{2}\int \frac{dt}{\sqrt{t}}$$
$$= -\sqrt{t} + C = -\sqrt{1-x^2} + C$$

This gives us

$$\int \sin^{-1}x\,dx = x\sin^{-1}x + \sqrt{1-x^2} + C$$

which you should check by differentiation.

3. Put $f(x) = \sin x$, $G(x) = \cos x$. Then $F(x) = -\cos x$, $g(x) = -\sin x$, and

$$\int \sin x \cos x\,dx = -\cos^2 x - \int \sin x \cos x\,dx$$

It looks at first that we have simply transformed the integral into itself, accomplishing nothing, but if we look carefully we see that we can now solve for the integral as an unknown. Writing I for $\int \sin x \cos x\,dx$, we have

$$I = -\cos^2 x - I$$

so $2I = -\cos^2 x$, and $I = -\frac{1}{2}\cos^2 x$; thus $\int \sin x \cos x\,dx = -\frac{1}{2}\cos^2 x + C$. This method is also useful in connection with the exponential and logarithmic functions. See Solved Exercise 5.

4. Let $u = x$ and $v = e^x$, so $dv = e^x\,dx$. Thus, using integration by parts,

$$\int xe^x\,dx = \int u\,dv = uv - \int v\,du$$
$$= xe^x - \int e^x\,dx = xe^x - e^x + C$$

5. Let $u = \sin x$ and $v = e^x$, so $dv = e^x\,dx$ and

$$\int e^x \sin x\,dx = e^x \sin x - \int e^x \cos x\,dx \qquad \textbf{(1)}$$

Repeating:

$$\int e^x \cos x\,dx = e^x \cos x + \int e^x \sin x\,dx \qquad \textbf{(2)}$$

where this time $u = \cos x$ and $v = e^x$. Substituting (2) in (1), we get

$$\int e^x \sin x\,dx = e^x \sin x - e^x \cos x - \int e^x \sin x\,dx$$

So, after transposing and dividing by 2, we have

$$\int e^x \sin x\,dx = \frac{1}{2}e^x(\sin x - \cos x) + C$$

6. Let $u = \sin 2x$ and $dv = \cos x\,dx$. Then $du = 2\cos 2x\,dx$, $v = \sin x$, and so

$$\int_0^{\pi/2} \sin 2x \cos x\,dx$$
$$= uv\Big|_0^{\pi/2} - \int_0^{\pi/2} v\,du$$
$$= \sin 2x \sin x\Big|_0^{\pi/2} - 2\int_0^{\pi/2} \sin x \cos 2x\,dx$$
$$= 0 - 2\int_0^{\pi/2} \sin x \cos 2x\,dx$$

Now integrate by parts again, with $u = \cos 2x$ and $dv = \sin x\,dx$. Then $du = -2\sin 2x\,dx$, $v = -\cos x$, and so

$$\int_0^{\pi/2} \sin x \cos 2x\,dx$$
$$= -\cos x \cos 2x\Big|_0^{\pi/2} - 2\int_0^{\pi/2} \sin 2x \cos x\,dx$$
$$= -(0-1) - 2\int_0^{\pi/2} \sin 2x \cos x\,dx$$

Writing I for $\int_0^{\pi/2} \sin 2x \cos x\,dx$ and combining the results of both integrations, we have

$$I = 0 - 2(1 - 2I) = -2 + 4I$$

So $3I = 2$ and $I = \frac{2}{3}$.

7. Let $u = \sqrt{x}$. Then $du/dx = 1/2\sqrt{x}$, $dx = 2\sqrt{x}\,du$, and

$$\int_0^{\pi^2} \sin \sqrt{x}\,dx = \int_0^{\pi} \sin u \cdot 2\sqrt{x}\,du$$
$$= 2\int_0^{\pi} u \sin u\,du$$

By Worked Example 2, this is

$$2(-u \cos u + \sin u)\Big|_0^{\pi}$$
$$= 2(-\pi \cos \pi + \sin \pi - 0) = 2\pi$$

8. Let $u = \ln x$, so $x = e^u$ and $du = (1/x)\,dx$. Then $\int \sin(\ln x)\,dx = \int (\sin u)e^u\,du$, which was evaluated in Solved Exercise 5. Hence

$$\int_1^e \sin(\ln x)\,dx = \int_0^1 e^u \sin u\,du$$
$$= \frac{1}{2}e^u(\sin u - \cos u)\Big|_0^1$$
$$= [\tfrac{1}{2}e^1(\sin 1 - \cos 1)] - [\tfrac{1}{2}e^0(\sin 0 - \cos 0)]$$
$$= \frac{e}{2}\left(\sin 1 - \cos 1 + \frac{1}{e}\right)$$

8

9. The nth bend occurs between $x = (2n - 2)\pi$ and $(2n - 1)\pi$. The area under this bend can be evaluated using integration by parts (Worked Example 2):

$$\int_{(2n-2)\pi}^{(2n-1)\pi} x \sin x \, dx = -x \cos x + \sin x \Big|_{(2n-2)\pi}^{(2n-1)\pi}$$

$$= -(2n - 1)\pi \cos[(2n - 1)\pi] + \sin[(2n - 1)\pi]$$
$$+ (2n - 2)\pi \cos(2n - 2)\pi - \sin(2n - 2)\pi$$

$$= -(2n - 1)\pi(-1) + 0 + (2n - 2)\pi(1) - 0$$

$$= (2n - 1)\pi + (2n - 2)\pi = (4n - 3)\pi$$

Thus the areas under successive bends are π, 5π, 9π, 13π, and so forth.

10. If $y = \sqrt{\sqrt{x} + 1}$, then $y^2 = \sqrt{x} + 1$, $\sqrt{x} = y^2 - 1$, and $x = (y^2 - 1)^2$. Thus we have

$$\int \sqrt{\sqrt{x} + 1} \, dx$$

$$= xy - \int x \, dy$$

$$= x\sqrt{\sqrt{x} + 1} - \int (y^4 - 2y^2 + 1) \, dy$$

$$= x\sqrt{\sqrt{x} + 1} - \tfrac{1}{5}y^5 + \tfrac{2}{3}y^3 - y + C$$

$$= x\sqrt{\sqrt{x} + 1} - \tfrac{1}{5}(\sqrt{x} + 1)^{5/2}$$
$$+ \tfrac{2}{3}(\sqrt{x} + 1)^{3/2} - (\sqrt{x} + 1)^{1/2} + C$$

11. Note that $g(1) = 2$ and $g(2) = 34$, so $f(2) = 1$ and $f(34) = 2$. Then

$$\int_2^{34} f(x) \, dx = 2 \cdot 34 - 1 \cdot 2 - \int_1^2 (y^5 + y) \, dy$$

$$= 66 - (\tfrac{1}{6}y^6 + \tfrac{1}{2}y^2) \Big|_1^2$$

$$= 66 - \left(\frac{64}{6} + \frac{4}{2}\right) + \left(\frac{1}{6} + \frac{1}{2}\right)$$

$$= 54$$

12. Refer to Fig. 8-3-4. Here R_1 is the region under the graph $y = f(x)$ on $[a, b]$, so its area is $\int_a^b f(x) \, dx$. Further, R_2 (if we flip the figure) is the region under the graph $x = g(y)$ on $[f(a), f(b)]$, so its area is $\int_{f(a)}^{f(b)} g(y) \, dy$. Finally, R_3 is a rectangle with area $af(a)$. Together R_1, R_2, and R_3 form a rectangle with area $bf(b)$, so we have

12. (*cont'd*)

$$\int_a^b f(x) \, dx + \int_{f(a)}^{f(b)} g(y) \, dy + af(a) = bf(b)$$

or

$$\int_a^b f(x) \, dx = bf(b) - af(a) - \int_{f(a)}^{f(b)} g(y) \, dy$$

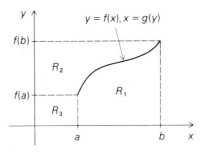

Fig. 8-3-4 The relation between the integrals of a function and its inverse.

CHAPTER 9

Solved Exercises for Section 9-1

1. Here $x_0 = 0$, $v_0 = 1$, and $\omega = 1$, so $x = x_0 \cos \omega t + (v_0/\omega) \sin \omega t = \sin t$.

2. (a) Here $y_0 = 1$, $v_0 = -1$, and $\omega = 3 = \sqrt{9}$ (using x in place of t and y in place of x), so

$$y = y_0 \cos \omega x + \frac{v_0}{\omega} \sin \omega x$$

$$= \cos 3x - \tfrac{1}{3} \sin 3x$$

   (b) The polar coordinates of $(1, -\tfrac{1}{3})$ are given by $\alpha = \sqrt{1 + \tfrac{1}{9}} = \sqrt{10}/3 \approx 1.05$ and $\theta = \tan^{-1}(-\tfrac{1}{3}) \approx -0.32$ (or $-18°$). Hence $y = \alpha \cos(\omega x - \theta)$ becomes $y = 1.05 \cos(3x + 0.32)$, which is sketched in Fig. 9-1-10. Here $\theta/\omega = -0.1$ and $2\pi/\omega = 2.1$.

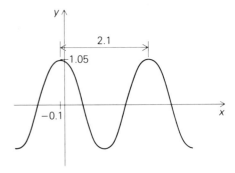

Fig. 9-1-10 The graph of $y = (1.05)\cos(3x + 0.32)$.

3. Let $P(t)$ denote the population. Since the rate of increase is 5%, $P'(t) = 0.05\,P(t)$, so $P(t) = P(0)e^{0.05t}$. In order for $P(t)$ to be $2P(0)$, we should have $2 = e^{0.05t}$; that is, $0.05t = \ln 2$ or $t = 20 \ln 2$. Using $\ln 2 = 0.6931$, $t \approx 13.862$ years.

4. (a) Let $f(t)$ denote the amount of radium at time t. We have $f'(t) = -0.000428 f(t)$ (minus since it is decay), so $f(t) = f(0)e^{-0.000428t}$. If $f(t) = \frac{1}{2} f(0)$, then $\frac{1}{2} = e^{-0.000428t}$; that is, $e^{0.000428t} = 2$; or $0.000428t = \ln 2$. Hence, $t = (\ln 2)/0.000428 \approx 1620$ years.

   (b) If $f'(t) = -\gamma f(t)$, then $f(t) = f(0)e^{-\gamma t}$. If we are to have $f(t) = \frac{1}{2}f(0)$, this means $\frac{1}{2} = e^{-\gamma t}$; that is, $e^{\gamma t} = 2$; that is, $t = (1/\gamma)\ln 2$.

5. 2^{29} cents, or \$5,368,709.12.

6. By definition,
$$\sinh(x+y) = \frac{e^{x+y} - e^{-x-y}}{2}$$
$$= \frac{e^x e^y - e^{-x}e^{-y}}{2}$$

Using the hint, we plug in $e^x = \cosh x + \sinh x$ and $e^{-x} = \cosh x - \sinh x$ to get
$$\sinh(x+y) = \tfrac{1}{2}[(\cosh x + \sinh x)(\cosh y + \sinh y)$$
$$- (\cosh x - \sinh x)(\cosh y - \sinh y)]$$

Expanding,
$$\sinh(x+y) = \tfrac{1}{2}(\cosh x \cosh y + \cosh x \sinh y$$
$$+ \sinh x \cosh y + \sinh x \sinh y$$
$$- \cosh x \cosh y + \cosh x \sinh y$$
$$+ \sinh x \cosh y - \sinh x \sinh y)$$
$$= \cosh x \sinh y + \sinh x \cosh y$$

7. $\tanh x = \sinh x / \cosh x$, so $(d/dx)\tanh x = (\cosh x \cdot \cosh x - \sinh x \cdot \sinh x)/\cosh^2 x = 1 - \tanh^2 x$. From $\cosh^2 x - \sinh^2 x = 1$ we get $1 - \sinh^2 x / \cosh^2 x = 1/\cosh^2 x$; that is, $1 - \tanh^2 x = \operatorname{sech}^2 x$.

8. By formula (20), with $\omega = 3$, $x_0 = 1$, and $v_0 = 1$, we have $x(t) = \cosh 3t + \frac{1}{3}\sinh 3t$ as our solution.

9. $(d/dx)\cosh x = \sinh x$ vanishes only if $x = 0$ (since $e^x = e^{-x}$ exactly when $e^{2x} = 1$; that is, $x = 0$) and $(d^2/dx^2)\cosh x = \cosh x$, which is 1 at $x = 0$, so $\cosh x$ has a minimum at $x = 0$.

10. By formula (21), $(d/dx)\sinh^{-1}x = 1/\sqrt{1+x^2}$, so by the chain rule,
$$\frac{d}{dx}\sinh^{-1}(3\tanh 3x)$$
$$= \frac{1}{\sqrt{1 + 9\tanh^2 3x}} \cdot 3\frac{d}{dx}\tanh 3x$$
$$= \frac{1}{\sqrt{1 + 9\tanh^2 3x}} \cdot 3 \cdot 3 \cdot \operatorname{sech}^2 3x$$
$$= \frac{9\operatorname{sech}^2 3x}{\sqrt{1 + 9\tanh^2 3x}}$$

11. Since
$$\frac{1}{1-x^2} = \frac{1}{2}\left(\frac{1}{1-x} + \frac{1}{1+x}\right)$$
an antiderivative is
$$\frac{1}{2}\ln|1-x| + \frac{1}{2}\ln|1+x|$$
$$= \frac{1}{2}\ln\left|\frac{1+x}{1-x}\right| \qquad |x| \neq 1$$

12. $\sinh^{-1}5 = \ln(5 + \sqrt{5^2 + 1}) = \ln(5 + \sqrt{26}) = \ln(10.100) \approx 2.31$.

13.
$$\int \frac{dx}{\sqrt{x^2+4}} = \frac{1}{2}\int \frac{dx}{[\sqrt{1+(x/2)^2}\,]}$$
$$= \frac{1}{2}\int \frac{2du}{\sqrt{1+u^2}} \quad (u = x/2)$$
$$= \sinh^{-1}u + C$$
$$= \sinh^{-1}(x/2) + C$$

Solved Exercises for Section 9-2

1. $y\,dy = \cos 2x\,dx$, so $y^2/2 = \frac{1}{2}\sin 2x + C$. Since $y = 1$ when $x = 0$, $C = 1$. Thus $y^2 = \sin 2x + 1$ or $y = \sqrt{\sin 2x + 1}$. (We take the $+$ square root since $y = +1$ when $x = 0$.)

2. $y\,dy = x\,dx/(1+x^2)$ so $y^2/2 = \frac{1}{2}\ln(1+x^2) + C$ where the integration was done by substitution. Thus $y^2 = \ln(1+x^2) + 2C$. Since $y = -1$ when $x = 0$, $C = \frac{1}{2}$. Since y is negative near $x = 0$, we choose the negative root: $y = -\sqrt{1 + \ln(1+x^2)}$.

3. Here the trick is to notice that the right-hand side factors: $y' = (x^2 - 1)(y^2 + 1)$. Thus $dy/(1+y^2) = (x^2 - 1)dx$ and integrating, $\tan^{-1}y = (x^3/3) - x + C$. Since $y(0) = 0$, $C = 0$. Hence $y = \tan[(x^3/3) - x]$.

4. (a) If we differentiate we have

$$y = kx^2 \quad \text{so} \quad k = y/x^2$$

$$y' = 2kx = 2\left(\frac{y}{x^2}\right)x = 2y/x$$

Thus any parabola $y = kx^2$ satisfies the equation $y' = 2y/x$.

(b) The slope of a line orthogonal to a line of slope m is $-1/m$, so the equation satisfied by the orthogonal trajectories is $y' = -x/2y$. This equation is separable:

$$2y\,dy = -x\,dx$$

$$y^2 = -\frac{x^2}{2} + C$$

or

$$y^2 + \frac{x^2}{2} = C$$

This is a family of ellipses. (See Fig. 9-2-12 and Section R-4.)

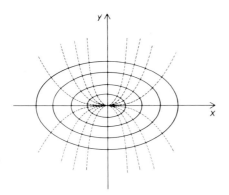

Fig. 9-2-12 The orthogonal trajectories of the family of parabolas is the family of ellipses.

5. Since mass times acceleration is force and since acceleration is the time derivative of velocity, we have the equation $m\,(dv/dt) = mg - \gamma v$; that is, $[m/(mg - \gamma v)]\,dv = dt$. Integrating, and assuming $mg - \gamma v > 0$, we obtain $(-m/\gamma) \ln(mg - \gamma v) = t + C$. Since we can measure the moment of release as $t = 0$ at which time $v = 0$, $C = (-m/\gamma) \ln(mg)$. Thus $\ln(mg - \gamma v) = -(\gamma/m)t + \ln(mg)$ and exponentiating,

$$mg - \gamma v = mg\,e^{-\gamma t/m}$$

$$v = \frac{mg}{\gamma}(1 - e^{-\gamma t/m})$$

5. (cont'd)

Note that as $t \to \infty$, $e^{-\gamma t/m} \to 0$ and so $v \to mg/\gamma$, the *terminal velocity*. See Fig. 9-2-13. For small t the velocity is approximately gt, what it would be if there were no air resistance. As t increases, the air resistance slows the velocity and a terminal velocity is approached.

Fig. 9-2-13 The velocity of an object moving in a resisting medium.

6. Here the slope at (x, y) is $-x/y$. We draw small line segments with these slopes at a number of selected locations to produce Fig. 9-2-14.

The equation is separable:

$$y\,dy = -x\,dx$$

$$\frac{y^2}{2} = -\frac{x^2}{2} + C$$

$$y^2 + x^2 = 2C$$

Thus any solution is a circle and the solutions taken together form a family of circles. This is consistent with the direction field.

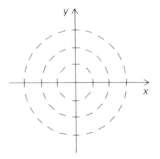

Fig. 9-2-14 The direction field for $y' = -x/y$.

7. The recursive procedure is summarized to the right. It is helpful to carefully record the data in a table as you proceed, as in the following. Here $x_0 = 0$, $y_0 = 0$, $a = \pi/4$, $n = 10$; thus $h = a/n = \pi/40 = 0.0785398$.

Thus $y(\pi/4) \approx .970263$.

	x_i	y_i	$x_{i+1} = x_i + h$	$y_{i+1} = y_i +$ $[x_i + (\cos y_i)]h$
$i = 0$	0 START 0		.0785398	.0785398
$i = 1$	.0785398	.0785398	.157080	.163006
$i = 2$	.157080	.163006	.235619	.252841
$i = 3$	.235619	.252841	.314159	.347390
$i = 4$	.314159	.347390	.392699	.445912
$i = 5$	.392699	.445912	.471239	.547614
$i = 6$	.471239	.547614	.549779	.651680
$i = 7$	.549779	.651680	.628318	.757304
$i = 8$	.628319	.757304	.706858	.863726
$i = 9$	.706858	.863726	.785398	.970263
$i = 10$	.785398	.970263	STOP	

8. The integrating factor is $\exp(-\int P(x)\,dx) = \exp(-\int -dx) = \exp(x)$. Thus

$$e^x(y' + y) = 1$$

$$\frac{d}{dx}(e^x y) = 1$$

$$e^x y = x + C$$

Since $y(0) = 1$, $C = 1$, so $y = (x + 1)e^{-x}$.

9. The integrating factor is $\exp(\int \tan x\, dx) = \exp(-\ln \cos x) = 1/\cos x$. (This is valid only if $\cos x > 0$, but since our initial condition is $x = 0$ where $\cos x = 1$, this is justified.) Thus

$$\frac{1}{\cos x}[y' + (\tan x)y] = \cos x$$

$$\frac{d}{dx}\left[\frac{y}{\cos x}\right] = \cos x$$

$$\frac{y}{\cos x} = \sin x + C$$

Since $y = 0$ when $x = 0$, $C = 0$. Thus $y = \cos x \sin x + 1$. It may be verified that this solution is valid for all x.

10. Let $y(t)$ denote the amount of pollutant in liters in the lake at time t. The amount of pollutant in one liter of lake water is thus

$$\frac{y(t)}{4 \times 10^7} = 10^{-7} \times y(t)/4$$

10. (cont'd)

The rate of change of $y(t)$ is the rate at which pollutant flows out; that is, $-10.67 \times 10^{-7} \times y(t)/4$ liters per second plus the rate at which it flows in; that is, 0.67 liter per second. Thus

$$y' = \frac{-10.67 \times 10^{-7}}{4} y + 0.67$$

The solution of $dy/dt = ay + b$, $y(0) = 0$ is found using the integrating factor e^{-at}:

$$e^{-at}(y' - ay) = be^{-at}$$

$$\frac{d}{dt}(e^{-at}y) = be^{-at}$$

$$e^{-at}y = -\frac{b}{a}e^{-at} + C$$

Since $y = 0$ at $t = 0$, $C = b/a$. Thus

$$e^{-at}y = \frac{b}{a}(1 - e^{-at})$$

$$y = \frac{b}{a}(e^{at} - 1)$$

In our case $a = -\dfrac{10.67 \times 10^{-7}}{4}$, $b = 0.67$, and so $y = (1 - e^{-2.67 \times 10^{-7}t})(2.51 \times 10^6)$ liters. For t small, y is relatively small; but for larger t, y approaches the (steady state) catastrophic value of 2.51×10^6 liters, that is, the lake is well over half pollutant. (See Exercise 14 to find out how long this takes.)

9

CHAPTER 10

Solved Exercises for Section 10-1

1. We choose the family P_x of planes such that P_0 contains the base of the cone and P_x is at distance x above P_0. Then the cone lies between P_0 and P_h, and the plane section by P_x is a circular region with radius $[(h-x)/h]r$ and area $\pi[(h-x)/h]^2 r^2$. (See Fig. 10-1-19.) The volume is

$$\int_0^h A(x)\,dx = \int_0^h \pi \frac{(h-x)^2}{h^2} r^2\,dx$$

$$= \frac{\pi r^2}{h^2}\int_0^h (h^2 - 2xh + x^2)\,dx$$

$$= \frac{\pi r^2}{h^2}\left(h^2 x - hx^2 + \frac{x^3}{3}\right)\Bigg|_0^h$$

$$= \frac{\pi r^2}{h^2}\left(h^3 - h^3 + \frac{h^3}{3}\right) = \frac{1}{3}\pi r^2 h$$

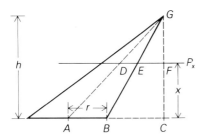

Fig. 10-1-19 $|DE| : |AB| = |GE| : |GB| = |GF| : |GC|$ by similar triangles. But $|AB| = r$, $|GC| = h$, and $|GF| = h - x$, so $|DE| = [(h-x)/h]r$.

2. The middle piece lies between the planes $P_{-r/3}$ and $P_{r/3}$ of Worked Example 2, and the area function is $A(x) = \pi(r^2 - x^2)$ as before. So the volume of the middle piece is

$$\int_{-r/3}^{r/3} \pi(r^2 - x^2)\,dx = \pi\left(r^2 x - \frac{x^3}{3}\right)\Bigg|_{-r/3}^{r/3}$$

$$= \pi\left(\frac{r^3}{3} - \frac{r^3}{81} + \frac{r^3}{3} - \frac{r^3}{81}\right)$$

$$= \frac{52}{81}\pi r^3$$

This leaves a volume of $\left(\frac{4}{3} - \frac{52}{81}\right)\pi r^3 = \frac{56}{81}\pi r^3$ to be divided between the two outside pieces. Since they are congruent, each of them has volume $\frac{28}{81}\pi r^3$. (You check this by computing $\int_{r/3}^r \pi(r^2 - x^2)\,dx$.)

3. The region is sketched in Fig. 10-1-20. Its volume is

$$\pi \int_0^1 (\sqrt{x})^2\,dx = \pi \int_0^1 x\,dx = \frac{\pi x^2}{2}\Bigg|_0^1 = \frac{\pi}{2}$$

Note that if this figure is rotated 90° in the xy plane, it fits right into the hollow of the bowl in Worked Example 5. Thus the volumes of the two solids (which happen, by chance, to be equal) should add up to the volume of a right circular cylinder with base radius 1 and height 1. This checks, since $\pi/2 + \pi/2 = \pi$.

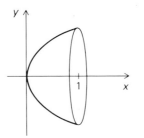

Fig. 10-1-20 The solid of revolution obtained by revolving the graph of $y = \sqrt{x}$ about the x axis.

4. The solid is sketched in Fig. 10-1-21. It has the form of a hollowed-out cone. The volume is that of the cone minus that of the hole. The cone is obtained by revolving the region under the graph of x on $[0,1]$ about the x axis, so its volume is

$$\pi \int_0^{\pi/2} x^2\,dx = \frac{\pi^4}{24} \quad \text{(compare Solved Exercise 1)}$$

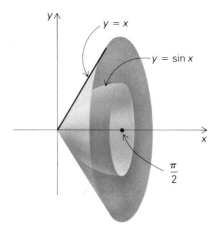

Fig. 10-1-21 The region between the graphs of $\sin x$ and x is revolved about the x axis.

4. (*cont'd*)

The hole is obtained by revolving the region under the graph of $\sin x$ on $[0, \pi/2]$ about the x axis, so its volume is

$$\pi \int_0^{\pi/2} \sin^2 x \, dx = \pi \int_0^{\pi/2} \frac{1 - \cos 2x}{2} \, dx$$

$$\text{(since } \cos 2x = 1 - 2 \sin^2 x \text{)}$$

$$= \pi \left(\frac{x}{2} - \frac{1}{4} \sin 2x \right) \Big|_0^{\pi/2}$$

$$= \pi \left(\frac{\pi}{4} - 0 - 0 + 0 \right) = \frac{\pi^2}{4}$$

Thus the volume of our solid is $\pi^4/24 - \pi^2/4 \approx 1.59$.

5. The solid is sketched in Fig. 10-1-22. Its volume is

$$2\pi \int_0^{\pi} x \sin x \, dx = 2\pi(-x \cos x + \sin x) \Big|_0^{\pi}$$

$$= 2\pi(\pi) = 2\pi^2$$

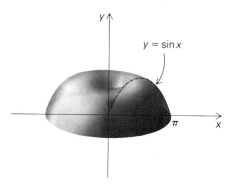

Fig. 10-1-22 The region under the graph of $\sin x$ is revolved about the y axis.

6. The doughnut-shaped solid is shown in Fig. 10-1-23. We observe that if the solid is sliced in half by a plane through the origin perpendicular to the y axis, the top half is the solid obtained by revolving about the y axis the region under the semicircle $y = \sqrt{1 - (x - 4)^2}$ on the interval $[3, 5]$.

The volume of that solid is

$$2\pi \int_3^5 x\sqrt{1 - (x - 4)^2} \, dx$$

$$= 2\pi \int_{-1}^1 (u + 4) \sqrt{1 - u^2} \, du \quad (u = x - 4)$$

$$= 2\pi \int_{-1}^1 \sqrt{1 - u^2} \, u \, du + 8\pi \int_{-1}^1 \sqrt{1 - u^2} \, du$$

6. (*cont'd*)

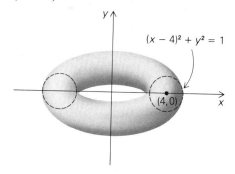

Fig. 10-1-23 The disk $(x - 4)^2 + y^2 \leq 1$ is revolved about the y axis.

Now $\int_{-1}^1 \sqrt{1 - u^2} \, u \, du = 0$ because the function $f(u) = \sqrt{1 - u^2} \, u$ satisfies $f(-u) = -f(u)$, so that $\int_{-1}^0 f(u) \, du$ is exactly the negative of $\int_0^1 f(u) \, du$. To see this analytically, let $v = -u$. Then

$$\int_{-1}^0 f(u) \, du = \int_1^0 f(-v)(-dv) = \int_0^1 f(-v) \, dv$$

$$= \int_0^1 -f(v) \, dv = -\int_0^1 f(v) \, dv$$

On the other hand, $\int_{-1}^1 \sqrt{1 - u^2} \, du$ is just the area of a semicircular region of radius 1—that is, $\pi/2$—so the volume of the upper half of the doughnut is $8\pi \cdot (\pi/2) = 4\pi^2$, and the volume of the entire doughnut is twice that, or $8\pi^2$. Notice that this is equal to the area π of the rotated disk times the circumference 8π of the circle traced out by its center $(4, 0)$.

Solved Exercises for Section 10-2

1. By the formula for average values, $\overline{\sqrt{1 - x^2}}_{[-1, 1]}$ $= (\int_{-1}^1 \sqrt{1 - x^2} \, dx)/2$. But $\int_{-1}^1 \sqrt{1 - x^2} \, dx$ is the area of the upper semicircle of $x^2 + y^2 = 1$, which is $\frac{1}{2}\pi$, so $\overline{\sqrt{1 - x^2}}_{[-1, 1]} = \pi/4 \approx 0.785$.

2. $\overline{x^2 \sin(x^3)}_{[0, \pi]}$

$$= \frac{1}{\pi} \int_0^{\pi} x^2 \sin x^3 \, dx$$

$$= \frac{1}{\pi} \int_0^{\pi^3} \sin u \, \frac{du}{3} \quad \text{(substituting } u = x^3\text{)}$$

$$= \frac{1}{3\pi} \int_0^{\pi^3} \sin u \, du = \frac{1}{3\pi}(-\cos u) \Big|_0^{\pi^3}$$

$$= \frac{1}{3\pi}(1 - \cos \pi^3) \approx 0.0088$$

10

3. Let f be continuous on $[a, b]$, and define $F(t) = \int_a^t f(t)\,dt$. By the fundamental theorem of calculus (alternative version), $F'(t) = f(t)$ for t in (a, b). One can easily show that $F(t)$ is continuous at a and b. By the mean value theorem for derivatives, there is some t_0 in (a, b) such that

$$F'(t_0) = \frac{F(b) - F(a)}{b - a}$$

Substituting for F and F' in terms of f, we have

$$f(t_0) = \frac{\displaystyle\int_a^b f(t)\,dt - \int_a^a f(t)\,dt}{b - a} = \frac{\displaystyle\int_a^b f(t)\,dt}{b - a}$$

$$= \overline{f(t)}_{[a,b]}$$

which establishes the mean value theorem for integrals.

4. Let m be the mass at each point. By formula (5), the center of mass is located at

$$\frac{m(1 \cdot 1 + 1 \cdot 2 + \cdots 1 \cdot n)}{m(1 + \cdots + 1)} = \frac{\displaystyle\sum_{i=1}^n i}{n}$$

$$= \frac{1}{n} \cdot \frac{n(n+1)}{2} \qquad \text{(see p. 176)}$$

$$= \frac{n+1}{2}$$

Notice that the center of mass is located halfway between the end masses and is independent of the value of m.

5. We take the vertices of the square to be $(0, 0)$, $(1, 0)$, $(1, 1)$, and $(0, 1)$. (See Fig. 10-2-13.) The center is at $(\frac{1}{2}, \frac{1}{2})$ and the center of mass is located by formula (6):

$$\bar{x} = \frac{1 \cdot 0 + 2 \cdot 1 + 3 \cdot 1 + 4 \cdot 0}{1 + 2 + 3 + 4} = \frac{1}{2}$$

$$\bar{y} = \frac{1 \cdot 0 + 2 \cdot 0 + 3 \cdot 1 + 4 \cdot 1}{1 + 2 + 3 + 4} = \frac{7}{10}$$

It is located $\frac{2}{10}$ unit above the center of the square.

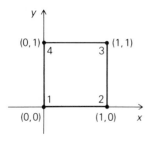

Fig. 10-2-13 The center of mass of these four weighted points is located at $(\frac{1}{2}, \frac{7}{10})$.

6. We take the region under the graph of $\sqrt{1 - x^2}$ on $[-1, 1]$. Since the y axis is an axis of symmetry, the center of mass must lie on this axis; that is, $\bar{x} = 0$. (You can also calculate $\int_{-1}^1 x\sqrt{1 - x^2}\,dx$ and find it to be zero.) Also, by equation (7),

$$\bar{y} = \frac{\dfrac{1}{2}\displaystyle\int_{-1}^1 (1 - x^2)\,dx}{\displaystyle\int_{-1}^1 \sqrt{1 - x^2}\,dx}$$

The denominator is the area $\pi/2$ of the semicircle. The numerator is $\frac{1}{2}\int_{-1}^1 (1 - x^2)\,dx = \frac{1}{2}[x - x^3/3]\,|_{-1}^1 = \frac{2}{3}$, so

$$\bar{y} = \frac{\frac{2}{3}}{\pi/2} = \frac{4}{3\pi} \approx 0.42$$

and the center of mass is located at $(0, 4/3\pi)$ (Fig. 10-2-14).

Fig. 10-2-14 The center of mass of the semicircular region is located at $(0, 4\pi/3)$.

7. The center of mass of the disk is at $(0, 0)$, since the x and y axes are both axes of symmetry. For the region under the graph of $\sin x$ on $[2\pi, 3\pi]$, the line $x = \frac{5}{2}\pi$ is an axis of symmetry. To find the y coordinate of the center of mass, we use formula (5) to obtain

$$\frac{\dfrac{1}{2}\displaystyle\int_{2\pi}^{3\pi} \sin^2 x\,dx}{\displaystyle\int_{2\pi}^{3\pi} \sin x\,dx} = \frac{\frac{1}{2} \cdot \pi/2}{2} = \frac{\pi}{8} \approx 0.393$$

(Notice that this region is more "bottom heavy" than the semicircular region.)

By the consolidation principle, the center of mass of the total figure is the same as one consisting of two points: one at $(0, 0)$ with mass $\rho\pi$ and one at $(\frac{5}{2}\pi, \pi/8)$ with mass $\rho 2$. The center of mass is, therefore, at $(\bar{x}, \bar{y})$, where

$$\bar{x} = \frac{\rho\pi \cdot 0 + \rho 2 \cdot \frac{5}{2}\pi}{\rho\pi + \rho 2} = \frac{5\pi}{2 + \pi}$$

and

$$\bar{y} = \frac{\rho\pi \cdot 0 + \rho 2 \cdot \pi/8}{\rho\pi + \rho 2} = \frac{\pi/4}{2 + \pi}$$

Hence $(\bar{x}, \bar{y})$ is approximately $(3.06, 0.15)$ (see Fig. 10-2-15).

7. (*cont'd*)

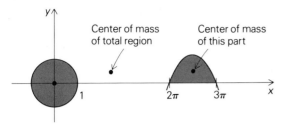

Fig. 10-2-15 The center of mass is found by the consolidation principle.

8. As we did when computing areas, we think of the region under the graph of f on $[a, b]$ as being composed of "infinitely many rectangles of infinitesimal width." The rectangle at x with width dx has area $f(x)\, dx$ and mass $\rho f(x)\, dx$; its center of mass is located at $(x, \frac{1}{2}f(x))$. (See Fig. 10-2-16.)

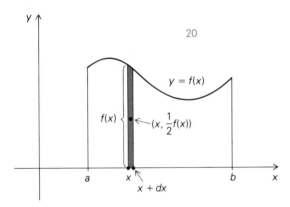

Fig. 10-2-16 The "infinitesimal rectangle" has mass $\rho f(x)\, dx$ and center of mass at $(x,\ f(x)/2)$.

Now we apply the consolidation principle, but instead of summing, as we did on pp. 428 and 429, we integrate:

$$\bar{x} = \frac{\displaystyle\int_a^b x\rho f(x)\, dx}{\displaystyle\int_a^b \rho f(x)\, dx} = \frac{\displaystyle\int_a^b x f(x)\, dx}{\displaystyle\int_a^b f(x)\, dx}$$

$$\bar{y} = \frac{\displaystyle\int_a^b \tfrac{1}{2}f(x)\rho f(x)\, dx}{\displaystyle\int_a^b \rho f(x)\, dx} = \frac{\tfrac{1}{2}\displaystyle\int_a^b [f(x)]^2\, dx}{\displaystyle\int_a^b f(x)\, dx}$$

Solved Exercises for Section 10-3

1. Let E be the energy content. By our formula for work, we have $\Delta E = \int_a^b F\, dx$, so $dE/dx = F$. By the chain rule, $P = dE/dt$ is equal to $dE/dx \cdot dx/dt = F \cdot v$. (In pushing a child on a swing, it is most effective to exert your force at the bottom of the swing, when the velocity is greatest.)

2. The total energy required to pump out the top h meters of water is

$$90g\pi \int_0^h (10 - x)^2 x\, dx$$

$$= 90g\pi \left(100\,\frac{h^2}{2} - 20\,\frac{h^3}{3} + \frac{h^4}{4} \right)$$

$$\approx 2770h^2 \left(50 - \frac{20}{3}h + \frac{h^2}{4} \right)$$

At the end of 6 minutes ($\frac{1}{10}$ hour), the pump has produced 10^4 joules of energy, so the water level is h meters from the top, where h is the solution of

$$2770h^2 \left(50 - \frac{20}{3}h + \frac{h^2}{4} \right) = 10^4$$

or

$$50h^2 - \frac{20}{3}h^3 + \frac{h^4}{4} = 3.61$$

Solving this numerically (see Solved Exercise 8, p. A-6 and Solved Exercise 6(d), p. A-23) gives $h \approx 0.27$ meter.

At the end of t hours, the total energy output is $10^5 t$ joules, so

$$90g\pi \int_0^h (10 - x)^2 x\, dx = 10^5 t$$

where h is the amount pumped out at time t. Differentiating both sides with respect to t gives

$$90g\pi (10 - h)^2 h\,\frac{dh}{dt} = 10^5$$

or

$$\frac{dh}{dt} = \frac{10^4}{9g\pi (10 - h)^2 h}$$

When $h = 0.27$, this is 1.41 meters per hour.

10

3. On June 21, we have $T = 0$, so $\sin D = \sin \alpha$, or $D = \alpha$. Also, we have $\cos l = \sin \alpha$ and $\sin l = \cos \alpha$, $\tan l = \cot \alpha$, so

$$E = \sqrt{\sin^2 \alpha - \sin^2 \alpha}$$
$$\qquad + \cos \alpha \sin \alpha \cos^{-1}(-\tan l \cot l)$$
$$= \cos \alpha \sin \alpha \cos^{-1}(-1)$$
$$= \frac{\pi}{2}\sin(2\alpha)$$

Evaluating at $\alpha = 23.5°$, we find $E \approx 1.15$, which is about 1.25 times the energy received at the equator on June 21. This excess is due, of course, to the long day at the Arctic Circle.

4. (a) We must sum E from $T = 0$ to 364; thus we get, for the annual energy,

$$F = \sum_{T=0}^{364} \{\sqrt{\cos^2 l - \sin^2 D}$$
$$\qquad + \sin l \sin D \cos^{-1}(-\tan l \tan D)\}$$

Expressing $\sin D$ in terms of T, we obtain

$$F = \sum_{T=0}^{364} \left\{ \sqrt{\cos^2 l - \sin^2 \alpha \cos^2\left(\frac{2\pi T}{365}\right)} \right.$$
$$\qquad + \sin l \sin \alpha \cos\left(\frac{2\pi T}{365}\right) \times$$
$$\qquad \left. \cos^{-1}\left[\frac{-\tan l \sin \alpha \cos(2\pi T/365)}{\sqrt{1 - \sin^2 \alpha \cos^2(2\pi T/365)}}\right]\right\}$$

(b) The sum in part (a) may be considered as a Riemann sum (see p. 187) and hence an approximation to the integral

$$\int_0^{365} \left\{ \sqrt{\cos^2 l - \sin^2 \alpha \cos^2\left(\frac{2\pi T}{365}\right)} \right.$$
$$\qquad + \sin l \sin \alpha \cos\left(\frac{2\pi T}{365}\right) \times$$
$$\qquad \left. \cos^{-1}\left[\frac{-\tan l \sin \alpha \cos(2\pi T/365)}{\sqrt{1 - \sin^2 \alpha \cos^2(2\pi T/365)}}\right]\right\} dT$$

No antiderivative for this integrand is evident. We will learn in the next chapter how to simplify this integral somewhat, but the result will be an "elliptic integral" (see Solved Exercise 2 in Section 11-2), which cannot be evaluated in terms of algebraic, trigonometric, or exponential functions. Fortunately, elliptic integrals occur in so many different applications of calculus that they are as well tabulated as the more familiar sines and cosines.

CHAPTER 11

Solved Exercises for Section 11-1

1. $\displaystyle\int (\sin^2 x + \sin^3 x \cos^2 x)\, dx$

$$= \int \sin^2 x\, dx + \int \sin^3 x \cos^2 x\, dx$$
$$= \int \left(\frac{1 - \cos 2x}{2}\right) dx$$
$$\qquad + \int (1 - \cos^2 x)\cos^2 x \sin x\, dx$$
$$= \frac{x}{2} - \frac{\sin 2x}{4} - \int (1 - u^2)u^2\, du \quad (u = \cos x)$$
$$= \frac{x}{2} - \frac{\sin 2x}{4} - \frac{\cos^3 x}{3} + \frac{\cos^5 x}{5} + C$$

2. $\displaystyle\int \sin^2 x \cos^2 x\, dx$

$$= \int \left(\frac{1 - \cos 2x}{2}\right)\left(\frac{1 + \cos 2x}{2}\right) dx$$
$$= \frac{1}{4}\int (1 - \cos^2 2x)\, dx$$
$$= \frac{x}{4} - \frac{1}{4}\int \cos^2 2x\, dx$$
$$= \frac{x}{4} - \frac{1}{4}\int \frac{1 + \cos 4x}{2}\, dx$$
$$= \frac{x}{4} - \frac{x}{8} - \frac{1}{8}\int \cos 4x\, dx$$
$$= \frac{x}{8} - \frac{\sin 4x}{32} + C$$

3. $\displaystyle\int \sin x \cos 2x\, dx = \frac{1}{2}\int (\sin 3x - \sin x)\, dx$

(see product formula (2a), p. 446)

$$= -\frac{\cos 3x}{6} + \frac{\cos x}{2} + C$$

4. $\displaystyle\int \cos 3x \cos 5x\, dx = \frac{1}{2}\int (\cos 8x + \cos 2x)\, dx$

$$= \frac{\sin 8x}{16} + \frac{\sin 2x}{4} + C$$

5. (a) $\int \dfrac{1}{1 + x^2} \, dx = \tan^{-1} x + C$ (see p. 345).

   (b) Let $x = \tan \theta$, so $dx = \sec^2 \theta \, d\theta$ and $1 + x^2 = \sec^2 \theta$. Thus

$$\int \frac{1}{1 + x^2} \, dx = \int \frac{\sec^2 \theta}{\sec^2 \theta} \, d\theta$$

$$= \int d\theta = \theta + C$$

$$= \tan^{-1} x + C$$

   The answers agree.

6. (a) Let $u = 4 - x^2$ so $du = -2x \, dx$. Thus

$$\int \frac{x}{\sqrt{4 - x^2}} \, dx = -\frac{1}{2} \int \frac{du}{\sqrt{u}}$$

$$= -\sqrt{u} + C$$

$$= -\sqrt{4 - x^2} + C$$

   (no trigonometric function appears)

   (b) $\int \dfrac{1}{\sqrt{x^2 - 4}} \, dx = \int \dfrac{du}{\sqrt{u^2 - 1}} \quad \left(u = \dfrac{x}{2} \right)$

$$= \cosh^{-1} u + C \quad \text{(see p. 390)}$$

$$= \cosh^{-1} \left(\frac{x}{2} \right) + C$$

$$= \ln \left(\frac{x}{2} + \sqrt{\frac{x^2}{4} - 1} \right) + C$$

7. To evaluate $\int (x^2/\sqrt{4 - x^2}) \, dx$, let $x = 2 \sin \theta$; $dx = 2 \cos \theta \, d\theta$; and $\sqrt{4 - x^2} = 2 \cos \theta$. Thus

$$\int \frac{x^2}{\sqrt{4 - x^2}} \, dx = \int \frac{4 \sin^2 \theta}{2 \cos \theta} \cdot 2 \cos\theta \, d\theta$$

$$= 4 \int \sin^2 \theta \, d\theta$$

$$= 4 \int \frac{1 - \cos 2\theta}{2} \, d\theta$$

$$= 2\theta - 2 \sin \theta \cos \theta + C$$

(See Worked Example 5.) From Fig. 11-1-3 we get

$$\int \frac{x^2}{\sqrt{4 - x^2}} \, dx$$

$$= 2 \sin^{-1} \left(\frac{x}{2} \right) - 2 \left(\frac{x}{2} \right) \left(\frac{\sqrt{4 - x^2}}{2} \right) + C$$

$$= 2 \sin^{-1} \frac{x}{2} - \frac{1}{2} x \sqrt{4 - x^2} + C$$

Fig. 11-1-3 Geometry of the substitution $x = 2 \sin \theta$.

8. $\int \dfrac{dx}{x^2 + x + 1} = \int \dfrac{dx}{(x + \frac{1}{2})^2 + \frac{3}{4}}$

$$= \int \frac{du}{u^2 + \frac{3}{4}} \quad (u = x + \tfrac{1}{2})$$

$$= \frac{1}{\sqrt{3/4}} \tan^{-1} \left(\frac{u}{\sqrt{3/4}} \right) + C$$

$$= \frac{2}{\sqrt{3}} \tan^{-1} \left(\frac{2x + 1}{\sqrt{3}} \right) + C$$

9. $\int \dfrac{dx}{\sqrt{x^2 + x + 1}}$

$$= \int \frac{du}{\sqrt{u^2 + \frac{3}{4}}} \quad (u = x + \tfrac{1}{2})$$

$$= \ln |u + \sqrt{u^2 + \tfrac{3}{4}}| + C$$

$$= \ln |x + \tfrac{1}{2} + \sqrt{(x + \tfrac{1}{2})^2 + \tfrac{3}{4}}| + C$$

$$= \ln |x + \tfrac{1}{2} + \sqrt{x^2 + x + 1}| + C$$

10. Let $u = x - 1$ so $du = dx$ and $x = u + 1$. Then

$$\int \frac{x^3 + 2x + 1}{(x - 1)^5} \, dx$$

$$= \int \frac{(u + 1)^3 + 2(u + 1) + 1}{u^5} \, du$$

$$= \int \frac{u^3 + 3u^2 + 5u + 4}{u^5} \, du$$

$$= \int \left(\frac{1}{u^2} + \frac{3}{u^3} + \frac{5}{u^4} + \frac{4}{u^5} \right) du$$

$$= -\frac{1}{u} - \frac{3}{2u^2} - \frac{5}{3u^3} - \frac{4}{4u^4} + C$$

$$= -\left[\frac{1}{x - 1} + \frac{3}{2(x - 1)^2} + \frac{5}{3(x - 1)^3} + \frac{1}{(x - 1)^4} \right] + C$$

11. The denominator factors as $(x - 1)(x^2 + x + 1)$, so

$$\frac{1}{x^3 - 1} = \frac{a_1}{x - 1} + \frac{A_1 x + B_1}{x^2 + x + 1}$$

   Thus $1 = a_1(x^2 + x + 1) + (x - 1)(A_1 x + B_1)$. We substitute values for x:

   $x = 1$: $1 = 3a_1$ so $a_1 = \frac{1}{3}$

   $x = 0$: $1 = \frac{1}{3} - B_1$ so $B_1 = -\frac{2}{3}$

   Comparing the x^2 terms, we get $0 = a_1 + A_1$, so $A_1 = -\frac{1}{3}$. Hence

$$\frac{1}{x^3 - 1} = \frac{1}{3} \left(\frac{1}{x - 1} - \frac{x + 2}{x^2 + x + 1} \right)$$

11

11. (*cont'd*)

Now

$$\int \frac{1}{x-1}\,dx = \ln|x-1| + C$$

and

$$\int \frac{x+2}{x^2+x+1}\,dx$$

$$= \frac{1}{2}\int \frac{2x+1}{x^2+x+1}\,dx + \frac{3}{2}\int \frac{dx}{\left(x+\frac{1}{2}\right)^2 + \frac{3}{4}}$$

$$= \frac{1}{2}\ln|x^2+x+1| + \frac{3}{2}\cdot\sqrt{\frac{4}{3}}\,\tan^{-1}\left(\frac{x+\frac{1}{2}}{\sqrt{3/4}}\right)$$

$$= \frac{1}{2}\ln|x^2+x+1| + \sqrt{3}\,\tan^{-1}\left(\frac{2x+1}{\sqrt{3}}\right)$$

Thus

$$\int \frac{dx}{x^3-1} = \frac{1}{3}\ln|x-1| - \frac{1}{6}\ln|x^2+x+1|$$

$$- \frac{1}{\sqrt{3}}\tan^{-1}\left(\frac{2x+1}{\sqrt{3}}\right)$$

$$= \frac{1}{3}\left[\frac{1}{2}\ln\left|\frac{(x-1)^2}{x^2+x+1}\right|\right.$$

$$\left. - \sqrt{3}\,\tan^{-1}\left(\frac{2x+1}{\sqrt{3}}\right)\right] + C$$

Observe that the innocuous-looking integrand $1/(x^3-1)$ has brought forth both logarithmic and trigonometric functions.

12. The denominator factors as $(x - \sqrt{2})^2(x + \sqrt{2})^2$, so

$$\frac{x^2}{(x^2-2)^2} = \frac{a_1}{x-\sqrt{2}} + \frac{a_2}{(x-\sqrt{2})^2}$$

$$+ \frac{b_1}{x+\sqrt{2}} + \frac{b_2}{(x+\sqrt{2})^2}$$

Thus

$$x^2 = a_1(x-\sqrt{2})(x+\sqrt{2})^2 + a_2(x+\sqrt{2})^2$$

$$+ b_1(x+\sqrt{2})(x-\sqrt{2})^2 + b_2(x-\sqrt{2})^2$$

We substitute values for x:

$$x = \sqrt{2}: \quad 2 = 8a_2 \quad \text{so } a_2 = \frac{1}{4}$$

$$x = -\sqrt{2}: \quad 2 = 8b_2 \quad \text{so } b_2 = \frac{1}{4}$$

12. (*cont'd*)

Therefore

$$x^2 = a_1(x^2-2)(x+\sqrt{2}) + \frac{1}{4}(x^2 + 2\sqrt{2x} + 2)$$

$$+ b_1(x^2-2)(x-\sqrt{2})$$

$$+ \frac{1}{4}(x^2 - 2\sqrt{2}\,x + 2)$$

$$= (a_1 + b_1)x^3 + (\sqrt{2}\,a_1 + \frac{1}{2} - \sqrt{2}\,b_1)x^2$$

$$+ (-2a_1 - 2b_1)x - 2\sqrt{2}\,a_1 + 1 + 2\sqrt{2}\,b_1$$

and so

$$a_1 + b_1 = 0 \quad \text{and} \quad \sqrt{2}\,a_1 + \frac{1}{2} - \sqrt{2}\,b_1 = 1$$

Thus

$$a_1 = \frac{1}{4\sqrt{2}}, \quad b_1 = -\frac{1}{4\sqrt{2}}$$

Hence

$$\frac{x^2}{(x^2-2)^2} = \frac{1}{4\sqrt{2}\,(x-\sqrt{2})} + \frac{1}{4(x-\sqrt{2})^2}$$

$$- \frac{1}{4\sqrt{2}\,(x+\sqrt{2})} + \frac{1}{4(x+\sqrt{2})^2}$$

and so

$$\int \frac{x^2}{(x^2-2)^2}\,dx$$

$$= \frac{1}{4\sqrt{2}}\ln\left|\frac{x-\sqrt{2}}{x+\sqrt{2}}\right| - \frac{1}{4(x-\sqrt{2})}$$

$$- \frac{1}{4(x+\sqrt{2})} + C$$

$$= \frac{1}{4\sqrt{2}}\ln\left|\frac{x-\sqrt{2}}{x+\sqrt{2}}\right| - \frac{x}{2(x^2-2)} + C$$

Solved Exercises for Section 11-2

1. If $f(x) = x^n$, then $f'(x) = nx^{n-1}$ and $\sqrt{1 + f'(x)^2} = \sqrt{1 + n^2 x^{2n-2}}$, so

$$L = \int_a^b \sqrt{1 + n^2 x^{2n-2}}\,dx$$

Introducing $u = n^{1/(n-1)}x$ as a new variable gives

$$L = \frac{1}{n^{1/(n-1)}}\int_{n^{1/(n-1)}a}^{n^{1/(n-1)}b} \sqrt{1 + u^{2n-2}}\,du$$

The two simplest cases for which we can evaluate the integral are $2n - 2 = 0$ ($n = 1$ gives a straight line) and $2n - 2 = 1$ ($n = \frac{3}{2}$). In Worked Example 3 we saw how to find the integral in the case $2n - 2 = 2$—that is, $n = 2$, which is a parabola.

2. We get

$$f(x) = \sqrt{1 - k^2 x^2}$$
$$f'(x) = -k^2 x / \sqrt{1 - k^2 x^2}$$

and

$$\sqrt{1 + f'(x)^2} = \sqrt{[1 + (k^4 - k^2)x^2]/(1 - k^2 x^2)}$$

so

$$L = \int_0^b \sqrt{\frac{1 + (k^4 - k^2)x^2}{1 - k^2 x^2}} \, dx$$

It turns out that the antiderivative

$$\int \sqrt{\frac{1 + (k^4 - k^2)x^2}{1 - k^2 x^2}} \, dx$$

cannot be expressed (unless $k^2 = 0$ or 1) in terms of algebraic, trigonometric, or exponential functions. It is a new kind of function called an *elliptic* function. (See Review Problem 18 for more examples of such functions.)

3. If, instead of the entire sphere, we take the band obtained by restricting x to $[a, b]$ $(-1 \le a \le b \le 1)$, the area is $2\pi \int_a^b r \, dx = 2\pi r (b - a)$. Thus the area obtained by slicing a sphere by two parallel planes and taking the middle piece is equal to $2\pi r$ times the distance between the planes, regardless of where the two planes are located (see Fig. 11-2-15). (Why doesn't the "longer" band around the middle have more area?)

Fig. 11-2-15 Bands of equal width have equal area.

4. (a) In case $f(x) = mx + q$, the surface is a frustum of a cone, as in Fig. 11-2-16. By the formula on p. 464, the area is

$$\pi(r_1 + r_2)s = \pi(a + b)\sqrt{1 + m^2} \, (b - a)$$
$$= \pi\sqrt{1 + m^2} \, (b^2 - a^2)$$
$$= \int_a^b 2\pi\sqrt{1 + m^2} \, x \, dx$$
$$= 2\pi \int_a^b x\sqrt{1 + f'(x)^2} \, dx \qquad (3)$$

By the same argument as we used in the case of revolution about the x axis, we extend formula (3) to piecewise linear f (the surface is again a conoid) and then to general f.

4. (*cont'd*)

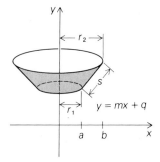

Fig. 11-2-16 A line segment revolved around the y axis becomes a frustum of a cone.

(b) If $f(x) = x^2$, $f'(x) = 2x$ and $\sqrt{1 + f'(x)^2} = \sqrt{1 + 4x^2}$. Then

$$A = 2\pi \int_1^2 x\sqrt{1 + 4x^2} \, dx = \frac{\pi}{4} \int_5^{17} u^{1/2} \, du$$
$$(u = 1 + 4x^2, \, du = 8x \, dx)$$
$$= \frac{\pi}{4} \left[\frac{2}{3} u^{3/2} \right]_5^{17}$$
$$= \frac{\pi}{6} (17^{3/2} - 5^{3/2}) \approx 30.85$$

Solved Exercises for Section 11-3

1. (a) We begin by making a table:

t	-2	-1	$-\frac{1}{2}$	0	$\frac{1}{2}$	1	2
$x = t^3 - t$	-6	0	$\frac{3}{8}$	0	$-\frac{3}{8}$	0	6
$y = t^2$	4	1	$\frac{1}{4}$	0	$\frac{1}{4}$	1	4

These points are plotted in Fig. 11-3-13. The number next to each point is the corresponding value of t. Notice that the point $(0, 1)$ occurs for $t = -1$ and $t = 1$.

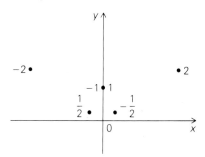

Fig. 11-3-13 Some points on the curve $(t^3 - t, \, t^2)$.

1. (*cont'd*)

   (b) We plot x and y against t in Fig. 11-3-14. From the graph of x against t, we conclude that as t goes from $-\infty$ to ∞, the point comes in from the left, reverses direction for a while, and then goes out to the right. From the graph of y against t, we see that the point descends for $t < 0$, reaching bottom at $y = 0$ when $t = 0$, and then ascends for $t > 0$. Putting this information together with the points we have plotted, we sketch the curve in Fig. 11-3-15.

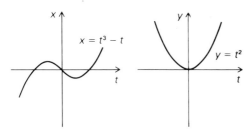

   Fig. 11-3-14   The graphs of x and y plotted separately against t.

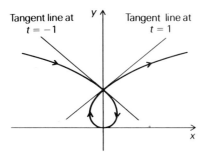

   Fig. 11-3-15   The parametric curve $(t^3 - t,\ t^2)$.

   (c) The slope of the tangent line at time t is

   $$\frac{dy}{dx} = \frac{dy/dt}{dx/dt} = \frac{2t}{3t^2 - 1}$$

   When $t = -1$, the slope is -1; when $t = 1$, the slope is 1. (See Fig. 11-3-15.)

   (d) We can eliminate t by solving the second equation for t to get $t = \pm\sqrt{y}$ and substituting in the first to get $x = \pm(y^{3/2} - y^{1/2})$. To obtain an equation without fractional powers, we square both sides. The result is $x^2 = y(y - 1)^2$, or $x^2 = y^3 - 2y^2 + y$. In this form, it is not so easy to predict the behavior of the curve, particularly at the "double point" $(0, 1)$.

2. The situation is close to that of a circle, but x ranges from $-\frac{1}{2}$ to $\frac{1}{2}$ and y ranges from $-\frac{1}{3}$ to $\frac{1}{3}$, so we try $x = \frac{1}{2}\cos t$, $y = \frac{1}{3}\sin t$. Then

   $$4x^2 + 9y^2 = \cos^2 t + \sin^2 t = 1$$

   so this parametrization works. As t goes from zero to 2π, the point moves once around the ellipse.

3. (a) Eliminating the parameter t, we have $y = x^{2/3}$. The graph has a cusp at the origin, as in Fig. 11-3-16. (See Section 6-3.)

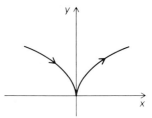

   Fig. 11-3-16   The curve $(t^3,\ t^2)$ has a cusp at the origin.

   (b) When $t = 1$, we have $x = t^3 = 1$, $y = t^2 = 1$, $dx/dt = 3t^2 = 3$, and $dy/dt = 2t = 2$, so the tangent line is given by

   $$x = 3(t - 1) + 1 \qquad y = 2(t - 1) + 1$$

   It has slope $\frac{2}{3}$. (You can also see this by differentiating $y = x^{2/3}$ and setting $x = 1$.)

   (c) When $t = 0$, we have $dx/dt = 0$ and $dy/dt = 0$, so the tangent line is not defined.

4. The tangent line is vertical when $dx/dt = 0$ and horizontal when $dy/dt = 0$. (If both are zero, there is no tangent line.)

   We have $dx/dt = -3\sin 3t$, which is zero when $t = 0, \pi/3, 2\pi/3, \pi, 4\pi/3, 5\pi/3$ (the curve repeats itself when t reaches 2π) and $dy/dt = \cos t$, which is zero when $t = \pi/2$ or $3\pi/2$. We make a table:

t	0	$\dfrac{\pi}{3}$	$\dfrac{\pi}{2}$	$\dfrac{2\pi}{3}$	π	$\dfrac{4\pi}{3}$	$\dfrac{3\pi}{2}$	$\dfrac{5\pi}{3}$
$x = \cos 3t$	1	-1	0	1	-1	1	0	-1
$y = \sin t$	0	$\dfrac{\sqrt{3}}{2}$	1	$\dfrac{\sqrt{3}}{2}$	0	$-\dfrac{\sqrt{3}}{2}$	-1	$-\dfrac{\sqrt{3}}{2}$
Tangent	ver	ver	hor	ver	ver	ver	hor	ver

4. *(cont'd)*

Using the fact that $\sqrt{3}/2 \approx 0.866$, we sketch this information in Fig. 11-3-17. Connecting these points in the proper order with a smooth curve, we obtain Fig. 11-3-18. This curve is an example of a *Lissajous figure* (see Review Problems 19 and 20 at the end of the chapter).

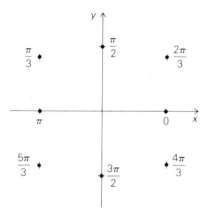

Fig. 11-3-17 Points on the curve (cos 3*t*, sin *t*) with horizontal or vertical tangent.

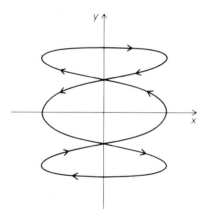

Fig. 11-3-18 The curve (cos 3*t*, sin *t*) is an example of a Lissajous figure.

5. We have $dx/dt = \cos 2t$ and $dy/dt = -2\cos t \sin t$, so

$$L = \int_0^\pi \sqrt{\cos^2 2t + 4\cos^2 t \sin^2 t}\; dt$$

$$= \int_0^\pi \sqrt{\cos^2 2t + \sin^2 2t}\; dt$$

$$= \int_0^\pi 1\; dt = \pi$$

6. We know that $g'(t)^2 \le f'(t)^2 + g'(t)^2$, so $g'(t) \le \sqrt{f'(t)^2 + g'(t)^2}$. Integrating from 0 to 1, we have

$$\int_0^1 g'(t)dt \le \int_0^1 \sqrt{f'(t)^2 + g'(t)^2}\; dt$$

By the fundamental theorem of calculus, the left-hand side is $g(1) - g(0) = a - 0 = a$; the right-hand side is the length L of the curve, so we have $a \le L$. If $a = L$, the integrands must be equal; that is, $g'(t) = \sqrt{f'(t)^2 + g'(t)^2}$, which is possible only if $f'(t)$ is identically zero, that is, $f(t)$ is constant. Since $f(0) = f(1) = 0$, we must have $f(t)$ identically zero; that is, the point (x, y) stays on the y axis.

We have shown that the shortest curve between the points $(0, 0)$ and $(0, a)$ is the straight line segment which joins them.

7. The length is

$$L = \int_0^{2\pi} \sqrt{r^2 + \left(\frac{dr}{d\theta}\right)^2}\; d\theta$$

$$= \int_0^{2\pi} \sqrt{(1 + \cos\theta)^2 + \sin^2\theta}\; d\theta$$

$$= \int_0^{2\pi} \sqrt{2 + 2\cos\theta}\; d\theta$$

This can be simplified by the half-angle formula $\cos^2(\theta/2) = (1 + \cos\theta)/2$, so

$$L = \int_0^{2\pi} 2\cos\frac{\theta}{2}\; d\theta = 0$$

Something is wrong here! We forgot that $\cos(\theta/2)$ can be negative, while the square root $\sqrt{2 + 2\cos\theta}$ must be positive; i.e.

$$\sqrt{2 + 2\cos\theta} = \sqrt{4\cos^2(\theta/2)} = 2|\cos(\theta/2)|.$$

The correct evaluation of L is as follows:

$$L = \int_0^{2\pi} 2\left|\cos\frac{\theta}{2}\right|\; d\theta$$

$$= \int_0^\pi 2\cos\frac{\theta}{2}\; d\theta - \int_\pi^{2\pi} 2\cos\frac{\theta}{2}\; d\theta$$

since $\cos(\theta/2) > 0$ on $(0, \pi)$ and $\cos(\theta/2) < 0$ on $(\pi, 2\pi)$. Thus

$$L = 4\sin\frac{\theta}{2}\Big|_0^\pi - 4\sin\frac{\theta}{2}\Big|_\pi^{2\pi}$$

$$= 4(1 - 0) - 4(0 - 1) = 8$$

11

8. The area dA of the shaded triangle in Fig. 11-3-6 is $\frac{1}{2}(r\,d\theta)r = \frac{1}{2}r^2\,d\theta$, so the area inside the curve is

$$\int_\alpha^\beta dA = \frac{1}{2}\int_\alpha^\beta r^2\,d\theta$$

as before.

9. The area enclosed is defined by $r = 1 + \cos\theta$ and the full range $0 \le \theta \le 2\pi$, so

$$A = \frac{1}{2}\int_0^{2\pi}(1 + \cos\theta)^2\,d\theta$$

$$= \frac{1}{2}\int_0^{2\pi}(1 + 2\cos\theta + \cos^2\theta)\,d\theta$$

Again using the half-angle formula,

$$A = \frac{1}{2}\int_0^{2\pi}\left(\frac{3}{2} + 2\cos\theta + \frac{\cos 2\theta}{2}\right)d\theta$$

$$= \frac{1}{2}\left[\frac{3\theta}{2} + 2\sin\theta + \frac{\sin 2\theta}{4}\right]_0^{2\pi} = \frac{3\pi}{2}$$

10. Suppose $r = f(\theta)$ and $r = g(\theta)$ are the two curves with $f(\theta) \ge g(\theta)$. We are required to find a formula for the shaded area in Fig. 11-3-19. The area is just the difference between the areas for f and g; that is,

$$A = \frac{1}{2}\int_\alpha^\beta [f(\theta)^2 - g(\theta)^2]\,d\theta$$

Fig. 11-3-19 The area of the shaded region is $\frac{1}{2}\int_\alpha^\beta [f(\theta)^2 - g(\theta)^2]\,d\theta$.

CHAPTER 12

Solved Exercises for Section 12-1

1. (a) The terms a_1 through a_6 are -1, $\frac{1}{2}$, $-\frac{1}{3}$, $\frac{1}{4}$, $-\frac{1}{5}$, $\frac{1}{6}$. They are plotted in Fig. 12-1-6. As n gets larger, the number $(-1)^n/n$ seems to get closer to zero. Therefore we guess that $\lim_{n\to\infty}(-1)^n/n = 0$.

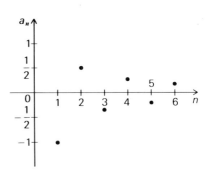

Fig. 12-1-6 The sequence $a_n = (-1)^n/n$ plotted in two ways.

(b) We have, for a_1 through a_6, $-\frac{1}{2}$, $\frac{2}{3}$, $-\frac{3}{4}$, $\frac{4}{5}$, $-\frac{5}{6}$, $\frac{6}{7}$. They are plotted in Fig. 12-1-7. In this case, the numbers a_n do not approach any particular number. (Some of them are getting near to 1, others to -1.) We guess that the sequence does not have a limit.

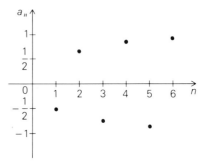

Fig. 12-1-7 The sequence $a_n = (-1)^n n/(n+1)$ plotted in two ways.

2. Using a calculator we find:

n	$\sqrt[n]{n}$
1	1
5	1.37973
10	1.25893
50	1.08138
100	1.04713
500	1.01251
1000	1.00693
5000	1.00170
10,000	1.00092

Thus it appears that $\lim_{n\to\infty} \sqrt[n]{n} = 1$. This is in fact true. (The easiest proof of this uses logarithms and l'Hôpital's rule; see Section 12-3.)

3. Looking at Worked Example 5 we observe that $a_n = 1 - 1/2^n$. This may be proved by using induction as follows. The formula is true for a_0 through a_3 by inspection; if it is true for k, then $a_{k+1} = \frac{1}{2}(1 + a_k) = \frac{1}{2}(1 + 1 - 1/2^k) = \frac{1}{2}(2 - 1/2^k) = 1 - 1/2^{k+1}$, so it is true for $k + 1$.

Now, given $\varepsilon > 0$, we wish to make $|a_n - 1| < \varepsilon$; that is, $|1 - (1/2^n) - 1| = 1/2^n < \varepsilon$. But the proof of Theorem 1, with $r = \frac{1}{2}$, shows that $1/2^n = (\frac{1}{2})^n < \varepsilon$ for N sufficiently large, so $\lim_{n\to\infty} a_n = 1$.

4. Given $\varepsilon > 0$, we must find an N such that $|a_n - c| < \varepsilon$ for all $n > N$. But $|a_n - c| = |c - c| = 0$, which is less than ε for all n. Thus we may choose N any way we wish, say $N = 1$. (The definition of limit *permits* N to grow as ε becomes small, but it does not *require* it.)

5. $\lim_{n\to\infty} (8n^2 + 2)/(3n^2 + n) =$
$\lim_{n\to\infty} (8 + 2/n^2)/(3 + 1/n) = \frac{8}{3}$

6. $\lim_{n\to\infty} [\pi + (\frac{2}{3})^n]^3 = \{\lim_{n\to\infty} [\pi + (\frac{2}{3})^n]\}^3 = \pi^3$.
(We use Theorem 1 and Theorem 2(5) with the continuous function $f(x) = x^3$.)

7. (a) Since $a_n \to 0$ as $n \to \infty$, for every $\varepsilon > 0$ there is an N such that $|a_n| < \varepsilon$ if $n > N$. Since $|b_n| \le |a_n|$, we also have $|b_n| < \varepsilon$ if $n > N$. Thus $b_n \to 0$ as $n \to \infty$.

7. *(cont'd)*
(b) (i) $|(\sin n)/n| \le 1/n$ since $|\sin n| \le 1$. Thus, with $a_n = 1/n$ and $b_n = (\sin n)/n$, part (a) gives us $b_n \to 0$ as $n \to \infty$.

(ii) Notice that
$$\left(\frac{1+n}{2n}\right)^n \le \left(\frac{\frac{1}{2}n + n}{2n}\right)^n$$
if $n \ge 2$, since then $1 \le n/2$. Thus if $n \ge 2$,
$$\left(\frac{1+n}{2n}\right)^n \le \left(\frac{\frac{1}{2}n + n}{2n}\right)^n = \left(\frac{\frac{3}{2}}{2}\right)^n$$
$$= \left(\frac{3}{4}\right)^n$$
Thus if $b_n = [(1 + n)/2n]^n$, $a_n = (\frac{3}{4})^n$, part (a) and Theorem 1 give $b_n \to 0$ as $n \to \infty$.

8. No; there is no single number to which a_n is close for large n because a_n is alternately $+1$ and -1.

9. We prove that $a_{n+1} > a_n$ by induction on n. For $n = 0$ it is true since $a_1 = \sqrt{3} > a_0 = 0$. Assume it is true for $n - 1$; that is, $a_n > a_{n-1}$. Then
$$a_{n+1} - a_n = \sqrt{3 + a_n} - \sqrt{3 + a_{n-1}}$$
$$= (\sqrt{3 + a_n} - \sqrt{3 + a_{n-1}})\left(\frac{\sqrt{3 + a_n} + \sqrt{3 + a_{n-1}}}{\sqrt{3 + a_n} + \sqrt{3 + a_{n-1}}}\right)$$
$$= \frac{a_n - a_{n-1}}{\sqrt{3 + a_n} + \sqrt{3 + a_{n-1}}} > 0$$
since $a_n - a_{n-1} > 0$. Thus $a_n > a_{n-1}$ for all n, and so a_n is increasing.

Likewise we can prove that $a_n < 5$ by induction. For $n = 0$ it is true. Assume that it is true for $n - 1$. Then
$$a_n = \sqrt{3 + a_{n-1}} < \sqrt{3 + 5} = \sqrt{8} < 5$$
so $a_n < 5$ also. (In fact, we can prove $a_n < 3$ the same way.)

Theorem 3 then guarantees that a_n converges to a limit; call it l. From $a_{n+1} = \sqrt{3 + a_n}$ we get, using Theorem 2(5) and continuity of $f(x) = \sqrt{3 + x}$,
$$\lim_{n\to\infty} a_{n+1} = \lim_{n\to\infty} \sqrt{3 + a_n} = \sqrt{3 + \lim_{n\to\infty} a_n}$$
Hence $l = \sqrt{3 + l}$ and $l^2 - l - 3 = 0$, so $l = [1 \pm \sqrt{1 - 4(1)(-3)}]/2 = (1 \pm \sqrt{13})/2$. Since $a_n \ge 0$, we get $l \ge 0$ (see Problem 13), and so l must be $(1 + \sqrt{13})/2$.

12

10. (a) By division, we have

$$\frac{b^{n+1} - a^{n+1}}{b - a}$$

$$= (b^n + b^{n-1} a + b^{n-2} a^2 + \cdots + a^n)$$

Since $0 \le a < b$, the right side is

$$< b^n + b^{n-1} \cdot b + b^{n-2} \cdot b^2 + \cdots + b^n$$
$$= (n + 1)b^n$$

since there are $n + 1$ terms.

Cross-multiplication gives

$$b^{n+1} - a^{n+1} < (n + 1)b^n(b - a)$$
$$= (n + 1)b^{n+1} - b^n a(n + 1)$$

Hence $b^n a(n + 1) - nb^{n+1} < a^{n+1}$, so
$b^n[(n + 1)a - nb] < a^{n+1}$.

(b) If $a = 1 + 1/(n + 1)$ and $b = 1 + 1/n$,
we get

$$\left(1 + \frac{1}{n}\right)^n \times$$

$$\left[(n + 1)\left(1 + \frac{1}{n+1}\right) - n\left(1 + \frac{1}{n}\right)\right]$$

$$< \left(1 + \frac{1}{n+1}\right)^{n+1}$$

Since the term in brackets [ ] is 1, we get

$$\left(1 + \frac{1}{n}\right)^n < \left(1 + \frac{1}{n+1}\right)^{n+1}$$

Hence $a_n < a_{n+1}$.

(c) If we set $a = 1$ and $b = 1 + (1/2n)$ in the
inequality of part (a), we get

$$\left(1 + \frac{1}{2n}\right)^n \left[(n + 1) - n\left(1 + \frac{1}{2n}\right)\right] < 1$$

Hence

$$\left(1 + \frac{1}{2n}\right)^n \left[(n + 1) - \left(n + \frac{1}{2}\right)\right] < 1$$

Thus

$$\left(1 + \frac{1}{2n}\right)^n < 2$$

$$\left(1 + \frac{1}{2n}\right)^{2n} < 4$$

(d) By part (b), $a_{n+1} < a_n$ and by part (c),
$a_{2n} < 4$, for any n. To show that $a_n < 4$,
notice that, by part (b),

$$a_n < a_{n+1} < a_{n+2} < \cdots < a_{n+n} = a_{2n}$$

which is less than 4 by part (c).

(e) Since a_n is increasing and bounded above,
it converges by Theorem 3.

11. (a) Directly,

$$S_n = \frac{1}{n^2} \sum_{i=1}^{n} (n + i)$$

$$= \frac{1}{n^2}\left[n^2 + \frac{n(n + 1)}{2}\right]$$

$$= \left(1 + \frac{1}{2} + \frac{1}{2n}\right) \rightarrow \frac{3}{2}$$

as $n \rightarrow \infty$. (We used $\sum_{i=1}^{n} i = n(n + 1)/2$
from p. 176.)

(b) $S_n = \sum_{i=1}^{n} (1 + i/n)(1/n)$ is the Riemann sum
for $f(x) = 1 + x$ with $a = 0$, $b = 1$, $t_0 = 0$, $t_1 = 1/n$, $t_2 = 2/n, \ldots, t_n = 1$, width
$\Delta t_i = 1/n$ and $c_i = t_i$. Hence as $n \rightarrow \infty$,

$$S_n \rightarrow \int_0^1 (1 + x)\, dx = \left. \frac{(1 + x)^2}{2} \right|_0^1 = \frac{3}{2}$$

12. Let $t_0, t_1, \ldots, t_n$ be a partition of $[a, b]$ with
maximum width $\rightarrow 0$. Let c_i belong to $[t_{i-1}, t_i]$.
Then

$$\int_a^b f(x)\, dx = \lim_{n \to \infty} \sum_{i=1}^{n} f(c_i)\, \Delta t_i$$

$$\int_a^b g(x)\, dx = \lim_{n \to \infty} \sum_{i=1}^{n} g(c_i)\, \Delta t_i$$

and

$$\int_a^b [f(x) + g(x)]\, dx = \lim_{n \to \infty} \sum_{i=1}^{n} [f(c_i) + g(c_i)]\, \Delta t_i$$

But, by the summation rules and limit laws, this
last expression equals

$$\lim_{n \to \infty} \sum_{i=1}^{n} f(c_i)\, \Delta t_i + \lim_{n \to \infty} \sum_{i=1}^{n} g(c_i)\, \Delta t_i$$

$$= \int_a^b f(x)\, dx + \int_a^b g(x)\, dx$$

Solved Exercises for Section 12-2

1. Following the hint, we have

$$\sqrt{x^2 + 1} - x = (\sqrt{x^2 + 1} - x)\frac{\sqrt{x^2 + 1} + x}{\sqrt{x^2 + 1} + x}$$

$$= \frac{x^2 + 1 - x^2}{\sqrt{x^2 + 1} + x} = \frac{1}{\sqrt{x^2 + 1} + x}$$

As $x \rightarrow \infty$, the denominator becomes arbitrarily
large, so we have $\lim_{x \to \infty} (\sqrt{x^2 + 1} - x) = 0$. For
a geometric interpretation, see Fig. 12-2-8.

1. (*cont'd*)

Fig. 12-2-8 As the length x goes to ∞, the difference $\sqrt{x^2+1} - x$ between the lengths of the hypotenuse and the long leg goes to zero.

2. We find $\lim\limits_{x\to+\infty} x/\sqrt{x^2+1} = \lim\limits_{x\to+\infty} 1/\sqrt{1+1/x^2} = 1$ and $\lim\limits_{x\to-\infty} x/\sqrt{x^2+1} = \lim\limits_{x\to-\infty} -\sqrt{x^2}/\sqrt{x^2+1} = \lim\limits_{x\to-\infty} -1/\sqrt{1+1/x^2} = -1$ (in the second limit we may take $x < 0$, so $x = -\sqrt{x^2}$). Hence the horizontal asymptotes are the lines $y = \pm 1$. See Fig. 12-2-9.

Fig. 12-2-9 The curve $y = x/\sqrt{x^2+1}$ has the lines $y = -1$ and $y = 1$ as horizontal asymptotes.

3. First of all, we note that $f(x) = e^{kx}$ is a decreasing, positive function. Given $\varepsilon > 0$, we wish to find A such that $x > A$ implies $e^{kx} < \varepsilon$. Taking logarithms of the last inequality gives $kx < \ln \varepsilon$, or $x > (\ln \varepsilon)/k$. So we may let $A = (\ln \varepsilon)/k$. (If ε is small, $\ln \varepsilon$ is a large negative number.)

4. (a) Let $\varepsilon > 0$ be given. We have to find $\delta > 0$ such that $|x^2 - a^2| < \varepsilon$ if $|x - a| < \delta$. We note that $x^2 - a^2 = (x-a)^2 + 2a(x-a)$ as in Worked Example 4. If $|x - a| < \delta$, we have $|x^2 - a^2| = |(x-a)^2 + 2|a|(x-a)| \leq |x-a|^2 + 2|a||x-a| < \delta^2 + 2|a|\delta$. Thus we want δ to be such that $\delta^2 + 2|a|\delta \leq \varepsilon$. We may choose δ such that $\delta \leq 1$ (so $\delta^2 \leq \delta$) and $\delta \leq \varepsilon/(1 + 2|a|)$. This choice (or a smaller one) guarantees that $\delta^2 + 2|a|\delta \leq (1 + 2|a|)\delta \leq \varepsilon$, and so $|x^2 - a^2| < \varepsilon$ if $|x - a| < \delta$.

(b) Given $\varepsilon > 0$, we want to choose $\delta > 0$ such that $|x^3 + 2x^2 + 2 - 47| < \varepsilon$ if $|x - 3| < \delta$. Write

$x^3 + 2x^2 - 45$
$= [(x-3)+3]^3 + 2[(x-3)+3]^2 - 45$
$= (x-3)^3 + 9(x-3)^2 + 27(x-3) + 27$
$\quad + 2(x-3)^2 + 12(x-3) + 18 - 45$
$= (x-3)^3 + 11(x-3)^2 + 39(x-3)$

4. (*cont'd*)

We must choose δ so that $\delta^3 + 11\delta^2 + 39\delta < \varepsilon$. For instance, choose $\delta < 1$ (so $\delta^3 < \delta$, $\delta^2 < \delta$) and also less than $\varepsilon/(1 + 11 + 39) = \varepsilon/51$. Then if $|x - 3| < \delta$, we have

$|x^3 + 2x^2 + 2 - 47|$
$\leq |x-3|^3 + 11|x-3|^2 + 39|x-3|$
$< \delta^3 + 11\delta^2 + 39\delta \leq \varepsilon$

so the statement $\lim\limits_{x\to 3} x^3 + 2x^2 + 2 = 47$ is verified.

5. (a) Given $\varepsilon > 0$ there is a $\delta > 0$ such that $|f(x)| < \varepsilon$ if $|x - x_0| < \delta$, by the assumption that $\lim\limits_{x\to x_0} f(x) = 0$. Given $\varepsilon > 0$, this same δ also gives $|g(x)| < \varepsilon$ if $|x - x_0| < \delta$ since $|g(x)| \leq |f(x)|$. Hence g has limit zero as $x \to x_0$ as well.

(b) Let $g(x) = x \sin(1/x)$ and $f(x) = x$. Then $|g(x)| \leq |x|$ for all $x \neq 0$, since $|\sin(1/x)| \leq 1$, so part (a) applies. (We know that $\lim\limits_{x\to 0} x = 0$ by the x rule.)

6. For $x \neq 0$ we have the identity

$$\frac{(3+x)^2 - 9}{x} = \frac{9 + 6x + x^2 - 9}{x} = 6 + x$$

The limit as $x \to 0$ is thus 6.

7. We find

$$f(x) = \frac{1}{x^2 - 5x + 6} = \frac{1}{(x-3)(x-2)}$$

Thus $f(x) \to \pm\infty$ as $x \to 3\pm$ and $f(x) \to \mp\infty$ as $x \to 2\pm$, so $x = 3$ and $x = 2$ are vertical asymptotes (Fig. 12-2-10).

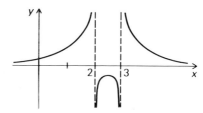

Fig. 12-2-10 The lines $x = 2$ and $x = 3$ are vertical asymptotes for

$$y = \frac{1}{(x-3)(x-2)}$$

12

8. (a) We have

$$f'(x) = \begin{cases} -1 & \text{for } x < 0 \\ 1 & \text{for } x > 0 \end{cases}$$

while $f'(0)$ is not defined. The graph of f' is sketched in Fig. 12-2-11.

Fig. 12-2-11 The graph of $(d/dx)\,|x|$.

(b) Inspecting the graph, we find that $\lim\limits_{x \to 0^-} f'(x) = -1$, while $\lim\limits_{x \to 0^+} f'(x) = 1$.

(c) The limit $\lim\limits_{x \to 0} f'(x)$ does not exist, since $f'(x)$ is not near any single number for both positive and negative small values of x.

Solved Exercises for Section 12-3

1. $\lim\limits_{x \to 0} \dfrac{\sin x - x}{\tan x - x} = \lim\limits_{x \to 0} \dfrac{\cos x - 1}{\sec^2 x - 1}$ (still in $\frac{0}{0}$ form)

$$= \lim_{x \to 0} \frac{-\sin x}{2 \sec x (\sec x \tan x)}$$

$$= \lim_{x \to 0} \frac{-\cos x}{2 \sec^2 x} = -\frac{1}{2}$$

2. No. For example, $\lim\limits_{x \to 0} (x^2 + 1)/x = \infty$, which is not equal to $\lim\limits_{x \to 0} 2x/1 = 0$. It is necessary to assume that the limit of the numerator is zero as well.

3. This is of the form 1^∞, again indeterminate. We have $x^{1/(x-1)} = e^{(\ln x)/(x-1)}$; applying L'Hôpital's rule gives

$$\lim_{x \to 1} \frac{\ln x}{x - 1} = \lim_{x \to 1} \frac{1/x}{1} = 1$$

so

$$\lim_{x \to 1} x^{1/(x-1)} = \lim_{x \to 1} e^{(\ln x)/(x-1)} = e^{\lim\limits_{x \to 1} (\ln x)/(x-1)} = e^1 = e$$

If we put $x = 1 + (1/n)$, then $x \to 1$ when $n \to \infty$; we have $1/(x - 1) = n$, so the limit we just calculated is $\lim\limits_{n \to \infty} (1 + 1/n)^n$. Thus L'Hôpital's rule gives another proof of the famous formula for e. (Actually, the proof on p. 332 is just L'Hôpital's rule in disguise.)

4. We may apply the comparison test with $g(x) = 1/(1 + x)^2$ and $f(x) = (\sin x)/(1 + x)^2$, since $|\sin x| \le 1$. To show that $\int_0^\infty dx/(1 + x)^2$ is convergent, we can compare $1/(1 + x)^2$ with $1/x^2$ on $[1, \infty)$, as in Worked Example 8, or we can evaluate the integral explicitly.

5. We write $\int_{-\infty}^\infty dx/(1 + x^2) = \int_{-\infty}^0 dx/(1 + x^2) + \int_0^\infty dx/(1 + x^2)$. To evaluate these integrals, we use the formula $\int dx/(1 + x^2) = \tan^{-1} x$. Then $\int_{-\infty}^0 dx/(1 + x^2) = \lim\limits_{a \to -\infty} (\tan^{-1} 0 - \tan^{-1} a) = 0 - \lim\limits_{a \to -\infty} \tan^{-1} a = -(-\pi/2) = \pi/2$. (See Fig. 6-2-8, p. 292 for the horizontal asymptotes of $y = \tan^{-1} x$.) Similarly, we have

$$\int_0^\infty \frac{dx}{1 + x^2} = \lim_{b \to \infty} (\tan^{-1} b - \tan^{-1} 0) = \frac{\pi}{2}$$

so

$$\int_{-\infty}^\infty \frac{dx}{1 + x^2} = \frac{\pi}{2} + \frac{\pi}{2} = \pi .$$

6. We use the comparison test in the reverse direction, comparing $1/\sqrt{1 + x^2}$ with $1/x$. In fact, for $x \ge 1$, we have $1/\sqrt{1 + x^2} \ge 1/\sqrt{x^2 + x^2} = 1/\sqrt{2}\,x$. If $\int_1^\infty dx/\sqrt{1 + x^2}$ were convergent, we could conclude from the comparison test that $\int_1^\infty dx/\sqrt{2}\,x$ converged too. But $\int_1^b dx/\sqrt{2}\,x = (1/\sqrt{2})\ln b$, and this diverges as $b \to \infty$.

7. We know that $\int \ln x\, dx = x \ln x - x + C$, so

$$\int_0^1 \ln x\, dx = \lim_{p \to 0+} (1 \ln 1 - 1 - p \ln p + p)$$

$$= 0 - 1 - 0 + 0 = -1$$

($\lim\limits_{p \to 0+} p \ln p = 0$ by Worked Example 4.)

8. This integral is improper at both ends; we write it as $I_1 + I_2$, where $I_1 = \int_0^1 (e^{-x}/\sqrt{x})\, dx$ and $I_2 = \int_1^\infty (e^{-x}/\sqrt{x})\, dx$ and apply the comparison test to each term. On $[0, 1]$, we have $e^{-x} \le 1$, so $e^{-x}/\sqrt{x} \le 1/\sqrt{x}$. Since $\int_0^1 dx/\sqrt{x}$ is convergent (Worked Example 9), so is I_1. On $[1, \infty)$, we have $1/\sqrt{x} \le 1$, so $e^{-x}/\sqrt{x} \le e^{-x}$. But $\int_1^\infty e^{-x}\, dx$ is convergent (Exercise 5(a)), so I_2 is convergent as well.

Solved Exercises for Section 12-4

1. Suppose that $\lim_{n\to\infty} x_n = \bar{x}$. Taking limits on both sides of the equation $x_{n+1} = x_n - f(x_n)/f'(x_n)$, we obtain $\bar{x} = \bar{x} - \lim_{n\to\infty} f(x_n)/f'(x_n)$, or $\lim_{n\to\infty} f(x_n)/f'(x_n) = 0$. Now let $a_n = f(x_n)/f'(x_n)$. Then we have $\lim_{n\to\infty} a_n = 0$, while $f(x_n) = a_n f'(x_n)$. Taking limits as $n \to \infty$ and using the continuity of f and f', we find

$$\lim_{n\to\infty} f(x_n) = \lim_{n\to\infty} a_n \lim_{n\to\infty} f'(x_n)$$

$$f(\bar{x}) = 0 \cdot f'(\bar{x}) = 0$$

2. The iteration formula is

$$x_{n+1} = x_n - \frac{x_n^5 - x_n^4 - x_n + 2}{5x_n^4 - 4x_n^3 - 1}$$

$$= \frac{4x_n^5 - 3x_n^4 - 2}{5x^4 - 4x^3 - 1}$$

For the purpose of convenient calculation, we may write this as

$$x_{n+1} = \frac{(4x_n - 3)x_n^4 - 2}{(5x_n - 4)x_n^3 - 1}$$

Starting at $x_0 = 1$, we find that the denominator is undefined, so we can go no further. (Can you interpret this difficulty geometrically?)

Starting at $x_0 = 2$, we get

$x_1 = 1.659574468$

$x_2 = 1.372968569$

$x_3 = 1.068606737$

$x_4 = -0.5293374382$

$x_5 = 169.5250382$

The iteration process seems to have sent us out on a wild goose chase. To see what has gone wrong, we look at the graph of $f(x) = x^5 - x^4 - x + 2$. (See Fig. 12-4-6.) There is a "bowl" near $x_0 = 2$; Newton's method attempts to take us down to a nonexistent root.

Finally, we start with $x_0 = -2$. The iteration gives

$x_0 = -2$ $f(x_0) = -44$

$x_1 = -1.603603604$ $f(x_1) = -13.61361361$

$x_2 = -1.323252501$ $f(x_2) = -3.799819057$

$x_3 = -1.162229582$ $f(x_3) = -0.782974790$

$x_4 = -1.107866357$ $f(x_4) = -0.067490713$

$x_5 = -1.102228599$ $f(x_5) = -0.000663267$

$x_6 = -1.102172085$ $f(x_6) = -0.000000061$

$x_7 = -1.102172080$ $f(x_7) = -0.000000003$

2. *(cont'd)*

Since the numbers in the $f(x)$ column appear to be converging to zero and those in the x column are converging, we obtain a root to be (approximately) -1.10217208. Since $f(x)$ is negative at this value (where $f(x) = -0.000000003$) and positive at -1.10217207 (where $f(x) = 0.000000115$), we can conclude, by the intermediate value theorem, that the root is between these two values.

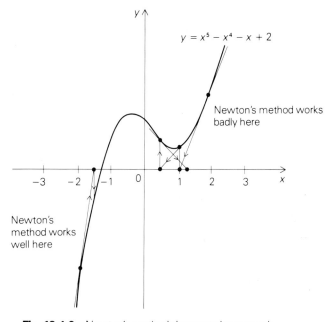

Fig. 12-4-6 Newton's method does not always work.

3. By formula (5), p. 531,

$$\int_0^1 f(x)\,dx \approx \tfrac{1}{30}[f(0) + 4f(0.1) + 2f(0.2) +$$

$$4f(0.3) + 2f(0.4) + 4f(0.5) +$$

$$2f(0.6) + 4f(0.7) + 2f(0.8) +$$

$$4f(0.9) + f(1.0)]$$

Inserting the given values of f and evaluating on a calculator, we get

$$\int_0^1 f(x)\,dx \approx \tfrac{1}{30}(49.042) = 1.63473$$

This result will be quite accurate unless the fourth derivative of f is very large.

12

4. Let $f(x) = e^{-x^2}$, $a = 2$, and $b = 4$. The error in the trapezoidal rule is $\leq [(b - a)/12] M_2 (\Delta x)^2$, where M_2 is the maximum of $|f''(x)|$ on $[a, b]$. We compute

$$f'(x) = -2xe^{-x^2}$$

$$f''(x) = -2e^{-x^2} + 4x^2 e^{-x^2} = 2(2x^2 - 1)e^{-x^2}$$

Now $f'''(x) = (12x - 8x^3) e^{-x^2} = 4x(3 - 2x^2)e^{-x^2} < 0$ on $[2, 4]$ so $f''(x)$ is decreasing. Also, $f''(x) > 0$ on $[2, 4]$, so $|f''(x)| = f''(x) \leq f''(2) = 2(2 \cdot 4 - 1)e^{-4} = M_2$, so the error is

$$\leq \frac{b - a}{12} M_2 (\Delta x)^2 = \frac{1}{6} \cdot 14e^{-4}(\Delta x)^2$$

If this is to be less than 10^{-6}, we should choose Δx so that

$$\tfrac{1}{6} \cdot 14e^{-4}(\Delta x)^2 < 10^{-6}$$

$$(\Delta x)^2 < e^4 10^{-6} \cdot \tfrac{3}{7} = 0.0000234$$

$$\Delta x < 0.0048$$

That is, we should take at least $n = (b - a)/\Delta x = 416$ divisions.

For Simpson's rule, the error is at most $[(b - a)/180] M_4 (\Delta x)^4$. Here

$$f'''(x) = (12x - 8x^3)e^{-x^2}$$

and

$$f''''(x) = 4(4x^4 - 12x^2 + 3)e^{-x^2}$$

On $[2, 4]$, we find that $4x^4 - 12x^2 + 3$ is increasing and e^{-x^2} is decreasing, so

$$|f''''(x)| \leq 4(4 \cdot 4^4 - 12 \cdot 4^2 + 3)e^{-4}$$
$$= 61.17 = M_4$$

Thus $[(b - a)/180] M_4 (\Delta x)^4 = \frac{1}{90} \cdot 61.17(\Delta x)^4 = 0.68(\Delta x)^4$. Hence if we are to have error less than 10^{-6}, it suffices to have

$$0.68(\Delta x)^4 \leq 10^{-6}$$

$$\Delta x \leq 0.035$$

Thus we should take at least $n = (b - a)/\Delta x = 57$ divisions. (If we had estimated f'' and f'''' more carefully, we would have obtained slightly smaller values of n.)

CHAPTER 13

Solved Exercises for Section 13-1

1. Although the number of time intervals being considered is infinite, the sum of their lengths is finite, so Achilles can overtake the tortoise in a finite time. The word *forever* in the sense of infinitely many terms is confused with "forever" in the sense of the time in the problem, resulting in the apparent paradox.

2. (a) $\sum_{n=0}^{\infty} 1/(6^{n/2}) = 1 + 1/\sqrt{6} + 1/\sqrt{6}^2 + \cdots = a/(1 - r)$, where $a = 1$ and $r = 1/\sqrt{6}$, so the sum is $1/(1 - 1/\sqrt{6}) = (6 + \sqrt{6})/5$. (Note that the index here is n instead of i.)

  (b) $\sum_{i=1}^{\infty} 1/5^i = 1/5 + 1/5^2 + \cdots = (1/5)/(1 - 1/5) = 1/4$. (We may also think of this as the series $\sum_{i=0}^{\infty} 1/5^i$ with the first term removed. The sum is thus $1/[1 - (1/5)] - 1 = 1/4$.)

3. The series is $\sum_{i=0}^{\infty} [2^i - (\tfrac{1}{2})^i]$. If it were convergent, we could add to it the convergent series $\sum_{i=0}^{\infty} (\tfrac{1}{2})^i$, and the result would have to converge by the sum rule. But the sum of the resulting series equals $\sum_{i=0}^{\infty} [2^i - (\tfrac{1}{2})^i + (\tfrac{1}{2})^i] = \sum_{i=0}^{\infty} 2^i$, which diverges because $2 > 1$, so the original series must itself be divergent.

4. Here $|a_i| = i/\sqrt{1 + i} = \sqrt{i}/\sqrt{1/i + 1} \to \infty$ as $i \to \infty$. Thus a_i does not tend to zero, so the series diverges.

5. This series is $\tfrac{1}{2} + \tfrac{1}{3} + \tfrac{1}{4} + \cdots$, which is the harmonic series with the first term missing; therefore this series diverges too.

6. The sum of the first three terms is

$$-\frac{1}{3^2 \cdot 1} + \frac{1}{3^3 \cdot 2} - \frac{1}{3^4 \cdot 3}$$

$$= -\frac{1}{9} + \frac{1}{54} - \frac{1}{243} = -\frac{47}{486} \approx -0.0967$$

To estimate

$$\left| \sum_{i=1}^{\infty} \frac{(-1)^i}{3^{i+1} i} - (-0.0967) \right|$$

we first observe that, for every series, $\sum_{i=1}^{\infty} a_i - \sum_{i=1}^{n} a_i = \sum_{i=n+1}^{\infty} a_i$. Thus we have

$$\left| \sum_{i=1}^{\infty} \frac{(-1)^i}{3^{i+1} i} - (-0.0967) \right| = \left| \sum_{i=4}^{\infty} \frac{(-1)^i}{3^{i+1} i} \right|$$

$$\leq \sum_{i=4}^{\infty} \frac{1}{3^{i+1} i} \quad \text{(by Theorem 2)}$$

$$\leq \sum_{i=4}^{\infty} \frac{1}{3^{i+1}} \quad \text{(since } i > 1\text{)}$$

$$= \frac{1}{3^5}\left(1 + \frac{1}{3} + \frac{1}{3^2} + \cdots\right)$$

$$= \frac{1}{3^5}\left(\frac{1}{1 - \frac{1}{3}}\right) = \frac{1}{3^5} \cdot \frac{3}{2} = \frac{1}{162} \approx 0.0062$$

Thus the error is no more than 0.0062. We may conclude that $\sum_{i=1}^{\infty} (-1)^i/(3^{i+1} i)$ lies somewhere in the interval $[-0.0967 - 0.0062, -0.0967 + 0.0062] = [-0.103, -0.090]$.

7. The term 2^i in the denominator "dominates" the term i for large i, so we apply Theorem 4 with $a_i = 1/(2^i - i)$ and $b_i = 1/2^i$. We have

$$\lim_{i \to \infty} \frac{a_i}{b_i} = \lim_{i \to \infty} \frac{1}{1 - i/2^i} = \frac{1}{1 - 0} = 1$$

($\lim_{i \to \infty} i/2^i = 0$ by L'Hôpital's rule). Since $\Sigma_{i=1}^{\infty} 1/(2^i)$ converges, so does $\Sigma_{i=1}^{\infty} 1/(2^i - i)$.

Solved Exercises for Section 13-2

1. An example is $1 - \frac{1}{2} + \frac{1}{3} - \frac{1}{4} + \cdots$, which is convergent since it is alternating (see Worked Example 1). The series of absolute values is $1 + \frac{1}{2} + \frac{1}{3} + \frac{1}{4} + \cdots$, which is the divergent harmonic series. A series which is convergent but not absolutely convergent is often called *conditionally convergent*.

2. The terms alternate in sign since $(-1)^i = 1$ if i is even and $(-1)^i = -1$ if i is odd. The absolute values, $1/(1 + i)^2$, are decreasing and converge to zero. Thus the series is alternating, so it converges.

3. The terms alternate in sign and tend to zero, but the absolute values are not monotonically decreasing. Thus the series is not an alternating one and Theorem 5 does not apply. If we group the terms by twos, we find that the series becomes

$(\frac{2}{2} - \frac{1}{2}) + (\frac{2}{3} - \frac{1}{3}) + (\frac{2}{4} - \frac{1}{4}) + (\frac{2}{5} - \frac{1}{5}) + \cdots$
$= \frac{1}{2} + \frac{1}{3} + \frac{1}{4} + \frac{1}{5} + \cdots$

which diverges. (Notice that the nth partial sum of the "grouped" series is the $2n$th partial sum of the original series.)

4. We consider the integral

$$\int_2^{\infty} \frac{1}{x(\ln x)^p} \, dx = \lim_{b \to \infty} \int_2^b (\ln x)^{-p} \frac{1}{x} \, dx$$
$$= \lim_{b \to \infty} \frac{(\ln x)^{-p+1}}{-p+1} \Big|_2^b$$
$$= \frac{1}{-p+1} \lim_{b \to \infty} [(\ln b)^{-p+1} - (\ln 2)^{-p+1}]$$

The limit is finite if $p = 2$ and infinite if $p = \frac{1}{2}$, so the integral converges if $p = 2$ and diverges if $p = \frac{1}{2}$. It follows that $\Sigma_{m=2}^{\infty} 1/(m\sqrt{\ln m})$ diverges and that $\Sigma_{m=2}^{\infty} 1/m(\ln m)^2$ converges.

5. (a) Let $a_j = (j^2 + 2j)/(j^4 - 3j^2 + 10)$ and $b_j = j^2/j^4 = 1/j^2$. Then

$$\lim_{j \to \infty} \frac{a_j}{b_j} = \lim_{j \to \infty} \frac{1 + 2/j}{1 - 3/j^2 + 10/j^4} = 1$$

Since $\Sigma_{j=1}^{\infty} b_j$ converges, so does $\Sigma_{j=1}^{\infty} a_j$.

5. (cont'd)

(b) Take $a_n = (3n + \sqrt{n})/(2n^{3/2} + 2)$ and $b_n = n/(n^{3/2}) = 1/\sqrt{n}$. Then

$$\lim_{n \to \infty} \frac{3 + (1/\sqrt{n})}{2 + (2/n^{3/2})} = \frac{3}{2}$$

Since $\Sigma_{n=1}^{\infty} b_n$ diverges, so does $\Sigma_{n=1}^{\infty} a_n$.

6. (a) Just as in the proof of formula (2) we have

$$\sum_{n=N+1}^{\infty} \frac{1}{n^p} \leq \int_N^{\infty} \frac{1}{x^p} \, dx$$

The left side is the error:

$$\sum_{n=N+1}^{\infty} \frac{1}{n^p} = \sum_{n=1}^{\infty} \frac{1}{n^p} - \sum_{n=1}^{N} \frac{1}{n^p}$$

Thus the error is no greater than

$$\int_N^{\infty} \frac{1}{x^p} \, dx = \frac{1}{(p-1)N^{p-1}}$$

(b) By part (a) the error in stopping at N terms is at most $1/N$. To have error $< 0.05 = \frac{1}{20}$, we must take 20 terms (note that 100 terms are needed to get two decimal places!). We find:

$$1 = 1$$
$$1 + \tfrac{1}{4} = 1.25$$
$$1 + \tfrac{1}{4} + \tfrac{1}{9} = 1.36$$
$$1 + \tfrac{1}{4} + \tfrac{1}{9} + \tfrac{1}{16} = 1.42$$
$$1 + \tfrac{1}{4} + \tfrac{1}{9} + \tfrac{1}{16} + \tfrac{1}{25} = 1.46$$

and so forth, obtaining 1.49, 1.51, 1.53, 1.54, 1.55, 1.56, ... Finally, $1 + \frac{1}{4} + \cdots + \frac{1}{400} = 1.596 \ldots$. (Notice the "slowness" of the convergence.) We may compare this with the exact value $\pi^2/6 = 1.6449 \ldots$.

7. (a) Here $a_n = 1/n!$, so

$$\frac{a_n}{a_{n-1}} = \frac{1/n(n-1) \cdots 3 \cdot 2 \cdot 1}{1/(n-1)(n-2) \cdots 3 \cdot 2 \cdot 1} = \frac{1}{n}$$

Thus $|a_n/a_{n-1}| = 1/n \to 0 < 1$, so we have convergence.

(b) Here $a_j = b^j/j!$, so

$$\frac{a_j}{a_{j-1}} = \frac{b^j/j!}{b^{j-1}/(j-1)!} = \frac{b}{j}$$

Thus $|a_j/a_{j-1}| = b/j \to 0$, so we have convergence. In this example, note that the numerator b^j and the denominator $j!$ tend to infinity, but the denominator does so much faster. In fact, since the series converges, $b^j/j! \to 0$ as $j \to \infty$.

13

8. (a) $\Sigma_{n=1}^{\infty} a_n - \Sigma_{n=1}^{N} a_n = \Sigma_{n=N+1}^{\infty} a_n$. As in the proof of the ratio test, $|a_{N+1}| < |a_N| r$, and in general $|a_{N+j}| < |a_N| r^j$, so $\Sigma_{j=1}^{\infty} |a_{N+j}| \leq |a_N| r/(1-r)$ by the formula for the sum of a geometric series and the comparison test. Hence the error is no greater than $|a_N|/(1-r)$.

   (b) Here $|a_n/a_{n-1}| = 1/n$, which is $< \frac{1}{5}$ if $n > 4 = N$. Hence the error is no more than $a_4/5(1 - \frac{1}{5}) = 1/4 \cdot 4! = 1/96 < 0.0105$. The error becomes small very quickly if N is increased.

9. (a) Let $l = \lim_{n\to\infty} (|a_n|^{1/n})$ and let $r = (1+l)/2$ be the midpoint of 1 and l, so $l < r < 1$. From the definition of limit, there is an N such that $(|a_n|)^{1/n} < r < 1$ if $n > N$. Hence

$$|a_n| < r^n \quad \text{if } n > N$$

   Thus, by direct comparison of $\Sigma_{n=N+1}^{\infty} |a_n|$ with the geometric series $\Sigma_{n=N+1}^{\infty} r^n$, which converges since $r < 1$, $\Sigma_{n=N+1}^{\infty} |a_n|$ converges. Since we have neglected only finitely many terms, the given series converges.

   (b) Here $a_n = 1/n^n$, so $|a_n|^{1/n} = 1/n$ converges to $0 < 1$, as $n \to \infty$; by part (a), the series converges.

Solved Exercises for Section 13-3

1. (a) We have $a_i = 1/i$, so

$$r = \lim_{i\to\infty} \left| \frac{a_i}{a_{i-1}} \right| = \lim_{i\to\infty} \left(\frac{i-1}{i} \right) = 1;$$

   the series converges for $|x| < 1$ and diverges for $|x| > 1$. When $x = 1$, $\Sigma_{i=1}^{\infty} x^i/i$ is the divergent harmonic series; for $x = -1$, the series is alternating, so it converges.

   (b) We have $a_i = 1/i^2$, so

$$r = \lim_{i\to\infty} (i-1)^2/i^2 = 1$$

   and the radius of convergence is again 1. This time, when $x = 1$ we get the p series $\Sigma_{i=1}^{\infty} 1/i^2$, which converges since $p = 2 > 1$. The series for $x = -1$, $\Sigma_{i=1}^{\infty} (-1)^i/i^2$, is also convergent.

2. Here $a_i = 1/i!$, so $|a_i/a_{i-1}| = (i-1)!/i! = 1/i \to 0$ as $i \to \infty$. Thus $r = 0$, so the series converges for all x.

3. (a) $|a_i x^i|^{1/i} = |a_i|^{1/i} |x| \to \rho |x| < 1$ if $|x| < 1/\rho$. Hence, by the root test, $\Sigma_{i=1}^{\infty} a_i x^i$ converges.

   (b) For $\Sigma_{i=1}^{\infty} x^i/(2 + 1/i)^i$, we can use part (a), $\rho = \lim_{i\to\infty} |a_i|^{1/i} = \lim_{i\to\infty} (1/(2 + 1/i)^i)^{1/i} = \lim_{i\to\infty} \{1/[2 + (1/i)]\} = \frac{1}{2}$, so the series converges if $|x| < 2$. The same answer may be obtained from Theorem 8.

4. (a) $1/(1-x) = 1 + x + x^2 + \cdots$, which converges for $|x| < 1$. Thus $\int dx/(1-x) = (x + x^2/2 + x^3/3 + \cdots) + C$, also valid for $|x| < 1$ by Theorem 9.

   (b) Since $1/(1-x) = 1 + x + x^2 + \cdots$ if $|x| < 1$, $1/x = 1/[1 - (1-x)] = 1 + (1-x) + (1-x)^2 + \cdots$ if $|1-x| < 1$. Thus $\ln x = \int dx/x = -(1-x) - (1-x)^2/2 - (1-x)^3/3 - \cdots$ if $|1-x| < 1$. (The constant of integration is chosen to make $\ln 1 = 0$.)

5. Since $f(x) = \Sigma_{i=0}^{\infty} x^i/i!$, we get

$$r = \lim_{i\to\infty} \frac{(i-1)!}{i!} = \lim_{i\to\infty} \frac{1}{i} = 0$$

   so f converges for all x. Then $f'(x) = \Sigma_{i=1}^{\infty} i x^{i-1}/i! = \Sigma_{i=1}^{\infty} x^{i-1}/(i-1)! = \Sigma_{i=0}^{\infty} x^i/i! = f(x)$. By the uniqueness of the solution of the differential equation $f'(x) = f(x)$ (see p. 379), $f(x)$ must be ce^x for some c. Since $f(0) = 1$, c must be 1, and so $f(x) = e^x$.

6. We get

$$\sin^2 x = \left(x - \frac{x^3}{3!} + \frac{x^5}{5!} - \cdots \right) \left(x - \frac{x^3}{3!} + \frac{x^5}{5!} - \cdots \right)$$

$$= x^2 - \frac{x^4}{3} + \left(\frac{2}{5!} + \frac{1}{(3!)^2} \right) x^6 - \cdots$$

$$= x^2 - \frac{x^4}{3} + \frac{2}{45} x^6 - \cdots$$

7. Let

$$f(x) = 1 + x + x^2 + \cdots = \frac{1}{1-x}$$

$$g(x) = -x^2 - x^3 - \cdots = \frac{-x^2}{1-x}$$

   The radius of convergence of $f(x)$ and $g(x)$ are both 1, but $f(x) + g(x) = 1 + x$; the radius of convergence of this series is ∞.

8. We differentiate $f(x)$ three times:

$$f(x) = \frac{1}{1+x^2} \qquad f(1) = \frac{1}{2} \qquad a_0 = f(1) = \frac{1}{2}$$

$$f'(x) = \frac{-2x}{(1+x^2)^2} \qquad f'(1) = -\frac{1}{2} \qquad a_1 = f'(1) = -\frac{1}{2}$$

$$f''(x) = \frac{6x^2 - 2}{(1+x^2)^3} \qquad f''(1) = \frac{1}{2} \qquad a_2 = \frac{f''(1)}{2!} = \frac{1}{4}$$

$$f'''(x) = \frac{-24x^3 + 24x}{(1+x^2)^4} \qquad f'''(1) = 0 \qquad a_3 = \frac{f'''(1)}{3!} = 0$$

so the Taylor series begins

$$\frac{1}{1+x^2} = \frac{1}{2} - \frac{1}{2}(x-1) + \frac{1}{4}(x-1)^2$$
$$+ 0 \cdot (x-1)^3 + \cdots$$

9. (a) We expand $1/(1+x^2)$ as a geometric series:

$$\frac{1}{1+x^2} = \frac{1}{1-(-x^2)} = 1 + (-x^2) + (-x^2)^2$$
$$+ (-x^2)^3 + \cdots = 1 - x^2 + x^4 - x^6 + \cdots$$

which is valid if $|-x^2| < 1$; that is, if $|x| <$ 1. By Theorem 12, this is the Maclaurin series of $f(x) = 1/(1+x^2)$.

(b) We find that $f'''''(0)/5!$ is the coefficient of x^5. Hence, as this coefficient is zero, $f'''''(0) = 0$. Likewise, $f''''''(0)/6!$ is the coefficient of x^6; thus $f''''''(0) = -6!$. This is *much* easier than calculating the sixth derivative of $f(x)$.

(c) Integrating from zero to x (justified by Theorem 9) gives

$$\int_0^x \frac{dt}{1+t^2} = x - \frac{x^3}{3} + \frac{x^5}{5} - \frac{x^7}{7} + \cdots$$

But we know that the integral of $1/(1+t^2)$ is $\tan^{-1} t$, so

$$\tan^{-1} x = x - \frac{x^3}{3} + \frac{x^5}{5} - \frac{x^7}{7} + \cdots \text{ for } |x| < 1.$$

(d) If we set $x = 1$, we get, since $\tan^{-1} 1 = \pi/4$, Euler's formula:

$$\frac{\pi}{4} = 1 - \frac{1}{3} + \frac{1}{5} - \frac{1}{7} + \cdots$$

But this is not quite justified, since the series for $\tan^{-1} x$ is valid only for $|x| < 1$. (It is plausible, though, since $1 - \frac{1}{3} + \frac{1}{5} - \frac{1}{7}$, being an alternating series, converges.) To justify Euler's formula, we may use the finite form of the geometric series expansion:

9. (cont'd)

$$\frac{1}{1+t^2} = 1 - t^2 + t^4 + \cdots + (-1)^n t^{2n}$$
$$+ (-1)^{n+1} \frac{t^{2n+2}}{1+t^2}$$

Integrating from 0 to 1, we have

$$\frac{\pi}{4} = \tan^{-1} 1 = 1 - \frac{1}{3} + \frac{1}{5} - \cdots + \frac{(-1)^n}{2n+1}$$
$$+ (-1)^{n+1} \int_0^1 \frac{t^{2n+2}}{1+t^2} dt$$

We will be done if we can show that the last term goes to zero as $n \to \infty$. We have

$$0 \le \int_0^1 \frac{t^{2n+2}}{1+t^2} dt \le \int_0^1 t^{2n+2} dt = \frac{1}{2n+3}$$

Since $\lim_{n\to\infty} 1/(2n+3) = 0$, the limit of

$$(-1)^{n+1} \int_0^1 \frac{t^{n+2}}{1+t^2} dt$$

is zero as well.

10. The binomial theorem, with $\alpha = \frac{1}{2}$ and x^2 in place of x, gives

$$(1+x^2)^{1/2} = 1 + \frac{1}{2}x^2 + \frac{\frac{1}{2}(\frac{1}{2}-1)}{2!}x^4$$
$$+ \frac{\frac{1}{2}(\frac{1}{2}-1)(\frac{1}{2}-2)}{3!}x^6 + \cdots$$

$$= 1 + \frac{1}{2}x^2 - \frac{1}{8}x^4 + \frac{1}{16}x^6 - \cdots$$

valid for $|x| < 1$.

11. Substituting $-x^2$ for x in the series for e^x gives

$$e^{-x^2} = 1 - x^2 + \frac{x^4}{2!} - \frac{x^6}{3!} + \frac{x^8}{4!} - \cdots$$

Integrating term by term gives

$$\int_0^x e^{-t^2} dt = x - \frac{x^3}{3} + \frac{x^5}{10} - \frac{x^7}{42} + \cdots$$

and so

$$\int_0^1 e^{-x^2} dx = 1 - \frac{1}{3} + \frac{1}{10} - \frac{1}{42} + \frac{1}{216} - \frac{1}{1320} + \cdots$$

This is an alternating series, so the error is no greater than the first omitted term. To have accuracy 0.001 we should include $\frac{1}{216}$. Thus within 0.001,

$$\int_0^1 e^{-x^2} dx \approx 1 - \frac{1}{3} + \frac{1}{10} - \frac{1}{42} + \frac{1}{216} \approx 0.747$$

13

11. (*cont'd*)

This method has an advantage over the methods in Section 12-4: to increase accuracy, we need only add on another term. Rules like Simpson's, on the other hand, require us to start over. Of course, if we have numerical data, or a function with an unknown or complicated series, using Simpson's rule may be necessary. (See also Review Problem 25 for Chapter 12.)

12. With $f(x) = \sin x$, $x_0 = \pi/4$, we have

$$f(x) = \sin x \qquad f(x_0) = \frac{1}{\sqrt{2}}$$

$$f'(x) = \cos x \qquad f'(x_0) = \frac{1}{\sqrt{2}}$$

$$f''(x) = -\sin x \qquad f''(x_0) = -\frac{1}{\sqrt{2}}$$

$$f'''(x) = -\cos x \qquad f'''(x_0) = -\frac{1}{\sqrt{2}}$$

$$f''''(x) = \sin x \qquad f''''(x_0) = \frac{1}{\sqrt{2}}$$

and so on. We have

$$R_n(x) = \frac{f^{(n+1)}(c)(x - x_0)^{n+1}}{(n+1)!}$$

for c between $\pi/4$ and $\pi/4 + 0.06$. Since $f^{(n+1)}(c)$ has absolute value less than 1, we have $|R_n(x)| \leq (0.06)^{n+1}/(n+1)!$. To make $|R_n(x)|$ less than 0.0001, it suffices to choose $n = 2$. The second-order approximation to $\sin x$ is

$$\frac{1}{\sqrt{2}} + \frac{1}{\sqrt{2}}\left(x - \frac{\pi}{4}\right) - \frac{1}{2\sqrt{2}}\left(x - \frac{\pi}{4}\right)^2$$

Evaluating at $x = \pi/4 + 0.06$ gives 0.7483.

If we had used the Maclaurin polynomial of degree n, the error estimate would have been $|R_n(x)| \leq (\pi/4 + 0.06)^{n+1}/(n+1)!$. To make $|R_n(x)|$ less than 0.0001 would have required $n = 6$.

CHAPTER 14

Solved Exercises for Section 14-1

1. (a) $a_1(3, 1) + a_2(6, 2) + a_3(-1, 1) = (3a_1 + 6a_2 - a_3, a_1 + 2a_2 + a_3)$; for this to equal $(5, 6)$, we must solve the equations

    $$3a_1 + 6a_2 - a_3 = 5$$
    $$a_1 + 2a_2 + a_3 = 6$$

    A solution of these equations is $a_1 = 0$, $a_2 = \frac{11}{8}$, $a_3 = \frac{13}{4}$. (Part (b) explains where this solution came from.)

    (b) We can rewrite the equations as

    $$6a_2 - a_3 = 5 - 3a_1$$

    and

    $$2a_2 + a_3 = 6 - a_1$$

    We may choose a_1 at will, obtaining a pair of linear equations in a_2 and a_3 which always have a solution, since the lines in the (a_2, a_3) plane which they represent have different slopes—that is, are not parallel and so intersect. The choice $a_1 = 0$ led to the equations $6a_2 - a_3 = 5$ and $2a_2 + a_3 = 6$, with the solution as given in part (a). The choice $a_1 = 6$, for instance, leads to the new solution $a_1 = 6$, $a_2 = -\frac{13}{8}$, $a_3 = \frac{13}{4}$, so the solution is not unique.

    (c) We notice that the sums $3 + 1 = 4$, $6 + 2 = 8$, and $-1 + 1 = 0$ of the components in each of the ordered pairs are even. Thus the sum $4a_1 + 8a_2$ of the components of $a_1(3, 1) + a_2(6, 2) + a_3(-1, 1)$ is even if a_1, a_2, and a_3 are integers. But $5 + 6 = 11$ is odd, so there is no solution with a_1, a_2, a_3 integers.

2. The pair (x, y) equals $(x, 0) + (0, y)$, which is equal to $x(1, 0) + y(0, 1)$.

3. We think of the molecule $N_x H_y$ (x atoms of nitrogen, y atoms of hydrogen) as represented by the ordered pair (x, y). Then the chemical equation given is equivalent to $2(1, 2) + (0, 2) = 2(1, 3)$. Indeed, both sides are equal to $(2, 6)$.

4. The solution is contained in Worked Example 1(c). If the coordinates of P are (x_1, y_1) and those of Q are (x_2, y_2), then the coordinates of the point $P - Q$ are $(x_1 - x_2, y_1 - y_2)$.

5. The vector $\mathbf{v}$ corresponds to the ordered pair $(2, -3) - (-1, 0) = (3, -3)$; $\mathbf{w}$ corresponds to $(1, 1) - (2, 0) = (-1, 1)$, so $\mathbf{v} + \mathbf{w}$ corresponds to $(3, -3) + (-1, 1) = (2, -2)$.

6. The vector $8\mathbf{v}$ corresponds to $8(3, -3) = (24, -24)$. If this vector is represented by the directed line segment from $(5, 6)$ to Q, $Q - (5, 6) = (24, -24)$, or $Q = (5, 6) + (24, -24) = (29, -18)$.

7. The vector **v** − **w** is such that (**v** − **w**) + **w** = **v**. It follows from the "triangle" construction of vector sums that if we draw **v** and **w** with a common tail, **v** − **w** is represented by the directed line segment from the head of **w** to the head of **v**. (See Fig. 14-1-28.)

Fig. 14-1-28 Geometric interpretation of vector subtraction.

8. We wish to reverse the process by which points are plotted if their coordinates are given. Thus the first thing we wish to do is to drop a perpendicular from P to the xy plane. This line will be parallel to the z axis, hence vertical in the figure, but we have no way of telling from Fig. 14-1-21 where this line will intersect the plane. The point P might be in the plane, above it, or below it. In fact, we may choose any point on the vertical line through P to be the point of intersection with the xy plane. We will make two very different choices, which will lead to two sets of coordinates for P (and two three-dimensional "interpretations" of Fig. 14-1-21).

In Fig. 14-1-29 we show the two choices, together with lines in the xy plane which are parallel to the axes. In Fig. 14-1-29(a), the point P lies below the xy plane; its coordinates are $(3, 7, -2)$. In Fig. 14-1-29(b), the point P lies above the xy plane; its coordinates are $(10, 12, 3)$.

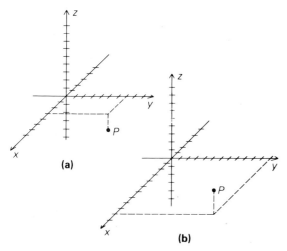

Fig. 14-1-29 Two different spatial "interpretations" of the point P.

8. (*cont'd*)

If you did this problem on your own, you will probably have obtained still other coordinates. Whatever they are, they should satisfy (approximately, since there will be errors in measurement) the equations

$$\frac{x-3}{7} = \frac{y-7}{5} = \frac{z+2}{5}$$

You can find an explanation for these equations in Section 14-2.

9. The vector $-2\mathbf{v}$ is twice as long as **v** but points in the opposite direction. See Fig. 14-1-30.

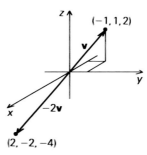

Fig. 14-1-30 Multiplying $(-1, 1, 2)$ by -2.

10. $\mathbf{v} - \frac{1}{3}\mathbf{w} = \frac{1}{3}(3\mathbf{v} - \mathbf{w})$. But vectors which are multiples of one another are parallel.

11. $\mathbf{v} = e\mathbf{i} + \pi\mathbf{j} - \sqrt{3}\,\mathbf{k}$.

12. (a) The displacement vector from $(1, 2, 3)$ is $\frac{1}{60}(10\mathbf{i}, 6\mathbf{j}, \mathbf{k}) = \frac{1}{6}\mathbf{i} + \frac{1}{10}\mathbf{j} + \frac{1}{60}\mathbf{k}$, so the new position is $(1, 2, 3) + (\frac{1}{6}, \frac{1}{10}, \frac{1}{60}) = (1\frac{1}{6}, 2\frac{1}{10}, 3\frac{1}{60})$.

(b) After t seconds ($= t/3600$ hours), the displacement vector is $(t/3600)(10\mathbf{i} + 6\mathbf{j} + \mathbf{k}) = (t/360)\mathbf{i} + (t/600)\mathbf{j} + (t/3600)\mathbf{k}$. The increase in altitude is the z component $t/3600$. This will equal 10 meters ($= \frac{1}{100}$ kilometer) when $t/3600 = 1/100$—that is, when $t = 36$ seconds.

13. The force **v** should be chosen so that $(\mathbf{i} + \mathbf{k}) + (\mathbf{j} + \mathbf{k}) + \mathbf{v} = 0$; that is, $\mathbf{v} = -(\mathbf{i} + \mathbf{k}) - (\mathbf{j} + \mathbf{k}) = -\mathbf{i} - \mathbf{j} - 2\mathbf{k}$.

14

Solved Exercises for Section 14-2

1. Refer to Fig. 14-2-11. The vector joining O to P is $\mathbf{j}$, and the vector joining O to Q is $\mathbf{i} + \mathbf{j} + \mathbf{k}$. The vector joining P to Q is the difference $\mathbf{i} + \mathbf{j} + \mathbf{k} - \mathbf{j} = \mathbf{i} + \mathbf{k}$; the vector joining P to the midpoint R of PQ (and hence of the face $PSQT$) is one-half of this, that is, $\frac{1}{2}(\mathbf{i} + \mathbf{k})$; the vector joining O to R is then $\mathbf{j} + \frac{1}{2}(\mathbf{i} + \mathbf{k})$, and the vector joining O to the point U one-third of the way from O to R is $\frac{1}{3}[\mathbf{j} + \frac{1}{2}(\mathbf{i} + \mathbf{k})] = \frac{1}{3}\mathbf{j} + \frac{1}{6}\mathbf{i} + \frac{1}{6}\mathbf{k} = \frac{1}{6}\mathbf{i} + \frac{1}{3}\mathbf{j} + \frac{1}{6}\mathbf{k}$. It follows that the coordinates of U are $(\frac{1}{6}, \frac{1}{3}, \frac{1}{6})$.

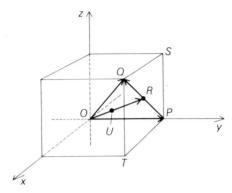

Fig. 14-2-11 The point U is one-third of the way from O to the midpoint of the face $PSQT$.

2. Refer to Fig. 14-2-12. Let $PQRS$ be the quadrilateral, $\mathbf{v}$ the vector from P to S, $\mathbf{w}$ from S to R, $\mathbf{t}$ from P to Q, and $\mathbf{u}$ from Q to R. The vector $\mathbf{a}$ from the midpoint J of PS to the midpoint K of SR satisfies $\frac{1}{2}\mathbf{v} + \mathbf{a} = \mathbf{v} + \frac{1}{2}\mathbf{w}$; solving for $\mathbf{a}$ gives $\mathbf{a} = \frac{1}{2}\mathbf{v} + \frac{1}{2}\mathbf{v} = \frac{1}{2}(\mathbf{v} + \mathbf{w})$. Similarly, the vector $\mathbf{b}$ from the midpoint M of PQ to the midpoint L of QR satisfies $\frac{1}{2}\mathbf{t} + \mathbf{b} = \mathbf{t} + \frac{1}{2}\mathbf{u}$, so $\mathbf{b} = \frac{1}{2}\mathbf{t} + \frac{1}{2}\mathbf{u} = \frac{1}{2}(\mathbf{t} + \mathbf{u})$.

   To show that $JKLM$ is a parallelogram, it suffices to show that the vectors $\mathbf{a}$ and $\mathbf{b}$ are equal. But $\mathbf{v} + \mathbf{w} = \mathbf{t} + \mathbf{u}$, since both sides are equal to the vector from P to R, so

$$\mathbf{a} = \tfrac{1}{2}(\mathbf{v} + \mathbf{w}) = \tfrac{1}{2}(\mathbf{t} + \mathbf{u}) = \mathbf{b}$$

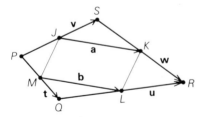

Fig. 14-2-12 The figure obtained by joining the midpoints of successive sides of $PQRS$ is the parallelogram $JKLM$.

3. If the lines intersect, there must be numbers t_1 and t_2 such that $(t_1, -6t_1 + 1, 2t_1 - 8) = (3t_2 + 1, 2t_2, 0)$; that is,

$$t_1 = 3t_2 + 1$$
$$-6t_1 + 1 = 2t_2$$
$$2t_1 - 8 = 0$$

From the third equation we have $t_1 = 4$. The first equation then becomes $4 = 3t_2 + 1$, or $t_2 = 1$. We must check whether these values satisfy the middle equation:

$$-6t_1 + 1 \stackrel{?}{=} 2t_2$$
$$-6 \cdot 4 + 1 \stackrel{?}{=} 2 \cdot 1$$
$$-24 + 1 \stackrel{?}{=} 2$$

The answer is no; the lines do not intersect.

4. The *line* through $(1, 1, 1)$ and $(2, 1, 2)$ is given by $R = (1 - t)(1, 1, 1) + t(2, 1, 2) = (1 + t, 1, 1 + t)$, as t takes on all real values. The point R lies between $(1, 1, 1)$ and $(2, 1, 2)$ only when $0 \le t \le 1$, so the line *segment* is described by $R = (1 + t, 1, 1 + t), 0 \le t \le 1$.

5. We have $\|\mathbf{i} + \mathbf{j} + c\mathbf{k}\| = \sqrt{1 + 1 + c^2} = \sqrt{2 + c^2}$. This equals 4 when $2 + c^2 = 16$, or $c = \pm\sqrt{14}$.

6. We find that $\mathbf{u} - \mathbf{v} = (\mathbf{u} - \mathbf{w}) + (\mathbf{w} - \mathbf{v})$, so the result follows from the triangle inequality with $\mathbf{u}$ replaced by $\mathbf{u} - \mathbf{w}$ and $\mathbf{v}$ replaced by $\mathbf{w} - \mathbf{v}$. Geometrically, we are considering the shaded triangle in Fig. 14-2-13.

Fig. 14-2-13 Illustrating the inequality $\|\mathbf{u} - \mathbf{v}\| \le \|\mathbf{u} - \mathbf{w}\| + \|\mathbf{w} - \mathbf{v}\|$

7. The diagram illustrating the sum $\mathbf{u} + \mathbf{v} = \mathbf{w}$ must be an equilateral triangle, as in Fig. 14-2-14. Knowing this, we may take $\mathbf{w} = \mathbf{i}$, $\mathbf{u} = \frac{1}{2}\mathbf{i} + (\sqrt{3}/2)\mathbf{j}$, $\mathbf{v} = \frac{1}{2}\mathbf{i} - (\sqrt{3}/2)\mathbf{j}$. Check that $\|\mathbf{u}\| = \|\mathbf{v}\| = \|\mathbf{w}\| = 1$ and that $\mathbf{u} + \mathbf{v} = \mathbf{w}$.

Fig. 14-2-14 The vectors $\mathbf{u}$, $\mathbf{v}$, and $\mathbf{w}$ are represented by the sides of an equilateral triangle.

8. (a) By the distance formula, the distance is

$$\sqrt{(t-3)^2 + (t-0)^2 + (t-0)^2}$$
$$= \sqrt{t^2 - 6t + 9 + t^2 + t^2} = \sqrt{3t^2 - 6t + 9}$$

(b) The distance is shortest when $3t^2 - 6t + 9$ is least—when $(d/dt)(3t^2 - 6t + 9) = 6t - 6 = 0$; that is, when $t = 1$.

(c) For $t = 1$, the distance is $\sqrt{6}$.

9. If $\mathbf{v} = a\mathbf{i} + b\mathbf{j} + c\mathbf{k}$, both sides equal $a^2 + b^2 + c^2$. Alternatively, use formula (3) with $\theta = 0$.

10. By Worked Example 10, the inner product of these two vectors is zero; therefore they are perpendicular, that is, $\theta = \pi/2$.

11. With $\mathbf{u} = \mathbf{i} - 2\mathbf{j}$ and $\mathbf{v} = \mathbf{i} + \mathbf{j}$, the orthogonal projection of $\mathbf{v}$ on $\mathbf{u}$ is

$$\frac{\mathbf{u} \cdot \mathbf{v}}{\mathbf{u} \cdot \mathbf{u}} \mathbf{u} = \frac{1-2}{1+4}(\mathbf{i} - 2\mathbf{j}) = -\frac{1}{5}(\mathbf{i} - 2\mathbf{j})$$

(See Fig. 14-2-15.)

Orthogonal projection of **v** on **u**

Fig. 14-2-15 The orthogonal projection of **v** on **u** equals $-\frac{1}{5}\mathbf{u}$.

12. The line has parametric equations $x = t$, $y = t$, $z = 2t$. It meets the plane when $x + y + 2z = t + t + 4t = 5$; that is, when $t = \frac{5}{6}$. The point of intersection is $P_1 = (\frac{5}{6}, \frac{5}{6}, \frac{5}{3})$.

   Since a normal to the plane is $\mathbf{n} = \mathbf{i} + \mathbf{j} + 2\mathbf{k}$, which is the same as the direction vector of this line, we see that the line is perpendicular to the plane at P_1. If P is also in the plane, consideration of the right triangle OP_1P shows that $\overrightarrow{OP_1}$ must be shorter than $\overrightarrow{OP}$ (see Fig. 14-2-16). Thus the distance from the origin to the plane is the length of $\overrightarrow{OP_1}$:

$$\sqrt{\frac{25}{36} + \frac{25}{36} + \frac{25}{9}} = \frac{\sqrt{150}}{6} = \frac{5\sqrt{6}}{6}$$

12. (*cont'd*)

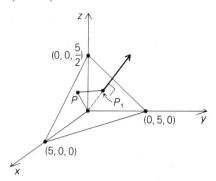

Fig. 14-2-16 The vector from O to the closest point P_1 on a plane is perpendicular to the plane.

13. The general equation of a plane is of the form $Ax + By + Cz + D = 0$. Since the points $(1, 1, 1)$, $(2, 0, 0)$, and $(1, 1, 0)$ lie on the plane, the coefficients A, B, C, D satisfy the three equations:

$$A + B + C + D = 0$$
$$2A \qquad\quad + D = 0$$
$$A + B \qquad + D = 0$$

Proceeding by elimination, we reduce this system to the form

$2A + D = 0$ (second equation)

$2B + D = 0$ (twice the third equation minus the second)

$C = 0$ (first equation minus the third)

Since the numbers A, B, C, and D are determined only up to a common factor, we can fix the value of one of them and then the others will be determined uniquely. If we let $D = -2$, then $A = 1$, $B = 1$, $C = 0$. Thus $x + y - 2 = 0$ is an equation of the plane that contains the given points. (You may go back and verify that the given points actually satisfy this equation.)

14

Solved Exercises for Section 14-3

1. We find that $\mathbf{i} \times (\mathbf{i} \times \mathbf{j}) = \mathbf{i} \times \mathbf{k} = -\mathbf{j}$, while $(\mathbf{i} \times \mathbf{i}) \times \mathbf{j} = \mathbf{0} \times \mathbf{j} = \mathbf{0}$, so the two expressions are *not* equal. This example means that the cross-product is *not associative*—one cannot move parentheses as in ordinary multiplication.

2. The normal to the plane is perpendicular to any vector which joins two points in the plane, so it is perpendicular to $\mathbf{v}_1 = (1,1,1) - (2,0,0) = (-1,1,1) = -\mathbf{i} + \mathbf{j} + \mathbf{k}$ and $\mathbf{v}_2 = (1,1,1) - (1,1,0) = (0,0,1) = \mathbf{k}$. A vector perpendicular to $\mathbf{v}_1$ and $\mathbf{v}_2$ is $\mathbf{v}_1 \times \mathbf{v}_2 = (-\mathbf{i} + \mathbf{j} + \mathbf{k}) \times \mathbf{k} = -\mathbf{i} \times \mathbf{k} + \mathbf{j} \times \mathbf{k} + \mathbf{k} \times \mathbf{k} = \mathbf{j} + \mathbf{i} + 0 = \mathbf{i} + \mathbf{j}$, so the equation of the plane has the form $x + y + D = 0$. Since $(1,1,1)$ lies in the plane, $2 + D = 0$, and the equation is $x + y - 2 = 0$. (In Solved Exercise 13 of Section 14-2, we got this result by solving a system of simultaneous equations. Here the cross-product does the solving for us.)

3. Let $\mathbf{v}_i = a_i \mathbf{i} + b_i \mathbf{j} + c_i \mathbf{k}$ $(i = 1, 2, 3)$. Then

$$\mathbf{v}_2 + \mathbf{v}_3 = (a_2 + a_3)\mathbf{i} + (b_2 + b_3)\mathbf{j} + (c_2 + c_3)\mathbf{k}$$

and

$$\mathbf{v}_1 \times (\mathbf{v}_2 + \mathbf{v}_3) = [b_1(c_2 + c_3) - c_1(b_2 + b_3)]\mathbf{i} +$$
$$[c_1(a_2 + a_3) - a_1(c_2 + c_3)]\mathbf{j} +$$
$$[a_1(b_2 + b_3) - b_1(a_2 + a_3)]\mathbf{k}$$
$$= (b_1 c_2 - c_1 b_2)\mathbf{i} + (c_1 a_2 - a_1 c_2)\mathbf{j} + (a_1 b_2 - b_1 a_2)\mathbf{k}$$
$$+ (b_1 c_3 - c_1 b_3)\mathbf{i} + (c_1 a_3 - a_1 c_3)\mathbf{j} + (a_1 b_3 - a_3 b_1)\mathbf{k}$$
$$= \mathbf{v}_1 \times \mathbf{v}_2 + \mathbf{v}_1 \times \mathbf{v}_3$$

4.
$$\begin{vmatrix} -\pi & 3 \\ \dfrac{\pi}{2} & 6 \end{vmatrix} = -6\pi - 3\pi/2 = -15\pi/2.$$

5.
$$\begin{vmatrix} a & b \\ c & d \end{vmatrix} = ad - bc \text{ and}$$

$$-\begin{vmatrix} b & a \\ d & c \end{vmatrix} = -(bc - ad) = ad - bc.$$

6. The volume is the absolute value of

$$\begin{vmatrix} 1 & 0 & 3 \\ 2 & 1 & -2 \\ 5 & 0 & 4 \end{vmatrix}$$

If we expand this by minors of the second column, the only nonzero term is

$$\begin{vmatrix} 1 & 3 \\ 5 & 4 \end{vmatrix} (1) = -11$$

so the volume equals 11.

7. Let P be the plane spanned by $\mathbf{v}_1$ and $\mathbf{v}_2$. (We assume that $\mathbf{v}_1$ and $\mathbf{v}_2$ are not parallel; otherwise $\mathbf{v}_1 \times \mathbf{v}_2 = \mathbf{0}$, and the determinant is zero.) Then $(\mathbf{v}_1, \mathbf{v}_2, \mathbf{v}_3)$ is right-handed [left-handed] if and only if $\mathbf{v}_3$ lies on the same [opposite] side of P as $\mathbf{v}_1 \times \mathbf{v}_2$; that is, if and only if $(\mathbf{v}_1 \times \mathbf{v}_2) \cdot \mathbf{v}_3$ is positive [negative]. But $(\mathbf{v}_1 \times \mathbf{v}_2) \cdot \mathbf{v}_3$ is the following determinant of

7. (*cont'd*)

$$\begin{vmatrix} a_1 & b_1 & c_1 \\ a_2 & b_2 & c_2 \\ a_3 & b_3 & c_3 \end{vmatrix}$$

If the determinant is zero (but $\mathbf{v}_1 \times \mathbf{v}_2$ is not zero), then $\mathbf{v}_3$ must lie *in* the plane P. In general, we may say that the determinant is zero when the vectors $\mathbf{v}_1$, $\mathbf{v}_2$, and $\mathbf{v}_3$ fail to span a solid parallelepiped, but instead lie in a plane, lie on a line, or are all zero. Such triples of vectors are said to be *linearly dependent* and are neither right-handed nor left-handed.

8. The expansion by minors of the second column is

$$-\begin{vmatrix} a_2 & c_2 \\ a_3 & c_3 \end{vmatrix} b_1 + \begin{vmatrix} a_1 & c_1 \\ a_3 & c_3 \end{vmatrix} b_2 - \begin{vmatrix} a_1 & c_1 \\ a_2 & c_2 \end{vmatrix} b_3$$

Expanding the 2×2 determinants gives

$$-(a_2 c_3 - c_2 a_3) b_1 + (a_1 c_3 - c_1 a_3) b_2$$
$$- (a_1 c_2 - c_1 a_2) b_3 =$$
$$-b_1 a_2 c_3 + b_1 c_2 a_3 + a_1 b_2 c_3$$
$$- c_1 b_2 a_3 - a_1 c_2 b_3 + c_1 a_2 b_3$$

Collecting the terms in a_3, b_3, and c_3, we get

$$(b_1 c_2 - c_1 b_2) a_3 - (a_1 c_2 - c_1 a_2) b_3$$
$$+ (a_1 b_2 - b_1 a_2) c_3$$
$$= \begin{vmatrix} b_1 & c_1 \\ b_2 & c_2 \end{vmatrix} a_3 - \begin{vmatrix} a_1 & c_1 \\ a_2 & c_2 \end{vmatrix} b_3 + \begin{vmatrix} a_1 & b_1 \\ a_2 & b_2 \end{vmatrix} c_3$$
$$= \begin{vmatrix} a_1 & b_1 & c_1 \\ a_2 & b_2 & c_2 \\ a_3 & b_3 & c_3 \end{vmatrix}$$

Solved Exercises for Section 15-1

1. The graph of $f(x, y) = 3x$ is the plane $z = 3x$. It contains the y axis and is shown in Fig. 15-1-12.

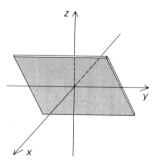

Fig. 15-1-12   The graph of $z = 3x$.

2. The surface $x^2 + y^2 + z^2 - 8 = 1$ (that is, $x^2 + y^2 + z^2 = 9$) is the set of points (x, y, z) whose distance from the origin is $\sqrt{9} = 3$; it is a sphere with radius 3. (See Fig. 15-1-13.)

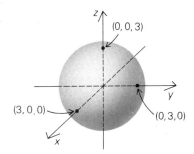

Fig. 15-1-13 The level surface of $x^2 + y^2 + z^2 - 8$ with value 1 is a sphere of radius 3.

3. The intersection of the plane $z = c$ and the graph of f consists of the points (x, y, c) in space such that $f(x, y) = c$. This set has the same shape as the level curve with value c, but it is moved from the xy plane up to the plane $z = c$. See Fig. 15-1-14.

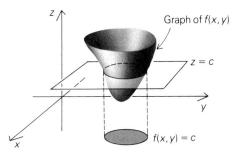

Fig. 15-1-14 The level curve of $f(x, y)$ with value c is obtained by finding the intersection of the graph of f with the plane $z = c$ and moving it down to the (x, y) plane.

4. The variable z does not occur in the equation, so the surface is a cylinder parallel to the z axis. Its cross-section is the plane curve $x^2 + y^2 = 25$, which is a circle of radius 5, so the surface is a right circular cylinder, as shown in Fig. 15-1-15.

4. (*cont'd*)

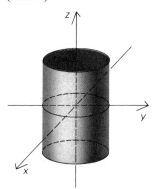

Fig. 15-1-15 The graph $x^2 + y^2 = 25$ is a right circular cylinder.

5. Completing the square, we write $z = x^2 + y^2 - 4x - 6y + 13$ as

$$z = x^2 - 4x + \quad y^2 - 6y \quad + 13$$
$$= x^2 - 4x + 4 + y^2 - 6y + 9 + 13 - 13$$
$$= (x - 2)^2 + (y - 3)^2$$

The level surface for value c is thus the circle $(x - 2)^2 + (y - 3)^2 = c$ with center $(2, 3)$ and radius $\sqrt{c}$. Comparing this result with Worked Example 6, we find that the surface is again a paraboloid of revolution, with its axis shifted to the line $(x, y) = (2, 3)$. (See Fig. 15-1-16.)

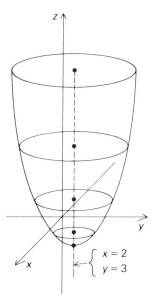

Fig. 15-1-16 The graph $z = x^2 + y^2 - 4x - 6y + 13$ is a shifted paraboloid of revolution.

15

Solved Exercises for Section 15-2

1. First, let z be constant. Then we get $x^2/9 + y^2/16 = 1 - z^2$. This is an ellipse centered at the origin if $-1 < z < 1$. If $z = 1$, we just get a point, $x = 0$, $y = 0$. Likewise, $(0, 0, -1)$ is on the surface. If $|z| > 1$ there are no (x, y) satisfying the equation.

   Setting $x = $ constant or $y = $ constant, we also get ellipses. We must have $|x| \leq 3$ and, likewise, $|y| \leq 4$. The surface, shaped like a stepped-on football, is easiest to draw if the intersections with the three coordinate planes are drawn first. (See Fig. 15-2-15.)

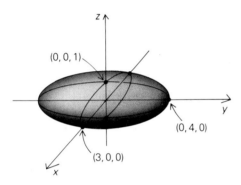

   Fig. 15-2-15 The surface $(x^2/9) + (y^2/16) + z^2 = 1$ is an ellipsoid.

2. If z is a constant, then $x^2 + y^2 = 4 + z^2$ is a circle. Thus in any plane parallel to the xy plane we get a circle. Our job of drawing the surface is simplified if we note right away that the surface is rotationally invariant about the z axis (since z depends only on $r^2 = x^2 + y^2$). Thus we can draw the curve traced by the surface in the yz plane (or xz plane) and revolve it about the z axis. Setting $x = 0$ we get $y^2 - z^2 = 4$, a hyperbola. Hence we get the surface shown in Fig. 15-2-16, a one-sheeted hyperboloid. Since this surface is symmetric about the z axis, it is also called a hyperboloid of revolution.

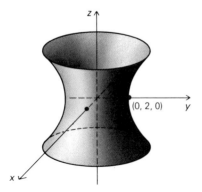

   Fig. 15-2-16 The surface $x^2 + y^2 - z^2 = 4$ is a one-sheeted hyperboloid of revolution.

2. (cont'd)

   The hyperboloid of one sheet has the property that it is *ruled*: that is, the surface is composed of straight lines (see Review Problem 23). It is therefore easy to make with string models and is useful architecturally. Figure 15-2-17 shows a practical application of this property. (See "Mathematical Games" by M. Gardner, *Scientific American*, September 1977, pp. 24–42, for more information.)

   Fig. 15-2-17 Hyperboloidal gears transmit motion to a skew shaft. (From M. Gardner, "Mathematical Games," *Scientific American*, September 1977. Copyright © 1977 by Scientific American, Inc. All rights reserved.)

3. (a) $r = \sqrt{x^2 + y^2} = \sqrt{2^2 + (-3)^2} = \sqrt{13}$ and $\theta = \tan^{-1}(-\frac{3}{2}) \approx -0.983 \approx -56.31°$; $z = 6$. See Fig. 15-2-18(a).

   (b) $x = r \cos \theta = 2 \cos 3\pi/4 = 2 \cdot (-\sqrt{2}/2) = -\sqrt{2}$ and $y = r \sin \theta = 2 \sin (3\pi/4) = 2 \cdot (\sqrt{2}/2) = \sqrt{2}$; $z = 1$. See Fig. 15-2-18(b).

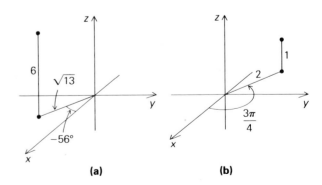

   Fig. 15-2-18 Two points in cylindrical coordinates.

4. The curve $r = \cos 2\theta$ in the xy plane is a four-leafed rose (see p. 277). Thus in cylindrical coordinates we obtain a vertical cylinder with the four-leafed rose as a base, as shown in Fig. 15-2-19.

Fig. 15-2-19 The surface $r = \cos 2\theta$ is a cylinder with a four-leafed rose as its base.

5. Increasing θ by π is a rotation through $180°$ about the z axis. Switching z to $-z$ reflects in the xy plane. See Fig. 15-2-20. Combining the two operations results in reflection through the origin.

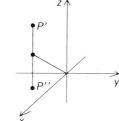

Fig. 15-2-20 The effect of replacing (r, θ, z) by $(r, \theta + \pi, -z)$ is to replace P by $-P$.

6. If we set $y = 0$ and take $x \geq 0$, then $r = x$ and $r = f(z)$ becomes $x = f(z)$; the remaining points satisfying $r = f(z)$ are then obtained by revolving the graph $x = f(z)$ about the z axis; note that $r = c$, $z = d$ is a circle centered on the z axis. Thus we have a surface of revolution with the z axis as axis of symmetry.

7. (a) $\rho = \sqrt{x^2 + y^2 + z^2} = \sqrt{2^2 + (-3)^2 + 6^2}$
$= \sqrt{49} = 7$

$\theta = \tan^{-1}(y/x) = \tan^{-1}(-3/2) \approx -0.983$
$\approx -56.31°$

$\phi = \cos^{-1}(z/\rho) = \cos^{-1}(6/7) \approx 0.541$
$\approx 31.0°$

See Fig. 15-2-21(a) (the point is the same as in Solved Exercise 3(a)).

(b) $x = \rho \sin \phi \cos \theta = 1 \sin (\pi/4) \cos (-\pi/2)$
$= (\sqrt{2}/2) \cdot 0 = 0$

$y = \rho \sin \phi \sin \theta = 1 \sin (\pi/4) \sin (-\pi/2)$
$= (\sqrt{2}/2)(-1) = -\sqrt{2}/2$

$z = \rho \cos \phi = 1 \cos (\pi/4) = \sqrt{2}/2$

See Fig. 15-2-21(b).

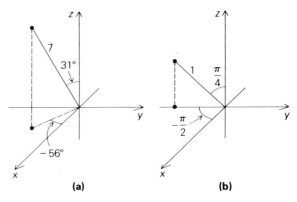

(a)   (b)

Fig. 15-2-21 Two points in spherical coordinates.

8. In spherical coordinates, ρ is the distance from the point (x, y, z) to the origin. Thus $\rho = 3$ consists of all points a distance 3 from the origin—that is, a sphere of radius 3 centered at the origin.

9. Increasing θ by π has the effect of rotating about the z axis through an angle of $180°$.

10. The key feature of the surface $\rho = f(\phi)$ is that its equation does not involve θ and hence is independent of rotations about the z axis; thus it is a surface of revolution. If we put $y = 0$, then $\rho = \sqrt{x^2 + z^2}$ and $\phi = \cos^{-1}(z/\sqrt{x^2 + z^2})$. Thus the surface $\rho = f(\phi)$ is obtained by revolving the curve in the xz plane given by

$$\sqrt{x^2 + z^2} = f\left(\cos^{-1}\left(\frac{z}{\sqrt{x^2 + z^2}}\right)\right)$$

about the z-axis.

15

Solved Exercises for Section 15-3

1. Since $x = \cos t$ and $y/2 = \sin t$, the point $(x, y, 0)$ satisfies $x^2 + y^2/4 = 1$, so the curve lies over this ellipse in the xy plane. As t increases from zero to 2π, the projection in the xy plane goes once around the same ellipse as in Worked Example 1 (Fig. 15-3-2), only now it starts at $(1, 0, 0)$ at $t = 0$ and proceeds counterclockwise since x behaves like $\cos t$ and y like $2\sin t$. Meanwhile, z increases steadily with t according to the formula $z = 2t$. The net result is a helix winding around the z axis, much like that of Worked Example 3 but no longer circular. It now lies on a cylinder of elliptical cross-section (see Fig. 15-3-14).

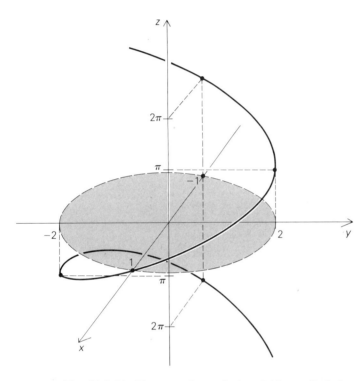

Fig. 15-3-14 The curve $(\cos t, 2\sin t, 2t)$ is an elliptical helix.

2. We find

$$(x, y, z) = (3t^3 + 2, t^3 - 8, 4t^3 + 3)$$
$$= (2, -8, 3) + t^3(3, 1, 4)$$

So the curve is a straight line through $(2, -8, 3)$ in the direction $(3, 1, 4)$.

3. Let $\boldsymbol{\sigma}_1(t) = f_1(t)\mathbf{i} + g_1(t)\mathbf{j} + h_1(t)\mathbf{k}$ and $\boldsymbol{\sigma}_2(t) = f_2(t)\mathbf{i} + g_2(t)\mathbf{j} + h_2(t)\mathbf{k}$. Therefore, $\boldsymbol{\sigma}_1(t)\cdot\boldsymbol{\sigma}_2(t) = f_1(t)f_2(t) + g_1(t)g_2(t) + h_1(t)h_2(t)$, so by the sum and product rules for real-valued functions, we have

$$\frac{d}{dt}[\boldsymbol{\sigma}_1(t)\cdot\boldsymbol{\sigma}_2(t)] = [f_1'(t)f_2(t) + f_1(t)f_2'(t)] +$$

$$[g_1'(t)g_2(t) + g_1(t)g_2'(t)] + [h_1'(t)h_2(t) + h_1(t)h_2'(t)]$$

Regrouping terms, we can rewrite this as

$$[f_1'(t)f_2(t) + g_1'(t)g_2(t) + h_1'(t)h_2(t)]$$
$$+ [f_1(t)f_2'(t) + g_1(t)g_2'(t) + h_1(t)h_2'(t)]$$
$$= [f_1'(t)\mathbf{i} + g_1'(t)\mathbf{j} + h_1'(t)\mathbf{k}]$$
$$\qquad\qquad \cdot [f_2(t)\mathbf{i} + g_2(t)\mathbf{j} + h_2(t)\mathbf{k}]$$
$$+ [f_1(t)\mathbf{i} + g_1(t)\mathbf{j} + h_1(t)\mathbf{k}]$$
$$\qquad\qquad \cdot [f_2'(t)\mathbf{i} + g_2'(t)\mathbf{j} + h_2'(t)\mathbf{k}]$$
$$= \boldsymbol{\sigma}_1'(t)\cdot\boldsymbol{\sigma}_2(t) + \boldsymbol{\sigma}_1(t)\cdot\boldsymbol{\sigma}_2'(t)$$

so the dot product rule is proved.

4. We have

$$\boldsymbol{\sigma}(t_0 + \Delta t) - \boldsymbol{\sigma}(t_0) = f(t_0 + \Delta t)\mathbf{i} + g(t_0 + \Delta t)\mathbf{j} +$$
$$h(t_0 + \Delta t)\mathbf{k} - f(t_0)\mathbf{i} - g(t_0)\mathbf{j} - h(t_0)\mathbf{k}$$
$$= [f(t_0 + \Delta t) - f(t_0)]\mathbf{i} +$$
$$[g(t_0 + \Delta t) - g(t_0)]\mathbf{j} + [h(t_0 + \Delta t) - h(t_0)]\mathbf{k}$$

so

$$\lim_{\Delta t\to 0}\frac{1}{\Delta t}[\boldsymbol{\sigma}(t_0 + \Delta t) - \boldsymbol{\sigma}(t_0)]$$

$$= \lim_{\Delta t\to 0}\left[\frac{f(t_0 + \Delta t) - f(t_0)}{\Delta t}\mathbf{i} +\right.$$

$$\left.\frac{g(t_0 + \Delta t) - g(t_0)}{\Delta t}\mathbf{j} + \frac{h(t_0 + \Delta t) - h(t_0)}{\Delta t}\mathbf{k}\right]$$

Taking the limit componentwise, and using the formula for the derivative of a real-valued function as a limit of difference quotients, we get

$$f'(t_0)\mathbf{i} + g'(t_0)\mathbf{j} + h'(t_0)\mathbf{k}$$

which is just $\boldsymbol{\sigma}'(t_0)$.

5. Let $\boldsymbol{\sigma}(t) = f(t)\mathbf{i} + g(t)\mathbf{j} + h(t)\mathbf{k}$. Then if $\boldsymbol{\sigma}'(t) = -\boldsymbol{\sigma}(t)$, we have $f'(t)\mathbf{i} + g'(t)\mathbf{j} + h'(t)\mathbf{k} = -f(t)\mathbf{i} - g(t)\mathbf{j} - h(t)\mathbf{k}$, so $f'(t) = -f(t)$, $g'(t) = -g(t)$, and $h'(t) = -h(t)$. Thus $f(t)$, $g(t)$, and $h(t)$ are all solutions of the differential equation $p'(t) = -p(t)$. But we know (see p. 379) that any solution of this equation is of the form $p(t) = p(0)e^{-t}$. Thus we have $f(t) = f(0)e^{-t}$, $g(t) = g(0)e^{-t}$, and $h(t) = h(0)e^{-t}$, so

$$\boldsymbol{\sigma}(t) = f(0)e^{-t}\mathbf{i} + g(0)e^{-t}\mathbf{j} + h(0)e^{-t}\mathbf{k}$$
$$= e^{-t}\boldsymbol{\sigma}(0)$$

5. (*cont'd*)

As $t \to \infty$, e^{-t} approaches zero, so $\boldsymbol{\sigma}(t) = e^{-t}\boldsymbol{\sigma}(0)$ approaches the zero vector along the line through $\boldsymbol{\sigma}(0)$. (See Fig. 15-3-15.)

Fig. 15-3-15 As $t \to \infty$, the solution of the equation $\boldsymbol{\sigma}'(t) = -\boldsymbol{\sigma}(t)$ approaches the zero vector.

6. The velocity vector is $\mathbf{v} = -\sin t\mathbf{i} + \cos t\mathbf{j} + \mathbf{k}$, so the speed is

$$v = \|\mathbf{v}\| = \sqrt{(-\sin t)^2 + (\cos t)^2 + 1}$$
$$= \sqrt{\sin^2 t + \cos^2 t + 1} = \sqrt{2}$$

7. Let $(f(t), g(t), h(t))$ be the parametric curve traced out by the particle, so that the velocity vector is $\boldsymbol{\sigma}'(t) = f'(t)\mathbf{i} + g'(t)\mathbf{j} + h'(t)\mathbf{k}$. The acceleration $\boldsymbol{\sigma}''(t)$ is equal to $-\mathbf{k}$, so we must have $f''(t) = 0$, $g''(t) = 0$, and $h''(t) = -1$. It follows that $f'(t)$ and $g'(t)$ are constant functions, and $h'(t)$ is a linear function with slope -1. Since $\boldsymbol{\sigma}'(0) = \mathbf{i} + \mathbf{j}$, we must have $\boldsymbol{\sigma}'(t) = \mathbf{i} + \mathbf{j} - t\mathbf{k}$. Integrating again and using the initial position $(0, 0, 1)$, we find that $(f(t), g(t), h(t)) = (t, t, 1 - \frac{1}{2}t^2)$. The particle drops below the plane $z = 0$ when $1 - \frac{1}{2}t^2 = 0$; that is, $t = \sqrt{2}$. At that time, the position is $(\sqrt{2}, \sqrt{2}, 0)$. The path traveled by the particle is a parabola in the plane $y = x$. (See Fig. 15-3-16.)

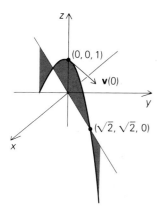

Fig. 15-3-16 The path of the parabola with initial position $(0, 0, 1)$, initial velocity $\mathbf{i} + \mathbf{j}$, and constant acceleration $-\mathbf{k}$ is a parabola in the plane $x = y$.

8. $\boldsymbol{\sigma}(t) = (\cos t, \sin t, t^2)$

$$\mathbf{v} = (-\sin t, \cos t, 2t)$$
$$\|\mathbf{v}\| = \sqrt{1 + 4t^2} = 2\sqrt{t^2 + (\tfrac{1}{2})^2}$$

so the arc length is

$$L = \int_0^\pi 2\sqrt{t^2 + (\tfrac{1}{2})^2}\, dt$$
$$= 2 \cdot \tfrac{1}{2}[t\sqrt{t^2 + (\tfrac{1}{2})^2}$$
$$+ (\tfrac{1}{2})^2 \ln(t + \sqrt{t^2 + (\tfrac{1}{2})^2})]\Big|_{t=0}^\pi$$
$$= \pi\sqrt{\pi^2 + \tfrac{1}{4}} + \tfrac{1}{4}\ln(\pi + \sqrt{\pi^2 + \tfrac{1}{4}}) - \tfrac{1}{4}\ln(\sqrt{\tfrac{1}{4}})$$
$$= \frac{\pi}{2}\sqrt{1 + 4\pi^2} + \frac{1}{4}\ln(2\pi + \sqrt{1 + 4\pi^2})$$
$$\approx 10.63$$

9. $\boldsymbol{\sigma}(t) = (e^t, t, e^t)$

$$\mathbf{v} = (e^t, 1, e^t)$$
$$\|\mathbf{v}\| = \sqrt{1 + 2e^{2t}}$$

So $L = \int_0^1 \sqrt{1 + 2e^{2t}}\, dt$. One way to do this integral is to put $u = \sqrt{1 + 2e^{2t}}$, which leads to

$$\int \sqrt{1 + 2e^{2t}}\, dt = \int u\,\frac{u\, du}{u^2 - 1}$$
$$= \int \left[1 + \frac{1}{2}\left(\frac{1}{u-1}\right) - \frac{1}{2}\left(\frac{1}{u+1}\right)\right] du$$
$$= u + \frac{1}{2}\ln(u - 1) - \frac{1}{2}\ln(u + 1)$$
$$= \sqrt{1 + 2e^{2t}} + \frac{1}{2}\ln\frac{\sqrt{1 + 2e^{2t}} - 1}{\sqrt{1 + 2e^{2t}} + 1}$$

(This result may be checked by differentiation.) To find L, we evaluate the last expression at $t = 0$ and $t = 1$ to obtain

$$L = \sqrt{1 + 2e^2} + \frac{1}{2}\ln\frac{\sqrt{1 + 2e^2} - 1}{\sqrt{1 + 2e^2} + 1} - \sqrt{3} - \frac{1}{2}\ln\frac{\sqrt{3} - 1}{\sqrt{3} + 1} \approx 2.64.$$

15

10. We use the rules of differentiation for vector functions:

$$\frac{d}{dt}(m\mathbf{r} \times \mathbf{v}) = m\frac{d\mathbf{r}}{dt} \times \mathbf{v} + m\mathbf{r} \times \frac{d\mathbf{v}}{dt}$$

$$= m\mathbf{v} \times \mathbf{v} + m\mathbf{r} \times \mathbf{a}$$

$$= \mathbf{0} + \mathbf{r} \times m\mathbf{a} = \mathbf{r} \times \mathbf{F}$$

If $\mathbf{F}$ is parallel to $\mathbf{r}$, then this last cross-product is $\mathbf{0}$. Thus $m\mathbf{r} \times \mathbf{v}$ must be a constant vector. It represents the angular momentum, a quantity which measures the tendency of a spinning body, for example a flywheel, to keep spinning. The magnitude of $m\mathbf{r} \times \mathbf{v}$ measures the amount of angular momentum, and the direction is along the axis of spin. If the derivative above is zero, it means that angular momentum is conserved; both its magnitude and its direction are preserved. This is the case for our model of planetary motion in which the sun is regarded as fixed and the gravitational force

$$\mathbf{F} = -\frac{GmM}{\|r\|^3}\mathbf{r}$$

is parallel to the vector $\mathbf{r}$ from the sun to the planet. The actual situation is a bit more complicated than this: in fact, both the sun and the planet move around their common center of mass. However, the mass M of the sun is so much greater than the mass m of the planet that this center of mass is very close to the center of the sun, and our approximation is quite good. Things would be more complicated, for example, in a double-star system where the masses were more nearly the same and the center of gravity somewhere between. What is conserved is the total angular momentum of the whole system, taking both stars into account.

11. Let $\mathbf{l} = m\mathbf{r} \times \mathbf{v}$ be the angular momentum vector. Then certainly $\mathbf{r} \cdot \mathbf{l} = 0$. We argued above that $d\mathbf{l}/dt = \mathbf{0}$, so $\mathbf{l}$ is a constant vector. Since $\mathbf{r}$ satisfies $\mathbf{r} \cdot \mathbf{l} = 0$, the planet always lies in the plane through the sun with normal vector $\mathbf{l}$.

12. The curvature is defined as $\|d\mathbf{T}/ds\|$. We must use the chain rule to express this in terms of the original parametrization t. First of all, we have $\mathbf{T} = \mathbf{v}/\|\mathbf{v}\| = \mathbf{v}/(\mathbf{v} \cdot \mathbf{v})^{1/2}$, so

$$\frac{d\mathbf{T}}{ds} = \frac{d\mathbf{T}}{dt}\frac{dt}{ds} = \frac{1}{(ds/dt)}\frac{d\mathbf{T}}{dt}$$

12. (*cont'd*)

Now $ds/dt = \|\mathbf{v}\|$ and

$$\frac{d\mathbf{T}}{dt} = \frac{d}{dt}\left(\frac{\mathbf{v}}{(\mathbf{v} \cdot \mathbf{v})^{1/2}}\right)$$

$$= \left(\frac{d}{dt}(\mathbf{v} \cdot \mathbf{v})^{-1/2}\right)\mathbf{v} + (\mathbf{v} \cdot \mathbf{v})^{-1/2}\frac{d\mathbf{v}}{dt}$$

$$= -\frac{1}{2}(\mathbf{v} \cdot \mathbf{v})^{-3/2}2\left(\mathbf{v} \cdot \frac{d\mathbf{v}}{dt}\right)\mathbf{v} + (\mathbf{v} \cdot \mathbf{v})^{-1/2}\frac{d\mathbf{v}}{dt}$$

$$= (\mathbf{v} \cdot \mathbf{v})^{-3/2}\left[-\left(\mathbf{v} \cdot \frac{d\mathbf{v}}{dt}\right)\mathbf{v} + (\mathbf{v} \cdot \mathbf{v})\frac{d\mathbf{v}}{dt}\right]$$

so

$$\left\|\frac{d\mathbf{T}}{dt}\right\|^2 = (\mathbf{v} \cdot \mathbf{v})^{-3}\left[-\left(\mathbf{v} \cdot \frac{d\mathbf{v}}{dt}\right)\mathbf{v} + (\mathbf{v} \cdot \mathbf{v})\frac{d\mathbf{v}}{dt}\right]$$

$$\cdot\left[-\left(\mathbf{v} \cdot \frac{d\mathbf{v}}{dt}\right)\mathbf{v} + (\mathbf{v} \cdot \mathbf{v})\frac{d\mathbf{v}}{dt}\right]$$

$$= (\mathbf{v} \cdot \mathbf{v})^{-3}\left[\left(\mathbf{v} \cdot \frac{d\mathbf{v}}{dt}\right)^2(\mathbf{v} \cdot \mathbf{v}) - \right.$$

$$\left. 2\left(\mathbf{v} \cdot \frac{d\mathbf{v}}{dt}\right)^2(\mathbf{v} \cdot \mathbf{v}) + (\mathbf{v} \cdot \mathbf{v})^2\left(\frac{d\mathbf{v}}{dt} \cdot \frac{d\mathbf{v}}{dt}\right)\right]$$

$$= (\mathbf{v} \cdot \mathbf{v})^{-2}\left[(\mathbf{v} \cdot \mathbf{v})\left(\frac{d\mathbf{v}}{dt} \cdot \frac{d\mathbf{v}}{dt}\right) - \left(\mathbf{v} \cdot \frac{d\mathbf{v}}{dt}\right)^2\right]$$

$$= \|\mathbf{v}\|^{-4}\left\|\mathbf{v} \times \frac{d\mathbf{v}}{dt}\right\|^2$$

(See Problem 23, Section 14-3.) Thus

$$k = \left\|\frac{d\mathbf{T}}{ds}\right\| = \frac{1}{\|\mathbf{v}\|}\|\mathbf{v}\|^{-2}\left\|\mathbf{v} \times \frac{d\mathbf{v}}{dt}\right\|$$

$$= \frac{\|\mathbf{v} \times d\mathbf{v}/dt\|}{\|\mathbf{v}\|^3}$$

which is formula (2).

13. We may suppose that the curve is parametrized by arc length. Then $\mathbf{v} = \mathbf{T}$ and $\|d\mathbf{T}/ds\| = 0$, so $\mathbf{v}$ is constant and the curve is a straight line. (If $f'(s)$, $g'(s)$, and $h'(s)$ are constants, then $f(s)$, $g(s)$, and $h(s)$ must be linear functions.)

CHAPTER 16

Solved Exercises for Section 16-1

1. The graph of $h(t) = f(t, y_0)$ is obtained as the section of the graph of f by the plane $y = y_0$.

2. We have $f(1, y, 3) = \sin(y/3)$; differentiating gives $\frac{1}{3}\cos(y/3)$ and setting $y = 2$ gives $\frac{1}{3}\cos(\frac{2}{3})$.

3. Since f_x is just the derivative holding y constant,

$$f_x(x, y) = \lim_{\Delta x \to 0}\frac{f(x + \Delta x, y) - f(x, y)}{\Delta x}$$

4. Holding r constant, we get

$$\frac{\partial h}{\partial s} = 2rs \sin (r^2 + s^2) + rs^2 \cdot 2s \cdot \cos (r^2 + s^2)$$

$$= 2rs \left[\sin (r^2 + s^2) + s^2 \cos (r^2 + s^2) \right].$$

5. Proceeding directly north means increasing the latitude x. Thus we seek

$$\frac{\partial T}{\partial x} = -(0.0003) \cdot 2xy$$

$$= -(0.0003) \cdot 2 \cdot (55.7) \cdot (120.2) \approx -4.017$$

So the temperature drops as we proceed north from Dawson Creek, at the rate of 4.017° C per degree of latitude.

6. We find $\partial u / \partial x = -yz \sin (xz) + \sin (yz)$ and $\partial u / \partial z = -xy \sin (xz) + xy \cos (yz)$. Thus

$$\frac{\partial^2 u}{\partial z \, \partial x} = -y \sin (xz) - xyz \cos (xz) + y \cos (yz)$$

Differentiation of $\partial u / \partial z$ with respect to x yields

$$\frac{\partial^2 u}{\partial x \, \partial z} = -y \sin (xz) - xyz \cos (xz) + y \cos (yz)$$

which shows that $\partial^2 u / \partial x \, \partial z = \partial^2 u / \partial z \, \partial x$.

7. Here $x_0 = 1$, $y_0 = 2$, and $f(1, 2) = \frac{5}{2}$. Also,

$$f_x (x, y) = \frac{2x \cdot xy - (x^2 + y^2) y}{(xy)^2}$$

$$= \frac{x^2 y - y^3}{(xy)^2} = \frac{x^2 - y^2}{x^2 y}$$

which is $-\frac{3}{2}$ at $(1, 2)$.

Similarly,

$$f_y (x, y) = \frac{y^2 x - x^3}{(xy)^2} = \frac{y^2 - x^2}{xy^2}$$

which is $\frac{3}{4}$ at $(1, 2)$

Thus the tangent plane is

$$z = -\frac{3}{2} (x - 1) + \frac{3}{4} (y - 2) + \frac{5}{2}$$

i.e., $4z = -6x + 3y + 10$

8. Since a normal vector is $-f_x (x_0, y_0) \mathbf{i} - f_y (x_0, y_0) \mathbf{j} + \mathbf{k}$, a unit normal is obtained by normalizing:

$$\mathbf{n} = \frac{[-f_x (x_0, y_0) \mathbf{i} - f_y (x_0, y_0) \mathbf{j} + \mathbf{k}]}{\sqrt{[f_x (x_0, y_0)]^2 + [f_y (x_0, y_0)]^2 + 1}}$$

9. Let $f(x, y) = \sin x \cos y$ and $x_0 = 0$, $y_0 = \pi$. Then if $x = 0.01$ and $y = 0.99\pi$,

$$f(x, y) \approx f_x (0, \pi)(x - 0) + f_y (0, \pi)(y - \pi)$$
$$+ f(0, \pi)$$
$$= -1(0.01) + 0 + 0$$
$$= -0.01$$

(The value of $f(x, y)$ computed on our calculator is -0.009994899.)

10. Let $\varepsilon > 0$ be given and let $f(x, y) = x$ and $l = x_0$. We seek a number > 0 such that $|f(x, y) - l| < \varepsilon$ whenever $d((x, y), (x_0, y_0)) < \delta$; that is, $|x - x_0| < \varepsilon$ whenever $\sqrt{(x - x_0)^2 + (y - y_0)^2} < \delta$. However, note that $|x - x_0| = \sqrt{(x - x_0)^2} \le \sqrt{(x - x_0)^2 + (y - y_0)^2}$, so if we choose $\delta = \varepsilon$, $d((x, y), (x_0, y_0)) < \delta$ will imply $|x - x_0| < \varepsilon$.

11. $(\partial / \partial x) \sqrt{x^2 + y^2} = x / \sqrt{x^2 + y^2}$ if $(x, y) \ne (0, 0)$. Thus

$$\lim_{(x, y) \to (0, 0)} (\partial / \partial x) \sqrt{x^2 + y^2} = \lim_{(x, y) \to (0, 0)} x / \sqrt{x^2 + y^2}$$

If we approach $(0, 0)$ on the y axis—that is, along points $(0, y)$—we get zero. Thus the limit, if it exists, is zero. On the other hand, if we approach $(0, 0)$ along the positive x axis, we have $y = 0$ and $x > 0$; then $x / \sqrt{x^2 + y^2} = 1$ because $x / \sqrt{x^2} = 1$, so the limit is 1. Since we are getting different answers, the full limit cannot exist.

Solved Exercises for Section 16-2

1. Substituting the formulas for x, y, and z in the formula for u gives

$$u = e^t \cdot e^{t \sin t} = e^{t(1 + \sin t)}$$

so

$$\frac{du}{dt} = [t \cos t + (1 + \sin t)] e^{t(1 + \sin t)}$$

The chain rule says that this should equal

$$\frac{\partial u}{\partial x} \frac{dx}{dt} + \frac{\partial u}{\partial y} \frac{dy}{dt} + \frac{\partial u}{\partial z} \frac{dz}{dt}$$

$$= e^{yz} e^t + xze^{yz} \cdot 1 + xye^{yz} \cos t$$

$$= e^{t \sin t} e^t + e^t \sin t \, e^{t \sin t} + e^t \cdot t \cdot e^{t \sin t} \cos t$$

$$= e^{t(1 + \sin t)}(1 + \sin t + t \cos t)$$

which it does.

2. If $z = xy$, then $\partial z / \partial x = y$ and $\partial z / \partial y = x$, so the chain rule gives $dz / dt = y(dx / dt) + x(dy / dt)$, which is precisely the product rule for functions of one variable.

16

3. According to the cited example, $z = e^{-y} \cos t + \sin(y + t^2)$. Where is the x of the chain rule? We may let $x = t$, so that $z = e^{-y} \cos x + \sin(y + x^2)$. When $(x, y) = (2, 3)$, we have $\partial z/\partial y = -e^{-3} \cos 2 + \cos 7$ by Worked Example 1 of Section 16-1 and hence $\partial z/\partial x = -e^{-3} \sin 2 + 4 \cos 7$ by Worked Example 2 of Section 16-1. Also, we are given that $dy/dt = 5$, while dx/dt must be 1, since it is dt/dt. By the chain rule,

$$\frac{dz}{dt} = \frac{\partial z}{\partial x}\frac{dx}{dt} + \frac{\partial z}{\partial y}\frac{dy}{dt}$$

$$= (-e^{-3} \sin 2 + 4 \cos 7)1 + (-e^{-3} \cos 2 + \cos 7)5$$

$$= -e^{-3}(\sin 2 + 5 \cos 2) + 9 \cos 7$$

$$\approx 6.843$$

4. By definition of the gradient,

$$\nabla\left(\frac{1}{r}\right) = \frac{\partial}{\partial x}\left(\frac{1}{r}\right)\mathbf{i} + \frac{\partial}{\partial y}\left(\frac{1}{r}\right)\mathbf{j} + \frac{\partial}{\partial z}\left(\frac{1}{r}\right)\mathbf{k}$$

But

$$\frac{\partial}{\partial x}\left(\frac{1}{r}\right) = \frac{\partial}{\partial x}\left(\frac{1}{\sqrt{x^2 + y^2 + z^2}}\right)$$

$$= -\frac{x}{(x^2 + y^2 + z^2)^{3/2}} = -\frac{x}{r^3}$$

and, similarly,

$$\frac{\partial}{\partial y}\left(\frac{1}{r}\right) = -\frac{y}{r^3}, \qquad \frac{\partial}{\partial z}\left(\frac{1}{r}\right) = -\frac{z}{r^3}$$

Thus

$$\nabla\left(\frac{1}{r}\right) = -\frac{x}{r^3}\mathbf{i} - \frac{y}{r^3}\mathbf{j} - \frac{z}{r^3}\mathbf{k}$$

$$= -\frac{1}{r^3}(x\mathbf{i} + y\mathbf{j} + z\mathbf{k})$$

$$= -\frac{\mathbf{r}}{r^3}$$

as required. Finally,

$$\left\|\nabla\left(\frac{1}{r}\right)\right\| = \left|-\frac{1}{r^3}\right| \|\mathbf{r}\| = \frac{r}{r^3} = \frac{1}{r^2}$$

$$= \frac{1}{x^2 + y^2 + z^2}$$

5. (a) If we sketched $\Phi(x, y) = 3y\mathbf{i} - 3x\mathbf{j}$ itself, the vectors at different base points would overlap. Instead we sketch $\frac{1}{6}\Phi(x, y) = \frac{1}{2}y\mathbf{i} - \frac{1}{2}x\mathbf{j}$ in Fig. 16-2-8.

(b) If $\Phi(x, y) = 3y\mathbf{i} - 3x\mathbf{j}$ were the gradient of a function $z = f(x, y)$, we would have $\partial z/\partial x = 3y$ and $\partial z/\partial y = -3x$. By the equality of mixed partial derivatives (since $3y$ and $-3x$ are continuous), $\partial^2 z/\partial x\,\partial y = -3$ and $\partial^2 z/\partial y\,\partial x = 3$ would have to be equal. But $3 \neq -3$, so our vector field must not be a gradient.

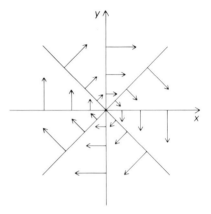

Fig. 16-2-8 The vector field $3y\mathbf{i} - 3x\mathbf{j}$ is not the gradient of a function.

6. If $f(\boldsymbol{\sigma}(t))$ is always equal to 2, the derivative $(d/dt)f(\boldsymbol{\sigma}(t))$ is zero. By the chain rule, $0 = \nabla f(\boldsymbol{\sigma}(t)) \cdot \boldsymbol{\sigma}'(t)$, so the gradient vector $\nabla f(\boldsymbol{\sigma}(t))$ and the velocity vector $\boldsymbol{\sigma}'(t)$ are perpendicular at all points of the curve.

7. This is the rate of change of f along the y axis, but in the negative direction. It equals $-f_y$.

8. We know $r = \sqrt{x^2 + y^2 + z^2}$, so

$$\nabla r = \frac{\partial r}{\partial x}\mathbf{i} + \frac{\partial r}{\partial y}\mathbf{j} + \frac{\partial r}{\partial z}\mathbf{k} = \frac{x}{r}\mathbf{i} + \frac{y}{r}\mathbf{j} + \frac{z}{r}\mathbf{k}$$

since $\partial r/\partial x = \frac{1}{2} \cdot 2x/\sqrt{x^2 + y^2 + z^2} = x/r$ and so forth. Thus

$$\nabla r = \frac{1}{r}(x\mathbf{i} + y\mathbf{j} + z\mathbf{k}) = \frac{\mathbf{r}}{r}$$

Thus r is increasing fastest in the direction of $\mathbf{r}/r$, which is a unit vector pointing outward from the origin. This makes sense since r is the distance from the origin.

9. The direction is

$$\nabla f(0,1) = \frac{\partial f}{\partial x}\mathbf{i} + \frac{\partial f}{\partial y}\mathbf{j} \quad \text{at } (0,1)$$
$$= 2x\mathbf{i} - 2y\mathbf{j} \quad \text{at } (0,1)$$
$$= -2\mathbf{j}$$

Thus one should head toward the origin along the y axis. The graph of f, sketched in Fig. 16-2-9, illustrates this.

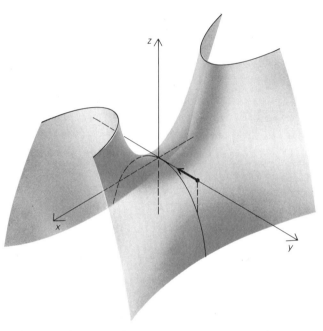

Fig. 16-2-9 Starting from $(0,1)$, moving in along the y axis makes the graph rise the steepest.

10. In order to cool the fastest, the captain should proceed in the direction in which T is decreasing the fastest; that is, in the direction $-\nabla T(1,1,1)$. But

$$\nabla T = \frac{\partial T}{\partial x}\mathbf{i} + \frac{\partial T}{\partial y}\mathbf{j} + \frac{\partial T}{\partial z}\mathbf{k}$$
$$= -e^{-x}\mathbf{i} - 2e^{-2y}\mathbf{j} + 3e^{3z}\mathbf{k}$$

Thus

$$-\nabla T(1,1,1) = e^{-1}\mathbf{i} + 2e^{-2}\mathbf{j} - 3e^{3}\mathbf{k}$$

is the direction required.

11. Let $f(x,y,z) = \sin(xy) - e^{z}$ so the surface is $f(x,y,z) = 0$. A normal is $\nabla f = y\cos(xy)\mathbf{i} + x\cos(xy)\mathbf{j} - e^{z}\mathbf{k}$. At $(1, \pi/2, 0)$ we get $-\mathbf{k}$. Thus $-\mathbf{k}$ (or $\mathbf{k}$) is the required unit normal. (It already has length 1, so there is no need to normalize.)

12. With $f(x,y,z) = z - g(x,y)$,

$$\nabla f(x,y,z) = f_x(x,y,z)\mathbf{i} + f_y(x,y,z)\mathbf{j} + f_z(x,y,z)\mathbf{k}$$
$$= -g_x(x,y)\mathbf{i} - g_y(x,y)\mathbf{j} + \mathbf{k}$$

This is exactly the normal to the tangent plane at (x,y) to the graph of g. (See p. 693).

13. By Solved Exercise 4, $\nabla(1/r) = -(\mathbf{r}/r^3)$. Therefore we can choose $V = -GMm/r$ to give $\mathbf{F} = -\nabla V$. The vector $\mathbf{F}$ points toward the origin. The level surfaces of V are $1/r = c$—that is, $r = 1/c$, a sphere. Clearly $\mathbf{F}$ is orthogonal to these surfaces.

14. If $\partial z/\partial y = 0$ but $\partial z/\partial x \neq 0$, where $z = F(x,y)$, then the gradient $\nabla F(x,y) = (\partial z/\partial x)\mathbf{i} + (\partial z/\partial y)\mathbf{j}$ is proportional to $\mathbf{i}$. The tangent line to the level curve $F(x,y) = 0$, being perpendicular to the gradient, must be vertical, so the level curve cannot be the graph of a function which is differentiable at that point.

For example, the equation $z = F(x,y) = x - y^3 = 0$ defines implicitly the function $y = f(x) = \sqrt[3]{x}$. We have $\partial z/\partial x = 1$ and $\partial z/\partial y = -3y^2$. So $\partial z/\partial y$ vanishes when $y = 0$; this is just the point on the graph $y = \sqrt[3]{x}$ where the cube root function is not differentiable.

15. Let $z = F(x,y) = x^y - 2$. Then $\partial z/\partial x = yx^{y-1}$ and $\partial z/\partial y = x^y \ln x$, so the relation is

$$yx^{y-1}\frac{dx}{dt} + x^y \ln x \frac{dy}{dt} = 0$$

Using the fact that $x^y = 2$, we can simplify this to

$$\frac{y}{x}\frac{dx}{dt} + \ln x \frac{dy}{dt} = 0$$

Solved Exercises for Section 16-3

1. We minimize the square of the distance: $d^2 = (x-1)^2 + (y-2)^2 + (z)^2$. Substituting $z^2 = x^2 + y^2$, we have the problem of minimizing

$$f(x,y) = (x-1)^2 + (y-2)^2 + x^2 + y^2$$
$$= 2x^2 + 2y^2 - 2x - 4y + 5$$

Now

$$f_x(x,y) = 4x - 2 \quad \text{and} \quad f_y(x,y) = 4y - 4$$

Thus the critical point, obtained by setting these equal to zero, is $x = \frac{1}{2}$, $y = 1$. This is the minimum point. The minimum distance is

$$d = \sqrt{(\tfrac{1}{2}-1)^2 + (1-2)^2 + (\tfrac{1}{2})^2}$$
$$= \sqrt{\tfrac{1}{4} + 1 + \tfrac{1}{4}} = \sqrt{3/2} \approx 1.225$$

16

2. Let x and y be the lengths of the sides of the base. Since the volume of the box is to be 256, the height must be $256/xy$. Two of the sides have area $x(256/xy)$, two sides have area $y(256/xy)$, and the base has area xy, so the total surface area is $A = 2x(256/xy) + 2y(256/xy) + xy = 512/y + 512/x + xy$. To minimize A, we must have

$$0 = \frac{\partial A}{\partial x} = -\frac{512}{x^2} + y \qquad 0 = \frac{\partial A}{\partial y} = -\frac{512}{y^2} + x$$

The first equation gives $y = 512/x^2$; substituting this into the second equation gives $0 = -512(x^2/512)^2 + x = -x^4/512 + x$. Discarding the extraneous root $x = 0$, we have $x^3/512 = 1$, or $x = \sqrt[3]{512} = 8$. Now $y = 512/x^2 = 8$, and the height is $256/xy = 4$, so the optimal box has a square base and is half as high as it is wide. (We have really shown only that the point $(8, 8)$ is a critical point for f, but if there is any minimum point this must be it.)

3. Since $z = 2(x^2 + y^2)e^{-x^2-y^2}$, we have

$$\frac{\partial z}{\partial x} = 4x(e^{-x^2-y^2}) + 2(x^2 + y^2)e^{-x^2-y^2}(-2x)$$

$$= 4x(e^{-x^2-y^2})(1 - x^2 - y^2)$$

and

$$\frac{\partial z}{\partial y} = 4y(e^{-x^2-y^2})(1 - x^2 - y^2)$$

These vanish when $x = y = 0$ or when $x^2 + y^2 = 1$.

4. Let $A = 2$, $B = -\frac{1}{2}$, $C = 1$. Then $A > 0$ and $AC - B^2 = 2 - \frac{1}{4} > 0$, so $2x^2 - xy + y^2$ has a minimum at $(0, 0)$. Hence f has a minimum value (of 3) at $(0, 0)$ as well.

5. We compute:

$$\frac{\partial z}{\partial x} = 2x \cos(x + 2y) - (x^2 + y^2) \sin(x + 2y)$$

$$\frac{\partial z}{\partial y} = 2y \cos(x + 2y) - 2(x^2 + y^2) \sin(x + 2y)$$

These vanish at $(0, 0)$ so $(0, 0)$ is a critical point.

5. (*cont'd*)
The second derivatives are:

$$\frac{\partial^2 z}{\partial x^2} = 2 \cos(x + 2y) - 4x \sin(x + 2y) - (x^2 + y^2) \cos(x + 2y)$$

$$\frac{\partial^2 z}{\partial x \, \partial y} = -4x \sin(x + 2y) - 2y \sin(x + 2y) - 2(x^2 + y^2) \cos(x + 2y)$$

$$\frac{\partial^2 z}{\partial y^2} = 2 \cos(x + 2y) - 8y \sin(x + 2y) - 4(x^2 + y^2) \cos(x + 2y)$$

Thus $A = 2$, $B = 0$, and $C = 2$, so $A > 0$, $AC - B^2 > 0$, and $(0, 0)$ is a local minimum.

6. Here $\partial z/\partial x = 2x/(x^2 + y^2 + 1)$ and $\partial z/\partial y = 2y/(x^2 + y^2 + 1)$, so $(0, 0)$ is the only critical point. We compute:

$$\frac{\partial^2 z}{\partial x^2} = \frac{-2x^2 + 2y^2 + 2}{(x^2 + y^2 + 1)^2}$$

$$\frac{\partial^2 z}{\partial y^2} = \frac{-2y^2 + 2x^2 + 2}{(x^2 + y^2 + 1)^2}$$

$$\frac{\partial^2 z}{\partial x \, \partial y} = \frac{-4xy}{(x^2 + y^2 + 1)^2}$$

Hence

$$A = C = 2 \quad \text{and} \quad B = 0$$

so $A > 0$, $AC - B^2 > 0$, and $(0, 0)$ is a local minimum.

7. The level curves of output are of the form $AK^\alpha L^{1-\alpha} = c$, or, solving for L,

$$L = \left(\frac{c}{A}\right)^{1/(1-\alpha)} K^{\alpha/(\alpha-1)}$$

Since $\alpha/(\alpha - 1) < 0$, these curves do look like those in Fig. 16-3-10. The partial derivatives of Q are

$$\frac{\partial Q}{\partial K} = \alpha A K^{\alpha-1} L^{1-\alpha}$$

and

$$\frac{\partial Q}{\partial L} = (1 - \alpha) A K^\alpha L^{-\alpha}$$

so there are no critical points except on the axes, where $Q = 0$. Thus the maximum must lie on the budget line $pL + qK = B$. The method of Lagrange multipliers gives the equations

$$\alpha A K^{\alpha-1} L^{1-\alpha} = \lambda q$$

$$(1 - \alpha) A K^\alpha L^{-\alpha} = \lambda p$$

$$pL + qK = B$$

7. (*cont'd*)

Eliminating λ from the first two equations gives

$$\alpha p L = (1 - \alpha) q K$$

and from the third equation we obtain

$$K = \frac{\alpha B}{q} \quad \text{and} \quad L = \frac{(1 - \alpha)B}{p}$$

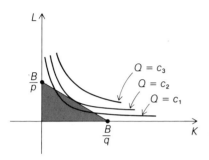

Fig. 16-3-10 What is the largest value of Q in the shaded triangle?

8. We extremize $f(x, y) = x^2 + y^2$ subject to the constraint $g(x, y) = x^6 + y^6 = 1$. The Lagrange multiplier equations are

$$2x = 6\lambda x^5$$
$$2y = 6\lambda y^5$$
$$x^6 + y^6 = 1$$

If we rewrite the first two of these equations as

$$x(6\lambda x^4 - 2) = 0$$
$$y(6\lambda y^4 - 2) = 0$$

we find the solutions $(0, \pm 1)$ and $(\pm 1, 0)$ with $\lambda = \frac{1}{3}$. If x and y are both nonzero, we have $x^4 = 1/3\lambda = y^4$, so $x = \pm y$, and we get the further solutions $(\pm \sqrt[6]{\frac{1}{2}}, \pm \sqrt[6]{\frac{1}{2}})$, with $\lambda = 2^{2/3}/3$.

To tell which points are maxima and minima, we compute; $f(0, \pm 1) = f(\pm 1, 0) = 1$, while $f(\pm \sqrt[6]{\frac{1}{2}}, \pm \sqrt[6]{\frac{1}{2}}) = 2\sqrt[3]{\frac{1}{2}} = 2^{2/3} > 1$, so the points $(0, \pm 1)$ and $(\pm 1, 0)$ are closest to the origin, while $(\pm \sqrt[6]{\frac{1}{2}}, \pm \sqrt[6]{\frac{1}{2}})$ are farthest (see Fig. 16-3-12).

8. (*cont'd*)

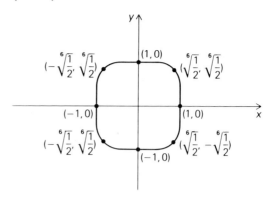

Fig. 16-3-12 Extreme points of $x^2 + y^2$ on the curve $x^6 + y^6 = 1$.

Solved Exercises for Section 16-4

1. We have $b_i = 1/m$ for $i = 1, \ldots, m$, so $AB = \sum_{i=1}^{m} a_i \cdot (1/m) = (1/m) \sum_{i=1}^{m} a_i$; it is just the average of the entries of A.

2. (a) We have

$$u = \begin{bmatrix} a & b & c \end{bmatrix} \begin{bmatrix} x \\ y \\ z \end{bmatrix} + d$$

and

$$v = \begin{bmatrix} e & f & g \end{bmatrix} \begin{bmatrix} x \\ y \\ z \end{bmatrix} + h$$

(b) The derivative matrix is

$$\frac{\partial(u, v)}{\partial(x, y, z)} = \begin{bmatrix} a & b & c \\ e & f & g \end{bmatrix}$$

all of whose entries are constants.

3. (a) Substituting for x, y, and z their expressions in t, we get

$$u = a(mt + n) + b(pt + q) + c(rt + s) + d$$
$$v = e(mt + n) + f(pt + q) + g(rt + s) + h$$

We can find the derivative matrix without multiplying out:

$$\frac{\partial(u, v)}{\partial(t)} = \begin{bmatrix} \dfrac{du}{dt} \\ \dfrac{dv}{dt} \end{bmatrix} = \begin{bmatrix} am + bp + cr \\ em + fp + gr \end{bmatrix}$$

16

3. (*cont'd*)

(b) The entries of $\partial(u, v)/\partial(t)$ are

$$[a \quad b \quad c] \begin{bmatrix} m \\ p \\ r \end{bmatrix} \quad \text{and} \quad [e \quad f \quad g] \begin{bmatrix} m \\ p \\ r \end{bmatrix}$$

Notice that they are obtained by multiplying the rows of $\partial(u, v)/\partial(x, y, z)$ by the (unique) column of $\partial(x, y, z)/\partial(t)$.

4. We find

$$\frac{\partial(x, y)}{\partial(r, \theta)} = \begin{bmatrix} \cos \theta & -r \sin \theta \\ \sin \theta & r \cos \theta \end{bmatrix}$$

When $r = 5$ and $\theta = \pi/6$, this becomes

$$\begin{bmatrix} \dfrac{\sqrt{3}}{2} & -\dfrac{5}{2} \\ \dfrac{1}{2} & \dfrac{5\sqrt{3}}{2} \end{bmatrix}$$

5. Let the unknown matrix B be

$$\begin{bmatrix} a & b \\ c & d \end{bmatrix}$$

Then

$$AB = \begin{bmatrix} 1 & 2 \\ 0 & 1 \end{bmatrix} \begin{bmatrix} a & b \\ c & d \end{bmatrix} = \begin{bmatrix} a + 2c & b + 2d \\ c & d \end{bmatrix}$$

If this equals I, then a, b, c, and d must satisfy the equations $a + 2c = 1$, $b + 2d = 0$, $c = 0$, $d = 1$. Solving for a and b, we get $a = 1$ and $b = -2$, so

$$B = \begin{bmatrix} 1 & -2 \\ 0 & 1 \end{bmatrix}$$

6. Let

$$A = \begin{bmatrix} a & b \\ c & d \end{bmatrix} \quad \text{and} \quad B = \begin{bmatrix} e & f \\ g & h \end{bmatrix}$$

Then $|A| = ad - bc$, $|B| = eh - gf$, and

$$\begin{aligned} |AB| &= \begin{vmatrix} ae + bg & af + bh \\ ce + dg & cf + dh \end{vmatrix} \\ &= (ae + bg)(cf + dh) - (ce + dg)(af + bh) \\ &= aecf + aedh + bgcf + bgdh - aecf - adgf \\ &\quad - cebh - bgdh \\ &= aedh + bgcf - adgf - cebh \\ &= (ad - bc)(eh - gf) \\ &= |A| \cdot |B| \end{aligned}$$

7. The result of Solved Exercise 3(b) may be written as

$$\frac{\partial(u, v)}{\partial(t)} = \frac{\partial(u, v)}{\partial(x, y, z)} \frac{\partial(x, y, z)}{\partial(t)}$$

8. By Worked Example 6,

$$\frac{\partial z}{\partial r} = \frac{\partial z}{\partial x} \cos \theta + \frac{\partial z}{\partial y} \sin \theta \tag{7}$$

Then

$$\frac{\partial^2 z}{\partial r^2} = \frac{\partial}{\partial r} \left[\frac{\partial z}{\partial x} \right] \cos \theta + \frac{\partial}{\partial r} \left[\frac{\partial z}{\partial y} \right] \sin \theta$$

Applying equation (7) with $\partial z/\partial x$ and $\partial z/\partial y$ replacing z, we get

$$\begin{aligned} \frac{\partial^2 z}{\partial r^2} &= \left(\frac{\partial^2 z}{\partial x^2} \cos \theta + \frac{\partial^2 z}{\partial y \, \partial x} \sin \theta \right) \cos \theta \\ &\quad + \left(\frac{\partial^2 z}{\partial x \, \partial y} \cos \theta + \frac{\partial^2 z}{\partial y^2} \sin \theta \right) \sin \theta \\ &= \frac{\partial^2 z}{\partial x^2} \cos^2 \theta + 2 \frac{\partial^2 z}{\partial y \, \partial x} \sin \theta \cos \theta + \frac{\partial^2 z}{\partial y^2} \sin^2 \theta \\ &= \frac{1}{x^2 + y^2} \left[x^2 \frac{\partial^2 z}{\partial x^2} + 2xy \frac{\partial^2 z}{\partial x \, \partial y} + y^2 \frac{\partial^2 z}{\partial y^2} \right] \end{aligned}$$

9. By the chain rule

$$\frac{\partial(t, s)}{\partial(u, v)} = \frac{\partial(t, s)}{\partial(x, y)} \frac{\partial(x, y)}{\partial(u, v)}$$

In this example,

$$\frac{\partial(x, y)}{\partial(u, v)} = \begin{bmatrix} 1 & -2 \\ 1 & 3 \end{bmatrix}$$

so

$$\begin{aligned} \frac{\partial(t, s)}{\partial(u, v)} &= \frac{\partial(t, s)}{\partial(x, y)} \begin{bmatrix} 1 & -2 \\ 1 & 3 \end{bmatrix} \\ &= \begin{bmatrix} \dfrac{\partial t}{\partial x} & \dfrac{\partial t}{\partial y} \\ \dfrac{\partial s}{\partial x} & \dfrac{\partial s}{\partial y} \end{bmatrix} \begin{bmatrix} 1 & -2 \\ 1 & 3 \end{bmatrix} \\ &= \begin{bmatrix} \dfrac{\partial t}{\partial x} + \dfrac{\partial t}{\partial y} & 3\dfrac{\partial t}{\partial y} - 2\dfrac{\partial t}{\partial x} \\ \dfrac{\partial s}{\partial x} + \dfrac{\partial s}{\partial y} & 3\dfrac{\partial s}{\partial y} - 2\dfrac{\partial s}{\partial x} \end{bmatrix} \end{aligned}$$

Solved Exercises for Section 17-1

1. The constant function $h(x, y) = 12 \geq f(x, y)$ has integral $12 \times 4 = 48$, so we get only the crude estimate $\int\int_D f(x, y)\, dx\, dy \leq 48$. To get a better one, divide D into four pieces:

$$D_1 = [0, 1] \times [1, 2]$$

$$D_2 = [0, 1] \times [2, 3]$$

$$D_3 = [1, 2] \times [1, 2]$$

$$D_4 = [1, 2] \times [2, 3]$$

Let h be the piecewise constant function given by taking the maximum value of f on each subrectangle (evaluated at the upper right-hand corner); that is,

$$h(x, y) = \begin{cases} 2 & \text{on } D_1 \\ 3 & \text{on } D_2 \\ 8 & \text{on } D_3 \\ 12 & \text{on } D_4 \end{cases}$$

The integral of h is $2 \times 1 + 3 \times 1 + 8 \times 1 + 12 \times 1 = 25$. Since $h \geq f$ we get

$$\int\int_D f(x, y)\, dx\, dy \leq 25$$

2. If ρ_1 is a piecewise constant function, then the integral of ρ_1 over D is the mass of the plate since mass = density × area. If ρ_1 is a piecewise constant density function with $\rho_1 \leq \rho$, then $\int\int_D \rho_1(x, y)\, dx\, dy \leq m$, where m is the mass of the plate with density ρ, since a lower density gives a smaller mass. Likewise, if ρ_2 is piecewise constant and $\rho_2 \geq \rho$, then $m \leq \int\int_D \rho_2(x, y)\, dx\, dy$. Thus the mass m lies to the right of L and to the left of U, so it must equal the integral $\int\int_D \rho(x, y)\, dx\, dy$.

3.
$$\int\int_D \sin(x + y)\, dx\, dy$$

$$= \int_0^{2\pi} \left[\int_0^{\pi} \sin(x + y)\, dx \right] dy$$

$$= \int_0^{2\pi} \left[-\cos(x + y) \Big|_{x=0}^{\pi} \right] dy$$

$$= \int_0^{2\pi} [\cos y - \cos(y + \pi)]\, dy$$

$$= [\sin y - \sin(y + \pi)] \Big|_{y=0}^{2\pi} = 0$$

4. The total mass is

$$\int\int_D \rho(x, y)\, dx\, dy = \int_0^1 \int_1^2 ye^{xy}\, dx\, dy$$

$$= \int_0^1 \left(e^{xy} \Big|_{x=1}^{2} \right) dy$$

$$= \int_0^1 (e^{2y} - e^y)\, dy$$

$$= \left(\frac{e^{2y}}{2} - e^y \right) \Big|_{y=0}^{1}$$

$$= \frac{e^2}{2} - e + \frac{1}{2} \approx 1.4762 \text{ grams}$$

5. The volume is

$$\int_{-1}^1 \int_0^3 (x^2 + y^2)\, dx\, dy$$

$$= \int_{-1}^1 \left(\frac{x^3}{3} + y^2 x \Big|_{x=0}^{3} \right) dy = \int_{-1}^1 (9 + 3y^2)\, dy$$

$$= 9y + y^3 \Big|_{-1}^{1} = 20$$

6. Here y ranges from x^3 to x^2, while x goes from 0 to 1. Hence the region is as shown in Fig. 17-1-20. The integral is

$$\int_0^1 \left(\frac{xy^2}{2} \Big|_{y=x^3}^{x^2} \right) dx = \int_0^1 \left(\frac{x^5}{2} - \frac{x^7}{2} \right) dx$$

$$= \left(\frac{x^6}{12} - \frac{x^8}{16} \right) \Big|_{x=0}^{1}$$

$$= \frac{1}{12} - \frac{1}{16} = \frac{1}{48}$$

Fig. 17-1-20 The region of integration for

$$\int_0^1 \int_{x^3}^{x^2} xy\, dy\, dx.$$

17

7. The region D is shown in Fig. 17-1-21. $\underline{D\text{ is a}}$ type 1 region with $\phi_1(x) = 0$, $\phi_2(x) = \sqrt{1 - x^2}$ and a type 2 region with $\psi_1(y) = 0$, $\psi_2(y) = \sqrt{1 - y^2}$. Thus

$$\int_0^1 \int_0^{\sqrt{1-x^2}} \sqrt{1 - y^2}\, dy\, dx$$

$$= \int_0^1 \int_0^{\sqrt{1-y^2}} \sqrt{1 - y^2}\, dx\, dy$$

$$= \int_0^1 \left[\sqrt{1 - y^2}\, x \,\Big|_{x=0}^{\sqrt{1-y^2}} \right] dy$$

$$= \int_0^1 (1 - y^2)\, dy$$

$$= y - \frac{y^3}{3} \,\Big|_0^1 = \frac{2}{3}$$

Evaluating the integral in the original order requires considerably more computation!

Fig. 17-1-21   The region of integration for $\int_0^1 \int_0^{\sqrt{1-x^2}} \sqrt{1 - y^2}\, dy\, dx$.

8. The region is sketched in Fig. 17-1-22. Thus the volume is

$$\iint_D (x^2 + y^2)\, dx\, dy$$

$$= \int_0^{5/2} \int_0^{(10-4y)/3} (x^2 + y^2)\, dx\, dy$$

$$= \int_0^{5/2} \left[\frac{(10 - 4y)^3}{3^4} + \frac{y^2(10 - 4y)}{3} \right] dy$$

$$= -\frac{(10 - 4y)^4}{3^4 \cdot 4 \cdot 4} + \frac{10y^3}{9} - \frac{y^4}{3} \,\Big|_0^{5/2}$$

$$= \frac{10^4}{3^4 \cdot 2^4} + \frac{10 \cdot 5^3}{3^2 \cdot 2^3} - \frac{5^4}{3 \cdot 2^4} = \frac{15625}{1296} \approx 12.056$$

(The volume can also be written as $5^6/(3^4 \cdot 2^4)$. Can any reader explain this simple factorization?)

8. (*cont'd*)

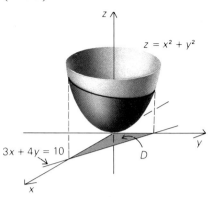

Fig. 17-1-22   The volume of the shaded region in space is $\iint_D (x^2 + y^2)\, dx\, dy$.

9. First we compute

$$\iint_D f(x, y)\, dx\, dy$$

$$= \int_0^\pi \int_0^\pi x \sin^2(xy)\, dx\, dy$$

$$= \int_0^\pi \left(\int_0^\pi \frac{1 - \cos(2xy)}{2}\, x\, dy \right) dx$$

$$= \int_0^\pi \left(\frac{y}{2} - \frac{\sin 2(xy)}{4x} \right) x \,\Big|_{y=0}^\pi dx$$

$$= \int_0^\pi \left(\frac{\pi x}{2} - \frac{\sin(2\pi x)}{4} \right) dx$$

$$= \left(\frac{\pi x^2}{4} + \frac{\cos(2\pi x)}{8\pi} \right) \Big|_0^\pi$$

$$= \frac{\pi^3}{4} + \frac{[\cos(2\pi^2) - 1]}{8\pi}$$

Thus the average value of f is

$$\frac{\pi^3/4 + [\cos(2\pi^2) - 1]/8\pi}{\pi^2}$$

$$= \frac{\pi}{4} + \frac{\cos(2\pi^2) - 1}{8\pi^3} \approx 0.7839$$

10. Writing z as a function of x and y, we have $z = g(x, y) = \pm\sqrt{f(x)^2 - y^2}$. The domain of this function consists of those (x, y) for which $-f(x) \le y \le f(x)$, so the surface in question lies over the region D (of type 1) defined by $a \le x \le b$, $-f(x) \le y \le f(x)$. (See Fig. 17-1-23.)

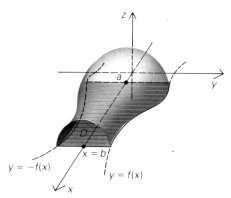

Fig. 17-1-23 Finding the surface area when $y = f(x)$ is rotated about the x axis.

The partial derivatives of g are

$$g_x(x, y) = f'(x)f(x)/\sqrt{f(x)^2 - y^2}$$

and

$$g_y(x, y) = -y/\sqrt{f(x)^2 - y^2},$$

so the surface area integrand is

$$\sqrt{1 + \frac{f'(x)^2 f(x)^2}{f(x)^2 - y^2} + \frac{y^2}{f(x)^2 - y^2}}$$

$$= \sqrt{\frac{f(x)^2 - y^2 + f'(x)^2 f(x)^2 + y^2}{f(x)^2 - y^2}}$$

$$= f(x)\sqrt{\frac{1 + f'(x)^2}{f(x)^2 - y^2}}$$

and the area is

$$A = 2 \int_a^b \int_{-f(x)}^{f(x)} f(x)\sqrt{\frac{1 + f'(x)^2}{f(x)^2 - y^2}}\, dy\, dx$$

(The factor of 2 occurs since half the surface lies below the xy plane.) We can carry out the integration over y:

$$A = 2\int_a^b f(x)\sqrt{1 + f'(x)^2}\left[\int_{-f(x)}^{f(x)} \frac{1}{\sqrt{f(x)^2 - y^2}}\, dy\right] dx$$

$$= 2\int_a^b f(x)\sqrt{1 + f'(x)^2}\left[\sin^{-1}\left(\frac{y}{f(x)}\right)\Big|_{y=-f(x)}^{f(x)}\right] dx$$

$$= 2\int_a^b f(x)\sqrt{1 + f'(x)^2}\left(\frac{\pi}{2} + \frac{\pi}{2}\right) dx$$

$$= 2\pi\int_a^b f(x)\sqrt{1 + f'(x)^2}\, dx$$

This is the formula for the area of a surface of revolution on p. 467.

Solved Exercises for Section 17-2

1. $\displaystyle\iiint_W (x + 2y + 3z)^2\, dy\, dz\, dx$

$$= \int_0^1 \int_0^{1/3} \int_{-1/2}^0 (x + 2y + 3z)^2\, dy\, dz\, dx$$

$$= \int_0^1 \int_0^{1/3} \left[\frac{(x + 2y + 3z)^3}{6}\Big|_{y=-1/2}^0\right] dz\, dx$$

$$= \int_0^1 \int_0^{1/3} \frac{1}{6}\left[(x + 3z)^3 - (x + 3z - 1)^3\right] dz\, dx$$

$$= \int_0^1 \frac{1}{6}\left[\frac{(x + 3z)^4}{12} - \frac{(x + 3z - 1)^4}{12}\Big|_{z=0}^{1/3}\right] dx$$

$$= \int_0^1 \frac{1}{72}\left[(x + 1)^4 + (x - 1)^4 - 2x^4\right] dx$$

$$= \frac{1}{72}\frac{1}{5}\left[(x + 1)^5 + (x - 1)^5 - 2x^5\right]_{x=0}^1 = \frac{1}{12}$$

2. Since density $\cdot$ volume = mass, the mass of the volume $dx\, dy\, dz$ is $\rho(x, y, z)\, dx\, dy\, dz$. Adding, we get $\iiint_W \rho(x, y, z)\, dx\, dy\, dz$ = total mass of W as in Solved Exercise 2, Section 17-1.

3. $\displaystyle\int_0^1 \int_0^1 \int_0^1 e^{x+y+z}\, dx\, dy\, dz$

$$= \int_0^1 \int_0^1 \left(e^{x+y+z}\Big|_{x=0}^1\right) dy\, dz$$

$$= \int_0^1 \int_0^1 (e^{1+y+z} - e^{y+z})\, dy\, dz$$

$$= \int_0^1 [e^{1+y+z} - e^{y+z}]_{y=0}^1\, dz$$

$$= \int_0^1 [e^{2+z} - 2e^{1+z} + e^z]\, dz$$

$$= [e^{2+z} - 2e^{1+z} + e^z]_0^1$$

$$= e^3 - 3e^2 + 3e - 1 = (e - 1)^3$$

4. W is of type II, so it can be expressed as the set of all (x, y, z) such that

$$a \le z \le b, \qquad \phi_1(z) \le y \le \phi_2(z)$$
$$\rho_1(z, y) \le x \le \rho_2(z, y)$$

Then

$$\iiint_W f(x, y, z)\, dx\, dy\, dz$$

$$= \int_a^b \int_{\phi_1(z)}^{\phi_2(z)} \int_{\rho_1(z,y)}^{\rho_2(z,y)} f(x, y, z)\, dx\, dy\, dz$$

W may also be expressed as the set of all (x, y, z) such that

$$c \le y \le d, \qquad \psi_1(y) \le z \le \psi_2(y)$$
$$\rho_1(z, y) \le x \le \rho_2(z, y)$$

Then

17

4. (cont'd)

$$\iiint_W f(x, y, z)\, dx\, dy\, dz$$

$$= \int_c^d \int_{\psi_1(y)}^{\psi_2(y)} \int_{\rho_1(z,y)}^{\rho_2(z,y)} f(x, y, z)\, dx\, dz\, dy$$

5. W can be expressed as the set of (x, y, z) with $\rho_1(z, y) = 0 \le x \le (z - y^2)^{1/2} = \rho_2(z, y)$ and (z, y) in D, where D is the subset of the yz plane with $0 \le z \le 2$ and $0 \le y \le z^{1/2}$ (see Fig. 17-2-12). Therefore

$$\iiint_W x\, dx\, dy\, dz$$

$$= \iint_D \left(\int_{\rho_1(z,y)}^{\rho_2(z,y)} x\, dx \right) dy\, dz$$

$$= \int_0^2 \left[\int_0^{z^{1/2}} \left(\int_0^{(z - y^2)^{1/2}} x\, dx \right) dy \right] dz$$

$$= \int_0^2 \int_0^{z^{1/2}} \left(\frac{z - y^2}{2} \right) dy\, dz$$

$$= \frac{1}{2} \int_0^2 \left(z^{3/2} - \frac{z^{3/2}}{3} \right) dz$$

$$= \frac{1}{2} \int_0^2 \frac{2}{3} z^{3/2}\, dz$$

$$= \left[\frac{2}{15} z^{5/2} \right]_0^2 = \frac{2}{15} 2^{5/2} = \frac{8\sqrt{2}}{15}$$

which agrees with our other answer.

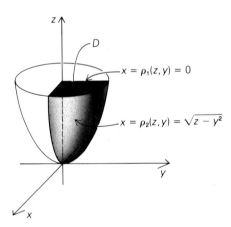

Fig. 17-2-12 W as a type II region.

6.

$$\int_0^1 \int_0^x \int_{x^2 + y^2}^1 dz\, dy\, dx$$

$$= \int_0^1 \int_0^x (1 - x^2 - y^2)\, dy\, dx$$

$$= \int_0^1 \left(x - x^3 - \frac{x^3}{3} \right) dx$$

$$= \frac{1}{2} - \frac{1}{4} - \frac{1}{12} = \frac{1}{6}$$

The region is sketched in Fig. 17-2-13.

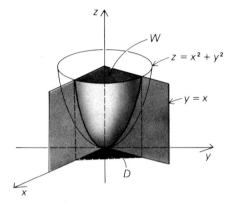

Fig. 17-2-13 The region W (type I) for Solved Exercise 6.

7. (a) Let c be the constant of proportionality. Then $T = c(x^2 + y^2 + z^2)$ and the average temperature is $\overline{T} = \frac{1}{8} \iiint_W T\, dx\, dy\, dz$, since the volume of the cube is 8. Thus

$$\overline{T} = \frac{c}{8} \int_{-1}^1 \int_{-1}^1 \int_{-1}^1 (x^2 + y^2 + z^2)\, dx\, dy\, dz$$

The triple integral is the sum of the integrals of x^2, y^2, and z^2. Since x, y, and z enter symmetrically into the description of the cube, the three integrals will be equal, so

$$\overline{T} = \frac{3c}{8} \int_{-1}^1 \int_{-1}^1 \int_{-1}^1 z^2\, dx\, dy\, dz$$

$$= \frac{3c}{8} \int_{-1}^1 z^2 \left(\int_{-1}^1 \int_{-1}^1 dx\, dy \right) dz$$

The inner integral is equal to the area of the square $[-1, 1] \times [-1, 1]$. The area of that square is 4, so

$$\overline{T} = \frac{3c}{8} \int_{-1}^1 4z^2\, dz = \frac{3c}{2} \left(\frac{z^3}{3} \right) \Big|_{-1}^1 = c$$

(b) The temperature is equal to the average temperature, where $c(x^2 + y^2 + z^2) = c$; that is, on the sphere $x^2 + y^2 + z^2 = 1$, which is inscribed in the cube W.

8. According to formula (8),

$$V(x_1, y_1, z_1)$$

$$= G \iiint_W \frac{\rho(x, y, z)\, dx\, dy\, dz}{\sqrt{(x - x_1)^2 + (y - y_1)^2 + (z - z_1)^2}}$$

$$= G\rho \iiint_W \frac{dx\, dy\, dz}{\sqrt{(x - x_1)^2 + (y - y_1)^2 + (z - z_1)^2}}$$

$$= G\,[\rho \text{ volume }(W)] \times$$

$$\frac{\displaystyle\iiint_W \frac{dx\, dy\, dz}{\sqrt{(x - x_1)^2 + (y - y_1)^2 + (z - z_1)^2}}}{\text{volume }(W)}$$

$$= GM \left(\overline{\frac{1}{r}}\right)$$

as required.

9. In polar coordinates, D is described by the set of points (r, θ) such that $1 \le r \le 2,\ 0 \le \theta \le \pi/2$. Hence

$$\iint_D \ln(x^2 + y^2)\, dx\, dy$$

$$= \int_{\theta=0}^{\pi/2} \int_{r=1}^{2} \ln(r^2)\, r\, dr\, d\theta$$

$$= \int_0^{\pi/2} \int_1^2 2(\ln r) \cdot r\, dr\, d\theta$$

$$= \int_0^{\pi/2} \left(\frac{r^2}{2}(2\ln r - 1) \Big|_{r=1}^{2} \right) d\theta \quad \text{(by parts)}$$

$$= \int_0^{\pi/2} \left(4\ln 2 - \frac{3}{2} \right) d\theta = \frac{\pi}{2}\left(4\ln 2 - \frac{3}{2} \right)$$

10. In spherical coordinates, W is described by

$$0 \le \rho \le 1 \qquad 0 \le \phi \le \pi \qquad 0 \le \theta \le 2\pi$$

Hence

$$\iiint_W \exp\left[(x^2 + y^2 + z^2)^{3/2}\right] dx\, dy\, dz$$

$$= \int_0^{2\pi} \int_0^{\pi} \int_0^{1} \exp(\rho^3) \cdot \rho^2 \sin\phi\, d\rho\, d\phi\, d\theta$$

$$= \frac{1}{3} \int_0^{2\pi} \int_0^{\pi} \left(\exp(\rho^3) \Big|_0^1 \right) \sin\phi\, d\phi\, d\theta$$

$$= \frac{1}{3} \int_0^{2\pi} \int_0^{\pi} (e - 1)\sin\phi\, d\phi\, d\theta$$

$$= \frac{1}{3}(e - 1) \int_0^{2\pi} \left((-\cos\phi) \Big|_{\phi=0}^{\pi} \right) d\theta$$

$$= \frac{1}{3}(e - 1) \int_0^{2\pi} 2\, d\theta = \frac{2}{3}(e - 1)(2\pi - 0)$$

$$= \frac{4\pi}{3}(e - 1)$$

11. Before evaluating the integral in formula (8), we make some observations which will simplify the computation. Since G and the density are constants, we may ignore them at first. Since the attracting body W is symmetric with respect to all rotations about the origin, the potential function $V(x_1, y_1, z_1)$ must itself be symmetric—thus $V(x_1, y_1, z_1)$ depends only on the distance $R = \sqrt{x_1^2 + y_1^2 + z_1^2}$ from the origin. Our computation will be simplest if we look at the point $(0, 0, R)$ on the z axis (see Fig. 17-2-14). Thus our integral is

$$V(0, 0, R) = \iiint_W \frac{dx\, dy\, dz}{\sqrt{x^2 + y^2 + (z - R)^2}}$$

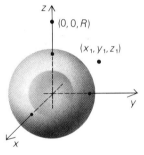

Fig. 17-2-14 The gravitational potential at (x_1, y_1, z_1) is the same as at $(0, 0, R)$, where $R = \sqrt{x_1^2 + y_1^2 + z_1^2}$.

In spherical coordinates, W is described by the inequalities $\rho_1 \le \rho \le \rho_2,\ 0 \le \theta \le 2\pi$, and $0 \le \phi \le \pi$, so by formula (11),

$$V(0, 0, R) =$$

$$\int_{\rho_1}^{\rho_2} \int_0^{\pi} \int_0^{2\pi} \frac{\rho^2 \sin\phi\, d\theta\, d\phi\, d\rho}{\sqrt{\rho^2 \sin^2\phi (\cos^2\theta + \sin^2\theta) + (\rho\cos\phi - R)^2}}$$

Replacing $\cos^2\theta + \sin^2\theta$ by 1, the integrand no longer involves θ, so we may integrate over θ to get

$$V(0, 0, R)$$

$$= 2\pi \int_{\rho_1}^{\rho_2} \int_0^{\pi} \frac{\rho^2 \sin\phi\, d\phi\, d\rho}{\sqrt{\rho^2 \sin^2\phi\,(\rho\cos\phi - R)^2}}$$

$$= 2\pi \int_{\rho_1}^{\rho_2} \rho^2 \left(\int_0^{\pi} \frac{\sin\phi\, d\phi}{\sqrt{\rho^2 - 2R\rho\cos\phi + R^2}} \right) d\rho$$

17

11. (*cont'd*)

The inner integral is easily evaluated by the substitution $u = -2R\rho \cos \phi$: it becomes

$$\frac{1}{2R\rho} \int_{-2R\rho}^{2R\rho} (\rho^2 + u + R^2)^{-1/2} \, du$$

$$= \frac{2}{2R\rho} (\rho^2 + u + R^2)^{1/2} \Big|_{-2R\rho}^{2R\rho}$$

$$= \frac{1}{R\rho} [(\rho^2 + 2R\rho + R^2)^{1/2} - (\rho^2 - 2R\rho + R^2)^{1/2}]$$

$$= \frac{1}{R\rho} \{ [(\rho + R)^2]^{1/2} - [(\rho - R)^2]^{1/2} \}$$

$$= \frac{1}{R\rho} (\rho + R - |\rho - R|)$$

The expression $\rho + R$ is always positive, but $\rho - R$ may not be, so we must keep the absolute value sign. (Here we have used the formula $\sqrt{x^2} = |x|$.) Substituting into the formula for V, we get

$$V(0, 0, R) = 2\pi \int_{\rho_1}^{\rho_2} \frac{\rho^2}{R\rho} (\rho + R - |\rho - R|) \, d\rho$$

$$= \frac{2\pi}{R} \int_{\rho_1}^{\rho_2} \rho(\rho + R - |\rho - R|) \, d\rho$$

We will now consider two possibilities for R, corresponding to the gravitational potential for objects outside and inside the hollow ball W.

If $R \geq \rho_2$ (that is, (x_1, y_1, z_1) is outside W), then $|\rho - R| = R - \rho$ for all ρ in $[\rho_1, \rho_2]$, so

$$V(0, 0, R) = \frac{2\pi}{R} \int_{\rho_1}^{\rho_2} \rho[\rho + R - (R - \rho)] \, d\rho$$

$$= \frac{4\pi}{R} \int_{\rho_1}^{\rho_2} \rho^2 \, d\rho$$

$$= \frac{1}{R} \frac{4\pi}{3} (\rho_2^3 - \rho_1^3)$$

The factor $(4\pi/3)(\rho_2^3 - \rho_1^3)$ is just the volume of W. Putting back the constants G and the mass density, we find that the gravitational potential is GM/R, where M is the mass of W. Thus V is just as it would be if all the mass of W were concentrated at the central point (see Solved Exercise 13, Section 16-2).

If $R \leq \rho_1$ (that is, (x_1, y_1, z_1) is inside the hole), then $|\rho - R| = \rho - R$ for ρ in $[\rho_1, \rho_2]$, and

$$V(0, 0, R) = \frac{2\pi}{R} \int_{\rho_1}^{\rho_2} \rho[\rho + R - (\rho - R)] \, d\rho$$

$$= 4\pi \int_{\rho_1}^{\rho_2} \rho \, d\rho = 2\pi(\rho_2^2 - \rho_1^2)$$

11. (*cont'd*)

The result is independent of R, so the potential V is constant inside the hole. Since the gravitational force is minus the gradient of V, we conclude that there is no gravitational force inside a uniform hollow planet!

We leave it to you (Problem 13) to compute $V(0, 0, R)$ for the case $\rho_1 < R < \rho_2$.

12. We have already shown that

$$\lim_{a \to \infty} \int\int_{D_a} e^{-(x^2+y^2)} \, dx \, dy$$

exists. Thus it suffices to show that

$$\lim_{a \to \infty} \left(\int\int_{R_a} e^{-(x^2+y^2)} \, dx \, dy - \int\int_{D_a} e^{-(x^2+y^2)} \, dx \, dy \right)$$

equals zero. The limit equals

$$\lim_{a \to \infty} \int\int_{C_a} e^{-(x^2+y^2)} \, dx \, dy$$

where C_a is the region between R_a and D_a (see Fig. 17-2-15). In the region C_a, $\sqrt{x^2 + y^2} \geq a$ (the radius of D_a), so $e^{-(x^2+y^2)} \leq e^{-a^2}$. Thus

$$0 \leq \int\int_{C_a} e^{-(x^2+y^2)} \, dx \, dy \leq \int\int_{C_a} e^{-a^2} \, dx \, dy$$

$$= e^{-a^2} \text{ area } (C_a)$$

$$= e^{-a^2}(4a^2 - \pi a^2)$$

$$= (4 - \pi)a^2 e^{-a^2}$$

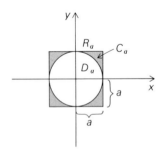

Fig. 17-2-15 C_a is the shaded region between R_a and D_a.

Thus it is enough to show that

$$\lim_{a \to \infty} a^2 e^{-a^2} = 0$$

12. (*cont'd*)

But, by L'Hôpital's rule (see Section 12-3),

$$\lim_{a\to\infty} a^2 e^{-a^2} = \lim_{a\to\infty} \left(\frac{a^2}{e^{a^2}}\right) \qquad (\tfrac{\infty}{\infty}\text{ form})$$

$$= \lim_{a\to\infty} \left(\frac{2a}{2ae^{a^2}}\right)$$

(differentiating numerator and denominator)

$$= \lim_{a\to\infty} \left(\frac{1}{e^{a^2}}\right) = 0$$

as required.

Solved Exercises for Section 18-1

1. Since the kinetic energy is zero at the beginning and the end of the process, the change in energy is zero and the sum of the work is zero. The two sources of work are you and the force field; since the work done by the force field is -5.28, the work done by you must be 5.28.

2. Let the path be given by $(x, y, z) = \boldsymbol{\sigma}(t)$, where $\boldsymbol{\sigma}(t_1) = (x_1, y_1, z_1)$ and $\boldsymbol{\sigma}(t_2) = (x_2, y_2, z_2)$. Think of x, y, and z as functions of t. Then $\boldsymbol{\sigma}'(t) = (dx/dt)\mathbf{i} + (dy/dt)\mathbf{j} + (dz/dt)\mathbf{k}$, and

$$W = \int_{t_1}^{t_2} -\frac{1}{(x^2+y^2+z^2)^{3/2}}(x\mathbf{i}+y\mathbf{j}+z\mathbf{k})\cdot$$
$$\left(\frac{dx}{dt}\mathbf{i}+\frac{dy}{dt}\mathbf{j}+\frac{dz}{dt}\mathbf{k}\right)dt$$

$$= -\int_{t_1}^{t_2} \frac{x(dx/dt)+y(dy/dt)+z(dz/dt)}{(x^2+y^2+z^2)^{3/2}}\,dt$$

$$= \int_{t_1}^{t_2} -\frac{\tfrac{1}{2}(d/dt)(x^2+y^2+z^2)}{(x^2+y^2+z^2)^{3/2}}\,dt$$

$$= \int_{t_1}^{t_2} \frac{d}{dt}(x^2+y^2+z^2)^{-1/2}\,dt$$

$$= (x^2+y^2+z^2)^{-1/2}\Big|_{t_1}^{t_2}$$

$$= (x_2^2+y_2^2+z_2^2)^{-1/2} - (x_1^2+y_1^2+z_1^2)^{-1/2}$$

$$= \frac{1}{R_2} - \frac{1}{R_1}$$

So the work done by the gravitational field when a particle moves from (x_1, y_1, z_1) to (x_2, y_2, z_2) is $1/R_2 - 1/R_1$. Notice that in this case the work is independent of the path taken between the two points.

3. We write $C = C_1 + C_2 + C_3$, where C_1 is the line segment from $(1, 0, 0)$ to $(0, 1, 0)$, C_2 is the line segment from $(0, 1, 0)$ to $(0, 0, 1)$, and C_3 is the line segment from $(0, 0, 1)$ to $(1, 0, 0)$. Parametrizations on $[0, 1]$ for these segments are

C_1: $(x, y, z) = (1-t, t, 0)$
so $dx = -dt$, $dy = dt$, $dz = 0$

C_2: $(x, y, z) = (0, 1-t, t)$
so $dx = 0$, $dy = -dt$, $dz = dt$

C_3: $(x, y, z) = (t, 0, 1-t)$
so $dx = dt$, $dy = 0$, $dz = -dt$

Then

$$\int_C \sin(\pi x)\,dy - \cos(\pi y)\,dz$$

$$= \sum_{i=1}^{3} \int_{C_i} \sin(\pi x)\,dy - \cos(\pi y)\,dz$$

$$= \int_0^1 \sin[\pi(1-t)]\,dt - \cos(\pi t)\cdot 0$$

$$\quad + \int_0^1 \sin(\pi\cdot 0)(-dt) - \cos[\pi(1-t)]\,dt$$

$$\quad + \int_0^1 \sin\pi t\cdot 0 - \cos(\pi\cdot 0)(-dt)$$

$$= \int_0^1 \sin[\pi(1-t)]\,dt - \int_0^1 \cos[\pi(1-t)]\,dt$$

$$\quad + \int_0^1 dt$$

$$= -\frac{1}{\pi}\{-\cos[\pi(1-t)]\}\Big|_0^1$$

$$\quad - \left(-\frac{1}{\pi}\right)\{\sin[\pi(1-t)]\}\Big|_0^1 + 1$$

$$= \frac{1}{\pi}(\cos 0 - \cos\pi) + \frac{1}{\pi}(\sin 0 - \sin\pi) + 1$$

$$= \frac{1}{\pi}[1-(-1)] + \frac{1}{\pi}(0-0) + 1 = \frac{2}{\pi} + 1$$

4. Since $\int_C \boldsymbol{\Phi}(\mathbf{r})\cdot d\mathbf{r} = \int_{t_1}^{t_2} \boldsymbol{\Phi}(\boldsymbol{\sigma}(t))\cdot\boldsymbol{\sigma}'(t)\,dt$, this integral will be zero because $\boldsymbol{\Phi}(\boldsymbol{\sigma}(t))\cdot\boldsymbol{\sigma}'(t) = 0$, as $\boldsymbol{\Phi}$ and $\boldsymbol{\sigma}'$ are orthogonal.

18

5. In this case we have

$$\frac{dx}{d\theta} = -3\cos^2\theta\sin\theta \qquad \frac{dy}{d\theta} = 3\sin^2\theta\cos\theta$$

$$\frac{dz}{d\theta} = 1$$

and so, substituting $x = \cos^3\theta$, $y = \sin^3\theta$, $z = \theta$ and replacing dx by $(dx/d\theta)\,d\theta$ and so on, we get

$$\int_C \sin z\,dx + \cos z\,dy - (xy)^{1/3}\,dz$$

$$= \int_0^{7\pi/2} (-3\cos^2\theta\sin^2\theta + 3\sin^2\theta\cos^2\theta$$

$$- \cos\theta\sin\theta)\,d\theta$$

$$= -\int_0^{7\pi/2}\cos\theta\sin\theta\,d\theta = -[\tfrac{1}{2}\sin^2\theta]_0^{7\pi/2} = -\tfrac{1}{2}$$

6. Formula (13) is applicable:

$$\int_C f_x\,dx + f_y\,dy = f(\boldsymbol{\sigma}(1)) - f(\boldsymbol{\sigma}(0))$$

$$= f(e, 1, \ln 2) - f(1, 0, 0) = e^3$$

7. If $a\,dx + b\,dy$ is an exact differential, then by definition $a = f_x$ and $b = f_y$. Thus

$$a_y = f_{xy} \quad \text{and} \quad b_x = f_{yx}$$

These are equal by Theorem 1 of Section 16-1.

8. (a) Choose the closed path C to be the perimeter of the unit square. As in Worked Example 5, we get

$$\int_C x^2\,dx + xy\,dy = \int_0^1 t^2\,dt + \int_0^1 1\cdot 0 + t\,dt$$

$$+ \int_0^1 (1-t)^2\,(-dt) + 0 + \int_0^1 0\,dt$$

$$= \tfrac{1}{3} + 0 + \tfrac{1}{2} - \tfrac{1}{3} + 0 + 0 = \tfrac{1}{2}$$

Since this is not zero, $x^2\,dx + xy\,dy$ is not exact.

(b) Here $a(x, y) = x^2$ and $b(x, y) = xy$, so

$$a_y(x, y) = 0 \qquad b_x(x, y) = y$$

Since these are not equal, $x^2\,dx + xy\,dy$ cannot be exact by Solved Exercise 7.

Solved Exercises for Section 18-2

1. If $\boldsymbol{\Phi}(x, y) = P\mathbf{i} + Q\mathbf{j}$, then $\boldsymbol{\Phi}(\mathbf{r})\cdot d\mathbf{r} = P\,dx + Q\,dy$. Now we apply Green's theorem to the case where D is the disk of radius r, $Q = -x$, and $P = y$, so $\partial Q/\partial x - \partial P/\partial y = -2$. Thus the integral is $\iint_D (-2)\,dx\,dy = (-2)$ (area of D) $= -2\pi r^2$.

2. By Green's theorem

$$\int_C PQ\,dx + PQ\,dy$$

$$= \iint_D \left[\frac{\partial}{\partial x}(PQ) - \frac{\partial}{\partial y}(PQ) \right] dx\,dy$$

$$= \iint_D \left(\frac{\partial P}{\partial x}Q + P\frac{\partial Q}{\partial x} - \frac{\partial P}{\partial y}Q - P\frac{\partial Q}{\partial y} \right) dx\,dy$$

$$= \iint_D \left[Q\left(\frac{\partial P}{\partial x} - \frac{\partial P}{\partial y} \right) + P\left(\frac{\partial Q}{\partial x} - \frac{\partial Q}{\partial y} \right) \right] dx\,dy$$

3. We could, of course, evaluate the integral directly, but it is easier to use Green's theorem. Let D be the unit square (bounded by the lines $x = 0$, $y = 0$, $x = 1$, and $y = 1$). Then

$$\int_C (y^4 + x^3)\,dx + 2x^6\,dy$$

$$= \iint_D \left[\frac{\partial}{\partial x}2x^6 - \frac{\partial}{\partial y}(y^4 + x^3) \right] dx\,dy$$

$$= \iint_D (12x^5 - 4y^3)\,dx\,dy$$

$$= \int_0^1 \left[\int_0^1 (12x^5 - 4y^3)\,dx \right] dy$$

$$= \int_0^1 (2 - 4y^3)\,dy = 1$$

4. Here $P = x$ and $Q = xy$. We have $\partial P/\partial y = 0$ and $\partial Q/\partial x = y$, so $\partial P/\partial y \neq \partial Q/\partial x$. Hence $x\mathbf{i} + xy\mathbf{j}$ is not a gradient and no such f can exist.

5. (a) The curve C goes from $(1, 0)$ to $(e, 1)$. Since $\partial(2x\cos y)/\partial y = -2x\sin y = \partial(-x^2\sin y)/\partial x$, the integrand is exact by Theorem 5. Thus we can replace C by any curve having the same endpoints, in particular by the polygonal path from $(1, 0)$ to $(e, 0)$ to $(e, 1)$. Thus the line integral must be equal to

$$\int_1^e 2t\cos 0\,dt + \int_0^1 -e^2\sin t\,dt$$

$$= (e^2 - 1) + e^2(\cos 1 - 1)$$

$$= e^2\cos 1 - 1$$

5. *(cont'd)*

Alternatively, using the antiderivative $u = f(x, y) = x^2 \cos y$, which may be found by the method of Worked Example 4,

$$\int_C 2x \cos y \, dx - x^2 \sin y \, dy$$

$$= \int_C \nabla f(\mathbf{r}) \cdot d\mathbf{r} = f(e, 1) - f(1, 0)$$

$$= e^2 \cos 1 - 1$$

(b) The solution is $x^2 \cos y = C$. Since $y = 0$ when $x = 3$ we have $x^2 \cos y = 9$; that is, $y = \cos^{-1}(9/x^2)$.

6. By the quotient rule for derivatives,

$$\frac{\partial P}{\partial y} = \frac{-(x^2 + y^2) + y(2y)}{(x^2 + y^2)^2} = \frac{y^2 - x^2}{(x^2 + y^2)^2}$$

and

$$\frac{\partial Q}{\partial x} = \frac{(x^2 + y^2) - x(2x)}{(x^2 + y^2)^2} = \frac{y^2 - x^2}{(x^2 + y^2)^2}$$

so $\partial P / \partial y = \partial Q / \partial x$. If $(x, y) = (\cos t, \sin t)$, $0 \le t \le 2\pi$, is the unit circle, then

$$\int_C P \, dx + Q \, dy$$

$$= \int_0^{2\pi} -\frac{\sin t}{1} \cdot (-\sin t \, dt) + \frac{\cos t}{1} \cos t \, dt$$

$$= \int_0^{2\pi} dt = 2\pi$$

Thus $P \, dx + Q \, dy$ is not exact, or its integral around this closed curve would be zero. This does not (surprisingly?) contradict Theorem 5, since P and Q are not defined on the whole plane (they are not defined at $(0, 0)$).

7. The function f is the integral of $P \, dx + Q \, dy$ along any path from $(0, 0)$ to (x, y). (In the proof of Theorem 5, we used paths parallel to the axes.) The expression $\int_0^1 xa(tx, ty) \, dt + yb(tx, ty) \, dt$ is just the integral of $P \, dx + Q \, dy$ along the path (tx, ty), $0 \le t \le 1$; that is, the straight line joining $(0, 0)$ to (x, y).

8. Let us write out the components. Since $\nabla f = (\partial f / \partial x, \partial f / \partial y, \partial f / \partial z)$ we have, by definition,

$$\nabla \times \nabla f = \begin{vmatrix} \mathbf{i} & \mathbf{j} & \mathbf{k} \\ \dfrac{\partial}{\partial x} & \dfrac{\partial}{\partial y} & \dfrac{\partial}{\partial z} \\ \dfrac{\partial f}{\partial x} & \dfrac{\partial f}{\partial y} & \dfrac{\partial f}{\partial z} \end{vmatrix}$$

$$= \left(\frac{\partial^2 f}{\partial y \, \partial z} - \frac{\partial^2 f}{\partial z \, \partial y} \right) \mathbf{i} +$$

$$\left(\frac{\partial^2 f}{\partial z \, \partial x} - \frac{\partial^2 f}{\partial x \, \partial z} \right) \mathbf{j} + \left(\frac{\partial^2 f}{\partial x \, \partial y} - \frac{\partial^2 f}{\partial y \, \partial x} \right) \mathbf{k}$$

Each component is zero because of the symmetry property of mixed partial derivatives; hence the result follows.

9. Refer to Fig. 18-2-20; C is the triangle in question and S is the region it bounds. By Stokes' theorem,

$$\int_C \mathbf{F} = \int_S (\nabla \times \mathbf{F}) \cdot \mathbf{n} \, dA$$

Now

$$\nabla \times \mathbf{F} = \begin{vmatrix} \dfrac{\partial}{\partial x} & \dfrac{\partial}{\partial y} & \dfrac{\partial}{\partial z} \\ x^2 & y^2 & -z \\ \mathbf{i} & \mathbf{j} & \mathbf{k} \end{vmatrix}$$

$$= 0\mathbf{i} + 0\mathbf{j} + 0\mathbf{k} = \mathbf{O}$$

Therefore the integral of $\mathbf{F}$ around C is zero.

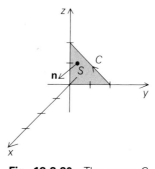

Fig. 18-2-20 The curve C of integration for Solved Exercise 9.

18

10. By the divergence theorem, the flux of **V** across *S* equals

$$\iiint_W \text{div } \mathbf{V} \, dx \, dy \, dz$$

where *W* is the ball $x^2 + y^2 + z^2 \leq 1$. But

$$\text{div } \mathbf{V}\,(x, y, z) = \frac{\partial}{\partial x}(2x) + \frac{\partial}{\partial y}(2y) + \frac{\partial}{\partial z}(2z)$$
$$= 6$$

Thus the flux is

$$6 \times \text{volume }(W) = 6 \cdot \frac{4}{3}\pi = 8\pi$$

11. Let *W* be a region in space with boundary surface *S*. By the divergence theorem,

$$\left\{\begin{array}{c} \text{Flux of } \mathbf{E} \\ \text{across } S \end{array}\right\} = \iint_S \mathbf{E} \cdot \mathbf{n} \, dA$$

$$= \iiint_W \text{div } \mathbf{E} \, dx \, dy \, dz$$

$$= \iiint_W \rho(x, y, z) dx \, dy \, dz$$

since div $\mathbf{E} = \rho$ by assumption. But since ρ is the charge per unit volume

$$Q = \iiint_W \rho \, dx \, dy \, dz$$

is the total charge inside *S*.

Appendix B

APPENDIX TO CHAPTER 13:
SERIES, COMPLEX NUMBERS, AND e^{ix}

Complex numbers provide a square root for -1.

This section provides a brief introduction to the algebra and geometry of complex numbers; i.e., numbers of the form $a + b\sqrt{-1}$. We point out the utility of complex numbers by examining the series expansions for $\sin x$, $\cos x$, and e^x derived in Section 13-3. This leads directly to Euler's formula relating the numbers 0, 1, e, π, and $\sqrt{-1}$: $e^{\pi\sqrt{-1}} + 1 = 0$.

HYPERBOLIC, EXPONENTIAL, AND TRIGONOMETRIC FUNCTIONS

If we compare the three power series,

$$\sin x = x - \frac{x^3}{3!} + \frac{x^5}{5} - \cdots \qquad (1)$$

$$\cos x = 1 - \frac{x^2}{2!} + \frac{x^4}{4!} - \cdots \qquad (2)$$

$$e^x = 1 + x + \frac{x^2}{2!} + \frac{x^3}{3!} + \cdots \qquad (3)$$

it looks as if $\sin x$ and $\cos x$ are almost the "odd and even parts" of e^x. We may write the series

$$e^{-x} = 1 - x + \frac{x^2}{2!} - \frac{x^3}{3!} + \cdots \qquad (4)$$

Combining (3) and (4) by subtraction and addition gives

$$\frac{e^x - e^{-x}}{2} = x + \frac{x^3}{3!} + \frac{x^5}{5!} + \cdots \qquad (5)$$

and

$$\frac{e^x + e^{-x}}{2} = 1 + \frac{x^2}{2!} + \frac{x^4}{4!} + \cdots \qquad (6)$$

The functions on the left-hand sides of (5) and (6) are the hyperbolic functions $\sinh x$ and $\cosh x$. (See Section 9-1.) Thus we see that, except for the alternating signs, the trigonometric functions $\sin x$ and $\cos x$ have the same Maclaurin series as $\sinh x$ and $\cosh x$.

On the other hand, adding (1) and (2) gives

$$\sin x + \cos x = 1 + x - \frac{x^2}{2!} - \frac{x^3}{3!} + \frac{x^4}{4!} + \cdots$$

which, except for the signs, is the series for e^x.

Can we get rid of these sign differences by an appropriate change of variable? We have already done all that we can by changing x to $-x$. Let us try changing x to ax, where a is some constant. We have, for example,

$$\cosh ax = 1 + a^2\frac{x^2}{2!} + a^4\frac{x^4}{4!} + a^6\frac{x^6}{6!} + \cdots$$

This would become the series for $\cos x$ if we have $a^2 = a^6 = a^{10} = \cdots = -1$ and $a^4 = a^8 = a^{12} = \cdots = 1$. In fact, all these equations would follow from the one relation $a^2 = -1$.

We know that the square of any real number is positive, so that the equation $a^2 = -1$ has no real solutions. Let us stretch our imagination, for a moment, and pretend that there is such a number; we will denote it by the letter i, for imaginary. Then we would have $\cosh ix = \cos x$.

Worked Example 1   What is the relation between $\sinh ix$ and $\sin x$?

Solution   Since $i^2 = -1$, we have $i^3 = -i$, $i^4 = -(i \cdot i) = 1$, $i^5 = i$, $i^6 = -1$, etc., so substituting ix for x in (5) gives

$$\sinh ix = ix - i\frac{x^3}{3!} + i\frac{x^5}{5!} - i\frac{x^7}{7!} + \cdots$$

Comparing this with (1), we find that $\sinh ix = i \sin x$.

The sum of the series in (5) and (6) is the series in (3), i.e., we have $e^x = \cosh x + \sinh x$. Substituting ix for x, we find

$$e^{ix} = \cosh ix + \sinh ix$$

or

$$e^{ix} = \cos x + i \sin x \qquad (7)$$

Formula (7) is called *Euler's formula*. Substituting π for x, we find that

$$e^{i\pi} = -1$$

and adding 1 to both sides gives

$$e^{i\pi} + 1 = 0 \qquad (8)$$

Since there is no real number having the property $i^2 = -1$, all of the calculations above belong so far to mathematical "science-fiction." In the next subsection, though, we will see how to construct a number system in which -1 does have a square root; in this new system, all the calculations which we have done above will be completely justified.

Solved Exercises*   1. Using formula (7), express the sine and cosine functions in terms of exponentials.

2. Find $e^{i(\pi/2)}$ and $e^{2\pi i}$.

Exercises   1. Using a trigonometric identity, show that $e^{ix}e^{-ix} = 1$.

2. Find $e^{(3\pi/2)i}$ and $e^{-i\pi}$.

COMPLEX NUMBERS

If we attempt to solve $x^2 + x + 1 = 0$ by the quadratic formula, we get

$$x = \frac{-1 \pm \sqrt{1^2 - 4 \cdot 1 \cdot 1}}{2 \cdot 1} = \frac{-1 \pm \sqrt{-3}}{2} = -\frac{1}{2} \pm \frac{\sqrt{3}}{2}(\sqrt{-1})$$

It is tempting to try to make some sense of the symbol $i = \sqrt{-1}$. This cannot be an ordinary number because the square of any real number is ≥ 0. Historically, square roots of negative numbers were deemed merely to be symbols on paper and not to have any real existence (whatever that means) and therefore to be "imaginary."† (Actually, the true history is a little more complicated in that imaginary numbers were not really taken seriously until the cubic and quartic equations were discussed in the 16th century.‡ A proper way to define square roots of negative numbers was finally obtained through the work of Girolamo Cardano around 1545 and Bombelli in 1572, but it was only with the work of L. Euler around 1747 that their importance was realized. A way to understand imaginaries in terms of real numbers was discovered by Wallis, Wessel, Argand, Gauss, Hamilton, and others in the early 19th century.)

To describe these new numbers, let us note that what we wish to define is the general *complex number*

$$a + b\sqrt{-1} = a + bi$$

where a and b are ordinary real numbers. The pair (a, b) of real numbers is nothing more than a point in the plane, so we can say that $a + ib$ is just another symbol for the point (a, b) in the xy plane. If we multiply formally, remembering that we desire $i^2 = -1$, we get

$$(a + bi)(c + di) = ac + adi + bci + bdi^2$$
$$= (ac - bd) + (ad + bc)i$$

another number of the same type.

Thus we can proceed as follows:

Fig. 13-A-1 A complex number is just a point (a, b) in the plane.

Definition A *complex number* is a point (a, b) in the xy plane. Complex numbers are added and multiplied as follows:

$$(a, b) + (c, d) = (a + c, b + d)$$
$$(a, b)(c, d) = (ac - bd, ad + bc)$$

The point $(0, 1)$ is denoted by the symbol i, so that $i^2 = (-1, 0)$ (using $a = 0$, $c = 0$, $b = 1$, $d = 1$ in the definition of multiplication). The x axis is called the *real axis* and the y axis the *imaginary axis*. (See Fig. 13-A-1.)

It is convenient to denote the point $(a, 0)$ just by a since we are thinking of points on the real axis as ordinary real numbers. Thus, in this notation, $i^2 = -1$. Also,

$$(a, b) = (a, 0) + (0, b) = (a, 0) + (b, 0)(0, 1)$$

as is seen from the definition of multiplication. Replacing (a, b) and $(b, 0)$ by a and b, and $(0, 1)$ by i, we see that

$$(a, b) = a + bi$$

Since two points in the plane are equal if and only if their coordinates are equal, we see that

$$a + ib = c + id \quad \text{if and only if} \quad a = c \text{ and } b = d$$

Thus, if $a + ib = 0$, both a and b must be zero.

We see now that sense can indeed be made of the symbol $a + ib$, where $i^2 = -1$. The notation $a + ib$ is much easier to work with than ordered pairs. In fact, we now revert to the old notation $a + ib$ and dispense with ordered pairs in our calculations. However, the geometric picture of plotting $a + ib$ as the point (a, b) in the plane is very useful and will be kept.

It can be verified, although we shall not do it, that the usual laws of algebra hold for complex numbers. For example, if we denote complex numbers by single letters such as $z = a + ib$, $w = c + id$, and $u = e + if$, we have

$$z(w + u) = zw + zu$$

etc.

Worked Example 2   Factor $x^2 + x + 3$.

Solution   By the quadratic equation, the roots of $x^2 + x + 3 = 0$ are $(-1 \pm \sqrt{1 - 12})/2 = (-1/2) \pm (\sqrt{11}/2)i$. Thus, we have $x^2 + x + 3 = (x + (1/2) - (\sqrt{11}/2)i) \cdot (x + (1/2) + (\sqrt{11}/2)i)$. (You may check by multiplying out.)

Solved Exercises

3. Simplify $(3 + 4i)(8 + 2i)$.

4. (a) Show that if $z = a + ib \neq 0$, $1/z = (a - ib)/(a^2 + b^2) = a/(a^2 + b^2) - ib/(a^2 + b^2)$ is a complex number whose product with z equals 1; thus, $1/z$ is the inverse of z, and we can divide by nonzero complex numbers.

   (b) Write $1/(3 + 4i)$ in the form $a + bi$.

5. Find $\sqrt{i}$.

Exercises

3. Plot the following complex numbers as points in the xy plane.

   (a) $4 + 2i$ (b) $-1 + i$ (c) $3i$ (d) $-(2 + i)$ (e) $-\frac{2}{3}i$
   (f) $3 + 7i$ (g) $.1 + .2i$ (h) $0 + 1.5i$ (i) $5i - 7$

4. Simplify the following and plot your result.

   (a) $(1 + 2i) - 3(5 - 2i)$ (b) $(4 - 3i)(8 + i) + (5 - i)$

   (c) $(2 + i)^2$ (d) $\dfrac{1}{1 + i}$ (e) $\dfrac{1}{5 - 3i}$

   (f) $\dfrac{2i}{1 - i}$ (g) $\dfrac{(1 + i)(3 - 2i)}{8 + i}$ (h) $\dfrac{(2 + 2i) + 6i}{(1 + 2i)(-4i)}$

5. Find:

   (a) $\sqrt{3 - i}$ (b) $\sqrt{1 + 2i}$

   (c) $\sqrt{1 + \sqrt{i}}$ (see Solved Exercise 5)

6. Give a geometric interpretation for the addition of the following pairs of complex numbers. [*Hint*: Plot both along with their sums and consider the resulting "diagram."]

   (a) $1 + \frac{1}{2}i,\ 3 - i$ (b) $-8 - 2i,\ 5 - i$ (c) $-3 + 4i,\ 6i$ (d) $7,\ 4i$

7. Can you now describe a method for adding two complex numbers geometrically? (See Exercise 6.)

FURTHER MANIPULATIONS WITH COMPLEX NUMBERS

We saw above how to make sense of $a + ib$, where $i = \sqrt{-1}$, and that these numbers can be manipulated by the usual rules of algebra.

Some further terminology is useful:

Definition   If $z = a + ib$ is a complex number, then

   (i) a is called the *real part* of z;

   (ii) b is called the *imaginary part* of z (note that the imaginary part is itself a real number);

   (iii) $a - ib$ is called the *complex conjugate* of z and is denoted $\bar{z}$;

   (iv) $r = \sqrt{a^2 + b^2}$ is called the *length* or *absolute value* of z and is denoted $|z|$.

   (v) θ defined by $a = r \cos \theta$ and $b = r \sin \theta$ is called the *argument* of z.

These notions are illustrated in Fig. 13-A-2. Note that the real and imaginary parts are simply the x and y coordinates, the complex conjugate is the reflection in the x axis, and the absolute value is (by Pythagoras' theorem) the length of the line joining the origin and z. The argument of z is the angle this line makes with the x axis. Thus, (r, θ) are simply the polar coordinates of the point (a, b).

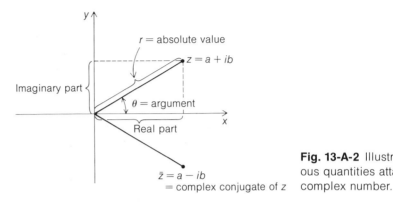

Fig. 13-A-2 Illustrating various quantities attached to a complex number.

The terminology above has come to play an important role in manipulations with complex numbers. For example, notice that

$$z \cdot \bar{z} = (a + ib)(a - ib) = a^2 + b^2 = |z|^2$$

so that

$$\frac{1}{z} = \frac{\bar{z}}{|z|^2}$$

(This reproduces the result of Solved Exercise 4 above.) Notice that we can remember this by: $1/z = (1/z)(\bar{z}/\bar{z}) = \bar{z}/z\bar{z} = \bar{z}/|z|^2$

Worked Example 3   Find the absolute value and argument of $1 + i$.

Solution   The real part is 1, and the imaginary part is 1. Thus the absolute value is $\sqrt{1^2 + 1^2} = \sqrt{2}$, and the argument is $\tan^{-1}(1) = \pi/4$.

Some useful properties are:

(i) $\overline{z_1 z_2} = \bar{z}_1 \cdot \bar{z}_2$, $\overline{z_1/z_2} = \bar{z}_1/\bar{z}_2$

(ii) z is real if and only if $z = \bar{z}$

(iii) $|z_1 z_2| = |z_1| \cdot |z_2|$, $|z_1/z_2| = |z_1|/|z_2|$

and

(iv) $|z_1 + z_2| \le |z_1| + |z_2|$ (triangle inequality)

The proofs of these properties are left to the Solved Exercises and Exercises.

Solved Exercises   6. Find the real part of $1/i$, $1/(1 + i)$, and $(8 + 2i)/(1 - i)$.

7. (a) Prove properties (i) and (ii) of complex conjugation.

   (b) Express $\overline{(1 + i)^{100}}$ without a bar.

8. Given $z = a + ib$, construct iz geometrically and discuss.

Exercises   8. Find the imaginary part of each of the following numbers:

(a) $\dfrac{1+i}{i}$

(b) $\dfrac{2-3i}{1+3i}$

(c) $\dfrac{10+5i}{(1+2i)^2}$

(d) $(1-8i)\left(2+\dfrac{1}{4}i\right)^{-1}$

(e) $\dfrac{(\frac{1}{2}+\frac{3}{5}i)}{(\frac{7}{8}-i)}$

(f) $\dfrac{\frac{3}{4}i}{(\frac{9}{4}+\frac{1}{5}i)}$

9. Write the solutions of the following equations in the form $a+bi$, where a and b are real numbers and $i=\sqrt{-1}$.

(a) $x^2+3=0$

(b) $x^2-2x+5=0$

(c) $x^2+\frac{1}{3}x+\frac{1}{2}=0$

(d) $x^3+2x^2+2x+1=0$

(e) $x^2-7x-1=0$

(f) $x^3-3x^2+3x-1=0$

10. Prove property (iii) of complex conjugation.

11. Prove property (iv) of complex conjugation.

12. Find the complex conjugate of each of the following complex numbers (answers should be given in the form $a+bi$). Sketch.

(a) $5+2i$

(b) $1-bi$

(c) $\sqrt{3}+\dfrac{1}{2}i$

(d) $\dfrac{1}{i}$

(e) $\dfrac{2-i}{3i}$

(f) $i(1+i)$

(g) $\dfrac{3-5i}{4+8i}$

(h) $\dfrac{1}{2i}\left(\dfrac{1+i}{1-i}\right)$

(i) 3

(j) $\dfrac{10+i}{7+4i}$

(k) $\sqrt{2i}$

13. Find the absolute value and argument of each of the following complex numbers. Plot.

(a) $-1-i$

(b) $7+2i$

(c) 2

(d) $4i$

(e) $\frac{1}{2}-\frac{2}{3}i$

(f) $3-2i$

(g) $-5+7i$

(h) $-10+\frac{1}{2}i$

(i) $-8-2i$

(j) $5+5i$

(k) $1.2+.7i$

(l) $50+10i$

14. Find $|(1+i)(2-i)(\sqrt{2}\,i)|$.

15. If $z=x+iy$, express x and y in terms of z and $\bar{z}$.

EULER'S FORMULA REVISITED

Using the algebra of complex numbers, we can define $f(z)$ when f is a rational function and z is a complex number.

Worked Example 4   If $f(x)=(1+x)/(1-x)$ and $z=1+i$, express $f(z)$ in the form $a+bi$.

Solution   Substituting $1+i$ for x, we have

$$f(1+i)=\frac{1+1+i}{1-(1+i)}=\frac{2+i}{-i}=-1-\frac{2}{i}=-1+2i$$

How can we define a transcendental function of complex numbers, like e^z? One way is to use the series, writing

$$e^z=1+z+\frac{z^2}{2!}+\frac{z^3}{3!}+\cdots=\sum_{n=0}^{\infty}\frac{z^n}{n!}$$

To make sense of this, we would have to define the limit of a sequence of complex numbers, so that the sum of the infinite series could be taken as the limit of its sequence of partial sums. Fortunately, this is possible, and in fact the whole theory of infinite series carries over to the complex numbers. This approach would take us too far afield,* though, and we prefer to take the direct approach of *defining e^{ix}*, for x real, by Euler's formula

$$e^{ix} = \cos x + i \sin x \qquad (9)$$

Because $e^{x+y} = e^x e^y$, we expect the same to be true of e^{ix}. In fact, it is.

Worked Example 5 Show that $e^{i(x+y)} = e^{ix} e^{iy}$. $\qquad (10)$

Solution The right-hand side is

$$(\cos x + i \sin x)(\cos y + i \sin y)$$
$$= \cos x \cos y - \sin x \sin y + i(\sin x \cos y + \sin y \cos x)$$
$$= \cos (x+y) + i \sin (x+y) = e^{i(x+y)}$$

by (9) and the addition formulae for sin and cos.

This proof may be summarized by saying that (10) is nothing more than a convenient way of expressing the trigonometric addition formula. This is why the use of e^{ix} is so convenient. The laws of exponents are easier to manipulate than the trigonometric identities.

As an illustration of the power of this method we refer to Solved Exercise 11 and to the paragraphs which follow.

Solved Exercises
9. (a) Calculate $\overline{e^{i\theta}}$ and $|e^{i\theta}|$. (b) Calculate $e^{i\pi/2}$ and $e^{i\pi}$.

10. Give a definition of e^z for $z = x + iy$.

11. Prove that

$$1 + \cos \theta + \cos 2\theta + \cdots + \cos n\theta = \frac{1}{2} + \frac{1}{2}\left(\frac{\cos n\theta - \cos (n+1)\theta}{1 - \cos \theta}\right)$$

[*Hint*: Consider $1 + e^{i\theta} + e^{2i\theta} + \cdots + e^{ni\theta}$.]

Exercises
16. If $z = x + iy$ with x and y real, what is $|e^z|$ and the argument of e^z?

17. Write the following in the form $a + bi$:
  (a) $e^{i\pi/3}$ (b) $e^{i\pi/4}$ (c) $e^{i3\pi/4}$ (d) e^{1+2i} (e) $e^{1+\pi i/2}$ (f) $e^{(1-\pi/6)i}$

18. Prove that $e^{i(\theta+3\pi/2)} = -ie^{-i\theta}$.

19. Prove that

$$\sin \theta + \sin 2\theta + \cdots + \sin n\theta = \left(\cot \frac{\theta}{2}\right)\left(\frac{1}{2} + \frac{1}{2}\left(\frac{\sin n\theta - \sin (n+1)\theta}{\sin \theta}\right)\right)$$

DE MOIVRE'S FORMULA†

Let us push our analysis of e^{ix} a little further. Notice that $e^{i\theta} = \cos \theta + i \sin \theta$ represents a point on the unit circle with argument θ. As θ ranges from 0 to

*See J. Marsden, Basic Complex Analysis, *Freeman, 1972, for a thorough treatment of complex series.

†Abraham DeMoivre (1667-1754), of French descent, worked in England around the time of Newton.

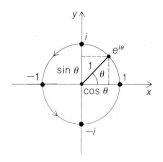

Fig. 13-A-3 As θ goes from 0 to 2π, the point $e^{i\theta}$ goes once around the unit circle in the complex plane.

2π, this point moves once around the circle (Fig. 13-A-3). (This is the same basic geometric picture we used to introduce the trigonometric functions in Section 6-1.)

Recall that if $z = x + iy$, and r, θ are the polar coordinates of (x, y), then $x = r \cos \theta$ and $y = r \sin \theta$. Thus

$$z = r \cos \theta + ir \sin \theta = r(\cos \theta + i \sin \theta) = re^{i\theta}$$

Hence we arrive at:

POLAR REPRESENTATION OF COMPLEX NUMBERS

If $z = x + iy$ and if (r, θ) are the polar coordinates of (x, y), i.e., the absolute value and argument of z, then

$$z = re^{i\theta}$$

This representation is very convenient for algebraic manipulations. For example,

$$\text{if } z_1 = r_1 e^{i\theta_1}, \quad z_2 = r_2 e^{i\theta_2}, \qquad \text{then } z_1 z_2 = r_1 r_2 e^{i(\theta_1 + \theta_2)}$$

which shows how the absolute value and arguments behave when we take products; i.e., it shows that $|z_1 z_2| = |z_1||z_2|$ (see also (iii), p. B-6) and that the argument of $z_1 z_2$ is the sum of the arguments of z_1 and z_2.

Let us also note that if $z = re^{i\theta}$, then $z^n = r^n e^{in\theta}$. Thus if we wish to solve $z^n = w$ and $w = \rho e^{i\phi}$ we must have $r^n = \rho$, i.e., $r = \sqrt[n]{\rho}$ (remember r, ρ are nonnegative) and $e^{in\theta} = e^{i\phi}$, i.e., $e^{i(n\theta - \phi)} = 1$, i.e., $n\theta = \phi + 2\pi k$ for an integer k (this is because $e^{it} = 1$ exactly when t is a multiple of 2π — see Fig. 13-A-3). Thus $\theta = \dfrac{\phi}{n} + \dfrac{2\pi k}{n}$.

Note that when $k = n$, $\theta = \dfrac{\phi}{n} + 2\pi$, so $e^{i\theta} = e^{i\phi/n}$. Thus we get the same value for $e^{i\theta}$ when $k = 0$ and $k = n$. Thus we need take only $k = 0, 1, 2, \ldots, n - 1$. Hence we get the following formula for the nth roots of a complex number:

DE MOIVRE'S FORMULA

The numbers z such that $z^n = w = \rho e^{i\phi}$, i.e., the nth roots of w, are given by

$$\sqrt[n]{\rho}\, e^{i(\phi/n + 2\pi k/n)}, \qquad k = 0, 1, 2, \ldots, n - 1$$

For example, the 9th roots of 1 are the complex numbers $e^{i2\pi k/9}$, $k = 0, 1, \ldots, 8$, which are 9 points equally spaced around the unit circle. See Fig. 13-A-4.

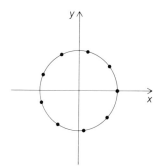

Fig. 13-A-4 The 9th roots of 1.

It is shown in more advanced books* that any nth degree polynomial $a_0 + a_1z + \cdots + a_nz^n$ has at least one complex root z_1 and, as a consequence, one shows that the polynomial can be completely factored,

$$a_0 + a_1z + \cdots + a_nz^n = (z - z_1) \cdots (z - z_n)$$

giving n roots $z_1, z_2, \ldots, z_n$ (some possibly repeated). For example,

$$z^2 + z + 1 = \left(z + \frac{1 + \sqrt{3}i}{2}\right)\left(z + \frac{1 - \sqrt{3}i}{2}\right)$$

although $z^2 + z + 1$ cannot be factored using only real numbers.

Solved Exercises

12. Redo Solved Exercise 8 using the polar representation.

13. Find the 4th roots of $1 + i$.

14. Give a geometric interpretation of multiplication by $(1 + i)$.

Exercises

20. (a) Prove that $(\cos \theta + i \sin \theta)^n = \cos n\theta + i \sin n\theta$, if n is any real number.

(b) Find the real part of

$$\left(\frac{1}{\sqrt{2}} + \frac{i}{\sqrt{2}}\right)^3$$

and the imaginary part of

$$\left(\frac{1}{2} + i\frac{\sqrt{3}}{2}\right)^9$$

21. (a) Find the 5th roots of $(1/2) - (i(\sqrt{3}/2))$ and $1 + 2i$. Sketch.

(b) Find the 4th roots of i and $\sqrt{i}$. Sketch.

(c) Find the 6th roots of $\sqrt{5} + 3i$ and $3 + \sqrt{5}i$. Sketch.

(d) Find the 3rd roots of $1/7$ and $i/7$. Sketch.

22. Find the polar representation (i.e., $z = re^{i\theta}$) of the following complex numbers.

(a) $1 + i$ (b) $\frac{1}{i}$ (c) $(2 + i)^{-1}$ (d) $\sqrt{3}$

(e) $7 - 3i$ (f) $4 + i^3$ (g) $-\frac{1}{2} - 3i$ (h) $\frac{(2 + 5i)}{(1 - i)}$

(i) $(3 + 4i)^2$ (j) $-1 + \frac{1}{3}i$

23. (a) Give a geometric interpretation of division by $1 - i$.

(b) Give a geometric interpretation of multiplication by an arbitrary complex number $z = re^{i\theta}$. (What happens if we divide?)

24. (a) Prove that if $z^6 = 1$ and $z^{10} = 1$, then $z = \pm 1$.

(b) Suppose we know that $z^7 = 1$ and $z^{41} = 1$. What can we say about z?

25. Let $z = re^{i\theta}$. Prove that $\bar{z} = re^{-i\theta}$.

---

*See any text in complex variables, such as J. Marsden, op. cit. The theorem referred to is called the "fundamental theorem of algebra." It was first proved by Gauss in his doctoral thesis of 1831.

26. (a) Let $f(z) = az^3 + bz^2 + cz + d$, where a, b, c, and d are real numbers. Prove that $f(\bar{z}) = \overline{f(z)}$.

    (b) Does equality still hold if a, b, c, and d are allowed to be arbitrary complex numbers?

27. Factor the following polynomials, where z is complex. [*Hint*: Find the roots.]

    (a) $z^2 + 2z + i$ (b) $z^2 + 2iz - 4$ (c) $iz^2 - z + 1 + i$

    (d) $3z^2 + z - e^{i\pi/3}$ (e) $z^3 - 1$

28. (a) Write $\tan i\theta$ in the form $a + bi$ where a and b are real functions of θ.

    (b) Write $\tan i\theta$ in the form $re^{i\phi}$.

Problems for the Appendix to Chapter 13

1. Let $z = f(t)$ be a *complex valued* function of the *real* variable t. If $z = x + iy = g(t) + ih(t)$, where g and h are real valued, we *define* $dz/dt = f'(t)$ to be $(dx/dt) + i(dy/dt) = g(t) + ih(t)$.

    (a) Show that $(d/dt)(Ce^{i\omega t}) = i\omega Ce^{i\omega t}$, if C is any complex number and ω is any real number.

    (b) Show that $z = Ce^{i\omega t}$ satisfies the *spring equation* (see Section 9-1) $z'' + \omega^2 z = 0$.

    (c) Show that $z = De^{-i\omega t}$ also satisfies the spring equation.

    (d) Find C and D such that $Ce^{i\omega t} + De^{-i\omega t} = f(t)$ satisfies $f(0) = A$, $f'(0) = B$. Express the resulting function $f(t)$ in terms of sines and cosines.

    (e) Compare the result of (d) with results in Section 9-1.

2. (a) Find λ such that the function $x = e^{\lambda t}$ satisfies the equation $x'' - 2x' + 2x = 0$; $x' = dx/dt$.

    (b) Express the function $e^{\lambda t} + e^{-\lambda t}$ in terms of sines, cosines, and *real* exponentials.

    (c) Show that the function in (b) satisfies the differential equation in (a).

3. Let z_1 and z_2 be nonzero complex numbers. Find an algebraic relation between z_1 and z_2 which is equivalent to the fact that the lines from the origin through z_1 and z_2 are perpendicular.

4. Let $w = f(z) = (1 + (z/2))/(1 - (z/2))$.

    (a) Show that if the real part of z is 0, then $|w| = 1$.

    (b) Are all points on the circle $|w| = 1$ in the range of f? [*Hint*: Solve for z in terms of w.]

5. (a) Show that, if $z^n = 1$, n a positive integer, then either $z = 1$ or $z^{n-1} + z^{n-2} + \cdots + z + 1 = 0$.

    (b) Show that, if $z^{n-1} + z^{n-2} + \cdots + z + 1 = 0$, then $z^n = 1$.

    (c) Find all the roots of the equation $z^3 + z^2 + z + 1 = 0$.

6. Describe the motion in the complex plane, as the real number t goes from $-\infty$ to ∞, of the point $z = e^{i\omega t}$, when

    (a) $\omega = i$ (b) $\omega = 1 + i$

    (c) $\omega = -i$ (d) $\omega = -1 - i$

    (e) $\omega = 0$ (f) $\omega = 1$

    (g) $\omega = -1$

7. Describe the motion in the complex plane, as the real number t varies, of the point $z = 93{,}000{,}000 \times e^{2\pi i(t/365)} + 1{,}000{,}000 e^{2\pi i(t/29)}$. Does this remind you of some astronomical phenomenon?

8. (a) Find *all* complex numbers z for which $e^z = -1$.

    (b) How might you define $\ln(-1)$? What is the difficulty here?

9. What is the relation between e^z and $e^{\bar{z}}$?

Solutions to Solved Exercises in the Appendix to Chapter 13

1. Substituting $-x$ for x in (7) and using the symmetry properties of cosine and sine, we obtain

$$e^{-ix} = \cos x - i \sin x$$

Adding this equation to (7) and dividing by 2 gives

$$\cos x = \frac{e^{ix} + e^{-ix}}{2}$$

while subtracting the equations and dividing by $2i$ gives

$$\sin x = \frac{e^{ix} - e^{-ix}}{2i}$$

2. Using formula (7), we have

$$e^{i(\pi/2)} = \cos \frac{\pi}{2} + i \sin \frac{\pi}{2} = i$$

and

$$e^{2\pi i} = \cos 2\pi + i \sin 2\pi = 1$$

3. $(3 + 4i)(8 + 2i) = 3 \cdot 8 + 3 \cdot 2i + 4 \cdot 8i + 2 \cdot 4i^2$

$$= 24 + 6i + 32i - 8$$

$$= 16 + 38i$$

4. (a) $\left(\dfrac{a - ib}{a^2 + b^2}\right)(a + ib)$

$= \left(\dfrac{1}{a^2 + b^2}\right)(a - ib)(a + ib)$

$= \left(\dfrac{1}{a^2 + b^2}\right)(a^2 + aib - iba - b^2 i^2)$

$= \left(\dfrac{1}{a^2 + b^2}\right)(a^2 + b^2)$

$= 1$

Hence

$z\left(\dfrac{a - ib}{a^2 + b^2}\right) = 1,$

so $(a - ib)/(a^2 + b^2)$ can be denoted $1/z$. Note that $z \neq 0$ means that not both a and b are zero, so $a^2 + b^2 \neq 0$ and division by the real number $a^2 + b^2$ is legitimate.

(b) $1/(3 + 4i) = (3 - 4i)/(3^2 + 4^2) = (3/25) - (4/25)i$ by the formula in (a).

5. We seek a number $z = a + ib$ such that $z^2 = i$. But $z^2 = a^2 - b^2 + 2abi$, so we must choose $a^2 - b^2 = 0$ and $2ab = 1$. Hence $a = \pm b$, so $b = \pm(1/\sqrt{2})$. Thus there are two numbers whose square is i, namely, $\pm[(1/\sqrt{2}) + (i/\sqrt{2})]$

i.e., $\sqrt{i} = \pm\dfrac{1}{\sqrt{2}}(1 + i) = \pm\dfrac{\sqrt{2}}{2}(1 + i)$

6. $1/i = (1/i)(-i/-i) = -i/1 = -i$, so the real part of $1/i = -i$ is zero. $1/(1 + i) = (1 - i)/((1 + i)(1 - i)) = (1 - i)/2$, so the real part of $1/(1 + i)$ is $1/2$.

$\dfrac{8 + 2i}{1 - i} = \dfrac{(8 + 2i)}{(1 - i)}\dfrac{1 + i}{1 + i} = \dfrac{8 + 10i - 2}{2}$

$= \dfrac{6 + 10i}{2} = 3 + 5i$

so the real part of $(8 + 2i)/(1 - i)$ is 3.

7. (a) Let $z_1 = a + ib$ and $z_2 = c + id$ so $\bar{z}_1 = a - ib$, $\bar{z}_2 = c - id$, $z_1 z_2 = (ac - bd) + (ad + bc)i$ and $\overline{z_1 z_2} = (ac - bd) - (ad + bc)i$. But $\bar{z}_1 \cdot \bar{z}_2 = (a - ib)(c - id) = (ac - bd) - ibc - aib = \overline{z_1 z_2}$. For the quotient, $z_2 \cdot z_1/z_2 = z_1$ so $\bar{z}_2 \cdot (\overline{z_1/z_2}) = \bar{z}_1$. Dividing by $\bar{z}_2$ gives the result.

(b) Since the complex conjugate of a product is the product of the complex conjugates (proved in (a)), we similarly have $\overline{z_1 z_2 z_3} = \overline{z_1} \overline{z_2} \overline{z_3} = \bar{z}_1 \bar{z}_2 \bar{z}_3$ and so on for any number of factors. Thus $\overline{z^n} = \bar{z}^n$, and hence $\overline{(1 + i)^{100}} = \overline{(1 + i)}^{100} = (1 - i)^{100}$.

8. If $z = a + ib$, $iz = ai - b = -b + ia$. Thus in the plane, $z = (a, b)$ and $iz = (-b, a)$. This point $(-b, a)$ is on the line perpendicular to the line Oz since the slopes are negative reciprocals. See Fig. 13-A-5 at the top of the next column. Since iz has the same length as z, we can say that iz is obtained from z by a rotation through 90°.

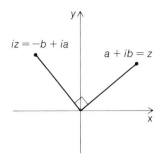

Fig. 13-A-5 The number iz is obtained from z by a 90° rotation about the origin.

9. (a) $e^{i\theta} = \cos\theta + i\sin\theta$, so by definition of the complex conjugate we should change the sign of the imaginary part:

$\overline{e^{i\theta}} = \cos\theta - i\sin\theta = \cos(-\theta) + i\sin(-\theta) = e^{-i\theta}$

since $\cos(-\theta) = \cos\theta$ and $\sin(-\theta) = -\sin\theta$. $|e^{i\theta}| = \sqrt{\cos^2\theta + \sin^2\theta} = 1$ using the general definition $|z| = \sqrt{a^2 + b^2}$ where $z = a + ib$.

(b) $e^{i\pi/2} = \cos(\pi/2) + i\sin(\pi/2) = i$ and $e^{i\pi} = \cos\pi + i\sin\pi = -1$.

10. We would like to have $e^{z_1 + z_2} = e^{z_1} e^{z_2}$ for any complex numbers, so we should define $e^{x + iy} = e^x \cdot e^{iy}$, i.e., $e^{x + iy} = e^x(\cos y + i\sin y)$. With this definition, $e^{z_1 + z_2} = e^{z_1} e^{z_2}$ can then be proved for all z_1 and z_2.

11. Since $\cos n\theta$ is the real part of $e^{in\theta}$, we are led to consider $1 + e^{i\theta} + e^{i2\theta} + \cdots + e^{in\theta}$. Recalling that $1 + r + \cdots + r^n = (1 - r^{n+1})/(1 - r)$, we get

$1 + e^{i\theta} + e^{i2\theta} + \cdots + e^{in\theta} = \dfrac{1 - e^{i(n+1)\theta}}{1 - e^{i\theta}}$

$= \dfrac{1 - e^{i(n+1)\theta}}{1 - e^{i\theta}} \cdot \dfrac{1 - e^{-i\theta}}{1 - e^{-i\theta}}$

$= \dfrac{1 - e^{-i\theta} - e^{i(n+1)\theta} + e^{in\theta}}{2 - (e^{i\theta} + e^{-i\theta})}$

$= \dfrac{1 - e^{-i\theta} - e^{i(n+1)\theta} + e^{in\theta}}{2(1 - \cos\theta)}$

Taking the real part of both sides gives the result.

12. Since $i = e^{i\pi/2}$, $iz = re^{i(\theta + \pi/2)}$ if $z = re^{i\theta}$. Thus iz has the same magnitude as z but its argument is increased by $\pi/2$. Hence iz is z rotated by 90°, in agreement with the solution to Solved Exercise 8.

13. $1 + i = \sqrt{2}\, e^{i\pi/4}$, since $1 + i$ has $r = \sqrt{2}$ and $\theta = \pi/4$. Hence the fourth roots are, according to DeMoivre's formula,

$\sqrt[8]{2}\, e^{i((\pi/16) + (\pi k/2))}, \qquad k = 0, 1, 2, 3$

i.e.,

$\sqrt[8]{2}\, e^{i\pi/16}, \quad \sqrt[8]{2}\, e^{i\pi 9/16}, \quad \sqrt[8]{2}\, e^{i\pi \cdot 17/16}, \quad \text{and} \quad \sqrt[8]{2}\, e^{i\pi \cdot 25/16}$

14. Since $(1 + i) = \sqrt{2}\, e^{i\pi/4}$, multiplication of a complex number z by $(1 + i)$ rotates z through an angle $\pi/4 = 45°$ and multiplies its length by $\sqrt{2}$.

APPENDIX TO CHAPTER 15: ROTATIONS AND THE SUNSHINE FORMULA

Rotations described in terms of the cross-product are used to derive the sunshine formula.

The purpose of this section is to derive the sunshine formula, which has been stated and used in Sections 6-3 and 10-3. Before we begin the actual derivation, we will study some properties of rotations in preparation for the description of the earth rotating on its axis. The cross-product, introduced in Section 14-3, will be used extensively here.

ROTATIONS

Fig. 15-A-1 If **r** rotates about **l**, its tip describes a circle.

Consider two unit vectors **l** and **r** in space with the same base point. If we rotate **r** about the axis through **l**, then the tip of **r** describes a circle (Fig. 15-A-1). (Imagine **l** and **r** glued rigidly at their base points and then spun about the axis through **l**.) Assume that the rotation is at a uniform rate counterclockwise (when viewed from the tip of **l**), making a complete revolution in T units of time. The vector **r** now is a vector function of time, so we may write $\mathbf{r} = \boldsymbol{\sigma}(t)$. Our first aim is to find a convenient formula for $\boldsymbol{\sigma}(t)$ in terms of its starting position $\mathbf{r}_0 = \boldsymbol{\sigma}(0)$.

Let λ denote the angle between **l** and $\mathbf{r}_0$; we can assume that $\lambda \neq 0$ and $\lambda \neq \pi$, i.e., **l** and $\mathbf{r}_0$ are not parallel, for otherwise **r** would not rotate. Construct the unit vector $\mathbf{m}_0$ as shown in Fig. 15-A-2. From this figure we see that

Fig. 15-A-2 Vector $\mathbf{m}_0$ is in the plane of $\mathbf{r}_0$ and **l**, is orthogonal to **l**, and makes an angle of $(\pi/2) - \lambda$ with $\mathbf{r}_0$.

$$\mathbf{r}_0 = \cos \lambda \, \mathbf{l} + \sin \lambda \, \mathbf{m}_0 \tag{1}$$

(In fact, formula (1) can be taken as the algebraic definition of $\mathbf{m}_0$; $\mathbf{m}_0 = (1/\sin \lambda)\mathbf{r}_0 - (\cos \lambda/\sin \lambda)\mathbf{l}$. We assumed that $\lambda \neq 0, \pi$, so $\sin \lambda \neq 0$.)

Now add to this figure the unit vector $\mathbf{n}_0 = \mathbf{l} \times \mathbf{m}_0$. (See Fig. 15-A-3.) (The triple $(\mathbf{l}, \mathbf{m}_0, \mathbf{n}_0)$ consists of three mutually orthogonal unit vectors, just like $(\mathbf{i}, \mathbf{j}, \mathbf{k})$.)

Fig. 15-A-3
The triple $(\mathbf{l}, \mathbf{m}_0, \mathbf{n}_0)$ is a right-handed orthogonal set of unit vectors.

Worked Example 1 Let $\mathbf{l} = (1/\sqrt{3})(\mathbf{i} + \mathbf{j} + \mathbf{k})$ and $\mathbf{r}_0 = \mathbf{k}$. Find $\mathbf{m}_0$ and $\mathbf{n}_0$.

Solution The angle between **l** and $\mathbf{r}_0$ is given by $\cos \lambda = \mathbf{l} \cdot \mathbf{r}_0 = 1/\sqrt{3}$. Thus $\sin \lambda = \sqrt{1 - \cos^2 \lambda} = \sqrt{\frac{2}{3}}$, and so from formula (1) we get

$$\mathbf{m}_0 = \frac{1}{\sin \lambda}\boldsymbol{\sigma}(0) - \frac{\cos \lambda}{\sin \lambda}\mathbf{l}$$

$$= \sqrt{\frac{3}{2}}\mathbf{k} - \frac{1}{\sqrt{3}}\sqrt{\frac{3}{2}} \cdot \frac{1}{\sqrt{3}}(\mathbf{i}+\mathbf{j}+\mathbf{k})$$

$$= \frac{2}{\sqrt{6}}\mathbf{k} - \frac{1}{\sqrt{6}}(\mathbf{i}+\mathbf{j})$$

and

$$\mathbf{n}_0 = \mathbf{l} \times \mathbf{m}_0 = \begin{vmatrix} \mathbf{i} & \mathbf{j} & \mathbf{k} \\ \frac{1}{\sqrt{3}} & \frac{1}{\sqrt{3}} & \frac{1}{\sqrt{3}} \\ -\frac{1}{\sqrt{6}} & -\frac{1}{\sqrt{2}} & \frac{2}{\sqrt{6}} \end{vmatrix} = \frac{1}{\sqrt{2}}\mathbf{i} - \frac{1}{\sqrt{2}}\mathbf{j}$$

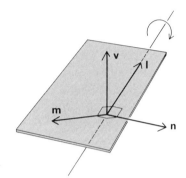

Fig. 15-A-4 The three vectors **v**, **m**, and **n** all rotate about **l**.

Return to Fig. 15-A-3 and rotate the whole picture about the axis **l**. Now **m** and **n** will vary with time as well. Since the angle λ remains constant, formula (1) applied after time t to **r** and **l** gives

$$\mathbf{m} = \frac{1}{\sin \lambda}\mathbf{r} - \frac{\cos \lambda}{\sin \lambda}\mathbf{l} \qquad (1')$$

(See Fig. 15-A-4.)

On the other hand, since **m** is perpendicular to **l**, it rotates in a circle in the plane of $\mathbf{m}_0$ and $\mathbf{n}_0$. It goes through an angle 2π in time T, so it goes through an angle $2\pi t/T$ in t units of time, and so

$$\mathbf{m} = \cos\left(\frac{2\pi t}{T}\right)\mathbf{m}_0 + \sin\left(\frac{2\pi t}{T}\right)\mathbf{n}_0$$

Inserting this in formula (1') gives

$$\mathbf{r} = \boldsymbol{\sigma}(t) = \cos \lambda\, \mathbf{l} + \sin \lambda \cos\left(\frac{2\pi t}{T}\right)\mathbf{m}_0 + \sin \lambda \sin\left(\frac{2\pi t}{T}\right)\mathbf{n}_0 \qquad (2)$$

This formula expresses explicitly how **r** changes in time as it is rotated about **l** in terms of the basic trihedral $(\mathbf{l}, \mathbf{m}_0, \mathbf{n}_0)$.

Worked Example 2 Express the function $\boldsymbol{\sigma}(t)$ explicitly in terms of **l**, $\mathbf{r}_0$, and T.

Solution We have $\cos \lambda = \mathbf{l} \cdot \mathbf{r}_0$ and $\sin \lambda = \|\mathbf{l} \times \mathbf{r}_0\|$. Furthermore $\mathbf{n}_0$ is a unit vector perpendicular to both **l** and $\mathbf{r}_0$, so we must have

$$\mathbf{n}_0 = \frac{\mathbf{l} \times \mathbf{r}_0}{\|\mathbf{l} \times \mathbf{r}_0\|}$$

Thus $\sin \lambda\, \mathbf{n}_0 = \mathbf{l} \times \mathbf{r}_0$. Finally, from formula (1), $\sin \lambda\, \mathbf{m} = \mathbf{r}_0 - \cos \lambda\, \mathbf{l} = \mathbf{r}_0 - (\mathbf{r}_0 \cdot \mathbf{l})\mathbf{l}$. Substituting all this into formula (2), we get

$$\mathbf{r} = (\mathbf{r}_0 \cdot \mathbf{l})\mathbf{l} + \cos\left(\frac{2\pi t}{T}\right)[\mathbf{r}_0 - (\mathbf{r}_0 \cdot \mathbf{l})\mathbf{l}] + \sin\left(\frac{2\pi t}{T}\right)(\mathbf{l} \times \mathbf{r}_0)$$

Solved Exercises   1. Show by a direct geometric argument that the speed of the tip of **r** is $2\pi/T \sin \lambda$. Verify that equation (2) gives the same formula.

Exercises   1. (a) Let $\mathbf{l} = (\mathbf{j} + \mathbf{k})/\sqrt{2}$ and $\mathbf{r}_0 = (\mathbf{i} = \mathbf{j})/\sqrt{2}$. Find $\mathbf{m}_0$ and $\mathbf{n}_0$.

   (b) Find $\mathbf{r} = \boldsymbol{\sigma}(t)$ if $T = 24$.

   (c) Find the equation of the line tangent to $\boldsymbol{\sigma}(t)$ at $t = 12$.

2. Verify from formula (2) that $\boldsymbol{\sigma}(T/2) \cdot \mathbf{n} = 0$. Also, show this geometrically. For what values of t is $\boldsymbol{\sigma}(t) \cdot \mathbf{n} = 0$?

THE MOTION OF THE EARTH ABOUT THE SUN

Now we apply our study of rotations to the motion of the earth about the sun, incorporating the rotation of the earth about its own axis as well. We will use a simplified model of the earth-sun system, in which the sun is fixed at the origin of our coordinate system and the earth moves at uniform speed around a circle centered at the sun. Let **u** be a unit vector pointing *from* the sun *to* the earth; we have $\mathbf{u} = \cos(2\pi t/T_y)\,\mathbf{i} + \sin(2\pi t/T_y)\,\mathbf{j}$, where T_y is the length of a year (t and T_y measured in the same units). See Fig. 15-A-5. Notice that the unit vector pointing from the earth to the sun is $-\mathbf{u}$ and that we have oriented our axes so that $\mathbf{u} = \mathbf{i}$ when $T = 0$.

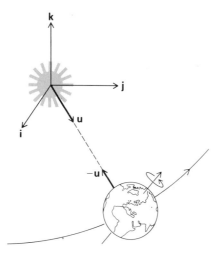

Fig. 15-A-5 The unit vector **u** points from the sun to the earth at time *t*.

Fig. 15-A-6 At $t = 0$, the earth's axis is tilted toward the sun.

Next we wish to take into account the rotation of the earth. The earth rotates about an axis which we represent by a unit vector **l** pointing from the center of the earth to the North Pole. We will assume that **l** is fixed* with respect to **i**, **j**, and **k**; astronomical measurements show that the inclination of **l** (the angle between **l** and **k**) is presently about 23.5°. We will denote this angle by α. If we measure time so that the first day of summer in the northern hemisphere occurs when $t = 0$, then the axis **l** must tilt in the direction $-\mathbf{i}$, and so we must have $\mathbf{l} = \cos \alpha\,\mathbf{k} - \sin \alpha\,\mathbf{i}$. (See Fig. 15-A-6.)

*Actually, the axis **l** is known to rotate about **k** once every 21,000 years. This phenomenon, called precession or wobble, is due to the irregular shape of the earth and may play a role in long-term climatic changes, such as ice ages. See pp. 130–134 of The Weather Machine by Nigel Calder, Viking, 1974.

Now let **r** be the unit vector at time t from the center of the earth to a fixed point P on the earth's surface. Notice that if **r** is located with its base point at P, then it represents the local *vertical* direction. We will assume that P is chosen so that at $t = 0$, it is noon at the point P; then **r** lies in the plane of **l** and **i** and makes an angle of less than 90° with $-\mathbf{i}$. Referring to Fig. 15-A-7, we introduce the unit vector $\mathbf{m}_0 = -(\sin \alpha)\mathbf{k} - (\cos \alpha)\mathbf{i}$ orthogonal to **l**. We then have $\mathbf{r}_0 = (\cos \lambda)\mathbf{l} + (\sin \lambda) \mathbf{m}_0$ where λ is the angle between **l** and $\mathbf{r}_0$. Since $\lambda = 90° - l$, where l is the latitude of the point P, we get $\mathbf{r}_0 = (\sin l)\mathbf{l} + (\cos l) \mathbf{m}_0$. As in the previous subsection, let $\mathbf{n}_0 = \mathbf{l} \times \mathbf{m}_0$.

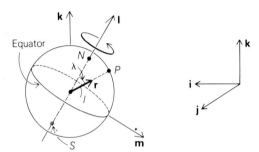

Fig. 15-A-7 The vector **r** is the vector from the center of the earth to a fixed location P. The latitude of P is l and the colatitude is $\lambda = 90° - l$. The vector $\mathbf{m}_0$ is a unit vector in the plane of the equator (orthogonal to **l**) and in the plane of **l** and $\mathbf{r}_0$.

Worked Example 3 Prove that $\mathbf{n}_0 = \mathbf{l} \times \mathbf{m}_0 = -\mathbf{j}$.

Solution Geometrically, $\mathbf{l} \times \mathbf{m}_0$ is a unit vector orthogonal to **l** and $\mathbf{m}_0$ pointing in the sense given by the right-hand rule. But **l** and $\mathbf{m}_0$ are both in the **ik** plane, so $\mathbf{l} \times \mathbf{m}_0$ points orthogonal to it in the direction $-\mathbf{j}$ (see Fig. 15-A-7).

Algebraically, $\mathbf{l} = (\cos \alpha)\mathbf{k} - (\sin \alpha)\mathbf{i}$ and $\mathbf{m}_0 = -(\sin \alpha)\mathbf{k} - (\cos \alpha)\mathbf{i}$, so

$$\mathbf{l} \times \mathbf{m}_0 = \begin{vmatrix} \mathbf{i} & \mathbf{j} & \mathbf{k} \\ -\sin \alpha & 0 & \cos \alpha \\ -\cos \alpha & 0 & -\sin \alpha \end{vmatrix} = -\mathbf{j}(\sin^2 \alpha + \cos^2 \alpha) = -\mathbf{j}$$

Now we apply formula (2) of the previous subsection, namely,

$$\mathbf{r} = \cos \lambda \mathbf{l} + \sin \lambda \cos \left(\frac{2\pi t}{T_d}\right) \mathbf{m}_0 + \sin \lambda \sin \left(\frac{2\pi t}{T_d}\right) \mathbf{n}_0$$

where T_d is the length of time it takes for the earth to rotate once about its axis (with respect to the "fixed stars"—i.e., our **i**, **j**, **k** vectors).* Substituting the expressions derived above for λ, **l**, $\mathbf{m}_0$, and $\mathbf{n}_0$, we get

$$\mathbf{r} = \sin l(\cos \alpha \, \mathbf{k} - \sin \alpha \, \mathbf{i}) + \cos l \cos \left(\frac{2\pi t}{T_d}\right)(-\sin \alpha \, \mathbf{k} - \cos \alpha \, \mathbf{i}) - \cos l \sin \left(\frac{2\pi t}{T_d}\right)\mathbf{j}$$

Hence

$$\mathbf{r} = -\left[\sin l \sin \alpha + \cos l \cos \alpha \cos \left(\frac{2\pi t}{T_d}\right)\right]\mathbf{i} - \cos l \sin \left(\frac{2\pi t}{T_d}\right)\mathbf{j}$$
$$+ \left[\sin l \cos \alpha - \cos l \sin \alpha \cos \left(\frac{2\pi t}{T_d}\right)\right]\mathbf{k} \qquad (3)$$

---

*T_d *is called the length of the* sidereal day. *It differs from the ordinary, or* solar, *day by about 1 part in 365. (Can you explain why?) In fact,* $T_d \approx 23.93$ *hours.*

Solved Exercise   2. What is the speed (in kilometers per hour) of a point on the equator due to the rotation of the earth? A point at latitude 60°? (The radius of the earth is 6371 kilometers.)

Exercises   3. If the earth rotated in the opposite direction about the sun, would T_d be longer or shorter than 24 hours?

4. Show by a direct geometric construction that $\mathbf{r} = \boldsymbol{\sigma}(T_d/4) = -\cos l\mathbf{j}$. Does this formula agree with formula (3)?

DERIVATION OF THE SUNSHINE FORMULA

With formula (3) at our disposal, we are now ready to derive the sunshine formula. The intensity of light on a portion of the earth's surface (or at the top of the atmosphere) is proportional to sin A, where A is the angle of elevation of the sun above the horizon (see Fig. 15-A-8). (At night sin A is negative, and the intensity then is of course zero.)

Fig. 15-A-8 The intensity of sunlight is proportional to sin A. The ratio of area 1 to area 2 is sin A.

Thus we want to compute sin A. From Fig. 15-A-9 we see that

$$\sin A = -\mathbf{u} \cdot \mathbf{r}$$

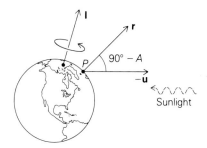

Fig. 15-A-9 The geometry for the formula sin A = cos $(90° - A)$ = $-\mathbf{u} \cdot \mathbf{r}$.

Substituting $\mathbf{u} = \cos(2\pi t/T_y)\mathbf{i} + \sin(2\pi t/T_y)\mathbf{j}$ and formula (3) into this formula for sin A and taking the dot product gives

$$\sin A = \cos\left(\frac{2\pi t}{T_y}\right)\left[\sin l \sin\alpha + \cos l \cos\alpha \cos\left(\frac{2\pi t}{T_d}\right)\right]$$

$$+ \sin\left(\frac{2\pi t}{T_y}\right)\left[\cos l \sin\left(\frac{2\pi t}{T_d}\right)\right]$$

$$\hspace{8cm}(4)$$

$$= \cos\left(\frac{2\pi t}{T_y}\right)\sin l \sin\alpha$$

$$+ \cos l\left[\cos\left(\frac{2\pi t}{T_y}\right)\cos\alpha\cos\left(\frac{2\pi t}{T_d}\right) + \sin\left(\frac{2\pi t}{T_y}\right)\sin\left(\frac{2\pi t}{T_d}\right)\right]$$

Worked Example 4 Set $t = 0$ in formula (4). For what l is $\sin A = 0$? Interpret your result.

Solution With $t = 0$ we get

$$\sin A = \sin l \sin \alpha + \cos l \cos \alpha = \cos (l - \alpha)$$

This is zero when $l - \alpha = \pm\pi/2$. Now $\sin A = 0$ corresponds to the sun on the horizon (sunrise or sunset), when $A = 0$ or π. Thus at $t = 0$ this occurs when $l = \alpha \pm (\pi/2)$. The case $l = \alpha + (\pi/2)$ is impossible, since l lies between $-\pi/2$ and $\pi/2$. The case $l = \alpha - (\pi/2)$ corresponds to a point on the Antarctic Circle; indeed at $t = 0$ (corresponding to noon on the first day of northern summer) the sun is just on the horizon at the Antarctic Circle.

Our next goal is to describe the variation of $\sin A$ with time *on a particular day*. For this purpose, the time variable t is not very convenient; it will be better to measure time from noon on the day in question.

To simplify our calculations, we will assume that the expressions $\cos (2\pi t/T_y)$ and $\sin (2\pi t/T_y)$ are constant over the course of any particular day—since T_y is 365 times as large as the change in t, this is a reasonable approximation. On the nth day (measured from June 21), we may replace $2\pi t/T_y$ by $2\pi n/365$, and formula (4) gives

$$\sin A = (\sin l)P + (\cos l)\left[Q \cos \left(\frac{2\pi t}{T_d}\right) + R \sin \left(\frac{2\pi t}{T_d}\right)\right] \qquad (5)$$

where $P = \cos (2\pi n/365) \sin \alpha$, $Q = \cos (2\pi n/365) \cos \alpha$, and $R = \sin (2\pi n/365)$.

We will write the expression $Q \cos (2\pi t/T_d) + R \sin (2\pi t/T_d)$ as $U \cos [2\pi(t - t_n)/T_d]$, where t_n is the time of noon on the nth day. To find U, we use the addition formula to expand the cosine:

$$U \cos \left(\frac{2\pi t}{T_d} - \frac{2\pi t_n}{T_d}\right) = U\left[\cos \left(\frac{2\pi t}{T_d}\right) \cos \left(\frac{2\pi t_n}{T_d}\right) + \sin \left(\frac{2\pi t}{T_d}\right) \sin \left(\frac{2\pi t_n}{T_d}\right)\right]$$

Setting this equal to $Q \cos (2\pi t/T_d) + R \sin (2\pi t/T_d)$ and comparing coefficients of $\cos 2\pi t/T_d$ and $\sin 2\pi t/T_d$ gives

$$U \cos \frac{2\pi t_n}{T_d} = Q \qquad \text{and} \qquad U \sin \frac{2\pi t_n}{T_d} = R$$

Squaring the two equations and adding gives

$$U^2 = Q^2 + R^2 \qquad \text{or} \qquad U = \sqrt{Q^2 + R^2}*$$

while dividing the second equation by the first gives $\tan (2\pi t_n/T_d) = R/Q$. We are interested mainly in the formula for U; substituting for Q and R gives

$$U = \sqrt{\cos^2 \left(\frac{2\pi n}{365}\right) \cos^2 \alpha + \sin^2 \left(\frac{2\pi n}{365}\right)}$$

$$= \sqrt{\cos^2 \left(\frac{2\pi n}{365}\right)(1 - \sin^2 \alpha) + \sin^2 \left(\frac{2\pi n}{365}\right)}$$

$$= \sqrt{1 - \cos^2 \left(\frac{2\pi n}{365}\right) \sin^2 \alpha}$$

We take the positive square root because $\sin A$ should have a local maximum when $t = t_n$.

Letting τ be the time from noon on the nth day, measured in hours, so that $(t - t_n)/T_d = \tau/24$, we may substitute into formula (5) to obtain the final formula:

$$\sin A = \sin l \cos\left(\frac{2\pi n}{365}\right)\sin \alpha$$

$$+ \cos l \sqrt{1 - \cos^2\left(\frac{2\pi n}{365}\right)\sin^2 \alpha}\cos\left(\frac{2\pi\tau}{24}\right) \tag{6}$$

which is identical (after some changes in notation) with formula (4) on p. 303.

Solved Exercise   3. How high is the sun in the sky in Edinburgh (latitude 56°) at 2 PM on February 1?

Exercise   5. Derive an "exact" formula for the time of sunset from formula (4). Make a suitable approximation to derive formula (6) on p. 304. Discuss the physical meaning of this approximation.

Problems for the Appendix to Chapter 15

1. Why does formula (6) for $\sin A$ not depend on the radius of the earth? The distance of the earth from the sun?

2. How high is the sun in the sky in Paris at 3 PM on January 15?

3. How much solar energy (relative to a summer day at the equator) does Paris receive on January 15?

4. How would your answer in Problem 3 change if the earth were to roll to a tilt of 32° instead of 23.5°?

Solutions to Solved Exercises in the Appendix to Chapter 15

1. The tip of $\mathbf{r}$ sweeps out a circle of radius $\sin \lambda$, so it covers a distance $2\pi \sin \lambda$ in time T. Its speed is therefore $(2\pi \sin \lambda)/T$ (Fig. 15-A-10).

Fig. 15-A-10 The tip of $\mathbf{r}$ sweeps out a circle of radius $\sin \lambda$.

From formula (2) we find the velocity vector to be

$$\frac{d\mathbf{r}}{dt} = -\sin \lambda \cdot \frac{2\pi}{T}\sin\left(\frac{2\pi t}{T}\right)\mathbf{m}_0$$

$$+ \sin \lambda \cdot \frac{2\pi}{T}\cos\left(\frac{2\pi t}{T}\right)\mathbf{n}_0$$

and its length is (since $\mathbf{m}_0$ and $\mathbf{n}_0$ are unit orthogonal vectors)

$$\left\|\frac{d\mathbf{r}}{dt}\right\| = \sqrt{\sin^2 \lambda \cdot \left(\frac{2\pi}{T}\right)^2\sin^2\left(\frac{2\pi t}{T}\right) + \sin^2 \lambda \cdot \left(\frac{2\pi}{T}\right)^2\cos^2\left(\frac{2\pi t}{T}\right)}$$

$$= \sin \lambda \cdot \left(\frac{2\pi}{T}\right)$$

as above.

2. From Solved Exercise 1, the speed is $s = (2\pi R/T_d) \cdot \sin \lambda = (2\pi R/T_d)\cos l$, where R is the radius of the earth and l is the latitude. (The factor R is inserted since $\mathbf{r}$ is a *unit* vector; the actual vector from the earth's center to P is $R\mathbf{r}$.) Using $T_d = 23.93$ hours and $R = 6371$ kilometers, we get $s = 1673 \cos l$ kilometers per hour. At the equator $l = 0$ so the speed is 1673 kilometers per hour; at $l = 60°$, $s = 836.4$ kilometers per hour.

3. We plug into formula (6): $\alpha = 23.5°$, $l = 90 - 56 = 34°$, $n =$ number of days after June 21 $= 225$, and $\tau = 2$ hours. We get

$$\sin A = 0.5196$$

$$\text{so } A = 31.3°.$$

Appendix C: Answers

ANSWERS TO ORIENTATION QUIZZES

Answers to Quiz A on p. 4

1. $-\dfrac{4}{3}$ **2.** $x > -\dfrac{2}{3}$ **3.** $x < -\dfrac{1}{3}$ and $x > 1$ **4.** $x = -2 \pm \dfrac{\sqrt{38}}{2}$

5.

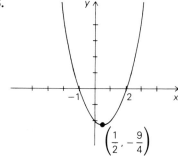

$\left(\dfrac{1}{2}, -\dfrac{9}{4}\right)$

6. $g(2) = \dfrac{4}{3}$. The domain of g is all x such that $x \neq 0, \dfrac{1}{2}$.   **7.** $x > -10$ **8.** At the points $(1, 1)$ and $(2, 4)$

9.

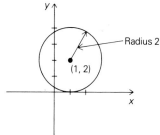

Radius 2

$(1, 2)$

10. 2

Answers to Quiz B on p. 4

1. $\dfrac{11}{4} = 2\dfrac{3}{4}$ **2.** $x(x + 3)$ **3.** 10   **4.** 4 kilograms   **5.** $(2x - 1)/2x$ (or $1 - 1/2x$)   **6.** x^7

7. $-6, -4, 0, \dfrac{1}{2}, 8$ **8.** $-\dfrac{3}{7}$ **9.** $\dfrac{1}{3}$ **10.** $x + 4$

Answers to Quiz C on p. 5

1. $(-2, -3)$ **2.** $\dfrac{8}{7}$ **3.** $150°$ **4.** $\dfrac{4}{5}$ **5.** $\sqrt{5}$ **6.** 8π centimeters   **7.** 14 meters

8. 18π cubic centimeters   **9.** $\dfrac{1}{2}$ **10.** 2

ANSWERS TO ODD-NUMBERED PROBLEMS

(Some solutions requiring proof are not included.)

CHAPTER 1

Section 1-1

1. (a) 2, 0   (b) 8, 0   (c) $-10, -10$ (d) 0, 10   (e) 30, 16   (f) 5, 6
(velocities in meters/sec; accelerations in meters/sec^2)

3. (a) $y = 2x - 1$ (b) $y = -4x + 4$ (c) $y = -3x + 1$ (d) $y = -3x + 5$

5. (a) $8x + 3$ (b) $2x - 2$ (c) $-2x$ (d) $-2x$ (e) $-4x + 5$ (f) -1

7. (a) 6   (b) 0   (c) -35 (d) 0   (e) -5 (f) 0   (g) -1

9. (a) Mileage (in miles/gallon)   (b) Fuel consumption (in gallons/mile)   (c) Fuel efficiency (in dollars per gallon)   (d) Operating cost (in dollars per mile). [These answers are subject to variable wording.]

11. (a) $\frac{5}{2}$ (b) -17 (c) $y = -x - 4$ (d) 4   (e) 29/2   **13.** 1

15. Two lines if $\bar{y} < \bar{x}^2$; one if $\bar{y} = \bar{x}^2$; none if $\bar{y} > \bar{x}^2$

17. Graph is a parabola opening upward if second derivative is positive; a parabola opening downward if it is negative; a straight line if it is zero.

19.   **21.** (a) $f'(x)$ $\begin{cases} < 0 \text{ for } x < -\frac{3}{4} \\ = 0 \text{ for } x = -\frac{3}{4} \\ > 0 \text{ for } x > -\frac{3}{4} \end{cases}$ (b)

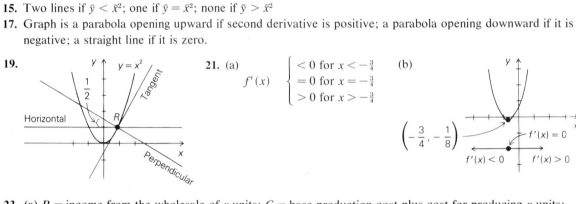

23. (a) R = income from the wholesale of x units; C = base production cost plus cost for producing x units; P = amount of income left from the sale of x units after accounting for the initial investment spent on producing x units.
(b) 2000 units   (c) 3000 units   (d) The profit made on each calculator sold

25. 104.94 mph   (b) $.008x + 1.1$ = rate of change of stopping distance with respect to speed.

27. $y = -\frac{5}{8}x + \frac{3}{4}$ **29.** (a) $y = 15x + 20$ (b) 15   (c) 170 ppm

Section 1-2

1. (a) 6   (b) $\frac{1}{5}$ (c) Does not exist   (d) 0   (e) Does not exist   (f) 0

3. (a) $2x - 1$ (b) $9x^2 + 1$ (c) $\frac{1}{2}$ (for $x \neq 0$)   (d) $\frac{-1}{x^2}$ $(x \neq 0)$ (e) $\frac{1 - x^2}{(1 + x^2)^2}$

5. (a) 2   (b) 4   (c) 7   (d) 22

7. (a) $\dfrac{dA}{dx} = 44x; \dfrac{d^2A}{dx^2} = 44$ (b) $\dfrac{dA}{dr} = \dfrac{3}{2}\pi r; \dfrac{d^2A}{dr^2} = \dfrac{3}{2}\pi$ (c) $\dfrac{dA}{dy} = \left(120 - \dfrac{25}{2}\pi\right)y; \dfrac{d^2A}{dy^2} = 120 - \dfrac{25}{2}\pi$

(d) $\dfrac{dA}{dx} = 26x; \dfrac{d^2A}{dx^2} = 26$

9. $\displaystyle\lim_{\Delta t \to 0} \dfrac{p(t + \Delta t) - p(t)}{\Delta t}$

11. (a) $\dfrac{dz}{dx} = 32x - 116, \dfrac{dy}{dx} = 2, \dfrac{dz}{dy} = 8y - 2, \dfrac{d^2z}{dx^2} = 32, \dfrac{d^2y}{dx^2} = 0$ (b) $\dfrac{dz}{dx} = -6x - 11, \dfrac{dy}{dx} = 1, \dfrac{dz}{dy} = -6y + 1,$

$\dfrac{d^2z}{dx^2} = -6, \dfrac{d^2y}{dx^2} = 0$ (c) $\dfrac{dz}{dx} = 20x + 40, \dfrac{dy}{dx} = 4x + 8, \dfrac{dz}{dy} = 5, \dfrac{d^2z}{dx^2} = 20, \dfrac{d^2y}{dx^2} = 4$

13. $13; -29; -0.2$

15. (a) 0   (b) 30   (c) 0   (d) -12 (Derivative at the midpoint is the same in each case.)

17. The exact values are in parentheses. (a) 4.08 (4.0804) (b) 39600 (39601) (c) 24.99 (24.990001)
(d) 1.004 (1.004004)

19. (a) $h(2.002) \approx 4.992$, $h(1.98) \approx 5.08$ (b) $f(5.005) \approx -5.055$, $f(5.01) \approx -5.11$
(c) $g(3.02) \approx -41.86$, $g(3.0002) \approx -41.0086$ (d) $f(9.989) \approx 443.076$, $f(10.021) \approx 445.764$

21. Too large by a factor $4(\Delta x)^2$, independent of the sign of Δx

$g'(3) = -16$; $g''(3) = -8$

23. (a) $f(T) = 0$, $g(T) = 0$
(b) One way the block might melt is to become a very thin sheet of ice of area A just before melting. In this case, $\lim_{t \to T} f(t) = A$, yet $f(T) = 0$, and $\lim_{t \to T} g(t) = 0$.
(c) $\lim_{t \to T} [f(t) \cdot g(t)] = 0$ (d) $[\lim_{t \to T} f(t)][\lim_{t \to T} g(t)] = A \cdot 0 = 0 = \lim_{t \to T} [f(t)g(t)]$

25. (a) Because the thermometer breaks
(b) The temperature of the flame, or doesn't exist if the temperature scale has too small a range.

(c)

(d)

Section 1-3

1. (a) $f'(r) = -30r^5 + 20r^3 - 26r$ (b) $g'(s) = 7s^6 + 78s^5 - 54s^2 + 3s$ (c) $h'(t) = 7t^6 - 5t^4 + 27t^2 - 9$

3. (a) $\dfrac{4x^9 + 7x^6 + x^4 - 2x}{(x^3 + 1)^2}$ (b) $\dfrac{-8(4x^2 + 21x - 1)}{(x^2 - 1)^4(x + 7)^3}$ (c) $\dfrac{-15x^6 - 9x^4 + 120x^3 + 30x^2 + 2}{(3x^4 + 2)^2}$

(d) $\dfrac{600(x - 1)^{299}}{(x + 1)^{301}}$ (e) $\dfrac{-8x}{(x^2 - 2)^2}$

5. (a) $y = \dfrac{-21}{2}x$ (b) $y = \dfrac{-25}{2}x + \dfrac{17}{2}$ (c) $y = \dfrac{-4}{8^8}x + \dfrac{9}{8^8}$

7. $\dfrac{d}{dx}\left(\dfrac{1}{ax^2 + bx + c}\right) = 0$ only at $x = \dfrac{-b}{2a}$. Any perfect square; $(x - 2)^2$ is an example.

9. -6250 **11.** Consider the first-order approximation.   **13.** $60 - x + \dfrac{x^2}{60}$ mi/hr

15. (a) $\frac{1}{2}$ (b) $(1 \pm \sqrt{46})/3$ (c) 0, 2   **17.** $(fg)'(x) = 1/(1 - x)^2 = f'(x)g'(x)$

19. (a) $\dfrac{d^2}{dx^2}x^n = n(n - 1)x^{n-2}$; $\dfrac{d^3}{dx^3}x^n = n(n - 1)(n - 2)x^{n-3}$ (b) $\dfrac{d^r}{dx^r}x^n = n(n - 1)\cdots(n - r + 1)x^{n-r}$
(c) $f'gh + fg'h + fgh'$

21. $1024x(1 + x^2)^7(1 + (1 + x^2)^8)^7(1 + (1 + (1 + x^2)^8)^8)^7$ **23.** (a) 7 (b) $\dfrac{-3}{16}$ (c) -1

25. Marginal revenue $= 8 - \dfrac{8x + 100}{x + 300} - \dfrac{2300x}{(x + 300)^2}$. Marginal profit $= 3 - \dfrac{8x + 100}{x + 300} - \dfrac{2300x}{(x + 300)^2} + 0.02x$.

Section 1-4

1. (a) $\dfrac{3}{2}x^2 + C$ (b) $\dfrac{1}{3}(t+1)^3 + C$ (c) $\dfrac{-1}{x} - \dfrac{1}{2x^2} + C$ (d) $\dfrac{3}{5}x^5 + x^4 + C$

3. (a) $\dfrac{-6x^2}{(x^3-1)^2}$ (b) $\dfrac{x^3+1}{x^3-1} + C$ 5. $G(t) = \displaystyle\int g(t)\,dt + C$. Constant is determined by $G(0)$.

7. (a) $t^3 + t^2 + t + C$ (b) $\dfrac{-1}{t} - \dfrac{1}{3t^3} - \dfrac{1}{4t^4} + C$ (c) $\dfrac{-1}{8(8t+1)} + C$ (d) $\dfrac{u^5}{5} - 3u^2 + C$

9. $\frac{3}{2}x^2 + x - \frac{3}{2}$ 11. 4 seconds

Chapter 1 — Review

1. (a) 6   (b) $2x + 9$ (c) $3x^2 + 2x$ (d) $\dfrac{-6}{x^2}$ (e) $26x(x^2+1)^{12}$

3. (a) $9x^2 - 2x - \dfrac{11}{2}$ (b) $6x^5 - 10x^4 - 12x^3 + 12x^2 + 8x$ (c) $\dfrac{x^3 + 12x^2 - 6x - 8}{(x+4)^3}$

   (d) $6x^2 - 10x + 1 + 3\dfrac{x^3 + 12x^2 - 6x - 8}{(x+4)(x^3 - x^2 - 2x)}$ (e) $4(6x^2 - 2x - 1)$

5. Reverse direction for $c < 0$ at $t = \pm\sqrt{\dfrac{-c}{3}}$; c does not affect acceleration.

7. (a) $y = -6x$ (b) $y = \dfrac{4}{7}x - \dfrac{4}{7}$ (c) $y = -5x + 3$ (d) $y = \dfrac{216}{361}x - \dfrac{413}{361}$ (e) $y = \dfrac{3}{32}x + \dfrac{11}{32}$

9. (a) 2   (b) 3   (c) 1   (d) -192; note that $[(h-2)^6 - 64]/h = [f(-2+h) - f(-2)]/h$ where $f(x) = x^6$.
   (e) Limit doesn't exist (or is $+\infty$).

11.

x_0	-3	-2	-1	0	1	2	3
$\displaystyle\lim_{x\to x_0} f(x)$	1	0	does not exist	does not exist	does not exist	does not exist	2

13. (a) $\dfrac{s^2(\pi s^4 - 4s + 3\pi)}{(\pi s^3 - 1)^2}$ (b) $20u^3$ (c) $8\pi r$ (d) $3s^2$ ($s = $ length of a side.)

15. (a) $f'(x) = \dfrac{-(x^2 - 2ax - c - 2ab)}{(x^2 + 2bx + c)^2}$; $f''(x) = \dfrac{2[x^3 - 3ax^2 - (6ab + 3c) + ac - 2bc - 4ab^2]}{(x^2 + 2bx + c)^3}$

   (b) $g'(t) = \dfrac{t^3(2t^4 - 3t^3 - 5t + 12)}{(t^3 - 1)^2}$; $g''(t) = \dfrac{2t^2(t^7 - 2t^4 - 9t^3 + 10t - 18)}{(t^3 - 1)^3}$

   (c) $h'(r) = 13r^{12} - 4\sqrt{2}r^3 + \dfrac{r^2 - 3}{(r^2 + 3)^2}$; $h''(r) = 156r^{11} - 12\sqrt{2}r^2 - \dfrac{2r(r^2 - 9)}{(r^2 + 3)^3}$

   (d) $h'(x) = 2(x - 2)^3(3x^2 - 2x + 4)$; $h''(x) = 2(x - 2)^2(15x^2 - 20x + 16)$

   (e) $q'(s) = \dfrac{s^4(5 - 16s + 10s^2)}{(1 - s)^2}$; $q''(s) = \dfrac{2s^3(10 - 45s + 54s^2 - 20s^3)}{(1 - s)^3}$

   (f) $f'(z) = \dfrac{ad - bc}{(cz + d)^2}$; $f''(z) = \dfrac{-2c(ad - bc)}{(cz + d)^3}$

17. (a) $2.45t^2 + 15t + C$ (b) $\dfrac{s^6}{6} + \dfrac{4s^5}{5} + 9s + C$ (c) $\dfrac{-1}{z} - \dfrac{2}{z^2} + C$ (d) $x^2(x^2 + 1) + C$

   (e) $\dfrac{1}{1-x} + C$ (f) $\dfrac{ax^3}{3} + \dfrac{bx^2}{2} + cx + C$ (g) $\dfrac{t^5}{t^2 + t} + C$

19. $y = 2(\sqrt{2} - 1)x$ 21. (c) If $\deg f(x) = 0$, then $f'(x) = 0$; this polynomial has no degree.

23. $\dfrac{\sqrt{3}}{2}s$ ($s = $ length of a side)   25. $fg'' + 2f'g' + f''g$

27. (a) $\dfrac{dy}{dx} = 2x + 3$; $\dfrac{dz}{dy} = 2y + 4$ (b) $\dfrac{dz}{dx} = 4x^3 + 18x^2 + 30x + 18$ (c) Same as part (b)

29. $\dfrac{dz}{dx} = n[f(x)]^{n-1}$ $\dfrac{df}{dx} = \dfrac{dz}{dy} \cdot \dfrac{dy}{dx}$ 31. Focal point is $\left(0, \dfrac{1}{4a}\right)$.

CHAPTER 2

Section 2-1

1. $f(x) = \begin{cases} 5x - 6 & -\frac{1}{2} < x < 2 \\ x + \dfrac{4}{x} & \text{otherwise} \end{cases}$ (This is not the only possible answer.)

3. $\lim\limits_{x \to 2-} h(x) = \lim\limits_{x \to 2+} h(x) = h(2)$; yes   **5.** $f(x) = \begin{cases} \dfrac{x^2 - 4}{x - 2} & x \ne 2 \\ 4 & x = 2 \end{cases}$

7. (a) $(-5, 1.01)$ (b) $(-5, 1.00001)$

    (c) $(-2 - \sqrt{3 + c}, -2 + \sqrt{3 + c})$ for $c > 6$; interval shrinks toward $(-5, 1)$ as c gets nearer to 6.

9. Suppose $f(x_0) \ne 0$, say $f(x_0) > 0$; use continuity to find an open interval containing x_0 but not containing an x_1 with $f(x_1) < 0$. No; counter examples:

    (1) $f(x) = \begin{cases} c & \text{if } x > x_0 \\ 0 & \text{if } x = x_0 \quad (c \ne 0) \\ -c & \text{if } x < x_0 \end{cases}$

    satisfies the other conditions but is discontinuous at x_0;

    (2) $f(x) = -x^2$ and $x_0 = 0$ satisfy the other conditions but in any open interval containing x_0, there does not exist an x_2 with $f(x_2) > 0$.

11. Reducing $f(x)$ to a quadratic in x^4 yields roots given by $x^4 = \dfrac{-3 \pm \sqrt{13}}{2}$ of which only $x^4 = \dfrac{-3 + \sqrt{13}}{2}$ yields real roots. Alternatively, look at $f(-1), f(0), f(1)$ and use the intermediate value theorem to show a root occurs between -1 and 0, and another root between 0 and 1.

13. (a) Easiest to use limits   (b) Apply the intermediate value theorem (second version) to $f - g$. The whole graph of f stays above that of g.

15. (a) $n \ge 7$ (b) $n \ge 12$ (c) $n \ge 14$ (d) $n \ge 1$ **17.** $\sqrt{7} \approx 2.65$ $(a = 2, b = 3, n \ge 7)$

19. No, because $\lim\limits_{x \to x_0-} f(x) \ne \lim\limits_{x \to x_0+} f(x)$

21. (a) No; $l = bh + ak + hk > 2$ for $h = k = 10^{-21}$, $a = b \ge 10^{21}$

    (b) 19th; $|l| \le |h| + |k| + |h||k| \le \max |h| + \max |k| + \max |h| \cdot \max |k|$

Section 2-2

1. (a) Increasing   (b) Neither   (c) Neither   (d) Increasing   (e) Decreasing

3. (a) $\pm \sqrt{2}$ (b) 1   (c) None   (d) None

5. Never   **7.** (a) Increasing   (b) Decreasing   (c) Decreasing   (d) Increasing

9. January, February, March; turning point at $t = 3$; at the end of March, inflation rate begins to increase and this trend continues for the remainder of that year.

11. $a(x^2 - 4x + 3)$, $a > 0$

13. (a)   (b)   (c)

    (d)   (e)   (f)

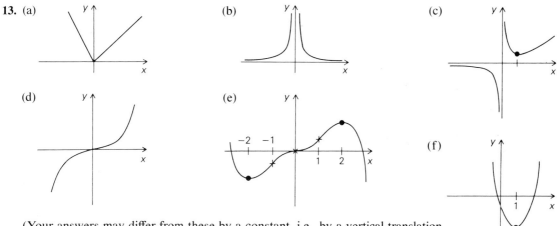

(Your answers may differ from these by a constant, i.e., by a vertical translation. See Section 1-4. $\times$ denotes an inflection point, and $\bullet$ denotes a turning point.)

15. (a) f increasing, f' increasing (b) f increasing, f' decreasing (c) f decreasing, f' increasing (d) f decreasing, f' decreasing

17. Inequalities are preserved upon multiplication by positive numbers. **19.** Turning point at $x = \dfrac{-b}{2a}$

21. $b^2 \le 3ac$ **23.** (a) $500 - 100N$ (b) If $N < N^* = 5$, then $\dfrac{dT}{dN} > 0$ implies $\dfrac{d}{dN}\left(\dfrac{I}{N}\right) < 0$.

25. $\dfrac{x^3}{3} - x$

Section 2-3

1. (a) $x = 0$ (local min.) (b) $x = 0$ (local min.) (c) $x = 0$ (local min.) (d) None (e) None

3. (a) 0 (b) 0 (c) $\dfrac{\pm\sqrt{3}}{3}$

5.

	Local max.	Local min.	Inflection points	Interval(s) on which f is increasing	Interval(s) on which f is decreasing	Interval(s) on which f is concave upward	Interval(s) on which f is concave downward
(a)	None	0	None	$(0, \infty)$	$(-\infty, 0)$	$(-\infty, \infty)$	Nowhere
(b)	$\frac{1}{2}$	None	None	$(-\infty, 0)\,;\left(0, \frac{1}{2}\right)$	$\left(\frac{1}{2}, 1\right)\,;(1, \infty)$	$(-\infty, 0)\,;(1, \infty)$	$(0, 1)$
(c)	-2	$\frac{2}{3}$	$\frac{-2}{3}$	$(-\infty, -2)\,;\left(\frac{2}{3}, \infty\right)$	$\left(-2, \frac{2}{3}\right)$	$\left(\frac{-2}{3}, \infty\right)$	$\left(-\infty, \frac{-2}{3}\right)$
(d)	$\frac{\pm\sqrt{2}}{2}$	0	$\frac{\pm\sqrt{6}}{6}$	$\left(-\infty, \frac{-\sqrt{2}}{2}\right)\,;\left(0, \frac{\sqrt{2}}{2}\right)$	$\left(\frac{-\sqrt{2}}{2}, 0\right)\,;\left(\frac{\sqrt{2}}{2}, \infty\right)$	$\left(\frac{-\sqrt{6}}{6}, \frac{\sqrt{6}}{6}\right)$	$\left(-\infty, \frac{-\sqrt{6}}{6}\right)\,;\left(\frac{\sqrt{6}}{6}, \infty\right)$
(e)	0	2	None	$(-\infty, 0)\,;(2, \infty)$	$(0, 2)$	$(2, \infty)$	$(-\infty, 2)$

7. $f(x) = \dfrac{x^4}{12} - \dfrac{x^3}{2} + x^2$

9. (a) (b)

• Turning point

× Inflection point

(c) 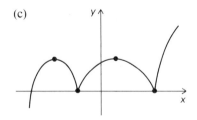 (d) Cannot exist

11. No; counterexample: $\begin{aligned} f(x) &= (x - x_0)(x - a) \\ g(x) &= (x - x_0)(x - b) \end{aligned}$ $b < x_0 < a$

13. The graph of f cannot lie both above and below its tangent at x_0. **15.** 28 ft

17. (a) Show $A'\left(\frac{1}{2}\right) = 0$ and $A''\left(\frac{1}{2}\right) < 0$. (b) Show $A\left(\frac{1}{2}\right) = \frac{1}{2}$ sq mi. (c) y satisfies $A\left(\frac{1}{2}\right) = \frac{1}{2}y$.

Section 2-4

1. (a) Odd   (b) Odd   (c) Neither   (d) Even   (e) Odd   **3.** (a) B (b) A (c) D (d) C

5. (a)

(b)

(c)

(d)

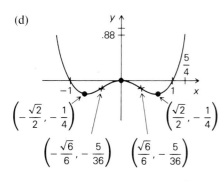

7. f has an inflection point at $x = \dfrac{-b}{3a}$ and nowhere else. The graph of g is a translation of the graph of f in such a way that g has an inflection point at $x = 0$. Hence symmetries for g about $(0, 0) = (0, g(0))$ translate to symmetries for f about its inflection point $\left(\dfrac{-b}{3a}, f\!\left(\dfrac{-b}{3a}\right)\right)$.

9. $e(x) = \frac{1}{2}[f(x) + f(-x)]$ and $o(x) = \frac{1}{2}[f(x) - f(-x)]$

11. (a)

(b)

(c)

(d)

(e)

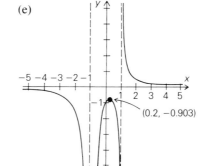

● Turning point
✕ Inflection point
■ Point where f''' equals zero

13.

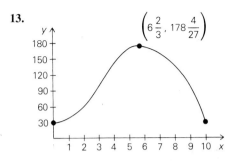

$\left(6\frac{2}{3}, 178\frac{4}{27}\right)$

15. General shape of the graph is unaffected by the reduction which consists only of a scaling by a factor $\frac{1}{a}$ and a translation by $\frac{-b}{4a}$.

17.

Type II₁

Type II₂

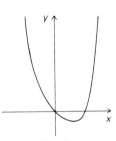

Type II₃

Chapter 2 — Review

1.

	(a)	(b)	(c)	(d)	(e)
Increasing	$(-\infty, \infty)$	$(-\infty, 0)$; $\left(\frac{12}{5}, \infty\right)$	Nowhere	$(-\infty, 0)$	Nowhere
Decreasing	Nowhere	$\left(0, \frac{12}{5}\right)$	$(-\infty, -3)$; $(-3, \infty)$	$(0, \infty)$	$\left(-\infty, \frac{-1}{\sqrt[3]{2}}\right)$; $\left(\frac{-1}{\sqrt[3]{2}}, \infty\right)$

3. (a)

(b)

$\left(\frac{9}{5}, -12.6\right)$

$\left(\frac{12}{5}, -19.9\right)$

(c)

(d)

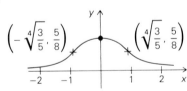

$\left(-\sqrt[4]{\frac{3}{5}}, \frac{5}{8}\right)$ $\left(\sqrt[4]{\frac{3}{5}}, \frac{5}{8}\right)$

(e)

$x = -\frac{1}{\sqrt[3]{2}}$

(0, 1) $\left(\frac{1}{\sqrt[3]{4}}, \frac{1}{2}\right)$

$x = -\frac{1}{2}$

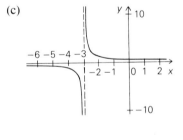

5. (a) False   (b) False   (c) False   (d) True   (e) False   (f) True   (g) True   (h) True   (i) True

7. (a)

(b)

(c)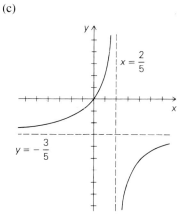

(d) $\left(-\sqrt{\frac{2}{3}}, 2.0887\right)$

$\left(\sqrt{\frac{2}{3}}, -0.0887\right)$

	Continuous	Differentiable	Monotonic	Concave upward	Concave downward	Critical points	Endpoints	Turning points	Local maxima	Local minima	Inflection points
(a)	$(-\infty, \infty)$	$(-\infty, \infty)$	$(-\infty, 0)$ $\left(0, \frac{4}{21}\right)$ $\left(\frac{4}{21}, \infty\right)$	$\left(-\infty, \frac{2}{21}\right)$	$\left(\frac{2}{21}, \infty\right)$	$0, \frac{2}{21}, \frac{4}{21}$	None	$0, \frac{4}{21}$	$\frac{4}{21}$	0	$\frac{2}{21}$
(b)	$(-\infty, \infty)$	$(-\infty, \infty)$	$\left(-\infty, \frac{-1}{4}\right)$ $\left(\frac{-1}{4}, \infty\right)$	$(-\infty, \infty)$	Nowhere	$\frac{-1}{4}$	None	$\frac{-1}{4}$	None	$\frac{-1}{4}$	None
(c)	$\left(-\infty, \frac{2}{5}\right)$ $\left(\frac{2}{5}, \infty\right)$	$\left(-\infty, \frac{2}{5}\right)$ $\left(\frac{2}{5}, \infty\right)$	$\left(-\infty, \frac{2}{5}\right)$ $\left(\frac{2}{5}, \infty\right)$	$\left(-\infty, \frac{2}{5}\right)$	$\left(\frac{2}{5}, \infty\right)$	None	None	None	None	None	None
(d)	$[-1, 2]$	$[-1, 2]$	$\left(-1, -\sqrt{\frac{2}{3}}\right)$ $\left(-\sqrt{\frac{2}{3}}, \sqrt{\frac{2}{3}}\right)$ $\left(\sqrt{\frac{2}{3}}, 2\right)$	$(0, 2)$	$(-1, 0)$	$0, \pm\sqrt{\frac{2}{3}}$	$-1, 2$	$\pm\sqrt{\frac{2}{3}}$	$-\sqrt{\frac{2}{3}}$	$\sqrt{\frac{2}{3}}$	0

9. Yes

11. (a) $c > \dfrac{3\sqrt{3}}{8}$

(b) $c < -\dfrac{3\sqrt{3}}{8}$; obtain graph by reflecting the graph of (a) about the y-axis.

(c) For (a), $c > -f'\left(\dfrac{\sqrt{3}}{3}\right)$; for (b), $c < -f'\left(\dfrac{-\sqrt{3}}{3}\right)$ where $\dfrac{\pm\sqrt{3}}{3}$ are the inflection points of f.

13. (a) 5000   (b) 45,000   (c) 15,500   **15.** $\dfrac{1}{50}$; 2 hr

17. No; let $x = a + h$, $y = b + k$, $a = b = 10^3$, $h = 10^{-13}$, $k = 9(10^{-13})$, then $l = 2ah + 2bk + h^2 + k^2$ has zeroes only in its first 8 decimal places.

19. $P_{\max} = P(\sqrt{\tfrac{1}{4} + \alpha}) = \dfrac{36\sqrt{\tfrac{1}{4} + \alpha}}{2\alpha + \tfrac{1}{2} + \sqrt{\tfrac{1}{4} + \alpha}} \approx 16.51$ where $\alpha = \left(\dfrac{\pi}{10}\right)^2$

21. (a)   (b) (graph not to scale)

CHAPTER 3

Section 3-1

1.

	Critical points	Endpoints	Singular points	Maximum points	Minimum points	Maximum value	Minimum value
(a)	$\dfrac{-1}{7}$	± 1	None	1	$\dfrac{-1}{7}$	13	$\dfrac{27}{7}$
(b)	None	None	None	None	None	None	None
(c)	$\dfrac{-1}{7}$	-4	None	-4	$\dfrac{-1}{7}$	108	$\dfrac{27}{7}$
(d)	$-2 \pm \sqrt{2}$	None	None	None	None	None	None
(e)	$-2 + 2\sqrt{2}$	6	None	6	$-2 + 2\sqrt{2}$	367	$47 - 32\sqrt{2}$
(f)	$-2 + 2\sqrt{2}$	-2	None	-2	$-2 + 2\sqrt{2}$	47	$47 - 32\sqrt{2}$
(g)	$-2 + 2\sqrt{2}$	± 4	None	4	$-2 + 2\sqrt{2}$	119	$47 - 32\sqrt{2}$

3.

	Critical points	Singular points	Endpoints	Maximum points	Minimum points
(a)	1.7	1	$-1, 2$	1	-1
(b)	1	0	-1	None	1
(c)	1.5, 3.5	None	None	3.5	1.5
(d)	None	1, 2	0, 3	1	0

5. (a) -6 (b) -4 (c) $-\dfrac{1}{2}$ (d) $-\dfrac{1001}{101}$ (e) none

7. If and only if f and g have at least one maximum point in common.

9.

Maximum points	Minimum point	Maximum value	Minimum value
22, 23	1000	506	$-955,000$

11. Suppose f attains a maximum at an interior point x_0, then the conditions $f(x) > f(x_0) + f'(x_0)(x - x_0)$ and $f'(x_0) = 0$ show x_0 is a local minimum, a contradiction.

Section 3-2

1. $\dfrac{1}{n}(a_1 + a_2 + \cdots + a_n)$ **3.** 4 and 32   **5.** 2000 rabbits

7.

		Length of a side of square area	Radius of circular area	Total area
(a)	Maximum	0 ft	$\dfrac{500}{\pi}$ ft	$\dfrac{250{,}000}{\pi}$ sq ft
(b)	Minimum	$\dfrac{1000}{\pi + 4}$ ft	$\dfrac{500}{\pi + 4}$ ft	$\dfrac{250{,}000}{\pi + 4}$ sq ft

9. $l = 24$, $w = h = 12$

11. With light sources at $x = \pm p$, $\pm 3p$, $p > 0$, the total
illumination $L(x) = \dfrac{1}{(x + 3p)^2} + \dfrac{1}{(x + p)^2} + \dfrac{1}{(x - p)^2} + \dfrac{1}{(x - 3p)^2}$
is least at x_1 and x_2, $-3p < x_1 < -2p$ and $p < x_2 < 3p$ (see fig.).
Using a symmetry argument without doing any calculations may
lead to an erroneous guess that minimum occurs at 0. Though 0 is a
local minimum, x_1 and x_2 are in fact the global minima on $[-3p, 3p]$.

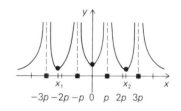

13. A semicircle of radius $\dfrac{500}{\pi}$ meters; maximum area $= \dfrac{125{,}000}{\pi}$ sq meters   **15.** $\dfrac{16}{0.22} \approx 72.72$

Section 3-3

1. $f(x) = 2g(x) + C$, where C is a constant.   **3.** Apply the mean value theorem to f on $[-1, 1]$.

5. (a) $\dfrac{2}{5}x^5 + \dfrac{8}{5}$ (b) $4x - \dfrac{x^2}{2} - 5$ (c) $\dfrac{x^5}{5} + \dfrac{x^4}{4} + \dfrac{x^3}{3} + \dfrac{13}{60}$ (d) $-\dfrac{1}{4x^4} + \dfrac{13}{4}$

7. (a) Apply the mean value theorem to f on $[-1, 0]$.   (b) Do the same to this f on $[-1, 1]$.

9. Apply Rolle's theorem (or the mean value theorem) to f to show f' vanishes at 2 distinct points, then apply it again to f'.

Chapter 3 — Review

1. The radius r and height h are given by $r^3 + 20pr^2 - \dfrac{V}{2\pi} = 0$, $h = \dfrac{V}{\pi r^2}$.

3. (a)

(b)

(c)

(d)

(e)

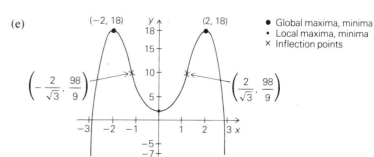

- Global maxima, minima
- Local maxima, minima
× Inflection points

5. 50 sq cm

7. Let $f(x) = (x-a_1)^4 + \cdots + (x-a_n)^4$ and show $f''(x) > 0$ for all x. Existence of a unique global minimum for f can be established by showing $f'(x) = 0$ has a unique solution.

9. (a) $t = \dfrac{-1}{3}$ (b) $t = \dfrac{-2}{3}$ **11.** Show that $\dfrac{dw}{dR} = 0$ and $\dfrac{d^2w}{dR^2} < 0$ at $R = R_i$.

13. Use the hypothesis about the tangent lines to show $\dfrac{d}{dx}\left[\dfrac{f(x)}{x}\right] = 0$ for all $x > 0$ and so $\dfrac{f(x)}{x} = m$

($m = $ constant).

15. 6 cm $\times$ 6 cm $\times$ 18 cm   **17.** (a) \$20/day   (b) \$1600   **19.** (a) Above average   (b) 92.16%

21. 83.544 cu in $< V(12) - V(11.8) < 86.4$ cu in

CHAPTER 4

11. Show $\quad -\dfrac{dw}{dR} = 0$ and

Section 4-1

1. (a) 14948   (b) 124   (c) $(n+1)^4 - 1$ (d) 0   **3.** 1515 grams

5. (a) $(n+1)^3 - 1 = 3\sum_{i=1}^{n} i^2 + 3\sum_{i=1}^{n} i + \sum_{i=1}^{n} 1$ and so $\sum_{i=1}^{n} i^2 = \dfrac{1}{3}\left[(n+1)^3 - 1 - 3\dfrac{n(n+1)}{2} - n\right]$

(b) $\dfrac{1}{6}[2(n^3 - m^3) + 3(n^2 + m^2) + (n - m)]$ (c) $\dfrac{n^2(n+1)^2}{4}$

7. (a) 24   (b) 21   (c) $\int_1^{10} g(t)dt \le 24$ (d) $\int_1^{10} 2g(t)dt \le 48$; $\int_1^{10} -g(t)dt \ge -24$

9. (b) $\int_a^b g(t)dt = \int_a^b f(t)dt + k(b-a)$

11. (a)   (b) 4.5

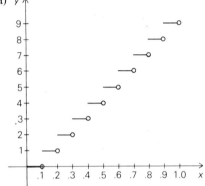

13. $\frac{1}{2}$ **17.** For $1 \le t \le 2, -3 \le t^3 - 4 \le 8$,   so $-3\int_1^2 dt \le \int_1^2 (t^3 - 4)dt \le 4\int_1^2 dt$.   **19.** $\frac{1}{6}$

Section 4-2

1. (a) 20   (b) 18.6   (c) $\dfrac{-268}{3}$ (d) $\dfrac{28}{3}\pi$ (e) $\dfrac{-78}{5}$

3. (a) $\dfrac{3}{(t^4 + t^3 + 1)^6}$ (b) $-t^2(1+t)^5$ (c) $\dfrac{-t^4}{(t^2 + 1)^3}$

5. (a)

(b) $\dfrac{14}{3}$ (c) $\dfrac{14}{3}$ (d) $F(t) = \begin{cases} \dfrac{t^3}{3} & 0 \le t \le 1 \\[2mm] t - \dfrac{2}{3} & 1 \le t < 5 \\[2mm] \dfrac{(t-6)^3 + 14}{3} & 5 \le t \le 6 \end{cases}$

(e) $F'(t) = f(t)$

7. 120,997.8   **9.** Apply the fundamental theorem of calculus.   **11.** (a) $\dfrac{1}{2}$ (b) $\dfrac{-193}{48}$

13. $\left(\dfrac{1}{n+2}\right)\left(x - \dfrac{1}{n+1}\right)(1+x)^{n+1} + C$ **15.** $\int_a^b t\,dt = (b^2 - a^2)/2$

17. (a) $-20t$ (b) $10(\Delta t)[2T + (\Delta t)]$ tons (in thousands)   (c) 450,000 tons

Section 4-3

1. (a) $\dfrac{22}{5}$ (b) 4   (c) $\dfrac{25}{3}$ (d) $\dfrac{26}{5}$ (e) 14   (f) $\dfrac{1}{2}$

(a)

(b)

(c)

(d)

(e)

(f)
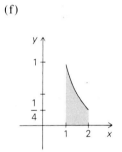

3. (a) $\dfrac{1}{2}bh$; same as from geometry   (b) $\dfrac{1}{2}bh$ (c) yes, $A = \int_0^a \dfrac{h}{a}x\,dx + \int_a^b \left(\dfrac{h}{a-b}x - \dfrac{bh}{a-b}\right)dx$

5. (a) $\dfrac{3}{10}$ (b) $\dfrac{67}{10}$ (c) $\dfrac{2437}{15}$ (d) $\dfrac{141}{80}$ (e) $\dfrac{129}{160}$ **7.** $\dfrac{16}{3}$ **11.** $x = \dfrac{8}{5}$ **13.** $\dfrac{112}{3}$

15. 15 min 43 sec

Chapter 4 — Review

1. (a) $\dfrac{21}{5}$ (b) $\dfrac{1777}{252}$ (c) 328,351   (d) $4n + 2$ (e) 379,250   **3.** (a) $\int_{x_1}^{x_2} \dfrac{dx}{f(x)}$ (b) $\dfrac{10^{3(n-1)} - 1}{n-1}$

5. 4   **7.** (a) $\dfrac{-718}{3}$ (b) $\dfrac{2}{3}$ (c) $\dfrac{-23}{36}$ (d) $\dfrac{7}{2}$

9. $\dfrac{40}{3}$

11. (a) 22 liters   (b) 16 liters   (c) $9\dfrac{13}{27}$ liters

13. (a) Upper sum ≈ 3.2399; lower sum ≈ 3.0399 (b) Average ≈ 3.1399; exact integral $= \pi$

15.

	Upper sums	Lower sums
(a)	$S > -\frac{3}{2}$	None
(b)	None	Any number

17. (b) $D(x) = 12.6 + 0.0045\, x^2$ billion cubic feet   (c) 12.8205 billion cubic ft

19. (b) $T \approx 5.2$ (c) ≈ 433 feet   (d) ≈ 0.4

21. (a) Second day   (b) Decreases by $4 \cdot 10^3$ bacteria/cm³.

CHAPTER 5

Section 5-1

1. $g(y) = $ the cost, in dollars, of y pounds of beans.

3. (a) f^{-1} is concave downward on $[f(a), f(b)]$.   (b) f^{-1} is concave upward on $[f(b), f(a)]$.

(a)   (b)

5.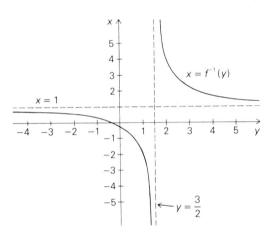

7. (a) $f^{-1}(y) = y^2 + 3;\ y \geq 0$ (b) $\dfrac{1}{2\sqrt{x-3}}$ **9.** (a) All $x \neq 0$ (b) Increasing on $(-\infty, \infty)$

11. (a) $\dfrac{3}{2\sqrt{x}} + \dfrac{1}{x^2}$ (b) $\dfrac{3 - x - 5x^3}{2\sqrt{x}\,(3 + x + x^3)^2}$ (c) $\dfrac{2}{(x^2 + 2)^{3/2}}$ (d) $\dfrac{1}{2\sqrt{x}\,(1 + \sqrt{x})^2}$ (e) $\dfrac{6 - 5x^4}{6\sqrt{x}\,(x^4 + 2)^{4/3}}$

13. (a) $\dfrac{1}{2}x^2 + \dfrac{2}{3}x^{3/2} + \dfrac{3}{4}x^{4/3} + C$ (b) $\dfrac{2}{3}(x - 1)^{3/2} + C$ (c) $\dfrac{8}{5}x^{5/2} + C$

17. (a)

Case $a = 1,\ q = \dfrac{1}{2}$

(b) $\dfrac{-2qx}{(a^2 + x^2)^{3/2}}$ (c) $\pm a\sqrt{3}$ **19.** $y'(x) \approx -6x + 18$ **21.** $\dfrac{\sqrt{2}}{48}$ sec/lb

23. (a) $y = \dfrac{50x}{x - 50}$ (b) $\dfrac{-2500}{(x - 50)^2}$ (c) $(0, 0)$ and $(100, 100)$

Section 5-2

1. (a) $h(x) = f(g(x))$ (b) $h(x) = k(l(x))$; $k(u) = m(n(u))$ **3.** $h'(x) = \dfrac{-x^2\,(x^3+4)\,(x^3-2)}{(x^3+1)^{2/3}\,(x^6+8)^{4/3}}$

5. (a) $3(x+4)\sqrt{x^2+8x+2}$ (b) $80(6\sqrt{x}-1)x^{3/2}(4\sqrt{x}-1)^{3/2}$ (c) $\dfrac{(x^2+1)(3x^{1/6}+2) - 12x^2(x^{1/6}+1)}{12x^{2/3}(x^2+1)^{3/2}(\sqrt{x}+\sqrt[3]{x})^{1/2}}$

(d) $\dfrac{2x^{3/2}(x^3+8+\sqrt{x}) - (x^2+2)(6x^{5/2}+1)}{2\sqrt{x}\,\sqrt{x^2+2}\,(x^3+8+\sqrt{x})^2}$ (e) $\dfrac{-48\sqrt{10}}{115}$

7. $(f\circ g)''(x) = f'(g(x))\cdot g''(x) + f''(g(x))\cdot [g'(x)]^2$

9. (a) The circle of radius $\dfrac{2}{3}$ centered at $\left(0, \dfrac{4}{3}\right)$ (b) 0 (c) $\left(\dfrac{\sqrt{2}}{3}, \dfrac{4-\sqrt{2}}{3}\right)$ or $\left(\dfrac{-\sqrt{2}}{3}, \dfrac{4+\sqrt{2}}{3}\right)$

11. (a) $2rx\,(1+x^2)^{r-1}$ (b) $\dfrac{3}{4}(1+x^2)^{2/3}+C$ **13.** $\dfrac{-41}{8}$ **15.** (a) $\dfrac{3}{4}$ (b) $y = \left(\dfrac{x}{8}\right)^4 + \dfrac{x}{4}$

17. Use the chain rule and the inverse function rule.   **21.** $\frac{9}{8}$ ft/sec   **23.** $\sqrt{\dfrac{\sqrt{5}}{2}}$

25. (a) In t hours, $x(t)$ miles had been traveled which used $A(x(t))$ gallons of gas.

(b) The statement is reasonable and is expressed by $\dfrac{d}{dt}[A(x(t))] = \dfrac{dA}{dx}\cdot\dfrac{dx}{dt}$

Chapter 5 — Review

1. (a) $\dfrac{5}{3}x^{2/3}$ (b) $\dfrac{(x^2+2)\sqrt{x}}{2(1+x^2)^{3/2}}$ (c) $\dfrac{3}{2}\sqrt{\dfrac{1+2\sqrt{x}}{x}}$ (d) $\dfrac{y(4-y^3)}{2(1-y^3)^{3/2}}$ **3.** $\dfrac{1}{2\sqrt{x}}f'(\sqrt{x})$

5. (a) $x^4+x^3+x^2+x+C$ (b) $10x+C$ (c) $\dfrac{1}{x}+\dfrac{1}{x^2}+\dfrac{1}{x^3}+\dfrac{1}{x^4}+C$ (d) $\dfrac{10}{21}x^{7/5}+C$

7. (a) $[-1,1]$ (b) $[5,9]$ (c) $-\frac{1}{3}$

9. (a) $\displaystyle\int\dfrac{dx}{2x^{3/4}(x^{1/4}+1)^2} = \dfrac{x^{1/4}-1}{x^{1/4}+1}+C$ (b) $\displaystyle\int\dfrac{4dx}{3(3x+1)^{4/3}(x-1)^{2/3}} = \sqrt[3]{\dfrac{x-1}{3x+1}}+C$

(c) $\displaystyle\int\dfrac{-2x\,dx}{(x^2-1)\sqrt{x^4-1}} = \sqrt{\dfrac{x^2+1}{x^2-1}}+C$ (d) $\displaystyle\int\dfrac{(3x^{5/2}+1)dx}{2\sqrt{x^4+2x^{3/2}+x}} = \sqrt{x^3+2\sqrt{x}+1}+C$

11. $\dfrac{d^2}{dy^2}[g(y)]^2 = \dfrac{2[f'(g(y)) - g(y)f''(g(y))]}{[f'(g(y))]^3}$ **13.** -1 **15.** $\dfrac{73\sqrt{130}}{26} \approx 32$ mi/hr

17. (a) $\sqrt{13}$ (b) 2 **19.** 79.20 lb/sec

21. (a) $R(x(13)) - R(x(12)) \approx \dfrac{d}{dt}(R(x(t)))\big|_{t=12}$ (b) $R(x(13)) - R(x(12)) \approx R'(x(12))x'(12)$ by the chain

rule, where $x'(12) = $ slope of tangent line to $x(t)$ at $t=12$ and $R'(x(12)) = $ slope of tangent line to $R(x)$ at $x = x(12)$.

Chapter 5 — Supplement

1. (a) $-10\sin(10x)$ (b) $15(1+2x)\cos[15(x+x^2)]$ (c) $\dfrac{-x\sin\sqrt{1+x^2}}{\sqrt{1+x^2}}$ (d) $\dfrac{(1+\sqrt{x})\cos\sqrt{x} - \sin\sqrt{x}}{2\sqrt{x}(1+\sqrt{x})^2}$

(e) $-\sin x\, e^{\cos x}$ (f) $\dfrac{e^x(e^x+x^2)\cos(e^x) - (e^x+2x)\sin(e^x)}{(e^x+x^2)^2}$ (g) $\dfrac{-e^x\sin\sqrt{1+e^x}}{2\sqrt{1+e^x}}$

(h) $5(\tan x+x^2)^4(\sec^2 x+2x)$ (i) $\dfrac{3}{\sqrt{1-9x^2}}$ (j) $\dfrac{x\sin^{-1}(x)}{\sqrt{1+x^2}} + \dfrac{\sqrt{1-x^4}}{1-x^2}$

3. (a) $\displaystyle\int x\sin(6x)\,dx = \dfrac{\sin(6x)}{36} - \dfrac{x\cos(6x)}{6} + C$ (b) $\displaystyle\int \ln x\,dx = x\ln x - x + C$

(c) $\displaystyle\int\dfrac{xe^x}{(x+1)^2}\,dx = \dfrac{e^x}{x+1}+C$ (d) $\displaystyle\int e^{2x}\cos x\,dx = \dfrac{e^{2x}}{5}(2\cos x + \sin x)+C$

5. (a) $\frac{1}{3}e^3(e^{12}-1)$ (b) $\frac{3}{2}+\sin 1 - \sin 2 + e(1-e)$ (c) $\ln(2)+\frac{7}{8}$ (d) $\sec x + (x^3/3) + C$

(e) $\dfrac{e^3}{2}(\sin 3 + \cos 3) - \dfrac{1}{2}$

7. Show that the derivative of the expression on the right-hand side equals the integrand.

9. (a) $\dfrac{\tan^{-1}(\sqrt{30})}{\sqrt{30}}$ (b) $\ln\left(\dfrac{3}{2}\right)$ (c) $\dfrac{1}{2}\ln\left(\dfrac{3}{2}\right)$

11. (a) $\dfrac{d}{dx}[\cot^{-1}(x)] = \dfrac{-1}{1+x^2}$ (b) $\dfrac{-3\cos(3x)}{[1+(\ln(\sin 3x))^2]\sin(3x)}$

13. Use the facts $\psi'(x) = \varphi(x)$ and $\psi''(x) = \dfrac{1}{\cos x}$.

15. (a) Use the chain rule to show the desired equation, conclude that $\ln(xx_0) - \ln x - \ln x_0 = C$, and determine C by letting $x = 1$.   (b) Apply the equation for $\ln xy$ to the product $e^x e^y$.

   (c) Show $\dfrac{d}{dx}\left(\dfrac{e^{x+x_0}}{e^x e^{x_0}}\right) = 0$ for a given fixed x_0 as in part (a) to conclude $\dfrac{e^{x+x_0}}{e^x e^{x_0}} = C = 1$ by letting $x = 0$.

CHAPTER 6

Section 6-1

1. (a) $\left(\sqrt{2}, \dfrac{-\pi}{4}\right)$ (b) $\left(2, \dfrac{\pi}{2}\right)$ (c) $\left(\dfrac{\sqrt{197}}{2}, \tan^{-1}14\right)$ (d) $\left(13, \pi + \tan^{-1}\dfrac{5}{12}\right)$ (e) $\left(\sqrt{73}, \pi - \tan^{-1}\dfrac{8}{3}\right)$

   (f) $\left(\dfrac{3\sqrt{2}}{4}, \dfrac{\pi}{4}\right)$ (g) $\left(4\sqrt{2}, \dfrac{3\pi}{4}\right)$ (h) $(\sqrt{226}, \tan^{-1}15)$ (i) $\left(\sqrt{370}, \tan^{-1}\dfrac{3}{19}\right)$

   (j) $\left(\sqrt{61}, \pi + \tan^{-1}\dfrac{6}{5}\right)$ (k) $\left(\dfrac{3\sqrt{10}}{10}, \tan^{-1}3\right)$ (l) $\left(\dfrac{\sqrt{10}}{2}, \pi - \tan^{-1}\dfrac{1}{3}\right)$

3. $\dfrac{475}{9}\pi$ meters   **5.** 10,722.534 ft   **9.** $\dfrac{2}{\sqrt{2+\sqrt{3}}}$

11.

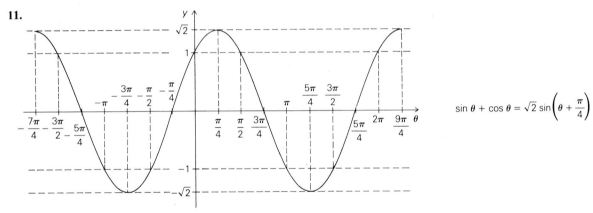

$$\sin\theta + \cos\theta = \sqrt{2}\sin\left(\theta + \dfrac{\pi}{4}\right)$$

13.

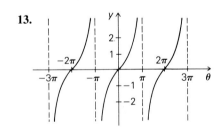

15. $\theta = \left(\dfrac{2n+1}{2}\right)\pi$ $(n = \text{integer})$

17.

$(x^2+y^2)^2 = y(3x^2 - y^2)$

As $\sin^3\theta = \cos\left[3\left(\theta - \dfrac{\pi}{6}\right)\right]$, its graph can be obtained by rotating the graph of Solved Exercise 20 through an angle of $\dfrac{\pi}{6}$.

19.

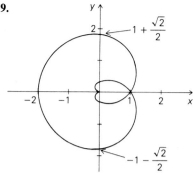

$$2(x^2 + y^2)^{3/2} - 4(x^2 + y^2) + (x^2 + y^2)^{1/2} + x = 0$$

21.

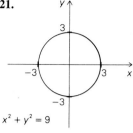

$$x^2 + y^2 = 9$$

23.

$$(x + 1)^2 + y^2 = 0$$

25.

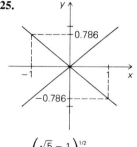

$$y = \pm\left(\frac{\sqrt{5} - 1}{2}\right)^{1/2} x = \pm 0.786x$$

27. $r = 1$ **29.** $r^2(2 + \sin 2\theta) = 2$ **31.** $r\sin\theta = \dfrac{1}{1 - r^2\cos^2\theta}$ **33.** $r(\sin\theta - \cos\theta) = 1$

43. (a) $\sqrt{3}(10^{10})$ cm/sec   (b) Use the fact that if $\sin\theta_1 = \sin\theta_2$ for θ_1 and θ_2 in $[0, \pi/2]$, then $\theta_1 = \theta_2$.
(c) $\approx 20.7°$

45. $A = 0.25$; $\omega = 400$

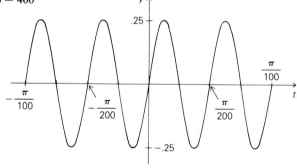

$$y(t) = 0.25\sin(400\,t)$$

47. $Ll\sin\theta$ **49.** $(-.000141779, .000141779)$

51. (b) Maxima of I are those t^* for which $\sin(311t^* + \theta) = 1$ and so $I_{\max} = I(t^*) = r(1)$.

(c) Period $= \dfrac{2\pi}{311}$; frequency $= \dfrac{311}{2\pi}$ (d) $\dfrac{-\tan^{-1}(2)}{311} \approx -.00356$

53. 2403 meters and 2368 meters

Section 6-2

1. (a) $\dfrac{1}{2\sqrt{x}} - 3\sin 3x$ (b) $\dfrac{-\sin x}{2\sqrt{\cos x}}$ (c) $\dfrac{3(x^2 + 2) - 2x\sqrt{1 - 9x^2}\,\sin^{-1} 3x}{(x^2 + 2)^2\sqrt{1 - 9x^2}}$

(d) $\dfrac{3}{2}(x^2\cos^{-1} x + \tan x)^{1/2}\left(2x\cos^{-1} x - \dfrac{x^2}{\sqrt{1 - x^2}} + \sec^2 x\right)$

(e) $\dfrac{3\sin 3\theta\cos 3\theta - \theta - \cos\theta\sqrt{\cos^2 3\theta + \theta^2}}{[1 + (\sin\theta + \sqrt{\cos^2 3\theta + \theta^2})^2]\sqrt{\cos^2 3\theta + \theta^2}}$ (f) $\dfrac{2\sqrt{1 - r^2} - 1}{(\sin r)\sqrt{1 - r^2}} - \dfrac{(r^2 + \sqrt{1 - r^2})(\sin r + r\cos r)}{r^2\sin^2 r}$

3. (a) $\dfrac{4\sin^{-1}(2x)}{\sqrt{1-4x^2}}+2x$ (b) $\dfrac{\tan 2x\,\sec 2x}{\sqrt{9-\sec^2 2x}}$, defined for $\frac{1}{3}\le|\cos x|\le 1$ (c) $\dfrac{-\theta}{(1-\theta^2)^{3/2}}$

(d) $v\sec\left(\dfrac{1}{v^2+1}\right)\left[\dfrac{\sec^2\sqrt{v^2+1}}{\sqrt{v^2+1}}-\dfrac{2}{(v^2+1)^2}\tan\sqrt{v^2+1}\cdot\tan\left(\dfrac{1}{v^2+1}\right)\right]$ (e) $\dfrac{-2\theta}{1+(\theta^2+1)^2}$ (f) 1

5. (a) 1

7. (a) The domain is $[-2,-\sqrt{2}]$, $[\sqrt{2},2]$ (b)

$$f'(x)=\dfrac{-2x}{\sqrt{1-(x^2-3)^2}}$$

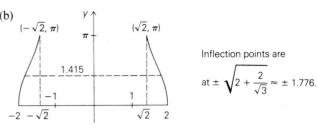

Inflection points are

at $\pm\sqrt{2+\dfrac{2}{\sqrt{3}}}\approx\pm1.776$.

9. (a) $\dfrac{dy}{dx}=-\left(\dfrac{y-\sqrt{1-x^2y^2}}{x-\sqrt{1-x^2y^2}}\right)$ (b) $\dfrac{dy}{dt}=-\left(\dfrac{y-\sqrt{1-x^2y^2}}{x-\sqrt{1-x^2y^2}}\right)\cdot\dfrac{1+t^2}{(1-t^2)^2}$

(c) $\dfrac{dx}{dt}=-\left(\dfrac{x-\sqrt{1-x^2y^2}}{y-\sqrt{1-x^2y^2}}\right)\dfrac{1}{\sqrt{1-t^2}}$ (d) $\dfrac{dy}{dt}=-\left(\dfrac{y-\sqrt{1-x^2y^2}}{x-\sqrt{1-x^2y^2}}\right)(3t^2+2)$

13. (a) $4\sin^{-1}x+C$ (b) $\dfrac{1}{2}\sin^{-1}(2x)+C$ (c) $\sin^{-1}\left(\dfrac{x}{2}\right)+C$ (d) $\dfrac{-\cos 2x}{2}+\dfrac{2}{3}x^{3/2}+C$

15. $\dfrac{5\pi}{2}$ meters per second   **17.** For example, $\tan^{-1}x$ **19.** $x=\left(\dfrac{2n+1}{12}\right)\pi$, for n any integer

23. (a) $-6000k\,(k\cos\theta-\sin\theta)(k\sin\theta+\cos\theta)^{-2}$ (b) $\theta=\tan^{-1}k$

Section 6-3

1. 2.2 meters per second

3. (a)

On $[0,2\pi]$

Zeros at 0.934, $2\pi-0.934$

Maxima at 0, π, 2π

Minima at $\dfrac{2\pi}{3}$, $\dfrac{4\pi}{3}$

Inflection at 0.936, 2.57, $2\pi-2.57$, $2\pi-0.963$

(b)

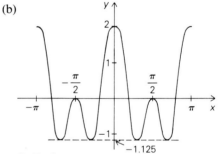

On $[0,\pi]$:

Zeros at 0.523, $\pi-0.523$

Maxima at 0, $\dfrac{\pi}{2}$, π

Minima at 0.912, $\pi-0.912$

Inflection at 0.557, 1.086, $\pi-1.086$, $\pi-0.557$

(c)

(d)

(e)

5. (a)

(−2.5, 2.261)

(b)

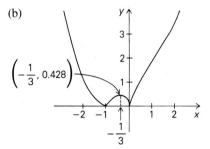

$\left(-\dfrac{1}{3}, 0.428\right)$

Inflection points: (−1.15, 0.197), (0.483, 1.027)

7. (a) 3   (b) $\frac{1}{3}$ (c) 1.877   (d) 0

9.

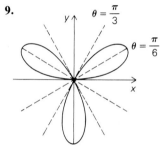

$\theta = \dfrac{\pi}{3}$

$\theta = \dfrac{\pi}{6}$

13. (a) max. height $= \dfrac{v_0^2 \sin^2 \alpha}{19.6}$;   range $= R = \dfrac{v_0^2 \sin 2\alpha}{9.8}$

15. $T = \dfrac{365}{2\pi} \cos^{-1}\left(\dfrac{\pm .327}{\sqrt{\tan^2 l + .0169}}\right)$ for $l > 16.7°$

l	T	Day
16.7°	0	June 21
20°	32.66	July 23
30°	57.23	Aug. 16
40°	68.28	Aug. 28
50°	75.20	Sept. 4
60°	80.24	Sept. 9
70°	84.33	Sept. 13
80°	87.90	Sept. 17
90°	91.25	Sept. 20

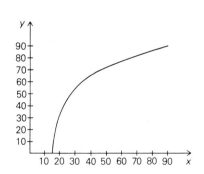

17. $S = 5{:}07{:}47$ P.M.

Chapter 6 — Review

1. (a) $f'(\theta) = 2\theta + \dfrac{\sin\theta - \theta\cos\theta}{\sin^2\theta}$; $f'(2) \approx 6.106$ (b) $g'(x) = \sec x\tan x - \dfrac{\cos(x+1) - x\sin(x+1)}{x^2\cos^2(x+1)}$

   (c) $h'(y) = 3y^2(1 + 2y\sec^2(y^3)) + 2\tan(y^3)$ (d) $x'(\theta) = \dfrac{2}{7}\left(\sin\dfrac{\theta}{2}\right)^{-3/7}\cos\left(\dfrac{\theta}{2}\right) + 9\theta^8 + \dfrac{11}{2\sqrt{\theta}}$

   (e) $y'(x) = -4x^3(2x^4 - 7)\sin(x^8 - 7x^4 - 10)$, $y'(0) = 0$

3. (a) $\dfrac{df}{dx} = 2[x + \cos(2x+1)]$; $\dfrac{df}{dt} = 6t^2[t^3 + 1 + \cos(2t^3 + 3)]$

   (b) $\dfrac{dg}{dr} = \dfrac{-2}{r^3} + \dfrac{2}{3}r(r^2 + 4)^{-2/3}$; $\dfrac{dg}{d\theta} = \dfrac{4\cos 2\theta}{3\sin^3 2\theta}[\sin^4 2\theta(\sin^2 2\theta + 4)^{-2/3} - 3]$

5. $\dfrac{150}{121} \approx 1.24$ meters per second

7. (a) $\sin x - \dfrac{1}{2}\cos 2x + \dfrac{1}{3}\sin 3x + C$ (b) $0.1446x^{8.3} + 0.4517x^{1.107} + 0.1x + C$ (c) $ax^n + C$

   (d) $2x^7 + \dfrac{1}{21}x^3 - 11x^2 + x + C$ (e) $-\cos(x^3) + x^2 + C$ (f) $\dfrac{x^3}{3} + 4\cos\left(\dfrac{1}{x} + 1\right) - 2x + C$

9.

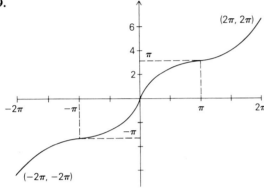

11. $x + 1 \geq \cos x$ for all $x \geq 0$.   **13.** $f(x) = A\sin 2x + B\cos 2x$ for any constants A and B

15. Concave up on $\left(\dfrac{2n+1}{2}\pi, (n+1)\pi\right)$, and concave down on $\left(n\pi, \left(\dfrac{2n+1}{2}\right)\pi\right)$ for all $n = 0, \pm 1, \pm 2, \cdots$.

    Inflection points at $x = \dfrac{n\pi}{2}$

19. (a)

   (b)

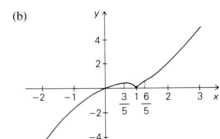

21. (b) $\dfrac{\pi}{2}$ (c) $\dfrac{\pi}{2}x - x\sin^{-1}x - \sqrt{1-x^2} + C = x\cos^{-1}x - \sqrt{1-x^2} + C$ (d) $x\sin^{-1}(3x) + \dfrac{\sqrt{1-9x^2}}{3} + C$

23. θ is max. when 13.856 ft away; θ is changing at 0 rad/sec.

25. (a) 13.75°/hr   (b) 21 min 49 sec   **27.** (a) No   (b) Yes

29. (a) $\dfrac{d\delta}{d\rho} = \dfrac{n\cos\rho}{\sqrt{1 - n^2\sin^2\rho}} - \dfrac{n\cos(A-\rho)}{\sqrt{1 - n^2\sin^2(A-\rho)}}$

31. (a) $\dfrac{dx}{dt} = \pi - \pi\cos\pi t;\ \dfrac{dy}{dt} = \pi\sin\pi t;\ \dfrac{d^2x}{dt^2} = \pi^2\sin\pi t;\ \dfrac{d^2y}{dt^2} = \pi^2\cos\pi t$ (b) 2π

33. (a) $(1-e^2)\left(x + \dfrac{le}{1-e^2}\right)^2 + y^2 = \dfrac{l^2}{1-e^2}$

CHAPTER 7

Section 7-1

1. (a) 2   (b) $2^{16/15}$ (c) 1   (d) 3   (e) -3 **3.** Yes   **5.** $b = \dfrac{3065}{1750};\ M = \dfrac{(1750)^2}{3065}$

7. Let $w = \log_{a^n} x$, then $(a^n)^w = x$ and so $nw = \log_a x;\ a > 0$ and $a \neq 1$.

9. $[1 + (b-1)]^n = 1 + n(b-1) + \binom{n}{2}(b-1)^2 + \cdots + n(b-1)^{n-1} + (b-1)^n \geq 1 + n(b-1)$.

11.

	Domain	Range
(a)	$(-\infty, -1);\ (3, \infty)$	$(-\infty, \infty)$
(b)	$(-\tfrac{1}{2}, \infty)$	$(-\infty, \infty)$
(c)	$(-1, 1)$	$(-\infty, 0)$

13. $g(y) = 2^y + 1;$ domain $= (-\infty, \infty)$

15. (a)

Domain $= (-\infty, \infty)$
Range $= [1, \infty)$

(b)

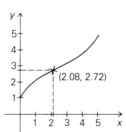

$\times$ = Inflection point

Domain $= [0, \infty)$
Range $= [1, \infty)$

(c)

Domain $= (-\infty, 0);\ (0, \infty)$
Range $= (0, 1);\ (1, \infty)$

17. (a) $(x^{1/2} + y^{1/2})(x^{1/2} - 2y^{1/2})$ (b) $(x^{1/2} + y^{1/2})(x^{1/2} - y^{1/2})$ (c) $(x^{1/3} + y^{1/3})(x^{2/3} + y^{2/3})$
   (d) $(x^{1/4} - 2)(x^{1/4} + 2)(x^{1/2} + 2)$ (e) $(x^{1/2} + 3^{1/2})^2$

19. (a) $(-5, -1);\ (-1, 1);\ (1, \infty)$ (b) $(-1, 1)$ (c) $(-5, -1);\ (-1, 1);\ (1, \infty)$
Domain of $f =$ domain of h and $f = h$.

21. Let $r = e^{\omega t}$, $\omega =$ constant, then $r' = \omega e^{\omega t} = \omega r$ and so the spiral seems to be expanding (or shrinking)
uniformly in all radial directions at a rate proportional to the radius. For $r = \theta = \omega t$, $r' = \omega$ and so
the radius increases (or decreases) uniformly at a constant rate independent of the radius and the angle.

23. (a) $S(I) = k\log_{10}(1) = 0$ (b) $S(2I) < 0;\ S\left(\dfrac{I}{2}\right) > 0$ (c) $-k\log_{10}(8) \leq S(I') \leq k\log_{10}(8)$

25. (a) 60 dB   (b) Increase the intensity from I to $10I$.   (c) Yes

27. (a) Yes   (b) Yes   (c) $10^{-5.5}$ moles/liter

Section 7-2

1. (a) $2(2x-1)e^{4x}$ (b) 6^x

3. (a) $\left(\dfrac{1}{x}+\ln x\right)e^x x^{e^x}$ (b) $\dfrac{-[2x^2 \sec^2 (x^2)\ln x + \tan (x^2)]}{x^{1+\tan (x^2)}}$ (c) $(1+\ln x)x^x \cos (x^x)$

    (d) $\left[2x \tan (x^2)\ln x + \dfrac{1}{x}\right]\sec (x^2)$ (e) $\dfrac{35(x+e^x)\cos x + 12(1+e^x)(1+\sin x)}{42(1+\sin x)^{1/6}(x+e^x)^{5/7}}$

5. $x=-1$ and those x satisfying $xe^x = \dfrac{(2n+1)}{2}\pi$; $n=0, \pm 1, \pm 2, \pm 3$.

7.

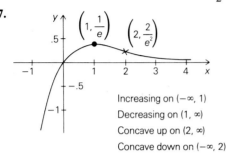

Increasing on $(-\infty, 1)$
Decreasing on $(1, \infty)$
Concave up on $(2, \infty)$
Concave down on $(-\infty, 2)$

9. (a) $\dfrac{-2\tan (2x)}{\ln \left(\frac{5}{3}\right)}$ (b) $\dfrac{x+1}{x(x+\ln x)}$ (c) $\dfrac{3}{2}x^{\sqrt{x}}\dfrac{(2+\ln x)}{\sqrt{x}}$ (d) $\dfrac{3}{2}x^{x/2}(1+\ln x)$ (e) $\dfrac{1}{x}+1+\ln x$

11. (a) $[f(x)+f'(x)]e^x + g'(x)$ (b) $[f'(x)+2x]e^{f(x)+x^2}$ (c) $[f(x)g'(x)+f'(x)]e^{g(x)}$

    (d) $[e^x + g'(x)]f'(e^x + g(x))$ (e) $\left[g(x)\dfrac{f'(x)}{f(x)}+g'(x)\ln f(x)\right]f(x)^{g(x)}$

13. (a) $\dfrac{3^x}{\ln 3}+C$ (b) $\dfrac{x^4}{4}+C$ (c) $x-3\ln |x+3|+C$; $x\neq -3$

    (d) $x\log_2 x - \dfrac{x}{\ln 2}+\ln |x|+C$; $x\neq 0$ (e) $\dfrac{x^2}{2}+10x+82\ln |x-8|+C$; $x\neq 8$

15. (a) Show $b^x = f(x) \approx f(0)+f'(0)(x-0)$.   (b) For $x=0.01$, $1+x\ln 2$ approximates 2^x to within $2.5(10^{-5})$ while for $x=0.0001$, the approximation is accurate to within 10^{-9}.

    (c) $e \approx \left(1+\dfrac{1}{n}\right)^n$

17. $\ln \left(\dfrac{1}{2}\right) = \lim_{h\to 0}\dfrac{1-2^h}{h2^h}$

19. (a) $\approx 7.25\%$

    (b) To see $1¢$ difference, deposit $\$P$ where $P \geq .01 \bigg/ \left[e^{.07}-\left(1+\dfrac{.07}{525600}\right)^{525600}\right] \approx \$1{,}072{,}653.$

    (The numerical answer will depend on your calculator, as substantial round-off errors can occur. An HP-33 gave $P \geq \$97.94$; a TI-58 gave $P \geq \$140{,}421$.)

21. $\approx 50.74\%$ **23.** $\$119.72$ **25.** ≈ 2881 years   **29.** (a) $\approx \$7278.37/\text{yr}$ (b) $\approx 25.92\%$

Chapter 7 — Review

1. (a) $\dfrac{x}{x+3}+\ln (x+3)$ (b) $[1+3x(x+2)^2]e^{(x+2)^3}$ (c) $-3\sin (3x+1)\cos [\cos (3x+1)]$

    (d) $\dfrac{\sin^2 [(\cos 2x)+1]-1}{2\sin (2x)\sin [(\cos 2x)+1]}$ (e) $\dfrac{1}{x\ln 2}$

3. (a) $\dfrac{-y}{x}$ (b) $\dfrac{y(y-3)}{x[1+(3-y)\ln x]}$ (c) $\dfrac{-1}{x}[y+3+\csc (xy+3x)]$

5. (a) $\dfrac{1}{3}e^{3x}+C$ (b) $\sin x + \dfrac{\ln x}{3}+C$ (c) $x+\ln x+C$

    (d) $\dfrac{\sin 2x}{2}-\dfrac{\cos 6x}{6}+C$ (e) $\dfrac{\tan^{-1}(\sqrt{2}x)}{\sqrt{2}}+C$

7. (a) 441.92 meters   (b) $10\int_0^{10} e^{-.07t} \approx 71.92 \approx 16\%$ of the distance. In the second ten seconds, neglecting $10\int_{10}^{20} e^{-.07t} \approx 35.71$ would give an overall error of 4% in the total distance traveled.

9. (a) e^{10} (b) e^{-12}

11. (a) $\frac{1}{2}$ (b) $\frac{3}{16}$

13.

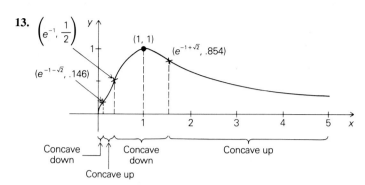

15. (a) $(\ln x)^{x-1} + (\ln x)^x \ln (\ln x)$ (b) $\left[\dfrac{7}{2(x+3)} + \dfrac{5}{3(x+8)} - \dfrac{12x}{11(x^2+1)}\right]\left[\dfrac{(x+3)^{7/2}(x+8)^{5/3}}{(x^2+1)^{6/11}}\right]$

17. For existence, apply the mean value theorem to $f(t) = e^t$ on $[0, x]$.
For the deduction, use the fact $0 < c < x$ implies $1 < e^c < e^x$.

19. (a) $r \ge 6.932\%$ (b) $r \ge 6.992\%$

21. $D = \dfrac{1}{\ln 10} \ln \left(\dfrac{I_0}{I}\right)$ 23. \$19,875.72

25. (a) $\dfrac{dP}{dt} = \dfrac{a^2 A}{e^{at}(b + Ae^{-at})^2}$ where $A = \left(\dfrac{a}{P_0}\right) - b$ (b) As $a > 0$, $\lim\limits_{t \to \infty} e^{-at} = 0$ and so $\lim\limits_{t \to \infty} \dfrac{a}{b + Ae^{-at}} = \dfrac{a}{b}.$

CHAPTER 8

Section 8-1

1. (a) $2a^3 + \dfrac{3}{2}a^2 + 2a$ (b) 6 (c) $\dfrac{2^{18}}{17}$ (d) $\dfrac{10}{9}$ (e) 193.6

3. (a) 0 (b) $\dfrac{5}{6}$

(c) $\displaystyle\int_{\pi/2}^x f(t)g(t)\, dt = \begin{cases} -\cos x & 0 < x \le 2 \\ \cos 2 - 2 \cos x & 2 \le x \le \pi \end{cases}$

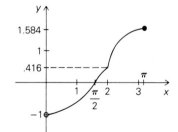

5. $1 + \ln (2) - \ln (e + 1)$ 7. (a) Show $\dfrac{d}{dx}(\sqrt{x^2 - 1} - \sec^{-1} x + C) = \dfrac{\sqrt{x^2 - 1}}{x}.$ (b) $\dfrac{\sqrt{5}}{2} - \sec^{-1}\left(\dfrac{3}{2}\right)$

9. $\frac{15}{32}$ 11. $\frac{16}{3} - \pi$ 13. 12 15. (b) \$45,231.46 17. (a) $\int_0^{10} 260e^{(0.1)t}\, dt$ (b) \$4,467.53 (c) 12

19. (a) As $D(t)$ is continuous on $[t_1, t_2]$, it has an antiderivative $N(t)$ which is the number of divorces t years after 1920. By the fundamental theorem of calculus, $N(t_2) - N(t_1) = \int_{t_1}^{t_2} D(t)\, dt.$ (b) 3.5×10^5

21. \$1510.72

Section 8-2

1. (a) $\cos\dfrac{1}{x} + C$ (b) $\dfrac{1}{3}(t^2 + 1)^{3/2} + C$ (c) $\dfrac{2}{15}(3t - 2)(t + 1)^{3/2} + C$ (d) $\sin \theta - \dfrac{1}{3}\sin^3\theta + C$

(e) $-\ln|\cos x| + C$

3. (a) $-\cos(\ln t) + C$ (b) $\ln | \ln x | + C$ (c) $e^{\tan x} + \tan x + C$ (d) $-\dfrac{3}{4}\left(3 + \dfrac{1}{x}\right)^{4/3} + C$

(e) $\frac{1}{2}\ln (1 + e^{2s}) + C$

5. (a) $\dfrac{2}{3}(\sqrt{17} - \sqrt{7})$ (b) 0 (c) $2\sqrt{2}$ (d) $\dfrac{1}{2}$

7. (a) $\dfrac{1}{3}(2\sqrt{2} - 1)$ (b) $\dfrac{\pi}{4} + 4 - \tan^{-1}(3)$ (c) $\dfrac{\pi - 2}{8}$ (d) $\ln 2$ (e) 1

9. (a) $\dfrac{1}{a}F(ax + b) + C$ where $F(u) = \int f(u)\, du$ (b) $\dfrac{1}{63}$ 11. $\dfrac{1}{2}$ 13. πab

Section 8-3

1. (a) $\dfrac{1}{25}(5x\sin 5x + \cos 5x) + C$ (b) $\dfrac{1}{4}(x^3-4)^{1/3}(x^3+12)+C$ (c) $x\ln(10x)-x+C$

(d) $\dfrac{x^3}{9}(3\ln x - 1)+C$ (e) $\dfrac{x^2}{5}[2\sin(\ln x)-\cos(\ln x)]+C$

3. $\dfrac{(x^n+1)^{m+1}[(m+1)x^n-1]}{n(m+1)(m+2)}+C$ **5.** (a) $\dfrac{\pi-2}{4}$ (b) 1 (c) $e-2$

7. $\dfrac{1}{a^2}[\sin(2\pi a)-2\pi a\cos(2\pi a)]$; the integral tends to zero.   **9.** (b) $\dfrac{5e^{3\pi/10}-3}{34}$

11. (a) For the case $\phi(a)\ge b$, $a\ge \phi^{-1}(b)$ and $\phi(x)\ge\phi(\phi^{-1}(b))=b$ on $[\phi^{-1}(b),a]$ as ϕ^{-1} and ϕ are increasing on $[0,\infty)$. By the formula for integrating inverse functions, $\int_0^b \phi^{-1}(y)\,dy = b\phi^{-1}(b)-\int_0^{\phi^{-1}(b)}\phi(x)\,dx$, and so $\int_0^b\phi^{-1}(y)\,dy+\int_0^a\phi(x)\,dx = b\phi^{-1}(b)+\int_{\phi^{-1}(b)}^a\phi(x)\,dx \ge b\phi^{-1}(b)+\int_{\phi^{-1}(b)}^a b\,dx = b\phi^{-1}(b)+b[a-\phi^{-1}(b)]=ab.$

(b)

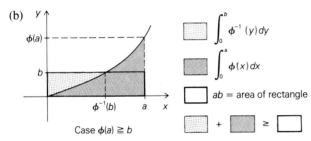

Case $\phi(a)\ge b$

$\square$ $\int_0^b \phi^{-1}(y)\,dy$

$\blacksquare$ $\int_0^a \phi(x)\,dx$

$\square$ ab = area of rectangle

$\square + \blacksquare \ge \square$

(c) Let $\phi(x)=x^{p-1}$ $(p>1)$, then $\phi^{-1}(y)=y^{1/(p-1)}=y^{q-1}$ since $\dfrac{1}{p}+\dfrac{1}{q}=1$ implies $p+q=pq$, $q=\dfrac{p}{p-1}$, and so $q-1=\dfrac{1}{p-1}$. As $\phi'(x)>0$ on $[0,\infty)$, part (a) gives $ab\le \int_0^a x^{p-1}\,dx + \int_0^b y^{q-1}\,dy = \dfrac{a^p}{p}+\dfrac{b^q}{q}.$

13. $1-\dfrac{40401}{e^{200}}\approx 1$ kg

15. $b_m = \begin{cases} \dfrac{-A}{m\pi} & \text{if } m=2k \\[2mm] \dfrac{A}{m\pi} & \text{if } m=2k-1 \qquad k=\text{non-zero integer} \\[2mm] 0 & \text{if } m=0 \end{cases}$

17. $Q(T)=EC\left[1-\dfrac{R}{2L}e^{-RT/2L}\left(T+\dfrac{2L}{R}\right)\right]$ coulombs

Chapter 8 — Review

1. (a) $\dfrac{x^2}{2}-\cos x + C$ (b) $\dfrac{x^2}{2}+\sin^{-1}x+C$ (c) $e^x-\dfrac{x^3}{3}-\ln|x|+\sin x + C$ (d) $\dfrac{x^4}{4}+\sin x + C$

(e) $\dfrac{8}{5}t^5-5\sin t + C$

3. $e^{\pi/2}+\dfrac{\pi^4}{64}+\dfrac{\pi^2}{4}+3\pi \approx 18.225$

5. (a) $-\dfrac{1}{3}\cos^3 x + C$ (b) $\dfrac{1}{3}e^{x^3}+C$ (c) $\dfrac{1}{3}\ln|3x+4|+C$ (d) $\dfrac{x^2}{2}\left[(\ln x)^2-\ln x + \dfrac{1}{2}\right]+C$

(e) $\dfrac{3}{4}[2x\sin(2x)+\cos(2x)]+C$

7. $\displaystyle\int \dfrac{x\,dx}{\sqrt{1-x^2}}=-\sqrt{1-x^2}+C$; $\displaystyle\int\dfrac{x^2\,dx}{\sqrt{1-x^2}}=\dfrac{1}{2}(\sin^{-1}x-x\sqrt{1-x^2})+C$

9. (a) $\dfrac{399}{4}$ (b) $16\pi^2$ (c) 6   (d) $2 + \dfrac{\pi}{8}$ (e) $\dfrac{2 - \sqrt{2}}{2}$ (f) 1   (g) $\ln 2$

(a)

(b)

(c)

(d)

(e)

(f)

(g)

11. (a) $\sin 1 - \sin \dfrac{1}{2}$ (b) $\dfrac{1}{6}\sin^2\left(\dfrac{\pi^3}{8}\right)$ (c) $\dfrac{1}{2}[(x^2 + 1)\tan^{-1}(x) - x] + C$ (d) $-\ln|\cos(e^x)| + C$

13. (a) $\dfrac{1}{36}(6x - 1)e^{6x} + C$ (b) $(x^2 - 2)\sin x + 2x \cos x + C$ (c) $\dfrac{x^3}{9}(3 \ln x - 1) + C$

(d) $\dfrac{x^4}{16}(4 \ln x - 1) + C$ (e) $\dfrac{x^{n+1}}{(n+1)^2}[(n+1)\ln x - 1] + C$ (f) $x \tan^{-1}x - \dfrac{1}{2}\ln(1 + x^2) + C$

(g) $2(\sqrt{x} - 1)e^{\sqrt{x}} + C$ (h) $\sqrt{x}(\ln x - 2) + C$

15. (a) The quadratic equation is $m^2 + (n + 2)m + (n + 1)^2 = 0$.

(b) The discriminant $b^2 - 4ac = -3n^2 - 4n < 0$ if $n > 0$.

(c) For example, $f(x) = 1/\sqrt{x}$, $g(x) = x^{(-3 \pm \sqrt{5})/2}$.

17. $\dfrac{2}{3\pi}\left[\cos\left(\dfrac{\pi x}{2}\right) + \sin\left(\dfrac{\pi x}{2}\right)\sin(\pi x)\right] + C$

19. (b) $\dfrac{1}{2}(x - \sin x \cos x) + C$ (c) $\dfrac{3}{8}x - \dfrac{3}{8}\sin x \cos x - \dfrac{1}{4}\sin^3 x \cos x + C$

21. (a) $\dfrac{1}{2}\sin^2(3x) + C$ (b) $-\dfrac{1}{3}(5 - x^2)^{3/2} + C$ (c) $\dfrac{2}{5}(x - 2)(x + 3)^{3/2} + C$ (d) $\dfrac{1}{2}e^{x^2} + C$

(e) $\dfrac{1}{2}e^{x^2}(x^2 - 1) + C$ (f) $\dfrac{2}{105}(x + 1)^{3/2}(15x^2 - 12x + 8) + C$

23. $\sqrt{6} - \sqrt{2}$ **25.** (a) $\tan^{-1}(e^x) + C$ (b) $\dfrac{6}{7}x^{7/6} - \dfrac{1}{3}x^3 + C$ (c) $\dfrac{1}{2}\tan^2 x + C$

27. (a) $f(g(x_0))\,g'(x_0)$ (b) $\dfrac{2}{x}$ **29.** $f(h(x))\,h'(x) - f(g(x))\,g'(x)$

CHAPTER 9

Section 9-1

1. (a) $-\cos(2t)$ (b) e^{3t} (c) $5\cos(5t) + \dfrac{5}{2}\sin(5t) = \dfrac{5\sqrt{5}}{2}\cos\left[5t - \tan^{-1}\left(\dfrac{1}{2}\right)\right]$

(d) $5\cos(5t) + (5 - \sqrt{2})\sin(5t) = \sqrt{52 - 10\sqrt{2}}\,\cos\left[5t - \tan^{-1}\left(\dfrac{5 - \sqrt{2}}{5}\right)\right]$

(a)

(b)

(c)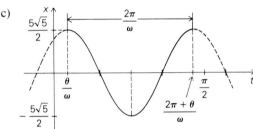

Phase shift $= \dfrac{\theta}{\omega} = \dfrac{\tan^{-1}\left(\dfrac{1}{2}\right)}{5}$

Period $= \dfrac{2\pi}{\omega} = \dfrac{2\pi}{5}$

(d)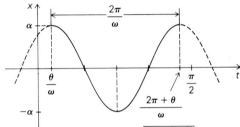

Amplitude $= \alpha = \sqrt{52 - 10\sqrt{2}}$

Phase shift $= \dfrac{\theta}{\omega} = \dfrac{1}{5}\tan^{-1}\left(\dfrac{5 - \sqrt{2}}{5}\right)$

Period $= \dfrac{2\pi}{\omega} = \dfrac{2\pi}{5}$

3. (a) 4   (b) $\cos(2t) + \dfrac{1}{2}\sin(2t) = \dfrac{\sqrt{5}}{2}\cos\left[2t - \tan^{-1}\left(\dfrac{1}{2}\right)\right]$

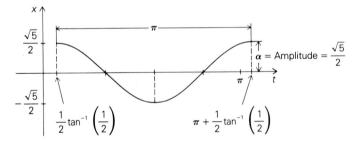

5. $(0.45)\dfrac{\ln 10}{\ln 2} \approx 1.4948676$ billion years

7. (a)

(b)

9. (a) $\cosh(9t) - \frac{1}{9}\sinh(9t)$ (b) $\cos(9t) - \frac{1}{9}\sin(9t)$ **11.** Apply Worked Example 6.

13. Reduce the problem to that of solving $y'(t) = -c\,y(t)$ by letting $y(t) = h - x(t)$; $K = h$.

15. (a) $\displaystyle\int \frac{dx}{1-4x^2} = \begin{cases} \frac{1}{2}\tanh^{-1}(2x) + C & |x| < \frac{1}{2} \\ \frac{1}{2}\coth^{-1}(2x) + C & |x| > \frac{1}{2} \end{cases}$

(b) $-\frac{1}{2}\coth(2x) + 3\ln|x| + C$; $x \neq 0$ (c) $-\operatorname{sech}^{-1}(2x) + C$; $0 < x < \frac{1}{2}$

19. (a) To show that x is a solution, show $x''(t) = -\omega^2 x + f(t)$ using the fact $y''(t) = -\omega^2 y + f(t)$. To show that any solution has the given form, let x be a solution and show $z = x - y$ satisfies $z''(t) + \omega^2 z(t) = 0$.

(b) $\dfrac{k}{\omega^2} - \dfrac{\sin(\omega t)}{\omega} + \dfrac{\omega^2 - k}{\omega^2}\cos(\omega t)$ (c) $t + \dfrac{2}{\omega}\sin(\omega t) - \cos(\omega t)$

21. (a) $y'(t) = P'(t)\,y_0 e^{P(t)} = p(t)\,y$ (b) $y(t) = e^{t^2/2}$

Section 9-2

1. (a) $y = \sqrt{x^2 + 1}$ (b) $y = -2x$ (c) y is given implicitly by $e^y(y-1) = \frac{1}{2}\ln|x^2 + 1|$

(d) $y = \alpha + \beta\,e^{3x^2/2}$; α, β are constants (e) $y = 2(\sin x - 1) + Ce^{-\sin x}$; $C = \text{constant}$

3. 61

5. (a) $\dfrac{dy}{dx} = \dfrac{-x}{4y}$ (b) $\dfrac{dy}{dx} = \dfrac{4y}{x}$; $y = \pm Kx^4$, $K = \text{constant}$

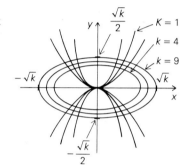

7. $a(t)$ is given implicitly by $\dfrac{-(\ln a - \ln a_0)}{(2b_0 - a_0)^2} - \dfrac{\left(\dfrac{1}{a} - \dfrac{1}{a_0}\right)}{2b_0 - a_0} + \dfrac{\ln|2b_0 - a_0 + a| - \ln(2b_0)}{(2b_0 - a_0)^2} = \dfrac{k}{2}t$

9. $y = \dfrac{1}{9}(3x - 1) + C\,e^{-3x}$; $C = \text{constant}$

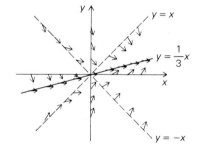

11. $y = e^{x^2/2}(1 + \int_0^x e^{-t^2/2}dt)$

13. (a) $\dfrac{dw}{dy} = (1-n)y^{-n} = (1-n)\dfrac{y^{1-n}}{y} = \dfrac{(1-n)}{y}w$ (b) $y = \dfrac{\pm 1}{x\sqrt{C - x^2}}$; $C = \text{constant}$

15. $s(t) = [5 - 4.99\,e^{-20(10^{-8})t}](10^6)$ kg; $t = \dfrac{-(10^8)}{20}\ln\left(\dfrac{.5}{4.99}\right) \approx 1.15(10^7)$ sec

17. $\displaystyle\int h(y)dy = -\int \frac{1}{g(x)}dx$ **19.** $y(1) \approx 0.6$

21. $y\left(\dfrac{\pi}{4}\right) \approx \begin{cases} 0.970263 & \text{using 10-step} \\ 0.975745 & \text{using 20-step} \end{cases}$ Though the 20-step should yield a more accurate result, we are probably safe to conclude $y\left(\dfrac{\pi}{4}\right) \approx 0.975 \pm .005$.

Chapter 9 — Review

1. (a) e^{3t} (b) $\dfrac{\sqrt{3}}{3}\sin(\sqrt{3}\,t)$ (c) $\dfrac{1}{3}(4e^{3t}-1)$ (d) e^t

3.

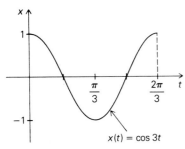

$x(t)=\cos 3t$

5. 27 minutes   **7.** (a) e^{-4t} (b) $\cos t - \sin t$ (c) $\cos(\sqrt{6}\,t)+\left(\dfrac{6-\cos\sqrt{6}}{\sin\sqrt{6}}\right)\sin(\sqrt{6}\,t)$

9. $y=\dfrac{2}{3\left(x+\frac{2}{3}\right)}$

11. (a) $\approx 5.127\%$ (b) Change the formula to $M(t)=M(0)e^{\alpha t/365}$ where $\alpha=\ln(1.05)=0.0488$.

13. $f(t)=-3\cos(\sqrt{5}\,t)+\dfrac{4}{\sqrt{5}}\sin(\sqrt{5}\,t)$

$\qquad =\sqrt{\dfrac{61}{5}}\,\cos\left[\sqrt{5}\,t-\pi+\tan^{-1}\left(\dfrac{4}{3\sqrt{5}}\right)\right]$

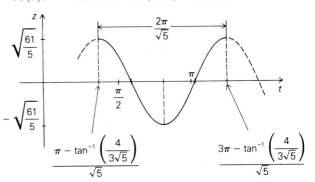

15. (a) $\dfrac{-4600\ln(2)}{\ln(.88)}\approx 24{,}942.446$ years

$\quad$ (b) $\dfrac{4600\ln(.01)}{\ln(.88)}\approx 165,714.02$ years (for 1%); $\dfrac{4600\ln(.001)}{\ln(.88)}\approx 248,571.04$ years $\left(\text{for }\dfrac{1}{1000}\right)$

17. 15.18 minutes; yes

21. (a) $y^2+9x^2=C$ (Solution is an ellipse if $C>0$, the point $(0,0)$ if $C=0$, and no solution if $C<0$.)

Case $C>0$

$\quad$ (b) $|y|^9=C\,|x|$; $\ C=$ constant

23. $y(1)=\begin{cases}\left(\dfrac{11}{10}\right)^{10}\approx 2.5937424 & \text{using 10-step}\\[2mm]\left(\dfrac{21}{20}\right)^{20}\approx 2.6532977 & \text{using 20-step}\\[2mm]e\ \ \approx 2.7182818 & \text{exact solution}\end{cases}$

The 10-step Euler method has an error of approximately 4.58%, nearly twice that of the 20-step (2.39%).

25. The solution "blows up" at $x=1$.

9. (a) $\cosh(9t) - \frac{1}{9}\sinh(9t)$ (b) $\cos(9t) - \frac{1}{9}\sin(9t)$ **11.** Apply Worked Example 6.

13. Reduce the problem to that of solving $y'(t) = -c\,y(t)$ by letting $y(t) = h - x(t)$; $K = h$.

15. (a) $\displaystyle\int \frac{dx}{1 - 4x^2} = \begin{cases} \frac{1}{2}\tanh^{-1}(2x) + C & |x| < \frac{1}{2} \\ \frac{1}{2}\coth^{-1}(2x) + C & |x| > \frac{1}{2} \end{cases}$

　　(b) $-\frac{1}{2}\coth(2x) + 3\ln|x| + C$; $x \neq 0$　(c) $-\operatorname{sech}^{-1}(2x) + C$; $0 < x < \frac{1}{2}$

19. (a) To show that x is a solution, show $x''(t) = -\omega^2 x + f(t)$ using the fact $y''(t) = -\omega^2 y + f(t)$. To show
　　that any solution has the given form, let x be a solution and show $z = x - y$ satisfies $z''(t) + \omega^2 z(t) = 0$.

　　(b) $\dfrac{k}{\omega^2} - \dfrac{\sin(\omega t)}{\omega} + \dfrac{\omega^2 - k}{\omega^2}\cos(\omega t)$　(c) $t + \dfrac{2}{\omega}\sin(\omega t) - \cos(\omega t)$

21. (a) $y'(t) = P'(t)\,y_0 e^{P(t)} = p(t)\,y$　(b) $y(t) = e^{t^2/2}$

Section 9-2

1. (a) $y = \sqrt{x^2 + 1}$　(b) $y = -2x$　(c) y is given implicitly by $e^y(y - 1) = \frac{1}{2}\ln|x^2 + 1|$
　　(d) $y = \alpha + \beta\,e^{3x^2/2}$; α, β are constants　(e) $y = 2(\sin x - 1) + Ce^{-\sin x}$; $C = $ constant

3. 61

5. (a) $\dfrac{dy}{dx} = \dfrac{-x}{4y}$　(b) $\dfrac{dy}{dx} = \dfrac{4y}{x}$; $y = \pm Kx^4$, $K = $ constant

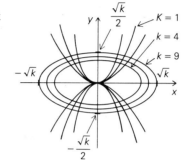

7. $a(t)$ is given implicitly by $\dfrac{-(\ln a - \ln a_0)}{(2b_0 - a_0)^2} - \dfrac{\left(\frac{1}{a} - \frac{1}{a_0}\right)}{2b_0 - a_0} + \dfrac{\ln|2b_0 - a_0 + a| - \ln(2b_0)}{(2b_0 - a_0)^2} = \dfrac{k}{2}t$

9. $y = \dfrac{1}{9}(3x - 1) + C\,e^{-3x}$; $C = $ constant

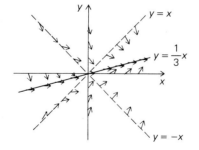

11. $y = e^{x^2/2}\left(1 + \int_0^x e^{-t^2/2}\,dt\right)$

13. (a) $\dfrac{dw}{dy} = (1 - n)y^{-n} = (1 - n)\dfrac{y^{1-n}}{y} = \dfrac{(1 - n)}{y}w$　(b) $y = \dfrac{\pm 1}{x\sqrt{C - x^2}}$; $C = $ constant

15. $s(t) = [5 - 4.99\,e^{-20(10^{-8})t}](10^6)$ kg; $t = \dfrac{-(10^8)}{20}\ln\left(\dfrac{.5}{4.99}\right) \approx 1.15(10^7)$ sec

17. $\displaystyle\int h(y)\,dy = -\int \dfrac{1}{g(x)}\,dx$　**19.** $y(1) \approx 0.6$

21. $y\!\left(\dfrac{\pi}{4}\right) \approx \begin{cases} 0.970263 & \text{using 10-step} \\ 0.975745 & \text{using 20-step} \end{cases}$　Though the 20-step should yield a more accurate result, we are
probably safe to conclude $y\!\left(\dfrac{\pi}{4}\right) \approx 0.975 \pm .005$.

Chapter 9 — Review

1. (a) e^{3t} (b) $\dfrac{\sqrt{3}}{3} \sin(\sqrt{3}\, t)$ (c) $\dfrac{1}{3}(4e^{3t} - 1)$ (d) e^t

3.

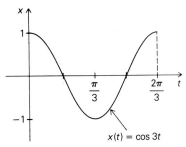

$x(t) = \cos 3t$

5. 27 minutes   **7.** (a) e^{-4t} (b) $\cos t - \sin t$ (c) $\cos(\sqrt{6}\, t) + \left(\dfrac{6 - \cos \sqrt{6}}{\sin \sqrt{6}}\right) \sin(\sqrt{6}\, t)$

9. $y = \dfrac{2}{3\left(x + \frac{2}{3}\right)}$

11. (a) $\approx 5.127\%$ (b) Change the formula to $M(t) = M(0)e^{\alpha t/365}$ where $\alpha = \ln(1.05) = 0.0488$.

13. $f(t) = -3 \cos(\sqrt{5}\, t) + \dfrac{4}{\sqrt{5}} \sin(\sqrt{5}\, t)$

$\qquad = \sqrt{\dfrac{61}{5}} \cos\left[\sqrt{5}\, t - \pi + \tan^{-1}\!\left(\dfrac{4}{3\sqrt{5}}\right)\right]$

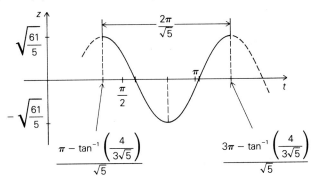

15. (a) $\dfrac{-4600 \ln(2)}{\ln(.88)} \approx 24{,}942.446$ years

(b) $\dfrac{4600 \ln(.01)}{\ln(.88)} \approx 165{,}714.02$ years (for 1%); $\dfrac{4600 \ln(.001)}{\ln(.88)} \approx 248{,}571.04$ years $\left(\text{for } \dfrac{1}{1000}\right)$

17. 15.18 minutes; yes

21. (a) $y^2 + 9x^2 = C$ (Solution is an ellipse if $C > 0$, the point $(0, 0)$ if $C = 0$, and no solution if $C < 0$.)

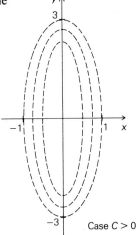

Case $C > 0$

(b) $|y|^9 = C |x|$; $C = $ constant

23. $y(1) = \begin{cases} \left(\dfrac{11}{10}\right)^{10} \approx 2.5937424 & \text{using 10-step} \\[2mm] \left(\dfrac{21}{20}\right)^{20} \approx 2.6532977 & \text{using 20-step} \\[2mm] e \qquad\;\; \approx 2.7182818 & \text{exact solution} \end{cases}$

The 10-step Euler method has an error of approximately 4.58%, nearly twice that of the 20-step (2.39%).

25. The solution "blows up" at $x = 1$.

CHAPTER 10

Section 10-1

1. (a) $4\pi r^2 h + \dfrac{\pi}{3}h^3$ (b) $4\pi r^2$ (the surface area of a sphere of radius r)

3. (a) $3(x-6)^5 - 10(x-6)^3 + 15(x-6) + 10{,}633 = 0$ (b) $x \approx 0.7466$ ft

5. (a) $\dfrac{71\pi}{105}$ (b) $\dfrac{4}{3}\pi r^3$ (c) $\dfrac{\pi^4}{12} - 2\pi$

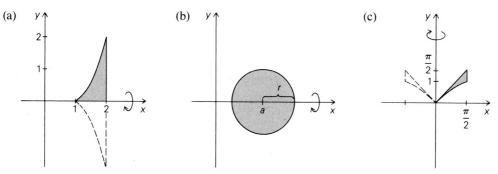

7. (a) $\pi \int_a^b [g(x)^2 - f(x)^2]\, dx$ (b) $2\pi \int_a^b x[g(x) - f(x)]\, dx$

	About x-axis	About y-axis	About x-axis	About y-axis
(a)	$\dfrac{86\pi}{15}$	$\dfrac{20\pi}{3}$		
(b)	$\dfrac{\pi^2}{8}$	$\dfrac{\pi}{2}\left(\dfrac{\pi}{2} - 1\right)$		
(c)	$\dfrac{83\pi}{3}$	$\dfrac{\pi}{3}(17 + 4\sqrt{2} - 6\sqrt{3})$		

11. $\dfrac{4\sqrt{3}}{3}$ **13.** $\dfrac{\pi^2}{4}(R^2 - r^2)(R - r)$ **15.** Volume of hemisphere + volume of cone = volume of cylinder.

Section 10-2

1. (a) $\dfrac{1}{2}$ (b) $9 + \sqrt{3}$ (c) $\dfrac{-\pi}{2}$ (d) $\dfrac{\sqrt{2}}{2} + \dfrac{8\sqrt{2}}{\pi} - \dfrac{\pi\sqrt{2}}{2} - \dfrac{12}{\pi}$ (e) $\dfrac{\pi}{2} - 1$ (f) $\dfrac{1}{e-1}$

3. (a) Apply the product rule to $a(x) = \dfrac{1}{x}\int_0^x f(t)\, dt$.

(b) The average of f over $[0, x]$ is $\begin{cases} \text{increasing} & \text{if } f(x) > a(x) \\ \text{decreasing} & \text{if } f(x) < a(x) \\ \text{at a critical point} & \text{if } f(x) = a(x) \end{cases}$

(c) The single strike-out represents a much larger proportion of the total times at bat near the beginning of the season than toward the end.

5. Show $f'(\bar{x}) = 0$ and $f''(\bar{x}) > 0$.

7. $(\bar{x}, \bar{y}) = \begin{cases} \left(\dfrac{-4}{21}, 0\right) & \text{if disc centers at } (0, 0) \text{ and hole centers at } (1, 0) \\ \left(\dfrac{4}{21}, 0\right) & \text{if disc centers at } (0, 0) \text{ and hole centers at } (-1, 0) \end{cases}$

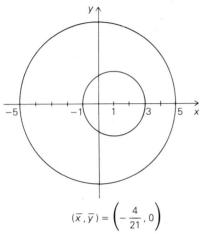

$$(\bar{x}, \bar{y}) = \left(-\frac{4}{21}, 0\right)$$

9. Show that $\dfrac{d^2\bar{x}}{dt^2}$, the acceleration of the center of mass, is zero.

11. There are three ways to do this: (1) consider the region as an infinite collection of infinitesimal strips; (2) prove the formula for the case where f and g are piecewise constant; (3) use formula (7) and the consolidation principle, following the method used for the area between graphs in Section 4-3.

13. *Hint:* Write $f(b) - f(a) = \int_a^b f'(x)\, dx$.

Section 10-3

1. $360 + \dfrac{96}{\pi}$ joules   3. 98 watts

5. (a) $4.1895(10^7)$ joules   (b) ≈ 6.62 m/sec   (c) $E = 1.257(10^8)$ joules; $v \approx 6.62$ m/sec; namely, the energy increased three-fold while the velocity remained unchanged.

7. (a) The seasons would be less pronounced.   (b) Hard to say for sure, but the weather presumably would become violent since the poles would vary from tropical to frigid as the year changed.

Chapter 10 — Review

1. (a) $2\pi^2$ (b) $\dfrac{\pi^2}{2}$ 3. (a) $\pi \int_a^b \rho(x)[f(x)]^2\, dx$ (b) $\dfrac{14\pi}{45}$

5. (a) For $0 \le t \le 1$, $\dfrac{1}{\sqrt{2}} \le \dfrac{1}{\sqrt{t^3 + 1}} \le 1$; integrate over $[0, 1]$.   (b) Use part (a) noting that $\dfrac{1}{\sqrt{2}} = \sin\left(\dfrac{\pi}{4}\right)$

and $1 = \sin\left(\dfrac{\pi}{2}\right)$; apply the intermediate value theorem to $\sin\theta$.

7. (a) $\left[\dfrac{1}{b-a}\displaystyle\sum_{j=1}^{n}\left\{\left[k_j - \dfrac{\sum_{i=1}^{n} k_i(t_i - t_{i-1})}{b-a}\right]^2 (t_j - t_{j-1})\right\}\right]^{1/2}$

(b) $\left[\dfrac{1}{n}\displaystyle\sum_{j=1}^{n}\left(k_j - \dfrac{1}{n}\sum_{i=1}^{n} k_i\right)^2\right]^{1/2}$ (c) Show $k_j = \mu$ for $j = 1, \cdots, n$ where $\mu = \dfrac{\sum_{i=1}^{n} k_i(t_i - t_{i-1})}{b-a}$

(d) $\sigma = \left[\dfrac{1}{n}\displaystyle\sum_{j=1}^{n}(a_j - \mu)^2\right]^{1/2}$ where $\mu = \dfrac{1}{n}\sum_{i=1}^{n} a_i$ (e) All the numbers in the list are the same.

9. *Hint:* Suppose $f(a) < 0$ and $f(b) > 0$. To prove there is a c in $[a, b]$ such that $f(c) = 0$, consider extending f beyond $[a, b]$ (by constant functions) to a function whose integral is zero.

11. $\dfrac{35}{12}$ meter²   13. $\dfrac{3}{98}$ m/sec

17. (a) The force exerted on a vertical rectangular slab is $\left(\int_0^{f(x)} \rho g h\, dh \right) dx = \tfrac{1}{2}\rho g [f(x)]^2\, dx$.

    (b) The force F is $\dfrac{\rho g}{2\pi}$ times the volume of the solid of revolution obtained by rotating the dam's face about its top edge.   (c) $653\tfrac{1}{3}(10^7)$ newtons

19. (a) Equator   (b) Pole   (c) Equator

CHAPTER 11

Section 11-1

1. (a) $\tfrac{1}{2}\ln\,(x^2 + 3) + C$ (b) $\sqrt{x^2 - 1} + C$ (c) $\tfrac{2}{3}(x - 2)\sqrt{x + 1} + C$ (d) $\tfrac{1}{16}x^4(4\ln x - 1) + C$
   (e) $\tfrac{2}{3}(1 + \sin x)^{3/2} + C$ (f) $\tfrac{1}{4}(e^x + 1)^4 + C$

3. (a) $\dfrac{5}{4} - \dfrac{3\pi}{8}$ (b) $2 + \dfrac{1}{3}\ln 3 + \dfrac{2\sqrt{3}}{3}\left(\tan^{-1}\dfrac{5\sqrt{3}}{3} - \tan^{-1} 3\sqrt{3}\right)$ (c) $\dfrac{1}{110}\left[5\ln\dfrac{76}{29} + 7\sqrt{5}\ln\dfrac{47 - \sqrt{5}}{47 + \sqrt{5}}\right]$

    (d) $(4 - \pi)/16$ (e) $\tfrac{1}{4}(5 + \ln 6)^4 - \tfrac{1}{4}(5 + \ln 3)^4$ (f) $e - 1$

5. Use the chain rule and the fundamental theorem of calculus.   7. $1, 0, \tfrac{1}{2}, 0, \tfrac{3}{8}, 0, \tfrac{5}{16}$ 9. $\pi \ln\left(\tfrac{225}{176}\right)$

11. $\bar{x} = \dfrac{\sqrt{5} - \sqrt{2}}{\ln\left(\dfrac{2 + \sqrt{5}}{1 + \sqrt{2}}\right)} - 1$ $\bar{y} = \dfrac{4\tan^{-1}2 - \pi}{8\ln\left(\dfrac{2 + \sqrt{5}}{1 + \sqrt{2}}\right)}$ 13. 125 15. $\sqrt{3}, \dfrac{9\sqrt{2}}{4}$

17. (b) $\left(\dfrac{3}{8}t - 3t\cos t + 3\sin t - \dfrac{3}{32}\sin 4t\right)^{113}$

    (c) Zeros of $s'(t)$ for $t > 1$ are $\cos \pi, 2\pi, 3\pi, \cdots$ Maxima at $(2k + 1)\pi$, $k = 1, 2, \cdots$

19. (a) $\ln\dfrac{x^2}{1 + x^2} - \dfrac{1}{2}\tan^{-1}x - \dfrac{1}{2}\dfrac{x}{x^2 + 1} + C$ (b) $\ln\sqrt{\dfrac{5}{2}}$ (c) $\dfrac{1}{12}\ln 3 + \dfrac{\sqrt{3}}{6}\tan^{-1}\left(\dfrac{4}{\sqrt{3}}x - \dfrac{1}{\sqrt{3}}\right)$

    (d) $\dfrac{1}{3}\ln\,|\tan x + 1\,| - \dfrac{1}{6}\ln\,|\tan^2 x - \tan x + 1\,| + \sqrt{3}\tan^{-1}\dfrac{2}{\sqrt{3}}\left(\tan x - \dfrac{1}{2}\right) + C$

Section 11-2

1. $\left(\dfrac{1}{27a^2}\right)[(4 + 9a^2 + 9a^2b)^{3/2} - (4 + 9a^2b)^{3/2}]$ Changing c has no effect.   3. $\int_0^1\sqrt{1 + (\cos x - x\sin x)^2}\, dx$

5. Use $0 \le \sin^2(\sqrt{3}x) \le 1$ to estimate $1 \le \sqrt{1 + 3\sin^2(\sqrt{3}x)} \le 2$, and so $2\pi \le \int_0^{2\pi}\sqrt{1 + 3\sin^2(\sqrt{3}x)}\, dx \le 4\pi$.

7. The length for the graph of f is no more than $(.9)\sqrt{2}$. The graph of g is at least as long as the distance between its endpoints.

9. $\tfrac{1}{4}(2b\sqrt{1 + 4b^2} + \ln\,(2b + \sqrt{1 + 4b^2}))$ 11. $2\pi[\sqrt{2} + \log\,(1 + \sqrt{2})]$

13. $\pi\left[e\sqrt{1 + e^2} - \sqrt{2} + \log\left(\dfrac{e + \sqrt{1 + e^2}}{1 + \sqrt{2}}\right)\right]$ 15. (a) $A = 2\pi\int_a^b t\sqrt{1 + f'(t)^2}\, dt$

17. $n > 2\pi\sqrt{17} = 25.9$. Take $n = 40$ so that $[0, 1]$ splits in intervals of length 0.025. Upper sum $= 3.7563383$; lower sum $= 3.1260868$. $A \approx 3.44$ sq units (to within $\tfrac{1}{3}$ sq unit).

19. (a) Area of cap $= 2\pi R(1 - \cos \theta)$, where $\sin\,(\theta/\theta) = C/\pi S$ and $R = S/2\theta$.
    (b) $4\pi R(1 - \cos \theta) + CH$ (c) \$7973.52

Section 11-3

1. (a) $y = \cos\,(\sqrt{x})$; horizontal tangent at $x = n^2\pi^2$, $n = 1, 2, 3, \ldots$ No vertical tangent; $y(0) = 1$,
    $\left.\dfrac{dy}{dx}\right|_{x=0} = -\dfrac{1}{4}$.   (b) $y = 2\sin 2x$; horizontal tangent at $x = \dfrac{2n + 1}{4}\pi$, $n = 0, \pm 1, \pm 2, \ldots$,   no vertical tangent.   (c) $x = 1 - 2y^2$; $-1 \le y \le 1$; vertical tangent at $x = 1$; no horizontal tangent.

(a)

(b)

(c)

3. (a) (i) $x = (\sqrt{2}/2)\cos t$ $\qquad y = \sin t$
  (ii) $x = (\sqrt{2}/2)\sin t$ $\qquad y = \cos t$

(b) (i) $x = t$ $\qquad y = 1/(4t)$
  (ii) $x = t/2$ $\qquad y = 1/(2t)$

(c) (i) $x = t$ $\qquad y = 3t - 2$
  (ii) $x = 1 + t$ $\qquad y = 1 + 3t$

(d) (i) $x = t$ $\qquad y = t^3 + 1$
  (ii) $x = t^3$ $\qquad y = t^9 + 1$

(e) (i) $x = (\sqrt{3}/3)\cosh t$ $\qquad y = \sinh t$
  (ii) $x = (\sqrt{3}/3)\sec t$ $\quad . \; y = \tan t$

(f) (i) $x = t$ $\qquad y = \cos 2t$
  (ii) $x = t/2$ $\qquad y = \cos t$

(g) (i) $x = \frac{1}{2}(\cosh t - 1)$ $\quad y = \frac{1}{2}\sinh t$
  (ii) $x = \frac{1}{2}(\sec t - 1)$ $\quad y = \frac{1}{2}\tan t$

5. (a)
$$x = f(t) = \begin{cases} 1 + t, & 0 \le t \le 1 \\ 2t, & 1 \le t \le 2 \\ 2 + t, & 2 \le t \le 3 \\ 11 - 2t, & 3 \le t \le 4 \\ 11 - 2t, & 4 \le t \le 5 \end{cases} \qquad y = g(t) = \begin{cases} 1 + t, & 0 \le t \le 1 \\ 2, & 1 \le t \le 2 \\ 4 - t, & 2 \le t \le 3 \\ 4 - t, & 3 \le t \le 4 \\ t - 4, & 4 \le t \le 5 \end{cases}$$

(b) $2 + 2\sqrt{2} + 2\sqrt{5}$ $\qquad$ (c) $12\pi(1 + \sqrt{2} + \sqrt{5})$

7. If s is the arc length, $s(x) = \dfrac{8}{27}\left(1 + \dfrac{9x}{4}\right)^{3/2} - \dfrac{8}{27}$. Use arc length as parameter and write $s = vt$, $v = $ constant speed. Obtain $v = (13^{3/2} - 8)/27$. The position at 1:30 P.M. is $(x, x^{3/2})$ where $x = \dfrac{1}{9}\left[\dfrac{3}{2}\,13^{3/2} - 4\right]^{2/3} - \dfrac{4}{9} \approx 1.376$, $y \approx 1.614$.

9. (a)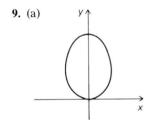

$A = 2 - \dfrac{\pi}{2}$

$L = \int_0^{\pi/2}\sqrt{\sec^4(\theta/2) + 4\tan^2(\theta/2)}\ d\theta$

(b)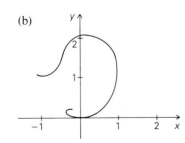

$A = \dfrac{7\pi^3}{96} - \dfrac{1}{2}\cos\left(\dfrac{9\pi^2}{16}\right) + \dfrac{1}{2}\cos\left(\dfrac{\pi^2}{16}\right) + \dfrac{1}{2}\int_{-\pi/4}^{3\pi/4}\sin^2(\theta^2)\ d\theta$

$L = \int_{-\pi/4}^{3\pi/4}\sqrt{(\theta + \sin(\theta^2))^2 + (1 + 2\theta\cos(\theta^2))^2}\ d\theta$

(c)

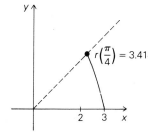

$$L = \int_0^{\pi/4} \sqrt{\sec^4 \theta + 4 \sec \theta + 4} \, d\theta$$

$$A = \frac{1 + \pi}{2} + 2 \ln (1 + \sqrt{2})$$

11. (a)

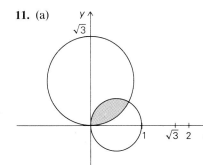

$$A = \frac{5\pi}{24} - \frac{\sqrt{3}}{4}$$

$$L = \frac{2 + \sqrt{3}}{6}$$

(b)

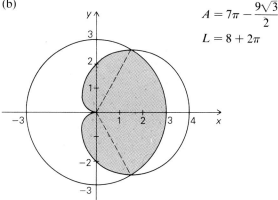

$$A = 7\pi - \frac{9\sqrt{3}}{2}$$

$$L = 8 + 2\pi$$

(c)

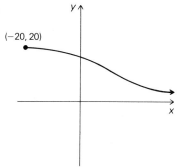

$$A = \pi$$

$$L = 8 + 2\pi$$

13. If $v = (f'(t)^2 + g'(t)^2)^{1/2}$, then $v'(t) = \frac{1}{v}[f'(t)f''(t) + g'(t)g''(t)]$.

This is 0 by hypothesis, so v is constant.

15. (a) Using string on the map of The United States in the *National Geographic Atlas of the World*, p. 25, (scale 1:7,800,000), the coastline of Maine was estimated at 338 miles.
(b) Using string on the map of Maine in the *State Farm Road Atlas*, Rand McNally, 1974, p. 47, (scale 1 in. = 20 mi) gives 688 miles.   (c) It would probably be longer.
(d) The measurement will depend on the definition and on the scale (and therefore the available detail) of the maps used.   (e) From the *World Almanac and Book of Facts — 1974*, Newspaper Enterprise Assoc., New York, 1973, p. 744, we have coastline: 228 miles; shoreline: 3,478 miles.

17. (d) Average rate of vertical ascent is $(y(T) - y(0))/T = \frac{1}{T} \int_0^T y'(t) \, dt$.

Average speed is $(s(T) - s(0))/T = \frac{1}{T} \int_0^T s'(t) \, dt$.

19.

$(-20, 20)$

Chapter 11 – Review

1. (a) $\tan^{-1}(x+2) + C$ (b) $\tan\theta + \frac{2}{3}\tan^3\theta + \frac{1}{5}\tan^5\theta + C$ (c) $2\sin\sqrt{x} - 2\sqrt{x}\cos\sqrt{x} + C$

(d) $\frac{1}{a}\tan\left(\frac{ax}{2}\right) + C$ (e) $-\frac{1}{2a}\cot\left(\frac{ax}{2}\right) - \frac{1}{6a}\cot^3\left(\frac{ax}{2}\right) + C$

3. (a) $\frac{1}{6a}\ln\left[\frac{(x-a)^3}{x^3-a^3}\right] + \frac{\sqrt{3}}{3a}\tan^{-1}\left(\frac{2\sqrt{3}}{3a}x + \frac{\sqrt{3}}{3}\right) + C$ where $a = \sqrt[3]{9}$

(b) $\frac{1}{2}x^2\ln x - \frac{1}{4}x^2 + 5x\ln x - 5x + C$ (c) $2(\sqrt{x}-1)e^{\sqrt{x}} + C$

(d) $-\frac{1}{15}(1-x)^{3/2}(3x^2+2) + C$ (e) $x - \ln(1+e^x) + C$

5. Use the chain rule and the fundamental theorem of calculus to check the integral. Then compute that

$\ln\left|\tan\left(\frac{x}{2}+\frac{\pi}{4}\right)\right| - \ln|\sec x + \tan x| = \ln(1) = 0$ and so the answers are the same.

7. Some results are listed here; you may be able to obtain others.

(a) $A = \begin{cases} \dfrac{1}{n+1}(b^{n+1} - a^{n+1}) & n \neq -1 \\ \ln(b/a) & n = -1 \end{cases}$

(b) $n=0$: $L = b-a$ $n=1$: $L = \sqrt{2}(b-a)$ $n=2$: see Worked Example 3 of Section 11-2

for $n = (2k+3)/(2k+2)$, $k = 0, 1, 2, 3, \ldots$ $L = \left\{\dfrac{n^{1/(1-n)}}{n-1}(1+n^2x^{2n-2})^{3/2}\sum_{j=0}^{k}\binom{k}{j}\dfrac{(-1)^{k-j}}{2j+3}(1+n^2x^{2n-2})^j\right\}\Big|_{x=a}^{x=b}$

$n=\frac{3}{2}$: $L = \frac{1}{27}[(4+9b)^{3/2} - (4+9a)^{3/2}]$ See also Solved Exercise 1, Section 11-2.

(c) $V_x = \begin{cases} \dfrac{\pi}{2n+1}(b^{2n+1} - a^{2n+1}), & n \neq -\frac{1}{2} \\ \pi\ln(b/a), & n = -\frac{1}{2} \end{cases}$ $V_y = \begin{cases} \dfrac{2\pi}{n+2}(b^{n+2} - a^{n+2}), & n \neq -2 \\ 2\pi\ln(b/a), & n = -2 \end{cases}$

(d) $n=0$; $A_x = 2\pi(b-a)$

$n=1$, $A_x = \sqrt{2}\pi(b^2 - a^2)$

$n=2$; $A_x = \dfrac{\pi}{32}[(1+8x^2)2x\sqrt{1+4x^2} - \ln(2x+\sqrt{1+4x^2})]|_{x=a}^{x=b}$

$n=3$; $A_x = \dfrac{\pi}{27}(1+9x^4)^{3/2}|_{x=a}^{x=b}$

$n = (2k+3)/(2k+1)$; $k = 0, 1, 2, 3, \ldots$

$A_x = \dfrac{2\pi}{n-1}n^{(1+n)/(1-n)}(1+n^2x^{2n-2})^{3/2}\sum_{j=0}^{k}\binom{k}{j}\dfrac{(-1)^{k-j}}{2j+3}(1+n^2x^{2n-2})^j$

$n=0$; $A_y = \pi(b^2 - a^2)$
$n=1$; $A_y = \sqrt{2}\pi(b^2 - a^2)$

$n=2$; $A_y = \dfrac{\pi}{6}[(1+4b^2)^{3/2} - (1+4a^2)^{3/2}]$

$n = (k+2)/(k+1)$; $k = 0, 1, 2, 3, \ldots$; $A_y = \dfrac{2\pi}{n-1}n^{2/(1-n)}(1+n^2x^{2n-2})^{3/2}\sum_{j=0}^{k}\binom{k}{j}\dfrac{(-1)^{k-j}}{2j+3}(1+n^2x^{2n-2})^j$

9. Some results are listed here. You may be able to obtain others.

(a) $n = 0, 1, 2, 3, \ldots$; $A = \sum_{j=0}^{n}\binom{k}{j}\dfrac{1}{2j+1}(b^{2j+1} - a^{2j+1})$

$n = -1$; $A = \tan^{-1}b - \tan^{-1}a$

If k is any other negative integer, obtain the formula

(1) $\displaystyle\int\frac{dx}{(1+x^2)^{k+1}} = \frac{1}{2k}\frac{x}{(1+x^2)^k} + \left(1 - \frac{1}{2k}\right)\int\frac{dx}{(1+x^2)^k}$

and use it repeatedly to reduce to the case of $n = -1$. For example,

$n = -2$; $A = \dfrac{1}{2}\left[\dfrac{b}{1+b^2} - \dfrac{a}{1+a^2} + \tan^{-1}b - \tan^{-1}a\right]$.

$n = -\dfrac{1}{2}$; $A = \ln[(b + \sqrt{1+b^2})/(a + \sqrt{1+a^2})]$

$$n = -\frac{3}{2}; \quad A = \frac{b}{\sqrt{1+b^2}} - \frac{a}{\sqrt{1+a^2}}$$

$$n = -\frac{2k+3}{2}; \quad k = 1, 2, 3, \ldots; \quad A = \frac{x}{\sqrt{1+x^2}} \sum_{j=0}^{k} \binom{k}{j} \frac{(-1)^j}{2j+1} \left(\frac{x^2}{1+x^2}\right)^j \Big|_{x=a}^{x=b}$$

$$n = \frac{2k-1}{2}; \quad k = 1, 2, 3, \ldots$$

$$A = \left\{ x\sqrt{1+x^2}\, \frac{(2k)!}{(k!)^2} \frac{1}{2^{2k}} \sum_{j=0}^{k-1} \frac{(j!)^2}{(2j+1)!}\, 2^{2j}(1+x^2)^j + \frac{(2k)!}{2^{2k}(k!)^2} \ln\,(x + \sqrt{1+x^2}) \right\} \Big|_a^b$$

(b) $n = 0; \quad L = b - a$

$n = 1; \quad L = \{x\sqrt{1+4x^2} + \frac{1}{2}\ln\,(2x + \sqrt{1+4x^2})\}|_{x=a}^{x=b}$

(See Worked Example 3, Section 11-2.)

(c) The results of part (a) may be used to obtain V_x when n is any integral multiple of $\frac{1}{4}$.

$$V_y = \begin{cases} \dfrac{\pi}{n+1}[(1+b^2)^{n+1} - (1+a^2)^{n+1}], & n \neq -1 \\[2mm] \pi \ln\left(\dfrac{1+b^2}{1+a^2}\right), & n = -1 \end{cases}$$

(d) $n = 0; \quad A = 2\pi(b - a)$

$n = 1; \quad A = \left\{ 2\pi\left[x\sqrt{1+4x^2} + \frac{1}{2}\ln\,(2x + \sqrt{1+4x^2})\right] + \frac{\pi}{32}[(1+8x^2)2x\sqrt{1+4x^2} - \ln\,(2x + \sqrt{1+4x^2})] \right\} \Big|_a^b$

$n = 0; \quad A_y = \pi\sqrt{2}\,(b - a)$

$n = 1; \quad A_y = \dfrac{\pi}{6}[(1 + 4b^2)^{3/2} - (1 + 4a^2)^{3/2}]$

11. $x + 3y = 3\sqrt{2}$

13. $\theta = t, \quad r = \csc^3 t$

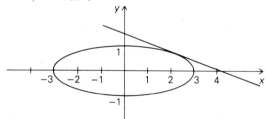

15. (a) $\pi^5/320$ (b) $3\pi/8$ (c) $\frac{1}{2}$

17. (a) $2\pi \int_\alpha^\beta f(\theta) \sin\theta \sqrt{f(\theta)^2 + f'(\theta)^2}\, d\theta$ (b) $2\pi \int_{-\pi/4}^{\pi/4} \cos 2\theta \sin\theta \sqrt{\cos^2 2\theta + 4\sin^2 2\theta}\, d\theta$

19. (a) $m = 1, \quad n = 1$ $m = 1, \quad n = 2$

$m = 1, n = 3$ $m = 1, n = 4$

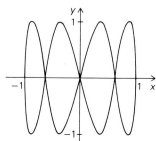

(b) The curve consists of n vertical loops.

(c) $m = 2$, $n = 1$ $m = 2$, $n = 2$

 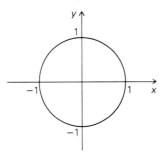

$m = 2$, $n = 3$ $m = 2$, $n = 5$

 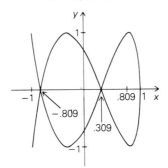

The case $m = 2$, $n = 4$ is the same as $m = 1$, $n = 2$.

(d) $m = 3$, $n = 4$ $m = 3$, $n = 5$

 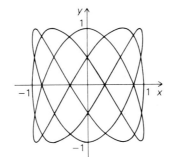

21. (a) $\dfrac{1}{k_2} \ln \left| \dfrac{N_0}{k_1 N_0 - k_2} \left(k_1 - \dfrac{k_2}{N(t)} \right) \right|$ (b) $N(t) = \dfrac{2k_1 k_2 \pm \sqrt{4 k_1^2 k_2^2 - 4 \left[k_1^2 - \left(k_1 - \dfrac{k_2}{N_0} \right)^2 e^{2k_2 t} \right] (k_2^2)}}{2 \left[k_1^2 - \left(k_1 - \dfrac{k_2}{N_0} \right)^2 \right]}$

23. $\int_t^{t+h} r^2 \, \theta \, ds = kh$ for all t.

CHAPTER 12

Section 12-1

1. (a) $0, -1, 4 - 2\sqrt{2}, 9 - 2\sqrt{3}, 12, 25 - 2\sqrt{5}$ (b) $1, 0, \dfrac{-1}{2}, \dfrac{1}{3}, \dfrac{-1}{8}, \dfrac{1}{30}$ (c) $\dfrac{1}{7}, \dfrac{1}{14}, \dfrac{1}{21}, \dfrac{1}{28}, \dfrac{1}{35}, \dfrac{1}{42}$

3. (a) Compare with $\dfrac{2}{n+1}$.   (b) Compare with $\dfrac{1}{n^2}$ (see Solved Exercise 7a).

(c) Use Theorems 1 and 2, and Solved Exercise 4.

5. (a) 0   (b) 0   (c) 2   (d) 6

7. (a) $\lim\limits_{n \to \infty} a_n = \infty$ means for every $M > 0$, there exists $N > 0$ such that $a_n > M$ whenever $n > N$.

(b) Pick $N > 4M + \sqrt{16M^2 + M - 1}$, then $n > N$ implies $\dfrac{1 + n^2}{1 + 8n} > M$.

11. $a_n = \dfrac{1}{2^n}$; $\lim\limits_{n \to \infty} a_n = 0$

13. (a) Suppose $\lim\limits_{n \to \infty} b_n < L$, then there exists N such that $b_n - a_n < 0$ for $n > N$,

contradicting the hypothesis $b_n > a_n$.

(b) Given $\epsilon > 0$, there exist N_1 such that $n > N_1$ implies $|a_n - L| < \epsilon$ and N_2 such that $n > N_2$ implies

$|c_n - L| < \epsilon$. As $a_n < b_n < c_n$, $n > N_3 = \max(N_1, N_2)$ implies $|b_n - L| < \epsilon$.

15. (a) $a_n = \begin{cases} 1.0000000 & \text{if } n = 1 \\ \sin(a_{n-1}) & \text{if } n = 2, 3, \ldots \end{cases}$
  (b) $\lim\limits_{n \to \infty} a_n = 0$

$\sin(a_1) \approx .841$
$\sin(a_2) \approx .746$
$\sin(a_3) \approx .678$

$a_2 = \sin(a_1)$ $a_1 = 1$ $a_3 = \sin(a_2)$

17. (a) $1 + \cfrac{1}{1 + \cfrac{1}{1 + \cfrac{1}{1 + \cfrac{1}{1 + \frac{1}{1 + \frac{1}{2}}}}}} \cong 2.414414$
  (b) $\lim\limits_{n \to \infty} \dfrac{1}{f^{(n)}(2)} \cong 2.73205$

19. (a) 3.55
  (b) 22 and $18 + \sqrt{2}$

21. $S_n \to \int_0^1 (x + x^2)\, dx$ as $n \to \infty$.
  23. $\dfrac{565}{16}$

Section 12-2

1. (a) $\frac{7}{9}$ (b) 3 (c) $\frac{3}{2}$ (d) 0 (e) 1

3. (a) Does not exist (b) $\frac{3}{5}$ (c) Does not exist (d) Does not exist (e) Does not exist

5.

	Horizontal asymptotes	Vertical asymptotes
(a)	$y = 0$	$x = \pm 1$
(b)	$y = 0$	$x = 0, \pm 2$
(c)	$y = \pm 1$	none

7. Let $N = \deg f$, $M = \deg g$, and consider the cases $N < M$, $N = M$, and $N > M$. The cases $l = \pm\infty$ occur

if $N > M$. If $N - M$ is even, then $\lim\limits_{x \to +\infty} \dfrac{f(x)}{g(x)} = \lim\limits_{x \to -\infty} \dfrac{f(x)}{g(x)}$. If $N - M$ is odd, then $\lim\limits_{x \to -\infty} \dfrac{f(x)}{g(x)} = -\lim\limits_{x \to +\infty} \dfrac{f(x)}{g(x)}$.

9. Choose $\delta < \min\left(1, \dfrac{\epsilon}{39}\right)$.

11. (a) $\lim\limits_{x \to \infty} f(x) = -\infty$ if given any number B, there exists a number A such that $f(x) < B$ whenever $x > A$.

(b)

13. (a)

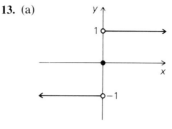

$\lim\limits_{x \to 0} f(x)$ does not exist.

(b)

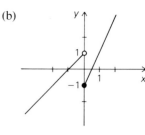

$\lim\limits_{x \to 0} g(x)$ does not exist.

15. 4 amperes

17. The result $\lim\limits_{n \to \infty} \dfrac{nr}{1 + (n-1)r} = 1$ can be obtained by letting $x = \dfrac{1}{n}$ and use the fact that $x \to 0$ as $n \to \infty$

to conclude $\lim\limits_{n \to \infty} \dfrac{nr}{1 + (n-1)r} = \lim\limits_{x \to 0} \dfrac{r}{x + (1-x)r} = 1$.

Section 12-3

1. (a) 0   (b) $\frac{3}{5}$ (c) Does not exist   (d) 0   **3.** (a) 0   (b) 0   (c) -1 (d) $\frac{1}{120}$

5.

    7. (a) Converges   (b) Diverges   (c) Converges   (d) Converges

9. (a) The singularity at $x = 0$ where $\dfrac{1}{x^2}$ is unbounded must be taken into account. The integral actually diverges. Also, the integral of a positive function, if it exists, should be a positive number.

(b) The fundamental theorem of calculus cannot be used since the denominator vanishes in the interval of integration. The integral actually diverges.

11. $\dfrac{4}{3} \ln 7 - \dfrac{\sqrt{3}}{3} \pi + \dfrac{2\sqrt{3}}{3} \tan^{-1}\left(\dfrac{5\sqrt{3}}{3}\right)$ **13.** $6\sqrt{3} \cong 10.39$ hr

15. (a) $\lim\limits_{b \to \infty} 2\pi \displaystyle\int_1^b \frac{1}{x}\sqrt{1 + \frac{1}{x^4}}\, dx > \lim\limits_{b \to \infty} 2\pi \displaystyle\int_1^b \frac{1}{x}\, dx = \lim\limits_{b \to \infty} 2\pi \ln b = \infty$ (b) $\lim\limits_{b \to \infty} \pi \displaystyle\int_1^b \frac{1}{x^2}\, dx = \pi$

(c) The amount of paint to coat an area is proportional to the area only if the thickness of the coat is constant.

(d) $r > 1$ (e) $r > \dfrac{1}{2}$; $V(r) = \dfrac{\pi}{2r-1}$

17. $2\pi(\sqrt{2} + \ln(1 + \sqrt{2}))$ [*Hint:* Evaluate $\int \sqrt{1 + w^2}\, dw$ by the method of Worked Example 3, Section 11-2.]

21. (a) $\mu = 104.5$ (c) Theoretical frequency ≈ 22.117155;   error $\approx 3.84\%$.

Section 12-4

1. (a) $x \approx 0.523$ (b) 7.724 and -0.247 **3.** $x \approx 4.49341$

5. $n = 78$ for trapezoidal rule; $n = 8$ for Simpson's rule   **7.** 2.1922

9. (b) 0.543   (c) (1) 0.258   (2) 0.259   (3) 0.257   **11.** 4

13. $|x_6 - \sqrt{2}| < \dfrac{1}{2^{127}} \approx 5.88\,(10^{-39})$ and so $x_6 = \sqrt{2}$ with 38-digit accuracy. ($x_5 = \sqrt{2}$ with only 18-digit accuracy, and so x_6 is the first that is 20-digit accurate.)

15. $T \approx 15{,}813.606$ sec

17. (a) $A = 2\pi \int_0^H f(x)\sqrt{1 + [f'(x)]^2}\, dx$ (b) $L = 2 \int_0^H \sqrt{1 + [f'(x)]^2}\, dx$

(c) Use the trapezoidal rule to get

$$A \approx 4\pi\left\{f(0)\sqrt{1+f'(0)^2}+f(H)\sqrt{1+f'(H)^2}+2\sum_{i=1}^{n-1}\left[f(4i)\sqrt{1+f'(4i)^2}\right]\right.$$

$$L \approx 4\left[\sqrt{1+f'(0)^2}+\sqrt{1+f'(H)^2}+2\sum_{i=1}^{n-1}\sqrt{1+f'(4i)^2}\right]$$

where $n=\dfrac{H}{4}$. Approximate $f'(4i) \approx \dfrac{f(4[i+1])-f(4i)}{4}$.

(d) $A \approx 2\pi\displaystyle\sum_{i=0}^{n-1}\left\{\sqrt{1+(m_i)^2}\,[8(2i+1)m_i+4b_i]\right\}$

where $m_i=\dfrac{f(4(i+1))-f(4i)}{4}$ and $b_i=(i+1)f(4i)-if(4(i+1))$

Chapter 12 — Review

1. (a) 1   (b) 1   (c) $\frac{3}{8}$ (d) 5   **3.** (a) $\tan 3$ (b) $\sec^2(3)$ (c) 0   (d) 0

5. (a) Take $\delta = \min\left(1, \dfrac{\epsilon}{5}\right)$.   (b) Take $n > \dfrac{1+\sqrt{1+\epsilon(1-3\epsilon)}}{3\epsilon}$ where $0 < \epsilon < \dfrac{1}{3}$.

7. Compute $a_0, \cdots, a_5$ and guess that $a_n \to \sqrt{2}$. To prove this, show $|a_n^2 - 2| < \dfrac{1}{4^n}$ which tends to 0 and so

$\displaystyle\lim_{n\to\infty} a_n^2 = 2$. As $a_n^2 \,\epsilon\,[1, \infty)$ and as the square root function is continuous on $[1, \infty)$, $\displaystyle\lim_{n\to\infty} a_n = \lim_{n\to\infty}\sqrt{a_n^2}$

$= \sqrt{2}$.

9. (a) 0   (b) e^2 (c) 0   (d) $\sin(\sqrt{2})$

11. (a)

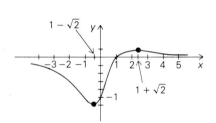

Horizontal asymptote: $y = 0$
Vertical asymptote: none

(b)

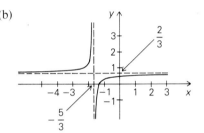

Horizontal asymptote: $y = \dfrac{2}{3}$

Vertical asymptote: $x = -\dfrac{5}{3}$

13. (b) $f'''(x_0) = \displaystyle\lim_{h\to 0}\dfrac{f(x_0+2h)-3f(x_0+h)+3f(x_0)-f(x_0-h)}{h^3}$ **15.** $\displaystyle\lim_{n\to\infty} S_n = \int_0^2 [2-\cos(\pi x)]\,dx$

17. (a) 1   (b) 1   (c) $\frac{1}{4}$ (d) 2

19. (a) Converges to 1   (b) Diverges   (c) Diverges   (d) Converges to 0   (e) Converges to 2

21. (a) $-\infty$ (does not exist)   (b) 0   (c) 1   (d) 0   (e) e^8

23. Integration by parts gives $\displaystyle\int_0^b \dfrac{\sin x}{1+x}\,dx = \dfrac{\cos b}{1+b}-1-\int_0^b \dfrac{\cos x}{(1+x)^2}\,dx$; then compare $\dfrac{\cos x}{(1+x)^2}$

with $g(x) = \begin{cases} 1 & 0 \le x \le 1 \\ \dfrac{1}{x^2} & x > 1 \end{cases}$

25. (a) ≈ 0.84270165 (b) .0002 (Answer may vary.)   (c) ≈ 0.94356671

(d)

Part	Evaluate	From table	From Simpson's rule	% error	Accuracy
(a)	erf(1)	0.842664	0.84270165	$< .0045\%$	good
(c)	erf(10)	1.0000	0.94356671	$> 5.6\%$	poor

The accuracy of the approximation to erf(t) via Simpson's rule decreases as $t \to \infty$.

27. $-1/2\pi$

29. (a)

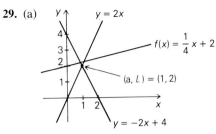

(c) Given $\epsilon > 0$, choose $h = \dfrac{1}{2}\dfrac{\epsilon}{m}$; then Lipschitz condition implies $|f(x) - L| < \epsilon$.

CHAPTER 13

Section 13-1

1. (a) Diverges   (b) Converges   (c) Converges   (d) Converges   (e) Converges

3. (a) Diverges   (b) Converges   (c) Converges   (d) Diverges   (e) Diverges   (f) Diverges

5. Use fact $\displaystyle\sum_{i=n+1}^{\infty} ar^i = \sum_{i=0}^{\infty} ar^i - \sum_{i=0}^{n} ar^i = \dfrac{a}{1-r} - a\left(\dfrac{1-r^{n+1}}{1-r}\right)$

7. (a) $\left(\dfrac{2}{3}\right)^{2n} h$ on nth bounce   (b) $2\left(\dfrac{2}{3}\right)^n \sqrt{\dfrac{2h}{g}}$ on nth bounce   (c) $5\sqrt{\dfrac{2h}{g}} < \infty$ (d) $\dfrac{13}{5}h$

(e) g is lower on the moon, so the times would be longer, but the total distance traveled would remain the same.

9. (b) $b_n = \dfrac{-1}{n};\ \displaystyle\sum_{n=1}^{\infty} \dfrac{1}{n(n+1)} = 1$

11. (a) Show $\displaystyle\sum_{k=1}^{n-1} A_k s_k = \sum_{k=1}^{n-1} a_k s_k + \sum_{k=1}^{n-1} A_k s_{k+1}$ (b) $\bar{a} = \dfrac{\displaystyle\sum_{i=1}^{n+1} s_i a_i}{\displaystyle\sum_{i=1}^{n} s_i} = \dfrac{\displaystyle\sum_{i=1}^{n} s^* a_i + s_{n+1} a_{n+1}}{\displaystyle\sum_{i=1}^{n} s^*}$

13. Compare with geometric series $\displaystyle\sum_{n=0}^{\infty} \left(\dfrac{3}{4}\right)^n$.

Section 13-2

1. (a) Converges   (b) Diverges   (c) Converges   (d) Converges

3. (a) Divergent   (b) Absolutely convergent   (c) Convergent; not absolutely convergent

     (d) Divergent   (e) Convergent; not absolutely convergent

5. (a) 2.98   (b) 0.90   (c) 11.55   (d) 2.30

7. (b) 1.0823; 9 terms; saves the work of summing 6 additional terms.   **9.** $p > 1$

11. (b)

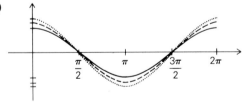

$S_1(r)$ ——— $S_1(0) = 1$

$S_2(r)$ - - - - $S_2(0) = \dfrac{82}{81} \approx 1.012$

$S_3(r)$ ············· $S_3(0) = \dfrac{51331}{50625} \approx 1.014$

Section 13-3

1. (a) $R = \infty$ (b) $R = \infty$ (c) $R = 2$ (d) $R = 1$

3. (a) $\sum_{n=0}^{\infty} x^n$ for $|x| < 1$ (b) $\sum_{n=0}^{\infty} (-1)^n x^n$ for $|x| < 1$ (c) $2\sum_{n=0}^{\infty} x^{2n+1}$ for $|x| < 1$

     (d) $\sum_{n=0}^{\infty} x^{2n}$ for $|x| < 1$ (e) $\sum_{n=0}^{\infty} x^{2n}$ for $|x| < 1$

5. (a) $\dfrac{1}{2} - \dfrac{x^2}{4!} + \dfrac{x^4}{6!} - \dfrac{x^6}{8!} + \cdots$ (b) $1 + \dfrac{1}{2}x^2 + \dfrac{5}{24}x^4 + \dfrac{61}{720}x^6 + \cdots$ (c) $1 + x - \dfrac{x^3}{3!} + \dfrac{x^5}{5!} - \cdots$

(d) $1 - \dfrac{2}{3}x + \dfrac{2}{9}x^2 - \dfrac{22}{81}x^3 + \dfrac{38}{243}x^4 - \dfrac{134}{729}x^5 + \cdots$ (e) $-1 + \dfrac{9}{2}x^2 - \dfrac{75}{8}x^4 + \dfrac{245}{16}x^6 - \cdots$

7. (a) $1 - \dfrac{x^2}{3!} + \dfrac{x^4}{5!} - \dfrac{x^6}{7!} + \cdots$ (b) $f'(0) = 0$; $f''(0) = \dfrac{-1}{3}$; $f'''(0) = 0$

9. 0.401 (Use Taylor's formula with remainder.)

13. $\displaystyle\int_1^x \ln t\, dt = \sum_{n=2}^{\infty} (-1)^n \dfrac{(x-1)^n}{n(n-1)}$; $\; x \ln x = (x-1) + \displaystyle\sum_{n=2}^{\infty} (-1)^n \dfrac{(x-1)^n}{n(n-1)}$; $\; \displaystyle\int_1^x \ln t\, dt = x \ln x + 1 - x$

15. (a) $\dfrac{1}{\sqrt{1-x^2}} = 1 + \displaystyle\sum_{n=1}^{\infty} \dfrac{(n-\frac{1}{2})(n-\frac{3}{2})\cdots(\frac{1}{2})}{n!} x^{2n}$ for $|x| < 1$

$\sin^{-1}(x) = x + \displaystyle\sum_{n=1}^{\infty} \dfrac{(n-\frac{1}{2})(n-\frac{3}{2})\cdots(\frac{1}{2})}{n!}\, \dfrac{x^{2n+1}}{2n+1}$ for $|x| < 1$

(b) $\sin^{-1}(\sin x) = x + 0x^2 + 0x^3 + \cdots$ (c) $\sin^{-1}(x) = x + \frac{1}{6}x^3 + \frac{3}{40}x^5 + \cdots$

(d) $g(x) = x - x^3 + 3x^5 + \cdots$

17. 3rd order $= 3.0365891$; 4th, 5th, 6th order each $= 3.0365889$

19. (a) $f(x) = \displaystyle\sum_{n=2}^{\infty} \dfrac{x^n}{n!}$ (b) $f(x) = e^x - x - 1$ **27.** Between 6.25 and 8.125 miles

31. (a) $-\sec \theta + \theta \tan \theta = \dfrac{1}{\cos \theta}(\theta \sin \theta - 1)$ where $\dfrac{1}{\cos \theta} = 1 + \dfrac{\theta^2}{2} + \dfrac{5}{24}\theta^4 + \dfrac{61}{720}\theta^6 + \dfrac{277}{8064}\theta^8 + \cdots$

(b) $I = \frac{32}{3}$; $\; \bar{y} \approx 14.84211$ (exact formula); $\; \bar{y} \approx 14.83522$ (finite series formula); error $< .007$
(percent error $< .05\%$)

Chapter 13 — Review

1. Converges   (b) Converges   (c) Diverges   (d) Converges   (e) Converges
3. (a) False   (b) False   (c) False   (d) True   (e) False   (f) False   (g) False   (h) True
(i) True
5. (a) Converges   (b) Converges   (c) Converges   (d) Diverges   (e) Converges
7. $B_1 = \dfrac{-1}{2}$, $B_2 = \dfrac{1}{6}$, $B_3 = 0$ **9.** (a) 1   (b) 1   **11.** (a) 0   (b) $\dfrac{10!}{4!}e$
13. (a) Use fact $\displaystyle\sum_{n=1}^{\infty} a_n(x^2)^n$ converges for $|x^2| < R$.   (b) $\dfrac{2}{\sqrt{\pi}}$
15. (b) $\lim\limits_{i \to \infty} b_i = \Sigma_{i=0}^{\infty} a_i = f(1)$; radius of convergence of g is < 1 (c) e
17. (a) 0.81   (b) 0.64   (c) 4.23   **21.** (b) $\Sigma_{n=0}^{\infty} (-1)^n A_{2n+1}$ **23.** 0.659178 with error $< 2(10^{-4})$
25. (b) $J_0(2.404) \approx \displaystyle\sum_{k=0}^{5} \dfrac{(-1)^k}{(k!)^2}(1.202)^{2k} = .0004972 \approx 0$ with error $< 5(10^{-4})$

CHAPTER 14

Section 14-1

1. (a) $(-3, -9, -15)$ (b) $(11, 0, 11)$ (c) $(0, 0, 0)$
5. (a) $4\mathbf{i} + 3\mathbf{j} + 6\mathbf{k}$ (b) $9\mathbf{i} + 12\mathbf{j} + 15\mathbf{k}$ (c) $-2\mathbf{i} + 2\mathbf{j} - 2\mathbf{k}$ (d) $26\mathbf{i} + 16\mathbf{j} + 38\mathbf{k}$ (e) $2\mathbf{i} + 5\mathbf{j} + 4\mathbf{k}$
7. The set of all points in the plane with integer coordinates

9. (a) $a = \frac{1}{2}$; $b = -\frac{1}{2}$ (b) $a = \frac{1}{2}$; $b = \frac{1}{2}$ (c) $a = 5$; $b = 2$

11. (a) $P_{-1} = (-1, -1, -1)$; $P_0 = (1, 0, 0)$; $P_1 = (3, 1, 1)$; $P_2 = (5, 2, 2)$

(b)   (c) The straight line through $(1, 0, 0)$ and parallel to the vector $(2, 1, 1)$

13. (a) With $r = 0$, $(0, 0) = r(1, 1)$. But there exists no r satisfying $(1, 1) = r(0, 0)$. So $(0, 0)$ is parallel to $(1, 1)$, but $(1, 1)$ is not parallel to $(0, 0)$. Hence this is not a good definition.

(b) Let $u = (m, n)$, then $x = \dfrac{md - nc}{ad - bc}$ and $y = \dfrac{na - bm}{ad - bc}$ are the desired numbers since $ad \neq bc$.

15. $F_{(1, 1, 1)} = (4.9, 4.9, 4.9)$; $F_{(-1, -1, 1)} = (-4.9, -4.9, 4.9)$

17. (a) $(-40, 100)$ (b)   **19.**

Section 14-2

1. Let $\mathbf{v}, \mathbf{w}$ be the vectors describing the sides of the rhombus and show $(\mathbf{v} + \mathbf{w}) \cdot (\mathbf{v} - \mathbf{w}) = 0$ using $\|\mathbf{v}\| = \|\mathbf{w}\|$.

3. (a) The xy-plane   (b) The x-axis   (c) The unit sphere centered at the origin
(d) The infinitely long circular cylinder of radius 1 centered around the z-axis.

5. (a) 3   (b) -2 (c) $2\sqrt{3}$ (d) 3

7. $(a_1 - a_2)x + (b_1 - b_2)y = \frac{1}{2}(a_1^2 - a_2^2 + b_1^2 - b_2^2)$ where $P_1 = (a_1, b_1)$ and $P_2 = (a_2, b_2)$

9. (a) $(\frac{42}{25}, \frac{56}{25}) = \frac{14}{25}(3, 4)$ (b) $\frac{7}{25}$ hr   (c) $\frac{2}{5}$ nautical mile

11. (a) L_1 and L_2: parallel and not intersecting
L_1 and L_3: intersect at $(0, -1, -1)$
L_1 and L_4: not parallel and not intersecting
L_2 and L_3: not parallel and not intersecting
L_2 and L_4: not parallel and not intersecting
L_3 and L_4: parallel and not intersecting

(b) L_1 and L_2: $2x - y - z - 2 = 0$
L_1 and L_3: $x - y - 1 = 0$
L_3 and L_4: $7x + 11y - 2z + 9 = 0$

(c) L_1: $\left(\dfrac{2}{3}, \dfrac{-1}{3}, \dfrac{-1}{3}\right)$; $x + y + z = 0$

L_2: $\left(\dfrac{2}{3}, \dfrac{-7}{3}, \dfrac{5}{3}\right)$; $x + y + z = 0$

L_3: $\left(\dfrac{10}{83}, \dfrac{-73}{83}, \dfrac{7}{83}\right)$; $x + y + 9z = 0$

L_4: $\left(\dfrac{-91}{83}, \dfrac{-8}{83}, \dfrac{11}{83}\right)$; $x + y + 9z = 0$

17. $\mathbf{F} = \mathbf{F_1} + \mathbf{F_2} = -\frac{1}{2}(\mathbf{i} + \mathbf{j}) + \frac{1}{2}(\mathbf{i} - \mathbf{j})$

19. (a) s and t satisfying $s = t\sqrt{a^2 + b^2 + c^2}$

(d) L_1 and L_2 have same direction cosines: $\mu = \lambda = \nu = \dfrac{\sqrt{3}}{3}$ and same direction angles:

$\alpha = \beta = \gamma = \cos^{-1}\left(\dfrac{\sqrt{3}}{3}\right) \approx 0.955$. L_3 and L_4 have same direction cosines: $\mu = \lambda = \dfrac{\sqrt{83}}{83}$, $\nu = \dfrac{9\sqrt{83}}{83}$;

and same direction angles: $\alpha = \beta = \cos^{-1}\left(\frac{\sqrt{83}}{83}\right) \approx 1.46$, $\gamma = \cos^{-1}\left(\frac{9\sqrt{83}}{83}\right) \approx 0.156$.

(e) $L = (0, 0, 0) + t(1, 1, 1)$

21. (a) $\frac{6}{\sqrt{2}}\mathbf{i} + \frac{6}{\sqrt{2}}\mathbf{j}$ (b) 1.249 radians (c) $36/\sqrt{2}$

Section 14-3

1. (a) $\mathbf{k} + \mathbf{j}$ (b) $\mathbf{k} + \mathbf{j}$ (c) $2\mathbf{i} - 2\mathbf{j} + 4\mathbf{k}$ **3.** (a) -2 (b) 0 (c) ac (d) 3 (e) -6

7. $\frac{\sqrt{2}}{6}(\mathbf{i} - \mathbf{j} + 4\mathbf{k})$ **9.** 12

13. *Hint:* (a, b) is a vector normal to line l_1: $ax + by = c$. So the line through P and perpendicular to l_1 is l_2: $(x, y) + t(a, b)$. Find the point of intersection Q of l_1 and l_2, then evaluate $\|P - Q\|$.

19. Expand by minors along the second row. Rearrange the terms and combine them as a sum of two 3×3 determinants, one the original array and the other λ times a determinant with two identical rows. Then use Problem 18.

21. (a) Subtract 4 times the first row from the second; subtract 7 times the first row from the third; reduce by minor of the first column; evaluate a 2×2 determinant. (b) 10

(c)

Problem (3d)	3
(3e)	-6
(4c)	0
(4d)	0
(4e)	10

Chapter 14 — Review

1. (a) $(-1, -2, 17)$ (b) $(10, 0, 5)$ (c) $2\mathbf{i} + 9\mathbf{j} + 5\mathbf{k}$ (d) $-4\mathbf{i} + 7\mathbf{j} - 11\mathbf{k}$ (e) 6

3. (a) $y = z = 0$ (b) $x^2 + y^2 + z^2 = 4$ (c) $l(t) = (1, 1, 2) + t(1, 1, 1) = t(2, 2, 3) + (1 - t)(1, 1, 2)$

(d) $x - y = 0$ (e) $l(t) = (0, 0, 3) + t(1, -1, 0)$

5. $\theta = \frac{\pi}{2} - \cos^{-1}\left(\frac{|(\mathbf{u} \times \mathbf{v}) \cdot \mathbf{w}|}{\|\mathbf{u} \times \mathbf{v}\| \|\mathbf{w}\|}\right)$ **7.** $\left(1, \frac{1}{6}\right)$ **9.** $3x + 5y + z = 12$

13. (a) The set of all points with integer coordinates lying strictly between the rays $y = \frac{1}{2}x$ and $y = 2x$, $x > 0$ and on the rays $y = \frac{1}{2}x + \frac{3n}{2}$, $n = 1, 2, 3, \cdots$

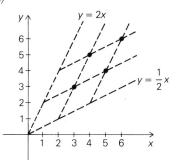

(b) The set of all points which are the intersections of the lines $y = \frac{1}{2}x + \frac{3n}{2}$ and the lines $y = 2x + 3m$; $n, m = 0, \pm 1, \pm 2, \cdots$

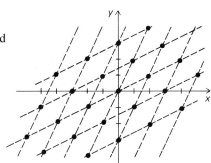

(c) The region strictly between the rays $y = \frac{1}{2}x$ and $y = 2x$; $x > 0$

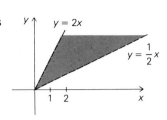

(d) The whole plane

15. (a) The set of all points on the ray $y = 2x$, $x > 0$ which are of the form $(r + \pi s)(1, 2)$; $r, s = 1, 2, 3, \cdots$

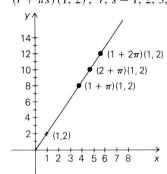

(b) The set of all points on the line $y = 2x$ which are of the form $(r + \pi s)(1, 2)$; $r, s = 0, \pm 1, \pm 2, \cdots$

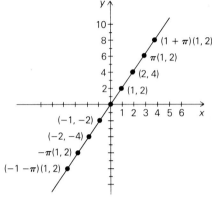

(c) The half line $y = 2x$, $x > 0$

(d) The line $y = 2x$

17. ≈ 2.1627528 miles; no   19. $(\frac{2}{3}, \frac{1}{3}, \frac{2}{3})$ and $(-\frac{2}{3}, -\frac{1}{3}, -\frac{2}{3})$

21. ± 1 (Its absolute value is the volume of a unit cube.)   25. (a) $V = \frac{1}{6} \begin{vmatrix} a_1 & a_2 & a_3 \\ b_1 & b_2 & b_3 \\ c_1 & c_2 & c_3 \end{vmatrix}$ (b) $\frac{1}{3}$

CHAPTER 15

Section 15-1

1. (a) All points except those on the unit circle centered at the origin; $f(1, 1) = 2$; $f(-1, 1) = 0$

(b) All points except the origin; $f(1, 1) = 1$; $f(-1, 1) = -1$

(c) All points except those on the hyperbola $x^2 - 4y^2 = 1$; $f(1, 0.5) = -1$; $f(0.1, 0.5) = \frac{-1}{1.99} \approx -.5025$

(d) All points except those on the lines $x = (2n - 1)\pi$, $n =$ integer; $f\left(0, \frac{\pi}{4}\right) = \frac{-\sqrt{2}}{4}$; $f\left(\frac{\pi}{4}, \pi\right) = \frac{\pi(2 - \sqrt{2})}{2}$

3. (a) A paraboloid

(b) A parabolic cylinder

(c) A paraboloid

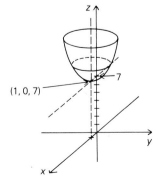

5. (a) For $c \geq 1$, level curve is a circle of radius $\sqrt{c-1}$ centered at $(0,0)$; there is no level curve for $c < 1$.

(b) For $c \leq 1$, level curve is a circle of radius $\sqrt{1-c}$ centered at $(0,0)$; there is no level curve for $c > 1$.

(c) For $c > 0$ $[c < 0]$ level curve is a hyperbola in the obtuse [acute] angles between the lines $x = 0$ and $y = -x$. For $c = 0$, the level curve is these two lines.

(d) For each c, level curve consists of (at most 3) straight lines of the form $x = x_0$ where x_0 is a root of the cubic $x^3 - x - c = 0$.

7. (a)

(b)

(c)

9. (a)

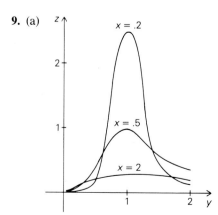

Section 15-2

1. (a) A hyperbolic cylinder about the y-axis; its intersection with the xz-plane is the hyperbola $x^2 - z^2 = 1$, $y = 0$.

(b) A hyperboloid of one sheet

(c) An ellipsoid

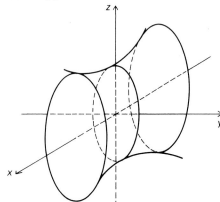

3. (a) For $z = c$, $c = \pm 1, \pm 2, \pm 3$, it is the circle of radius $|c|$ centered at $(0, 0, c)$.

(b)
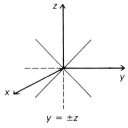

For $x = 0$ For $y = 0$

$y = \pm z$ $x = \pm z$

(d) The surface in the cited exercise is the upper half (*nappe*) of this cone.

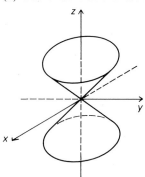

(e) The ellipse $\dfrac{x^2}{\left(\dfrac{ad}{c}\right)^2} + \dfrac{y^2}{\left(\dfrac{bd}{c}\right)^2} = 1$, for $z = d =$ constant.   (f) For $x = 0$, $z = \pm \dfrac{c}{b} y$; for $y = 0$, $z = \pm \dfrac{c}{a} x$.

5.

	Cylindrical coordinates	Spherical coordinates
(a)	$\left(3, \dfrac{\pi}{2}, 4\right)$	$\left(5, \dfrac{\pi}{2}, \cos^{-1}\dfrac{4}{5}\right)$
(b)	$\left(\sqrt{3}, \pi + \tan^{-1}\left(\dfrac{-1}{\sqrt{2}}\right), 0\right)$	$\left(\sqrt{3}, \pi + \tan^{-1}\left(\dfrac{-1}{\sqrt{2}}\right), \dfrac{\pi}{2}\right)$
(c)	$(0, \theta, 0)$	$(0, \theta, \phi)$
(d)	$(\sqrt{2}, \pi, 1)$	$\left(\sqrt{2}, \pi, \dfrac{\pi}{4}\right)$
(e)	$\left(4, \dfrac{7\pi}{6}, 3\right)$	$\left(5, \dfrac{7\pi}{6}, \cos^{-1}\dfrac{3}{5}\right)$

7.

	Cartesian coordinates	Cylindrical coordinates
(a)	$(0, 0, -1)$	$(0, \theta, -1)$
(b)	$(0, -1, \sqrt{3})$	$\left(1, \dfrac{-\pi}{2}, \sqrt{3}\right)$
(c)	$(0, 0, 0)$	$(0, \theta, 0)$
(d)	$(0, 0, -2)$	$(0, \theta, -2)$
(e)	$\left(\dfrac{1}{2}, 0, \dfrac{-\sqrt{3}}{2}\right)$	$\left(\dfrac{1}{2}, 0, \dfrac{-\sqrt{3}}{2}\right)$

9.

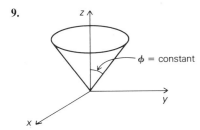

11. $\left(\rho, \theta, \phi + \dfrac{\pi}{2}\right)$ is obtained by rotating the point (ρ, θ, ϕ) by 90° clockwise about the origin in the plane determined by the z-axis and the line joining the origin to (ρ, θ, ϕ).

15. (a) $(x + 1)^2 + \dfrac{(y - 2)^2}{4} + (z - 1)^2 = 1$

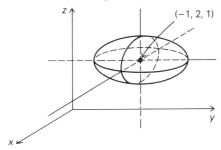

An ellipsoid centered at $(-1, 2, 1)$

(b) $\dfrac{8(x + 1)^2}{21} + \dfrac{4(y + \frac{3}{2})^2}{7} - \dfrac{16(z + 1)^2}{21} = -1$, a hyperboloid of two sheets centered at $\left(-1, \dfrac{-3}{2}, -1\right)$ (figure similar to that on page 655).

17. Place the cartesian coordinate system (x, y, z) as indicated in the figure. Then the air space inside the tank is given in cylindrical coordinates by $0 \leq r \leq 10, \dfrac{-\pi}{2} \leq \theta \leq \dfrac{\pi}{2}, 0 \leq z \leq 16$.

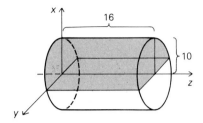

19. Place cartesian coordinates as in the figure. Then the cartridge is given in cylindrical coordinates by $\frac{5}{16} \leq r \leq \frac{9}{4}, 0 \leq \theta \leq 2\pi, 0 \leq z \leq \frac{28}{5}$.

Section 15-3

1. (a) An elliptical helix, much like that of Solved Exercise 1, except that it starts at $(4, 0, 0)$, goes around once, and stops at $(4, 0, 2\pi)$. (b)

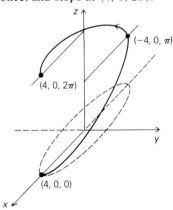

3. (a) Centered at $(0, 0, \cos \phi)$ of radius $\sin \phi$

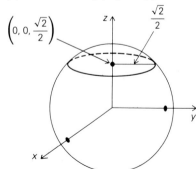

(b) Centered at $(0, 0, 0)$ of radius 1 and lying on the plane determined by the z-axis and θ fixed.

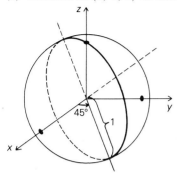

5.

Problem	Velocity vector	Acceleration vector
1 (a)	$(-4 \sin t, 2 \cos t, 1)$	$(-4 \cos t, -2 \sin t, 0)$
1 (b)	$\left(1, \dfrac{-1}{t^2}, 1\right)$	$\left(0, \dfrac{2}{t^3}, 0\right)$
2 (a)	$(1, -1, 2t)$	$(0, 0, 2)$
2 (b)	$(\sinh t, \cosh t, 1)$	$(\cosh t, \sinh t, 0)$

7. Curve $\boldsymbol{\sigma}(t) = (-\cos t + c_1, -\sin t + c_2, dt + c_3)$ is a generalized circular helix which starts at $(c_1 - 1, c_2, c_3)$ and winds clockwise about the line $x = c_1$, $y = c_2$, parallel to the z-axis, at a rate d.

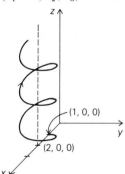

$\boldsymbol{\sigma}(t) = (2 - \cos t, -\sin t, dt)$

If $d = 0$, $\boldsymbol{\sigma}(t)$ is a circle on the plane $z = c_3$ centered at (c_1, c_2, c_3) with radius 1.

9. ≈ 6046 sec/cycle

11. (a) $\boldsymbol{\sigma}$ lies in the plane perpendicular to $\mathbf{u}$.　　(b) $\boldsymbol{\sigma}$ lies in a plane parallel to the one in (a).
　　(c) $\boldsymbol{\sigma}$ lies on a cone with axis $\mathbf{u}$ and vertex angle $2 \cos^{-1} b$.

13. (b) $\boldsymbol{\sigma}(t) = (a_1 f(t) + a_2 h(t), b_2 f(t) + b_2 h(t), c_1 f(t) + c_2 h(t))$ where $f(t) = \sin t$, $h(t) = \cos t$ and $a_i, b_i, c_i, i = 1, 2$, are constants.

15. The set of points (x, y, z) with $x = x_0$, $y = y_0$, $z = \cos^{-1}(x_0) + 2n\pi$, $n = 0, \pm 1, \pm 2, \cdots$; 2π

17. (a) $\theta = \tan^{-1}\left(\dfrac{v^2}{rg}\right)$; the bank angle θ is such that $\tan \theta$ is proportional to the square of the speed v and inversely proportional to the radius r.　　(b) The bank angle θ depends on position and is given by (a) where r is the radius of curvature of the ellipse.

19. $\dfrac{5}{2}$　　23. $\mathbf{B} = \left(\dfrac{\sqrt{2}}{2} \sin t, \dfrac{-\sqrt{2}}{2} \cos t, \dfrac{\sqrt{2}}{2}\right)$; $\tau = \dfrac{1}{2}$

Chapter 15 — Review

1. (a) The plane $x + y - z = 0$ determined by the z-axis and the line $x + y = 0$, $z = 0$

line $x + y = 0$, $z = 0$

(b) Right circular cone about the z-axis with vertex at the origin and opening downward

(c) An elliptical helix starting at $(0, 0, 1)$ and winding counterclockwise about the y-axis; it lies on the elliptic cylinder $\left(\dfrac{x}{3}\right)^2 + z^2 = 1$.

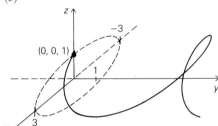

$(0, 0, 1)$

(d) An elliptical cone

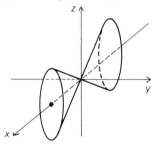

3. (a) Ellipse   (b) Hyperbola
(c)

$x = 0$ $x = 1$ $x = 2$ $x = 3$

$\dfrac{\sqrt{3}}{2}$ $z = -y$ $z = y$ $\dfrac{\sqrt{5}}{2}$

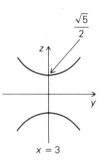

(a degenerate hyperbola)

5.   Elliptical paraboloid $\begin{cases} \text{(a) vertex at } (0, 0, -1) \text{ opening upward} \\ \text{(b) vertex at } (0, 0, 0) \text{ opening upward} \\ \text{(c) vertex at } (0, 0, 1) \text{ opening downward} \\ \text{(d) vertex at } (0, 0, 0) \text{ opening downward} \end{cases}$

The figures look like the paraboloid of revolution, distorted horizontally.

(a)

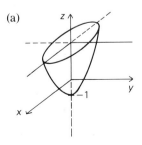

7. (a) $\mathbf{v} = (e^t, \cos t, -\sin t)$; $\mathbf{a} = (e^t, -\sin t, -\cos t)$ (b) $\mathbf{v} = \left(\dfrac{2t}{(1+t^2)^2}, 1, 0\right)$; $\mathbf{a} = \left(\dfrac{2-6t^2}{(1+t^2)^3}, 0, 0\right)$

(c) $\mathbf{v} = \left(e^t + \dfrac{2t}{(1+t^2)^2}, \cos t + 1, -\sin t\right)$; $\mathbf{a} = \left(e^t + \dfrac{2-6t^2}{(1+t^2)^3}, -\sin t, -\cos t\right)$

9. $x = t + 1$, $y = \dfrac{t-1}{2}$, $z = \dfrac{t-2}{3}$

11.

	Level curves	Values of c at which sudden changes occur	Level curve for $c = -1$	Level curve for $c = 0$	Level curve for $c = 1$
(a)	For each value of c, a straight line	None			
(b)	Hyperbola through $(\pm\sqrt{c}, 0)$ if $c > 0$; lines $y = \pm x$ if $c = 0$; hyperbola through $(0, \pm\sqrt{c})$ if $c < 0$	0			
(c)	Hyperbola through $(0, \pm\sqrt{c})$ if $c > 0$; lines $y = \pm x$ if $c = 0$; hyperbola through $(\pm\sqrt{c}, 0)$ if $c < 0$	0			
(d)	Circles centered at $(0, 0)$ of radius $\sqrt{c}$ if $c > 0$; the point $(0, 0)$ if $c = 0$; no level curves if $c < 0$	None	None		
(e)	Hyperbola if $|c| > 0$; the x and y axes if $c = 0$	0			
(f)	For each value of c, a parabola with vertex at $(0, c)$ opening upward	None			

13. (a) Circular hyperboloid intersecting at z-axis   if $c < 0$
Right circular cones about the z-axis   if $c = 0$
Circular hyperboloid intersecting at y-axis   if $c > 0$
(b) (i) $c = -4$; (ii) $c = -1$; (iii) $c = 0$; (iv) $c = 1$; (v) $c = 4$
(c) Shape changes from circular to elliptic.

15. (a) The terrain is sloping, elevating at a gradual rate.   (b) The terrain is steep, elevating at a less gradual rate.   (c) Indicates that the terrain for that distance is steep mountains. No, elevation is not really a function of position at such points since there is more than one height (output) for a given position (input). No, this is not a good direction from which to climb — too steep!

(d)

(e)

17. $3(x^2 + y^2) = z^2 + 1$

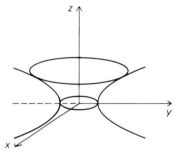

19. (a) $\left(\frac{1}{2}, -\frac{\sqrt{3}}{2}, 0\right)$ (b) $\sqrt{5\pi/3}$ (c) $\mathbf{v} = (-2t \sin (t^2), 2t \cos (t^2), 0)$; speed $= 2t$

   (d) $\mathbf{v} = (\sqrt{5\pi}, \sqrt{5\pi/3}, 0)$; speed $= 2\sqrt{5\pi/3}$ (e) $\sqrt{5\pi/3} + 3/2\sqrt{5\pi}$

21. (a) $\boldsymbol{\sigma}(0) = \mathbf{u}_1$; $\boldsymbol{\sigma}\left(\frac{\pi}{2}\right) = \mathbf{u}_2$ (b) The plane spanned by $\mathbf{u}_1$ and $\mathbf{u}_2$, and the unit sphere (Other answers are possible.)   (c) $\cos^{-1}(\mathbf{u}_1 \cdot \mathbf{u}_2)$ (the angle between $\mathbf{u}_1$ and $\mathbf{u}_2$)   (d) Same as (c)

   (e) $\boldsymbol{\sigma}_1(t) = \mathbf{u}_1 \cos \omega t + [1 - (\mathbf{u}_1 \cdot \mathbf{u}_2)^2]^{-1/2}[\mathbf{u}_2 - (\mathbf{u}_1 \cdot \mathbf{u}_2)\mathbf{u}_1] \sin \omega t$, where $\omega = (2/\pi) \cos^{-1}(\mathbf{u}_1 \cdot \mathbf{u}_2)$.

23. (a) The equations are $x_0 a + y_0 b = z_0$ and $a^2 + b^2 = 1$.   (b) No. (See Fig. 15-2-9 for a similar surface.)   (c) Every hyperboloid of one sheet is ruled; no hyperboloid of two sheets is ruled.

25. Rod: $0 \le r \le 5$, $0 \le \theta \le 2\pi$, $0 \le z \le 15$
   Winding: $5 \le r \le 6.2$, $0 \le \theta \le 2\pi$, $0 \le z \le 15$,
   where the z-axis of the cartesian coordinate system runs through the center of the solenoid, which sits on the xy-plane.

CHAPTER 16

Section 16-1

1. (a) $f_x = y$, $f_x(1, 1) = 1$; $f_y = x$, $f_y(1, 1) = 1$

   (b) $f_x = \frac{1}{y}$, $f_x(1, 1) = 1$; $f_y = \frac{-x}{y^2}$, $f_y(1, 1) = -1$

   (c) $f_x = \frac{1}{1 + (x - 3y^2)^2}$, $f_x(1, 0) = \frac{1}{2}$; $f_y = \frac{-6y}{1 + (x - 3y^2)^2}$, $f_y(1, 0) = 0$

   (d) $f_x = yz$, $f_y = xz$, $f_z = xy$; $f_x(1, 1, 1) = f_y(1, 1, 1) = f_z(1, 1, 1) = 1$

   (e) $f_x = \frac{x}{\sqrt{x^2 + y^2 + z^2}}$, $f_x(3, 0, 4) = \frac{3}{5}$; $f_y = \frac{y}{\sqrt{x^2 + y^2 + z^2}}$, $f_y(3, 0, 4) = 0$;

   $f_z = \frac{z}{\sqrt{x^2 + y^2 + z^2}}$, $f_z(3, 0, 4) = \frac{4}{5}$

3. (a) 2   (b) $z = 1 + y^2$, a parabola opening upward with vertex at $(0, 1)$ (c) $\mathbf{j} + 2\mathbf{k}$

5. (a) $2x + 6y - z = 4$ (b) $2x - 6y + z = 4$ (c) $z = 1$

7. (a) $\dfrac{-\mathbf{i}+\mathbf{j}+\mathbf{k}}{\sqrt{3}}$ (b) $\dfrac{-4\mathbf{i}+8\mathbf{j}+\mathbf{k}}{9}$ (c) $\dfrac{-\mathbf{i}-\mathbf{j}+\mathbf{k}}{\sqrt{3}}$

9. (a) $\dfrac{\partial u}{\partial x}=6x$; $\dfrac{\partial u}{\partial y}=4y$ (b) $\dfrac{\partial u}{\partial x}=(2x-3y)\cos\,(x^2-3xy)$; $\dfrac{\partial u}{\partial y}=-3x\cos\,(x^2-3xy)$

   (c) $\dfrac{\partial u}{\partial x}=\dfrac{2+7y}{3y}$; $\dfrac{\partial u}{\partial y}=\dfrac{-2x}{3y^2}$ (d) $\dfrac{\partial u}{\partial x}=2xy^3(1+xy)e^{2xy}$; $\dfrac{\partial u}{\partial y}=(3+2xy)x^2y^2e^{2xy}$

11. (b) $\partial h/\partial r$ is negative, $\partial h/\partial s$ is positive.   13. $1+\Delta v-\dfrac{\Delta a}{6}$

15. $f_{xy}=2x+2y$; $f_{yz}=2z$; $f_{zx}=0$; $f_{xyz}=0$

19. (b)

21. (a) $g'(t_0)=2e^{2t_0}-\sin\,(2t_0)$

23. (a) $f_x=\dfrac{x^4y+4x^2y^3-y^5}{(x^2+y^2)^2}$, $(x,\,y)\neq(0,\,0)$; $f_y=\dfrac{x^5-4x^3y^2-xy^4}{(x^2+y^2)^2}$, $(x,\,y)\neq(0,\,0)$

   (b) $f(x,\,0)=f(0,\,y)=0$ (c) Use the limit definition of f_x and f_y at $(0,\,0)$ and the result of part (b).

25. (a) \$27   (b) \$47; $\dfrac{\partial c}{\partial z}$ is the change in cost due to a one-unit increase in production of the most

expensive skate, the levels of production of the lower-priced units being held fixed.

Section 16-2

1. (a) $\dfrac{d}{dt}f(\boldsymbol{\sigma}(t))=(e^{2t}-e^{-2t})[1+\ln\,(e^{2t}+e^{-2t})]$ (b) $\dfrac{d}{dt}f(\boldsymbol{\sigma}(t))=-\sin\,t+\sin\,(2t)+3t^2$

   (c) $\dfrac{d}{dt}f(\boldsymbol{\sigma}(t))=(1+4t^2)e^{2t^2}$

3. (a) $x[1+\ln\,(x^2+y^2)]\mathbf{i}+y[1+\ln\,(x^2+y^2)]\mathbf{j}$ (b) $\mathbf{i}+2y\mathbf{j}+3z^2\mathbf{k}$ (c) $(2x^2+1)e^{x^2+y^2}\mathbf{i}+2xye^{x^2+y^2}\mathbf{j}$

5. (b) Use parts (e) and (a) of Problem 4. Solved Exercise 4 yields the same result.   (c) $\dfrac{-2\mathbf{r}}{r^4}$, $(r\neq0)$

7. (a) $\dfrac{-14}{\sqrt{3}}$ (b) $\dfrac{218}{\sqrt{3}}e^{-10101}$ (c) $\dfrac{\pi}{8}-\dfrac{1}{2}$ 9. (a) $\mathbf{n}=\dfrac{1}{\sqrt{2}}(\mathbf{i}+\mathbf{j})$ (b) $\mathbf{n}=\dfrac{x_0}{\sqrt{x_0^2+y_0^2}}\mathbf{i}+\dfrac{y_0}{\sqrt{x_0^2+y_0^2}}\mathbf{j}$

11. (a) $\dfrac{1}{\sqrt{5}}(\mathbf{i}+2\mathbf{j})$ (b) $\dfrac{1}{\sqrt{5}}(\mathbf{i}-2\mathbf{j})$ (c) $\sin\,(1)\,\mathbf{i}+\cos\,(1)\,\mathbf{j}$

13. $\nabla f(1,\,1)=\left(\dfrac{9-2\sqrt{5}}{\sqrt{2}}\right)\mathbf{i}+\left(\dfrac{2\sqrt{5}-3}{\sqrt{2}}\right)\mathbf{j}$; directional derivative $=\dfrac{3+2\sqrt{5}}{\sqrt{10}}$

15. (a) $x^x(1+\ln\,x)$; use the fact $x^x=f(\boldsymbol{\sigma}(x))$ where $f(y,\,z)=y^z$ and $\boldsymbol{\sigma}(x)=(x,\,x)$

   (b) $x^x(1+\ln\,x)$; use the fact $\ln\,(x^x)=x\ln\,x$ and differentiate.

19. (b) First expression $=\dfrac{\partial}{\partial x}f(g(\zeta),\,h(t+\Delta t))\cdot g'(\zeta)\,\Delta t$; Second expression $=\dfrac{\partial}{\partial y}f(g(t),\,h(\eta))\cdot h'(\eta)\,\Delta t$

where ζ and η lie in the interval $(t,\,t+\Delta t)$

   (c) As $\Delta t\to0$, $t+\Delta t\to t$, $\zeta\to t$, $\eta\to t$ and $\dfrac{dz}{dt}=\dfrac{\partial}{\partial x}f(g(t),\,h(t))\cdot g'(t)+\dfrac{\partial}{\partial y}f(g(t),\,h(t))\cdot h'(t)$

by the continuity of the partials.

21. (a) $\dfrac{\partial z}{\partial x}=100y-4x$ (b) $\dfrac{\partial z}{\partial y}(5,\,3)=464=$ the expected change in production per \$1 million increase

in capital investment with a fixed number of staff when the labor force numbered 5000 and the capital

investment totaled \$3 million.

23. (a) $\nabla V(x, y) = \dfrac{\lambda}{2\pi\epsilon_0}\left(\dfrac{x + x_0}{r_2^2} - \dfrac{(x - x_0)}{r_1^2}, \dfrac{y}{r_2^2} - \dfrac{y}{r_1^2}\right)$

(b) $\dfrac{\partial^2 V}{\partial x^2} = \dfrac{\lambda}{2\pi\epsilon_0}\left[\dfrac{r_2^2 - 2(x + x_0)^2}{r_2^4} + \dfrac{2(x - x_0)^2 - r_1^2}{r_1^4}\right]$; $\dfrac{\partial^2 V}{\partial y^2} = \dfrac{\lambda}{2\pi\epsilon_0}\left[\dfrac{r_2^2 - 2y^2}{r_2^4} + \dfrac{2y^2 - r_1^2}{r_1^4}\right]$;

Use the facts $r_2^2 = (x + x_0)^2 + y^2$ and $r_1^2 = (x - x_0)^2 + y^2$ to show $\dfrac{\partial^2 V}{\partial x^2} + \dfrac{\partial^2 V}{\partial y^2} = 0$

25. (a) $P = \dfrac{RT}{V - \beta} - \dfrac{\alpha}{V^2}$ is a function of T and V; $T = \left(P + \dfrac{\alpha}{V^2}\right)\left(\dfrac{V - \beta}{R}\right)$ is a function of P and V;

V can be obtained as a solution of the cubic equation $PV^3 - (\beta P + RT)V^2 + \alpha V - \alpha\beta = 0$, hence V can be expressed in terms of its coefficients which are functions of P and T and so V itself is a function of P and T.

(b) $\dfrac{\partial T}{\partial P} = \dfrac{V - \beta}{R}$ = the expected change in temperature per 1 unit increase in pressure holding volume

constant; $\dfrac{\partial P}{\partial V} = \dfrac{2\alpha}{V^3} - \dfrac{RT}{(V - \beta)^2}$ = the expected change in pressure per 1 unit increase in volume holding

temperature constant; $\dfrac{\partial V}{\partial T} = \dfrac{-R}{(V - \beta)\left[\dfrac{2\alpha}{V^3} - \dfrac{RT}{(V - \beta)^2}\right]}$ = the expected change in volume per 1 unit increase

in temperature holding pressure constant (obtained by differentiating $P = \dfrac{RT}{V - \beta} - \dfrac{\alpha}{V^2}$ implicitly).

Section 16-3

1. (a) $(0, 0)$ is a saddle point.   (b) $(0, 0)$ is a saddle point.   (c) $(0, 0)$ is a saddle point; (x, y) satisfy-

ing $xy = \dfrac{\pi}{2}(2n + 1)$ where n = even is a local maximum; (x, y) satisfying $xy = \dfrac{\pi}{2}(2n + 1)$ where n = odd

is a local minimum. [The second derivative test fails but maxima and minima may be determined from the facts that $|\sin(xy)| \le 1$ and $\ln(2 + \sin xy)$ is an increasing function of xy.]

(d) $(3, 7)$ is a local minimum.

3. $(0, 0)$ is a saddle point.

5. (a) $(0, 0)$ is a saddle point.   (b) $(0, 0)$ is a local minimum for all C such that $|C| < 2$ and a global minimum for $|C| = 2$; $(0, 0)$ is a saddle point for all C such that $|C| > 2$; behavior changes qualitatively at $C = 2$.

7. (a) Minimum distance = 1 at $(\pm 1, 0, 0)$.   (b) Minimum distance = $\dfrac{\sqrt{20397}}{90}$ at $\left(\dfrac{-7\sqrt{41}}{45}, \dfrac{\sqrt{41}}{6}, \dfrac{28}{45}\right)$

and $\left(\dfrac{7\sqrt{41}}{45}, \dfrac{-\sqrt{41}}{6}, \dfrac{28}{45}\right)$.   (c) Minimum distance = $\sqrt{3}$ at $(1, 1, 1)$, $(-1, -1, 1)$, $(1, -1, -1)$ and

$(-1, 1, -1)$.

9. (a) $y = \dfrac{4}{7}x + 1$ (b) $y = \dfrac{3}{2}x + \dfrac{1}{6}$

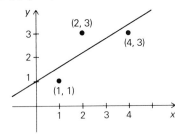

15. Local minimum at $(0, 0)$ **17.** Minimum at $\left(\dfrac{1}{2}, \dfrac{1}{2}\right)$ with minimum value = $\dfrac{1}{2}$

19. Aluminum: $\dfrac{20}{3}\sqrt[3]{3}$ tons; iron: $10\sqrt[3]{3}$ tons; magnesium: $5\sqrt[3]{3}$ tons

21. (a) They are parallel.   (b) $1/3\sqrt{3}$, $-1/3\sqrt{3}$

23. (a) $g(x, y) = Ax^2 + 2Bxy + Cy^2$ has minimum values of 0 [maximum values] along the line $x + By/A = 0$ if $A > 0$ $[A < 0]$.

(b)

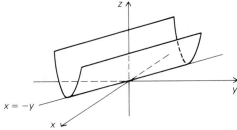

(c) $g(x, y) = Cy^2$, a cylindrical parabola

27. (a) $\dfrac{T_{l-1}}{T_l} = \left(\dfrac{p_l^2}{p_{l-1} p_{l+1}}\right)^{1-(1/n)}$ if $1 \le l \le k$ (b) $p_1 = (p_0 p_4)^{3/4}\left(\dfrac{T_0^3 T_2}{T_1 T_3^3}\right)^{n/4(n-1)}$

31. At production level $(x, y, z) = \left(\dfrac{10^5}{4}, \dfrac{10^5}{4}, \dfrac{10^5}{2}\right)$

Section 16-4

1. (a) $\begin{bmatrix} a & b \\ 0 & 0 \end{bmatrix}$ (b) $\begin{bmatrix} 0 \\ b \end{bmatrix}$ (c) Not defined.

3. $B = \begin{bmatrix} 5 & -2 \\ -2 & 1 \end{bmatrix}$ (a) $(x, y) = (1, 0)$ (b) $(x, y) = (0, 0)$

7. (a) $\dfrac{\partial(x, y)}{\partial(t, s)} = \begin{bmatrix} 1 & 1 \\ 1 & -1 \end{bmatrix}$; $\dfrac{\partial(u, v)}{\partial(x, y)} = \begin{bmatrix} 2x & 2y \\ 2x & -2y \end{bmatrix}$ (b) $\dfrac{\partial(u, v)}{\partial(t, s)} = \begin{bmatrix} 4t & 4s \\ 4s & 4t \end{bmatrix}$

(c) $\dfrac{\partial(u, v)}{\partial(x, y)} \cdot \dfrac{\partial(x, y)}{\partial(t, s)} = \dfrac{\partial(u, v)}{\partial(t, s)}$

9. (a) $\dfrac{\partial(u, v)}{\partial(x, y)} = \begin{bmatrix} 1 & 0 \\ 0 & -1 \end{bmatrix}$; $\dfrac{\partial(x, y)}{\partial(t, s)} = \begin{bmatrix} s & t \\ s & t \end{bmatrix}$; $\dfrac{\partial(u, v)}{\partial(t, s)} = \begin{bmatrix} s & t \\ -s & -t \end{bmatrix}$

(b) $\dfrac{\partial(u, v, w)}{\partial(x, y, z)} = \begin{bmatrix} y & x & 0 \\ z & 0 & x \\ z & 0 & x \end{bmatrix}$; $\dfrac{\partial(x, y, z)}{\partial(t, s)} = \begin{bmatrix} 2t & 2s \\ 2t & -2s \\ 2s & 2t \end{bmatrix}$; $\dfrac{\partial(u, v, w)}{\partial(t, s)} = \begin{bmatrix} 4t^3 & -4s^3 \\ 6t^2 s + 2s^3 & 2t^3 + 6ts^2 \\ 6t^2 s + 2s^3 & 2t^3 + 6ts^2 \end{bmatrix}$

11. (a) r (b) r (c) $-r^2 \sin \phi$

(d) $\left|\dfrac{\partial(x, y, z)}{\partial(u, \phi, \theta)}\right| = \begin{vmatrix} a \sin \phi \cos \theta & au \cos \theta \cos \phi & -au \sin \phi \sin \theta \\ b \sin \phi \sin \theta & bu \sin \theta \cos \phi & bu \sin \phi \cos \theta \\ c \cos \phi & -cu \sin \phi & 0 \end{vmatrix}$

Expand by minors of the third row.

17. Show that $R_{\theta_2} R_{\theta_1} = R_{\theta_1} R_{\theta_2}$ where $R_{\theta_i} = \begin{bmatrix} \cos \theta_i & \sin \theta_i \\ -\sin \theta_i & \cos \theta_i \end{bmatrix}$ and $(R_{\theta_i} R_{\theta_j})\begin{bmatrix} x \\ y \end{bmatrix} = R_{\theta_i}\left(R_{\theta_j}\begin{bmatrix} x \\ y \end{bmatrix}\right)$

Chapter 16 — Review

1. (a) $f_x = \dfrac{-[y \sin (xy) + \cos (xy)]}{e^x + \cos (xy)}$; $f_y = \dfrac{-x \sin (xy)}{e^x + \cos (xy)}$; $f_z = 0$ (b) $g_x = \dfrac{\pi \cos (\pi x)}{1 + y^2}$; $g_y = \dfrac{-2y \sin (\pi x)}{(1 + y^2)^2}$

(c) $h_x = 2x e^{-x^2 - 2y^2} \sin(e^{-x^2 - 2y^2})$; $h_y = 4y e^{-x^2 - 2y^2} \sin(e^{-x^2 - 2y^2})$; $h_z = 0$

3. (a) $(-\frac{1}{2}, 0, 0)$ (b) $(\pi, 0)$ (c) $(0, 0, 0)$ **7.** $t \approx .85$ sec

9. (a) Maximum value $= \frac{1}{2}$; minimum value $= -\frac{1}{2}$ (b) Maximum value $= 4$; minimum value $= 1$

11. The only critical points are $\left(\dfrac{2n - 1}{2}, 0\right)$, $n = 0, \pm 1, \pm 2, \cdots$ which lie on the x-axis; maximum when $n = $ odd and minimum when $n = $ even.

13. (a) $\mathbf{n}_1 = 2\mathbf{i} + 2\mathbf{j} + 4\mathbf{k}$; P_1: $x + y + 2z = 6$; $\mathbf{n}_2 = 4\mathbf{i} + 6\mathbf{j} + 4\mathbf{k}$; P_2: $2x + 3y + 2z = 9$

(b) $\theta = \cos^{-1}\left(\dfrac{9}{\sqrt{102}}\right) \approx .47$ radians   (c) $\left(3, 0, \dfrac{3}{2}\right) + t\left(-2, 1, \dfrac{1}{2}\right)$

15. (a) Use the equality of the mixed partials of f.   (b) Similar to part (a).   (c) No, suppose there is and apply part (a) to get a contradiction.

17. (a) $x^2 e^{xy^2}$ (b) Does not exist   (c) $\ln(1 + x^2 + y^2)$ (d) Does not exist

19. $\dfrac{\partial z}{\partial x} = \dfrac{4(1+y)\,e^{-2x-2y+2xy}}{(e^{-2x-2y} - e^{2xy})^2};\quad \dfrac{\partial z}{\partial y} = \dfrac{4(1+x)\,e^{-2x-2y+2xy}}{(e^{-2x-2y} - e^{2xy})^2}$

23. $\left(n, \dfrac{2n+1}{2}\right)$ is a saddle point for $n = $ odd and is a maximum for $n = $ even.

25. Let $z = f(x, y)$
   (a) $\lim\limits_{y \to 0} f(0, y) = -1$ and $\lim\limits_{x \to 0} f(x, 0) = 1$
   (b) Along $y = mx$ (any m), $\lim\limits_{\substack{y=mx \\ x \to 0}} f(x, y) = 0$; along $x = \pm y^2$, $\lim\limits_{\substack{x=\pm y^2 \\ y \to 0}} f(x, y) = \pm 1$

27. (a) $f_x = \dfrac{x^4 + 6x^2y^2 - 3y^4}{(x^2+y^2)^2}$ and $f_y = \dfrac{-8x^3y}{(x^2+y^2)^2}$ for $(x, y) \neq (0, 0)$; $f_x(0, 0) = -3$ and $f_y(0, 0) = 0$

29. $D_2/D_1 = \left(\dfrac{l_1 Q_2^2}{l_2 Q_1^2}\right)^{1/6}$ **31.** (a) $-3.97\,\epsilon_0$ (b) $\Delta C^* = \dfrac{\partial C^*}{\partial R}(.02) + \dfrac{\partial C^*}{\partial h}(.15)$

33. (a) $u_x = \dfrac{-k^2\,\text{sech}^2\,\alpha}{1 - kt(2u-k)\,\text{sech}^2\,\alpha};\ u_t = \dfrac{ku(u-k)\,\text{sech}^2\,\alpha}{1 - kt(2u-k)\,\text{sech}^2\,\alpha}$ where $\alpha = kx - u(u-k)t$

   (b) First substitute the expression for kx to yield $\text{sech}^2\,[kx - u(u-k)t] = \dfrac{u(2k-u)}{k^2}$ using the fact $\text{sech}^2\,\theta = 1 - \tanh^2\theta$; then substitute the expression for t into u_t.

35. $y = \dfrac{C}{x}$, ($C = $ constant) and wetted perimeter $= 2y + x$ has a minimum at $x = \sqrt{2C}$, namely, at

$$2y = \dfrac{2C}{\sqrt{2C}} = x$$

CHAPTER 17

Section 17-1

1. (a) $\dfrac{1}{36}(e^{36} - e^9 - 9e^4 + 9e)$ (b) $\dfrac{15}{2}\left(\dfrac{3}{2} + \ln\dfrac{3}{2}\right)$ (c) $\dfrac{1}{3}\dfrac{e^{-\cos 1} - 1}{\cos 1} + \dfrac{1}{3}(1 - e^{-1})$

3. (a) $\dfrac{1}{3}$ (b) $\dfrac{\pi^6}{768} + \dfrac{\pi}{2} - 1$ (c) $\dfrac{13}{2}$ **5.** 148π **7.** $\dfrac{1888}{15}$ **9.** (a) $\dfrac{\pi}{4}$ (b) $\dfrac{11}{12}$

11. $\dfrac{\pi^4}{3} + 4\pi^2 \approx \$71.95$ **13.** (a) $\tfrac{4}{5}(2 + 9\sqrt[3]{4})$ (b) $\tfrac{1}{45}(1056 - 49\sqrt{7} - 169\sqrt{13})$

15. (a) $\tfrac{2}{3}$ (b) $a^2 + a + \tfrac{2}{3}$ (c) $\tfrac{2}{3}a^2$ (d) $\tfrac{1}{2}a^2$ **17.** $2\int_a^b \int_{-f(x)}^{f(x)} \sqrt{[f(x)]^2 - y^2}\,dydx = 2\pi \int_a^b [f(x)]^2\,dx$

19. $(0, 0)$ **21.** 8 **23.** (a) 3.1 (b) π **25.** $(8a^3/3n)\tan(\pi/2n)$ **27.** (c) $\dfrac{\rho\omega^2}{24}\,ab(a^2 + b^2)$

Section 17-2

1. (a) $\dfrac{2}{3}$ (b) 0 (c) $\dfrac{2\pi}{7}$ **3.** (a) $\sqrt{2}\,\pi$ (b) $\dfrac{4}{3}\pi\left(1 - \dfrac{\sqrt{2}}{2}\right)$ (c) $\dfrac{16}{3}a^3$

5. (b) $\int_0^1 \int_z^1 \int_y^1 f(x, y, z)\,dx\,dy\,dz$

7. The double integral of W over the base of the box, times the height of the box.

11. (a) $\left(\dfrac{1}{2}, \dfrac{1}{2}, \dfrac{1}{2}\right)$ (b) $\left(\dfrac{3}{4}, 3, \dfrac{8}{5}\right)$ **13.** $2\pi\rho G\left(\rho_2^2 - \dfrac{1}{3}R^2 - \dfrac{2}{3R}\rho_1^3\right)$

15. (c) $\underset{S}{\iiint} f(x, y, z)\,dx\,dy\,dz = \underset{S^*}{\iiint} h(u, v, w)\left|\dfrac{\partial(x, y, z)}{\partial(u, v, w)}\right|\,du\,dv\,dw$ **17.** $\dfrac{32}{3}$, 64π

19. (a) $\bar{x} = 2$, $\bar{y} = 3$, $\bar{z} = (200 - 8\pi)/(40 - \pi) \approx 4.74$ (b) $\bar{x} = 0$, $\bar{y} = 0$, $\bar{z} \approx 7.55$

Chapter 17 — Review

1. (a) $960 - \sin(11) + \sin(10) + \sin(7) - \sin(6) \approx 961.4$
   (b) $\dfrac{75}{4} - \sin(4) + \sin(3) + \sin(2) - \sin(1) \approx 19.7$ (c) $\dfrac{15}{2}$

3. (a) $\dfrac{10}{3}$ (b) 36π (c) $\dfrac{4}{3}\pi abc$ (d) $\dfrac{13}{192}\pi$ **7.** (a) $\pi \ln (\sec 1 + \tan 1)$ (b) $\dfrac{124{,}352}{315} \approx 394.77$

9. (a) $\dfrac{10}{3}$ (b) $\dfrac{25}{3}$ **11.** (b) The average of the values at the vertices   **13.** $(0, 0, -0.203)$

17. $-GmM \ln r$, where r is the distance from the center of the disc, G is the gravitational constant, M is the mass of the disc, and m is the mass of the point (valid for $r \ge$ radius of the disc).

CHAPTER 18

Section 18-1

1. (a) 1   (b) -1 (c) 0   **3.** (a) 7/6   (b) 0   (c) 0   **5.** (a) $2\pi^2$ (b) 0   (c) 0   (d) 0

9. (a) 0   (b) $\pi\sqrt{2}$ (c) $(5^{3/2} - 1)/12$ (d) 2   (e) $52\sqrt{14}$ (f) $\frac{16}{3} - 2\sqrt{3}$ **11.** No

Section 18-2

1. 67/6   **7.** 24π **9.** $x^2 + x^2 \cos y + C$

11. (a) The derivatives are transferred from Q to P, introducing a minus sign.   **15.** Use Green's theorem.

17. Both sides equal zero.   **19.** $3\pi a^2$ **21.** No

23. (a) Not exact   (b) $4xy - x^4 + y^4 = C$ (c) $x^2 y^2 + 2x^3 y = C$ (d) Not exact   (e) Not exact

25. 2

Chapter 18 — Review

1. (a) $\frac{10}{3} - 2\cos 2 + 2\sin 2 - 2\sin 1$ (b) π (c) -32 **3.** $\frac{1}{3}(e - 1)$

5. (a) $\cos 3 - \cos 5 + [\ln 2]^2 + \frac{44}{3}\ln 2 - \frac{83}{18}$ (b) 0   **7.** $e^3 - e - \ln 2$

9. (a) $\frac{1}{4}\{(y_2 - y_1)(x_1 + x_2) + (x_2 - x_1)(y_1 + y_2) + (y_3 - y_2)(x_3 + x_2) + (x_3 - x_2)(y_3 + y_2) + (y_1 - y_3)(x_1 + x_3)$
$+ (x_1 - x_3)(y_1 + y_3)\}$ (b) $\frac{1}{4}\sum_{i=1}^{n}\{(y_{i+1} - y_i)(x_i + x_{i+1}) + (x_{i+1} - x_i)(y_i + y_{i+1})\}$
where $y_{n+1} = y_1$ and $x_{n+1} = x_1$. (Other forms of the answer are also possible.)

11. (b) $f(x, y) = x \tan y + C$ (d) $f(x, y) = x^3 y^2 + C$

13. (a) $f(x, y) = xe^y \sin x + C$ (b) $f(x, y, z) = y \sin x + x \sin z + z \sin y + C$ (c) Not exact

     (d) $f(x, y) = \frac{1}{2}\exp(x^2 + y^2) + C$

15. (a) Not exact   (b) Not exact   (c) Not exact   (d) $-xy - 3x^3 + x + 2y^2 = C$

     (e) $x^2(y - 1) + y = C$

17. (a) $\nabla \times \mathbf{F} = 2zx^3 y\, \mathbf{i} - 3zx^2 y^2\, \mathbf{j} - 2\mathbf{k}$

     $\nabla \cdot \mathbf{F} = x^3 y^2$

     (b) 2π (c) $1/12$

19. $\int_S (\nabla \times \mathbf{F}) \cdot \mathbf{n}\, dA$ where C is the boundary of S.

21. A surface in space with a hole cut in it; the inner hole is traversed in the opposite sense to the outer one.

25. (b) $\dfrac{d}{dt}\iiint_W \rho\, dx\, dy\, dz = \iint_{\partial W} \rho\, \mathbf{V} \cdot \mathbf{n}\, dA$

Index